Sample Preparation in Biological Mass Spectrometry

Alexander R. Ivanov · Alexander V. Lazarev
Editors

Sample Preparation in Biological Mass Spectrometry

Springer

Editors
Alexander R. Ivanov
Harvard School of Public Health
Harvard University
Department of Genetics and Complex Diseases
Huntington Avenue, SPH-1 665
02115 Boston Massachusetts
USA
aivanov@hsph.harvard.edu

Alexander V. Lazarev
Pressure BioSciences, Incorporated
Research and Development
Norfolk Avenue 14
02375 South Easton
Massachusetts
USA
alazarev@pressurebiosciences.com

ISBN 978-94-007-0758-0 e-ISBN 978-94-007-0828-0
DOI 10.1007/978-94-007-0828-0
Springer Dordrecht Heidelberg London New York

Library of Congress Control Number: 2011928676

Printed on acid-free paper

Springer is part of Springer Science+Business Media (www.springer.com)

Preface

Materialistic Dualism of Sample Preparation: Boring but Important!

On two occasions I have been asked,—"Pray, Mr. Babbage, if you put into the machine wrong figures, will the right answers come out?" . . . I am not able rightly to apprehend the kind of confusion of ideas that could provoke such a question.

Charles Babbage, *Passages from the Life of a Philosopher.*

During the past several decades tremendous investments have been made to develop analytical technologies in biology, chemistry and in other related disciplines. Biological samples in particular have received a lot of attention fueled by the increased focus in biomedical research. It seems that analytical technologies are approaching the limits of sensitivity and resolution relevant to biomedical research; computer hardware and data analysis algorithms are becoming ever more powerful every year, allowing individual researchers to complete in a month or two projects only large institutions could have dreamt about a couple of decades ago. It is commonly overlooked, however, that most analytical technologies cannot accept biological samples as they exist in vivo, therefore, elaborate manipulations must precede the actual analysis step to transform samples into a form suitable as input for a given analytical instrument.

Applications of biological mass spectrometry (MS) span across several different areas including proteomics, genomics, metabolomics, pharmacokinetics, tissue imaging, etc. Novel applications continue to be developed in areas such as environmental monitoring and ecology. Moreover, mass spectrometry is finally entering into clinical diagnostics and epigenetics. The task of describing sample preparation for biological MS is even more confusing because there is no commonly agreed upon definition of the very concept of "sample preparation". For some, sample preparation stops at cell or tissue lysis and extraction of the analytes, while others consider steps such as proteolytic digestion and sample/matrix spotting for matrix-assisted ionization experiments, analyte enrichment, sample derivatization, isotopic labeling and other analogous manipulations as components of an all-encompassing sample preparation routine.

Sample preparation has most commonly been defined as "…the process of converting a representative sample into a form suitable for chemical analysis…" (Harris, D., 2002). In other words, the goal and philosophy of sample preparation in biological mass spectrometry is to reveal the actual multicomponent molecular structure of a biological specimen assessable by mass spectrometry tools. In this volume we continue to adhere to the above definitions of sample preparation as inclusive of all of the steps preceding mass spectrometry analysis, while compiling a collection of practical examples of classical, popular and innovative sample manipulation methods, leading to successful fulfillment of the analytical goals. We have also included several chapters that give the reader a perspective on the transition from sample preparation to analysis and an overview of steps involving separation and analysis of a broad variety of relevant analytes.

Given that good sample quality is a foundation of analytical success, the importance of highly reproducible and thorough sample preparation practice cannot be overstated. Unfortunately, sample preparation techniques still used in many laboratories today remain very conservative, sometimes, without much exaggeration, based on principles (e.g., mortar and pestle) utilized by early researchers at the onset of human civilization. Such conservatism, perhaps, also acknowledges the importance of hands-on expertise in sample preparation as an art, where small nuances can make a considerable difference. Consequently, numerous volumes have been written about analytical applications, such as methods for separation and detection of molecules of biological relevance. Sample preparation, however, has remained poorly covered in the literature. We believe that there are several reasons for that trend: sample preparation is often looked at as a "boring" step which does not justify any efforts beyond a paragraph or two in a "Methods" section; it is believed not to be conducive to innovation, or to elicit the enthusiasm of funding agencies. Moreover, healthy common sense frequently tells researchers "if it ain't broke, don't fix it", leading young investigators to continue routinely using sample preparation methods which "worked perfectly well" years ago for their advisors, without ever questioning whether those methods satisfy the demands imposed by the newest instruments blessed with high sensitivity, throughput and resolution, which did not exist even in the wildest dreams of their predecessors. However, as the analytical instrumentation is beginning to reach the plateau of its capabilities relevant to biomedical research, it is clear that "traditional" sample preparation is becoming an increasingly important bottleneck in any modern analytical workflow. Therefore, attention to sample preparation methods in recent years has started to rise considerably, both in academia and in the life science industry.

The recent surge of publications in the peer-reviewed literature related to development and optimization of biological sample preparation methods has been quite evident (Fig. 1). For those of us who dedicated nearly two decades of research to optimization of biological sample preparation, it is very satisfying to see such heightened attention to this topic; it is an indicator that the research community is beginning to pay attention to the "boring" steps, finally accepting the famous

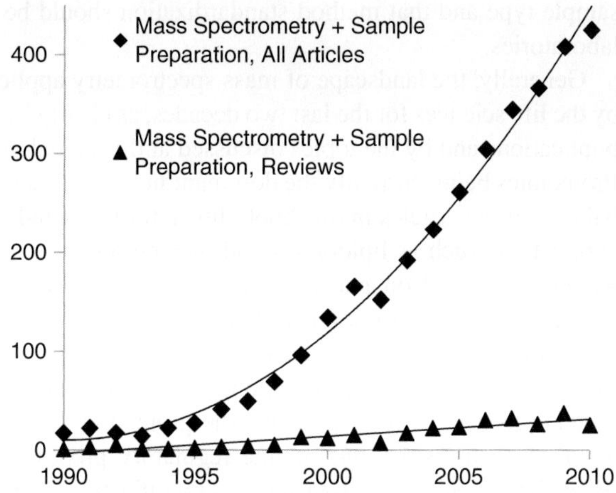

Fig. 1 PubMed Search with keywords ("Sample Preparation" and "Mass spectrometry") in the title and/or abstract. The linear trend of increase in the number of review publications does not match an exponential growth observed in the number of published experimental articles

computer science axiom "Garbage In, Garbage Out" (http://en.wikipedia.org/wiki/Garbage_In,_Garbage_Out). However, only a few volumes are available to date which compile and summarize the information from public domain literature on this subject (Rosenfeld, 2004; Smejkal and Lazarev, 2006; Posch, 2008a, b; Von Hagen, 2008).

Note that the resurgence of attention to the sample preparation problems with respect to mass spectrometry analysis is associated with some of the latest breakthroughs in analytical technologies, mainly, the broad commercial availability of high resolution accurate mass instruments and new ionization and ion fragmentation techniques which enable label-free quantitation, leading, in turn, to the necessity of improving reproducible and quantitative recovery of analytes from samples. Moreover, recent trends to use Multiple Reaction Monitoring (MRM) assays for quantitative analysis of known protein biomarkers have increased the demand for quantitative analyte recovery and the efficiency of proteolytic digestion. The rapid progress in mass spectrometry instrument development leads us to believe that the level of interest to systematic, effectual, and reproducible sample preparation approaches will continue to grow in the future.

Separation and fractionation are perhaps other confusing fields that may or may not be considered part of the sample preparation workflow. In reality, the boundaries between pre-fractionation and analytical separation steps do not exist. Moreover, mass spectrometers may also be considered as the instruments carrying out separation of ionized analyte species in gas phase. Despite the invasion of digital technologies into our daily lives, we all still live in an analog world. Therefore, we tend to think of all the steps in the analytical process as a continuum of overlapping transformations, which leads us to believe that an adequate coverage of transitions between well-defined areas of knowledge is well justified in this book. One of the predominant thoughts which threads through practically every chapter of this book, is that sample preparation requires both optimization for each analytical method or sample type and that method standardization should be applicable across multiple laboratories.

Generally, the landscape of mass spectrometry applications has been dominated by the life sciences for the last two decades, as clearly illustrated by the frequency of publications and by the topics discussed at the annual mass spectrometry meetings. Proteomics being currently the dominant field in life science mass spectrometry; the balance of the articles in this book illustrates this trend as well. However, emerging applications such as lipidomics and metabolomics are getting stronger every year. Research focused on phosphorylation, glycosylation and other post-translational modifications is becoming mainstream as well. Proteomics, as we know it, is also changing. After nearly two decades of frustrating attempts to detect specific disease biomarkers in blood plasma among abundant protein indicators of global fundamental physiological phenomena, the proteomics community is finally shifting its attention to investigations of fine regulatory processes in tissues and subcellular organelles, specific post-translational modifications of proteins, protein–protein and protein–lipid interactions, and other important markers of physiological states and diseases. Such a shift from the traditional protein mapping workflow to targeted, hypothesis-driven proteomic experiments, requires drastic re-thinking of the sample preparation philosophy.

While a large proportion of the proteomics laboratories have switched to LC-MS/MS workflows during the last decade, polyacrylamide gel-based separation and in-gel digestion still remain important applications. However, the goal of the in-gel digestion workflow has been transformed to accommodate the gel step as a pre-fractionation technique followed by multi-dimensional chromatography. For obvious reasons, quantitative approaches in proteomics require more rigorous attention to quantitative recovery of the proteins during sample lysis and dissolution, as well as quantitative extraction of peptides after in-gel digestion.

The novel trends of using alternative approaches to proteomics sample preparation have only emerged in recent years, both in the extraction and digestion parts of the workflow. A constant search for completeness of proteolytic digestion has prompted exploration of physical methods such as vigorous agitation, ultrasound, microwave, and high hydrostatic pressure in the attempts to improve or accelerate enzymatic reactions (Capelo et al., 2009; Lill, et al., 2007, as well as Chapters 5, 8, and 30 herein). A variety of alternative reduction and alkylation schemes have been proposed over the years to replace the traditional iodoacetamide/DTT method. Such approaches bring certain advantages, namely the ability to simultaneously reduce and alkylate proteins without concern about quenching the reducing or the alkylating reagents (Gorg et al., 2009; Granvogl et al., 2007 and Chapters 1 and 2 of this book). Approaches to conduct reduction and alkylation in buffers compatible with downstream labeling techniques such as ICAT or iTRAQ are also discussed extensively in Chapters 27 and 28 of this volume. A variety of applications with immobilized enzymes resulting in higher recovery of peptides and faster digestion due to the increase in the apparent enzyme-to-substrate ratio, are also discussed. Over a decade of attempts to automate tryptic digestion have not yet resulted in a commonly accepted commercial platform, although attempts to optimize protein digestion using immobilized trypsin continue to be reported (Massolini and Calleri, 2005; Krenkova and Svec, 2009; Josic and Clifton, 2007; Kim J. et al., 2010).

In other areas of mass spectrometry analysis it is important to note the recent attention to metabolomics and the advances in pharmaceutical analysis and drug discovery applications which are also driven by changes in the mass spectrometry industry. We are also excited to acknowledge the efforts that several groups, multi-laboratory consortia, and organizations have dedicated to this goal. It is important to mention the on-going HUPO (Human Proteome Organization), CPTAC (Clinical Proteomic Technology Assessment for Cancer) and, ABRF (Association of Biomolecular Resource Facilities) sPRG (Proteomics Standards Research Group) initiatives for improvement and standardization of sample preparation. Numerous inter-laboratory groups in biopharmaceutical analysis and drug metabolism research also dedicate their time and effort to standardize and validate the sample handling methods. However, with respect to standardization, the world of sample preparation for biological mass spectrometry is still far from perfect. While the problem with mass spectrometry data portability between laboratories is frequently attributed to the differences in analytical and data analysis platforms, it is no longer a secret that significant variability originates in the sample handling methods as well.

We believe that this volume may add to the joint efforts of many laboratories to improve preanalytical sample preparation practices, eventually leading to better quality data and to better return on investment from the latest analytical instruments. Written as a compilation of hands-on methods exemplified by the work of several recognized leaders in the field, this book may serve as a guide for selection of optimal sample preparation strategies to solve particular biological problems. We hope that this book will help the reader to embrace the idea that sample preparation efforts need more standardization. We would also like to encourage the industrial community and government funding agencies to stimulate and support efforts in this field, which continues to present business opportunities as the research field evolves. There is a plethora of problems in biological sample preparation that have to be solved as soon as possible in order to take full advantage of modern analytical instrumentation and "to expand the knowledge base in medical and associated sciences" as well as "to foster fundamental creative discoveries, innovative research strategies, and their applications as a basis for ultimately protecting and improving human health" as the NIH mission states among its other goals.

The aim of this book, *Sample Preparation in Biological Mass Spectrometry*, is to provide the researcher with important sample preparation strategies for a wide variety of analyte molecules, specimens, methods, and biological applications requiring mass spectrometric analysis as the detection end-point. In this volume we have compiled contributions from several laboratories that are employing mass spectrometry for biological analysis. Our goal is to give the reader several successful examples of sample preparation, development and optimization, leading to success in the analytical steps and proper conclusions made at the end of the day. This book is structured as a compilation of contributed chapters ranging from protocols to research articles and reviews. We deliberately present a diversity of sample preparation applications by not restricting the individual contributions to conform to an unified standard, therefore, you will find comprehensive reviews and short protocols arranged in sections by application, or by sample type, as opposed to by the type of chapter. The main philosophy of this volume is that sample preparation methods have to be optimized and validated for every project, for every sample type and for every downstream analytical technique. There is no universal recipe for biological sample preparation. What we have attempted to do is to present examples of successful optimization of methods to prepare samples for biomedical analysis. Such examples should serve as a good starting point in your efforts. We sincerely hope that this book will help you in your research!

Finally, we must express our sincere gratitude for the cooperation and hard work of all contributing authors who made this book possible. We are also indebted to our family members and colleagues for their patience and encouragement.

Boston, Massachusetts — Alexander R. Ivanov
South Easton, Massachusetts — Alexander V. Lazarev

References

Babbage, C. (1864). Passages from the Life of a Philosopher. (London, Pickering & Chatto Publishers), (1994), ISBN: 978-1851960408.

Capelo, J.L., Carreira, R., Diniz, M., Fernandes, L., Galesio, M., Lodeiro, C., Santos, H.M., and Vale, G. (2009). Overview on modern approaches to speed up protein identification workflows relying on enzymatic cleavage and mass spectrometry-based techniques. Anal Chim Acta *650*, 151–159.

Gorg, A., Drews, O., Luck, C., Weiland, F., and Weiss, W. (2009). 2-DE with IPGs. Electrophoresis *30 Suppl 1*, S122–132.

Granvogl, B., Ploscher, M., and Eichacker, L.A. (2007). Sample preparation by in-gel digestion for mass spectrometry-based proteomics. Anal Bioanal Chem *389*, 991–1002.

Harris, D. (2002). Quantitative Chemical Analysis, Sixth Edition (New York, NY, W. H. Freeman and Co.), ISBN: 978-0-7167-4464-1.

Josic, D., and Clifton, J.G. (2007). Use of monolithic supports in proteomics technology. J Chromatogr A *1144*, 2–13.

Kim, J., Kim, B.C., Lopez-Ferrer, D., Petritis, K., and Smith, R.D. (2010). Nanobiocatalysis for protein digestion in proteomic analysis. Proteomics *10*, 687–699.

Krenkova, J., and Svec, F. (2009). Less common applications of monoliths: IV. Recent developments in immobilized enzyme reactors for proteomics and biotechnology. J Sep Sci *32*, 706–718.

Lill, J.R., Ingle, E.S., Liu, P.S., Pham, V., and Sandoval, W.N. (2007). Microwave-assisted proteomics. Mass Spectrom Rev *26*, 657–671.

Massolini, G., and Calleri, E. (2005). Immobilized trypsin systems coupled on-line to separation methods: recent developments and analytical applications. J Sep Sci *28*, 7–21.

Posch, A. (2008a). 2D PAGE: Sample Preparation and Fractionation: Volume 1 Series: Methods in Molecular Biology. (Totowa, NJ, Humana Press), 459p, ISBN: 978-1-58829-722-8.

Posch, A. (Ed.) (2008b) 2D PAGE: Sample Preparation and Fractionation; Volume 2; Series: Methods in Molecular Biology (Totowa, NJ, Humana Press), 340p, ISBN: 978-1-60327-209-4.

Rosenfeld, J.M. (Ed.) (2004). Sample Preparation for Hyphenated Analytical Techniques (Hoboken, NJ, Wiley-Blackwell), ISBN: 978-0-8493-2374-6.

Smejkal, G.B., and Lazarev, A. (Ed.) (2006). Separation Methods in Proteomics (Farmington, CT, CRC Press/Taylor and Francis Group), ISBN: 978-0-8247-2699-5.

von Hagen, J. (Ed.) (2008). Proteomics Sample Preparation (London, Wiley), ISBN:978-3-527-31796-7.

Contents

Contributors

Guillaume O. Adelmant Department of Cancer Biology, Harvard Medical School, Boston, MA, USA, Guillame_Adelmant@dfci.harvard.edu

Ganesh Kumar Agrawal Research Laboratory for Biotechnology and Biochemistry (RLABB), Kathmandu, Nepal, gkagrawal123@gmail.com

Dikran Aivazian Biogen Idec, San Diego, CA, USA, aivazian@gmail.com; dikran.aivazian@biogenidec.com

Liaqat Ali Department of Medical Biochemistry, Göteborg University, Gothenburg, Sweden, liaqat.ali@medkem.gu.se

Wasfi Alrawashdeh Molecular Oncology and Imaging Centre, Barts Cancer Institute, John Vane Science Centre, Queen Mary University of London, London, UK, W.Alrawashdeh@qmul.ac.uk

Michail A. Alterman Tumor Vaccines and Biotechnology Branch, Division of Cellular and Gene Therapies, Center for Biologics Evaluation and Research, Food and Drug Administration, Bethesda, MD, USA, michail.alterman@fda.hhs.gov

Roland S. Annan Proteomics and Biological Mass Spectrometry Laboratory, GlaxoSmithKline, Collegeville, PA, USA, roland.s.annan@gsk.com

David Arnott Protein Chemistry Department, Genentech, Inc., South San Francisco, CA, USA, arnott@gene.com

Keith Ashman Centro Nacional de Investigaciones Oncológicas (CNIO), Madrid, Spain, kashman@cnio.es

Erin Baker Biological Science Division, Pacific Northwest National Laboratory, Richland, WA, USA, erin.baker@pnl.gov

Bryan A. Ballif Department of Biology and Vermont Genetics Network Proteomics Facility, University of Vermont, Burlington, VT, USA, bballif@uvm.edu

Robert F. Balshaw NCE CECR Centre of Excellence for the Prevention of Organ Failure, Vancouver, BC, Canada; Department of Statistics, University of British Columbia (UBC), Vancouver, BC, Canada, robert.balshaw@hli.ubc.ca

Vibeke Barkholt Enzyme and Protein Chemistry, Department of Systems Biology, Technical University of Denmark, Lyngby, Denmark, vb@bio.dtu.dk

Miklós Békés Apoptosis and Cell Death Research, Sanford|Burnham Medical Research Institute, La Jolla, CA, USA, mbekes@sanfordburnham.org

Mikhail E. Belov Biological Science Division, Pacific Northwest National Laboratory, Richland, WA, USA, mikhail.belov@pnl.gov

Matthew J. Berberich Laboratory of Neurotoxicology, National Institute of Mental Health, National Institutes of Health, Bethesda, MD, USA, matthew.berberich@yale.edu

Axel P. Bergman Immunity and Infection Research Centre, Vancouver Coastal Health Research Institute, Vancouver, BC, Canada, axelbergman@gmail.com

Laurence V. Bindschedler The BioCentre, University of Reading, Reading, UK; Department of Chemistry, University of Reading, Reading, UK; School of Biological Sciences/BioCentre, University of Reading, Reading, Berkshire, UK, l.v.bindschedler@reading.ac.uk

Anders Laurell Blom Møller Enzyme and Protein Chemistry, Department of Systems Biology, Technical University of Denmark, Lyngby, Denmark, albm@bio.dtu.dk

Antonio Boccia Biogen Idec, San Diego, CA, USA, aboccia22@yahoo.com; antonio.boccia@biogenidec.com

Eric Bonneil Caprion Proteomics, Montréal, QC, Canada; Present address: Institute for Research in Immunology and Cancer, Université de Montréal, Montréal, QC, Canada, eric.bonneil@umontreal.ca

Christoph H. Borchers Department of Biochemistry and Microbiology, University of Victoria, Victoria, BC, V8W 3P6; University of Victoria-Genome BC Proteomics Centre, #3101-4464 Markham Street, Vancouver Island Technology Park, Victoria, BC, V8Z 7X8, Canada, christoph@proteincentre.com

Laurence M. Brill NCI Cancer Center Proteomics Facility, Sanford|Burnham Medical Research Institute, La Jolla, CA, USA, lbrill@burnham.org

Sylvain Brunet Caprion Proteomics, Montréal, QC, Canada, sbrunet@caprion.com

Alex Buko Biogen Idec, Cambridge, MA, USA, alex.buko@biogenidec.com

Andres Buonanno Section on Molecular Neurobiology, Eunice Kennedy Shriver National Institute of Child Health and Human Development, National Institutes of Health, Bethesda, MD, USA, buonanno@mail.nih.gov

Alexander G. Camenzind University of Victoria-Genome BC Proteomics Centre, Victoria, BC, Canada, alex@proteincentre.com

María Isabel Ruppen Cañás Centro Nacional de Investigaciones Oncológicas (CNIO), Madrid, Spain, miruppen@cnio.es

Job D. Cardoza Blais Proteomics Center, Dana-Farber Cancer Institute, Harvard Medical School, Boston, MA, USA, job_cardoza@dfci.harvard.edu

Sebastien Carpentier Laboratory of Tropical Crop Improvement, Department of Biosystems, K.U. Leuven, Heverlee, Belgium, Sebastien.Carpentier@biw.kuleuven.be

Elena Chernokalskaya EMD Millipore, Danvers, MA, USA, elena.chernokalskaya@merckgroup.com

Feixia Chu Department of Molecular, Cellular & Biomedical Sciences, University of New Hampshire, Durham, NH, USA, fvy2@unh.edu

Martin Clynes National Institute for Cellular Biotechnology, Dublin City University, Dublin 9, Ireland, martin.clynes@dcu.ie

Gabriela V. Cohen Freue Department of Statistics, University of British Columbia (UBC), Vancouver, BC, Canada; NCE CECR Centre of Excellence for the Prevention of Organ Failure, Vancouver, BC, Canada, Gabriela.Cohen-Freue@hli.ubc.ca

Ben Colson Biogen Idec, San Diego, CA, USA, benjamincolson@gmail.com

Darwin L. Conwell Division of Gastroenterology, Hepatology and Endoscopy, Center for Pancreatic Disease, Brigham and Women's Hospital, Boston, MA, USA; Department of Medicine, Harvard Medical School, Boston, MA, USA, dconwell@partners.org

Rainer Cramer The BioCentre, University of Reading, Reading, UK; Department of Chemistry, Harborne Building, University of Reading, Reading, UK, r.k.cramer@reading.ac.uk

Tatjana Crnogorac-Jurcevic Molecular Oncology and Imaging Centre, Barts Cancer Institute, John Vane Science Centre, Queen Mary University of London, London, UK, T.C.Jurcevic@qmul.ac.uk

Robert J. Cybulski Division of Bacteriology, US Army Medical Research Institute of Infectious Diseases, Fort Detrick, MD, USA, robert.cybulski@us.army.mil

Lan Dai Bioinformatics Program, Medical School, University of Michigan, Ann Arbor, MI, USA, lan.dai2@pfizer.com

Mark Daniels Biogen Idec, San Diego, CA, USA, daniels.mark@yahoo.com

Biplab Das Department of Drug Metabolism and Pharmacokinetics, Biogen Idec, Cambridge, MA, USA, biplab.das@biogenidec.com

Philip Doran Department of Biological Chemistry, David Geffen School of Medicine, University of California, Los Angeles, CA, USA, Philip-doran@yahoo.co.uk

Vladimir M. Doroshenko Science and Engineering Services Inc., Columbia, MD, USA, dorosh@apmaldi.com

Alan A. Doucette Dalhousie University, Halifax, NS, Canada, alan.doucette@dal.ca

Paul Dowling National Institute for Cellular Biotechnology, Dublin City University, Glasnevin, Dublin 9, Ireland, paul.dowling@dcu.ie

Michael J. Dunn School of Medicine and Medical Science, UCD Conway Institute of Biomolecular and Biomedical Research, University College Dublin, Dublin 4, Ireland, Michael.Dunn@ucd.ie

Ravi C. Dwivedi Department of Internal Medicine, Manitoba Centre for Proteomics and Systems Biology, University of Manitoba, Winnipeg, Canada, dwivedi@cc.umanitoba.ca

Monica H. Elliott University of Victoria-Genome BC Proteomics Centre, Victoria, BC, Canada, monica@proteincentre.com

Mari Enoksson Apoptosis and Cell Death Research, Sanford|Burnham Medical Research Institute, La Jolla, CA, USA, enoksson@sanfordburnham.org

Scott B. Ficarro Department of Cancer Biology, Harvard Medical School, Boston, MA, USA, Scott_Ficarro@dfci.harvard.edu

Christine Finnie Enzyme and Protein Chemistry, Department of Systems Biology, Technical University of Denmark, Lyngby, Denmark, csf@bio.dtu.dk

Anik Forest Caprion Proteomics, Montréal, QC, Canada, aforest@caprion.com

Isabelle Fournier MALDI Imaging Team, Laboratoire de Spectrométrie de Masse Biologique Fondamentale et Appliquée (FABMS), EA 4550, Université Nord de France, F-59655, Villeneuve d'Ascq Cedex, France, isabelle.fournier@univ-lille1.fr

Emily A. Freeman Department of Genetics and Complex Diseases, Harvard School of Public Health, Boston, MA, USA; The Harvard School of Public Health Proteomics Resource, Boston, MA, USA, efreeman@hsph.harvard.edu

Steve French Biogen Idec, Cambridge, MA, USA, Steve.French@biogenidec.com

David B. Friedman Proteomics Laboratory, Mass Spectrometry Research Center, Vanderbilt University School of Medicine, Nashville, TN, USA, david.friedman@vanderbilt.edu

Fernando García Martínez Centro Nacional de Investigaciones Oncológicas (CNIO), Madrid, Spain, fgarcia@cnio.es

Craig A. Gelfand BD Diagnostics, Preanalytical Systems, Franklin Lakes, NJ, USA, craig_gelfand@bd.com

Melkamu Getie-Kebtie Tumor Vaccines and Biotechnology Branch, Division of Cellular and Gene Therapies, Center for Biologics Evaluation and Research, Food and Drug Administration, Bethesda, MD, USA, Melkamu.Getiekebtie@fda.hhs.gov

Tapasree Goswami Department of Cell Biology, Harvard Medical School, Boston, MA, USA, tapasree.goswami@cantab.net

Vera Gross Pressure BioSciences Incorporated, South Easton, MA, USA, vgross@pressurebiosciences.com

William S. Hancock The Barnett Institute of Chemical and Biological Sciences, Northeastern University, Boston, MA, USA, wi.hancock@neu.edu

Zhiqi Hao Thermo Fisher Scientific, San Jose, CA, USA, zhiqi.hao@thermofisher.com

Darryl B. Hardie University of Victoria-Genome BC Proteomics Centre, Victoria, BC, Canada, darryl@proteincentre.com

Stuart Hassard London BioScience Innovation Centre, 2 Royal College Street, London, NW1 ONH, stuart@deltadot.com

Marina Hincapie The Barnett Institute of Chemical and Biological Sciences, Northeastern University, Boston, MA, USA, m.hincapie@neu.edu

Misato Hirano Division of Integrative Physiology, Department of Physiology, Jichi Medical University School of Medicine, Shimotsuke, Tochigi, Japan, misa-hirano@jichi.ac.jp

Kim K. Hixson Environmental Molecular Sciences Laboratory, Pacific Northwest National Laboratory, Richland, WA, USA, kim.hixson@pnl.gov

Steffan Ho Biogen Idec, San Diego, CA, USA, steffan.ho@pfizer.com

Marc J. Horn Prospect Biosystems, Inc., Newark, NJ, USA, mhorn@prospectbiosys.com

Michael J. Huddleston Proteomics and Biological Mass Spectrometry Laboratory, GlaxoSmithKline, Collegeville, PA, USA, Michael.J.Huddleston@gsk.com

Joseph P.M. Hui Caprion Proteomics, Montréal, QC, Canada; Present address: Institute for Marine Biosciences NRC, Halifax, NS, Canada, joseph.hui@nrc.ca

Yehia Ibrahim Biological Science Division, Pacific Northwest National Laboratory, Richland, WA, USA, yehia.ibrahim@pnl.gov

Samah Issa Department of Medical Biochemistry, Göteborg University, Gothenburg, Sweden; School of Chemistry, National University of Ireland, Galway, Ireland, samah.issa@medkem.gu.se

Alexander R. Ivanov HSPH Proteomics Resource, Department of Genetics and Complex Diseases, Harvard School of Public Health, Boston, MA, USA, aivanov@hsph.harvard.edu

Shelley N. Jackson Intramural Research Program, National Institute on Drug Abuse, National Institutes of Health, Baltimore, MD, USA, shelley.jackson@nih.hhs.gov

Michel Jaquinod Caprion Proteomics, Montréal, QC, Canada; Present address: Laboratoire d'Etude de la Dynamique des Protéomes, CEA, Grenoble, France, Michel.Jaquinod@cea.fr

Oliver A.H. Jones School of Engineering and Computing Sciences, University of Durham, Christopherson Building, Durham, UK, o.jones@gmail.com

Ingrid Joseph Biogen Idec, San Diego, CA, USA, dribj1@yahoo.com

Vivek Joshi EMD Millipore, Danvers, MA, USA, vivek.joshi@merckgroup.com

Niclas G. Karlsson Department of Medical Biochemistry, Göteborg University, Gothenburg, Sweden; School of Chemistry, National University of Ireland, Galway, Ireland, niclas.karlsson@medkem.gu.se

Diarmuid T. Kenny Department of Medical Biochemistry, Göteborg University, Gothenburg, Sweden; School of Chemistry, National University of Ireland, Galway, Ireland, diarmuid.kenny@medkem.gu.se

Paul A. Keown Division of Nephrology, University of British Columbia (UBC), Vancouver, BC, Canada, keown@interchange.ubc.ca

Andre Kopoyan Department of Biochemistry and Molecular Pharmacology, University of Massachusetts Medical School, Worcester, MA, USA, Andre.Kopoyan@umassmed.edu

Olga A. Koroleva School of Biological Sciences/BioCentre, University of Reading, Reading, Berkshire, UK; Current address: Pursuit Dynamics plc, Huntingdon, Cambridgeshire, UK, olgakoroleva@pursuitdynamics.com

Jeffrey A. Kowalak Laboratory of Neurotoxicology, National Institute of Mental Health, National Institutes of Health, Bethesda, MD, USA, jkowalak@mail.nih.gov

Oleg V. Krokhin Department of Internal Medicine, Manitoba Centre for Proteomics and Systems Biology, University of Manitoba, Winnipeg, Canada, krokhino@cc.umanitoba.ca

Ruwan Kurulugama Biological Science Division, Pacific Northwest National Laboratory, Richland, WA, USA, ruwan_kurulugama@yahoo.com

Xianyin Lai Department of Cellular and Integrative Physiology, Indiana University School of Medicine, Biotechnology Research and Training Center, Indianapolis, IN, USA, xlai@iupui.edu

WenKui Lan Prospect Biosystems, Inc., Newark, NJ, USA, wlan@prospectbiosys.com

Iulia M. Lazar Department of Biological Sciences, Virginia Polytechnic Institute and State University, Blacksburg, VA, USA, malazar@vt.edu

Alexander V. Lazarev Pressure BioSciences Incorporated, South Easton, MA, USA, alazarev@pressurebiosciences.com

Richard S. Lee Department of Urology, Harvard Medical School, Children's Hospital Boston, Boston, MA, USA, Richard.Lee@childrens.harvard.edu

Chen Li Department of Chemistry, University of Michigan, Ann Arbor, MI, USA, chen.li@regeneron.com

Alexandria Lipka deltaDOT QSTP-LLC, Qatar Science & Technology Park, Doha, State of Qatar, a.lipka@deltadot-qatar.com

Peter S. Liu Protein Chemistry Department, Genentech, Inc., South San Francisco, CA, USA, liu.peter@gene.com

Joseph A. Loo Department of Chemistry and Biochemistry, University of California, Los Angeles, CA, USA; Department of Biological Chemistry, David Geffen School of Medicine, University of California, Los Angeles, CA, USA, jloo@chem.ucla.edu

Daniel Lopez-Ferrer Pacific Northwest National Laboratory, Biological Science Division, Richland, WA; Caprion Proteomics U.S., LLC, 1455 Adams Drive, Menlo Park, CA, USA, dlopezferrer@caprion.com

David M. Lubman Bioinformatics Program, Medical School, University of Michigan, Ann Arbor, MI, USA; Department of Chemistry, University of Michigan, Ann Arbor, MI, USA; Department of Surgery, Medical School, University of Michigan Medical Center, Ann Arbor, MI, USA, dmlubman@umich.edu

Jose L. Luque-Garcia Department Analytical Chemistry, University Complutense of Madrid. Av. Complutense s/n, Madrid, Spain, jlluque@quim.ucm.es

Mahon L. Maguire BHF Magnetic Resonance Unit, Department of Cardiovascular Medicine, Wellcome Trust Centre for Human Genetics, University of Oxford, Oxford, UK, mahon.maguire@gmail.com

Anthony J. Makusky Laboratory of Neurotoxicology, National Institute of Mental Health, National Institutes of Health, Bethesda, MD, USA, makuskya@mail.nih.gov

Xiuli Mao Virginia Bioinformatics Institute, Blacksburg, VA, USA, xlmao80@vbi.vt.edu

Sanford P. Markey Laboratory of Neurotoxicology, National Institute of Mental Health, National Institutes of Health, Bethesda, MD, USA, markeys@mail.nih.gov

Nicky Marlin deltaDOT Ltd, London, UK, n.marlin@deltadot.com

Brian Martin Laboratory of Neurotoxicology, National Institute of Mental Health, National Institutes of Health, Bethesda, MD, USA, bm35s@nih.gov

Jarrod A. Marto Department of Cancer Biology and Blais Proteomics Center, Dana-Farber Cancer Institute, Department of Biological Chemistry and Molecular Pharmacology, Harvard Medical School, Boston, MA, USA, jarrod_marto@dfci.harvard.edu

Yoshinori Masuo Laboratory of Neuroscience, Department of Biology, Toho University, Funabashi Chiba, Japan, yoshinori.masuo@bio.sci.toho-u.ac.jp

Ciara A. McManus School of Medicine and Medical Science, UCD Conway Institute of Biomolecular and Biomedical Research, University College Dublin, Dublin 4, Ireland, ciara.mcmanus@nibrt.ie

Bruce M. McManus NCE CECR Centre of Excellence for the Prevention of Organ Failure, Vancouver, BC, Canada; The James Hogg iCAPTURE Research Centre, University of British Columbia and Providence Health Care, Vancouver, BC, Canada; Institute of Heart and Lung Health, Vancouver, BC, Canada; Department of Pathology and Laboratory Medicine, University of British Columbia (UBC), Vancouver, BC, Canada, bruce.mcmanus@hli.ubc.ca

W. Robert McMaster Vancouver Coastal Health Research Institute, Vancouver, BC, Canada; Department of Medical Genetics, University of British Columbia (UBC), Vancouver, BC, Canada, robm@interchange.ubc.ca

Dean E. McNulty Proteomics and Biological Mass Spectrometry Laboratory, GlaxoSmithKline, Collegeville, PA, USA, Dean.E.McNulty@gsk.com

Céline Mériaux CNRS, MALDI Imaging Team, Laboratoire de Neuroimmunologie et Neurochimie Evolutives, Université Nord de France, Villeneuve d'Ascq Cedex, France, celine1.meriaux@etudiant.univ-lille1.fr

Paula Meleady National Institute for Cellular Biotechnology, Dublin City University, Dublin 9, Ireland, paula.meleady@dcu.ie

Hans-Peter Mock Department of Physiology and Cell Biology, Applied Biochemistry, Leibniz Institute of Plant Genetics and Crop Plant Research (IPK), Gatersleben, Germany, mock@ipk-gatersleben.de

Patricia Molina Protein Chemistry Department, Genentech, Inc., South San Francisco, CA, USA, molina.patricia@gene.com

Khatereh Motamedchaboki NCI Cancer Center Proteomics Facility, Sanford|Burnham Medical Research Institute, La Jolla, CA, USA, motamed@sanfordburnham.org

Alice Mui Department of Surgery, University of British Columbia (UBC), Vancouver, BC, Canada, mui@interchange.ubc.ca

Judit M. Nagy Department of Chemistry, Imperial College London, London, UK, j.nagy@imperial.ac.uk

Hyung Wook Nam Molecular Pharmacology and Experimental Therapeutics (MPET), Mayo Clinic, Rochester, MN, USA, Nam.Hyung@mayo.edu

Raymond T. Ng NCE CECR Centre of Excellence for the Prevention of Organ Failure, Vancouver, BC, Canada; Department of Computer Science, University of British Columbia (UBC), Vancouver, BC, Canada, rng@cs.ubc.ca

Jeremy L. Norris Protein Discovery, Inc., Knoxville, TN, USA, Jeremy@proteindiscovery.com

Tetsuo Ogawa School of Medicine, Showa University, Shinagawa, Tokyo, Japan, t.ogawa@med.showa-u.ac.jp

Rachel R. Ogorzalek Loo Department of Biological Chemistry, Molecular Biology Institute, UCLA-DOE Joint Institute for Genomics & Proteomics, University of California-Los Angeles, Los Angeles, CA, USA, rloo@mednet.ucla.edu

Leanne B. Ohlund University of Victoria-Genome BC Proteomics Centre, Victoria, BC, Canada, leanne@proteincentre.com

Berk Oktem Science and Engineering Services Inc., Columbia, MD, USA, oktem@apmaldi.com

Gilbert S. Omenn Center for Computational Medicine and Bioinformatics, University of Michigan, Ann Arbor, MI, 48109-2218, USA, gomenn@umich.edu

Carol E. Parker University of Victoria-Genome BC Proteomics Centre, Victoria, BC, Canada, carol@proteincentre.com

Joao A. Paulo Division of Gastroenterology, Hepatology and Endoscopy, Center for Pancreatic Disease, Brigham and Women's Hospital and Department of Medicine, Harvard Medical School, Boston, MA, USA; Department of Pathology, Children's Hospital Boston, Harvard Medical School, Boston, MA, USA, Joao.Paulo@childrens.harvard.edu

Natalia Penner Department of Drug Metabolism and Pharmacokinetics, Biogen Idec, Cambridge, MA, USA, natasha.penner@biogenidec.com

Fiona Pereira Department of Chemistry, Imperial College London, London, UK, fiona@Deltadot.com

Lilian Phu Protein Chemistry Department, Genentech, Inc., South San Francisco, CA, USA, phu.lilian@gene.com

Tatiana Plavina Biogen Idec, Inc., Boston, MA, USA, tatiana.plavina@biogenidec.com

George O. Poinar Jr Department of Zoology, Oregon State University, Corvallis, OR, USA, poinarg@science.oregonstate.edu

Bradford S. Powell Division of Bacteriology, US Army Medical Research Institute of Infectious Diseases, Fort Detrick, MD, USA, bpowell@mris.com

Chandra Prakash Department of Drug Metabolism and Pharmacokinetics, Biogen Idec, Cambridge, MA, USA, chandra.prakash@biogenidec.com

Randeep Rakwal Laboratory of Neuroscience, Department of Biology, Toho University, Funabashi, Chiba, Japan; Research Laboratory for Biotechnology and Biochemistry (RLABB), Kathmandu, Nepal; School of Medicine, Showa University, Tokyo, Japan, plantproteomics@gmail.com

Jane Razumovskaya Science and Engineering Services Inc., Columbia, MD, USA, jrazumovski@apmaldi.com

Jennifer D. Reid University of Victoria-Genome BC Proteomics Centre, Victoria, BC, Canada, jreid@proteincentre.com

Pier Giorgio Righetti Department of Chemistry, Materials, and Chemical Engineering, Politecnico di Milano, Milan, Italy, piergiorgio.righetti@polimi.it

Lee D. Roberts Harvard Medical School, Cardiovascular Research Center, Massachusetts General Hospital, Charlestown, USA, ldroberts@partners.org

Lisa Robertson NCE CECR Centre of Excellence for the Prevention of Organ Failure, Vancouver, BC, Canada, lisa.robertson@hli.ubc.ca

Aisling A. Robinson School of Medicine and Medical Science, UCD Conway Institute of Biomolecular and Biomedical Research, University College Dublin, Dublin 4, Ireland, aisling.robinson@ucd.ie

Ilyana Romanovsky Pressure BioSciences Incorporated, South Easton, MA, USA, ilyanar@gmail.com

Michel Salzet MALDI Imaging Team, Laboratoire de Spectrométrie de Masse Biologique Fondamentale et Appliquée (FABMS), EA 4550, Université Nord de France, F-59655, Villeneuve d'Ascq Cedex, France, michel.salzet@univ-lille1.fr

Wendy N. Sandoval Protein Chemistry Department, Genentech, Inc., South San Francisco, CA, USA, sandoval.wendy@gene.com

Mayu Sasaki PROOF Centre of Excellence, Vancouver, BC, Canada, mayu.sasaki@hli.ubc.ca

Philip Serwer Department of Biochemistry, The University of Texas Health Science Center at San Antonio, San Antonio, TX, USA, serwer@uthscsa.edu

Junko Shibato Laboratory of Neuroscience, Department of Biology, Toho University, Funabashi Chiba, Japan, rjunko@nifty.com

Mototada Shichiri Health Research Institute (HRI), National Institute of Advanced Industrial Science and Technology (AIST), Ikeda, Osaka, Japan, mototada-shichiri@aist.go.jp

Seiji Shioda School of Medicine, Showa University, Shinagawa, Tokyo, Japan, shioda@med.showa-u.ac.jp

Timothy W. Sikorski Blais Proteomics Center, Dana-Farber Cancer Institute, Harvard Medical School, Boston, MA, USA, tsikorsk@fas.harvard.edu

Gary B. Smejkal Harvard Medical School, Boston, MA, USA, gary_smejkal@hms.harvard.edu

Igor P. Smirnov Institute of Physical-Chemical Medicine, Moscow, Russian Federation; A.N. Belozersky Institute of Physico-Chemical Biology, Moscow State University, Moscow, Russian Federation, smirnov_i@hotmail.com

Derek S. Smith University of Victoria-Genome BC Proteomics Centre, Victoria, BC, Canada, derek@proteincentre.com

Celia J. Smith The BioCentre, University of Reading, Reading, UK, c.j.smith@reading.ac.uk

Richard D. Smith Biological Science Division, Pacific Northwest National Laboratory, Richland, WA, USA, rds@pnl.gov

Melissa A. Sondej Department of Chemistry and Biochemistry, University of California, Los Angeles, CA, USA, masondej@ucla.edu

Hideaki Soya Laboratory of Sports Biochemistry, Institute of Health & Sports Science, University of Tsukuba, Tsukuba, Ibaraki, Japan, hsoya@taiiku.tsukuba.ac.jp

Vic Spicer Department of Physics and Astronomy, University of Manitoba, Winnipeg, Canada, vspicer@cc.umanitoba.ca

Hanno Steen Department of Pathology, Harvard Medical School, Children's Hospital Boston, Boston, MA, USA; Proteomics Center at Children's Hospital Boston, Boston, MA, USA, Hanno.Steen@childrens.harvard.edu

Appavu K. Sundaram Science and Engineering Services Inc., Columbia, MD, USA, asundaram@apmaldi.com

Birte Svensson Enzyme and Protein Chemistry, Department of Systems Biology, Technical University of Denmark, Lyngby, Denmark, bis@bio.dtu.dk

Sun W. Tam Department of Biochemistry and Molecular Pharmacology, University of Massachusetts Medical School, Worcester, MA, USA, sunny.tam@umassmed.edu

Shigeru Tamogami Laboratory of Biologically Active Compounds, Department of Biological Production, Akita Prefectural University, Akita, Japan, tamo_chem@akita-pu.ac.jp

Pierre Thibault Caprion Proteomics, Montréal, QC, Canada; Present address: Institute for Research in Immunology and Cancer, Université de Montréal, Montréal, QC, Canada, pierre.thibault@umontreal.ca

Ali Tiss The BioCentre, University of Reading, Reading, UK; Department of Chemistry, University of Reading, Reading, UK, ali.tiss@dasmaninstitute.org

Diem Tran Biogen Idec, San Diego, CA, USA, sdwtran@yahoo.com; diem.tran@biogenidec.com

Fumihiko Urano Program in Gene Function and Expression, University of Massachusetts Medical School, Worcester, MA, USA, Fumihiko.Urano@umassmed.edu

Ali R. Vaezzadeh Department of Urology, Harvard Medical School, Children's Hospital Boston, Boston, MA, USA, Alireza.Vaezzadeh@childrens.harvard.edu

Annelies Vertommen Laboratory of Tropical Crop Improvement, Department of Biosystems, K.U. Leuven, Heverlee, Belgium, Annelies.Vertommen@biw.kuleuven.be

Rosa Viner Thermo Fisher Scientific, San Jose, CA, USA, rosa.viner@thermofisher.com

Detlef Vullhorst Section on Molecular Neurobiology, Eunice Kennedy Shriver National Institute of Child Health and Human Development, National Institutes of Health, Bethesda, MD, USA, vullhord@mail.nih.gov

Ina-Beate Wanner Semel Institute for Neuroscience and Human Behavior, Intellectual and Developmental Disability Research Center (IDDRC), David Geffen School of Medicine, University of California, Los Angeles, CA, USA, IWanner@mednet.ucla.edu

Jing Wei Biogen Idec, San Diego, CA, USA, jwei@jadebio.com

Susan T. Weintraub Department of Biochemistry, The University of Texas Health Science Center at San Antonio, San Antonio, TX, USA, weintraub@uthscsa.edu

Maxence Wisztorski MALDI Imaging Team, Laboratoire de Spectrométrie de Masse Biologique Fondamentale et Appliquée (FABMS), EA 4550, Université Nord de France, F-59655, Villeneuve d'Ascq Cedex, France, maxence.wisztorski@univ-lille1.fr

Katja Witzel Department of Physiology and Cell Biology, Applied Biochemistry, Leibniz Institute of Plant Genetics and Crop Plant Research (IPK), Gatersleben, Germany; Institute of Plant Nutrition and Soil Science, Christian Albrechts University, Kiel, Germany, kwitzel@plantnutrition.uni-kiel.de

Frank A. Witzmann Department of Cellular and Integrative Physiology, Biotechnology Research and Training Center, Indiana University School of Medicine, Indianapolis, IN, USA, fwitzman@iupui.edu

Amina S. Woods Intramural Research Program, National Institute on Drug Abuse, National Institutes of Health, Baltimore, MD, USA, awoods@intra.nida.nih.gov

Caroline Woodward Department of Drug Metabolism and Pharmacokinetics, Biogen Idec, Cambridge, MA, USA, caroline.woodward@biogenidec.com

Yasukazu Yoshida Health Research Institute (HRI), National Institute of Advanced Industrial Science and Technology (AIST), Ikeda, Osaka, Japan, yoshida-ya@aist.go.jp

Vlad Zabrouskov Thermo Fisher Scientific, San Jose, CA, USA, vlad.zabrouskov@thermofisher.com

Yi Zhang Department of Biological Chemistry and Molecular Pharmacology, Harvard Medical School, Boston, MA, USA, yi_zhang@dfci.harvard.edu

Terry Zhang Thermo Fisher Scientific, San Jose, CA, USA, terry.zhang@thermofisher.com

Part I
Traditional and Improved Techniques in Sample Preparation for Proteomics

Chapter 1
Gel-Based and Gel-Free Sample Preparation for LC-MS/MS Analysis

Xianyin Lai and Frank A. Witzmann

Abstract To understand the biological processes in health and disease, it would be useful to profile all the nucleic acids, glycans, lipids, proteins, and metabolites in a particular system. Unfortunately, no single analytical approach exists that is capable of analyzing all of these elements simultaneously. Rather, each unique sample type requires distinct methods for preparation and analysis: genomics for nucleic acids, glycomics for glycans, lipidomics for lipids, proteomics for proteins, and metabolomics for metabolites. In proteomic analyses, unique experimental designs frequently apply various degrees of fractionation, diverse gel-based or gel-free separation techniques, and a host of mass spectrometric analytical approaches. These issues are discussed in the introduction to encourage the broad community of scientists to develop a specific method for each unique experiment. At the same time, this chapter describes several special sample preparation methods for LC-MS/MS using electrospray ionization and a linear ion-trap mass spectrometer. The methods are optimized, reproducible, and as simple as possible, and they are capable of generating excellent results. However, optimization for each unique experiment is always a good idea.

Keywords Gel-based · Gel-free · LC-MS/MS · Sample preparation · Tandem mass spectrometry

1.1 Introduction

Sample preparation is one of the early and most important steps in biomedical research where living systems in health and disease are studied. The "life processes" and cell activities involve millions of ions and molecules that comprise a living

X. Lai (✉)
Department of Cellular and Integrative Physiology, Indiana University School of Medicine, Biotechnology Research and Training Center, Indianapolis, IN 46202, USA
e-mail: xlai@iupui.edu

A.R. Ivanov, A.V. Lazarev (eds.), *Sample Preparation in Biological Mass Spectrometry*, DOI 10.1007/978-94-007-0828-0_1,

organism. A typical mammalian cell includes inorganic ions (3%), nucleic acids (5%), glycans (6%), lipids (16%), proteins (60%), and miscellaneous metabolites (10%) (Prescher and Bertozzi, 2005). Techniques for the study of inorganic ions are well-established, but those for organic molecules remain under development or are emerging, and they face great challenges. To achieve a fundamental yet comprehensive understanding of the biological processes in living systems in health and disease, ideally, researchers need a "moleculome" that profiles all the nucleic acids, glycans, lipids, proteins, and metabolites in the system. However, there presently exists no such technique now that can analyze them simultaneously. Currently, researchers apply unique sample preparation and analytical methods for each: genomics for nucleic acids, glycomics for glycans, lipidomics for lipids, proteomics for proteins, and metabolomics for metabolites.

In proteomic research, methods for sample preparation depend on the experimental design (Fig. 1.1). To develop a superior method, the following issues should be considered. First, researchers must ask the following questions. What kind of sample will be analyzed? Will the sample be fractionated? Is a gel-based or gel-free approach used? What kind of protein analytical approach (e.g. mass spectrometry) will be used? Second, different strategies have advantages and disadvantages. Researchers must compare these for each particular sample. Third, the following steps should be considerable factors in the choice of methods that are appropriate to the entire experiment. While some publications in periodicals and books focus on sample preparation methods (Bodzon-Kulakowska et al., 2007; Lipton and Pasa-Tolic, 2009), it remains difficult for scientists to decide which method(s) should be used. Nevertheless, sample preparation methods constitute a step critical to achieving reliable results. Sample preparation issues listed in Fig. 1.1 are discussed to encourage the broad community of scientists to develop a specific method for each unique experiment.

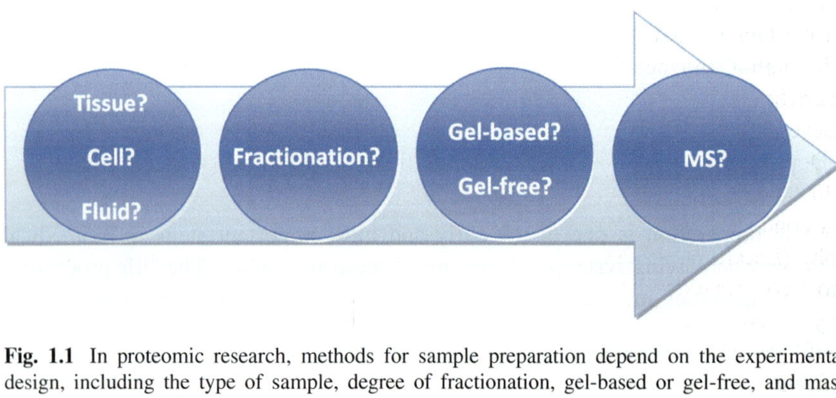

Fig. 1.1 In proteomic research, methods for sample preparation depend on the experimental design, including the type of sample, degree of fractionation, gel-based or gel-free, and mass spectrometry applied

1.1.1 The Source of Sample

Proteomics is "the study of proteins, how they're modified, when and where they're expressed, how they're involved in metabolic pathways and how they interact with one another" (Wilkins et al., 1996). Proteins can originate from any organism, such as human, murine, plants, bacteria, and viruses. In biomedical research, proteomics principally attempts to examine all proteins expressed or present in tissues, cells, and/or fluids. Accordingly, the ultimate goal of sample preparation is to recover for analysis all proteins in a sample, while effectively eliminating all other components, under reasonable conditions. Each sample type has its own characteristic composition. Sometimes, the same approach can be applied to samples from different sources. However, most samples require unique preparation methods. That decision depends on the experimental design and goals. For example, bile and urine have numerous dissolved salts that are incompatible with many protein analytical procedures. In this case, a general method for desalting and extraction of proteins can also be used for these two body fluid samples. If protein depletion or fractionation is not employed, plasma, serum and saliva require no special preparation. On the contrary, in brain tissue, lipids are particularly abundant, while in healthy liver they are not. Therefore, unique preparation methods for brain and liver tissue are required. Proteins in CSF (cerebrospinal fluid) are low in abundance and require concentration prior to analysis, a procedure not generally needed for most other samples.

1.1.2 Fractionation and Enrichment

Current proteomic techniques cannot analyze all constituent sample proteins in a single analysis. Several strategies have been used to fractionate samples to enrich the proteins in which researchers are interested, such as membrane proteins, mitochondrial proteins, glycoproteins, etc. First, enrichment of proteins is conducted according to their abundance. Many methods have been reported that deplete the high-abundance proteins in plasma/serum (Pernemalm et al., 2009). Second, enrichment of proteins is often conducted based on their cellular localization. Several papers have reviewed sample preparation methods for membrane proteins (Dormeyer et al., 2008), mitochondrial proteins (Chaiyarit and Thongboonkerd, 2009), Golgi proteins (Takatalo et al., 2006), etc. Third, enrichment of proteins is conducted according to size by methods such as size-exclusion chromatography (Lecchi et al., 2003). Fourth, enrichment of proteins is conducted according to a combination of factors such as size, charge, and structure of ionized proteins, by methods such as strong-cation-exchange chromatography (Essader et al., 2005). Fifth, enrichment of proteins can also be conducted based on post-translational modification, such as glycoproteins (Geng et al., 2001) and phosphoproteins (Delom and Chevet, 2006), a process aided by the proteins' binding affinities to specific ligands such as immobilized metal ion affinity chromatography (IMAC) for phosphoproteins and lectins for glycoproteins. Bear in mind that the appropriate fractionation method typically depends on the purpose of the analysis.

1.1.3 Gel-Based or Gel-Free Protein Analysis

Since the development of liquid chromatography – tandem mass spectrometric (LC-MS/MS) application for proteomics, many protein analysis strategies have changed from gel-based to gel-free (Lambert et al., 2005). These two approaches are vastly different. Sample preparation for proteins to be loaded onto a gel (or immobilized pH-gradient (IPG) strip) is quite different from sample preparation for LC-MS/MS. For example, more complex buffers are required for gel-based than gel-free separations. A more specific example is that protocols for preparation of tissue and cell samples for two-dimensional gel electrophoresis (2-DE) published earlier by one of the authors (Witzmann, 2005) are different from the protocols of same samples for LC-MS/MS in this chapter. The focus of this chapter is to describe differences between preparation of gel-based and gel-free samples for mass spectrometric analysis.

1.1.4 Mass Spectrometry

For protein identification using mass spectrometry, several issues must be considered. First, different mass spectrometers require different sample preparation methods. MALDI-TOF requires matrix for solid phase ionization, the LTQ ion trap (linear trap) is limited to the analysis of proteolytic peptides, while the LTQ-Orbitrap (linear ion trap + orbitrap) is capable of analyzing both peptides and intact proteins. Obviously, researchers must prepare samples differently for unique mass spectrometers. Second, various applications of the same instrument, such as electrospray ionization (ESI) and nanospray ESI (NSI) in the LTQ, may require distinct sample preparation methods. In NSI, the amount of sample and reagents must be adjusted from the ESI-specific procedures. Furthermore, additional procedures, such as ZipTip desalting, may be required in NSI (Du et al., 2009), compared to ESI. These procedures are described in this chapter. Third, proteomics strategies such as bottom-up and top-down approaches using the LTQ-Orbitrap require specific sample preparation methods. In the bottom-up approach, proteins are proteolytically cleaved to produce peptides for analysis; the top-down approach analyzes intact proteins.

The most important issue for a researcher to consider is the development and optimization of a sample preparation method appropriate for the sample types and mass spectrometry platform at hand, rather than simply repeating a previously published method. This chapter does not attempt to address all the methods used in proteomic research. Alternatively, it describes several special sample preparation methods for LC-MS/MS using the LTQ and an ESI mode. All the protocols listed in this chapter have been used successfully in the authors' laboratory and most of them have been published elsewhere. They reflect rigorous optimization to render the methods reproducible yet simple, enabling the generation of reliable results. However, optimization for each unique experiment is always a good idea. Bear in mind that researchers should always improve sample preparation methods with the development of new analytical techniques.

1.2 Materials

1.2.1 Equipment

1. OneTouch 2D gel spotpicker (The Gel Company, San Francisco, CA, Cat. #:P2D1.5).
2. Shake microplate incubator shaker (Boekel Scientific, Feasterville, PA, Model #: 130000).
3. Centrifugal evaporator concentrator (Jouan, Inc., Winchester, VA, Model #: RC10.10).
4. Centrifuge (Eppendorf, Hamburg, Germany, Model #: 5417 C).
5. Ultrafree-MC centrifugal filter units with microporous membrane (Millipore, Billerica, MA, Cat. #: UFC30GVNB).
6. DUALL®, Glass Pestle and Tube, with Screw-Cap (Kimble/Kontes, Vineland, NJ, Cat. #: 885350-0021).
7. Beckman TL-100 ultracentrifuge (Beckman Coulter, Inc., Fullerton, CA, Cat. #: TL-100).
8. Vortex-Genie 2 (Scientific Industries, Inc., Bohemia, NY, Cat. #: G560).
9. 1.5 and 2.0 mL Biopur® Safe-Lock micro test tubes (Eppendorf, Hamburg, Germany, Model #: 022600028 and 022600044).

1.2.2 Reagents and Solutions

1. HPLC grade H_2O (EMD Chemicals, Gibbstown, NJ, Cat. #: WX0008-1).
2. 100 mM NH_4HCO_3 (790.6 mg/100 mL H_2O, Sigma, St. Louis, MO, Cat. #: A6141).
3. Acetonitrile (ACN) (EMD Chemicals, Cat. #: AX0145-1).
4. 10 mM DL-Dithiothreitol (DTT) in 100 mM NH_4HCO_3 (38.5 mg/25 mL of 100 mM NH_4HCO_3, Sigma, Cat. #: D5545).
5. 55 mM Iodoacetamide in 100 mM NH_4HCO_3 (254.3 mg/25 mL of 100 mM NH_4HCO_3, Sigma, Cat. #: I1149).
6. *100 ng/μL trypsin.* Add 200 μL of H_2O into 20 μg trypsin vial (Porcine, Modified Sequencing Grade, Princeton Separations, Freehold, NJ, Cat. #: EN-151).
7. *6 ng/μL trypsin for gel-based sample digestion.* Take 23 μL of 100 ng/μL trypsin solution and place in a 1.5 mL tube. Add 350 μL of H_2O to this tube and vortex it.

8. *20 ng/μL trypsin for gel-free sample digestion.* Take 200 μL of 100 ng/μL trypsin solution and place in a 1.5 mL tube. Add 800 μL of H_2O and vortex it.
9. Formic acid (FA) (EMD Chemicals, Cat. #: 11670-1).
10. 5% ACN in 0.1% FA (50 μL ACN + 950 μL of 0.1% FA).
11. 4 M urea (NH_2CONH_2) (2.4240 g/ 10 mL of water, Sigma, Cat. #: U5378).
12. Lysis buffer (8 M urea, 10 mM DTT solution freshly prepared, 4.848 g of urea, 15.4 mg of DTT in 10 mL of H_2O).
13. Bio-Rad protein assay dye reagent concentrate (Bio-Rad Laboratories, Hercules, CA, Cat. #: 500-0006).
14. 2-D Quant Kit (GE Life Sciences, Piscataway, NJ, Cat. #: 80-6483-56).
15. ProXPRESSION™ Biomarker Manual Enrichment Kit (PerkinElmer, Wellesley, MA, Cat. #: BMK102).
16. D-Tube™Dialyzers (EMD Biosciences, San Diego, CA, Cat. #: 71506-3).
17. ProteoPrep® 20 Plasma Immunodepletion Kit Single (Sigma, Cat. #: PROT20S).
18. *Reduction/alkylation cocktail.* Add 3.9 mL of ACN into a 15 mL BD Falcon™ conical tube (BD, Franklin Lakes, NJ, Cat. #: 352096). Add 20 μL of triethylphosphine $(C_2H_5)_3P$, Sigma, Cat. #: 245275) and 80 μL of iodoethanol (ICH_2CH_2OH, Sigma, Cat. #: 176850) to the tube. Mix to get 0.5% triethylphosphine and 2% iodoethanol in ACN (see Note 1).

1.3 Methods

1.3.1 Gel-Based Sample Preparation

1.3.1.1 Gel Cutting

1. Place 3 gels stained with Coomassie Blue dye on a light box (see Note 2).
2. Cut corresponding gel spots from each gel (For example, ID: 1019 in Fig.1.2) using a spot picker.
3. Place the gel plugs into a 96-well plate.

1.3.1.2 Destaining

1. Add 50 μL of 100 mM NH_4HCO_3 into each well followed by 50 μL of 100% ACN into each well. Shake the plate for 10 min (Boekel Jitterbug Microplate Incubator Shaker, Feasterville, PA). Remove the entire 100 μL of solution.
2. Repeat Step 1.
3. Add 50 μL of 100% ACN to each well. Shake the plate for 5 min. Remove the entire 50 μL of solution and discard.
4. Allow the ACN to evaporate for 10 min (see Note 3).

1.3.1.3 Reduction/Alkylation

1. Add 50 μL of DTT in 100 mM NH_4HCO_3 to each well. Shake the plate for 30 min. Do not remove the solution.

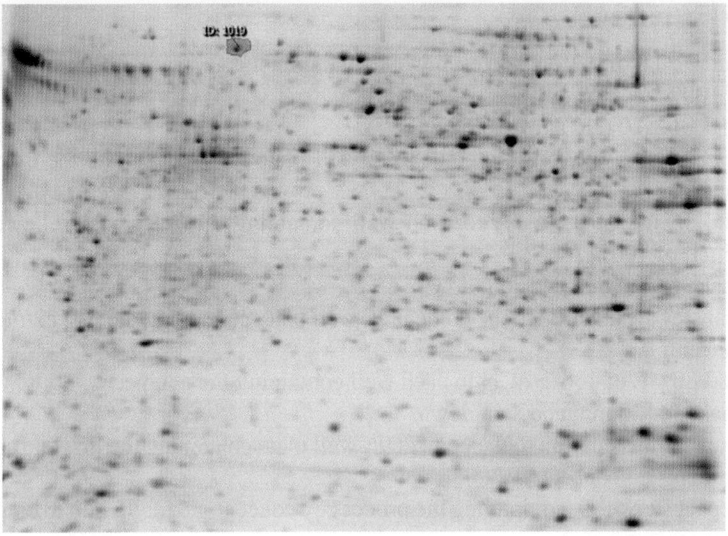

Fig. 1.2 An example of 2D gel stained with Coomassie Blue dye. The spot (ID: 1019) was cut using an OneTouch 2D gel spotpicker

2. Add 50 μL of iodoacetamide in 100 mM NH_4HCO_3 to each well. Shake the plate for 20 min. Do not remove the solution.
3. Add 100 μL of 100% ACN to each well. Wait for 5 min, then remove all 200 μL of solution.

1.3.1.4 Wash/Dehydration

1. Add 50 μL of 100 mM NH_4HCO_3 to each well. Shake the plate for 10 min. Do not remove the solution.
2. Add 50 μL of 100% ACN to each well. Shake the plate for 5 min. Remove the 100 μL of solution.
3. Add 50 μL of 100% ACN to each well. Shake the plate for 5 min and remove the solution.
4. Allow the ACN to evaporate for 20 min.

1.3.1.5 Digestion

1. Add 25 μL of diluted trypsin solution (6 ng/μL) to each well of the 96-well plate.
2. Shake the covered plate containing gel plugs for 3 h (settings: Mix = 7, Temp = 37°C).
3. Spin down the covered plate using the centrifugal evaporator concentrator to get the solution to the bottom of each well.
4. Transfer the trypsin/peptide solution (~25 μl) to a new conical 96-well plate. Concentrate the transferred solution using the SpeedVac centrifugal concentrator system.

1.3.1.6 Extraction

1. Add 21 μL of 0.2% FA to each well containing the gel pieces followed by 9 μL of 100% ACN.
2. Shake for 20 min and spin down.
3. Transfer the solution to a conical 96-well plate and concentrate the extracted solution.
4. Add 15 μL of 0.2% FA to each well containing the gel pieces followed by 15 μL of 100% ACN.
5. Shake for 20 min and spin down.
6. Transfer the solution to the conical 96-well plate containing the first extract and concentrate.
7. Add 30 μL of 100% ACN to each well containing the gel pieces.
8. Shake for 20 min and spin down.
9. Transfer the solution to the conical 96-well plate containing the first and second extracts and concentrate.
10. Repeat steps 1–9, completing the process by concentrate for 45 min to complete dryness.

1.3.1.7 Reconstitution

1. Add 58 μL of 5% ACN in 0.1% FA to each well.
2. Shake for 20 min.
3. Transfer the solution from the each well to an Ultrafree-MC Centrifugal filter (see Note 4).
4. Centrifuge at 3,500×*g* for 2 min using Eppendorf centrifuge.
5. Transfer 50 μL of each sample into a sample vial and place the vials on the autosampler tray.
6. Forty μL of each sample is loaded onto the column and analyzed by LTQ (see Notes 5 and 6).

1.3.2 Gel-Free Sample Preparation

1.3.2.1 Tissue Sample

1. This protocol is designed to extract proteins from liver for general proteomic experiments. Optimization is required for other tissues.
2. While keeping the liver on ice, cut and weigh 250 mg of tissue. For best results, it is recommended that, prior to excision, the in situ liver is first perfused with ice cold physiological saline to remove blood. If this is not possible, tissue pieces cut from unperfused livers should be washed thoroughly to remove residual blood.
3. Place the liver tissue in a 50 mL beaker and add 2 mL (8 volumes) of lysis buffer (see Section 1.2.2, Step 12) (at room temperature).

4. Thoroughly mince the liver tissue with needle-nose surgical scissors.
5. Transfer the buffer/minced tissue solution to a ground-glass homogenizer tube (see Note 7).
6. Use the ground-glass homogenizer to grind up material by twisting with an up/down motion until no solid tissue remains.
7. Pour homogenized tissue solution into a 2 mL microcentrifuge tube, labeled with the sample name or identification number.
8. Shake for 60 min at 37°C. This step enables complete solubilization without the risk of protein carbamylation (which occurs only at temperatures >40°C).
9. Spin sample at 100,000×g for 20 min at 22°C using a Beckman TL-100 ultracentrifuge to remove insoluble materials.
10. Fully solubilized samples are then stored at –45 or –80°C for later analysis.
11. Protein concentration is determined by the Bradford Protein Assay using Bio-Rad protein assay dye reagent concentrate.
12. The sample is ready for Reduction/Alkylation (see Section 1.3.2.6) and digestion (see Section 1.3.2.7).

1.3.2.2 Cell Culture Samples

1. This protocol is designed to generally extract cell proteins from cultured cells (Witzmann et al., 2002). This should be applicable to most mammalian cell types and has been used successfully in primary rat hepatocytes, rat type II alveolar cells, primary rat macrophages, lymphocytes, mouse distal tubule cells, calu-3 lung cells, endothelial cells, and human caco-2/HT29-MTX co-cultures.
2. Add 400 μL of lysis/solubilization buffer (see Section 1.2.2, Step 12) to each well of ~1 × 10^6 cells/well (6 well plate).
3. Shake for 60 min at 37°C and spin down using the centrifugal evaporator concentrator.
4. Remove entire volume from each well and place in a 2.0 mL tube (see Note 8).
5. Determine protein concentration by the Bradford Protein Assay using Bio-Rad protein assay dye reagent concentrate.
6. The sample is ready for Reduction/Alkylation (1.3.2.6) and digestion (1.3.2.7).

1.3.2.3 Plasma/Serum

1. This protocol is designed to enrich high molecular weight proteins bound to albumin and other plasma/serum carrier proteins.
2. Enrich serum carrier protein-bound biomarkers according to the ProXPRESSION Biomarker Manual Enrichment user guide.
3. Determine the protein concentration by the Bradford Protein Assay using Bio-Rad protein assay dye reagent concentrate.
4. The sample is ready for Reduction/Alkylation (1.3.2.6) and digestion (1.3.2.7) (see Note 9).

1.3.2.4 Urine/Bile/Liver Cyst Fluid

1. This protocol is a general approach used to obtain proteins from body fluids such as in urine, bile, and liver cyst fluid.
2. Place the collected urine/bile/ liver cyst fluid in D-Tube^{TM} Dialyzers and dialyze overnight at 4°C against H_2O (volume dependent on sample required).
3. Determine the protein concentration by the Bradford Protein Assay using Bio-Rad protein assay dye reagent concentrate.
4. The sample is ready for Reduction/Alkylation (1.3.2.6) and digestion (X3.2.7) (see Note 10).

1.3.2.5 Renal Cyst Fluid

1. This protocol is designed to prepare renal cyst fluid proteins.
2. Renal cyst fluid contains an excessive abundance of albumin and other high-abundance plasma proteins.
3. Protein concentration is determined by the Bradford Protein Assay using Bio-Rad protein assay dye reagent concentrate.
4. Deplete the high-abundance plasma proteins using the ProteoPrep® 20 plasma immunodepletion kit according the ProteoPrep 20 User Guide.
5. Determine the protein concentration using 2-D Quant Kit assay.
6. The sample is ready for Reduction/Alkylation (1.3.2.6) and digestion (1.3.2.7)(see Note 11).

1.3.2.6 Reduction/Alkylation

1. Place 100 μg of protein from each sample into a 2.0 mL tube.
2. Concentrate the solution to dryness using the centrifugal evaporator concentrator.
3. Reconstitute the sample by adding 200 μL of 4 M urea solution.
4. Add 200 μL of reduction/alkylation cocktail to the sample.
5. Incubate the mixture for 1 h at 37°C.
6. Concentrate the reaction mixture 5 h to total dryness using the centrifugal evaporator concentrator to get rid of the organic solvents that affect the trypsin activity (see Note 12).

1.3.2.7 Digestion

1. Add 100 μL of the 100 mM NH_4HCO_3 solution to the dried sample tube. Vortex to dissolve the dried pellets completely.
2. Add 150 μL of 20 ng/μL trysin solution to the sample and vortex it.
3. Incubate the mixture for 3 h at 37ºC.
4. Add another 150 μL of 20 ng/μL trypsin solution to the sample.
5. Incubate the mixture for another 3 h at 37ºC.
6. Transfer the solution from the each tube to Ultrafree-MC Centrifugal filters and centrifuge them at 3500 rcf for 2 min using Eppendorf centrifuge.

7. Take 120 μL of the sample into a sample vial and place on the autosampler tray (see Note 13).
8. One hundred microliters of each sample is loaded onto the column and analyzed by LTQ mass spectrometer.

Notes

1. Triethylphosphine is highly flammable (spontaneously flammable in air) and causes burns. Read the material safety data sheet (MSDS) carefully before using it. Prepare this cocktail very quickly on ice in a hood.
2. Commonly, three replicate gels are used for one spot. For a more intensely stained spot, one gel plug is usually sufficient. However, most spots require three gel plugs.
3. Before this step, Steps 1–3 can be repeated to eliminate most of Coomassie brilliant blue stain.
4. C18 ZipTip treatment significantly improves MALDI-MS data because it concentrates and desalts the sample (Huynh et al., 2009). However, for this protocol, C18 ZipTip treatment for LC-MS/MS analysis is not recommended.
5. Tryptic peptides were analyzed by tandem mass spectrometry, identified by SEQUEST, and validated by TPP (Keller et al., 2002; Nesvizhskii et al., 2003). Based on ID confidence, sequence coverage, number of unique peptide, number of total peptide, pI, and MW (Fig. 1.3), and observed pI and MW, accurate protein identification is straightforward. The scores of protein and peptides are summarized in Table 1.1.
6. This gel-based sample preparation method was published in the Proteomics Journal (Richardson et al., 2009).
7. Ensure that ALL minced tissue pieces are transferred to the tube (none left behind in the beaker).

1 IPI00323357 1.0000

coverage: 47.7% num unique peps: 34 tot num peps: 90 | share of spectrum id's: 62.02% | subsumed entries: 1

>IPI:IPI00323357 Gene_Symbol=Hspa8 Heat shock cognate 71 kDa protein E:ENSMUSP00000015800 Tr:Q3KQJ4 Sw:P63017

	weight	peptide sequence	nsp adj prob	init prob	ntt	nsp	total	pep grp ind
	wt-0.99	2_FEELNADLFR	0.9998	0.9990	2	31.01	2	
	wt-0.98	2_IINEPTAAAIAYGLDK	0.9998	0.9990	2	31.02	1	a-2
	wt-0.99	2_IINEPTAAAIAYGLDKK	0.9998	0.9990	2	31.01	1	g-2
	wt-0.98	2_NQVAMNPTNTVFDAK	0.9998	0.9990	2	31.01	1	
	wt-0.99	2_NSLESYAFNMK	0.9998	0.9990	2	31.00	1	
	wt-1.00	2_SFYPEEVSSMVLTK	0.9998	0.9990	2	30.99	2	
	wt-0.99	2_SINPDEAVAYGAAVQAAILSGDK	0.9998	0.9990	2	31.00	1	b-2
	wt-0.99	2_STAGDTHLGGEDFDNR	0.9998	0.9990	2	31.01	3	c-2
	wt-0.97	2_TTPSYVAFTDTER	0.9998	0.9990	2	31.03	1	e-2
*	wt-1.00	2_TVTNAVVTVPAYFNDSQR	0.9998	0.9990	2	30.99	1	d-2
	wt-0.97	2_VEIIANDQGNR	0.9998	0.9990	2	31.03	8	

Fig. 1.3 The spot (ID: 1019) was cut and prepared according to current protocol, analyzed by LTQ, identified by SEQUEST, and validated by TPP. It has high confidence (100%), coverage (47.7%), number of unique peptide (34), and number of total peptide (90). Its theoretical pI (5.37) and MW (70.827 kDa) fit the 2D gel position

Table 1.1 The scores of protein and peptides from spot (ID: 1019) were generated by TPP, which also provides protein name (Heat shock cognate 71 kDa protein), ID (IPI00323357), confidence (100%), coverage (47.7%), number of unique peptide (34), number of total peptide (90), theoretical pI (5.37), and MW (70.827 kDa)

Weight	Peptide sequence	nsp adj prob	init prob	ntt	nsp	Total	pep grp ind
wt-0.99	2_FEELNADLFR	0.9998	0.9990	2	31.01	2	
wt-0.98	2_IINEPTAAAIAYGLDK	0.9998	0.9990	2	31.02	1	a-2
wt-0.99	2_IINEPTAAAIAYGLDKK	0.9998	0.9990	2	31.01	1	g-2
wt-0.98	2_NQVAMNPTNTVFDAK	0.9998	0.9990	2	31.01	1	
wt-0.99	2_NSLESYAFNMK	0.9998	0.9990	2	31.00	1	
wt-1.00	2_SFYPEEVSSMVLTK	0.9998	0.9990	2	30.99	2	
wt-0.99	2_SINPDEAVAYGAAVQAAILSGDK	0.9998	0.9990	2	31.00	1	b-2
wt-0.99	2_STAGDTHLGGEDFDNR	0.9998	0.9990	2	31.01	3	c-2
wt-0.97	2_TTPSYVAFTDTER	0.9998	0.9990	2	31.03	1	e-2
wt-1.00	2_TVTNAVVTVPAYFNDSQR	0.9998	0.9990	2	30.99	1	d-2
wt-0.97	2_VEIIANDQGNR	0.9998	0.9990	2	31.03	8	
wt-0.98	3_GPAVGIDLGTTYSCVGVFQHGK	0.9998	0.9990	2	31.01	1	
wt-0.98	3_IINEPTAAAIAYGLDK	0.9998	0.9990	2	31.02	1	a-3
wt-0.99	3_QTQTFTTYSDNQPGVLIQVYEGER	0.9998	0.9990	2	31.01	1	
wt-0.99	3_SINPDEAVAYGAAVQAAILSGDK	0.9998	0.9990	2	31.00	3	b-3
wt-0.99	3_STAGDTHLGGEDFDNR	0.9998	0.9990	2	31.01	4	c-3
wt-1.00	3_TVTNAVVTVPAYFNDSQR	0.9998	0.9990	2	30.99	1	d-3
wt-0.97	3_TTPSYVAFTDTER	0.9996	0.9976	2	31.03	2	e-3
wt-0.99	2_DAGTIAGLNVLR	0.9993	0.9963	2	31.00	1	
wt-0.99	2_FELTGIPPAPR	0.9993	0.9959	2	31.00	2	
wt-0.99	2_LLQDFFNGK	0.9979	0.9881	2	31.02	1	
wt-0.97	2_ITITNDKGR	0.9975	0.9862	2	31.03	6	
wt-0.99	3_SQIHDIVLVGGSTR	0.9972	0.9848	2	31.02	3	
wt-0.99	2_VQVEYKGETK	0.9966	0.9812	2	31.02	5	f-2
wt-0.99	2_VCNPIITK	0.9965	0.9805	2	31.02	18	
wt-1.00	3_QATKDAGTIAGLNVLR	0.9959	0.9776	2	31.01	1	
wt-0.99	2_GTLDPVEK	0.9918	0.9560	2	31.04	2	

Table 1.1 (continued)

Weight	Peptide sequence	nsp adj prob	init prob	ntt	nsp	Total	pep grp ind
wt-0.99	2_LSKEDIER	0.9913	0.9531	2	31.05	5	
wt-0.99	3_VQVEYKGETK	0.9912	0.9529	2	31.05	5	f-3
wt-0.99	3_IINEPTAAAIAYGLDKK	0.9900	0.9465	2	31.06	1	g-3
wt-0.99	2_EIAEAYLGK	0.9865	0.9290	2	31.07	1	
wt-0.98	3_NQVAMNPTNTVFDAKR	0.9719	0.8606	2	31.15	1	
wt-0.99	3_DNNLLGKFELTGIPPAPR	0.8310	0.5310	2	31.47	1	
wt-0.99	3_ELEKVCNPIITK	0.7656	0.4656	2	31.53	2	

Note regarding weight: If a peptide is present in multiple sequence entries, this weight indicates the contribution of this peptide to each of those sequence entries. nsp adj prob (number of sibling peptides adjusted probability). init prob: original or initial probability assigned to this peptide by PeptideProphet (without regard to corresponding protein). ntt (number of tolerable (tryptic) termini): This number indicates whether the peptide sequence exhibits zero, one, or two expected cleavage termini. nsp bin: This is the bin (ranging from 0 to 7) reflecting the number of sibling peptides after discretization. total: number of instances that this peptide was identified in the underlying dataset(s). pep grp ind (peptide group indicator): If a peptide was identified from spectra acquired on ions of different precursor charge states, these different identifications of the same peptide are considered independent evidence for that peptide identity.

8. Witzmann et al. (2002) discovered that the recovery of the cells from monolayer cell culture by scraping results in the introduction of significant variability between the samples. Therefore, the direct addition of lysis buffer is applied in this protocol.
9. This sample preparation method for serum carrier protein-bound proteins has been published in the Electrophoresis Journal (Lai et al., 2009a).
10. This sample preparation method for liver cyst fluid has been published in the Proteomics Journal (Lai et al., 2009b).
11. This sample preparation method for renal cyst fluid has been published in the Proteomics – Clinical Applications Journal (Lai et al., 2008).
12. This reduction/alkylation method was modified from a method developed by a research group in Eli Lilly and Company (Hale et al., 2004).
13. Each sample provides sufficient material for 3 technical replicates (3 injections).

Acknowledgments The authors gratefully acknowledge the technical assistance of Mrs. Heather Ringham. Much of this work was supported by NIH-NIAAA R21 AA016-217-01 and AFOSR Grant FA9550-06-1-0083.

References

Bodzon-Kulakowska, A., Bierczynska-Krzysik, A., Dylag, T., Drabik, A., Suder, P., Noga, M., Jarzebinska, J., and Silberring, J. (2007). Methods for samples preparation in proteomic research. J Chromatogr B Analyt Technol Biomed Life Sci *849*, 1–31.

Chaiyarit, S., and Thongboonkerd, V. (2009). Comparative analyses of cell disruption methods for mitochondrial isolation in high-throughput proteomics study. Anal Biochem *394*, 249–258.

Delom, F., and Chevet, E. (2006). Phosphoprotein analysis: From proteins to proteomes. Proteome Sci *4*, 15.

Dormeyer, W., van Hoof, D., Mummery, C.L., Krijgsveld, J., and Heck, A.J. (2008). A practical guide for the identification of membrane and plasma membrane proteins in human embryonic stem cells and human embryonal carcinoma cells. Proteomics *8*, 4036–4053.

Du, J., Xu, H., Wei, N., Wakim, B., Halligan, B., Pritchard, K.A., Jr., and Shi, Y. (2009). Identification of proteins interacting with GTP cyclohydrolase I. Biochem Biophys Res Commun *385*, 143–147.

Essader, A.S., Cargile, B.J., Bundy, J.L., and Stephenson, J.L., Jr. (2005). A comparison of immobilized pH gradient isoelectric focusing and strong-cation-exchange chromatography as a first dimension in shotgun proteomics. Proteomics *5*, 24–34.

Geng, M., Zhang, X., Bina, M., and Regnier, F. (2001). Proteomics of glycoproteins based on affinity selection of glycopeptides from tryptic digests. J Chromatogr B Biomed Sci Appl *752*, 293–306.

Hale, J.E., Butler, J.P., Gelfanova, V., You, J.S., and Knierman, M.D. (2004). A simplified procedure for the reduction and alkylation of cysteine residues in proteins prior to proteolytic digestion and mass spectral analysis. Anal Biochem *333*, 174–181.

Huynh, M.L., Russell, P., and Walsh, B. (2009). Tryptic digestion of in-gel proteins for mass spectrometry analysis. Methods Mol Biol *519*, 507–513.

Keller, A., Nesvizhskii, A.I., Kolker, E., and Aebersold, R. (2002). Empirical statistical model to estimate the accuracy of peptide identifications made by MS/MS and database search. Anal Chem *74*, 5383–5392.

Lai, X.Y., Bacalla, R.L., Blazer-Yost, B.L., Hong, D., Mason, S.B., and Witzmann, F.A. (2008). Characterization of the renal cyst fluid proteome in autosomal dominant polycystic kidney disease (ADPKD) patients. Proteomics Clin Appl *2*, 1140–1152.

Lai, X., Liangpunsakul, S., Crabb, D.W., Ringham, H.N., and Witzmann, F.A. (2009a). A proteomic workflow for discovery of serum carrier protein-bound biomarker candidates of alcohol abuse using LC-MS/MS. Electrophoresis *30*, 2207–2214.

Lai, X.Y., Blazer-Yost, B.L., Gattone, V.H., Muchatuta, M.N., and Witzmann, F.A. (2009b). Protein composition of liver cyst fluid from the BALB/c-cpk/+ mouse model of autosomal recessive polycystic kidney disease. Proteomics *9*, 3775–3782.

Lambert, J.P., Ethier, M., Smith, J.C., and Figeys, D. (2005). Proteomics: From gel based to gel free. Anal Chem *77*, 3771–3787.

Lecchi, P., Gupte, A.R., Perez, R.E., Stockert, L.V., and Abramson, F.P. (2003). Size-exclusion chromatography in multidimensional separation schemes for proteome analysis. J Biochem Biophys Methods *56*, 141–152.

Lipton, M.S., and Pasa-Tolic, L. (2009). Mass spectrometry of proteins and peptides: Methods and protocols. Methods Mol Biol *492*, 1–474.

Nesvizhskii, A.I., Keller, A., Kolker, E., and Aebersold, R. (2003). A statistical model for identifying proteins by tandem mass spectrometry. Anal Chem *75*, 4646–4658.

Pernemalm, M., Lewensohn, R., and Lehtio, J. (2009). Affinity prefractionation for MS-based plasma proteomics. Proteomics *9*, 1420–1427.

Prescher, J.A., and Bertozzi, C.R. (2005). Chemistry in living systems. Nat Chem Biol *1*, 13–21.

Richardson, M.R., Lai, X., Dixon, J.L., Sturek, M., and Witzmann, F.A. (2009). Diabetic dyslipidemia and exercise alter the plasma low-density lipoproteome in Yucatan pigs. Proteomics *9*, 2468–2483.

Takatalo, M.S., Kouvonen, P., Corthals, G., Nyman, T.A., and Ronnholm, R.H. (2006). Identification of new Golgi complex specific proteins by direct organelle proteomic analysis. Proteomics *6*, 3502–3508.

Wilkins, M.R., Pasquali, C., Appel, R.D., Ou, K., Golaz, O., Sanchez, J.C., Yan, J.X., Gooley, A.A., Hughes, G., Humphery-Smith, I., *et al.* (1996). From proteins to proteomes: Large scale protein identification by two-dimensional electrophoresis and amino acid analysis. Biotechnology *14*, 61–65.

Witzmann, F.A. (2005). Preparation of Mammalian Tissue Samples for Two-Dimensional Electrophoresis. In The Proteomics Protocols Handbook, J.M. Walker, ed. (Totowa, NJ, Humana Press), pp. 31–35.

Witzmann, F.A., Clack, J.W., Geiss, K., Hussain, S., Juhl, M.J., Rice, C.M., and Wang, C. (2002). Proteomic evaluation of cell preparation methods in primary hepatocyte cell culture. Electrophoresis *23*, 2223–2232.

Chapter 2
Manipulating the Mass Spectrometric Properties of Peptides through Selective Chemical Modification

David Arnott, Peter S. Liu, Patricia Molina, Lilian Phu, and Wendy N. Sandoval

Abstract Mass spectrometry is routinely applied to the detection of chemically modified peptides, and researchers deliberately induce chemical modifications in peptides for a variety of reasons. One motivation for doing so is to manipulate the behavior of the peptide in the mass spectrometer itself. Ionization efficiency, for example, can be selectively enhanced or suppressed in MALDI MS, and the charge state distribution altered in ESI. Addition of fixed or localized charges can enhance or suppress entire ion series upon CID, with dramatic effects in MALDI, and more subtle changes in ESI. Newer activation techniques such as ECD and ETD can be favorably combined with ESI and charge derivatization to simplify product ion spectra and map sites of posttranslational modification. Peptides containing specific amino acids and PTMs can be targeted for selective detection or recognized amid a complex background of other peptides by virtue of characteristic product ions or neutral losses conferred by a well-chosen chemical modification. The mass spectrometrist thus possesses a rich set of chemical modifications that can be applied to achieve specific goals in peptide and protein analysis.

Keywords Tandem mass spectrometry · Peptide sequencing · Posttranslational modification · Derivatization

2.1 Introduction

Mass spectrometry is among the most powerful and commonly applied technologies for the characterization of peptides and their structural modifications. Such modifications can originate in vivo, appear as artifacts of sample handling, or as the deliberate result of chemical manipulation. Modifications of the latter sort are performed routinely for a variety of reasons. These include long-established aspects of protein sample preparation such as reduction of disulfide bonds and alkylation

---

D. Arnott (✉)
Protein Chemistry Department, Genentech, Inc., South San Francisco, CA 94080, USA
e-mail: arnott@gene.com

A.R. Ivanov, A.V. Lazarev (eds.), *Sample Preparation in Biological Mass Spectrometry*, DOI 10.1007/978-94-007-0828-0_2,

of free cysteine residues (Cleland, 1964; Dickens, 1933), derivatization to alter problematic physical characteristics, for example by the beta elimination of phosphate from phosphoserine and phosphothreonine, the introduction of affinity tags or reactive groups for purification or crosslinking (Wong, 1991), and as a way of incorporating a stable isotope for quantification via isotope dilution (Corthals and Rose, 2007). A less appreciated motivation for chemical derivatization of proteins and peptides lies in altering their behavior in the mass spectrometer itself, and it is this class of modifications that is reviewed here. Altering the physical or chemical properties of peptides can be performed with the objectives of enhancing the ionization of compounds of interest, directing the fragmentation of peptides in tandem mass spectrometry, and enabling the recognition or selective detection of particular amino acids and the peptides that contain them.

2.2 Modifications That Enhance Ionization

Derivatization of compounds that would not otherwise be desorbed or ionized has long been practiced by mass spectrometrists. Carbohydrates, for example, are still frequently derivatized prior to mass spectrometry (Lamari et al., 2003), and some of the first peptides to be sequenced by tandem mass spectrometry were permethylated and introduced to the mass spectrometer by gas chromatography (Biemann and Martin, 1987). With the advent of "soft" ionization techniques such as fast atom bombardment (FAB), matrix-assisted laser desorption ionization (MALDI), and electrospray ionization (ESI), peptides and proteins can be analyzed on many mass spectrometry platforms directly. Chemical modification of peptides to enhance or modulate the ionization characteristics of peptides has nevertheless continued as an active area of research. It was recognized, for example, that peptides containing the amino acid arginine produced higher abundance ions in MALDI spectra than peptides containing lysine as their most basic residue (Krause et al., 1999). Consequently, spectra of peptide mixtures produced by tryptic digestion of proteins discriminate against the roughly half of such peptides that terminate at lysine. Several laboratories subsequently reported that conversion of lysine to homoarginine via guanidination reaction with O-methylisourea enhanced the ion abundances of these peptides and thereby increased protein sequence coverage in peptide mass mapping experiments (Beardsley et al., 2000; Brancia et al., 2000; Hale et al., 2000). An optimized protocol for this reaction, which produces a 42 Da mass shift, and efficiently modifies the ε-amine of lysine but not the free amine at peptide N-termini except where the N-terminal amino acid is glycine, followed shortly thereafter (Beardsley and Reilly, 2002). These results are attributable to the propensity for analytes to accept a proton in the MALDI process; arginine has by far the highest proton affinity of the amino acids at approximately 245 kcal/mol vs. 229 kcal/mol for lysine (Wu and Fenselau, 1992). Increasing the gas phase basicity of a peptide, as the guanidination reaction accomplishes, should thus increase the abundance of its MALDI-produced peak in a mass spectrum. By the same token, ionization can be suppressed by reducing a peptide's basicity, for example by acetylation of its N-terminus and any lysine residues it contains. These effects are illustrated in Fig. 2.1a: the MALDI time-of-flight spectrum of the lysine-terminated peptide LHQLAFDTYQEFEEAYIPK in an equimolar mixture of unmodified, guanidinated, and doubly acetylated forms. It is evident that the guanidinated version is at

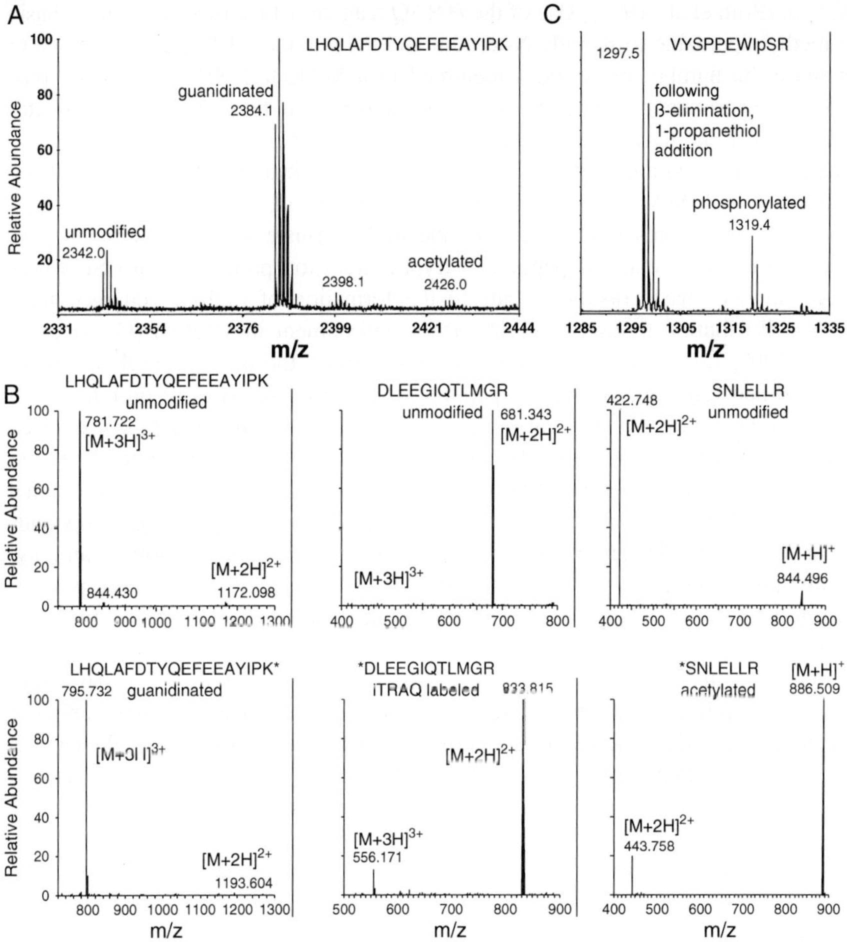

Fig. 2.1 The effect of charge-derivatization on ionization efficiency and charge state. (**a**) MALDI: Equimolar amounts of a lysine-terminated peptide in its unmodified, Lys-guanidinated, and doubly-acetylated forms were mixed and analyzed by MALDI-TOF (4800; Applied Biosystems; α-hyroxy-4-cinnamic acid matrix; with reflectron and delayed-extraction). (**b**) ESI: Full MS scans of three peptides in their unmodified (*top panels*) and Lys-guanidinated, iTRAQ labeled, and acetylated forms, respectively (*bottom panels*) were acquired in the ICR cell of a hybrid linear ion trap-Fourier transform mass spectrometer (LTQ-FT; Thermo Electron). Sites of modification are indicated with ∗. (**c**) MALDI: Equimolar amounts of the peptide VYSPPEWISR in its Ser-phosphorylated state and as the product of β-elimination and Michael addition of 1-propanethiol. The *underlined* proline is a stable isotope analog of the natural amino acid ($^{13}C_5{}^{15}N_1$)

least four fold more abundant than the unmodified peptide, and that the acetylated peptide is effectively suppressed. Similar or greater enhancements in ionization efficiency have been claimed through the conversion of peptide primary amines to more basic groups, for instance by reaction with ethyl picolinimidate (Kim et al., 2009), trifluoroethylisothiocyanate (Bartlet-Jones et al., 1994), or through a variety of other charge-fixing chemistries, extensively reviewed by Roth, Huang, Sadagopan, and Watson (Roth et al., 1998). Use of the iTRAQ reagent, which incorporates the basic N-methylpiperazine to peptide N-termini and the ε-amine of lysine was shown to increase the number of peptides identified in a MALDI-TOF/TOF shotgun proteomics experiment compared to label-free samples, a finding attributed to a greater number of lysine-terminated peptides detected (Ernoult et al., 2008).

Most derivatization studies aiming to enhance ionization efficiency have been performed with MALDI in mind, with relatively few addressing electrospray ionization as well. This may be due to the high intrinsic ionization efficiency of the most common class of peptides – tryptic digestion products – in ESI, where protonation of basic residues leads to the formation of multiply charged ions, with a distribution thought to reflect their environment in solution (Guevremont et al., 1992; Loo et al., 1989). Because free amines, the most general, and readily modified functional groups of peptides, are already the sites of protonation in electrospray, relatively little gain might be expected from the attachment of fixed charges or the like. Ionization efficiency is the result of multiple factors, however, and is influenced by peptide hydrophobicity, solvent composition, and additional species present such as salts, surfactants, and other ion suppressing compounds (Nishikaze and Takayama, 2006). An amine-reactive reagent employing an alkyl chain and quaternary amine to confer greater hydrophobicity and a fixed positive charge was shown to produce measurable, and in some cases dramatic improvements in peptide ionization efficiency, although the effect was variable and not readily predicted from peptide sequence (Mirzaei and Regnier, 2006). Treatment of a whole yeast lysate with the iTRAQ reagent was reported to increase the number of lysine-terminated tryptic peptides identified by database search to equivalence with arginine-terminated peptides (Ross et al., 2004). Even where ionization efficiency is not greatly changed, the distribution of charge states observed can be altered by derivatization. In Fig. 2.1b, for example, the ESI spectra of three peptides are shown in their unmodified forms, and as modified by guanidination, the iTRAQ reagent, and acetylation. Guanidination of a lysine-terminated peptide produced almost no change; the peptide is exclusively triply charged before modification and remains so with the lysine converted to homoarginine (no quadruply charged species is present). Addition of the high proton affinity iTRAQ reagent to an arginine-containing peptide shifted a modest, but noticeable fraction of the peptide from the doubly to triply charged form. Acetylation, which blocks the peptide N-terminus, has a much more dramatic effect, shifting the charge state of an arginine-containing peptide from being almost exclusively doubly charged to a predominantly singly charged state. The ability to alter the charge state of peptides has utility for discriminating among peptide classes in proteomics experiments, and by making a peptide more amenable to fragmentation by collision-induced dissociation (where doubly charged ions are ideal, and high charge states problematic) or electron-capture or electron-transfer dissociation (where highly charged peptides are desirable) (Good et al., 2007).

Phosphopeptides represent a special case in that they are more acidic and hydrophilic than their nonphosphorylated counterparts, and are often more readily detected as negative ions than in the positive ion mode (Janek et al., 2001). The precise contribution of the phosphate to ionization efficiency in the positive ion mode has received little systematic study, but is generally considered deleterious and has been shown in some cases to be severely so (Gao and Wang, 2007). Several chemical modification approaches have been reported to address this problem. The picolinamidination reaction already mentioned was reported to enhance the signal of phosphopeptides in MALDI by two orders of magnitude (Kim et al., 2009), and similar gains have been claimed for the derivatization of carbodiimide-activated peptide carboxylic acids with 1-(2-pyrimidyl) piperazine in both MALDI and ESI (Xu et al., 2008). A more radical approach is available for phosphoserine and phosphothreonine: removal of the phosphate altogether by β-elimination, generally followed by Michael addition of a nucleophile to the resulting dehydrated amino acid (Jaffe et al., 1998; Molloy and Andrews, 2001). This has the benefit of removing the negatively charged and hydrophilic phosphate and replacing it with a neutral and typically hydrophobic group. Elimination of phosphate from the peptide VYSPPEWIpSR and addition of 1-propanethiol (Fig. 2.1c) resulted in an approximately three fold increase in peak height. Even greater improvements have been reported; a survey of N- and S-nucleophiles for the addition reaction found that a 10-fold or greater enhancement of signal could be achieved with 2-phenylethanethiol in MALDI-TOF analysis (Klemm et al., 2004).

2.3 Modifications That Direct the Fragmentation of Peptides in Tandem Mass Spectrometry

Tandem mass spectrometry of peptides produces sequence-specific product ions that can be used to sequence the peptides or match them to entries in a protein database. The dissociation of peptides occurs, to a first approximation, randomly along the peptide backbone producing fragments containing the N-terminus (a-, b-, c-ions) or the C-terminus (x-, y-, and z-ions) where the particular fragments observed depend on both the peptide sequence and the means used to activate the peptide. Collision-induced dissociation (CID) at high energies – keV lab frame – produces spectra that can contain any of these ion series including side chain fragmentation products d- and w-ions that can distinguish leucine and isoleucine. Neutral losses of water, ammonia, and other side chain-associated groups are also common. Low energy CID, as typically performed in triple quadrupole, ion trap, and hybrid quadrupole-time-of-flight instruments, predominantly produces b- and y-ions and associated neutral losses, with a-ions and internal ion series (the result of multiple fragmentation events for a single ion) commonly generated in quadrupole collision cells but unusual in ion trap resonant excitation. The peptides's charge state and the location of basic amino acids within the peptide sequence can influence the ions that are observed and their relative abundances. Consequently, product ion spectra can be difficult to interpret, and CID will not always provide complete sequence coverage or the information necessary to localize posttranslational modifications. Features of peptide fragmentation and interpretation of CID spectra are detailed at length in several excellent reviews (Kinter and Sherman, 2000; Staudenmann and James, 2001).

One of the principal uses of chemical modification has been to assist the sequencing or identification of peptides through CID by enhancing or suppressing a chosen ion series. It was recognized early on, for instance, that high energy CID spectra are very informative due to their rich repertoire of products but difficult to interpret for the same reason. Upon attachment of a fixed charge at the peptide's N-terminus, however, N-terminal sequence ions dominate the product ion spectrum – at least for singly-charged precursors (Roth et al., 1998; Stults et al., 1993). Likewise, early results from low energy CID indicated that product ions with an N-terminal proline were unusually prominent, leading to the use of a t-butyl protected proline NHS ester as a peptide N-terminal label that enhanced b- and a- ion series (Krishnamurthy et al., 1989). Since those early observations many approaches to simplifying CID spectra through judicious derivatization have been employed. The majority, which include those just mentioned as well as those based on quaternary amine or phosphonium derivatives (Roth et al., 1998) and high proton affinity derivatives such as the methylpiperazine found in the iTRAQ reagent (Ross et al., 2004) are targeted to the N-terminal amine. Fewer target the peptide's C-terminus, because the chemistry involved in activating the carboxyl group is more complicated and difficult to apply to low abundance peptides, but some have taken this route as well (English et al., 2009; Wagner et al., 1991; Xu et al., 2008). A notable c-terminal modification is the guanidination of lysine, which enhances the y-ion series of peptides terminating at lysine, and can be performed in conjunction with other amine-reactive reagents that would otherwise label both the N-terminus and the ε-amine of lysine. A third strategy is to modify peptides so as to suppress, rather than enhance, a particular ion series. This can be accomplished by incorporation of a negatively charged group such as a sulfonic acid derivative (Keough et al., 1999). When applied to a peptide's N-terminus, for example, the N-terminal ion series are effectively suppressed, and C-terminal ion series (usually y-ions) dominate the product ion spectrum. Peptides containing lysine are typically guanidinated prior to sulfonation (Keough et al., 2002; Wang et al., 2004).

The effects of these strategies for directing ion series in CID are illustrated in Fig. 2.2. In the top panel, the product ion spectrum of the unmodified, lysine-terminated peptide LHQLAFDTYQEFEEAYIPK is shown as acquired on a MALDI-TOF/TOF instrument. The spectrum is predominantly composed of b-ions, unsurprisingly in light of the basic histidine residue near the N-terminus, but also contains a short series of y-ions. Guanidination of the peptide's C-terminal lysine dramatically increases the number and abundance of the y-ions so that they are now the dominant ion series, although all of the b-ions are still present (middle panel). Finally, the doubly modified peptide with its lysine guanidinated and a sulfonic acid incorporated at the N-terminus via reaction with 4-sulfophenylisothiocyanate yields

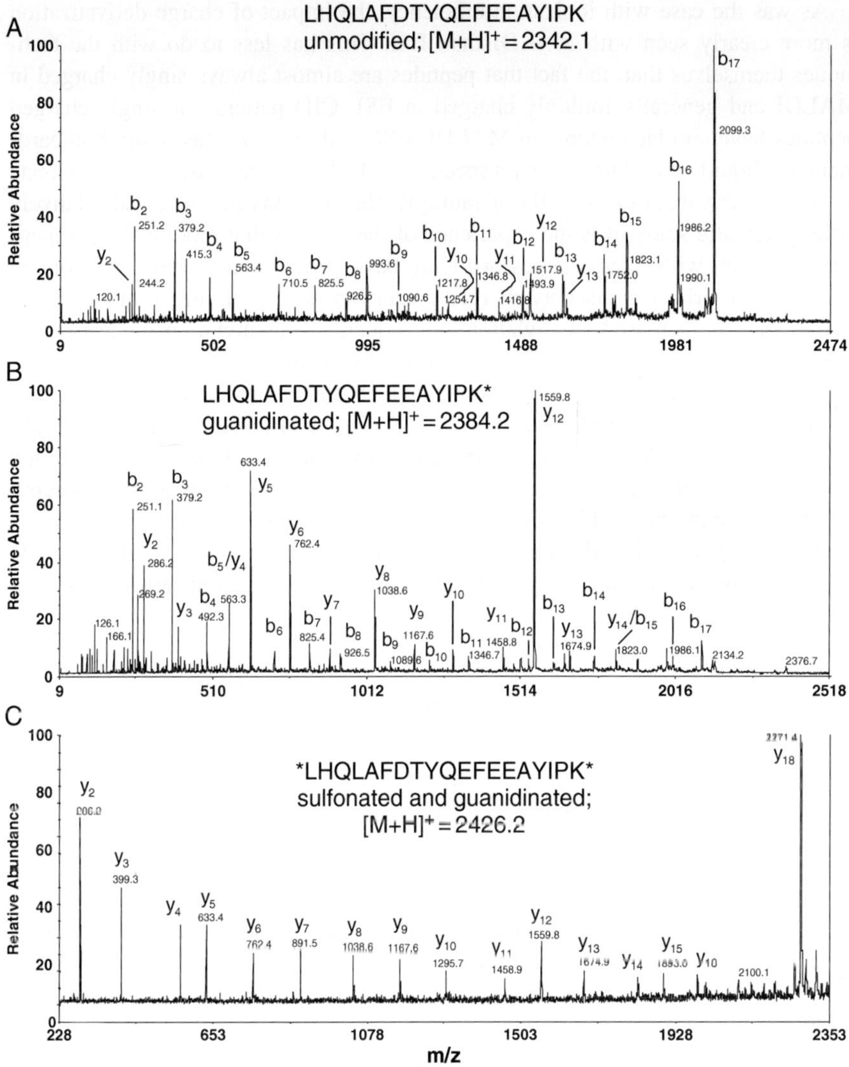

Fig. 2.2 The influence of charge-derivatization on the product ion series of peptide LHQLAFDTYQEFEEAYIPK acquired by MALDI-CID-TOF/TOF (4800; Applied Biosystems). (**a**) Unmodified peptide. (**b**) Guanidinated at C-terminal Lys. (**c**) Guanidinated at C-terminal Lys and sulfonated at the N-terminal amine

a CID spectrum devoid of b-ions and containing a nearly complete y-ion series, allowing the sequence to be read with relative ease (bottom panel). Interestingly, the most information-rich spectrum was obtained with guanidination alone, but the b-ion-suppressed spectrum resulting from guanidination and sulfonation was much more readily interpreted. Chemical modification strategies can therefore be tailored to the priorities of the scientist and the samples at hand.

As was the case with ionization efficiency, the impact of charge derivatization is more clearly seen with MALDI than ESI. This has less to do with the techniques themselves than the fact that peptides are almost always singly charged in MALDI and generally multiply charged in ESI. CID patterns of singly charged peptides have similar features in MALDI, ESI, and the older fast atom bombardment and liquid secondary ion mass spectra (Roth et al., 1998). Although the effects of charge derivatization on CID of multiply charged ions has received relatively little systematic study, it is the experience of the authors that modification of peptide N-termini with fixed charges or highly basic groups can moderately promote b- and a- ion series from doubly charged precursors, with declining influence as the charge state is increased. Acetylation and reductive dimethylation modestly suppress N-terminal ion series, and sulfonation does so somewhat more emphatically. In Fig. 2.3, for example, the same peptide which underwent such striking changes upon modification in MALDI-TOF/TOF MS (Fig. 2.2) undergoes relatively subtle changes in ESI MS/MS from unmodified (top panel) to Lys-guanidinated and N-terminally sulfonated (bottom panel). In the lower mass regions of these spectra, which were both obtained by resonant excitation of doubly charged precursors in a linear ion trap, there is indeed a shift from b- and y-ions in the unmodified peptide to y-ions alone in the modified peptide, but at higher masses, both series are present

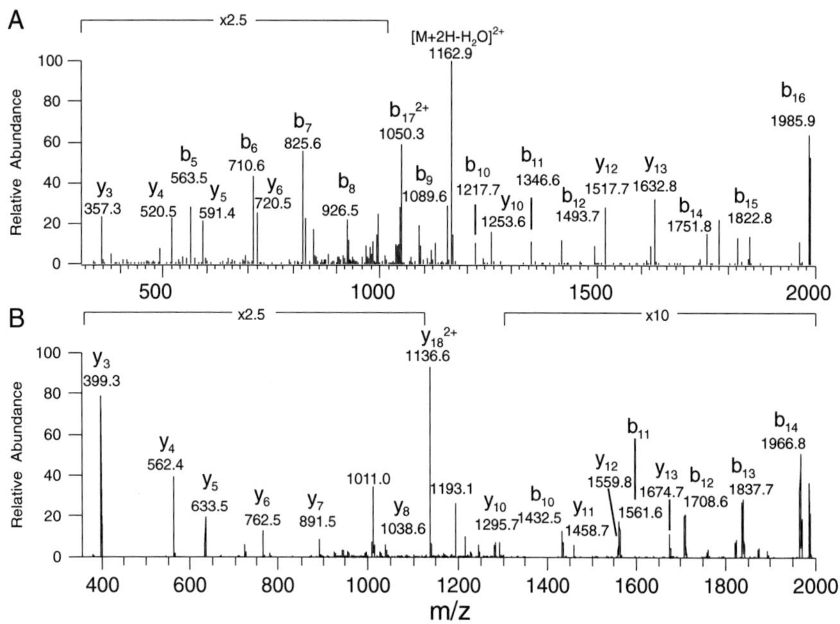

Fig. 2.3 The influence of charge-derivatization on the product ion series of the doubly charged peptide LHQLAFDTYQEFEEAYIPK acquired by ion trap ESI-CID-MS/MS (LTQ; Thermo Fisher). (**a**) Unmodified peptide. (**b**) Guanidinated at C-terminal Lys and sulfonated at the N-terminal amine

and indeed b-ions account for some of the most prominent peaks. This behavior can be attributed to the mechanism of low energy CID, which invokes a "mobile proton" required for charge-induced bond cleavage (Lee et al., 2004; McCormack et al., 1993; Wang et al., 2004). The more highly charged a peptide, the less influence any single site of charge localization can exert. This mechanism also explains some of the differences between charge-fixing modifications like the quaternary amine or phosphonium-conferring reagents (which cannot liberate a mobile proton) and those that localize charge with a basic but protonated functionality. A peptide whose only charge is provided by the former should only undergo charge-remote fragmentation, usually requiring high energy CID for production of informative sequence-specific fragments. In the latter case and for all peptides whose charge is contributed to by protonation in addition to charge-fixing modifications, the typical features of low energy CID should apply. Caution must be exercised in applying simplistic generalizations, however. Singly charged trimethylphosphonium derivatives, for example, produce primarily a-ions, apparently through charge remote mechanisms even in low energy CID (Sadagopan and Watson, 2000), and the "fixed charge" trimethylammonium containing tags can generate a "mobile proton" from dissociation of the tag itself, thus enabling both charge-remote and charge-induced fragmentation (He and Reilly, 2008). Therefore, it seems that derivatization of peptides for CID in ESI is of less utility for directing fragmentation than in MALDI, but can be useful in particular cases, and may be performed for other reasons such as quantitation via introduction of isotope-coded tags (Corthals and Rose, 2007). It should be noted that trypsin has long been the enzyme of choice in proteomic sample preparation, and that the vast majority of peptides analyzed therefore contain a basic groups at their N- and C-termini. The advantages of fragmentation-directing chemical modifications might well be more apparent in non-tryptic peptides.

If the case for charge derivatization in ESI-CID-MS/MS is less than compelling, the outlook is brighter for other activation methods such as electron capture dissociation (ECD) or electron transfer dissociation (ETD). In contrast to CID, particularly low energy CID, these methods are thought to involve non-ergodic, charge remote fragmentation pathways and produce product ion spectra with distinct features such as the preservation of labile modifications and an "even" distribution of sequence specific products, primarily c- and z-ions (Syka et al., 2004; Zubarev et al., 2000). Also in contrast to CID, these techniques perform best with highly charged precursors (Swaney et al., 2008). Under these circumstances, charge derivatization of peptides can be beneficial, both for increasing the peptide's charge state, and simplifying the product ion spectra. Digesting proteins with endoproteinase Lys-N, which places two basic groups (N-terminal amine and lysine ε-amine) at the peptide's N-terminus, has been shown to result in ETD product ion spectra dominated by c-ions, facilitating interpretation (Taouatas et al., 2008). The same effect can be achieved by chemical modification of peptide N-termini with a phosphonium group, and this strategy has been taken to increase the sequence coverage for O-phosphorylated and O-glycosylated peptides (Chamot-Rooke et al., 2007). Likewise, charge derivatives of carboxylic acids including the peptide C-terminus can enhance the z-ion series and improve the spectral quality of peptides that are acidic or would not otherwise bear a charge at their C-terminus (English et al., 2009). N-terminal sulfonation serves to suppress N-terminal ions just as in CID, but in favor of z-ions rather than y-ions. This approach has been taken with a hybrid activation method combining ETD with supplemental CID and found useful for sequencing proline-rich sequences, nontryptic peptides, and phosphorylation site mapping (Madsen and Brodbelt, 2009). Interestingly, the dominance of z-type ions was observed even when a strongly basic residue like lysine or arginine was not present at the peptide C-terminus.

A fortuitous benefit of N-terminal peptide modification is the appearance of product ions that establish the peptide's N-terminal amino acid. Under ordinary circumstances, CID spectra of unmodified peptides rarely contain the b_1, a_1, or y^{n-1} ion, although the b_2 and a_2 ions are frequently prominent. Consequently, the order (and sometimes identity) of the first two residues of the peptide cannot be directly established. This phenomenon has been attributed to various causes, including the stability of the b_2 ion, and a mechanism of peptide fragmentation that invokes attack by the carbonyl group of the N-terminally adjacent residue (Arnott et al., 1994; Yalcin et al., 1995). Whatever the case, it is clear that many of the N-terminal derivatives can produce these b_1 or a_1 ions upon low energy CID. Phenylisothiocyanate (Brancia et al., 2001) and a series of other isothiocyanates (Wang et al., 2009) have been shown to produce prominent b_1 ions; likewise acetylation and other modifications that form a new amide bond (Ross et al., 2004). An interesting contrast is found in the reaction of divinylsulfone with peptide N-termini to form a thiomorpholine 1,1-dioxide, which upon CID preferentially produces a_1 ions (Boja et al., 2004). Finally, upon reductive dimethylation, the a_1 ions in peptide CID are frequently the base peaks of the spectra. Besides being useful for de novo sequence interpretation, the ability to easily define a peptide's N-terminal amino acid can restrict the results of automated database searching by a factor of 20, improving the specificity of such searches (Brancia et al., 2001).

2.4 Modifications That Enable Recognition or Selective Detection of Particular Amino Acids

In addition to the wholesale direction of peptide dissociation patterns, certain modifications can change the fragmentation behavior associated with specific amino acids. One of the oldest motivations for chemically modifying peptides is to disambiguate lysine and glutamine (Hunt et al., 1986). These amino acids have the same nominal masses and are not easily distinguished without high resolution mass spectrometers. Acetylation, or indeed, most labels of free amines will produce a mass shift for product ion series that contain lysine along with the peptide N-terminus, allowing the number of lysines in the peptide to be counted directly as well as giving lysine and glutamine distinct residue masses. Likewise, methyl (or other) esterification of carboxylic acids yields a count of a peptide's aspartic and glutamic acid residues and imparts a mass shift to all product ions containing the C-terminus. These tricks, and the related tactic of enzymatically incorporating ^{18}O

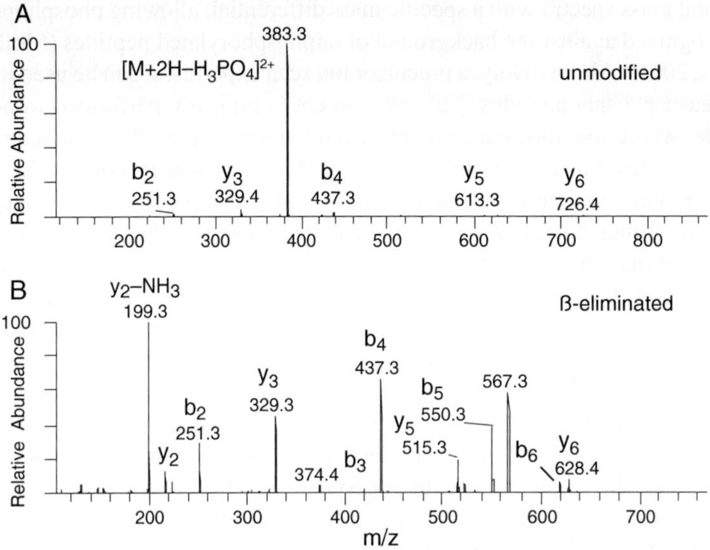

Fig. 2.4 Low energy CID of the phosphopeptide HLADLpSK acquired in a linear ion trap. (**a**) Before and (**b**) after β-elimination of the phosphate by treatment with 20 mM $Ba(OH)_2$ for 60 min. at 60°C

into the peptide's C-terminal carboxylate to mark all product ions containing the C-terminus (Takao et al., 1991), are all valuable tools for facilitating de novo sequence interpretation (Hunt et al., 1981).

In addition to their ionization properties described previously, phosphopeptides present unique challenges and opportunities for tandem mass spectrometry. Whereas phosphotyrosine-containing peptides fragment with broadly similar results to their unphosphorylated counterparts, peptides containing phosphoserine and phosphothreonine are often dominated by peaks resulting from the neutral loss of phosphoric acid from their precursors (DeGnore and Qin, 1998). This phenomenon is attributable to the lability of the phosphoester bond and the stepwise activation of peptides through multiple low energy collisions. In Fig. 2.4a, for example, most of the ion current in the CID spectrum of the doubly charged peptide HLADLpSK, obtained by low energy resonant excitation in a linear ion trap, is carried by this neutral loss; no b- or y- series product ion exceeds 4% of the base peak's abundance. The lack or low abundance of sequence-specific product ions poses a challenge for both manual interpretation and computational techniques for peptide identification and phosphorylation site localization. A simple way to rectify the situation is to remove the phosphate by β-elimination. In the example given, conversion of the phosphoserine to dehydroalanine yields a peptide whose product ion spectrum is much more readily interpreted (Fig. 2.4b).

Following β-elimination of phosphate with Michael addition of appropriate nucleophiles can be used to achieve further benefits. If two nucleophiles are incorporated in the reaction, each phosphopeptide will appear as a pair of peaks in the subsequent mass spectra with a specific mass differential, allowing phosphopeptides to be recognized against the background of unphosphorylated peptides (Molloy and Andrews, 2001). Alternatively, a precursor ion scan experiment can be used to selectively detect phosphopeptides. This experiment is typically performed in negative ion mode, where the appearance of phosphate anion at m/z 79 is characteristic of phosphopeptides, but peptide CID is preferably performed in positive ion mode, requiring either a second experiment (Annan et al., 2001) or sacrifice of duty-cycle time by switching polarities repeatedly within an experiment (Williamson et al., 2006). β-elimination and Michael addition provide an alternative, whereby the negatively charged phosphate group is replaced with a 2-dimethylaminoethanesulfoxide group that produces a strong sulfenic acid derivative at m/z 122 upon CID in positive ion mode. This ion can thus be the target of a precursor ion scan that is followed by data-dependent product ion scans in a single experiment (Steen and Mann, 2002).

The strategy of using characteristic product ions – typically immonium ions – to identify or selectively detect peptides containing a specific amino acid is not limited to phosphoserine and phosphothreonine. Several of the naturally occurring amino acids can be targeted in this way; proline, histidine, tyrosine and phenylalanine produce particularly strong immonium ions, for example. Other amino acids that are not as readily targeted can be chemically modified so as to produce abundant, specific immonium ions. One such example is cysteine, which when converted to pyridylethylcysteine (peCys) by reaction with 4-vinylpyridine produces a prominent immonium ion at m/z 106.1 (Moritz et al., 1996). The selective detection of cysteine-containing peptides can be valuable for several reasons. Important families of proteins such as cytokines often have evolutionary conservation of their cysteines, even when there is little other sequence identity among family members. Cysteine patterns are also conserved in monoclonal antibodies, and mark the borders of four of the six hypervariable sequence regions that confer the antibody's specificity for its antigen, making detection of cysteine-containing peptides particularly desirable when sequencing antibodies or attempting to identify a specific monoclonal antibody (Johnson and Wu, 2000). The selective detection of peCys-containing peptides in a tryptic digest of a monoclonal antibody is illustrated in Fig. 2.5. The top panel is the total ion chromatogram of full mass range scans acquired during LC-MS on a triple quadrupole-linear ion trap hybrid mass spectrometer (QTRAP 4000; Applied Biosystems). The second panel represents a precursor ion scan, preformed in the same experiment, revealing peptides that fragment to produce the peCys immonium ion at m/z 106.1. The product ion of one of the peptides revealed by this scan (triggered automatically by a peak in the precursor ion scan) is displayed in the bottom panel. Every prominent peak in the precursor ion chromatogram corresponds to a cysteine-containing peptide, and ten of the 16 cysteines in the antibody sequence were found among these peptides (Table 2.1). A similar strategy has been employed to detect methionine-containing peptides. S-alkylation with phenacylbromide at low pH is selective for methionine and produces a phenacyl sulfonium fixed charge derivative that fragments readily under low energy CID with a characteristic and very prominent neutral loss. This modification was employed in a neutral loss scan experiment to selectively detect methionine-containing peptides from proteolytic digests, and relative quantitation of peptides between samples was demonstrated through the use of a deuterated analog of the reagent (Reid et al.,

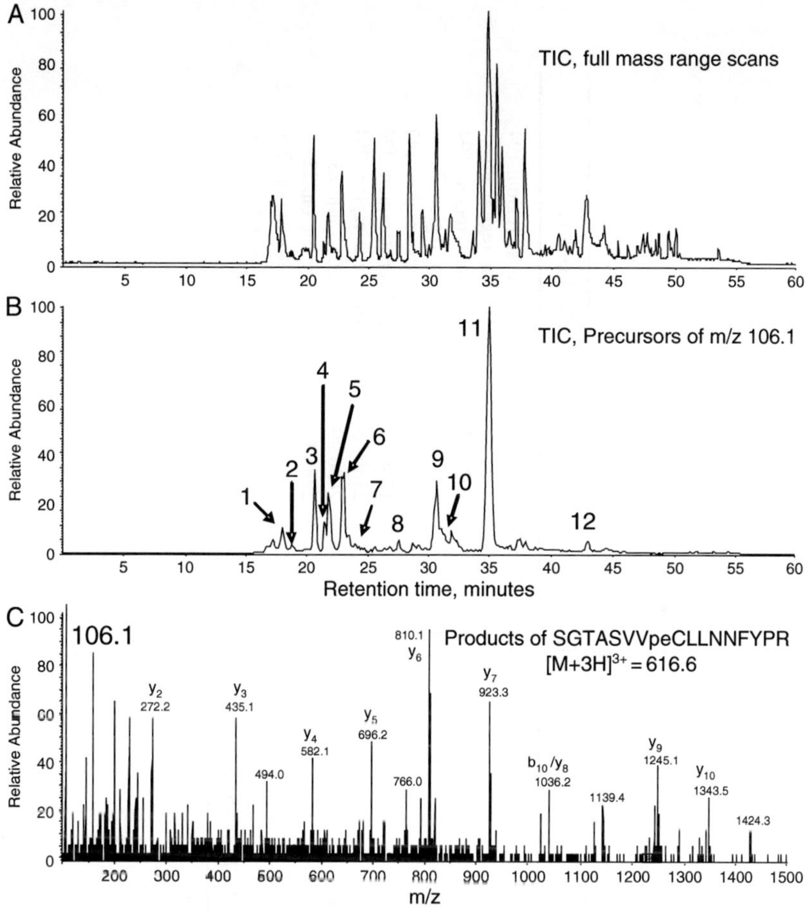

Fig. 2.5 Selective detection of cysteine-containing peptides in the total trypsin digest of a monoclonal antibody following reduction with dithiothreitol and alkylation with 4-vinylpyridine. (**a**) Total ion chromatogram of full mass range scans acquired in a linear ion trap (QTRAP4000, Applied Biosystems). (**b**) Total ion chromatogram for precursor ion scans of m/z 106.1. Numbered peaks correspond to Cys-containing tryptic peptides listed in Table 2.1. (**c**) Product ion spectrum of the peptide corresponding to peak 12 triggered by the precursor ion scan and acquired in the linear ion trap. Note the prominent peCys immonium ion at m/z 106.1

2005). The branched peptide chains generated by tryptic digestion of ubiquitylated proteins can also be recognized by a characteristic neutral loss from sulfonated peptides in both MALDI and ESI MS/MS (Wang and Cotter, 2005; Wang et al., 2006).

2.5 Conclusions

The mass spectrometrist possesses a rich set of chemical modifications, some of which are summarized in Table 2.2, which can be applied to achieve specific goals in peptide and protein analysis. Ionization efficiency can be selectively enhanced or

Table 2.1 Cysteine mapping by precursor ion scan for pyridylethyCys

#	Peptide sequence	MH^+ obs.	MH^+ calc.	Cys location:residue
1	HKVYACEVTHQGLSSPVTK	2188.1	2188.109	LC const. region: 194
2	AEDTAVYYCSR	1381.7	1381.597	HC var. region: 96
3	VYACEVTHQGLSSPVTK	1922.8	1922.956	LC const. region: 194
4	LSCAASGFNIK	1214.7	1214.611	HC var. region: 22
5	STSGGTAALGCLVK	1368.6	1368.707	HC const. region: 147
6	NVQSLTCLVK	1208.5	1208.658	HC const. region: 370
7	TPEVTCVVVDVSHEDPEVK	2186.0	2186.056	HC const. region: 264
8	DIQMTQSPSSLSASVGDRVTITCR	2655.9	2656.294	LC var. region: 23
9	LSCAASGFNIKDTYIHWVR	2284.9	2285.141	HC var. region: 22
10	THTCPPCPAPELLGGPSVFLFPPKPK	2941.2	2939.523	HC const. region: 229, 232
11	SGTASVVCLLNNFYPR	1844.7	1844.924	LC const. region: 134
12	TVAAPSVFIFPPSDEQLKSGTASVVCLLNNFYPR	3772.6	3771.933	LC const. region: 134

Table 2.2 Selected chemical derivatizations

Name	Reactive groups	Reaction	Reaction conditions, briefly	References
Acetylation	N-terminus, lysine ε-amine	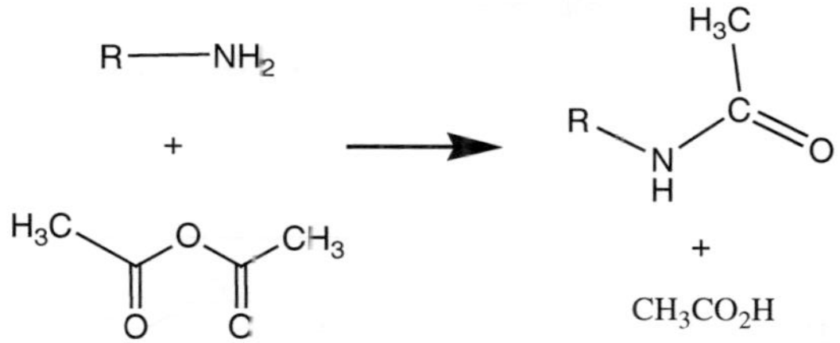	Peptides in 200 mM NH_4HCO_3 pH 8.5, 1% v/v acetic anhydride. 30 min at 20°C. Use NH_4OAc pH 6 for n-terminal selectivity	Che and Fricker, (2002)
iTRAQ	N-terminus, lysine ε-amine	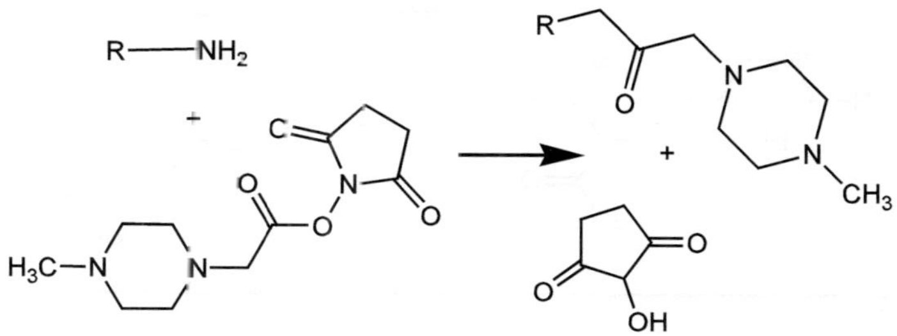	Peptides in 250 mM Et_3NHHCO_3, 75% v/v EtOH, 1% w/v iTRAQ reagent. 30 min at 20°C	Ross et al., (2004)

Table 2.2 (continued)

Name	Reactive groups	Reaction	Reaction conditions, briefly	References
Guanidination	Lysine ε-amine; N-terminal glycine	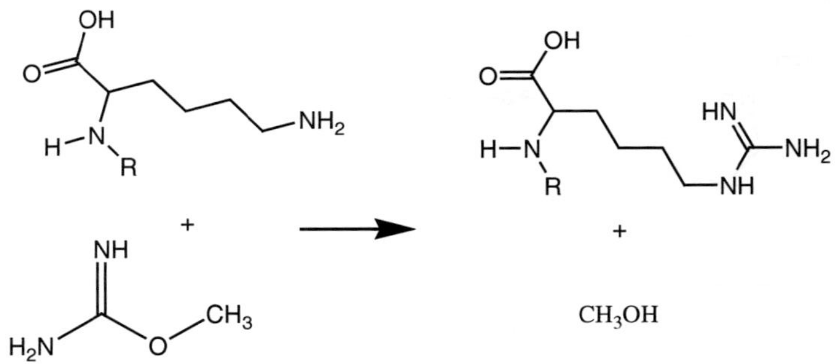	Peptides in 250 mM Na_2CO_3 pH 11, 0.5 M O-methylisourea. 30 min at 65°C	Beardsley and Reilly, (2002) and Keough et al. (2002)
Sulfonation	N-terminus, lysine ε-amine	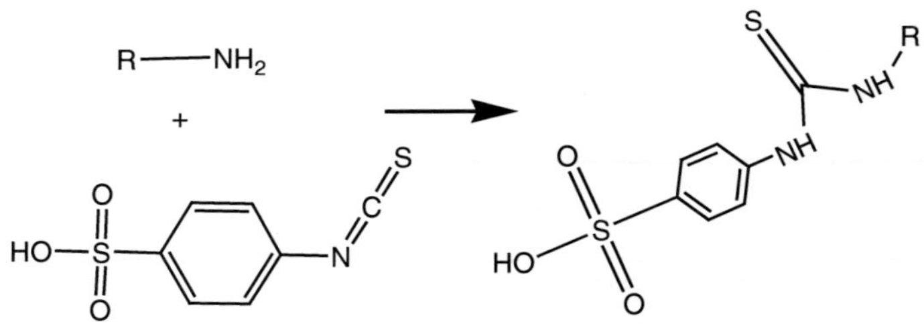	Peptides in 20 mM $NaHCO_3$ pH 9.5, 10 μg/μL sulfophenyl isothiocyanate. 30 min at 55°C	Wang et al. (2004)

Table 2.2 (continued)

Name	Reactive groups	Reaction	Reaction conditions, briefly	References
Reductive methylation	N-terminus, lysine ε-amine	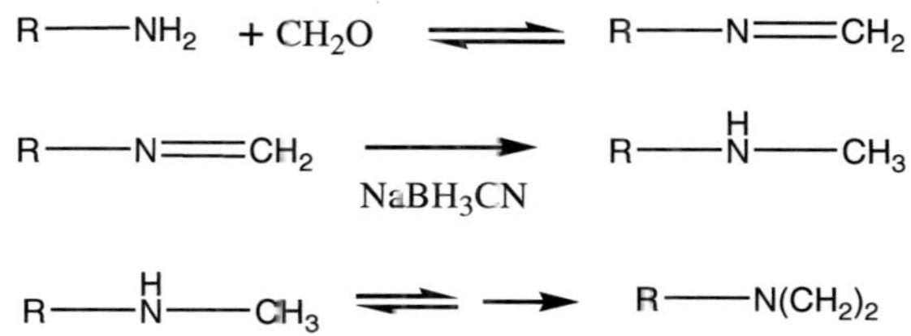	Peptides in 50 mM HEPES pH 7, 20 mM formaldehyde, 10 mM $NaBH_3CN$. 2 h at 20°C.	Hsu et al., (2005; Jentoft and Dearborn, (1983)
Methyl esterification	Aspartic and glutamic acids, C-terminus	$R\text{-}CO_2H \xrightarrow[HCl]{M_eOH} R\text{-}CO_2M_e$	Prepare methanolic HCl by adding 15 µL acetyl chloride to 85 µL anhydrous MeOH. Add to dried peptides. 4 h at 37°C.	(Kim et al., 2005)
β-elimination of phosphate	pSer, pThr, O-linked sugars	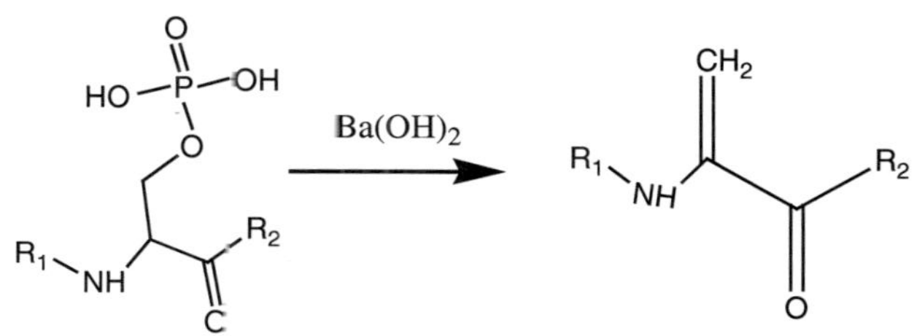	Peptides in 20 mM $Ba(OH)_2$. 1 h at 60°C.	Klemm et al. (2004)

Table 2.2 (continued)

Name	Reactive groups	Reaction	Reaction conditions, briefly	References
β-elimination/ Michael Addition	pSer, pThr, O-linked sugars	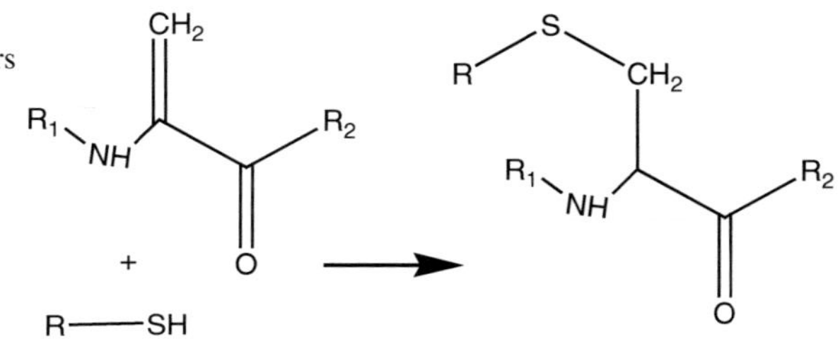	Peptides in 20 mM $Ba(OH)_2$, 50 mM 1-propanethiol, 42% v/v EtOH. 1 h at 60°C.	Klemm et al. (2004)

suppressed in MALDI MS, and the charge state distribution altered in ESI. Addition of fixed or localized charges can enhance or suppress entire ion series upon CID, with dramatic effects in MALDI, and more subtle changes in ESI. Newer activation techniques such as ECD and ETD can be favorably combined with ESI and charge derivatization to simplify product ion spectra and map sites of posttranslational modification. Peptides containing specific amino acids and PTMs can be targeted for selective detection or recognized amid a complex background of other peptides by virtue of characteristic product ions or neutral losses conferred by a well-chosen chemical modification. Problematic classes of peptides that might otherwise be difficult to detect, identify, sequence, or otherwise characterize by mass spectrometry are often excellent candidates for chemical manipulation.

References

Annan, R.S., Huddleston, M.J., Verma, R., Deshaies, R.J., and Carr, S.A. (2001). A multidimensional electrospray MS-based approach to phosphopeptide mapping. Anal Chem *73*, 393–404.

Arnott, D., Kottmeier, D., Yates, N., Shabanowitz, J., and Hunt, D.F. (1994). Fragmentation of Multiply Protonated Peptides Under Low Energy Conditions. Paper presented at: Proceedings of the 42nd ASMS Conference on Mass Spectrometry and Allied Topics (Chicago, IL).

Bartlet-Jones, M., Jeffery, W.A., Hansen, H.F., and Pappin, D.J. (1994). Peptide ladder sequencing by mass spectrometry using a novel, volatile degradation reagent. Rapid Commun Mass Spectrom *8*, 737–742.

Beardsley, R.L., Karty, J.A., and Reilly, J.P. (2000). Enhancing the intensities of lysine-terminated tryptic peptide ions in matrix-assisted laser desorption/ionization mass spectrometry. Rapid Commun Mass Spectrom *14*, 2147–2153.

Beardsley, R.L., and Reilly, J.P. (2002). Optimization of guanidination procedures for MALDI mass mapping. Anal Chem *74*, 1884–1890.

Biemann, K., and Martin, S.A. (1987). Mass spectrometric determination of the amino acid sequence of peptides and proteins. Mass Spectrom Rev *6*, 1–75.

Boja, E.S., Sokoloski, E.A., and Fales, H.M. (2004). Divinyl sulfone as a postdigestion modifier for enhancing the a(1) Ion in MS/MS and postsource decay: potential applications in proteomics. Anal Chem *76*, 3958–3970.

Brancia, F.L., Butt, A., Beynon, R.J., Hubbard, S.J., Gaskell, S.J., and Oliver, S.G. (2001). A combination of chemical derivatisation and improved bioinformatic tools optimises protein identification for proteomics. Electrophoresis *22*, 552–559.

Brancia, F.L., Oliver, S.G., and Gaskell, S.J. (2000). Improved matrix-assisted laser desorption/ionization mass spectrometric analysis of tryptic hydrolysates of proteins following guanidination of lysine-containing peptides. Rapid Commun Mass Spectrom *14*, 2070–2073.

Chamot-Rooke, J., van der Rest, G., Dalleu, A., Bay, S., and Lemoine, J. (2007). The combination of electron capture dissociation and fixed charge derivatization increases sequence coverage for O-glycosylated and O-phosphorylated peptides. J Am Soc Mass Spectrom *18*, 1405–1413.

Che, F.Y., and Fricker, L.D. (2002). Quantitation of neuropeptides in Cpe(fat)/Cpe(fat) mice using differential isotopic tags and mass spectrometry. Anal Chem *74*, 3190–3198.

Cleland, W.W. (1964). Dithiothreitol, a new protective reagent for Sh groups. Biochemistry *3*, 480–482.

Corthals, G.L., and Rose, K. (2007). Quantitation in Proteomics. In Proteome Research: Concepts, Technology and Application, M.R. Wilkins, R.D. Appel, K.L. Williams, and D.F. Hochstrasser, eds. (Berlin, Heidelberg, Springer), pp. 69–93.

DeGnore, J.P., and Qin, J. (1998). Fragmentation of phosphopeptides in an ion trap mass spectrometer. J Am Soc Mass Spectrom *9*, 1175–1188.

Dickens, F. (1933). Interaction of halogenacetates and SH compounds: The reaction of halogenacetic acids with glutathione and cysteine. The mechanism of iodoacetate poisoning of glyoxalase. Biochem J *27*, 1141–1151.

English, M., Udeshi, N., Shabanowitz, J., and Hunt, D.F. (2009). Evaluation of ETD Fragmentation-Enhancing Peptide Charge Modification Strategies Amenable to Complex Samples and Direct Use With HPLC-MS. Paper presented at: Proceedings of the 57th ASMS Conference on Mass Spectrometry and Allied Topics (Philadelphia, PA).

Ernoult, E., Gamelin, E., and Guette, C. (2008). Improved proteome coverage by using iTRAQ labelling and peptide OFFGEL fractionation. Proteome Science *6*, 27–39.

Gao, Y., and Wang, Y. (2007). A method to determine the ionization efficiency change of peptides caused by phosphorylation. J Am Soc Mass Spectrom *18*, 1973–1976.

Good, D.M., Wirtala, M., McAlister, G.C., and Coon, J.J. (2007). Performance characteristics of electron transfer dissociation mass spectrometry. Mol Cell Proteomics *6*, 1942–1951.

Guevremont, R., Siu, K.W.M., Le Blanc, J.C.Y., and Berman, S.S. (1992). Are the electrospray mass spectra of proteins related to their aqueous solution chemistry? J Am Soc Mass Spectrom *3*, 216–224.

Hale, J.E., Butler, J.P., Knierman, M.D., and Becker, G.W. (2000). Increased sensitivity of tryptic peptide detection by MALDI-TOF mass spectrometry is achieved by conversion of lysine to homoarginine. Anal Biochem *287*, 110–117.

He, Y., and Reilly, J.P. (2008). Does a charge tag really provide a fixed charge? Angew Chem Int Ed Engl *47*, 2463–2465.

Hsu, J.L., Huang, S.Y., Shiea, J.T., Huang, W.Y., and Chen, S.H. (2005). Beyond quantitative proteomics: signal enhancement of the a1 ion as a mass tag for peptide sequencing using dimethyl labeling. J Proteome Res *4*, 101–108.

Hunt, D.F., Bone, W.M., Shabanowitz, J., Rhodes, J., and Ballard, J.M. (1981). Sequence analysis of oligopeptides by secondary ion/collision activated dissociation mass spectrometry. Anal Chem *53*, 1704–1706.

Hunt, D.F., Yates, J.R., 3rd, Shabanowitz, J., Winston, S., and Hauer, C.R. (1986). Protein sequencing by tandem mass spectrometry. Proc Natl Acad Sci USA *83*, 6233–6237.

Jaffe, H., Veeranna, and Pant, H.C. (1998). Characterization of serine and threonine phosphorylation sites in beta-elimination/ethanethiol addition-modified proteins by electrospray tandem mass spectrometry and database searching. Biochemistry *37*, 16211–16224.

Janek, K., Wenschuh, H., Bienert, M., and Krause, E. (2001). Phosphopeptide analysis by positive and negative ion matrix-assisted laser desorption/ionization mass spectrometry. Rapid Commun Mass Spectrom *15*, 1593–1599.

Jentoft, N., and Dearborn, D.G. (1983). Protein labeling by reductive alkylation. Methods Enzymol *91*, 570–579.

Johnson, G., and Wu, T.T. (2000). Kabat Database and its applications: 30 years after the first variability plot. Nucleic Acids Res *28*, 214–218.

Keough, T., Lacey, M.P., and Youngquist, R.S. (2002). Solid-phase derivatization of tryptic peptides for rapid protein identification by matrix-assisted laser desorption/ionization mass spectrometry. Rapid Commun Mass Spectrom *16*, 1003–1015.

Keough, T., Youngquist, R.S., and Lacey, M.P. (1999). A method for high-sensitivity peptide sequencing using postsource decay matrix-assisted laser desorption ionization mass spectrometry. Proc Natl Acad Sci USA *96*, 7131–7136.

Kim, T.Y., Brun, Y.V., and Reilly, J.P. (2005). Effects of tryptic peptide esterification in MALDI mass spectrometry. Anal Chem *77*, 4185–4193.

Kim, J.S., Cui, E., and Kim, H.J. (2009). Picolinamidination of phosphopeptides for MALDI-TOF-TOF mass spectrometric sequencing with enhanced sensitivity. J Am Soc Mass Spectrom *20*, 1751–1758.

Kinter, M., and Sherman, N.E. (2000). Protein Sequencing and Identification Using Tandem Mass Spectrometry (New York, NY, Wiley).

Klemm, C., Schroder, S., Gluckmann, M., Beyermann, M., and Krause, E. (2004). Derivatization of phosphorylated peptides with S- and N-nucleophiles for enhanced ionization efficiency in matrix-assisted laser desorption/ionization mass spectrometry. Rapid Commun Mass Spectrom *18*, 2697–2705.

Krause, E., Wenschuh, H., and Jungblut, P.R. (1999). The dominance of arginine-containing peptides in MALDI-derived tryptic mass fingerprints of proteins. Anal Chem *71*, 4160–4165.

Krishnamurthy, T., Szafraniec, L., Hunt, D.F., Shabanowitz, J., Yates, J.R., 3rd, Hauer, C.R., Carmichael, W.W., Skulberg, O., Codd, G.A., and Missler, S. (1989). Structural characterization of toxic cyclic peptides from blue-green algae by tandem mass spectrometry. Proc Natl Acad Sci USA *86*, 770–774.

Lamari, F.N., Kuhn, R., and Karamanos, N.K. (2003). Derivatization of carbohydrates for chromatographic, electrophoretic and mass spectrometric structure analysis. J Chromatogr B *793*, 15–36.

Lee, Y.H., Han, H., Chang, S.B., and Lee, S.W. (2004). Isotope-coded N-terminal sulfonation of peptides allows quantitative proteomic analysis with increased de novo peptide sequencing capability. Rapid Commun Mass Spectrom *18*, 3019–3027.

Loo, J.A., Udseth, H.R., and Smith, R.D. (1989). Peptide and protein analysis by electrospray ionization-mass spectrometry and capillary electrophoresis-mass spectrometry. Anal Biochem *179*, 404–412.

Madsen, J.A., and Brodbelt, J.S. (2009). Simplifying fragmentation patterns of multiply charged peptides by N-terminal derivatization and electron transfer collision activated dissociation. Anal Chem *81*, 3645–3653.

McCormack, A.L., Somogyi, A., Dongre, A., and Wysocki, V. (1993). Fragmentation of protonated peptides: Surface-induced dissociation in conjunction with a quantum mechanical approach. Anal Chem *65*, 2859–2872.

Mirzaei, H., and Regnier, F. (2006). Enhancing electrospray ionization efficiency of peptides by derivatization. Anal Chem *78*, 4175–4183.

Molloy, M.P., and Andrews, P.C. (2001). Phosphopeptide derivatization signatures to identify serine and threonine phosphorylated peptides by mass spectrometry. Anal Chem *73*, 5387–5394.

Moritz, R.L., Eddes, J.S., Reid, G.E., and Simpson, R.J. (1996). S-pyridylethylation of intact polyacrylamide gels and in situ digestion of electrophoretically separated proteins: A rapid mass spectrometric method for identifying cysteine-containing peptides. Electrophoresis *17*, 907–917.

Nishikaze, T., and Takayama, M. (2006). Cooperative effect of factors governing molecular ion yields in desorption/ionization mass spectrometry. Rapid Commun Mass Spectrom *20*, 376–382.

Reid, G.E., Roberts, K.D., Simpson, R.J., and O'Hair, R.A. (2005). Selective identification and quantitative analysis of methionine containing peptides by charge derivatization and tandem mass spectrometry. J Am Soc Mass Spectrom *16*, 1131–1150.

Ross, P.L., Huang, Y.N., Marchese, J.N., Williamson, B., Parker, K., Hattan, S., Khainovski, N., Pillai, S., Dey, S., Daniels, S., *et al.* (2004). Multiplexed Protein Quantitation in Saccharomyces cerevisiae using amine-reactive Isobaric tagging reagents. Mol Cell Proteomics *3*, 1154–1169.

Roth, K.D., Huang, Z.H., Sadagopan, N., and Watson, J.T. (1998). Charge derivatization of peptides for analysis by mass spectrometry. Mass Spectrom Rev *17*, 255–274.

Sadagopan, N., and Watson, J.T. (2000). Investigation of the tris(trimethoxyphenyl)phosphonium acetyl charged derivatives of peptides by electrospray ionization mass spectrometry and tandem mass spectrometry. J Am Soc Mass Spectrom *11*, 107–119.

Staudenmann, W., and James, P. (2001). Interpreting Peptide Tandem Mass-Spectrometry Fragmentation Spectra. In Proteome Research: Mass Spectrometry, P. James, ed. (Berlin, Springer), pp. 143–166.

Steen, H., and Mann, M. (2002). A new derivatization strategy for the analysis of phosphopeptides by precursor ion scanning in positive ion mode. J Am Soc Mass Spectrom *13*, 996–1003.

Stults, J.T., Lai, J., McCune, S., and Wetzel, R. (1993). Simplification of high-energy collision spectra of peptides by amino-terminal derivatization. Anal Chem *65*, 1703–1708.

Swaney, D.L., McAlister, G.C., and Coon, J.J. (2008). Decision tree-driven tandem mass spectrometry for shotgun proteomics. Nat Methods *5*, 959–964.

Syka, J.E., Coon, J.J., Schroeder, M.J., Shabanowitz, J., and Hunt, D.F. (2004). Peptide and protein sequence analysis by electron transfer dissociation mass spectrometry. Proc Natl Acad Sci USA *101*, 9528–9533.

Takao, T., Hori, H., Okamoto, K., Harada, A., Kamachi, M., and Shimonishi, Y. (1991). Facile assignment of sequence ions of a peptide labelled with 18O at the carboxyl terminus. Rapid Commun Mass Spectrom *5*, 312–315.

Taouatas, N., Drugan, M.M., Heck, A.J.R., and Mohammed, S. (2008). Straightforward ladder sequencing of peptides using a Lys-N metalloendopeptidase. Nat Methods *5*, 405–407.

Wagner, D.S., Salari, A., Gage, D.A., Leykam, J., Fetter, J., Hollingsworth, R., and Watson, J.T. (1991). Derivatization of peptides to enhance ionization efficiency and control fragmentation during analysis by fast atom bombardment tandem mass spectrometry. Biol Mass Spectrom *20*, 419–425.

Wang, D., and Cotter, R.J. (2005). Approach for determining protein ubiquitination sites by MALDI-TOF mass spectrometry. Anal Chem *77*, 1458–1466.

Wang, D., Fang, S., and Wohlhueter, R.M. (2009). N-terminal derivatization of peptides with isothiocyanate analogues promoting Edman-type cleavage and enhancing sensitivity in electrospray ionization tandem mass spectrometry analysis. Anal Chem *81*, 1893–1900.

Wang, D., Kalb, S.R., and Cotter, R.J. (2004). Improved procedures for N-terminal sulfonation of peptides for matrix-assisted laser desorption/ionization post-source decay peptide sequencing. Rapid Commun Mass Spectrom *18*, 96–102.

Wang, D., Kalume, D., Pickart, C., Pandey, A., and Cotter, R.J. (2006). Identification of protein ubiquitylation by electrospray ionization tandem mass spectrometric analysis of sulfonated tryptic peptides. Anal Chem *78*, 3681–3687.

Williamson, B.L., Marchese, J., and Morrice, N.A. (2006). Automated identification and quantification of protein phosphorylation sites by LC/MS on a hybrid triple quadrupole linear ion trap mass spectrometer. Mol Cell Proteomics *5*, 337–346.

Wong, S.S. (1991). Chemistry of Protein Conjugation and Cross-Linking (Boca Raton, FL, CRC Press).

Wu, Z., and Fenselau, C. (1992). Proton affinity of arginine measured by the kinetic approach. Rapid Commun Mass Spectrom *6*, 403–405.

Xu, Y., Zhang, L., Lu, H., and Yang, P. (2008). Mass spectrometry analysis of phosphopeptides after peptide carboxy group derivatization. Anal Chem *80*, 8324–8328.

Yalcin, T., Khouw, C., Csizmadia, I.G., Peterson, M.R., and Harrison, A.G. (1995). Why are B ions stable species in peptide spectra? J Am Soc Mass Spectrom *6*, 1165–1174.

Zubarev, R.A., Horn, D.M., Fridriksson, E.K., Kelleher, N.L., Kruger, N.A., Lewis, M.A., Carpenter, B.K., and McLafferty, F.W. (2000). Electron capture dissociation for structural characterization of multiply charged protein cations. Anal Chem *72*, 563–573.

Chapter 3
Sample Preparation for 2D Electrophoresis and CE-Based Proteomics

Judit M. Nagy, Alexandria Lipka, Fiona Pereira, Nicky Marlin, and Stuart Hassard

Abstract This chapter describes sample preparation for gel-based proteomics, capillary electrophoresis and capillary electrophoresis coupled with mass spectrometry. Two dimensional gel electrophoresis (2D-PAGE) remains one of the most commonly used separation techniques for complex protein mixtures. Stringent and consistent sample preparation is essential for reproducible and high quality 2D-PAGE analysis of complex protein mixtures. The goal of this step is to produce homogeneous sample, free from contaminants and with consistent chemical parameters (pH, salt concentration etc) that will allow the user to generate reproducible results. Capillary electrophoresis (CE) separates analytes from a complex mixture with high resolution based on differential migration through a liquid filled capillary in a strong electric field. CE is easily automated and lends itself well to microfluidic device technology. The evolution of these microfluidic devices allows the potential for the hyphenation of CE to other separation and identification methods like gel chromatography, high performance liquid chromatography (HPLC) and mass spectrometry.

Keywords Capillary electrophoresis · Consistent chemical parameters · Proteomics

3.1 Introduction

Two dimensional gel electrophoresis (2D-PAGE) remains one of the most commonly used separation techniques for complex protein mixtures. 2D-PAGE separates proteins according to charge (*pI*) by isoelectric focusing (IEF) in the first dimension and according to size (Mr) by SDS-PAGE in the second dimension. It has a unique capacity for resolving complex protein mixtures making the simultaneous analysis of hundreds or even thousands of gene products possible. With the introduction of immobilized pH gradients 2D-PAGE has become the workhorse of

---

S. Hassard (✉)
London BioScience Innovation Centre, 2 Royal College Street, London, NW1 ONH
e-mail: stuart@deltadot.com

A.R. Ivanov, A.V. Lazarev (eds.), *Sample Preparation in Biological Mass Spectrometry*, DOI 10.1007/978-94-007-0828-0_3,

protein separation and a quick method for obtaining protein expression patterns. Up to 10,000 distinct protein and peptide spots can be separated on one gel (Gorg et al., 2004; Klose and Kobalz, 1995). Never the less, 2D-PAGE has a bad reputation because it is limited to the visualization of those proteins which are soluble in the 2D-PAGE buffer and have a *pI* and molecular weight within the separation ranges of the 1st and 2nd dimension gels. 2D-PAGE is time consuming, does not lend itself to automation, and, as a result of this, reproducibility is poor. The dynamic range of current visualization methods is limited, which means that only the more abundant proteins can be seen, and the low abundant proteins cannot be identified. Further drawbacks are that 2D-PAGE is not quantitative due to the staining technique used for the visualisation of protein spots, and each step requires lengthy optimisation and user intervention. In addition, sample handling can easily introduce artefacts into the 2D-PAGE pattern, such as oxidation, side-chain modifications and spot elimination, if protein enrichment or fractionation is used. In spite of these limitations 2D-PAGE is still among the most commonly used separation methods because it is relatively cheap and gives visual results immediately. Currently there is no single method, which can provide qualitative and quantitative information of all protein components of a complex mixture. For the reasons discussed above, in our laboratories, we combined 2D-PAGE with capillary electrophoresis (CE) for quality control of protein mixtures prior to separation by 2D-PAGE and to use the complementary advantages of the two techniques. In both separation techniques it has been proven that, as long as the sample preparation is standardized and the running conditions optimized for the specific sample, the results are highly reproducible. The recently completed 2D-PAGE reproducibility study carried out by the Industrial Advisory Board (IAB) of the Human Proteome Organisation (HUPO) and several academic laboratories came to the conclusion, that using identical samples and following the standard operating procedures rigorously, the laboratories produced 2D images which fell into the 95% confidence level when compared to the master gel (http://www.fixingproteomics.org/).

3.2 Sample Preparation for Gel-Based Proteomics

3.2.1 Introduction

Stringent and consistent sample preparation is essential for reproducible and high quality 2D-PAGE analysis of complex protein mixtures. Considering the entire 2D-PAGE process, sample preparation is the most laborious and time-consuming step, and the most crucial as errors here effect all downstream processes, yet cannot be detected until the process is over and resources committed. Uniform sample preparation (as far as that is possible across the diverse spectrum of proteomics) is also the key to standardisation across the proteomics arena, an important and long-term aim of HUPO. The goal of this step is to produce homogeneous sample, free from contaminants and with consistent chemical parameters (pH, salt concentration etc) that will allow the user to generate reproducible results.

The dual purposes of sample preparation are to reduce (or completely remove) the number of contaminants in order to enrich for all desired proteins of interest and to present the sample to the separation media in an optimal state. The first step of sample preparation for 2D-PAGE requires cell lysis or tissue disruption followed by purification and solubilisation of the proteins. There is no one common protocol which can be applied to all samples when it comes to 2D-PAGE sample preparation because of the wide dynamic range of proteins in complex cell lysates (Dunn and Bradd, 1993). Optimal sample preparation for 2D-PAGE would be no intervention at all however this is very rarely an option. Thus, the methods employed must strive to be as unobtrusive as possible to avoid damage or irreversible protein cross-linking, or sample loss and yield proteins of interest at detectable levels, which may involve the removal of abundant proteins or non-relevant classes.

3.2.2 Sample Type

The manner in which a sample is prepared is also dependant on the target data set of any given experiment. Sample preparation will be dictated by the set of proteins the experiment is concerned with, for instance, the preparation of a pure total protein extract will require a different protocol from another one which involves enriching the sample for low abundant proteins. Similarly, sample preparations from urine or blood will vary significantly to those for bacterial cell lysates. It is also important to note that not all samples are suitable for 2D-PAGE because it is limited to the visualization of those proteins which are soluble in the 2D-PAGE buffers and have a *pI* and molecular weight within the separation ranges of the 1st and 2nd dimension gels.

3.2.3 Purification

Protein samples for 2D-PAGE must be free from contaminants such as cellular debris and DNA, and must not contain high salt concentrations or ionic detergents, such as sodium dodecyl sulfate (SDS). These contaminants will disturb the IEF stage and hinder the ability of the protein mixture to separate efficiently, causing spot streaking in the gel and compromising reproducibility (Weiss and Gorg, 2009). Assiduous sample preparation, in turn, ensures that sample interference in mass spectrometry is minimal and ultimately facilitates more efficient data collection.

Protein precipitation is routinely applied to remove various contaminants from biological samples (Englard and Seifter, 1990). Whether common solutions or commercially available kits are used, the methods employed select for proteins while any contaminants remain in solution. However, proteins exhibit marked different precipitation characteristics so while various classes of proteins may be selected for, this is not the technique to use if an entire protein profile is sought. A typical technique is trichloroacetic acid (TCA) precipitation, where TCA is added to a final concentration of 10–20% and the solution precipitated on ice for 30 min. Following centrifugation, the pellet is washed with either acetone or ethanol to remove residual

TCA. Other precipitation methods include variations of acetone precipitation, TCA in acetone precipitation (Cilia et al., 2009), ammonium sulfate precipitation (Jiang et al., 2004), ammonium acetate in methanol with phenol extraction (Rodrigues et al., 2009). There are also a wide range of commercial kits available such as; ProteoHook™ (Proteopure); ReadyPrep 2-D Cleanup Kit (Bio-Rad); ProteoPrep® Protein Precipitation Kit and ProteoPrep® Sample Extraction Kit (Sigma-Aldrich); and Qproteome FFPE Tissue 2D-PAGE Kit (Qiagen).

The presence of salt in samples decreases the efficiency of IEF focusing, as the salt ions result in high IPG strip conductivity. Their removal is therefore vital and can be achieved through a variety of techniques ranging from TCA precipitation, dialysis (either by tubing, microdialysers or spin columns; (Canas et al., 2007), desalting columns, fast protein liquid chromatography (FPLC, (Freeman and Hemby, 2004); gel filtration, solid phase extraction (SPE, reviewed in (Bodzon-Kulakowska et al., 2007) and high-throughput nickel and glutathione disks (Chambers, 2002). There are also commercial kits such as Princeton Centri-Spin columns, the Millipore Amicon range or Sartorius Vivapure columns. The removal of detergents, particularly SDS, which forms negatively charged complexes with proteins and results in severely weakened isoelectric focusing, can also be achieved with protein precipitation, dialysis, gel filtration chromatography, ion exchange chromatography, hydrophobic adsorption chromatography, ceramic hydroxyapatite chromatography (Dong et al., 1997), affinity bead chromatography, SPE (Hannam et al., 1998) and commercial kits such as the ProteoSpin™ Detergent Clean-Up Kit (Norgen Biotek) and the Detergent-OUT™ series by GBiosciences.

For the removal of nucleic acids from a sample, the most common method is precipitation with TCA, generally preceded by treatment with DNase and RNase and followed by ultracentrifugation (Gorg et al., 2004). However, these methods have now been found to be less efficient than a very simple two phase liquid-liquid extraction in chloroform/phenol/isoamyl alcohol performed at a pH value of 9.5 or higher, a procedure commonly used to eliminate protein contaminants from DNA samples (Antonioli et al., 2009).

Other contaminants such as polysaccharides can be removed with TCA, acetone, ammonium sulphate or phenol ammonium acetate precipitation and centrifugation and the commonly used 2-D Clean-Up Kit from GE Healthcare. Plant phenols may interact with proteins, causing horizontal gel streaking and can be removed by TCA precipitation and extraction with ice-cold acetone or by binding to polyvinyl polypyrrolidone (Granier, 1988).

The removal of lipids is very important when working with plasma samples and can be done by TCA/acetone precipitation or precipitation using acetonitrile with the addition of 1% trifluoroacetic acid (TFA).

High abundant proteins generally hinder the separation and the detection of lower abundance proteins. For example, albumin, which constitutes up to 60% of the total protein content of plasma (Pieper et al., 2003), can be a problem because it masks other proteins with similar isoelectric points and/or molecular weights. Although there are several albumin removal kits available (ProteoSeek™ Albumin/IgG Removal Kit (Thermo-Pierce); ProteoSpin™ Abundant Serum Protein Depletion Kit (Norgen Biotek) there is also the potential to remove proteins other than albumin when using them. For the purification of low abundant proteins, delipidation by centrifugation, IgG removal with Protein Sepharose and human serum albumin (HSA) depletion with sodium chloride/ethanol precipitation have been found to be an efficient and reproducible method (Fountoulakis et al., 2004; Fu et al., 2007).

Any combination of precipitation protocols can be used to remove salts, detergents, nucleic acids, lipids, polysaccharides and various other molecules from biological samples, and each sample preparation regime will encompass a range of the techniques described.

3.2.4 Solubilisation

Complex protein mixtures for 2D-PAGE commonly contain proteins, that in their native state, form aggregates or associate with other molecules forming insoluble complexes. Samples for 2D-PAGE must be completely soluble, free from aggregates and complexes, denatured and reduced to produce gels free of streaks and blurred spots (Shaw and Riederer, 2003). Solubilisation techniques for protein samples have been well documented, although these techniques need to be compatible with 2D-PAGE. In order to solubilise samples for 2D-PAGE, it is necessary to break any secondary structure bonds, prevent aggregation without compromising the protein's integrity, and keep the sample in a soluble state long enough for analysis. Solubilisation techniques may be helped by agitation or ultracentrifugation, but temperature shifts should be avoided in order to maintain constant and reproducible results (Gorg et al., 1991).

Sample solubilisation can broadly be described as the prevention of disulfide bonds and non-covalent interactions, and is usually completed in a buffer containing chemicals with different levels of action. Surfactants like sulfobetaines (SB 3–10, SB 3–12 and ASB-14; (Chevallet et al., 1998; Luche et al., 2003) and treatment with reducing agents [β-mercaptoethanol (BME), dithiothreitol (DTT) and dithioerythritol (DTE); (Rabilloud, 1996)] can be used, again depending on the target protein groups. However, since DTT and DTE are charged and migrate out of the pH gradient during IEF, they result in the loss of solubility of some proteins thus, tributyl phosphine (TBP) is sometimes used instead as it has been found to enhance protein solubility (Herbert et al., 1998). Alternatively, tris (2-carboxyethyl) phosphine (TCEP) is used in the saturation labelling procedure in DIGE and can be considered.

Neutral detergents such as octyl-glucoside, dodecyl-maltoside, Triton X-100 and zwitterionic detergents such as CHAPS; 3-[(3-cholamidopropyl)-dimethylammonio] propanesulfonate and CHAPSO; 3-([3-cholamidopropyl] dimethylammonio)-2-hydroxyl-1-propanesulfonate can be used for very stable species, such as membrane proteins. Chaotropic chemicals such as urea and thiourea are commonly used and while urea disrupts hydrogen bonds, leading to protein unfolding, thiourea is more efficient at breaking hydrophobic interactions. However, its usefulness is somewhat limited due to its poor solubility in water (Rabilloud et al., 1997). Currently the best solution for solubilization of hydrophobic proteins is a combination of 5–7 M urea and 2 M thiourea, in conjunction with 1–4% appropriate detergents (Ahmed, 2009). It should also be noted that samples containing urea should never be heated as an increase in temperature causes urea to hydrolyze to isocyanate, modifying proteins by carbamylation (Dunn and Bradd, 1993). Some

Table 3.1 Commonly used 2D-PAGE buffer components and suitability

Buffer component	Category	Applications
Tris	Buffer agent	Suitable for use in 1st and 2nd dimension
Glycerol	Buffer agent	Suitable for use in 1st and 2nd dimension
Chaotropes Disrupt intramolecular interactions		
Urea	Denaturing, disrupts hydrogen bonds	Should never be heated, elevated temperatures cause urea to hydrolyze to isocyanate, modifying proteins by carbamylation
Thiourea	More efficient than urea	Poorly soluble in H_20
Neutral detergents Useful in membrane protein purification as can be easily extracted from final sample		
Triton X-100	Common detergent	Used for IEF, concentrations between 1 and 4% (v/v). Also used for cell permeabilization and protein solubilization including high molecular mass proteins
Zwitterionic detergents Protect native state of proteins without altering native charge of protein		
CHAPS	Solubilising, non-denaturing agent	Work particularly well for purifying membrane proteins
CHAPSO	Solubilising, non-denaturing agent	Higher solubility than CHAPS due to more polar head group
Sulfobetaines	Synthetic zwitterionic non-detergents	They may improve the yield of membrane proteins when used with detergents and reportedly prevent aggregation of denatured proteins
Aminosulfobetaines	Synthetic zwitterionic detergents	Work well in conjunction with CHAPS to offer full solubility; optimal solubility achieved in urea-thiourea mixtures and not in urea alone
Anionic detergent		
SDS	Denatures proteins	Not suitable in 1st dimension, can disturb IEF because of charge, causing streaking
Reducing agents		
BME	Disrupts disulfide bonds	Should not be used if samples are to be, silver stained or labelled using DIGE; can interfere with cysteine labelling
DTT	Disrupts disulfide bonds	Charged, migrates out of pH gradient during IEF resulting in loss of solubility of some proteins; contains thiol groups, not suitable for DIGE
DTE	Same as DTT	Same as DTT
TBP	Odourless, alternative to DTT and DTE	Can enhance protein solubility of basic proteins

Table 3.1 (continued)

Buffer component	Category	Applications
TCEP	Odourless, alternative to DTT and DTE	Can be used in DIGE
Carrier ampholytes		
Pharmalyte	pH range approximating pH range of IEF separation to be performed	Should not be used if samples are to be labelled using DIGE due to presence of amines
IPG Buffer	pH range of buffer should correspond to pH range of IEF separation to be performed.	Should not be used if samples are to be labelled using DIGE due to presence of amines

proteins tend to precipitate at their isoelectric point. To prevent this one may add salt to the protein solution. However, as soon as one performs isoelectric focusing, the salt will migrate away from protein promoting precipitation. To solve this problem, carrier ampholytes are often included in the buffer.

Although SDS is a fairly common detergent, it is negatively charged and should be avoided as it interferes with the first dimension IEF separation, forming negatively charged complexes. SDS is chosen for the second dimension separation because it binds to most proteins in a constant fashion (about 1.4 g of SDS per gram of protein) and also masks any charge of the protein by forming these large anionic complexes. SDS also disrupts most of hydrogen bonds, blocks many hydrophobic interactions and partially unfolds the protein molecules minimising differences based on secondary or tertiary structure. Sample solubilisation can be achieved using these conventional reagents (see Table 3.1) or commercially available kits, (ZOOM® 2D Protein Solubilizer Kit (Invitrogen); Bio-X Hydrophobic Protein Solubilizer for 2D Electrophoresis (InfiniteBio Inc.); FOCUS™ Protein Reduction-Alkylation Kit (GBiosciences), ProteoPrep® Reduction & Alkylation Kit (Sigma-Aldrich) and Protein Fraction Enrichment Kit (ProFEK) from ITSI Biosciences) with specific techniques applicable to a range of protein analyses. There is no one protocol which can be applied to all samples and the manner in which a mixture is solubilised depends greatly upon the nature of the sample, the types of proteins present in the sample and the overall experimental parameters.

3.2.5 Labelling

The general practice in the PAGE is to stain the proteins post electrophoresis. Relatively new techniques allow labelling of proteins in a sample prior to 2D-PAGE. Difference gel electrophoresis (DIGE) allows differentially labelling of all proteins within a cellular extract before 2D-PAGE (Unlu et al., 1997). This allows different protein samples to be compared on the same gel, significantly limiting any inter-gel variation discrepancies. Fluorescent dyes were chosen for their accessibility and were designed based on four principles (i) each matched set of dyes must react with the same amino acid residues, (ii) they must preserve the charge of the target amino acid, (iii) the dyes must have similar molecular masses, and (iv) they must have distinct fluorescent characteristics (Minden et al., 2009). Subsequently, the commercial development of these dyes was developed in collaboration with Amersham Biosciences (now GE Healthcare).

Two types of DIGE dyes have been developed; a lysine reactive or minimal dye and a cysteine reactive or saturation dye. Lysine reactive minimal dyes (Cy2, Cy3 and Cy5) have an NHS-ester reactive group and covalently attach to the epsilon amino group of lysine of proteins via an amide linkage. To ensure that only a single lysine is labelled, the ratio of dye to protein is specifically designed to ensure the dyes are limiting in the reaction and only cover approximately 1–2% of the available proteins and then only on a single lysine per protein. Cysteine reactive saturation dyes have a maleimide reactive group, and are designed to covalently bond to the thiol group of cysteine residues of proteins via a thioether linkage. To achieve maximum labelling of cysteine residues, the protocol uses a high fluor to protein labelling ratio.

The lysine amino acid in proteins carries an intrinsic single positive charge at neutral or acidic pH. CyDye DIGE Fluor minimal dyes also carry a single positive charge which, when coupled to the lysine, replaces the lysine's single positive charge with its own, ensuring that the *pI* of the labelled protein does not significantly alter compared to the same unlabelled protein. The CyDye DIGE Fluor saturation dyes have a net neutral charge and therefore do not alter the charge on a protein after labelling. They are also matched for molecular weight ensuring that the same protein from a standard and test sample labelled with the different dyes will overlay when separated on a single gel.

After 2D-PAGE, the gel is scanned using the excitation wavelength of each dye, allowing for the relative quantification of all proteins detected. DIGE also has the ability to minimize any discrepancies resulting from sample preparation due to the internal standard, which is the pooled aliquot of all samples and included on each gel. This mixture is made once, labelled (usually with Cy2) and divided between all gels. Two additional samples are labelled with Cy3 and Cy5, respectively, and these samples are, in turn, compared to the internal standard (i.e. Cy2:Cy3 and Cy2:Cy5) in order to give a comprehensive quantitative measurement for each labelled protein. These ratios can then be normalized for all gel set in an experiment. Current software packages available for the analysis of DIGE labelled proteins include Nonlinear Dynamics' Progenesis SameSpots v.3.0, GE Healthcare's DeCyder v.6.5 and Syngene's Dymension 3 programs (Kang et al. 1998)

3.2.6 Gel Loading

The amount of protein loaded onto the IEF strip is dependent upon many factors such as the sample (complex or pure protein) the length and pH range of strip used, the 2D-PAGE gel used and the staining regime. Typically 50–300 μg for complex samples with many proteins, and 5–50 μg for pure protein samples are loaded in the first dimension step. The concentration of a protein sample can be determined, classically, by a Bradford Assay, which is a colourimetric assay and uses Coomassie and a standard curve (usually BSA) to determine protein concentration. Commercially available protein assay kits, such as GE Healthcare's 2-D Quant Kit, are also widely used.

As mentioned previously, one of the main goals of 2D-PAGE is high-quality, reproducible data. The quality of the sample preparation is one of the key factors implicated in this reproducibility and is vital for reliable and accurate result. Applications for screening the quality of samples for 2D-PAGE are not widely used. However, methods such as capillary electrophoresis would facilitate the entire 2D-PAGE sample preparation process as they could save time, energy, costs and valuable samples.

3.3 Sample Preparation for Capillary Electrophoresis

3.3.1 Introduction

Capillary electrophoresis (CE) separates analytes from a complex mixture with high resolution based on differential migration through a liquid filled capillary in a strong electric field (Beale, 1998; Camilleri, 1993; Hjerten, 1967; Knox, 1994; Marina and Torre, 1994). The capillaries typically have an outer diameter of 375 μm and internal diameters ranging from 25 to 100 μm depending on the separation parameters. CE can be performed in several modes, which are a function of the separation matrix employed to obtain separation. Capillary Gel Electrophoresis (CGE) can be used to separate proteins that vary in molecular weight while Micellar Electrokinetic Capillary chromatography (MECC) is employed in the analysis of samples that differ in hydrophobicity (Knox, 1994). Separation modes like capillary zone electrophoresis (CZE) and MECC are dependent on an electrically driven bulk flow of solution known as the electro-osmotic flow (EOF) (Engelhardt et al., 1993; Monnig and Kennedy, 1994; Watzig and Gunter, 2003). These techniques can be used to analyse complex mixtures of molecules independently and also sequentially.

CGE is generally seen to be the CE technique most closely associated to the type of analyses typically performed by 1D gel electrophoresis (Karger et al., 1989). Capillary gel electrophoresis has advantages over slab gel electrophoresis in that the small diameter of the capillaries allows better heat dissipation, thereby minimizing band broadening due to Joule heating (Cetin and Li, 2008). This also means very strong electric fields can be applied, reducing run time and diffusion. Significantly smaller sample aliquots are needed in CE with injected plug volumes at the nanoliter scale and high sensitivity may be obtained especially when using fluorescent markers (Beale, 1998; Engelhardt et al., 1993; Monnig and Kennedy, 1994). CE is easily automated and lends itself well to microfluidic device technology. The evolution of these microfluidic devices allows the potential for the hyphenation of CE to other separation and identification methods like gel chromatography, high performance liquid chromatography (HPLC, (Zhang et al., 2001) and mass spectrometry (MS; Gasper et al., 2008; Schmitt-Kopplin and Frommberger, 2003). The coupling of CE to MS has been a benefit to proteomics, but there are still several problems that need to be addressed.

Capillary electrophoresis is well suited for analysis of complex multi-component samples. However, the analysis of analytes present in matrices such as plasma, serum, whole blood, tissue homogenates, saliva or urine requires well-designed sample preparation procedures. While a factor in CGE, this is particularly relevant to CZE as the direct injection of complex protein samples in to the capillary will lead to reduction of EOF due to protein adsorption to the capillary wall. Proteins can precipitate and adsorb irreversibly to the silanol groups on the inner wall of the capillary. This causes a local disturbance in the EOF around the area of adsorption, which leads to a deterioration of separation efficiency. Once the capillary is coated with denatured proteins, it becomes a source of secondary interaction with analytes resulting in severe tailing of analyte peaks (Ghosal, 2003). Loading highly complex mixtures of proteins into a gel filled column may lead to column blockage, loss of resolution and elevated baselines. It is simpler to fractionate the sample prior to analysis and focus on achieving resolution in specific mass ranges for superior results. It is clear that some useful means of sample preparation is required to reduce the sample complexity. Methods for sample preparation include protein precipitation, liquid-liquid extraction, solid-phase extraction and membrane methods such as dialysis and ultracentrifugation (Hernández-Borges et al., 2007; Mikus and Marakova, 2009; Puig et al., 2008).

As with any analysis technique sample preparation is often a complex operation that causes the most variability in analytical results. Ideally, sample manipulation prior to CE analysis should be as simple as possible and target-oriented. Objectives for sample preparation for analysis of components in biological fluids are removal of macromolecular components, mainly high abundance proteins, removal of salts, concentration of the analyte of interest and matching the sample buffer to the running CE buffer. Automation of this procedure is driven by the need to make it faster, safer and less prone to human error. While advances have been made in the automation of sample preparation for CE, it still remains the step, which limits the throughput.

3.3.2 Sample Types

Categorisation of sample type, based on sample preparation, can be seen as Venn diagram (Fig. 3.1). The first distinction (or set) can be made based on the mode of capillary electrophoresis employed; the second distinction can be the differentiation between sample types, such as analysis of protein/peptide in biological fluid or removal of protein from biological matrices to assess for drugs or other biomarkers; and the third set the level of analysis required, from basic detection level analysis

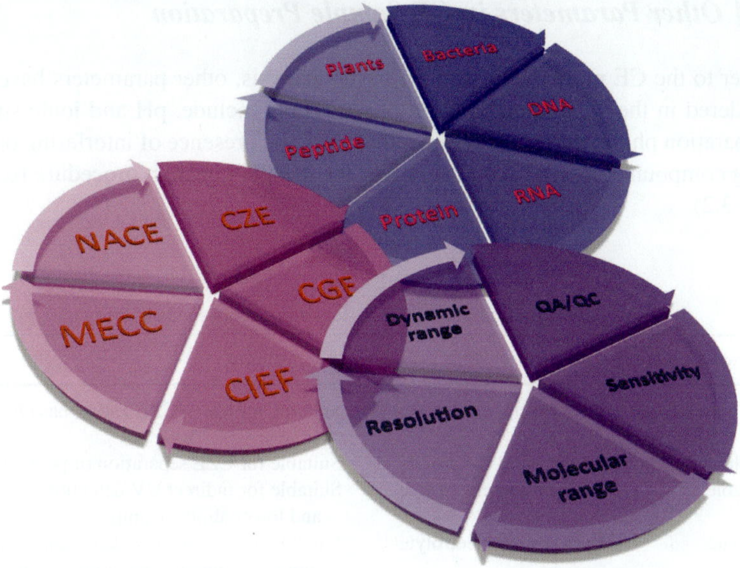

Fig. 3.1 Venn diagram of mode of capillary electrophoresis employed (*pink*), the sample types (*blue*) and the level of analysis required (*purple*)

to very high resolution QA/QC applications. The intersection represents the optimal experimental parameters for a defined analysis.

3.3.3 Purification

Prior to CE several other techniques may be used to reduce sample complexity. One of the commonly used methods for sample clean up is off-line solid phase extraction. This technique works on the differential adsorption of the sample to stationary and mobile phases, where macromolecules can be isolated based on their surface chemistry or morphology and has been reviewed in detail by several publications (Bergkvist et al., 2002; Ekstrom et al., 2002; Mader et al., 2002; Morita and Sawada, 1993; Tamai et al., 1991). Dialysis is another clean up or separation technique that utilises mass flux through a semi-permeable membrane separating two chambers. A molecular mass cut-off dialysis membrane provides molecular separation based on molecular hydrodynamic dimensions. The major drawback of dialysis is that this technique does not recover analytes quantitatively and the sample is usually diluted in the acceptor fluid. It is therefore necessary to concentrate the sample post dialysis. Sample filtration or centrifugation is also frequently used to isolate liquid from solid particles. Centrifugation may be used to isolate proteins precipitated out of solution using organic solvents, inorganic salts or strong acids. This technique is simple and can be used to concentrate a complex protein mixture and exchange to a desired buffer.

3.3.4 Other Parameters in CE Sample Preparation

Further to the CE mode being applied to the analysis, other parameters have to be considered in the sample preparation phase. These include, pH and ionic strength of separation phase, detection system employed, the presence of interfering or competing compounds such as reductants, and the sample injection procedure (see also Table 3.2).

Table 3.2 Commonly used CE buffer components and suitability

Buffer component	Category	Applications
Tris Borate EDTA	Background electrolyte	Suitable for use in CZE and as base buffer in CGE
Phosphate Buffer	Background electrolyte	Suitable for CZE separation of peptides
Imidazole	Background electrolyte	Suitable for indirect UV detection of cations and lower aliphatic amines
Chromate	Background electrolyte	Suitable for indirect UV detection of organic/inorganic anions when used in conjunction with CTAB to reverse EOF
Sodium Tetraborate	Background electrolyte	Suitable for CZE separations
Chaotropes Disrupt intra-molecular interactions		
Urea	Denaturing, disrupts hydrogen bonds	Should never be heated, elevated temperatures cause urea to hydrolyze to isocyanate, modifying proteins by carbamylation
Thiourea	More efficient than urea	Poorly soluble in H_20, absorbs UV, can be used in sample as a denaturant and as an EOF marker
Detergents Solubilising and micelle formation		
CHAPS	Solubilising, non-denaturing agent	Integral part of many background electrolytes
CTAB	Cationic detergent	Used in CZE to reverse EOF, can be used to form micelles above critical micelle concentration (CMC)
SDS	Anionic detergent, denatures proteins and forms micelles	Most commonly used surfactant in protein molecular weight separations. When used above CMC, forms micelles allowing separation of hydrophobic compounds
Reducing agents		
BME	Disrupts disulfide bonds	Typically used for denaturation prior to CGE, can interfere with cysteine labelling

Table 3.2 (continued)

Buffer component	Category	Applications
DTT	Disrupts disulfide bonds	Charged, migrates out of pH gradient during IEF resulting in loss of solubility of some proteins; contains thiol groups, not suitable for DIGE
TCEP	Odourless, alternative to DTT and DTE	Does not oxidise, used to maintain protein in denatured state
Additives		
Glycerol	Additive	Used to increase intrinsic viscosity of sieving matrix, improving separation efficiency. High levels should be avoided as this affects sample loading onto capillary
Cellulose based polymers	Sieving matrix	Hydroxy ethyl cellulose (HEC), hydroxy propyl methyl cellulose (HPMC), methyl cellulose (MC) etc. are all used to form a non-cross linked, UV transparent sieving matrix in protein and DNA gel based separations
Polyvinyl-pyrrilidone	EOF control coating	Used at 2–10% prior to loading DNA sieving matrix
Polyacrylamide	Coating and sieving matrix	Linear form used as coating/separating matrix; cross-linked form used as sieving matrix in DNA and protein separations. Cannot be used at high concentration for UV detection at 214 nm and 256 nm
Dextran Sulphate and Polybrene	EOF control coating	Successive multiple ionic layers are generated using alternating layers of these compounds

- pH: Charged-based CE separations of proteins depend on their overall surface charge which depend on their isoelectric point and can be controlled by choosing the appropriate pH for sample and running buffer. This ensures that the sample moves into the capillary and past the detector. Often the pH of the buffer can be chosen such that the charge on the wall of the capillary is similar to that of the protein in that buffer, reducing electrostatic protein adhesion to its surface.
- Injection technique: The sample injection technique is also a good example of how sample preparation can affect downstream processes. Electrokinetic injections can introduce a molecular weight-based bias, as smaller molecules will migrate preferentially to the injection interface. When CE protocols require electrokinetic injection, desalting is required because salt ions compete with the analyte ions so only a relatively small amount of sample is introduced into the capillary. Higher levels of salt in the sample plug also lead to current instability during separation. Protein and nucleic acid samples are frequently desalted by procedures based on ultrafiltration, dialysis or centrifugation. Pressure injections,

which show no such biases, are therefore more reproducible principally with differently charged analytes of small size.

- Reductant: When separating proteins under reduced conditions, care should be taken to ensure that the reductant used is compatible with the separation buffer and that it does not interfere with the detection system. DTT and BME degrade within a few days, even stored cold or frozen and will oxidize and derivatize cysteine residues, such that proteins become modified. These derivatizations may interfere with the overall mobility of the proteins, but can be removed prior to analysis with addition of freshly prepared reducing reagent. As with 2D-PAGE, TCEP is a convenient replacement for BME or DTT as it is an odourless, thiol-free reducing agent, resistant to oxidation (Herbert et al., 1998). All three reducing agents absorb UV at 214 nm, a common detection wavelength in label-free CE systems and should not be added to the separation matrix unless absolutely necessary. The ordered structure of proteins can also be eliminated using denaturants like guanidine-HCl and urea. Urea, however, modifies proteins by forming carbamoylate derivatives of lysine and other residues at temperatures above 30°C rendering a protein unsuitable for many enzymatic digests and leading to a more populated CE electropherogram. Carbamylation is detrimental to further protein characterization as the isocyanic acid reacts with the amino terminus blocking the protein to N-terminal sequencing. Typically, high levels of urea are required, between 6 and 8 M solutions, to denature proteins. These levels of saturation can lead to crystallization of the salts within the instrument leaving deposits on electrodes and other exposed surfaces. Often this leads to problems with the current stability and overall performance of the instrument. Samples prepared in urea should be maintain at a suitable temperature prior to and during separation, the urea should be freshly prepared, the capillary used should be cleaned between every separation and during the rinse steps the outlet end of the capillary should be immersed in water.

3.3.5 Labelling for CE Applications

Depending on the detection system used, the sample may require further preparative steps to derivatize it prior to addition of a fluorescent label. It is important to ensure that the fluorophor to sample ratio is appropriate to prevent a dye-induced change in sample migration behaviour. When using multiple dyes within a single separation, care should be taken in choosing dyes with similar charge and morphology, as this will allow unbiased comparison between samples. Label-free systems tend to require simpler systems as they require no chemical conditioning prior to imaging. They do however have a significantly lower limit of detection than labelled systems and must use other strategies to regain the signal to noise advantages of fluorescent labels such increasing the path length with a bubble chamber topology, the use of conductivity interface stacking techniques analogous to those used in gels (Mikus et al., 2009, Puig et al., 2008) multipixel detection (Culbertson and Jorgenson, 1998, Pereira et al., 2009) and amperometric detection (Xu et al., 1996).

3.4 Sample Preparation for Capillary Electrophoresis-Mass Spectrometry

3.4.1 Introduction

As mentioned previously, there are multiple modes of separation possible within CE, the two most predominant being CZE and CGE. CZE is more routinely coupled to MS detection due to its inherent buffer composition being more compatible for effective ionization. Although there are multiple ionization modes for MS, CZE is most commonly linked to Electrospray Ionization-Mass Spectrometry (ESI-MS) for selective detection using a nanoflow interface device (Ohnesorge et al., 2005).

ESI-MS takes ions from the liquid phase and coverts them into the gas phase, making it an ideal ionization method for coupling to liquid-based separation techniques. Effective coupling is achieved by forcing the liquid phase through a charged electrospray needle, generating a fine spray of charged droplets. The solvent from the droplet is evaporated so the droplet shrinks until eventually only the charged analyte molecules exist in the gas phase. Electrospray is a soft method of ionization, allowing both singly or multiply charged analyte molecules to form, which is why ESI-MS is an important technique in biological analysis and is particularly suited to the studies of large biomolecules such as proteins.

3.4.2 Interface/Coupling

In coupling CE to MS a number of considerations must be taken into account. Firstly, a completed electrical circuit must be present at all times for electrophoresis to occur, while the circuit must be completed in close proximity to the sprayer. The low flow-rates generated by electrophoresis (compared to pump flow HPLC) also need to be considered. Capillary electrophoretic flow-rates are very low (20–200 nl/min) while ionization spray flow required is often nano- to micro-flow (Monton and Terabe, 2005). Often a sheath liquid is employed to aid transportation and to ensure a stable flow-rate for the electrospray.

A number of different interfaces are possible.

- Liquid-junction approach (Lee et al., 1988) has been used to couple CZE to MS for peptide analysis, but gave rise to unsatisfactory ionization.
- Sheathless coupling, which uses an electrically conductive coated capillary tip to ensure a completed circuit (Vitali et al., 1994). This approach gives greater sensitivity of detection, however the capillary tip was easily damaged and required frequent replacement.
- Sheath-flow coupling (more commonly employed by over 75% of CE-MS instruments (Paces et al., 2003) first employed by Smith and Udseth (1988) provides stable ionization and is more compatible with nano-flow ionspray interfaces already used for MS coupling LC techniques. Examples of currently available CE-MS configurations are the Agilent 7100 CE system, which can be interfaced

directly to any of the Agilent 6000 Series MS systems, or the Beckman Coulter P/ACE MDQ CE system, which has an external detector adapter to allow simple interfacing with mass spectrometry. However a key drawback to this technique is the dilution factor introduced by the sheath-liquid, and the resulting loss in sensitivity. It is believed that although the sample is diluted, it actually occurs to a lesser extent than expected, probably due to incomplete mixing in the Taylor cone (Pelzing and Neususs, 2005). The sheath flow rate should therefore be adjusted to ensure optimum stability with minimum dilution. It has also been found that the radius of the jet emerging from the Taylor cone increases with decreasing conductivity and flow rate. As such, high-conductivity and low flow rates are also desirable, as smaller droplets are formed (Schmitt-Kopplin and Frommberger, 2003). The choice of sheath liquid has a direct effect on the ionization efficiency and the ionization state of the analytes, therefore should be considered carefully. The sheath liquid must be conductive in order to complete the electrical circuit, but the ionic strength should not be so high that arcing or electrical discharges are observed in the ESI. The counter ion employed in the sheath liquid should ideally be the same as that present in the run buffer, to ensure that problems do not arise from moving ionic boundaries. The liquid should be highly volatile to aid evaporation (>50% organic solvent) and normally contain an acid or base to enhance ionization of the given analyte. Acetonitrile or methanol are commonly mixed with water, and acidified with a small addition of acid such as acetic acid.

Another consideration in coupling the electrophoretic capillary to the MS detector is the increase in capillary length often required in order to reach the electrospray source. This in turn weakens the field strength applied over the separation, and as such the voltage should be increased to compensate. In some instances the separation voltage cannot be increased beyond a limit of the CE instrument, and as such the total field strength may be weaker that with a conventional CE-UV separation.

3.4.3 CE Buffer Considerations

When considering CE separations to be coupled to MS detection, a number of constraints must be observed. Firstly, the buffer employed must be suitable for electrophoresis of the sample, but more importantly, the running buffer must be volatile. Ammonium acetate and ammonium formate are therefore ideal buffers to be used, however standalone CE methods frequently employ non-volatile buffers such as sodium phosphate or borate. These buffers have been reported for use in CE-MS (Engelhardt et al., 1993, Monnig and Kennedy, 1994, Karger et al., 1989, Watzig and Gunter, 2003 and Cetin and Li 2008). However, they have been used at very dilute concentrations resulting in lower sensitivity MS data. Buffer modifiers such as organic solvents are ideal as they increase the volatility of the liquid phase, reducing the viscosity and surface tension, as well as introducing an extra factor for resolution by CE.

If the salt concentration is too high then Joule heating within the capillary can also be induced. Therefore, sample desalting is often beneficial for stable CE and CE-MS separations to be performed. The separation of biomolecules by CE-MS may therefore require existing CE methods to be reviewed and modified to allow successful coupling, and new methods being developed must take the buffer constraints into consideration.

3.4.4 Dilution Effects/Sample Stacking

Injection volumes within CE are extremely low (in the nL range) and are directly related to the capillary dimension and buffer viscosity. Therefore, when using UV detection alone, the concentration sensitivity can be somewhat poor at times as it is dependent upon the amount of solution loaded into the capillary (Monton and Terabe, 2005). When analysing proteins and peptides, often only low concentrations of sample are available. Therefore, it is often necessary to increase the sample loading onto the capillary in order to achieve the desired sensitivity. This is often achieved by using a concentration step during sample loading, termed "stacking". In order to achieve effective stacking, the sample matrix must differ significantly from the electrophoretic buffer, for example; in terms of conductivity, pH or salt concentration. The aim is to achieve a decrease in the electrophoretic mobility of the analyte as it passes from the sample matrix into the running buffer, thereby generating a focussing effect at the interface between these phases. Ionic strength stacking is routinely used, where the field strength should be very high in the injected sample relative to the electrophoresis buffer. In this instance the analyte migrates rapidly through the sample buffer to the solvent interface before the separation takes place. Low-conductivity buffers that do not generate much joule heating, such as borate, are important for this technique. Using ionic strength stacking, the sample is commonly dissolved into the separation buffer, but at 10x lower ionic strength. This can also be achieved by adding a small plug of water before injecting the sample. Stacking can increase the peak height by approximately 3–10 times (Shihabi, 1999). It is also possible to use electrokinetic injection to create a stacking effect while introducing the sample to the capillary; however, this technique can introduce "bias" in the sample loading. Therefore, these results may not be quantitative with respect to all of the sample components. It should be noted however, that this effect is often beneficial or indeed desirable if the analyte of interest is only a small percentage of the total sample and has a high electrophoretic mobility (Landers, 2007).

References

Ahmed, F.E. (2009). Sample preparation and fractionation for proteome analysis and cancer biomarker discovery by mass spectrometry. J Sep Sci *32*, 771–798.

Antonioli, P., Bachi, A., Fasoli, E., and Righetti, P.G. (2009). Efficient removal of DNA from proteomic samples prior to two-dimensional map analysis. J Chromatogr A *1216*, 3606–3612.

Beale, S.C. (1998). Capillary electrophoresis. Anal Chem *70*, 279R–300R.

Bergkvist, J., Ekstrom, S., Wallman, L., Lofgren, M., Marko-Varga, G., Nilsson, J., and Laurell, T. (2002). Improved chip design for integrated solid-phase microextraction in on-line proteomic sample preparation. Proteomics *2*, 422–429.

Bodzon-Kulakowska, A., Bierczynska-Krzysik, A., Dylag, T., Drabik, A., Suder, P., Noga, M., Jarzebinska, J., and Silberring, J. (2007). Methods for samples preparation in proteomic research. J Chromatogr B Anal Technol Biomed Life Sci *849*, 1–31.

Camilleri, P. (1993). Capillary Electrophoresis: Theory and Practice (Boca Raton, FL, CRC Press).

Canas, B., Pineiro, C., Calvo, E., Lopez-Ferrer, D., and Gallardo, J.M. (2007). Trends in sample preparation for classical and second generation proteomics. J Chromatogr A *1153*, 235–258.

Cetin, B., and Li, D. (2008). Effect of Joule heating on electrokinetic transport. Electrophoresis *29*, 994–1005.

Chambers, S.P. (2002). High-throughput protein expression for the post-genomic era. Drug Discov Today *7*, 759–765.

Chevallet, M., Santoni, V., Poinas, A., Rouquie, D., Fuchs, A., Kieffer, S., Rossignol, M., Lunardi, J., Garin, J., and Rabilloud, T. (1998). New zwitterionic detergents improve the analysis of membrane proteins by two-dimensional electrophoresis. Electrophoresis *19*, 1901–1909.

Cilia, M., Fish, T., Yang, X., McLaughlin, M., Thannhauser, T.W., and Gray, S. (2009). A comparison of protein extraction methods suitable for gel-based proteomic studies of aphid proteins. J Biomol Tech *20*, 201–215.

Dong, M., Baggetto, L.G., Falson, P., Le Maire, M., and Penin, F. (1997). Complete removal and exchange of sodium dodecyl sulfate bound to soluble and membrane proteins and restoration of their activities, using ceramic hydroxyapatite chromatography. Anal Biochem *247*, 333–341.

Dunn, M.J., and Bradd, S.J. (1993). Separation and analysis of membrane proteins by SDS-polyacrylamide gel electrophoresis. Methods Mol Biol *19*, 203–210.

Ekstrom, S., Malmstrom, J., Wallman, L., Lofgren, M., Nilsson, J., Laurell, T., and Marko-Varga, G. (2002). On-chip microextraction for proteomic sample preparation of in-gel digests. Proteomics *2*, 413–421.

Engelhardt, H., Beck, W., Kohr, J., and Schmitt, T. (1993). Capillary electrophoresis – methods and scope. Angew Chem Int Ed *32*, 629–649.

Englard, S., and Seifter, S. (1990). Precipitation techniques. Methods Enzymol *182*, 285–300.

Fountoulakis, M., Juranville, J.F., Jiang, L., Avila, D., Roder, D., Jakob, P., Berndt, P., Evers, S., and Langen, H. (2004). Depletion of the high-abundance plasma proteins. Amino Acids *27*, 249–259.

Freeman, W.M., and Hemby, S.E. (2004). Proteomics for protein expression profiling in neuroscience. Neurochem Res *29*, 1065–1081.

Fu, Q., Bovenkamp, D.E., and Van Eyk, J.E. (2007). A rapid, economical, and reproducible method for human serum delipidation and albumin and IgG removal for proteomic analysis. Methods Mol Biol *357*, 365–371.

Gasper, A., Englmann, M., Fekete, A., Harir, M., and Schmitt-Kopplin, P. (2008). Trends in CE-MS 2005–2006. Electrophoresis *29*, 66–79.

Ghosal, S. (2003). The effect of wall interactions in capillary-zone electrophoresis. J Fluid Mech *491*: 285–300.

Gorg, A., Postel, W., Friedrich, C., Kuick, R., Strahler, J.R., and Hanash, S.M. (1991). Temperature-dependent spot positional variability in two-dimensional polypeptide patterns. Electrophoresis *12*, 653–658.

Gorg, A., Weiss, W., and Dunn, M.J. (2004). Current two-dimensional electrophoresis technology for proteomics. Proteomics *4*, 3665–3685.

Granier, F. (1988). Extraction of plant proteins for two-dimensional electrophoresis. Electrophoresis *9*, 712–718.

Hannam, C., Lange, G.L., and Mellors, A. (1998). Synthesis of a radiolabeled zwitterionic detergent and its use in protein purification. Anal Biochem *258*, 246–250.

Herbert, B.R., Molloy, M.P., Gooley, A.A., Walsh, B.J., Bryson, W.G., and Williams, K.L. (1998). Improved protein solubility in two-dimensional electrophoresis using tributyl phosphine as reducing agent. Electrophoresis *19*, 845–851.

Hernández-Borges, J., Borges-Miquel, T.M., Rodríguez-Delgado, M.Á., and Cifuentes, A. (2007). Sample treatments prior to capillary electrophoresis-mass spectrometry. J Chromatogr A *1153*, 214–226.

Hjerten, S. (1967). Free zone electrophoresis. Chromatogr Rev *9*, 122 –219.

Jiang, L., He, L., and Fountoulakis, M. (2004). Comparison of protein precipitation methods for sample preparation prior to proteomic analysis. J Chromatogr A *1023*, 317–320.

Kang, Y., Techanukul, T., Mantalaris, A., and Nagy, J. (2009). Comparison of three commercially available DIGE analysis software packages using conditioned media samples. J Proteome Res *8*, 1077–1084.

Karger, B.L., Cohen, A.S., and Guttman, A. (1989). High-performance capillary electrophoresis in the biological sciences. J Chromatogr *492*, 585–614.

Klose, J., and Kobalz, U. (1995). Two-dimensional electrophoresis of proteins: An updated protocol and implications for a functional analysis of the genome. Electrophoresis *16*, 1034–1059.

Knox, J. (1994). Terminology and nominclature in capillary electroseparation systems. J Chromatogr A *680*, 3–13.

Lee, E.D., Muck, W., Henion, J.D., and Covey, T.R. (1988). On-line capillary zone electrophoresis-ion spray tandem mass spectrometry for the determination of dynorphins. J Chromatogr *458*, 313–321.

Luche, S., Santoni, V., and Rabilloud, T. (2003). Evaluation of nonionic and zwitterionic detergents as membrane protein solubilizers in two-dimensional electrophoresis. Proteomics *3*, 249–253.

Mader, R.M., Rizovski, B., and Steger, G.G. (2002). On-line solid-phase extraction and determination of paclitaxel in human plasma. J Chromatogr B *769*, 357–361.

Marina, M.L., and Torre, M. (1994). Capillary electrophoresis. Talanta *41*, 1411–1433.

Mikus, P., and Marakova, K. (2009). Advanced CE for chiral analysis of drugs, metabolites, and biomarkers in biological samples. Electrophoresis *30*, 2773–2802.

Minden, J.S., Dowd, S.R., Meyer, H.E., and Stuhler, K. 2009: Difference gel electrophoresis. Electrophoresis *30*(Suppl 1): S156–161.

Monnig, C.A., and Kennedy, R.T. (1994). Capillary electrophoresis. Anal Chem *66*, R280–R314.

Monton, M.R.N., and Terabe, S. (2005). Recent developments in capillary electrophoresis-mass spectrometry of proteins and peptides. Anal Sci *21*, 5–13.

Morita, I., and Sawada, J. (1993). Capillary electrophoresis with online sample pretreatment for the analysis of biological samples with direct-injection. J Chromatogr *641*, 375–381.

Ohnesorge, J., Neususs, C., and Watzig, H. (2005). Quantitation in capillary electrophoresis-mass spectrometry. Electrophoresis *26*, 3973–3987.

Paces, M., Kosek, J., Marek, M., Tallarek, U., and Seidel-Morgenstern, A. (2003). Mathematical modelling of adsorption and transport processes in capillary electrochromatography: Open-tubular geometry. Electrophoresis *24*, 380–389.

Pelzing, M., and Neususs, C. (2005). Separation techniques hyphenated to electrospray-tandem mass spectrometry in proteomics: Capillary electrophoresis versus nanoliquid chromatography. Electrophoresis *26*, 2717–2728.

Pieper, R., Su, Q., Gatlin, C.L., Huang, S.T., Anderson, N.L., and Steiner, S. (2003). Multi-component immunoaffinity subtraction chromatography: An innovative step towards a comprehensive survey of the human plasma proteome. Proteomics *3*, 422–432.

Puig, P., Borrull, F., Calull, M., and Aguilar, C. (2008). Sorbent preconcentration procedures coupled to capillary electrophoresis for environmental and biological applications. Anal Chimica Acta *616*, 1–18.

Rabilloud, T. (1996). Solubilization of proteins for electrophoretic analyses. Electrophoresis *17*, 813–829.

Rabilloud, T., Adessi, C., Giraudel, A., and Lunardi, J. (1997). Improvement of the solubilization of proteins in two-dimensional electrophoresis with immobilized pH gradients. Electrophoresis *18*, 307–316.

Rodrigues, S.P., Ventura, J.A., Zingali, R.B., and Fernandes, P.M. (2009). Evaluation of sample preparation methods for the analysis of papaya leaf proteins through two-dimensional gel electrophoresis. Phytochem Anal *20*, 456–464.

Schmitt-Kopplin, P., and Frommberger, M. (2003). Capillary electrophoresis – mass spectrometry: 15 years of developments and applications. Electrophoresis *24*, 3837–3867.

Shaw, M.M., and Riederer, B.M. (2003). Sample preparation for two-dimensional gel electrophoresis. Proteomics *3*, 1408–1417.

Shihabi, Z.K. (1999). Field amplified injection in the presence of salts for capillary electrophoresis. J Chromatogr A *853*, 3–9.

Smith, R.D., and Udseth, H.R. (1988). Capillary zone electrophoresis-MS. Nature *331*, 639–640.

Tamai, G., Edani, M., and Imai, H. (1991). Determination of ketoprofen enantiomers in plasma by soild-phase extraction and column switching high-perfomrnace liquid-chromatography. Anal Sci *7*, 29–32.

Unlu, M., Morgan, M.E., and Minden, J.S. (1997). Difference gel electrophoresis: A single gel method for detecting changes in protein extracts. Electrophoresis *18*, 2071–2077.

Watzig, H., and Gunter, S. (2003). Capillary electrophoresis – A high performance analytical separation technique. Clin Chem Lab Med *41*, 724–736.

Weiss, W., and Gorg, A. (2009). High-resolution two-dimensional electrophoresis. Methods Mol Biol *564*, 13–32.

Zhang, X.M., Hu, H.L., Xu, S.Y., Yang, X.H., and Zhang, J. (2001). Comprehensive two-dimensional capillary LC and CE for resolution of neutral components in traditional Chinese medicines. J Sep Sci *24*, 385–391.

Chapter 4
Filtration as a Sample Preparation Technique Prior to Mass Spectrometry: Selecting the Right Filtration Device

Vivek Joshi and Elena Chernokalskaya

Abstract Filtration is one of the simplest forms of sample preparation prior to any analysis including mass spectrometry. Being the simplest, filtration is also one of the most neglected steps during optimization of sample preparation which can lead to problems in downstream analysis. This report provides guidelines towards selection of the right filtration device for the application. The factors which decide choice of filter are sample size, chemical characteristics of sample, particle load and downstream analytical technique. This report also provides useful information regarding extractables and analyte binding and ways to minimize both of these.

Keywords Analyte binding · Extractables · Filter selection · Microfiltration · Ultrafiltration

4.1 Introduction

Mass spectrometry, MS is one of the most commonly used analytical techniques used for quantitation (Yang et al., 2005) as well as identification and structure determination of biological molecules (Liu and Hop, 2005). In bio-analytical arena, mass spectrometry is used as a detector for either analysis of small molecular weight analytes such as active pharmaceuticals (Zurbonsen et al., 2004), pollutants (González et al., 2006) or large molecular weight analytes such as proteins (Heuvel and Heck, 2004), peptides (Machtejevas et al., 2007) or oligonucleotides (Willems et al., 2004).

Sample preparation prior to analysis helps to bring a sample to a format that is compatible with the analytical technique. Sample preparation also helps to reduce complexity of the sample by selectively removing interfering impurities from the matrix and thereby concentrating the analyte prior to analysis. A typical sample for mass spectrometry should be free of particulates and dissolved in a MS compatible solvent. Since biological samples are very complex in nature, multistep sample preparation is usually required before mass spectrometry analysis (Brandt et al.,

V. Joshi (✉)
EMD Millipore, Danvers, MA 01923, USA
e-mail: vivek.joshi@merckgroup.com

A.R. Ivanov, A.V. Lazarev (eds.), *Sample Preparation in Biological Mass Spectrometry*, DOI 10.1007/978-94-007-0828-0_4,

2005). In order to reduce the sample complexity further, chromatography is often coupled with mass spectrometry. In this situation, the sample requirement for mass spectrometry are similar to that for High Performance Liquid Chromatography, HPLC, which again means that the sample is particle free and completely soluble in the solvent compatible with the HPLC system.

One of the simple sample preparation techniques that can be used prior to mass spectrometry is use of membrane filtration (Tsao et al., 2004; Machtejevas et al., 2007). Microfiltration and ultrafiltration, are two common filtration techniques that can be used as a sample preparation step. Microfiltration refers to use of membrane filters of 0.1–0.45 μm pore size which allow removal of particulate impurities prior to analysis. For samples containing higher load of particles or larger particles, a pre-filter membrane with larger pore size can be used followed by final filtration.

Ultrafiltration is very commonly used for preparation of protein samples prior to analysis. It typically deals with removal of soluble impurities based on their molecular weight (size) from the sample. Typically the pore size of these membranes range from 0.1 to 0.005 μm and are referenced by nominal molecular weight cut off, NMWCO (200–5 kDa) of the membrane. These membranes remove proteinaceous impurities larger than the NMWCO of the membrane used thereby reducing complexity of the sample. If the protein under study has tendency to aggregate in certain buffers, careful consideration needs to be given to buffer prior to sample preparation, as the membrane can inadvertently also remove portions of the sample. Ultrafiltration membranes also remove particulate impurities during the process. On the other hand, UF can be used to concentrate molecules larger than NMWCO and remove smaller contaminants that may interfere with the downstream analysis.

For complex samples containing particulate impurities, filtration alone can not provide the sample necessary for analysis, but forms an integral part of an overall sample preparation strategy which involves other sample preparation techniques like extraction, centrifugation, depletion. This combination helps reduce sample complexity (Chéneau et al., 2007) and also brings it to a form suitable for analysis.

Since filtration is a very simple sample preparation technique, it is one of the most neglected steps in sample prep optimization. Not enough attention is paid to filtration and the choice of filtration devices and materials which can lead to inconsistency and even failure of the downstream analysis. In this report we will provide some guidelines for selection of appropriate filter prior to chromatography and mass spectrometric analysis of biological samples.

4.2 Selection of an Appropriate Filter

Following parameters are important for selection of an appropriate filtration device in order to achieve fast and efficient sample preparation.

1. Sample Size
2. Chemical nature of the sample
3. Particle load and viscosity
4. Downstream analytical technique

4.2.1 Sample Size

Filters are typically available in various sizes (4–50 mm), which essentially implies the size of the membrane/filtration area available for sample filtration. As expected, the larger the filtration area, the easier it is to filter the sample and the faster will be the filtration speed, so it might appear that it's best to choose the largest filter for sample preparation. But the fact is, larger sized filters also have higher sample hold up volume which can cause loss of precious sample volume. Larger membrane area also means that there is a higher potential for analyte losses due to the binding to the filter as well as higher amount of extractable impurities contaminating the sample. Extractables are soluble impurities that leach out of the filter device as the sample is being filtered through. Thus there is an ideal filter size based on the sample volume which provides optimum filtration performance and minimizes the risk of extractables and analyte binding. Table 4.1 shows the recommended correlation between the sample size and the filter size.

When preparing large number of samples, using syringe filters or ultrafilters which filter one sample at a time can be very time consuming. In such situations, filtration in multi-well format is preferred. Use of multi-well plates enables filtration of 96 or 384 samples directly into a collection plate, which can be analyzed directly on an LC-MS system. Multi-well plates are also compatible with various automation formats available in the market thereby making the operation as hands off as possible. Multi-well plates are commercially available from various vendors with number of membrane/pore size/pre-filter combinations that suit various application needs. These products are typically volume restricted (Max. about 1.5 mL), but for MS-based applications, larger sample volume is typically not required. Use of filter plates allows for sample preparation of multiple samples at the same time, but these are not very cost effective when only a few samples are to be processed.

Table 4.1 Sample volume and appropriate filter size

Sample volume (mL)	Filter size (mm)	Hold up volume (μL)	Filtration area (cm^2)
< 1	4	10	0.1
1–10	13	25	0.65
10–100	25	100	3.6
10–100	33	80	4.5

4.2.2 Chemical Nature of a Sample

It is important to consider chemical nature of the sample prior to using the filters as various membranes used in filters can have different chemical compatibility. Using a filter where the membrane or the housing material is not compatible with the sample can lead to issues such as incomplete filtration of sample, extraction of impurities into the sample and in the worst case even complete membrane disintegration. Table 4.2 provides information on chemical compatibility of various

Table 4.2 Chemical compatibility of various membranes and housing materials

	Polar organic solvents			Aqueous basic solutions			Acid	Salt solution	Surfactants	
	ACN	MeOH	EtOH	3N NaOH	NH_4OH	Na_2CO_3 solution	1N HCl	Brine	SDS	Tween 20
Housing materials										
HDPE	E	E	E	E	E	P	E	E	P	P
PP	E	E	E	E	P	E	E	E	N	N
Membranes										
PTFE	E	E	E	E	E	E	E	E	N	E
PVDF	G	E	E	E	E	E	E	E	N	P
Nylon	E	G	E	E	P	E	E	E	N	P
PES	P	E	E	E	P	N	E	N	N	N

E = Excellent, **G** = Good, **P** = Poor, **N** = No Data

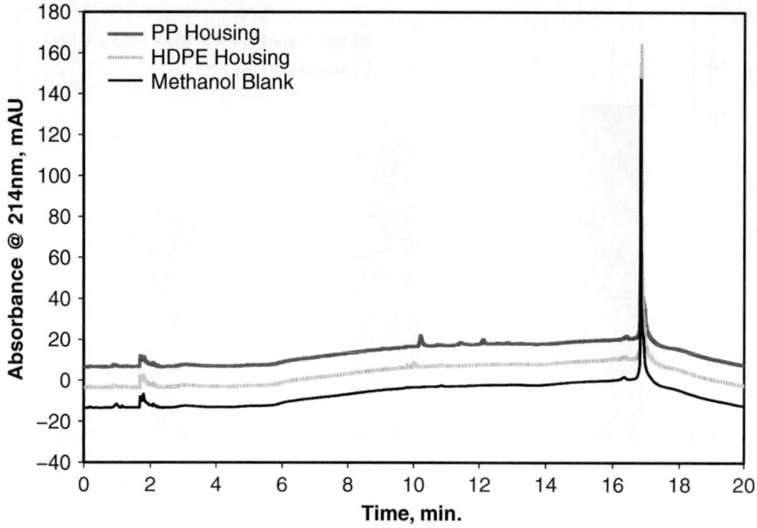

Fig. 4.1 Low extractables from filter housing. Methanol was filtered through syringe filters without any membrane and filtrate collected and analyzed by reversed-phase HPLC using a C18 column. Water and Acetonitrile were used as mobile phase components and UV absorbance was measured at 214 nm

membranes that are commonly present in syringe filters and housing materials commonly present in syringe filters as well as multi-well plates.

The chromatogram in the Fig. 4.1 also shows UV extractables profile of High Density Polyethylene, HDPE and Polypropylene, PP housing materials used commonly in syringe filters. Very low amount of extractables shown in this chromatogram coupled with their broad chemical compatibility (Table 4.1) indicate that these filter housings do not contribute significant impurities to the sample. Even with the low level of extractables, minor contaminating peaks resulting from leachables might interfere with the detection of low abundance analytes.

4.2.3 Particle Load and Viscosity

Presence of particles in the sample affects filtration speed. Sometimes single membrane filters (0.2 or 0.45 μm) get clogged before sufficient volume of sample can pass through. In such cases it is recommended to use syringe filters which contain a pre-filter in it. These filters typically contain one or more layers of pre-filter material followed by the final membrane filter. The pre-filter material helps trap the larger particulates thereby protecting the final membrane from fouling and allowing higher volume of sample to be filtered before the filter clogs. Figure 4.2 shows benefit of using a filter with a pre-filter for filtration of high particulate samples. Using

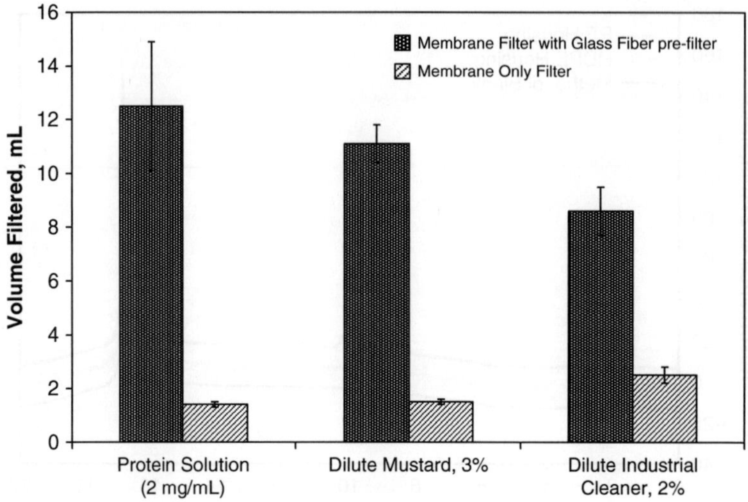

Fig. 4.2 Effect of pre-filter on the filtration efficiency. Various solutions were filtered through a membrane only and a membrane combined with a glass fiber pre-filter containing syringe filters till the filter clogged. Results were reported as average of 3 measurements

multi layered filters allow for filtration of 4–6 times higher volume than with the membrane only filter.

Most commonly used pre-filter material in these multi membrane filters is glass fiber and it is important to remember that presence of glass fibers in the syringe filter device can lead to higher level of extractables or analyte binding. It can be especially critical when working with samples containing low levels of proteins/peptides/oligonucleotides as analytes as they all have a strong tendency to bind to glass fiber. In such cases it might be better to filter sample through a filter which contains an inert/low protein binding filter and a pre-filter.

4.2.4 Downstream Analytical Technique

Finally, it is critical to choose the syringe filter based on the downstream analytical technique. As discussed above, mass spectrometry is frequently coupled with a separation technique such as HPLC, UHPLC (Ultra High Performance Liquid Chromatography) or Capillary Electrophoresis, CE. HPLC requires sample filtration through 0.45 μm filter prior to injection onto the system. On the other hand, the newer more sensitive and faster analytical techniques such as UHPLC which offer significant benefits because of its higher speed and higher peak capacity also pose more stringent requirements on sample quality. This new LC technique requires filtration of sample through a 0.2 μm filter. Thus, depending on the analytical technique used, a 0.45 or a 0.2 μm filter is required for sample preparation.

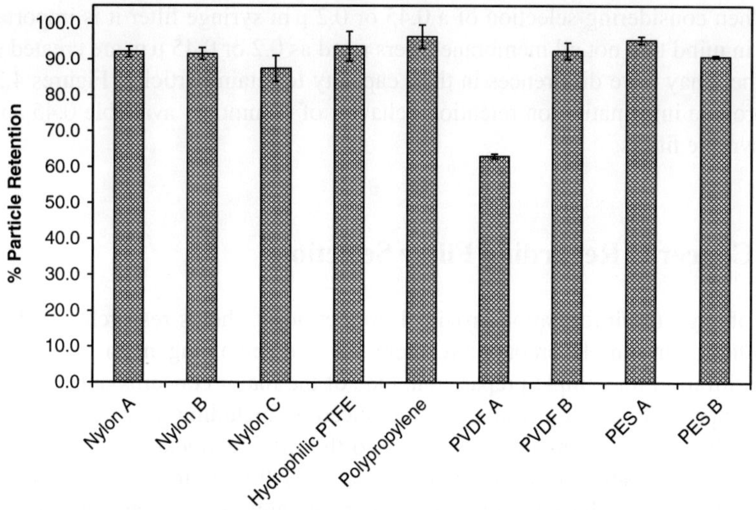

Fig. 4.3 Particle retention efficiency of various 0.45 µm syringe filters. A 0.005% suspension of 0.5 µm latex particles in Triton-X 100 was filtered through various syringe filters and UV absorbance of filtrate determined

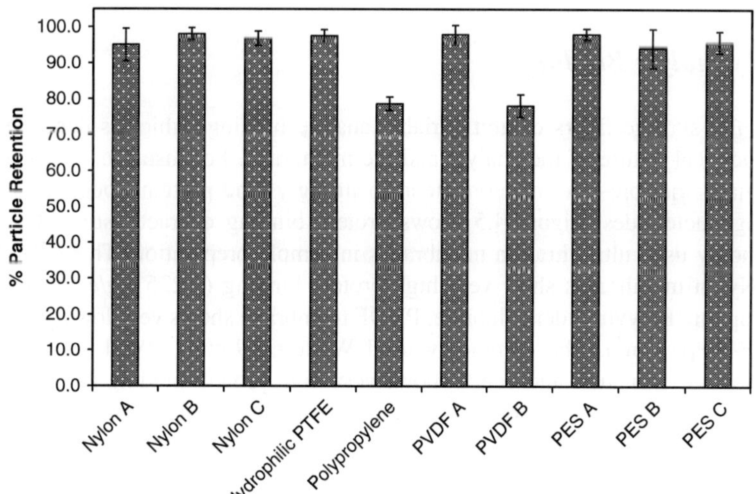

Fig. 4.4 Retention efficiency of 0.2 µm syringe filters. A 0.005% suspension of 0.3 µm latex particles in Triton-X 100 was filtered through various syringe filters and UV absorbance of filtrate determined

When considering selection of a 0.45 or 0.2 μm syringe filter it is important to keep in mind that not all membrane filters rated as 0.2 or 0.45 μm are created equal and they may have differences in their capacity to retain particles. Figures 4.3 and 4.4 provide information on retention behavior of commonly available 0.45 and 0.2 μm syringe filters.

4.3 Concerns Regarding Filter Selection

Even though the information provided in Section 4.2 helps researchers select the right filter for sample preparation, there are two remaining main concerns with using filtration for sample preparation. One of the major concerns during selection of appropriate filter for any analytical application, including mass spectrometry, is analyte loss due to non-specific binding to the filter, extractables, and leachables from the filter itself. These concerns are not only limited to the choice of a filter device; any surface that comes in contact with the sample has a potential to introduce extractable impurities into the sample as well as to result in analyte binding.

The discussion below provides some information on analyte binding and extractables as it pertains to various filter devices and how its impact on analysis can be minimized.

4.3.1 Analyte Binding

Typically syringe filters exhibit variable analyte binding, which is dependent on the chemical nature of the analyte and the membrane. For instance, as stated earlier, glass fiber pre-filter materials tend to strongly bind proteins, peptides as well as oligonucleotides. Figure 4.5 shows protein binding characteristics of various commonly used ultrafiltration membranes in sample preparation. This data shows that Nylon membranes show very high protein binding (∼225 μg/cm^2), whereas Hydrophilic Polyvinylidene fluoride, PVDF membrane shows very low protein biding (∼15 μg/cm^2). The commonly used Western Blotting PVDF membrane is hydrophobic in nature and hence shows very high protein binding (BSA Binding capacity, ∼200 μg/cm^2) as opposed to the hydrophilic PVDF membrane.

Multiple secondary interactions such as hydrogen bonding, electrostatic interactions and hydrophobic interactions are responsible for binding of the analyte to the membrane surface. Analyte binding to the membrane surface is also dependent on concentration of an analyte. Since there are limited numbers of binding sites on the membrane surface, at higher concentration the significance of analyte binding is diminished. To the contrary at low analyte concentration, a larger fraction of sample can bind to the membrane surface before the membrane surface is saturated. Thus, analyte binding can become a much bigger problem at low concentration than at higher concentration. Figures 4.6 and 4.7 show drug binding to Nylon and Hydrophilic Polytetrafluoroethylene, PTFE membrane filters. It is very clear

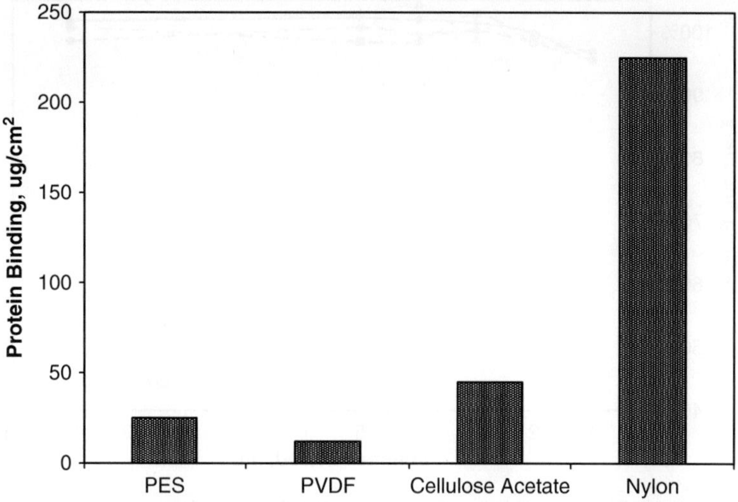

Fig. 4.5 Protein binding characteristics of various membranes. A comparative study of protein binding was done on various 0.2 μm membranes. 1 mg/ mL solution of ^{125}I labeled IgG was filtered through 13 mm membrane discs. After incubation, bound protein was determined

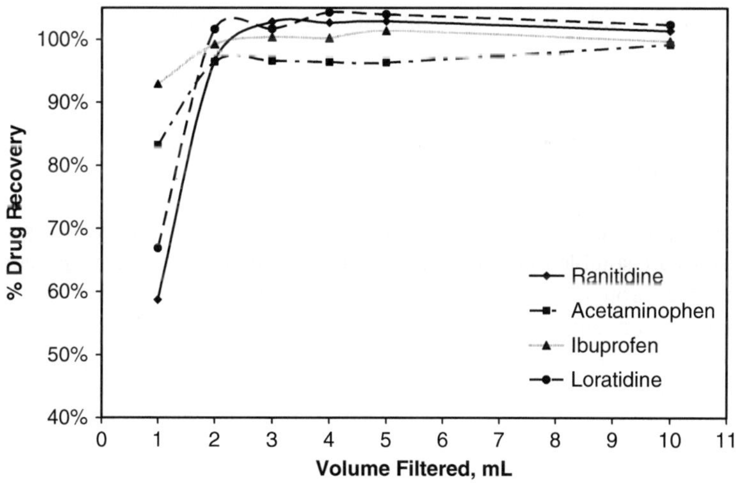

Fig. 4.6 Drug binding characteristics of nylon syringe filters. This test was carried out using a drug dissolution testing protocol. Commercially available oral formulations (tablets) of the drugs tested were dissolved in their respective dissolution media. At the end of the dissolution test, samples were filtered and various fractions of the filtrate were collected and analyzed using UV-visible spectrophotometry

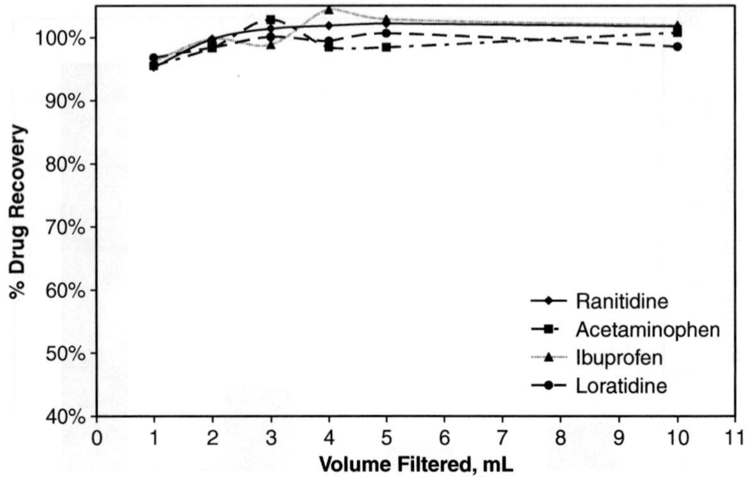

Fig. 4.7 Drug binding characteristics of hydrophilic PTFE syringe filters

from these graphs that Nylon membranes show very high and variable drug binding depending on the analyte studied. On the other hand, Hydrophilic PTFE membranes demonstrate very little analyte binding and should be preferred for high sensitivity analyses.

4.3.2 Extractables/Leachables

Presence of extractables and leachables can affect downstream analysis of samples. Extractables are soluble impurities that leach out of the filter device as the sample is being filtered through it. If the filter is not chemically compatible with the sample, it can lead to high amount of extractable impurities. Presence of these extractable impurities leads to higher background during sample analysis, thereby reducing the sensitivity of detection and quantitation. If these extractable impurities co-elute with an analyte of interest in downstream LC-MS analysis that leads to incorrect quantitation of the analyte. In cases where the extractable impurities are well separated from the analyte of interest, that can lead to appearance of HPLC detected peaks which are difficult to explain and also vary from sample to sample. There are several examples of extractable impurities as seen by HPLC–UV analysis shown in Figs. 4.8 and 4.9.

Even though extractables can pose a problem in downstream analysis, these can be reduced significantly by just washing the membrane filter with the sample or a solvent prior to filtration. Figure 4.10 shows the effect of pre-rinsing of the filter on extractables level. Just by rinsing the membrane with 1 mL of solvent as in this case, the extractables were reduced to undetectable level by LC–UV.

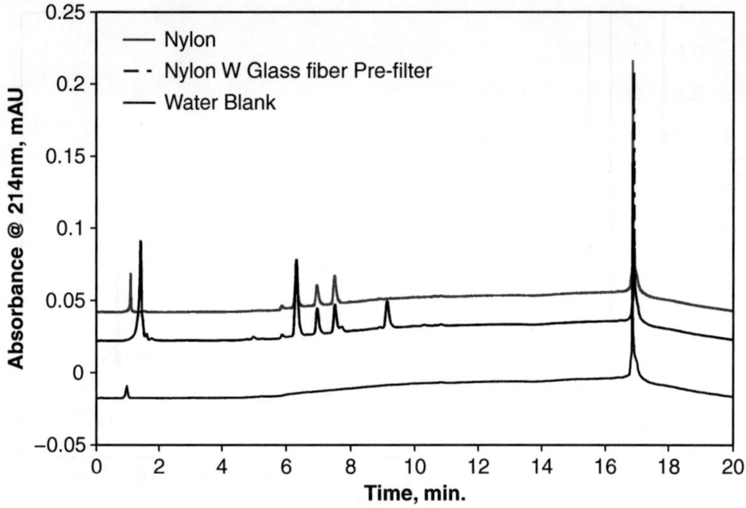

Fig. 4.8 UV extractables from nylon syringe filters. The extraction solvent was filtered using an appropriate syringe filter and the first milliliter of the filtrate was collected and analyzed with reversed-phase HPLC using a C18 column and water and Acetonitrile as mobile phase components. A UV detector set at 214 nm was used

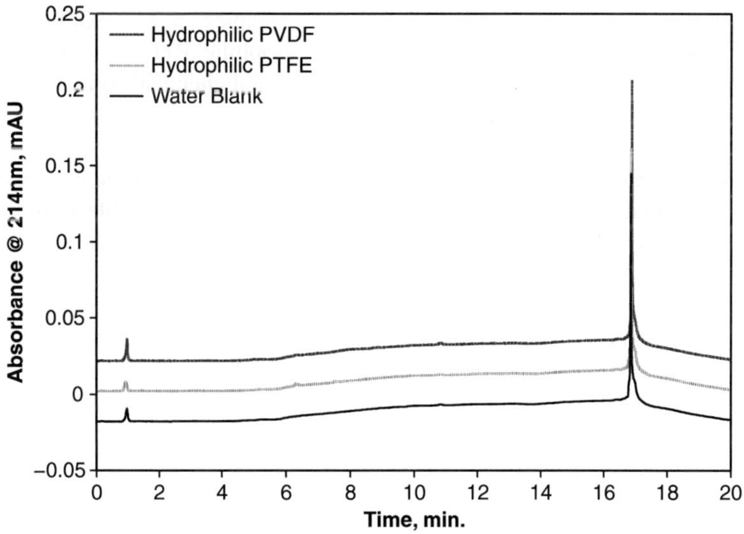

Fig. 4.9 UV extractables from hydrophilic PTFE and PVDF syringe filters

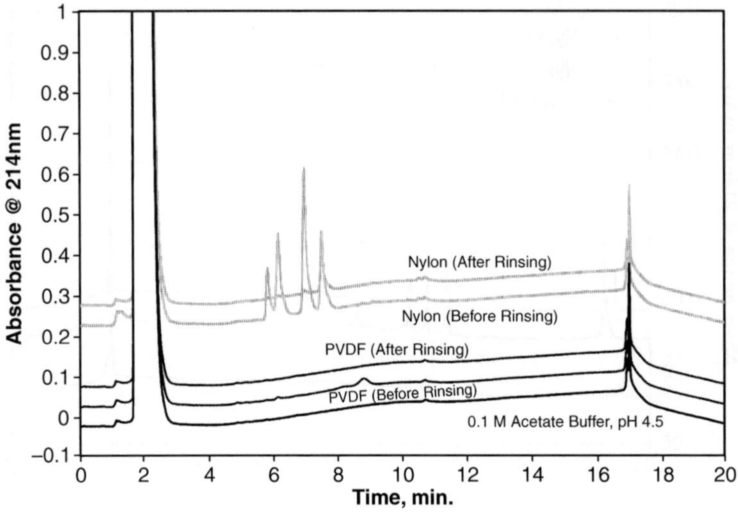

Fig. 4.10 Effect of rinsing on reduction in extractables Various syringe filters were rinsed with 1.0 mL of extracting solvent and the second milliliter of the filtrate was collected and analyzed using reversed-phase HPLC similar to Fig. 4.8

For MS analysis it is important to also look at extractables as seen by mass spectrometry. Figures 4.11 and 4.12 provide examples of MS extractables for two different types of syringe filters, viz. Hydrophilic PTFE and Hydrophilic Polypropylene. Both these membranes have very good chemical compatibility and low extractables by LC-UV analysis, but when the same membranes are tested for extractables by mass spectrometry, the Polypropylene membrane shows much higher level of extractables as compared with the Hydrophilic Teflon membrane.

We compared MS extractables profile of syringe filters available from various mass spec. solvents. Eight commonly used mass spec. solvents were used in this study (Milli-Q® Water, Acetonitrile, Methanol, 0.1% Formic Acid in water, 0.1% Trifluoroacetic Acid (TFA) in water, 0.1% Ammonia in water, 5% Tetrahydrofuran (THF) in water and 5% Isopropyl alcohol (IPA) in water). Syringe filters containing Hydrophilic PTFE, Polypropylene and Nylon membranes were used in this study. Table 4.3 given below shows comparison of MS extractables of various syringe filters commercially available. This data also reinforces superiority of Hydrophilic PTFE membrane from extractables perspective.

All this extractables data clearly shows that different membranes show different levels of extractables, but overall, hydrophilic PTFE membrane shows the least amount of extractables in LC-UV as well as MS.

In contrast with microfiltration membranes, UF membranes typically contain additives such as glycerin or Poly ethylene glycol, PEG, as a humectant (Wright et al., 2005) and these additives contaminate the sample as the sample is filtered through them. These additives can affect downstream analysis and hence it is

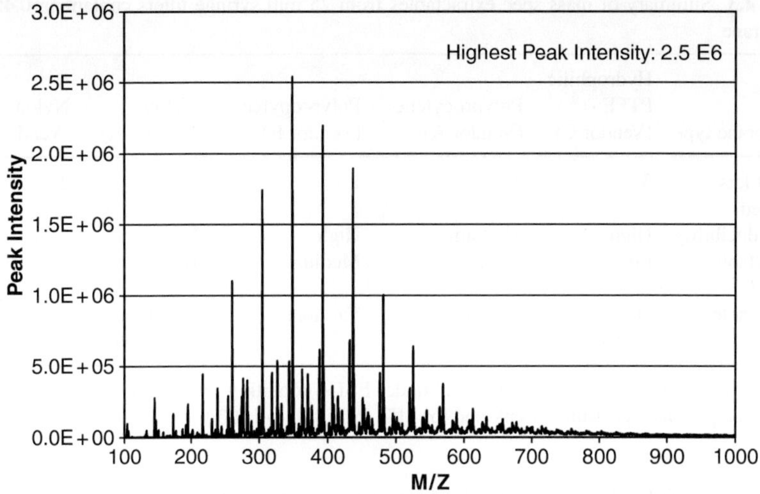

Fig. 4.11 Extractable profile acquired for a polypropylene syringe filters (0.45 μm) using mass spectrometry. Acetonitrile was filtered through the syringe filters and first milliliter of the filtrate was collected. This filtrate was infused into API 2000 at a flow rate of 20 μL/min for 15 min. Total ion current and av. Mass spectra were measured using ESI + mode and M/Z range of 100–1000

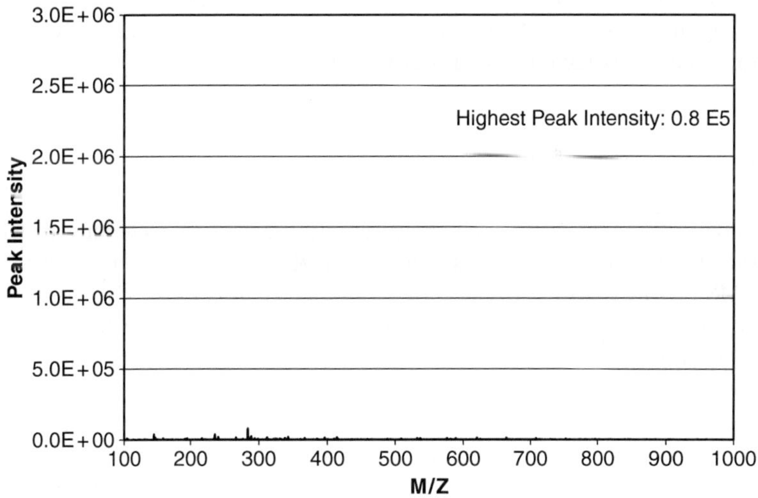

Fig. 4.12 Extractable profile of hydrophilic Teflon filters (0.45 μm) using mass spectrometry

important to have knowledge about them before selecting the ultrafiltration device. Especially, the presence of PEG in ultrafiltration membranes shows a very distinct pattern of peaks separated by 44 amu in mass spectrum. To reduce interference from these contaminants, the filters can be washed multiple times with a solvent (buffer, water etc.) that may be present in the sample.

Table 4.3 Summary of mass spec extractables from 25 mm syringe filters containing 0.45 μm membrane

Membrane type	Hydrophilic PTFE (Vendor C)	Polypropylene (Vendor A)	Polypropylene (Vendor B)	Nylon (Vendor A)	Nylon (Vendor B)
No. of lots tested	3	2	2	2	2
Reproducibility	High	Medium	High	Low	Low
Extractable level	Low	High	Medium	High	High
Extractable type	100–400 (M/Z)	Polymeric	Variable	Polymeric-variable	Polymeric-variable

MS infusion experiments were carried out under ESI + condition for 15 min at 20 μL/min flow rate. Average mass spectrum was plotted for M/Z from 100 to 1000

As a general rule of thumb, the best solvent for pre-rinsing of the filter is the sample itself. Using the sample not only removes the extractable impurities, but also saturates the filter with the analyte of interest thereby significantly reducing analyte binding. Volume of solvent needed to pre-rinse the filter varies depending on the filter size, but typically for a 25 or 33 mm syringe filter, pre-rinsing with 1–3 mL of sample is sufficient. If sufficient amount of sample is not available for pre-rinsing, then a solvent, buffer or water in which the sample is dissolved can be used for reducing extractable impurities from filters.

4.4 Summary

In summary, filtration is a very simple sample preparation technique that helps removal of particulate material as well as soluble impurities from samples prior to LC-MS or MS analysis. Because of its simplicity, it is most often neglected in the process of optimization of sample preparation. In order for filtration to succeed, choice of filtration device needs to be made appropriately based on various sample characteristics such as sample size, chemical nature of the sample, particle load and intended downstream analytical technique. Also, two important factors to take into account while selecting a filtration device are the extractables and analyte binding characteristics of the device and how that affects the downstream analysis.

References

Brandt, I., Vriendt, K., Devreese, B., Beeumen, J., Dongen, W., Augustyns, K., Meester, I., Scharpé, S., and Lambeir, A. (2005). Search for substrates for prolyl oligopeptidase in porcine brain. Peptides *26*, 2536–2546.

Chéneau, E., Henri, J., Pirotais, Y., Abjean, J., Roudaut, B., Sanders, P., and Laurentie, M. (2007). Liquid chromatography–electrospray tandem mass spectrometric method for quantification of monensin in plasma and edible tissues of chicken used in pharmacokinetic studies: Applying a total error approach. J Chromatogr B *850*, 15–23.

González, A., Rodríguez-Velasco, M., Ben-Gigirey, B., and Botana, L. (2006). First evidence of spirolides in Spanish shellfish. Toxicon *48*, 1068–1074.

Heuvel, R.H., and Heck, A.J. (2004). Native protein mass spectrometry: From intact oligomers to functional machineries. Curr Opin Chem Biol *8*, 519–526.

Liu, D.Q., and Hop, C. (2005). Strategies for characterization of drug metabolites using liquid chromatography–tandem mass spectrometry in conjunction with chemical derivatization and on-line H/D exchange approaches. J Pharm Biomed Anal *37*, 1–18.

Machtejevas, E., Andrecht, S., Lubda, D., and Unger, K. (2007). Monolithic silica columns of various format in automated sample clean-up/multidimensional liquid chromatography/mass spectrometry for peptidomics. J Chromatogr A *1144*, 97–101.

Tsao, R., Yang, R., Young, J., Zhu, H., and Manolis, T. (2004). Separation of geometric isomers of native lutein diesters in marigold (*Tagetes erecta* L.) by high-performance liquid chromatography–mass spectrometry. J Chromatogr A *1045*, 65–70.

Willems, A., Deforce, D., Lambert, W., Peteghem, C., and Bocxlaer, J. (2004). Rapid characterization of oligonucleotides by capillary liquid chromatography–nano electrospray quadrupole time-of-flight mass spectrometry. J Chromatogr A *1052*, 93–101.

Wright, S., Pellegrino, J., and Amy, G. (2005). Humectant release from membrane materials. J Membr Sci *246*, 227–234.

Yang, A., Sun, L., Musson, D., and Zhao, J. (2005). Application of a novel ultra-low elution volume 96-well solid-phase extraction method to the LC/MS/MS determination of simvastatin and simvastatin acid in human plasma. J Pharm Biomed Anal *38*, 521–527.

Zurbonsen, K., Bressolle, F., Solassol, I., Aragon, P., Culine, S., and Pinguet, F. (2004). Simultaneous determination of dexamethasone and 6-hydroxydexamethasone in urine using solid-phase extraction and liquid chromatography: Applications to in vivo measurement of cytochrome P450 3A4 activity. J Chromatogr B *804*, 421–429.

Chapter 5
Pressure-Assisted Lysis of Mammalian Cell Cultures Prior to Proteomic Analysis

Emily A. Freeman and Alexander R. Ivanov

Abstract Mass spectrometry-based proteomic analysis of specimens derived from mammalian cell cultures and tissues provides informative views into the complex molecular mechanisms of cellular processes. Good control, efficiency, and reproducibility of protein extraction from cells and tissues are essential aspects for downstream proteomic assays, as well as for other diverse biological and basic research applications. However, common cell lysis techniques suffer from multiple limitations including the lack of reproducibility in protein extraction, covalent alteration of the sample molecules, and misrepresentation of subpopulations of cellular proteins, among others. In this work we have examined the effect of altering both the lysis buffer and the mechanism of lysis in order to enhance proteomic coverage from a cell culture sample. Ultra-high hydrostatic pressure cycling and the addition of fluorinated alcohol in the lysis buffer have been tested due to their proposed individual abilities to solubilize and change conformations of proteins in order to facilitate downstream enzymatic digestion and LC-MS/MS-based proteomic analysis. The resulting data shows that both changes in the cell lysis method demonstrate enrichment for different classes of proteins and protein classes, including membrane- and organelle-associated proteins, meanwhile increasing the total number of recovered proteins and peptides when compared to the conventional lysis methods.

Keywords Cell lysis · Fluoroalcohols · Hexafluoroisopropanol · Pressure-cycling technology · Protein extraction

5.1 Introduction

Both the quality and quantity of proteomics samples provide the groundwork for a successful analysis; in many experiments utilizing tissue cultured cells, a sufficient quantity of sample protein is much less difficult to achieve than a sample that

---

A.R. Ivanov (✉)
HSPH Proteomics Resource, Department of Genetics and Complex Diseases, Harvard School of Public Health, Boston, MA 02115, USA
e-mail: aivanov@hsph.harvard.edu

A.R. Ivanov, A.V. Lazarev (eds.), *Sample Preparation in Biological Mass Spectrometry*, DOI 10.1007/978-94-007-0828-0_5,

properly represents the cellular proteome at a given time. Given the complexity of the proteome in mammalian cell cultures, and of the mammalian cell itself, it is not surprising that huge portions of extremely important and specialized protein types are misrepresented in cell culture-derived proteomic profiles (Annan et al., 2001; Griffin and Schnitzer, 2010; Santoni et al., 2000). In this body of work we have investigated alternative methods for the isolation of proteins from mammalian cell culture using variations in physical extraction methods, as well as lysis buffer composition.

Unlike other strains used in cell culture, such as yeast and bacterial species, most mammalian tissue cultures are composed of cells that are relatively simple to lyse. Techniques often employed for this task are homogenization, sonication, detergent-based, and shearing force-based methods. While they all will physically disrupt the cell, these techniques also present problems with reproducibility, efficacy, and bias towards subsets of protein families, especially a failure to isolate and identify membrane and integral proteins. In essence, data generated using these protocols can be misrepresentative of the actual proteome of the cell culture under investigation, providing a critical hindrance to meaningful progression a study relying on proteomic profiling.

Similar to the techniques for cell lysis, there tends to be minimal variation in lysis buffers that are traditionally used. Often these buffers contain denaturants such as urea, detergents such as CHAPS, sodium dodecyl sulfate (SDS), or Triton X-100, and reductants such as 2-mercaptoethanol (BME) or dithiothreitol (DTT). Others contain varying concentration of salts to take advantage of the cell lysis by osmotic shock. Recent studies have evaluated the use of alcohols and organic solvents such as methanol in order to facilitate membrane protein solubilization and improve cell lysis (Blonder et al., 2006). Fluoroalcohols have received attention in the last decade corresponding to their unique influence on protein conformations and structure, mainly their ability to unfold the three-dimensional conformation (Gast et al., 2001; Redeby et al., 2004, 2006; Redeby and Emmer, 2005; Sirangelo et al., 2003). We investigated the use of fluoroalcohols in cell lysis by using buffers that did and did not contain hexafluoroisopropanol (HFIP) in addition to the more traditional reagents, and evaluated resulting protein extracts for the presence of proteins of specific cellular localization and functional categories.

Hydrostatic pressure has been examined as a factor influencing biochemistry in multiple ways, such as unicellular growth inhibition and changes in chemical reaction equilibria (Gross and Jaenicke, 1994). This principle has fueled the concept of using alternating hydrostatic pressure, referred to as Pressure Cycling Technology (PCT), in the enzymatic proteolysis of protein samples (Freeman and Ivanov, 2010; Lopez-Ferrer et al., 2008). In addition, recent work has examined the use of PCT as an aid to disrupt tissue samples prior to downstream MS-based proteomics (Gross et al., 2008; Ringham et al., 2007; Smejkal et al., 2006; Smejkal et al., 2007). In this study we intended to optimize conditions for cellular lysis by alternating hydrostatic pressure in the presence of various solvents and buffers and understand the respective contribution of physical and chemical factors to protein extraction efficiency in proteomic sample preparation. Our experimental purpose was to utilize technologies currently available to both streamline the protein isolation process, as well as minimize sample to sample variability. Here we have compared the use of conventional protocols and those utilizing PCT. Application of pressure-assistance and additions of fluoroalcohols to the conventional sonication- and urea-based lysis buffer resulted in significant improvement of throughput and reproducibility of sample preparation. Superior extraction rates for cytosolic, nuclear, mitochondrial, ribosomal, and membrane-associated proteins, as well as for proteins related to vital gene ontology (GO) processes (RNA splicing, chromatin assembly, organelle organization among others) were observed in pressure-assisted and organic solvent-assisted sample preparation. There was also enrichment in proteins which contained putative transmembrane domains (TMDs). It is our hypothesis that modifications of the lysis procedure, such as the ones which we have tested here, will function to improve recovery of proteins from defined subcellular localizations due to the characteristic hydrophobic properties of proteins associated with membranes and organelles.

5.2 Methods

5.2.1 Lysis of HepG2 Cells

The conventional lysis buffer was formulated according to Table 5.1. The HFIP buffer was of the same formulation in 30% aqueous HFIP instead of water. Human hepatocellular carcinoma HepG2 cells (ATCC catalog # HB-8065) were grown in 10% FBS-supplemented MEM to 90% confluence in three individual 10 cm culture dishes, and were treated as follows: plate A received conventional lysis buffer and was treated with sonication lysis; plate B received conventional lysis buffer and was treated with PCT-assisted cell lysis; plate C received 30% HFIP lysis buffer and sonication lysis. The cell layers were first rinsed with cold PBS, then exposed to

Table 5.1 The composition of the conventional lysis buffer

Conventional buffer
8 M urea
50 mM NaCl
50 mM NH_4HCO_3 (pH ~ 8.0)
Protease inhibitors (1 tablet per 10 mL)
1 mM NaF
1 mM β-glycerophosphate
1 mM sodium orthovanadate
10 mM sodium pyrophosphate
1 mM PMSF
2 mM CaCl2
5 mM TCEP
[a]3275 μL H_20

[a]The 30% HFIP buffer substituted HFIP for the water portion of the conventional buffer

1.0 mL of the indicated hypotonic lysis buffer. All plates were exposed to the lysis buffers for 10 min at 4°C, when cells were then scraped into appropriate collection containers. Plates A and C were scraped into 15-mL conical tubes, and plate B was scraped into a 1.4 mL PULSE Tube (Pressure BioSciences, Inc.).

The 15-mL conical tube samples were lysed further at osmotic shock conditions with assistance of sonication in an ultrasonic bath on ice for 15 min, then vortexed for 1 min. The PULSE Tube sample was treated for 30 min in a Barocycler NEP2320 (Pressure BioSciences, Inc.) using a program that consisted of 20 pressure cycles of 99 s at 241 MPa and 5 s at atmospheric pressure. The PULSE Tube sample was then transferred to a 15-mL conical tube and vortexed for 1 min. All lysate samples were centrifuged at 3000×*g* for 5 min (4°C) to spin down cell debris. The clarified portions of the lysate supernatant were transferred to clean conical tubes and used for subsequent analyses.

5.2.2 Tryptic Digestion of Cell Lysates

All lysate samples were initially subjected to a detergent compatible protein concentration assay (BioRad, Hercules, CA) to measure approximate protein concentrations in each sample type. The total volumes of all samples were brought to 2.5 mL with 50 mM ammonium bicarbonate (NH_4HCO_3, pH 8.5), and was supplemented with Tris[2-carboxyethyl] phosphine (TCEP) to sustain its concentration at 5 mM. The lysates were vortexed and incubated for 15 min (60°C) to reduce remaining disulfide bonds, then cooled to room temperature. Iodoacetamide (IAA) was added to 10 mM; the lysates were vortexed and incubated for 30 min at room temperature in the dark. To quench excessive IAA, TCEP was added to 5 mM and incubated for 15 min in the dark (37°C). The lysates were diluted fivefold with 25 mM NH_4HCO_3 (pH 8.5). Six 20 μL aliquots from each of the resulting lysates were transferred to clean 0.5 mL Eppendorf tubes and subjected to tryptic digestion. Each tube of lysate was estimated to contain 20 μg of total protein based upon initial concentration measurements, and each tube received 300 ng of trypsin (0.3 μg in 100 μL). The samples were incubated overnight (37°C), then 10% formic acid was added to halt the enzymatic reaction. Samples were concentrated to 5–10 μL and resuspended to a total volume of 50 μL in 2% formic acid, 2% acetonitrile. Samples were centrifuged for 15 min at 10,000×*g*, after which no sediment was observed. Samples were transferred to polypropylene autosampler tubes before subsequent analysis on LC-MS/MS in duplicate.

5.2.3 Protein Identification by 1D Nano-LC Tandem Mass Spectrometry

A CTC Autosampler (LEAP Technologies, NC) was equipped with two 10-port Valco valves and a 20 μL injection loop. A 2D LC system (Eksigent, CA) was used to deliver the flow rate of 3 μL/min during sample loading and 250 nL/min during nanoflow-LC separation. The following in-house-packed columns were used: a C18 solid phase extraction "trapping" column (250 μm internal diameter (i.d.) × 10 mm) and a nano-LC capillary column (100 μm i.d. × 15 cm, 8 μm i.d. pulled tip (New Objective)), both packed with a Magic C18AQ, 3 μm, 200 Å (Michrom Bioresources) stationary phase. A protein digest (10 μL) was injected onto the trapping column connected on-line with the nano-LC column through the 10-port Valco valve. The sample was cleaned and concentrated using the trapping column, eluted onto and separated on the nano-LC column with a linear gradient of acetonitrile in 0.1% formic acid. The LC MS/MS solvents were Solvent A: 2% acetonitrile in aqueous 0.1% formic acid; and Solvent B: 5% isopropanol 85% acetonitrile in aqueous 0.1% formic acid. An 85-min long LC gradient program was used for analysis of the lysates and included the following elution conditions: 2%B for 1 min; 2–35%B in 60 min; 35–90%B in 10 min; 90%B for 2 min; and 90–2%B in 2 min. The eluent was introduced into LTQ Orbitrap mass spectrometer (Thermo Electron, CA) equipped with a nanoelectrospray source (New Objective, MA) by nanoelectrospray. The source voltage was set to 2.2 kV and the temperature of the heated capillary was set to 160°C. For each scan cycle one full MS scan was acquired in the Orbitrap mass analyzer at 60,000 mass resolution, 6×10^5 AGC target and a 1200 ms maximum ion accumulation time was followed by 7 MS/MS scans acquired for the 7 most intense ions in the *m/z* range 350–1700 amu. The LTQ mass analyzer was set for a 30,000 AGC target and 100 ms maximum accumulation time, a 2.2 Da isolation width, and 30 ms activation at 35% normalized collision energy. Dynamic exclusion was enabled for 45 s for each of the 250 ions that had been already selected for fragmentation to exclude them from repeated fragmentation. The HepG2 samples were prepared in triplicate and each digest was analyzed by LC MS/MS in at least two replicates.

5.2.4 LC-MS/MS Peptide Identification and Quantitative Analysis

For all LC-MS/MS analyses, the MS data .raw files acquired by the LTQ Orbitrap mass spectrometer and Xcalibur (version 2.0.6; Thermo Electron, CA) were analyzed using the Sorcerer IDA2 search engine (version 3.5 RC2; Sage-N Research, Thermo Electron, CA) using the SEQUEST-Sorcerer database search algorithm (version 4.0.4). The search was performed against a concatenated FASTA database containing 114,356 sequences in total, and was comprised of 57,049 proteins from the human (25 *Homo sapiens*) UniProt KB database downloaded from EMBL-EBI on October 23rd of 2008, the 129 common contaminants from our in-house database, and reverse sequences. Methionine, histidine, and tryptophan oxidation [+15.994915 atomic mass units (amu)] and cysteine alkylation [+57.021464 amu with iodoacetamide derivative] were set as differential modifications. No static modifications or other differential posttranslational modifications were employed in the database search. A peptide mass tolerance equal to 30 ppm and a fragment ion mass tolerance equal to 0.8 amu were used in all searches. The monoisotopic mass type, semitryptic peptide termini, and up to 2 missed cleavages were used in all searches. The database search results were filtered and validated using PeptideProphet and ProteinProphet software platforms of the Trans-Proteomic Pipeline (TPP, version 4.0.2., Institute for Systems Biology, WA). The FPR was calculated as the number of peptide matches from a "reverse" database divided by the total number of "forward" protein matches, and converting this to a percentage, similar to the method suggested by Joshua Elias and Steven Gygi (Elias and Gygi, 2007). The filtering was performed to bring the FPR to <0.5% for all datasets. The human FASTA protein sequence database used for searching HepG2 data was processed using the TMHMM Server v. 2.0 (Center for Biological Sequence Analysis, Technical University of Denmark, Lyngby, Denmark) to predict the number of trans-membrane domains (TMDs) for each protein.

5.2.5 Gene Ontology (GO) Term Enrichment Analysis

For comparative functional and subcellular loci characterization of proteins recovered using alternative lysis conditions, we applied the GeneGo MetaCore^{TM} software suite (Version 6.2, GeneGo Inc., St. Joseph, MI). LC-MS/MS protein identification results corresponding to profiling experiments of HepG2 lysates were processed using Peptide Prophet, as described above, and the resulting pepXML files were combined and filtered to FDR <1% using an in-house script. The script removed redundant entries and combined indistinguishable proteins (by identified peptides) into protein groups. The resulting .txt file assembled unique protein accession numbers with corresponding spectral counts detected in each protocol. The merged results were further filtered to remove protein groups detected with a spectral count value lower than 3, common contaminants, and protein groups detected in fewer than two experiments. The resulting false positive rate for protein identification was brought to 1% or lower. The lists of UniProt accession numbers corresponding to proteins recovered in individual protocols were uploaded to the MetaCore suite and analyzed using the GO Localization and GO Processes enrichment module. In instances when MS detected peptides led to identification of a protein group comprised of multiple accession numbers indistinguishable by MS data, we utilized only one of those accession numbers for the downstream bioinformatics analysis. The resulting lists of GO terms and corresponding p-values reflecting the coverage of a term by detected proteins were exported and analyzed using Excel. Pair-wise comparison of GO term enrichment results for sample preparation protocols were performed elucidating the differences between corresponding negative log transformed p-values for each GO term. Seventy of the most differently enriched GO terms were plotted as bar-charts per comparative analysis. The sets of filtered proteins and protein groups subjected to GO enrichment analysis were used to build Venn diagrams and identify overlapping and uniquely detected proteins per pair of compared protocols.

5.3 Results and Discussion

5.3.1 Pressure- and Fluorinated Alcohol-Assisted Cell Lysis

We chose to examine the effects of pressure on protein recovery during the lysis of human hepatocellular carcinoma HepG2 cells; these results were compared to those generated using more traditional sonication lysis techniques. We also compared the use of a more traditional buffer to the use of a buffer containing fluorinated alcohol (HFIP): the "conventional" buffer containing 8 M urea was used in comparison to a 30% HFIP version of the same buffer. In addition, we hypothesized that pressure would function to change protein conformations and accelerate the solubilization of less soluble proteins without mechanical lysis. We also assumed that the ability of HFIP to rapidly denature and unfold proteins, disassemble protein-complexes, deactivate enzymes including kinases, phosphatases and proteases, in cooperation with its solubilizing properties, would be beneficial in improving efficiency and reproducibility of cell lysis and protein extraction.

5.3.2 Improvement of Protein Recovery and the Depth of LC-MS Proteomic Profiling When Using Alternative Lysis Buffers and Pressure Cycling

Each LC-MS/MS injection used protein extracts derived from approximately 10,000 cells, based upon cell counts using a hemocytometer just prior to lysis. The use of PCT for cell lysis resulted in a 23% increase of the total peptide identification rate (8979 spectral counts) when compared to the conventional sonication lysis method (7294 spectral counts) (Fig. 5.1a). When an alternative lysis buffer containing 30% HFIP was used, peptide spectral matches increased by approximately 51% at ambient pressure (11038) (Fig. 5.1a). This trend was also demonstrated by the number of unique peptides identified (Fig. 5.1b), showing the same relative pattern between the three protocols. The combined searches of all samples provided the following numbers of unique peptides per protocol: Conventional = 1720, PCT-assisted = 2365, and HFIP-assisted = 3389.

The total number of identified proteins (Fig. 5.1c) and protein sequence coverage (not shown) were consistent with the results from the peptide identification and spectral count results. The conventional urea-based buffer had the lowest number of total human proteins identified (834) of the lysis conditions tested (30% HFIP = 1120, PCT-assisted = 912, Fig. 5.1c). Similar to the results observed using spectral counts (Fig. 5.1a), the addition of pressure alone to cell lysis showed a slight increase when compared to the conventional protocol, and the addition of 30% HFIP increased the total proteins identified by nearly 34% when compared to

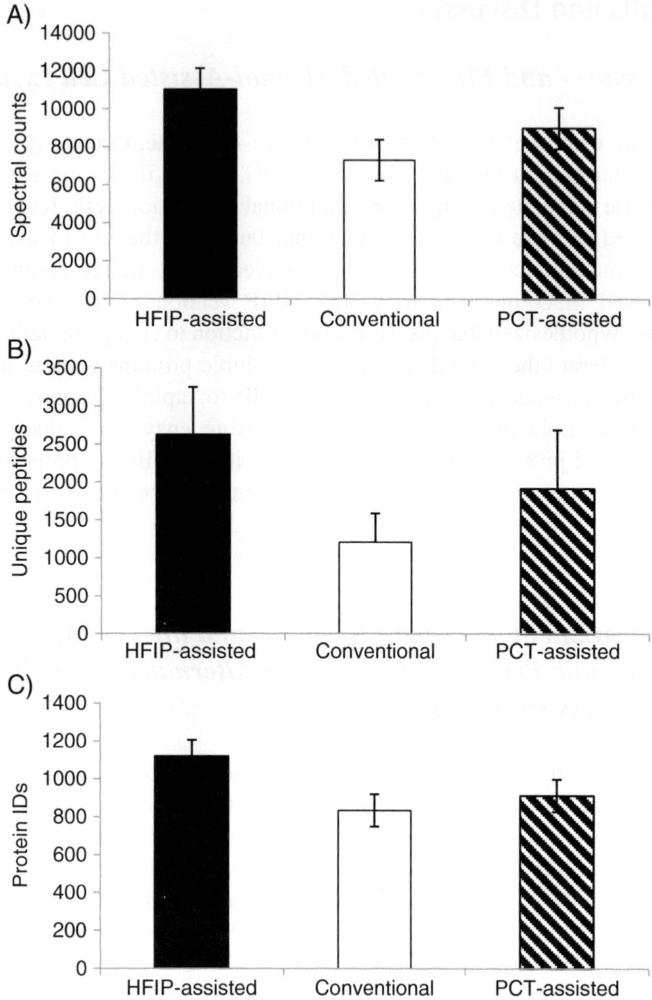

Fig. 5.1 Results from analysis of HepG2 lysate generated using three different lysis methods; conventional sonication, HFIP-assisted, and PCT-assisted. (**a**) Spectral counts generated using the three methods. (**b**) Total unique peptides identified. (**c**) Total proteins identified

the conventional protocol. In addition to the increased numbers of recovered and identified proteins, we also observed better reproducibility in sample quantities and protein identification results when utilizing the described changes in lysis buffer and method. This aspect of improved pre-analytical reproducibility is an immensely important feature of sample preparation techniques when attempting to create a proteomic profile for research or diagnostic purposes. Overall, these results demonstrate the benefit of changes in both the lysis buffer used for cellular disruption, and the physical method of cell lysis.

5.3.3 Uniqueness, Functionality, and Localization of Recovered Proteins Using Different Lysis Techniques

The proteomic profiling results of the discussed experiments were used to evaluate alterations in protein localizations and functional associations recovered using the different lysis methods. We have indicated that observed differences in populations of detected proteins cannot be attributed to analytical variability because the overlap in the filtered and validated results of the profiling experiments was always greater than 90% between technical LC-MS/MS replicates. To evaluate the cell lysis protocols and their resulting recovery of proteins corresponding to different functional and subcellular localization categories, we conducted comparisons for unique protein identifications. The proteins were evaluated and categorized using the Metacore software suite from GeneGo, an on-line program that can reveal highly-recovered protein networks and show functional relationships for identified proteins.

The comparison of conventional and PCT-assisted lysis shows that the majority of proteins (669, 62%) were identified in both, with the lesser remaining proteins to be uniquely identified using the two protocols, yet more unique identifications were made in the PCT-assisted form ("Conventional" = 165 IDs, PCT-assisted = 243 IDs) (Fig. 5.2a). Comparative GO (gene ontology) term enrichment analysis between PCT-assisted and ultrasonic lysis illustrated that high pressure allowed for selective moderate enrichment of proteins forming multi-molecular complexes, ribosomal, nuclear and cytosolic proteins (Fig. 5.2b). In contrast, sonication lysis enabled significant enrichment for mostly mitochondrial proteins in addition to some enrichment for other organelle-associated protein groups (Fig. 5.2b). Comparative analysis for GO cellular processes shows that PCT-lysis enriched for protein production, RNA splicing, RNA processing, nuclear complex, proteasomal, lysosomal, and mobilization pathway proteins, while sonication lysis showed preference for metabolic process- and some chromatin-associated proteins. These results support our hypothesis that pressure-cycling could aid the disruption of protein complexes, thus making them more susceptible to tryptic digestion and identification with LC-MS/MS.

The use of a 30% HFIP lysis buffer also showed that the majority of all identified proteins (708, 56%) were identified using both lysis buffers (Fig. 5.3a). The HFIP-assisted protocol had almost 4× as many unique proteins (412) as did the conventional buffer version (126), highlighting the potential benefits of adding the fluoroalcohol to the lysis solution. As for the previous comparison, the resulting proteins were examined on the GeneGO platform; the HFIP-assisted lysis showed enrichment for ribonucleoprotein complex-, ribosome-, cytosol-, organelle-, and membrane-associated proteins. The conventional buffer mostly demonstrated the enrichment of cytosolic proteins, as well as some recovery of membrane-associated and mitochondrial proteins (Fig. 5.3b). The HFIP-assisted buffer also showed enhanced recovery of RNA processing, RNA splicing, translation, ribonucleoprotein biogenesis, and protein mobilization pathways, while the conventional buffer had comparatively more enrichment for metabolic and catabolic proteins.

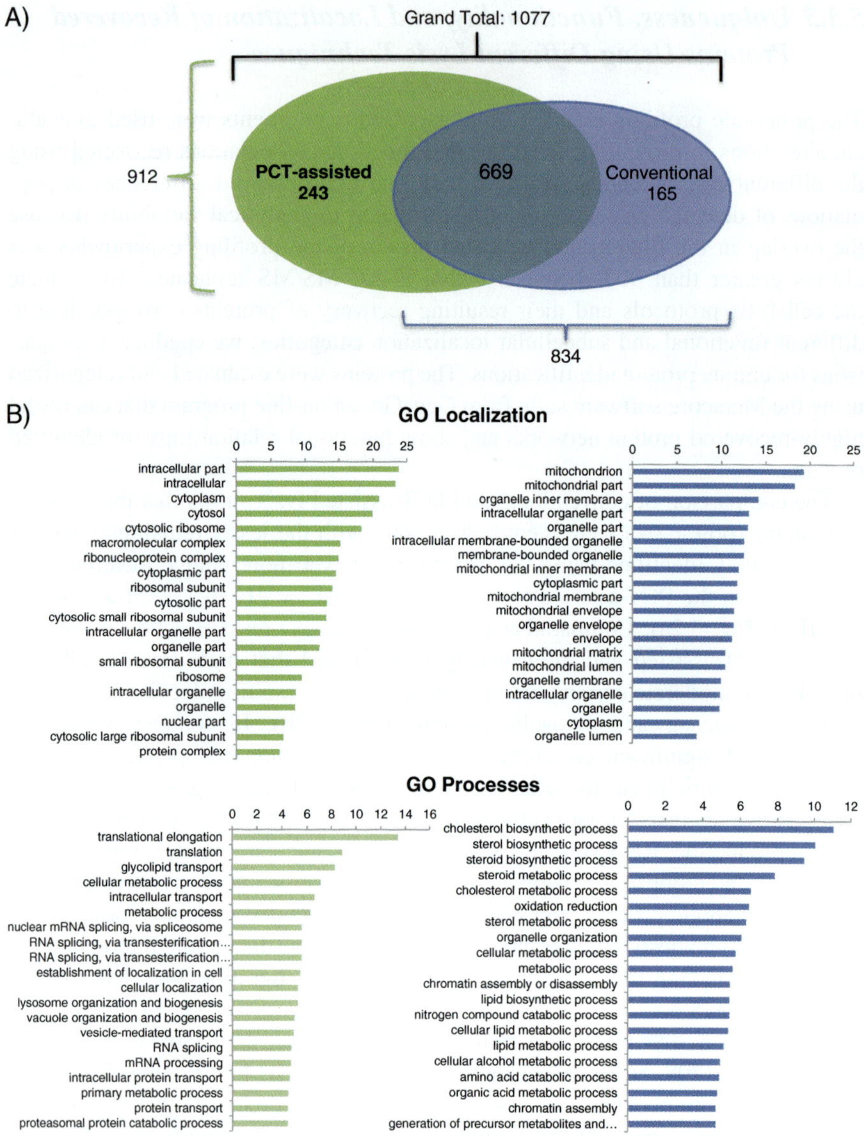

Fig. 5.2 Unique proteins identified using PCT-assisted lysis. (**a**) The overlap of protein identifications between PCT-assisted cell lysis, and that of conventional (sonication) lysis. (**b**) The GO term enrichment analysis reveals enrichment of specialized protein groups and associated functions per each method

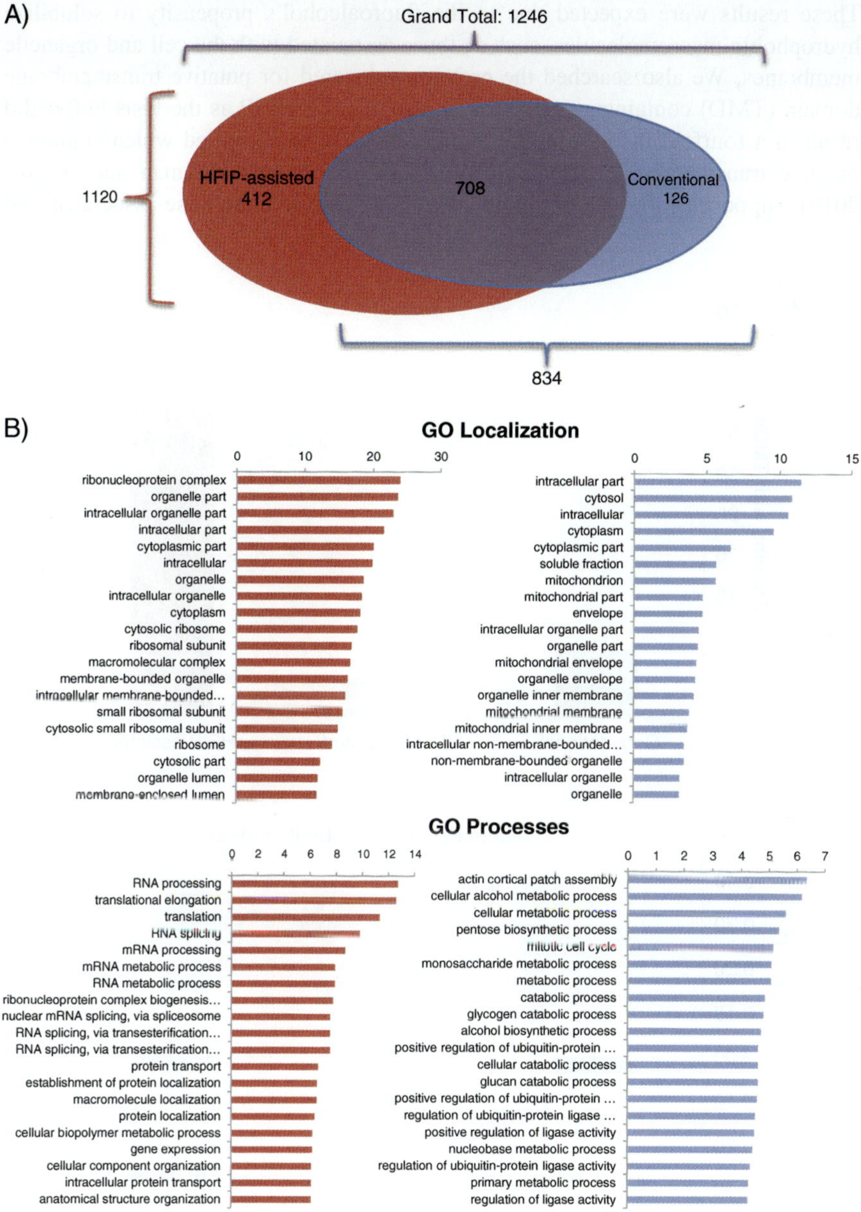

Fig. 5.3 Unique proteins identified using HFIP-assisted lysis. (**a**) The overlap of protein identifications between HFIP-assisted cell lysis, and that of conventional (sonication) lysis. (**b**) The GO term enrichment analysis reveals enrichment of specialized protein groups and associated functions per each method

These results were expected due to the fluoroalcohol's propensity to solubilize hydrophobic macromolecules, such as those associated with the cell and organelle membranes. We also searched the proteins recovered for putative transmembrane domain (TMD) containing proteins. The use of 30% HFIP as the lysis buffer did result in a fourfold increase in the number of proteins identified which contained putative transmembrane domains (TMDs) (Fig. 5.4a and Freeman and Ivanov, 2010), supporting the use of fluoroalcohols to extract membrane-associated and

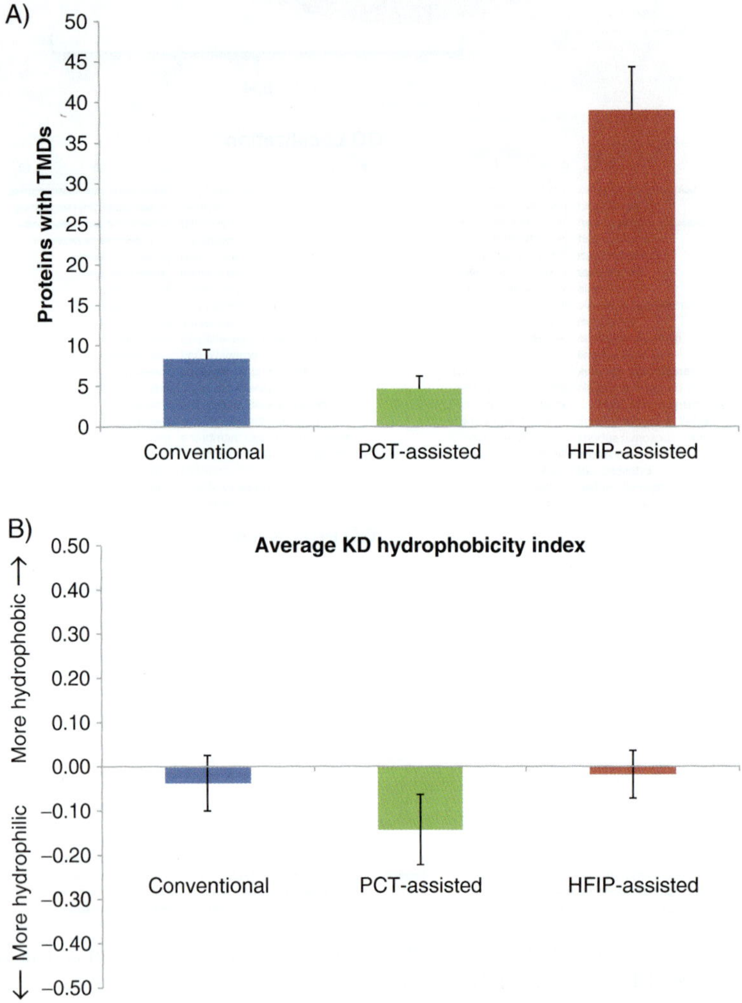

Fig. 5.4 Additional properties of the unique protein identifications made using the three lysis methods. (**a**) Total proteins containing putative TMDs identified using each of the methods. (**b**) Relative average Kyte–Doolittle hydrophobicity indexes for the total human proteins identified from each method

other hydrophobic proteins. PCT-assisted lysis showed less pronounced enrichment for TMD-containing proteins on its own. The peptides acquired from the lysates were also given calculated Kyte–Doolittle hydrophobicity values (Kyte and Doolittle, 1982), and then these values were averaged per protocol to examine changes between lysis buffers or PCT-assistance in recovering more or less hydrophobic proteins and peptides. Contrary to the results observed from recovered protein data, the result showed that there were no major differences in the hydrophobicity of tryptic peptides derived from proteins extracted by the discussed methods (Fig. 5.4b), regardless of the method or buffer used (Freeman and Ivanov, 2010).

5.4 Concluding Remarks

Our analysis on specific alterations in the method of human cell lysis has shown that both of the tested major categorical changes, the physical lysis mechanism and the lysis buffer, provide significant changes in the protein families and characters which are recovered. Simple alterations in the constituents of lysis buffers, such as the ones we have presented here with the addition of a fluoroalcohol, can provide enhanced recovery and downstream identification of "difficult" proteins, such as those that are membrane-associated and/or hydrophobic in nature. Similarly, the use of ultra-high alternating hydrostatic pressure was shown to increase the recovery of unique proteins from the complex cell lysate mixture. It is therefore suggested that the independent but cooperative effects of these two lysis components are able to achieve better recovery of complex-forming and organellar proteins. We therefore hypothesize that these methods could very likely provide essential mechanisms to isolate proteins that compose regulatory and survival pathways in healthy and disease-model cells, providing information that is critical to the age of discovery-driven proteomics. We predict that the design of our experiments can provide a predictive window into the enrichment patterns for specific protein sets that would also be observed when using larger proteomic samples and similar lysis methods, depending upon the cell type of origin. Relatively simple changes in very traditional procedures for cell lysis could provide novel and valuable data for the era of fast-paced proteomics research.

Acknowledgements We would like to acknowledge the experimental and analytical contributions of Yelena Margolin for this project. We are also thankful for fruitful discussions and experimental support provided by Vera Gross and Alexander Lazarev of Pressure BioSciences. We thank the Department of Genetics and Complex Diseases at the Harvard School Public Health for funding support.

References

Annan, R.S., Huddleston, M.J., Verma, R., Deshaies, R.J., and Carr, S.A. (2001). A multidimensional electrospray MS-based approach to phosphopeptide mapping. Anal Chem *73*, 393–404.

Blonder, J., Chan, K.C., Issaq, H.J., and Veenstra, T.D. (2006). Identification of membrane proteins from mammalian cell/tissue using methanol-facilitated solubilization and tryptic digestion coupled with 2D-LC-MS/MS. Nat Protoc *1*, 2784–2790.

Elias, J.E., and Gygi, S.P. (2007). Target-decoy search strategy for increased confidence in large-scale protein identifications by mass spectrometry. Nat Methods *4*, 207–214.

Freeman, E., and Ivanov, A.R. (2010). Proteomics Under Pressure: Development of Essential Sample Preparation Techniques in Proteomics Using Ultra-High Hydrostatic Pressure (Boston, MA, Harvard School of Public Health).

Gast, K., Siemer, A., Zirwer, D., and Damaschun, G. (2001). Fluoroalcohol-induced structural changes of proteins: Some aspects of cosolvent-protein interactions. Eur Biophys J *30*, 273–283.

Griffin, N.M., and Schnitzer, J.E. (2010). Overcoming key technological challenges in using mass spectrometry for mapping cell surfaces in tissues. Mol Cell Proteomics *10*(2), R110.000935.

Gross, V., Carlson, G., Kwan, A.T., Smejkal, G., Freeman, E., Ivanov, A.R., and Lazarev, A. (2008). Tissue fractionation by hydrostatic pressure cycling technology: The unified sample preparation technique for systems biology studies. J Biomol Tech *19*, 189–199.

Gross, M., and Jaenicke, R. (1994). Proteins under pressure. The influence of high hydrostatic pressure on structure, function and assembly of proteins and protein complexes. Eur J Biochem *221*, 617–630.

Kyte, J., and Doolittle, R.F. (1982). A simple method for displaying the hydropathic character of a protein. J Mol Biol *157*, 105–132.

Lopez-Ferrer, D., Petritis, K., Hixson, K.K., Heibeck, T.H., Moore, R.J., Belov, M.E., Camp, D.G., 2nd, and Smith, R.D. (2008). Application of pressurized solvents for ultrafast trypsin hydrolysis in proteomics: Proteomics on the fly. J Proteome Res *7*, 3276–3281.

Redeby, T., Carr, H., Bjork, M., and Emmer, A. (2006). A screening procedure for the solubilization of chloroplast membrane proteins from the marine green macroalga Ulva lactuca using RP-HPLC-MALDI-MS. Int J Biol Macromol *39*, 29–36.

Redeby, T., and Emmer, A. (2005). Membrane protein and peptide sample handling for MS analysis using a structured MALDI target. Anal Bioanal Chem *381*, 225–232.

Redeby, T., Roeraade, J., and Emmer, A. (2004). Simple fabrication of a structured matrix-assisted laser desorption/ionization target coating for increased sensitivity in mass spectrometric analysis of membrane proteins. Rapid Commun Mass Spectrom *18*, 1161–1166.

Ringham, H., Bell, R.L., Smejkal, G.B., Behnke, J., and Witzmann, F.A. (2007). Application of pressure cycling technology to tissue sample preparation for 2-DE. Electrophoresis *28*, 1022–1024.

Santoni, V., Molloy, M., and Rabilloud, T. (2000). Membrane proteins and proteomics: Un amour impossible? Electrophoresis *21*, 1054–1070.

Sirangelo, I., Dal Piaz, F., Malmo, C., Casillo, M., Birolo, L., Pucci, P., Marino, G., and Irace, G. (2003). Hexafluoroisopropanol and acid destabilized forms of apomyoglobin exhibit structural differences. Biochemistry *42*, 312–319.

Smejkal, G.B., Robinson, M.H., Lawrence, N.P., Tao, F., Saravis, C.A., and Schumacher, R.T. (2006). Increased protein yields from Escherichia coli using pressure-cycling technology. J Biomol Tech *17*, 173–175.

Smejkal, G.B., Witzmann, F.A., Ringham, H., Small, D., Chase, S.F., Behnke, J., and Ting, E. (2007). Sample preparation for two-dimensional gel electrophoresis using pressure cycling technology. Anal Biochem *363*, 309–311.

Chapter 6
Multiplexed Preparation of Biological Samples for Mass Spectrometry Using Gel Electrophoresis

Jeremy L. Norris and Alan A. Doucette

Abstract In spite of the perception that gel electrophoresis has limited compatibility with modern mass spectrometry, polyacrylamide gel electrophoresis remain the gold standard for high resolution protein separation. Recent work has sought to capitalize on both the advantages of gel electrophoresis and mass spectrometry as a tool for protein separation and analysis, combining the two technologies into integrated workflows that have unique capabilities. This chapter discusses recent advances of this technology and applications for its use in mass spectrometry assays.

Keywords Electrophoresis · Fractionation · Proteomics · Quantitation · Sample preparation

6.1 Introduction

The preparation of proteins for mass spectrometry analysis is limited to only a few commonly used separation tools. One can broadly classify these tools as solution-based (e.g. SPE, HPLC, membrane ultrafiltration) or gel-based approaches. Polyacrylamide gel electrophoresis (PAGE) has long been the method of choice for high resolution protein separation. The advantages of gel-based separations are numerous, but include efficiency, scalability, and low cost. Beyond these advantages, PAGE separations use high concentrations of SDS to fully solubilize the protein and impart uniform charge to the molecule. This removes the bias of protein solubility, which often is of concern in solution-based approaches. Moreover, SDS PAGE separation is linear over a wide molecular weight range, a benefit in a proteomics workflow.

Despite the simplicity of the approach, as well as the high resolution separations achievable with SDS PAGE, gel-based approaches have long held an unfavorable reputation. Once comprehensive MS-based proteome analysis first began to develop,

---

J.L. Norris (✉)
Protein Discovery, Inc., Knoxville, TN 37902, USA
e-mail: Jeremy@proteindiscovery.com

A.R. Ivanov, A.V. Lazarev (eds.), *Sample Preparation in Biological Mass Spectrometry*, DOI 10.1007/978-94-007-0828-0_6,

HPLC approaches for separation of peptides from globally digested proteomes quickly became adopted as the favored method. Not only were such solution workflows more readily automated, incorporating a single solution-phase digestion over multiple in-gel digestions, but these also seemed to provide improved proteome coverage. The classic method for protein analysis involves proteome separation and visualization through 2D gel electrophoresis, with spot excision and in-gel digestion for coupling to MS. Using this classic approach, it was found that the strategy is strongly biased to high abundance proteins (Gygi et al., 2000). Most recently, however, a variant on the classic gel-to-MS workflow was established, which removed this bias.

The "GeLC" proteome workflow is a simple extension of Mann & Shevchenko's robust in-gel digestion strategy which enables bottom-up MS analysis of SDS PAGE-separated proteins (Shevchenko et al., 2006, 1996). Here, the entire gel lane from a one-dimensional SDS PAGE separation is subject to in-gel digestion and MS analysis. Thus, in this workflow it is no longer a requirement that a protein be detected (i.e. visualized through staining) prior to MS analysis. Given that mass spectrometry is a more sensitive detector than conventional silver or Coomassie staining, it is therefore a logical approach to analyze all regions of the gel band, despite an apparent lack of protein as determined through staining. It has since been proven that gel-based separation workflows have the capacity to generate reliable MS data for low abundance proteins. Recent studies incorporating GeLC as a sample preparation platform for bottom-up MS have shown nearly comprehensive proteome coverage (de Godoy et al., 2008). Furthermore, recent comparisons of the GeLC approach to other methods of proteome fractionation (e.g. cation exchange, isoelectric focusing, etc.) at both the protein and peptides level demonstrate the GeLC approach to have superior performance (Fang et al., 2010; Piersma et al., 2010).

Since it is difficult to argue the merits of SDS PAGE as a reliable platform for biological sample preparation ahead of mass spectrometry, one might still have reservations in choosing such an approach over solution-based alternatives. For one, the GeLC strategy is limited to bottom-up MS analysis, in that it relies on in-gel digestion to achieve efficient recovery of peptides from the gel. The in-gel digestion process itself is often considered a cumbersome, labor-intensive approach. What is therefore desired is an approach which yields reliable separation of proteome samples based on protein molecular weight, but avoids the necessity of in-gel digestion. This chapter highlights our recent work to develop a simple, easy to use, preparative scale electrophoresis approach that can be used in conjunction with mass spectrometry. The core technology, Gel Eluted Liquid Fraction Entrapment Electrophoresis (GELFrEE) was originally developed and published by Tran et al. in 2008 and has been subsequently commercialized in two formats, an 8-channel and a 96-channel platform as described below (Tran and Doucette, 2008). The technology utilizes concepts from continuous tube gel electrophoresis, but incorporates a unique process of trapping eluted fractions in a defined liquid volume, which permits significant concentration even for high mass fractions. Additionally, the approach incorporates important improvements in gel and buffer chemistry which provide for fast, reproducible, and high resolution separations in a format that is directly compatible with enzymatic digestion and high performance nano-liquid chromatography tandem mass spectrometry (nLC/MS/MS). The workflow provides unique advantages over other techniques of sample preparation. The design, use, and advantages of these approaches for various sample preparation tasks will be reviewed.

6.2 Multiplexed Preparative Electrophoresis Devices

Specific design requirements must be considered for preparative devices that differ significantly from those of an analytical gel. The key considerations to mention include loading capacity, speed of separation, ease of use, reproducibility, and recovery. Taking all of these design elements under consideration, a format was developed that allowed for each of these requirements to be met or exceeded.

6.2.1 8-channel Electrophoretic Separations

Experiments are performed using the Gelfree® 8100 Fractionation System (Protein Discovery, Inc.) which is comprised of two components: a controller power supply and a single use gel cartridge. The electrophoretic controller provides for application of constant voltage up to 240 V. The cartridge (Fig. 6.1) has five separate chambers: (a) the cathode buffer reservoir, (b) the sample introduction chamber, (c) the tube gel, (d) the sample collection chamber, and (e) the anode reservoir. The tube gel is comprised of polyacrylamide gels cast inside 8 mm O.D. borosilicate glass tubes. A 3.5 kDa molecular weight cut off membrane separates the cathode buffer chamber from the sample introduction chamber, as well as the sample collection chamber from the anode buffer chamber. These membranes permit electrophoretic current to pass but trap molecules greater than 3.5 kDa. The sample is

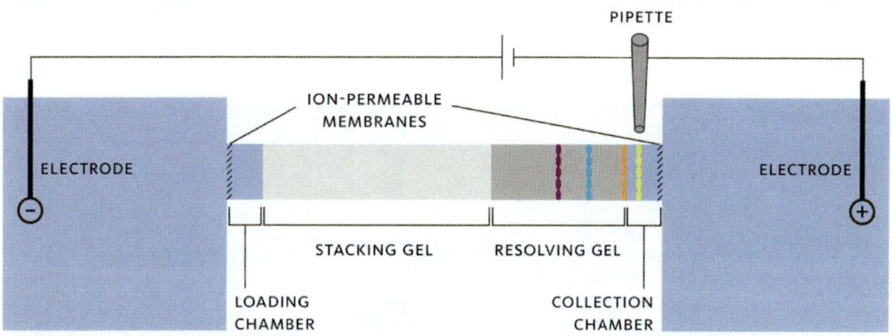

Fig. 6.1 Schematic representation of GELFREE 8100 continuous elution tube gel electrophoresis device. The sample is placed into the Loading Chamber and electrophoretically separated through the tube gel, finally eluting into the Collection Chamber. Protein samples are sequentially collected using a pipettor

introduced into the device through the sample introduction chamber. The sample is mixed with a sample buffer provided with the cartridge kit, adjusting the volume to completely fill the sample introduction chamber and buffering the sample pH. The sample buffer contains reducing agent and detergent to aid in the preparation and electrophoretic separation of the sample. The remaining 3 chambers, the cathode reservoir, anode reservoir, and sample collection chamber are filled with electrophoresis running buffer. During the separation, the proteins contained in the sample are electrophoretically driven from the sample introduction chamber on the cathode side of the cartridge toward the anode. The separation is carried out in a polyacrylamide gel in two phases. To compensate for the large volume of sample introduced into the sample chamber and to maximize resolution, a stacking gel is employed. In this region, the sample is focused into a tight band prior to migration into the resolving region of the gel. The separation of the sample into discrete fractions of molecular weight is accomplished in the resolving gel. As samples are eluted from the end of the resolving gel, they are emptied into a collection chamber of fixed volume bound on one side by the tube gel and on the other side with the ion-permeable membrane. The membranes are specifically chosen to allow buffer ions to pass, but not proteins. The entrapped sample is removed from the collection chamber using a pipette and the fraction can be stored for subsequent analysis.

6.2.2 96-channel Electrophoretic Separations

For applications that necessitate higher numbers of samples be prepared using the GELFrEE technique, a different device has been developed, permitting the simultaneous preparation of as many as 96 samples is a single experiment. The system, called the Passport® 1200 Sample Prep System, consists of three components: a power supply, control software, and a single use cartridge. Figure 6.2 shows a CAD drawing of the 96-well GELFrEE cartridge. The design utilizes the principles from the 8-channel GELFrEE device above. In this case however, the tube gels are turned vertically and formatted in an 8 × 12 format at 9 mm pitch. This format

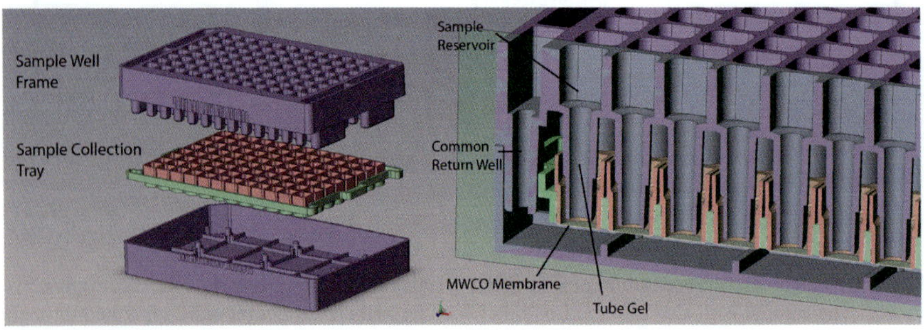

Fig. 6.2 Schematic of 96-channel tube gel electrophoresis device

integrates directly with liquid handling robotics, making it possible to fully automate the preparation of many more samples. In order to accommodate 96 GELFrEE separation chambers and fit within the confines of a standard 96 well microtiter plate footprint, several modifications are necessary. To save space the number of independent counter electrodes is reduced. The design incorporates a common return area below the molecular weight cutoff membranes to complete the circuit. This design saves space and allows for the necessary format to interface with existing laboratory equipment. The cartridge is constructed entirely from molded biocompatible plastic, and has three components: the base, sample collection tray, and sample well frame. Tube gels are cast directly into the cartridge well frame. The tube gel is filled allowing for a sample loading volume of at least 100 μL. The sample collection tray has a membrane spanning the bottom of the reservoir. The user fills the sample collection tray to the desired volume prior to loading the sample. The collection chamber can accommodate 35–200 μL of buffer. When the cartridge well frame is seated onto the sample collection tray, the tube gels are immersed into the buffer, completing a circuit. Once assembled, the sample is loaded into the well frame directly on top of the gel. The area above the sample is filled with buffer to control the pH of the separation. As the voltage is applied, the analytes migrate through the tube gel and elute into the sample collection tray below. The sample collection chamber traps the sample between the end of the tube gel and a 500 Da molecular weight cut-off membrane. Once the desired molecular fraction has been collected, the voltage is turned off, the well frame is removed, and the sample can be transferred to a clean 96 well plate for analysis. The preparation time for 96 samples is approximately 1 h.

Current is applied to the cartridge using a customized, 96-channel electrophoretic controller (Harkins et al., 2008). The cartridge is loaded into the instrument through a mechanical drawer. Once inside the instrument, an array of electrodes is lowered into position, immersing the electrodes into the wells. The experiment is defined in the control software using a graphical user interface. The system is controlled by the application of constant current to each well, passing a predetermined number of coulombs of charge for each well. Each well is independently controllable allowing for rapid method development by systematically varying time, applied potential, loading conditions, or buffer conditions. Once a method has been optimized, the system is used to accomplish parallel processing of samples for high throughput. Methods may be stored and recalled for use at a later time.

6.3 The Use and Application of the GELFrEE Technique

The principles of separation in GELFrEE are identical to those that drive SDS PAGE separation. The only difference of the GELFrEE experiment is that the voltage application is not terminated, forcing the proteins to migrate from the end of the gel "column". In doing so, proteins are separated in time, with smaller proteins migrating from the gel column and being collected first. Since proteins are required to migrate through the entire length of the gel column, a relatively short length of resolving gel is sufficient to resolve proteins over a broad molecular weight

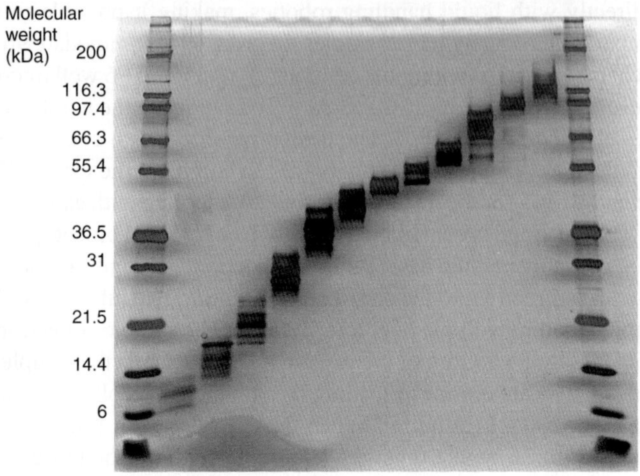

Fig. 6.3 One-dimensional gel electrophoresis of Gelfree 8100 fractionated sample. Two hundred micrograms of lysate from *S. cerevisiae* were separated, using tube gel electrophoresis, into 12 fractions. The fractions collected from that separation are visualized using 1D gel separation

range (~10 kDa to >200 kDa). The use of a short resolving gel also enables these broad mass separations to be complete in approximately 90 min of applied voltage. Typically, up to 20 discrete fractions are collected over this mass range. Figure 6.3 demonstrates the separation power of the approach with a complex cell lysate. A sample of yeast, *S. cerevisiae*, was lysed in the presence of protease inhibitors to create a complex sample for demonstration of the GELFrEE technique. A total of 200 μg of yeast lysate was loaded into the Gelfree 8100 device and the separation was carried out, collecting 12 fractions spanning the mass range from ~3.5 to 150 kDa. The resulting fractions were prepared and analyzed using a 1D tris-glycine gel (Invitrogen, Carlsbad, CA) and silver stained to visualize the separation. The collection chamber effectively retains the solubilized proteins until such time as the fraction is removed from the GELFrEE device. Protein yields are nearly quantitative over the mass range ~10 to >200 kDa with the cartridge (Tran and Doucette, 2008). A general workflow for proteome analysis using GELFrEE as a sample preparation device is outlined in Fig. 6.4. As can be seen from the workflow, simple to carry out and amenable to both bottom-up and top-down MS.

6.3.1 Efficient Removal of SDS

The high recovery of protein from GELFrEE is partly attributed to the fact that SDS is present in solution within the collection chamber, which keeps proteins fully dissolved in the running buffer. However, the presence of SDS is also a limiting factor in an MS detection workflow if it not properly removed, given the incompatibility of this ionic detergent to electrospray ionization (Beavis and Chait, 1990; Ikonomou

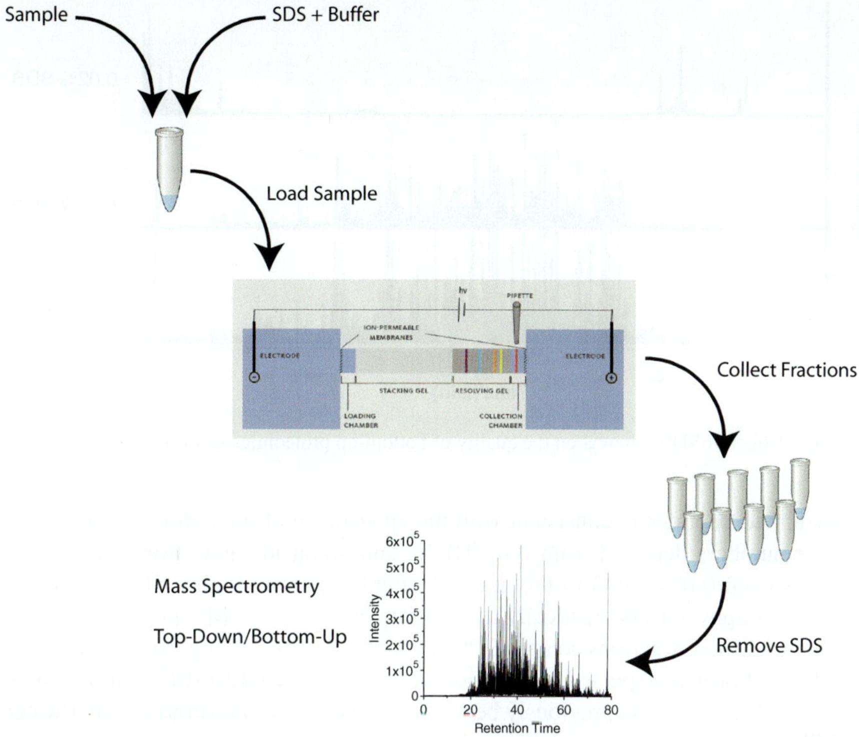

Fig. 6.4 Sample Preparation Workflow of the GELFREE™ 8100 Sample Fractionation Station

et al., 1990). While it is possible to operate GELFrEE under native conditions, the benefits of mass-based separation imparted by using SDS make this the favored approach. Perhaps the most infamous contaminant in MS, SDS is often avoided from the proteome workflow. This is unfortunate since SDS is not only a common and effective solubilizing agent, but it is also needed in SDS PAGE for molecular weight separation. The method of SDS removal is critical to the success of the workflow, noting that there are several options available, as long as the percent SDS is reduced to levels that are tolerable for LC/MS. Coupling GELFrEE to MS requires an effective strategy for removal of SDS from the sample buffer prior to analysis. While it is agreed that SDS can cause severe signal suppression in ESI, there is some debate as to what level mass spectrometry will tolerate SDS. Most literature cites the suppression effects of SDS in direct infusion ESI experiments. However, such findings have little bearing on a proteome experiment, as these inevitably couple LC to MS. In doing so, the tolerance threshold of SDS is actually related to the chromatography, as opposed to the ionization. Considering an LC/MS "bottom-up" proteomic experiment, we have found that SDS can be tolerated up to levels of 0.01% in the injected sample without causing any ill effects (Fig. 6.5 – base peak chromatograms showing no SDS, 0.01% SDS, and 0.02% SDS). The number

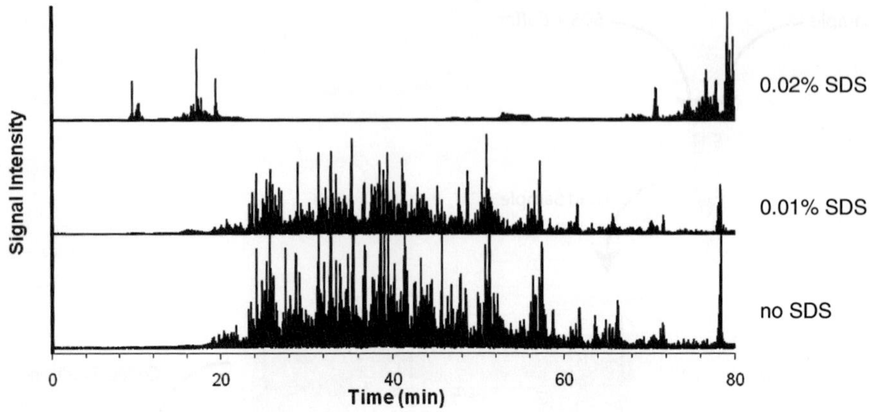

Fig. 6.5 Effect of SDS removal on the quality of bottom-up proteomics analysis

of identified peptides is consistent with the appearance of the chromatogram, with similar numbers detected from 0 to 0.01%, and a considerable drop in identified proteins from 0.02% SDS in the injected sample. At even higher SDS levels, the chromatography suffers drastically and consequently so does MS analysis.

The GELFrEE fractions contain SDS at concentrations ranging from 0.1 to 2.0%. The level of protein expected to be collected from a single GELFrEE fraction can be as low as 1–5 μg. It has previously been shown that the chloroform/methanol/water (CMW) precipitation protocol (Wessel and Flugge, 1984) is an effective means of coupling SDS-containing protein samples to MALDI MS (Puchades et al., 1999). The method was therefore applied to GELFrEE fractions for coupling to LC/MS/MS (Tran and Doucette, 2008). With nearly a 1000 fold reduction of SDS, protein fractions are readily analyzed. The CMW protocol has the desirable feature of being rapid, economical, and easily scalable. With some practice by the user, consistent yields on the order of 50–70% are achievable. The method also enables top-down analysis (Section 6.3.3). Alternative precipitation protocols have also been explored. Acetone precipitation, when combined with 1 or 2 additional washing steps, will reduce SDS to below threshold levels, even at initial level of SDS in the sample of 2%. From triplicate analysis, a comparison of GELFrEE fractions analyzed by bottom-up MS through acetone and CMW precipitation revealed 70% of identified proteins to be in common, with a slightly greater number of proteins (9%) observed through acetone precipitation (Botelho et al., 2010). The presence of SDS in the sample does not prevent MS analysis, and simple protocols are available for effective removal of the interfering detergent prior to MS.

While removal of SDS from GELFrEE fractions is generally not a limiting step in the proteome workflow, it does present added challenges. Most particularly, with any precipitation step, there is concern for protein loss. Protein losses are most likely due to incomplete resolubilization of the sample. Any solubilizing additive added at this stage must be compatible with mass spectrometry, or at the very least easily removed (such as through protein trapping on an HPLC guard column). The addition of urea, thiourea, or mass spectrometry compatible detergents (e.g. cleavable surfactants) can be considered to improve protein solubilization to achieve near quantitative recovery.

6.3.2 Bottom-Up Proteomics

As a broad mass fractionation strategy, the resolution of GELFrEE is not sufficient to fully resolve discrete proteins from a complex mixture. However, GELFrEE resolution is more than sufficient to simplify a complex mixture ahead of LC/MS analysis. This strategy of fractionation of complex mixtures has been shown effective using a number of differing separation strategies. The advantage of fractionation of complex mixtures includes an increase in sensitivity, proteome coverage, and dynamic range of the analysis. In the case of GELFrEE, the proteins detected by MS follow a predicted pattern of increasing molecular weight for the larger fractions. Of the detected proteins, two thirds are confined to 1 or 2 fractions. For other proteins, detection in multiple fractions is typically a consequence of trace levels of that protein diffusing over multiple fractions or protein degradation due to residual protease activity after extraction of the protein from the native environment. Degradation of the proteome will enevitable cause fragments of proteins to be detected in molecular weight fractions other than the molecular weight fraction predicted by the genome.

Figure 6.6 shows the results of the analysis of 100 μg of yeast, *S. cerevisiae*, fractionated into 16 fractions using the Gelfree 8100 Fractionation System. The 16 fractions were processed to remove SDS, digested using trypsin, and analyzed by LC/MS/MS on an LTQ linear ion trap equipped with a nanospray source. The peptides were separated on 150 mm × 180 μm i.d. Biobasic C18 column (Thermo Scientific, San Jose, CA) using a gradient from 2 to 40% acetonitrile. Peptides were

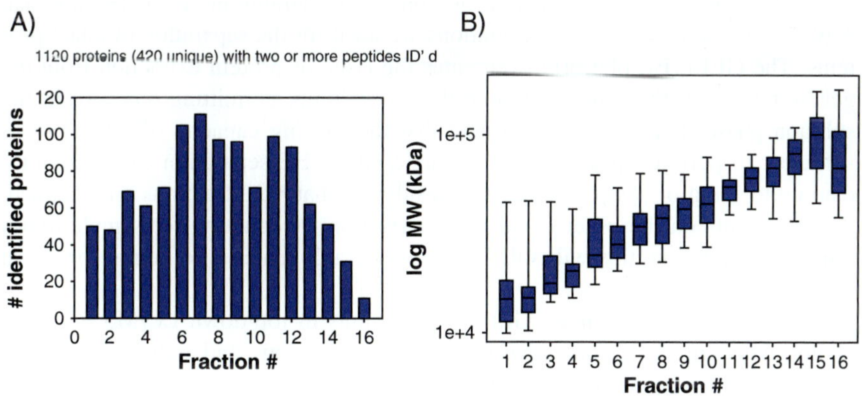

Fig. 6.6 Results of bottom-up analysis of *S. cerevisiae* after GELFREE fractionation. (**a**) Plot shows the distribution of 1120 protein with two or more peptides identified per protein. (**b**) A whisker box plot of the protein molecular weight versus fraction number for all identified proteins shows a strong correlation between median protein size and fraction number. Adapted with permission from Tran, J.C., Doucette, A.A. (2009). Multiplexed Size Seperation of Intact Proteins in Solution Phase for Mass Spectrometry. *Anal Chem.*, 2009, 81 (15), pp 6201–6209

identified using Sequest; peptides with charge +1, +2, and +3 were accepted with X_{corr} scores greater than 1.9, 2.2, and 3.75 respectively. Proteins were identified with a false positive rate of 1%. Figure 6.6a shows the distribution of peptide identification in each fraction of the 16 GELFrEE fraction collected and analyzed. As expected the majority of the protein mass is found in the middle fractions, resulting in greater numbers of proteins identified from these fractions. Analysis of the predicted molecular weight of the proteins identified shows a similarity in trend with the gel image in Fig. 6.3. Each subsequent fraction contains proteins having a higher average molecular weight. In total, the analysis yielded 1110 proteins identified (4891 peptides) from the combined 16 fractions, 428 of those being unique proteins (Tran, 2008).

6.3.3 Top-Down Proteomics

The development of high-throughput solution-based separation platforms for proteome analysis was almost completely driven by bottom-up MS analysis. Single and multidimensional modes of peptide separation, including ion-exchange and reversed-phase chromatography, as well as chromatofocusing, isoelectric focusing and other techniques, can all achieve very high resolution peptide separation. More importantly, these techniques generally have high recovery for all peptides within the sample. The same cannot be said when these techniques are applied to the separation of intact proteins. Given that most solubilizing additives (detergents in particular) must be avoided during chromatographic separation, unbiased recovery for all protein components of a sample becomes problematic. As an example, the recovery of high molecular weight proteins, as well as extremely hydrophobic proteins is compromised in reversed-phase chromatography. In other words, the separation of proteins in solution is considerably more challenging than peptide separation. Gel-based separations are ideal for the separation of intact proteins. The GELFrEE platform overcomes the issue of protein extraction from the gel by migrating the sample through the gel column, permitting recovery in the solution phase. It is important to note that the loading capacity of GELFrEE is sufficient to permit top-down MS. A single GELFrEE separation can accommodate up to 1 mg of sample material. Considering that ~20 fractions are obtained, this represents on the order of 50 μg per fraction. Accounting for sample losses during precipitation, this will still leave considerable material to permit top-down analysis.

The coupling of low mass GELFrEE fractions to top-down LC/MS has been demonstrated in the comparative study of asynchronous and mitotic cells from human cell lines (Lee et al., 2009). HeLa S3 cells were grown in culture, lysed and fractionated using GELFrEE, and analyzed using a top-down approach. The cell lysate was fractionated into 16 fraction of 150 μl each containing approximate 10 μg of protein. Five of the fractions spanning the mass range 5–25 kDa were subject to chloroform/methanol/water precipitation to remove residual SDS from the fractions. The proteins were resuspended in the HPLC starting solvents (95% water,

5% ACN, 0.1% FA), and subjected to LC/MS/MS. The fractions were separated on a 100 mm × 75 μm i.d. C_4 or PLRP-S analytical column (New Objective, Woburn, MA). The samples were analyzed with a 12T or 14.5T LTQ-FT-Ultra mass spectrometer (Thermo Fisher Scientific, San Jose, CA). Data dependent, top 2 or top 3 MS/MS acquisition runs with a 10 m/z isolation window were used. Data was processed and analyzed using ProSightPC 2.0. Analysis of a HeLa cell nuclear extract resulted in detection of 286 distinct protein forms with 225 total protein identifications. This corresponds to the identification of 35 unique protein accession numbers. A total of 10–20 unique protein accession numbers were identified in each GELFrEE/LC/MS/MS injection. The coupling of GELFrEE to reversed-phase HPLC therefore constitutes a robust 2D separation platform for solubilized proteins. This coupling resulted in the identification of several intact proteins within the predicted mass range, including measuring relative changes in protein phosphorylation when comparing asynchronous and mitotically arrested cells.

6.3.4 High-Throughput Peptide Quantitation

One application where the number of samples requires the throughput of the 96-well design of the Passport® 1200 Sample Prep Station is the quantitation of peptides using LC/MS/MS (Harkins et al., 2008). Peptide quantitation is becoming more necessary and widespread due to the growth in development and production of biological drugs and the need for quantitative proteomics assays. Peptides are important in drug development, as they represented over $8B in pharmaceutical revenue in 2005. The analysis of analogues in the discovery phase demands generic high-throughput LC-MS/MS procedures. The high sensitivity (pg/ml) and specificity of nanoLC-MS/MS makes it an attractive choice for the quantitation of endogenous peptides. Quantification of peptides from complex sample matrices at high rates using LC-MS/MS requires efficient sample preparation to remove interferences that cause instrument contamination and ion suppression. However, traditional sample preparation methods such as protein precipitation, though inexpensive, often provide poor selectivity, requiring long chromatographic runs. The high relative expense of mass spectrometers, long chromatographic cycles, and serial analysis make it difficult for nanoLC-MS/MS to compete with multiplexed immunoaffinity assays in a clinical setting. Electrophoretic enrichment combined with nanoLC-MS/MS was thus evaluated for sample quality, chromatographic throughput, and matrix interference removal.

For this application, native gel electrophoresis was used to prepare endogenous plasma peptides for analysis by mass spectrometry. Rather than collecting discrete ranges of protein molecular weight, this application requires only collection of the lowest molecular weight species in the sample. During this mode of operation, the sample is electrophoretically transferred through the gel, and the separation is purposefully stopped only after the peptides of interest have eluted from the gel. By using this approach, the larger molecular weight species are effectively depleted from the sample, preventing ion suppression effects and column overloading due to the contaminating protein species. In addition to high molecular weight depletion of the sample, neutral species are readily removed from the sample during electrophoresis. Under native conditions, the pH of the sample dictates the charge carrying species in the sample. The result is that under a given polarity, cations and anions move in opposite directions while neutral species do not migrate at all. By choosing the polarity of the instrument appropriately, either anionic or cationic species can be selectively enriched from the sample. This feature allows for the removal of some well known ion suppression agents.

Phospholipids are known ion suppression agents and the extent of the effect of phospholipids has been studies by numerous groups (Little et al., 2006). The benefits of electrophoretic sample preparation for quantitative nano LC/MS/MS assays are demonstrated using the example of angiotensin I. The angiotensin I assay was performed using a 0.075 × 50 mm C18 reversed-phase column coupled to a triple quarupole mass spectrometer. Figure 6.7 demonstrates the negative effect that phospholipids have on quantitative assays. The endogenous peptide, angiotensin I, coelutes with a series of phospholipids that were simultaneously monitored during a nanoLC/MS/MS analysis of a protein precipitated sample. These compounds

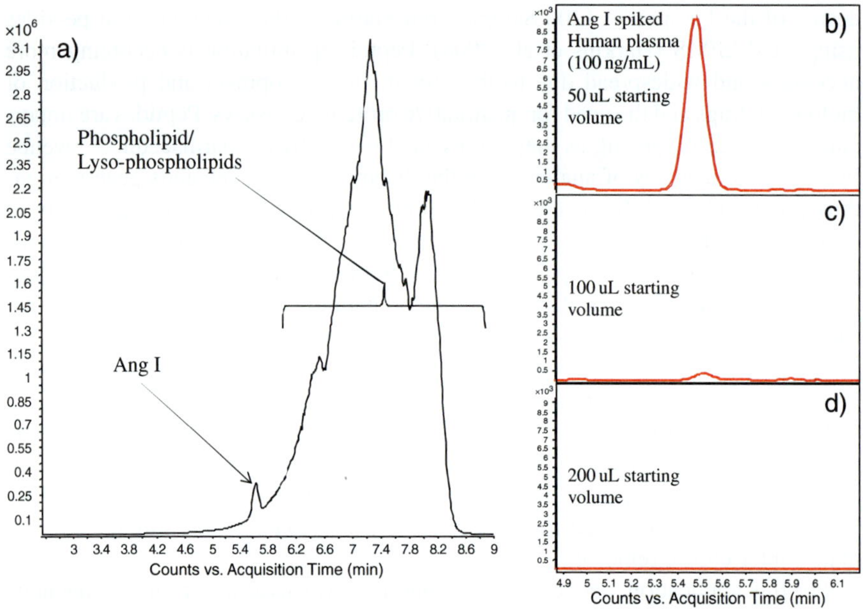

Fig. 6.7 The effect of protein precipitation on angiotensin I spiked (100 ng/mL) into human plasma. Clinical samples are contaminated with many chemical species that interfere with, and can often prevent, the analysis of peptides at endogenous levels. Panel (**a**) above shows a chromatogram of angiotensin I spiked human plasma prepared for analysis by protein precipitation. Co-elution of lipid contaminants with the analyte of interest limits sensitivity of the assay. Attempts to measure endogenous levels of peptides by increasing the starting volume of plasma also increase the level of ion suppression, ultimately decreasing sensitivity (**b–d**)

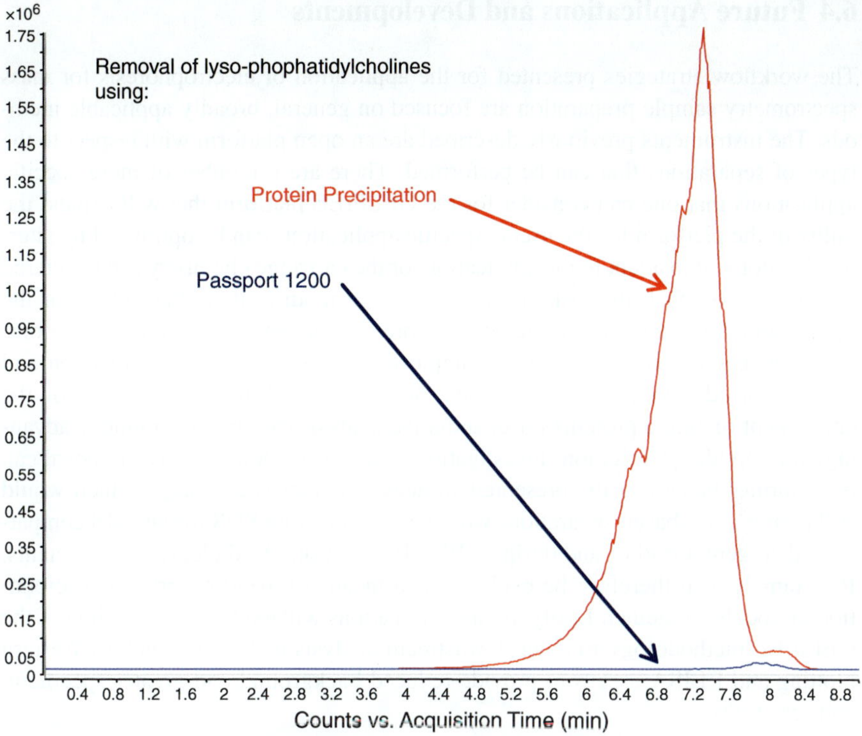

Fig. 6.8 Comparison of phospholipid contamination after protein precipitation and preparation by Passport 1200. The chromatogram demonstrates the removal of greater than 99% of the contaminating phospholipids from human plasma samples prepared using Passport 1200 sample preparation. The *red trace highlights* a representative example, a lyso-phosphatidylcholine (MS/MS transition 496 → 186). The *blue trace* is representative of the same plasma sample processed using Passport 1200

limit the sensitivity of the assay. Furthermore, an attempt to increase the sensitivity of the analysis by starting the analysis with more plasma results in enrichment of the contaminants and a loss in sensitivity for the analyte of interest. Enrichment of the peptides using electrophoresis provides a means to exclude the phospholipids components efficiently. Figure 6.8 shows an example of a single phospholipid in a protein precipitated sample compared to the same volume of plasma processed using electrophoresis. Greater than 99% of the phospholipids monitored were removed from the sample. Cleanup at of this type enabled the analysis of angiotensin I at endogenous levels directly from 10 μL of human plasma using nanoLC/MS/MS. The lower limit of detection was determined to be below 1 ng/mL. In one example, the endogenous level was determined to be 6.3 ng/mL, corresponding to only 48 amol of angiotensin I in the 10 μL aliquot of plasma analyzed. This level of sensitivity makes possible the measurement of numerous endogenous peptides.

6.4 Future Applications and Developments

The workflow strategies presented for the application of electrophoresis for mass spectrometry sample preparation are focused on general, broadly applicable methods. The instruments previously described are an open platform with respect to the types of separations that can be performed. There are a number of more specific applications that one can consider for the GELFrEE platform that will expand the utility of the platform for the users. Specific applications can be optimized by altering the protocol to adapt to the application, or the cartridge chemistry can be altered to better accomplish the task at hand. Cartridge modifications that allow for the optimized fractionation and enrichment of proteins of specific mass ranges, accomplish native separations of protein complexes, and allow for the purification and recovery of other biopolymers are under development. Protocols that focus on the enrichment of single proteins for quantitative analysis can provide a unique advantage for peptide quantitation. Investigation of novel methods for SDS removal can offer further benefit to the presented strategy. An alternate strategy which would still permit mass-based separations would be to substitute SDS for an "MS compatible" detergent ahead of and during GELFrEE fraction. Acid cleavable surfactants, for example, can therefore be explored as a means of avoiding protein precipitation protocols altogether. Lastly, future applications will explore the coupling of the GELFrEE methodology to other downstream analysis techniques such as Western blotting and ELISA assays to provide reduced background and greater specificity and sensitivity.

6.5 Conclusions

This review provides an overview of recent work and available products that combine gel electrophoresis with LC/MS analysis in a novel way. The implementations of PAGE separation described here are preparative in nature, allowing for the separation of large aliquots of sample. Furthermore, the design permits parallel separation of either 8 or 96 samples to maximize throughput of the analysis. SDS PAGE electrophoresis is an ideal method for the preparation of samples for proteomics analysis due to the lack of bias against hydrophobic proteins and high recovery of the proteins of interest due to the high capacity for SDS to solubilized proteins. The GELFrEE device, being a molecular weight-based separation platform, is orthogonal to other modes of separation including charge (pI) and partially to hydrophobicity. Strategies for the removal of SDS from these samples are now numerous and have been shown effective. Results demonstrate that the technique provides superior recovery of proteins relative to competitive techniques, and the coupling of GELFrEE to two or even three orthogonal dimensions of separation can impart additional separation if required.

References

Beavis, R.C., and Chait, B.T. (1990). Rapid, sensitive analysis of protein mixtures by mass spectrometry. Proc Natl Acad Sci USA *87*, 6873–6877.

Botelho, D., Wall, M.J., Vieira, D.B., Fitzsimmons, S., Liu, F., and Doucette, A. (2010). Top-down and bottom-up proteomics of SDS-containing solutions following mass-based separation. J Proteome Res *9*, 2863–2870.

de Godoy, L.M., Olsen, J.V., Cox, J., Nielsen, M.L., Hubner, N.C., Frohlich, F., Walther, T.C., and Mann, M. (2008). Comprehensive mass-spectrometry-based proteome quantification of haploid versus diploid yeast. Nature *455*, 1251–1254.

Fang, Y., Robinson, D.P., and Foster, L.J. (2010). Quantitative analysis of proteome coverage and recovery rates for upstream fractionation methods in proteomics. J Proteome Res *9*, 1902–1912.

Gygi, S.P., Corthals, G.L., Zhang, Y., Rochon, Y., and Aebersold, R. (2000). Evaluation of two-dimensional gel electrophoresis-based proteome analysis technology. Proc Natl Acad Sci USA *97*, 9390–9395.

Harkins, J.B.t., Katz, B.B., Pastor, S.J., Osucha, P., Hafeman, D.G., Witkowski, C.E., 2nd, and Norris, J.L. (2008). Parallel electrophoretic depletion, fractionation, concentration, and desalting of 96 complex biological samples for mass spectrometry. Anal Chem *80*, 2734–2743.

Ikonomou, M.G., Blades, A.T., and Kebarle, P. (1990). Investigations of the electrospray interface for liquid chromatography mass spectrometry. Anal Chem *62*, 957–967.

Lee, J.E., Kellie, J.F., Tran, J.C., Tipton, J.D., Catherman, A.D., Thomas, H.M., Ahlf, D.R., Durbin, K.R., Vellaichamy, A., Ntai, I., *et al.* (2009). A robust two-dimensional separation for top-down tandem mass spectrometry of the low-mass proteome. J Am Soc Mass Spectrom *20*, 2183–2191.

Little, J.L., Wempe, M.F., and Buchanan, C.M. (2006). Liquid chromatography-mass spectrometry/mass spectrometry method development for drug metabolism studies: Examining lipid matrix ionization effects in plasma. J Chromatogr B Anal Technol Biomed Life Sci *833*, 219–230.

Piersma, S.R., Fiedler, U., Span, S., Lingnau, A., Pham, T.V., Hoffmann, S., Kubbutat, M.H., and Jimenez, C.R. (2010). Workflow comparison for label-free, quantitative secretome proteomics for cancer biomarker discovery: method evaluation, differential analysis, and verification in serum. J Proteome Res *9*, 1913–1922.

Puchades, M., Westman, A., Blennow, K., and Davidsson, P. (1999). Removal of sodium dodecyl sulfate from protein samples prior to matrix-assisted laser desorption/ionization mass spectrometry. Rapid Commun Mass Spectrom *13*, 344–349.

Shevchenko, A., Tomas, H., Havlis, J., Olsen, J.V., and Mann, M. (2006). In-gel digestion for mass spectrometric characterization of proteins and proteomes. Nat Protoc *1*, 2856–2860.

Shevchenko, A., Wilm, M., Vorm, O., and Mann, M. (1996). Mass spectrometric sequencing of proteins silver-stained polyacrylamide gels. Anal Chem *68*, 850–858.

Tran, J.C. (2008). Development of Effective Proteomic Separations for Biological Analysis Using Mass Spectrometry. Thesis (Ph.D) – Dalhousie University, Halifax, Canada. In Chemistry (Halifax, NS, Dalhousie University), p. 177.

Tran, J.C., and Doucette, A.A. (2008). Gel-eluted liquid fraction entrapment electrophoresis: An electrophoretic method for broad molecular weight range proteome separation. Anal Chem *80*, 1568–1573.

Tran, J.C., and Doucette, A.A. (2009). Multiplexed Size Seperation of Intact Proteins in Solution Phase for Mass Spectrometry. Anal Chem *81*, pp 6201–6209.

Wessel, D., and Flugge, U.I. (1984). A method for the quantitative recovery of protein in dilute solution in the presence of detergents and lipids. Anal Biochem *138*, 141–143.

Part II
Methods for Improved Proteolytic Digestion

Chapter 7
Development of an On-Bead Digestion Procedure for Immunoprecipitated Proteins

Matthew J. Berberich, Jeffrey A. Kowalak, Anthony J. Makusky, Brian Martin, Detlef Vullhorst, Andres Buonanno, and Sanford P. Markey

Abstract Combining antibody coated magnetic bead affinity purification as a separation tool with mass spectrometry as a protein identification tool has been a powerful method to determine the composition of interacting proteins and protein complexes. Receptor proteins and insoluble complexes offer added challenges, as they may contain hydrophobic membrane proteins. When strong detergents such as SDS are required to release the complexes from the magnetic beads or prevent their aggregation and precipitation, the subsequent required removal of anionic detergents appears to cause significant sample losses prior to MS analysis. We describe a procedure in which protein denaturation and trypsin digestion are performed directly on the affinity bound complex on the magnetic beads, circumventing detergents and reducing sample loss prior to LC/MS/MS analysis.

Keywords On-bead digestion · Trypsin · Antibody pull-down · Immunoprecipitation · Protein complex · SCX fractionation · Post-synaptic density · ErbB4

7.1 Introduction

The identification of interacting proteins in subcellular complexes in biological systems can provide insight into the structure of complexes as well as their function. The combination of antibody-coated magnetic beads to isolate protein complexes, with mass spectrometry (MS) based proteomics to determine the identity of the constituent proteins, has been an effective strategy to characterize many complexes (reviewed in Gingras et al., 2007). The typical workflow for these experiments involves solubilizing or suspending the insoluble protein complex, trapping and purifying the complex in single or tandem affinity steps, elution and recovery of

S.P. Markey (✉)
Laboratory of Neurotoxicology, National Institute of Mental Health, National Institutes of Health, Bethesda, MD 20892, USA
e-mail: markeys@mail.nih.gov

A.R. Ivanov, A.V. Lazarev (eds.), *Sample Preparation in Biological Mass Spectrometry*, DOI 10.1007/978-94-007-0828-0_7,

the complex, gel based separation of the constituent proteins, and in-gel trypsin digestion followed by LC-MS/MS analysis of the released peptides (Aebersold and Mann, 2003).

However, we have encountered several limitations in applying this affinity purification-mass spectrometry based workflow, particularly when sample quantities are limited and/or contain membrane proteins. Detergents required to solubilize these proteins are problematic for ESI-MS, however the necessity of detergent removal often results in substantial losses when small quantities or membrane protein-containing samples are being handled. While the use of in-gel trypsin digestion has become a standard procedure in proteomics, the peptide extraction efficiency can be poor and variable, especially for hydrophobic peptides. Finally, when the bait protein in the immunopurification is present at low levels in an overwhelming background, the mass spectrometric identification of the bait protein may be suppressed, requiring further strategies to provide evidence that the immunoaffinity procedure was specific and successful.

Our goal was to develop a method to identify affinity-purified protein complexes by mass spectrometry without the use of SDS as a solubilization detergent and without the use of SDS-PAGE as a separation tool. We decided to use mass spectrometry-compatible detergents to solubilize the antibody-protein complex, and then perform all steps in solution and directly on the antibody-coated magnetic beads. Many synthetic and enzymatic reactions are routinely performed directly on surface resins or magnetic beads, thus we suspected that others had not probed direct digestion of materials for mass spectrometric analysis because of anticipated high background from affinity substrates. We reasoned that the beads would be coated in layers, with the outermost being loosely bound and the innermost being tightly associated with covalently bound antibodies. This led to a step-wise approach to direct digestion.

This antibody complex fractionation procedure consists of three distinct steps. First, an initial round of heating, reduction, and alkylation of the affinity bound sample is used to separate protein components that are lightly bound and to collect these as a fraction. Proteins of this fraction may be either contaminants, non-specific interactors e.g., heat shock proteins, or secondary-tertiary interactors. A second fraction is collected from proteolytic digestion using conditions similar to those employed in conventional solution phase digestion procedures. This step produces a very complex peptide mixture requiring 2D LC/MS/MS for analysis. A third fraction is collected by proteolytically stripping the beads, using digestion in the presence of urea to ensure that remaining proteins (or fragments thereof) are effectively solubilized and removed from the magnetic beads.

We tested this procedure with two types of affinity purifications previously performed in our work, and typical of many immunoprecipitations. The first example is a cardiac tissue extract containing the over-expressed bait protein ErbB4, and the second is an affinity-bound insoluble PSD-95 complex isolated from rat brain. While significant quantities of antibody and Protein A peptides were identified in these analyses, this background did not preclude our ability to detect ErbB4 in the first example, or identify proteins previously seen in the usual 1-D PAGE followed by LC/MS (geLC/MS) analysis of SDS eluted PSD-95 complexes. Of particular interest, the peptides that were observed when the bait protein was stripped from the affinity column contained the amino acid segment that was used to generate the polyclonal ErbB4 antibody. Our results suggest that others may find that stepwise digestions of immunoaffinity separated complexes provide a useful probe of complexes directly on-bead, and that the resulting antibody and matrix peptide background spectra are consistent and readily subtracted from those derived from captured targets.

7.2 Experimental Section

Ammonium bicarbonate, dithiothreitol (DTT), 2-iodoacetoamide (IAA) and other chemicals were purchased from Sigma (St Louis MO) and HPLC solvents were purchased from Honeywell Burdick and Jackson (Morristown, NJ).

7.2.1 Sample Isolation and Immunoprecipitation

For each preparative immunoprecipitation (IP), 1.2 ml of Protein A coated magnetic Dynabeads (Invitrogen, Carlsbad, CA) were washed 3 times with IP wash buffer (10 mM Tris-HCl, pH 7.5, 150 mM NaCl, 0.1% NP-40) and incubated with 72 μg of polyclonal affinity purified anti-ErbB4 antibody for 12 h. After incubation, beads were washed with PBS (3×) and antibody was cross linked with 0.1 mM DSS (disuccinyl suberate). The reaction was stopped after 30 min with 1 M Tris-HCl, pH 7.5, and the beads were washed again 3 times with PBS.

Whole heart was dissected from a $ErbB4^{MHCerbB4}$ (Tidcombe et al., 2003) and homogenized in 20 volumes of IP lysis buffer (10 mM Tris-HCl, pH 7.5, 150 mM NaCl, 1% NP-40) with an Ultra-Turrax T 25 basic S1 homogenizer (IKA Works, Inc., Germany). The homogenate was centrifuged at 20,000 g to remove cell debris (SS-34 rotor, RC5B Sorvall centrifuge, DuPont Instruments, Manalapan, NJ). The supernatant was removed, and the sample was diluted to a total volume of 24 ml with lysis buffer. To remove proteins which bind nonspecifically to substrates not containing any antibody, pre-clearing steps were performed with 5 ml of Sepharose 4B beads (Sigma), which had been washed 3 times with IP lysis buffer. The homogenate was incubated with the washed beads for 30 min, then centrifuged (2500 rpm), and supernatant was added to another batch of IP lysis buffer-washed Sepharose beads. This process was repeated 3 times. The pre-cleared homogenate was then added to the anti-ErbB4-bound beads and incubated for 3 h at 4°C. The beads were washed 3 times with IP wash buffer, collected on a magetic rack, and the supernatant was discarded. The experiment was performed in triplicate.

PSD-95 affinity purified samples were prepared as previously described (Dosemeci et al., 2007). Brains from adult Sprague-Dawley rats were custom collected by Pel Freez Biologicals (Rogers, AR), frozen immediately in liquid nitrogen, and shipped on dry ice. Brains were thawed by a 1-min immersion in isotonic sucrose at 37°C and dissected on ice to remove white matter, and cerebral cortices were then rapidly homogenized. A synaptosomal fraction was obtained and treated with 0.5% Triton X-100. Detergent-insoluble pellets were collected, and a PSD-enriched fraction was separated by sucrose density centrifugation. This fraction was extracted once more with 0.5% Triton X-100, 75 mM KCl. Dynabeads (M-450 coated with goat anti-mouse IgG) were obtained from Dynal (Oslo, Norway) and were further coated with a monoclonal PSD-95 antibody (MA1–046; ABR-Affinity BioReagents, Inc., Golden, CO). PSD fraction was resuspended in 10 ml of 2% BSA (of 99% purity to minimize contamination of the preparation by the blocker), 0.01% Tween 20 in phosphate-buffered saline using a Kontes Micro Ultrasonic Cell Disrupter, operated at minimum power. Sonication (3 × 1 min with ~3 min cooling intervals in ice) was done at the lowest power setting (<5 watts) taking care not to warm the samples. Resuspended PSD fraction was incubated with antibody-coated magnetic beads (2×10^8 beads) in a final volume of 10 ml for 2 h at 4°C with continuous rotation. Unbound material was removed, and the beads were washed three times with 2% BSA, 0.01% Tween 20 in phosphate-buffered saline and five times with 0.01% Tween 20 in phosphate-buffered saline for a total of ~3 h. The beads were then resuspended in 20% glycerol and stored at –20°C prior to the antibody complex fractionation procedure.

7.2.2 *Antibody Complex Fractionation*

7.2.2.1 Initial Heating-Reduction-Alkylation

The magnetic beads were resuspended in 500 μl of IP wash buffer and heated at 60°C for 45 min. Disulfides in the sample were reduced by adding DTT (6 μl of 45 mM) at 60°C for 30 min. After reduction, the sample was alkylated by adding 10 μl of 200 mM iodoacetamide (IAA) and incubating in the dark for 30 min at room temperature. The magnetic beads were removed on a magnetic rack, and the supernatant was removed and digested overnight at 37°C with 20 μg porcine trypsin (Promega, Madison, WI). The digested peptide mixture was desalted on MacroSpin Vydac Silica C18 columns (The Nest Group, Southborough, MA) using elution buffer containing 80% acetonitrile and 0.1% TFA, dried and resuspended in 15 μl of reversed-phase Buffer A (95:5:0.1 (v:v:v) water: acetonitrile: formic acid). 10 μl of the resulting sample was analyzed by LC-MS/MS.

7.2.2.2 On-Bead Trypsin Digestion

The beads were again resuspended in IP wash buffer, and heating, reduction, alkylation and trypsin digestion were performed as described in 7.3.1. The digested peptide mixture from this step was cleaned (desalted) by MacroSpin Columns (as above), dried and resuspended in 15 μl of reverse phase Buffer A (95:5:0.1 (v:v:v) water: acetonitrile: formic acid) and 10 μl was analyzed by LC-MS/MS. The gradient used for this sample was extended from 45 to 65 min to permit better separation and detection of peptides.

Due to the increased complexity of the PSD sample, strong cation exchange chromatography (SCX) was used to separate the peptides prior to LC-MS/MS. After the pull-down was completed, the sample was resuspended in 100 μl of SCX Buffer A (10 mM ammonium formate, 20% acetonitrile, 0.05% formic acid) for strong cation exchange chromatography using a Shimadzu LC system (SCL-10A system controller, LC-10AD pump, FRC-10A fraction collector, SPD-10A UV detector monitoring absorbance at 220 and 280 nm). The sample was loaded onto a 4.6 × 200 mm PolySULFOETHYL A column (PolyLC, Columbia, MD), and peptides were separated by a 36 min stepwise gradient run to 500 mM ammonium formate, 20% acetonitrile, 2.0% formic acid (SCX Buffer B), with Buffer B concentration of 0, 1, 3, 5, 7, 10, 15, 20, 30, 50, 70, and 100% with elution steps of 3 min. Fractions were collected every 2 min, dried in a centrifugal evaporator and resuspended in ddH_2O. As above, the process was repeated three times to remove residual ammonium formate prior to LC-MS/MS.

7.2.2.3 Urea + On-Bead Trypsin Digestion

After the above treatment, the beads were resuspended in 8 M urea and heated, reduced, and alkylated as described before. The sample was diluted to < 2 M urea with 100 mM ammonium bicarbonate, pH 8.0, and digested overnight with trypsin at 37°C. The sample was dried in a centrifugal evaporator and resuspended in 100 ul Nanopure ddH_2O (Millipore, Billerica, MA). This lyophilization step was repeated three times to remove residual volatile ammonium bicarbonate. Strong cation exchange chromatography was then performed as described above. The flow-through fractions were not subjected to LC-MS/MS

7.2.3 Mass Spectrometry, Protein Identification, and Post Processing

Peptides were separated and analyzed by LC-MS/MS. The LC was an Eksigent NanoLC-2D system with the mobile phases consisting of 95:5:01 (v:v:v) water: acetonitrile:formic acid (Buffer A) and 80:20:0.1 (v:v:v) acetonitrile: water: formic acid (Buffer B). A Spark-Holland autosampler with capillary flow was used to load the sample onto a Peptide Captrap (0.5 × 2 mm, PLRP-S 5 μm 100 Å) (Michrom Bioresources, Inc., Auburn, CA) at 8 μl/min, followed by the application of a reversed-phase gradient to a BetaBasic, C18 PicoFrit Column, 75 mm ID × 360 mm OD; 15 micron tip (New Objective, Inc., Woburn, MA). The gradient was programmed linearly from 5% B to 45% B over 45 min. For samples analyzed using an extended gradient this separation was applied over 65 min.

The MS/MS system was an LTQ-ETD (ThermoFischer Scientific, Waltham, MA). ESI voltage was 1.5 kV, isolation width was 2.00, ion accumulation time was 50 ms. Survey scans were taken from 400 to 2000 m/z, followed by MS/MS on the 5 most intense ions using CID. Normalized collision energy was 35.0 and the minimum precursor ion intensity required was 5000.0.

Following data acquisition, the *.raw data files were searched using Mascot (version 2.2.03) via the Mascot Daemon (Matrix Science, Inc., Boston, MA). The files were searched against the Swissprot database (v57.12 portion of UniProt v15.12), with carbamidomethyl cysteine as a fixed modification (+ 57 Da) and oxidized methionine (+16 Da) as a variable modification. Subsequent searches were performed restricting the taxonomy to Mammalia and considering protein *N*-acetylation (+ 42 Da) as well as pyro-Glutamic acid formation (from Gln), and pyro-carbamidomethyl Cys (–17 Da, from *N*-terminal camCys) as variable modifications. The peptide tolerance was 1.2 Da and the MS/MS tolerance was 0.8 Da. Post processing was performed using the MassSieve software program (Slotta et al., 2010) which filtered for ion scores that were greater than or equal to the identity score, and applied the rules of parsimony (Yang et al., 2004).

7.3 Results and Discussion

7.3.1 General Procedure of On-Bead Digestion

The aim of the antibody complex fractionation procedure is to separate out the components of an immunoprecipitated protein complex without the use of SDS or the use of gel electrophoresis. The main components of this complex consisted of the Protein A coat of the magnetic bead, a cross-linked IgG antibody that can bind the receptor, proteins which are specific binders to the antigen receptor, and a variety of non-specific binding proteins (left, Fig. 7.1). The separation of these components is achieved through three steps (right, Fig. 7.1). Step 1 in the procedure involves an initial sequence of heating, reduction with DTT and alkylation with IAA. This will allow for the removal of proteins which may interact very weakly, proteins that may be the result of non-specific interactions, as well as contaminants such as keratins. Proteins detected in this initial step may be the result of specific or non-specific interactions. The determination of which protein identifications represent biologically valid interactions are based on the peptide and peptide hit enrichment seen in subsequent steps. The second step of the antibody complex fractionation procedure is to add trypsin directly to the antibody-coated magnetic bead protein complex and digest overnight. For extremely complex mixtures, it is necessary to perform offline fractionation prior to LC-MS/MS. While it is expected that peptides resulting from the antibody chains will be present, the vast majority of peptides should be from potentially biologically relevant protein interactions as well as peptides from the used bait protein. In the final step of the antibody complex fractionation procedure following the removal of the peptide supernatant, the beads are resuspended in 8 M urea to ensure that there are no protein components remaining on the magnetic beads. While this step will result in the release of a large amount of the antibody chains and Protein A, there may still be peptides from the bait protein detected.

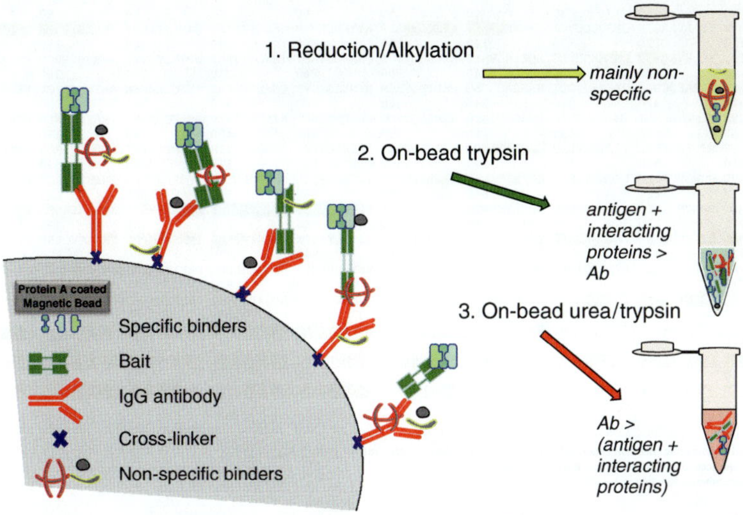

Fig. 7.1 Illustration describing the antibody complex fractionation procedure. This procedure consists of 3 steps, with the first consisting of an initial round of heating-reduction-alkylation to eliminate weak antibody complex interactions and to aid in the removal on non-specific binders (1). Following this, in solution trypsin digestion is performed directly on the magnetic beads, resulting in the disruption of stronger interactions. Peptides resulting from the bait protein, as well as peptides from interacting proteins, will dominate this fraction compared to peptides generated from the antibody (2). The final step involves resuspending the beads in urea and performing another round of in solution trypsin digestion. The majority of these peptides will result from antibody fragments and will ensure that all of the protein as been effectively removed from the magnetic beads (3)

7.3.2 Analysis of ErbB4 Complexes in Heart

To determine if the antibody complex fractionation procedure can be applied to identify immunoprecipitated complexes, we used a biological system where the bait protein is overexpressed. ErbB4 is a receptor tyrosine kinase involved in synaptic function and plasticity (Buonanno and Fischbach, 2001; Mei and Xiong, 2008), and is essential during embryonic development for trabiculation of the heart (Gassmann et al., 1995). In order to circumvent embryonic lethality, an ErbB4 knockout mouse was developed that harbors a transgenic myosin heavy chain promoter construct driving expression of the receptor specifically in the heart (Tidcombe et al., 2003). These mice transgenically expressing ErbB4 in heart, denoted ErbB4^{MHCerbB4}, live to adulthood. We have initially used these mice to study the ErbB4 proteome in heart because compared to the diversity of protein seen in the brain, the protein composition in heart is less complex. A polyclonal antibody against the intracellular component of ErbB4 was successfully used for immunoprecipitation of ErbB4 (Vullhorst et al., 2009). Using heart tissue from ErbB4 heart transgenic mice, we

```
        10         20         30         40         50         60         70         80         90        100
MKPATGLWVW VSLLVAAGTV QPSDSQSVCA GTENKLSSLS DLEQQYRALR KYYENCEVVM GNLEITSIEH NRDLSFLRSV REVTGYVLVA LNQFRYLPLE
       110        120        130        140        150        160        170        180        190        200
NLRIIRGTKL YEDRYALAIF LNYRKDGNFG LQELGLKNLT EILNGGVYVD QNKFLCYADT IHWQDIVRNP WPSNLTLVST NGSSGCGRCH KSCTGRCWGP
       210        220        230        240        250        260        270        280        290        300
TENHCQTLTR TVCAEQCDGR CYGPYVSDCC HRECAGGCSG PKDTDCFACM NFNDSGACVT QCPQTFVYNP TTFQLEHNFN AKYTYGAFCV KKCPHNFVVD
       310        320        330        340        350        360        370        380        390        400
SSSCVRACPS SKMEVEENGI KMCKPCTDIC PKACDGIGTG SLMSAQTVDS SNIDKFINCT KINGNLIFLV TGIHGDPYNA IEAIDPEKLN VFRTVREITG
       410        420        430        440        450        460        470        480        490        500
FLNIQSWPPN MTDFSVFSNL VTIGGRVLYS GLSLLILKQQ GITSLQFQSL KEISAGNIYI TDNSNLCYYH TINWTTLFST INQRIVIRDN RKAENCTAEG
       510        520        530        540        550        560        570        580        590        600
MVCNHLCSSD GCWGPGPDQC LSCRRFSRGR ICIESCNLYD GEFREFENGS ICVECDPQCE KMEDGLLTCH GPGPDNCTKC SHFKDGPNCV EKCPDGLQGA
       610        620        630        640        650        660        670        680        690        700
NSFIFKYADP DRECHPCHPN CTQGCNGPTS HDCIYYPWTG HSTLPQHART PLIAAGVIGG LFILVIVGLT FAVYVRRKSI KKKRALRRFL ETELVEPLTP
       710        720        730        740        750        760        770        780        790        800
SGTAPNQAQL RILKETELKR VKVLGSGAFG TVYKGIWVPE GETVKIPVAI KILNETTGPK ANVEFMDEAL IMASMDHPHL VRLLGVCLSP TIQLVTQLMP
       810        820        830        840        850        860        870        880        890        900
HGCLLEYVHE HKDNIGSQLL LNWCVQIAKG MMYLEERRLV HRDLAARNVL VKSPNHVKIT DFGLARLLEG DEKEYNADGG KMPIKWMALE CIHYRKFTHQ
       910        920        930        940        950        960        970        980        990       1000
SDVWSYGVTI WELMTFGGKP YDGIPTREIP DLLEKGERLP QPPICTIDVY MVMVKCWMID ADSRPKFKEL AAEFSRMARD PQRYLVIQGD DRMKLPSPND
      1010       1020       1030       1040       1050       1060       1070
SKFFQNLLDE EDLEDMMDAE EYLVPQAFNI PPPIYTSRAR IDSNRSEIGH SPPPAYTPMS GNQFVYRDGG FAAEQGVSVP YRAPTSTIPE APVAQGATAE
      1110       1120       1130       1140       1150       1160       1170       1180       1190       1200
IFDDSCCNGT LRKPVAPHVQ EDSSTQRYSA DPTVFAPERS PRGELDEEGY MTPMRDKPKQ EYLNPVEENP FVSRRKNGDL QALDNPEYHN ASNGPPKAED
      1210       1220       1230       1240       1250       1260       1270       1280       1290       1300
EYVNEPLYLN TFANTLGKAE YLKNNILSMP EKAKKAFDNP DYWNHSLPPR STLQHPDYLQ EYSTKYFYKQ NGRIRPIVAE NPEYLSEFSL KPGTVLPPPP

YRHRNTVV
```

Underline = Immunogenic fragment generated in Buonanno lab (against mouse ErbB4 = 1036-1239)
Yellow = Sequences identified in on-bead trypsin digestion (Steps 2 and 3)
Blue = transmembrane domain (652–675)

Fig. 7.2 Sequence coverage of ErbB4 from antibody complex fractionation procedure. The peptides identified from ErbB4 in our over-expressed cardiac tissue model cover over 30% of the sequence (considering peptides with 6 or more residues and up to 4000 Da in mass). The coverage increases to 40% if one removes potential sites of glycosylation from consideration. Observed peptides covered most of the immunogenic fragment used to generate the polyclonal antibody (*underline*, 1036–1239), as well as an isoform specific peptide (*SEIGHSPPPAYTPMSG*, (1046–1061)). Yellow, sequences identified in on-bead trypsin digestion (Steps 2 and 3); Blue, transmembrane domain (652–675)

applied the antibody complex fractionation procedure in triplicate. As can be seen in Fig. 7.2, over 30% of the ErbB4 sequence was detected in Steps 2 and 3 using the antibody complex fractionation procedure. Peptide hit distributions are located in Table 7.1. Peptides were detected in both the extracellular and intracellular regions of the protein. Of particular note is that the peptides seen cover most of the immunogenic component used to generate the antibody.

One concern about the digestion of the antibody protein complex directly on the magnetic beads is that relative to the amount of bait protein, the amount of antibody and Protein A is extremely high and will impact the ability to detect the bait protein. While the amount of peptide hits from Protein A and the various antibody chains was high relative to the peptide hits detected from ErbB4, we were nevertheless able to detect the bait protein due to the distinct retention times in chromatography and consistent elution of chemical background peptides (Table 7.2). In addition, while the number of peptide hits from Protein A and the various antibody chains was high, these correspond to less than 20 peptides facilitating their recognition and subtraction, making the use of exclusion lists feasible. Furthermore, the initial round of heating, reduction and alkylation was effective at removing keratins from the immunoprecipitated sample, increasing the ability to detect peptides from the bait protein (Fig. 7.3).

Table 7.1 Peptide hit distribution of ErbB4 in biological triplicate runs of antibody complex fractionation procedure

	Peptide hits – Trypsin digestion					
	On-bead			Urea treated on-bead		
Sequence	ErbB4 IP I	ErbB4 IP II	ErbB4 IP III	ErbB4 IP I	ErbB4 IP II	ErbB4 IP III
AEDEYVNEPLYLNTFANTLGK		1	1		1	2
ANVEFMDEALIMASMDHPHLVR	1					
APTSTIPEAPVAQGATAEIFDDSCCNGTLR	1					
CPDGLQGANSFIFK		2	4			
DGGFAAEQGVSVPYR	7		7	3	2	2
EVTGYVLVALNQFR	1		2	1		
FLETELVEPLTPSGTAPNQAQLR	6					1
FTHQSDVWSYGVTIWELMTFGGKPYDGIPTR	3					
ICIESCNLYDGEFR	3	5	8			
IRPIVAENPEYLSEFSLKPGTVLPPPPYR		1				
KPVAPHVQEDSSTQR			1			
LSSLSDLEQQYR	2	3	11	1	2	2
NGDLQALDNPEYHNASNGPPK			1			
QEYLNPVEENPFVSR			1		2	
SEIGHSPPPAYTPMSGNQFVYR	2					
STLQHPDYLQEYSTK		2	2			
VLGSGAFGTVYK		5	7			
VLYSGLSLLILK			1	1		
YALAIFLNYR	2					
YLVIQGDDR		4	9			
YSADPTVFAPER	3	5	11	3		2
GELDEEGYMTPMR						1

Table 7.2 Overview of number of peptides and peptide hits from three different preparations of ErbB4 immuno-captured on magnetic beads. The numbers of selected background immunoglobulin and Protein A, as well as targeted ErbB4 peptides identified by LC/MS/MS are summarized for the three fractions collected after (1) initial heating-reduction-alkylation, (2) on-bead trypsin digestion, and (3) urea treated on-bead trypsin digestion. Numbers in parentheses are results from analysis of beads containing bound Ab without application of any tissue extract. Numbers in parenthesis are from fraction that only contained antibody

ErbB4 IP – Heart

		Initial heating-reduction-alkylation		On bead trypsin digestion		Urea + on bead trypsin digestion	
Protein name	Description	Peptides	Peptide Hits	Peptides	Peptide hits	Peptides	Peptide hits
ERBB4_HUMAN	Receptor tyrosine-protein kinase erbB-4;			11, 9, 14	31, 28, 66	5, 4, 6	9, 7, 10
SPA_STAAU	Immunoglobulin G-binding protein A;			14, 14, 16 (1)	171, 97, 134 (1)	14, 20, 17 (19)	280, 100, 92 (112)
IGHG_RABIT	Ig gamma chain C region;	1, 0, 0	3, 0, 0	6, 3, 3 (1)	53, 30, 5 (1)	7, 4, 4 (7)	545, 82, 74 (92)
KAC4_RABIT	Ig kappa-b4 chain C region;	0, 3, 2 (2)	0, 8, 12 (7)	3, 3, 3 (2)	75, 30, 53 (11)	4, 4, 4 (4)	188, 82, 74 (93)
KACB_RABIT	Ig kappa-b4 chain C region;					0, 1, 1	0, 2, 2
LAC_RABIT	Ig lambda chain C region;			0, 1, 0	1, 0, 0	1, 2, 1 (1)	16, 10, 5 (4)
Others (Total)		20, 57, 33 (16)	464, 170, 222 (66)	53, 51, 85 (14)	588, 222, 424 (53)	55, 112, 93 (80)	1259, 401, 326 (424)
Total proteins (Parsimony)		14, 17, 18 (9)		18, 17, 34 (8)		20, 54, 50 (37)	

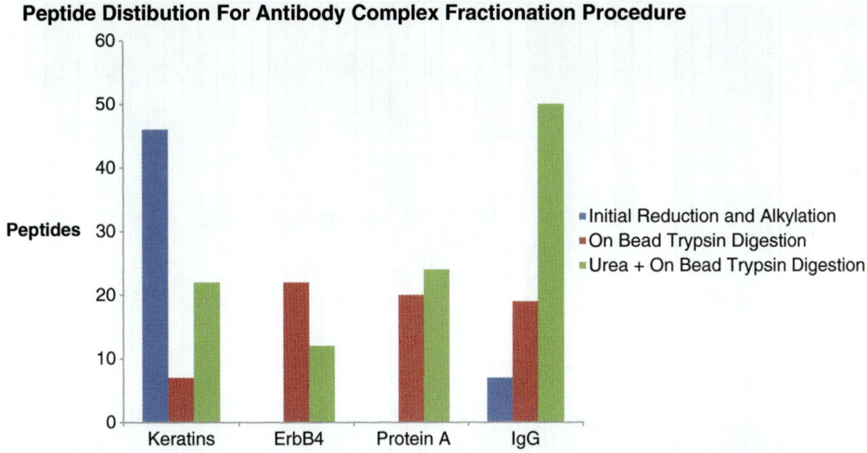

Fig. 7.3 Peptide distribution of keratins, ErbB4, Protein A and Ig peptides found in steps of antibody complex fractionation procedure. The application of an initial heating-reduction-alkylation step removes contaminants such as keratin. The next step in the procedure produces peptides that result from ErbB4 as well Protein A and IgG. The final treatment with urea produces a dramatic increase in the number of Protein A and Ig peptides, but ErbB4 peptides are still identified

7.3.3 Analysis of PSD95 Complexes in Brain

To show that the on-bead antibody complex fractionation procedure was applicable to more complex biological samples in which the bait protein is not overexpressed, the procedure was applied to PSD-95 affinity purified samples isolated from rat brain. As seen in earlier experiments, Step 1 in the antibody-complex fractionation procedure is dominated by peptides from proteins that may be considered contaminants, including keratins and serum albumin (Table 7.3), but also contains at least one protein associated with the PSD95 complex (BAIP2_RAT). Step 2 of the antibody complex fractionation procedure resulted in the identification of the bait protein PSD-95, as well as hundreds of others. To determine how consistent our list was, we compared it to previous results from our lab (Dosemeci et al., 2007), as well as from other labs (Fernandez et al., 2009). As can be seen in Table 7.3, there is an excellent correlation between the proteins identified using the on-bead fractionation procedure and proteins that were seen in previous studies using PSD-95 affinity purification followed by GeLC-MS. In Step 3 of the antibody complex fractionation procedure, several Ig proteins were identified but there were no proteins that were unique to this step of the procedure, indicating that the majority of the PSD-95 protein complex was removed in Step 2. The beads used for this experiment were not coated with Protein A, but an antibody was not cross-linked to the beads, showing that the antibody complex fractionation procedure can be applied to a wide variety of conditions.

Table 7.3 Application of antibody complex fractionation procedure to PSD-95 immunopurified sample and comparison of ID with published components of PSD

Protein name	Peptides	PepHits	Parsimony type	Mass	Description	Dosemeci Rank	Fernandez Affinity
Initial heating reduction-alkylation							
ALBU_BOVIN	22	54	DIFFERENTIABLE	71,244.19	(P02769) Serum albumin;		
K1C10_HUMAN	6	8	DIFFERENTIABLE	59,702.99	(P13645) Keratin, type I cytoskeletal 10;		
TRYP_PIG	3	53	DIFFERENTIABLE	25,078.06	(P00761) Trypsin;		
BAIP2_RAT	2	4	SUPERSET	59,373.96	(Q6GMN2) Brain-specific angiogenesis inhibitor 1-associated protein 2	16	+
K2C1_HUMAN	2	10	SUPERSET	66,170.07	(P04264) Keratin, type II cytoskeletal 1;		
On bead trypsin digestion							
ACTB_RAT	5	12	SUBSET	41,737	(P60711) Actin, cytoplasmic 1 subset of ACTA_RAT	8	–
ACTN1_RAT*	1	2	EQUIVALENT	10,3465.79	(Q9Z1P2) Alpha-actinin-1	47 = Q6GMN8_RAT	+
ADT1_RAT	4	12	DISCRETE	33,196.32	(Q05962) ADP/ATP translocase 1	24	–
AINX_RAT	7	16	DIFFERENTIABLE	56,252.7	(P23565) Alpha-internexin	67	+
ANS1B_RAT	8	44	DISCRETE	140,317.54	(P0C6S7)Ankyrin repeat and sterile alpha motif domain-containing protein 1B	11 = Q8BZM2_MOUSE	+
BAIP2_RAT	18	80	DISCRETE	59,373.96	(Q6GMN2) Brain-specific angiogenesis inhibitor 1-associated protein 2	16	+
BEGIN_RAT	6	13	DISCRETE	67,522.01	(O88881) Brain-enriched guanylate kinase-associated protein	obs, unranked	+
BSN_RAT	19	65	DISCRETE	420,277.68	(O88778) Protein bassoon	73	–
CALM_RAT	2	7	DISCRETE	16,826.83	(P62161) Calmodulin	–	–
CCG8_RAT	2	3	DISCRETE	43,641.18	(Q8VHW5) Voltage-dependent calcium channel gamma-8 subunit	46	+
CNKR2_RAT	4	13	DISCRETE	118,457.28	(Q9Z1T4) Connector enhancer of kinase suppressor of ras 2	obs, unranked	–
CTNA2_MOUSE	6	10	DISCRETE	106,018.12	(Q61301) Catenin alpha-2	20 = Q5R416_PONPY	–
CTRO_MOUSE	5	12	DISCRETE	237,294.42	(P49025) Citron Rho-interacting kinase	44 = Q9QX19_RAT	–
CYLD_RAT	4	8	DISCRETE	108,297.82	(Q66H62) Probable ubiquitin carboxyl-terminal hydrolase CYLD	33 = Q66H62_RAT	–
DLG1_RAT	7	15	DIFFERENTIABLE	100,908.24	(Q62696) Disks large homolog 1	39 = DLG1_HUMAN	+
DLG2_RAT	17	66	DIFFERENTIABLE	95331.45	(Q63622) Disks large homolog 2	6	+
DLG3_RAT	7	17	DIFFERENTIABLE	94,051.41	(Q62936) Disks large homolog 3	40	+
DLG4_RAT	23	189	DIFFERENTIABLE	80757.89	(P31016) Disks large homolog 4	2	+
DLGP1_RAT	14	39	DIFFERENTIABLE	111,363.37	(P97836) Disks large-associated protein 1	23	+
DLGP2_RAT	11	38	DIFFERENTIABLE	120,271.95	(P97837) Disks large-associated protein 2	28	+
DLGP3_MOUSE	12	31	DISCRETE	106,776.89	(Q6PFD5) Disks large-associated protein 3	25	+
DLGP4_RAT	10	32	DIFFERENTIABLE	108,880.68	(P97839) Disks large-associated protein 4	22	+
E41L1_RAT	2	8	DISCRETE	98,866.31	(Q9WTP0) Band 4.1-like protein 1	obs, unranked	–
ERC2_RAT	5	11	DIFFERENTIABLE	110,721.05	(Q8K3M6) ERC protein 2	71	–
FA81A_MOUSE	4	20	DISCRETE	42,105.01	(Q8TBF8) Protein FAM81A	31 = Q8TBF8_HUMAN	+
G3P_RAT	2	6	DIFFERENTIABLE	36,090.27	(P04797) Glyceraldehyde-3-phosphate dehydrogenase	48	+
GLNA_RAT	2	5	DISCRETE	42,981.58	(P09606) Glutamine synthetase	21	+
GRIA1_RAT	8	20	DIFFERENTIABLE	102,312.53	(P19490) Glutamate receptor 1	18	+
GRIA2_RAT	15	68	DIFFERENTIABLE	99,252.39	(P19491) Glutamate receptor 2	7	+
GRIA3_RAT	8	32	DIFFERENTIABLE	100,878.84	(P19492) Glutamate receptor 3	12	+
GRIA4_RAT	3	9	DIFFERENTIABLE	101,491.54	(P19493) Glutamate receptor 4	32	+
GRIK2_RAT	2	8	DISCRETE	103,088.61	(P42260) Glutamate receptor, ionotropic kainate 2	obs, unranked	+
HOME1_RAT	9	34	DISCRETE	41,393.87	(Q9Z214) Homer protein homolog 1	13	–
HSP7C_RAT	5	21	DIFFERENTIABLE	70,871	(P63018) Heat shock cognate 71 kDa protein	37	+
IQEC1_MOUSE	9	20	DIFFERENTIABLE	108,803.53	(Q8R0S2) IQ motif and SEC7 domain-containing protein 1	17 = Q6DN90_HUMAN	+

Table 7.3 (continued)

IQEC2_MOUSE	23	82	DIFFERENTIABLE	162,663.23	(Q5DU25) IQ motif and SEC7 domain-containing protein 2	10	+
KALRN_RAT	3	3	DISCRETE	339,511.53	(P97924) Kalirin	49 = HAPIP_RAT	+
KCC2A_RAT	14	154	DIFFERENTIABLE	54,650.76	(P11275) Calcium/calmodulin-dependent protein kinase type II alpha chain	1	+
KCC2B_RAT	9	82	DIFFERENTIABLE	61,104.9	(P08413) Calcium/calmodulin-dependent protein kinase type II beta chain	obs, unranked	+
KCC2D_RAT	7	49	DIFFERENTIABLE	60,669.79	(P15791) Calcium/calmodulin-dependent protein kinase type II delta chain	obs, unranked	+
KCC2G_RAT	10	71	DIFFERENTIABLE	59,685.32	(P11730) Calcium/calmodulin-dependent protein kinase type II gamma chain	obs, unranked	–
KPCG_RAT	4	9	DISCRETE	79,551.54	(P63319) Protein kinase C gamma type	34	–
LPHN1_RAT	2	4	DISCRETE	168,562.19	(O88917) Latrophilin-1	obs,unranked	–
LRRC7_RAT	19	49	DISCRETE	168,689.62	(P70587) Leucine-rich repeat-containing protein 7	19	–
LRTM1_HUMAN	0	0	N/A	38,171	(Q9HBL6) Leucine-rich repeat and transmembrane domain-containing protein 1	41	+
MBP_RAT	2	15	DISCRETE	21,545.99	(P02688) Myelin basic protein S	–	–
MYH10_RAT*	1	1	DISCRETE	229,793.37	(Q9JLT0) Myosin-10	27	–
MYO5A_RAT	2	7	DISCRETE	213,283.46	(Q9QYF3) Myosin-Va	–	–
NEB1_RAT	6	10	DIFFERENTIABLE	122,888.74	(O35867) Neurabin-1	obs, unranked	–
NFL_RAT	3	8	DISCRETE	61,355.06	(P19527) Neurofilament light polypeptide	137	+
NLGN3_RAT	5	21	DIFFERENTIABLE	94,456.34	(Q62889) Neuroligin-3	30	+
NMDE1_RAT	6	20	DISCRETE	166,674.87	(Q00959) Glutamate [NMDA] receptor subunit epsilon-1	38 = O08948_RAT	+
NMDE2_RAT	13	42	DISCRETE	167,675.75	(Q00960) Glutamate [NMDA] receptor subunit epsilon-2	26	+
NMDZ1_RAT	10	40	DISCRETE	106,126.06	(P35439) Glutamate [NMDA] receptor subunit zeta-1	45 = Q62648_RAT	+
PCLO_RAT	8	23	DISCRETE	554,200.64	(Q9JKS6) Protein piccolo	–	–
PLEC1_RAT	9	12	DISCRETE	535,152.26	(P30427) Plectin-1	36	–
PP1A_RAT*	1	2	EQUIVALENT	37,512	(P62138) Serine/threonine-protein phosphatase PP1-alpha catalytic subunit	29	–
Q63128_RAT	0	0	N/A	109,037	(Q63128_RAT) CPG2 protein	50	–
SHAN1_RAT	27	72	DISCRETE	226,394.42	(Q9WV48) SH3 and multiple ankyrin repeat domains protein 1	14	+
SHAN2_RAT	12	56	DISCRETE	159,099.2	(Q9QX74) SH3 and multiple ankyrin repeat domains protein 2	15	–
SHAN3_RAT	20	52	DISCRETE	193,824.1	(Q9JLU4) SH3 and multiple ankyrin repeat domains protein 3	9	–
SNIP_RAT	7	13	DISCRETE	130,006.77	(Q9QXY2) p130Cas-associated protein	35	–
SRBS2_RAT	3	8	DISCRETE	61,355.06	(Q3UTJ2) Sorbin and SH3 domain-containing protein 2	–	–
SYGP1_RAT	28	185	DISCRETE	145,600.4	(Q9QUH6) Ras GTPase-activating protein SynGAP	4	+
SYN2_RAT	3	14	DISCRETE	63,701.88	(Q63537) Synapsin-2	80	+
T132B_HUMAN*	1	1	DISCRETE	119,477	Transmembrane protein 132B	42 = Q8NA73_HUMAN	–
TBA4A_RAT	14	70	DIFFERENTIABLE	49,924	(Q5XIF6) Tubulin alpha-4A chain	5 = Q5XIF6_RAT	–
TBB5_RAT	12	78	DIFFERENTIABLE	50,095.14	(P69897) Tubulin beta-5 chain	3	–
VDAC1_RAT	5	13	DISCRETE	30,850.62	(Q9Z2L0) Voltage-dependent anion-selective channel protein 1	43	+
VDAC2_RAT	4	9	DISCRETE	32,352.86	(P81155) Voltage-dependent anion-selective channel protein 2	78	+
Urea treated on-bead digestion [a]							
ALBU_BOVIN	35	220	DIFFERENTIABLE	71,244.19	(P02769) Serum albumin		
ANS1B_RAT	3	3	DISCRETE	140,317.54	(P0C6S7) Ankyrin repeat and sterile alpha motif domain-containing protein 1B	11	+
BAIP2_RAT	2	2	DISCRETE	59,373.96	(Q6GMN2) Brain-specific angiogenesis inhibitor 1-associated protein 2	16	+
BSN_RAT	4	5	DISCRETE	420,277.68	(O88778) Protein bassoon	69	–

Table 7.3 (continued)

DLG4_RAT	11	17	DISCRETE	80,757.89	(P31016) Disks large homolog 4	2	+
DLGP1_RAT	3	5	DISCRETE	111,363.37	(P97836) Disks large-associated protein 1	23	+
GRIA1_RAT	2	2	DISCRETE	102,312.53	(P19490) Glutamate receptor 1	18	+
GRIA2_RAT	5	7	DIFFERENTIABLE	99,252.39	(P19491) Glutamate receptor 2	7	+
GRIA3_RAT	3	3	DIFFERENTIABLE	100,878.84	(P19492) Glutamate receptor 3	12	+
HOME1_RAT	3	5	DISCRETE	41,393.87	(Q9Z214) Homer protein homolog 1	13	–
IGG2B_MOUSE	8	23	DISCRETE	44,972.46	(P01867) Ig gamma-2B chain C region		
IGKC_MOUSE	4	9	DISCRETE	11,941.62	(P01837) Ig kappa chain C region		
IQEC1_MOUSE	2	2	DIFFERENTIABLE	108,803.53	(Q8R0S2) IQ motif and SEC7 domain-containing protein 1	17	+
IQEC2_MOUSE	6	7	DIFFERENTIABLE	162,663.23	(Q5DU25) IQ motif and SEC7 domain-containing protein 2	10	+
K1C10_RAT	5	10	DISCRETE	56,698.57	(Q6IFW6) Keratin, type I cytoskeletal 10		
KCC2A_RAT	9	28	DIFFERENTIABLE	54,650.76	(P11275) Calcium/calmodulin-dependent protein kinase type II alpha chain	1	+
KCC2G_RAT	3	9	DIFFERENTIABLE	59,685.32	(P11730) Calcium/calmodulin-dependent protein kinase type II gamma chain	obs, unranked	
MBP_RAT	2	3	DISCRETE	21,545.99	(P02688) Myelin basic protein S	–	–
PLEC1_RAT	2	3	DISCRETE	535,152.26	(P30427) Plectin-1	36	–
SYGP1_RAT	6	11	DIFFERENTIABLE	145,600.4	(Q9QUH6) Ras GTPase-activating protein SynGAP	4	+
TBA1B_RAT	4	7	SUPERSET	50,152	(Q6P9V9) Tubulin alpha-1B chain	–	–

Data from MassSieve using rat parsimony list; also included single peptide entries (marked with *) that match Dosemeci list of AP top 50; shading:green = on-bead matches Dosemeci and Fernandez identifications; yellow = on-bead differs from both Dosemeci and Fernandez; gray = found in Dosemeci top 50, but not on-bead; white = observed on-bead and Dosemeci, but not Fernandez); omitted from 'On bead trypsin digestion' portion of Table: Ig, keratin, and tryptic peptides

[a]data from MassSieve rat parsimony with >2 peptides/protein; mouse data used for IGG and IQEC proteins; mammal data used for albumin

7.4 Conclusions

We have described an antibody complex fractionation and analysis procedure that allows the identification of immunoprecipitated protein complexes without elution and subsequent protein fractionation. This avoids the requirement for strong detergents that necessitate buffer exchange or gel prior to LC/MS analysis. We were able to readily identify the bait protein, and the detected peptides that may give insight into the epitope that is recognized by the antibody. The presence of an initial step of heating, reduction, and alkylation assists in the removal of contaminants and non-specific binders. The final urea treatment ensures that the entire protein complex is solubilized and is removed from the beads. We believe that the antibody complex fractionation procedure can be applied to a wide variety of buffer conditions, enzymes, and sample amounts. *During the editing of this chapter, a thorough review and discussion regarding alternative on-bead digestion procedures for immunocaptured proteins was published (Holzmann et al., 2011).*

Acknowledgments This work was supported by the Intramural Research Programs of the National 24 Institute of Mental Health (MH000274). Dr. Ayse Dosemeci (NINDS) kindly provided immunoaffinity purified post-synaptic densities; her advice and assistance was most helpful. M.B., D.V. and A.B. and part of the work was supported by the Eunice Kennedy Shriver National Institute of Child Health and Human Development.

References

Aebersold, R., and Mann, M. (2003). Mass spectrometry-based proteomics. Nature *422*, 198–207.

Buonanno, A., and Fischbach, G.D. (2001). Neuregulin and Erbb receptor signaling pathways in the nervous system. Curr Opin Neurobiol *11*, 287–296.

Dosemeci, A., Makusky, A.J., Jankowska-Stephens, E., Yang, X., Slotta, D.J., and Markey, S.P. (2007). Composition of the synaptic Psd-95 complex. Mol Cell Proteomics *6*, 1749–1760.

Fernandez, E., Collins, M.O., Uren, R.T., Kopanitsa, M.V., Komiyama, N.H., Croning, M.D., Zografos, L., Armstrong, J.D., Choudhary, J.S., and Grant, S.G. (2009). Targeted tandem affinity purification of Psd-95 recovers core postsynaptic complexes and schizophrenia susceptibility proteins. Mol Syst Biol *5*, 269.

Gassmann, M., Casagranda, F., Orioli, D., Simon, H., Lai, C., Klein, R., and Lemke, G. (1995). Aberrant neural and cardiac development in mice lacking the Erbb4 neuregulin receptor. Nature *378*, 390–394.

Gingras, A.C., Gstaiger, M., Raught, B., and Aebersold, R. (2007). Analysis of protein complexes using mass spectrometry. Nat Rev Mol Cell Biol *8*, 645–654.

Holzmann, J., Fuchs, J., Pichler, P., Peters, J.-M., and Mechtler, K. (2011). Lesson from the stoichiometry determination of the cohesin complex: a short protease mediated elution increases the recovery from cross-linked antibody-conjugated beads. J Proteome Res *10*, 780–789.

Mei, L., and Xiong, W.C. (2008). Neuregulin 1 in neural development, synaptic plasticity and schizophrenia. Nat Rev Neurosci *9*, 437–452.

Slotta, D.J., McFarland, M.A., and Markey, S.P. (2010). Masssieve: panning Ms/Ms peptide data for proteins. Proteomics *10*, 3035–3039.

Tidcombe, H., Jackson-Fisher, A., Mathers, K., Stern, D.F., Gassmann, M., and Golding, J.P. (2003). Neural and mammary gland defects in Erbb4 knockout mice genetically rescued from embryonic lethality. Proc Natl Acad Sci USA *100*, 8281–8286.

Vullhorst, D., Neddens, J., Karavanova, I., Tricoire, L., Petralia, R.S., McBain, C.J., and Buonanno, A. (2009). Selective expression of Erbb4 in interneurons, but not pyramidal cells, of the rodent hippocampus. J Neurosci *29*, 12255–12264.

Yang, X., Dondeti, V., Dezube, R., Maynard, D.M., Geer, L.Y., Epstein, J., Chen, X., Markey, S.P., and Kowalak, J.A. (2004). Dbparser: web-based software for shotgun proteomic data analyses. J Proteome Res *3*, 1002–1008.

Chapter 8
Ultra-Fast Sample Preparation for High-Throughput Proteomics

Daniel Lopez-Ferrer, Kim K. Hixson, Mikhail E. Belov, and Richard D. Smith

Abstract Sample preparation oftentimes can be the *Achilles Heel* of any analytical process, and in the field of proteomics, preparing samples for mass spectrometric analysis is no exception. Current goals, concerning proteomic sample preparation on a large scale, include efforts toward improving reproducibility, reducing the time of processing and ultimately the automation of the entire workflow. This chapter reviews an array of recent approaches applied to bottom-up proteomics sample preparation to reduce the processing time down from hours to minutes. The current state-of-the-art approaches in the field use different energy inputs such as microwave, ultrasound or pressure to perform the four basic steps in sample preparation: protein extraction, denaturation, reduction/alkylation, and digestion. No single energy input for enhancement of proteome sample preparation has become the universal gold standard. Instead, a combination of different energy inputs tends to produce the best results. This chapter further describes the future trends in the field such as the hyphenation of sample preparation with downstream detection and analysis systems. Finally, a detailed protocol describing the combined use of both pressure cycling technology and ultrasonic energy inputs to hasten proteomic sample preparation is presented.

Keywords Proteomics · Enzymatic digestion · High pressure and mass spectrometry

Abbreviations

ESI	Electrospray ionization
FDR	False discovery rate
HIFU	High intensity focused ultrasound
HPP	High pressure processing
IAM	Iodoacetamide

D. Lopez-Ferrer (✉)
Pacific Northwest National Laboratory, Biological Science Division, Richland, WA;
Caprion Proteomics U.S., LLC, 1455 Adams Drive, Menlo Park, CA 94025, USA
e-mail: dlopezferrer@caprion.com

A.R. Ivanov, A.V. Lazarev (eds.), *Sample Preparation in Biological Mass Spectrometry*, DOI 10.1007/978-94-007-0828-0_8,

IT	Ion trap
MAPED	Microwave-assisted protein enzymatic digestion
MS	Mass spectrometry
MS/MS	Tandem mass spectrometry
MW	Molecular weight
NA	Not assigned
PCT	Pressure cycling technology
RP	Reversed phase
TCEP	Tris(2-carboxyethyl) phosphine
TOF	Time-of-flight

8.1 Introduction

We are entering an age where systems biology approaches, including proteomics, transcriptomics, metabolomics, and other – omics techniques are being measured with greater depth and frequency. These large interconnected databases, relating networks and macromolecular associations, represent the burgeoning industrialization of biology. At the heart of understanding the inner workings of life, identification and characterization of proteins is paramount. Proteins are the fundamental molecular machines of cells. They are essential to the structure, communication, metabolism, replication, catabolism, and incorporation or synthesis of all necessary chemicals needed to sustain life. The genome, which is essentially static, may only be able to provide us with a propensity towards a certain cellular state. The proteome, however, is dynamic, actually revealing what is occurring in the cell at a specific time under a specific condition. To this end, the rapid study of the proteome in a short time frame is crucial to solving most present day biological challenges, and from a clinical perspective it is vital in the development of rapid and accurate diagnosis and prognosis of disease (Qian et al., 2009).

Currently most high-throughput proteomic strategies for rapid comprehensive protein profiling employ ultra-high performance and multidimensional peptide chromatography systems (HPLC) coupled with high resolution mass spectrometers (MS) for detection and identification of a previously digested protein sample (i.e., shotgun proteomics) (Aebersold and Mann, 2003; Domon and Aebersold, 2006; Washburn et al., 2001; Wolters et al., 2001). The first and most important step during this type of proteome analysis is the sample preparation. If this primary step fails by being irreproducible (e.g., incomplete protein digestion) or low in quality (i.e., inaccurate determination of peptide concentration or is contaminated with polymers, salts, or surfactants), the downstream peptide separation, detection, or data analysis will fail. In addition, with proteome experiments constantly evolving to be larger and more complex, it has become increasingly necessary to optimize the time-consuming sample preparation strategies conventionally used in today's high-throughput proteomic laboratories. The typical proteome sample preparation process incorporates cell lysis, protein denaturation, reduction of disulfide bonds, prevention of the reformation of disulfide bonds by alkylation with iodoacetamide, digestion with a protease (trypsin being the most commonly used), and, finally, desalting/clean-up.

Typically, the largest bottleneck in sample preparation is the time required for the proteolytic digestion of the protein sample. A traditional proteome digestion calls for several hours of incubation time to produce a thorough and reproducible digest. Here we highlight recent advances in the acceleration of protein digestions down to just a few minutes using an array of different energy inputs to accelerate the kinetics of the digestion reaction. Additionally, we present a protocol that provides a rapid and thorough digestion of proteins using modified sequencing grade trypsin under high hydrostatic pressure.

8.2 Analysis Workflow

Life comes in many forms (e.g., bacteria, animals, fungi, plants) that can be found in vastly different environments (i.e., growth media, soil, bodies of water, and other organisms). For this reason, there is no single universal proteome sample procedure for the initial protein extraction. After protein extraction, most workflows follow similar paths consisting of protein solubilization, protein reduction, alkylation, and protein digestion (Lopez-Ferrer et al., 2004; Manza et al., 2005; Wang et al., 2005; Wisniewski et al., 2009). A detailed schematic of a traditional proteome sample preparation workflow in comparison to new improved workflow implementations is illustrated in Fig. 8.1.

8.2.1 Protein Extraction and Solubilization

The first step of sample processing in proteomic experiments is to obtain high quality and adequate protein amounts from the sample to be studied. The protein extraction varies significantly depending on where the sample comes from, the relative abundance of the proteins of interest as well as the presence of high abundance proteins that may obscure dissolution, detection and analysis of lower abundance proteins.

Apart from working with body fluids, ground water, and secreted protein extracts, most proteome analyses target some sort of cell mass that requires an approach to lyse, homogenize, or solubilize cells in order to extract the proteins. During this process, it is critical to keep the samples ice cold and to use protease inhibitors in the lysis medium if the sample is particularly susceptible to endogenous proteases. Not doing so may increase the likelihood of undesirable side reactions, which could translate into irreproducible results. The most commonly used techniques for cell lysis/tissue disruption are: (a) manual homogenization, (b) vortexing, grinding, or bead beating with beads such as zirconia/silica beads, (c) freeze/thaw using liquid nitrogen or dry ice/ethanol (alongside manual grinding if working with plant tissues or yeast) and (d) sonication or (e) pressure. The combination of these approaches with the use of detergents is also becoming very popular, but this use needs to be compatible with the overall analytical strategy (i.e., detergents typically need to be removed prior to mass spectrometric analysis). In addition to the cell lysis methods

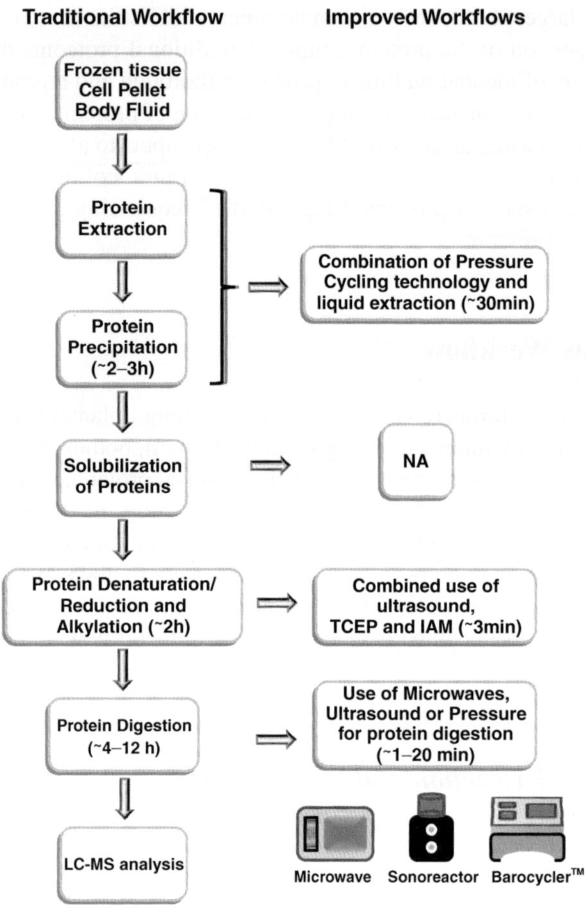

Fig. 8.1 Proteomics workflow. Proteins are extracted from different biological materials, precipitated to eliminate salts and other possible artifacts. Then the proteins are solubilized and completely denatured. The proteins are then digested and submitted for LC-MS analysis

listed above, pressure cycling technology in recent years has also become available as a means to provide effective and efficient lysis of many types of cells and tissues.

Usually the next step is to isolate proteins from most other cellular components (e.g., DNA, RNA, lipids, salts). The most popular technique is to precipitate proteins with either trichloroacetic acid (TCA) or TCA/cold-acetone precipitation (Isaacson et al., 2006). Other protocols have also been very successful, such as the original phenol-chloroform extraction method published by Chomczynsky and Sacchi in the late 1980s (Chomczynski and Sacchi, 1987). Phenol protein extraction followed by protein purification using an ammonium acetate precipitation also works well to extract and purify proteins, especially those coming from difficult samples (e.g., plant leaves containing high concentrations of polyphenols and carbohydrates) (Isaacson et al., 2006). Other recent developments in the realm of protein extraction include the work by Gross et al. consisting in the combined use of fluorinated alcohols and aliphatic hydrocarbons with pressure cycling technology (PCT). The authors demonstrated that efficient protein extraction from a sample was obtainable and furthermore after lysis this emulsion could be separated into a pellet and two liquid phases (Gross et al., 2008). DNA, RNA and their associated proteins were found in the pellet at the bottom of the tube, while proteins and small molecules remained in the middle phase, and the upper layer contained the lipids. Using this strategy, most proteins could be easily extracted by isolating the middle layer and drying off the volatile solvent.

Many protein extraction methods leave the protein as a precipitate which then requires solubilization in order to carry out the downstream reduction, alkylation and digestion. The chosen solubilization method will need to be compatible with subsequent analytical methods used in the protein analysis (i.e., LC-MS). Common chaotropes used include urea, thiourea, and guanidine-HCl. These chaotropes denature the proteins and prevent protein aggregation by disrupting hydrogen bonds as well as intra- and intermolecular forces. Depending on which kinds of proteins are present in the sample, solubilization may be more or less challenging and may require different solvent mixtures such as organic-aqueous solvents like TFE (Wang et al., 2005), MeOH (Blonder et al., 2006) or ACN (Russell et al., 2001) in concentrations varying from 20 to 60%. When using high concentrations of organic solvent, care must be given since this strategy may also precipitate out some other proteins. Another strategy is to use detergents like the non-denaturing surfactant CHAPS, which due to its zwitterionic behavior can be removed using a strong cation exchange (SCX) solid phase extraction (SPE) clean-up after protein digestion (Hixson et al., 2006).

8.2.2 Reduction and Alkylation Strategies

Proteins naturally contain intra- and intermolecular covalent disulfide linkages (bridging cysteines with other cysteine residues) which serve the purpose of providing protein stability and an appropriate confirmation and activity in vivo. To elucidate the protein sequence using tandem mass spectrometry it is necessary to reduce these bonds so that the protein can fully unfold for a complete digestion and accurate detection of the cysteine-containing peptide sequences. Typical reducing agents used in proteome sample preparation schemes include dithiothreitol (DTT), dithioerythritol (DTE) or β-mercaptoethanol. These reagents, although widely used, have a very pungent odor associated with them. In the case of DTT and DTE, they have to be made up fresh because they lose their reducing ability relatively quickly. More recently, phosphine derivatives like Tris(2-carboxyethyl) phosphine (TCEP) and tributylphosphine (TBP) have become more popular to use for reduction since these chemicals are stable and can be kept in neutralized solutions for many months at room temperature. In addition, these reagents can be used in conjunction with alkylating agents such as iodoacetamide (IAM) (Lopez-Ferrer et al., 2008a). Reduction reactions are commonly achieved in 45 min at 60°C.

After the disulfide reduction, the free sulphydryl groups need to be blocked to protect them from undesired side reactions by the use of an alkylating agent. Alkylating reagents, such iodoacetic acid (IAA), IAM and acrylamide can be used for this purpose. Traditionally IAM is the most used reagent, the main reason being that it was the most compatible with 2-DE (one of the earliest methods of choice for proteome analyses) because it does not change the protein pI. IAM to covalently block the sulfhydryl groups, thus preventing the reformation of disulfide bonds is referred to specifically as "carbamidomethilation" (or more generally "alkylation") and can be carried out in about an hour. Since IAM is photosensitive, alkylations using this chemical are carried out in the dark to limit any unwanted side reactions.

In recent years, ultrasonic energy or microwaves have been demonstrated to be effective at accelerating reduction and alkylation steps down to just a few minutes (Rial-Otero et al., 2007). Capelo and coworkers used indirect ultrasound energy by means of a sonoreactor which is $\sim$30 times less powerful than an ultrasonic probe, but $\sim$50 times more powerful than a regular bench top ultrasonic bath. The authors show that both steps, reduction and alkylation, could be effectively accelerated to just 3 min for each step. Lopez-Ferrer and coworkers have recently demonstrated that the two processes of reduction and alkylation can be coupled by using TCEP and IAM at the same time, reducing the time for complete reduction and alkylation to only 3 min total (Lopez-Ferrer et al., 2008a). In addition, the authors showed that by reducing the time of exposure to the ultrasonic energy, little or no carbamylation occurred. More recently, Basile and coworkers, showed that these reactions could also be accelerated by the use of microwave energy (Hauser and Basile, 2008; Hauser et al., 2008).

8.2.3 Enzymatic Digestion

Typical protease digestion protocols consist of the addition of a specific protease (i.e., one that cleaves a polypeptide or a protein at specific amino acid residues). For proteomic purposes several sequencing grade proteases are available. Trypsin, a serine protease that cleaves exclusively after arginine (Arg) and lysine (Lys), is by far the most widely used because it is economical (compared with other proteases), has been modified to be very robust, and is well characterized. Additionally, Arg and Lys are typically spread fairly evenly throughout most protein sequences, thus a trypsin digest usually provides an array of detectable peptides for each protein present (Kiser et al., 2009). Other proteases commonly used in proteomic research but with less frequency include endoproteinase Lys-C (cleaves at lysine (Boyne et al., 2009)), *S. aureus* V8 (also known as endoproteinase Glu-C, and cleaves at glutamic acid (Glu) and aspartic acid (Asp) residues (Xiang et al., 2004)), and proteinase K (Wu et al., 2003; Wu and Yates, 2003) (cleaves aromatic, and aliphatic residues, but with less specificity for the latter). As an additional note, cyanogen bromide (although not an enzyme) has been commonly employed to cleave proteins at methionine residues, especially in applications that are focused on proteome characterizations within membranes (van Montfort et al., 2002).

In proteomic experiments, enzyme digestions are typically the most time-consuming step within the sample preparation process because it is often recommended to digest a sample for many hours to ensure complete digestion. It is well known that the enzyme activity increases with temperature, only up to a point where the temperature becomes high enough to denature the protease enzyme resulting in a dramatic decrease in activity. Some researchers have investigated the use of thermo stable trypsin. Other pursuits at enhancing trypsin activity include efforts to modify the pH or the buffer composition, such as adding organic solvents or salts like $CaCl_2$ to stabilize the enzyme-substrate complex (Russell et al., 2001; Wang et al., 2005). Despite these efforts, digestion times still usually require at least 30 min to be effective and complete (Havlis et al., 2003). Unconventional methods for enhancement of enzyme kinetics such as microwave, ultrasound and, more recently, pressure have been explored in the last 5 years as an alternative way for applying energy to a digestion reaction in order to accelerate the enzymatic hydrolysis reaction.

8.2.3.1 Microwave Irradiation

For many years now, researchers have utilized microwaves to hasten reaction rates in their protein chemistry and proteomics experiments (Lill et al., 2007; Zhong et al., 2004). The instrumentation used ranges from domestic microwave ovens to sophisticated laboratory microwave equipment. In both cases, it has been shown that the effect of microwave radiation applied to proteases can dramatically accelerate their enzyme kinetics. At this point, the field is divided by those who believe that the improvement of the reaction kinetics is due solely to the thermal effect produced by the efficient absorption of the radiation through the medium, and those who believe that the main reason for enhanced enzyme kinetics is through both thermal and non-thermal effects derived from both energy transfer from the electromagnetic field and vibration modes of the molecules altering their dipoles. The latter hypothesis was demonstrated when Pramanik and coworkers fixed the temperature of a trypsin digestion reaction at 37°C while controlling the level of microwave irradiation (Pramanik et al., 2002). Their results indicate that heat is not the main player driving the reaction. Microwave assisted protein digestion (MAPED) has been applied in proteomics for in-solution and in-gel digestion of proteins. MAPED has been optimized by using different organic co-solvents as well as by using immobilized trypsin immobilized on magnetic micro- or nanospheres.

Microwaves have also been used to accelerate acid hydrolysis of peptides in order to generate peptide ladders, where the difference between the detected ladder masses is the mass of the amino acid that has been cleaved. This methodology provides a very accurate peptide sequence and is excellent for protein characterization. Its application for high-throughput proteomics is more limited because the absence of specificity on the peptide cleavage makes it extremely challenging to interpret data resulting from a complex protein mixture digest (Zhong et al., 2004). Recently, Fenselau and coworkers have also explored what seems to be a very promising approach that utilizes low concentrations of formic or acetic acid to produce specific chemical cleavages in proteins (Swatkoski et al., 2007a, b, 2008). It has been

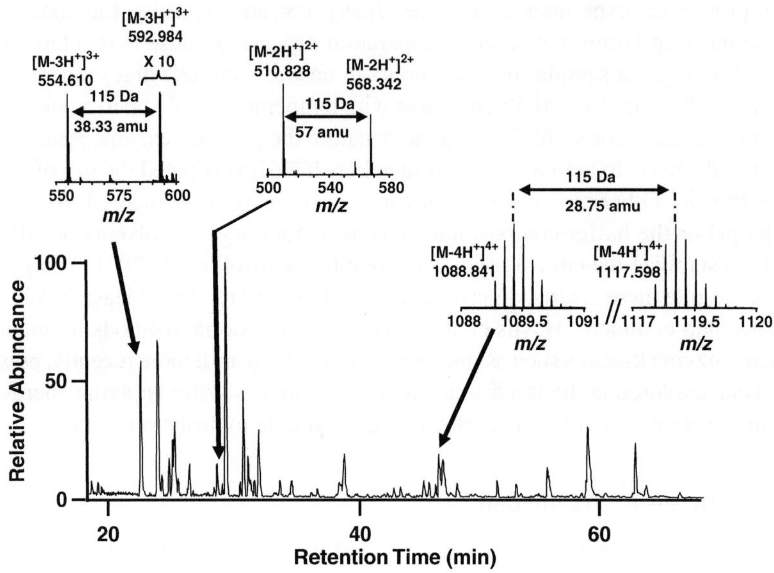

Fig. 8.2 Chromatographic profile of a microwave D-digestion of four proteins. Detailed mass spectra show pairs of signals 115 Da apart. This indicates that cleavage was occurring before and after aspartic acid residues. Also shown are a variety of peptide masses ranging from 1000 up to 4500 Da

demonstrated that these chemical cleavages under certain conditions can be very specific, preferring amino acid sites before and after Asp. This approach is particularly interesting because Asp residues are not as common as Arg or Lys, providing a completely different map of the proteins that can yield a better characterization of the protein primary structure. Additionally, because it cleaves before and after the Asp residues it is very common to see doublets of a certain peptide in the mass spectrum, which improves the forthcoming protein identification. Figure 8.2 shows the digestion of four protein standards using this methodology. Basile and coworkers have taken this method a step further by developing an on-line acid digestion system (Hauser and Basile, 2008; Hauser et al., 2008).

8.2.3.2 Ultrasound

The use of ultrasound energy for accelerating enzyme kinetics in proteomics was introduced for the first time by Lopez-Ferrer and coworkers in 2005 (Lopez-Ferrer et al., 2005). The study demonstrated that acceleration of a trypsin digestion was feasible for both in-gel digestions of single proteins and for in-solution digestions of either single proteins or complex global proteomes. The authors were able to apply ultrasonic energy to small volumes (less than 50 μL) using an ultrasonic probe. It is hypothesized that the effectiveness of this method is based upon the formation of localized high pressure and temperature zones in which the enzyme-substrate kinetics is enchanced. This process is called cavitation and it is visible when transient bubbles appear in the solution during the ultrasonic irradiation. Lopez-Ferrer and coworkers further demonstrated that the reaction time could be shortened even further if immobilized enzymes were used (Lopez-Ferrer et al., 2009). By using immobilized enzymes, the enzyme:substrate ratio increases dramatically and also there is a better mass transfer achievable (Kim et al., 2009) which altogether aids in a significantly more efficient digestion in the presence of ultrasound. Most recently, a protocol was developed where the use of high intensity focused ultrasound (HIFU) was applied to perform quantitative proteomic experiments using an $^{16}O/^{18}O$ labeling strategy. In this application, the enzyme is used not in the digestion of the proteins but in the post-digestion exchange of oxygen atoms at the C-termini. This hydrolysis/oxygen exchange reaction creates a 2 or 4 Da shift depending on the degree of reaction completion per peptide. These mass shifts can then be used to identify peptide abundance changes in a simultaneous mass spectrometric analysis of two mixed samples (namely, one that is labeled with ^{16}O added in equal mass to one that is labeled with ^{18}O).

8.2.3.3 High Pressure

For a number of years, MAPED and HIFU have been used in proteomic sample processing schemes and have proven useful for increasing the kinetics of proteolytic digestions. In HIFU, along with an increase in heat, short-lived micro-environments of high pressure are also experienced by the sample.

In food and biotechnology industries, hydrostatic pressure have been used to speed-up or inhibit certain reactions (e.g., enhancement of the curd formation in the manufacturing of cheese, inactivation of food pathogens orinhibition of enzymes). Evidence also exists to suggest that low to moderate pressures (e.g., 100–400 MPa) can activate enzyme activity (Cano et al., 1997), whereas higher pressures tend to inactivate the same enzymes (Hernandez and Cano, 1998). Lopez-Ferrer et al. demonstrated that proteome samples subjected to increased pressures (i.e., up to ~20,000 psi) could effectively hasten trypsin digestion (Lopez-Ferrer et al., 2008b). It was found that under these high pressures, globular sample proteins can become more denatured (the thermodynamic understanding of this process is currently unknown), allowing for the modified sequencing grade trypsin (trypsin that has been reductively methylated and enhanced by N-tosyl-L-phenylalanyl chloromethyl ketone (TPCK) treatment) to rapidly digest the proteins.

8.3 Future Trends

Considering the current and growing demand for proteomic analyses, especially those that utilize shotgun proteomic methodologies, the development of in-solution methods that enhance and hasten the digestions carried out during proteome sample preparation still needs to be optimized, automated and integrated with mass spectrometric analyses. The uses of microfluidic devices that can presently integrate different steps in the proteome workflow are currently restricted to specialized labs.

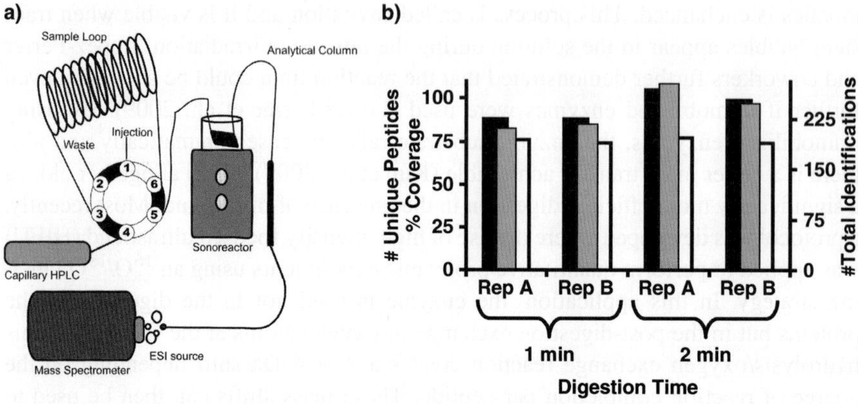

Fig. 8.3 (**a**) Schematic of a sonoreactor being used in an on-line set-up. (**b**) Complete digestion of protein samples in the presence of the low pH compatible protease pepsin can be achieved in 1–2 min as is indicated in (**b**), where is shown the level of proteolysis obtained by using the on-line sonoreactor setup with a bovine serum albumin sample digested with pepsin

However, we have proposed two simple alternatives that can be easily implemented into every proteomic facility based on the use of either ultrasounds or pressure. Ultrasound probes are usually commonplace in most molecular biology laboratories and these same probes can be used for effective digestion and/or ^{18}O labeling using trypsin to produce peptides in a matter of minutes. The direct hyphenation with LC-MS is also feasible as it is shown in Fig. 8.3. In this case, digestion of 1 pmol of bovine serum albumin was performed with pepsin (being that pepsin is already compatible with the low pH solvents used for the RP separations), and the flow rate was modulated in order to allow the sample to be under the ultrasonic field for 1 or 2 min. As it is shown in Fig. 8.3b, the digestion is extremely efficient providing a proteome coverage of over 70% in less than a minute.

Initial demonstrations of enhanced digestions using pressure conducted by Lopez-Ferrer et al. were carried out using a Barocylcer^{TM}, an instrument capable of subjecting samples to a temperature and pressure controlled environment which can be cycled between ambient pressure and 35,000 psi. This instrument, originally marketed for cell lysis conducted in enclosed disposable sample containers (Smejkal et al., 2006, 2007), is currently being adopted by many laboratories as the way to control enzymatic digestion of protein samples. Moreover, Lopez-Ferrer et al., recently demonstrated that high-pressure digestion could potentially be automated by using a modified high pressure HPLC equipment (Lopez-Ferrer et al., 2008c). Since high-pressure is integral to an HPLC system, protein samples added with trypsin to a sample loop can be pressurized briefly followed by direct injection onto a reversed phase C18 column. This set-up demonstrated the proof of concept for future automated sample preparation devices for high-throughput proteome sample preparation. With developments such as these, perhaps in the near future we will be able to routinely obtain a full proteome analyses (from whole cells to MS data) or

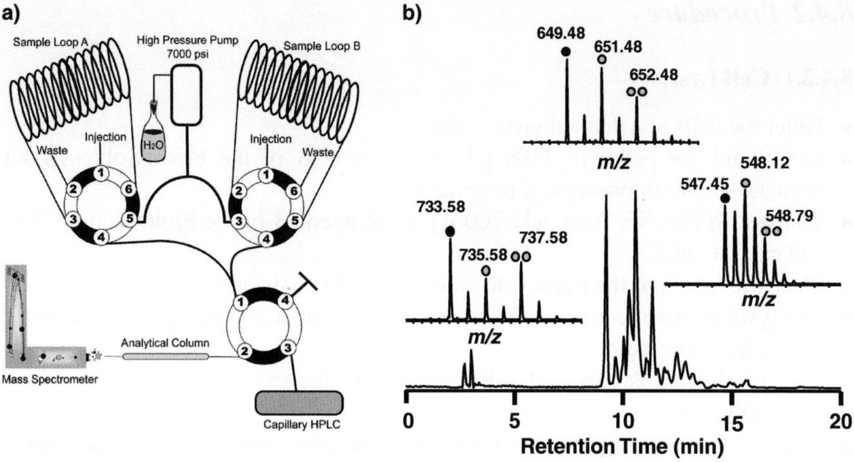

Fig. 8.4 Schematic showing the valve diagram and coupling to an LC-MS system for an on-line ^{18}O differential proteomics digestion system. (**b**) Chromatographic profile of a BSA digestion either with ^{18}O or regular water. Detailed mass spectra of different peptide pairs is shown for +2 and +3 ions

even comparative (using automated pressure assisted digestions in an on-line system with $^{18}O/^{16}O$ labeling) proteome analyses in just a manner of minutes (Fig. 8.4).

8.4 Protocol for Rapid Sample Preparation of Peptides for Shotgun Proteomics

8.4.1 Materials

8.4.1.1 Reagents

Sequencing grade trypsin was obtained from Promega (Madison, WI). Iodoacetamide (IAM), ammonium bicarbonate, formic acid, HPLC grade solvents, ^{18}O water, pepsin and bovine serum albumin were purchased from Sigma-Aldrich (St. Louis, MO). Tris[2-carboxyethyl]phosphine (TCEP) was purchased from Pierce (Rockford IL, USA). ProteoSolve-SB kit was obtained from Pressure BioSciences (South Easton, MA, USA).

8.4.1.2 Equipment

Barocycler^{TM} NEP-3229 instrument, disposable polypropylene PULSE tubes FT-500, PCT MicroTube^{TM} and a PCT MicroTube^{TM} adapter kit were obtained from Pressure BioSciences (South Easton, MA, USA). A UTR200 Sonoreactor was purchase from Hischler (Teltow, Germany)

8.4.2 Procedure

8.4.2.1 Cell Lysis

- Pellet the cells at a defined growth phase.
- Resuspend the pellet in 1100 μL of Reagent A of the ProteoSolve-SB kit, sonicate in the sonoreactor, if necessary.
- Transfer to PULSE Tube, add 200 μL of Reagent B of the ProteoSolve-SB kit and cap the tube.
- Place the tube into the Barocycler and initiate the lysis program.
- The lysis program consist of 20 cycles of pressure at 35 kpsi for 30 s and 5 s at atmospheric pressure.
- Transfer the sample to a 2 mL microcentrifuge tube and centrifuge for 10–15 min at 15,000 rpm.
- Transfer the bottom phase to a new reaction tube (protein content) and dry it down in the speedvac.

8.4.2.2 Reduction and Alkylation

- Resuspend the sample in 100 μL of 8 M urea, sonicate the pellet into solution, if necessary. (Do not vortex)
- Measure the protein concentration at this point (Sapan et al., 1999)
- Add 1 μL of Bond Breaker TCEP solution (Final concentration 5 mM)
- Add 4 μL of 500 mM IAM solution (Final concentration 20 mM)
- Sonicate for 3 min in the sonoreactor at 50% power

8.4.2.3 Protein Digestion

- Dilute the sample 4-fold to 400 μL with 50 mM ammonium bicarbonate.
- Add a 1:50 mass ratio of trypsin to the sample protein.
- Divide the sample between 4 and 100-μL Barocycler^{TM} tubes.
- Place into the Barocycler^{TM} and initiate the digestion program.
- The digestion program consist of 10 cycles of pressure at 20 kpsi for 20 s and 5 s at atmospheric pressure.
- Evaporate the samples down to ~150 mL in a SpeedVac concentrator.
- Measure and adjust the concentration of the peptide sample to optimum concentration for LC-MS system used.

Acknowledgments Portions of this work were supported by the NIH National Center for Research Resources (NCRR, RR018522), NIH National Cancer Institute (R21 CA12619-01), and the Pacific Northwest National Laboratory's (PNNL) Laboratory Directed Research and Development Program. This research was enabled in part by capabilities developed under support from the U.S. Department of Energy (DOE) Office of Biological and Environmental Research and the NCRR, and was conducted in the Environmental Molecular Sciences Laboratory, a DOE national scientific user facility located at the Pacific Northwest National Laboratory (PNNL) in Richland, WA. PNNL is a multiprogram national laboratory operated by Battelle for the DOE under Contract No. DE-AC05-76RLO 1830.

References

Aebersold, R., and Mann, M. (2003). Mass spectrometry-based proteomics. Nature *422*, 198–207.

Blonder, J., Chan, K.C., Issaq, H.J., and Veenstra, T.D. (2006). Identification of membrane proteins from mammalian cell/tissue using methanol-facilitated solubilization and tryptic digestion coupled with 2D-LC-MS/MS. Nat Protoc *1*, 2784–2790.

Boyne, M.T., Garcia, B.A., Li, M., Zamdborg, L., Wenger, C.D., Babai, S., and Kelleher, N.L. (2009). Tandem mass spectrometry with ultrahigh mass accuracy clarifies peptide identification by database retrieval. J Proteome Res *8*, 374–379.

Cano, M.P., Hernandez, A., and DeAncos, B. (1997). High pressure and temperature effects on enzyme inactivation in strawberry and orange products. J Food Sci *62*(1), 85–88.

Chomczynski, P., and Sacchi, N. (1987). Single-step method of RNA isolation by acid guanidinium thiocyanate-phenol-chloroform extraction. Anal Biochem *162*, 156–159.

Domon, B., and Aebersold, R. (2006). Mass spectrometry and protein analysis. Science *312*, 212–217.

Gross, V., Carlson, G., Kwan, A.T., Smejkal, G., Freeman, E., Ivanov, A.R., and Lazarev, A. (2008). Tissue fractionation by hydrostatic pressure cycling technology: the unified sample preparation technique for systems biology studies. J Biomol Tech *19*, 189–199.

Hauser, N.J., and Basile, F. (2008). Online microwave D-cleavage LC-ESI-MS/MS of intact proteins: Site-specific cleavages at aspartic acid residues and disulfide bonds. J Proteome Res *7*, 1012–1026.

Hauser, N.J., Han, H., McLuckey, S.A., and Basile, F. (2008). Electron transfer dissociation of peptides generated by microwave D-cleavage digestion of proteins. J Proteome Res *7*, 1867–1872.

Havlis, J., Thomas, H., Sebela, M., and Shevchenko, A. (2003). Fast-response proteomics by accelerated in-gel digestion of proteins. Anal Chem *75*, 1300–1306.

Hernandez, A., and Cano, M.P. (1998). High-pressure and temperature effects on enzyme inactivation in tomato puree. J Agric Food Chem *46*, 266–270.

Hixson, K.K., Adkins, J.N., Baker, S.E., Moore, R.J., Chromy, B.A., Smith, R.D., McCutchen-Maloney, S.L., and Lipton, M.S. (2006). Biomarker candidate identification in Yersinia pestis using organism-wide semiquantitative proteomics. J Proteome Res *5*, 3008–3017.

Isaacson, T., Damasceno, C.M., Saravanan, R.S., He, Y., Catala, C., Saladie, M., and Rose, J.K. (2006). Sample extraction techniques for enhanced proteomic analysis of plant tissues. Nat Protoc *1*, 769–774.

Kim, B.C., Lopez-Ferrer, D., Lee, S.M., Ahn, H.K., Nair, S., Kim, S.H., Kim, B.S., Petritis, K., Camp, D.G., Grate, J.W., *et al.* (2009). Highly stable trypsin-aggregate coatings on polymer nanofibers for repeated protein digestion. Proteomics *9*, 1893–1900.

Kiser, J.Z., Post, M., Wang, B., and Miyagi, M. (2009). Streptomyces erythraeus trypsin for proteomics applications. J Proteome Res *8*, 1810–1817.

Lill, J.R., Ingle, E.S., Liu, P.S., Pham, V., and Sandoval, W.N. (2007). Microwave-assisted proteomics. Mass Spectrom Rev *26*, 657–671.

Lopez-Ferrer, D., Capelo, J.L., and Vazquez, J. (2005). Ultra fast trypsin digestion of proteins by high intensity focused ultrasound. J Proteome Res *4*, 1569–1574.

Lopez-Ferrer, D., Heibeck, T.H., Petritis, K., Hixson, K.K., Qian, W., Monroe, M.E., Mayampurath, A., Moore, R.J., Belov, M.E., Camp, D.G., 2nd, *et al.* (2008a). Rapid sample processing for LC-MS-based quantitative proteomics using high intensity focused ultrasound. J Proteome Res *7*, 3860–3867.

Lopez-Ferrer, D., Hixson, K.K., Smallwood, H.S., Squier, T.C., Petritis, K., and Smith, R.D. (2009). Evaluation of a high-intensity focused ultrasound-immobilized trypsin digestion and 18O-labeling method for quantitative proteomics. Anal Chem *81*, 6272–6277.

Lopez-Ferrer, D., Martinez-Bartolome, S., Villar, M., Campillos, M., Martin-Maroto, F., and Vazquez, J. (2004). Statistical model for large-scale peptide identification in databases from tandem mass spectra using SEQUEST. Anal Chem *76*, 6853–6860.

Lopez-Ferrer, D., Petritis, K., Hixson, K.K., Heibeck, T.H., Moore, R.J., Belov, M.E., Camp, D.G., 2nd, and Smith, R.D. (2008b). Application of pressurized solvents for ultrafast trypsin hydrolysis in proteomics: Proteomics on the fly. J Proteome Res *7*, 3276–3281.

Lopez-Ferrer, D., Petritis, K., Lourette, N.M., Clowers, B., Hixson, K.K., Heibeck, T., Prior, D.C., Pasa-Tolic, L., Camp, D.G., 2nd, Belov, M.E., *et al.* (2008c). On-line digestion system for protein characterization and proteome analysis. Anal Chem *80*, 8930–8936.

Manza, L.L., Stamer, S.L., Ham, A.J., Codreanu, S.G., and Liebler, D.C. (2005). Sample preparation and digestion for proteomic analyses using spin filters. Proteomics *5*, 1742–1745.

Pramanik, B.N., Mirza, U.A., Ing, Y.H., Liu, Y.H., Bartner, P.L., Weber, P.C., and Bose, A.K. (2002). Microwave-enhanced enzyme reaction for protein mapping by mass spectrometry: A new approach to protein digestion in minutes. Protein Sci *11*, 2676–2687.

Qian, W.J., Liu, T., Petyuk, V.A., Gritsenko, M.A., Petritis, B.O., Polpitiya, A.D., Kaushal, A., Xiao, W., Finnerty, C.C., Jeschke, M.G., *et al.* (2009). Large-scale multiplexed quantitative discovery proteomics enabled by the use of an (18)O-labeled "universal" reference sample. J Proteome Res *8*, 290–299.

Rial-Otero, R., Carreira, R.J., Cordeiro, F.M., Moro, A.J., Fernandes, L., Moura, I., and Capelo, J.L. (2007). Sonoreactor-based technology for fast high-throughput proteolytic digestion of proteins. J Proteome Res *6*, 909–912.

Russell, W.K., Park, Z.Y., and Russell, D.H. (2001). Proteolysis in mixed organic-aqueous solvent systems: Applications for peptide mass mapping using mass spectrometry. Anal Chem *73*, 2682–2685.

Sapan, C.V., Lundblad, R.L., and Price, N.C. (1999). Colorimetric protein assay techniques. Biotechnol Appl Biochem *29*(Pt 2), 99–108.

Smejkal, G.B., Robinson, M.H., Lawrence, N.P., Tao, F., Saravis, C.A., and Schumacher, R.T. (2006). Increased protein yields from Escherichia coli using pressure-cycling technology. J Biomol Tech *17*, 173–175.

Smejkal, G.B., Witzmann, F.A., Ringham, H., Small, D., Chase, S.F., Behnke, J., and Ting, E. (2007). Sample preparation for two-dimensional gel electrophoresis using pressure cycling technology. Anal Biochem *363*, 309–311.

Swatkoski, S., Gutierrez, P., Ginter, J., Petrov, A., Dinman, J.D., Edwards, N., and Fenselau, C. (2007a). Integration of residue-specific acid cleavage into proteomic workflows. J Proteome Res *6*, 4525–4527.

Swatkoski, S., Gutierrez, P., Wynne, C., Petrov, A., Dinman, J.D., Edwards, N., and Fenselau, C. (2008). Evaluation of microwave-accelerated residue-specific acid cleavage for proteomic applications. J Proteome Res *7*, 579–586.

Swatkoski, S., Russell, S., Edwards, N., and Fenselau, C. (2007b). Analysis of a model virus using residue-specific chemical cleavage and MALDI-TOF mass spectrometry. Anal Chem *79*, 654–658.

van Montfort, B.A., Doeven, M.K., Canas, B., Veenhoff, L.M., Poolman, B., and Robillard, G.T. (2002). Combined in-gel tryptic digestion and CNBr cleavage for the generation of peptide maps of an integral membrane protein with MALDI-TOF mass spectrometry. Biochim Biophys Acta *1555*, 111–115.

Wang, H., Qian, W.J., Mottaz, H.M., Clauss, T.R., Anderson, D.J., Moore, R.J., Camp, D.G., 2nd, Khan, A.H., Sforza, D.M., Pallavicini, M., *et al.* (2005). Development and evaluation of a micro- and nanoscale proteomic sample preparation method. J Proteome Res *4*, 2397–2403.

Washburn, M.P., Wolters, D., and Yates, J.R., 3rd (2001). Large-scale analysis of the yeast proteome by multidimensional protein identification technology. Nat Biotechnol *19*, 242–247.

Wisniewski, J.R., Zougman, A., Nagaraj, N., and Mann, M. (2009). Universal sample preparation method for proteome analysis. Nat Methods *6*, 359–362.

Wolters, D.A., Washburn, M.P., and Yates, J.R., 3rd (2001). An automated multidimensional protein identification technology for shotgun proteomics. Anal Chem *73*, 5683–5690.

Wu, C.C., MacCoss, M.J., Howell, K.E., and Yates, J.R., 3rd (2003). A method for the comprehensive proteomic analysis of membrane proteins. Nat Biotechnol *21*, 532–538.

Wu, C.C., and Yates, J.R., 3rd (2003). The application of mass spectrometry to membrane proteomics. Nat Biotechnol *21*, 262–267.

Xiang, R., Shi, Y., Dillon, D.A., Negin, B., Horvath, C., and Wilkins, J.A. (2004). 2D LC/MS analysis of membrane proteins from breast cancer cell lines MCF7 and BT474. J Proteome Res *3*, 1278–1283.

Zhong, H., Zhang, Y., Wen, Z., and Li, L. (2004). Protein sequencing by mass analysis of polypeptide ladders after controlled protein hydrolysis. Nat Biotechnol *22*, 1291–1296.

Chapter 9
Exploring the Capabilities of the Protein Identification by Unconventional Sample Preparation Approaches: LC/MALDI/On-Target Digestion Approach and High Pressure-Assisted In-Gel Tryptic Digestion

Melkamu Getie-Kebtie and Michail A. Alterman

Abstract The rapid and comprehensive separation, identification, and characterization of proteins from complex biological samples are formidable challenges that the growing field of proteomics faces. A robust and efficient digest of separated proteins is central to the development of any quantitative and/or qualitative proteomic approach. We describe here two unconventional approaches to a trypsin digestion of biological samples. One of those approaches integrates separation of proteins using RP-HPLC with on-target proteolytic digestion of the proteins for subsequent MALDI-MS analysis. This approach allows the combined information from peptide mass fingerprinting (PMF), MS/MS peptide fragment fingerprinting (PPF) and whole protein MS to increase confidence in protein identification and structural analysis of proteins. The other approach is based on the application of high-pressure cycling technology (PCT) for in-gel trypsin digest of 1D electrophoretically separated proteins. That approach was used in our laboratory to develop a label-free mass spectrometry based method for a relative quantification of proteins after SDS PAGE separation, including co-migrating proteins.

Keywords LC · MALDI · PCT · Tryptic Digestion · On-target

9.1 Introduction

Proteomics, as we know it today, was shaped by three technologies: protein/peptide separation, mass spectrometry and bioinformatics. Mass spectrometry and bioinformatics aspects of proteomics are not directly related to the topic of this book and this chapter, though we should point out that, analytical and computational power alone

M.A. Alterman (✉)
Tumor Vaccines and Biotechnology Branch, Division of Cellular and Gene Therapies,
Center for Biologics Evaluation and Research, Food and Drug Administration, Bethesda,
MD 20892, USA
e-mail: michail.alterman@fda.hhs.gov

A.R. Ivanov, A.V. Lazarev (eds.), *Sample Preparation in Biological Mass Spectrometry*, DOI 10.1007/978-94-007-0828-0_9,

is not sufficient to ensure success of a proteomics project. Major factors influencing the outcome of a proteomic project are:

- Carefully defined objective (e.g. global proteomics vs targeted proteomics)
- Appropriate choice of a sample preparation and pre-fractionation/separation approach
- Multiplicity and diversity of protein species
- Dynamic range of protein expression.

A high level of expertise in biological sample preparation and handling has about the same, if not higher, importance for successful experiments in bio-mass spectrometry as the presence of considerable mass spectrometry experience; both are required.

Proteomics researchers have to take into account difficulties particular to different strategies of the sample handling specific for different MS techniques, i.e. ESI relies largely on HPLC, while MALDI ionization brings to the research community sample separation strategies independent of HPLC. MALDI TOF instruments have been used successfully for analyzing mixtures of proteins and peptides, most notably, tryptic or similar digests. Sample preparation becomes particularly important when one moves from qualitative to quantitative proteomics experiments.

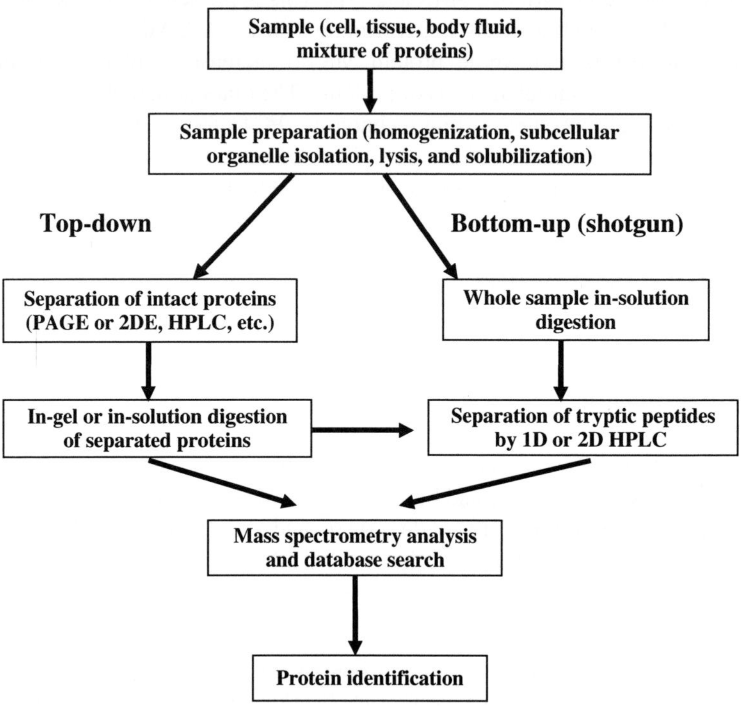

Fig. 9.1 Protein identification strategies

The strategies that protein chemists employ to approach this challenge can be broadly divided into two broad categories "top-down" and "bottom-up" (Fig. 9.1). The most popular bottom-up approach, multidimensional protein identification technology (MudPIT), relies heavily on liquid chromatography (LC) to reduce complexity of sample peptide mixture prior to MS analysis. Typically, this technique involves digestion of the whole biological sample followed by separation of the peptide mixture by a combination of various LC methods prior to MS analysis. Top-down methods, in turn, rely on intact protein analysis which starts with separation of proteins and is followed by MS analysis performed either on the level of whole proteins or tryptic peptides.

Our laboratory is dealing with a wide variety of biological samples ranging from vaccine samples to cell cultures and subcellular organelles and we have to consider a number of sample preparation-specific difficulties such as limited sample amounts, heterogeneous cellular compositions in samples, the fact that many proteins of interest are hydrophobic membrane proteins, subcellular organelle isolation and protein solubilization, and, in particular, tryptic digest conditions for reliable quantitative experiments.

Here we present two approaches to tryptic digestion coupled with two different separation techniques: first one involves an on-target tryptic digest preceded by an RP HPLC separation of proteins, and second is an enhanced high-pressure-assisted in-gel tryptic digestion.

9.2 Materials

1. BSA Lyophilized Powder (Lot No. 126K7405), cytochrome c from bovine heart, insulin from bovine pancreas, β-lactoglobulin from bovine milk, catalase from bovine liver, glyceraldehyde-3-phosphate dehydrogenase from rabbit muscle, myoglobin from equine skeletal muscle, glucose oxidase from Aspergillus niger, α-cyano-4-hydroxycinnamic acid, acetonitrile, methanol, ammonium bicarbonate, iodoacetamide, DTT, EZ Blue, Protein Extraction Reagent 4, Protease Inhibitor Cocktail, and tributylphosphine (all from Sigma-Aldrich, St. Louis, MO).
2. BSA ampules, 2 mg/mL, Lot No. IG115177 and Ferritin from horse spleen (Pierce Biotechnology, Rockford, IL).
3. BSA Standard grade Powder (Lot No. BAH63-717) (Equitech-Bio, Kerrville, TX).
4. Human cytochrome P450 1A2, human cytochrome P450 2E1, Solubilizer I, NuPAGE 4–12% Bis–Tris gel, NuPAGE MES SDS Running Buffer (20X), NuPAGE reducing agent (10X), NuPAGE antioxidant, and LDS Loading (Life Technologies, Carlsbad, CA).
5. Electrophoretically homogeneous rat cytochrome P450 2B1 was isolated as described previously (Alterman et al., 1995).
6. Trifluoroacetic acid (Fluka, Milwaukee, WI).
7. Sequencing grade modified porcine trypsin (Promega, Madison, WI).

8. C4 polymeric reversed-phase (RP) column (5 μm, 300 Å, 150 μm × 50 mm) purchased from VYDAC (Alltech associates, Inc., Deerfield, IL).
9. MRC5 cell (ATCC, Cat. No. CCL-171) extract was prepared as described under Section "Preparation of MRC5 Cell Extract." (Page 147).
10. The Tempo LC MALDI and the 4800 MALDI TOF/TOF instruments were from Applied Biosystems (Foster City, CA).
11. The Pressure Cycling Technology (PCT) instrument, Barocycler NEP-3229, and PCT Microtubes were from Pressure Biosciences, Inc. (South Easton, MA).

9.3 Methods

9.3.1 On-Target Tryptic Digestion

The applicability of the on-target digestion in protein identification platforms was studied by depositing trypsin on proteins immobilized on a MALDI target by either direct spotting of a single protein solution (Section 9.3.1.1) or after chromatographic separation of proteins from a simple protein mixture as well as a more complex biological sample (Section 9.3.1.2). The following sections describe the detailed procedure employed in this study.

9.3.1.1 On-Target Digestion of Proteins After Direct Spotting

1. *Sample preparation:* A 1 mg/mL solution of each of the following proteins was prepared in 25 mM ammonium bicarbonate (ABC) buffer. The proteins were BSA from different sources (described under Section 9.2), horse cytochrome C, human cytochrome P450s (recombinant CYP1A2 and CYP2E1), and rat cytochrome P450 2B1.
2. *Preparation of trypsin*: 20 μg of trypsin was reconstituted in 100 μL of 0.1 N HCl, aliquoted, and stored at –20°C until used. Immediately before use an aliquot of this solution was diluted four times with 25 mM ABC.
3. *Direct spotting:* A 0.5 μL aliquot of each protein solution was manually spotted on wells of a 384-well MALDI plate and allowed to dry in open air. All the spots dried within a maximum of 10 min. To investigate whether application of a proteolytic enzyme prior to or following spotting the sample material had any influence on its digestion efficiency, half of the wells contained a pre-deposited trypsin. To half of the spots that didn't contain pre-deposited trypsin, a 0.5 μL aliquot of trypsin was added after deposition of sample proteins.
4. *Reduction/alkylation:* The effect of reduction/alkylation on the effectiveness of on-target digestion was evaluated by using BSA (Equitech-Bio, protease free) as a model protein. A 2 mg/mL solution of BSA in 25 mM ABC was denatured with an equal volume of acetonitrile. BSA was first reduced by incubation with 100 mM DTT for 1 h at room temperature and then alkylated with 500 mM iodoacetamide at room temperature for 1 h protected from light. The alkylated BSA was diluted with 25 mM ABC to a final concentration of 0.25 μg/mL.

Table 9.1 Number of tryptic peptides matched to their corresponding proteins from on-target tryptic digestion of different samples of BSA and cytochrome P450 by peptide fragment fingerprinting

	BSA					
Sample type	Sigma	Pierce	Equitech	CYP1B2	CYP1A2	CYP2E1
Control (– trypsin)	0	0	0	0	0	0
Experiment (+ trypsin)	7	9	6	9	9	19

Aliquots of 0.5 μL were spotted on a 384-well MALDI plate and digested with trypsin, as described above. A non-reduced, non-alkylated BSA solution of the same concentration was used as a control.

5. *Deposition of matrix:* In all cases, a 0.5 μL aliquot of CHCA matrix solution (5 mg/mL) in 0.1% TFA/70% acetonitrile was deposited prior to MS analysis.

Table 9.1 summarizes the number of peptides identified by MS/MS fragmentation, i.e., peptide fragment fingerprinting (PFF), from BSA of different sources and the different cytochrome P450's. This study demonstrated that variable number of peptides was obtained by on-target digestion of the proteins, whereas the control (undigested) samples did not show any peptide peaks. This observation clearly indicates that the on-target tryptic digestion approach is a viable strategy for rapid protein digestion, which can be achieved in a period of 10 min. The results presented in this table were obtained by depositing trypsin on dried-out proteins pre-deposited on MALDI target. To assess the influence of the order of trypsin addition on the digestion efficiency, we performed an experiment reversing this sequence, i.e., trypsin was first pre-deposited and dried out on a MALDI target before the deposition of the protein under study. This experiment proved that the order of trypsin addition has no significant effect on on-target tryptic digestion. We also evaluated the influence of reduction/alkylation of proteins on the on-target tryptic digestion approach using BSA as a model protein. Our result demonstrated that reduction/alkylation significantly improved the protein digest quality as exhibited by the increased number of peptides identified as compared to 16 peptides identified from the non-reduced, non-alkylated sample. The sequence coverage also increased from 23 to 74%.

9.3.1.2 On-Target Digestion of Proteins After Chromatographic Separation

The success of the on-target tryptic digestion following direct spotting of individual proteins on a MALDI target led us to evaluation of the applicability of this technique for digestion of protein mixtures using a Tempo LC MALDI instrument. Figure 9.2 presents the scheme of the Tempo LC MALDI/on-target digestion experiment. This experiment was first performed with a simple mixture of eight proteins and then with protein extracts of MRC5 cells. The detailed procedure and observed results are discussed in the following sections.

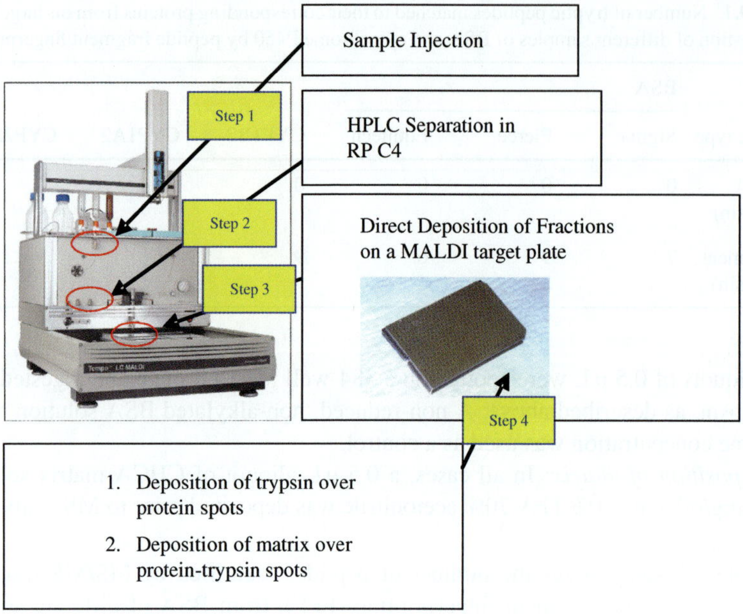

Fig. 9.2 Scheme for protein RP HPLC separation/on-target tryptic digestion

The Tempo LC MALDI instrument is a hybrid of an HPLC system and a MALDI target fraction collector. The spotting device directly deposits samples eluted from the HPLC system onto a MALDI target plate for subsequent analysis by a mass spectrometer. This automated device enables highly efficient protein digestion by providing a means to enhance protein concentration through precise positioning of liquid deposition into a very small area of the MALDI target surface. This improves sensitivity for analysis of proteins present in low abundance. In addition, while most solution-phase digestion experiments are performed over a period of hours and are subjected to high degree of autolysis of the enzymes used, the on-target digestion is performed in a few minutes (10 min on the average) with minimal occurrence of autolysis. Furthermore, this approach has the advantage that the collection of the proteins, enzymatic digestion, and MS analysis are all integrated into one MALDI plate without any sample transfer in order to avoid unnecessary contamination or sample loss.

Preparation of a Simple Protein Mixture

- A 1 mg/mL solution of each of the following proteins was prepared in a solvent containing acetonitrile (30%), 2-propanol (5%), TFA (0.1%), and distilled water (64.9%). The proteins were bovine cytochrome c, bovine insulin, bovine β-lactoglobulin, bovine catalase, rabbit GAPDH, equine myoglobin, *A. niger* glucose oxidase, and equine ferritin.

- Equal volumes of each protein solution were mixed and the volume was adjusted with the same solvent so that the final concentration of each protein becomes 100 μg/mL.

Preparation of MRC5 Cell Extract

- MRC5 cells were grown to confluence in a 75 cm^2 flask in Eagles's Minimal Essential medium with Earle's BSS, 2 mM L-glutamine, 1.0 mM sodium pyruvate, 0.1 mM nonessential amino acids, 50 IU/mL penicillin, 50 IU/mL streptomycin and 10% heat-inactivated (56°C for 30 min.) bovine fetal serum.
- The medium was removed from the flask and the monolayer was washed three times with 10 mL of Dulbecco's PBS without calcium and magnesium.
- Two milliliters of Protein Extraction Reagent 4 containing 1% Protease Inhibitor Cocktail and 5 mM tributylphosphine were applied to the monolayer and rocked for approximately 1 h at room temperature.
- The lysate was then further disrupted with assistance of PCT (Pressure Cycling Technology, in a Barocycler NEP 3229 system, Pressure BioSciences, Inc., South Easton, MA) for 20 pressure cycles, with each cycle consisting of 20 s at 35,000 psi followed by 10 s at atmospheric pressure at room temperature.
- The lysate was then centrifuged at 10,000*g* for 10 min.
- The supernatant was separated and alkylated with 2.8% dimethyl acrylamide for 30 min at room temperature on a rotary shaker.
- Residual dimethyl acrylamide was quenched with 32.6 μL of 2 M DTT.

Reversed Phase HPLC Separation of Proteins

- This experiment was performed by off-line coupling of a Tempo LC MALDI and a 4800 MALDI TOF/TOF instruments from Applied Biosystems (Foster City, CA) as illustrated in Fig. 9.1.
- A 2 μL aliquot of the mixture of the eight proteins or MRC5 cell extract prepared as described above was loaded into a C4 polymeric reversed phase column.
- The proteins were separated using a nonlinear gradient of two mobile phases at a flow rate of 3 μL/min. The compositions of the mobile phases were:

  Mobile Phase A: 0.1% TFA in 2% acetonitrile
  Mobile Phase B: 0.1% TFA in 98% acetonitrile

- The gradient profile used for solvent B was as follows: 5% for 2 min; 5–95% in 12 min; 95% for 5 min; 95–50% in 2 min; 50% for 2 min; 50–5% in 2 min, and 5% for 9 min, with a total run time of 35 min.
- The fractions (0.6 μL each) were deposited on a MALDI target plate without a matrix.
- When the spots dried out, 0.6 μL of a trypsin solution in 25 mM ABC (50 ng/μL, pH ∼ 8.0) was deposited manually on each spot.
- Digestion took place at ambient temperature while the droplets dried out. This took on average about 10 min.

- An aliquot of 0.6 μL of CHCA in 70% ACN containing 0.1% TFA (5 mg/mL) was then deposited over each spot using the Tempo LC MALDI instrument and air-dried.
- The spots were then analyzed using the 4800 MALDI TOF/TOF instrument.

Intact Protein Molecular Weight Determination

- The molecular weights of the intact proteins were determined after chromatographic separation of proteins from the simple eight protein mixture, containing cytochrome c, insulin, β-lactoglobulin, catalase, GAPDH, glucose oxidase, myoglobin, and ferritin. The intact molecular weight information was used to confirm protein identifications obtained from the protein digest.
- A 2 μL aliquot of the mixture of the eight proteins was loaded into a C4 polymeric reversed phase column. The proteins were separated using a nonlinear gradient of two mobile phases at a flow rate of 3 μL/min. Composition of the mobile phases and the gradient profile for solvent B were as described above.
- The fractions (0.6 μL each) were deposited on a 384-well MALDI target plate premixed with equal volumes of a 5 mg/mL solution of CHCA in 0.1% TFA/70% acetonitrile.
- The droplets were air-dried.
- The molecular weights of the proteins contained in each spot were determined using the 4800 MALDI TO/TOF MS in linear detector mode.

By coupling HPLC with linear mode MALDI TOF, we were able to detect seven of the eight proteins included in the mixture, i.e., cytochrome c, insulin, β-lactoglobulin, catalase, GAPDH, myoglobin, and ferritin (Table 9.2). The identity of these proteins, including glucose oxidase (no intact mass detected), was further confirmed with on-target digestion of the chromatographic fractions deposited on a MALDI target at the same retention times as the spots from which the intact proteins

Table 9.2 Proteins and their respective numbers of peptides identified from RP HPLC separation/on-target tryptic digestion of the mixture of 8 proteins

	Protein MW (Da)			
Protein name	Observed	Theoretical	No. of peptides matched	% Seq. coverage
Insulin	5,644	5733	–	–
Cyt C	12,142	11,565	10	63
β-LG	18,230	19,870	9	56
GADPH	35,207	35,666	9	36
Catalase	57,289	57,550	19	49
Myoglobin	16,859	16,941	12	82
Glucose oxidase	Not detected	65,597	11	36
Ferritin LC	19,890	19,818	6	34

were detected. It was observed that some of the spots contained multiple proteins co-eluted from the column. However, this protein overlap did not create a problem for reliable identification of the proteins as there were enough peptides detected to achieve sufficient sequence coverage. A summary of the proteins identified, along with the number of peptides matched to each protein, and their sequence coverage is presented in Table 9.2. The sequence coverage ranged from 34% (for ferritin) to 82% (for myoglobin).

As a next step, we examined the applicability of the HPLC/MALDI/on-target digestion approach for identification of proteins in complex biological systems. The protein extract of MRC5 cell line was subjected to HPLC fractionation and subsequent on-target tryptic digestion using the Tempo LC MALDI instrument and protein identification using the Mascot search engine. A total of about 50 proteins were successfully identified using this technique (see Table 9.3).

Table 9.3 List of proteins identified from the extract of MRC5 cell culture

Protein group	Acc. no.	MW (kDa)	No. pep.	% SC
Glutathione *S*-transferase Mu 1	P09488	25.6	6	31
Heterogeneous nuclear ribonucleoproteins	P22626	37.4	4	19
Methyltransferase-like protein 2	Q96IZ6	43.5	7	23
Macrophage inflammatory protein-2-alpha precursor	P19875	11.4	3	42
Heat-shock protein	P04792	22.8	9	45
Profilin-1	P07737	14.9	4	39
Beta-defensin 108 precursor	Q8NET1	8.3	4	64
Cofilin	P23528	18.4	2	15
Calreticulin precursor	P27797	48.1	5	19
Putative protein 15E1.2	O43716	15.1	4	38
Small proline-rich protein 4	Q96PI1	8.8	5	45
Macrophage migration inhibitory factor	P14174	12.3	2	17
Calumenin precursor	O43852	37.1	3	9
Cytochrome P450 26B1	Q9NR63	57.5	2	6
Serine/threonine-proteinkinase 24	Q9Y6E0	49.3	4	7
Galectin-1	P09382	14.6	4	34
Nucleophosmin	P06748	32.6	2	11
40S ribosomal protein	P62269	17.7	5	26
SH3 domain-binding glutamicacid-rich-like protein 3	Q9H299	10.4	2	20
Ras GTPase-activating protein1	P20936	116.3	10	12
Retinol dehydrogenase 12	Q96NR8	35.0	7	18
Spindlin-like protein 3	Q99865	29.1	5	24
Hexokinase D	P35557	52.2	7	18
Suppressor of cytokine signaling 1	O15524	23.5	5	29
Tropomyosin	P09493	32.7	3	10
Fructose-bisphosphate aldolase A	P04075	39.3	6	25
Collagen-binding protein2 precursor	P50454	46.4	11	49
Cytochrome P450 4F2	P78329	59.8	7	16
Kininogen precursor	P01042	71.9	8	16
Carbonylreductase [NADPH] 1	P16152	30.2	5	29
Histone	O60814	13.8	3	26

Table 9.3 (continued)

Protein group	Acc. no.	MW (kDa)	No. pep.	% SC
Glyceraldehyde-3-phosphate dehydrogenase	P04406	35.9	4	21
Vimentin	P08670	53.5	13	35
Alphaenolase	P06733	47.0	7	20
Myosin light polypeptide 6	P60660	16.8	5	34
Peripherin	P41219	53.8	4	8
StAR-related lipid transfer protein 5	Q9NSY2	23.8	6	30
Pyruvate kinase	P14618	57.8	9	23
Transgelin	Q01995	22.5	8	36
Glutathione *S*-transferase P	P09211	23.2	3	22
Heat shock cognate 71 kDa protein	P11142	70.9	13	26
Protein disulfide-isomerase precursor	P07237	57.1	3	8
Receptor activity-modifying protein 3 precursor	O60896	16.5	3	33
Tubulin beta-2 chain	P07437	49.6	11	36
Annexin A2	P07355	38.4	9	33
Actin	P63261	41.8	12	51
DNA-binding death effector domain-containing protein 2	Q8WXF8	36.2	4	19
14-3-3 protein	P62258	29.2	5	27
Annexin A5	P08758	35.8	16	64

MW – protein molecular weight; SC – sequence coverage

In conclusion, the method described above integrates separation of proteins using RP-HPLC with on-target proteolytic digestion of the proteins for subsequent MALDI-MS analysis. This method has been proven effective for a number of soluble and membrane proteins as well as an MRC5 cell lysate. The rapid protein separation by RP HPLC and direct deposition of eluted proteins fractions on a MALDI target plate followed by tryptic digestion on the target combined with off-line interface with the MALDI MS is a unique platform for rapid protein identification with improved sequence coverage. This simple and robust approach also significantly reduces the sample handling and potential loss in large-scale proteomics experiments. This approach allows the combined information from peptide mass fingerprinting (PMF), MS/MS peptide fragment fingerprinting (PPF) and intact protein MS to increase confidence in protein identification and structural analysis of proteins on MALDI-TOF instruments.

9.3.2 Pressure-Assisted Tryptic Digestion

Tryptic digestion of proteins continues to be a fundamental part of any proteomics workflow. A number of enhanced digestion protocols have been developed in recent years (Harris and Reilly, 2002, Warscheid and Fenselau, 2004). Nonetheless, a need still exists for new digestion approaches that meet the demands of qualitative and quantitative proteomics. Here we describe an application of pressure cycling technology (PCT) for tryptic in-gel digestion of proteins. Our results demonstrated that high pressure-assisted in-gel tryptic digestion compares with the standard in-gel tryptic digestion with respect to number of detected peaks; number of identified peptides, and sequence coverage. In all three parameters, PCT digestion performed at least as well, and sometimes better, than the standard digestion approach. This approach also reduced the extended time (about 16 h) required by the standard in-gel digestion to 15 min. We have also found that this high pressure-assisted, label-free approach can be used for the rapid, relative quantitation of proteins after 1DE separation (manuscript in preparation).

The digestion efficiency of the pressure-assisted in-gel tryptic digestion approach was compared with that of the conventional overnight in-gel digestion approach, which is currently used as a standard technique in many proteomics labs. Although this protocol focuses on in-gel tryptic digestion, the pressure-assisted approach is shown to be effective for in-solution protein digestion as well (López-Ferrer et al., 2008).

9.3.2.1 Sample Preparation

1. A 3 mg/mL solution of each of the three proteins (Fig. 9.3) was prepared in 50 mM ammonium bicarbonate containing 0.1% RapiGest (Waters, Milford, MA) and stored at –80°C until used. A 1 μL aliquot of this solution was diluted with 12 μL of distilled water prior to mixing with LDS loading buffer (Life Technologies, Carlsbad, CA).
2. LDS Loading Buffer containing a reducing agent was prepared by mixing 1 part of Reducing Agent (10X) with 2.5 parts of LDS Loading Buffer (4X).
3. A mixture of the sample and loading buffer was prepared by mixing 13 parts of the sample with 7 parts of the loading buffer and heated at 95°C for 3 min to denature the proteins. The sample was then cooled down to room temperature for subsequent loading on NuPAGE 4–12% Bis–Tris gel (Life Technologies, Carlsbad, CA).

9.3.2.2 SDS PAGE

1. Sample loading: A 20 μL aliquot of each of the denatured and reduced protein was loaded on NuPAGE 4–12% Bis–Tris gel in duplicate.
2. Running buffer: The running buffers used both for the inner and outer chambers were NuPAGE MES SDS Running Buffer, except the inner chamber contained NuPAGE antioxidant.
3. Running condition: Electrophoresis was performed at a constant voltage of 200 V for 35 min using electrophoretic apparatus from Life Technologies (Carlsbad, CA).
4. Staining and destaining: The gels were staining with EZ Blue colloidal Coomassie Blue stain (Sigma-Aldrich, St. Louis, MO) with gentle shaking for an hour. Gels were de-stained by repetitive rinsing in distilled water.

9.3.2.3 In-Gel Tryptic Digestion

1. Excision of bands: The protein bands of interest were excised from the gels and the gel pieces were placed in 1.5 mL Eppendorf-style microcentrifuge tubes pre-washed with acetonitrile.
2. Washing/Destaining: 100 μL of 50 mM ammonium bicarbonate was added to each tube and the sample was incubated for 20 min at 40°C. The samples were then centrifuged at 10,000 rpm for 1 min and the supernatant was removed. Next, the gel pieces were dehydrated by addition of 100 μL of acetonitrile and incubation for 20 min at 40°C. The acetonitrile was removed after brief centrifugation as above. These steps were repeated until the gel pieces were visually destained.
3. Reduction: 100 μL of 10 mM DTT was added to each sample and incubated at 60°C for 30 min to reduce the disulfide bonds of proteins. The sample was then centrifuged at 10,000 rpm for 1 min and the supernatant was removed.
4. Alkylation: 100 μL of 50 mM IAA was added and the sample was incubated at 60°C for 30 min, protected from light. The IAA was then removed after centrifugation. Before proceeding to the following steps, the gel pieces were dehydrated by addition of 100 μL of acetonitrile.

After this step, the following downstream procedures were employed for standard or pressure-assisted digestion, as appropriate.

9.3.2.4 Conventional Tryptic Digestion

1. Fifty microliters of 20 ng/μL trypsin in 50 mM ABC was added to each sample and incubated at 37°C overnight (~16 h).
2. To extract the peptides 50 μL acetonitrile was added to the tubes and the sample was incubated at 37°C for 1 h.
3. The samples were then centrifuged and the supernatants were withdrawn for analysis.
4. A 0.7 μL aliquot of the digest was spotted on a MALDI target plate in triplicate and the MS and MS/MS spectra was collected using a 4800 MALDI TOF/TOF instrument.

9.3.2.5 Pressure-Assisted Tryptic Digestion

1. The gel pieces were washed, destained, reduced, and alkylated as described above, then transferred into PCT Microtubes specially designed to contract and expand upon application of alternating high and low hydrostatic pressure (Pressure Biosciences, So. Easton, MA).
2. Fifty microliters of 20 ng/μL trypsin in 50 mM ABC was then added to the Microtubes. Samples were treated with 45 cycles of hydrostatic pressure between high (35,000 psi) and low (atmospheric pressure) at 37°C in a Barocycler NEP-3229 instrument (Pressure BioSciences). The high pressure was maintained for 55 s and the low pressure for 5 s in each cycle.
3. Peptides were then extracted from the gel by addition of 50 μL of acetonitrile and application of 10 cycles of high and low pressure as described above (step 2 of Section 9.3.2.5).

4. A 0.7 μL aliquot of the digest was spotted on a MALDI target plate in triplicate and the MS and MS/MS spectra were collected using a 4800 MALDI TOF/TOF instrument.

Figure 9.3 shows the MS spectra of GlOx and PyKin in-gel digest obtained using the pressure assisted and the conventional tryptic digestion approach. For all the proteins studied, we observed a high degree of similarity among the mass spectra

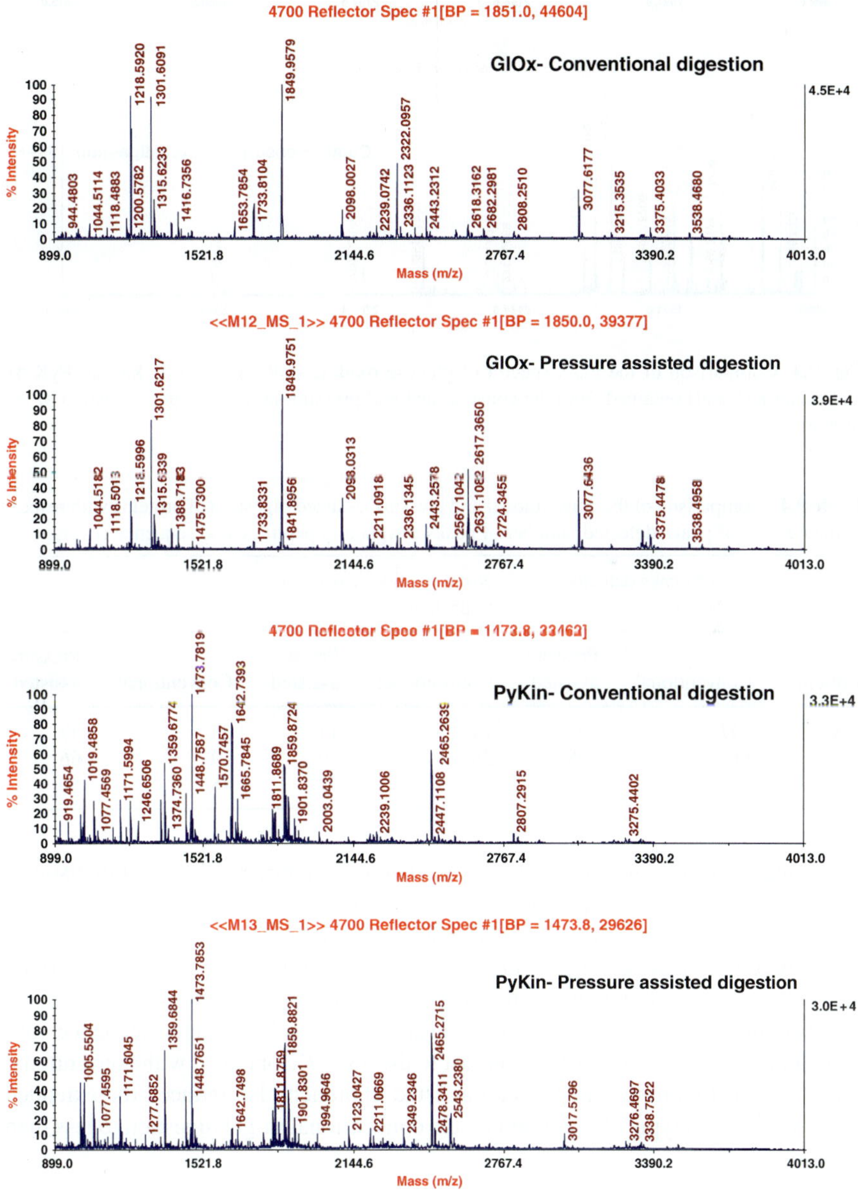

Fig. 9.3 (continued)

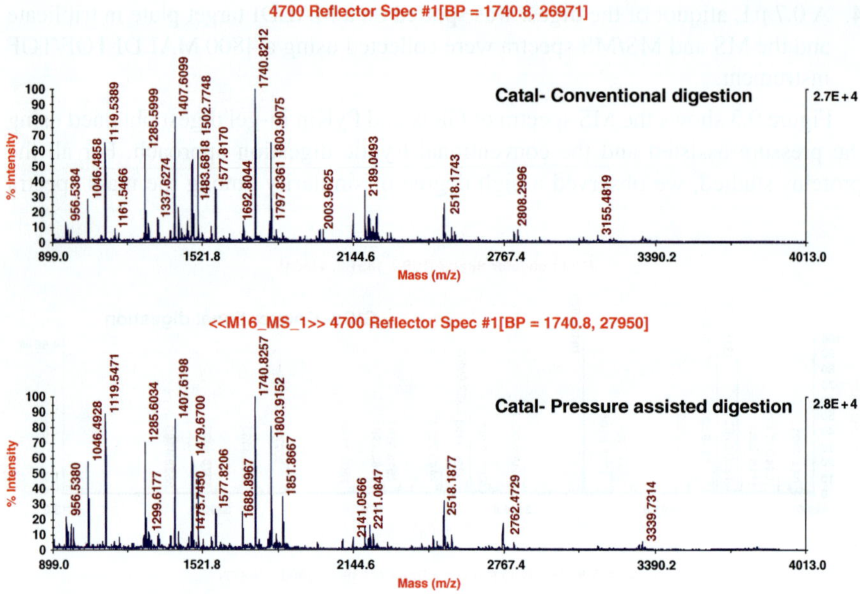

Fig. 9.3 Comparison of the mass spectra of glucose oxidase (GlOx), pyruvate kinase (PyKin), and catalase (Catal) obtained from the conventional and pressure-assisted tryptic digestion of the proteins

Table 9.4 Comparison of the conventional and pressure-assisted digestion of proteins with respect to the number of peaks detected, number of peaks assigned to peptides, and sequence coverage

	No. of peaks detected, *S/N > 20		No of peptides assigned to protein		% Sequence Coverage	
Protein	Conventional	Pressure assisted	Conventional	Pressure assisted	Conventional	Pressure assisted
GlOx	97	92	14	17	38	49
PyKin	112	128	25	30	59	66
Catal	86	80	18	21	45	50

generated under the two conditions of digestion. A summary of the comparison of the two tryptic digestion approaches with respect to the number of peaks detected, the number of peptides identified, and sequence coverage, is presented in Table 9.4. In all three parameters, the pressure-assisted tryptic digestion approach performed at least as well, and sometimes better, than the conventional digestion approach. For example, pressure-assisted tryptic digestion approach presented sequence coverage ranging from 49 to 66% in contrast to the 38–59% obtained with conventional digestion. Furthermore, the pressure-assisted approach also reduced the extended time (about 16 h) typically required by the conventional in-gel digestion to less than 30 min.

References

Alterman, M.A., Chaurasia, C.S., Lu, P., and Hanzlik R.P. (1995). Heteroatom substitution shifts regioselectivity of lauric acid metabolism from omega-hydroxylation to (omega-1)-oxidation. Biochem Biophys Res Commun *214*, 1089–1094.

Harris, W.A., and Reilly, J.P. (2002) On-probe digestion of bacterial proteins for MALDI-MS. Anal Chem *74*, 4410–4416.

López-Ferrer, D., Petritis, K., Hixson, K.K., Heibeck, T.H., Moore, R.J., Belov, M.E., Camp, D.G., 2nd, and Smith, R.D. (2008). Application of pressurized solvents for ultrafast trypsin hydrolysis in proteomics: proteomics on the fly. J Proteome Res *7*(8), 3276–3281.

Warscheid, B., and Fenselau, C. (2004). A targeted proteomics approach to the rapid identification of bacterial cell mixtures by matrix-assisted laser desorption/ionization mass spectrometry. Proteomics *4*, 2877–2892.

Part III
Methods for Tissue, Cell, and Organelle Preparation and Analysis

Chapter 10
Sample Preparation of Formalin-Fixed Paraffin-Embedded (FFPE) Tissue for Proteomic Analyses

Diem Tran, Mark Daniels, Ben Colson, Dikran Aivazian, Antonio Boccia, Ingrid Joseph, Steffan Ho, Steve French, Alex Buko, and Jing Wei

Abstract A sample preparation procedure has been developed to prepare Formalin-Fixed Paraffin-Embedded (FFPE) tissues for proteomic profiling. Protocols have been established for processing FFPE sections on slides, including deparaffinization, rehydration, protein extraction, and enzymatic digestion, all using in-house developed and optimized buffers and techniques. These methods are practical and allow for parallel sample processing. The efficiency of protein extraction from FFPE tissue is comparable to the results from fresh or frozen tissue. The prepared samples can be directly coupled to a wide range of proteomic technologies, including isobaric labeling and on-line multidimensional chromatography. Comprehensive proteomic profiling of prepared FFPE tissue was performed using a fully automated tandem multidimensional liquid chromatography (LC) system coupled with on-line electrospray ionization (ESI) tandem mass spectrometry (MS): LC/LC-MS/MS. Routinely around 4,000 unique proteins of high confidence could be identified from 100 μg of extracted protein. The method presented here will be valuable for others who plan to analyze FFPE tissues.

Keywords FFPE · iTRAQ · Mass Spectrometry · Multidimensional Chromatography · Proteomics

10.1 Introduction

Formalin fixed paraffin embedded (FFPE) tissues have been used for decades to evaluate tissue histology and diagnose disease (Blonder and Veenstra, 2009). FFPE tissue blocks are routinely preserved and stored after pathological diagnosis, and such archived material provides an ample resource for research and clinical proteomics. High-throughput proteomic studies of archival FFPE tissues have the potential to be a powerful tool for examining the clinical course of disease (Fowler et al., 2007). However, proteomic analysis of FFPE tissue has proven especially

J. Wei (✉)
Biogen Idec, San Diego, CA 92122, USA
e-mail: jwei@jadebio.com

A.R. Ivanov, A.V. Lazarev (eds.), *Sample Preparation in Biological Mass Spectrometry*, DOI 10.1007/978-94-007-0828-0_10,

difficult due to: (1) The poor yield of protein extraction caused by formaldehyde-induced inter-molecular and intra-molecular cross-linking (Bedino, 2003); (2) The low percentage of identifiable and quantifiable peptides; and (3) The labor intensive process of sample preparation. Considerable efforts have been made by many researchers to improve this process (Addis et al., 2009; Belief et al., 2008; Crockett et al., 2005; Guo et al., 2007; Hood et al., 2006; Hood et al., 2005; Jain et al., 2008; Negishi et al., 2009; Nirmalan et al., 2009; Nishimura et al., 2009; Palmer-Toy et al., 2005; Patel et al., 2008; Sprung et al., 2009). The aim of our study was to develop an efficient, robust, standardized, and facile technique to process FFPE tissues for mass spectrometry (MS) based proteomic analysis.

Effective sample preparation protocols are crucial components of a successful proteomics analysis. As a process science, a successful platform must comprise a set of seamlessly integrated technologies, from sample preparation and chromatography, to MS instrumentation and bioinformatics. Samples prepared with the presently described method can be directly applied with (or is compatible with) a wide range of proteomic technologies. In this example, comprehensive proteomic profiling of processed FFPE samples was accomplished using a fully automated tandem multidimensional separation system coupled with on-line ESI-tandem MS– (LC/LC-MS/MS), which was modified from a previously developed multi-dimensional proteome platform named On-line Three-Dimensional (3D) LC-MS/MS(Wei et al., 2005). Additionally, customized bioinformatics software packages have been developed for optimal analysis of these large (multi-GB) datasets. The efficiency of protein extraction, the total number of protein identifications and the overall reproducibility of FFPE extraction was tested and compared to that of fresh-frozen tissue. LC/LC-MS/MS analyses of typical sample loads of 100 μg extracted and processed FFPE protein yielded thousands of high-confidence protein identifications, and was similar to what was obtained from frozen tissue. Finally, to further evaluate the new FFPE protocol, prepared samples were coupled with isobaric and Isotopic Mass Tagging reagents such as Isotope Tags for Relative and Absolute Quantitation (iTRAQ) of AB SCIEX and Tandem Mass Tag (TMT) of Thermo Fisher Scientific, which represent some of the most advanced technologies for proteomics quantitation. These reagents can be directly applied to prepared samples using conventional manufacturer-developed protocols. Overall, this FFPE proteomics platform was found to be sensitive and reproducible. The sample preparation protocol presented here will be valuable for investigators who plan to analyze FFPE tissues.

10.2 Materials and Experimental Procedures

Instrumentation: LTQ mass spectrometer equipped with a nano-spray source from Thermo Fisher Scientific (San Jose, CA), Agilent 1100 series HPLC from Agilent (Palo Alto, CA), in-line microfilter assemblies from Upchurch Scientific (Oak Habor, WA)

Software: Spectrum Mill program by Agilent (Palo Alto, CA)

Reagents: Proteomics grade endoproteinases Lys-C and Trypsin from Roche Diagnostics, Germany; Rapigest from Waters, Milford, MA; 1 M HEPES buffer, 0.5 M tris(2-carboxyethyl)phosphine (TCEP) at neutral pH, and H_2O and Acetonitrile from Thermo Fisher Scientific, Pittsburgh, PA; DC Protein Assay from Biorad, Hercules, CA; Glycoprotein Deglycosylation Kit from EMD Chemicals, Gibbstown, NJ; Zorbax SB-C18 reversed-phase material from Agilent (Palo Alto, CA); polysulfoethyl A strong cation exchange material from PolyLC Inc. (Columbia, MD); all SDS PAGE supplies and reagents, SilverXpress silver staining kit from Life Technologies (Carlsbad, CA), and iTRAQ reagents from AB SCIEX (Foster City, CA); all other chemicals, unless otherwise noted, obtained from Sigma-Aldrich(St. Louis, MO).

Buffers: Digestion buffer (150 mM NaCl, 50 mM HEPES, 1 mM EDTA, pH 7.4). Extraction buffer (150 mM NaCl, 50 mM Hepes, 1 mM EDTA, 4% Rapigest, pH 7.4).

Tools: Surgical blades (size 11) and glass microscope slides from Thermo Fisher Scientific (Pittsburgh, PA).

10.2.1 Sample Preparation

Freshly harvested mouse xenograft melanoma tumor ($\sim$1 cm^3) was divided in half. One half of the tumor tissue was used for generating FFPE sections following standard histology procedures and the other half was frozen. Subsequent sample preparation of the frozen tissue was identical to that of FFPE samples with the omission of step 1 and 5 in section 10.2.1.3:

10.2.1.1 Prepare Sections on Slides

To generate unstained sections, cut the tissue block into 10 μm sections and place three sections onto a slide. Incubate at 37°C to secure the sections on the slides.

10.2.1.2 Deparaffinization

1. Transfer the slide to a staining dish containing fresh xylene. Incubate for 10 min at room temperature.
2. Repeat Step 1 twice, using fresh xylene each time.
3. Transfer the slide to a staining dish containing fresh 100% ethanol. Incubate for 10 min at room temperature.
4. Repeat Step 3 using fresh 100% ethanol.
5. Transfer the slide to a staining dish containing fresh 96% ethanol. Incubate for 10 min at room temperature.
6. Repeat Step 5 using fresh 96% ethanol.
7. Transfer the slide to a staining dish containing fresh 70% ethanol. Incubate for 10 min at room temperature.
8. Repeat Step 7 using fresh 70% ethanol.
9. Remove the slide and air dry.

10.2.1.3 Protein Extraction

1. Dip the tip of a surgical blade into a 1.5 mL tube containing Extraction buffer. Scrape the tissue off of the slide with the blade and transfer to a 1.5 mL tube containing 90 μL Extraction buffer.
2. Heat sample at 95°C for 20 min.
3. Centrifuge the tube at 10,000 rcf for 1 min in a microcentrifuge to recombine condensation.
4. Dilute sample with 7 volumes of Digestion buffer. Vortex.
5. Incubate the sample at 80°C for 2 h with agitation at 750 rpm.
6. Place sample on ice.
7. Perform a protein quantitation using a Bio-Rad DC Protein Assay kit.
8. Save 3 μg of protein for a 1D SDS-PAGE analysis.
9. Estimate the protein yield: *Protein yield (μg)* = *~0.3 μg/mm²/μm* × *section area (mm²)* × *thickness (μm)*. Using the estimated protein yield to calculate the amount of deglycosylase, dephosphorylase and endopeptidase enzymes to use in the digestion steps below.

10.2.1.4 Protein Digestion

1. Deglycosylate the proteins by adding 1 μL of each deglycosylase enzyme in the 5-enzyme degylcosylation kit for every 1 mg of sample protein. Incubate for 3 h at 37°C. (optional)
2. Dephosphorylate the protein by adding 1 μL of protein phosphatase for every 500 μg of sample protein. Incubate for 30 min at 30°C. (optional)
3. Resuspend lyophilized trypsin to 1 mg/mL with cold Digestion buffer and keep on ice until added to the samples.
4. Add trypsin to protein sample at a ratio of 1:30 (w/w Incubate overnight at 37°C with agitation at 500 rpm.
5. Perform a protein quantitation after trypsin digestion using a Bio-Rad DC Protein Assay kit. The protein quantity determined before digestion will be artificially lower because the protein extraction from FFPE tissue is fully completed during the digestion process.
6. Add 0.5 M TCEP for a final concentration of 1 mM. Incubate for 60 min at 37°C.
7. During incubation prepare iodoacetamide (IAM) solution at 100 mg/mL in Digestion buffer.
8. Add the IAM solution to a final concentration of 0.4 mg/mL. Incubate at 37°C in the dark for 30 min.
9. Add an additional aliquot of trypsin to achieve 1:60 (w/w) ratio (based on the protein yield after step 5) to the sample. Incubate for 4 h at 37°C.
10. Perform protein quantitation after Trypsin digestion using a Bio-Rad DC Protein Assay kit.
11. Remove 3 μg aliquot of digested sample to run a 1D SDS-PAGE gel.
12. Flash freeze the remaining sample for future use.
13. Run a 1D SDS-PAGE gel using 3 μg of digested and 3 μg of undigested sample to ensure complete digestion.

14. Stain using the SilverXpress silver staining protocol provided by the vendor.
15. Scan the gel for record keeping.

10.2.1.5 iTRAQ Labeling (Optional)

Aliquot and dry the desired amount of digested protein. Perform iTRAQ labeling according to the vendor protocol.

10.2.1.6 Prepare the Sample for LC/LC-MS/MS Analysis

Remove Rapigest according to the vendor protocol.

10.3 LC/LC-MS/MS Analysis

10.3.1 Column and Buffer Preparation

The peptide samples were analyzed by an LC/LC-MS/MS system consisting of an Agilent 1100 series HPLC driving a 3D column: a reversed-phase (RP), a strong cation exchange (SCX), and a second reversed-phase microcapillary column (RP1-SCX-RP2), and an LTQ mass spectrometer equipped with a nano-spray source. Each column was packed individually in-house at 1000 psi using a pressure bomb designed in our lab and machined at University of California, San Diego. A 3D column was constructed with RP1, SCX and RP2 coupled in series by in-line micro-filter assemblies and packed with three LC phases. The first microcapillary, RP1, (200 μm i.d. x 365 μm o.d. x 30 cm) was packed with Zorbax SB-C18 reversed-phase material (5 μm). The second microcapillary, SCX, (200 μm i.d. × 365 μm o.d. x 10 cm) was packed with polysulfoethyl A strong cation exchange material (5 μm, PolyLC Inc., Columbia). The third microcapillary tip, RP2, (200 μm i.d. x 365 μm o.d. x 15 cm) was packed with Zorbax SB-C18 reversed-phase particles (5 μm). The column was then connected to the HPLC pump through RP1 and coupled to the LTQ through RP2.

Without desalting, 100 μg of peptide mixture from each digested protein sample was directly loaded onto the RP1 section of the microcapillary column using a pressure bomb. The LC separation was carried out in a fully automated manner using four buffer solutions: buffer A (100% H_2O, 0.25% formic acid), buffer B (100% ACN, 0.25% formic acid), buffer C (100 mM ammonium acetate in H_2O, 0.25% formic acid), and buffer D (1 M ammonium acetate in H_2O, 1.25% formic acid, 2.5% acetic acid).

10.3.2 Chromatography

The chromatography applied here was modified from a 3D LC-MS method described earlier (Wei et al., 2005) to allow for easier optimization for different sample types. The method is a 2D method with 3D column configuration: (1) a C18 column for peptide trapping and clean-up, (2) a strong-ion exchange column for salt separation, and (3) a C18 column for reverse-phased separation. An RP gradient of

0–40% B over 150 min with a post-split flow rate of 200 nL/min was applied to the first 14 salt steps to elute a fraction of the absorbed peptides from the RP1 region, which were retained on the SCX phase. Then a salt step of 10 min with a post-split flow rate of 1 μL/min was used to subfractionate peptides from the SCX phase onto the RP2 region. The peptides on RP2 were then separated using the 0–40% B RP gradient of 150 min with a post-split flow rate of 150 nL/min. The peptides eluted from RP2 were then directly analyzed by the LTQ mass spectrometer. When the elution from RP2 was completed, another salt step with higher salt concentrations was applied to transfer another subfraction from the SCX region to RP2, followed by the same RP gradient for RP2 separation. When this series of 18 salt steps (0, 10, 30, 40, 45, 50, 55, 60, 62.5, 65, 67.5, 70, 75, 80, 85, 90, 95, 100 mM ammonium acetate pH 3.0) were completed within the 0–40% B RP gradient, a higher RP gradient of 0–60% B was applied for the next 14 salt steps (110, 115, 120, 125, 130, 140, 150, 160, 170, 180, 190, 210, 250, 300 mM ammonium acetate pH 3.0). For the last set of 11 salt steps (350, 400, 450, 500, 550, 600, 650, 725, 900, 1100, and 2000 mM ammonium acetate), the RP gradient of 0–80% B over 150 min.

10.3.3 Mass Spectrometry Analysis for Label-Free Peptides

The total mass spectral scan range was divided into three segments to utilize the gas-phase fractionation power of the instrument. The LTQ mass spectrometer was set to divide the full MS scan into three smaller sections covering a total range of 400–2000 *m*/*z* (400–700, 700–1200, 1200–2000). Each of the MS scans were followed by 10 MS/MS scans of the 10 most intense ions from the preceding MS scan using collision-induced dissociation (CID). The normalized collision energy was set to 35%. Dynamic exclusion was enabled with a repeat count of two and a 30 s duration window.

10.3.4 Mass Spectrometry Analysis for Labeled Peptides

The chromatography method remained the same as for label-free peptide analysis while the mass spectrometer method was modified. Each of the MS scans was followed by 10 MS/MS scans of the 5 most intense ions from the preceding MS scan, sequentially using collision-induced dissociation (CID) and pulse Q dissociation (PQD) for each selected peptide ions. Collision energy was typically set to 35% for both CID and PQD.

10.4 Data Analysis

Spectral data were extracted and, using the Spectrum Mill program, searched against the IPI Human Database with concatenated reversed sequences. Sequences for enzymes used in sample preparation and common protein contaminants were also included in the database. Spectra were searched for fully tryptic peptide matches with up to two missed cleavages. Variable modification of methionine oxidation and fixed modification of cysteine carbamidomethylation were included as database search parameters. For iTRAQ labeled samples a fixed modification for the mass of the isobaric tags is added to the N-termini of peptides and to lysine residues. Determination of valid proteins was achieved by setting a peptide score cutoff for each observed charge state such that no more than 0.1% of peptide matches above the score could come from matches to the reversed protein sequences. iTRAQ isotopic purity values for each lot of reagent were supplied by the manufacturer and purity correction was applied using code from i-Tracker for implementation of Cramer's rule (Shadforth et al., 2005). iTRAQ intensities for proteins are calculated by summing the corrected iTRAQ intensities of all valid peptides for each protein. Normalization of iTRAQ intensities for each sample in a multiplexed run is achieved by applying a global linear correction to each sample determined by comparison to the sample with the highest summed iTRAQ intensity.

10.5 Results and Discussion

Several FFPE sample preparation protocols were evaluated based on the efficiency of protein extraction and reversal of formalin cross-linking, as well as subsequent suitability for proteolytic digestion, overall reproducibility, robustness, and the quality of resulting proteomic analysis. The tools to evaluate these criteria included Bio-Rad DC Protein Assay, 1D SDS-PAGE with silver staining, online LC/LC-MS/MS, and downstream bioinformatic analysis. As a quality control benchmark the results of each evaluation step (Fig. 10.1) were compared with those from matched frozen tissue, which has a well established sample preparation protocol.

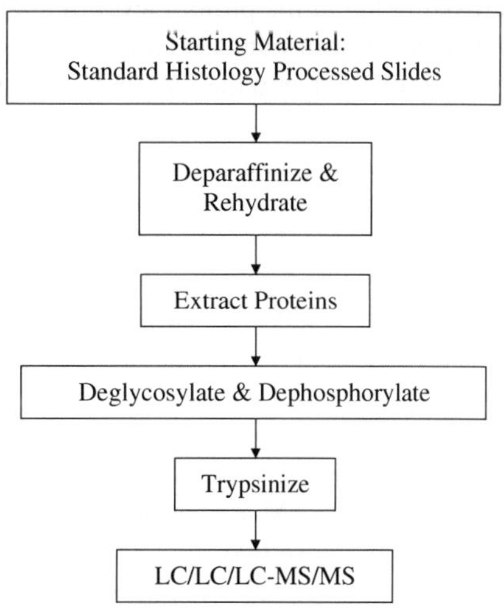

Fig. 10.1 FFPE sample preparation and analysis flowchart

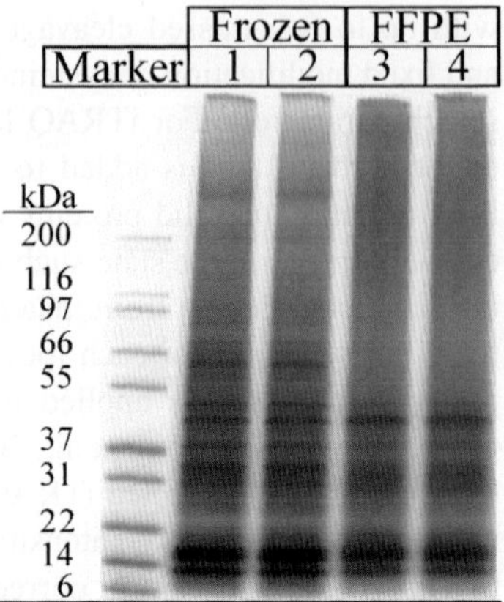

Protein yield from 6 sections (μg)	511	576	684	655

Fig. 10.2 SDS-PAGE analysis of combined proteins extracted from Frozen and FFPE sections of mouse xenograft melanoma tumor. Six sections were extracted for each sample. *Lane M*: molecular weight marker. *Lanes 1* and 2: frozen tissue in duplicate. *Lanes 3* and *4*: FFPE tissue in duplicate

The efficiency of the protein extraction was evaluated first at the protein level. The total protein yield from six extracted FFPE sections was between 500 and 700 μg (lanes 3 and 4, Fig. 10.2), which was very similar to the yield from frozen tissue (lanes 1 and 2, Fig. 10.2), and all replicates analyzed were within that range of concentration. Protein yield can be estimated based on the sizes of tissue sections, as outlined in the Sample Preparation section (3.9), and is used to estimate the number of FFPE sections to combine for profiling and the amount of digestion enzymes to add.

Several sets of initial extraction conditions gave rise to "smearing" in the 1D gels of protein extracted from FFPE, whereby no clear bands could be detected. The smearing was a consequence of the majority of extracted proteins remaining formalin cross-linked. The optimized protocol is more efficient at reversing the cross-linking at the protein level than all other protocols tested (Fig. 10.2). Clearly resolved protein bands could now be observed, particularly in the lower part of the gel, likely due to the fact that smaller proteins have proportionately fewer cross-linked sites. Un-reversed cross-linking in larger proteins is unlikely to be detrimental as those proteins will have many free peptides liberated by digestion.

Extracted protein from FFPE tissue was digested with trypsin using the standard digestion protocol (see Sample Preparation section 4), and the results analyzed by 1D silver stained gel (Fig. 10.3). The protocol led to complete proteolytic digestion and the results were similar between FFPE and frozen tissue. The final quantity of extracted material was determined after trypsinization. In the protocol described, deglycosylation and dephosphorylation are optional. In our lab, both steps have been integrated into the routine analysis for biomarker discovery where the interest is in changes of protein expression levels but not those of post-translational modifications (PTMs). PTM are best studied with specialized protocols for their enrichment. By skipping these two steps in the described protocol, preservation of glycans and phosphates on proteins could be potentially achieved with FFPE samples.

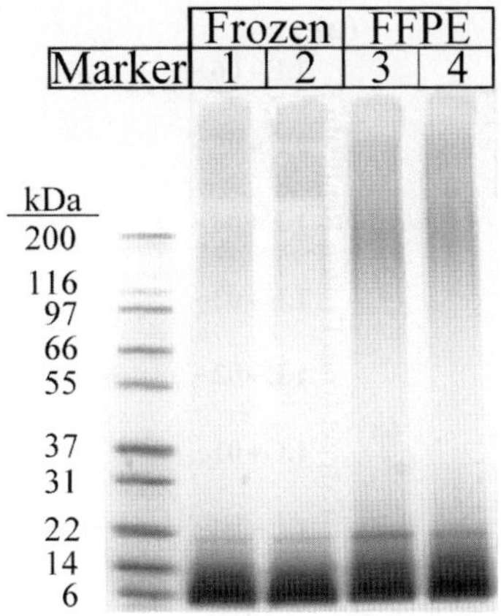

Fig. 10.3 SDS-PAGE analysis of digested proteins from Frozen and FFPE sections of mouse xenograft melanoma tumor. *Lane M*: molecular weight marker. *Lanes 1* and *2*: frozen digested protein in duplicate. *Lanes 3* and *4*: FFPE digested protein in duplicate. The digestion was complete and results were comparable between FFPE and frozen tissue

This FFPE sample preparation protocol was developed for seamless integration with downstream mass spectrometry-based technologies. As a result, prepared samples can be directly coupled to wide range of proteomic technologies. For direct analysis, 100 μg of a label-free protein digest of extracted FFPE tissue was loaded on a 3D column and analyzed using the method described in the experimental LC/LC-MS/MS analysis (Section 10.3). Approximately two million spectra were collected per sample, representing over 100,000 unique peptides, and over 4,000 unique proteins were identified from each single analysis. The range of identifications was comparable between FFPE and frozen tissue, and the variability was lower than between biologically different samples. Data analysis was performed as described above with a false discovery rate set at 0.1%.

Peptides prepared from FFPE tissue were also successfully labeled and analyzed with iTRAQ reagents for multiplex proteomic analysis. iTRAQ labeling is both an advanced tool for quantitative proteomics analysis and a useful indicator of effective reversal of formalin cross-linking, as the amide reactive group of the iTRAQ label can only be attached to lysine in the absence of bound formalin. 100 μg each of two FFPE and two frozen samples were labeled using 4-plex iTRAQ reagent. After Rapigest removal, half of the multiplexed sample was loaded on a 3D column and analyzed using the method described in the experimental LC/LC-MS/MS analysis (Section 10.3). The results of this multiplex experiment demonstrate that reproducibility between replicates of FFPE sample runs is very high. More than 2500 protein were quantifiable with iTRAQ following this protocol. The R^2 value is 0.99 when comparing two FFPE samples prepared in parallel (Fig. 10.4a), which is comparable to the R^2 value of the frozen duplicates (Fig. 10.4b). The overall proteome profiles are similar between FFPE and frozen tissues, however, the level of a few of the detected proteins was different between FFPE and frozen tissues (Fig. 10.4c). The variability of these protein levels is likely attributable to the formalin fixation.

The protocol is quite reproducible, provided it is applied to fairly homogeneous samples. However, when dealing with clinical samples, their intrinsic heterogeneity

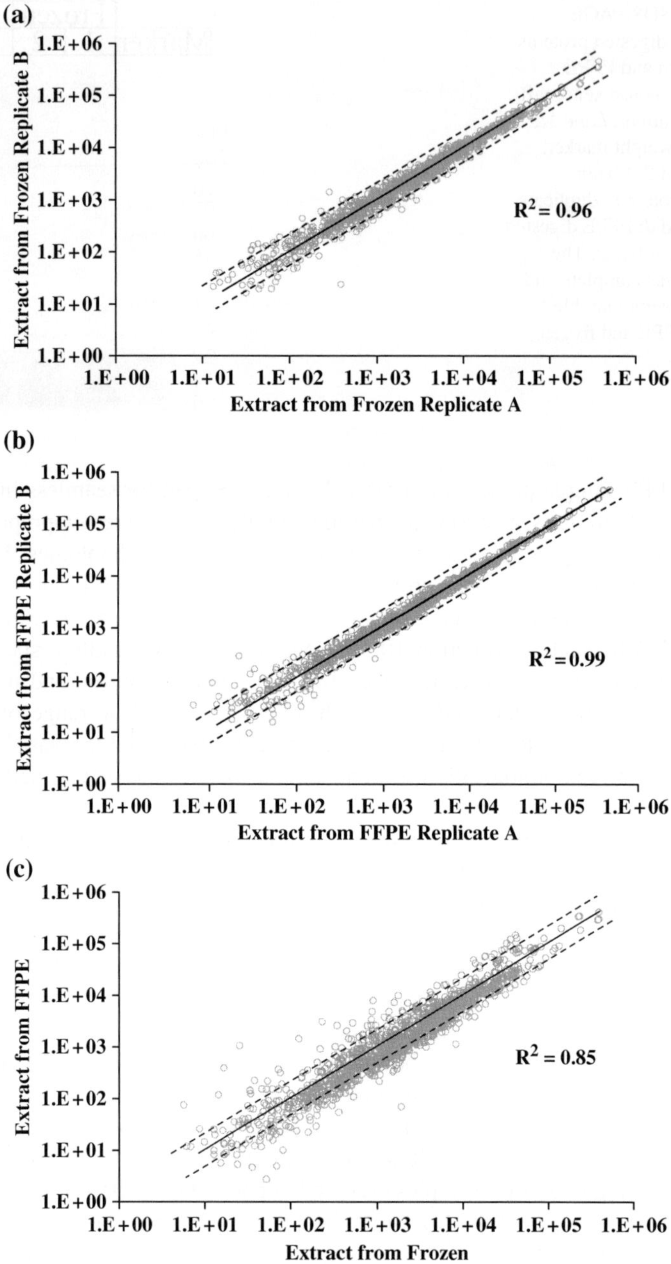

Fig. 10.4 Comparisons of normalized iTRAQ intensities for all quantified proteins in a representative multiplexed run of xenograft tumor protein extracts (2556 proteins). *Solid lines* represents linear regression with R squared shown on the plot. *Dotted lines* represent intensity ratios of two fold change. The iTRAQ intensities have been linearly normalized such that the linear regression equation for each plot is $y = x$. Data is plotted in log base 10 scale. (**a**) Replicate extracts from the frozen tumor half. (**b**) Replicate extracts from the FFPE processed tumor half. (**c**) Matched extracts from frozen and FFPE processed tumor halves

must be accounted for. At Biogen Idec this FFPE tissue protocol has been successfully applied to clinical samples for profiling of pre- and post-treatment tumor biopsy proteomes. From this study, we have gained experience and a better understanding of the sources of FFPE sample heterogeneity. The main sources of such heterogeneity are blood proteins, extra-cellular matrix contamination, and intrinsic tumor tissue diversity. These variations could, and should be reduced during and after sample collection. For instance, dissecting the tissue to remove blood vessels prior to fixation or micro-dissection of the FFPE sections would be beneficial. These efforts will help to minimize background interference and lead to increased identification of truly differentially expressed or regulated proteins.

Gaining access to the proteome of FFPE samples is an important step that, when combined with well controlled sample collection and carefully considered downstream applications, can lead to successful discovery proteomics outcomes. The FFPE tissue protocol presented here is a facile, robust, efficient, and reproducible means to prepare these samples for proteomic applications. This optimized protocol reverses cross-linking, efficiently extracts protein, and allows for complete proteome digestion leading to protein identification equivalent to what can be achieved with frozen tissues. From a typical sample load of 100 μg, each single analysis yields thousands of high-confidence protein identifications which can be used to construct detailed proteome maps. The protocol and proteomics platform described herein are fully compatible with isobaric multiplexing technologies and can be utilized as a foundation for quantitative proteomic discovery and validation in FFPE samples.

References

Addis, M.F., Tanca, A., Pagnozzi, D., Crobu, S., Fanciulli, G., Cossu Rocca, P., and Uzzau, S. (2009). Generation of high-quality protein extracts from formalin-fixed, paraffin-embedded tissues. Proteomics *9*, 3815–3823.

Bedino, J.H. (2003). Embalming chemistry: glutaraldehyde versus formaldehyde (Springfield, OH, The Champion Company *649*, 19.)

Belief, V., Boissiere, F., Bibeau, F., Desmetz, C., Berthe, M.L., Rochaix, P., Maudelonde, T., Mange, A., and Solassol, J. (2008). Proteomic analysis of RCL2 paraffin-embedded tissues. J Cell Mol Med *12*, 2027–2036.

Blonder, J., and Veenstra, T.D. (2009). Clinical proteomic applications of formalin-fixed paraffin-embedded tissues. Clin Lab Med *29*, 101–113.

Crockett, D.K., Lin, Z., Vaughn, C.P., Lim, M.S., and Elenitoba-Johnson, K.S. (2005). Identification of proteins from formalin-fixed paraffin-embedded cells by LC-MS/MS. Laboratory investigation. J Tech Meth Pathol *85*, 1405–1415.

Fowler, C.B., Cunningham, R.E., O'Leary, T.J., and Mason, J.T. (2007). 'Tissue surrogates' as a model for archival formalin-fixed paraffin-embedded tissues. Laboratory investigation. J Tech Meth Pathol *87*, 836–846.

Guo, T., Wang, W., Rudnick, P.A., Song, T., Li, J., Zhuang, Z., Weil, R.J., DeVoe, D.L., Lee, C.S., and Balgley, B.M. (2007). Proteome analysis of microdissected formalin-fixed and paraffin-embedded tissue specimens. J Histochem Cytochem *55*, 763–772.

Hood, B.L., Conrads, T.P., and Veenstra, T.D. (2006). Unravelling the proteome of formalin-fixed paraffin-embedded tissue. Brief Funct Genomic Proteomic *5*, 169–175.

Hood, B.L., Darfler, M.M., Guiel, T.G., Furusato, B., Lucas, D.A., Ringeisen, B.R., Sesterhenn, I.A., Conrads, T.P., Veenstra, T.D., and Krizman, D.B. (2005). Proteomic analysis of formalin-fixed prostate cancer tissue. Mol Cell Proteomic *4*, 1741–1753.

Jain, M.R., Liu, T., Hu, J., Darfler, M., Fitzhugh, V., Rinaggio, J., and Li, H. (2008). Quantitative proteomic analysis of formalin fixed paraffin embedded oral HPV lesions from HIV patients. Open Proteomics J *1*, 40–45.

Negishi, A., Masuda, M., Ono, M., Honda, K., Shitashige, M., Satow, R., Sakuma, T., Kuwabara, H., Nakanishi, Y., Kanai, Y., et al. (2009). Quantitative proteomics using formalin-fixed paraffin-embedded tissues of oral squamous cell carcinoma. Cancer Sci *100*, 1605–1611.

Nirmalan, N.J., Harnden, P., Selby, P.J., and Banks, R.E. (2009). Development and validation of a novel protein extraction methodology for quantitation of protein expression in formalin-fixed paraffin-embedded tissues using western blotting. J Pathol *217*, 497–506.

Nishimura, T., Nomura, M., Tojo, H., Hamasaki, H., Fukuda, T., Fujii, K., Mikami, S., Bando, Y., and Kato, H. (2009). Proteomic analysis of laser-microdissected paraffin-embedded tissues: (2) MRM assay for stage-related proteins upon non-metastatic lung adenocarcinoma. J Proteomics. *73*(6), 1100–1110.

Palmer-Toy, D.E., Krastins, B., Sarracino, D.A., Nadol, J.B., Jr., and Merchant, S.N. (2005). Efficient method for the proteomic analysis of fixed and embedded tissues. J Proteom Res *4*, 2404–2411.

Patel, V., Hood, B.L., Molinolo, A.A., Lee, N.H., Conrads, T.P., Braisted, J.C., Krizman, D.B., Veenstra, T.D., and Gutkind, J.S. (2008). Proteomic analysis of laser-captured paraffin-embedded tissues: a molecular portrait of head and neck cancer progression. Clin Cancer Res *14*, 1002–1014.

Shadforth, I.P., Dunkley, T.P., Lilley, K.S., and Bessant, C. (2005). i-Tracker: for quantitative proteomics using iTRAQ. BMC Genomic *6*, 145.

Sprung, R.W., Jr., Brock, J.W., Tanksley, J.P., Li, M., Washington, M.K., Slebos, R.J., and Liebler, D.C. (2009). Equivalence of protein inventories obtained from formalin-fixed paraffin-embedded and frozen tissue in multidimensional liquid chromatography-tandem mass spectrometry shotgun proteomic analysis. Mol Cell Proteomic *8*, 1988–1998.

Wei, J., Sun, J., Yu, W., Jones, A., Oeller, P., Keller, M., Woodnutt, G., and Short, J.M. (2005). Global proteome discovery using an online three-dimensional LC-MS/MS. J Proteom Res *4*, 801–808.

Chapter 11
Brain Proteomics: Sample Preparation Techniques for the Analysis of Rat Brain Samples Using Mass Spectrometry

Yoshinori Masuo, Misato Hirano, Junko Shibato, Hyung Wook Nam, Isabelle Fournier, Céline Mériaux, Maxence Wisztorski, Michel Salzet, Hideaki Soya, Ganesh Kumar Agrawal, Tetsuo Ogawa, Seiji Shioda, and Randeep Rakwal

Abstract Rat brain proteomics is progressing gradually from gel-based approaches to gel-free systems and the emerging technique of MALDI imaging is making its mark on direct analysis of proteins in the brain sections. In this protocol chapter, we provide the details on the sample preparation methods for brain proteins analysis by mass spectrometry. Sample preparation methods include first grinding the brain and/or its regions in liquid nitrogen, followed by extracting proteins efficiently with various extraction buffers. The two protocols described here, are the modified TCA/acetone extraction method and extraction using Tris-buffered saline with addition of Tween 20. The soluble proteins obtained from the powdered brain samples can be directly analyzed after digestion with trypsin by gel-free approach to enhance the proteome coverage. Finally, we present a detailed protocol for protein identification in brain sections using the emerging technology of MALDI imaging.

Keywords 2-DGE · MALDI imaging · Protein extraction · Proteome · Rat brain

11.1 Introduction

Proteomics, a systematic study of proteins present in a cell, tissue, organ, or organism at a particular moment during the life cycle, is one of the key research areas in functional genomics (Wilkins et al., 1995; Aebersold and Mann, 2003; Steinberg et al., 2003; Fountoulakis, 2004; Garbis et al., 2005; Fountoulakis and Kossida, 2006; Cox and Mann, 2007; Klose, 2009). Although, the DNA-based microarray technology has become an increasingly popular research field for profiling genome-wide expression of genes at the mRNA level, it is "proteins, not genes" that are

R. Rakwal (✉)
Laboratory of Neuroscience, Department of Biology, Toho University, Funabashi, Chiba 274-8510, Japan; Research Laboratory for Biotechnology and Biochemistry (RLABB), GPO Box 13265, Kathmandu, Nepal; School of Medicine, Showa University, Tokyo 142-8555, Japan
e-mail: plantproteomics@gmail.com

A.R. Ivanov, A.V. Lazarev (eds.), *Sample Preparation in Biological Mass Spectrometry*, DOI 10.1007/978-94-007-0828-0_11,

directly responsible for the cell functions and phenotype. Moreover, these powerful genomic tools do not provide any direct information on protein levels and their state of modification mainly due to post-translational regulation, which results in a lack of correlation between mRNA and protein abundance. This difference is just one of the examples of many types of information, such as the total number of genes in a given genome, protein function, localization and compartmentalization, and protein–protein interactions, which we cannot obtain from the study of genes alone. Hence the need to study proteins marks a new urgency. Other than the vast wealth of genomic data available to the proteomics researcher, a series of tremendous technological developments in immobilized pH gradient (IPG) based two-dimensional gel electrophoresis (2-DGE), mass spectrometry (MS), staining and scanning methods, chromatography, bioinformatics, and protein chips, have all fuelled the proteomics revolution, resulting in rapid accumulation of large data sets for many organisms, such as yeast, humans (mammals) and even plants (Agrawal and Rakwal, 2008).

Proteomics along with genomics and metabolomics (commonly referred to as the "omics" approaches) is being increasingly applied to the study of animal models of stress and human diseases, specifically the disorders of the brain (Fountoulakis, 2004; Garbis et al., 2005; Fountoulakis and Kossida, 2006). Animal models are genetically homogenous, and their environment can be easily controlled allowing for more interventions than in clinical cases. In addition, since the models usually have simpler nervous systems, their behaviors can be more easily interpreted (Sagvolden et al., 2005). One such animal model is the rat. Genome sequence of the Brown Norway (BN) rat, one of the laboratory rat (*Rattus norvegicus*), was deciphered by Rat Genome Sequencing Project Consortium (Gibbs et al., 2004). Moreover, rat databases for genes and proteins have been provided by several organizations. Out of these databases, Rat Genome Database (RGD, http://rgd.mcw.edu) is one of major resources for rat researchers. In the RGD, more than 36,317 genes have been submitted until 2007 (Dwinell et al., 2009).

In this chapter, we will discuss the rat brain as a sample in proteomics approaches and as an example of proteomics studies of psychiatric and neurological disorders, including depression (Tafet and Bernardini, 2003; Mu et al., 2008; Schmidt et al., 2008). As human population is exposed to an increasing levels of physical, chemical, biological, and mental stress, the study of stress and understanding its underlying mechanisms is critical for "human health and happiness." Our team, previously the Mental Stress Team at the Health Technology Research Center of Advanced Industrial Science and Technology institute (AIST), and now at Laboratory of Neuroscience, Toho University (Japan) has been primarily studying regional brainomics – we define it as the study of brain at the omics level – from rat models of attention deficit hyperactivity disorder (ADHD), and the effects of physical/mental stresses (such as coffee aroma, gamma ray, sake intake, depression, exercise and neurotoxins such as 6-hydroxydopamine (6-OHDA) to rat brains (Hirano et al., 2006, 2007, 2008; Seo et al., 2008; Masuo et al., 2009). In this chapter we present the techniques and tools we have developed during our research in collaboration with other scientists working on investigation of the brain proteome.

Starting with dissection of whole rat brain and regions, we will move towards preparation of brain powders to be stored in liquid nitrogen for further analyses, and to protein extraction for gel-based [1-DGE (term for sodium dodecyl sulfate-polyacrylamide gel electrophoresis; SDS-PAGE) and 2-DGE] and gel-free approaches. We would like to emphasize that individual researchers have used different sample preparation protocols for studying each of the "omes": genome (by DNA microarrays), transcriptome (by mRNA microarrays), proteome (using mostly 2-DGE and MS), and metabolome (large-scale metabolite analysis), making inter-omic comparisons difficult. When starting "brainomics" in early 2005, we thought that if we could use the same single sample for analyzing the genes, proteins and metabolites simultaneously, it would make inter-omic comparisons much easier and more efficient. To address this problem, we have developed a novel sample preparation system, which utilizes the use of liquid nitrogen-based grinding of tissues of interest for parallel investigation of the genes, proteins and metabolites from the same sample (Hirano et al., 2006, 2007, 2008; Seo et al., 2008; Masuo et al., 2009). With all the techniques in place and optimized for all of the analyte groups mentioned above, we are currently investigating selected genes/proteins at the functional level, and moving into the systems biology era for integration of omics data, in understanding stress and providing ways for practical applications. In this report, we will concentrate on the proteomics part, and describe each sample preparation protocol that we are utilizing for revealing the identities of the proteins in the rat brain proteome. Recently, we have also initiated collaboration with the leading group of Prof. Michel Salzet at the University of Lille (France) for applying matrix-assisted laser desorption/ionization (MALDI) imaging (Wisztorski et al., 2008; http://www.maldi-imaging.com) for unraveling the effects of 6-OHDA on rat brain proteins. Therefore, the methodology for this exciting and emerging advance in brain proteomics technology is presented in the later part of this chapter.

11.2 Dissection of Brain and Brain Regions

Adult male and female Wistar rats (Clea, Japan) were generally used in our experiments, or as otherwise mentioned in individual references. It should be noted that the experiments were carried out in accordance with regional legal regulations.

1. As specified, the rats were decapitated and the whole brains rapidly removed from female or male rats.
2. Ten brain regions are dissected out along with the pituitary gland following the method of Glowinski and Iversen (1966) with some modifications (Hirano et al., 2006). The Fig. 11.1 shows the representative rat whole brain and the 11 dissected regions. For the midbrain, we made a frontal section between the anterior and posterior edges of the *substantia nigra* in order to investigate alterations in this region rich in dopaminergic neurons. Moreover, studies using animal models of human diseases have to investigate changes in specific regions including cortex, thalamus, and hippocampus, because specific information related to each region is important and can be lost when analyzing the whole brain (reviewed in Williams et al., 2004; Fountoulakis and Kossida, 2006).
3. Each sample is immediately weighed, flash-frozen in liquid nitrogen and stored at –80°C.

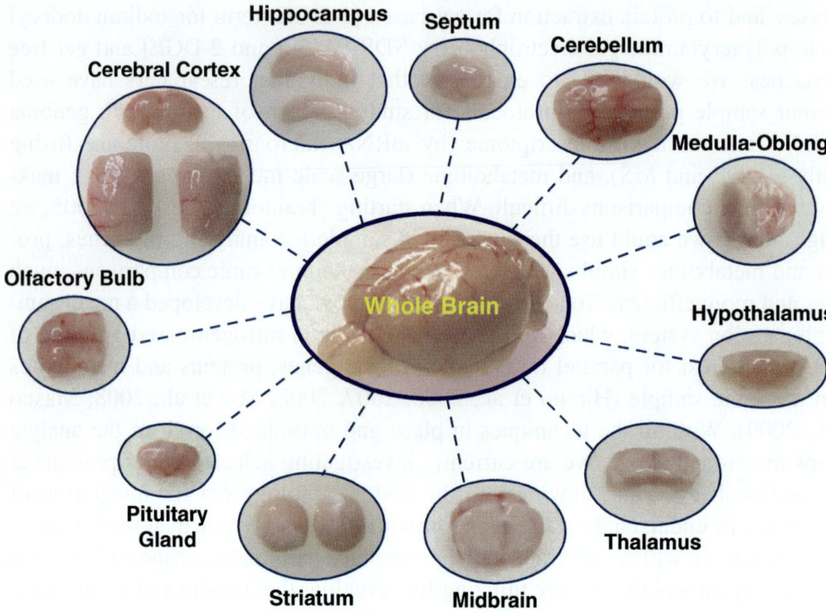

Fig. 11.1 Rat whole brain and its regions. The whole brain (*center*) was removed, each region was dissected out along with pituitary gland: whole brain (1.7 g), olfactory bulb (81 mg), cerebral cortex (553 mg, frontal and other cortices), hippocampus (65 mg), septum (38 mg), cerebellum (249 mg), medulla-oblongata (219 mg), hypothalamus (100 mg), thalamus (97 mg), midbrain (95 mg), striatum (90 mg) and pituitary gland (11 mg)

11.3 Preparation of Powdered Brain/Regions

1. The frozen whole brains and brain regions are individually placed in liquid nitrogen, and ground thoroughly to a very fine powder with a mortar and pestle (Fig. 11.2). Extreme care should be taken to grind the brain/regions very finely, as this step is critical for downstream analyses, because the quality of the powder affects subsequent extraction, separation, and corresponding protein identification.
2. Individual replicates can be ground separately or pooled, depending on the experimental goals. For whole brains, biological replicates are ground separately due to their large size. This tissue powder is transferred in pre-chilled sterile microfuge tubes or cyrotubes and stored at –80°C until further preparation and analysis. It should be noted that this "powder" of brain or individual brain regions can be separated into aliquots and used for proteome, transcriptome and metabolome analyses, one of the key point in preparing samples by grinding in liquid nitrogen. Moreover, finely powdered sample tissue facilitates good extraction of its respective molecular components.

Fig. 11.2 The brain grinder's protocol. Details as mentioned in the figure

11.4 Protein Extraction for Gel-Based Proteomics

Previously, some regions of the rat brain, such as the thalamus (Paulson et al., 2004), hippocampus (Fountoulakis et al., 2005), striatum (Yeom et al., 2005) and frontal cortex (Kim et al., 2005), were homogenized or sonicated directly in chloroform/methanol/water (Paulson et al., 2004), isoelectric focusing (IEF) resuspension buffer (Fountoulakis et al., 2005; Yeom et al., 2005), or Tris-HCl buffer (Kim et al., 2005) to extract proteins. Direct extraction of proteins usually resulted in a much lower yield of proteins. Many research groups use quite different protein extraction protocols for rat brains and their sub-regions, however, it would be extremely desirable to establish a standardized protein extraction protocol to achieve better reproducibility between laboratories. It must be emphasized that we do not criticize any particular method but aim to improve the resolution of 2-DGE by employing an innovative trichloroacetic acid (TCA)/acetone extraction buffer (AEB), termed TCAAEB, for precipitating proteins coupled with a modified lysis and re-solubilization lysis buffer (LRB, see below LB-TT, i.e., lysis buffer containing thiourea and Tris). The efficiency of protein extraction by LB depends on two factors: (i) removal of contaminants before addition of LB; and (ii) means to maximize the exposure of the sample surface to the extraction buffer. Here, we have used two protocols of protein extraction for gel-based proteomics. One is an extraction method using TCAAEB protocol (Hirano et al., 2006). Though TCAAEB protocol yielded greater amount of protein, it bears the risk of selective loss of proteins, present in aqueous proteome fraction, but may be carried away with the acetone/TCA. The second extraction protocol used Tris-buffered saline containing 0.1% Tween 20 (TBS-T20). The obtained total protein pellet after extraction with these buffers was solubilized into lysis buffer [LRB or LB-TT; 7 M urea, 2 M thiourea, 4% (w/v) CHAPS, 18 mM Tris-HCl (pH 8.0), 14 mM Trizma base, two EDTA-free proteinase inhibitor cocktail tablets in a final volume of 100 mL buffer, 0.2% (v/v) Triton X-100 (R), containing ampholyte (1% w/v) and 50 mM dithiothreitol (DTT)] (Hirano et al., 2008) for gel-based proteomics (Fig. 11.3). After a suitable image analysis procedure, protein spots/bands from 2- and 1-D (SDS-PAGE) gels were excised, digested with trypsin and processed for MS (nanoESI-LC-MS/MS) (for details see Hirano et al., 2008). Moreover, as good sample preparation – extraction of a maximum number of proteins from a given cell, tissue, organ or organism – is the first and most crucial step in protein separation and identification, we focused on establishing a new protein extraction protocol for rat brain proteomics that could theoretically be applied to any organ/tissue of the rat.

11.4.1 TCA/Acetone Precipitation

This protocol, also commonly used for extracting proteins from a variety of tissue samples, form mammals to plants, is composed of two-steps (Hirano et al., 2008). This protocol is simple but very efficient.

1. The frozen tissue powder (50–100 mg) of the rat brain is transferred to sterile tubes containing cold TCAAEB [acetone containing 10% (w/v) TCA, and 0.07% mercaptoethanol (ME)], and the proteins are precipitated for 1 h at –20°C, followed by centrifugation at 15,000 rpm for 15 min at 4°C.
2. The supernatant is decanted, and the pellet washed twice with the chilled wash buffer (acetone containing 0.07% ME, 2 mM EDTA, and one EDTA-free proteinase inhibitor cocktail tablet in a final volume of 100 mL buffer), followed by removal of all the acetone for at least 24 h at –80°C. Alternatively, the precipitated pellets can be lyophilized to remove remaining acetone, and stored until used for further analysis.
3. The protein pellet is solubilized with LB-TT, incubated for 30 min at room temperature with occasional vortexing and sonication (30 s × 5 times using a BRANSON 200 Sonicator), and centrifuged at 15,000 rpm for 15 min at 20°C.

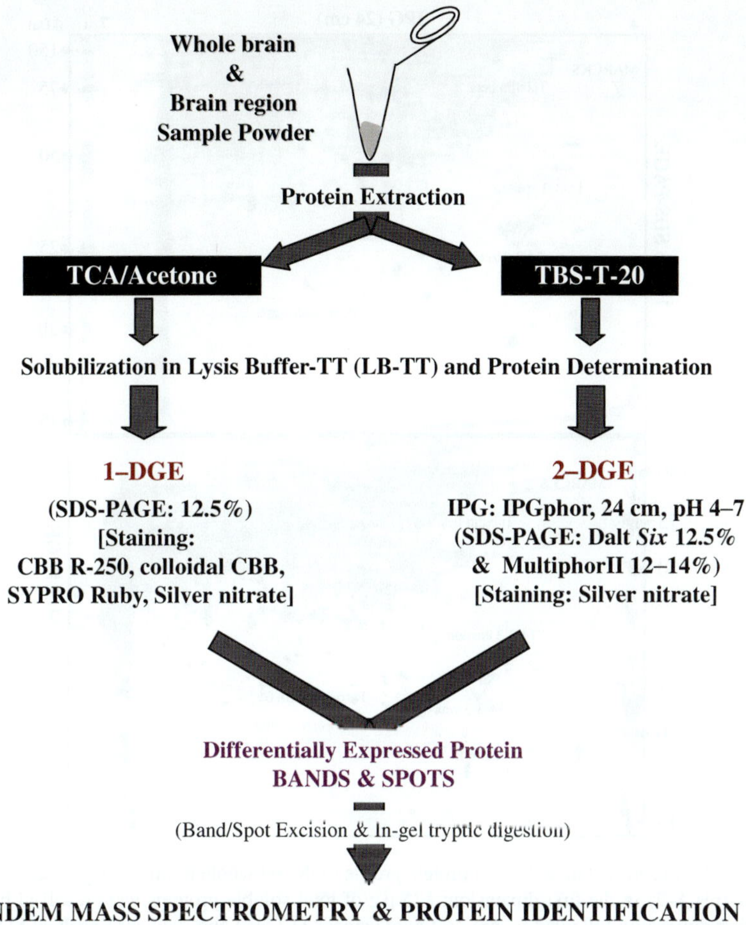

Fig. 11.3 A schematic presentation of a strategy for gel-based proteomics in rat brain. Total soluble protein extracted TCA/acetone or TBS-T20 can be processed either through 1-D (SDS-PAGE) and 2-DGE approaches and stained by appropriate staining reagents. Differentially expressed protein bands and spots are digested by trypsin and identified utilizing tandem mass spectrometry

4. The protein solubilized in LB-TT should be stored at –80°C until analysis for protein concentration [Coomassie Plus^{TM} (PIERCE, Rockford, IL, USA) protein assay kit] followed by SDS-PAGE or 2-DGE.
5. Prior to 2-DGE, further purification/clean up of the solubilized protein samples should be performed using the 2-D Clean-Up Kit (GE Healthcare Bio-Sciences AB, Uppsala, Sweden); the purified pellet was resolubilized in LB-TT as above. For details on the 2-DGE protocol, readers are referred to the paper of Hirano and co-workers (Hirano et al., 2008). In Fig. 11.4, we show a representative 2-D gel images using both homogenous and gradient pre-cast gels.

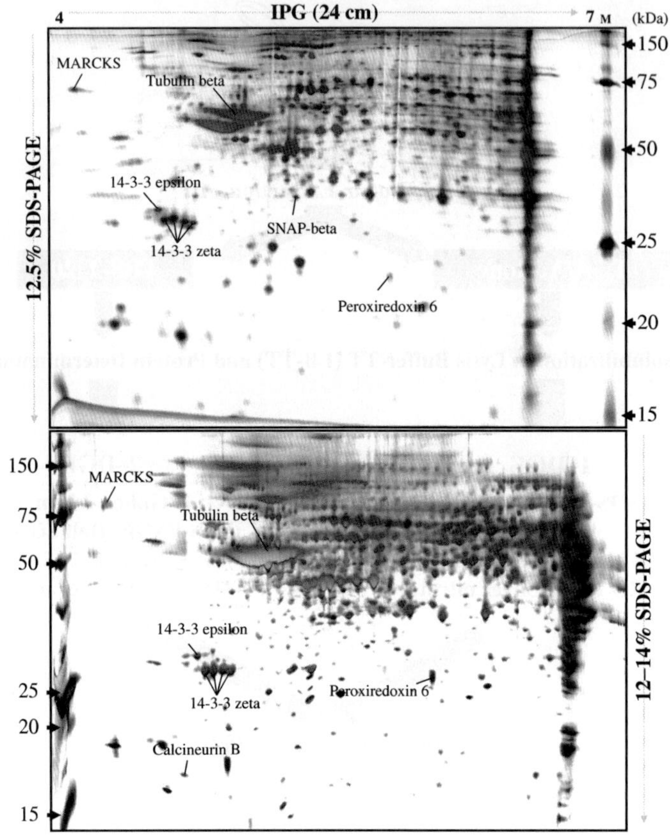

Fig. 11.4 A representative 2-D gel protein profile of the rat whole brain using pre-cast 24 cm IPG strips (pH 4–7) in the first dimension; 125 μg of total soluble proteins were resolved by IEF in the first dimension followed by pre-cast homogenous (12.5%) and gradient ExcelGel SDS-PAGE (12–14%) in the second dimension. Proteins were visualized by staining with silver nitrate. Some of the identified proteins are marked by *arrows*

11.4.2 TBS-T20

1. The frozen tissue powder (50–100 mg) of rat brain is transferred to sterile tubes and cold TBS-TT is added followed by vortexing and occasional sonication (30 s × 5 times using a BRANSON 200 Sonicator) to disperse the tissue powder.
2. SDS sample buffer (SDS-SB) [62 mM Tris (pH 6.8), 10% (v/v) glycerol, 2.5% (w/v) SDS, and 5% (v/v) ME] is added to the TBS-TT extracted protein solution to extract more soluble protein. The vortexing and sonication step is repeated followed by heat treatment (95°C for 5 min) in a heat block, and cooling to room temperature.
3. Insoluble fraction is precipitated and discarded by centrifugation at 15,000 rpm for 10 min at 4°C.

4. The supernatant including soluble proteins is transferred to sterile tubes following by clean-up of proteins using either the 2-D Clean-Up kit (see Section 11.4.1) or the ProteoExtract® protein precipitation kit commercialized by Calbiochem. The cleaned-up pellet is resolubilized into LB-TT (LRB) and the concentration was determined as mentioned in Section 11.4.1. We have used this soluble protein fraction for both SDS-PAGE and 2-DGE (Hirano et al., unpublished results).

11.4.3 1-DGE and Shotgun Analysis

1. The total soluble protein (described in Sections 11.4.1 and 11.4.2) is mixed with 2.5X SDS-SB containing the bromophenol blue dye.
2. A defined protein amount (20–30 μg)/volume (20–30 μL) is loaded into the well and SDS-PAGE was performed (Hirano et al., unpublished results).
3. The separated proteins are stained with Coomassie brilliant blue (CBB), colloidal CBB, silver nitrate, or appropriate fluorescent satin, followed by band excision of interest, trypsin digested and analyzed by nanoLC-ESI-MS/MS. A representative SDS-PAGE image of SYPRO Ruby (Molecular Probes) – stained gel and some of the identified proteins from a few brain regions of the male rat are shown in Fig. 11.5. This is a simple and efficient way to obtain separation of proteins on the basis of their molecular weights. When combined with the MS/MS, this method increases protein coverage compared to direct MS of protein samples.

11.5 Protein Extraction for Gel-Free Proteomics

This involves utilizing complex protein samples in solution, digestion with trypsin and analysis by tandem MS. We will give two examples of the methods we are using for a gel-free approach.

11.5.1 MudPIT

Multidimensional protein identification technology (MudPIT; Washburn et al., 2001) is a gel-free proteomics technique for separation and identification of complex protein and peptide mixtures by LC-MS/MS, based on 2-dimensional chromatography separation prior to ESI-MS. This technology has several advantages: (i) it is a fast technique which allows the identification of thousands of proteins a day, (ii) it allows the detection of very high and very low molecular weight proteins, and (iii) it allows the separation of hydrophobic proteins and peptides as well as those with basic or acidic isoelectric points (Mathy and Sluse, 2008), typically absent on 2-D gels. Complex protein and peptide mixtures can be separated by exploiting each peptide's unique physical properties of charge and hydrophobicity. 2-D chromatographic separation have become an effective means of resolving complicated peptide mixtures with an increased loading capacity, and have expanded the

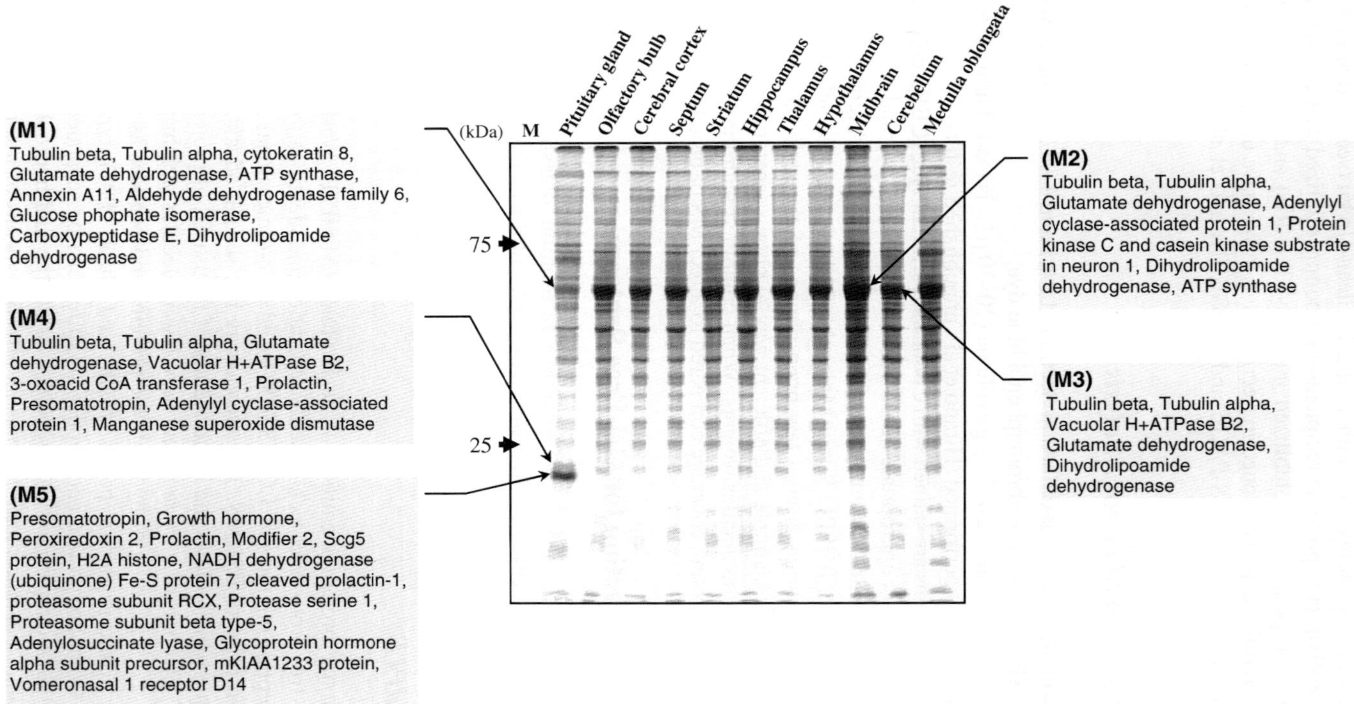

Fig. 11.5 Representative 1-D gel image of male rat brain proteins (10 regions and pituitary gland as shown in Fig. 11.1). Total soluble protein (20 μg) was separated by 12.5% SDS-PAGE and visualized by fluorescent staining reagent, SYPRO Ruby. Bands M1–M5 are marked by *arrows* and the identified proteins from each band are named below. Molecular mass standard are marked in *left side* (M; Precision Plus Protein Standards, Bio-Rad)

dynamic range of identified proteins through a sequential salt gradient elution steps to displace peptides from SCX column followed by MS/MS analysis. Charged peptides are trapped by the SCX resin and uncharged peptides are flowing through onto the reversed-phase (RP) portion of the column that comprises the second chromatographic dimension. Uncharged peptides are eluted to an analytical column from the RP column by LC gradient separation and introduced into a mass spectrometer. After the flow-through run, charged peptides are eluted to the RP column by multiple salt gradient fractions and introduced to the mass spectrometer via electrospray ionization in the same manner as the eluent of the flow-through run. It is critical to optimize the conditions for both salt (SCX) and RP gradient elutions for achieving satisfactory performance of the method. For the salt gradient, ammonium acetate or ammonium bicarbonate (AMBIC) are typically used. The reversed-phase gradient separation usually applies C18 stationary phases and water-acetonitrile linear gradients in the presence of a counter ions generated by addition of formic or acetic acid.

We have performed MudPIT analysis of the frontal cortex, striatum, and midbrain in ADHD model rats (Hirano et al., unpublished data). Sample preparation was performed using the protocol reported by Kang et al. with modifications (Kang et al., 2005; Hirano et al., manuscript to be published elsewhere; Fig. 11.6a):

1. The finely ground tissue powder is solubilized in lysis buffer containing 7.0 M urea, 2.0 M thiourea and 0.25 M AMBIC, and the soluble fraction was separated from the insoluble fraction, mostly comprised of cell debris, membrane fragments, and lipid structures, by ultracentrifugation.

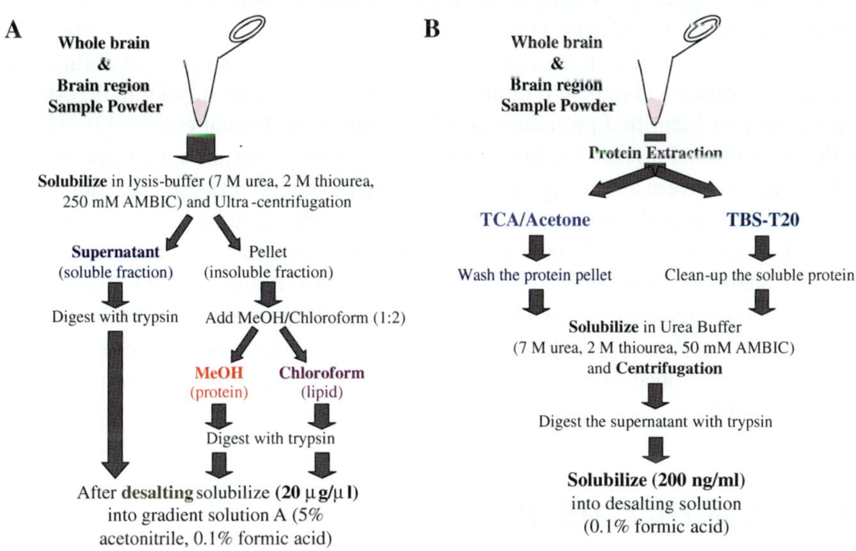

Fig. 11.6 The gel-free approach for protein identification. MudPIT workflow is shown in **a**, while the nano1D-LC-MS/MS workflow is presented in **b**. Details are in the text

2. To the insoluble fraction methanol: chloroform (1:2) is added and the mixture was separated by centrifugation into methanol layer that included membrane proteins and chloroform layer that includes lipid.
3. The supernatant, membrane fraction and lipid fraction is digested by sequencing grade trypsin in solution followed by reduction and alkylation utilizing dithiothreitol and iodoacetamide.
4. The peptide mixture is re-solubilized in solvent A containing 0.1% aqueous formic acid and 5.0% acetonitrile after complete evaporation and loaded onto the SCX column of first dimension using an auto-sampler.
5. After the flow-through run, peptides eluted by increasing the concentration of AMBIC solution to 13, 21, 28, 35, 50, 100 mM and 1 M are analyzed.
6. The number of detected peptides was more than 1000 in each fraction including the flow-through run. However, some peptides were repeatedly detected in multiple salt fractions. The fewest number of proteins was identified in the flow-through run and nearly 400 in each of the four fractions including the highest salt concentration. Some abundant proteins, such as creatine kinase and tubulin, were also repeatedly identified in multiple fractions.
7. In the final step of the analysis, acquired mass spectra are searched using bioinformatics software and databases, such as Mascot and NCBI. The typical parameters used in the MASCOT MS/MS ion search were: maximum of two trypsin miss cleavage, fixed modification cysteine carbamidomethylation, variable modification methionine oxidation and lysine ubiquitinylation, peptide mass tolerance ± 1.2 Da, fragment mass tolerance ± 1.0 Da, and threshold ($p < 0.05$).

These results suggest possibility of decreased chance for detecting other low abundance peptide ions and low content proteins due to the repeated detection of abundant protein species in multiple salt fractions. To eliminate this possibility, we may need to remove abundant proteins, such as tubulin and actin, prior to digestion. The number of identified proteins was 1266/1082 in the frontal cortex, 1361/1378 in the striatum and 1314/1543 in the midbrain of control/treated rats, respectively. In this assay we were analyzing the differentially expressed proteins. We found that abundances of several proteins were changed, as was detected by both by 2-DGE in MudPIT analyses, although more changes in protein abundances were detected by MudPIT analysis. These results suggest that both 2-DGE and MudPIT approaches are complementary, agreeing with a previous paper by Koller et al. (2002). Taking advantage of method's usefulness for identification of hydrophobic proteins and peptides as mentioned above, we also analyzed protein components contained in insoluble pellets.

11.5.2 Nano-1DLC-MS/MS

Nano-1DLC is also essential to separate complex samples such as protein and peptide efficiently. Protein extraction for nano-1DLC-MS/MS is based on the

TCAAEB and TBS-T20 protocols (Sections 11.4.1 and 11.4.2) similar to gel-based proteomics. In this example, we present the analysis of whole rat brains under stressed (leading to depression) conditions (Hirano et al., unpublished data). Sample preparation is performed as follows (Fig. 11.6b).

1. Protein pellet is solubilized in lysis buffer containing 7.0 M urea, 2.0 M thiourea, and 0.05 M AMBIC, followed by sonication for 30 min.
2. Solubilized samples are centrifuged at 15,000 rpm (maximum speed) for more than 20 min at 4°C in order to precipitate insoluble particles, such as insoluble proteins. Contamination with insoluble particles might cause the clogging of the connecting LC capillaries and columns.
3. The supernatant (containing 20 μg of soluble protein) is digested by sequencing grade trypsin in-solution followed by reduction and alkylation utilizing dithiothreitol and iodoacetamide. (Kang et al., 2005; Hirano et al., unpublished data)
4. The dried peptides are solubilized in 100 μL of desalting solution containing 0.1% formic acid.
5. One microliter of the peptides (200 ng) is injected on a LC-MS-IT-TOF (Shimadzu, Japan) for MS analysis using an ESI source. On-line capillary nano-flow LC (LC-20AD and LC-20ADnano, Shimadzu) included a reversed-phase trap column (L-column Micro, 0.3 × 5 mm, CERI, Tokyo, Japan) and a monolithic C18 fused-silica capillary column (PicoFrit column, 0.075 × 100 mm).
6. Sample is loaded onto a peptide trap for concentration and desalting prior to final separation on the C18 column using a linear acetonitrile gradient ranging from 2 to 95% solvent B [H_2O/acetonitrile/formic acid, 5/95/0.1 (v/v)] in solvent A [H_2O/acetonitrile/formic acid, 98/2/0.1 (v/v)] for a duration of 35 min.
7. The mass/charge (*m/z*) ratios of eluted peptides and fragmented ions from the spray tip are analyzed in the data-dependent positive ion acquisition mode on LC-MS/MS.
8. Dynamic exclusion is used with the following parameters: repeat count (3) on MS1, repeat count (10) in MS2, repeat duration (0 min), exclusion list size (71), and exclusion duration (3.0 min). Following each full scan (400–1500 *m/z*), three data-dependent triggered MS/MS scans (50–1500 *m/z*) for the most intense parent ions were acquired. The heated fused-silica PicoTip emitter was held at ion spray voltage of 2.5 kV and a flow rate of 300 nL/min.
9. Acquired LC-MS/MS data are submitted to the MASCOT server (www.matrixscience.com) for querying all MS/MS ion search against the MS protein sequence databases (MSDB, Rat Protein Database, updated at 20060831 and National Center for Biotechnology Information, NCBInr, Rat Protein Database, updated at 20071108). The typical parameters used in the MASCOT MS/MS ion search were: maximum of one trypsin missed cleavage, fixed modification – cysteine carbamidomethylation; variable modification – methionine oxidation, peptide mass tolerance ± 0.05 Da, threshold ($p < 0.05$), minimum ion counts (0), and fragment mass tolerance ± 0.05 Da.

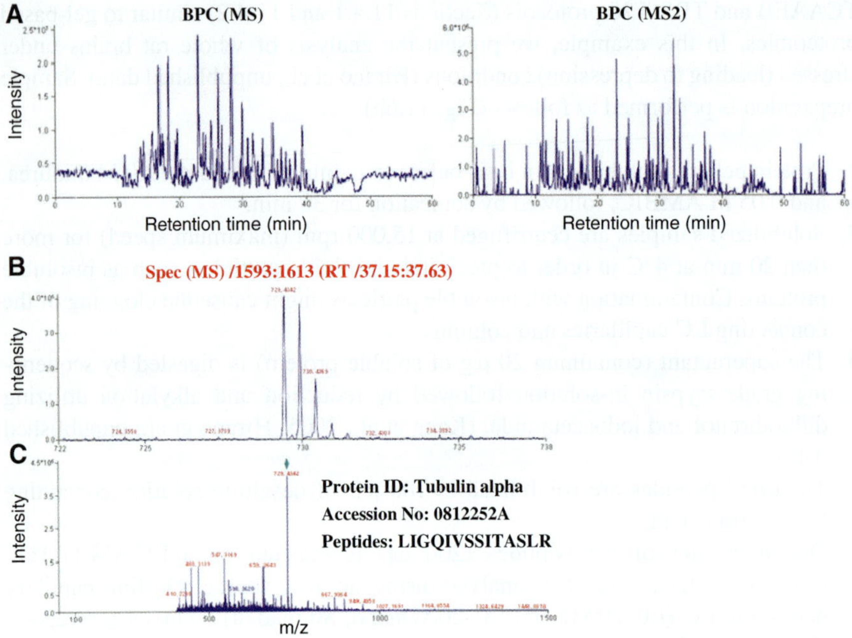

Fig. 11.7 A representative protein ID showing the MS/MS spectra of the tubulin peptide identified from rat whole brain. (**a**) The base peak chromatograms for MS and MS2 are shown. (**b**) and (**c**) MS spectrum view of the identified protein tubulin alpha

Using this approach, we have examined global protein profiles in the whole brain of stressed rats. We detected 67 proteins in the Mascot search results using by MSDB database. Fractionation of proteins based on their physical characteristics such as molecular weight prior to digestion will be needed to increase the number of identified proteins using this method. A representative peptide sequence and a spectrum of tubulin alpha chain peptide is shown in Fig. 11.7.

11.6 Sample Preparation for MALDI-Imaging

MALDI-Mass Spectrometry imaging (MSI) is an emerging technique for unraveling the proteome in sections of desired tissues. This technique is gaining prominence in the field of brain proteomics. This methodological part is composed according to the MSI workflow (Fig. 11.8) (Wisztorski et al., 2008). Here, we will primarily focus on tissue sections preparation prior to MALDI-MSI, which is an extremely important part of the methods. In particular, snap-frozen tissues require specific tissue treatment (washing procedures) prior to any other steps. One of the procedures is a fixation step of tissues using ethanol (Chaurand et al., 2008). The second procedure is used to remove small organic compounds (namely lipids) that could hamper detection of peptides in the low *m/z* range by use of organic solvents (Lemaire et al., 2006b). While these steps are independent from each other, it is

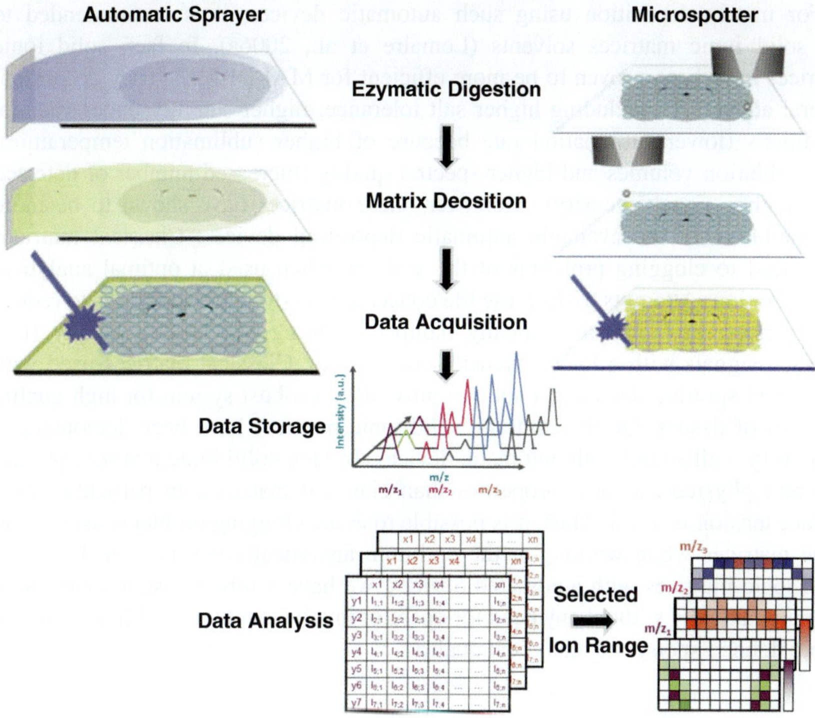

Fig. 11.8 Schematic representation of the strategies for in situ digestion using an automatic sprayer or microspotter devices

strongly recommended to perform *in situ* digestion and identification of proteins (Franck et al., 2009b). Because *in situ* digestion and/or matrix deposition must be performed with precision, good reproducibility and prevention of molecular delocalization, it is recommended to use an automated device. Two types of instruments are commonly used (Franck et al., 2009a):

– An automatic microspotter delivering picoliters of solution to the locations with predefined coordinates and allowing to cover the entire surface of the tissue section with an array of spot spaced by approximately 200 μm (Franck et al., 2009a).
– An automatic sprayer able to pulverize a solution to form very small droplets (~25 μm size in average) above the tissue section. In this case the tissue section is fully covered by the solution.

The main advantages of these devices are to limit delocalization of molecules potentially induced by the solvents while still permitting optimal extraction of molecules, allowing to maintain enzyme activity or incorporation of analytes into matrix crystal lattice during matrix crystallization. These devices currently represent the best compromise to minimize lateral diffusion of analytes while keeping optimal enzymatic digestion and analyte incorporation efficiency.

For matrix deposition using such automatic devices, it is recommended to use solid ionic matrices solvents (Lemaire et al., 2006a). In fact, solid ionic matrices have been proven to be more efficient for MALDI-MSI. They do present several advantages including higher salt tolerance, higher stability under vacuum conditions (lower sublimation rate because of higher sublimation temperature), lower ablation volumes and higher spectral quality (increased number of detected ions, higher signal intensity). Moreover, these matrices have shown to be more compatible with the available automatic deposition devices. Classical matrices often lead to clogging problems of the systems when used at optimal analytical concentration. Attempts to decrease the concentration of matrix solutions to reduce the clogging problems dramatically hamper resulting MS spectra quality (less detected signals with a lower intensity on average). Classical matrices used with automated spotting devices do not currently offer a robust system for high quality analyses of tissues. On the contrary, solid ionic matrices have been demonstrated to be very well suited with automatic devices. In fact, solid ionic matrices present different physico-chemical properties than classical matrices. In particular, their surface tension is lower. Thus, it is possible to avoid clogging problems using solid ionic matrices while working at the optimum analytical concentration. By using solid ionic matrices with automated devices, we have a robust system that can be run automatically without any need for operator involvement, giving higher printing (more defined spots) and analytical quality.

11.6.1 Preparation of Frozen Tissue Section

The following procedures were optimized and used for preparation of frozen tissue sections.

11.6.1.1 Snap-Frozen Tissue

1. The organ has to be dissected and rinsed with a saline solution suited for the considered tissue to remove blood and other tissue fragments from the surface. Alternative: Prior to sacrifice, the animal can be perfused with the saline solution to remove blood inside the organ.
2. Morphology of the organ needs to be carefully maintained. Thus, the tissue should not be placed in a tube or wrapped in an aluminium foil to avoid deformation of the organ (due to the adaptation to the outlines of the container).
3. Snap freezing procedure is applied for tissue conservation to maintain tissue morphology and to prevent ice crystal formation and cell damage. In fact, differences in the cooling rate of various parts of the organ or direct dipping of the organ into liquid nitrogen leads to the formation of cracks and fragmentation of the tissue. Therefore, the use of isopentane cooled at –45°C with dry ice is recommended. Freezing time is dependent on the size of the organ. It is preferable not to use any embedding media. For very small organs or surgically removed pieces, cutting without embedding material increases deformations and damages of the tissue sections. In such cases, a solution containing non-polymeric compounds,

namely 10% gelatin solution, helps to obtain high quality tissue sections. Tissue is embedded in 10% gelatin directly after dissection and frozen as previously described.

4. After snap freezing, tissue is stored at –80°C. We strongly recommend not to store any samples beyond a period of 6 months. Variation in the molecular profiles could be observed after storage over 6 months, if no sample stabilization procedure is used. Preferentially, tissues should be analyzed a few days or weeks after snap freezing.

11.6.1.2 Tissue Cryosection and Thaw Mounting

1. The use of cryopreservative solutions containing organic polymers such as the Optimal Cutting Temperature (O.C.T.) polymer should be restricted to the attachment of the tissue to the sample holder and not used for complete embedding of the tissue. Moreover, all parts of the cryostat in contact with the tissue must be cleaned to prevent any sample cross-contamination or carryover of the polymer. In case of contact between the tissue and cryopreservative solutions containing polymers, MS spectra will be dominated by polymer signals such as polyethylene glycol (PEG).
2. A tissue specimen should be placed in the cryostat for sufficient time before sectioning in order to warm the sample slowly to the cryostat temperature. If the tissue is too cold, poor quality sections are typically obtained.
3. 10 μm thickness tissue sections should be cut using a cryomicrotome at –20°C. Different tissue types may need other temperature settings. The thickness of 10 μm is optimal for brain samples.
4. Collect the tissue sections onto Indium Tin Oxide (ITO) glass slides pre-cooled to –20°C. Transfer is performed by applying the cooled ITO slide onto the section. Adhesion of the frozen sections to the glass slides is obtained by placing heated fingers under the slide or by incubating the slide at room temperature. This transfer procedure, contrary to the classical thaw mounting; prevents formation of ice crystals at the surface of the cryostat microtome cutting plate.
5. Mounted sections are stored in a sealed container at –80°C until used.

11.6.1.3 Pre-analysis Treatment

Tissue Fixation

1. A closed container stored at –80°C should be warmed at room temperature in a vacuum desiccator to prevent water condensation at the surface of the frozen slide.
2. After complete drying, the ITO slide has to be washed. Washing steps are optional and dependent on the molecules to be analyzed. Careful washing is crucial for conserving spatial localization of molecules.
3. For analysis of small molecules like lipids or drugs, no washing steps were used. For peptides or proteins washing procedures were generally used. Washing is performed by immersing the glass slide gently into ice-cold 75% ethanol for 30 s.

No agitation or shaking is needed. This step washes out salts, cells fragments or residual fluids.

4. Take the slide out and remove the excess of liquid around the section. A stream of nitrogen over the surface could help to remove excess of ethanol.
5. The ITO glass slide should be then placed in a vacuum desiccator to complete the drying of the tissue. The time of drying is dependent on the size of the section.
6. Optional: The second bath of fresh ice-cold 75% ethanol for 30 s followed by a complete drying under vacuum desiccator can be performed.
7. After complete drying, the sample has to be dipped into cold 95% ethanol for 30 s. No agitation or shaking is needed. This step prevents degradation of the proteome by dehydration and fixation of the tissue.
8. The slide is completely dried as described in Steps 4 and 5 above.

Removal of Lipids

1. After complete drying, immerse the glass slide gently into an ice-cold chloroform for 30 s. No agitation or shake is needed. This step removes lipids (especially phospholipids) present in high concentration in the tissue (components of cell membranes), which may cause signal suppression in MS spectra (Lemaire et al., 2006b).
2. Take the slide out and place it in the vacuum desiccator for complete drying of the tissue.

11.6.2 Preparation of FFPE Tissue Sections

11.6.2.1 FFPE Tissue Section

1. 10 μm thick FFPE tissue sections should be cut using a microtome at room temperature. Paraffin blocks can be cooled down to –20°C prior to preparation to facilitate tissue sectioning.
2. Sections have to be transferred onto a conductive ITO-glass slide on top of a water droplet.
3. Glass slide should be warmed up (30–40°C) on a hot plate to let the sections unfold.
4. Excess of water has to be removed and glass slide has to be stored in an incubator at 30°C for 20 min for good adherence. Subsequently, glass slides with FFPE tissue sections can be stored for several months at room temperature.

11.6.2.2 FFPE Tissue Deparaffinization

1. After complete drying, the glass slide has to be gently dipped into a bath of xylene for 5 min. This procedure should be repeated twice. No agitation or shaking is needed.
2. The slide has to be then washed in stepwise immersion, for 5 min each, in 100% ethanol twice, 95% ethanol, 75% ethanol and 30% ethanol for rehydration of tissue sections.

3. The ITO glass slide has to be placed in the vacuum desiccator for complete drying of the sections.

11.6.3 Matrix Deposition

11.6.3.1 Matrix Deposition for Protein Analysis

Using a Microspotter

1. An ITO slide after the washing step as described for frozen tissue samples has to be used.
2. On each defined spot, 20 nL of sinapinic acid/aniline (SA/ANI) solution should be applied. Five droplets of 100 pL should be deposited at each spot per cycle, 40 iterations are necessary to deposit the total required volume.
3. Check matrix coverage using an optical microscope.
4. A rapid MS analysis on one spot is recommended to verify that a sufficient amount of matrix is deposited. Increase of number of iterations may improve MSI when signal intensity appears to be low.

Using an Automatic Sprayer

1. The use of an ITO slide after the washing step for frozen tissue is recommended.
2. A method with different steps of spraying, incubation and drying phase is required. The ImagePrep (Bruker Daltonics) method for SA/ANI deposition is based on the default SA method provided with the ImagePrep. Optimization is necessary for each of the different types of tissue. Briefly, the spray time has to be around 2 s (depending on the surface of the tissue section). An incubation time of 30 s (except for the initialization phase: 10 s) allows an efficient extraction of proteins. A particular attention is required to set correctly the drying time to achieve complete crystallization on the tissue section. If the time is too short, the section will be too wet and a delocalization of molecules will be observed. The minimum drying time is around 45 s.
3. Check the matrix coverage using an optical microscope
4. A rapid MS analysis at one position can be performed to check that a sufficient amount of matrix has been deposited. If not, several cycles of the last phase of deposition can be performed again, which may improve MSI when signals intensity is too low.

11.6.3.2 For Peptides Analysis

Using a Microspotter

1. The use of an ITO slide after washing step as described in the procedure for frozen tissues or digestion for FFPE or frozen tissues is suggested.
2. On each defined spot, 20 nL of HCCA/ANI solution has to be applied. Five droplets of 100 pL have to be deposited at each spot per cycle, then 40 iterations

are necessary to deposit the required total volume. For slides after digestion, the matrix has to be deposited with the same array than the one used for trypsin deposition (for further details see, Franck et al., 2009b).

3. Check matrix coverage using an optical microscope
4. A rapid MS analysis on one spot is recommended to verify that a sufficient amount of matrix is deposited. Increase of iterations number may improve MSI when signal intensity appears to low.

Using an Automatic Sprayer

1. The use of an ITO slide after the washing step for frozen tissue or digestion for FFPE or frozen tissues as described above is recommended.
2. The ImagePrep method for HCCA/ANI deposition is based on the default HCCA method provided with the ImagePrep. Optimization is required for each type of tissue. Briefly, the spray time is around 2 s. An incubation time of 20 s (except for initialization phase: 10 s) typically allows for effective extraction of proteins. Particular attention has to be drawn to correctly set the drying time for complete crystallization on the tissue section. If the time is too short, the section will be too wet and a delocalization of molecules will be observed. The minimum drying time is recommended to be around 120 s.
3. Check matrix coverage using an optical microscope
4. A rapid MS analysis at one position can be performed to check out that a sufficient amount of matrix has been deposited. If not, several cycles of the last phase of deposition can be repeated, which may improve MSI when signals intensity is too low.

11.6.4 Mass Spectrometry Analysis

11.6.4.1 MALDI MSI Experiment

Mass Spectrometry Analysis for Proteins MSI

1. 0.5 μL of protein calibration solution was deposited near to the location of the tissue section and mixed with 0.5 μL of SA/ANI solution.
2. The mass spectrometer was calibrated with a calibration spot.
3. Using FlexImaging, an area of interest was selected on the tissue after definition of the training points.
4. The distance between each measurement point was set. Distance between measurement points depends on the method used for matrix deposition:

   – With Chip-1000 (Shimadzu) deposition, the spots are generally spaced by 250 μm (center to center). It is possible to define the same raster than has been defined during matrix deposition. Due to the size of the spot it is possible to accumulate spectra at different position in the same spot. This increases statistics and reduces spot-to-spot variability.

- With ImagePrep deposition, distance between two measurements can be chosen by the user. Generally the recommended resolution is around 100 μm.

5. In FlexControl, the adequate methods for proteins analysis were set in positive linear mode and a total of 500 spectra were acquired at each position at the laser frequency of 200 Hz.
6. The images were saved and reconstructed using FlexImaging 2.1.

Mass Spectrometry Analysis for Peptide MSI

1. 0.5 μL of the peptide calibration solution was deposited near to the location of the tissue section and mixed with 0.5 μL of the HCCA/ANI solution.
2. The mass spectrometer was calibrated with the calibration spot.
3. Using FlexImaging, an area of interest was selected on the tissue after definition of the teaching points.
4. The distance between each measurement point was set. Distance between measurement points is dependent of the method used for matrix deposition

   - With Chip-1000, deposition spots are generally spaced by 250 μm center to center. It is possible to define the same raster than for matrix deposition. Due to the size of the spots spectra can be accumulated at different positions in the same spot.
   - With ImagePrep deposition, distance between two measurements is chosen by users. Generally the resolution is suggested to be around 100 μm.

5. In FlexControl, the adequate methods for peptides analysis was set in the positive reflector mode and a total of 500 spectra were acquired at each position at a laser frequency of 200 Hz. However, the negative reflectron mode can also be used for specific classes of peptides.
6. The images were saved and reconstructed using FlexImaging 2.1.

Two examples of the MALDI MSI experiments are presented in Figs. 11.9 and 11.10.

11.7 Concluding Remarks

We have presented methods for sample preparation developed in our quest to analyze the rat brain proteomes. These involve the whole brain or its regions to be dissected and ground in liquid nitrogen for good extraction of proteins using two modified protocols, TCAAEB and TBS-T20, which give excellent protein profiles on both 2-DGE and 1-DGE for subsequent analysis by MS/MS. Moreover, we have also provided a method for sample preparation for gel-free proteomics using the same powdered samples. Finally, we have shown how to prepare brain tissue sections for identifying proteins by MALDI imaging. We hope that researchers investigating the brain samples will find this chapter useful in utilizing the sample

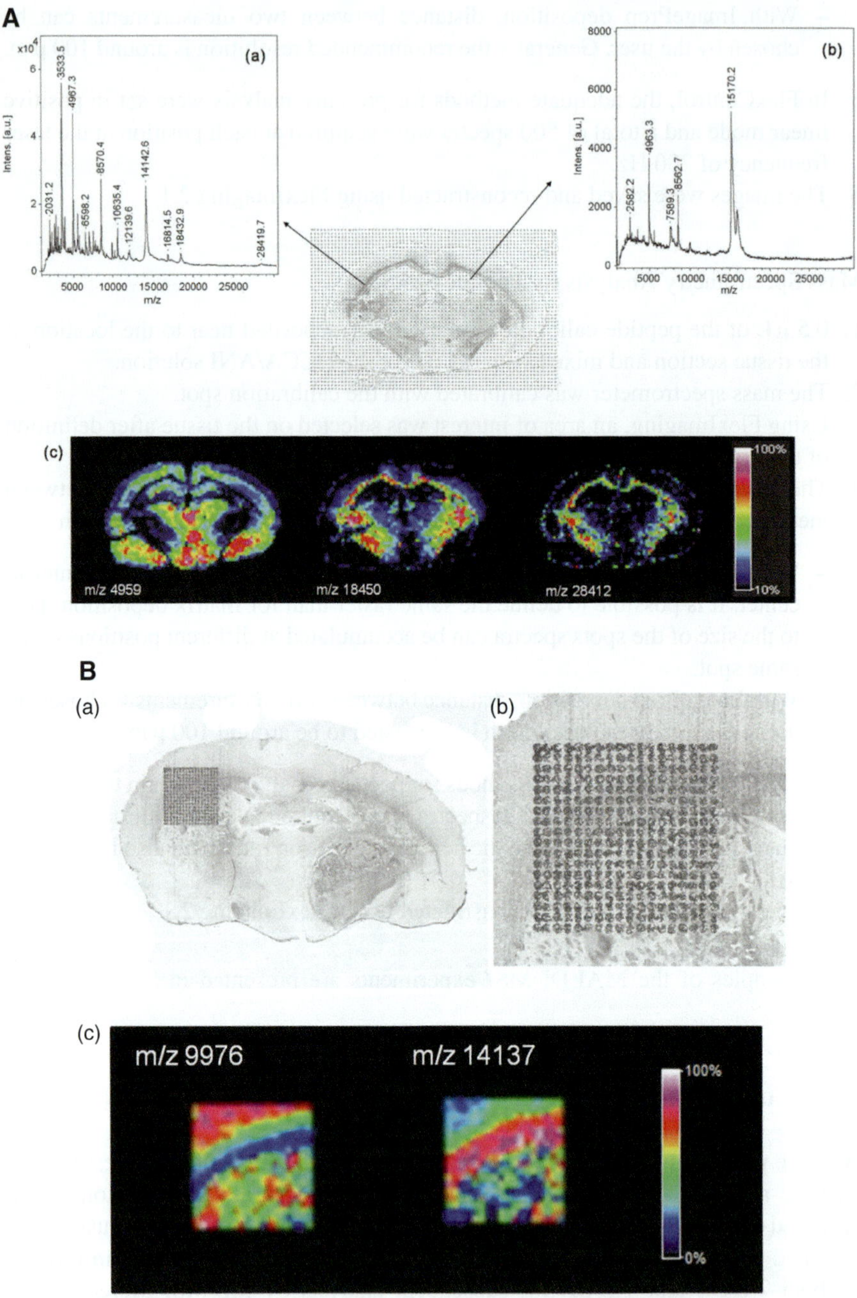

Fig. 11.9 **A:** (**a, b**) MALDI MS spectra recorded in two different locations of a rat brain tissue section after automated microspotting of the matrix solution (SA/ANI) and (**c**) reconstructed molecular images for *m/z* 4959.3, *m/z* 18,450.7 and *m/z* 28,412. **B:** (**a**) Optical image of spots obtained after SA/ANI deposition using the CHIP-1000 with a spatial resolution of 150 μm. (**b**) Zooming of region containing spots of SA/ANI. (**c**) Reconstructed molecular images for *m/z* 9976 and *m/z* 14137. With permission from Analytical Chemistry (Franck et al., 2009a)

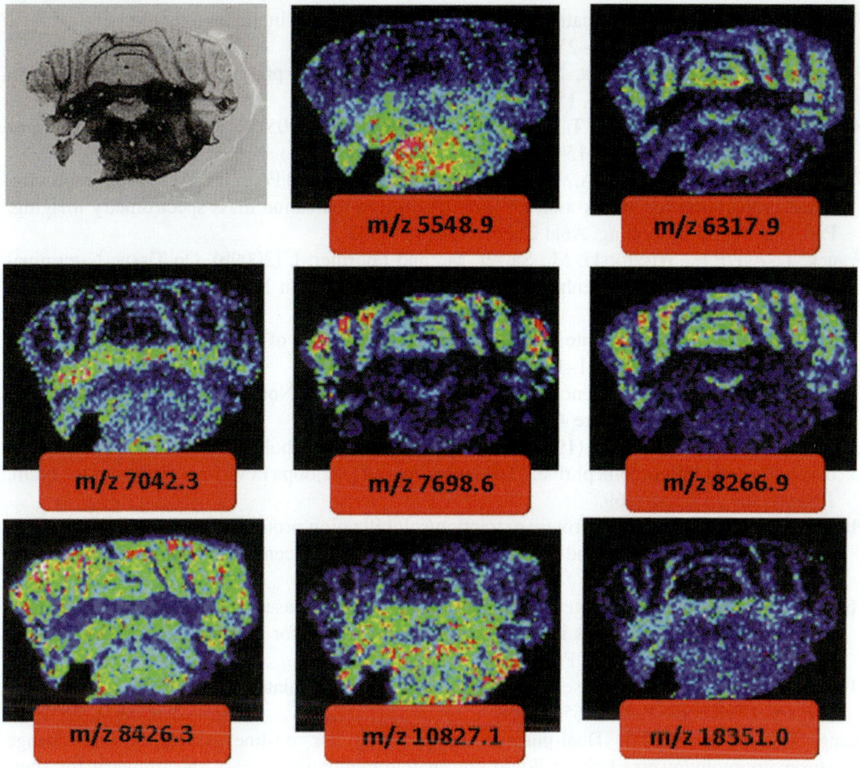

Fig. 11.10 Reconstructed molecular MALDI images for different *m/z* from a rat brain tissue section after automatic spray of the matrix solution (SA/ANI)

preparation methods for use in both gel-based and gel-free proteomics approaches. In the end, the challenge remains as to effectively and exhaustively investigate the protein components of each brain region, to generate high resolution gel-based reference maps for comparative proteomics and MS-based identification of specific proteomes and their quantification, and to give brain proteomics a functional dimension by studying post-translational modifications. This will require a lot of hard work and patience and most importantly, reproducible techniques along with good planning.

References

Aebersold, R., and Mann, M. (2003). Mass spectrometry-based proteomics. Nature *422*, 198–207.

Agrawal, G.K., and Rakwal, R. (eds.) (2008). Plant Proteomics: Technologies, Strategies, and Applications (Hoboken, NJ, Wiley).

Chaurand, P., *et al.* (2008). Imaging mass spectrometry of intact proteins from alcohol-preserved tissue specimens: Bypassing formalin fixation. J Proteome Res *7*, 3543–3555.

Cox, J., and Mann, M. (2007). Is proteomics the new genomics? Cell *130*, 395–398.

Dwinell, M.R., *et al.* (2009). The Rat Genome Database 2009: Variation, ontologies and pathways. Nucleic Acids Res *37*, D744–D749.

Fountoulakis, M. (2004). Application of proteomics technologies in the investigation of the brain. Mass Spectrom Rev *23*, 231–258.

Fountoulakis, M., and Kossida, S. (2006). Proteomics-driven progress in neurodegeneration research. Electrophoresis *27*, 1556–1573.

Fountoulakis, M., Tsangaris, G.T., Maris, A., and Lubec, G. (2005). The rat brain hippocampus proteome. J Chrom B *819*, 115–129.

Franck, J., Arafah, K., Barnes, A., Wisztorski, M., Salzet, M., and Fournier, I. (2009a). Improving tissue preparation for matrix-assisted laser desorption ionization mass spectrometry imaging. Part 1: using microspotting. Anal Chem *81*, 8193–8202.

Franck, J., Ayed, E., Wisztorski, M., Salzet, M., and Fournier, I. (2009b). On Tissue N-terminal peptide derivatizations for enhancing protein identification in MALDI-MSI strategies. Anal Chem *81*, 8305–8317.

Garbis, S., Lubec, G., and Fountoulakis, M. (2005). Limitations of current proteomics technologies. J Chromatogr A *1077*, 1–18.

Gibbs, R.A., *et al.* (2004). Genome sequence of the Brown Norway rat yields insights into mammalian evolution. Nature *428*, 493–521.

Glowinski, J., and Iversen, L.L. (1966). Regional studies of catecholamines in the rat brain. I. The disposition of [3H]norepinephrine, [3H]dopamine and [3H]dopa in various regions of the brain. J Neurochem *13*, 655–669.

Hirano, M., *et al.* (2006). New protein extraction/solubilization protocol for gel-based proteomics of rat (female) whole brain and brain regions for investigating central nervous system disorders. Mol Cells *22*, 119–125.

Hirano, M., *et al.* (2008). Proteomics- and transcriptomics-based screening of differentially expressed proteins and genes in brain of Wig rat: A model for attention deficit hyperactivity disorder (ADHD) research. J Proteome Res *7*, 2471–2489.

Hirano, M., *et al.* (2007). Gel-based proteomics of unilateral irradiated striatum after gamma knife surgery. J Proteome Res *6*, 2656–2668.

Kang, D., *et al.* (2005). Dual-purpose sample trap for on-line strong cation-exchange chromatography/reversed-phase liquid chromatography/tandem mass spectrometry for shotgun proteomics. Application to the human Jurkat T-cell proteome. J Chromatogr A *1070*, 193–200.

Kim, S.-Y., *et al.* (2005). Proteomic analysis of phosphotyrosyl proteins in morphine-dependent rat brains. Mol Brain Res *133*, 58–70.

Klose, J. (2009). From 2-D electrophoresis to proteomics. Electrophoresis *30*, S142–S149.

Koller, A., Washburn, M.P., *et al.* (2002). Proteomic survey of metabolic pathways in rice. Proc Natl Acad Sci USA *99*, 11969–11974

Lemaire, R., Tabet, J.C., Ducoroy, P., Hendra, J.B., Salzet, M., and Fournier, I. (2006a). Solid ionic matrixes for direct tissue analysis and MALDI imaging. Anal Chem *78*, 809–819.

Lemaire, R., Wisztorski, M., Desmons, A., Tabet, J.C., Day, R., Salzet, M., and Fournier, I. (2006b). MALDI-MS direct tissue analysis of proteins: Improving signal sensitivity using organic treatments. Anal Chem *78*, 7145–7153.

Masuo, Y., *et al.* (2009). Omic analyses unravels global molecular changes in the brain and liver of a rat model for chronic Sake (Japanese alcoholic beverage) intake. Electrophoresis *30*, 1259–1275.

Mathy, G., and Sluse, F.E. (2008). Mitochondrial comparative proteomics: Strengths and pitfalls. Biochim Biophys Acta *1777*, 1072–1077.

Mu, J., *et al.* (2008). Proteomic analysis of a rat model of depression. Expert Rev Proteomics *5*, 315–320.

Paulson, L., *et al.* (2004). Comparative proteome analysis of thalamus in MK-801-treated rats. Proteomics *4*, 819–825.

Sagvolden, T., *et al.* (2005). Rodent models of attention-deficit/hyperactivity disorder. Biol Psychiatry *57*, 1239–1247.

Schmidt, M.V., *et al.* (2008). Chronic stress and individual vulnerability. Ann N Y Acad Sci *1148*, 174–183.

Seo, H.S., Hirano, M., Shibato, J., Rakwal, R., Hwang, I.K., and Masuo, Y. (2008). Effects of coffee bean aroma on the rat brain stressed by sleep deprivation: A selected transcript- and 2D gel-based proteome analyses. J Agric Food Chem *56*, 4665–4673.

Steinberg, T.H., *et al.* (2003). Global quantitative phosphoprotein analysis using multiplexed proteomics technology. Proteomics *3*, 1128–1144.

Tafet, G.E., and Bernardini, R. (2003). Psychoneuroendocrinological links between chronic stress and depression. Prog Neuropsychopharmacol Biol Psychiatry *27*, 893–903.

Washburn, M.P., Wolters, D., and Yates J.R. 3rd (2001). Large-scale analysis of the yeast proteome by multidimensional protein identification technology. Nat Biotechnol *19*, 242–247.

Wilkins, M.R., Sanchez, J.C., Gooley, A.A., Appel, R.D., Humphery-Smith, I., *et al.* (1995). Progress with proteome projects: Why all proteins expressed by a genome should be identified and how to do it. Biotechnol Genet Eng Rev *13*, S19–S50.

Williams, K., *et al.* (2004). Recent advances in neuroproteomics and potential application to studies of drug addiction. Neuropharmacology 47, 148–166.

Wisztorski, M., Croix, D., Macagno, E., Fournier, I., and Salzet, M. (2008). Molecular MALDI imaging: An emerging technology for neuroscience studies. Dev Neurobiol *68*, 845–858.

Yeom, M., Shim, I., Lee, H.-J., and Hahm, D.-H. (2005). Proteomic analysis of nicotine-associated protein expression in the striatum of repeated nicotine-treated rats. Biochem Biophys Res Commun *326*, 321–328.

Chapter 12
Subcellular Fractionation Using Step-Wise Density Extraction for Mass Spectrometry Analysis

WenKui Lan, Marc J. Horn, Fumihiko Urano, Andre Kopoyan, and Sun W. Tam

Abstract In recent years, significant technological advances in mass spectrometry instrumentation have moved towards higher sensitivity, resolution, and throughput. However, the ability to generate quantitative information on complex biological samples, especially for identification of low abundance proteins, subcellular compartment-specific protein expression and biomarker discovery, still depends upon the quality and the reproducibility of the sample preparation. Upstream fractionation of tissue or cell culture samples can play a key role in reducing the complexity of the sample. The ability to enrich selective fractions with proteins of interest can significantly expedite the identification of those proteins. We have developed a novel technology (Edge^{TM}) and instrumentation (Edge 200) platform for upstream proteomics sample preparation and fractionation, which uses stepwise density extraction of biological particles within a sample, based upon the densities of those particles.

Keywords Density extraction · Enrichment · Organelle · Sample preparation · Subcellular fractionation

12.1 Introduction

Efficient fractionation of biological particles from cell or tissue homogenates is crucial to future developments in the study and understanding of structural biology. Additionally, the fractionation and enrichment of low abundance proteins from organelles and other subcellular particles is central to biomarker discovery in pharmaceutical drug development. Current methods for isolation and characterization of functional low-abundance proteins specific to organellar compartments are complex and time-consuming, due to the heterogeneity of subcellular particles. Standard

W. Lan (✉)
Prospect Biosystems, Inc., Newark, NJ, USA
e-mail: wlan@prospectbiosys.com

A.R. Ivanov, A.V. Lazarev (eds.), *Sample Preparation in Biological Mass Spectrometry*, DOI 10.1007/978-94-007-0828-0_12,

methods for isolation of organelles typically involve multi-step centrifugations, including differential and density gradient ultracentrifugation, and may require up to 24–48 h to obtain acceptable separations. Additionally, these classical fractionation methods have not kept pace with the increased sensitivity of mass spectrometry in protein analysis.

Separation of particles can be accomplished by simple gravity sedimentation, wherein samples are allowed to sit and separation occurs due to the differences in the size and shape of the particles. Gravitational sedimentation has limited practical value for particles under a few micrometers in diameter due to the prohibitively long settling times. Sedimentation can be accelerated, however, by coupling with driving forces including centrifugal force, magnetic force, electric force and other forces.

The most common methods to separate biological particles have involved the use of centrifugation including, differential, rate-zonal and isopycnic centrifugation. These methods have been the subject of numerous reviews (Anderson et al., 1966; Hinton and Mullock, 1997; Huber et al., 2003).

Towards the goal of providing a method for the fractionation of biological particles and organelles quickly and efficiently, in a manner which is both reproducible and scalable, we have developed a technology and instrumentation platform which allows for subcellular fractionation by step-wise density-based extraction. The process, termed Edge^{TM} (*E*nhanced *d*ensity *g*radient *e*xtraction), relies upon the difference in density between subsets of particles in the sample and that of the extracting medium.

Based on the Stokes' equation, particles that are equal to or lower in density than the density of the extracting medium will float in the extracting medium while particles higher in density than the density of the medium will sediment towards the bottom of the extracting medium. When subjected to an external force, e.g., gravitational, centrifugal, magnetic, electrical, etc., the supernatant will contain particles with a density less than or equal to the density of the extracting medium ($\rho_1{}^1$), while the sedimented pellet will contain particles with a density greater than that of the extracting medium. The supernatant (s^1) can be isolated, and the remaining pellet can then be resuspended in an extracting medium where the density has been increased incrementally ($\rho_1{}^2$). When an external force is applied again, the resulting supernatant (s^2) will contain particles that have a density in the range greater than ($\rho_1{}^1$) and less than or equal to ($\rho_1{}^2$). The Edge process is shown pictorially in Fig. 12.1.

The remaining pellet can then be treated iteratively, as above, to obtain additional particles present in a sample depending upon their density. This process may be optionally repeated until a desired density of the medium is reached or until no pellet is produced when the external force is applied.

The Edge process is implemented on the Edge 200 Separation System. At the heart of the system is the Edge 200 (Fig. 12.2), an air-driven separation instrument which uses centrifugal force as the external force to accelerate the Edge separation process. Using "house air" or an air compressor, the Edge 200 rotor reaches speeds of 95,000–100,000 RPM in less than a minute, providing 120,000–150,000$\times g$. All separations are carried out at 3.5–5.5°C.

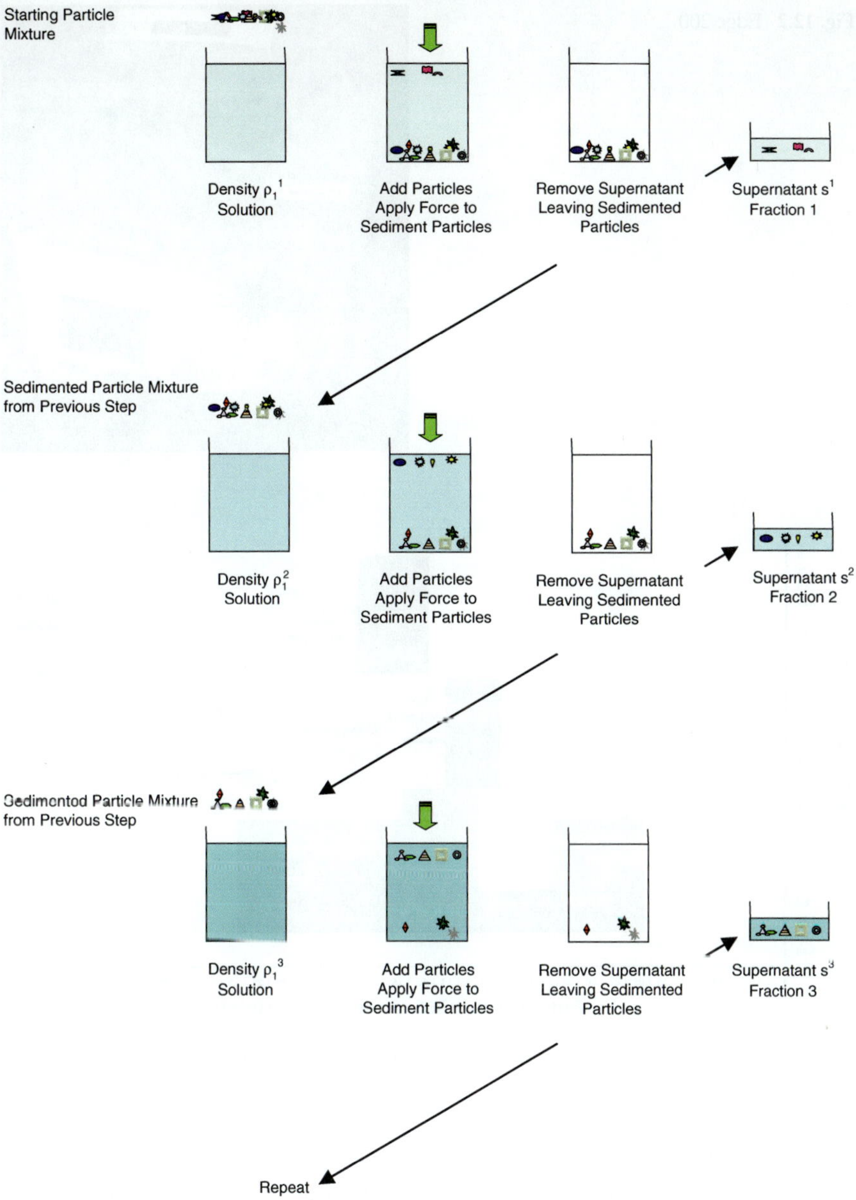

Fig. 12.1 Edge process of density-based extraction

Individual organelles display a broad range of heterogeneity in both size and density (Fig. 12.3). While many approaches (Cox and Emili, 2006; Guillemin et al., 2005; Hoffmann et al., 2005; Song et al., 2006; Srinivas et al., 2004) have been used in an attempt to isolate high purity organelles, the "concept of 'pure' organelles is

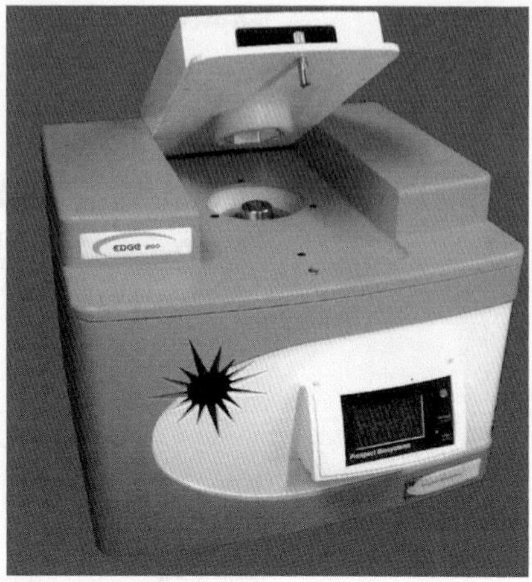

Fig. 12.2 Edge 200

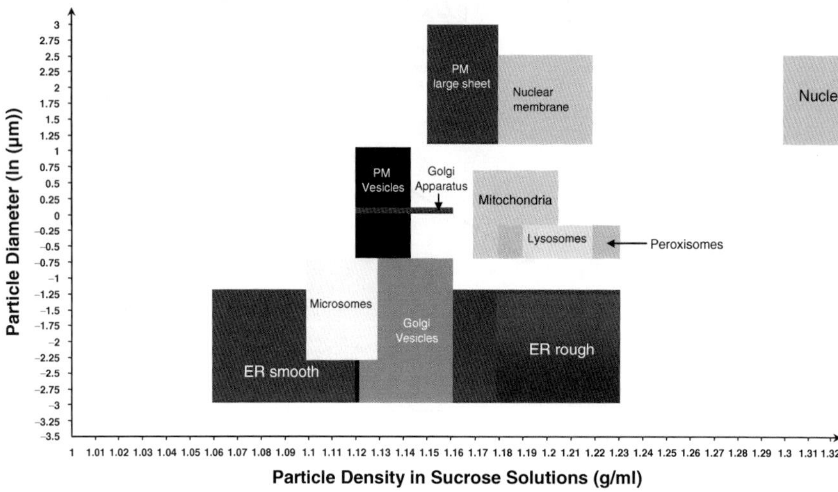

Fig. 12.3 Plot of biological particle density in sucrose solution vs. particle diameter in Ln (µm)

untenable" (Hu et al., 2007). The Edge fractionation process, on the other hand, is designed to complement downstream analytical techniques by providing rapid enrichment and fractionation of subcellular particles and the proteins therein, while also providing a direct correlation to the density of the particle. Additionally, the flexibility of the process to use extraction volumes smaller than the initial sample volume can produce significant enrichment and concentration of both the particles and their proteins.

The use of well-defined, accurate density media, e.g., GradiSpec^{TM} reagents, allows highly reproducible means to directly and rapidly compare changes, fraction by fraction, between differentially treated or expressed samples. The identification of small changes in overall difference between two samples often may be amplified by quantitating ratios of specific markers in pairs of identical fractions of the two samples.

12.2 Protocols

12.2.1 General Workflow

The general workflow involved in subcellular fractionation using step-wise density extraction is shown below (Fig. 12.4).

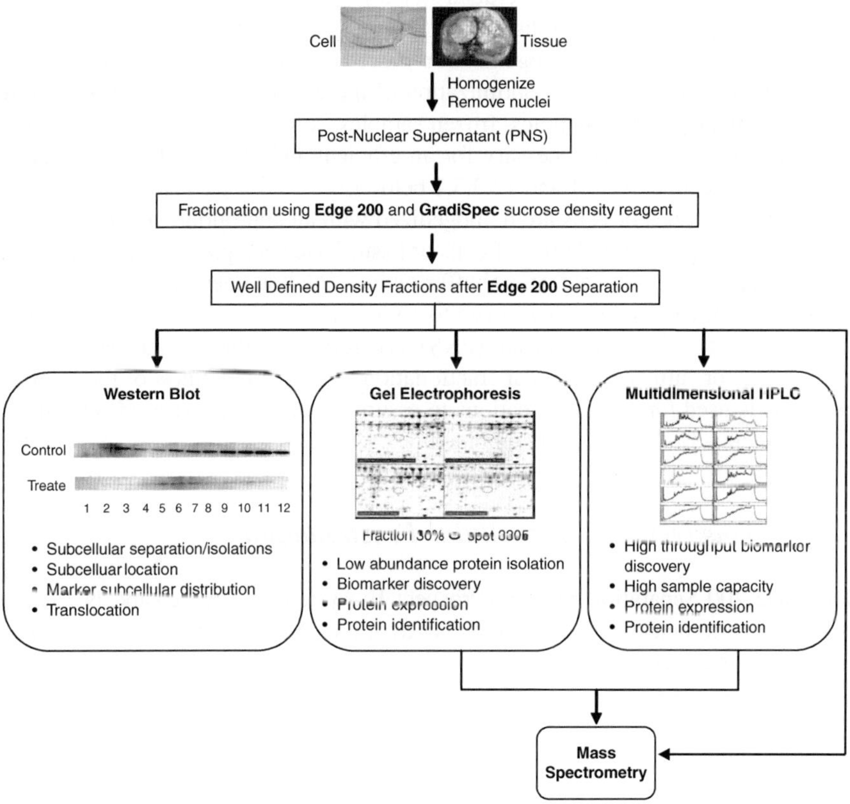

Fig. 12.4 General work flow of cells or tissues using Edge process, followed by gel or gel-free analytical techniques, e.g., western blot, 2DE, HPLC, etc., and mass spectrometry for final protein identification or quantitation. *Bulleted points* on the lower row of panels represented potential applications

12.2.2 Materials

Tissue or culture cells
Homogenization buffer with protease inhibitors
Sucrose density extraction media or GradiSpec reagent kit

12.2.3 Equipment

Dissecting tools
Homogenizer
Low-speed centrifuge
Edge 200 Separation System

12.2.4 Homogenization

The homogenization process is a crucial step for subcellular fractionation. The ideal homogenization process is to disrupt the cells and keep subcellular compartments intact. There are standard homogenization procedures which can be used (Graham, 1997, for example). However, optimization of these protocols based on the nature of the sample, e.g., fresh samples, frozen samples, soft tissues, hard tissues, different cultured cells, etc., is necessary for an efficient and effective homogenization process (see Sections 12.3.1 and 12.3.3.1 below).

Generally, tissues or cells are homogenized with an appropriate homogenizer in a specific homogenization buffer. If cells or tissue have been previously frozen, they are first thawed in homogenization buffer on ice. The homogenate is transferred to centrifuge tubes and spun at low speed to remove nuclei. The supernatant, which is termed the post nuclear supernatant (PNS), is carefully removed to a clean tube and kept on ice for further subcellular fractionation. A single repetition of the homogenization is used for most applications to assure maximum recovery of subcellular compartments. Both PNS's are combined.

12.2.5 Density-Based Subcellular Fractionation

The Edge 200 Separation System (Prospect Biosystems, Inc., Newark, NJ, USA) is used for the density-based subcellular fractionation. The fractionation method follows the manufacturer's procedure. Briefly:

1. Transfer the combined PNS's into a sample container (SteriLiner™, Prospect Biosystems, Inc.).
2. Insert the sample container into a sample rotor.
3. Place the rotor into the Edge 200, and automatically run the system's fractionation step: the rotor will accelerate to 100 K RPM, the maximum speed will be maintained for 2 min, and the rotor will then decelerate to rest.
4. Repeat Step 3 one more time when the rotor comes to rest. Remove the sample container from the rotor. At this point it was noted that almost all of the particles are pelleted on the vertical wall of the sample container.
5. Carefully remove the supernatant to a clean eppendorf tube.
   Note: The supernatant from the combined PNS's is referred to as the first fraction.

6. Add a specific volume of GradiSpec extraction medium reagent (Prospect Biosystems, Inc.) having a higher density than the PNS to the sample container. Vortex the container for 20–30 s or until all the pellet is resuspended in the density solution.
7. Repeat Steps 2, 3, 5 and 6 above to collect the desired fractions. For Step 6, GradiSpec extraction medium reagents with increasing sucrose density media (e.g., 10, 15, 20, 25, 30, 35, 40, 45, 50, 55 and 60%), as desired, can be used.

12.2.6 Gel Electrophoresis, Western Blot and Mass Spectrometry

Any secondary analysis before mass spectrometry, such as SDS-gel electrophoresis, two dimensional gel electrophoresis (2DE), western blots, HPLC, etc., can be used for subsequent separation or marker detection. The commonly accepted protocols should be used for each respective downstream separation and analysis method. Both MALDI-TOF and LC-MS/MS can be used for protein identification.

12.3 Applications and Results

12.3.1 Selective Fractionation and Enrichment

In the search for and isolation of low abundance proteins, up-stream fractionation of a tissue or cell culture sample plays a key role in reducing the sample complexity. The ability to enrich selective fractions with proteins of interest can significantly expedite the identification of those proteins.

12.3.1.1 Experimental Procedure

One frozen rat liver (NIEHS G822812-A, #10)(~5 g) was thawed in ice cold 1x homogenization buffer (20 mM HEPES, 10 mM KCl, 1 mM Na_2EDTA and 250 mM sucrose, pH 7.4) with protease inhibitor cocktail. The tissue was dissected into 2–3 mm^3 pieces and the liquid was carefully discarded. The tissue pieces were resuspended in five volumes of ice cold homogenization buffer and the suspension was transferred to a glass homogenizer (Kontes Dounce Tissue Grinder, 7 mL, Fisher Scientific). The tissue was homogenized on ice by doing 10 up and down strokes with a loose pestle, followed by 10 up and down strokes with a tight pestle. A portion of the homogenate (5 mL) was transferred to centrifuge tubes and centrifuged at $1000\times g$ for 10 min to remove nuclei. The PNS (3.5 mL) was fractionated using the Edge process into the following density fractions: 8.5, 10, 14, 18, 22, 26, 30, 34, 38, 42, 46, 50, 54 and 60%. Selected fractions were further separated by 2DE. Gel images were acquired on Typhoon imager (GE healthcare). Digitized gel images were analyzed and compared using Progenesis software (Nonlinear Dynamics). Selected anchor spots were used to optimize the comparison. Selected spots were excised and analyzed by MALDI-QIT-TOF (Shimadzu, Axima QIT).

12.3.1.2 Results

Figure 12.5 shows a comparison of the original PNS with density fractions 14, 26, 38 and 46% of the Edge separation. The image analysis clearly shows the enrichment of certain proteins in specific fractions. Additionally, proteins not visible in the PNS may be easily identified following fractionation. Enrichment also highlights the location of spot 1734, in two distinctly different fractions, 38 and 46%, to be the same protein, electron transfer flavoprotein-ubiquinone oxidoreductase precursor (Fig. 12.6, Table 12.1), a mitochondrial inner membrane protein, indicating possible heterogeneity of the mitochondria.

12.3.2 Pathway Biomarkers

Information exists regarding key proteins that are part of specific biochemical pathways. To the extent that such proteins are reflective of activity within those pathways, following their movement and location can be used to understand and manage disease as well as to identify additional proteins which may be markers of that pathway.

IRE1 is a signaling molecule activated by endoplasmic reticulum (ER) stress, a type of cell stress caused by accumulation of misfiled and unfolded proteins (Fonseca et al., 2009). Research has shown that ER stress plays a role in the pathology of diabetes (Fonseca et al., 2009). Edge-fractionated biological samples provide a method to quickly and efficiently track pathway biomarkers. Relative percentage distribution of the IRE1 marker within fractions is expected to differ between normal and treated (or diseased) states of pathologies involving the ER stress signaling pathway. Probing the distribution of known pathway biomarkers such as IRE1 within defined fractions, other differentially expressed proteins related to the specific pathway may be easily found.

12.3.2.1 Experimental Procedure

About 20 million INS-1 pancreatic beta cells were grown in 5 mM glucose (control), or stimulated with 1 mM thapsigargin, a chemical inducer of ER stress (treated). The frozen cells were resuspended in 1 mL of homogenization buffer (20 mM Tris/HCl, 1 mM Na_2EDTA and 250 mM sucrose, pH 7.4) with protease inhibitor cocktail and incubated on ice for 10 min. The cell suspension was then transferred to a glass Dounce homogenizer. The tube which contained the cell suspension was rinsed with 1 mL of homogenization buffer which was transferred to the Dounce homogenizer. Cells were homogenized using 20 up and down strokes with a tight pestle. Trypan blue was used to optimize and monitor the completion of the cell homogenization. One microliter of cell homogenate was mixed with one microliter of trypan blue, and the mixture was placed on a glass slide. Under a phase contrast microscope, the completion of the cell homogenization was examined, and homogenized further until >90% of cells were homogenized. The homogenate was transferred to centrifuge tubes and centrifuged at $1000\times g$ for 10 min to remove nuclei. The resulting

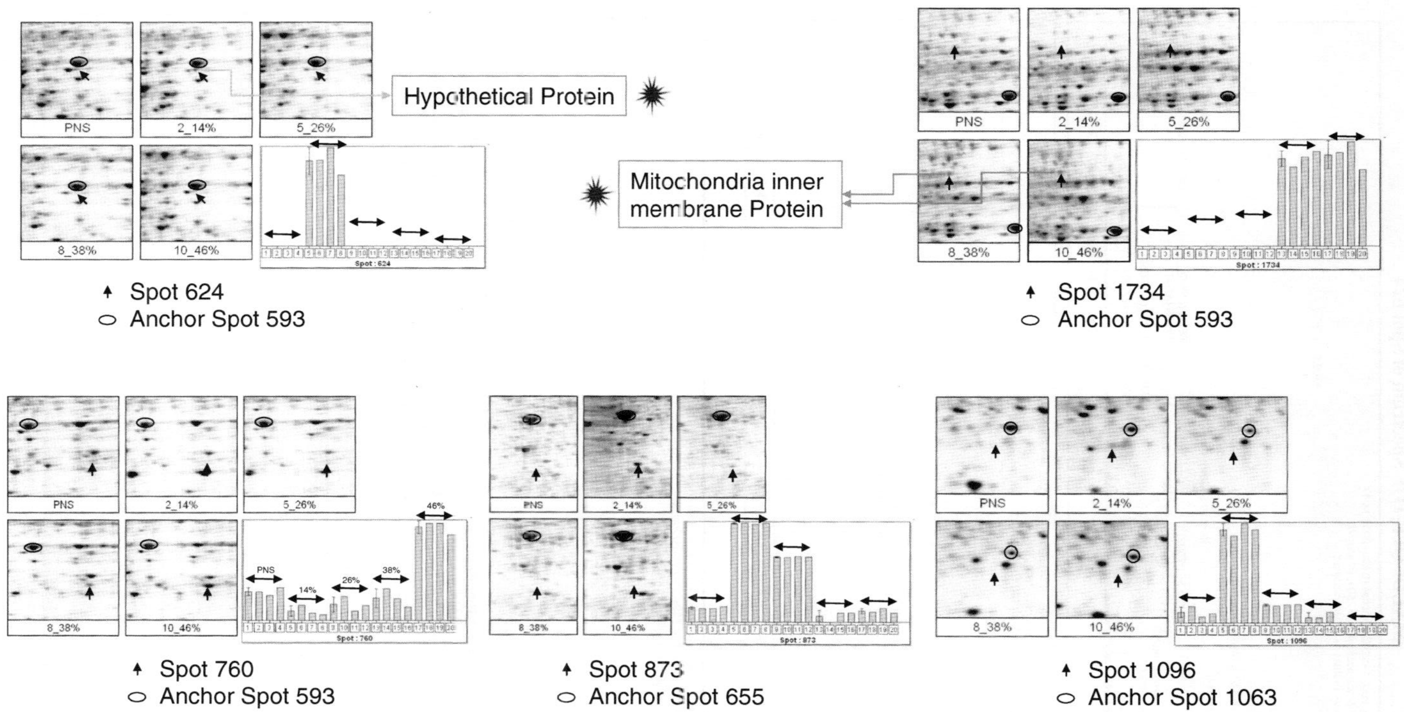

Fig. 12.5 2DE and image analyses of specific fractions of rat liver homogenate, following homogenate fractionation using Edge technology. Fractions: PNS (bars 1–4 in the histogram plot, the sample mixture before fractionation; 2_14% (fraction 2 in 14% sucrose reagent, bars 5–8 in the histogram plot); 5_26% (fraction 5 in 26% sucrose reagent, bars 9–12 in the histogram plot); 8_38% (fraction 8 in 38% sucrose reagent, bars 13–16 in the histogram plot) and 10_46% (fraction 10 in 46% sucrose reagent, bars 17–20 in the histogram plot) were selected for the 2DE analysis. Five anchor spots (two anchor spots shown) were chosen and identified by LC/MS/MS (data not shown) to increase gel overlay accuracy

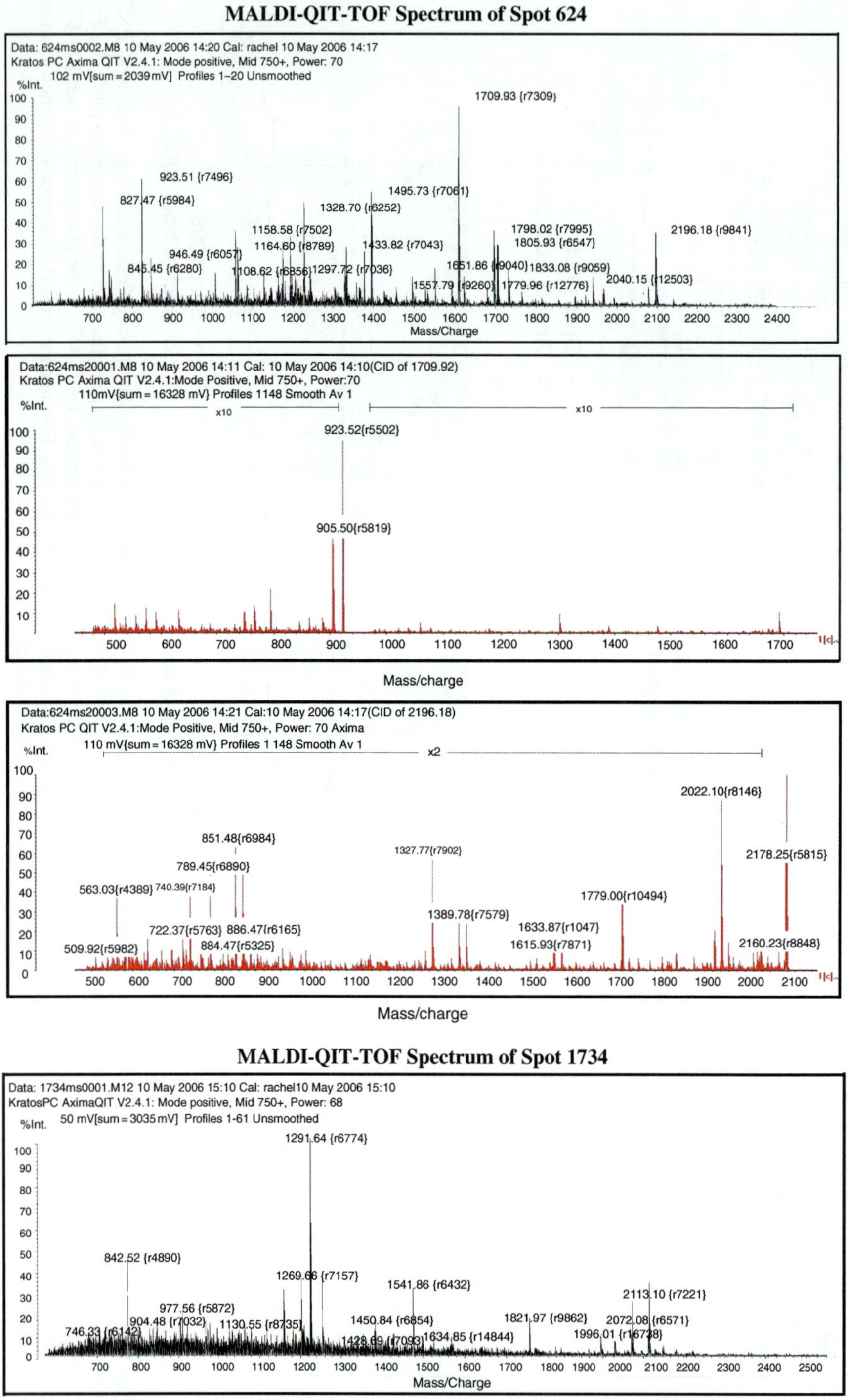

Fig. 12.6 MALDI-TOF spectra of protein spots 624 (from fraction 2 in 14% sucrose reagent) and 1734 (from fraction 8 in 46% sucrose reagent and fraction 10 in 46% sucrose reagent)

Table 12.1 Selected spot protein ID using MALDI-QIT-TOF

Spot No.	Protein
624	Predicted: Similar to Psmc6 protein (gi/34869622) (18 peptides matched)
1096	Ferritin light chain (gi/204123) (14 peptides matched)
760	Urate Oxidase (gi/56971244) (18 peptides matched)
873	40S ribosomal protein SA (p40) (34/67 kDa laminin receptor) (7 peptides matched)
1734	Electron transfer flavoprotein-ubiquinone oxidoreductase precursor (18 peptides matched)

PNS was fractionated using the Edge 200 Separation System into the following density fractions: 8.5, 10, 15, 20, 25, 30, 35, 40, 45, 50, 55 and 60%. Samples from each fraction were subjected to standard western blot analyses using anti-IRE1 as probing marker. Fractions showing significant relative expression changes of IRE1 were selected and analyzed by 2DE. Gel images were analyzed as described above for protein expression differences. Protein spots showing >twofold up- or down-regulation were excised, digested and identified by MALDI-TOF/TOF (Shimadzu, Axima QIT).

12.3.2.2 Results

Following western blot analysis, the relative percentage distribution of the IRE1 marker was calculated for all fractions. IRE1 is a trans-membrane protein kinase localized to the ER and is thought to be a central component of the ER stress signaling pathway (Ron and Walter, 2007). Both cell types gave similar-shaped total protein distribution profiles of IRE1 within the 13 fractions (Fig. 12.7). However, sub-location of IRE1 was shifted in the ER stressed cells compared with controlled cells. While the majority of the IRE1 in the controlled cells was located in higher density particles, the majority of the IRE1 in the stressed cells was located in lower density particles. This illustrated that separation by particle density using Edge 200

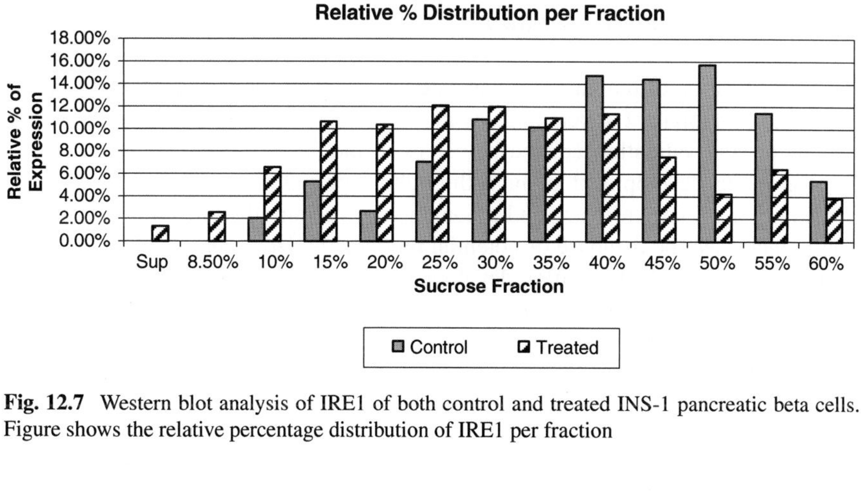

Fig. 12.7 Western blot analysis of IRE1 of both control and treated INS-1 pancreatic beta cells. Figure shows the relative percentage distribution of IRE1 per fraction

Table 12.2 IRE1 Expression changes vs. number of differentially expressed proteins

Fraction (%)	Relative change in IRE1 expression	Number of differentially expressed proteins
25	0.71	37
30	0.10	31
35	0.08	18
45	−0.93	31
50	−2.73	56

can demonstrate IRE1 sub-location change as an important function in managing ER stress. Such changes in location may arise from certain biological events, such as translocation, or from morphological differences between the subcellular compartments in controlled and stressed cells, which in turn may affect the density of those compartments.

The Edge density extraction technology provides fractionation and separation of subcellular compartments based on their individual density, as opposed to fractionation and separation of individual proteins. Thus, groups of markers may be found within the context of their biologically relevant environments, leading to the discovery and identification of pathway-related process markers or disease markers.

As such, following 2-dimensional electrophoresis of fractions 25, 30, 35, 45 and 50% of the control and treated samples, fractions that showed significant differences in IRE1 expression, gels were compared to find spots having large differences in expression. Table 12.2 shows that the 25 and 50% fractions contain the largest number of differentially expressed proteins from the group of fractions chosen.

Differentially expressed spots from fractions 25 to 50% were identified using mass spectrometry, and then those identities were searched for relationships with ER stress and diabetes using the PubMed database. The following table (Table 12.3), showing a representative sampling of proteins found, demonstrates the high degree of correlation between those proteins found in fractions containing the largest amount of IRE1 differential expression, and diabetes and ER stress.

12.3.3 Biomarker Evaluation

The identification of biomarkers is an essential element for early prediction of diseases and development of personalized medicine in the future. Biomarker discovery, evaluation and validation are the key steps in biomarker development processes. The on-going development of high-throughput proteomics, which includes high sensitivity mass spectrometry and automation of protein identification, has significantly increased the database of potential biomarkers. However, the evaluation and validation steps of the biomarker development process remain a bottleneck.

Rat models of high fat diet (HF) have shown that a diet high in fat has a profound impact on brain function (Wu et al., 2004). Phosphosynapsin (p-synapsin I) has been reported as a marker for synaptic dysfunction (Wu et al., 2008). To evaluate

Table 12.3 Protein ID of selected differentially expressed spots using MALDI-TOF/TOF

	Protein	Accession No.	Relative expression	ER stress pathway	Diabetes
Frx 25%	Expressed in non-metastatic cell 1, protein (NM23A) (nucleoside diphosphate kinase)	gi|19924089(11)[a]	Up	√	√
	N-ethylmaleimide sensitive fusion protein	gi|13489067(8)	Up	√	√
	Similar to eukaryotic translation initiation factor 5A	gi|27672956(14)	Up	Unknown	Unknown
	Eno 1 protein	gi|50926833(12)	Up	√	√
	Unnamed protein product	gi|56200(18)	Up	Unknown	Unknown
	Arginase 1	gi|8392920(9)	Up	√	√
	Guanine nucleotide-binding protein, beta-1 subunit	gi|13591874(19)	Up	√	√
	Expressed in non-metastatic cell 2	gi|55778652(13)	Up	X	X
	Signal sequence receptor 4	gi|8394364(8)	Up	√	√
	Unnamed protein product	gi|56082(12)	Down	Unknown	Unknown
	Aconitase 2, mitochondrial	gi|40538860(7)	Down	√	√
Frx 50%	Isovaleryl Coenzyme A dehydrogenase	gi|6981112(17)	Up	√	√
	Eef1g protein	gi|51261278(10)	Up	X	X
	Similar to prohibitin (BAP 32)	gi|34873234(16)	Up	Unknown	Unknown
	Unnamed protein product	gi|56200(15)	Up	Unknown	Unknown
	ATP synthase, H+ transporting, mitochondrial F1 complex, beta subunit	gi|54792127(9)	Up	√	√
	Mitochondrial aconitase	gi|10637996(15)	Up	√	√
	Glycerol-3-phosphate dehydrogenase 2	gi|6980978(13)	Up	√	X
	dnaK-type molecular chaperone grp 75 precursor	gi|2119726(7)	Down	√	√
	Glucose regulated protein, 58 kDa	gi|38382858(19)	Down	√	√
	Insulin I	gi|56488(17)	Down	√	√

[a]Number of peptides matched

the utility of p-synapsin I as a marker of synaptic dysfunction arising from high fat diet, brain samples of rats fed high fat diets were compared with those of rats fed a control diet.

12.3.3.1 Experimental Procedure

Twenty rats were separated into two groups of 10, one group fed with special high fat diet (60 kcal% fat, D12492, Research Diets, Inc.) for 6 weeks and the other group fed with regular low fat diet for the same time. Whole rat brains from each group were snap-frozen. One frozen brain (~1 g) was thawed in ice cold 1x homogenization buffer (20 mM HEPES, 10 mM KCl, 1 mM Na_2EDTA and 320 mM sucrose, pH 7.4) with protease inhibitor cocktail. The tissue was dissected into 2–3 mm^3 pieces and the liquid was carefully discarded. The tissue pieces were resuspended in five volumes of ice cold homogenization buffer and the suspension was transferred to a glass homogenizer. Using a loose pestle, the tissue was homogenized on ice by doing 10 up and down strokes with a loose pestle, followed by 10 up and down strokes with a tight pestle. The homogenate was transferred to centrifuge tubes and centrifuged at 1000×*g* for 10 min to remove nuclei. The PNS, was fractionated using the Edge process into the following density fractions: 10, 15, 20, 25, 30, 35, 40, 45, 50, 55 and 60%. All remaining rat brain tissues were fractionated using the same procedure as above.

All fractions from all samples were subjected to western blot analysis using the antibodies against rat synaptic function marker p-synapsin I. Results were analyzed using Prospect's algorithm (EDGE*TEST) giving clustered data for the marker.

12.3.3.2 Results

Fractions from each animal were analyzed by western blot analysis for p-synapsin I, and relative percentage distributions for the marker were calculated and plotted across all fractions for all animals. The histograms in Fig. 12.8 show the relative percentage distributions of p-synapsin I among eleven fractions of 10 individual control brains, and among eleven fractions of 10 individual brains from high fat diet animals.

Statistical analysis of changes in relative percentage distribution of the p-synapsin I marker between the control and high fat diet groups revealed that the ratio of the marker in the 10% vs. the 40% fraction was significantly different in each group, thus providing a possible means for evaluation.

The data from the western blot analysis above is presented in Table 12.4, showing a specific ratio (Fraction 10%/Fraction 40%) of p-synapsin I expression in control and HF diet rat brains, as well as the mean and SD (standard deviation) for each group.

Figure 12.9 is a graphical representation of the data in Table 12.4. Ratios (Fraction 10%/Fraction 40%) of the relative percentage distribution of p-synapsin I expression cluster into two specific groups, unique to either the high fat diet or control rat brain origin. Error bars indicate mean ±1 SD.

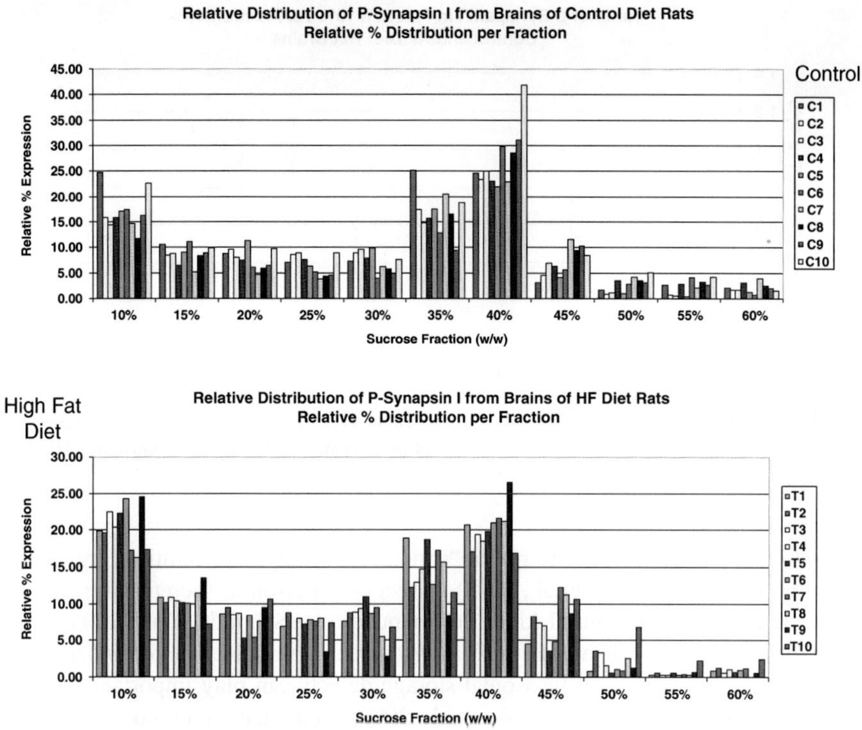

Fig. 12.8 Western blot analysis of p-synapsin I for both control and high fat diet rat brain specimens. Figure shows the relative percentage distribution of p-synapsin I per fraction

Table 12.4 Statistical analysis of p-synapsin I distribution data

	Ratio of p-synapsin I expression in Frx 10% vs. Frx 40%										Mean	SD
Control	1.01	0.68	0.58	0.69	0.78	0.58	0.64	0.41	0.52	0.54	0.64	0.165
HF	0.96	1.15	1.16	1.10	1.13	1.15	0.80	0.79	1.03	0.93	1.02	0.144

The evaluation method relies on the ratio between internal fractions of the sample, and thus data arising from the process is internally referenced. This aspect of the method makes analysis of the data independent of sample amount, and does not require internal or external standards.

12.4 Summary

Recently, there has been a tremendous growth in subcellular studies of the mammalian proteomes. In depth analysis of the adipocyte proteome by Mann and colleagues have yielded an extensive list of proteins across nuclei, mitochondria, membrane and cytosol (Adachi et al., 2007). Aebersold and colleagues have made

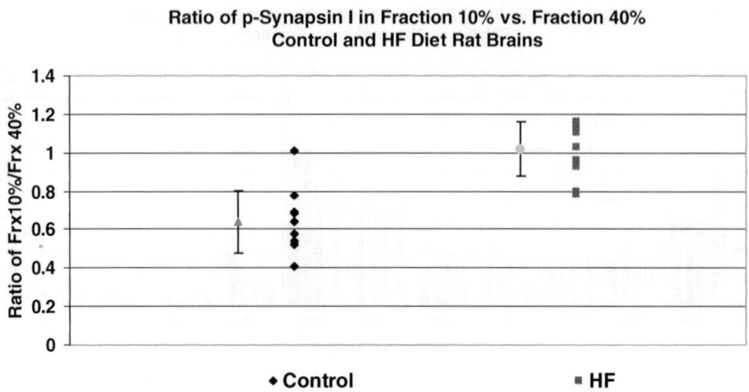

Fig. 12.9 Graphical representation of the data in Table 12.4. The graph shows the clustering of data points for the ratio of Frx 10% vs. Frx 40% for p-synapsin I expression in control and HF diet rat brains. *Error bars* indicate mean ±1 SD

a combination of approaches to reveal "concurrent" analysis of multiple organelles from the same lysate (Yan et al., 2008). Direct comparison of secretory lysosomes has differentiated two natural killer cell populations (Schmidt et al., 2008). The study of an individual organelle, such as lysosome, has allowed better understanding of many diseases since lysosomal storage defects can play important roles in cancer or Alzheimer disease (Lubke et al., 2009). Many have focused on correlating individual proteins with their specific organellar locations. Overall, the field of the organelle proteomic study has identified many previously unknown signaling intermediates.

Despite these advances, there still exists a fundamental need for a general, simple and robust sample preparation method for proteomic studies of complex biological materials. Quality sample preparation and subcellular fractionation are crucial initial steps in any successful downstream analysis including mass spectrometry. Analytical processes such as gel electrophoresis are limited by the amount of sample which may be analyzed at one time. Fractionation allows proteins of low abundance to be visualized within fractions, when they normally would be obscured by higher abundance proteins in the total sample, or not be visible at all due to the large dynamic range of protein concentration within the sample.

Edge technology is a non-denaturing front-end sample fractionation and separation methodology developed by Prospect Biosystems, comprising a stepwise extraction of a particle mixture using extraction media of increasing densities. It forms the basis of a broad-range subcellular fractionation and separation capability and an instrumentation platform, providing non-denatured, intact biological compartments in which a biological event has occurred. Groups of markers can be identified within biologically relevant environments allowing the user to follow the biology throughout the disease process. This rapid biology-focused fractionation and separation technology provides a well defined and reproducible method for biomarker discovery and proteomics analysis from complex tissue homogenate and cell lysate samples. Edge fractionation is compatible with and complements all downstream analyses including gel electrophoresis, HPLC, mass spectrometry, microarrays, etc. Since Edge technology fractionates intact biological compartments without denaturing those compartments, additional biological information about subcellular location may also be learned. Additionally, Edge fractionation followed by statistical analysis of the relative percentage distribution of a marker within specific subcellular compartments provides a fast, straightforward means for evaluation of the marker.

References

Adachi, J., Kumar, C., Zhang, Y., and Mann, M. (2007). In-depth analysis of the adipocyte proteome by mass spectrometry and bioinformatics. Mol Cell Proteomics *6*(7), 1257–1273.

Anderson, N.G., Harris, W.W., Barber, A.A., Rankin, C.T., Jr., and Candler, E.L. (1966). Separation of Subcellular Components and Viruses by Combined Rate- and Isopycnic-Zonal Centrifugation. In The Development of Zonal Centrifuges and Ancillary Systems for Tissue Fractionation and Analysis, National Cancer Institute, N.G. Anderson, ed. (Washington, D.C., U.S. Government Printing Office), Monograph 21, pp. 253–268.

Cox, B., and Emili, A. (2006). Tissue subcellular fractionation and protein extraction for use in mass-spectrometry-based proteomics. Nat Protoc *1*(4), 1872–1878.

Fonseca, S.G., Burcin, M., Gromada, J., and Urano, F. (2009). Endoplasmic reticulum stress in beta cells and development of diabetes. Curr Opin Pharmacol. *9*(6), 763–770.

Graham, J.M. (1997). Homogenization of Tissues and Cells. In Subcellular Fractionation, J.M. Graham, D. Rickwood, eds. (New York, NY, Oxford University Press), pp. 1–29.

Guillemin, I., Becker, M., Ociepka, K., Friauf, E., and Nothwang, H.G. (2005). A subcellular prefractionation protocol for minute amounts of mammalian cell cultures and tissue. Proteomics *5*(1) 35–45.

Hinton, R.H., and Mullock, B.M. (1997). Isolation of Subcellular Fractions. In Subcellular Fractionation, J.M. Graham, D. Rickwood, eds. (New York, NY, Oxford University Press), pp. 31–69.

Hoffmann, K., Blaudszun, J., Brunken, C., Höpker, W.-W., Tauber, R., and Steinhart, H. (2005). New application of a subcellular fractionation method to kidney and testis for the determination of conjugated linoleic acid in selected cell organelles of healthy and cancerous human tissues. Anal Bioanal Chem *381*(6), 1138–1144.

Hu, Z.-Z., Valencia, J.C., Huang, H., Chi., A., Shabanowitz, J., Hearing, V.J., Appella, E., and Wu, C. (2007). Comparative bioinformatics analyses and profiling of lysosome-related organelle proteomes. Int J Mass Spectrom *259*, 147.

Huber, L.A., Pfaller, K., and Vietor, I. (2003). Organelle proteomics: Implications for subcellular fractionation in proteomics. Circ Res *92*(9), 962–968.

Lubke, T., Lobel, P., and Sleat, D.E. (2009). Proteomics of the lysosome. Biochim Biophys Acta *1793*(4), 625–635.

Ron, D., and Walter, P. (2007). Signal integration in the endoplasmic reticulum unfolded protein response. Nat Rev Mol Cell Biol *8*(7), 519–529.

Schmidt, H., Gelhaus, C., Nebendahl, M., Lettau, M., Watzl, C., Kabelitz, D., Leippe, M., and Janssen, O. (2008). 2-D DIGE analyses of enriched secretory lysosomes reveal heterogeneous profiles of functionally relevant proteins in leukemic and activated human NK cells. Proteomics *8*(14), 2911–2925.

Song, Y., Hao, Y., Sun, A., Li, T., Li, W., Guo, L., Yan, Y., Geng, C., Chen, N., Zhong, F., Wei, H., Jiang, Y., and He, F. (2006). Sample preparation project for the subcellular proteome of mouse liver. Proteomics *6*(19), 5269–5277.

Srinivas, K.S., Chandrasekar, G., Srivastava, R., and Puvanakrishnan, R. (2004). A novel protocol for the subcellular fractionation of C3A hepatoma cells using sucrose density gradient centrifugation. J Biochem Biophys Methods *60*(1), 23–27.

Wu, A., Ying, Z., and Gomez-Pinilla, F. (2004). The interplay between oxidative stress and brain-derived neurotrophic factor modulates the outcome of a saturated fat diet on synaptic plasticity and cognition. Eur J Neurosci *19*(7), 1699–1707.

Wu, A., Ying, Z., and Gomez-Pinilla, F. (2008). Docosahexaenoic acid dietary supplementation enhances the effects of exercise on synaptic plasticity and cognition. Neuroscience *155*(3), 751–759.

Yan, W., Hwang, D., and Aebersold, R. (2008). Quantitative proteomic analysis to profile dynamic changes in the spatial distribution of cellular proteins. Methods Mol Biol, *432*, 389–401.

Part IV
2D Gel-Based Proteomics

Chapter 13
Two-Dimensional Polyacrylamide Gel Electrophoresis

Aisling A. Robinson, Ciara A. McManus, and Michael J. Dunn

Abstract Two-dimensional polyacrylamide gel electrophoresis (2-DE), when combined with mass spectrometry (MS) has the power to identify and quantify differential protein expression of large sample sets analysed in parallel. By separating proteins by isoelectric point (p*I*) and molecular weight (MW), it has the potential to detect 2,000–3,000 different protein spots from a single sample separation. With the development of technologies such as Difference In-Gel Electrophoresis (DIGE), which also allow the parallel addition of an internal standard within each gel, 2-DE has shown improved reproducibility and moreover improved protein detection sensitivity. Furthermore, 2-DE provides protein abundance information as well as p*I* and MW and so has the potential to resolve and visualise post-translational modifications such as phosphorylation, a major advantage over non gel-based separation techniques. In this chapter we review sample preparation procedures from prevention of proteolysis, including the potential of a novel protein stabilising device, to lysing of a given sample with an appropriate lysis buffer. We describe additional steps which may be taken in order to achieve a protein sample which will yield a highly reproducible 2-DE map with distinct protein spots. This chapter also discusses the basic methodology behind 2-DE and the various protein detection reagents which are available. Finally, we discuss how the 2-DE images can be quantified and differentially expressed proteins can be identified by MS.

Keywords 2-DE · DIGE · Protein preparation · Proteolysis · Protein spot digestion

Abbreviations

2-DE	Two dimensional electrophoresis
ACN	Acetonitrile
ANOVA	Analysis of variance
DIGE	Difference gel electrophoresis

A.A. Robinson (✉)
School of Medicine and Medical Science, UCD Conway Institute of Biomolecular and Biomedical Research, University College Dublin, Dublin 4, Ireland
e-mail: aisling.robinson@ucd.ie

A.R. Ivanov, A.V. Lazarev (eds.), *Sample Preparation in Biological Mass Spectrometry*, DOI 10.1007/978-94-007-0828-0_13,

DTT	Dithiothreitol
ECL	Enhanced chemiluminesence
ELISA	Enzyme linked immunosorbant assay
FDR	False discovery rates
IEF	Isoelectric focusing
IgG	Immunoglobulin
IPG	Immobilised pH gradients
LC	Liquid chromatography
LCM	Lasercapture microdissection
MALDI	Matrix assisted laser desorption/ionisation
MW	Molecular weight
PAGE	Polyacrylamide gel electrophoresis
PCA	Principle component analysis
p*I*	Isoelectric point
PMSF	Phenylmethylsulfonyl fluoride
PTM	Post translational modification
SDS	Sodium dodecyl sulphate
TCA	Tricholoracetic acid
TOF	Time of flight

13.1 Introduction

Global analysis of clinical and biological sample proteomes by two-dimensional electrophoresis (2-DE) has the potential to detect and quantify protein expression changes among large samples which can lead to the identification of diagnostic markers of disease as well as novel therapeutic targets within clinical studies. By comparing at a protein level, stimulated/dysfunctional to control/healthy states, specific proteins which may lead to an altered functional status can be detected. 2-DE was first described in the 1950s where it was used to separate serum in electrophoresis buffer in the first dimension using filter paper and a starch gel was used to further separate proteins in the second dimension (Smithies and Poulik, 1956). Separation of proteins by polyacrylamide gel electrophoresis (PAGE) in two dimensions was subsequently proposed in 1964 (Klose, 2009; Raymond, 1964). 2-DE as we now know it was first reported by O'Farrell in 1975 (O'Farrell, 1975), in which proteins are separated in the first dimension by isolectric focusing (IEF) according to their isoelectric point (p*I*), which is the pH at which the net charge of the protein is the same as the surrounding pH. Following IEF, the proteins are further separated in the second dimension according to their molecular weight (MW) utilising sodium dodecyl sulphate polyacryalmide gel electrophoresis (SDS-PAGE) in the second dimension (O'Farrell, 1975). A protein map of a given sample is produced with separation of up to 2,000–3,000 proteins (Jager et al., 2002). Less than 1 ng of protein per spot can subsequently be detected and quantified (Gorg et al., 2004). 2-DE in combination with mass spectrometry (MS) has major advances over non gel-based proteomic techniques: (i) in that it can separate and visualise a considerable number of proteins in a single run, (ii) and it can also provide valuable information on protein abundance together with p*I* and MW which thus can detect multiple protein isoforms resulting from post translational modifications (PTMs) such as phosphorylation or proteolysis. Other PTMs such as glycosylation can also be detected as distinct spot trains in the horizontal axis of the gel. However, 2-DE does have its limitations. The major drawback of this technique is the inability to resolve very hydrophobic proteins such as membrane associated proteins. Another difficulty arises with high abundance proteins within a sample which can often appear as large spots with a tendency to smear, thereby often masking lower abundance spots. Moreover, high molecular weight proteins can also be difficult to resolve by 2-DE.

In this chapter we will discuss the preparation of samples and subsequent highly reproducible separation of the sample proteins by 2-DE. We will also review how differential protein expression can be detected once samples are separated by 2-DE, and how particular spots/proteins of interest can be excised and identified by MS.

13.2 Sample Preparation and Solubilisation for 2-DE

13.2.1 Proteolytic Inhibition Prior to Sample Lysis

In order for quantitative proteomic studies to yield meaningful results, tissue or cell proteomes must as closely as possible represent their in vivo state following isolation. Proteolytic fragments of highly abundant proteins can mask lower MW proteins or peptides, while also affecting the abundance of the intact protein (Zhang et al., 2003). The effect of protein degradation can clearly be seen in studies using 2-DE (Grassl et al., 2009). Ideally, such artefacts should be prevented by preserving the native proteome during sample preparation. Snap-freezing in liquid nitrogen is currently the method of choice for maintaining cell and tissue proteome integrity with the addition of protease inhibitors during lysis and sample preparation. However, studies have shown that degradation can still occur particularly during IEF (Finnie and Svensson, 2002; Svensson et al., 2009). In addition, many protease inhibitors can interfere with labelling of proteins for 2-DE analysis as occurs with Difference In-Gel Electrophoresis (DIGE) (see Section 13.4.1.4).

An alternative method, which has recently been proposed uses heat to inactivate proteolytic enzymes and thus allows preservation of proteome integrity (Svensson et al., 2009). This novel technique, using the Stabilizor T1 device (Denator AB, Sweden), is based on rapid denaturation of proteins by heat, while positive pressure (or negative pressure, i.e. vacuum) is used to facilitate even heat distribution. It has been shown to effectively preserve yeast cell extracts (Grassl et al., 2009) as well as post-mortem brain tissue (Robinson et al., 2009) with no effect on protein solubility following treatment. The Stabilizor T1 can be used on freshly isolated cell or tissue samples or on samples which have been frozen in liquid nitrogen. If the sample being prepared is kept frozen before stabilisation, both methods were equally beneficial (Robinson et al., 2009). Cell or tissue samples, up to 5 mm in

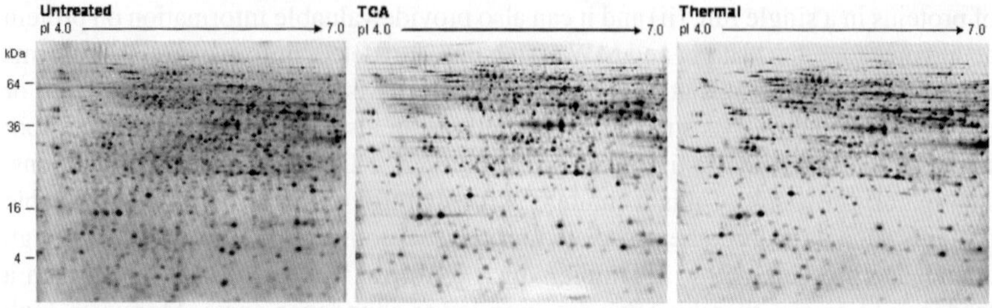

Fig. 13.1 Separation of yeast cells (*Saccharomyces cerevisiae*) by 2-DE. Evidence of a proteolysis cloud can be seen in the lower region of the gel in the untreated sample. Protein stabilisation and removal of the proteolysis cloud is clearly seen following TCA and thermal treatment with the Stabilizor T1 device (Denator AB) (Grassl et al., 2009)

depth and 30 mm in diameter are placed inside a teflon foil cartridge, which then enters the treatment chamber. A class 1 laser evaluates tissue size and determines an appropriate treatment time, where upon aluminium heat blocks press the sample and a vacuum seal (5–10 mbar) ensures that these blocks are in constant contact with the teflon encapsulated tissue. Fresh tissue samples are heated to 95°C, under vacuum, for 50 s, while frozen samples require longer times up to 100 s.

Svensson et al. have shown that phosphatase activity is decreased to background levels following thermal stabilisation and also that the phosphorylation status of a number of proteins was maintained following stabilisation up to 2 h post treatment (Svensson et al., 2009). 2-DE analysis showed yeast cell extracts to have a decrease in low MW proteins with a parallel increase in high MW proteins following thermal stabilisation, when compared to direct cell harvesting and freezing in liquid nitrogen. Specific analysis of a particular protein, enolase 1, supported the hypothesis that many of these low MW spots which disappeared following thermal stabilisation were indeed fragments of higher MW proteins (Grassl et al., 2009), see Fig. 13.1. Analysis of post-mortem tissue showed heart tissue proteome to be preserved equally by thermal stabilisation or snap freezing (Robinson et al., 2009). However, post-mortem brain tissue is known to be particularly susceptible to proteolytic activity (Harrison et al., 1995; Li et al., 2004; Vawter et al., 2006) and thermal stabilisation with the Stabilizor T1 demonstrated significant preservation of post-mortem brain tissue. A dramatic decrease in low MW proteins and a concurrent increase in abundance of high MW proteins was observed. Many of the low MW proteins were subsequently identified as protein fragments which showed reduced expression following stabilisation (Robinson et al., 2009).

13.2.2 Protein Lysis

A thorough reproducible sample preparation is vital for a comparative quantitative 2-DE analysis. As sample diversity is so large, optimisation of sample preparation protocols are necessary for each study and no universal standard procedure exists. Cell or tissue samples must first be disrupted which can be carried out by homogenisation, grinding with a pestle and mortar, sonication, freeze-thaw cycles or by detergent lysis. Samples with robust cell walls, such as yeast, can be homogenised using glass beads for example (Grassl et al., 2009). Softer tissue samples, such as heart tissue, which contain no cell walls, are routinely prepared by grinding while frozen with a pestle and mortar in liquid nitrogen (Robinson et al., 2009). A gentler cell disruption technique is applied in the preparation of intact organelles. However, recent studies have shown that mitochondria are best isolated by sonication (Chaiyarit and Thongboonkerd, 2009). The homogenised tissue or cell extract is then solubilised in a lysis buffer which will prevent protein aggregation, precipitation or modification which can ultimately result in sample loss. Lysis must also occur rapidly and efficiently to prevent the action of proteases and other enzymes (e.g. phosphatases and kinases) which can result in alterations in both p*I* and MW of proteins. If all these factors are taken into account during sample preparation for 2-DE, the presence of artifactual spots will be minimal.

13.2.3 Lysis Buffers

In order for proteins to be separated by their charge state or isoelectric point, and represented as a single spot on a 2-DE image, a typical 2-DE solubilisation or lysis buffer must contain: (i) a chaotropic or denaturing reagent, (ii) a detergent (surfactant) compatible with IEF, (iii) a reducing agent, (iv) protease inhibitors, (v) carrier ampholytes, and (vi) must be buffered to approximately pH 8 (O'Farrell, 1975).

13.2.3.1 Chaotropes

A chaotropic or denaturing agent disrupts hydrogen bonds and hydrophilic interactions which unfolds the secondary structure of a protein. Typically with 2-DE analysis, the neutral chaotrope urea (8–9 M) is used, as the intrinsic charge of the proteins remains unaltered. Adding the more potent chaotrope, thiourea, increases protein solubility (Rabilloud et al., 1997). However, as thiourea is poorly soluble in water, typically 2 M thiourea is combined with 7 M urea to achieve the best solubilisation. However, solutions containing urea should not be heated above 37°C as this will lead to degradation of urea to form ammonium isocyanate, resulting in carbamylation modification of lysine residues within proteins which can lead to alterations in p*I* (Gorg et al., 2004; McCarthy et al., 2003).

13.2.3.2 Detergents

Detergents disrupt hydrophobic interactions and so are necessary to efficiently solubilise complex biological samples by minimising protein aggregation and precipitation. Detergents are available in a number of charged states such as; negatively charged (e.g. SDS), uncharged (e.g. Triton X-100), or both positively and negatively charged (zwitterionic, e.g. CHAPS). SDS is known to solubilise proteins effectively, however its charge properties are incompatible with isoelectric focusing (IEF) in the first dimension of 2-DE. The negative charge of SDS masks the inherent charge of the proteins, so that they no longer focus at their p*I*, leading to horizontal streaking (Bodzon-Kulakowska et al., 2007; Luche et al., 2003). The zwitterionic detergent, CHAPS is routinely used in lysis buffers for downstream 2-DE analysis (Trott et al., 2009). However, if membrane proteins are to be resolved by 2-DE, alternative detergents are required. Sulphobetaines such as SB 3-10 or ASB 14 have been used to solubilise membrane proteins for 2-DE analyses (Molloy, 2000; Santoni et al., 2000).

13.2.3.3 Reducing Agent

The standard reducing agent, dithiothreitol (DTT) disrupts intra- and intermolecular disulfide bonds between cysteine residues within a protein to further unfold its complex structure. Once reduced, proteins must be prevented from re-oxidation before separation by 2-DE. Reducing agents are typically added in excess (>100 mM). Other reducing agents such as TCEP and TBP can also be used and which reducing agent best suits a particular sample must be optimised (Getz et al., 1999; Khoudoli et al., 2004).

13.2.3.4 Protease Inhibitors

There are several different types of proteases including serine-, cysteine-, and metallo- proteases among others which have specific inhibitors and should be included in lysis buffer preparation. Several commercially available protease inhibitor cocktails exist which account for many of the proteases present. Many of these cocktails include phenylmethylsulfonyl fluoride (PMSF), leupeptin, and EDTA which inhibit serine-, cysteine-, and metallo-proteases respectively.

13.2.3.5 Carrier Ampholytes

Carrier ampholytes are low molecular weight components with both amino and carboxyl groups. Carrier ampholytes of the appropriate pH range to match the IEF separation are added to the sample to improve solubility and also to act as a cyanate scavenger.

13.2.4 Sample Clean Up Prior to Electrophoresis

As some cellular components, such as DNA, and salts can interfere with 2-DE, care must be taken with solubilisation or lysis buffer conditions. In particular IEF is salt sensitive and thus salt ions can interfere with protein separation by causing protein precipitation during IEF as well as interfering with focusing of proteins during the first dimension (Bodzon-Kulakowska et al., 2007). Since salt can increase the conductivity of the gel, IEF in some instances can actually stop if the salt concentration is too high. Therefore it is recommended that salt ions should be removed if the concentration is >50 mM. There are various methods available for the removal of salt ions including precipitation of proteins with trichloroacetic acid (TCA) or ice cold acetone, or using commercially available spin columns. Alternatively, a commercial

2-D clean-up kit (GE Healthcare) has proven to be an efficient, easy and an effective way of getting rid of various impurities such as salts, lipids and nucleic acids and moreover aids with efficient IEF and minimises gel streaking.

The presence of highly abundant proteins in samples, such as albumin, can prove problematic since these can affect protein separation. Furthermore these abundant proteins can greatly affect the detection of lower abundance proteins by either preventing their loading onto the gel or masking their 2-D profile. Various commercial depletion kits are available to remove abundant proteins but it is important to note that the specificity of the kits cannot be guaranteed and thus there is the potential that other proteins will bind and also be removed. This will be discussed further below in 13.2.5.

13.2.5 Prefractionation Prior to 2-DE

The analysis of tissue or cell samples is made all the more difficult by the fact that 90% of the protein mass of a typical cell proteome is composed of only 10% of the 10,000–20,000 different protein species (Lilley et al., 2002). Removal of these highly abundant proteins is often carried out to try and achieve an in-depth analysis of a particular sample as is routinely done with serum samples. Other fractionation techniques used prior to 2-DE include isolation of individual organelles or tissue sections by chemical means or by laser capture microdissection (LCM), or enrichment of a particular component of the cell such as membrane proteins. Complex samples can also be fractionated in solution according to their *pI* and subsequently separated on zoom IPG IEF gels which will also be discussed in this section.

13.2.5.1 Removal of Abundant Proteins

Protein concentrations may vary over 10 orders of magnitude within biological samples (Anderson et al., 2009). Moreover, highly expressed proteins can mask other lower abundant species in 2-DE. Among these highly expressed proteins which are found in plasma, serum and cerebrospinal fluid are albumin, immunoglobulins (IgGs) and transthyretin (Bodzon-Kulakowska et al., 2007). Antibody based approaches to achieve immunodepletion of these high abundant proteins are frequently used. However, affinity based methods such as Cibracon blue which binds albumin, or protein A or G which have a strong affinity for IgGs, among other depletion methods have also been used and are reviewed in Kim et al. (Kim and Kim, 2007). Commercially available immunodepletion columns exist, such as the Multiple Affinity Removal System (MARS, Agilent Technologies), which is available in several formats that can deplete serum samples of the 6–14 most abundant proteins (MARS, Agilent Technologies) (Echan et al., 2005; Lei et al., 2008), see Fig. 13.2. However, immunodepletion can also result in removal of additionally bound proteins which may be of interest and must be considered (Righetti et al., 2005). Recently, a combinatorial ligand library based method has been developed and applied to biological samples (Boschetti and Righetti, 2008; Righetti et al., 2005). A diverse library of peptides, which act as bait and bind specific proteins,

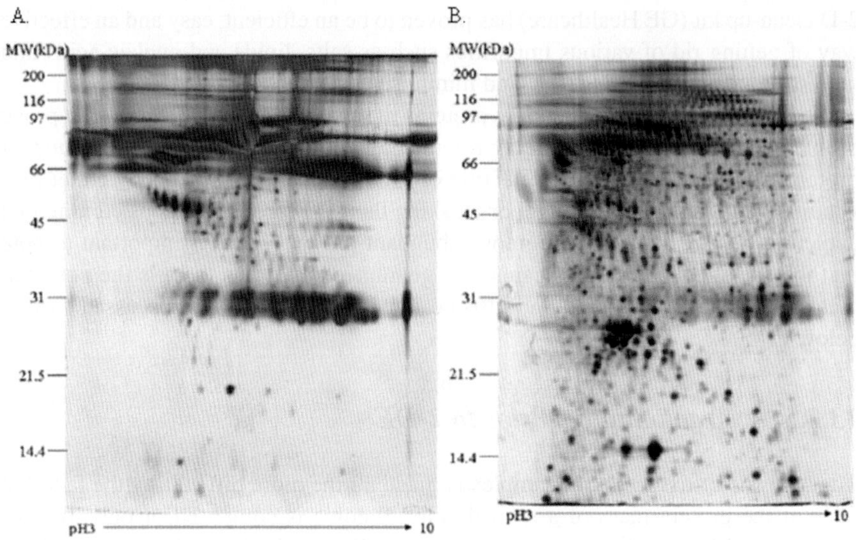

Fig. 13.2 Separation of serum by 2-DE following immunodepletion using the MARS 6 column (Agilent Technologies). Serum proteins which bound to the column (**a**) and serum which has been depleted (**b**) is shown (Lei et al., 2008)

are coupled to spherical porous beads, one peptide per bead. The complex proteome to be "equalised" is passed over these beads and a sampling of all proteins present is achieved with decreased expression of the most abundant proteins and an increase in those that are expressed at a low level. This particular procedure has had success with many complex biological samples including serum. Serum is of particular importance as it is thought to contain all proteins found within the body and will most likely reveal disease biomarkers (Anderson and Anderson, 2002). Using this method 3,300 proteins were detected within plasma (Righetti et al., 2006) which included > 1,500 unique proteins not reported previously.

13.2.5.2 Organelle Isolation

Cellular organelles can be isolated by many procedures with centrifugation being the most common. Nuclei are isolated in this way (Turck et al., 2004), while mitochondria require differential centrifugation (Chaiyarit and Thongboonkerd, 2009) or sucrose gradient centrifugation (Lopez et al., 2000b; Righetti et al., 2005). Commercial kits such as the ProteoExtract subcellular fractionation kit (Merck Biosciences), based on selective solubilisation using solutions of increasing strength, have been used to isolate individual cellular subproteomes, in particular the cytosolic, membrane, nuclear and cytoskeletal components (Abdolzade-Bavil et al., 2004; Murray et al., 2009). LCM allows for the enrichment of a particular set of proteins by selectively isolating the area of interest from sections of tissue using a laser and collecting the isolated tissue into a sample tube. For 2-DE analysis, frozen tissue samples are cut into sections at –20°C and stained to visualise the area of interest. LCM is of particular use when proteomes of malignant tumours or of atherosclerotic plaques are to be investigated (Bagnato et al., 2007; Craven and Banks, 2003; Domazet et al., 2008).

13.2.5.3 Membrane Isolation

The membrane proteome is of immense interest as it is the first point of contact within a cell before downstream signalling is initiated. However due to their hydrophobicity and the fact that they are embedded within lipid bilayers, analysis of cellular membrane proteins has proven to be a difficult task. Two-dimensional gel analysis of membrane proteins has also proven difficult due to poor solubilisation of these proteins in IEF compatible buffers, with resulting protein precipitation during the first dimension separation (Henningsen et al., 2002). Sulfobetaine detergents, such as SB 3-10 or ASB 14, have shown increased efficiency in the solubilisation of membrane proteins for 2-DE analysis (Molloy, 2000; Santoni et al., 2000). More recently, integral membrane proteins such as the sacroplasmic reticulum have been solubilised using the detergent DHPC (Babu et al., 2004). However it is still agreed that membrane proteins with multiple transmembrane domains are poorly represented using IEF based 2-DE (Peirce et al., 2004; Rabilloud et al., 2008). Replacement of the IEF with electrophoresis using charged (cationic or anionic) detergents has been suggested (Braun et al., 2007). It is possible to enrich for membrane proteins prior to 2-DE separation. Triton X-114 for example has successfully isolated the membrane proteome of cardiac ventricular tissue (Donoghue et al., 2008) and has been used to separate membrane and non membrane cellular components by 2-DE (Taylor et al., 2000).

13.3 Protein Separation by 2-DE

13.3.1 IPG Strips

The development of immobilized pH gradients (IPG) prompted major advances in IEF and commercial IPG strips are now readily available with quite an extensive choice of pH gradient. Today, IPG strips can be cast to cover any desired pH range between pH 2.5 and 12. Similarly, several lengths are also commercially available between 7 and 24 cm although longer strips of up to 54 cm have also been produced in the laboratory (Righetti and Drysdale, 1973). Commercially available IPG strips offer high reproducibility and ease of use in 2-DE. The development of rehydration and focusing manifolds for IEF has also aided in reproducibility and has reduced the level of experimental variation since up to 12 IPG strips can be focused simultaneously (Gorg et al., 1999).

Although, traditionally IEF for 2-DE is carried out using wide-range IPG gel strips spanning pH 3–10 or pH 4–7 (Trott et al., 2009), more complex samples have proven to be difficult to separate using these wide pH ranges and further in depth 2-DE analyses have used very narrow range IPGs (Westbrook et al., 2001). The use

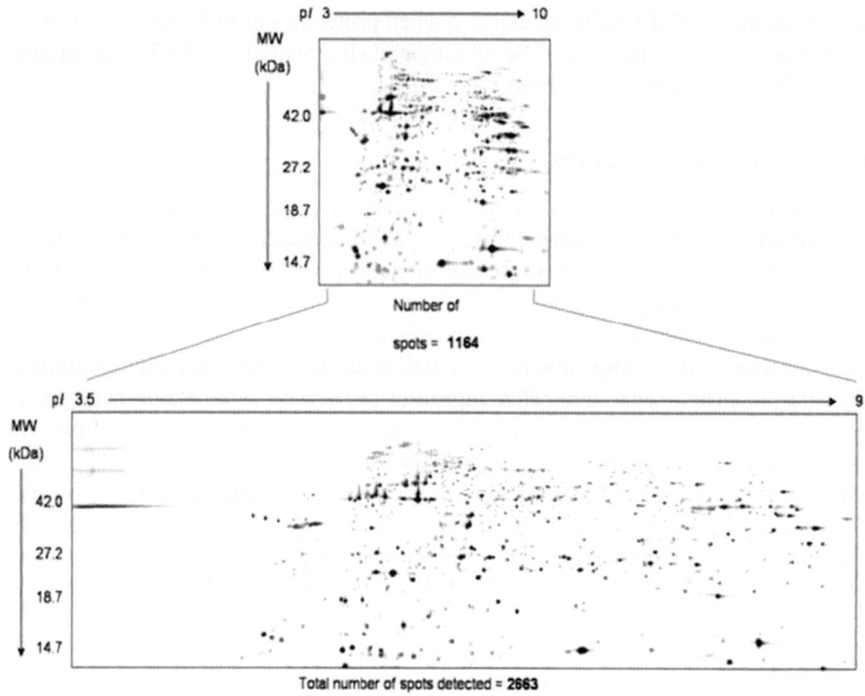

Fig. 13.3 Comparison of human heart left ventricle proteins separated using pH 3–10 IPG strips. Separation on a single IPG strip pH 3–10 detected 1,164 spots, while a combined image achieved from the separation of the tissue using narrow range IPG strips between pH 3.5–9, detected 2,663 protein spots (Westbrook et al., 2006)

of overlapping narrow IPGs, or "zoom-in gels" enables further pulling apart of the protein spots and allows for optimum resolution of proteins with each single spot representing a single protein. Furthermore, protein loadings can be increased significantly allowing less abundant proteins to be visualised and more importantly be detected by MS (Gorg et al., 2000). Previous studies have demonstrated a three-fold increase in the number of spots detected using this method in comparison to separation with IPG 3-10 (Wildgruber et al., 2000). In addition, Westbrook et al. showed not only an increase in spot number but upon MS analysis, an enhanced number of identifications in proteins species and isoforms (Westbrook et al., 2001, 2006), see Fig. 13.3. These very narrow IPGs can have a range as low as one pH unit (e.g. pH 3.5–4.5) or even ultra narrow ranges (e.g. pH 4.9–5.3). Each p*I* range can be investigated and then a comprehensive 2-D map can be established giving a thorough overview of proteins. However, broader range IPGs should be considered for first look 2-DE analyses in order to determine which pH region is of most interest.

An alternative approach is to pre-fractionate the protein sample according to p*I* in the liquid phase, using a device such as Zoom IEF Fractionator (Invitrogen) which first separates the sample in solution according to p*I*. Following this, each fraction can be analysed by 2-DE using a specific narrow IPG which corresponds to the pH range of each fraction. Prefactionation using the Zoom IEF Fractionator resulted in a 3.5 fold increase in spot detection from an *E. Coli* sample when compared to separation on broad range (pH 3–10) IPG strips (Richardson et al., 2008). Moreover, this fractionation step avoids significant sample loss which would occur if the sample was applied directly to the narrow range IPGs.

13.3.1.1 Sample Application to IPG Strips

Samples are usually applied to IPG strips by an "in-gel rehydration" approach, i.e. the strips are rehydrated with buffer (8 M urea, 0.5–4% nonionic or zwitterionic detergents (e.g. 2% CHAPS), a reductant (e.g. 0.4% DTT) and 0.5% IPG buffer of Pharmalyte 3–10) containing the appropriate amount of the solubilised protein sample (Rabilloud et al., 1994). However, this method is not recommended for samples with very alkaline or hydrophobic containing proteins or very high molecular weight proteins. In these instances, samples should be introduced onto the strips by a "cup loading" method. Here IPG strips are rehydrated overnight and the sample is then applied into disposable plastic cups which are placed on the surface of the IPG strip, usually at the anodic end.

Problems can still be encountered in the separation of basic proteins (pH 8.5–11). This is caused by DTT, the essential reducing agent, which has a weakly acidic p*K* value and hence migrates towards the anodic end during IEF (Pennington et al., 2004). The resulting protein streaking often renders this basic region of the gel unusable for further analysis. However, various technological improvements have been made. These include the addition of hydroxyethylene-disulfide or DeStreak reagent (GE Healthcare) which inhibits streaking caused by the depletion of DTT. Furthermore, "paper-bridge loading" has overcome the associated problem of protein precipitation at the sample cup during "cup loading". The technique originally reported by Sabounchi-Schutt et al. was further developed by Kane et al. specifically for basic protein separation (Kane et al., 2006; Sabounchi-Schutt et al., 2001). Instead of applying the sample to a cup, a paper strip is applied to the cathode end of the IPG strip and proteins are slowly loaded to the paper medium and using a stepwise voltage increase subsequently introduced into the gel strip. This method offers high resolution and allows significantly more sample to be applied in comparison to cup loading, which is clearly limited by the size of the cup, an important aspect for downstream MS analysis. Furthermore the paper bridges can be re-used for separation of the residual proteins using other pH ranges such as IPG 4–7 gels, crucial if the sample is limited in availability (Kane et al., 2006), see Fig. 13.4.

13.3.1.2 Running Conditions for IEF

Various factors must be taken into consideration when optimizing conditions for separation in the first dimension. Care must be taken not to under- or over-focus the strips as this can lead to horizontal streaking. Moreover, the latter can also cause water evaporation at the strip surface and can result in loss of proteins or

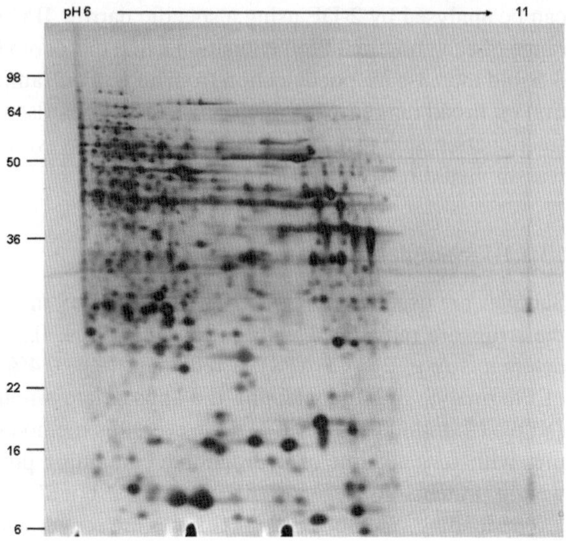

Fig. 13.4 2-DE map of the dorsolateral prefrontal cortex (dlPFC) region of the human brain. Protein extracts were separated on pH 6-11 IPG strips using the paper bridge loading technique. Proteins were subsequently separated on 12% SDS polyacrylamide gels and visualised by silver staining (McManus et al., 2010)

distorted spot patterns. Thus it is important to establish optimum focusing time, taking into account the pH and length of the IPG strip, for each protein sample type. The temperature during IEF has been shown to be crucial as it can greatly affect spot patterns and reproducibility. Gorg et al. has reported the optimum temperature to be 20°C (Gorg et al., 1991) and today instruments for running IPG strips such as IPGphors (GE Healthcare) now contain temperature controllers which maintain the temperature at 20°C throughout the entire run.

13.3.1.3 Equilibration of IPG Gel Strips

Following the first-dimension IEF separation, IPG strips are equilibrated for 10–15 min in a Tris-HCl buffer containing 2% SDS, 1% DTT, 6 M Urea and 30% glycerol (Gorg, 1987). This step crucially allows the separated proteins to be reduced and to interact with SDS, which helps with protein transfer to the second dimension. The urea and glycerol limit the electroendosmotic effect. A further incubation of 10–15 min in the same buffer containing 4% iodoacetamide instead of DTT is then preformed. The iodoacetamide is essential for alkylating the sulfhydryl groups of the proteins. Failure to do this can cause reoxidation during the second dimension leading to vertical spot streaking. In addition iodoacetamide helps to alkylate any free DTT. This equilibration step is also recommended for spot identification by MS.

13.3.2 Second Dimension Electrophoresis

Separation in the second SDS-PAGE dimension can be preformed on either horizontal or vertical systems (Gorg et al., 1995). Vertical systems are generally used for large scale analysis since they allow the parallel running of multiple gels (e.g. up to 20) and can therefore offer a higher throughput as well as maximizing the reproducibility. Horizontal setups make good use of the commercially available ready-made gels. Gels sizes are commonly around 20 × 25 cm^2 with a thickness of 1–1.5 mm. Laemmli's discontinuous buffer system is the most widely used buffer for separation in the second dimension (Laemmli, 1970). However modifications to Laemmli's buffer can be made or alternate buffers for different sample types, such as borate buffers for the separation of highly glycosylated proteins (Patton et al., 1991). Typically the second dimension is performed overnight at a constant power of 1–2 W per gel.

13.4 Protein Detection and Quantitation

13.4.1 Protein Spot Detection

In order to visualise and analyse quantitative changes in protein expression by 2-DE, each 2-D gel must be stained and digitally scanned. Several different protein stains exist which are compatible with proteomic studies. However, these differ in their sensitivity as well as their ability to detect protein spots at concentrations over several orders of magnitude (Riederer, 2008). Another important element of protein stains is the requirement for a high linear dynamic range in order that quantitative expression changes can be determined accurately. Protein concentration in biological samples can be greater than 10 orders of magnitude (Anderson and Anderson, 2002; Anderson et al., 2009). For proteomic studies, it is also vital that the staining used is compatible with mass spectrometry and will allow identification of the detected protein spots. As well as detecting protein spots on a 2-DE gel, protein modifications such as phosphorylation can also be visualised with certain specific protein stains. Before a gel is stained, it must first be fixed in a solution containing methanol and acetic acid (50 and 10% respectively for example). Subsequent staining should then be performed with gentle agitation so as to avoid uneven staining of the gel. The various staining methods currently available are discussed in detail below.

13.4.1.1 Coomassie Brilliant Blue R

Coomassie brilliant blue R protein stain was originally introduced in 1963 (Fazekas de St Groth et al., 1963). The detection limit is about 0.2–0.5 μg of protein/2-DE spot, and visualisation requires destaining as the entire gel is stained non-specifically by Coomassie staining in an overnight incubation. After destaining, specific proteins spots can be clearly visualised. Colloidal Coomassie staining has increased sensitivity with 8–10 ng of protein (Neuhoff et al., 1988). While Coomassie staining methods are easy to use and cost effective, their weaknesses include limited sensitivity and low linear dynamic range (10–30 fold changes detected), thus limiting their use in quantitative 2-DE analysis.

13.4.1.2 Silver Staining

Silver staining requires the binding of silver ions to proteins within the gel. Silver nitrate has a detection limit of approximately 2–5 ng of protein or 0.1 ng of protein/2-DE spot and a low dynamic range with 10–40 fold differences in protein concentration detected (Gorg et al., 2004; Merril et al., 1981). For further proteomic analysis by MS, glutaraldehyde and formaldehyde, used in fixing or sensitising the gel before staining, must be omitted from this multistep staining procedure (Yan et al., 2000). However, omission of these aldehydes, which result in protein crosslinking, reduces the sensitivity of this stain. A major limitation of this technique is the fact that the end point of silver nitrate staining is subjective and thus it is not ideal for quantitative analysis and this also hinders gel to gel reproducibility (Gorg et al., 2004). Despite these drawbacks, silver staining is routinely used in 2-DE analyses and staining can be achieved in approximately 5 h.

13.4.1.3 Fluorescent Staining

Fluorescent non-covalent dyes have great sensitivity and a linear quantitation range over several orders of magnitude making them superior to either Coomassie or silver stain (Lopez et al., 2000a). Taking only 1 h to efficiently stain a gel is also advantageous. One such example, Sypro Ruby protein stain can detect proteins in the range of 1–2 ng/2-DE protein spot and over three orders of magnitude (Berggren et al., 2000; Gorg et al., 2004). Sypro Ruby binds primary amines of lysines. Similar dyes which are more cost effective have emerged (Rabilloud et al., 2001) and appear to have great sensitivity and larger orders of magnitude while still allowing subsequent analysis by mass spectrometry. These include Deep Purple and Lightning Fast (Mackintosh et al., 2003; Riederer, 2008). However, Sypro Ruby is still the fluorescent dye of choice for protein detection following protein separation by 2-DE (Harris et al., 2007).

13.4.1.4 2D-DIGE

Two-Dimensional Difference in Gel Electrophoresis (2D-DIGE) was first described by Ünlü et al. in 1997 (Unlu et al., 1997) as a means of overcoming problems associated with 2-DE variability. In this approach, prior to electrophoresis, commercially available fluorescent dyes (GE Healthcare) are covalently bound to the protein samples. These cyanine dyes are highly sensitive and are both size- and charge-matched, yet they are spectrally distinct enabling high accuracy with no cross talk between dyes. Both minimal labelling and saturation labelling are feasible as both methods are found to minimally effect the molecular mass migration of proteins. Minimal labelling utilises Cy2, Cy3 and Cy5 dyes, binds a maximum of one dye molecule per protein and is lysine reactive. In contrast, saturation labelling Cy dyes react with cysteine residues and 100% of all reactive residues are dye-coupled. The protein samples and the internal standard (a pool of equal amounts of all samples in the study, routinely labelled with Cy2) are each labelled with a different dye. DIGE allows multiple samples to be run on the same gel and thus there are several advantages: (i) experimental variations in spot quantity are minimised due to the utilisation of internal standards; (ii) alterations in protein abundance can be detected, even small differences and also (iii) gel to gel variation is eliminated. Moreover, it is possible to compare these protein patterns to previously generated gels which have been stained with Sypro Ruby, Coomassie blue or silver nitrate techniques.

Two quality fluorescent gel imaging systems are currently available for DIGE image capture; laser scanning of gels in low-fluorescence glass plates and a CCD system that uses wide field illumination on open gels. Both offer sensitive, wide dynamic range images of 2D-DIGE gels. The images are scanned at different emission wavelengths, thereby enabling, multiple images corresponding to the differently labelled samples to be generated from a single 2-DE gel. Crucial to the use of DIGE is experimental design, focusing particularly on the number of biological replicates required (Karp et al., 2005). From a data analysis point of view, Karp et al. also suggest using a 2 dye design rather than a 3 dye approach with minimal labelling due to a bias in *p* value distribution (Karp et al., 2007). Cy2, utilised to label the internal standard produces a relatively weak signal in comparison to the other dyes and therefore using Cy3 and Cy5 only along with good statistical analysis will overcome this problem.

To date, 2D-DIGE has been used in various cells and tissues to greatly enhance our understanding of both normal and disease biological processes (English et al., 2009; Trott et al., 2009). It offers a sensitive and reproducible method while simultaneously controlling for differences in sample preparation and separation.

A CyDye immunoblotting methodology, combining DIGE and Enhanced Chemiluminesence (ECL) Plex labelling has recently been described by our laboratory which allows total proteome expression to be profiled in conjunction with the co-detection of specific immunolabelled protein(s) (McManus et al., 2009). Moreover this is a simple, rapid yet reproducible means of protein confirmation following software expression analysis and/or MS identification. Protein samples are pre-labelled with "minimal labelling" Cy3 and Cy5 reagents, and then separated by 1-D or 2-D gel electrophoresis. Subsequently, the proteins are transferred by electroblotting onto membranes and probed with a specific primary antibody of choice. The membranes are then incubated with wavelength specific ECL Plex fluor-labelled species-specific secondary antibodies (GE Healthcare). Thus, both the total protein expression profile and the specific immunoreactive protein(s) can be easily visualised. This technology offers significant advantages including; high sensitivity as a result of the CyDye labelling, decreased amount of sample required, reduces the number of replicate gels required and most notably allows specific protein spots of interest to be easily correlated with other analytical or micro-preparative 2-D gels for further differential expression analysis or mass spectrometry identification. Detection of multiple protein targets is also possible with the ECL Plex system since secondary antibodies labelled with different CyDyes are detected at specific wavelengths suggesting an added quantitative capability to this approach, see Fig. 13.5.

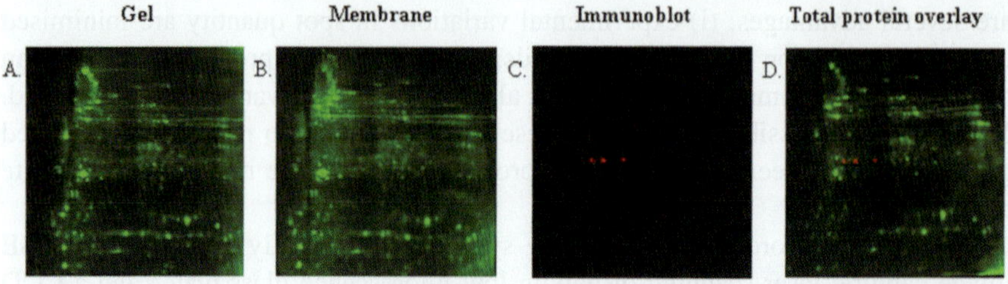

Fig. 13.5 ECL Plex 2-D CyDye blotting. Human heart (*left ventricle*) was labelled with Cy3, separated by 2-DE (**a**), transferred onto nitrocellulose membrane (**b**), and probed with Hsp27 primary antibody with a Cy5-conjugated secondary antibody (**c**). Total protein expression combined with specific Hsp27 protein detection was visualised at specific wavelengths for both CyDyes (**d**). Modified from (Donoghue et al., 2006)

13.4.1.5 Specific Staining for Post Translational Modifications (PTM)

A powerful attribute of 2-DE is the ability to detect PTMs due to the knowledge that most modifications introduce charged groups into the protein and so result in alterations in p*I*. This addition of charged groups also results in a change in mass of a protein, however this is often too small to detect by 1-DE. As these modifications can result in alteration of protein activities as well as interactions and subcellular location, their detection and quantitation is an important part of proteomic research. PTMs include phosphorylation and glycosylation, which are often considered to be the most important, in addition to acetylation, lipidation, sulfation among many others. Phosphorylation, addition of a phosphate group, of amino acid residues can be visualised on a 2-D gel as a horizontal series of well separated spots. While specific phosphorylation events such as the modification of tyrosine, serine or threonine residues can be visualised by probing a 2-D Western blot with specific antibodies, specific gel stains such as Pro-Q Diamond are also commercially available and which can detect all phosphorylation events (Steinberg et al., 2003). Pro-Q Diamond, which is compatible with subsequent MS analysis, can detect phosphorylated proteins at concentrations of 1–2 ng. However, non-phosphorylated proteins can also be detected, albeit to a lesser extent (Murray et al., 2004).

Glycosylation involves the incorporation of sugar moieties to a given protein. The specific stain, Pro-Q Emerald, can detect glycosylated proteins at concentrations as low as 5–18 ng (Hart et al., 2003). Lectins, labelled fluorescently or otherwise, can also be used to bind glycoproteins and allow for their detection following 2-DE.

13.4.2 2-DE Image Analysis and Quantitation

In order to detect differential protein expression changes between a set of 2-D gels, each 2D gel image must be compared with all the others in the experimental set. Before importing into software which will allow quantitation to be carried out, 2-DE images must be digitally scanned. Standard protein stains can be visualised using most digital flatbed scanners which record light transmitted through or reflected from the gel. However, DIGE labelled proteins separated by 2-DE require fluorescent scanners (e.g. Typhoon Imager 9400, GE Healthcare), where the gel is scanned at a specific excitation wavelength and the fluorescence emitted is measured. The images are subsequently imported into dedicated 2-DE computer software packages to (i) align the images, that allows for gel–gel variation which can occur during the 2-DE process, (ii) normalise the staining intensity and remove background staining, (iii) detect spots, which generally requires some manual editing as sections of the gel can have both horizontal and vertical streaking, and (iv) highlight differentially expressed spots with statistical confidence.

Several commercial 2-DE image analysis software packages are available such as DeCyder (GE Healthcare) or Progenesis Same Spots (Nonlinear Dynamics). Many other image analyses tools are also available and have been compared and contrasted to each other and to the two aforementioned (Kang et al., 2009; Morris et al., 2008; Raman et al., 2002; Rosengren et al., 2003). In most of these analyses software tools, the images are uploaded, spots detected and further subjected to spot editing before the "volumes" for each spot are computed. Typically a reference gel is chosen and all gels are subsequently matched to this. Progenesis Samespots has a particular user friendly setup, allowing the user to input vectors to align some spots manually at various sections of the gel and the programme automatically detects the remaining spots, adds remaining vectors and aligns the gel images. This particular step is quite laborious but care must be taken to align all images accurately. The software packages carry out statistical tests (Analysis of Variance (ANOVA)) based on spot size and intensity and the spots are then assigned *p*-values, fold changes between groups, and False Discovery Rates (FDR) or *q*-values. Principle Component Analysis (PCA) is also carried out by these bioinformatical tools which allows the user to see if any particular sample or gel is an outlier or not typical of the particular set. Differences in protein abundance can then be detected above a certain threshold, such as a p-value > 0.5.

13.4.3 2-DE Proteome Databases

Proteome databases allow experimental and mass spectrometric data to be uploaded in order to disseminate and make proteomic data publicly available. The Proteomics Identification Database (PRIDE) database is one of the main public repositories of mass spectrometry derived data (Vizcaino et al., 2009). Two-dimensional gel databases have recently become available, displaying annotated representative 2-DE reference maps for a variety of cell and tissue types. A list of the publicly available databases is found at http://www.expasy.org/ch2d/2d-index.html. This link highlights the World-2DPAGE Constellation which is composed of the established WORLD-2DPAGE list of 2-D PAGE database servers, the World-2DPAGE Portal, which displays the list of reference maps available, and the WORLD-2DPAGE Repository, which displays the publication links to each reference map and how the sample was prepared (Hoogland et al., 2008).

The SWISS-2DPAGE is the largest database at present, currently containing 36 different reference 2-DE proteome maps from 7 different species, including *S. Cerevisiae* and *E. Coli,* as well as many different human tissue and organs, such as kidney and liver (Hoogland et al., 2000). This database also annotates the protein identifications giving information on physiological and pathological functions as well as experimental data (MW and p*I*, amino acid composition and peptide masses), and citations. PROTICdb is a specific 2-DE database designed to store and analyse plant proteome data (Ferry-Dumazet et al., 2005). This database was created to record data on experimental setup, reliability of results and display protein spot identification but does not integrate MS results. More recently, a 2-DE database was set up for *E. Coli*, 2Dbase (Vijayendran et al., 2007). At an organelle level, a 2-DE reference map of mitochondrial proteins from human placenta has been produced (Rabilloud et al., 1998) and many other 2-DE reference maps are available on line.

13.5 Protein Identification by MS Analysis

13.5.1 Spot Cutting

Spots can be excised from 2-D gels manually or using a robotic spot cutter. When spots are excised care must be taken to remove the protein spot with minimal contamination from neighbouring spots or excess polyacrylamide gel, in order to facilitate an easy digestion while also taking care to limit the contaminating environmental factors.

For manual cutting the gel must be visibly stained (e.g. silver or Coomassie as detailed in Section 4.1). Fluorescently labelled spots such as DIGE gels can be post-stained to facilitate comparison with silver stained preparative gels. Manually, spots can be excised with a sterile scalpel, pipette tip or "spot cutter" in a laminar flow hood to prevent keratin contamination. A 1 mL pipette tip, with the tip removed approximately 1–3 mm from the top to obtain a suitable diameter for the gel spot can be used for larger gel spots, while 0.2 mL tips can be utilised for smaller spots. It is important to ensure that all tips are dust free. Commercially available spot and band cutters can be more convenient for excision of large numbers of spots. However, excising manually from a gel can be quite cumbersome especially if there are a significant number of spots to be excised. Firstly, the spots should be located and marked on a printed image of the 2D gel, subsequently located on the gel and finally the gel plugs transferred to a microtitre plate or fresh Eppendorf tube. Great caution must be taken when using this method since errors can occur quite easily.

Robotic spot pickers are directly linked to the image analysis software and thus can detect both visible and fluorescent stained gels. Automated spot pickers will excise the spots from the gel slab and transfer them directly to predefined wells in microtitre plates. This process can be achieved using a CCD camera or alternatively the pixel co-ordinates can be transferred from the image analysis system to the robotic spot picker. A major advantage to the use of robotic cutters is the fact that spot excision takes place in one unit eliminating gel, protein and plate handling and moreover greatly reducing environmental contamination.

13.5.2 Trypsin Digestion of Protein Spots

Once the gel plug has been successfully removed from the 2-D gel, the protein is then digested with a protease, typically trypsin and identified by MS. Various methods and adaptations for digesting the protein with trypsin are available, but these are ultimately based on the "in-gel digestion" technique reported by Wilm et al. in which the proteins are enzymatically digested in situ within the gel pieces, thereby allowing the resulting peptides to be extracted (Wilm et al., 1996).

The initial stage of digestion involves the removal of all components of electrophoresis as the buffer used is not compatible with mass spectrometry instruments. Prior to digestion, the gel piece is dehydrated with acetonitrile and dried under a vacuum with centrifugation. The proteins are then subjected to reduction with DTT and alkylation with iodoacetamide. These steps break the disulfide bonds between cysteine residues and prevent them from reforming. The gel pieces are then swelled in a solution (usually ammonium bicarbonate) containing trypsin, enabling the enzyme to enter into the gel pieces. The digestion can be for a minimum of 3–4 h but is generally carried out overnight at 37°C. The peptides produced can then be extracted as they are now small enough to diffuse efficiently through the polyacrylamide gel piece, whereas previously the intact protein was too large. This extraction is carried out with increasing concentrations of acetonitrile (30% and then 70%) until the gel plugs turn white in colour. The trypsinised gel pieces are centrifuged and the supernatant containing the peptides removed, dried under vacuum and resuspended in a buffer which must be compatible with the intended mass spectrometer (see Table 13.1).

Table 13.1 Trypsin digestion of 2-DE protein spots

Digestion	Protocol
Destain	1. 30 mM $K_3Fe(CN)_6$,100 mM $Na_2S_2O_3$ (1:1) shake 20 min 22°C
	2. ddH_2O shake 5 min 22°C × 3
Wash	3. 200 mM NH_4HCO_3 shake 10 min 37°C
	4. 200 mM NH_4HCO_3/ACN (2:3) shake 10 min 37°C
	5. 50 mM NH_4HCO_3 shake 10 min 37°C
	6. ACN shake 10 min 37°C
Reduction	7. 10 mM DTT in 100 mM NH_4HCO_3 shake 60 min at 56°C
Alkylation	8. 50 mM IAA in 100 mM NH_4HCO_3 shake 30 min at 22°C
Wash	9. 100 mM NH_4HCO_3 shake 15 min at 37°C
	10. 20 mM NH_4HCO_3/ACN (1:1) shake 15 min at 37°C
	11. ACN shake 10 min at 37°C
Digest	12. 0.2 μg Trypsin in 200 mM NH_4HCO_3 shake overnight at 37°C
Extraction	13. 30% ACN/0.2% TFA shake 10 min at 37°C
	14. 70% ACN/0.2% TFA shake 10 min at 37°C

13.5.3 MS Analysis

Typically 2-DE protein spots are identified by Matrix Assisted Laser Desorption/Ionisation Time of Flight (MALDI-TOF) MS but can also be identified by various liquid chromatography (LC)-MS/MS methods. The amino acid sequence of each peptide is determined and searched against a known protein sequence database (e.g. Uniprot) using a search engine such as MASCOT, X!Tandem or SEQUEST to identify the protein(s) present within the spot of interest.

13.6 Discussion

2-DE has the ability to reproducibly separate complex protein samples, separating 2,000–3,000 different protein spots on an individual gel (Gorg et al., 2004), and to quantify and identify protein expression changes. It has a strong advantage over non-gel based protein techniques, as it is able to provide MW and p*I* information of a particular protein, and so can therefore enable PTMs to be visualised.

However, when designing a 2-DE based proteomic study, a number of considerations must first be made during sample preparation which we have discussed in this chapter. As with all proteomic studies, great care and attention must be made to preserve protein integrity and prevent proteolysis during sample preparation. As 2-DE separates proteins in the first dimension according to their p*I*, the sample lysis buffer must be carefully chosen with low-ionic-strength and low salt concentration so as not to affect protein focusing and result in streaking of the proteins on the gel. As high abundant proteins can result in spot smearing, care must be taken with 2-DE analysis as this could result in masking of adjacent low abundance proteins. In this chapter we have also given an insight into the methodology behind 2-DE discussing separation by IEF using IPG strips of various pH ranges and also separation by MW in the second dimension. We have discussed how proteins can be visualised by various staining methods including Coomassie blue, silver or DIGE and how following detection, 2-DE protein maps can then be analysed and differential protein abundance quantified using commercially available software such as DyCyder and Progenesis Samespots. Finally, we have described how specific protein spots of interest can be identified by excising them from the 2-D gel, digesting with trypsin, subsequently analysing by MS and searching the identified peptides against various sequence databases. While protein identifications might match to the predicted p*I* and MW, most publications of proteomic studies will however require further validation of the differences detected by either Western blotting, Enzyme-linked immunosorbent assay (ELISA) or immunohistochemistry for example.

The use of 2-DE to successfully analyse various proteomes and differential protein abundance is well proven. When combined with MS, 2-DE remains the most popular method for proteomic analysis. Recent advancements with protein detection including DIGE has allowed for reproducible spot patterns for any given sample and the parallel detection of proteins of interest with greatly increased statistical confidence. While novel non-gel based quantitative proteomic techniques are emerging, there is no standard method which can currently replace 2-DE (Gorg et al., 2004; Penque, 2009).

References

Abdolzade-Bavil, A., Hayes, S., Goretzki, L., Kroger, M., Anders, J., and Hendriks, R. (2004). Convenient and versatile subcellular extraction procedure, that facilitates classical protein expression profiling and functional protein analysis. Proteomics *4*, 1397–1405.

Anderson, N.L., and Anderson, N.G. (2002). The human plasma proteome: History, character, and diagnostic prospects. Mol Cell Proteomics *1*, 845–867.

Anderson, N.L., Anderson, N.G., Pearson, T.W., Borchers, C.H., Paulovich, A.G., Patterson, S.D., Gillette, M., Aebersold, R., and Carr, S.A. (2009). A human proteome detection and quantitation project. Mol Cell Proteomics *8*, 883–886.

Babu, G.J., Wheeler, D., Alzate, O., and Periasamy, M. (2004). Solubilization of membrane proteins for two-dimensional gel electrophoresis: Identification of sarcoplasmic reticulum membrane proteins. Anal Biochem *325*, 121–125.

Bagnato, C., Thumar, J., Mayya, V., Hwang, S.I., Zebroski, H., Claffey, K.P., Haudenschild, C., Eng, J.K., Lundgren, D.H., and Han, D.K. (2007). Proteomics analysis of human coronary atherosclerotic plaque: A feasibility study of direct tissue proteomics by liquid chromatography and tandem mass spectrometry. Mol Cell Proteomics *6*, 1088–1102.

Berggren, K., Chernokalskaya, E., Steinberg, T.H., Kemper, C., Lopez, M.F., Diwu, Z., Haugland, R.P., and Patton, W.F. (2000). Background-free, high sensitivity staining of proteins in one- and two-dimensional sodium dodecyl sulfate-polyacrylamide gels using a luminescent ruthenium complex. Electrophoresis *21*, 2509–2521.

Bodzon-Kulakowska, A., Bierczynska-Krzysik, A., Dylag, T., Drabik, A., Suder, P., Noga, M., Jarzebinska, J., and Silberring, J. (2007). Methods for samples preparation in proteomic research. J Chromatogr B Anal Technol Biomed Life Sci *849*, 1–31.

Boschetti, E., and Righetti, P.G. (2008). The ProteoMiner in the proteomic arena: A non-depleting tool for discovering low-abundance species. J Proteomics *71*, 255–264.

Braun, R.J., Kinkl, N., Beer, M., and Ueffing, M. (2007). Two-dimensional electrophoresis of membrane proteins. Anal Bioanal Chem *389*, 1033–1045.

Chaiyarit, S., and Thongboonkerd, V. (2009). Comparative analyses of cell disruption methods for mitochondrial isolation in high-throughput proteomics study. Anal Biochem. 2009 Nov 15; *394*(2):249–258.

Craven, R.A., and Banks, R.E. (2003). Laser capture microdissection for proteome analysis. Curr Protoc Protein Sci (Chapter 22, Unit 22.3.1–22.3.27).

Domazet, B., Maclennan, G.T., Lopez-Beltran, A., Montironi, R., and Cheng, L. (2008). Laser capture microdissection in the genomic and proteomic era: Targeting the genetic basis of cancer. Int J Clin Exp Pathol *1*, 475–488.

Donoghue, P.M., Hughes, C., Vissers, J.P., Langridge, J.I., and Dunn, M.J. (2008). Nonionic detergent phase extraction for the proteomic analysis of heart membrane proteins using label-free LC-MS. Proteomics *8*, 3895–3905.

Donoghue, P.M., McManus, C.A., O'Donoghue, N.M., Pennington, S.R., and Dunn, M.J. (2006). CyDye immunoblotting for proteomics: Co-detection of specific immunoreactive and total protein profiles. Proteomics *6*, 6400–6404.

Echan, L.A., Tang, H.Y., Ali-Khan, N., Lee, K., and Speicher, D.W. (2005). Depletion of multiple high-abundance proteins improves protein profiling capacities of human serum and plasma. Proteomics *5*, 3292–3303.

English, J.A., Dicker, P., Focking, M., Dunn, M.J., and Cotter, D.R. (2009). 2-D DIGE analysis implicates cytoskeletal abnormalities in psychiatric disease. Proteomics *9*, 3368–3382.

Fazekas de St Groth, S., Webster, R.G., and Datyner, A. (1963). Two new staining procedures for quantitative estimation of proteins on electrophoretic strips. Biochim Biophys Acta *71*, 377–391.

Ferry-Dumazet, H., Houel, G., Montalent, P., Moreau, L., Langella, O., Negroni, L., Vincent, D., Lalanne, C., de Daruvar, A., Plomion, C., *et al.* (2005). PROTICdb: A web-based application to store, track, query, and compare plant proteome data. Proteomics *5*, 2069–2081.

Finnie, C., and Svensson, B. (2002). Proteolysis during the isoelectric focusing step of two-dimensional gel electrophoresis may be a common problem. Anal Biochem *311*, 182–186.

Getz, E.B., Xiao, M., Chakrabarty, T., Cooke, R., and Selvin, P.R. (1999). A comparison between the sulfhydryl reductants tris(2-carboxyethyl)phosphine and dithiothreitol for use in protein biochemistry. Anal Biochem *273*, 73–80.

Gorg, A., Boguth, G., Obermaier, C., Posch, A., and Weiss, W. (1995). Two-dimensional polyacrylamide gel electrophoresis with immobilized pH gradients in the first dimension (IPG-Dalt): The state of the art and the controversy of vertical versus horizontal systems. Electrophoresis *16*, 1079–1086.

Gorg, A., Obermaier, C., Boguth, G., Harder, A., Scheibe, B., Wildgruber, R., and Weiss, W. (2000). The current state of two-dimensional electrophoresis with immobilized pH gradients. Electrophoresis *21*, 1037–1053.

Gorg, A., Obermaier, C., Boguth, G., and Weiss, W. (1999). Recent developments in two-dimensional gel electrophoresis with immobilized pH gradients: Wide pH gradients up to pH 12, longer separation distances and simplified procedures. Electrophoresis *20*, 712–717.

Gorg, A., Postel, W., Friedrich, C., Kuick, R., Strahler, J.R., and Hanash, S.M. (1991). Temperature-dependent spot positional variability in two-dimensional polypeptide patterns. Electrophoresis *12*, 653–658.

Gorg, A., Postel, W., Weser, J., Gunther, S., Strahler, J.R., Hanash, S.M., Somerlot, L. (1987). Elimination of point streaking on silver stained two-dimensional gels by addition of iodoacetamide to the equilibration buffer. Electrophoresis *8*, 122–124.

Gorg, A., Weiss, W., and Dunn, M.J. (2004). Current two-dimensional electrophoresis technology for proteomics. Proteomics *4*, 3665–3685.

Grassl, J., Westbrook, J.A., Robinson, A., Boren, M., Dunn, M.J., and Clyne, R.K. (2009). Preserving the yeast proteome from sample degradation. Proteomics *9*, 4616–4626.

Harris, L.R., Churchward, M.A., Butt, R.H., and Coorssen, J.R. (2007). Assessing detection methods for gel-based proteomic analyses. J Proteome Res *6*, 1418–1425.

Harrison, P.J., Heath, P.R., Eastwood, S.L., Burnet, P.W., McDonald, B., and Pearson, R.C. (1995). The relative importance of premortem acidosis and postmortem interval for human brain gene expression studies: Selective mRNA vulnerability and comparison with their encoded proteins. Neurosci Lett *200*, 151–154.

Hart, C., Schulenberg, B., Steinberg, T.H., Leung, W.Y., and Patton, W.F. (2003). Detection of glycoproteins in polyacrylamide gels and on electroblots using Pro-Q Emerald 488 dye, a fluorescent periodate Schiff-base stain. Electrophoresis *24*, 588–598.

Henningsen, R., Gale, B.L., Straub, K.M., and DeNagel, D.C. (2002). Application of zwitterionic detergents to the solubilization of integral membrane proteins for two-dimensional gel electrophoresis and mass spectrometry. Proteomics 2, 1479–1488.

Hoogland, C., Mostaguir, K., Appel, R.D., and Lisacek, F. (2008). The World-2DPAGE Constellation to promote and publish gel-based proteomics data through the ExPASy server. J Proteomics *71*, 245–248.

Hoogland, C., Sanchez, J.C., Tonella, L., Binz, P.A., Bairoch, A., Hochstrasser, D.F., and Appel, R.D. (2000). The 1999 SWISS-2DPAGE database update. Nucleic Acids Res *28*, 286–288.

Jager, D., Jungblut, P.R., and Muller-Werdan, U. (2002). Separation and identification of human heart proteins. J Chromatogr B Anal Technol Biomed Life Sci *771*, 131–153.

Kane, L.A., Yung, C.K., Agnetti, G., Neverova, I., and Van Eyk, J.E. (2006). Optimization of paper bridge loading for 2-DE analysis in the basic pH region: Application to the mitochondrial subproteome. Proteomics *6*, 5683–5687.

Kang, Y., Techanukul, T., Mantalaris, A., and Nagy, J.M. (2009). Comparison of three commercially available DIGE analysis software packages: Minimal user intervention in gel-based proteomics. J Proteome Res *8*, 1077–1084.

Karp, N.A., McCormick, P.S., Russell, M.R., and Lilley, K.S. (2007). Experimental and statistical considerations to avoid false conclusions in proteomics studies using differential in-gel electrophoresis. Mol Cell Proteomics *6*, 1354–1364.

Karp, N.A., Spencer, M., Lindsay, H., O'Dell, K., and Lilley, K.S. (2005). Impact of replicate types on proteomic expression analysis. J Proteome Res *4*, 1867–1871.

Khoudoli, G.A., Porter, I.M., Blow, J.J., and Swedlow, J.R. (2004). Optimisation of the two-dimensional gel electrophoresis protocol using the taguchi approach. Proteome Sci *2*, 6.

Kim, M.R., and Kim, C.W. (2007). Human blood plasma preparation for two-dimensional gel electrophoresis. J Chromatogr B Anal Technol Biomed Life Sci *849*, 203–210.

Klose, J. (2009). From 2-D electrophoresis to proteomics. Electrophoresis *30*(Suppl 1), S142–149.

Laemmli, U.K. (1970). Cleavage of structural proteins during the assembly of the head of bacteriophage T4. Nature *227*, 680–685.

Lei, T., He, Q.Y., Wang, Y.L., Si, L.S., and Chiu, J.F. (2008). Heparin chromatography to deplete high-abundance proteins for serum proteomics. Clin Chim Acta *388*, 173–178.

Li, J.Z., Vawter, M.P., Walsh, D.M., Tomita, H., Evans, S.J., Choudary, P.V., Lopez, J.F., Avelar, A., Shokoohi, V., Chung, T., *et al.* (2004). Systematic changes in gene expression in postmortem human brains associated with tissue pH and terminal medical conditions. Hum Mol Genet *13*, 609–616.

Lilley, K.S., Razzaq, A., and Dupree, P. (2002). Two-dimensional gel electrophoresis: Recent advances in sample preparation, detection and quantitation. Curr Opin Chem Biol *6*, 46–50.

Lopez, M.F., Berggren, K., Chernokalskaya, E., Lazarev, A., Robinson, M., and Patton, W.F. (2000a). A comparison of silver stain and SYPRO ruby protein gel stain with respect to protein detection in two-dimensional gels and identification by peptide mass profiling. Electrophoresis *21*, 3673–3683.

Lopez, M.F., Kristal, B.S., Chernokalskaya, E., Lazarev, A., Shestopalov, A.I., Bogdanova, A., and Robinson, M. (2000b). High-throughput profiling of the mitochondrial proteome using affinity fractionation and automation. Electrophoresis *21*, 3427–3440.

Luche, S., Santoni, V., and Rabilloud, T. (2003). Evaluation of nonionic and zwitterionic detergents as membrane protein solubilizers in two-dimensional electrophoresis. Proteomics *3*, 249–253.

Mackintosh, J.A., Choi, H.Y., Bae, S.H., Veal, D.A., Bell, P.J., Ferrari, B.C., Van Dyk, D.D., Verrills, N.M., Paik, Y.K., and Karuso, P. (2003). A fluorescent natural product for ultra sensitive detection of proteins in one-dimensional and two-dimensional gel electrophoresis. Proteomics *3*, 2273–2288.

McCarthy, J., Hopwood, F., Oxley, D., Laver, M., Castagna, A., Righetti, P.G., Williams, K., and Herbert, B. (2003). Carbamylation of proteins in 2-D electrophoresis--myth or reality? J Proteome Res *2*, 239–242.

McManus, C.A., Donoghue, P.M., and Dunn, M.J. (2009). A fluorescent codetection system for immunoblotting and proteomics through ECL-Plex and CyDye labeling. Methods Mol Biol *536*, 515–526.

McManus, C.A., Polden, J., Cotter, D.R., and Dunn, M.J. (2010). Two-dimensional reference map for the basic proteome of the human dorsolateral prefrontal cortex (dlPFC) of the prefrontal lobe region of the brain. Proteomics. 2010 Jul; *10*(13):2551–2555.

Merril, C.R., Goldman, D., Sedman, S.A., and Ebert, M.H. (1981). Ultrasensitive stain for proteins in polyacrylamide gels shows regional variation in cerebrospinal fluid proteins. Science *211*, 1437–1438.

Molloy, M.P. (2000). Two-dimensional electrophoresis of membrane proteins using immobilized pH gradients. Anal Biochem *280*, 1–10.

Morris, J.S., Clark, B.N., and Gutstein, H.B. (2008). Pinnacle: A fast, automatic and accurate method for detecting and quantifying protein spots in 2-dimensional gel electrophoresis data. Bioinformatics *24*, 529–536.

Murray, C.I., Barrett, M., and Van Eyk, J.E. (2009). Assessment of ProteoExtract subcellular fractionation kit reveals limited and incomplete enrichment of nuclear subproteome from frozen liver and heart tissue. Proteomics *9*, 3934–3938.

Murray, J., Marusich, M.F., Capaldi, R.A., and Aggeler, R. (2004). Focused proteomics: Monoclonal antibody-based isolation of the oxidative phosphorylation machinery and detection of phosphoproteins using a fluorescent phosphoprotein gel stain. Electrophoresis *25*, 2520–2525.

Neuhoff, V., Arold, N., Taube, D., and Ehrhardt, W. (1988). Improved staining of proteins in polyacrylamide gels including isoelectric focusing gels with clear background at nanogram sensitivity using Coomassie Brilliant Blue G-250 and R-250. Electrophoresis *9*, 255–262.

O'Farrell, P.H. (1975). High resolution two-dimensional electrophoresis of proteins. J Biol Chem *250*, 4007–4021.

Patton, W.F., Chung-Welch, N., Lopez, M.F., Cambria, R.P., Utterback, B.L., and Skea, W.M. (1991). Tris-tricine and tris-borate buffer systems provide better estimates of human mesothelial cell intermediate filament protein molecular weights than the standard tris-glycine system. Anal Biochem *197*, 25–33.

Peirce, M.J., Wait, R., Begum, S., Saklatvala, J., and Cope, A.P. (2004). Expression profiling of lymphocyte plasma membrane proteins. Mol Cell Proteomics *3*, 56–65.

Pennington, K., McGregor, E., Beasley, C.L., Everall, I., Cotter, D., and Dunn, M.J. (2004). Optimization of the first dimension for separation by two-dimensional gel electrophoresis of basic proteins from human brain tissue. Proteomics *4*, 27–30.

Penque, D. (2009). Two-dimensional gel electrophoresis and mass spectrometry for biomarker discovery. Proteomics Clin Appl *3*, 155–172.

Rabilloud, T., Adessi, C., Giraudel, A., and Lunardi, J. (1997). Improvement of the solubilization of proteins in two-dimensional electrophoresis with immobilized pH gradients. Electrophoresis *18*, 307–316.

Rabilloud, T., Chevallet, M., Luche, S., and Lelong, C. (2008). Fully denaturing two-dimensional electrophoresis of membrane proteins: A critical update. Proteomics *8*, 3965–3973.

Rabilloud, T., Kieffer, S., Procaccio, V., Louwagie, M., Courchesne, P.L., Patterson, S.D., Martinez, P., Garin, J., and Lunardi, J. (1998). Two-dimensional electrophoresis of human placental mitochondria and protein identification by mass spectrometry: Toward a human mitochondrial proteome. Electrophoresis *19*, 1006–1014.

Rabilloud, T., Strub, J.M., Luche, S., van Dorsselaer, A., and Lunardi, J. (2001). A comparison between Sypro Ruby and ruthenium II tris (bathophenanthroline disulfonate) as fluorescent stains for protein detection in gels. Proteomics *1*, 699–704.

Rabilloud, T., Valette, C., and Lawrence, J.J. (1994). Sample application by in-gel rehydration improves the resolution of two-dimensional electrophoresis with immobilized pH gradients in the first dimension. Electrophoresis *15*, 1552–1558.

Raman, B., Cheung, A., and Marten, M.R. (2002). Quantitative comparison and evaluation of two commercially available, two-dimensional electrophoresis image analysis software packages, Z3 and Melanie. Electrophoresis *23*, 2194–2202.

Raymond, S. (1964). Acrylamide gel electrophoresis. Ann N Y Acad Sci *121*, 350–365.

Richardson, M.R., Liu, S., Ringham, H.N., Chan, V., and Witzmann, F.A. (2008). Sample complexity reduction for two-dimensional electrophoresis using solution isoelectric focusing prefractionation. Electrophoresis *29*, 2637–2644.

Riederer, B.M. (2008). Non-covalent and covalent protein labeling in two-dimensional gel electrophoresis. J Proteomics *71*, 231–244.

Righetti, P.G., Boschetti, E., Lomas, L., and Citterio, A. (2006). Protein equalizer technology: The quest for a "democratic proteome". Proteomics *6*, 3980–3992.

Righetti, P.G., Castagna, A., Antonioli, P., and Boschetti, E. (2005). Prefractionation techniques in proteome analysis: The mining tools of the third millennium. Electrophoresis *26*, 297–319.

Righetti, P.G., and Drysdale, J.W. (1973). Small-scale fractionation of proteins and nucleic acids by isoelectric focusing in polyacrylamide gels. Ann N Y Acad Sci *209*, 163–186.

Robinson, A.A., Westbrook, J.A., English, J.A., Boren, M., and Dunn, M.J. (2009). Assessing the use of thermal treatment to preserve the intact proteomes of post-mortem heart and brain tissue. Proteomics *9*, 4433–4444.

Rosengren, A.T., Salmi, J.M., Aittokallio, T., Westerholm, J., Lahesmaa, R., Nyman, T.A., and Nevalainen, O.S. (2003). Comparison of PDQuest and Progenesis software packages in the analysis of two-dimensional electrophoresis gels. Proteomics *3*, 1936–1946.

Sabounchi-Schutt, F., Astrom, J., Eklund, A., Grunewald, J., and Bjellqvist, B. (2001). Detection and identification of human bronchoalveolar lavage proteins using narrow-range immobilized pH gradient DryStrip and the paper bridge sample application method. Electrophoresis *22*, 1851–1860.

Santoni, V., Kieffer, S., Desclaux, D., Masson, F., and Rabilloud, T. (2000). Membrane proteomics: Use of additive main effects with multiplicative interaction model to classify plasma membrane proteins according to their solubility and electrophoretic properties. Electrophoresis *21*, 3329–3344.

Smithies, O., and Poulik, M.D. (1956). Two-dimensional electrophoresis of serum proteins. Nature *177*, 1033.

Steinberg, T.H., Agnew, B.J., Gee, K.R., Leung, W.Y., Goodman, T., Schulenberg, B., Hendrickson, J., Beechem, J.M., Haugland, R.P., and Patton, W.F. (2003). Global quantitative phosphoprotein analysis using multiplexed proteomics technology. Proteomics *3*, 1128–1144.

Svensson, M., Boren, M., Skold, K., Falth, M., Sjogren, B., Andersson, M., Svenningsson, P., and Andren, P.E. (2009). Heat stabilization of the tissue proteome: A new technology for improved proteomics. J Proteome Res *8*, 974–981.

Taylor, R.S., Wu, C.C., Hays, L.G., Eng, J.K., Yates, J.R., 3rd, and Howell, K.E. (2000). Proteomics of rat liver golgi complex: Minor proteins are identified through sequential fractionation. Electrophoresis *21*, 3441–3459.

Trott, D., McManus, C.A., Martin, J.L., Brennan, B., Dunn, M.J., and Rose, M.L. (2009). Effect of phosphorylated hsp27 on proliferation of human endothelial and smooth muscle cells. Proteomics *9*, 3383–3394.

Turck, N., Richert, S., Gendry, P., Stutzmann, J., Kedinger, M., Leize, E., Simon-Assmann, P., Van Dorsselaer, A., and Launay, J.F. (2004). Proteomic analysis of nuclear proteins from proliferative and differentiated human colonic intestinal epithelial cells. Proteomics *4*, 93–105.

Unlu, M., Morgan, M.E., and Minden, J.S. (1997). Difference gel electrophoresis: A single gel method for detecting changes in protein extracts. Electrophoresis *18*, 2071–2077.

Vawter, M.P., Tomita, H., Meng, F., Bolstad, B., Li, J., Evans, S., Choudary, P., Atz, M., Shao, L., Neal, C., *et al.* (2006). Mitochondrial-related gene expression changes are sensitive to agonal-pH state: Implications for brain disorders. Mol Psychiatry *11*, 615, 663–679.

Vijayendran, C., Burgemeister, S., Friehs, K., Niehaus, K., and Flaschel, E. (2007). 2DBase: 2D-PAGE database of Escherichia coli. Biochem Biophys Res Commun *363*, 822–827.

Vizcaino, J.A., Cote, R., Reisinger, F., Foster, J., Mueller, M., Rameseder, J., Hermjakob, H., and Martens, L. (2009). A guide to the PRIDE proteomics data repository. Proteomics. 2009 Sep; *9*(18):4276–4283.

Westbrook, J.A., Wheeler, J.X., Wait, R., Welson, S.Y., and Dunn, M.J. (2006). The human heart proteome: Two-dimensional maps using narrow-range immobilised pH gradients. Electrophoresis *27*, 1547–1555.

Westbrook, J.A., Yan, J.X., Wait, R., Welson, S.Y., and Dunn, M.J. (2001). Zooming-in on the proteome: Very narrow-range immobilised pH gradients reveal more protein species and isoforms. Electrophoresis *22*, 2865–2871.

Wildgruber, R., Harder, A., Obermaier, C., Boguth, G., Weiss, W., Fey, S.J., Larsen, P.M., and Gorg, A. (2000). Towards higher resolution: Two-dimensional electrophoresis of Saccharomyces cerevisiae proteins using overlapping narrow immobilized pH gradients. Electrophoresis *21*, 2610–2616.

Wilm, M., Shevchenko, A., Houthaeve, T., Breit, S., Schweigerer, L., Fotsis, T., and Mann, M. (1996). Femtomole sequencing of proteins from polyacrylamide gels by nano-electrospray mass spectrometry. Nature *379*, 466–469.

Yan, J.X., Wait, R., Berkelman, T., Harry, R.A., Westbrook, J.A., Wheeler, C.H., and Dunn, M.J. (2000). A modified silver staining protocol for visualization of proteins compatible with matrix-assisted laser desorption/ionization and electrospray ionization-mass spectrometry. Electrophoresis *21*, 3666–3672.

Zhang, H., Li, X.J., Martin, D.B., and Aebersold, R. (2003). Identification and quantification of N-linked glycoproteins using hydrazide chemistry, stable isotope labeling and mass spectrometry. Nat Biotechnol *21*, 660–666.

Chapter 14
Quantitative Intact Proteomic Strategies to Detect Changes in Protein Modification and Genomic Variation

David B. Friedman

Abstract Each of the major proteomics technology platforms, notably those that focus on peptides ("bottom-up" or "shotgun" MS strategies) and those that focus on intact proteins (2D gel-based, MALDI-imaging and "top-down" MS strategies) have unique strengths and limitations that in many cases can be complementary. These differences become even more evident in the realm of quantitative proteomics, where technical noise can play a large role in obscuring biological significance. This chapter focuses on the unique strengths that the 2D-gel based Difference Gel Electrophoresis (DIGE) quantitative approach can provide by enabling the interrogation of over 1000 resolved intact species from multiple biological replicates of multiple biological conditions with low technical noise and high statistical power. Examples will be highlighted where vital information on protein modifications or genomic variants were facilely obtained from global-scale analyses on intact protein forms.

Keywords 2D gel electrophoresis · Difference gel electrophoresis · Quantitative proteomics · Shotgun analysis

14.1 Introduction

No single proteomics technology platform is capable of accomplishing a true global analysis of the entire proteome. Whereas the LC-MS/MS strategies provide the greatest depth of coverage, they rely on surrogate peptides (generated by protease digestion of the starting protein sample) for protein identification, thereby loosing quantitative information on isoforms of the intact proteins (Elliott et al., 2009; Lottspeich, 2009). Technologies that retain this information, such as 2D gels and DIGE in particular, are limited especially in very hydrophobic proteins

D.B. Friedman (✉)
Proteomics Laboratory, Mass Spectrometry Research Center, Vanderbilt University School of Medicine, Nashville, TN 37232-8575, USA
e-mail: david.friedman@vanderbilt.edu

A.R. Ivanov, A.V. Lazarev (eds.), *Sample Preparation in Biological Mass Spectrometry*, DOI 10.1007/978-94-007-0828-0_14,

and proteins of extreme pH and MW, but can provide quantitative information on these biologically-relevant proteome changes, and on a global scale with very low technical noise and thus high statistical power (Karp and Lilley, 2005).

Proteins that are "difficult" to resolve by intact-protein methods are often readily identified using the LC-MS/MS shotgun approach because sufficient surrogate peptides can be obtained and mass analyzed (assuming that the proteins remain soluble during the various isolation, extraction, and digestion steps). The shotgun approach also provides for greater sensitivity and dynamic range, especially using multidimensional protein identification technology (MudPIT) approaches (Link et al., 1999; Washburn et al., 2001; Wolters et al., 2001). However, using peptide-based approaches it is more challenging to analyze modifications and/or perform quantitative analyses in a non-targeted discovery experiment, especially for experiments that require independent replicates across multiple experimental conditions.

Perhaps most often the decision of which proteomics platform to use can come down to what you can see and how well you can see it. This trade-off between sensitivity (depth of coverage) and statistical power (is it biologically significant?) is quickly becoming a major focus of investigation in quantitative proteomics studies (e.g. see (Anderle et al., 2004; Cho et al., 2007; Karp and Lilley, 2005, 2009)). It is often the case that sample fractionation and enrichment strategies are necessary to provide greater depth of coverage; but to achieve this often means introducing technical variation/noise into the experiment. This ultimately lowers the statistical power (and therefore biological insight) of the experiment unless a sufficient number of technical replicates are analyzed in addition to the requisite number of biological replicates. Analyzing unfractionated samples introduces the least amount of technical variation/noise, enabling a much higher statistical power, but at the cost of overall sensitivity/depth of coverage. Ultimately, the decision of which approach to utilize is best directed by the nature of the experimental question being asked, and also by the experimental instrumentation/expertise available. In the best case scenarios, a proteomics project will not be limited by the utilization of only one technology platform.

14.2 Difference Gel Electrophoresis (DIGE)

Sodium Dodecyl Sulfate polyacrylamide gel electrophoresis (SDS-PAGE), first described by Laemmli (1970), has long been the method of choice for resolving intact proteins (based on apparent molecular mass, commonly referred to as molecular weight, MW). For more complex mixtures, two-dimensional gel electrophoresis (2D gel, 2DE; (Klose, 1975; O'Farrell, 1975)) introduces a concentration phase prior to SDS-PAGE via isoelectric focusing (IEF), whereby proteins are separated as a function of isoelectric point, or charge. When proteins reach their isoelectric point (pI) in a pH gradient and cease to migrate in an electric field, they become "focused".

These two orthogonal separations, each based on physicochemical properties of intact proteins, are particularly powerful for visualizing protein isoforms that result from charged post-translational modification, such as phosphorylation and sulfation (which add charge), or acetylation (which neutralize charge). They are also useful in detecting splice variants and proteolytic cleavages, as well as genomic differences that are reflected in the proteome, all of which may result in protein species with altered MW and pI.

Proteins resolved by 2D-gels are of course directly amenable to bottom-up protein identification strategies after gel-excision and digestion of the target protein into peptides directly within the gel sections. Historically this platform has been used for cataloguing experiments, but more recently this endeavor has largely been replaced by the more sensitive and comprehensive LC-MS/MS and MudPIT shotgun proteomic strategies. Especially with the advent of DIGE technology, 2D-gel experiments have more recently been used mostly for differential expression studies on a global scale, where proteomes are compared between multiple experimental conditions.

Especially for these quantitative approaches, replicate gels are required to ensure that changes are biologically significant and not due to experimental/sample preparation variation and analytical gel-to-gel variation. Until recently, these strategies lacked the ability to easily quantify abundance changes due mainly to the inability to directly correlate migration patterns and protein staining between gel separations (analytical gel-to-gel variation). Stable isotope labeling, which is prevalent in LC-MS/MS studies, has also been used in gel-based proteomics, whereby different proteomes have been separately labeled with different stable isotopes (e.g. growing cells using ^{14}N vs. ^{15}N-labeled medium) prior to mixing and running together through the same 2DE separation (Vogt et al., 2003). In this case, abundance changes are monitored during the mass spectrometry stage, which must be performed on each resolved protein. Thus these types of studies have typically been limited to comparisons made within a single gel, where any statistical advantage is essentially lost.

Difference Gel Electrophoresis (DIGE) technology has more recently been used for direct quantification of abundance changes on a global scale without interference from gel-to-gel variation. This is done using spectrally-resolvable MW and charge-matched fluorescent dyes (Cy2, Cy3 and Cy5) to pre-label protein samples which are then multiplexed onto 2D gels. These dyes offer sub-nanogram detection limits and a linear dynamic range of *ca.* 4-orders of magnitude (Tonge et al., 2001; Von Eggeling et al., 2001), and when used with an internal standard methodology, enable a straight-forward comparison of global expression changes between multiple experimental conditions (Bengtsson et al., 2007; Friedman et al., 2006, 2007; Seike et al., 2004; Suehara et al., 2006), with each condition represented by independent (biological) replicates to provide high statistical power (Karp and Lilley, 2005).

The full statistical advantage of the DIGE approach is obtained when a Cy2-labeled internal standard (comprised of a mixture of all samples in the experiment) is co-resolved on a series of gels that each contains individual samples labeled with Cy3 or Cy5, thereby enabling sufficient independent (biological) replicates to be co-analyzed. Because these individual samples are multiplexed with an equal aliquot of the same Cy2-standard mixture, each resolved feature can be directly related to the cognate feature in the Cy2 standard mixture within that gel. These intra-gel ratios can then be normalized to all other ratios for that feature from the other samples across the other gels in the experiment with extremely low technical (analytical) noise and high statistical power (Alban et al., 2003; Friedman et al., 2004; Karp and Lilley, 2005). This approach is also directly amenable to multivariate statistical analyses such as principle component analysis (PCA) and hierarchical clustering, that can be extremely beneficial in visualizing the variation within a set of experimental samples, to help determine if the major source of variation is describing the biology or indicating unanticipated variation between samples (or introduced during sample preparation), in addition to pinpointing subsets of proteins that respond collectively to an experimental stimulus or classification (Friedman et al., 2006, 2007; Hatakeyama et al., 2006; Seike et al., 2004; Suehara et al., 2006).

The strengths of 2D gel proteome analysis lie mainly in the resolution and quantification of intact protein species, including charged post-translational modifications as well as proteolytic forms, and facile protein identification directly from the resolved intact protein by gel excision, in-gel digestion and MS. For quantitative studies, this platform provides the highest statistical power when using the DIGE technology with the mixed-sample internal standard method. Several studies have now been published that address the validation of the DIGE technique with respect to quantitative assessment (Kolkman et al., 2005), as well as same/same experiments that demonstrate low technical noise, obviating the need for technical replicates and enabling a statistically-powered experiment using a minimal number of independent biological replicates (e.g. 2-fold changes with $N = 4$ independent samples, (Karp and Lilley, 2005)). Total protein fluorescent stain detection systems can detect levels as low as 1 ng (e.g. 20 fmole of a 50 kDa protein), which is nearing the lower limit of detection in most mass spectrometers after in-gel digestion (Friedman et al., 2004).

14.3 DIGE and Statistical Power

Technical and biological replicates are important for ensuring the accuracy and biological significance of quantitative measurements. Technical replicates are necessary to control for variation in the analytical measurement. However, biological replicates are vital to assess whether or not changes in protein abundance/modification are descriptive of the biology rather than arising from unanticipated sources of experimental variation (e.g. inter-subject variation, sample preparation variation, analytical variation of the instrument). Technical replicates provide confidence in the result from the tested samples, but do not provide any confidence to the biological relevance. Determining whether any observed changes come from the biology rather than technical variation can only be assessed with independent biological replicates. In cases where technical noise has been demonstrated to be sufficiently low, then biological replicates can also serve as technical replicates, which is what is typically done with the DIGE platform.

"Statistical Power" is the ability to visualize an X-fold magnitude change (effect size) at a Y-% confidence interval (e.g. $p < 0.05$) and be "correct" (usually expressed as power of 0.8, or 80% of the time it is "correct"). Statistical power depends heavily on the analytical variation of the instrumentation making the measurement as well as the number of independent (biological) replicates. For example, the experimental noise of DIGE is extremely low owing to the internal standard experimental design, enabling statistically-powered experiments with very few biological replicates (Karp and Lilley, 2005), whereas LC/MS experiments have a relatively high degree of analytical variation with little-to-no internal standard methodologies being employed, often resulting in under-powered experiments unless sufficient biological and technical replicates are employed (Anderle et al., 2004; Cho et al., 2007).

Pooling independent (biological) replicates to either produce sufficient material or to minimize costs (the "economics of proteomics") should be done with extreme caution with respect to the statistical power of the resulting data. Pooling samples can be effective if the technical variation between samples is very low (e.g. sample preparation, analytical platforms with low noise), and disastrous if it is high (e.g. biological variation unrelated to experiment, analytical platforms with high noise). Even with sufficient technical replicates of the same samples to account for high technical variability of the sample preparation and analysis, performing the $N = 1$ experiment on pooled samples assumes that the averaging of populations is reflective of biological signal, and that the technical noise is low. In some cases it may be valid to create sub-pools from a larger experiment, but in these cases it is still essential to maintain some degree of individualization of samples to retain statistical power (Karp and Lilley, 2009).

In the spirit that no single technology platform is capable of providing all possible proteomic answers (Lottspeich, 2009), two recent case studies are presented below that highlight examples of where a quantitative intact proteomics (QIP) approach, in these cases Difference Gel Electrophoresis (DIGE), is used to discover polymorphisms that correlate with phenotype that otherwise would have gone undetected using peptide-based proteomics technologies

14.4 Case Study A: Delineation of a Carcinogenic *Helicobacter pylori* Proteome

This group had previously demonstrated that an in vivo-adapted *H. pylori* strain 7.13 reproducibly induces pre-malignant and malignant gastric lesions in 2 different rodent models, while cancer was not induced by the progenitor strain B128 (Franco et al., 2005, 2009). An initial study using DNA array technology failed to detect any significant differences between the strains, so a proteomics approach using the DIGE-MS platform was undertaken to define a subset of protein changes that correlated with the carcinogenic phenotype. Membrane and cytoplasmic fractions were analyzed from both genotypes, and each was produced as 4 independent (biological) replicates to control for unanticipated technical variation. The resulting 16 samples were co-resolved on 8 coordinated DIGE gels, whereby each gel contained two of the individual samples that were labeled with either Cy3 or Cy5 (half of the samples from each classification are labeled with Cy3 and the other

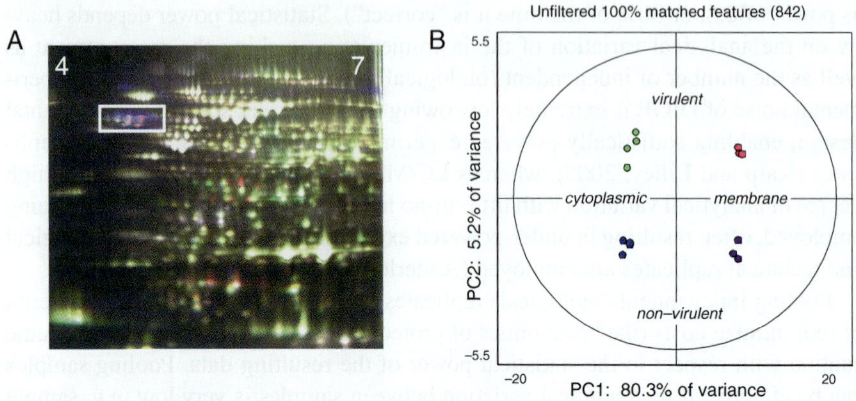

Fig. 14.1 (**a**) False-color overlay of a representative gel from the pH 4–7 range. Cy3- B128 cytoplasm (*green*) and Cy5- 7.13 cytoplasm (*red*), and Cy2 mixed-sample internal standard (*blue*) are shown for illustrative purposes only. 842 features were matched across the 8-gel set. *Boxed area* is highlighted in Fig. 2. (**b**) principle component analysis (PCA) for the 842 resolved features matched between all 8 pH 4–7 DIGE gels. PCA indicated that the major source of variation (*PC1*, 80.3%) distinguished the cytoplasmic and membrane samples, as expected. Of interest was the additional 5.2% of variance (PC2) that distinguished the carcinogenic strain 7.13 from the non-carcinogenic strain B128. Adapted from Franco et al., 2009

half with Cy5 to control for any labeling bias) along with an equal aliquot of a Cy2-labeled grand mixture of all 16 samples. Each of the 8 gels contained an aliquot of the same Cy2-labeled mixture, and each co-resolved intact protein form in the Cy2 channel served as a unique internal standard for the Cy3/Cy5 labeled signals from that feature arising from the individual samples co-resolved on that gel. This enabled the normalization of all intra-gels ratios for each of the 842 matched features separately and independently (Fig. 14.1a).

This strategy also enabled an extremely powerful multivariate statistical analysis of the dataset, allowing for the visualization of the variation in the experiment. For example, Principle Component Analysis (PCA) performed on the 842 matched features demonstrated that the majority of variation ($PC1 = 80\%$) was consistent with differences between cytoplasmic and membrane fractions, as expected. However, the second principle component (PC2), which described the second greatest source of variation between the samples, accounted for only an additional 5% variation, but this variation clustered the samples by virulent (7.13) vs. non-virulent (B128) strains. Thus, PCA analysis clearly indicated low technical noise amongst the independently derived samples, as well as highlighted a significant subpopulation of protein changes that were candidate markers for virulence (Fig. 14.1b).

One important set of differences observed in this significant subpopulation was in the expression of *H. pylori* strain-specific FlaA isoforms. FlaA encodes a flagellar protein that is expressed as at least four major isoforms that differ in charge, creating closely-related species resolved in the first dimension isoelectric focusing on the 2D gels (Fig. 14.2). Resolution of the individual samples on separate 2D gels

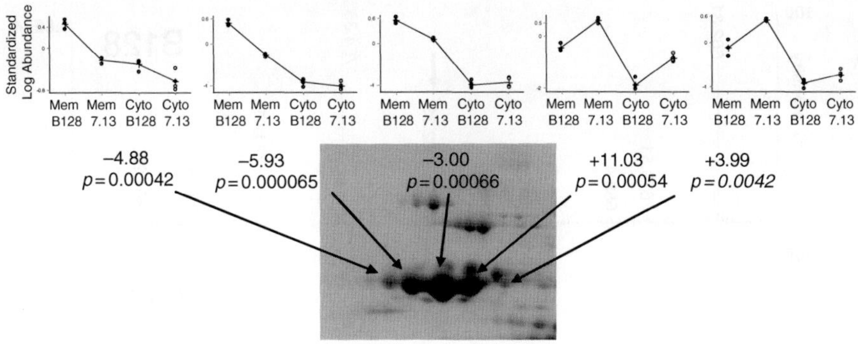

Fig. 14.2 DIGE analysis revealed global shift in FlaA isoforms between strain B128 and 7.13 when samples were mixed and co-resolved over a series of DIGE gels (each isoform was excised and identied by mass spectrometry, not shown). *Graphs* represent normalized abundance changes for the indicated isoforms. Values are mean fold-change for the indicated isoform in the membrane fractions between strain B128 and strain 7.13 with associated Student's t-test *p*-value. Adapted from Franco et al., 2009

revealed four isoforms that were all shifted with respect to the isoelectric point in the 7.13 strain relative to the non-carcinogenic progenitor strain B128, suggesting an amino acid substitution that altered the relative charge (pI) of all four isoforms. Mass spectrometry analysis clearly indicated the substitution of a cysteine residue for an arginine in strain 7.13 (Fig. 14.3) which could occur via a single basepair mutation. This was verified as the only mutation in flaA by genomic sequencing, and subsequent validation using motility assays and ultrastructural analysis by electron microscopy indicated a strong correlation between this mutation in FlaA and structural aberrations in the flaggela that compromised motility. This result was readily detected via DIGE MS because the mutation altered the pI of the resolved, intact protein forms, but otherwise did not affect the relative expression level of the flaA protein. As such, this alteration would most likely have gone undetected in a peptide-base, bottom-up LC-MS/MS only strategy.

14.5 Case Study B: Identification of Quantitative Trait Loci Underlying Proteome Variation in Human Lymphoblastoid Cells

The study of genetic determinants of complex phenotypes and diseases by quantitative trait loci (QTL) analysis has been around for over a quarter century, and much has been focused on transcript expression variation (Garge et al., 2010). In a novel approach, Garge et al. used a quantitative intact proteomics approach to investigate protein expression variation amongst 24 human lymphoblastoid cell lines as QTLs (peQTLs). Genomic changes such as single-nucleotide polymorphisms (SNPs) that alter a charged amino acid and thereby a protein's isoelectric point, or more gross changes in MW and pI that could result from chromosomal rearrangements, would

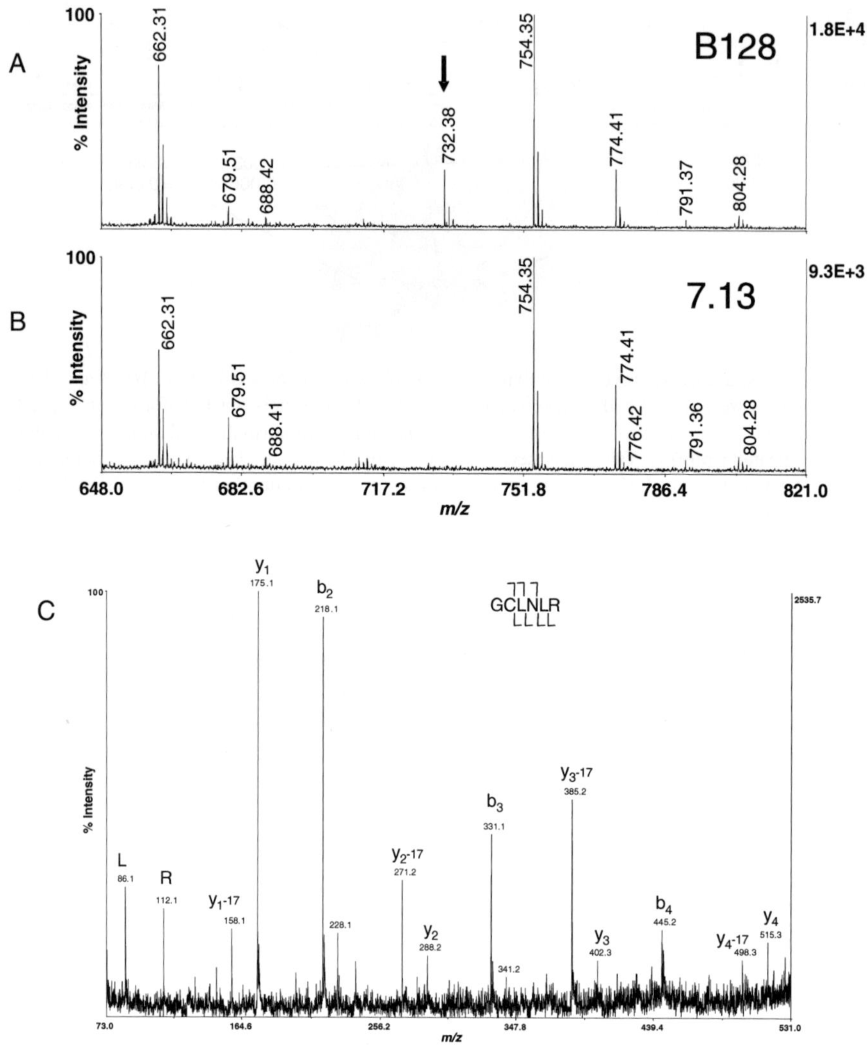

Fig. 14.3 Evidence for C296R mutation in strain 7.13. Expanded mass range of MALDI-TOF peptide mass maps from trypsin digestion of FlaA from strain B128 (**a**) and 7.13 (**b**) that were resolved on separate gels to isolate strain-specific forms. (**c**) MALDI-TOF/TOF fragmentation spectrum of the 7.13-specific peptide at m/z 732.38 (*black arrow* in panel A). The fragmentation pattern is consistent with the peptide containing the C296R mutation (with R296 making a novel trypsin cleavage site). Labeled b-ions and y-ions are denoted by cleavage brackets above and below the sequence, respectively. The cysteine was carbamidomethylated (mass shift of +57 Da). Adapted from Franco et al., 2009

also be directly visualized by DIGE along with other changes in expression or post-translational modification. This study then correlated these DIGE expression values with known genomic variants in these cell lines that were identified in the International Human HapMap repository (hapmap.ncbi.hlm.nih.gov/).

Experimental design was similar to that described in the Franco et al. study, using the Cy3 and Cy5 dyes to label individual samples that were then co-resolved in random pairs along with an aliquot of a Cy2-labeled mixed-sample internal standard (using a 12-gel matrix in this case). In an effort to define genetic components that lead to proteome variation, peQTLs were derived from the 544 expressed protein features that were matched across the 24 individual proteome maps, and correlated with both cis-only SNPs as well as with genome-wide association approaches, resulting in 24 peQTLs associated with individual protein phenotypes.

In several cases where a peQTL was associated with a genetic alteration, a putative causative SNP that contributed to variation in allelic protein expression levels was also found. For example, among the protein phenotypes were six variant forms of the protein L-Plastin (LCP1) that were all mapped to cis-acting SNPs that specified an amino acid substitution (lysine-to-glutamate) altering the resulting isoelectric point of the protein forms (and thereby the position on the 2D gel). Similar results were obtained for isoforms of CDNP2 and GLRX1 (glutaredoxin-1). These changes were readily accessible in the 2D DIGE analysis because the SNP specified a change in migration pattern of the intact protein form.

This work established that quantitative proteomic methods could be used for QTL discovery, which may be a significant step towards understanding complex phenotypes and diseases. Had this analysis been approached using the bottom-up, peptide-based proteomics platform, some of these phenotypes would almost certainly have gone undetected, unless the peptides that carried the altered amino acid(s) were not only analyzed during the mass spectrometry acquisition, but also accounted for in the database used to search the resulting data.

14.6 Conclusions

There are many points of comparison and contrast between the standard LC-MS/MS "shotgun" analyses and the 2D-gel based analyses, such as sample consumption, depth of proteome coverage, and quantitative statistical power. Both platforms have the ability to provide for resolution on hundreds to thousands of features, so the difference is really in what portion of the proteome is investigated. An interesting trade-off in this regard may well be between depth of coverage based on a few surrogate peptides vs. alterations of modifications/isoforms on intact proteins. But ultimately, one can conclude that these approaches are both of great value to a proteomics study and often provide complementary information for an overall richer analysis.

LC-MS/MS shotgun analyses achieves exceptional depth of coverage owing to the fact that the analysis is based on surrogate peptides, and that peptides are relatively homogeneous in their physiochemical properties and can be resolved over multiple chromatographic phases. Analysis of surrogate peptides also enables the identification of proteins that are otherwise refractory to 2D gels due to issues such as solubility, hydrophobicity, and extreme MW/pI of the intact proteins. However, important post-translational modifications and processed forms would only be detected if the important peptides are captured or directly targeted with this technology (because information on the intact protein is not retained), and it also becomes increasingly challenging to perform complex experimental comparisons that are relatively straight-forward by DIGE (Anderle et al., 2004; Cho et al., 2007; Karp and Lilley, 2005). If the specific alteration were known ahead of time, specific modifications can also be quantified using MRM-directed mass spectrometry approaches (Wolf-Yadlin et al., 2007) whereas they might be below the detection limits of 2D gels.

Whether these differences become a strength or a limitation ultimately depends on the experimental goals. For a global, discovery-based experiment these intact protein attributes can be used in a multivariate analysis (e.g., principle component analysis) to analyze the variation between samples and to assess if the variation is descriptive of the biology or a reflection of unanticipated variables. Furthermore, examples were provided of case studies wherein the phenotypic readout was an alteration in protein composition, not in abundance, that correlated with the biology. Importantly, these types of phenotypes would go undetected in a global-scale bottom-up, peptide-based strategy but are readily observable and quantifiable using a quantitative intact proteomics (QIP) approach.

References

Alban, A., David, S.O., Bjorkesten, L., Andersson, C., Sloge, E., Lewis, S., and Currie, I. (2003). A novel experimental design for comparative two-dimensional gel analysis: Two-dimensional difference gel electrophoresis incorporating a pooled internal standard. Proteomics *3*, 36–44.

Anderle, M., Roy, S., Lin, H., Becker, C., and Joho, K. (2004). Quantifying reproducibility for differential proteomics: Noise analysis for protein liquid chromatography-mass spectrometry of human serum. Bioinformatics *20*, 3575–3582. Epub 2004 Jul 3529.

Bengtsson, S., Krogh, M., Szigyarto, C.A., Uhlen, M., Schedvins, K., Silfversward, C., Linder, S., Auer, G., Alaiya, A., and James, P. (2007). Large-scale proteomics analysis of human ovarian cancer for biomarkers. J Proteome Res *6*, 1440–1450. Epub 2007 Feb 1422.

Cho, H., Smalley, D.M., Theodorescu, D., Ley, K., and Lee, J.K. (2007). Statistical identification of differentially labeled peptides from liquid chromatography tandem mass spectrometry. Proteomics *7*, 3681–3692.

Elliott, M.H., Smith, D.S., Parker, C.E., and Borchers, C. (2009). Current trends in quantitative proteomics. J Mass Spectrom *44*, 1637–1660.

Franco, A.T., Friedman, D.B., Nagy, T.A., Romero-Gallo, J., Krishna, U., Kendall, A., Israel, D.A., Tegtmeyer, N., Washington, M.K., and Peek, R.M., Jr. (2009). Delineation of a carcinogenic helicobacter pylori proteome. Mol Cell Proteomics *25*, 25.

Franco, A.T., Israel, D.A., Washington, M.K., Krishna, U., Fox, J.G., Rogers, A.B., Neish, A.S., Collier-Hyams, L., Perez-Perez, G.I., Hatakeyama, M., *et al.* (2005). Activation of beta-catenin by carcinogenic helicobacter pylori. Proc Natl Acad Sci USA *102*, 10646–10651. Epub 12005 Jul 10618.

Friedman, D.B., Hill, S., Keller, J.W., Merchant, N.B., Levy, S.E., Coffey, R.J., and Caprioli, R.M. (2004). Proteome analysis of human colon cancer by two-dimensional difference gel electrophoresis and mass spectrometry. Proteomics *4*, 793–811.

Friedman, D.B., Stauff, D.L., Pishchany, G., Whitwell, C.W., Torres, V.J., and Skaar, E.P. (2006). *Staphylococcus aureus* redirects central metabolism to increase iron availability. PLoS Pathogens *2*, e87.

Friedman, D.B., Wang, S.E., Whitwell, C.W., Caprioli, R.M., and Arteaga, C.L. (2007). Multivariable difference gel electrophoresis and mass spectrometry: A case study on TGF-beta and ErbB2 signaling. Mol Cell Proteomics *6*, 150–169.

Garge, N., Pan, H., Rowland, M.D., Cargile, B.J., Zhang, X., Cooley, P.C., Page, G.P., and Bunger, M.K. (2010). Identification of quantitative trait loci underlying proteome variation in human lymphoblastoid cells. Mol Cell Proteomics *9*, 1383–1399.

Hatakeyama, H., Kondo, T., Fujii, K., Nakanishi, Y., Kato, H., Fukuda, S., and Hirohashi, S. (2006). Protein clusters associated with carcinogenesis, histological differentiation and nodal metastasis in esophageal cancer. Proteomics *6*, 6300–6316.

Karp, N.A., and Lilley, K.S. (2005). Maximising sensitivity for detecting changes in protein expression: Experimental design using minimal cydyes. Proteomics *5*, 3105–3115.

Karp, N.A., and Lilley, K.S. (2009). Investigating sample pooling strategies for DIGE experiments to address biological variability. Proteomics *9*, 388–397.

Klose, J. (1975). Protein mapping by combined isoelectric focusing and electrophoresis of mouse tissues. A novel approach to testing for induced point mutations in mammals. Humangenetik *26*, 231–243.

Kolkman, A., Dirksen, E.H., Slijper, M., and Heck, A.J. (2005). Double standards in quantitative proteomics: Direct comparative assessment of difference in gel electrophoresis and metabolic stable isotope labeling. Mol Cell Proteomics *4*, 255–266. Epub 2005 Jan 2004.

Laemmli, U.K. (1970). Cleavage of structural proteins during the assembly of the head of bacteriophage T4. Nature *227*, 680–685.

Link, A.J., Eng, J., Schieltz, D.M., Carmack, E., Mize, G.J., Morris, D.R., Garvik, B.M., and Yates, J.R., 3rd. (1999). Direct analysis of protein complexes using mass spectrometry. Nat Biotechnol *17*, 676–682.

Lottspeich, F. (2009). Introduction to proteomics. Methods Mol Biol *564*, 3–10.

O'Farrell, P.H. (1975). High resolution two-dimensional electrophoresis of proteins. J Biol Chem *250*, 4007–4021.

Seike, M., Kondo, T., Fujii, K., Yamada, T., Gemma, A., Kudoh, S., and Hirohashi, S. (2004). Proteomic signature of human cancer cells. Proteomics *4*, 2776–2788.

Suehara, Y., Kondo, T., Fujii, K., Hasegawa, T., Kawai, A., Seki, K., Beppu, Y., Nishimura, T., Kurosawa, H., and Hirohashi, S. (2006). Proteomic signatures corresponding to histological classification and grading of soft-tissue sarcomas. Proteomics *6*, 4402–4409.

Tonge, R., Shaw, J., Middleton, B., Rowlinson, R., Rayner, S., Young, J., Pognan, F., Hawkins, E., Currie, I., and Davison, M. (2001). Validation and development of fluorescence two-dimensional differential gel electrophoresis proteomics technology. Proteomics *1*, 377–396.

Vogt, J.A., Schroer, K., Holzer, K., Hunzinger, C., Klemm, M., Biefang-Arndt, K., Schillo, S., Cahill, M.A., Schrattenholz, A., Matthies, H., *et al.* (2003). Protein abundance quantification in embryonic stem cells using incomplete metabolic labelling with 15 N amino acids, matrix-assisted laser desorption/ionisation time-of-flight mass spectrometry, and analysis of relative isotopologue abundances of peptides. Rapid Commun Mass Spectrom *17*, 1273–1282.

Von Eggeling, F., Gawriljuk, A., Fiedler, W., Ernst, G., Claussen, U., Klose, J., and Romer, I. (2001). Fluorescent dual colour 2D-protein gel electrophoresis for rapid detection of differences in protein pattern with standard image analysis software. Int J Mol Med *8*, 373–377.

Washburn, M.P., Wolters, D., and Yates, J.R., 3rd. (2001). Large-scale analysis of the yeast proteome by multidimensional protein identification technology. Nat Biotechnol *19*, 242–247.

Wolf-Yadlin, A., Hautaniemi, S., Lauffenburger, D.A., and White, F.M. (2007). Multiple reaction monitoring for robust quantitative proteomic analysis of cellular signaling networks. Proc Natl Acad Sci USA *104*, 5860–5865. Epub 2007 Mar 5826.

Wolters, D.A., Washburn, M.P., and Yates, J.R., 3rd. (2001). An automated multidimensional protein identification technology for shotgun proteomics. Anal Chem *73*, 5683–5690.

Chapter 15
Two-Dimensional Non-denaturing Gel Electrophoresis for Characterization of Proteins in Multi-molecular Particles by Mass Spectrometry

Susan T. Weintraub and Philip Serwer

Abstract Analysis of intact multi-molecular complexes by gel electrophoresis requires the use of non-denaturing conditions. Agarose is the matrix of choice for separation of large, intact complexes because gels with larger pore sizes can be more readily prepared and manipulated compared to acrylamide. In the absence of SDS (or any other denaturant), the separation is based on both the average electrical surface charge density (σ) and the effective radius (R_E). In order to separately measure σ and R_E, two dimensions of electrophoresis are required. Native 1-D and 2-D agarose gel electrophoresis (AGE) have been used for analysis of bacteriophages and related particles, clathrin coated vesicles, cross-linked microtubules, multi-enzyme complexes, protein/polysaccharide vaccines, and ribosomes. 2-D AGE has proved to be especially valuable for characterization of bacteriophages and related particles because not only is information obtained about particle σ and R_E, but also, the separated particles can be isolated from the agarose gel and examined by microscopy or subjected to DNA or protein analysis, as illustrated in this chapter for bacteriophage T3 proteins.

Keywords Non-denaturing agarose gel electrophoresis · Bacteriophage assembly · Protein analysis · Mass spectrometry

15.1 Introduction

Polyacrylamide gel electrophoresis (PAGE) is widely used for separation of proteins prior to analysis by mass spectrometry (MS). Applications include use of one-dimensional (1-D) and two-dimensional (2-D) separations under a variety of conditions, depending on the specific goals of the experiment. Most commonly, separations are conducted in the presence of sodium dodecyl sulfate (SDS) in order to

---

S.T. Weintraub (✉)
Department of Biochemistry, The University of Texas Health Science Center at San Antonio, San Antonio, TX 78229-3900, USA
e-mail: weintraub@uthscsa.edu

A.R. Ivanov, A.V. Lazarev (eds.), *Sample Preparation in Biological Mass Spectrometry*, DOI 10.1007/978-94-007-0828-0_15,

dissociate protein complexes and denature proteins. The SDS also coats the proteins and makes their average electrical surface charge density (σ) more uniform. This makes it possible to use a 1-D gel to separate proteins on the basis of molecular weight.

Analysis of intact multi-molecular complexes requires the use of non-denaturing conditions, including both PAGE and native agarose gel electrophoresis (AGE). Agarose is the matrix of choice for separation of large, intact complexes because gels with larger pore sizes can be more readily prepared and manipulated compared to polyacrylamide. In the absence of SDS (or any other denaturant), the separation is based on both σ and the effective radius (R_E) (reviewed in Serwer et al., 1989; Serwer and Griess, 1999). However, 1-D gel electrophoresis does not provide independent determination of σ and R_E. In order to separately measure σ and R_E, a second dimension of electrophoresis is required. Native 1-D and 2-D AGE have been used for analysis of bacteriophages and related particles (Casjens et al., 1992; Duda et al., 2009; Fang et al., 2008), clathrin coated vesicles (Kedersha and Rome, 1986), cross-linked microtubules (Serwer et al., 1989), multi-enzyme complexes (Easom et al., 1989; Serwer et al., 1989), protein/polysaccharide vaccines (Serwer and Hayes, 1986; Tietz, 2007), and ribosomes (Tokimatsu et al., 1981). 2-D AGE has proved to be especially valuable for characterization of bacteriophages and related particles (Casjens et al., 1992; Serwer et al., 1989) because not only is information obtained about particle σ and R_E, but also, the separated particles can be isolated from the agarose gel and examined by microscopy or subjected to DNA or protein analysis, as illustrated below for bacteriophage T3.

Initial fractionation of bacteriophages and related particles is usually accomplished by some type of ultracentrifugation. Buoyant density centrifugation in a cesium chloride density gradient is often the method of choice because double-stranded DNA phage particles, with typical densities in the 1.4–1.5 g/mL range, can be well separated from proteins (1.27–1.32 g/mL) and double-stranded DNA (~1.7 g/mL). In our studies, buoyant density centrifugation is usually preceded by centrifugation through a pre-formed step gradient of cesium chloride in order to separate the phages from the more slowly sedimenting proteins and lower density fragments of the envelope of the host cell (including outer membrane vesicles).

Interest in studying bacteriophages stems not only from their role in microbial communities (Srinivasiah et al., 2008; Weinbauer and Rassoulzadegan, 2004) and potential use for treatment of antibiotic-resistant infections (Skurnik et al., 2007; Deresinski, 2009), but also from use of their DNA packaging motors as models for other biological motors (Fang et al., 2008; Rao and Feiss, 2008; Sun et al., 2008). In addition, studies of phage infection have revealed many of the fundamentals of molecular biology (Campbell, 2003; Gottesman and Weisberg, 2004) and have been important in development of procedures used for determining the structure and structural dynamics of multi-molecular complexes [including conventional electron microscopy (Ackermann, 2009; Buhle et al., 1985) and cryo-electron microscopy (Johnson and Chiu, 2007; Zhou, 2008)].

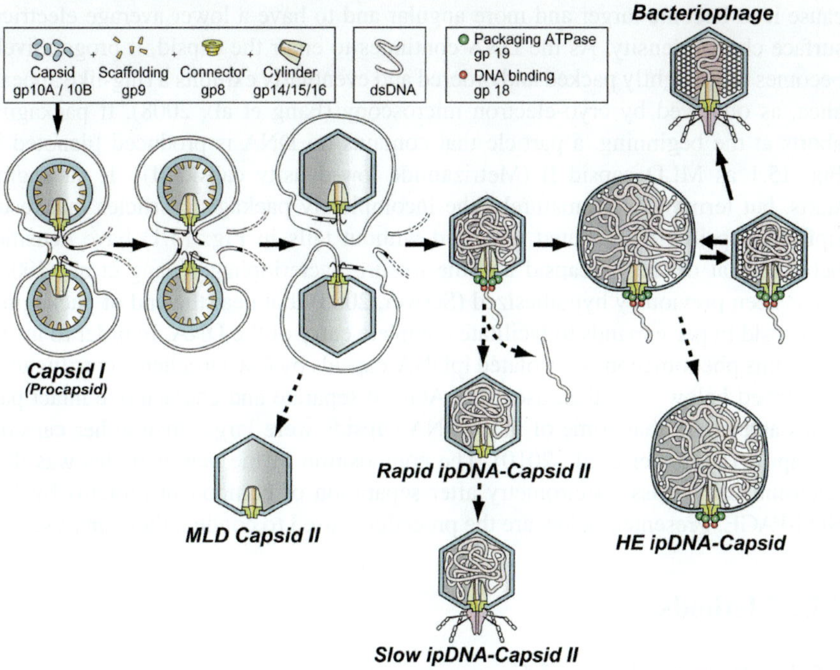

Fig. 15.1 The DNA packaging pathway of the related bacteriophages, T3 and T7. The steps indicated by *solid arrows* have been deduced from in vivo data. The *dashed arrows* show production of a variety of particles that have been detected, isolated, and characterized (Fang et al., 2008; Serwer et al., 2010). MLD, Metrizamide low density; ipDNA, incompletely packaged DNA; HE, hyper-expanded. Analysis of proteins in HE ipDNA-capsid particles is described at the end of this chapter

Bacteriophages T3 and T7 are short-tailed, double-stranded DNA bacterial viruses (podoviruses) that consist of an outer protein head (capsid) with icosahedral symmetry (that contains the double-stranded DNA genome) and a tail. There is a high degree of similarity between the proteins of T3 and T7. Details of T3 and T7 sequence and genetics have been reviewed by Molineux and co-workers (Pajunen et al., 2002). DNA packaging in bacteriophage T3/T7 is illustrated in Fig. 15.1. Structural proteins of T3/T7 are initially assembled into a round procapsid (also called capsid I). After assembly of capsid I, the bacteriophage DNA starts to be packaged into the capsid by the DNA packaging motor. This motor consists of a connector ring of 12 subunits of gene product (gp) 8 embedded in the head along with a co-axial multimer of packaging ATPase subunits (gp19). For some bacteriophages, including T3 and T7, the packaged DNA molecule is cleaved from an end-to-end concatemer of mature genomes.

DNA packaging is an energy-requiring process that is fueled by ATP hydrolysis that is catalyzed, at least in part, by the packaging ATPase. In concert with DNA packaging in T3/T7, the outer shell undergoes several major transitions which cause it to become larger and more angular and to have a lower average electrical surface charge density. As the DNA continues to enter the capsid, it progressively becomes more tightly packed and ordered and eventually exhibits a ring-like appearance, as observed by cryo-electron microscopy (Fang et al., 2008). If packaging aborts at the beginning, a particle that contains no DNA is produced [denoted in Fig. 15.1 as MLD capsid II (Metrizamide low-density capsid II)]. If packaging starts, but terminates prematurely, the incompletely packaged particles that result (ipDNA-capsid II; illustrated with and without tails in Fig. 15.1) have densities between that of the procapsid and the mature bacteriophage (Fang et al., 2008). It has been previously hypothesized (Serwer, 2003) that near the end of packaging, the capsid hyper-expands to facilitate complete entry of the DNA. In order to investigate this phenomenon, we isolated ipDNA-capsids by CsCl gradient centrifugation (described below). We then used 2-D AGE to separate and characterize intact particles and found that some of the ipDNA-capsids were larger than either capsid I or capsid II (Serwer et al., 2010). The composition of the new particles was then determined by mass spectrometry after separation of component proteins by 1-D SDS-PAGE. Presented below are the procedures used to conduct these analyses.

15.2 Methods

15.2.1 Preparation and Concentration of Bacteriophage-related Particles

The following protocol is based on published procedures (Fang et al., 2008) and is scaled for use with up to 12 L of aerated *Escherichia coli* BB/1 infected with phage T3 (multiplicity = 0.01), incubated at 30°C until spontaneous lysis.

Medium (2× LB broth): 20 g Bacto Tryptone, 10 g yeast extract, 5 g NaCl in 1 L of water
Add sterile solid NaCl to the lysate to a final concentration of 0.5 M.
Immediately remove debris by pelleting in six 1.0-L bottles at 4,000 rpm (Beckman JS4.2 rotor) for 15 min.
Add solid polyethylene glycol 8,000 to a final concentration of 9%.
Precipitate the capsids, ipDNA-capsids and bacteriophage particles by allowing this mixture to sit at 4°C for 2–4 days.
Decant the supernatant.
Compact the precipitate by centrifugation at 4,000 rpm for 45 min (JS4.2 rotor).
Remove residual supernatant and discard.
Resuspend the precipitated particles in buffer containing 0.5 M NaCl, 1 mM $MgCl_2$, 10 mM Tris-Cl, pH 7.4. Use 7 mL/L of original lysate.
Centrifuge at 10,000 rpm (JLA16.250 rotor), 4°C, for 10 min to remove large cell envelope components.
Remove supernatant.
Wash the pellet two more times with the same buffer and combine washes with the previous supernatant.
Add buffer to a volume of 160 mL.

Add solid polyethylene glycol 8,000 to the resuspended particles to a final concentration of 9%.

Precipitate at 4°C for ~16 h.

After resuspension (~17 mL final volume) and clarification as described above, add DNase I (2.5 μg/mL final concentration) and incubate at 30°C for 60 min. [Note: (1) This concentration of DNase reduces sample viscosity but does not completely degrade DNA. (2) Pour the CsCl step gradient (see Section 15.2.2.1) during the 60-min incubation.]

15.2.2 Density-Based Fractionation

15.2.2.1 Cesium Chloride Step Density Gradient

Prepare a cesium chloride step gradient consisting of the following steps: 1.226 g/mL (1.5 mL), 1.349 g/mL (1.0 mL); 1.446 g/mL (1.5 mL); 1.510 g/mL (1.0 mL); 1.728 g/mL (1.0 mL). Use a buffer containing 1 mM $MgCl_2$ and 10 mM Tris-Cl, pH 7.4, for the CsCl solutions. [Note: A step gradient of this size is appropriate for a 2- to 3-L bacteriophage lysate.]

Layer 5.6 mL of the DNase-digested mixture on top of the step gradient.

Centrifuge at 33,000 rpm (Beckman SW41 rotor), 18°C, for 3 h.

Illuminate the tube from the bottom using a fluorescent light source and photograph the light scattered from the particles in the gradient.

Remove each fraction (typically 0.2–0.4 mL) by pipetting from the top. [Note: In comparison to tube puncture, pipetting from the top reduces inter-fraction cross-contamination. In addition, pipetting from the top facilitates fraction selection because there is sufficient time for visual inspection of the bands during collection.]

15.2.2.2 Buoyant Density Centrifugation in a Cesium Chloride Density Gradient

For each fraction of interest, dilute the corresponding step gradient fraction with an equal volume of gradient buffer (1 mM $MgCl_2$ and 10 mM Tris-Cl, pH 7.4) and remove debris by spinning at 10,000 rpm (Beckman JLA16.250 rotor), 4°C, for 10 min. [Note: In some cases, it may be necessary to pool corresponding fractions from step gradients of replicate samples prior to buoyant density centrifugation in order to obtain a sufficient quantity of bacteriophage particles for subsequent analysis.]

Adjust the density of the solution to 1.340 g/mL and the volume to 5.2 mL by addition of the appropriate amount of solid CsCl and buffer.

Subject each fraction to buoyant density centrifugation at 42,000 rpm (Beckman SW50.1 rotor) for 20 h.

Illuminate the tube from the bottom using a fluorescent light source and photograph the light scattered from the particles in the gradient.

Remove each fraction by pipetting from the top.

15.2.3 Non-denaturing Agarose Gel Electrophoresis (AGE)

15.2.3.1 1-D AGE

Add 182 mL of water and the required amount of agarose (Seakem LE; Lonza, Rockland ME) to a flask to achieve the desired final agarose concentration.
Determine the weight of the flask and mixture.
Microwave at high power until the solution boils.
Vortex to mix and microwave to boiling again. [Note: This step is necessary to insure that the agarose is uniformly dispersed.]
Weigh and replace lost water as needed.
Add 18 mL of 10× electrophoresis buffer (10 mM $MgCl_2$, 0.9 M Tris-acetate, pH 8.4).
Mix by swirling.
Cool for 10 min in a 50°C water bath.
Pour the gel in an apparatus for horizontal, submerged gel electrophoresis.
Allow the gel to sit at room temperature for 45 min.
Submerge the gel under electrophoresis buffer.
Mix an aliquot of a density gradient fraction with 0.35× volume of electrophoresis buffer (1 mM $MgCl_2$, 90 mM Tris-acetate, pH 8.4). [Note: Aliquot volumes of 25 μL are typically appropriate.]
Add 0.11× volume of 50% sucrose in electrophoresis buffer.
Layer each solution in a well of the submerged gel. [Note: After loading, let the gel sit in the electrophoresis buffer for 60 min before applying the electrical potential in order to dialyze enough CsCl out of the sample so that increased ionic strength at the origin does not interfere with the initial movement of particles.]
Electrophorese at 2.0 V/cm, 25°C for 10 h. [Note: The electrophoresis buffer must be circulated during electrophoresis to prevent formation of pH gradients; circulation should be started 30 min after application of the electric field.]
Stain for DNA. [Note: Typical stains include ethidium bromide and GelStar (Lonza). For bacteriophage separations, electrophoresis buffer is used to dissolve the stain in order to help preserve particle integrity.]
Acquire a gel image using a device with the appropriate fluorescence detection.
Stain the gel for protein (e.g., with a visible stain like Coomassie blue).
Acquire a gel image using the appropriate device.

15.2.3.2 2-D AGE

Use samples prepared as described above for 1-D AGE. [Note: Bovine serum albumin can be added at this stage (final concentration of 100 μg/mL) to minimize sample binding to the agarose at the gel origin.]

Cast a submerged, horizontal 1.8% agarose gel in an apparatus in which the direction of electrophoresis can be changed by 90° without manipulating the gel (Aquebogue Machine Shop; Riverhead, NY). Block the location for the first-dimension segment (see below) with an appropriate sized piece of plastic when pouring this gel. [Note: A detailed schematic of the apparatus is shown in Serwer (1985).]

Embed a 0.3% agarose first dimensional gel within the blocked region of the 1.8% agarose gel, submerged beneath electrophoresis buffer.

Electrophorese at 2.0 V/cm for 4.0 h, circulating the buffer starting at 30 min after electrophoresis.

Change the direction of electrophoresis by 90° and conduct the second dimension at 1.8 V/cm for 10.0 h with buffer circulation.

Stain and acquire images for DNA and protein (see above).

15.2.4 Identification of Proteins in the 2-D AGE Gels

15.2.4.1 1-D SDS-PAGE

Excise individual plugs from regions of interest in the agarose gel using a surgical blade. [Note: Print a full-size image of the gel (DNA or protein stain, as appropriate) to use as a guide for determining plug locations.]

Transfer each plug to a 12 × 75 mm culture tube.

Re-image the gel to confirm the location of excised plugs.

To each gel plug, add 60 μL of buffer containing 20.5% glycerol, 4.8% SDS, 8 mM EDTA, 4.8% β-mercaptoethanol, 0.24 mg/mL bromophenol blue and 0.23 M Tris-Cl, pH 6.8.

Incubate at room temperature for 1.5 h with agitation.

Melt the agarose by incubating the tubes for 3 min at 85–90°C in a water bath.

Transfer tubes to a heating block and maintain temperature at ~80°C to prevent re-gelation of the agarose.

Load 45 μL of the melted agarose plug into a well of an SDS-PAGE gel (Criterion, Bio-Rad) using a pipette that has been pre-warmed to 37°C. [Note: A pipettor fitted with a polypropylene tip that has had ~3 mm of the small end cut off can be used without pre-warming.]

Electrophorese for a time sufficient to run the tracking dye the desired distance into the separating polyacrylamide gel. [Note: The total gel run distance is determined on the basis of expected sample complexity. For the example presented at the end of this chapter, this distance was approximately 2.5 cm.]

Stain the gel for total protein. [Note: A high-sensitivity stain like SYPRO Ruby (Invitrogen) can be used to obtain gel images, but post-staining with a visible stain that is MS-compatible (e.g., Coomassie blue) is highly recommended to facilitate gel excision.]

Acquire gel images using an appropriate scanner or imager, depending on the stain used.

15.2.4.2 In-Gel Digestion

Excision and Destaining of SDS-PAGE Slices

Excise the protein-containing region of each SDS-PAGE lane into slices using an X-acto knife fitted with a clean flat blade (13 mm). [Note: The number of slices excised is determined by the length of the gel run. Slices of ~3 mm are suitable for digestion in standard 0.6-mL polypropylene tubes. For a 2.5-cm total run, the protein-containing region is ~1.5 cm.]
Divide each slice into cubes (~1.5 mm per side).
Using flat forceps (Bio-Rad part number 165-4070) and the X-acto blade, transfer the cubes to a 0.6-mL polypropylene tube (Sarstedt) containing 400 μL destaining solution (40 mM NH_4CO_3/50% acetonitrile) [Note: There must be liquid in the tube to release the gel pieces from the forceps.]
Aspirate and discard the destaining solution using a 1-mL pipettor.
Add 500 μL destaining solution.
Incubate on an inversion mixer at room temperature for 20 min.
Aspirate and discard the destaining solution.
Repeat the above wash/aspirate steps.
Add 500 μL acetonitrile.
Invert the tube twice.
Aspirate and discard the acetonitrile.
Add 500 μL acetonitrile.
Incubate on an inversion mixer at room temperature for 20 min.
Aspirate and discard the acetonitrile.
Spin at low speed in a bench-top centrifuge.
Aspirate and discard the remaining 10–20 μL of acetonitrile.
The slice can be used immediately for in-gel digestion or it can be stored at room temperature for an extended period of time.

In-Gel Digestion

Add 10–15 μL (depending on the size of the slice) of trypsin (Promega, sequencing grade, item number V5111; 10 ng/μL in 40 mM NH_4CO_3/20% acetonitrile).
Incubate at 37°C overnight.
Add 100 μL 0.1% trifluoroacetic acid (TFA) to the digest and incubate at 30°C for 1 h.
Transfer the liquid above the gel cubes to a sample tube or autosampler vial.
Add 100 μL of 0.1% TFA/50% acetonitrile to the gel cubes.
Incubate at 30°C for 1 h.
Remove the liquid above the gel cubes and combine with the first 0.1% TFA extract.
Spin to dryness in a vacuum centrifuge.
Resuspend the digest in 10 μL 0.5% TFA. [Note: When samples are either of a low degree of complexity or contain low levels of protein, it is often

advantageous to combine the digests from all of the slices in a gel lane prior to analysis by the MS-based method of choice.]

15.3 Use of 2-D AGE and Mass Spectrometry to Characterize Unique Bacteriophage T3 Intermediates

As a part of our investigation of biological motors, we have been studying the dynamics of bacteriophage DNA packaging (illustrated in Fig. 15.1). Non-denaturing agarose gel electrophoresis and mass spectrometry have been essential techniques for detection and characterization of intermediate particles generated during this process. Results from analyses of capsid-containing particles of bacteriophage T3 LG114 (a deletion mutant that contains a shortened double-stranded DNA genome) are presented below.

The goal of the analysis was to elucidate the nature of T3 DNA packaging intermediates that appear to be hyper-expanded—that is, they contain DNA but the

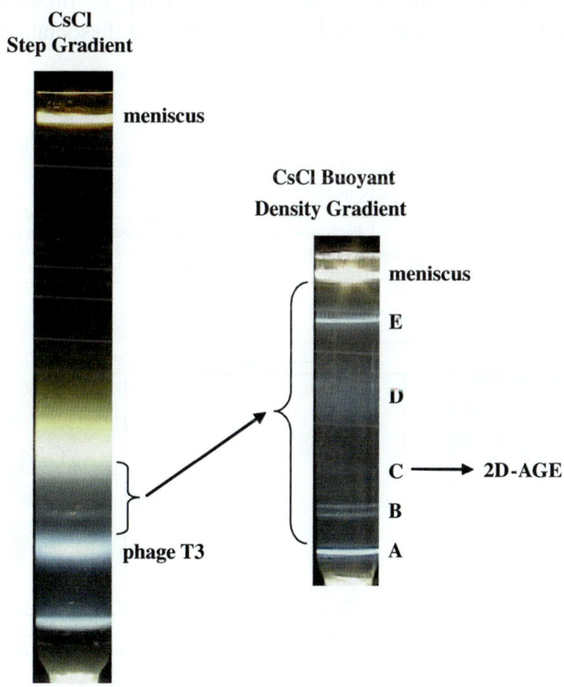

Fig. 15.2 Isolation of ipDNA-capsids of bacteriophage T3 LG114 by CsCl density gradient centrifugation. Particles in the step gradient fraction with densities between 1.25 and 1.44 g/mL were further separated by CsCl buoyant-density gradient centrifugation; A, residual intact phage T3; B, dimers of phage T3 and empty capsids; C, 1.3404 g/mL ipDNA-capsids, some of which are hyper-expanded; D, cell envelope fragments; E, empty capsids. The 1.3404-g/mL fraction was analyzed by 2-D agarose gel electrophoresis (2D-AGE; Fig. 15.3)

diameter of the capsid is larger than that of mature T3. For this study, phages and related particles (including ipDNA-capsids) were separated by a CsCl step gradient (see Section 15.2.2.1) and then the particles with densities between 1.25 and 1.44 g/mL (lower density than mature T3, as shown in Fig. 15.2) were subsequently separated with higher resolution by CsCl buoyant-density gradient centrifugation (see Section 15.2.2.2). The 1.3404-g/mL buoyant-density gradient fraction was added to the sample well of a 0.3% submerged agarose gel that was embedded within a 1.8% agarose gel, in preparation for 2-D AGE (see Section 15.2.3.2). After electrophoresis in the first dimension for 4 h at 2.0 V/cm, the electrical field was turned 90° and applied for 10 h at 1.8 V/cm for the second dimension. Images of the two dimensions of AGE separations (GelStar DNA stain) are shown in Fig. 15.3. (The stain penetrates the particles and binds to the packaged DNA.) As described in Section 15.2.4.1, the indicated gel plugs were excised, boiled in SDS sample buffer, and the proteins separated by 1D SDS-PAGE (1.5 cm protein-containing region). The proteins in the gel slices were digested in situ with trypsin (see Section 15.2.4.2), and the digests for each gel lane were combined and analyzed by capillary HPLC-electrospray ionization-tandem mass spectrometry (HPLC-ESI-MS/MS) on a Thermo Fisher LTQ mass spectrometer, as previously described (Fang et al., 2008). The uninterpreted collision-induced dissociation spectra were

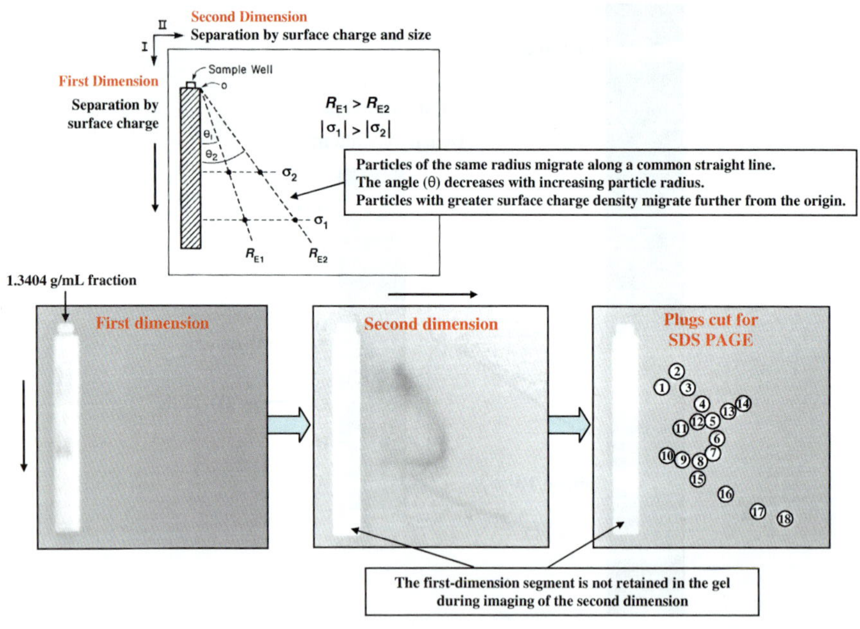

Fig. 15.3 Analysis of hyper-expanded assembly intermediates of bacteriophage T3 by 2-D, non-denaturing agarose gel electrophoresis. The particles separated in this process are all negatively charged and move toward the anode (shown as *down* in the first dimension and to the *right* in the second dimension). Proteins in the indicated gel plugs were subsequently separated by 1-D SDS-PAGE, digested in situ with trypsin, and the digests analyzed by HPLC-ESI-MS/MS. The results of the analyses are shown in Table 15.1

Table 15.1 Proteins identified in hyper-expanded assembly intermediates of bacteriophage T3

	Gel plug																	
Protein	1	2	3	4	5	6	7	8	9	10	11	12	13	14	15	16	17	18
gp10	4	35	26	12	12	15	17	11	6	4	4	9	12					3
gp8		15	7															
gp15		9																
gp16		4																

The values represent the number of tryptic peptides from the HPLC-ESI-MS/MS analysis assigned with ≥95% confidence.

searched against the NCBInr database using Mascot (Matrix Science, London, UK). Methionine oxidation was considered as a variable modification for all searches. Cross correlation of the Mascot results with X! Tandem and determination of peptide and protein identity probabilities were accomplished by Scaffold (Proteome Software).

Shown in Table 15.1 are the numbers of peptides assigned with ≥95% confidence to the indicated bacteriophage T3 proteins. Entries for trypsin, keratin, and the BSA that was added to the sample as a carrier for 2D-AGE are not shown. The results revealed that all of the arc-forming particles analyzed contained bacteriophage proteins, predominantly the high-abundance major head protein (gp10A/B). It was only possible to detect the minor T3 capsid proteins in one of the samples (plug 2), presumably due to the relatively low quantity of particles in the fraction that was analyzed. It is important to note that no host proteins were identified in the arc-forming particles, indicating that these particles are not vesicles derived from the *E. coli* host. These protein analyses, together with electron microscopy of particles analyzed (Serwer et al., 2010), provide the first evidence to support the existence of hyper-expanded capsids that are produced during bacteriophage assembly.

Acknowledgments This research was supported by grants from the National Institutes of Health (GM24365 and GM069757), the Welch Foundation (AQ-764), and The Robert J. Kleberg, Jr. and Helen C. Kleberg Foundation. MS analyses were conducted in the UTHSCSA Institutional Mass Spectrometry Laboratory. We thank Elena Wright, Kevin W. Hakala, and Curtis Johnson for their invaluable assistance, and the UTHSCSA Bioinformatics Center for help with computational aspects of the project.

References

Ackermann, H.W. (2009). Basic phage electron microscopy. Meth Mol Biol *501*, 113–126.

Buhle, E.L., Jr., Aebi, U., and Smith, P.R. (1985). Correlation of surface topography of metal-shadowed specimens with their negatively stained reconstructions. Ultramicroscopy *16*, 436–450.

Campbell, A. (2003). The future of bacteriophage biology. Nat Rev Genet *4*, 471–477.

Casjens, S., Wyckoff, E., Hayden, M., Sampson, L., Eppler, K., Randall, S., Moreno, E.T., and Serwer, P. (1992). Bacteriophage P22 portal protein is part of the gauge that regulates packing density of intravirion DNA. J Mol Biol *224*, 1055–1074.

Deresinski, S. (2009). Bacteriophage therapy: exploiting smaller fleas. Clin Infect Dis *48*, 1096–1101.

Duda, R.L., Ross, P.D., Cheng, N., Firek, B.A., Hendrix, R.W., Conway, J.F., and Steven, A.C. (2009). Structure and energetics of encapsidated DNA in bacteriophage HK97 studied by scanning calorimetry and cryo-electron microscopy. J Mol Biol *391*, 471–483.

Easom, R.A., DeBuysere, M.S., Olson, M.S., and Serwer, P. (1989). Size determination of multienzyme complexes using two-dimensional agarose gel electrophoresis. Proteins *5*, 224–232.

Fang, P.-A., Wright, E.T., Weintraub, S.T., Hakala, K., Wu, W., Serwer, P., and Jiang, W. (2008). Visualization of bacteriophage T3 capsids with DNA incompletely packaged in vivo. J Mol Biol *384*, 1384–1399.

Gottesman, M.E., and Weisberg, R.A. (2004). Little lambda, who made thee? Microbiol Mol Biol Rev *68*, 796–813.

Johnson, J.E., and Chiu, W. (2007). DNA packaging and delivery machines in tailed bacteriophages. Curr Opin Struct Biol *17*, 237–243.

Kedersha, N.L., and Rome, L.H. (1986). Preparative agarose gel electrophoresis for the purification of small organelles and particles. Anal Biochem *156*, 161–170.

Pajunen, M.I., Elizondo, M.R., Skurnik, M., Kieleczawa, J., and Molineux, I.J. (2002). Complete nucleotide sequence and likely recombinatorial origin of bacteriophage T3. J Mol Biol *319*, 1115–1132.

Rao, V.B., and Feiss, M. (2008). The bacteriophage DNA packaging motor. Annu Rev Genet *42*, 647–681.

Serwer, P. (1985). Two-dimensional agarose gel electrophoresis without gel manipulation. Anal Biochem *144*, 172–178.

Serwer, P. (2003). Models of bacteriophage DNA packaging motors. J Struct Biol *141*, 179–188.

Serwer, P., Easom, R.A., Hayes, S.J., and Olson, M.S. (1989). Rapid detection and characterization of multimolecular cellular constituents by two-dimensional agarose gel electrophoresis. TIBS *14*, 4–7.

Serwer, P., and Griess, G.A. (1999). Advances in the separation of bacteriophages and related particles. J Chromatogr B *722*, 179–190.

Serwer, P., and Hayes, S.J. (1986). Two-Dimensional Agarose Gel Electrophoresis. In Electrophoresis '86, M.J. Dunn, ed. (Weinheim, VCH), pp. 243–252.

Serwer, P., Wright, E.T., Hakala, K., Weintraub, S.T., Su, M., and Jiang, W. (2010). DNA packaging-associated hyper-capsid expansion of bacteriophage T3. J Mol Biol *397*, 361–374.

Skurnik, M., Pajunen, M., and Kiljunen, S. (2007). Biotechnological challenges of phage therapy. Biotechnol Lett *29*, 995–1003.

Srinivasiah, S., Bhavsar, J., Thapar, K., Liles, M., Schoenfeld, T., and Wommack, K.E. (2008). Phages across the biosphere: contrasts of viruses in soil and aquatic environments. Res Microbiol *159*, 349–357.

Sun, S., Kondabagil, K., Draper, B., Alam, T.I., Bowman, V.D., Zhang, Z., Hegde, S., Fokine, A., Rossmann, M.G., and Rao, V.B. (2008). The structure of the phage T4 DNA packaging motor suggests a mechanism dependent on electrostatic forces. Cell *135*, 1251–1262.

Tietz, D. (2007). Computer-assisted 2-D agarose electrophoresis of Haemophilus influenzae type B meningitis vaccines and analysis of polydisperse particle populations in the size range of viruses: A review. Electrophoresis *28*, 512–524.

Tokimatsu, H., Strycharz, W.A., and Dahlberg, A.E. (1981). Gel electrophoretic studies on ribosomal proteins L7/L12 and the Escherichia coli 50 S subunit. J Mol Biol *152*, 397–412.

Weinbauer, M.G., and Rassoulzadegan, F. (2004). Are viruses driving microbial diversification and diversity? Environ Microbiol *6*, 1–11.

Zhou, Z.H. (2008). Towards atomic resolution structural determination by single-particle cryo-electron microscopy. Curr Opin Struct Biol *18*, 218–228.

Part V
Sample Preparation and Analysis Techniques for Biological Fluids and Biomarker Discovery

Chapter 16
Preanalytical Variables for Plasma and Serum Proteome Analyses

Craig A. Gelfand and Gilbert S. Omenn

Abstract Proteomic biomarker discovery, confirmation, and validation studies depend absolutely on great care in the management and documentation of preanalytical sources of variation. This consideration is particularly important for such complex and dynamic specimens as plasma. Clues to the awareness of the investigator are clarity about whether the specimen is serum or plasma, reasons stated for the choice, specification of the anticoagulant in the case of plasma, and statement about whether enrichment, depletion, or protease inhibition methods were deployed. This chapter provides a systematic discussion of preanalytical variables affecting plasma and serum and offers practical guidance for reducing undesirable variation in results from these sources.

Keywords Plasma · Serum · Preanalytical variables · Protease inhibitors · Anticoagulants · Mass spectrometry · Reproducibility · Coefficient of variation · Biomarkers

16.1 Introduction

Advances in proteomics methods over the past decade, particularly high-resolution mass spectrometry (MS), have brought the potential of a new era for clinical diagnostics. The term "biomarkers" has become part of the modern lexicon, often replacing "analyte" or even "ligand." Biomarkers, as molecules that bear diagnostic value, imply a higher-tech approach to detection might be part of the diagnostic process. A great deal of effort in proteomics has been devoted to push the abilities of MS instruments and software to detect more protein biomarker candidates and build a link between abundance or molecular form (e.g., post-translational modifications) and disease conditions. Discussions of reproducibility in measurement methods, and in algorithms to identify peptides from MS spectra and proteins from the peptides,

C.A. Gelfand (✉)
BD Diagnostics, Preanalytical Systems, Franklin Lakes, NJ 07417, USA
e-mail: craig_gelfand@bd.com

A.R. Ivanov, A.V. Lazarev (eds.), *Sample Preparation in Biological Mass Spectrometry*, DOI 10.1007/978-94-007-0828-0_16,

have been increasingly prominent. It is generally agreed to require at least two tryptic peptides to gain sufficient confidence to claim protein identifications. Most reviewers and journals now have a routine expectation of a determination of the measurement's coefficient of variation (CV), with CVs becoming smaller (10% or less for measurement CV is routine, with better instrumentation), and MS getting to the point where CV is beginning to achieve performance expected for clinical diagnostic platforms.

By contrast, critical information about sample variables and numerous preanalytical variables has not been a major focus of research for most proteomics programs. Even basic information is lacking in many experimental annotations. "Preanalytical variables," simply defined, are all of the variables that can introduce error into the subsequent analytical processes. An understanding of preanalytical variables is a routine aspect of clinical chemistry. There are well-established reasons why, for example, serum and plasma are not interchangeable samples for certain clinical tests (Doumas et al., 1989; Jacobsen et al., 1994). The proteomics community has focused far more on detection technology and biomarker discovery, too often overlooking sample variables and sources of variability. This deficiency may account for a significant part of the very limited uptake of proteomic biomarkers into clinical chemistry labs.

If this is such a critical topic, how is it so easily overlooked? First, many proteomics investigators entered the field from protein biochemistry, as we did, typically studying individual proteins from vendors or those that were highly purified in the lab. Such basic clinical distinctions as the differences between serum and plasma may be unknown to the typical protein biochemist, so it is understandable that the complexity of sample variability might be naively overlooked. Second, project research aims and funding aspire to develop new biomarkers, better diagnostics, personalized therapies, and saving lives. Questions like "how reproducible is my blood sample?" seem trivial and subordinate in comparison. Grants written solely to understand sample reproducibility do not have nearly the excitement or good prospects for funding as seeking a new early cancer diagnostic, even though logically one would expect that better sample reproducibility would be a requirement for such diagnostics. The National Institute for Science and Technology (NIST) and the National Cancer Institute (NCI) have recognized this need; some large consortia are trying to define best practices with regard to methodologies (Barker et al., 2006; Paulovich et al., 2010; Rai et al., 2005; Rudnick et al., 2010) and disease states/sample types, e.g., (Hamacher et al., 2008; Hu et al., 2007; Lovestone et al., 2009).

To bridge the gap in development of diagnostics, the field needs to merge the world of proteomics with the world of clinical diagnostics. Aspects of this evolution of proteomics have been described previously (Vitzthum et al., 2005).

In this chapter, we describe sample acquisition and sample handling variables with respect to using blood in proteomics experiments. We detail some of the lessons learned, through experience and citations where possible, from projects at the interface between proteomics and clinical chemistry. In the end, the complexity of the proteome likely remains the most significant hurdle for improving the current state of "clinical proteomics," but an understanding of preanalytical variability can lead to mitigating strategies, which ultimately should accelerate biomarker development and validation.

16.2 Preanalytical Variables in Blood Collection

The importance of the topic of preanalytical variables in the collection and handling of blood is a much more widely recognized area in routine clinical diagnostics than in newer technology areas like proteomics-based diagnostics. Recently, preanalytical variability for clinical chemistry assays was expertly reviewed by Bowen et al. (2010). The focus of this chapter will be more aligned to variables that uniquely and perhaps more severely affect proteomics and mass spectrometry measurements.

In 2005, a list of variables in the sample collection and handling process was included in the first report from the HUPO Plasma Proteome Project (PPP) Sample Collection and Handling Committee (Rai et al., 2005; Table 16.1). At that time, there were few data to substantiate the effect of such variables upon proteomics measurements from plasma, but there was growing awareness that reproducibility would be compromised if variability between samples were ignored. In the sections that follow, several of these variables are described in detail, including discussion of available data.

In the five years since that initial publication, additional aspects of sample collection have come to our attention. Thus, Table 16.1 has been expanded. Logically, the more thoroughly reproducible all of these items can be made, the more reproducible the plasma proteomics data will be. The counter-argument, also logically, is that as this list grows, the complexity of the challenge of managing preanalytical variability also grows, putting a premium on thought and planning prior to initiating new studies.

The sections that follow take a look at some of these variables in greater detail, to help define not only the potential variability hiding within each topic, but ways in which that variability creeps into and alters proteomics measurements. The topics are arranged roughly in order of workflow, from collection to storage (Table 16.1).

16.2.1 *Sample and Collection Devices and Technique*

There are many variables included in the very first steps of phlebotomy and the nature of the container used to collect the blood sample. The needle gauge should be recorded and standardized, and the phlebotomy technique, including use of tourniquet and for how long, should be standardized. For example, variations in needle gauge can alter the sheer force the blood experiences during phlebotomy, and the speed of filling the tube, and thus could lead to irreproducible results, e.g., potentially hemolysis, if not controlled between subjects (Lippi et al., 2006). Important variables to consider are in vivo aspects of the sample, particularly the intrinsic enzymatic activity carried in the blood, and in vitro biochemistry such as choice of anticoagulants, induction of clotting ex vivo, and stabilizing additives.

Table 16.1 Preanalytical variables for blood collection and processing

Patient information
- Gender
- Age
- Diet
- Genetics
- Medical background
- Health background
- Special conditions: pregnancy, post/pre-menopausal, medications
- Social history – alcohol intake, smoking status
- Fasting or post-prandial sample

Venepuncture
- Needle gauge
- Blood collection set (details of)
- Length of line*

Phlebotomy
- Tourniquet technique (used? length of time?)
- Patient position: seated/standing/lying
- First tube versus last tube (or track tube order)
- Blood source: venepuncture or from existing line

Sample/collection device
- Tube or bag
- Glass or plastic
- Separator? If yes, gel or non-gel
- Intrinsic enzyme activity ex vivo*
- Plasma or serum
  - If plasma, nature of anticoagulant (EDTA, citrate, heparin)
  - If serum, clotting mechanism and/or clot activator
- Dilution of blood sample by additives*
- Other tube additives
  - Protease inhibitors
  - Stabilizers, excipients*

Sample processing
- Dwell between collection, processing, and analysis
  - Length of time
  - Temperature during dwell
- Separate blood from cells or left as whole blood
- If centrifuged
  - Speed and duration
  - Rotor: fixed angle, swing bucket, ...*
  - Temperature*
- Aliquot before analysis?
  - Handling/storage, time and temperature conditions
- Depletion of abundant proteins: which proteins, method, etc.*
- Fractionation into "Sub-omes": target, method, ...*

Shipping and storage
- Shipping between collection site and analysis lab?*
  - Shipped separated or whole blood*
  - Shipping temperature, duration in transit*
- Frozen storage?
- If frozen
  - Final target temperature
  - Rate of freezing
  - Rate of thawing *
- Elapsed time and temperature, prior to freezing
- Short versus long term storage
- Storage temperature
- Expiration dating
- Storage materials: glass versus plastic
  - Type of plastic
- Number of freeze/thaw cycles

Adapted from Rai et al. (2005)
*Addition beyond original citation

16.2.1.1 Intrinsic Enzymes: Intrinsically Damaging?

The presence and turnover rates of proteolytic enzymes need to be considered, and possibly reduced. Some of the enzymatic activity in a drawn blood sample is obvious, primarily the coagulation pathway, a cascade of proteases that ultimately cleave fibrinogen, resulting in clotting. This pathway is intentionally activated in serum samples, and must be inhibited by anticoagulants in plasma samples. While coagulation proteases are often referred to as being highly specific toward their targets, their specificity as proteases can be measured upon the plasma proteome more generally. O'Mullan et al. (2009) recently demonstrated that thrombin can exert extensive proteolytic damage upon the plasma proteome, nearly to the extent that trypsin does when intentionally introduced for bottom-up MS measurements. This observation is well-known (Blomback et al., 1972), but has renewed relevance and significance in modern proteomic analyses.

There are other proteolytic activities in blood, involved with particular signaling pathways. For example, dipeptidyl peptidase IV (DPP-IV) is part of the glucose metabolism regulation pathway, and very quickly cleaves glucagon-like-peptide 1 (GLP-1) in drawn blood (see, for example, Deacon and Holst, 2009). GLP-1 half-life in drawn blood samples is at most minutes. If this enzyme can cleave beyond its expected biological specificity, the damage could occur rapidly among other peptides. The more general extent of "non-specific" protease and peptidase activity and its effect on plasma proteomics is discussed below, in the protease inhibitor section (see 16.2.1.4).

Beyond proteolysis, esterase activity has been shown to degrade active forms of ghrelin, a peptide hormone involved in dietary responses (De Vriese et al., 2004). Classical observations showed that thrombin itself exerts weak esterase activity toward a broad range of substrates (Sherry et al., 1965). There are surely many other examples, e.g., phosphatases, kinases, glycolases, and lipases, which may be acting upon small molecule metabolites, thus altering metabolomics measurements, or possibly altering the post-translational modification state of proteins, and thus interfering with post-translational modification (PTM)-related studies and purification approaches.

16.2.1.2 Anticoagulants

Among the many choices of additives for blood samples, the most obvious are anticoagulants, included to inhibit the proteases involved in coagulation. There are three typical anticoagulant choices. Each can exert its own unique form of interference upon proteomics experiments, depending on the nature of the desired analysis format. For the purposes of this discussion, we can assume that their robustness as anticoagulants, in the "bulk" clinical definition of preventing clot formation, is not in question.

Heparin is a polymeric glucosaminoglycan, which will readily show up as a typical repeating building-block polymer in MS, making it far from ideal for many MS approaches. Citrate is a liquid reagent contained in the blood-drawing tube, and thus dilutes the plasma sample by some amount (typically 1:9 dilution), and needs to be accounted for when measuring protein concentrations. EDTA and citrate both chelate calcium, which inhibits coagulation, but chelation can also disrupt desired analytical approaches. For example, if the same sample is intended to be used for MS proteomics and an ELISA assay, an ELISA read-out enzyme relying on calcium will have greatly reduced activity in the presence of EDTA, and thus read-outs will be unreliable unless steps are built into the analytical process to remove the chelation. Removing or overcoming the interferences caused by chelators must be approached with some caution. For example, merely adding a calcium salt to saturate EDTA could lead to reactivation of the clotting enzymes. The effects of these proteases upon plasma proteins have been discussed above (see Yi et al., 2007). Also important, and more embarrassing for anyone that has had this happen, the sample can go all the way to the gelled fibrin network state of a clot. Clearly, when different analytes are to be measured, more than one type of tube may be necessary to meet the requirements of each.

16.2.1.3 Serum Versus Plasma

One might consider that use of serum could avoid interference, because exogenous anticoagulants are not part of serum samples. However, many tubes for drawing serum contain clot accelerators, either chemicals or glass that quickly activate intrinsic coagulation enzymes, or even extra enzyme to accelerate clot formation. Given the evidence of thrombin's degradation of plasma proteins, enzymatic acceleration of clot formation may further destabilize the plasma proteome. While this may not lead to a bad outcome for any particular experiment or format, it is one more consideration, and serves as a reminder that a sample called "serum" may not be sufficient to describe the exact biochemical nature of the sample, or, more dangerously, comparisons between "serum" samples. It is unacceptable for an author to use the terms "serum" and "plasma" interchangeably in referring to a specimen analyzed, or to fail to indicate which anticoagulant was used in collecting and preparing the plasma specimen.

There are examples of studies where analyte levels across common plasma or serum samples were compared. Differences in certain selected analytes can be detected easily as a function of sample type (Doumas et al., 1989; Hosnijeh et al., 2010; Jacobsen et al., 1994; Yoshimura et al., 2008) when using the same assay system, indicating either "matrix effect" (the nature of the sample influences the assay outcome) or differences in the sample induced by the different sample chemistries (Long, 1993; Wood, 1991). The "matrix effect" is well known, and manifests in certain tests, even clinical tests, requiring the use of *only* serum or *only* plasma as the input sample. Such effects for biomarker discovery studies may be less obvious, to the point of potentially obfuscating "real" biomarkers in the midst of sample-induced irreproducibility. Realizing the painstaking and extensive effort required to optimize assays for individual analytes, one can readily appreciate the challenge of optimizing simultaneously for dozens or hundreds or even thousands of proteins in proteomic studies.

16.2.1.4 Intrinsic Proteases and Protease Inhibitors

Another popular class of additives is protease inhibitors (PIs). Alone or in cocktails, PIs are very typical additives in cell lysates, preventing obvious and rampant damage upon cell disruption. Many commercial formats, formulations, and cocktails are available, and are cited in methods frequently throughout the literature by name or source, but generally without knowledge of the actual components of the proprietary cocktail. It has been shown that the extent of proteolytic damage to the proteome can be reduced by introduction of protease inhibitors immediately upon blood draw (Craft et al., 2009; Yi et al., 2007, 2008); BD and other companies have made specialty blood collection tubes for such purposes.

In creating these protease inhibitor tubes, work at BD focused first on observations of what intrinsic enzymatic degradation could be detected, as a function of time that the plasma sample dwelled at room temperature. The changes in peptide content as a function of time are either artifacts that need to be minimized, as described in our work (Yi et al., 2007), or perhaps may be enzymatic biomarkers that can be exploited. Given the ease with which peptide "laddering" (exopeptidase activity at both ends of parent peptides leading to sequentially shortened peptides) can be seen (Villanueva et al., 2006; Yi et al., 2007, 2008), it seems likely that any individual peptide, either intrinsic or created in vitro (e.g., by adding trypsin), could easily show up in multiple peaks, which would make accurate peptide biomarker quantitation very challenging. Protease inhibitors help minimize the intrinsic laddering, giving more reproducible results.

In either case, important initial observations in this work also included the fact that various common samples (e.g., serum and the three common plasma types) are dissimilar in peptide content. This may not be particularly surprising, because the samples are chemically different. However, these experiments point out an additional complication, in that the differences are manifest in an "as fast as possible" situation, where less than 30 minutes elapsed between initial phlebotomy and having the samples spun and prepared for MALDI MS. Even with ideal handling conditions, extensive protease and peptidase activity intrinsic to healthy human blood samples modifies the sample, and the changes increase with time until freezing. Because centrifugation is virtually a universal first step for processing blood, and it takes up to 20 minutes, there is no way to achieve any processing speed that can overcome fully the protease activity. Hence, there is interest in including protease inhibitors in the primary blood drawing tube, with chemical additives being a straightforward time-mitigating strategy, giving a more reproducible sample at the protein and intrinsic peptide level (Craft et al., 2009).

We sought better control over time effects by spiking plasma and serum samples with heavy-isotope synthetic peptides to see the degradation rate of these distinct peptides. Observations that the degradation rate of peptides varies with respect to both plasma sample type and peptide sequence (Yi et al., 2008), and that intrinsic peptidase activity varies even among healthy samples (Craft et al., 2009), make this aspect of preanalytical variability even more complicated, highlighting the potential value of protease inhibitors. Further, because spiked, labeled peptides are often used as controls, even the stability and thus the utility of controls becomes a preanalytical variable to be evaluated and managed. This fact has been overlooked to-date. The speed of degradation can be minutes (or faster) in a serum sample; thus, spiking even "immediately" before analysis might not be as accurate as intended by such a method.

As with any additive, potential interferences with analyses by protease inhibitors should be carefully considered prior to starting any large studies. Exposure to high levels of irreversible active-site inhibitors can lead to non-specific covalent modification of proteins beyond active sites (Conboy et al., 2008; Rai et al., 2005; Schuchard et al., 2005). Thus, the use of inhibitor formulations and choice of concentrations of inhibitors for particular purposes, such as stopping proteases released upon lysis of cultured cells, cannot safely be applied to other biological samples. Care must be taken to select and demonstrate non-interfering conditions or to use products specifically formulated for blood rather than cell lysates. Any processes requiring proteolytic activity downstream, e.g., introducing trypsin for bottom-up MS (Washburn et al., 2003) or using intrinsic protease activity as the actual biomarkers (Findeisen et al., 2008; Villanueva et al., 2008), could be altered by introduction of protease inhibitors. For the former, the in vitro added trypsin activity far exceeds intrinsic protease activity that needs to be inhibited; thus, the protease inhibitor cocktail developed specifically for blood at BD does not appear to alter traditional digestion protocols.

16.2.2 Sample Processing

There is a myriad of processing choices, in advance of another extensive set of analytical choices. We focus here on clinical processes, such as centrifugation and initial handling conditions, and then on some of the first proteomic process choices, namely protein depletion and sub-proteome isolation or enrichment.

16.2.2.1 Temperature

Temperature of handling has obvious relationship to underlying enzyme or biological processes. Logically, reducing temperature will slow down most enzymes, and might be considered to provide some protection from intrinsic plasma enzyme activity, including, for example, proteases, glycosylases, and phosphorylases.

However, a plasma sample, and of course the whole blood sample itself, is a complex mixture which may include significant cellular content. In particular, platelets can be cold-activated, leading to increased coagulation and complement activity in the sample (Suontaka et al., 2005). If cold exacerbates non-specific and potentially prolific thrombin-mediated proteolysis, then such cold activation may have an opposite effect upon protein stability from the protection expected. Cold-activation of platelets has significant variability among individuals and even within different samples from the same individual (Suontaka et al., 2005). Like many other issues in this chapter, the relatively simple measurements used to demonstrate cold-activation, in this case kallekrein-like activity measured with a chromogenic substrate, may underestimate the molecular damage that would be evident in proteomics experiments, particularly in MS-based assays.

16.2.2.2 Centrifugation

Obviously, the speed and duration of centrifugation of a blood sample alters the cellular content of the resulting plasma. One common oversight among proteomics scientists is that there is a distinct residual cellular content in plasma (and to a lesser extent in serum) which should not be ignored. This topic is intermingled with other storage considerations, and is discussed in more detail later.

16.2.2.3 Depletion of Highly Abundant Proteins

Given the wide dynamic range of proteins in plasma, it seems obvious that removal of abundant proteins could facilitate measurement of many proteins present in much lower concentrations. This need was recognized well before the advent of modern proteomics, with Cibacron Blue being an albumin removal tool utilized for decades (Angal and Dean, 1977). A number of companies provide depletion reagents and kits, facilitating removal of 6, 7, 20, or even 50 of the most abundant proteins. The need to remove abundant proteins depends on the detection method used, but the depletion process itself is yet another preanalytical variable. The nature of what is being depleted has been questioned, with some proposing that the "albuminome," or what binds to albumin and is thus depleted along with albumin, might itself have diagnostic value (Lowenthal et al., 2005). There have been many studies specifically evaluating depletion approaches (for example, Björhall et al., 2005; Seam et al., 2007); there is rather poor reproducibility and limited efficacy with present methods (Gundry et al., 2009; Roche et al., 2009).

The gains from depletion have also been questioned. For example, removal of 99% of albumin still leaves albumin one of the most abundant proteins, because it starts at 50% of the protein mass (and would still be 0.5%). Its depletion may introduce more variability than solving any particular dynamic range problem, especially because many peptides and proteins are bound to albumin. Studies have shown that the "next level" of abundant proteins is still so dominating as to continue to confound detection of the desired low-abundance proteins. These second-tier abundant proteins have a large number of post-translational modifications, which complicates problems caused by abundance alone. Using 2-D gels, many previously undetected spots can be seen after depletion of albumin and several other abundant proteins, but the additional spots are nearly all additional isoforms of already detected relatively abundant proteins, not previously undetected proteins. However, when additional fractionation was performed, many previously undetected lower-abundance proteins were identified (Tang et al., 2005).

16.2.2.4 Sub-omes

Another way of avoiding abundant proteins is to positively select subsets of proteins based on other chemical attributes. The peptidome, for example, is relatively easily extracted by size exclusion chromatography, mass cut-off filtration or other methods (Penno et al., 2009; Tammen et al., 2005; Yi et al., 2007; Zheng et al., 2006). However, the variability of preparing the peptides can be substantial, and may require parallel processing and pooling as a means to "average out" unavoidable processing variability (Craft et al., 2009; Penno et al., 2009).

There is growing interest in the glycoproteome, accessed by enrichment of glycosylated proteins by affinity to immobilized lectins (Plavina et al., 2007), and also having the advantage of providing several different specific lectin affinities for sub-fractionation. A complementary glycoprotein enrichment uses hydrazide chemistry (LAB 2, Table 16.1 in Omenn et al., 2005). Glycoprotein enrichment has the special benefit of depleting albumin, which is not glycosylated.

Certainly the simplification or enrichment provided by sub-fractionation can significantly benefit plasma proteomics studies, but the added steps almost certainly bring preanalytical processing variation, which, similar to variables discussed above, needs to be considered, tracked, and managed.

16.2.3 Shipping and Storage

Because the clinical collection sites are often remote to the proteomics lab, interim storage and shipping are unavoidable topics. Further, because many proteomics analytical processes are extremely detailed and time consuming, it is likely that sample collection can go much faster than analysis, so long-term storage also is likely, especially in the biomarker discovery applications of MS-based proteomics. Naturally, these aspects can introduce further preanalytical variability.

16.2.3.1 Freezing

Many publications discuss that cold or freezing can induce cellular damage or apoptosis in isolated cells and tissues (Clarke et al., 2004; Paasch et al., 2005; Salahudeen et al., 2003; Yiu et al., 2006). The underlying biochemistry of cold-induced ischemia (Bhaumik et al., 1995; van der Woude et al., 2004) is relevant. These publications do not deal directly with blood samples or blood cells, but the themes are similar: cold-induced apoptosis, caspase activation, and downstream pathway induction, any of which could happen in the cells remaining in plasma or serum samples. This implies that both refrigeration and freezing can induce protein changes in plasma, likely evident in both protein level and post-translational modification status. Diligence in examining stability with respect to particular biomarkers of interest should be included in experimental planning.

16.2.3.2 Shipping

Another aspect is the facilities and costs associated with preparation, storage and shipping of frozen samples. One recent study concluded that the cost of dry-ice shipping is becoming so high that inclusion of protease inhibitors, including the extra expense associated with their use, was more cost-effective, in addition to improving the reproducibility of the protein biomarker in question (Matheson et al., 2008). Smaller clinical facilities, where facilitating science or clinical trials may be an extremely infrequent activity, typically will not have the ability (or willingness) to purchase, store and use dry ice, or may not even have adequate freezer capabilities. This situation is even more common in more remote or rural field sites, which may enter into consideration in large trials/studies. Furthermore, shipping of dry ice is complicated by its status as a hazardous compound (US DOT 49 CFR Part 173.217). Similarly, clinical sites may not even have access to pipettors and small storage tubes, so it may be impossible to have aliquots made, and thus the sample is frozen "bulk," defeating the single freeze/thaw preferred approach. All of this means that diligence is required to thoroughly understand what happened to the sample before it arrived in the proteomics laboratory.

16.2.3.3 Freeze/Thaw Cycling, Long-Term Storage and Residual Cellular Content

The effects of freezing and thawing of samples are yet another complication. Inextricably entangled with freeze/thaw are the variable residual cellular content of plasma/serum and long-term sample storage (usually frozen). These three topics are interwoven in this section.

Freezing (and thus thawing) of samples is generally an unavoidable consideration, because so many studies involve sample collection at sites remote from the proteomics laboratory. Freezing and shipping of samples is commonplace, and accounts for the majority of sample handling in the biomarker discovery arena today. The growth of biorepositories and sample banks obviously requires long-term frozen storage and quality control, with associated standard procedures for assembly, storage, and provision of samples to users (Troyer, 2008). Studies specific to each facility and local processes should be undertaken by scientists maintaining such repositories.

It is generally accepted that repeated freeze/thaw cycles will increase damage to proteins in blood samples. For the Plasma Proteome Project, HUPO recommended minimizing freeze/thaw cycling (Omenn et al., 2005; Rai et al., 2005), especially by preparing aliquots from the start. Several studies have reported cold or freezing instability of proteins, generally in less complex specimens than plasma (Bhatnagar et al., 2007; Pikal-Cleland et al., 2002); for example, many antibody preparations must be stored refrigerated because freezing induces irreversible denaturation. For plasma samples, it is becoming routine to make many aliquots of fresh samples, so that experiments will always use only single freeze/thaw cycle samples. Several studies have carefully mapped time and storage effects for particular markers of interest. Flower et al. (2000) reported an average 17% increase in tumor necrosis factor alpha (TNFα) in samples after three freeze/thaw cycles of EDTA-plasma. Decreases of 10–15% were observed for other markers, depending on the nature of the sample storage conditions. Increases as a function of storage or handling, perhaps a less intuitive observation, are almost surely due to cell lysis or induced degranulation (such as from activated platelets, see below), releasing intracellular protein content into the plasma, or perhaps protein neogenesis in vitro, which is possible with induced apoptosis.

The dangers of freeze/thaw are exacerbated in a complex sample, such as blood, and warrant further consideration. Freezing lyses cells, and is typically used as a process step in intentional cell-lysis processes. Virtually all plasma samples are created by centrifugation to remove the cellular components of a whole blood sample, but that "removal" is not complete and may be overlooked. A "hard" spin, beyond normal clinical plasma preparation, to generate so-called platelet poor plasma, depletes more cells, but the clinical definition of platelet poor plasma allows for up to 10^3 platelets per microliter (per Clinical Laboratory Standards Institute (CLSI) guidelines), or 10 platelets per nanoliter. This may be "poor" by clinical definition, relating to, for example, certain kinds of platelet function testing where platelet-poor plasma is prepared and used as a diluent to normalize the platelet count from platelet-rich plasma. But consider that 10 platelets per nanoliter is actually in the range of femtomolar platelets, and platelets are cells, not mere molecules; the contents of platelet granules alone have an extensive proteomic content (Piersma et al., 2009). Freeze/thaw lysis and release of such intracellular content will contaminate the plasma; remember, the definition of both plasma and serum is the extracellular content of blood. Worse, proteases released upon freeze/thaw will now be active, so the very reason that inhibitors are frequently included when lysing cultured cells becomes a complication for blood. One freeze/thaw cycle of stored specimens is an inevitability, so understanding the potential issues is essential. Multiple freeze/thaws lyse more cells with each cycle, and thus reproducibility of the detectable proteome could certainly be compromised.

The link between residual cellular content and freeze/thaw stability also requires consideration of the nature of the centrifugation that created each plasma sample in a study. A plasma sample spun at $1300\times g$ surely has more residual cells than a plasma sample spun at $2500\times g$. Without any other knowledge, logic might suggest that we prefer the harder spin, to deplete more cells that could cause interference. Mixing and matching samples centrifuged under different conditions could result in artifacts in protein content or level.

16.2.4 Measurement Methods and Targets with Respect to Preanalytical Variables

The reader probably appreciates the importance of establishing reproducible assay methods. However, achieving such reproducibility can be a challenging goal even for the most savvy proteomics experiment designers. Sample variability and preanalytical error are inextricably connected to assay reproducibility, such that untangling the sources of variation can be difficult, if not impossible. Without a reproducible assay system, it is impossible to study sample variability, yet without a reproducible sample, it may be impossible to prove that the assay system itself is reproducible. This inevitable "Catch 22" situation is perhaps the most significant hurdle facing new biomarker discovery, partly accounting for the paucity of new biomarkers in recent years.

A tangible aspect of this problem is the drive to reduce the CV of mass spectrometry measurements. A discussion or statement of CV for measurements has become almost essential content in proteomics publications. We were asked to measure CV in our first papers using MALDI MS as the tool for assessing preanalytical variability of blood samples (Yi et al., 2007), and ultimately published a raw CV of less than 25%, and less than 7% with spectrum normalization for our intrinsic plasma peptide method. Yi et al. (2008) and others have continued to improve CVs.

Figures like 10% CV are a reasonable benchmark for the field at this time. As is obvious from peptidase activity time-course experiments (see above), a failure to account for degradation in standard samples can lead to exceptionally large CVs, which are entirely due to preanalytical instability, and therefore can never be fixed by having even perfectly reproducible MS spectra. Logically, the largest source of error defines the smallest possible size of the error bars of the resulting data. Thus, spending time refining only the detection step, through better instruments and/or methods, cannot overcome preanalytical variability. A reasonable goal for reducing preanalytical variability, then, would be to have the intrinsic variability of other processes mask *optimized* preanalytical variability, such that the latter truly can be ignored. The content of this chapter should make it clear that this goal is not close to being achieved.

The situation becomes more complicated with conflicting information about different measurement systems reporting the same analyte. MS, because of mass accuracy and the ability to have MS/MS to confirm protein identity, is a relatively incontrovertible measurement. However, that mass accuracy also makes MS the most susceptible to confounding effects like exopeptidase activity. By contrast, sandwich ELISA assays merely measure the presence of two epitopes on the same molecule, and if exopeptidase action has occurred, ELISA will not report it. This may not be an issue for many proteins, but for a peptide like GLP-1, where the difference between the biologically active and inactive signals is merely the presence or absence (via proteolytic cleavage) of the two amino-terminal residues, the intactness of the target molecule is an important property to measure, one that carries significant immunoassay challenges (Deacon and Holst, 2009). Worse still are Radioimmunoassay (RIA) and other single-antibody assays, where what is reported is literally just a single epitope; thus, extensive damage can befall a biomarker without altering the assay at all. Clearly, different assays may have different, unintended outcomes. Particularly for MS, one should not rely on a published report that used ELISA to demonstrate analyte stability, because what is measured in MS may still be highly unstable.

The opposite effect is also possible, in that using MS-based methods may rescue biomarkers for which antibody assays are insensitive to the required molecular state or when antibody assays simply fail, for a wide variety of reasons beyond the scope of this chapter. For example, Whiteaker et al. (2007) described attempts to order commercially-available antibodies to build assays to validate several dozen proteins implicated as biomarker candidates. In the end, they managed only one reproducible ELISA assay. However, they were able to use certain antibodies, especially anti-peptide antibodies, in preparation of samples for multiple reaction monitoring (MRM) in the MS, which gave them mass accuracy and, ultimately, useful validation data.

Finally, quantitative measurements are affected by sample concentration and preanalytical factors that ultimately alter concentrations. Generally, processing steps involving dilution or concentration steps are accounted for, often by using the original volume of plasma or serum input into the experiment as the benchmark for normalizing quantitative biomarker measurements. But preanalytical factors could be overlooked. For example, hematocrit, the basic clinical measurement of the red cell intracellular volume in a drawn blood sample, reflects an altered balance of intracellular and extracellular volume inside the primary blood tube, and thus might affect the way circulating concentration of a biomarker is calculated or normalized. Dilution of a primary blood sample with a liquid additive, for example the citrate formulation used in most citrate plasma tubes, needs to be accounted for. Recently, biological dilution has been evaluated, under the hypothesis that obese patients intrinsically carry a hemodilution effect from their increased circulating plasma volume (Grubb et al., 2009), thus reducing the concentration of biomarkers like prostate-specific antigen (PSA). Patients with extreme circulating lipid levels and lipemic plasma will have reduced aqueous phase.

16.3 Ways to Improve Preanalytical Reproducibility

When looked at with a magnifying glass, covering all of these sources of variability, it could easily seem that discovering or validating a new protein or peptide biomarker in blood is an impossible task. More than one proteomics scientist has wished they could go back to a previous setting where the binary nature of single-nucleotide polymorphisms (SNPs) in stable DNA was the only goal of their pharmacogenomics biomarker discovery studies. In reality, for successful proteomics, the primary requirement is nothing more than vigilant attention to detail. As usual, unknown sources of variability are the most painful once finally discovered. The "Monday morning quarterback" predicament, coming to the end of collecting a large clinical specimen set only to find that you missed some details or discover uncontrolled variables that have ruined your ability to detect biomarkers, is expensive and unpleasant. At the very least, we can all learn from the collective wisdom of our published literature, where many of the "mistakes" become evident through retrospection on processes and results. Of course, that requires full description (probably in Supplementary Material of publications) of details of preanalytical variables and insights about their associated effects.

16.3.1 "The Dry Run"

There are several obvious mitigating strategies. Essentially these are thought experiments, and thus inexpensive except for time. Small wetlab experiments can be suggested after learning from mistakes made by academic, clinical, and industry labs alike. The primary suggestion is to make a complete dress rehearsal of the entire process you intend to use, starting with phlebotomy and proceeding all the way through data collection and even data analysis. This can be done dry, or perhaps with a limited number of actual samples. This approach allows you to define your written standard operating procedures (SOPs) first, and put them to the test before committing the time and money toward an intended large study. The dress rehearsal may be a uniquely powerful way to discover and correct problems associated with studies involving multi-site sample collection, which is quite typical for consortium-style research projects, and especially when expensive and highly specialized analytical equipment is employed, usually meaning that scientists that will process the samples are nowhere near the blood collection site(s). Roles of each person involved in various aspects of the specimen collection, processing, storage, and handling can be specified and visualized.

The following are a few examples of actual hiccups, either by direct experience or through anecdote, all of which should be easily uncovered through dry or wet rehearsals focused on the preanalytical space:

- You define an SOP requiring centrifuging the primary blood tube soon after phlebotomy, only to find that some of your clinical collection sites do not have the capability to spin as fast as you need them to, and a few sites do not have centrifuges at all.
- You plan on overseeing sample collection and the entire study will be done within your one site. You define in your SOP a required time within which samples must be processed only to find your clinical centrifuge is in use when you show up to use it.
- Your SOP requires shipping samples from the hospital to a core proteomics lab on dry ice, but some of your clinical collaborators, who will recruit patients and draw the samples, are not equipped to handle dry ice, or, worse, don't have a freezer at all.
- You ask for plasma to be aliquoted into 500 μL portions, and shipped to the proteomics lab on dry ice. Some sites send you individual microfuge tubes, while others send you microtiter plates, and, for the latter, thawing one aliquot at a time for careful controlled use is not possible.
- You define plasma as the sample you will collect. When the samples arrive, you find that some collaborators sent you EDTA plasma while others sent you heparin plasma.
- Your SOP strictly controls freeze/thaw, such that only one thawing is allowed, but later one of your clinical collaborators tells you that they froze the sample because they didn't have time to aliquot it on the day they collected it, and they thawed it the next day to make the aliquots, but they put them back into the freezer right away.

These all may seem trivial, but they are potentially important variables, and could alter your measurements. Taking the extra effort before starting a full study can help detect unforeseen variables, so they can be either controlled or eliminated – or you might be eliminating your data, before you ever started.

16.3.2 Data Recording

The "dry run" approach leads to another recommendation: document everything, from the important to the minute and trivial – the vendor and part number of the needle, the lot number of the microfuge tubes for aliquots, the exact time the sample was drawn and the exact moment it was put into and taken from the centrifuge, the phlebotomist's name – everything. Consider Table 16.2, as well as Bowen et al. (2010) for a thorough list of topics. No SOP will be strictly adhered to, and even with planning and dry runs, there could still be variables missed during the planning stage. Having everyone in contact with the sample record's minute details might be easier, and could allow a retrospective understanding to help filter data in the event ambiguities start to emerge once enough samples are analyzed. This might not even be much extra work, because, when disease biomarker candidates are studied, you

Table 16.2 Details that can be planned in an SOP and recorded during acquisition and processing

Collection details	Processing details
Date/time of draw	– Room temperature of processing site
Phlebotomist name	– Centrifuge
SOP	• Operator/technician name
• Was phlebotomist trained on this SOP?	– Trained on this SOP?
(operator should certify yes or no)	– Version of SOP being used
• Version of SOP being used	• Time at start of spin
Patient information	• Condition of sample between draw and spin
• Time of last meal	(e.g., dwelled at r.t., 4° C, . . .)
• Orientation (e.g., sitting, standing)	• Duration
Room temperature of collection site	• Speed (G or RPM?)
Blood collection process:	• Internal temperature
• Needle/set used	• Make/model of centrifuge
– Type: straight, set ("butterfly")	• Make/model of rotor
• Length of tubing in the set	• Date of last calibration
• Needle gauge	– Aliquotting
• Vendor, product name, number, and lot	• Operator/technician name
number, expiration date	• Trained on this SOP?
• Number of sticks (e.g., was a second stick	• Version of SOP being used
required to find vein?)	• Time of aliquotting
• Tourniquet used? (if so, part #, etc.)	• Volume of aliquot
• Site disinfectant (part #, etc.).	• Container (vendor, part #, etc.)
Blood collection tube(s) used	• Condition of sample between spin and
• Vendor, product name, number, and lot	aliquotting (e.g., dwelled at r.t., 4°C, . . .)
number, expiration date	– Storage
Total volume of blood collected	• Time of placement into storage condition
• From the venepuncture event	• Target temperature
• In each tube	• Date of calibration of refrigerator/freezer
Draw order of all tubes used	• Method of freezing (e.g., flash freeze on dry
• Was a "waste tube" used to prime the blood	ice or directly placed in freezer)
collection device?	• Rate of cooling (if known)
Mixing of tube (to dissolve additive)	– Shipping
• Number of inversions	• Temperature
	• Method of maintaining temperature (wet
	ice, dry ice, gel pack, etc.)
	• Date/time of pick-up
	• Shipping vendor information,
	company, tracking number, etc.

expect as full and detailed a set of medical data about the subject as you can possibly get. Why should the preanalytical and analytical processes not receive that same level of attention? Consider making a form that will be filled out for each sample, with blanks for every type of data that you want collected. Having such information will be especially helpful if a replication study is performed, or a follow-on study.

16.3.3 Least Common Denominators

An important aspect that might also come from "dry runs" is more specific to projects where parallel processes might be employed, e.g., if multiple collection sites will do phlebotomy and initial sample processing, or you have a consortium of proteomics labs that will work on a subset of samples and pool their data afterward. For the "dry runs" have every participating site/lab participate, so that any differences in processing abilities may become evident. Then, think about the "least common denominator" (LCD) that all sites will be able to execute, so that every step can be performed as comparably and reproducibly as possible. For example, if one phlebotomy site cannot centrifuge the sample, and will need to ship it to you as whole blood, then consider having all of your collection sites do that. This LCD version of your SOP may not be "ideal," but at least all samples should be comparable, which may be the best that can be achieved. Or perhaps request duplicate samples, one adhering to the LCD version of your SOP and the other adhering to the ideal SOP, if it can be accommodated, so that you get a chance at both reproducibility/comparability from all sites and ideality from some subset.

16.3.4 Calibration and Control

Finally, do not overlook instrument calibration and process controls. When running a mass spectrometer, frequent checks of mass accuracy, generally using a mass calibration standard, are required to achieve accurate peptide and protein detection and identification. This should be a well-known concept to proteomics scientists. Most labs take the effort to calibrate their small volume pipettors. But, have you considered what other instruments might be involved in the preanalytical workflow? When was the last time the slow-speed centrifuge you use to spin down your blood sample was checked for having accurate RPMs? A critical parameter like reproducible levels of residual cell contamination might be at stake. Can you be certain a frozen sample actually stayed frozen during shipping? Some shipping services include dry ice refilling during transit, to ensure samples arrive still frozen, but unless samples arrive thawed, you can't tell what happened during shipping. A simple trick here is to ship an ice cube along with your samples, in a zip lock bag. A sample that thawed in route and was re-frozen will be evident with a frozen puddle in the baggie, distinctly not ice-cube shaped. There are also temperature-sensitive tapes, which are now used globally to ensure the cold-chain for vaccines delivered in sites far away from hospitals.

16.4 Conclusion

Preanalytical variables are sources of potential disaster for otherwise carefully conducted clinical and biological proteomics studies. Close attention to detail, noting the many variables in Table 16.1 and the practical guidance in our Improving Preanalytical Reproducibility section, can yield major benefits for the research investigators and their clinical collaborators.

References

Angal, S., and Dean, P.D. (1977). The effect of matrix on the binding of albumin to immobilized Cibacron Blue. Biochem J *167*, 301–303.

Barker, P.E., Wagner, P.D., Stein, S.E., Bunk, D.M., Srivastava, S., and Omenn, G.S. (2006). Standards for plasma and serum proteomics in early cancer detection: A needs assessment report from the National Institute of Standards and Technology--National Cancer Institute Standards, Methods, Assays, Reagents and Technologies Workshop, August 18–19, 2005. Clin Chem *52*, 1669–1674.

Bhatnagar, B.S., Bogner, R.H., and Pikal, M.J. (2007). Protein stability during freezing: separation of stresses and mechanisms of protein stabilization. Pharm Dev Technol *12*, 505–523.

Bhaumik, G., Srivastava, K.K., Selvamurthy, W., and Purkayastha, S.S. (1995). The role of free radicals in cold injuries. Int J Biometeorol *38*, 171–175.

Björhall, K., Miliotis, T., and Davidsson, P. (2005). Comparison of different depletion strategies for improved resolution in proteomic analysis of human serum samples. Proteomics *5*, 307–317.

Blomback, B., Hessel, B., Iwanaga, S., Reuterby, J., and Blomback, M. (1972). Primary structure of human fibrinogen and fibrin. J Biol Chem *247*, 1496–1512.

Bowen, R.A., Hortin, G.L., Csako, G., Otañez, O.H., and Remaley, A.T. (2010). Impact of blood collection devices on clinical chemistry assays. Clin Biochem *43*, 4–25.

Clarke, D.M., Baust, J.M., Van Buskirk, R.G., and Baust, J.G. (2004). Addition of anticancer agents enhances freezing-induced prostate cancer cell death: implications of mitochondrial involvement. Cryobiology *49*, 45–61.

Conboy, J.J., Wood, K.G., Lame, M.E., Durham, R.A., and Geoghegan, K.F. (2008). Modification of amyloid-[beta](1-40) by a protease inhibitor creates risk of error in mass spectrometric quantitation of amyloid-[beta](1-42). Anal Biochem *382*, 147–149.

Craft, D., Yi, J., and Gelfand, C. (2009). Time-dependent and sample-to-sample variations in human plasma peptidome are both minimized through use of protease inhibitors. Anal Lett *42*, 1398–1406.

Deacon, C.F., and Holst, J.J. (2009). Immunoassays for the incretin hormones GIP and GLP-1. Best Pract Res Clin Endocrinol Metab *23*, 425–432.

De Vriese, C., Gregoire, F., Lema-Kisoka, R., Waelbroeck, M., Robberecht, P., and Delporte, C. (2004). Ghrelin degradation by serum and tissue homogenates: identification of the cleavage sites. Endocrinology *145*, 4997–5005.

Doumas, B.T., Hause, L.L., Simuncak, D.M., and Breitenfeld, D. (1989). Differences between values for plasma and serum in tests performed in the Ektachem 700 XR Analyzer, and evaluation of "plasma separator tubes (PST)." Clin Chem *35*, 151–153.

Findeisen, P., Peccerella, T., Post, S., Wenz, F., and Neumaier, M. (2008). Spiking of serum specimens with exogenous reporter peptides for mass spectrometry based protease profiling as diagnostic tool. Rapid Commun Mass Spectrom *22*, 1223–1229.

Flower, L., Ahuja, R.H., Humphries, S.E., and Mohamed-Ali, V. (2000). Effects of sampler handling on the stability of interleukin 6, tumour necrosis factor-alpha and leptin. Cytokine *12*, 1712–1716.

Grubb, R.L., Black, A., Izmirlian, G., Hickey, T.P., Pinsky, P.F., Mabie, J.E., *et al.* (2009). Serum prostate-specific antigen hemodilution among obese men undergoing screening in the Prostate, Lung, Colorectal, and Ovarian Cancer Screening Trial. Cancer Epidemiol Biomark Prevent *18*, 748–751.

Gundry, R.L., White, M.Y., Nogee, J., Tchernyshyov, I., and Van Eyk, J.E. (2009). Assessment of albumin removal from an immunoaffinity spin column: critical implications for proteomic examination of the albuminome and albumin-depleted samples. Proteomics *9*, 2021–2028.

Hamacher, M., Stephan, C., Eisenacher, M., Hardt, T., Marcus, K., and Meyer, H.E. (2008). Maintaining standardization: an update of the HUPO Brain Proteome Project. Expert Rev Proteomics *5*, 165–173.

Hosnijeh, F.S., Krop, E.J.M., Portengen, Lt, Rabkin, C.S., Linseisen, J., Vineis, P., *et al.* (2010). Stability and reproducibility of simultaneously detected plasma and serum cytokine levels in asymptomatic subjects. Biomarkers *15*, 140–148.

Hu, S., Loo, J.A., and Wong, D.T. (2007). Human saliva proteome analysis. Ann N Y Acad Sci *1098*, 323–329.

Jacobsen, D.W., Gatautis, V.J., Green, R., Robinson, K., Savon, S.R., Secic, M., *et al.* (1994). Rapid HPLC determination of total homocysteine and other thiols in serum and plasma: sex differences and correlation with cobalamin and folate concentrations in healthy subjects. Clin Chem *40*, 873–881.

Lippi, G., Franchini, M., Montagnana, M., Salvagno, G.L., Poli, G., and Guidi, G.C. (2006). Quality and reliability of routine coagulation testing: can we trust that sample? Blood Coagul Fibrinol *17*, 513–519.

Long, T. (1993). Statistical power in the detection of matrix effects. Arch Pathol Lab Med *117*, 387–392.

Lovestone, S., Francis, P., Kloszewska, I., Mecocci, P., Simmons, A., Soininen, H., *et al.* (2009). AddNeuroMed – the European collaboration for the discovery of novel biomarkers for Alzheimer's disease. Ann N Y Acad Sci *1180*, 36–46.

Lowenthal, M.S., Mehta, A.I., Frogale, K., Bandle, R.W., Araujo, R.P., Hood, B.L., *et al.* (2005). Analysis of albumin-associated peptides and proteins from ovarian cancer patients. Clin Chem *51*, 1933–1945.

Matheson, L.A., Duong, T.T., Rosenberg, A.M., and Yeung, R.S.M. (2008). Assessment of sample collection and storage methods for multicenter immunologic research in children. J Immunol Methods *339*, 82–89.

Omenn, G.S., States, D.J., Adamski, M., Blackwell, T.W., Menon, R., Hermjakob, H., *et al.* (2005). Overview of the HUPO Plasma Proteome Project: results from the pilot phase with 35 collaborating laboratories and multiple analytical groups, generating a core dataset of 3020 proteins and a publicly-available database. Proteomics *5*, 3226–3245.

O'Mullan, P., Craft, D., Yi, J., and Gelfand, C. (2009). Thrombin induces broad spectrum proteolysis in human serum samples. Clin Chem Lab Med *47*, 685–693.

Paasch, U., Grunewald, S., Wuendrich, K., Jope, T., and Glander, H.-J. (2005). Immunomagnetic removal of cryo-damaged human spermatozoa. Asian J Androl *7*, 61–69.

Paulovich, A.G., Billheimer, D., Ham, A.-J.L., Vega-Montoto, L., Rudnick, P.A., Tabb, D.L., *et al.* (2010). Interlaboratory study characterizing a yeast performance standard for benchmarking LC-MS platform performance. Mol Cell Proteomics *9*, 242–254.

Penno, M.A., Ernst, M., and Peter, H. (2009). Optimal preparation methods for automated matrix-assisted laser desorption/ionization time-of-flight mass spectrometry profiling of low molecular weight proteins and peptides. Rapid Commun Mass Spectrom *23*, 2656–2662.

Piersma, S.R., Broxterman, H.J., Kapci, M., de Haas, R.R., Hoekman, K., Verheul, H.M.W., *et al.* (2009). Proteomics of the TRAP-induced platelet releasate. J Proteomics *72*, 91–109.

Pikal-Cleland, K., Cleland, J., Anchordoquy, T., and Carpenter, J. (2002). Effect of glycine on pH changes and protein stability during freeze-thawing in phosphate buffer systems. J Pharm Sci *91*, 1969–1979.

Plavina, T., Wakshull, E., Hancock, W.S., and Hincapie, M. (2007). Combination of abundant protein depletion and multi-lectin affinity chromatography (M-LAC) for plasma protein biomarker discovery. J Proteome Res *6*, 662–671.

Rai, A.J., Gelfand, C.A., Haywood, B.C., Warunek, D.J., Yi, J., Schuchard, M.D., *et al.* (2005). HUPO Plasma Proteome Project specimen collection and handling: towards the standardization of parameters for plasma proteome samples. Proteomics *5*, 3262–3277.

Roche, S., Tiers, L., Provansal, M., Seveno, M., Piva, M.-T., Jouin, P., *et al.* (2009). Depletion of one, six, twelve or twenty major blood proteins before proteomic analysis: the more the better? J Proteomics *72*, 945–951.

Rudnick, P.A., Clauser, K.R., Kilpatrick, L.E., Tchekhovskoi, D.V., Neta, P., Blonder, N., *et al.* (2010). Performance metrics for liquid chromatography-tandem mass spectrometry systems in proteomics analyses. Mol Cell Proteomics *9*, 225–241.

Salahudeen, A.K., Huang, H., Joshi, M., Moore, N.A., and Jenkins, J.K. (2003). Involvement of the mitochondrial pathway in cold storage and rewarming-associated apoptosis of human renal proximal tubular cells. Am J Transplant *3*, 273–280.

Schuchard, M.D., Mehigh, R.J., Cockrill, S.L., Lipscomb, G.T., Stephan, J.D., Wildsmith, J., *et al.* (2005). Artifactual isoform profile modification following treatment of human plasma or serum with protease inhibitor, monitored by 2-dimensional electrophoresis and mass spectrometry. BioTechniques *39*, 239–247.

Seam, N., Gonzales, D.A., Kern, S.J., Hortin, G.L., Hoehn, G.T., and Suffredini, A.F. (2007). Quality control of serum albumin depletion for proteomic analysis. Clin Chem *53*, 1915–1920.

Sherry, S., Alkjaersig, N., and Fletcher, A.P. (1965). Comparative activity of thrombin on substituted arginine and lysine esters. Am J Physiol *209*, 577–583.

Suontaka, A.M., Silveira, A., Söderström, T., and Blombäck, M. (2005). Occurrence of cold activation of transfusion plasma during storage at +4 degrees C. Vox Sang *88*, 172–180.

Tammen, H., Schulte, I., Hess, R., Menzel, C., Markus, K., Mohring, T., *et al.* (2005). Peptidomic analysis of human blood specimens: comparison between plasma specimens and serum by differential peptide display. Proteomics *5*, 3414–3422.

Tang, H.-Y., Ali-Khan, N., Echan, L.A., Levenkova, N., Rux, J.J., and Speicher, D.W. (2005). A novel four-dimensional strategy combining protein and peptide separation methods enables detection of low-abundance proteins in human plasma and serum proteomes. Proteomics *5*, 3329–3342.

Troyer, D. (2008). Biorepository standards and protocols for collecting, processing, and storing human tissues. Methods Mol Biol *441*, 192–220.

van der Woude, F., Schnuelle, P., and Yard, B. (2004). Preconditioning strategies to limit graft immunogenicity and cold ischemic organ injury. J Investig Med *52*, 323–329.

Villanueva, J., Nazarian, A., Lawlor, K., Yi, S.S., Robbins, R.J., and Tempst, P. (2008). A sequence-specific exopeptidase activity test (SSEAT) for "functional" biomarker discovery. Mol Cell Proteomics *7*, 509–518.

Villanueva, J., Shaffer, D.R., Philip, J., Chaparro, C.A., Erdjument-Bromage, H., Olshen, A.B., *et al.* (2006). Differential exoprotease activities confer tumor-specific serum peptidome patterns. J Clin Invest *116*, 271–284.

Vitzthum, F., Behrens, F., Anderson, N.L., and Shaw, J.H. (2005). Proteomics: from basic research to diagnostic application. A review of requirements & needs. J Proteome Res *4*, 1086–1097.

Washburn, M.P., Ulaszek, R.R., and Yates, J.R. (2003). Reproducibility of quantitative proteomic analyses of complex biological mixtures by multidimensional protein identification technology. Anal Chem *75*, 5054–5061.

Whiteaker, J.R., Zhang, H., Zhao, L., Wang, P., Kelly-Spratt, K.S., Ivey, R.G., *et al.* (2007). Integrated pipeline for mass spectrometry-based discovery and confirmation of biomarkers demonstrated in a mouse model of breast cancer. J Proteome Res *6*, 3962–3975.

Wood, W. (1991). "Matrix effects" in immunoassays. Scand J Clin Lab Invest Suppl *205*, 105–112.

Yi, J., Kim, C., and Gelfand, C.A. (2007). Inhibition of intrinsic proteolytic activities moderates preanalytical variability and instability of human plasma. J Proteome Res *6*, 1768–1781.

Yi, J., Liu, Z., Craft, D., O'Mullan, P., Ju, G., and Gelfand, C.A. (2008). Intrinsic peptidase activity causes a sequential multi-step reaction (SMSR) in digestion of human plasma peptides. J Proteome Res *7*, 5112–5118.

Yiu, W.-K., Cheng, S.W.K., and Sumpio, B.E. (2006). Vascular smooth muscle cell apoptosis induced by "supercooling" and rewarming. J Vasc Interv Radiol *17*, 1971–1977.

Yoshimura, T., Fujita, K., Kawakami, S., Takeda, K., Chan, S., Beligere, G., *et al.* (2008). Stability of pro-gastrin-releasing peptide in serum versus plasma. Tumor Biol *29*, 224–230.

Zheng, X., Baker, H., and Hancock, W.S. (2006). Analysis of the low molecular weight serum peptidome using ultrafiltration and a hybrid ion trap-Fourier transform mass spectrometer. J Chromatogr A *1120*, 173–184.

Chapter 17
Biomarker Discovery in Biological Fluids

Wasfi Alrawashdeh and Tatjana Crnogorac-Jurcevic

Abstract The quest for disease biomarkers that can be used for diagnostic, prognostic and selection and monitoring of treatment has been at the heart of translational research for decades. The recent advances in proteomic research and the ability to profile the proteomes of various biological samples, particularly biological fluids, have revolutionized this field. The initial optimism, however, was soon met with tremendous challenges related to the vast complexity and our limited understanding of the biological samples and the factors influencing their proteomic profiles, as well as the lack of clear pathway for biomarker discovery and development. In this chapter we will discuss biomarker discovery in biological fluids, primarily blood and urine, including the influence of various pre-analytical variables as well as the biomarker discovery pipeline.

Keywords Biological fluids · Biomarker discovery · Biomarker pipeline · Sample preparation

17.1 Introduction

A biomarker is a measurable biological indicator of a normal biological or pathological process or a pharmacological response to a therapeutic intervention (Group, 2001). It can be a gene, a protein, a metabolite or a pattern of expression of these molecules. In order to fulfil this function, an ideal biomarker should display high sensitivity (low false negative results) and specificity (low false positive results). The clinical assay for biomarker detection should ideally be optimised on easily obtainable biological samples and should be easy and cost-effective to perform. The clinical application of a good biomarker should also demonstrate improved patient outcome compared to current best practice. At present, biomarkers are used to aid

T. Crnogorac-Jurcevic (✉)
Molecular Oncology and Imaging Centre, Barts Cancer Institute, John Vane Science Centre, Queen Mary University of London, EC1M 6BQ London, UK
e-mail: T.C.Jurcevic@qmul.ac.uk

A.R. Ivanov, A.V. Lazarev (eds.), *Sample Preparation in Biological Mass Spectrometry*, DOI 10.1007/978-94-007-0828-0_17,

screening, diagnosis, staging, monitoring of disease recurrence and prognostication of various disease states. More recently, with the advent of targeted molecular therapies, biomarkers are being increasingly used for selection of patients for an individualised treatment and monitoring of response (Eneman et al., 2004; Nielsen et al., 2009). It is the early detection of diseases, however, particularly in the cancer field, where biomarkers are most studied and where they are likely to play a major role, as all cancers have better prognosis if detected at early stages where available treatment modalities are most effective. The positive impact of screening and early detection of cervical and breast cancers has already been demonstrated (Etzioni et al., 2003).

Prior to the emergence of the proteomics era in biomarker discovery, a large number of gene expression profiling studies was performed on multiple organs/tissues to identify markers related to various clinical parameters. However, such studies rely on availability of tissue specimens, which requires an invasive procedure. Proteins, on the other hand, are the primary determinants of the cellular phenotype, their qualitative and quantitative characteristics are only partially determined by their respective genes (Anderson and Seilhamer, 1997; Mann et al., 2004), and they can be detected in easily accessible specimens such as body fluids. The recent advances in proteomics enabled large scale profiling for unbiased discovery of biomarkers for various benign and malignant conditions. This resulted in a surge of publications describing hundreds of potential biomarkers. This early promise of proteomics in changing the face of clinical practice has so far been elusive, as many of the reported potential biomarkers started to display signs of weakness as early as during their initial validation (McLerran et al., 2008a, b; Rogers et al., 2003). Several factors contributed to this, starting with a lack of clear study design and suboptimal statistical power, almost complete lack of standardization of sample collection and handling, problems with reproducibility of sample processing, inherent differences between proteomic platforms, lack of consistent bioinformatics and data analysis procedures and finally a lack of clear pathway or pipeline for development of biomarkers. Many of these important issues have been addressed in recent excellent reviews (Anderson, 2005; Ludwig and Weinstein, 2005; Mischak et al., 2007; Rifai et al., 2006).

In this chapter we will review issues related to the collection and preparation of biological fluids for subsequent Mass Spectrometry (MS)-based biomarker discovery. In addition, we will briefly discuss the biomarker pipeline and address current problems and potential solutions in the discovery of biomarkers in translational research and their potential use within the realms of personalised medicine.

17.2 Body Fluids in Biomarker Discovery

Biological fluids are dynamic compartments that largely reflect the changes in organs/tissues they come in contact with. Therefore, they represent a rich source for biomarker discovery. Although blood and urine are the most studied, other body fluids have also been explored, including saliva, cerebrospinal fluid (CSF), tears, ascitic fluid, gastric and pancreatic juice, bile, prostatic secretions, nipple aspirate and endometrial fluid. Fluids in close proximity to the disease site often contain proteins shed or secreted as a result of the disease process (tissue leakage proteins), however, many of these, so called “proximal fluids”, are difficult to access and require invasive procedures for their collection (CSF, gastric fluid, pancreatic juice and bile). Blood (plasma/serum), urine and saliva are on the other hand, easily accessible and can be obtained in sufficiently large volumes.

17.2.1 Blood

Blood is the most sampled biofluid. Because it circulates throughout the whole body, it is reasonable to assume that it is possible, with the right approach and technology, to detect most (if not all) tissue proteins. The fact that most currently used biomarkers are tissue leakage proteins and the detection of several intracellular proteins in recent serum/plasma proteomic studies support this (Anderson et al., 2004; Chignard et al., 2006). In addition, the relative ease of collection makes blood a very attractive biofluid for biomarker discovery. However, its proteome also represents the most complex and comprehensive of all the body proteomes (Anderson and Anderson, 2002). In serum, 60–80 mg/mL of proteins comprise a vast mixture of structural and haemostatic proteins, hormones, immunoglobulins, enzymes, lipoproteins, cytokines as well as foreign proteins derived from various microorganisms (Omenn et al., 2005; Tirumalai et al., 2003). Such complexity of blood raises several challenges. The major one is its massive dynamic range, which spans at least 12 orders of magnitude (Hortin and Sviridov, 2010). This is further complicated by the fact that only 22 common proteins constitute almost 99% of plasma proteins, while the low abundance proteins, the potentially most useful pool of biomarkers, constitute the remaining 1% (Tirumalai et al., 2003). The level of complexity of the proteome extends beyond just the amount of a particular protein and includes various splice variants, post translational modifications (PTM) and degradation products. Taking all this into account, it has been estimated that plasma might contain more than 10 million different proteins (Anderson and Anderson, 2002). In addition, as the proteomic profile of blood is very sensitive to sample handling, processing and analytical procedures (see later), it becomes quickly apparent how enormous is the task of profiling such a complex proteome.

The liquid phase of blood can be processed in two different ways, to obtain either serum or plasma. Serum is prepared by allowing blood to clot in plain tubes with/without silica particles and with/without separator gel. The clot is subsequently removed by centrifugation. Plasma, on the other hand, is prepared by collecting blood in tubes containing anticoagulants: EDTA and citrate inhibit coagulation by chelating calcium, whereas the anticoagulant effect of heparin is mediated through activation of antithrombin III (Hirsh et al., 1995; Tammen et al., 2005). In order to analyse protein constituents of both plasma and serum, identify biological sources of intra- and inter-individual variation and create a global data repository, the Human Proteome Organisation (HUPO) initiated the Plasma Proteome Project in 2002.

In 2005, the results of this comprehensive evaluation from 35 collaborating laboratories were reported: 3020 individual proteins with two or more peptides were identified, and the use of plasma rather than serum was recommended (Omenn et al., 2005). The generation of serum involves ex vivo activation of the coagulation cascade and several proteases resulting in the formation of new peptides and fragments from several high abundance proteins. Many of these peptides are thus specific for serum and not present in plasma (>40% of all detected peptides in a study of Tammen et al. were serum-specific) (Hsieh et al., 2006; Tammen et al., 2005). Furthermore, the clotting process is sensitive to several factors such as time, temperature and medications, and therefore may be difficult to standardize. Nonetheless, large archives of serum samples are available and can be invaluable for biomarker discovery, particularly for retrospective and longitudinal validation studies. And while the recommendation for preferential use of plasma in proteomic studies still holds, a recent study has exploited the ex vivo generation of peptide fragments in serum and concluded that some of these are disease-specific secondary to the presence of a disease-associated proteases. These can potentially be used as surrogate markers for the detection and classification of diseases, as shown for prostate, bladder and breast cancers (Villanueva et al., 2006).

17.2.2 Urine

Urine can be obtained completely non-invasively and in large amounts, and has been used in medical practice for diagnostic purposes throughout centuries. The first biomarker (in today's meaning of the word), Bence Jones protein, was discovered in urine in 1848 (Jones, 1848). Today, urine is still an extremely attractive source for biomarker discovery, with the major advantages of a lower dynamic range and more stable proteome compared to plasma/serum (Papale et al., 2007; Schaub et al., 2004; Thongboonkerd, 2007). More than 1500 proteins have been described in urine, the majority of which are membrane, cytoplasmic and extracellular matrix derived proteins (Adachi et al., 2006).

Urine is an ultrafiltrate of plasma, and it is the ultimate destination of most of <60 kDa proteins (Anderson and Anderson, 2002; Lafitte et al., 2002); in fact, most of the cleavage products of potentially important disease-specific proteins might be more readily discovered in urine as they would be relatively quickly cleared by the kidneys (Banks et al., 2005). The glomerular filtration barrier allows free filtration of solutes and low molecular weight (LMW) proteins, whilst retaining high MW proteins (although abundant proteins in plasma such as albumin and globulins are also known to enter the lumen of nephron). Most of the filtered load is reabsorbed in the proximal tubules resulting in less than 150 mg of protein in the normal urine/day (Papale et al., 2007). Of these proteins, 49% are soluble and 48% are within debris and cells. The remaining 3% lie within exosomes (Zhou et al., 2006), which are 50–100 nm vesicles formed by fusion of the outer membranes of late endocytic vesicles with the apical plasma membranes of renal epithelial cells from all nephron segments. The proteome of these secreted vesicles currently comprises 295 proteins some of which appear to be promising biomarkers for renal diseases (Pisitkun et al., 2004). The final urinary proteome is a function of the plasma protein content, the integrity of the glomerular filtration barrier and the tubular reabsorption mechanisms, and the pathophysiological status of the lower urinary tract. Abnormalities within any of these can alter the composition of the urinary proteome and present biomarkers not only indicative of renal and lower urinary tract pathologies, but also of systemic diseases such as cancers of lung, ovary, breast and pancreas (Roy et al., 2004; Tantipaiboonwong et al., 2005; Weeks et al., 2008; Ye et al., 2006).

Despite all the advantages, the analysis of the urinary proteome is not without challenges. Urine contains large amounts of salts and other interfering compounds such as glycosaminoglycans, lipids and small charged molecules, and therefore requires extensive sample clean-up prior to most downstream proteomics applications, particularly in 2D-gel analysis (Oh et al., 2004; Thongboonkerd et al., 2006). The protein content in the urine is low and exhibits high degree of intra- and inter-individual variability (Papale et al., 2007; Thongboonkerd et al., 2006), which often requires concentration of the samples and some form of normalisation to enable comparability. Several physiological and environmental factors can also affect the urinary proteome, including fluid intake, smoking, dietary habits and exercise (Airoldi et al., 2009; Poortmans et al., 1996; Russo et al., 2002; Thongboonkerd et al., 2006).

17.2.3 Saliva

Saliva is another easily accessible and collectable body fluid. Although it "lacks the drama of blood, the sincerity of sweat and the emotional appeal of tears" (Mandel, 1990), it remains a complex mixture of secretions from major (parotid, submandibular and sublingual) and around 700 minor salivary glands, gingival fluid, nasal and bronchial secretions, blood, oral mucosal cells, oral microbial/viral flora and cellular and food debris (whole saliva) (Kaufman and Lamster, 2002). Per day, 500–1500 ml of saliva is produced and contains 3 g/L of proteins (0.3%) (in comparison, plasma contains 70 g/L proteins) (Aps and Martens, 2005; Humphrey and Williamson, 2001). The most common of these are enzymes (20%), immune response proteins (18%), and proteases and their inhibitors (10%) (Esser et al., 2008). Both the production rate and the composition of saliva are under neurological control and can be affected by several physiological and environmental factors such as age, gender, diet, medications, oral health status and exercise, in addition to circadian and circannual variations (peak secretory flow is in winter) (Aps and Martens, 2005; Esser et al., 2008; Humphrey and Williamson, 2001; Schipper et al., 2007). Several studies profiled different types of saliva specimens (stimulated and unstimulated, whole and gland-specific saliva) using a variety of proteomic platforms (Denny et al., 2008; Schipper et al., 2007). The "Human Salivary Proteome Consortium" recently reported the identification of over 1100 nonredundant proteins from parotid and submandibular/sublingual saliva specimens (Denny et al., 2008); combined with data from whole saliva, a catalogue of 1939 nonredundant salivary proteins was compiled. 597 of these proteins have also been observed in plasma (Yan et al., 2009b). The ability of saliva to reflect both local and systemic health status and its use in drug monitoring made it an attractive body fluid for proteomic studies. Several salivary proteins have been suggested as potential biomarkers for local pathologies such as oral squamous cell carcinoma (Hu et al., 2008) as well as systemic diseases such as ovarian and breast cancer (Streckfus and Bigler, 2005), type 2 diabetes (Rao et al., 2009) and Sjogren syndrome (Baldini et al., 2008). However, like other major biofluid proteomes, the salivary proteome is highly complex and dynamic (Oppenheim et al., 2007). Several PTMs occur both before and after protein secretion including glycosylation, phosphorylation, proteolytic degradation and formation of protein–protein complexes (Helmerhorst and Oppenheim, 2007). Furthermore, the salivary proteome is very unstable due to the presence of many proteases and also shows large inter- and intra-individual variability (Esser et al., 2008; Quintana et al., 2009; Schipper et al., 2007).

17.2.4 Gastrointestinal (GI) Fluids

Several GI fluids have been used for biomarker discovery, including bile (Chen et al., 2007a; Farina et al., 2009; Kristiansen et al., 2004), pancreatic juice (Gronborg et al., 2004) and gastric juice (Hsu et al., 2007). The collection of these fluids is invasive and requires endoscopic procedures or surgery, which cannot be readily performed and repeated. Only small amounts can usually be obtained and these samples are often contaminated with blood. In addition, bile and pancreatic juice can also cross contaminate each other (Zhou et al., 2007). Bile contains high levels of salts, bile acids, cholesterol, phospholipids, bile pigment and other interfering substances, and therefore requires extensive pre-processing. Due to different processing procedures and pathologies analysed (cholangiocarcinoma, cholelithiasis, biliary stenosis due to pancreatic cancer) as well as the small number of subjects used in the majority of reported studies, the number of proteins identified in bile varied from 80 to more than 200 (Guerrier et al., 2007; Zhou et al., 2005), with the overlap between the studies being less than 30% (Farina et al., 2009).

Pancreatic juice represents a good source of biomarkers for benign and malignant diseases of pancreas. In addition to inter-individual variability and high salt content, pancreatic juice contains high concentrations of zymogens and low protein content (Chen et al., 2006b, 2007b; Gronborg et al., 2004; Zhou et al., 2007). At present, around 170 proteins, one third of which are enzymes, have been described in pancreatic juice (Gronborg et al., 2004). Using SELDI (surface enhanced laser desorption ionisation) ProteinChip, Rosty et al. identified Hepatocarcinoma-Intestine-Pancreas/Pancreatic Associated Protein (HIP/PAP-I) as a potential biomarker of pancreatic adenocarcinoma and validated this finding using serum samples (Rosty et al., 2002). A homologous PAP2 protein was also recently identified using 1D gel and LC-MS/MS approach (Gronborg et al., 2004). Currently our understanding of GI fluid proteomes remains very limited and more studies exploring methods of collection and optimisation of sample preparation as well as use of analytical platforms are needed. Due to the invasive nature of GI fluids collection, however, it is difficult to imagine their widespread usage as a biomarker source in a routine clinical practice.

17.2.5 Cerebrospinal Fluid (CSF)

The majority of CSF is produced by the choroid plexus at a rate of 500 mL/day, with around 125 mL being present in the ventricular system at any time. It contains lipids, sugars, electrolytes and proteins. CSF can provide invaluable information about the physiology of the central nervous system (CNS) and blood/brain barrier as well as pathologies of the CNS (Brechlin et al., 2008; Castano et al., 2006; Hanrieder et al., 2009). The collection of CSF, however, is an invasive procedure requiring lumbar puncture, although occasionally it can also be collected through ventricular catheters that have been inserted for therapeutic or diagnostic purposes. The collection method of CSF has significant implications, as many CSF proteins show rostro-caudal concentration gradient. For example, blood-derived CSF proteins show 20% decrease in concentration between the 1st and 12th mL of CSF withdrawn from lumbar puncture (Reiber and Peter, 2001). Further challenges in CSF proteomics arise from the composition of the CSF itself: the protein content is low, 200–400 fold less than serum (You et al., 2005), with the majority of these proteins being transudates from plasma (80%), hence rich in albumin, transferrin and immunoglobulins, while the remaining are brain-derived proteins (Thompson, 1995). Due to the high salt content (Yuan and Desiderio, 2005), CSF also requires a significant sample pre-processing; inter- and intra-individual variations as well as age-related variations further complicate comparative proteomic studies (Hu et al., 2005; Zhang et al., 2005).

Published proteomics data on eleven different body fluids, including bronchoalveolar lavage, seminal and amniotic fluids, as well as synovial fluid, tears and milk have recently been compiled into a Sys-Body Fluid database (http://www.biosino.org/bodyfluid/database.jsp), where readers can access the associated proteins, their detailed annotations, description, sequence and gene ontology information.

17.3 Sample Preparation

Sample collection and preparation in body fluid proteomics can be a source of considerable bias resulting in false biomarker discovery or failure to validate many potentially promising biomarkers (Omenn et al., 2005). Since most of the proteomic methodological studies involving biofluids focused on plasma/serum and urine, these will be discussed in more details below.

17.3.1 Sample Collection and Handling

Steps to ensure comparability between sample sets starts as early as subject selection. A biomarker should be able to differentiate subjects with a particular disease from both healthy and other disease states, particularly those affecting the same organ or system; samples representative of all these groups of subjects should be included. Although the influence of age and sex on the overall proteomic profiles of serum and plasma have not been systematically explored, the expression of abundant and frequently detected plasma Apolipoproteins E and A1 correlated with the age of the subjects (Yan et al., 2009a). Several other studies have also demonstrated variation in specific proteins with age and sex (Groesbeck et al., 2008; Scully et al., 2009). In urine, aging has been shown to induce proteomic changes similar to chronic kidney diseases (Zurbig et al., 2009) and unmatched sex and age can result in differential expression of some proteins that may be misinterpreted in the context of biomarker discovery experiments (Munro et al., 2006). Several studies showed significant differences between male and female urine proteomes; the number of shared peaks in SELDI analysis of male and female samples by Papale et al. was <50% (Papale et al., 2007). Other disease-related clinical parameters should also be considered; Yan et al. found that biliary obstruction, often associated with benign and malignant pancreatic and biliary diseases, can increase the levels of Apolipoprotein A1, E and transthyretin irrespective of the underlying pathology (Yan et al., 2009a). Samples should also be matched in respect to other patient characteristics including co-morbidities, medications, nutritional habits, etc. These examples only serve to illustrate the significant effects of subject characteristics on proteomic profiling, many of which still remain to be systematically explored.

Blood collection conditions (in outpatient clinics, before or after surgery) can account for the differences between sets of samples, and if not fully assessed can also result in false biomarker discovery; for example prolactin, a putative ovarian cancer marker was shown not to differentiate between cases and controls if the collection conditions were accounted for, as elevation of its levels was induced during anaesthesia prior to surgical procedures (Thorpe et al., 2007).

Several sample handling variables have also been shown to alter proteomic profiles. The polymeric material in the collection tube may produce interfering signals (Drake et al., 2004) although this has not been confirmed by Baumann (Baumann et al., 2005). While EDTA and citrate plasma profiles appeared similar in one study (Tammen et al., 2005), other studies reported variations in plasma samples prepared using different anticoagulants (Banks et al., 2005) with EDTA samples being the most divergent (Hsieh et al., 2006). Inadequate removal of platelets can also alter the plasma profile, particularly for plasma samples when preparation was carried out at lower temperature (4°C), which can result in platelet activation and release of platelet-derived proteins (Tammen et al., 2005). Serum profiles have been shown to be altered by prolonged transport and incubation at room temperature, with some of the putative biomarkers for early detection of ovarian cancer such as ITIH4 (inter-alpha (globulin) inhibitor H4) and hemoglobin beta being particularly unstable, hence unlikely to be robust disease biomarkers (Hsieh et al., 2006; Timms et al., 2007). Several studies have demonstrated the effect of prolonged clotting time, depending on the subsequent analytical method, and particularly affected the peptidome, with formation of various fibrinogen-derived peptides (Baumann et al., 2005; Govorukhina et al., 2009). In one study, the effect of the prolonged incubation time at room temperature (22°C) on serum proteomic profiles approached the effect of biological variance, with complement component C3f, fibrin and kininogen fragments being sensitive indicators of sample handling-induced variability (West-Norager et al., 2007). In summary, it is recommended that plasma is processed within an hour at room temperature, and clotting for serum preparation should not be left for longer than 30–60 min.

Although the addition of protease inhibitors, which contain peptide and small molecule inhibitors, to plasma may have favourable effects in preventing degradation, they may also be a source of variations. The sulfonyl fluoride AEBSF, for example, can covalently bind to some proteins altering their isoelectric points (p*I*s) and the peptide protease inhibitor aprotinin can compete with endogenous peptides in MS (Rai et al., 2005; Schuchard et al., 2005; Zhou et al., 2005). While judicious use of protease inhibitors, immediately after blood collection, in top-down proteomics may be beneficial (Rai et al., 2005), it is not routinely recommended. In urine proteomics, the use of protease inhibitors is also not recommended, particularly for gel-based approaches (Havanapan and Thongboonkerd, 2009). However, they may still be required for proteinuric urine (Havanapan and Thongboonkerd, 2009; Thongboonkerd, 2007) and are essential to preserve the exosome-associated proteins (Zhou et al., 2006). Furthermore, several pathological conditions of the urinary tract my result in a disease-specific increased protease activity as seen in nephritic syndrome (Candiano et al., 2006; Magistroni et al., 2009) and this also has to be taken into account before the decision to use or omit protease inhibitors is made.

Timing of sample collection is also important. Although 24 h urine collection is preferable as it avoids the diurnal variation and accurately measures the urinary protein excretion, this is often not practical. Moreover, as in the case of 1st morning urine, it may suffer from contamination with bacterial overgrowth or bladder epithelial cells (Bottini et al., 2002). It was also shown that 1st void has a different profile to midstream urine, it is less stable, prone to bacterial contamination and often contains extra signals from urogenital tract secretions, particularly in females, where additional peaks with average masses consistent with alpha-defensins were seen (Schaub et al., 2004). Therefore, the second-morning or random midstream urine collection is recommended (Thongboonkerd, 2007). Following collection, urine samples are centrifuged at low speed to remove cells and debris; centrifugation also appears to delay bacterial overgrowth (Thongboonkerd and Saetun, 2007). It has been shown that prolonged incubation at room temperature can lead to bacterial overgrowth and changes in proteomic profiles which can occur as early as after 8–12 h, whilst keeping urine at 4°C delays bacterial growth to 20/48 h in uncentrifuged/centrifuged urine (Thongboonkerd and Saetun, 2007). The addition of preservatives such as boric acid or sodium azide to delay bacterial overgrowth should be considered, particularly if a prolonged time between sample collection and freezing is anticipated; however, if possible, freezing the urine specimens within 2 h is the most optimal (Papale et al., 2007; Schaub et al., 2004; Thongboonkerd and Saetun, 2007).

Urine samples should also be tested for general biochemical features including creatinine, pH, nitrites, glucose and protein concentration, as well as presence of blood and leucocytes. The presence of blood has been shown to alter the proteomic profile even after removal of cells by centrifugation due to the presence of plasma proteins (Schaub et al., 2004). Whilst the use of proteinuric urine is an integral part of biomarker discovery for renal disease, the use of such urine in the search for biomarkers for non-renal diseases is problematic. Proteinuria resulting from renal pathology has differential effects on the urine proteomic profile which will hamper the interpretation of the results (Lafitte et al., 2002). Although in many previous studies the urine pH was adjusted prior to processing, a recent report by Thongboonkerd et al. found that urine pH has no effect on gel-based urine proteomic profiles and therefore does not need adjustment (Thongboonkerd et al., 2009). However, pH may affect ion exchange chromatography; in a study of the urinary peptidome using magnetic bead (MB) fractionation and matrix-assisted laser desorption/ionization-time-of-flight (MALDI-TOF MS) it was found that pH adjustment to 7 is necessary for weak cation-exchange and metal ion affinity MB preparation (Fiedler et al., 2007).

17.3.2 Sample Storage

Storage temperature and time are additional important factors. Although many authors recommend storage at temperatures lower or equal to –20°C, some studies showed continuous modification of the serum profiles at –20°C, even during short term storage of 10 days, such samples for example, artificially differentiated multiple sclerosis patients from healthy people (Pieragostino et al., 2009). This time and temperature-related false biomarker discovery suggests that long-term storage at –80°C is a safer approach, although specific analytes may be altered even at such low temperature. This was demonstrated by significant changes in the intensities of 15 peaks after 10 months storage of serum at –80°C (compared to 62 differential peaks that appeared between samples stored at –20°C vs. –80°C) (Ahmad et al., 2009). Interestingly, some of these peaks were again attributable to fibrinogen, fibrinopeptides, kininogen and C3f, changes of which were previously reported to result from sampling-induced variability (West-Norager et al., 2007). Therefore, case and control samples should be collected in tandem and stored either at –80° or, preferably, in liquid nitrogen (Ahmad et al., 2009).

Storage of urine at –20°C can result in progressive decrease in urinary protein concentration compared to storage at –70°C, although not all urinary proteins are equally sensitive to storage conditions (for example, immunoglobulin G is much less stable than albumin) (Klasen et al., 1999; Schultz et al., 2000). Another study also reported that storage at –20°C can induce loss of proteins in the precipitate and vigorous shaking of the sample after thawing should be applied to restore the protein content to its baseline (Pisitkun et al., 2004). In addition, freezing at –80°C preserves the urinary exosome-associated proteins better than –20°C (Zhou et al., 2006).

It is generally believed that the minimal number of freeze-thaw cycles should be applied. Results from studies addressing this important issue have so far been variable. Hsieh et al. found no major effect of repeated freeze-thaw up to 10 cycles on the 100 SELDI peaks they evaluated within the LMW proteins of serum (Hsieh et al., 2006), whereas Baumann et al. found that best reproducibility is achieved with single thawing of frozen samples, again looking at LMW proteins in serum (Baumann et al., 2005). The larger MW proteins may be even more sensitive to freeze-thaw cycles; therefore it is probably best to aliquot the samples into small volumes and avoid more than one freeze-thaw. Urine on the other hand, appears to be more resistant to changes related to freezing as several studies have shown no significant effect on the urinary proteome following 4–5 freeze-thaw cycles (Papale et al., 2007; Schaub et al., 2004; Thongboonkerd, 2007).

17.3.3 Protein Recovery, Concentration and Clean Up

Several biofluids such as urine, CSF and saliva have low protein content and require a concentration step, particularly prior to gel-based proteomic approaches. In urine, several methods have been used to recover proteins. In their extensive assessment of 38 different protocols (including precipitation with various solvents, centrifugal filtration, lyophilisation and ultracentrifugation), Thongbonkerd et al. concluded that although 90% ethanol gave the best yield, different protocols are complementary and several may need to be combined in order to obtain a comprehensive coverage of the proteome (Thongboonkerd et al., 2006). In an earlier study, the same group demonstrated that acetone precipitated more acidic and hydrophilic proteins whereas ultracentrifugation precipitated more basic and hydrophobic proteins, with only albumin being commonly precipitated by both protocols (Thongboonkerd et al., 2002), suggesting again that combining several approaches increases the chances of better proteome coverage. In addition to whole protein recovery, exosomal subproteome can be isolated using high-speed centrifugation (200,000×g for 1 h) and resuspending the pellet in an appropriate isolation solution (Pisitkun et al., 2004). Salts and metabolites can interfere with some proteomic platforms, particularly in gel-based approaches, and their removal may be necessary. Whilst some protein recovery methods such as solid phase extraction (SPE) and centrifugal filtration with MW cut-off filters (MWCO) can desalt samples at the same time, other methods such as prolonged dialysis of samples can also be used. Performing dialysis prior to protein extraction by organic solvents appears to improve extraction efficiency (Thongboonkerd et al., 2002).

17.3.4 *Depletion of Highly Abundant Proteins, Fractionation and Enrichment*

Because of the large dynamic range of body fluids, particularly of plasma/serum, which is unmatched by the (limited) sensitivity of current proteomics platforms, and the low abundance of clinically useful biomarkers, it is prudent to attempt to reduce the complexity of initial sample mixture as highly abundant proteins compete with low abundant proteins and prevent their detection in MS. Therefore, depletion of the highly abundant proteins and fractionation of samples became standard practice in proteomics experiments. In addition, enrichment of certain cellular subproteomes of interest can also be performed (for review see Stasyk and Huber, 2004).

17.3.4.1 Depletion

Concentrating and "unmasking" lower abundance proteins by depletion of highly abundant proteins can be achieved by several approaches. These include dye-based methods such as Cibacron Blue (CB) F3-GA which binds albumin but also enzymes and other plasma proteins (Gianazza and Arnaud, 1982), protein A/G systems which bind immunoglobulins (Eliasson et al., 1988), ultrafiltration with MWCO filters and immunoaffinity (Bjorhall et al., 2005; Echan et al., 2005). Echan et al. compared depletion of high abundance proteins using dye-based affinity, immunoaffinity depletion of albumin and IgG and Multiple Affinity Removal System (MARS) columns, which deplete six highly abundant proteins (albumin, transferrin, haptoglobin, IgG, IgA, and alpha-1 antitrypsin). They reported 85% depletion of the total serum protein using MARS columns and lack of efficiency and specificity with the other methods. However, they also found that such depletion (although improving the gel resolution) resulted in only a modest increase in the number of newly identified spots on 2D gels, most of which were minor isoforms of relatively abundant proteins (Echan et al., 2005). Furthermore, the lowest abundance proteins detected, such as amyloid P and carboxypeptidase N (both around 30 μg/mL) indicated that depletion of six major proteins combined with 2D gel analysis is still not sensitive enough for the detection of potential biomarkers, which are in the ng/mL or pg/mL range (Echan et al., 2005). Another study comparing CB dye-based method with immunodepletion of six high abundance proteins also reported the limited specificity and efficiency of the dye-based system compared to 98% depletion efficiency of targeted proteins and minimal non-specific binding with the immunoaffinity method (Zolotarjova et al., 2005). Roche et al. compared the results of immunoaffinity depletion of one, six, 12 and 20 high abundance proteins from serum samples (Roche et al., 2009). Depletion of 12 proteins removed around 87% of the total protein which was significantly better than depleting one or six proteins, but similar to depletion of 20 proteins. In addition, the authors found that increasing the number of depleted proteins from 12 to 20 resulted in loss of SELDI signals, which could be due to inadvertent loss of non-targeted proteins or peptides (Roche et al., 2009). The Avian IgY (Immunoglobulin Yolk) immunoaffinity subtraction system has also been evaluated in several studies (Huang et al., 2005b; Liu et al., 2006b). Liu et al. used IgY-12 affinity columns to deplete 12 high abundance proteins and found that this method efficiently depleted approximately 95% of the total plasma proteins, but they reproducibly identified 38 (using LC-MS/MS) and 108 (using 2D LC-MS/MS) non-target proteins in the bound fraction (Liu et al., 2006b). In urine, depletion of high abundance proteins has not been routinely used in non-proteinuric urine. However, due to leak of large amounts of plasma proteins into proteinuric urine, there may be some merit in depleting these samples of high abundance proteins (Kushnir et al., 2009).

One of the drawbacks of depletion of high abundance proteins is the inevitable loss of other protein species, particularly with dye-based methods, and to a lesser extent with more specific antibody-based depletion methods. Albumin and other high abundance proteins are known to act as "molecular sponges" and non-specifically bind other proteins, protein fragments, hormones, and peptides including cytokines thus retaining them in the circulation and essentially amplifying their concentration; however, these may be lost during the depletion process (Granger et al., 2005; Manabe et al., 2003). Zhou and colleagues investigated the serum "interactome" by isolating six of the most highly abundant proteins using immunoaffinity/affinity chromatography (under mild buffer conditions to help preserve the protein–protein interactions) followed by digestion of bound proteins and peptides and subsequent microRPLC-MS/MS (Zhou et al., 2004). They identified 210 proteins, several of which (such as prostate-specific antigen (PSA) and pregnancy plasma protein) are biomarkers already in clinical use. In a study by Shen et al., depletion of albumin was associated with the reduction of 815 other proteins (Shen et al., 2005). The specificity of immunodepletion seems to depend not only on the antibody but also on the optimal buffers and depletion conditions (Echan et al., 2005; Zolotarjova et al., 2005). Supporting the same notion, results of another study showed that performing serum immunodepletion of albumin and IgG under partially denaturing conditions with (5%) acetonitrile (which disrupts the binding of LMW proteins to their carriers) minimized the loss of bound proteins and improved detection of lower abundance proteins (Huang et al., 2005a). On the other hand, the study of this "carrier protein-bound proteome" and the ratio of bound and unbound molecules in the "albuminome" could also potentially be of interest in biomarker discovery (Gundry et al., 2007) as it can comprise potential biomarkers for various pathologies as has been shown for alcohol abuse (Lai et al., 2009) and ovarian cancer (Lowenthal et al., 2005).

Due to their efficiency, specificity and reproducibility, immunoaffinity methods are the most commonly used approach for the depletion of high abundance proteins. Other available methods, although cheaper, are perhaps better used as fractionation rather than depletion methods with both fractions subsequently profiled to avoid missing potentially important biomarkers. In either case, it is clear that optimizing the depletion buffers and conditions is necessary to achieve efficient depletion with minimal loss of non-targeted proteins (Echan et al., 2005). While high abundance proteins vary according to the biofluids, immunodepletion systems are primarily designed to deplete high abundance proteins in serum/plasma.

To further improve proteomic coverage of other body fluids, depletion methods targeting high abundance proteins in those body fluids are required.

Finally, a chromatographic depletion has recently been suggested. Gao et al. fractionated liver tissue proteins with strong cation exchange/reversed-phase HPLC (micro-SCX/cRPLC). They applied a signal intensity of 0.1 absorbance unit as a cut-off, and defined peaks with higher signal intensities as high abundance proteins. The fractions containing such peaks were removed and the remaining fractions pooled before further analysis. Using this approach, 58 high abundance proteins were depleted, which resulted in a 2.7 fold increase in the number of identified proteins (from 451 to 1213) compared to non-depleted samples (Gao et al., 2008).

17.3.4.2 Fractionation and Enrichment

Even with successful depletion of the majority of high abundance plasma proteins (99% of plasma proteins), the remaining 1% has an estimated concentration dynamic range greater than 7 logs (Hoffman et al., 2007), which is still beyond the sensitivity of current proteomics platforms. Therefore, fractionation of plasma and other complex biological samples prior to proteomic analysis is critical in order to further reduce the sample complexity and allow higher sample load; both of which result in a substantial improvement in the detection of low abundance proteins.

Several fractionation and separation methods have been used in proteomics, including biomarker discovery. These methods either exploit the physicochemical characteristics of proteins/peptides such as charge and MW, or use affinity matrices to capture specific peptide/protein groups. These approaches can be broadly divided into electrophoretic and chromatographic methods. 2D electrophoresis (2DE) has been the cornerstone platform for proteomics experiments. It utilises isoelectric focusing (IEF) to separate proteins according to their p*I*s in the first dimension followed by SDS-PAGE to separate them according to relative MW in the second dimension. The major advantages of 2DE are its ability to resolve intact proteins, to detect PTMs and fragmentation of proteins as well as its quantitative ability. However, it has limited ability to resolve proteins with extremes of MW, p*I*s and hydrophobic proteins despite several improvements such as narrow pH strips, improved buffers and two-dimensional differential gel electrophoresis (2D-DIGE) technology. In addition, 2DE is labour-intensive and its dynamic range remains around 10^2–10^4 (Hoffman et al., 2007). Both electrophoretic separation principles (IEF, SDS-PAGE) can also be used separately. IEF can be performed on immobilized pH gradient (IPG) strips or in solution (such as microscale IEF and free-flow electrophoresis (FFE)). In FFE-IEF, liquid phase IEF is used to separate both proteins and peptides into separate fractions according to their p*I*s. Separation can be performed under both native and denaturing conditions (Wildgruber et al., 2008). This has several advantages. First, unlimited sample loads can be separated due to the continuous flow under matrix-free separation with less non-specific interactions, and second, the minimum loss of proteins due to precipitation permits high recovery rate of sample (Moritz et al., 2004). Finally, p*I* values can be useful in further validation of the MS data (Moritz et al., 2004; Nissum et al., 2007). Microscale

IEF separates proteins into several fractions separated by membrane partitions of defined pH. It is compatible with most downstream proteomic platforms, particularly 2DE, where fractionated samples can be directly applied to narrow range IPG strips (Zuo and Speicher, 2002). It still suffers some drawbacks including run-to-run variation, which can be reduced by using 2D-DIGE, and loss of the proteins in the membrane partitions (Tang et al., 2005). Capillary electrophoresis is another form of electrophoresis that has been used particularly in urinary proteomics. Despite the limited sample loading capability, it provides fast, low-cost, efficient and high-resolution separation of intact proteins and peptides, requires small analyte volumes, and can be coupled directly to MS (Ahmed, 2009; Fang et al., 2009). SDS-PAGE can also be used on its own or as the last protein separation step prior to MS; the gel is stained, bands are visualised, cut and digested prior to MS identification.

Chromatographic methods are based on fractionating proteins/peptides according to their interactions with the chromatographic surfaces. Ion-exchange chromatography (IEC) separates molecules according to their charge under nondenaturing conditions and is therefore most useful for soluble proteins, whereas hydrophobic interaction chromatography (i.e. RP) separates proteins based on the hydrophobicity profile and thus allows analysis of insoluble proteins. Size-exclusion chromatography (SEC) fractionates proteins based on their molecular mass; it is less commonly used because of similar separation parameter as SDS-PAGE and potential for significant protein loss (Lescuyer et al., 2004). Affinity chromatography has been used to separate proteins/peptides based on specific affinity interactions with the chromatographic surfaces, such as metal and heparin affinity (Jmeian and El Rassi, 2009; Lescuyer et al., 2004). Affinity chromatography can also be used to enrich for specific groups of proteins such as phosphoproteins (using immobilized metal ion affinity chromatography (IMAC), or phospho-specific antibodies), glycoproteins (using hydrazide chemistry or lectin affinity), ubiquitinated proteins, and cysteine containing proteins (using thioredoxin-sepharose affinity or covalent chromatography) (Azarkan et al., 2007; Stasyk and Huber, 2004). In both affinity and SPE chromatography the unbound fraction is washed away and the bound fraction can be either analysed directly (SELDI ProteinChips or functionalized MALDI targets) or eluted in one step prior to further analysis (Callesen et al., 2009). In liquid chromatography, on the other hand, almost all proteins bind to the stationary chromatography surface with varying affinity; subsequently they are eluted in separate fractions using the liquid phase gradient (pH, salt or detergents).

There is ample evidence in the literature that high dimensional fractionation, which includes one or more protein separation methods followed by trypsin digestion and subsequent analysis of resulting peptide mixtures, by either LC-MS/MS or LC/LC-MS/MS, is required for deep mining of the complex proteomes for exploration of low abundance proteins (Dayarathna et al., 2008; Hoffman et al., 2007; Jin et al., 2005; Nissum et al., 2007; Vasudev et al., 2008). The basic principle of multi-dimensional fractionation is to separate proteins and peptides according to various orthogonal physicochemical properties and/or affinity interactions, resulting in much less complex fractions. In addition to increasing the number of identified proteins and the dynamic range of detection, this multi-dimensionality also increases the confidence of protein identifications. Furthermore, the protein fractionation prior to digestion and MS identification combines the advantages of both bottom-up and top-down proteomic approaches. This however comes at a cost, namely low throughput, sample loss during fractionation procedures and high experimental cost. The power of combining several electrophoretic and chromatographic fractionation methods will be illustrated through selected reports of several recent studies.

Multidimensional protein identification technology (MudPIT) is a prime example of the enhanced proteome coverage in bottom-up proteomic approaches. It was developed in the Yates laboratory and it utilises biphasic columns (typically SCX followed by RP) to separate protein digests prior to MS/MS; this results in a high number of protein identifications, including low abundance proteins (Washburn et al., 2001). A combination of protein fractionation using chromatofocusing chromatography (ProteomLab PF 2D fractionation system) and MudPIT in analyses of proteome of a breast cancer cell line resulted in the identification of 1160 proteins from 1 mg of cell lysate (Chen et al., 2006a). In another study, serum immunodepleted with a MARS column was followed by RP fractionation of proteins using macroporous RP C18 column under high temperature (80°C) and optimised elution conditions, followed by SDS-PAGE or 2D LC-MS/MS (Martosella et al., 2005). The authors demonstrated reproducible separation of proteins with near complete recovery from the RP column. This approach resulted in increased identification of unique proteins, improved sequence coverage and detection of some low abundance proteins such as macrophage stimulatory protein (MSP), which is normally present at low ng/mL levels (Martosella et al., 2005). A similar approach was later applied with immunodepletion of 14 high abundance proteins (resulting in depletion of 94 and 92% of serum and plasma proteins, respectively). They were able to detect more than 20 low abundance proteins, such as insulin-like growth factor binding proteins 3, 4 and 6, interleukin–27 beta chain, etc, several of which were represented by multiple peptides (Zolotarjova et al., 2008). Pieper et al. used a 3D fractionation approach of serum proteins prior to 2DE. They first performed immunoaffinity subtraction chromatography to deplete nine high abundance proteins, followed by anion exchange chromatography (AEC) and SEC in the 2nd and 3rd dimension, respectively. 2DE Protein spots were subsequently identified using MALDI-TOF-MS or LC-MS/MS. A total of 620, 2100 and 3500 spots were identified with 1D, 2D and 3D pre-fractionation, respectively. While 2D pre-fractionation resulted in 157 protein identifications, 3D yielded 295, and both yielded 325 distinct proteins, many of them posttranslationally modified. With the 3D pre-fractionation the authors were able to identify several low abundance proteins in the sub ng/mL range in addition to cellular and tissue leakage proteins including several clinically used biomarkers such as C-reactive protein, interleukin-6, PSA, creatine kinase M and several cathepsins (Pieper et al., 2003). Barnea et al. evaluated several protein fractionation methods. Serum samples were depleted of albumin and IgG followed by strong cation exchange (SCX) chromatography, SDS-PAGE or liquid phase IEF. Proteins were subsequently digested and peptides separated using either RPLC or MudPIT followed by ESI-MS/MS. A total of 470 proteins were identified by all approaches, 47% of which by only one peptide. Although SCX fractionation resulted in the largest number of identified proteins, each method revealed a set of unique proteins emphasizing the complementary nature of these methods. In addition, the use of MudPIT instead of LC-MS/MS dramatically increased the number of uniquely identified proteins; however, this is not a high-throughput approach due to extensive MS time usage (Barnea et al., 2005). The enhanced coverage and dynamic range achieved by employing several fractionation procedures was also excellently illustrated by Liu et al. in their profiling of the plasma proteome of trauma patients (Liu et al., 2006a). They first immunodepleted plasma of 12 highly abundant proteins, which resulted in depletion of 94.4% of high abundance proteins; the immunodepleted plasma was then divided into two aliquots and subjected to either solid phase cysteinyl peptide or N- glycoprotein capture. All four fractions (Cys-peptides, non-Cys peptides, N-Glycopeptides and non-glycopeptides) were then fractionated by SCX chromatography and analysed by capillary LC-MS/MS. This approach resulted in the identification of 3654 proteins (1494 with more than one peptide); again, high and medium abundance proteins were represented with more peptides, while low abundance proteins were usually identified by only one peptide). Only 9% of proteins were detected in all four fractions, again emphasizing the complementary nature of different approaches. This strategy resulted in the detection of more than 100 low abundance proteins such as different cytokines and their receptors, and cathepsins, with an estimated dynamic detection range of 10^7 (Liu et al., 2006a). Serum samples from prostate cancer and benign prostatic hyperplasia were fractionated using anion displacement liquid chromatofocusing that separates proteins according to their p*I*s using a pH gradient (performed on Proteomelab PF 2D HPLC system, Beckman Coulter) followed by IEF gels and 2D-DIGE. This improved the resolution, increased the number of resolved spots and resulted in the identification of several low abundance proteins, including squamous cell carcinoma antigen 1 (SCCA1) (Qin et al., 2005). Another study used a similar approach employing AEC of MARS-depleted plasma samples prior to 2D-DIGE. Depletion and fractionation resulted in increased number of resolved spots from 290 to 1200, with 33 differentially expressed spots. Leucine-rich alpha-2-glycoprotein (LRG) was identified as a potential plasma biomarker of pancreatic cancer and its over-expression was confirmed by western blotting (Kakisaka et al., 2007). Moritz et al. used 2D fractionation of human plasma following MARS depletion. They employed FFE-IEF under nondenaturing conditions in the first dimension to collect 96 fractions followed by rapid (6 min) RP-HPLC of each FFE fraction. Using MS, they identified 78 proteins from 16% of the collected FFE pools (55 with > one peptide). In addition to low abundance proteins such as L-selectin, they also identified cancer-related protection peptide (CRISPP) that was co-eluted with the highly abundant alpha1-antitrypsin protein from the bound fraction of a MARS column, thus confirming a "protein sponge" hypothesis (Moritz et al., 2005). An even higher dimensional (4D) fractionation approach has been suggested and named "protein array pixelation" (Tang et al., 2005). In this approach, plasma and serum were immunodepleted using a MARS column prior to solution IEF and 1DE. Gel bands containing a group of proteins with known p*I* and MW (pixels) were cut and proteins digested with trypsin followed by RP-LC peptide separation prior to ESI-MS/MS. They identified 744 proteins

(185 with > one peptide) from plasma and 4377 proteins (752 with > one peptide) from serum. Several proteins were detected at ng/mL to high pg/mL levels such as metalloproteinase inhibitor 1, calcitonin and tumor necrosis factor alpha which provided an estimated dynamic range of 10^9 for this method (Tang et al., 2005).

Lastly, the use of peptide ligand libraries to reduce the complexity of proteomes has been increasingly investigated recently. In this approach, a large number of diverse short peptides (usually hexapeptides) are synthesized and attached to beads or other solid phase matrices. When a complex protein mixture is added, proteins bind to specific peptide(s) based on their affinity until those peptides are saturated leaving the rest of their protein partners in the solution. The bound proteins are subsequently eluted and analysed (Thulasiraman et al., 2005). If the number of peptides is large enough, the library will be able to capture most of the proteins within the mixture, resulting in reduced dynamic range of the biological sample by both diluting high abundance proteins and concentrating low abundance ones. However, this approach remains incompletely characterized and requires further research despite many studies demonstrating its advantages (Castagna et al., 2005; Guerrier et al., 2007; Sennels et al., 2007).

While many of the described approaches have not been used routinely for biomarker discovery, they illustrate a few important points: firstly, a combination of orthogonal multidimensional separation is essential for in depth mining of complex body fluid proteomes, and secondly, as most of the reported approaches largely detected abundant species, novel strategies are needed to be able to routinely access the pool of low abundance peptides and proteins.

The repertoire of issues and methods we have discussed in relation to sample collection and preparation are by no means exhaustive. And whilst selection of the best and most appropriate experimental protocols is essential, it is more important that samples are matched in as many variables as possible and the adopted protocols are applied similarly to all samples. Several minor variables can have a cumulative effect resulting in bias that may be difficult to identify by data analysis.

17.4 Biomarker Pipeline

The lack of stringent and appropriate study design, sample collection protocols, experimental procedures as well as data analysis and reporting in biomarker discovery has indicated the need for critical changes in proteome research if the goal of finding clinically useful protein biomarkers is to be achieved (Ransohoff, 2004). With this in mind, the adoption of a phased approach with appropriate guidelines in biomarker discovery, similar to the ones in drug development has been advocated by many authors in several recent reviews (Pepe et al., 2001; Ransohoff, 2007; Rifai et al., 2006; Zolg and Langen, 2004). In addition, new guidelines for proteomic experiments (MAIPE) (Taylor et al., 2007), data reporting (REMARK, STARD) (Bossuyt et al., 2003; McShane et al., 2006) as well as analytical validation of soluble biomarker assays (OMERACT 9 v2) (Maksymowych et al., 2009) have

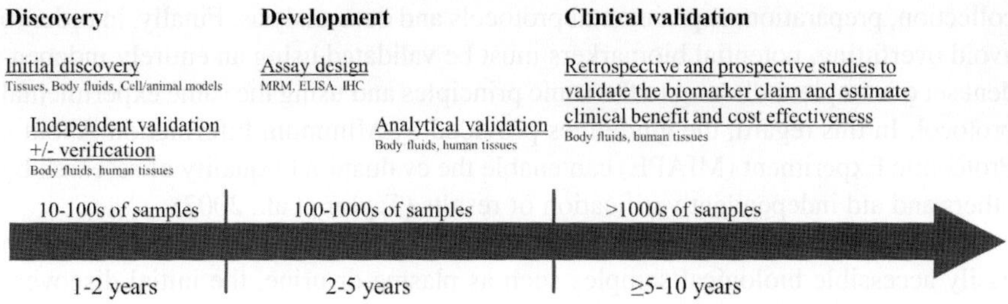

Fig. 17.1 Biomarker discovery pipeline. The three phases of biomarker discovery are illustrated including sub-phases, biological samples used and approximate timescale. (MRM: multiple reaction monitoring; ELISA: enzyme-linked immunosorbent assay; IHC: immunohistochemistry)

recently been published. A summary of the potentially most relevant findings will be discussed here. Irrespective of their intended use, candidate biomarkers must go through a rigorous pipeline, which is typically divided into three general phases: discovery and verification, development, and finally clinical validation (Fig. 17.1). In addition, several ethical and regulatory issues, and compliance with the food and drug administration (FDA) regulations for in vitro diagnostic devices have to be achieved before biomarkers can be brought to the market and employed in clinical practice (Gutman and Kessler, 2006).

17.4.1 Discovery

The process of biomarker development starts with the discovery phase, to identify biomarker(s) that can, reproducibly and reliably, differentiate one sample group from another with high sensitivity and specificity. There is a general consensus that a perfect biomarker, i.e. a single biomarker that shows a 100% sensitivity and specificity for a certain disease, probably does not exist (Mor et al., 2005; Srinivas et al., 2001). In addition, several studies have demonstrated the added value of using a panel of biomarkers for diagnosis of ovarian (Zhang et al., 2007), prostate (Landers et al., 2005) and other cancers (Cordero et al., 2008). Furthermore, clinically useful biomarkers in current practice such as PSA for prostate cancer, carcinoembryonic antigen (CEA) for colorectal cancer and carbohydrate antigen 19–9 (CA19.9) for pancreatic cancer have variable degrees of performance and are often used as part of decision making algorithms along with other clinical, laboratory and imaging parameters (Greene et al., 2009; Locker et al., 2006). It is inevitable that some of the potential biomarkers identified during the discovery phase will not survive through subsequent validation steps. To ensure the maximum confidence in the results from the discovery phase, to avoid bias and increase the chances of success of potential biomarkers, several principles should be followed in the discovery phase. These include an appropriate study design with adequate statistical power, rigorous documentation of demographic and clinical data, and careful selection of study subjects, as well as documentation and adherence to standard procedures regarding sample collection, preparation, experimental protocols and data analysis. Finally, in order to avoid overfitting, potential biomarkers must be validated using an entirely independent set of samples adhering to the same principles and using the same experimental protocol. In this regard, the guidelines provided by Minimum Information About a Proteomic Experiment (MIAPE) can enable the evaluation of quality of research by others and aid independent verification of results (Taylor et al., 2007).

Although the biomarker assay, in its final format, will likely be performed on easily accessible biological samples such as plasma or urine, the initial discovery stage can also be performed on tissue specimens and proximal fluids. Each of these strategies has its own merits and disadvantages and the choice of specimens for the initial discovery will depend on many factors, including the sample availability, expertise, technology and the intended clinical use of the biomarker(s). Tissue specimens have the advantage of four to five logs lower dynamic range compared to plasma (Hortin and Sviridov, 2010), hence, biomarker approaches based on tissue of origin or a proximal fluid, have the advantage of sampling concentrated proteins which is more likely to result in the detection of lower abundance proteins. At a later stage, one can search for selected candidates in more accessible body fluids in a targeted approach using sensitive methods such as multiple reaction monitoring (MRM) and antibody-based assays (Roessler et al., 2005; Rosty et al., 2002; Zhang and Chan, 2007). Using tissues also allows for enrichment of specific cell types such as cancer cells with techniques like laser microdissection (Johann et al., 2009). It is worth mentioning that important biomarkers can also be derived from the tumoral microenvironment and can be missed by profiling only cancer cells (Bijian et al., 2009; Paulitschke et al., 2009). The disadvantages of using tissues as starting material in the discovery phase include the usually invasive nature of obtaining such samples and the limited availability of specimens particularly for the less common diseases. Similarly, many proximal body fluids require invasive collection procedures, are often collected in small volumes and from limited number of subjects. In addition, the proteomic profiles of many of these fluids are poorly understood and the pre-analytical and analytical variables that are influencing them (as discussed previously) are still incompletely characterised. Furthermore, some of the biomarkers that can show good discriminatory power in tissues or proximal fluids may not display the same power in blood (Kawai et al., 2004; Ucar et al., 2004). In addition to human biofluids and tissue samples, animal models and cell cultures can also be used for biomarker discovery (Hingorani et al., 2003; Kuick et al., 2007; Paulitschke et al., 2009). Although these approaches rely on the ability of the animal/cell models to recapitulate the human disease, they have the advantage of minimal inter-individual variability and genetic and environmental influence on samples, in addition to more controlled sample handling approaches (Kuick et al., 2007; Rifai et al., 2006).

The prioritisation of potential biomarkers from a large number of differentially expressed proteins is pivotal. Informed selection based on statistical analysis, biomarker performance and biological knowledge is much more likely to result in a selection of successful candidate biomarkers compared to statistics alone. Availability of specific antibodies or assays may also favour the pursuit of some biomarkers. Of note, several recent reviews of 2-DE based proteomic experiments reported a group of highly abundant proteins commonly found to be differentially expressed regardless of the tissue/cell type, species or experimental objectives (Petrak et al., 2008; Wang et al., 2009). Many of these proteins are related to the cytoskeleton and stress response (such as enolase 1, several heat shock proteins, peroxiredoxins and actins) and therefore would unlikely be successful as biomarkers.

Regardless of the starting sample type and approach, the initial discovery must be followed by rigorous validation in an entirely independent set of samples. In addition, where samples used for discovery are of a different type to the samples that will be used clinically, verification of the potential biomarkers using the same sample type, which will be used for the development of the final clinical assay, is necessary. This requires a targeted approach to measure selected biomarkers. The high sensitivity and multiplexing ability of MRM lends itself well to the validation/verification process. MRM is a specific and sensitive MS-based method in which unique peptides and their fragment ions (corresponding to the putative biomarker) are chosen from the initial MS/MS data and targeted for quantification (Yocum and Chinnaiyan, 2009). Although relative quantitation of protein abundance between samples can be adequate in the initial discovery stage, absolute quantitiation is often mandatory to establish cut-off levels at which biomarkers display maximum sensitivity and specificity. Several quantitation methods such as inclusion of isotopically labelled peptides in MRM, or ICAT and iTRAQ can be employed. In addition, ELISA assays can also be used when already available. Where appropriate, immunohistochemistry can also be the assay of choice for both initial validation and subsequent clinical validation. This is particularly true for some prognostic and treatment selection biomarkers that can be detected directly in tissues obtained through surgery or biopsy.

In addition to specifically identified biomarkers, another biomarker discovery approach utilises MS peaks, usually generated by SELDI, to establish a pattern of peaks that can distinguish diseased from control states without identifying the corresponding proteins/peptides, the so-called pattern diagnostics (Veenstra et al., 2004). The publication of the flagship paper by Petricoin and others (Petricoin et al., 2002) that identified a pattern of peaks that can distinguish ovarian cancer patients from controls with 100% sensitivity and 95% specificity generated massive interest, but was soon followed by criticism. While SELDI-based approaches require minimal sample preparation and are high throughput (Petricoin and Liotta, 2004), concerns about the bias, lack of reproducibility and difficulties in protein identification, have been raised (Baggerly et al., 2005; Ransohoff, 2005). Furthermore, although the identity of MS peaks is not necessary for pattern diagnosis once the peak discriminating power has been validated, it may provide important clues into the biology of the disease process. Finally, most of the peaks, when identified, appeared to be derived from highly abundant proteins or their fragments and hence are likely resulting from systemic secondary effects of the disease, again raising questions about their specificity (Li et al., 2005; Xiao et al., 2005). At the same time, OvaCheck, a diagnostic tool based on serum pattern diagnostics for ovarian cancer was independently designed for screening purposes; illustrating well the potential danger of precipitate clinical implementation of insufficiently validated discovery data and technologies (Check, 2004).

17.4.2 Assay Development

The second phase of the biomarker pipeline involves the development and analytical validation of the biomarker(s) assay that will be used clinically which may or may not involve transferring to a different platform. Many protein assays in clinical practice involve antibody-based ELISA-type sandwich assays. These have the advantage of high throughput and sensitivity, cost effectiveness and ease of large-scale integration into clinical practice. However, ELISA relies on the availability of antibodies that recognise specific epitopes in the target protein. It also has limited multiplexing ability, and although some multiplexed assays do exist, optimising the condition of the assay for several proteins of different dynamic ranges and antibody affinities may be challenging, resulting in suboptimal measurement of analytes (Leng et al., 2008). Moreover, an ELISA-type assay may not be feasible particularly where the biomarker is a small peptide (Good et al., 2007). MRM- based assays on the other hand have large multiplexing capability (Kuzyk et al., 2009) and good sensitivity and reproducibility (Addona et al., 2009; Keshishian et al., 2007). Whilst MRM-based validation avoids the need for specific antibodies, including the potential time and cost involved in their development, it also does not require a platform switch (where it has been used in the discovery phase) and therefore bypasses many of the analytical validation requirements for a new assay. However, this approach has substantial economic implications, as it requires large capital investment in MS platforms within the health system, which is currently not feasible. Whilst MRM could be the assay of choice for biomarker measurements, ELISA will probably remain the more affordable assay for the foreseeable future.

The analytical validation of the clinical assay involves evaluation of the assay performance in relation to accuracy, precision, reproducibility and dynamic range. Furthermore, whilst the overall proteomic profile may be stable under certain conditions, specific analytes may not be. For example, the levels of soluble interleukin 2 and transferrin receptors were increased in EDTA plasma (De Jongh et al., 1997), whereas plasma levels of circulating vascular endothelial growth factor levels were increased by delayed processing time, room temperature and low centrifugal force (Hetland et al., 2008). Therefore, the stability of candidate biomarkers under various handling and storage conditions should be established. Possible confounding variables including subject demographics, physiological and environmental factors should also be determined. Importantly, reference intervals and the sensitivity and specificity of the test using various cut-off levels should be established. This is often achieved using ROC (Receiver Operating Characteristic) curves, a statistical method that measures the diagnostic accuracy of the biomarker by correlating false positive and true positive rates at various cut-off levels and allows selection of the optimal cut-off values (Soreide, 2009). Whilst samples in the discovery phase are often collected from a highly selected cohort of subjects to minimise biological variability, samples for this stage should be more representative of the targeted population for testing and their number should provide enough statistical power depending on the desired sensitivity and specificity. This relies on several factors including the intended use of the test, the number of biomarkers being assessed, the variability of the test results and the prevalence of the disease. While the stringent requirements of specificity in relation to screening the general population may not be needed once a diagnosis is made and the biomarker is used for follow up or stratification, high sensitivity is required for diagnostic test of life threatening conditions such as cancer (Vasan, 2006).

17.4.3 Clinical Validation

This last phase is perhaps the most variable in the biomarker discovery pipeline, because the details and extent of this phase including study design and subject selection relies heavily on the intended use of the biomarkers. In general however, it involves large number of subjects (thousands to millions) and multicentre collaboration to fully validate the clinical performance of the test and provide information about the benefits and potential harm as well as economic implications of adopting a particular biomarker. Therefore, before the start of this phase, a biomarker assay should have already demonstrated acceptable analytical performance and sufficient sensitivity and specificity. Validation of a biomarker that is intended for diagnostic purposes involves selection of subjects with the suspected diagnosis and others with clinically similar diseases and comparing the performance of the test with the gold standard clinical practice. Pepe et al. described five phases for biomarker discovery for early detection of cancer (Pepe et al., 2001). In addition to the initial discovery and assay development, the authors proposed three more phases that have distinct roles in the development of such biomarkers:

- Retrospective longitudinal repository studies primarily to evaluate the ability of the biomarker to detect pre-clinical disease.
- Retrospective screening studies to evaluate detection and false referral rates and assessing the feasibility of implementing the screening program.
- Cancer control studies to evaluate the benefits of the screening program including the impact of screening on cancer mortality and its cost effectiveness.

During the validation phase, further refinement of the biomarker(s) panel should also be achieved. This includes potential reduction in the number of biomarkers within the panel, comparison of different combinations of selected markers, establishment of possible physiological and pathological confounding factors and better understanding of the variability of the test result in the target population.

The cost of biomarker discovery can be enormous, particularly during this last phase. Many of the validation studies can be integrated into ongoing clinical trials (Taube et al., 2009). Biomarkers for diagnosis, prognosis and selection and monitoring of therapy (such as HER2 and EGFR) can be conveniently integrated into ongoing therapeutic and interventional clinical trials. In addition to lowering the cost and logistic requirements, these, often randomised controlled trials, offer an excellent study design and repositories of well-annotated biological specimens. This integration, however, has several implications in terms of study design, sample size and availability and allocation of specimens and therefore should be planned appropriately (Sargent and Allegra, 2002).

The complexity of the final validation stage of the biomarker pipeline can be demonstrated through the story of PSA. PSA was discovered in 1979 and was approved for detection of recurrence of prostate cancer by FDA in 1986. In 1994 it was also approved by the FDA for screening for prostate cancer. 15 years later, the debate regarding the benefits/potential harms of its use for screening is still ongoing due to the high rate of false positive results, and consequent overdiagnosis and overtreatment of screen-detected cancers, many of which are indolent (Heijnsdijk et al., 2009). Recently, the results of two clinical trials were published. Andriole et al. reported the results of the PLCO (Prostate, Lung, Colorectal, and Ovarian) cancer screening trial, which found that the rate of death from prostate cancer did not differ significantly between the screened and the control groups (Andriole et al., 2009). In the same issue of the New England Journal of Medicine, Schroder et al. published the data of the ERSPC (The European Randomized Study of Screening for Prostate Cancer) trial which showed that PSA screening reduced mortality from prostate cancer by 20%, but acknowledged the high risk of over-diagnosis (Schroder et al., 2009). Naturally, these two papers attracted a great deal of interest as well as critical comments regarding the methodology and subject selection (Catalona, 2009; Ojha et al., 2009; Preston, 2009). This example illustrates well the complexity of the validation of a biomarker and the strict requirements for well-designed studies for clinical validation.

17.5 Concluding Remarks

So, where are we currently? We have certainly moved from the point where biomarker discovery in body fluids was commonly referred to as a fishing expedition; however, the unfortunate reality is that no major, clinically useful biomarker discovery has been made recently, and the current speed of FDA approval (one biomarker/year) is not particularly promising. The reasons for this are becoming more obvious; we are still not able to technologically "conquer" the major challenges, i.e. complexity and large dynamic range combined with low abundance of useful biomarkers in the majority of biofluids. In addition, we have still not been able to quantify the extent of human and disease variation in addition to the vast pre-analytical variability that we are becoming aware of and that we hope to have illustrated throughout this chapter. Despite these enormous challenges, the concept of diagnosing/screening for a particular disease by a simple blood/urine test would be a dream-come true for both patients and clinicians. One thing is certain: the transition from discovery phase, successful clinical validation and finally clinical implementation of candidate biomarkers is not going to happen overnight. The recently learned lessons of the critical importance of standardisation of sample collection and handling, as well as rapid technological progress in addition to the vast potential clinical benefits of well validated biomarkers allow us to look to the future with optimism, albeit a cautious one.

References

Adachi, J., Kumar, C., Zhang, Y., Olsen, J.V., and Mann, M. (2006). The human urinary proteome contains more than 1500 proteins, including a large proportion of membrane proteins. Genome Biol *7*, R80.

Addona, T.A., Abbatiello, S.E., Schilling, B., Skates, S.J., Mani, D.R., Bunk, D.M., Spiegelman, C.H., Zimmerman, L.J., Ham, A.J., Keshishian, H., *et al.* (2009). Multi-site assessment of the precision and reproducibility of multiple reaction monitoring-based measurements of proteins in plasma. Nat Biotechnol *27*, 633–641.

Ahmad, S., Sundaramoorthy, E., Arora, R., Sen, S., Karthikeyan, G., and Sengupta, S. (2009). Progressive degradation of serum samples limits proteomic biomarker discovery. Anal Biochem *394*, 237–242.

Ahmed, F.E. (2009). The role of capillary electrophoresis-mass spectrometry to proteome analysis and biomarker discovery. J Chromatogr B Anal Technol Biomed Life Sci *877*, 1963–1981.

Airoldi, L., Magagnotti, C., Iannuzzi, A.R., Marelli, C., Bagnati, R., Pastorelli, R., Colombi, A., Santaguida, S., Chiabrando, C., Schiarea, S., *et al.* (2009). Effects of cigarette smoking on the human urinary proteome. Biochem Biophys Res Commun *381*, 397–402.

Anderson, N.L. (2005). The roles of multiple proteomic platforms in a pipeline for new diagnostics. Mol Cell Proteomics *4*, 1441–1444.

Anderson, N.L., and Anderson, N.G. (2002). The human plasma proteome: History, character, and diagnostic prospects. Mol Cell Proteomics *1*, 845–867.

Anderson, L., and Seilhamer, J. (1997). A comparison of selected mRNA and protein abundances in human liver. Electrophoresis *18*, 533–537.

Anderson, N.L., Polanski, M., Pieper, R., Gatlin, T., Tirumalai, R.S., Conrads, T.P., Veenstra, T.D., Adkins, J.N., Pounds, J.G., Fagan, R., *et al.* (2004). The human plasma proteome: A nonredundant list developed by combination of four separate sources. Mol Cell Proteomics *3*, 311–326.

Andriole, G.L., Crawford, E.D., Grubb, R.L., 3rd., Buys, S.S., Chia, D., Church, T.R., Fouad, M.N., Gelmann, E.P., Kvale, P.A., Reding, D.J., *et al.* (2009). Mortality results from a randomized prostate-cancer screening trial. N Engl J Med *360*, 1310–1319.

Aps, J.K., and Martens, L.C. (2005). Review: The physiology of saliva and transfer of drugs into saliva. Forensic Sci Int *150*, 119–131.

Azarkan, M., Huet, J., Baeyens-Volant, D., Looze, Y., and Vandenbussche, G. (2007). Affinity chromatography: A useful tool in proteomics studies. J Chromatogr B Anal Technol Biomed Life Sci *849*, 81–90.

Baggerly, K.A., Morris, J.S., Edmonson, S.R., and Coombes, K.R. (2005). Signal in noise: Evaluating reported reproducibility of serum proteomic tests for ovarian cancer. J Natl Cancer Inst *97*, 307–309.

Baldini, C., Giusti, L., Bazzichi, L., Lucacchini, A., and Bombardieri, S. (2008). Proteomic analysis of the saliva: A clue for understanding primary from secondary Sjogren's syndrome? Autoimmun Rev *7*, 185–191.

Banks, R.E., Stanley, A.J., Cairns, D.A., Barrett, J.H., Clarke, P., Thompson, D., and Selby, P.J. (2005). Influences of blood sample processing on low-molecular-weight proteome identified by surface-enhanced laser desorption/ionization mass spectrometry. Clin Chem *51*, 1637–1649.

Barnea, E., Sorkin, R., Ziv, T., Beer, I., and Admon, A. (2005). Evaluation of prefractionation methods as a preparatory step for multidimensional based chromatography of serum proteins. Proteomics *4*, 3367–3375.

Baumann, S., Ceglarek, U., Fiedler, G.M., Lembcke, J., Leichtle, A., and Thiery, J. (2005). Standardized approach to proteome profiling of human serum based on magnetic bead separation and matrix-assisted laser desorption/ionization time-of-flight mass spectrometry. Clin Chem *51*, 973–980.

Bijian, K., Mlynarek, A.M., Balys, R.L., Jie, S., Xu, Y., Hier, M.P., Black, M.J., Di Falco, M.R., LaBoissiere, S., and Alaoui-Jamali, M.A. (2009). Serum proteomic approach for the identification of serum biomarkers contributed by oral squamous cell carcinoma and host tissue microenvironment. J Proteome Res *8*, 2173–2185.

Bjorhall, K., Miliotis, T., and Davidsson, P. (2005). Comparison of different depletion strategies for improved resolution in proteomic analysis of human serum samples. Proteomics *5*, 307–317.

Bossuyt, P.M., Reitsma, J.B., Bruns, D.E., Gatsonis, C.A., Glasziou, P.P., Irwig, L.M., Lijmer, J.G., Moher, D., Rennie, D., and de Vet, H.C. (2003). Towards complete and accurate reporting of studies of diagnostic accuracy: The STARD initiative. BMJ *326*, 41–44.

Bottini, P.V., Ribeiro Alves, M.A., and Garlipp, C.R. (2002). Electrophoretic pattern of concentrated urine: Comparison between 24-hour collection and random samples. Am J Kidney Dis *39*, E2.

Brechlin, P., Jahn, O., Steinacker, P., Cepek, L., Kratzin, H., Lehnert, S., Jesse, S., Mollenhauer, B., Kretzschmar, H.A., Wiltfang, J., *et al.* (2008). Cerebrospinal fluid-optimized two-dimensional difference gel electrophoresis (2-D DIGE) facilitates the differential diagnosis of creutzfeldt-jakob disease. Proteomics *8*, 4357–4366.

Callesen, A.K., Madsen, J.S., Vach, W., Kruse, T.A., Mogensen, O., and Jensen, O.N. (2009). Serum protein profiling by solid phase extraction and mass spectrometry: A future diagnostics tool? Proteomics *9*, 1428–1441.

Candiano, G., Musante, L., Bruschi, M., Petretto, A., Santucci, L., Del Boccio, P., Pavone, B., Perfumo, F., Urbani, A., Scolari, F., *et al.* (2006). Repetitive fragmentation products of albumin and alpha1-antitrypsin in glomerular diseases associated with nephrotic syndrome. J Am Soc Nephrol *17*, 3139–3148.

Castagna, A., Cecconi, D., Sennels, L., Rappsilber, J., Guerrier, L., Fortis, F., Boschetti, E., Lomas, L., and Righetti, P.G. (2005). Exploring the hidden human urinary proteome via ligand library beads. J Proteome Res *4*, 1917–1930.

Castano, E.M., Roher, A.E., Esh, C.L., Kokjohn, T.A., and Beach, T. (2006). Comparative proteomics of cerebrospinal fluid in neuropathologically-confirmed Alzheimer's disease and non-demented elderly subjects. Neurol Res *28*, 155–163.

Catalona, W.J. (2009). Prostate-cancer screening. N Engl J Med *361*, 202; author reply 204–205.

Check, E. (2004). Proteomics and cancer: Running before we can walk? Nature *429*, 496–497.

Chen, B., Dong, J.Q., Chen, Y.J., Wang, J.M., Tian, J., Wang, C.B., and Zou, S.Q. (2007a). Two-dimensional electrophoresis for comparative proteomic analysis of human bile. Hepatobiliary Pancreat Dis Int *6*, 402–406.

Chen, E.I., Hewel, J., Felding-Habermann, B., and Yates, J.R., 3rd. (2006a). Large scale protein profiling by combination of protein fractionation and multidimensional protein identification technology (MudPIT). Mol Cell Proteomics *5*, 53–56.

Chen, R., Pan, S., Cooke, K., Moyes, K.W., Bronner, M.P., Goodlett, D.R., Aebersold, R., and Brentnall, T.A. (2007b). Comparison of pancreas juice proteins from cancer versus pancreatitis using quantitative proteomic analysis. Pancreas *34*, 70–79.

Chen, R., Pan, S., Yi, E.C., Donohoe, S., Bronner, M.P., Potter, J.D., Goodlett, D.R., Aebersold, R., and Brentnall, T.A. (2006b). Quantitative proteomic profiling of pancreatic cancer juice. Proteomics *6*, 3871–3879.

Chignard, N., Shang, S., Wang, H., Marrero, J., Brechot, C., Hanash, S., and Beretta, L. (2006). Cleavage of endoplasmic reticulum proteins in hepatocellular carcinoma: Detection of generated fragments in patient sera. Gastroenterology *130*, 2010–2022.

Cordero, O.J., De Chiara, L., Lemos-Gonzalez, Y., Paez de la Cadena, M., and Rodriguez-Berrocal, F.J. (2008). How the measurements of a few serum markers can be combined to enhance their clinical values in the management of cancer. Anticancer Res *28*, 2333–2341.

Dayarathna, M.K., Hancock, W.S., and Hincapie, M. (2008). A two step fractionation approach for plasma proteomics using immunodepletion of abundant proteins and multi-lectin affinity chromatography: Application to the analysis of obesity, diabetes, and hypertension diseases. J Sep Sci *31*, 1156–1166.

De Jongh, R., Vranken, J., Vundelinckx, G., Bosmans, E., Maes, M., and Heylen, R. (1997). The effects of anticoagulation and processing on assays of IL-6, sIL-6R, sIL-2R and soluble transferrin receptor. Cytokine *9*, 696–701.

Denny, P., Hagen, F.K., Hardt, M., Liao, L., Yan, W., Arellanno, M., Bassilian, S., Bedi, G.S., Boontheung, P., Cociorva, D., *et al.* (2008). The proteomes of human parotid and submandibular/sublingual gland salivas collected as the ductal secretions. J Proteome Res *7*, 1994–2006.

Drake, S.K., Bowen, R.A., Remaley, A.T., and Hortin, G.L. (2004). Potential interferences from blood collection tubes in mass spectrometric analyses of serum polypeptides. Clin Chem *50*, 2398–2401.

Echan, L.A., Tang, H.Y., Ali-Khan, N., Lee, K., and Speicher, D.W. (2005). Depletion of multiple high-abundance proteins improves protein profiling capacities of human serum and plasma. Proteomics *5*, 3292–3303.

Eliasson, M., Olsson, A., Palmcrantz, E., Wiberg, K., Inganas, M., Guss, B., Lindberg, M., and Uhlen, M. (1988). Chimeric IgG-binding receptors engineered from staphylococcal protein A and streptococcal protein G. J Biol Chem *263*, 4323–4327.

Eneman, J.D., Wood, M.E., and Muss, H.B. (2004). Selecting adjuvant endocrine therapy for breast cancer. Oncology (Williston Park) *18*, 1733–1744; discussion 1744–1735, 1748, 1751–1734.

Esser, D., Alvarez-Llamas, G., de Vries, M.P., Weening, D., Vonk, R.J., and Roelofsen, H. (2008). Sample stability and protein composition of saliva: Implications for its use as a diagnostic fluid. Biomark Insights *3*, 25–27.

Etzioni, R., Urban, N., Ramsey, S., McIntosh, M., Schwartz, S., Reid, B., Radich, J., Anderson, G., and Hartwell, L. (2003). The case for early detection. Nat Rev Cancer *3*, 243–252.

Fang, X., Balgley, B.M., and Lee, C.S. (2009). Recent advances in capillary electrophoresis-based proteomic techniques for biomarker discovery. Electrophoresis *30*, 3998–4007.

Farina, A., Dumonceau, J.M., Frossard, J.L., Hadengue, A., Hochstrasser, D.F., and Lescuyer, P. (2009). Proteomic analysis of human bile from malignant biliary stenosis induced by pancreatic cancer. J Proteome Res *8*, 159–169.

Fiedler, G.M., Baumann, S., Leichtle, A., Oltmann, A., Kase, J., Thiery, J., and Ceglarek, U. (2007). Standardized peptidome profiling of human urine by magnetic bead separation and matrix-assisted laser desorption/ionization time-of-flight mass spectrometry. Clin Chem *53*, 421–428.

Gao, M., Deng, C., Yu, W., Zhang, Y., Yang, P., and Zhang, X. (2008). Large scale depletion of the high-abundance proteins and analysis of middle- and low-abundance proteins in human liver proteome by multidimensional liquid chromatography. Proteomics *8*, 939–947.

Gianazza, E., and Arnaud, P. (1982). Chromatography of plasma proteins on immobilized cibacron blue F3-GA. Mechanism of the molecular interaction. Biochem J *203*, 637–641.

Good, D.M., Thongboonkerd, V., Novak, J., Bascands, J.L., Schanstra, J.P., Coon, J.J., Dominiczak, A., and Mischak, H. (2007). Body fluid proteomics for biomarker discovery: Lessons from the past hold the key to success in the future. J Proteome Res *6*, 4549–4555.

Govorukhina, N.I., de Vries, M., Reijmers, T.H., Horvatovich, P., van der Zee, A.G., and Bischoff, R. (2009). Influence of clotting time on the protein composition of serum samples based on LC-MS data. J Chromatogr B Analyt Technol Biomed Life Sci *877*, 1281–1291.

Granger, J., Siddiqui, J., Copeland, S., and Remick, D. (2005). Albumin depletion of human plasma also removes low abundance proteins including the cytokines. Proteomics *5*, 4713–4718.

Greene, K.L., Albertsen, P.C., Babaian, R.J., Carter, H.B., Gann, P.H., Han, M., Kuban, D.A., Sartor, A.O., Stanford, J.L., Zietman, A., *et al.* (2009). Prostate specific antigen best practice statement: 2009 update. J Urol *182*, 2232–2241.

Groesbeck, D., Kottgen, A., Parekh, R., Selvin, E., Schwartz, G.J., Coresh, J., and Furth, S. (2008). Age, gender, and race effects on cystatin C levels in US adolescents. Clin J Am Soc Nephrol *3*, 1777–1785.

Gronborg, M., Bunkenborg, J., Kristiansen, T.Z., Jensen, O.N., Yeo, C.J., Hruban, R.H., Maitra, A., Goggins, M.G., and Pandey, A. (2004). Comprehensive proteomic analysis of human pancreatic juice. J Proteome Res *3*, 1042–1055.

Group, B.D. (2001). Biomarkers and surrogate endpoints: Preferred definitions and conceptual framework. Clin Pharmacol Ther *69*, 89–95.

Guerrier, L., Claverol, S., Finzi, L., Paye, F., Fortis, F., Boschetti, E., and Housset, C. (2007). Contribution of solid-phase hexapeptide ligand libraries to the repertoire of human bile proteins. J Chromatogr A *1176*, 192–205.

Gundry, R.L., Fu, Q., Jelinek, C.A., Van Eyk, J.E., and Cotter, R.J. (2007). Investigation of an albumin-enriched fraction of human serum and its albuminome. Proteomics Clin Appl *1*, 73–88.

Gutman, S., and Kessler, L.G. (2006). The US food and drug administration perspective on cancer biomarker development. Nat Rev Cancer *6*, 565–571.

Hanrieder, J., Wetterhall, M., Enblad, P., Hillered, L., and Bergquist, J. (2009). Temporally resolved differential proteomic analysis of human ventricular CSF for monitoring traumatic brain injury biomarker candidates. J Neurosci Methods *177*, 469–478.

Havanapan, P.O., and Thongboonkerd, V. (2009). Are protease inhibitors required for gel-based proteomics of kidney and urine? J Proteome Res *8*, 3109–3117.

Heijnsdijk, E.A., der Kinderen, A., Wever, E.M., Draisma, G., Roobol, M.J., and de Koning, H.J. (2009). Overdetection, overtreatment and costs in prostate-specific antigen screening for prostate cancer. Br J Cancer *101*, 1833–1838.

Helmerhorst, E.J., and Oppenheim, F.G. (2007). Saliva: A dynamic proteome. J Dent Res *86*, 680–693.

Hetland, M.L., Christensen, I.J., Lottenburger, T., Johansen, J.S., Svendsen, M.N., Horslev-Petersen, K., Nielsen, L., and Nielsen, H.J. (2008). Circulating VEGF as a biological marker in patients with rheumatoid arthritis? Preanalytical and biological variability in healthy persons and in patients. Dis Markers *24*, 1–10.

Hingorani, S.R., Petricoin, E.F., Maitra, A., Rajapakse, V., King, C., Jacobetz, M.A., Ross, S., Conrads, T.P., Veenstra, T.D., Hitt, B.A., *et al.* (2003). Preinvasive and invasive ductal pancreatic cancer and its early detection in the mouse. Cancer Cell *4*, 437–450.

Hirsh, J., Raschke, R., Warkentin, T.E., Dalen, J.E., Deykin, D., and Poller, L. (1995). Heparin: Mechanism of action, pharmacokinetics, dosing considerations, monitoring, efficacy, and safety. Chest *108*, 258S–275S.

Hoffman, S.A., Joo, W.A., Echan, L.A., and Speicher, D.W. (2007). Higher dimensional (Hi-D) separation strategies dramatically improve the potential for cancer biomarker detection in serum and plasma. J Chromatogr B Anal Technol Biomed Life Sci *849*, 43–52.

Hortin, G.L., and Sviridov, D. (2010). The dynamic range problem in the analysis of the plasma proteome. J Proteomics *73*, 629–636.

Hsieh, S.Y., Chen, R.K., Pan, Y.H., and Lee, H.L. (2006). Systematical evaluation of the effects of sample collection procedures on low-molecular-weight serum/plasma proteome profiling. Proteomics *6*, 3189–3198.

Hsu, P.I., Chen, C.H., Hsieh, C.S., Chang, W.C., Lai, K.H., Lo, G.H., Hsu, P.N., Tsay, F.W., Chen, Y.S., Hsiao, M., *et al.* (2007). Alpha1-antitrypsin precursor in gastric juice is a novel biomarker for gastric cancer and ulcer. Clin Cancer Res *13*, 876–883.

Hu, S., Arellano, M., Boontheung, P., Wang, J., Zhou, H., Jiang, J., Elashoff, D., Wei, R., Loo, J.A., and Wong, D.T. (2008). Salivary proteomics for oral cancer biomarker discovery. Clin Cancer Res *14*, 6246–6252.

Hu, Y., Malone, J.P., Fagan, A.M., Townsend, R.R., and Holtzman, D.M. (2005). Comparative proteomic analysis of intra- and interindividual variation in human cerebrospinal fluid. Mol Cell Proteomics *4*, 2000–2009.

Huang, H.L., Stasyk, T., Morandell, S., Mogg, M., Schreiber, M., Feuerstein, I., Huck, C.W., Stecher, G., Bonn, G.K., and Huber, L.A. (2005a). Enrichment of low-abundant serum proteins by albumin/immunoglobulin G immunoaffinity depletion under partly denaturing conditions. Electrophoresis *26*, 2843–2849.

Huang, L., Harvie, G., Feitelson, J.S., Gramatikoff, K., Herold, D.A., Allen, D.L., Amunngama, R., Hagler, R.A., Pisano, M.R., Zhang, W.W., *et al.* (2005b). Immunoaffinity separation of plasma proteins by IgY microbeads: Meeting the needs of proteomic sample preparation and analysis. Proteomics *5*, 3314–3328.

Humphrey, S.P., and Williamson, R.T. (2001). A review of saliva: Normal composition, flow, and function. J Prosthet Dent *85*, 162–169.

Jin, W.H., Dai, J., Li, S.J., Xia, Q.C., Zou, H.F., and Zeng, R. (2005). Human plasma proteome analysis by multidimensional chromatography prefractionation and linear ion trap mass spectrometry identification. J Proteome Res *4*, 613–619.

Jmeian, Y., and El Rassi, Z. (2009). Liquid-phase-based separation systems for depletion, prefractionation and enrichment of proteins in biological fluids for in-depth proteomics analysis. Electrophoresis *30*, 249–261.

Johann, D.J., Rodriguez-Canales, J., Mukherjee, S., Prieto, D.A., Hanson, J.C., Emmert-Buck, M., and Blonder, J. (2009). Approaching solid tumor heterogeneity on a cellular basis by tissue proteomics using laser capture microdissection and biological mass spectrometry. J Proteome Res *8*, 2310–2318.

Jones, H.B. (1848). On a new substance occuring in the urine of a patient with mollities ossium. Philos Trans R Soc Lond *138*, 55–62.

Kakisaka, T., Kondo, T., Okano, T., Fujii, K., Honda, K., Endo, M., Tsuchida, A., Aoki, T., Itoi, T., Moriyasu, F., *et al.* (2007). Plasma proteomics of pancreatic cancer patients by multi-dimensional liquid chromatography and two-dimensional difference gel electrophoresis (2D-DIGE): Up-regulation of leucine-rich alpha-2-glycoprotein in pancreatic cancer. J Chromatogr B Anal Technol Biomed Life Sci *852*, 257–267.

Kaufman, E., and Lamster, I.B. (2002). The diagnostic applications of saliva – a review. Crit Rev Oral Biol Med *13*, 197–212.

Kawai, M., Uchiyama, K., Tani, M., Onishi, H., Kinoshita, H., Ueno, M., Hama, T., and Yamaue, H. (2004). Clinicopathological features of malignant intraductal papillary mucinous tumors of the pancreas: The differential diagnosis from benign entities. Arch Surg *139*, 188–192.

Keshishian, H., Addona, T., Burgess, M., Kuhn, E., and Carr, S.A. (2007). Quantitative, multiplexed assays for low abundance proteins in plasma by targeted mass spectrometry and stable isotope dilution. Mol Cell Proteomics *6*, 2212–2229.

Klasen, I.S., Reichert, L.J., de Kat Angelino, C.M., and Wetzels, J.F. (1999). Quantitative determination of low and high molecular weight proteins in human urine: Influence of temperature and storage time. Clin Chem *45*, 430–432.

Kristiansen, T.Z., Bunkenborg, J., Gronborg, M., Molina, H., Thuluvath, P.J., Argani, P., Goggins, M.G., Maitra, A., and Pandey, A. (2004). A proteomic analysis of human bile. Mol Cell Proteomics *3*, 715–728.

Kuick, R., Misek, D.E., Monsma, D.J., Webb, C.P., Wang, H., Peterson, K.J., Pisano, M., Omenn, G.S., and Hanash, S.M. (2007). Discovery of cancer biomarkers through the use of mouse models. Cancer Lett *249*, 40–48.

Kushnir, M.M., Mrozinski, P., Rockwood, A.L., and Crockett, D.K. (2009). A depletion strategy for improved detection of human proteins from urine. J Biomol Tech *20*, 101–108.

Kuzyk, M.A., Smith, D., Yang, J., Cross, T.J., Jackson, A.M., Hardie, D.B., Anderson, N.L., and Borchers, C.H. (2009). Multiple reaction monitoring-based, multiplexed, absolute quantitation of 45 proteins in human plasma. Mol Cell Proteomics *8*, 1860–1877.

Lafitte, D., Dussol, B., Andersen, S., Vazi, A., Dupuy, P., Jensen, O.N., Berland, Y., and Verdier, J.M. (2002). Optimized preparation of urine samples for two-dimensional electrophoresis and initial application to patient samples. Clin Biochem *35*, 581–589.

Lai, X., Liangpunsakul, S., Crabb, D.W., Ringham, H.N., and Witzmann, F.A. (2009). A proteomic workflow for discovery of serum carrier protein-bound biomarker candidates of alcohol abuse using LC-MS/MS. Electrophoresis *30*, 2207–2214.

Landers, K.A., Burger, M.J., Tebay, M.A., Purdie, D.M., Scells, B., Samaratunga, H., Lavin, M.F., and Gardiner, R.A. (2005). Use of multiple biomarkers for a molecular diagnosis of prostate cancer. Int J Cancer *114*, 950–956.

Leng, S.X., McElhaney, J.E., Walston, J.D., Xie, D., Fedarko, N.S., and Kuchel, G.A. (2008). ELISA and multiplex technologies for cytokine measurement in inflammation and aging research. J Gerontol A Biol Sci Med Sci *63*, 879–884.

Lescuyer, P., Hochstrasser, D.F., and Sanchez, J.C. (2004). Comprehensive proteome analysis by chromatographic protein prefractionation. Electrophoresis *25*, 1125–1135.

Li, J., Orlandi, R., White, C.N., Rosenzweig, J., Zhao, J., Seregni, E., Morelli, D., Yu, Y., Meng, X.Y., Zhang, Z., *et al.* (2005). Independent validation of candidate breast cancer serum biomarkers identified by mass spectrometry. Clin Chem *51*, 2229–2235.

Liu, T., Qian, W.J., Gritsenko, M.A., Xiao, W., Moldawer, L.L., Kaushal, A., Monroe, M.E., Varnum, S.M., Moore, R.J., Purvine, S.O., *et al.* (2006a). High dynamic range characterization of the trauma patient plasma proteome. Mol Cell Proteomics *5*, 1899–1913.

Liu, T., Qian, W.J., Mottaz, H.M., Gritsenko, M.A., Norbeck, A.D., Moore, R.J., Purvine, S.O., Camp, D.G., 2nd., and Smith, R.D. (2006b). Evaluation of multiprotein immunoaffinity subtraction for plasma proteomics and candidate biomarker discovery using mass spectrometry. Mol Cell Proteomics *5*, 2167–2174.

Locker, G.Y., Hamilton, S., Harris, J., Jessup, J.M., Kemeny, N., Macdonald, J.S., Somerfield, M.R., Hayes, D.F., and Bast, R.C., Jr. (2006). ASCO 2006 update of recommendations for the use of tumor markers in gastrointestinal cancer. J Clin Oncol *24*, 5313–5327.

Lowenthal, M.S., Mehta, A.I., Frogale, K., Bandle, R.W., Araujo, R.P., Hood, B.L., Veenstra, T.D., Conrads, T.P., Goldsmith, P., Fishman, D., *et al.* (2005). Analysis of albumin-associated peptides and proteins from ovarian cancer patients. Clin Chem *51*, 1933–1945.

Ludwig, J.A., and Weinstein, J.N. (2005). Biomarkers in cancer staging, prognosis and treatment selection. Nat Rev Cancer *5*, 845–856.

Magistroni, R., Ligabue, G., Lupo, V., Furci, L., Leonelli, M., Manganelli, L., Masellis, M., Gatti, V., Cavazzini, F., Tizzanini, W., *et al.* (2009). Proteomic analysis of urine from proteinuric patients shows a proteolitic activity directed against albumin. Nephrol Dial Transplant *24*, 1672–1681.

Maksymowych, W.P., Landewe, R., Tak, P.P., Ritchlin, C.J., Ostergaard, M., Mease, P.J., El-Gabalawy, H., Garnero, P., Gladman, D.D., Fitzgerald, O., *et al.* (2009). Reappraisal of OMERACT 8 draft validation criteria for a soluble biomarker reflecting structural damage endpoints in rheumatoid arthritis, psoriatic arthritis, and spondyloarthritis: The OMERACT 9 v2 criteria. J Rheumatol *36*, 1785–1791.

Manabe, T., Yamaguchi, N., Mukai, J., Hamada, O., and Tani, O. (2003). Detection of protein-protein interactions and a group of immunoglobulin G-associated minor proteins in human plasma by nondenaturing and denaturing two-dimensional gel electrophoresis. Proteomics *3*, 832–846.

Mandel, I.D. (1990). The diagnostic uses of saliva. J Oral Pathol Med *19*, 119–125.

Mann, K.G., Brummel-Ziedins, K., Undas, A., and Butenas, S. (2004). Does the genotype predict the phenotype? Evaluations of the hemostatic proteome. J Thromb Haemost *2*, 1727–1734.

Martosella, J., Zolotarjova, N., Liu, H., Nicol, G., and Boyes, B.E. (2005). Reversed-phase high-performance liquid chromatographic prefractionation of immunodepleted human serum proteins to enhance mass spectrometry identification of lower-abundant proteins. J Proteome Res *4*, 1522–1537.

McLerran, D., Grizzle, W.E., Feng, Z., Bigbee, W.L., Banez, L.L., Cazares, L.H., Chan, D.W., Diaz, J., Izbicka, E., Kagan, J., *et al.* (2008a). Analytical validation of serum proteomic profiling for diagnosis of prostate cancer: Sources of sample bias. Clin Chem *54*, 44–52.

McLerran, D., Grizzle, W.E., Feng, Z., Thompson, I.M., Bigbee, W.L., Cazares, L.H., Chan, D.W., Dahlgren, J., Diaz, J., Kagan, J., *et al.* (2008b). SELDI-TOF MS whole serum proteomic profiling with IMAC surface does not reliably detect prostate cancer. Clin Chem *54*, 53–60.

McShane, L.M., Altman, D.G., Sauerbrei, W., Taube, S.E., Gion, M., and Clark, G.M. (2006). REporting recommendations for tumor MARKer prognostic studies (REMARK). Breast Cancer Res Treat *100*, 229–235.

Mischak, H., Apweiler, R., Banks, R.E., Conaway, M., Coon, J., Dominiczak, A., Ehrich, J.H.H., Fliser, D., Girolami, M., Hermjakob, H., Hochstrasser, D., Jankowski, J., Julian, B.A., Kolch, W., Massy, Z.A., Neusuess, C., Novak, J., Peter, K., Rossing, K., Schanstra, J., Semmes, O.J., Theodorescu, D., Thongboonkerd, V., Weissinger, E.M., Van Eyk, J.E., Yamamoto., T. (2007). Clinical proteomics: A need to define the field and to begin to set adequate standards. Proteomics Clin Appl *1*, 148–156.

Mor, G., Visintin, I., Lai, Y., Zhao, H., Schwartz, P., Rutherford, T., Yue, L., Bray-Ward, P., and Ward, D.C. (2005). Serum protein markers for early detection of ovarian cancer. Proc Natl Acad Sci USA *102*, 7677–7682.

Moritz, R.L., Clippingdale, A.B., Kapp, E.A., Eddes, J.S., Ji, H., Gilbert, S., Connolly, L.M., and Simpson, R.J. (2005). Application of 2-D free-flow electrophoresis/RP-HPLC for proteomic analysis of human plasma depleted of multi high-abundance proteins. Proteomics *5*, 3402–3413.

Moritz, R.L., Ji, H., Schutz, F., Connolly, L.M., Kapp, E.A., Speed, T.P., and Simpson, R.J. (2004). A proteome strategy for fractionating proteins and peptides using continuous free-flow electrophoresis coupled off-line to reversed-phase high-performance liquid chromatography. Anal Chem *76*, 4811–4824.

Munro, N.P., Cairns, D.A., Clarke, P., Rogers, M., Stanley, A.J., Barrett, J.H., Harnden, P., Thompson, D., Eardley, I., Banks, R.E., *et al.* (2006). Urinary biomarker profiling in transitional cell carcinoma. Int J Cancer *119*, 2642–2650.

Nielsen, D.L., Andersson, M., and Kamby, C. (2009). HER2-targeted therapy in breast cancer. Monoclonal antibodies and tyrosine kinase inhibitors. Cancer Treat Rev *35*, 121–136.

Nissum, M., Kuhfuss, S., Hauptmann, M., Obermaier, C., Sukop, U., Wildgruber, R., Weber, G., Eckerskorn, C., and Malmstrom, J. (2007). Two-dimensional separation of human plasma proteins using iterative free-flow electrophoresis. Proteomics *7*, 4218–4227.

Oh, J., Pyo, J.H., Jo, E.H., Hwang, S.I., Kang, S.C., Jung, J.H., Park, E.K., Kim, S.Y., Choi, J.Y., and Lim, J. (2004). Establishment of a near-standard two-dimensional human urine proteomic map. Proteomics *4*, 3485–3497.

Ojha, R.P., Thertulien, R., and Fischbach, L.A. (2009). Prostate-cancer screening. N Engl J Med *361*, 203; author reply 204–205.

Omenn, G.S., States, D.J., Adamski, M., Blackwell, T.W., Menon, R., Hermjakob, H., Apweiler, R., Haab, B.B., Simpson, R.J., Eddes, J.S., *et al.* (2005). Overview of the HUPO plasma proteome project: Results from the pilot phase with 35 collaborating laboratories and multiple analytical groups, generating a core dataset of 3020 proteins and a publicly-available database. Proteomics *5*, 3226–3245.

Oppenheim, F.G., Salih, E., Siqueira, W.L., Zhang, W., and Helmerhorst, E.J. (2007). Salivary proteome and its genetic polymorphisms. Ann N Y Acad Sci *1098*, 22–50.

Papale, M., Pedicillo, M.C., Thatcher, B.J., Di Paolo, S., Lo Muzio, L., Bufo, P., Rocchetti, M.T., Centra, M., Ranieri, E., and Gesualdo, L. (2007). Urine profiling by SELDI-TOF/MS: Monitoring of the critical steps in sample collection, handling and analysis. J Chromatogr B Anal Technol Biomed Life Sci *856*, 205–213.

Paulitschke, V., Kunstfeld, R., Mohr, T., Slany, A., Micksche, M., Drach, J., Zielinski, C., Pehamberger, H., and Gerner, C. (2009). Entering a new era of rational biomarker discovery for early detection of melanoma metastases: Secretome analysis of associated stroma cells. J Proteome Res *8*, 2501–2510.

Pepe, M.S., Etzioni, R., Feng, Z., Potter, J.D., Thompson, M.L., Thornquist, M., Winget, M., and Yasui, Y. (2001). Phases of biomarker development for early detection of cancer. J Natl Cancer Inst *93*, 1054–1061.

Petrak, J., Ivanek, R., Toman, O., Cmejla, R., Cmejlova, J., Vyoral, D., Zivny, J., and Vulpe, C.D. (2008). Deja vu in proteomics. A hit parade of repeatedly identified differentially expressed proteins. Proteomics *8*, 1744–1749.

Petricoin, E.F., Ardekani, A.M., Hitt, B.A., Levine, P.J., Fusaro, V.A., Steinberg, S.M., Mills, G.B., Simone, C., Fishman, D.A., Kohn, E.C., *et al.* (2002). Use of proteomic patterns in serum to identify ovarian cancer. Lancet *359*, 572–577.

Petricoin, E.F., and Liotta, L.A. (2004). SELDI-TOF-based serum proteomic pattern diagnostics for early detection of cancer. Curr Opin Biotechnol *15*, 24–30.

Pieper, R., Gatlin, C.L., Makusky, A.J., Russo, P.S., Schatz, C.R., Miller, S.S., Su, Q., McGrath, A.M., Estock, M.A., Parmar, P.P., *et al.* (2003). The human serum proteome: Display of nearly 3700 chromatographically separated protein spots on two-dimensional electrophoresis gels and identification of 325 distinct proteins. Proteomics *3*, 1345–1364.

Pieragostino, D., Petrucci, F., Del Boccio, P., Mantini, D., Lugaresi, A., Tiberio, S., Onofrj, M., Gambi, D., Sacchetta, P., Di Ilio, C., *et al.* (2009). Pre-analytical factors in clinical proteomics investigations: Impact of ex vivo protein modifications for multiple sclerosis biomarker discovery. J Proteomics *73*, 579–592.

Pisitkun, T., Shen, R.F., and Knepper, M.A. (2004). Identification and proteomic profiling of exosomes in human urine. Proc Natl Acad Sci USA *101*, 13368–13373.

Poortmans, J.R., Geudvert, C., Schorokoff, K., and De Plaen, P. (1996). Postexercise proteinuria in childhood and adolescence. Int J Sports Med *17*, 448–451.

Preston, S.H. (2009). Prostate-cancer screening. N Engl J Med *361*, 202–203; author reply 204–206.

Qin, S., Ferdinand, A.S., Richie, J.P., O'Leary, M.P., Mok, S.C., and Liu, B.C. (2005). Chromatofocusing fractionation and two-dimensional difference gel electrophoresis for low abundance serum proteins. Proteomics *5*, 3183–3192.

Quintana, M., Palicki, O., Lucchi, G., Ducoroy, P., Chambon, C., Salles, C., and Morzel, M. (2009). Inter-individual variability of protein patterns in saliva of healthy adults. J Proteomics *72*, 822–830.

Rai, A.J., Gelfand, C.A., Haywood, B.C., Warunek, D.J., Yi, J., Schuchard, M.D., Mehigh, R.J., Cockrill, S.L., Scott, G.B., Tammen, H., *et al.* (2005). HUPO Plasma proteome project specimen collection and handling: Towards the standardization of parameters for plasma proteome samples. Proteomics *5*, 3262–3277.

Ransohoff, D.F. (2004). Rules of evidence for cancer molecular-marker discovery and validation. Nat Rev Cancer *4*, 309–314.

Ransohoff, D.F. (2005). Lessons from controversy: Ovarian cancer screening and serum proteomics. J Natl Cancer Inst *97*, 315–319.

Ransohoff, D.F. (2007). How to improve reliability and efficiency of research about molecular markers: Roles of phases, guidelines, and study design. J Clin Epidemiol *60*, 1205–1219.

Rao, P.V., Reddy, A.P., Lu, X., Dasari, S., Krishnaprasad, A., Biggs, E., Roberts, C.T., and Nagalla, S.R. (2009). Proteomic identification of salivary biomarkers of type-2 diabetes. J Proteome Res *8*, 239–245.

Reiber, H., and Peter, J.B. (2001). Cerebrospinal fluid analysis: Disease-related data patterns and evaluation programs. J Neurol Sci *184*, 101–122.

Rifai, N., Gillette, M.A., and Carr, S.A. (2006). Protein biomarker discovery and validation: The long and uncertain path to clinical utility. Nat Biotechnol *24*, 971–983.

Roche, S., Tiers, L., Provansal, M., Seveno, M., Piva, M.T., Jouin, P., and Lehmann, S. (2009). Depletion of one, six, twelve or twenty major blood proteins before proteomic analysis: The more the better? J Proteomics *72*, 945–951.

Roessler, M., Rollinger, W., Palme, S., Hagmann, M.L., Berndt, P., Engel, A.M., Schneidinger, B., Pfeffer, M., Andres, H., Karl, J., *et al.* (2005). Identification of nicotinamide N-methyltransferase as a novel serum tumor marker for colorectal cancer. Clin Cancer Res *11*, 6550–6557.

Rogers, M.A., Clarke, P., Noble, J., Munro, N.P., Paul, A., Selby, P.J., and Banks, R.E. (2003). Proteomic profiling of urinary proteins in renal cancer by surface enhanced laser desorption ionization and neural-network analysis: Identification of key issues affecting potential clinical utility. Cancer Res *63*, 6971–6983.

Rosty, C., Christa, L., Kuzdzal, S., Baldwin, W.M., Zahurak, M.L., Carnot, F., Chan, D.W., Canto, M., Lillemoe, K.D., Cameron, J.L., *et al.* (2002). Identification of hepatocarcinoma-intestine-pancreas/pancreatitis-associated protein I as a biomarker for pancreatic ductal adenocarcinoma by protein biochip technology. Cancer Res *62*, 1868–1875.

Roy, R., Wewer, U.M., Zurakowski, D., Pories, S.E., and Moses, M.A. (2004). ADAM 12 cleaves extracellular matrix proteins and correlates with cancer status and stage. J Biol Chem *279*, 51323–51330.

Russo, L.M., Bakris, G.L., and Comper, W.D. (2002). Renal handling of albumin: A critical review of basic concepts and perspective. Am J Kidney Dis *39*, 899–919.

Sargent, D., and Allegra, C. (2002). Issues in clinical trial design for tumor marker studies. Semin Oncol *29*, 222–230.

Schaub, S., Wilkins, J., Weiler, T., Sangster, K., Rush, D., and Nickerson, P. (2004). Urine protein profiling with surface-enhanced laser-desorption/ionization time-of-flight mass spectrometry. Kidney Int *65*, 323–332.

Schipper, R.G., Silletti, E., and Vingerhoeds, M.H. (2007). Saliva as research material: Biochemical, physicochemical and practical aspects. Arch Oral Biol *52*, 1114–1135.

Schroder, F.H., Hugosson, J., Roobol, M.J., Tammela, T.L., Ciatto, S., Nelen, V., Kwiatkowski, M., Lujan, M., Lilja, H., Zappa, M., *et al.* (2009). Screening and prostate-cancer mortality in a randomized European study. N Engl J Med *360*, 1320–1328.

Schuchard, M.D., Mehigh, R.J., Cockrill, S.L., Lipscomb, G.T., Stephan, J.D., Wildsmith, J., Valdes-Camin, R., Kappel, W.K., Rai, A.J., and Scott, G.B. (2005). Artifactual isoform profile modification following treatment of human plasma or serum with protease inhibitor, monitored by 2-dimensional electrophoresis and mass spectrometry. Biotechniques *39*, 239–247.

Schultz, C.J., Dalton, R.N., Turner, C., Neil, H.A., and Dunger, D.B. (2000). Freezing method affects the concentration and variability of urine proteins and the interpretation of data on microalbuminuria. The oxford regional prospective study group. Diabet Med *17*, 7–14.

Scully, P., Tighe, P., Gilmore, G.A., Wallace, J.M., Strain, J.J., McNulty, H., Ward, M., and Gilmore, W.S. (2009). The relationship between gender and age with monocyte tissue factor expression. J Thromb Thrombolysis *28*, 156–165.

Sennels, L., Salek, M., Lomas, L., Boschetti, E., Righetti, P.G., and Rappsilber, J. (2007). Proteomic analysis of human blood serum using peptide library beads. J Proteome Res *6*, 4055–4062.

Shen, Y., Kim, J., Strittmatter, E.F., Jacobs, J.M., Camp, D.G., 2nd., Fang, R., Tolie, N., Moore, R.J., and Smith, R.D. (2005). Characterization of the human blood plasma proteome. Proteomics *5*, 4034–4045.

Soreide, K. (2009). Receiver-operating characteristic curve analysis in diagnostic, prognostic and predictive biomarker research. J Clin Pathol *62*, 1–5.

Srinivas, P.R., Kramer, B.S., and Srivastava, S. (2001). Trends in biomarker research for cancer detection. Lancet Oncol *2*, 698–704.

Stasyk, T., and Huber, L.A. (2004). Zooming in: Fractionation strategies in proteomics. Proteomics *4*, 3704–3716.

Streckfus, C., and Bigler, L. (2005). The use of soluble, salivary c-erbB-2 for the detection and post-operative follow-up of breast cancer in women: The results of a 5-year translational research study. Adv Dent Res *18*, 17–24.

Tammen, H., Schulte, I., Hess, R., Menzel, C., Kellmann, M., Mohring, T., and Schulz-Knappe, P. (2005). Peptidomic analysis of human blood specimens: Comparison between plasma specimens and serum by differential peptide display. Proteomics *5*, 3414–3422.

Tang, H.Y., Ali-Khan, N., Echan, L.A., Levenkova, N., Rux, J.J., and Speicher, D.W. (2005). A novel four-dimensional strategy combining protein and peptide separation methods enables detection of low-abundance proteins in human plasma and serum proteomes. Proteomics *5*, 3329–3342.

Tantipaiboonwong, P., Sinchaikul, S., Sriyam, S., Phutrakul, S., and Chen, S.T. (2005). Different techniques for urinary protein analysis of normal and lung cancer patients. Proteomics *5*, 1140–1149.

Taube, S.E., Clark, G.M., Dancey, J.E., McShane, L.M., Sigman, C.C., and Gutman, S.I. (2009). A perspective on challenges and issues in biomarker development and drug and biomarker codevelopment. J Natl Cancer Inst *101*, 1453–1463.

Taylor, C.F., Paton, N.W., Lilley, K.S., Binz, P.A., Julian, R.K., Jr., Jones, A.R., Zhu, W., Apweiler, R., Aebersold, R., Deutsch, E.W., *et al.* (2007). The minimum information about a proteomics experiment (MIAPE). Nat Biotechnol *25*, 887–893.

Thompson, E.J. (1995). Cerebrospinal fluid. J Neurol Neurosurg Psychiatry *59*, 349–357.

Thongboonkerd, V. (2007). Practical points in urinary proteomics. J Proteome Res *6*, 3881–3890.

Thongboonkerd, V., Chutipongtanate, S., and Kanlaya, R. (2006). Systematic evaluation of sample preparation methods for gel-based human urinary proteomics: Quantity, quality, and variability. J Proteome Res *5*, 183–191.

Thongboonkerd, V., McLeish, K.R., Arthur, J.M., and Klein, J.B. (2002). Proteomic analysis of normal human urinary proteins isolated by acetone precipitation or ultracentrifugation. Kidney Int *62*, 1461–1469.

Thongboonkerd, V., Mungdee, S., and Chiangjong, W. (2009). Should urine pH be adjusted prior to gel-based proteome analysis? J Proteome Res *8*, 3206–3211.

Thongboonkerd, V., and Saetun, P. (2007). Bacterial overgrowth affects urinary proteome analysis: Recommendation for centrifugation, temperature, duration, and the use of preservatives during sample collection. J Proteome Res *6*, 4173–4181.

Thorpe, J.D., Duan, X., Forrest, R., Lowe, K., Brown, L., Segal, E., Nelson, B., Anderson, G.L., McIntosh, M., and Urban, N. (2007). Effects of blood collection conditions on ovarian cancer serum markers. PLoS One *2*, e1281.

Thulasiraman, V., Lin, S., Gheorghiu, L., Lathrop, J., Lomas, L., Hammond, D., and Boschetti, E. (2005). Reduction of the concentration difference of proteins in biological liquids using a library of combinatorial ligands. Electrophoresis *26*, 3561–3571.

Timms, J.F., Arslan-Low, E., Gentry-Maharaj, A., Luo, Z., T'Jampens, D., Podust, V.N., Ford, J., Fung, E.T., Gammerman, A., Jacobs, I., *et al.* (2007). Preanalytic influence of sample handling on SELDI-TOF serum protein profiles. Clin Chem *53*, 645–656.

Tirumalai, R.S., Chan, K.C., Prieto, D.A., Issaq, H.J., Conrads, T.P., and Veenstra, T.D. (2003). Characterization of the low molecular weight human serum proteome. Mol Cell Proteomics *2*, 1096–1103.

Ucar, T., Baykal, A., Akyuz, M., Dosemeci, L., and Toptas, B. (2004). Comparison of serum and cerebrospinal fluid protein S-100b levels after severe head injury and their prognostic importance. J Trauma *57*, 95–98.

Vasan, R.S. (2006). Biomarkers of cardiovascular disease: Molecular basis and practical considerations. Circulation *113*, 2335–2362.

Vasudev, N.S., Ferguson, R.E., Cairns, D.A., Stanley, A.J., Selby, P.J., and Banks, R.E. (2008). Serum biomarker discovery in renal cancer using 2-DE and prefractionation by immunodepletion and isoelectric focusing; increasing coverage or more of the same? Proteomics *8*, 5074–5085.

Veenstra, T.D., Prieto, D.A., and Conrads, T.P. (2004). Proteomic patterns for early cancer detection. Drug Discov Today *9*, 889–897.

Villanueva, J., Shaffer, D.R., Philip, J., Chaparro, C.A., Erdjument-Bromage, H., Olshen, A.B., Fleisher, M., Lilja, H., Brogi, E., Boyd, J., *et al.* (2006). Differential exoprotease activities confer tumor-specific serum peptidome patterns. J Clin Invest *116*, 271–284.

Wang, P., Bouwman, F.G., and Mariman, E.C. (2009). Generally detected proteins in comparative proteomics--a matter of cellular stress response? Proteomics *9*, 2955–2966.

Washburn, M.P., Wolters, D., and Yates, J.R., 3rd. (2001). Large-scale analysis of the yeast proteome by multidimensional protein identification technology. Nat Biotechnol *19*, 242–247.

Weeks, M., Hariharan, D., Petronijevic, L., Radon, TP., Whiteman, HJ., Kocher, HM., Timms, JF., Lemoine, NR., and Crnogorac-Jurcevic, T. (2008). Analysis of the urine proteome in patients with pancreatic ductal adenocarcinoma. Proteomics Clin Appl *2*, 1047–1057.

West-Norager, M., Kelstrup, C.D., Schou, C., Hogdall, E.V., Hogdall, C.K., and Heegaard, N.H. (2007). Unravelling in vitro variables of major importance for the outcome of mass spectrometry-based serum proteomics. J Chromatogr B Anal Technol Biomed Life Sci *847*, 30–37.

Wildgruber, R., Yi, J., Nissum, M., Eckerskorn, C., and Weber, G. (2008). Free-flow electrophoresis system for plasma proteomic applications. Methods Mol Biol *424*, 287–300.

Xiao, Z., Prieto, D., Conrads, T.P., Veenstra, T.D., and Issaq, H.J. (2005). Proteomic patterns: Their potential for disease diagnosis. Mol Cell Endocrinol *230*, 95–106.

Yan, L., Tonack, S., Smith, R., Dodd, S., Jenkins, R.E., Kitteringham, N., Greenhalf, W., Ghaneh, P., Neoptolemos, J.P., and Costello, E. (2009a). Confounding effect of obstructive jaundice in the interpretation of proteomic plasma profiling data for pancreatic cancer. J Proteome Res *8*, 142–148.

Yan, W., Apweiler, R., Balgley, B.M., Boontheung, P., Bundy, J.L., Cargile, B.J., Cole, S., Fang, X., Gonzalez-Begne, M., Griffin, T.J., *et al.* (2009b). Systematic comparison of the human saliva and plasma proteomes. Proteomics Clin Appl *3*, 116–134.

Ye, B., Skates, S., Mok, S.C., Horick, N.K., Rosenberg, H.F., Vitonis, A., Edwards, D., Sluss, P., Han, W.K., Berkowitz, R.S., *et al.* (2006). Proteomic-based discovery and characterization of glycosylated eosinophil-derived neurotoxin and COOH-terminal osteopontin fragments for ovarian cancer in urine. Clin Cancer Res *12*, 432–441.

Yocum, A.K., and Chinnaiyan, A.M. (2009). Current affairs in quantitative targeted proteomics: Multiple reaction monitoring-mass spectrometry. Brief Funct Genomic Proteomic *8*, 145–157.

You, J.S., Gelfanova, V., Knierman, M.D., Witzmann, F.A., Wang, M., and Hale, J.E. (2005). The impact of blood contamination on the proteome of cerebrospinal fluid. Proteomics *5*, 290–296.

Yuan, X., and Desiderio, D.M. (2005). Proteomics analysis of human cerebrospinal fluid. J Chromatogr B Anal Technol Biomed Life Sci *815*, 179–189.

Zhang, H., and Chan, D.W. (2007). Cancer biomarker discovery in plasma using a tissue-targeted proteomic approach. Cancer Epidemiol Biomarkers Prev *16*, 1915–1917.

Zhang, J., Goodlett, D.R., Peskind, E.R., Quinn, J.F., Zhou, Y., Wang, Q., Pan, C., Yi, E., Eng, J., Aebersold, R.H., *et al.* (2005). Quantitative proteomic analysis of age-related changes in human cerebrospinal fluid. Neurobiol Aging *26*, 207–227.

Zhang, Z., Yu, Y., Xu, F., Berchuck, A., van Haaften-Day, C., Havrilesky, L.J., de Bruijn, H.W., van der Zee, A.G., Woolas, R.P., Jacobs, I.J., *et al.* (2007). Combining multiple serum tumor markers improves detection of stage I epithelial ovarian cancer. Gynecol Oncol *107*, 526–531.

Zhou, H., Chen, B., Li, R.X., Sheng, Q.H., Li, S.J., Zhang, L., Li, L., Xia, Q.C., Wang, H.Y., and Zeng, R. (2005). Large-scale identification of human biliary proteins from a cholesterol stone patient using a proteomic approach. Rapid Commun Mass Spectrom *19*, 3569–3578.

Zhou, L., Lu, Z., Yang, A., Deng, R., Mai, C., Sang, X., Faber, K.N., and Lu, X. (2007). Comparative proteomic analysis of human pancreatic juice: Methodological study. Proteomics *7*, 1345–1355.

Zhou, M., Lucas, D.A., Chan, K.C., Issaq, H.J., Petricoin, E.F., 3rd., Liotta, L.A., Veenstra, T.D., and Conrads, T.P. (2004). An investigation into the human serum "interactome". Electrophoresis *25*, 1289–1298.

Zhou, H., Yuen, P.S., Pisitkun, T., Gonzales, P.A., Yasuda, H., Dear, J.W., Gross, P., Knepper, M.A., and Star, R.A. (2006). Collection, storage, preservation, and normalization of human urinary exosomes for biomarker discovery. Kidney Int *69*, 1471–1476.

Zolg, J.W., and Langen, H. (2004). How industry is approaching the search for new diagnostic markers and biomarkers. Mol Cell Proteomics *3*, 345–354.

Zolotarjova, N., Martosella, J., Nicol, G., Bailey, J., Boyes, B.E., and Barrett, W.C. (2005). Differences among techniques for high-abundant protein depletion. Proteomics *5*, 3304–3313.

Zolotarjova, N., Mrozinski, P., Chen, H., and Martosella, J. (2008). Combination of affinity depletion of abundant proteins and reversed-phase fractionation in proteomic analysis of human plasma/serum. J Chromatogr A *1189*, 332–338.

Zuo, X., and Speicher, D.W. (2002). Comprehensive analysis of complex proteomes using microscale solution isoelectrofocusing prior to narrow pH range two-dimensional electrophoresis. Proteomics *2*, 58–68.

Zurbig, P., Decramer, S., Dakna, M., Jantos, J., Good, D.M., Coon, J.J., Bandin, F., Mischak, H., Bascands, J.L., and Schanstra, J.P. (2009). The human urinary proteome reveals high similarity between kidney aging and chronic kidney disease. Proteomics *9*, 2108–2117.

Chapter 18
Sample Handling of Body Fluids for Proteomics

Joao A. Paulo, Ali R. Vaezzadeh, Darwin L. Conwell, Richard S. Lee, and Hanno Steen

Abstract This chapter provides an overview of different approaches that can be used for sample preparation of body fluids for proteomics. Sample collection, protein extraction, protease inhibitor supplementation, sample storage, and abundant protein depletion are presented here in the context of various human body fluids. We emphasize that the particular set of techniques chosen for such investigations tightly correlates with the fluid to be analyzed, as no consensus methods are appropriate for all body fluids. However, we stress the need for standardized methods for the individual body fluids which is paramount in obtaining reproducible and robust data when analyzing human body fluids. In addition, we provide examples of optimized sample handling techniques using a systemic (urine) and a proximal body fluid (pancreatic fluid).

Keywords Body fluids · Biomarkers · Sample handling · Urine · Pancreas · Pancreatitis · Kidney

18.1 Introduction

Standardization of the preanalytical phase, including sample collection, preparation, and storage must be established prior to sample collection and analysis so that proteomics can be extended to clinical use. Minimizing preanalytical disparities, analytical standardization and quality control measures are imperative for successful proteomic analysis of body fluids aimed at biomarker discovery in both a research setting and for translation into routine clinical testing. Human body fluids (such as

J.A. Paulo (✉)
Division of Gastroenterology, Hepatology and Endoscopy, Center for Pancreatic Disease, Brigham and Women's Hospital and Department of Medicine, Harvard Medical School, Boston, MA 02115, USA; Department of Pathology, Children's Hospital Boston, Harvard Medical School, Boston, MA 02115, USA
e-mail: Joao.Paulo@childrens.harvard.edu

A.R. Ivanov, A.V. Lazarev (eds.), *Sample Preparation in Biological Mass Spectrometry*, DOI 10.1007/978-94-007-0828-0_18,

Table 18.1 Common human body fluids

Fluid	Origin	Function
Amniotic fluid	Amniotic sac	Nourishing and protecting fetus
Aqueous humor	Eye	Maintains the intraocular pressure
Bile	Gall bladder/liver	Aids in digestion of lipids
Blood (plasma/serum)	Systemic	Maintenance of cells and tissues
Breast milk	Female breast	Nourishment
Bronchoalveolar lavage	Lungs	Exchange of gases from atmosphere into blood stream
Cerebrospinal fluid	Surrounds spine and brain	Cushion and maintenance
Gastroduodenal fluid	Stomach and duodenum	Digestion
Interstitial fluid	Surrounds extracellular space	Maintenance of cells and tissues
Lymph	Systemic/lymph node	Bathe cells with water and nutrients
Mucus/nasal lavage	Nasal cavity	Protection form infection/foreign agents
Pancreatic fluid	Pancreas	Digestion
Perilymph	Cochlea	Bathes spiral ganglion cell bodies of the auditory nerve
Pleural fluid	Pleural cavity, surrounds lungs	Aids in respiration
Saliva	Mouth	Digestion
Semen	Male gonads	Reproduction
Sweat	Skin	Thermoregulation
Synovial fluid	Synovial joints	Reduce friction between the articular cartilage
Tears	Eye	Clean and lubricate the eyes
Urine	Urinary system	Waste removal
Vaginal secretion	Vagina	Maintain vaginal moisture
Vitreous fluid	Eye	Maintains pressure between lens and retina

See references in text.

those listed in Table 18.1) may be a more attractive option than tissue biopsies for the diagnosis and prognosis of diseases as obtaining body fluid has the advantage of being less invasive, less costly, easier to collect and possibly easier to process. Body fluids are excellent sources of protein biomarkers because during their circulation, they come in contact with a variety of tissues and pick up proteins secreted or shed by tissues.

Despite the attractiveness of body fluids for biomarker discovery, the task of their quantitative analysis using proteomics technologies is challenging. Body fluids, such as plasma, are rich and complex reservoirs of proteins. Large differences in protein concentration (ranging from several milligrams to less than 1 pg/mL) and the high number of posttranslational modifications are among the challenges of body fluids proteomics. Other challenges are related to the development of standardized sample preparation and handling protocols. High variation in body fluids increases the necessity of standardized protocols and large sample cohorts to obtain statistically viable biological conclusions.

18.1.1 Proper Handling of Body Fluids for Proteomics Is Essential

Establishing clear and consistent sample collection and processing methodologies is an initial stage in the development of clinical proteomics assays. In translational research, there often are insufficiently standardized protocols in regards to specimen collection, storage and processing. In urine proteomics, for example, several articles have stressed the importance of standardized sample handling in reducing experimental variability (Barratt and Topham, 2007; Decramer et al., 2008; Hortin and Sviridov, 2007; Muller and Brenner, 2006; Munro et al., 2006; Thongboonkerd, 2007, 2008). Similarly, for blood and serum, there are ongoing efforts to optimize sample handling protocols (Barelli et al., 2007; Hsieh et al., 2006; Luque-Garcia and Neubert, 2007; Rai and Vitzthum, 2006). Significant changes in the proteomic profile may be introduced during sample preparation if a consensus methodology is not in place. In biomarker discovery, these procedural artifacts may significantly impair the experimentation and analysis. All clinical and sample variables cannot be entirely avoided; however, it should be kept to a minimum with standardized sample handling methodologies.

18.1.2 Systemic and Proximal Body Fluids

In broad terms, there are two types of body fluids, proximal and systemic. Many body fluids, such as urine, can be considered both. Both proximal and systemic fluids have their advantages and disadvantages. A proximal fluid is the immediate down-stream body fluid of a particular organ or disease. In contrast, systemic fluids, such as blood or urine, can represent the entire body and may provide a snapshot of the organism under a given set of conditions or disease. The majority of the proteins detected in a systemic fluid may not necessarily be directly related to the disease of interest (Issaq et al., 2007); and systemic fluid analysis may identify the response of non-primary organs to the disease. For example, children with obstructive sleep apnea may have changes in renal function that can be detected in the urinary proteome (Gozal et al., 2009).

The human blood/serum proteome originates from a variety of tissues as a result of secretion or shedding, and can reflect human physiological states which can be used for disease diagnosis and prognosis (Anderson et al., 2004). As blood comes in contact with all organs of the body, it is a source of many potential biomarkers. However, such biomarkers may be less specific than those obtained from proximal organ fluids, as the origin of suspected biomarkers may be difficult to attain. Like other body fluids, blood contains proteases, cells, and lipids, and their removal or deactivation may be necessary for protein analysis. The non-cellular fraction of blood represents proteins essential for circulatory functions, organ secretions that are released into circulation, and diseases such as myocardial infarctions (Omenn et al., 2005). Blood (both serum and plasma) has one of the highest dynamic ranges of any body fluid, with a protein concentration spanning over 10 orders of magnitude (Anderson and Anderson, 2002).

Urine can be considered both a systemic and proximal fluid. Specifically, urine represents the metabolic end product of blood, but it is also influenced by the status of the kidney and the lower urinary tract. It is estimated that up to 70% of the urinary proteome may originate from the urinary tract and kidney (Mischak et al., 2007). Urine is often considered the ideal body fluid source for diseases of the urogenital tract and kidney; however, at the same time, a large number of the proteins identified may be originating from other organ systems.

The potential advantage of analyzing a *proximal* body fluid is that it will increase the likelihood of biomarker discovery in the context of particular diseased organ (Rifai et al., 2006). For example, as cerebrospinal fluid (CSF) is in direct contact with the extracellular surface of the brain, its biochemical composition is altered by disorders related to the central nervous system. Synovial fluid, which is the fluid present in joints, has been used in the diagnosis of rheumatoid arthritis and similar inflammation. Synovial fluid reduces friction between cartilage and other tissues and may reflect that pathophysiological conditions of joints. Its investigation may result in effective markers of rheumatoid arthritis and osteoarthritis (De Ceuninck and Berenbaum, 2009; Wilson et al., 2009).

Although serum is typically considered to be a systemic fluid, in certain contexts it can be a proximal fluid. One particular example is of blood obtained directly from the coronary sinus of the heart, which is the collection of veins that collects blood from the myocardium of the heart. Blood obtained from this source may provide new insights into myocardial damage and ischemia (Gerszten and Wang, 2008). Other proximal fluids of interest include pleural fluid, which is the fluid between the layers of tissue surrounding the lungs. Abnormal accumulation of pleural fluid is a potential rich source of protein analysis for primary lung and pleural disease (Tyan et al., 2005a,b). Bronchoalveolar lavage (BAL) fluid is another potential source of proximal fluid for pulmonary disease, such as asthma. BAL fluid analysis demonstrates a great diversity of cellular origins and functions, and a comparative analysis of serum and bronchoalveolar lavage proteomes revealed that some proteins were more abundant in bronchoalveolar lavage than in plasma, suggesting that they are specifically produced in the airways (Noel-Georis et al., 2001; Wattiez and Falmagne, 2005). Nipple aspirates is another promising proximal fluid to identify biomarkers for breast cancer diagnosis and therapy. The analysis of tear fluid for specific eye disorders such as Sjogren's syndrome and dry eye syndrome is another example (Nguyen and Peck, 2009; Tomosugi et al., 2005).

Proteomic analysis of saliva may represent the proximal fluid of the upper airway and organs of the mouth and offers a rapid, simple and non-invasive method for short- and long-term monitoring of pathological disorders, such as oral squamous cell carcinoma (Wu et al., 2010) and periodontitis (Haigh et al., 2010). Pancreatic fluid is an excellent clinical specimen for identification of novel biomarkers of the upper gastrointestinal tract and pancreas, as its protein composition is of lower complexity compared to serum (Paulo et al., 2011). Pancreatic fluid primarily consists of fluid originating from the exocrine pancreas function, and may be an ideal fluid source for the identification of novel biomarkers of pancreatitis and pancreatic adenocarcinoma (Grote and Logsdon, 2007; Lohr and Faissner, 2004; Wandschneider et al., 2001).

18.1.3 Sample Accessibility

While systemic fluids are often more easily accessible, proximal fluids may provide a more immediate view of an organ's microenvironment at the expense of invasiveness. Although blood and urine can be accessed by relatively noninvasive means and allow for analysis of non-diseased patients for comparative proteomic studies, it may be difficult to obtain many other proximal body fluids from/for healthy controls. Fluid accessibility via noninvasive or minimally invasive methods is imperative when performing comparative analyses that require sampling of a corresponding body fluid from a healthy individual. While some fluids are easy to access, such as urine, saliva, and blood, other fluids may be more difficult to acquire, necessitate well-trained individuals, and have inherent associated risks. Such less accessible proximal fluids, including cerebrospinal fluid, pleural fluid and synovial fluid must often be acquired only from individuals who are suspected of having the corresponding disease. Appropriate controls are sorely needed; however, controls usually consist of those with clinical suspicion of the disease, but whose diagnosis was negative. For example, in the study of pancreatic fluid and bile, fluid collection via ERCP (endoscopic retrograde cholangiopancreatography) is not recommended on a healthy individual due to the significant risk of complications which may arise and result in acute pancreatitis, and is usually only performed if underlying gastrointestinal dysfunction is suspected (Crnogorac-Jurcevic et al., 2005; Lohr and Faissner, 2004). CSF collection involves a lumbar puncture, which may result in complications (Pendyala et al., 2009; Xiao et al., 2009). For such investigations, normal or healthy controls must be culled from patients who have a negative result for a particular test of that disease.

Proximal body fluids considered easily accessible include BAL fluid, urine (in specific diseases), nipple aspirates, tear fluid, and saliva. For example, BAL can be noninvasively obtained by the inhalation of a soluble hypotonic aerosol to induce expectoration (Beier et al., 2004). Collection of these body fluids is not associated with high risks to the patients, and normal controls may be more easily obtained. Overall, sample accessibility must be taken into consideration in the study design prior to undertaking any proteomic investigations.

18.1.4 Sample Variability

The composition of body fluids may be influenced by environmental factors, age, gender, sex, and confounding diseases, which may result in substantial subject-to-subject variability. Factors affecting composition of many body fluids, such as saliva, urine, pancreatic fluid, bile, and serum include circadian rhythm, health status, exercise, medications, and food intake. For example, in a study of the cerebrospinal fluid proteomes from 10 individuals, 38% of the identified proteins were unique to individual subjects, whereas only 6% were common among all ten subjects (Wenner et al., 2004). Similarly, studies of human breast milk demonstrate a relationship between a mother's nutrition and milk proteome composition (Cavaletto et al., 2004). At the same time, many of the common milk proteins, such as immunoglobulins, casein, and serum albumin, are often modified by posttranslational modifications or proteolytic processes (D'Auria et al., 2005).

Urine is a prime example of how confounding factors can influence protein content. Urine is the metabolic end product of blood which can be affected by renal function. It is estimated that 70% of proteins from urine originate from the kidney, whereas the remaining proteins are derived from plasma (Mischak et al., 2007). The protein concentration of urine is also dependent on stability, binding, and clearance of small proteins that are usually reabsorbed into the blood via normal renal function. Typically, it is thought that the highest concentration of protein in urine is usually in the morning; however, this may be contaminated by bacterial protein contamination, protein degradation in the bladder, and longer storage times in the bladder. As such, the second urine of the day may be a more useful and less variable specimen for proteomic analysis. When investigating and designing discovery based proteomics studies of human body fluids, sample variability must be taken into consideration. Large individual and biological variability requires a larger sample size due and further highlights the need for standardized collection methodologies.

18.1.5 Fluid Mixing

Along with patient and environmental variability, certain body fluids are composed of mixtures from different glands, each of which if isolated separately may have its own unique proteome. For example, human saliva is secreted from multiple salivary glands including parotid, submandibular, sublingual, and other minor glands lying beneath the oral mucosa. Although most proteins isolated from saliva originate from the salivary glands, some blood (Huang, 2004), oral tissue (Kojima et al., 2000) and bacterial (Macarthur and Jacques, 2003) proteins have also been readily identified in this fluid. Seminal fluid is another complex mixture consisting of spermatozoa suspended within secretions of male sex glands including the seminal vesicles and prostate. Similarly, cervical–vaginal fluid, which has been analyzed to identify antimicrobial peptides, is composed of a mixture of different secretions (Di Quinzio et al., 2007; Shaw et al., 2007; Tang et al., 2007). In addition, the study of gastrointestinal diseases, such as pancreatitis, often involve the collection of fluid which is to some degree a mixture of gastric fluid, duodenal fluid, pancreatic fluid and bile (Paulo et al., 2010a). Amniotic fluid is another example of natural mixing of body fluids. Amniotic fluid surrounds the developing fetus and plays a crucial role in normal development. Amniotic fluid is a mixture of fetal and maternal proteins and corresponding degradation products. The composition of this body fluid includes proteins exchanged between mother and fetus, proteins excreted by the fetus, and proteins from the maternal placenta. Along with the caveats of accessibility and variability, the innate mixing of different fluids may be unavoidable and must be considered throughout the body fluid analysis process.

18.2 Sample Preparation

No single sample preparation method is optimal for all body fluids. However, standardization of methods should be established at the clinical collection interface and the laboratory for each body fluid to avoid introducing potential confounders. Collection methods may vary depending on the location, or even the person who is acquiring the sample. Along with differences in collection methods, temporal and environmental (e.g. temperature of transport) differences in transporting the fluid from the clinic to the bench may result in the loss of reproducibility. Efforts should be made to standardize the protocols used to obtain the sample so as to increase the accuracy and reproducibility of the data.

18.2.1 Protein Extraction and Fractionation

The optimal method of protein extraction is highly dependent on the sample under investigation. Sample fractionation can be performed using a variety of methods including centrifugation, precipitation, liquid chromatography and electrophoresis. It is not uncommon, that for many body fluids, a large number of intracellular proteins are identified due to shedding of cells into the interstitial fluid, which is a common source of confounding proteins. At times, whole cells may be present, and can be removed via centrifugation prior to further sample processing. For body fluid proteomic analyses, proteins should be segregated from lipids, metabolites, and other non-proteinaceous substances. Most often, salt removal is accomplished via dialysis (spin, micro) (Manabe et al., 1999), ultrafiltration (Fountoulakis et al., 2004; Jiang et al., 2004), gel filtration/electrophoresis, precipitation with TCA or organic solvents, and/or solid-phase extraction.

18.2.1.1 Centrifugation

One of the simplest methods of protein enrichment is ultracentrifugation using sequentially increasing centrifugal forces, or sucrose gradients, which permits separation at various subcellular levels. In addition to isolating proteins from membranes, mitochondria, or nuclei, differential centrifugation can also investigate vesicles such as exosomes, which may be an alternative source of biomarkers. Exosomes are small vesicles secreted by cells and may be a mechanism for selective removal of many plasma membrane proteins (Simpson et al., 2009). Recently, a nano-LC-MS/MS analysis identified 295 proteins in urinary exosomes, including multiple proteins known to be involved in renal and systemic diseases. Urinary exosomes may be a rich source for disease biomarker discovery in urine (Gonzales et al., 2009; Pisitkun et al., 2004).

18.2.1.2 Precipitation

Several precipitation methods are available for protein isolation, these include acetone, trichloroacetic acid, ethanol, isopropanol, chloroform/methanol, and ammonium sulfate. The efficiency of protein precipitation varies among different organic solvents. For example, acetone appears to precipitate more acidic and hydrophilic proteins, whereas ultracentrifugation fractionates more basic, hydrophobic, and membrane proteins (Thongboonkerd et al., 2002). Alternatively, chloroform methanol extraction has been used to successfully extract hydrophobic proteins in human bile (Stark et al., 1999). For comparison, precipitation with 10% TCA/90% acetone/20 mM DTT is well-suited for salivary proteomics (Hu et al., 2005). Whereas, using 0.02% sodium bisulfate in ethanol and acetic acid (1:1, v/v) has been successful in protein extraction of sinonasal lavage, which has been used to investigate the effects of smoking on the nasal cavity proteins of smokers and non-smokers (Casado et al., 2005). Moreover, we have determined that trichloroacetic acid and acetone precipitation extracted the most proteins from pancreatic fluid and gastroduodenal fluid, respectively (Paulo et al., 2010b, c). Again, due to the intrinsic differences in the composition of the body fluid, such protein precipitation methodologies must be investigated for the particular fluid of interest and for the specific types of proteins that are to be isolated.

18.2.1.3 Depletion of Abundant Proteins

Due to sample complexity and the high dynamic range in some body fluids, depletion of abundant proteins may be necessary (Hortin et al., 2006). The depletion of proteins of relatively high abundance, such as those listed in Table 18.2, is advantageous for biomarker discovery. For example, albumin and immunoglobulins (main constituents of blood and urine) account for 80% of vitreous fluid proteins, which if

Table 18.2 Most abundant plasma proteins (Heide et al., 1997)

Protein	mg/mL
Albumin	50
IgG, IgA, IgM	12–15
Fibrinogen	3
α-, β-lipoproteins	2–6
α1-antitrypsin	2–5
α2-macroglobulin	2–4
Transferrin	2–3
α-1-acid glycoprotein	1
Hemopexin	1
α-lipoproteins	0.6–1.5
Haptoglobin	3
Ceruloplasmin	0.3

not depleted, may prevent the identification of lower abundant proteins (Koyama et al., 2003). Albumin is often depleted from serum or plasma using dye-based columns. Traditionally, Cibachrome Blue dye is used to deplete albumin (Thresher and Swaisgood, 1990). However, as this dye is nonspecific, monoclonal antibodies may be a better alternative (Bjorhall et al., 2005): anti-human serum albumin columns, for instance, have been shown to be more efficient and specific in the depletion of albumin from blood (Echan et al., 2005). Independently IgG is often removed with an immobilized protein A column.

In addition to albumin and IgG, there are other abundant proteins that are also often depleted to improve identification of low abundant proteins. For example, the most abundant proteins (human serum albumin, transferrin, IgG, IgA, IgM, fibrinogen, and hemoglobin) in CSF and blood are often immunodepleted to improve the identification of less abundant proteins (Hortin et al., 2006; Thouvenot et al., 2008). Multiple affinity removal columns that remove up to 7–14 of the most abundant proteins simultaneously are available (MARS – Multiple Affinity Removal System, Agilent) (Pieper et al., 2003; Bjorhall et al., 2005).

However, the disadvantage of removing albumin is that albumin functions as a carrier protein and its removal may result in the loss or co-depletion of other significant proteins (Zhou et al., 2004). Additionally, each additive step in the workflow may also lead to more sample loss and variability. In one study using blood, the depletion of albumin was associated with the additional reduction of 815 proteins possibly from co-depletion (Shen et al., 2005). Several other studies suggest that albumin binds to certain low-molecular weight proteins and peptides, which are lost while albumin is depleted (Decramer et al., 2008; Zolotarjova et al., 2005). To address the problem of co-depletion, addition of 5 to 20% acetonitrile may be useful in disrupting the binding of albumin to lower molecular weight proteins that may be otherwise lost during albumin depletion (Huang et al., 2005). Alternatively, albumin may be used to isolate small bound proteins, by specifically binding albumin under normal conditions to a solid matrix and releasing bound peptides for analysis via mass spectrometry (Lowenthal et al., 2005). Disrupting these protein–protein interactions by ultracentrifugation with detergents and/or chaotrophic agents can also help to avoid protein loss.

Currently, the vast majority of depletion methods have been developed specifically for serum. Adapting these methods to other body fluids, such as urine, can be challenging because of the inherent differences in each body fluid. The particular body fluid under investigation may contain a lower concentration of the high-abundant proteins of serum, or alternatively, other abundant proteins. The decision of whether to deplete proteins, and if so, with what modality, must be determined with the particular study and body fluid in mind.

18.2.2 Protease Inhibitors Supplementation

The use of protease inhibitors is assay-specific as their use may be detrimental to particular experiments. For example, certain protease inhibitors may interfere with bottom-up proteomic experiments (particularly for in-solution digests) where specific enzymatic digestion (e.g., with trypsin) is necessary. Protease inhibitors should not be added to peptidomics studies aiming to investigate the temporal changes of enzyme activity (Ivanov and Yatskin, 2005; Schulz-Knappe et al., 2005). Although, the addition of specific protease inhibitors (e.g. phenylmethylsulfonyl fluoride (PMSF), aminoethyl benzylsulfonyl fluoride (AEBSF), ethylene diamine tetraacetic acid (EDTA), pepstatin, benzamidine, leupeptin, aprotinin) may be recommended in particular situations, it should be done so with caution, as inhibitors may modify proteins, introduce adducts, and interfere with further peptide studies. For example, certain peptide inhibitors, such as high-concentration aprotinin, may interfere with MS analysis and several small molecule inhibitors, such as PMSF and AEBSF have been shown to form covalent bonds with proteins (Finnie and Svensson, 2002), thereby changing p*I* and electrophoretic mobility (Rai et al., 2005). In addition, many protease inhibitor cocktails contain small molecule or peptide inhibitors, which can interfere with subsequent peptide ionization (Marshall et al., 2003).

Conversely, the addition of protease inhibitors might be advantageous for sample preservation during other applications, such as in-gel tryptic digests, in which low molecular weight protease inhibitors could be separated from the proteins and/or inactivated during SDS-PAGE. Body fluid that are shipped off site or are at risk for fragmentation because of temperature fluctuations during transport may require protease inhibitors during transport. The use of protease inhibitors should be tested to determine their need for each particular application or clinical sample situation.

18.2.3 Storage Conditions and Protein Stability

Both long- and short-term storage of body fluids may have an impact on sample quality. For example, blood is typically incubated at room temperature to allow for the initiation of the coagulation cascade. After clotting, the blood can be centrifuged to separate the serum and plasma components. However, significant variation of serum composition can occur secondary to variations of clotting time and temperature exposure. Plasma may be more stable by the nature of its collection process. To obtain plasma, various anticoagulants, such as heparin, citrate and EDTA (Lundblad, 2004) are added in defined amounts. Currently, HUPO (Human Proteome Organization) suggests the use of plasma with EDTA over serum because of the stability of plasma. Studies have shown improved resolution, sensitivity and reproducibility of protein identifications via tandem mass spectrometry in plasma verses serum (Omenn et al., 2005).

In addition to natural inherent variables of samples, multiple freeze–thaw cycles can have a detrimental effect on proteome stability. Freeze–thaw cycles can induce precipitation, as has been shown in urine and saliva (Nurkka et al., 2003). Even though proteins from a particular fluid may be resistant to degradation or precipitation resulting from freeze–thaw cycles, it may be of additional benefit to store samples for extended periods in smaller aliquots to minimize, or essentially eliminate, the need for multiple sample thaws. Several studies have been performed to investigate the stability of urine, which has a lower complexity and relatively high thermostability compared to serum, under different storage conditions. In urine proteomics, for example, several recent reviews have stressed the importance of standardized sample storage conditions in reducing experimental variability (Barratt and Topham, 2007; Decramer et al., 2008; Hortin and Sviridov, 2007; Muller and Brenner, 2006; Munro et al., 2006; Thongboonkerd, 2007, 2008). For pancreatic fluid, such variations are especially pronounced due to its inherent high concentration of active proteolytic enzymes, and for which proteolysis is evident after 30 min at room temperature on SDS-PAGE gels. In addition, proteins in particular are subject to modifications, degradation, precipitation, adductions (e.g. acetylation, oxidation), which again can confound studies aimed to identify biomarkers. Storage conditions may vary with the body fluid under investigation, but a cautious and conservative approach should be considered until all variables are suitably investigated. Currently, we recommend minimal transport time, transport at low temperatures, separating clinical samples into experiment specific single use aliquots, and long term storage at –80°C.

18.2.4 Sample Analysis

There are numerous different methodologies that can be used to perform sample analysis. Each method has its advantages and disadvantages in the context of the particular body fluid analysis. Typically the down-stream needs of the workflow and complexity of the sample will determine the optimal method of sample analysis. We provide examples of different methodologies.

18.2.4.1 Protein Fractionation

The substantial complexity of body fluid proteomes requires multiple dimensions of separation in order to be resolved and comprehensively studied (as briefly mentioned in Section 2.1). Separation is usually based on different physicochemical properties of the species and can be performed at the protein level (prior to proteolytic digestion), at the peptide level (post-proteolytic digest), or both. A diverse range of separation methods, which can be used in tandem are available to reduce the complexity of a given body fluid sample.

To reduce sample complexity at the protein level, proteins can be fractionated by differential centrifugations, size exclusion chromatography or centrifugation, reversed-phase liquid chromatography, or gel electrophoresis. A combination of these methods can be utilized if further fractionation is necessary due to sample complexity.

GeLC-MS/MS is one of the most common methods of separation. In GeLC-MS/MS, one-dimensional SDS-PAGE gel electrophoresis is initially performed to fractionate proteins by electrophoretic mobility/molecular weight. The entire gel lane is cut into numerous gel slices and in-gel proteolytic digestion results in numerous peptide fractions, which are then submitted to secondary separation by standard reverse phase liquid chromatography prior to MS/MS analysis. GeLC-MS/MS can be advantageous in body fluid proteomics, because proteins are immobilized to the gel matrix which allows for cleaning of salts, lipids, and other substances (which are common in body fluids) that can interfere with mass spectrometry (Shevchenko et al., 1996, 2006). However, large scale studies involving multiple samples can be difficult.

18.2.4.2 2D-GE

Another commonly used protein separation method is two-dimensional gel electrophoresis (2D-GE). Proteins are first separated according to their isoelectric point followed by separation based on their electrophoretic mobility/molecular weight. Gel spots are excised and proteins can be proteolytically digested prior to mass spectrometry analysis. Although progress has been made to increase reproducibility, there remains still significant run-to-run variability, which is a deterrent for comparative proteomics experiments designed to detect differences between two samples. Additionally, interfering compounds and salts in many of the body fluids can make 2D-GE difficult. Furthermore, a large number of clinical samples may prohibit the efficiency of 2D-GE.

Many of the difficulties of 2D-GE have been overcome by the development of difference imaging gel electrophoresis (DiGE) (Seike et al., 2004; Tian et al., 2008; Walsh et al., 2009). In DIGE, two or three samples are labeled with spectrally distant isobaric fluorescent dyes (Cy2, Cy3, or Cy5) and run simultaneously on a single gel. Images for each wavelength are then merged for quantitative and qualitative assessment of differences without need of cross gel comparisons, thus eliminating gel-to-gel variation (Mujumdar et al., 1989; Unlu et al., 1997). Furthermore, gel spots corresponding to proteins that differ between images, or proteins of interest that overlap in both samples can be excised and analyzed by mass spectrometry.

18.2.4.3 Peptide Fractionation

Protocols typically would begin with in-solution digestion of the protein sample resulting in peptides which are fractionated in a single dimension (revered-phase, strong cation exchange, isoelectric focusing) (Manadas et al., 2009) or via MUDPIT (multidimensional protein identification technology) (Peng et al., 2003; Washburn et al., 2001).

As opposed to separating at the protein level, initial separation can occur at the peptide level to reduce sample complexity. Orthogonal methods of separation include high pH reversed-phase chromatography, strong cation exchange chromatography, and/or isoelectric focusing prior to mass spectrometric analysis. Several recent studies have compared the various fractionation methods mentioned above (Delmotte et al., 2007; Dwivedi et al., 2008; Elschenbroich et al., 2009; McDonald et al., 2002; Motoyama and Yates, 2008). However, such studies are performed to answer specific questions, and their advantages may not be applicable to all study designs.

18.2.4.4 SELDI

Surface-enhanced laser desorption/ionization (SELDI) (Hauskrecht et al., 2005; Koopmann et al., 2004; Ortsater et al., 2007; Rosty and Goggins, 2002; Scarlett et al., 2007; Verma et al., 2001) is commonly used to investigate body fluids prepared by the methods outlined above. A subfraction of proteins are isolated by adsorptive surfaces on a chip and the proteins or peptides that bind are analyzed by matrix-assisted laser desorption/ionization (MALDI). The SELDI chip contains a chromatographic coating on which sample components of a certain type (i.e., hydrophobic, or metal binding) are captured. Using this technique, the mass-to-charge peak pattern is identified; however, the corresponding proteins or peptides identifications are not. More recent advances have allowed for some protein identification when the ProteinChip (BioRAD) technology is coupled to tandem mass spectrometers.

18.2.4.5 Quantitative Proteomics

Although DiGE does overcome some of the limitations of 2D-GE, it still carries over biases against very small, very large and hydrophobic proteins, many of which are of great importance when attempting to discover biomarkers of human body fluids. For such analyses, other methods of quantitative proteomics may provide an unbiased approach to comprehensive proteome analysis. SILAC (stable isotope labeling in cell culture) is one of the original methods of labeling for quantitative proteomics. Unfortunately it is only applicable in cell culture or animal models. For cell culture, stable isotope labels may be introduced during cell growth to attain up to 100% labeling efficiency. Typically, two cell states are prepared, with one set of cells being grown in media with heavy isotope labeled arginine and/or lysine, combined, and chromatographically separated to assess relative differences and similarities in protein content. Currently, SILAC cannot be used with human body fluids. Human body fluids can only be labeled after they are collected.

Currently, all labeling methods occur at the peptide level. Among the simplest method of labeling is the incorporation of ^{18}O into the C-terminus of the peptides during proteolysis, e.g. with trypsin (Stewart et al., 2001). Similarly, peptides can be labeled with chemical tags. Chemical labeling can be performed in a multiplexed manner using isobaric tags (e.g., TMT (Thompson et al., 2003) or iTRAQ (Ross et al., 2004)). From two to eight disease states, time points, or samples can be chemically labeled and compared in a single experiment. After samples are differentially labeled, they are pooled, undergo separation as described above, and then further fractionation by liquid chromatography prior to tandem mass spectrometry analysis. The fragmentation of the attached tag generates low molecular mass reporter ions that can be used to relatively quantify the peptides and the proteins from which they originate. These quantitative methods allow slight changes in the proteome to be detected, which would otherwise not be if only qualitative (presence vs. absence) of proteins were investigated. Once, the protein of interest has been defined, directed quantitation methods such as isotope dilution (also known as AQUA for Absolute Quantitation) can be applied. Using this method, a synthetic heavy-isotope-labeled standard peptide is introduced into cell lysates at a known concentration and selected reaction monitoring is used to detect and quantitate the peptide of interest (Gerber et al., 2003, 2007; Kirkpatrick et al., 2005).

These stable isotope labeling remain the 'gold standard', but they have not been widely used for large-scale, multiplexed analyses owing to their relatively high cost, the limited availability of different mass-coded labels, and the frequent under-sampling associated with tandem mass spectrometry. Label-free methods for LC-MS represent attractive alternatives (Lundgren et al., 2010). They are based on the principle that the relative abundance of the same peptide in different samples can be estimated by the precursor ion signal intensity across consecutive LC-MS runs, given that the measurements are performed under identical conditions. In contrast to differential labeling, every biological specimen needs to be measured separately in a label-free experiment. Typically, peptide signals are detected at the MS1 level and distinguished from chemical noise through their characteristic isotopic pattern. These patterns are then tracked across the retention time dimension and used to reconstruct a chromatographic elution profile of the monoisotopic peptide mass. The total ion current of the peptide signal is then integrated, or the counts of spectra corresponding to a particular protein are recorded, and used as a quantitative measurement of the original peptide concentration. Label-free methods are very replicate dependent. To be statistically significant, chromatographic separation reproducibility must be very high. The exact alignment of the chromatograms and a highly reproducible ion count are mandatory. Not every chromatographic system is prepared for this performance, and the competition for the ionization may be problematic when very complex samples are separated. Advances in dedicated software are still needed to study the comparative behavior of both samples during the analysis in a trustable manner.

18.3 Examples

18.3.1 Urine

Urine is one of the ideal biological sources for prognosis and diagnosis of human diseases. Before the proteomics era, urine was analyzed by physicians for hundreds of years. Urine is produced by the kidney and was typically thought to be the filtered waste product of blood; however, it is now known that urine may contain many potential markers of systemic disease and the genitourinary tract. In a healthy normal human approximately 150–180 mg of proteins are daily excreted into human urine (Bramham et al., 2009). The majority of these proteins are derived from the kidney and the urogenital tract (70%) with the rest originating from plasma

(30%) (Gonzalez-Buitrago et al., 2007). A low amount of exosomes and apical membranes of renal tubular epithelial cells are also shed into urine. During normal kidney function, glomerular filtration results in removal of proteins with molecular weights greater than 40,000 Da. Smaller proteins (less than 15,000 Da) pass freely through the glomerular barriers, but are almost completely reabsorbed in the proximal tubules (99%). About two thirds of the filtered proteins is comprised of albumin, transferrin, some immunoglobulins and low molecular weight proteins (Barratt and Topham, 2007). Alteration of normal kidney function may lead to disruption of the normal transport mechanisms and subsequent changes in the urinary proteome.

Urine has several characteristics that make it a preferred clinical specimen as compared to other body fluids. First and foremost is the fact that urine is readily available, noninvasively obtained, and usually present in large volumes. This availability simplifies repeated sampling from the same individual and permits critical longitudinal studies. Urine has the specific advantage of being the proximal fluid (primary effluent) to the majority of urogenital organs. In contrast to samples such as blood, urine is exceptionally stable (Lee et al., 2008). It can sit at room temperature for hours or be stored at –20°C for long periods without any significant quantitative change to the general proteome. However, the proteins stability should not be interpreted in a generalized manner, since several studies have shown alteration in the peptidome (Fiedler et al., 2007) and the exosomes (Zhou et al., 2006) after long exposures to room temperature. In addition, it is unclear how different subproteomes would change under similar conditions.

Another advantage of urine is that it appears to be less complex than other most widely used body fluids. Although the urine proteome's dynamic range still spans over several orders of magnitude, its simplicity over serum makes it a rather attractive alternative (Fliser et al., 2007). Furthermore, urine contains a high amount of natural peptides and low-molecular weight proteins, which are soluble and can be directly analyzed. Capillary electrophoresis mass spectrometry (CE-MS) has been extensively used to study these compounds (Fliser et al., 2007). Finally, urine is a rich source of glycoproteins, where half of all proteins are thought to contain a glycan group (Wang et al., 2006). Glycosylation is a widespread and important post-translational modification and is involved in many biological processes such as immune response, signal transduction, inflammatory reaction and cell adhesion (Haltiwanger and Lowe, 2004).

Despite attractiveness of urine for biomarker discovery, some confounding effects have made its use in routine clinical applications challenging (Vaezzadeh et al., 2009). The composition of urine is highly variable. These changes are due to diet, hydration, circadian rhythm, metabolic and catabolic processes, exercise, diseases and other environmental factors. This variability has rendered the task of defining "normal" human urine proteome extremely intricate. The diurnal variation of the urinary proteome is also important to take into account (Bottini et al., 2002). Another major obstacle is the presence of high concentrations of interfering compounds such as urea, salts, metabolites and other charged compounds. Another significant drawback is the presence of a few proteins such as uromodulin, immunoglobulins and albumin in high abundance. These proteins can mask the detection of low-abundant proteins. Finally, the inconsistency of urine pH may result in greater variability in the composition of particular peptide fragments.

The advantages and challenges of urine as a source for biomarker discovery demonstrate the necessity of developing standard sample collection and preparation procedures. The initiation of Human Kidney and Urine Proteome Project (HKUPP) by Human Proteome Organization (HUPO) is an indication of the international effort of the scientific community towards this goal (Yamamoto et al., 2008). In this section, we provide a step-by-step practical review of current urinary proteomics sample handling challenges and offer recommendations based on our own experience and the literature.

18.3.1.1 Sample Collection and Storage

Urine can be collected at different time points during the day. However, Bottini et al. showed that the 24 h and first morning urine samples are at risk of contamination from bacteria or bladder epithelial cells (Bottini et al., 2002). Twenty-four hour urine samples should not be used for discovery based proteomics analysis because of their inherent nature of contamination in collection. The second-morning sample is recommended for urinary proteomics. Another aspect of the sample collection is when to collect during urination. Ideally a catheterized sample would be obtained. However, typically a voided sample is what is usually collected. It is suggested to use a midstream urine sample to decrease the risk of contamination, particularly for females (Schaub et al., 2004). Recommendations on methods to clean the genital area, to avoid contamination in the urine sample, can be found in the literature (Lifshitz and Kramer, 2000; Schaub et al., 2004; Vaillancourt et al., 2007).

In the past it was suggested to add the protease inhibitors for urine collection (Zhou et al., 2006). However, since urine is stagnated in the bladder for a long term, the proteases may not be as active as in other body fluids. Moreover, their addition might result in interference with protein digestion and analysis and down-stream mass spectrometry steps (Rai et al., 2005). Previous reports described increased bacterial activity in samples exposed to room temperature for >8 h or stored at 4°C for a >12 h (Thongboonkerd and Saetun, 2007). They suggest the addition of preservatives such as sodium azide (0.1–1 mM) or boric acid (200 mM) to retard bacterial growth. However, in our hands a sample, which was exposed to room temperature for 24 h did not show any noteworthy bacterial growth (Lee et al., 2008). Our search of the entire Swiss-Prot database using this sample resulted in identification of only three single-peptide protein identifications from *Escherichia coli*, where two of these three proteins had a high degree of homology to their human counterparts. In addition, our comparison of urine samples exposed at room temperature for 0, 4, 8 and 24 h showed that 97.5% of the top 50th percentile of the identified proteins were seen at all times (Lee et al., 2008). By performing label-free quantitation, we also showed that peptide spectral counts levels were stable at room temperature up to 24 h (Lee et al., 2008). Earlier reports of contamination may have been more a function of the voided collection methodology.

The pH of urine may vary largely (4.4–8.0) based on the dietary acid-base load, fruit and vegetables and meat intakes and also for pathological reasons (Welch et al., 2008). Despite the fact that it has been demonstrated that urine's pH has no effect on proteome analysis using 2D-GE (Thongboonkerd et al., 2009), sample pH can have a significant influence on the partitioning of proteins between sediments and the soluble portion of urine (Saetun et al., 2009). Further work has to be done to assess pH effects on posttranslational modifications and protease activity in samples with proteinuria. We found that while depleting albumin from urine using Hitrap Blue columns (GE Healthcare), the pH can have a significant effect on the efficiency of the depletion as well as the recovery (manuscript in preparation). Interestingly at low pH, the depletion was very effective; however, the recovery was not optimal. Since the reverse was true for basic pH, we systematically adjust the sample's pH to neutral in order to obtain high recovery and efficient depletion.

Samples are usually centrifuged at low speed or filtrated to remove cells and debris. Immediately after this step, urine samples are stored at –20°C or lower (we recommend –80°C). Repetitive freeze–thaw cycles have been reported to alter the proteome (Schaub et al., 2004). We experimented five freeze–thaw cycles and found that 98.3% of the top 50th percentile of the identified proteins remained unchanged (Lee et al., 2008). Using spectral count quantitation on the top 200 abundant peptides, we also found that freeze–thaw cycles did not have any significant effect on the peptide quantification. These results are confirmed by Fiedler et al. using MALDI to compare the peptidome of once frozen sample to three freeze–thaw cycles (Fiedler et al., 2007).

18.3.1.2 Sample Preparation

To date, there is no consensus on methods for concentrating, purifying and depleting urine samples for proteomics analysis. It is well known that using different methods will yield different proteins. In addition, variation in sample handling may significantly alter the urinary proteome and create artificial differences between comparable specimens.

Protein Concentration and Purification

Isolating and concentrating urinary proteins is an essential and important step as urine specimens usually have a very low protein concentration in non-nephrotic diseases, and in comparison to blood and tissue samples. Numerous methods have been employed including lyophilization, ultrafiltration, centrifugal filtration, reversed-phase extraction, and precipitation with organic solvents. Lyophilization was shown in several studies to afford the best quantitative yield since it has the least protein loss (Lee et al., 2008; Thongboonkerd et al., 2006). However, the most important drawback of this method is that it concentrates the proteins and also salts and other contaminations. Precipitation with organic solvents such as ethanol (Lee et al., 2008), acetone (Thongboonkerd et al., 2006) and acetic acid (Thongboonkerd et al.,

2006) favors isolation of mostly hydrophilic proteins. Proteins isolated with precipitation methods may still contain salt and lipids. Centrifugal filtration has a superior consistency compared with other methods but results in some protein loss (Lee et al., 2008). Reversed-phase trapping columns made of C4 or C18 resins are efficient tools to concentrate urinary proteins and peptides, respectively (Lee et al., 2008). In addition, the trapping columns can be used to for desalting of electrolytes and urea, but other interfering compounds, such as lipids, might remain in the sample.

We compared four different protein extraction methods to determine the most effective method: ethanol precipitation, lyophilization, microconcentrators and C4 trapping column. We found that out of the top 100 proteins identified, 89% were found in all four methods. It has been shown that a combination of different methods can result in obtaining a better coverage of the proteome. By combining 10 different protocols, Thongboonkerd and colleagues were able to increase the number of 2D-GE detected protein spots from around 100 in each method to 700 (Thongboonkerd et al., 2006). We believe that the choice of the method should be based on the technical ease, throughput, cost and compatibility with the down-stream protocols. While methods such as ethanol or dual-phase methanol/chloroform precipitation are efficient in enriching the urinary protein, the samples remain salty and are not suitable for protein digestion. In our hands, C4 purification was efficient to remove the salts but some other compounds remained in the samples which resulted in high current issues during isoelectric focusing, usually a sign of interference from charged molecules.

High-Abundant Protein Depletion

The dynamic range of protein concentrations in urine spans several orders of magnitude. Albumin, uromodulin (Tamm–Horsfall protein), transferrin, immunoglobulins and α1-antitrypsin constitute the majority of the high abundant proteins in the urinary proteome. Depletion of high abundant proteins may provide access to previously unreachable protein, particularly high potential biomarker proteins with low-abundance (Pieper et al., 2004; Zolotarjova et al., 2005). However, some of these high-abundant proteins may provide invaluable information in some diseases. A repetitive pattern of albumin fragments and some of its specific fragments have been associated with nephrotic diseases (Candiano et al., 2006) and type 2 diabetics (Mischak et al., 2004). Albumin oxidation has also been reported in patients with active focal segmental glomerulosclerosis (Musante et al., 2007). Further discussion concerning arguments for and against the depletion of abundant proteins is discussed in Section 18.2.1.

Recently, a different approach entitled "Proteominer" has been introduced (Boschetti et al., 2007). This method is based on combinatorial library of hexameric peptide ligands bound to polyacrylate beads. Any protein in the complex body fluid mixture can interact with one of the millions of the peptide ligands. This allows the equilibration of proteins concentrations in contrast to removing specific high-abundant species. This technology has been applied to the analysis of urine (Castagna et al., 2005). Castagna et al. reported that the use of the Proteominer in conjunction with nano-LC-MS/MS increased the number of identified proteins from 134 for the non-treated urine to 300 proteins. However, since the natural abundance of proteins is altered with this method, performing quantitative analysis might prove to be challenging.

18.3.2 Pancreatic Fluid

The search for evidence of early chronic pancreatitis eventually leads to an exhaustive lists of invasive (endoscopic) and non-invasive (radiologic) pancreatic imaging studies. If endoscopic and radiologic findings are absent or equivocal, patients are generally referred to a tertiary care center for further testing. Pancreas function testing is considered the non-histological gold standard to diagnose mid-to-late stage chronic pancreatitis (Chowdhury et al., 2005). Routine pancreatic biopsy is not recommended due to potential complications of bleeding and fistulae formation. Chronic inflammation in the pancreas will decrease acinar and duct cell secretory function. This degree of dysfunction can be determined by measurement of specific concentrations of cellular secretory components in pancreatic fluid after hormone stimulation. Thus, pancreatic dysfunction is a "surrogate" marker of chronic pancreatitis (DiMagno et al., 1973). Function testing is advantageous as it may detect abnormalities in secretion before the development of steatorrhea or radiologic abnormalities. Earlier chronic pancreatic disease detection is needed to assess the benefit of treatments that may retard or modify disease progression, and decrease the number of diagnostic tests which will ultimately lower the healthcare costs.

18.3.2.1 Sample Processing To Date

Several mass spectrometry-based proteomic analyses of pancreatic fluid have been reported to date (Chen et al., 2007a,b; Gronborg et al., 2004, 2006). The studies used different approaches in all aspects including: sample collection, method of collection, protein extraction, the use of protease inhibitors, and data analysis. Several of these proteomic investigations of pancreatic fluid have been performed using specimens collected surgically or via ERCP (Chen et al., 2005, 2007a,b; Cui et al., 2008; Gronborg et al., 2004, 2006; Ke et al., 2009; Li et al., 2008; Zhao et al., 2009). Although these specimens were not directly collected for proteomics, as they were considered excess specimens, both methods are highly invasive and associated with significant risks for the patients. In contrast, the ePFT (endoscopic pancreatic function test) collection method, which is much less invasive compared to ERCP and surgery, and permits the safe collection of 10-fold larger volumes of pancreatic fluid, making it a well-suited method for comprehensive proteome analysis (Conwell et al., 2002, 2003a,b; Stevens et al., 2004a,b; Wu and Conwell, 2009).

Once fluid is collected, it must be processed to efficiently extract proteins for analysis. Following a short centrifugation spin to remove any particulates, protein extraction is generally performed on the pancreatic fluid sample. Care must be taken to prevent proteolysis as it is prevalent due to the elevated amounts of pancreatic proteolytic enzymes. Although some publications investigating fluids of pancreatic origin do not specifically mention the protein precipitation method used (Chen et al., 2007a,b; Gronborg et al., 2004, 2006; Wandschneider et al., 2001), other publications have cited acetone precipitation for protein extraction (Ke et al., 2009; Zhou et al., 2007). In addition, there is data reported for and against the use of protease inhibitors in pancreatic fluid studies (Gronborg et al., 2006; Lohr and Faissner, 2004; Thongboonkerd, 2007; Wandschneider et al., 2001). Finally, mass spectrometry-based studies of pancreatic fluids investigating either pancreatic cancer (Chen et al., 2007b; Cui et al., 2008; Ke et al., 2009; Li et al., 2008; Zhao et al., 2009) or chronic pancreatitis have utilized GeLC-MS/MS (Gronborg et al., 2004; Paulo et al., 2010a), SELDI (Rosty and Goggins, 2002), and ICAT (Chen et al., 2007a). Proteins identified varied in number from 22 to over 170 based on the MS methodology utilized.

18.3.2.2 Sample Optimization

In efforts to standardize sample processing methodologies, we initiated a study using SDS-PAGE protein profiling to establish a robust protocol (Fig. 18.1) for pancreatic fluid analysis for future studies (Paulo et al., 2010b). These experiments were designed to maximize the integrity of our pancreatic fluid samples, as protein degradation was expected in this endogenously protease-rich body fluid. We have shown that a sample of pancreatic fluid that is incubated at 37°C shows progressive degradation overtime, most noticeably for time points beyond 1 h (Fig. 18.2). Similarly, we have shown that protein degradation was prevalent even if the sample was maintained at room temperature for longer than 1 h. However, no degradation was evident after 8 h if the samples were maintained on ice. Thus, to limit sample

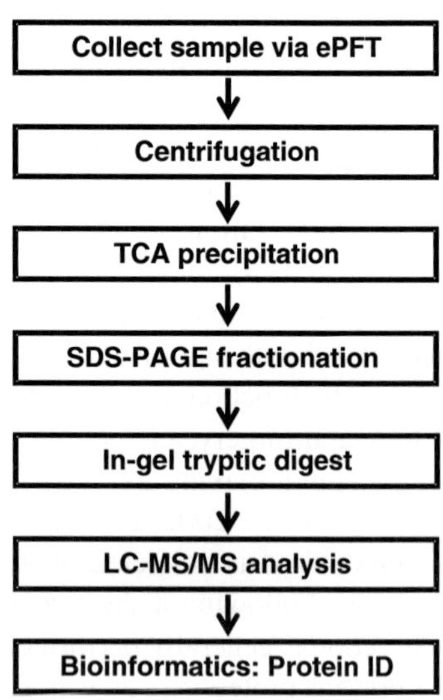

Fig. 18.1 General workflow of an ePFT-mass spectrometry-based proteomics experiment

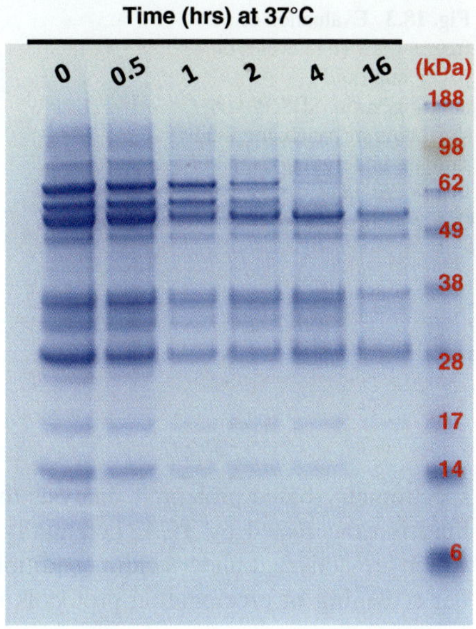

Fig. 18.2 Autodigestion of pancreatic fluid at 36°C over 16 h

degradation, we maintain our pancreatic fluid samples at 4°C and minimize its handling prior to precipitation or storage at –80°C. Such results support the importance of maintaining samples, which are rich in proteolytic enzymes, such as pancreatic fluid, at cold temperatures during preparation for subsequent proteomic analysis.

In addition, we examined the effects of multiple freeze–thaw cycles on the degree of protein degradation. As samples often require repeated analysis, their integrity following freeze–thaw cycles should be examined. To test this for pancreatic fluid, we subjected samples to repeated cycles of freezing at –80°C and thawing on ice for approximately 1 h with intermittent agitation. Following five freeze/thaw cycles, SDS-PAGE protein pattern analysis revealed only very slight differences in the protein bands (Fig. 18.3). Even though we show that pancreatic fluid proteins are resistant to degradation resulting from freeze–thaw cycles, it may be of additional benefit to store samples for extended periods in smaller aliquots to minimize, or essentially eliminate, the need for multiple sample thaws.

Protein extraction from pancreatic fluid was optimized. Both chemical and non-chemical protein extraction techniques were tested for maximal protein extraction with minimal degradation including vacuum centrifugation, 5 kDa molecular weight cut-off ultrafiltration, C4 trapping column, trifluoroacetic acid/acetonitrile precipitation, trichloroacetic acid (TCA) precipitation, ethanol precipitation (Fig. 18.4). Of the methods investigated, TCA precipitation maximized protein extraction. While the cold temperatures slow enzymatic reactions, TCA precipitation most efficiently and rapidly inactivated pancreatic enzymes as a result of the decrease in pH, and thus results in less overall protein degradation. However, this method did not apply to all fluids, including those of the upper gastrointestinal tract. We have shown by mass

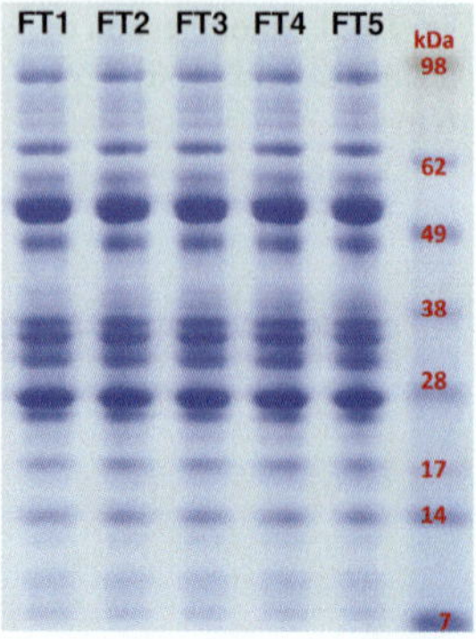

Fig. 18.3 Evaluation of freeze–thaw (FT) cycles on pancreatic fluid auto-digestion. SDS-PAGE separation of pancreatic fluid samples undergoing a total of five freeze–thaw cycles. FT, freeze–thaw cycle

spectrometry-based proteomic analysis that protein degradation in gastroduodenal fluid is exacerbated by TCA, potentially due to the presence of gastric enzymes which are activated under acidic conditions (Paulo et al., 2010c). We suggest similar screening of precipitation protocols for other body fluids to optimize protein extraction for proteomic analysis.

Furthermore, we have shown that the addition of protease inhibitors may be ineffective in the case of pancreatic fluid (at the manufacturer's recommended concentration), as there was no difference in the protein profile with or without their addition, as was depicted in Fig. 18.4. This is possibly the result of the unusually high protease concentration encountered in pancreatic fluid. There are conflicting reports in the literature, both for and against the use of protease inhibitors in pancreatic fluid studies (Gronborg et al., 2006; Lohr and Faissner, 2004;

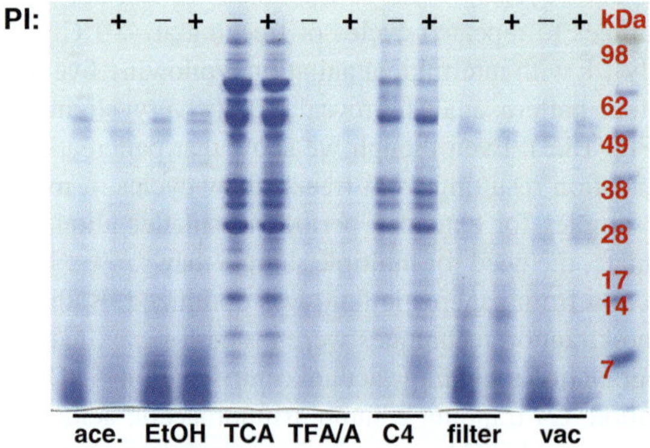

Fig. 18.4 Pancreatic fluid protein extraction methods. SDS-PAGE gel image of pancreatic fluid proteins extracted using the following techniques: *vac*, vacuum centrifugation; *filter*, 5 kDa molecular weight cut-off Centricon filtration; *C4*, C4 trapping column; *TFA/A*, trifluoroacetic acid/acetonitrile precipitation; *TCA*, trichloroacetic acid precipitation; *EtOH*, ethanol precipitation; *ace*, acetone precipitation; *PI*, protease inhibitors

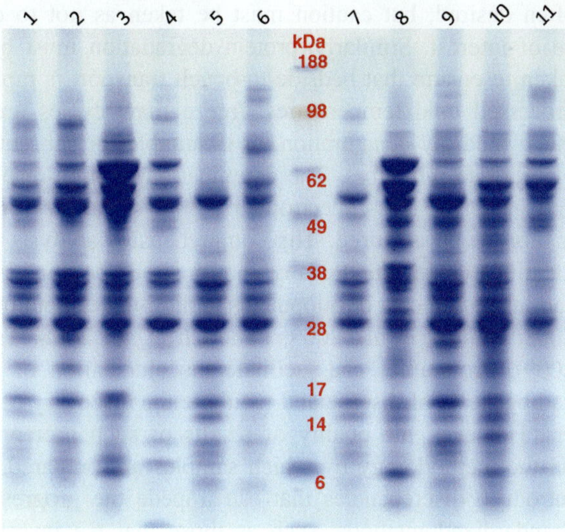

Fig. 18.5 Processing of pancreatic fluid from 11 patients using our optimized methodology. Samples were collected by secretin-stimulated ePFT and processed using our optimized methodology

Thongboonkerd, 2007; Wandschneider et al., 2001). Our results have shown that auto-digestion is minimized whether or not the sample is supplemented with the protease inhibitors, as long as samples are handled on ice and/or are TCA precipitated (Paulo et al., 2010b).

Such experiments (involving precipitation optimization, autodigestion, freeze–thaws, use of protease inhibitors) are vital, not only for pancreatic fluid, but for proteomics of any body fluids which may susceptible to degradation. We can apply our optimized methodology to pancreatic fluid from various individuals and produce robust and relatively reproducible SDS-PAGE protein banding patterns (Fig. 18.5). Additionally, using our sample preparation methodology coupled with GeLC-MS/MS, we have been able to consistently identify over 250 proteins from a single sample of ePFT-collected pancreatic fluid (manuscript in preparation). Our optimized sample preparation methods can be utilized in future studies investigating the proteome of pancreatic fluids, but can also be extended to studies of other body fluids.

18.4 Conclusions

Sample preparation is an indispensible step in the study of body fluid proteomics, as it has a tremendous impact on subsequent analyses. It is crucial to decide on a particular body fluid preparation strategy with the final goal in mind. No single method can be applied to all samples, and it is imperative to optimize procedures for a particular sample type. The depletion of highly abundant proteins is often desired, but caution must be taken as not to deplete proteins which may be of interest. Similarly, protein degradation must be minimized. Care must be taken to ensure that bedside-to-bench transport is rapid and efficient, samples are processed cold temperatures, freeze–thaw cycles are limited and proteases are inactivated, by denaturation, or if suitable for subsequent studies, by the addition of protease inhibitors. The combination of high-resolution separation techniques and powerful mass spectrometric analysis allows previously unattainable information to be acquired. The comprehensive analysis of protein mixtures and the identification of hundreds or thousands of proteins is possible for clinical applications with recent developments in high-throughput mass spectrometry (Han et al., 2008; Kentsis et al., 2009; Latterich et al., 2008; Tsangaris, 2009; Yates et al., 2009). Proteomics can facilitate the elucidation of proteins which regulate the pathogenesis of disease and facilitate the discovery of clinically-relevant biomarkers. However, the quality of results of such proteomic studies depends heavily on the methodology by which samples are prepared. Variations in methods may introduce discrepancies that can impede the progress of body fluid proteomics. Standardized methods, such as those that we investigate herein, can maximize protein extraction and minimize the heterogeneity of samples by reducing protein degradation.

The establishment and optimization of clear and consistent sample collection, handling, and processing methodologies is paramount to the development of clinical proteomics to create a platform upon which such assays can be further developed. Quantitative proteome profiling may be key, as a protein may be present, albeit at varying levels, both in the normal and in the diseased state, and without quantitative information, the value of that protein as a biomarker may be overlooked. In comparative proteomics, sample preparation is of utmost importance as minor differences in experimental and control samples are under investigation. Whereas the genome is relatively stable and identical in all cells, the proteome varies by organ, cell, subcellular location, temporally, and due to stimuli, such as changes in health, diet, and environment. Differences exist not solely in tissue/cell-specific protein content, but also in protein processing, such as posttranslational modifications and splice variants. Biological fluids are attractive for biomarker discovery, owing to several factors including ease of accessibility, avoidance of risky, costly and invasive biopsies, monitoring of the disease state by multiple samplings, and resulting in the development of relatively easily implemented prognostic/diagnostic tests (Good et al., 2007). Consideration must be taken that the samples remain stable during collection, transportation from site of collection to site of analysis, storage, and preanalytical preparation. To this end, one should meticulously document the conditions for sample preparation and handling and, thereby carefully track all preanalytical variables. Thus, although comprehensive proteomic analysis is a valuable tool for biomarker discovery and understanding the pathophysiology of disease, fluid-specific standardized methodology must be established for successful comprehensive proteomic analyses.

Protocol Box

Below are protocols, culled from a variety of sources including Ke et al. (2009), Lee et al. (2008), Thongboonkerd et al. (2006), Zhou et al. (2007), which have been mentioned throughout the chapter. Volumes are appropriate for pancreatic and gastroduodenal fluids (0.5–1.5 mg/mL), but may be adjusted for less protein rich body fluids, such as urine.

Acetone precipitation. Add four sample volumes (800 μL) of ice-cold 100% acetone to 200 μL of sample, vortex briefly, and incubate at –20°C for 3 h. Subsequently, centrifuge the samples at 20,000×*g* at 4°C for 30 min. Carefully aspirate the supernatant and air dry the pellets at 23°C.

Chloroform-methanol precipitation. Solubilize or reduce sample volume to 100 or 200 μL in water (or aqueous buffer). Add 1 mL of a chloroform/methanol (2:1, vol:vol) solvent. Vortex thoroughly and place on ice. Centrifuge at 4°C for 20 min at 14,000 rpm. Carefully extract the lower phase containing the chloroform with glass syringe. Add 500–1000 μL of MeOH to the supernatant and vortex thoroughly. Centrifuge at 4°C for 20 min at 14,000 rpm. Discard the supernatant and dry the pellet in speed-vac.

C4 column. A final concentration of 0.1% trifluoroacetic acid (TFA) and 2% acetonitrile (ACN) was added to 200 μL of pancreatic fluid. Loaded the samples onto a Michrom C4 trapping column which has been pre-equilibrated with 0.1% TFA/2% ACN (Buffer A). Wash the samples twice with 500 μL of Buffer A and elute with 0.1% TFA/90% ACN. Dry the eluted samples at 23°C in a vacuum centrifuge.

Ethanol precipitation. Add a total of 800 μL of 100% ethanol to a 200 μL aliquot of sample. Vortex thoroughly and place on ice for 30 min. Centrifuge at 14,000×*g* at 4°C for 30 min. Carefully aspirate the supernatant and air dry the pellets at 23°C.

Microfiltration. Load 200 μL of sample onto a Millipore 5 kDa Microcon centrifugal filter concentrators and centrifuge at 14,000×*g* at 4°C until the volume is at least one-tenth of the original sample volume. Wash the retained samples twice with 200 μL of water in the centrifugal filter concentrator. Dry the final 20 μL sample at 23°C in a vacuum centrifuge.

TFA/A precipitation. Add two sample volumes (400 μL) of ice-cold 0.1% trifluoroacetic acid (TFA) in 99.9% acetonitrile (ACN) to 200 μL of sample. Vortex thoroughly and place on ice for 30 min. Centrifuge at 14,000×*g* at 4°C for 30 min. Carefully aspirate the supernatant and air dry the pellets at 23°C.

TCA precipitation. Add 25 μL of ice-cold 100% trichloroacetic acid to 200 μL of sample. Vortex thoroughly and incubate at 4°C for 2 h. Centrifuge the sample at 20,000×*g* at 4°C for 30 min and carefully aspirate the supernatant. Add 1 mL of 100% ice-cold acetone to the pellets. Vortex briefly and incubate at –20°C for 1 h. Centrifuge the sample at 20,000×*g* at 4°C for 30 min. Gently wash the pellet twice with 100% ice-cold acetone. Allow the final pellets air dry at 23°C.

Vacuum centrifugation. Dry 200 μL aliquots of sample at 23°C in a vacuum centrifuge (SPD1010 Thermo Savant, Waltham, MA) for approximately 3–4 h.

References

Anderson, N.L., and Anderson, N.G. (2002). The human plasma proteome: History, character, and diagnostic prospects. Mol Cell Proteomics *1*, 845–867.

Anderson, N.L., Polanski, M., Pieper, R., Gatlin, T., Tirumalai, R.S., Conrads, T.P., Veenstra, T.D., Adkins, J.N., Pounds, J.G., Fagan, R., *et al.* (2004). The human plasma proteome: A nonredundant list developed by combination of four separate sources. Mol Cell Proteomics *3*, 311–326.

Barelli, S., Crettaz, D., Thadikkaran, L., Rubin, O., and Tissot, J.D. (2007). Plasma/serum proteomics: Pre-analytical issues. Expert Rev Proteomics *4*, 363–370.

Barratt, J., and Topham, P. (2007). Urine proteomics: The present and future of measuring urinary protein components in disease. CMAJ *177*, 361–368.

Beier, J., Beeh, K.M., Kornmann, O., and Buhl, R. (2004). Induced sputum methodology: Validity and reproducibility of total glutathione measurement in supernatant of healthy and asthmatic individuals. J Lab Clin Med *144*, 38–44.

Bjorhall, K., Miliotis, T., and Davidsson, P. (2005). Comparison of different depletion strategies for improved resolution in proteomic analysis of human serum samples. Proteomics *5*, 307–317.

Boschetti, E., Lomas, L., Citterio, A., and Righetti, P.G. (2007). Romancing the 'hidden proteome', Anno Domini two zero zero seven. J Chromatogr A *1153*, 277–290.

Bottini, P.V., Ribeiro Alves, M.A., and Garlipp, C.R. (2002). Electrophoretic pattern of concentrated urine: Comparison between 24-hour collection and random samples. Am J Kidney Dis *39*, E2.

Bramham, K., Mistry, H.D., Poston, L., Chappell, L.C., and Thompson, A.J. (2009). The non-invasive biopsy--will urinary proteomics make the renal tissue biopsy redundant? QJM *102*, 523–538.

Candiano, G., Musante, L., Bruschi, M., Petretto, A., Santucci, L., Del Boccio, P., Pavone, B., Perfumo, F., Urbani, A., Scolari, F., *et al.* (2006). Repetitive fragmentation products of albumin and alpha1-antitrypsin in glomerular diseases associated with nephrotic syndrome. J Am Soc Nephrol *17*, 3139–3148.

Casado, B., Pannell, L.K., Iadarola, P., and Baraniuk, J.N. (2005). Identification of human nasal mucous proteins using proteomics. Proteomics *5*, 2949–2959.

Castagna, A., Cecconi, D., Sennels, L., Rappsilber, J., Guerrier, L., Fortis, F., Boschetti, E., Lomas, L., and Righetti, P.G. (2005). Exploring the hidden human urinary proteome via ligand library beads. J Proteome Res *4*, 1917–1930.

Cavaletto, M., Giuffrida, M.G., and Conti, A. (2004). The proteomic approach to analysis of human milk fat globule membrane. Clin Chim Acta *347*, 41–48.

Chen, R., Brentnall, T.A., Pan, S., Cooke, K., Moyes, K.W., Lane, Z., Crispin, D.A., Goodlett, D.R., Aebersold, R., and Bronner, M.P. (2007a). Quantitative proteomics analysis reveals that proteins differentially expressed in chronic pancreatitis are also frequently involved in pancreatic cancer. Mol Cell Proteomics *6*, 1331–1342.

Chen, R., Pan, S., Cooke, K., Moyes, K.W., Bronner, M.P., Goodlett, D.R., Aebersold, R., and Brentnall, T.A. (2007b). Comparison of pancreas juice proteins from cancer versus pancreatitis using quantitative proteomic analysis. Pancreas *34*, 70–79.

Chen, R., Yi, E.C., Donohoe, S., Pan, S., Eng, J., Cooke, K., Crispin, D.A., Lane, Z., Goodlett, D.R., Bronner, M.P., *et al.* (2005). Pancreatic cancer proteome: the proteins that underlie invasion, metastasis, and immunologic escape. Gastroenterology *129*, 1187–1197.

Chowdhury, R., Bhutani, M.S., Mishra, G., Toskes, P.P., and Forsmark, C.E. (2005). Comparative analysis of direct pancreatic function testing versus morphological assessment by endoscopic ultrasonography for the evaluation of chronic unexplained abdominal pain of presumed pancreatic origin. Pancreas *31*, 63–68.

Conwell, D.L., Zuccaro, G., Morrow, J.B., Van Lente, F., O'Laughlin, C., Vargo, J.J., and Dumot, J.A. (2002). Analysis of duodenal drainage fluid after cholecystokinin (CCK) stimulation in healthy volunteers. Pancreas *25*, 350–354.

Conwell, D.L., Zuccaro, G., Jr., Vargo, J.J., Morrow, J.B., Obuchowski, N., Dumot, J.A., Trolli, P.A., Burton, A., O'Laughlin, C., and Van Lente, F. (2003a). An endoscopic pancreatic function test with cholecystokinin-octapeptide for the diagnosis of chronic pancreatitis. Clin Gastroenterol Hepatol *1*, 189–194.

Conwell, D.L., Zuccaro, G., Jr., Vargo, J.J., Trolli, P.A., Vanlente, F., Obuchowski, N., Dumot, J.A., and O'Laughlin, C. (2003b). An endoscopic pancreatic function test with synthetic porcine secretin for the evaluation of chronic abdominal pain and suspected chronic pancreatitis. Gastrointest Endosc *57*, 37–40.

Crnogorac-Jurcevic, T., Gangeswaran, R., Bhakta, V., Capurso, G., Lattimore, S., Akada, M., Sunamura, M., Prime, W., Campbell, F., Brentnall, T.A., *et al.* (2005). Proteomic analysis of chronic pancreatitis and pancreatic adenocarcinoma. Gastroenterology *129*, 1454–1463.

Cui, Y., Tian, M., Zong, M., Teng, M., Chen, Y., Lu, J., Jiang, J., Liu, X., and Han, J. (2008). Proteomic analysis of pancreatic ductal adenocarcinoma compared with normal adjacent pancreatic tissue and pancreatic benign cystadenoma. Pancreatology *9*, 89–98.

D'Auria, E., Agostoni, C., Giovannini, M., Riva, E., Zetterstrom, R., Fortin, R., Greppi, G.F., Bonizzi, L., and Roncada, P. (2005). Proteomic evaluation of milk from different mammalian species as a substitute for breast milk. Acta Paediatr *94*, 1708–1713.

De Ceuninck, F., and Berenbaum, F. (2009). Proteomics: Addressing the challenges of osteoarthritis. Drug Discov Today *14*, 661–667.

Decramer, S., Gonzalez de Peredo, A., Breuil, B., Mischak, H., Monsarrat, B., Bascands, J.L., and Schanstra, J.P. (2008). Urine in clinical proteomics. Mol Cell Proteomics *7*, 1850–1862.

Delmotte, N., Lasaosa, M., Tholey, A., Heinzle, E., and Huber, C.G. (2007). Two-dimensional reversed-phase x ion-pair reversed phase HPLC: An alternative approach to high-resolution peptide separation for shotgun proteome analysis. J Proteome Res *6*, 4363–4373.

Di Quinzio, M.K., Oliva, K., Holdsworth, S.J., Ayhan, M., Walker, S.P., Rice, G.E., Georgiou, H.M., and Permezel, M. (2007). Proteomic analysis and characterisation of human cervico-vaginal fluid proteins. Aust N Z J Obstet Gynaecol *47*, 9–15.

DiMagno, E.P., Go, V.L., and Summerskill, W.H. (1973). Relations between pancreatic enzyme outputs and malabsorption in severe pancreatic insufficiency. N Engl J Med *288*, 813–815.

Dwivedi, R.C., Spicer, V., Harder, M., Antonovici, M., Ens, W., Standing, K.G., Wilkins, J.A., and Krokhin, O.V. (2008). Practical implementation of 2D HPLC scheme with accurate peptide retention prediction in both dimensions for high-throughput bottom-up proteomics. Anal Chem *80*, 7036–7042.

Echan, L.A., Tang, H.Y., Ali-Khan, N., Lee, K., and Speicher, D.W. (2005). Depletion of multiple high-abundance proteins improves protein profiling capacities of human serum and plasma. Proteomics *5*, 3292–3303.

Elschenbroich, S., Ignatchenko, V., Sharma, P., Schmitt-Ulms, G., Gramolini, A.O., and Kislinger, T. (2009). Peptide separations by on-line MudPIT compared to isoelectric focusing in an off-gel format: Application to a membrane-enriched fraction from C2C12 mouse skeletal muscle cells. J Proteome Res *8*, 4860–4869.

Fiedler, G.M., Baumann, S., Leichtle, A., Oltmann, A., Kase, J., Thiery, J., and Ceglarek, U. (2007). Standardized peptidome profiling of human urine by magnetic bead separation and matrix-assisted laser desorption/ionization time-of-flight mass spectrometry. Clin Chem *53*, 421–428.

Finnie, C., and Svensson, B. (2002). Proteolysis during the isoelectric focusing step of two-dimensional gel electrophoresis may be a common problem. Anal Biochem *311*, 182–186.

Fliser, D., Novak, J., Thongboonkerd, V., Argiles, A., Jankowski, V., Girolami, M.A., Jankowski, J., and Mischak, H. (2007). Advances in urinary proteome analysis and biomarker discovery. J Am Soc Nephrol *18*, 1057–1071.

Fountoulakis, M., Juranville, J.F., Jiang, L., Avila, D., Roder, D., Jakob, P., Berndt, P., Evers, S., and Langen, H. (2004). Depletion of the high-abundance plasma proteins. Amino Acids *27*, 249–259.

Gerber, S.A., Kettenbach, A.N., Rush, J., and Gygi, S.P. (2007). The absolute quantification strategy: Application to phosphorylation profiling of human separase serine 1126. Methods Mol Biol *359*, 71–86.

Gerber, S.A., Rush, J., Stemman, O., Kirschner, M.W., and Gygi, S.P. (2003). Absolute quantification of proteins and phosphoproteins from cell lysates by tandem MS. Proc Natl Acad Sci USA *100*, 6940–6945.

Gerszten, R.E., and Wang, T.J. (2008). The search for new cardiovascular biomarkers. Nature *451*, 949–952.

Gonzales, P.A., Pisitkun, T., Hoffert, J.D., Tchapyjnikov, D., Star, R.A., Kleta, R., Wang, N.S., and Knepper, M.A. (2009). Large-scale proteomics and phosphoproteomics of urinary exosomes. JASN *20*, 363–379.

Gonzalez-Buitrago, J.M., Ferreira, L., and Lorenzo, I. (2007). Urinary proteomics. Clin Chim Acta *375*, 49–56.

Good, D.M., Thongboonkerd, V., Novak, J., Bascands, J.L., Schanstra, J.P., Coon, J.J., Dominiczak, A., and Mischak, H. (2007). Body fluid proteomics for biomarker discovery: Lessons from the past hold the key to success in the future. J Proteome Res *6*, 4549–4555.

Gozal, D., Jortani, S., Snow, A.B., Kheirandish-Gozal, L., Bhattacharjee, R., Kim, J., and Capdevila, O.S. (2009). Two-dimensional differential in-gel electrophoresis proteomic approaches reveal urine candidate biomarkers in pediatric obstructive sleep apnea. Am J Respir Crit Care Med *180*, 1253–1261.

Gronborg, M., Bunkenborg, J., Kristiansen, T.Z., Jensen, O.N., Yeo, C.J., Hruban, R.H., Maitra, A., Goggins, M.G., and Pandey, A. (2004). Comprehensive proteomic analysis of human pancreatic juice. J Proteome Res *3*, 1042–1055.

Gronborg, M., Kristiansen, T.Z., Iwahori, A., Chang, R., Reddy, R., Sato, N., Molina, H., Jensen, O.N., Hruban, R.H., Goggins, M.G., *et al.* (2006). Biomarker discovery from pancreatic cancer secretome using a differential proteomic approach. Mol Cell Proteomics *5*, 157–171.

Grote, T., and Logsdon, C.D. (2007). Progress on molecular markers of pancreatic cancer. Curr Opin Gastroenterol *23*, 508–514.

Haigh, B.J., Stewart, K.W., Whelan, J.R., Barnett, M.P., Smolenski, G.A., and Wheeler, T.T. (2010). Alterations in the salivary proteome associated with periodontitis. J Clin Periodontol *37*, 241–247.

Haltiwanger, R.S., and Lowe, J.B. (2004). Role of glycosylation in development. Annu Rev Biochem *73*, 491–537.

Han, X., Aslanian, A., and Yates, J.R., 3rd (2008). Mass spectrometry for proteomics. Curr Opin Chem Biol *12*, 483–490.

Hauskrecht, M., Pelikan, R., Malehorn, D.E., Bigbee, W.L., Lotze, M.T., Zeh, H.J., Whitcomb, D.C., and Lyons-Weiler, J. (2005). Feature selection for classification of SELDI-TOF-MS proteomic profiles. Appl Bioinformatics *4*, 227–246.

Heide, K., Haupt, H., and Schwick, H.G. (1997). Plasma Protein Fractionation, Vol. 3 (New York, NY, Academic).

Hortin, G.L., Jortani, S.A., Ritchie, J.C., Jr., Valdes, R., Jr., and Chan, D.W. (2006). Proteomics: A new diagnostic frontier. Clin Chem *52*, 1218–1222.

Hortin, G.L., and Sviridov, D. (2007). Diagnostic potential for urinary proteomics. Pharmacogenomics *8*, 237–255.

Hsieh, S.Y., Chen, R.K., Pan, Y.H., and Lee, H.L. (2006). Systematical evaluation of the effects of sample collection procedures on low-molecular-weight serum/plasma proteome profiling. Proteomics *6*, 3189–3198.

Hu, S., Xie, Y., Ramachandran, P., Ogorzalek Loo, R.R., Li, Y., Loo, J.A., and Wong, D.T. (2005). Large-scale identification of proteins in human salivary proteome by liquid chromatography/mass spectrometry and two-dimensional gel electrophoresis-mass spectrometry. Proteomics *5*, 1714–1728.

Huang, C.M. (2004). Comparative proteomic analysis of human whole saliva. Arch Oral Biol *49*, 951–962.

Huang, H.L., Stasyk, T., Morandell, S., Mogg, M., Schreiber, M., Feuerstein, I., Huck, C.W., Stecher, G., Bonn, G.K., and Huber, L.A. (2005). Enrichment of low-abundant serum proteins by albumin/immunoglobulin G immunoaffinity depletion under partly denaturing conditions. Electrophoresis *26*, 2843–2849.

Issaq, H.J., Xiao, Z., and Veenstra, T.D. (2007). Serum and plasma proteomics. Chem Rev *107*, 3601–3620.

Ivanov, V.T., and Yatskin, O.N. (2005). Peptidomics: A logical sequel to proteomics. Expert Rev Proteomics *2*, 463–473.

Jiang, L., He, L., and Fountoulakis, M. (2004). Comparison of protein precipitation methods for sample preparation prior to proteomic analysis. J Chromatogr A *1023*, 317–320.

Ke, E., Patel, B.B., Liu, T., Li, X.M., Haluszka, O., Hoffman, J.P., Ehya, H., Young, N.A., Watson, J.C., Weinberg, D.S., *et al.* (2009). Proteomic analyses of pancreatic cyst fluids. Pancreas *38*, 33–42.

Kentsis, A., Monigatti, F., Dorff, K., Campagne, F., Bachur, R., and Steen, H. (2009). Urine proteomics for profiling of human disease using high accuracy mass spectrometry. Proteomics Clin Appl *3*, 1052–1061.

Kirkpatrick, D.S., Gerber, S.A., and Gygi, S.P. (2005). The absolute quantification strategy: A general procedure for the quantification of proteins and post-translational modifications. Methods *35*, 265–273.

Kojima, T., Andersen, E., Sanchez, J.C., Wilkins, M.R., Hochstrasser, D.F., Pralong, W.F., and Cimasoni, G. (2000). Human gingival crevicular fluid contains MRP8 (S100A8) and MRP14 (S100A9), two calcium-binding proteins of the S100 family. J Dent Res *79*, 740–747.

Koopmann, J., Fedarko, N.S., Jain, A., Maitra, A., Iacobuzio-Donahue, C., Rahman, A., Hruban, R.H., Yeo, C.J., and Goggins, M. (2004). Evaluation of osteopontin as biomarker for pancreatic adenocarcinoma. Cancer Epidemiol Biomarkers Prev *13*, 487–491.

Koyama, R., Nakanishi, T., Ikeda, T., and Shimizu, A. (2003). Catalogue of soluble proteins in human vitreous humor by one-dimensional sodium dodecyl sulfate-polyacrylamide gel electrophoresis and electrospray ionization mass spectrometry including seven angiogenesis-regulating factors. J Chromatogr B Anal Technol Biomed Life Sci *792*, 5–21.

Latterich, M., Abramovitz, M., and Leyland-Jones, B. (2008). Proteomics: New technologies and clinical applications. Eur J Cancer *44*, 2737–2741.

Lee, R.S., Monigatti, F., Briscoe, A.C., Waldon, Z., Freeman, M.R., and Steen, H. (2008). Optimizing sample handling for urinary proteomics. J Proteome Res *7*, 4022–4030.

Li, C., Simeone, D.M., Brenner, D.E., Anderson, M.A., Shedden, K.A., Ruffin, M.T., and Lubman, D.M. (2008). Pancreatic cancer serum detection using a lectin/glyco-antibody array method. J Proteome Res *8*, 483–492.

Lifshitz, E., and Kramer, L. (2000). Outpatient urine culture: does collection technique matter? Arch Intern Med *160*, 2537–2540.

Lohr, M., and Faissner, R. (2004). Proteomics in pancreatic disease. Pancreatology *4*, 67–75.

Lowenthal, M.S., Mehta, A.I., Frogale, K., Bandle, R.W., Araujo, R.P., Hood, B.L., Veenstra, T.D., Conrads, T.P., Goldsmith, P., Fishman, D., *et al.* (2005). Analysis of albumin-associated peptides and proteins from ovarian cancer patients. Clin Chem *51*, 1933–1945.

Lundblad, R.L. (2004). Considerations for the use of blood plasma and serum for proteomic analysis. Internet J Genomics Proteomics *1*.

Lundgren, D.H., Hwang, S.I., Wu, L., and Han, D.K. (2010) Role of spectral counting in quantitative proteomics. Expert Rev Proteomics *7*, 39–53.

Luque-Garcia, J.L., and Neubert, T.A. (2007). Sample preparation for serum/plasma profiling and biomarker identification by mass spectrometry. J Chromatogr A *1153*, 259–276.

Macarthur, D.J., and Jacques, N.A. (2003). Proteome analysis of oral pathogens. J Dent Res *82*, 870–876.

Manabe, T., Miyamoto, H., Inoue, K., Nakatsu, M., and Arai, M. (1999). Separation of human cerebrospinal fluid proteins by capillary isoelectric focusing in the absence of denaturing agents. Electrophoresis *20*, 3677–3683.

Manadas, B., English, J.A., Wynne, K.J., Cotter, D.R., and Dunn, M.J. (2009). Comparative analysis of OFFGel, strong cation exchange with pH gradient, and RP at high pH for first-dimensional separation of peptides from a membrane-enriched protein fraction. Proteomics *9*, 5194–5198.

Marshall, J., Kupchak, P., Zhu, W., Yantha, J., Vrees, T., Furesz, S., Jacks, K., Smith, C., Kireeva, I., Zhang, R., *et al.* (2003). Processing of serum proteins underlies the mass spectral fingerprinting of myocardial infarction. J Proteome Res *2*, 361–372.

McDonald, W.H., Ohi, R., Miyamoto, D.T., Mitchison, T.J., and Yates, J.R. (2002). Comparison of three directly coupled HPLC MS/MS strategies for identification of proteins from complex mixtures: Single-dimension LC-MS/MS, 2-phase MudPIT, and 3-phase MudPIT. Int J Mass Spectrom *219*, 245–251.

Mischak, H., Kaiser, T., Walden, M., Hillmann, M., Wittke, S., Herrmann, A., Knueppel, S., Haller, H., and Fliser, D. (2004). Proteomic analysis for the assessment of diabetic renal damage in humans. Clin Sci (Lond) *107*, 485–495.

Mischak, H., Rolf, A., Banks, E., Mark, C., Joshua, C., Anna, D., Jochen, H.H.E., Danilo, F., Mark, G., Henning, H., *et al.* (2007). Clinical proteomics: a need to define the field and to begin to set adequate standards. Proteomics Clin Appl *1*, 148–156.

Motoyama, A., and Yates, J.R., 3rd (2008). Multidimensional LC separations in shotgun proteomics. Anal Chem *80*, 7187–7193.

Mujumdar, R.B., Ernst, L.A., Mujumdar, S.R., and Waggoner, A.S. (1989). Cyanine dye labeling reagents containing isothiocyanate groups. Cytometry *10*, 11–19.

Muller, H., and Brenner, H. (2006). Urine markers as possible tools for prostate cancer screening: Review of performance characteristics and practicality. Clin Chem *52*, 562–573.

Munro, N.P., Cairns, D.A., Clarke, P., Rogers, M., Stanley, A.J., Barrett, J.H., Harnden, P., Thompson, D., Eardley, I., Banks, R.E., *et al.* (2006). Urinary biomarker profiling in transitional cell carcinoma. Int J Cancer *119*, 2642–2650.

Musante, L., Candiano, G., Petretto, A., Bruschi, M., Dimasi, N., Caridi, G., Pavone, B., Del Boccio, P., Galliano, M., Urbani, A., *et al.* (2007). Active focal segmental glomerulosclerosis is associated with massive oxidation of plasma albumin. J Am Soc Nephrol *18*, 799–810.

Nguyen, C.Q., and Peck, A.B. (2009). Unraveling the pathophysiology of Sjogren syndrome-associated dry eye disease. Ocul Surf *7*, 11–27.

Noel-Georis, I., Bernard, A., Falmagne, P., and Wattiez, R. (2001). Proteomics as the tool to search for lung disease markers in bronchoalveolar lavage. Dis Markers *17*, 271–284.

Nurkka, A., Obiero, J., Kayhty, H., and Scott, J.A. (2003). Effects of sample collection and storage methods on antipneumococcal immunoglobulin A in saliva. Clin Diagn Lab Immunol *10*, 357–361.

Omenn, G.S., States, D.J., Adamski, M., Blackwell, T.W., Menon, R., Hermjakob, H., Apweiler, R., Haab, B.B., Simpson, R.J., Eddes, J.S., *et al.* (2005). Overview of the HUPO Plasma Proteome Project: Results from the pilot phase with 35 collaborating laboratories and multiple analytical groups, generating a core dataset of 3020 proteins and a publicly-available database. Proteomics *5*, 3226–3245.

Ortsater, H., Sundsten, T., Lin, J.M., and Bergsten, P. (2007). Evaluation of the SELDI-TOF MS technique for protein profiling of pancreatic islets exposed to glucose and oleate. Proteomics *7*, 3105–3115.

Paulo, J.A., Lee, L.S., Wu, B., Repas, K., Mortele, K.J., Banks, P.A., Steen, H., and Conwell, D.L. (2010a). Identification of pancreas-specific proteins in endoscopically (endoscopic pancreatic function test) collected pancreatic fluid with liquid chromatography-tandem mass spectrometry. Pancreas *39*, 889–896.

Paulo, J.A., Lee, L.S., Wu, B., Repas, K., Banks, P.A., Conwell, D.L., and Steen, H. (2010b). Optimized sample preparation of endoscopic collected pancreatic fluid for SDS-PAGE analysis. Electrophoresis *31*, 2377–2387.

Paulo, J.A., Lee, L.S., Wu, B., Repas, K., Banks, P.A., Conwell, D.L., and Steen, H. (2010c). Proteomic analysis of endoscopically (endoscopic pancreatic function test) collected gastroduodenal fluid using in-gel tryptic digestion followed by LC-MS/MS. Proteomics Clin Appl *4*, 715–725.

Paulo, J.A., Lee, L.S., Wu, B., Banks, P.A., Steen, H., and Conwell, D.L. (2011). Mass spectrometry-based proteomics of endoscopically collected pancreatic fluid in chronic pancreatitis research. Proteomics Clin Appl *5*, 109–120.

Pendyala, G., Trauger, S.A., Kalisiak, E., Ellis, R.J., Siuzdak, G., and Fox, H.S. (2009). Cerebrospinal fluid proteomics reveals potential pathogenic changes in the brains of SIV-infected monkeys. J Proteome Res *8*, 2253–2260.

Peng, J., Elias, J.E., Thoreen, C.C., Licklider, L.J., and Gygi, S.P. (2003). Evaluation of multidimensional chromatography coupled with tandem mass spectrometry (LC/LC-MS/MS) for large-scale protein analysis: The yeast proteome. J Proteome Res *2*, 43–50.

Pieper, R., Gatlin, C.L., McGrath, A.M., Makusky, A.J., Mondal, M., Seonarain, M., Field, E., Schatz, C.R., Estock, M.A., Ahmed, N., *et al.* (2004). Characterization of the human urinary proteome: A method for high-resolution display of urinary proteins on two-dimensional electrophoresis gels with a yield of nearly 1400 distinct protein spots. Proteomics *4*, 1159–1174.

Pieper, R., Su, Q., Gatlin, C.L., Huang, S.T., Anderson, N.L., and Steiner, S. (2003). Multi-component immunoaffinity subtraction chromatography: An innovative step towards a comprehensive survey of the human plasma proteome. Proteomics *3*, 422–432.

Pisitkun, T., Shen, R.F., and Knepper, M.A. (2004). Identification and proteomic profiling of exosomes in human urine. Proc Natl Acad Sci USA *101*, 13368–13373.

Rai, A.J., Gelfand, C.A., Haywood, B.C., Warunek, D.J., Yi, J., Schuchard, M.D., Mehigh, R.J., Cockrill, S.L., Scott, G.B., Tammen, H., *et al.* (2005). HUPO Plasma Proteome Project specimen collection and handling: Towards the standardization of parameters for plasma proteome samples. Proteomics *5*, 3262–3277.

Rai, A.J., and Vitzthum, F. (2006). Effects of preanalytical variables on peptide and protein measurements in human serum and plasma: Implications for clinical proteomics. Expert Rev Proteomics *3*, 409–426.

Rifai, N., Gillette, M.A., and Carr, S.A. (2006). Protein biomarker discovery and validation: The long and uncertain path to clinical utility. Nat Biotechnol *24*, 971–983.

Ross, P.L., Huang, Y.N., Marchese, J.N., Williamson, B., Parker, K., Hattan, S., Khainovski, N., Pillai, S., Dey, S., Daniels, S., *et al.* (2004). Multiplexed protein quantitation in Saccharomyces cerevisiae using amine-reactive isobaric tagging reagents. Mol Cell Proteomics *3*, 1154–1169.

Rosty, C., and Goggins, M. (2002). Early detection of pancreatic carcinoma. Hematol Oncol Clin North Am *16*, 37–52.

Saetun, P., Semangoen, T., and Thongboonkerd, V. (2009). Characterizations of urinary sediments precipitated after freezing and their effects on urinary protein and chemical analyses. Am J Physiol Renal Physiol *296*, F1346–1354.

Scarlett, C.J., Samra, J.S., Xue, A., Baxter, R.C., and Smith, R.C. (2007). Classification of pancreatic cystic lesions using SELDI-TOF mass spectrometry. ANZ J Surg *77*, 648–653.

Schaub, S., Wilkins, J., Weiler, T., Sangster, K., Rush, D., and Nickerson, P. (2004). Urine protein profiling with surface-enhanced laser-desorption/ionization time-of-flight mass spectrometry. Kidney Int *65*, 323–332.

Schulz-Knappe, P., Schrader, M., and Zucht, H.D. (2005). The peptidomics concept. Comb Chem High Throughput Screen *8*, 697–704.

Seike, M., Kondo, T., Fujii, K., Yamada, T., Gemma, A., Kudoh, S., and Hirohashi, S. (2004). Proteomic signature of human cancer cells. Proteomics *4*, 2776–2788.

Shaw, J.L., Smith, C.R., and Diamandis, E.P. (2007). Proteomic analysis of human cervico-vaginal fluid. J Proteome Res *6*, 2859–2865.

Shen, Y., Kim, J., Strittmatter, E.F., Jacobs, J.M., Camp, D.G., 2nd, Fang, R., Tolie, N., Moore, R.J., and Smith, R.D. (2005). Characterization of the human blood plasma proteome. Proteomics *5*, 4034–4045.

Shevchenko, A., Tomas, H., Havlis, J., Olsen, J.V., and Mann, M. (2006). In-gel digestion for mass spectrometric characterization of proteins and proteomes. Nat Protoc *1*, 2856–2860.

Shevchenko, A., Wilm, M., Vorm, O., and Mann, M. (1996). Mass spectrometric sequencing of proteins silver-stained polyacrylamide gels. Anal Chem *68*, 850–858.

Simpson, R.J., Lim, J.W., Moritz, R.L., and Mathivanan, S. (2009). Exosomes: Proteomic insights and diagnostic potential. Expert Rev Proteomics *6*, 267–283.

Stark, M., Jornvall, H., and Johansson, J. (1999). Isolation and characterization of hydrophobic polypeptides in human bile. Eur J Biochem *266*, 209–214.

Stevens, T., Conwell, D., Zuccaro, G., Van Lente, F., Khandwala, F., Hanaway, P., Vargo, J.J., and Dumot, J.A. (2004a). Analysis of pancreatic elastase-1 concentrations in duodenal aspirates from healthy subjects and patients with chronic pancreatitis. Dig Dis Sci *49*, 1405–1411.

Stevens, T., Conwell, D.L., Zuccaro, G., Van Lente, F., Khandwala, F., Purich, E., Vargo, J.J., Fein, S., Dumot, J.A., Trolli, P., *et al.* (2004b). Electrolyte composition of endoscopically collected duodenal drainage fluid after synthetic porcine secretin stimulation in healthy subjects. Gastrointest Endosc *60*, 351–355.

Stewart, II, Thomson, T., and Figeys, D. (2001). 18O labeling: A tool for proteomics. Rapid Commun Mass Spectrom *15*, 2456–2465.

Tang, L.J., De Seta, F., Odreman, F., Venge, P., Piva, C., Guaschino, S., and Garcia, R.C. (2007). Proteomic analysis of human cervical-vaginal fluids. J Proteome Res *6*, 2874–2883.

Thompson, A., Schafer, J., Kuhn, K., Kienle, S., Schwarz, J., Schmidt, G., Neumann, T., Johnstone, R., Mohammed, A.K., and Hamon, C. (2003). Tandem mass tags: A novel quantification strategy for comparative analysis of complex protein mixtures by MS/MS. Anal Chem *75*, 1895–1904.

Thongboonkerd, V. (2007). Proteomics of Human Body Fluids: Principles, Methods, and Applications (Totowa, NJ, Humana Press).

Thongboonkerd, V. (2008). Urinary proteomics: Towards biomarker discovery, diagnostics and prognostics. Mol Biosyst *4*, 810–815.

Thongboonkerd, V., Chutipongtanate, S., and Kanlaya, R. (2006). Systematic evaluation of sample preparation methods for gel-based human urinary proteomics: Quantity, quality, and variability. J Proteome Res *5*, 183–191.

Thongboonkerd, V., McLeish, K.R., Arthur, J.M., and Klein, J.B. (2002). Proteomic analysis of normal human urinary proteins isolated by acetone precipitation or ultracentrifugation. Kidney Int *62*, 1461–1469.

Thongboonkerd, V., Mungdee, S., and Chiangjong, W. (2009). Should urine pH be adjusted prior to gel-based proteome analysis? J Proteome Res *8*, 3206–3211.

Thongboonkerd, V., and Saetun, P. (2007). Bacterial overgrowth affects urinary proteome analysis: Recommendation for centrifugation, temperature, duration, and the use of preservatives during sample collection. J Proteome Res *6*, 4173–4181.

Thouvenot, E., Urbach, S., Dantec, C., Poncet, J., Seveno, M., Demettre, E., Jouin, P., Touchon, J., Bockaert, J., and Marin, P. (2008). Enhanced detection of CNS cell secretome in plasma protein-depleted cerebrospinal fluid. J Proteome Res *7*, 4409–4421.

Thresher, W.C., and Swaisgood, H.E. (1990). Characterization of specific interactions of coenzymes, regulatory nucleotides and cibacron blue with nucleotide binding domains of enzymes by analytical affinity chromatography. J Mol Recognit *3*, 220–228.

Tian, M., Cui, Y.Z., Song, G.H., Zong, M.J., Zhou, X.Y., Chen, Y., and Han, J.X. (2008). Proteomic analysis identifies MMP-9, DJ-1 and A1BG as overexpressed proteins in pancreatic juice from pancreatic ductal adenocarcinoma patients. BMC Cancer *8*, 241.

Tomosugi, N., Kitagawa, K., Takahashi, N., Sugai, S., and Ishikawa, I. (2005). Diagnostic potential of tear proteomic patterns in Sjogren's syndrome. J Proteome Res *4*, 820–825.

Tsangaris, G.T. (2009). From proteomics research to clinical practice. Expert Rev Proteomics *6*, 235–238.

Tyan, Y.C., Wu, H.Y., Lai, W.W., Su, W.C., and Liao, P.C. (2005a). Proteomic profiling of human pleural effusion using two-dimensional nano liquid chromatography tandem mass spectrometry. J Proteome Res *4*, 1274–1286.

Tyan, Y.C., Wu, H.Y., Su, W.C., Chen, P.W., and Liao, P.C. (2005b). Proteomic analysis of human pleural effusion. Proteomics *5*, 1062–1074.

Unlu, M., Morgan, M.E., and Minden, J.S. (1997). Difference gel electrophoresis: A single gel method for detecting changes in protein extracts. Electrophoresis *18*, 2071–2077.

Vaezzadeh, A.R., Steen, H., Freeman, M.R., and Lee, R.S. (2009). Proteomics and opportunities for clinical translation in urological disease. J Urol *182*, 835–843.

Vaillancourt, S., McGillivray, D., Zhang, X., and Kramer, M.S. (2007). To clean or not to clean: Effect on contamination rates in midstream urine collections in toilet-trained children. Pediatrics *119*, e1288–e1293.

Verma, M., Wright, G.L., Jr., Hanash, S.M., Gopal-Srivastava, R., and Srivastava, S. (2001). Proteomic approaches within the NCI early detection research network for the discovery and identification of cancer biomarkers. Ann N Y Acad Sci *945*, 103–115.

Walsh, N., O'Donovan, N., Kennedy, S., Henry, M., Meleady, P., Clynes, M., and Dowling, P. (2009). Identification of pancreatic cancer invasion-related proteins by proteomic analysis. Proteome Sci *7*, 3.

Wandschneider, S., Fehring, V., Jacobs-Emeis, S., Thiesen, H.J., and Lohr, M. (2001). Autoimmune pancreatic disease: preparation of pancreatic juice for proteome analysis. Electrophoresis *22*, 4383–4390.

Wang, L., Li, F., Sun, W., Wu, S., Wang, X., Zhang, L., Zheng, D., Wang, J., and Gao, Y. (2006). Concanavalin A captured glycoproteins in healthy human urine. Mol Cell Proteomics *5*, 560–562.

Washburn, M.P., Wolters, D., and Yates, J.R., 3rd (2001). Large-scale analysis of the yeast proteome by multidimensional protein identification technology. Nat Biotechnol *19*, 242–247.

Wattiez, R., and Falmagne, P. (2005). Proteomics of bronchoalveolar lavage fluid. J Chromatogr B Anal Technol Biomed Life Sci *815*, 169–178.

Welch, A.A., Mulligan, A., Bingham, S.A., and Khaw, K.T. (2008). Urine pH is an indicator of dietary acid-base load, fruit and vegetables and meat intakes: Results from the European Prospective Investigation into Cancer and Nutrition (EPIC)-Norfolk population study. Br J Nutr *99*, 1335–1343.

Wenner, B.R., Lovell, M.A., and Lynn, B.C. (2004). Proteomic analysis of human ventricular cerebrospinal fluid from neurologically normal, elderly subjects using two-dimensional LC-MS/MS. J Proteome Res *3*, 97–103.

Wilson, R., Whitelock, J.M., and Bateman, J.F. (2009). Proteomics makes progress in cartilage and arthritis research. Matrix Biol *28*, 121–128.

Wu, B., and Conwell, D.L. (2009). The endoscopic pancreatic function test. Am J Gastroenterol *104*, 2381–2383.

Wu, J.Y., Yi, C., Chung, H.R., Wang, D.J., Chang, W.C., Lee, S.Y., Lin, C.T., Yang, Y.C., and Yang, W.C. (2010). Potential biomarkers in saliva for oral squamous cell carcinoma. Oral Oncol *46*, 226–231.

Xiao, F., Chen, D., Lu, Y., Xiao, Z., Guan, L.F., Yuan, J., Wang, L., Xi, Z.Q., and Wang, X.F. (2009). Proteomic analysis of cerebrospinal fluid from patients with idiopathic temporal lobe epilepsy. Brain Res *1255*, 180–189.

Yamamoto, T., Langham, R.G., Ronco, P., Knepper, M.A., and Thongboonkerd, V. (2008). Towards standard protocols and guidelines for urine proteomics: A report on the Human Kidney and Urine Proteome Project (HKUPP) symposium and workshop, 6 October 2007, Seoul, Korea and 1 November 2007, San Francisco, CA, USA. Proteomics *8*, 2156–2159.

Yates, J.R., Ruse, C.I., and Nakorchevsky, A. (2009). Proteomics by mass spectrometry: Approaches, advances, and applications. Annu Rev Biomed Eng *11*, 49–79.

Zhao, Y., Lee, W.N., Lim, S., Go, V.L., Xiao, J., Cao, R., Zhang, H., Recker, R.R., and Xiao, G.G. (2009). Quantitative proteomics: Measuring protein synthesis using 15 N amino acid labeling in pancreatic cancer cells. Anal Chem *81*, 764–771.

Zhou, L., Lu, Z., Yang, A., Deng, R., Mai, C., Sang, X., Faber, K.N., and Lu, X. (2007). Comparative proteomic analysis of human pancreatic juice: methodological study. Proteomics *7*, 1345–1355.

Zhou, M., Lucas, D.A., Chan, K.C., Issaq, H.J., Petricoin, E.F., 3rd, Liotta, L.A., Veenstra, T.D., and Conrads, T.P. (2004). An investigation into the human serum 'interactome'. Electrophoresis *25*, 1289–1298.

Zolotarjova, N., Martosella, J., Nicol, G., Bailey, J., Boyes, B.E., and Barrett, W.C. (2005). Differences among techniques for high-abundant protein depletion. Proteomics *5*, 3304–3313.

Zhou, H., Yuen, P.S., Pisitkun, T., Gonzales, P.A., Yasuda, H., Dear, J.W., Gross, P., Knepper, M.A., and Star, R.A. (2006). Collection, storage, preservation, and normalization of human urinary exosomes for biomarker discovery. Kidney Int *69*, 1471–1476.

Part VI
Sample Preparation Methods in Plant Proteomics

Chapter 19
Quantitative Plant Proteomics Using Hydroponic Isotope Labeling of Entire Plants (HILEP)

Laurence V. Bindschedler, Celia J. Smith, and Rainer Cramer

Abstract Hydroponic cultivation is a soil-free technique commonly used in agriculture where plants are grown in a highly controlled environment leading to high crop yields. Hydroponic Isotope Labeling of Entire Plants (HILEP) is the most cost-effective isotope labeling method for quantitative plant proteomics, enabling the metabolic labeling of whole and mature plants with a stable isotope such as ^{15}N. Employing inorganic ^{15}N-containing salts as the sole nitrogen source, healthy plants can be easily grown and labeled in hydroponic solutions. Close to 100% ^{15}N-labeling of proteins can be achieved using HILEP. Moreover, hydroponic cultivation allows tight control of growth conditions. Plants grown in ^{14}N- and ^{15}N-hydroponic media are typically pooled straight after harvest, eliminating any bias due to subsequent sample preparation and analysis. The pooled ^{14}N-/^{15}N-protein extracts can be fractionated in any convenient way and digested with trypsin (or any other enzyme of choice). Peptides can then be analyzed by techniques such as liquid chromatography electrospray ionization tandem mass spectrometry (LC-ESI-MS/MS). Following protein identification, the spectra of ^{14}N/^{15}N-peptide pairs are typically compared and relative protein amounts are calculated from the ^{14}N/^{15}N-ion signal ratios. An increasing number of bioinformatics tools are now available for determining these ratios in a convenient way.

Keywords HILEP · Hydroponics · Mass spectrometry · Metabolic labeling · Plant proteomics

---

L.V. Bindschedler (✉)
The BioCentre, University of Reading, Reading, UK; Department of Chemistry, University of Reading, Reading, UK; School of Biological Sciences/BioCentre, University of Reading, Reading, Berkshire, UK
e-mail: l.v.bindschedler@reading.ac.uk

R. Cramer (✉)
The BioCentre, University of Reading, RG6 6AS, Reading, UK;
Department of Chemistry, Harborne Building, University of Reading, RG6 6AS, Reading, UK
e-mail: r.k.cramer@reading.ac.uk

A.R. Ivanov, A.V. Lazarev (eds.), *Sample Preparation in Biological Mass Spectrometry*, DOI 10.1007/978-94-007-0828-0_19,

19.1 Introduction

Despite the ever increasing demand for more accurate and sensitive analyses of whole proteomes, their quantitative analysis using mass spectrometry (MS) is still one of the major challenges in proteomics. The typically poor correlation between analyte concentration and signal intensity, which is due to variation in ionization efficiency in the presence of molecular competitors or contaminants, prevents straightforward quantitation using the recorded ion signal intensity. Unfortunately, the two widely used ionization techniques in proteomic MS, electrospray ionization (ESI) and matrix-assisted laser desorption/ionization (MALDI), are no exception to this limitation. However, most of the methods available to quantify entire proteomes are based on MS techniques involving the measurement of ion signal intensities.

Quantitative proteomic methods can be classified as either non-MS or MS-based methods as well as "label-free" or "labeling", of which the latter can be further subdivided into the various types of labeling approaches such as chemical and metabolic labeling.

19.1.1 Quantitative Proteomics Using Non-MS Methods

One of the most important non-MS approaches employs differential protein derivatization using fluorescent dyes such as CyDyes^{TM}. The protein samples to be compared are labeled using different dyes and are then pooled prior to protein separation by electrophoresis on the same gel, thus limiting quantitation errors that easily occur if samples are run on different gels (Marouga et al., 2005; Timms and Cramer, 2008; Unlu et al., 1997). This technique, called DIfference Gel Electrophoresis (DIGE), has the advantage of determining and then selectively quantifying the differentially expressed proteins (or protein isoforms), rather than identifying and quantifying the entire polypeptide content as it is the case in quantitative Multidimensional Protein Identification Technology (MudPIT) analyses. Although DIGE provides several advantages with respect to analyte separation, purification and enrichment as well as subsequent MS analysis time, it requires the use of expensive dyes and sophisticated image analysis software and is intrinsically limited by the disadvantages typical for gel electrophoretic separation of proteins.

There are also other quantitative gel-based methods available such as ^{35}S-labeling or quantitation of gel spots/bands by densitometry. However, DIGE has the inherent advantage of drastically reducing the technical bias of gel electrophoresis and is therefore often the method of choice for gel electrophoretic quantitation of proteins.

19.1.2 Quantitative Proteomics Using MS-Based Methods

Unlike DIGE where the intensity of a protein spot is used for quantitation, the MS-based proteomic quantitation methods mostly employ the so-called "bottom-up" strategies by identifying and quantifying proteins at the peptide level using proteolytic digests and their MS (or MS/MS) analysis. Proteins are typically digested with a protease, usually trypsin, prior to detection and characterization of the resultant peptides by MS or tandem mass spectrometry (MS/MS). The amount of protein is determined by the ion intensities of the identified peptides or their MS/MS products. A protein is usually identified and quantified by more than one peptide, enabling statistical analysis and the discrimination between protein isoforms by selecting only the peptides specific to a single protein isoform (an isoform being defined by a gene locus). However, isoform-specific quantitation is often limited by the absence or low intensity of isoform-specific peptide ion signals as well as the sequence-determined possible limitation of detecting only one or a few isoform-specific peptides.

19.1.2.1 Label-Free Methods

There are several quantitation methods available that directly exploit the recorded MS ion signal intensity without the use of any labeled internal standards or comparison between differentially labeled samples. These are called label-free methods. One such method, spectral counting (Ishihama et al., 2005), counts the number of recorded peptide spectra per protein and is based on the assumption that peptide ions from an abundant protein will be analyzed many more times than peptide ions from a protein that is less abundant. The exponentially modified Protein Abundance Index (emPAI) (Ishihama et al., 2005) is calculated from the number of spectral counts of each peptide identified in one protein. It takes into account the fact that generally more peptides are detected for larger proteins. The emPAI value is readily available when using the protein identification software Mascot (Matrix Science, London, UK).

Other label-free methods use the signal intensities of individual peptides rather than the spectral counts to compare the relative abundance of proteins between samples (Hughes et al., 2006; Palmblad et al., 2007b; Silva et al., 2005; Wang et al., 2006). However, this approach is only valid if the sets of samples to be compared are all relatively similar, as the presence of contaminants or other ion species will influence the intensity of an ion signal, due to ion suppression. For methods with LC separation, reproducibility of the chromatographic separation and software to align the LC peaks, as well as accurate mass measurement are all prerequisites for the success of this technique (Palmblad et al., 2007b; Silva et al., 2005).

19.1.2.2 Chemical Labeling

Other strategies that can be employed to circumvent the constraints of label-free quantitation include differential labeling of peptides with chemical tags, e.g. tags containing stable isotopes as reviewed in Ong and Mann (Ong and Mann, 2005). The advantage of isotope-labeled tags is that they are physically and chemically identical, with the exception of the labels' mass and/or isotope distribution. This means that it is possible to pool the samples after the labeling step and then determine the protein ratios by comparing ion intensities of the corresponding protein-derived light/heavy peptides or their fragment ions in the same mass spectrum. Thus, there is no requirement to align and match LC peaks from different runs, as is required for many of the label-free quantitation techniques. Chemical labeling with isotopes overcomes the problem of ion suppression encountered in label-free strategies and thus, reduces the need to run many MS technical replicates. The Isotope-Coded Affinity Tag (ICAT) approach, one of the first differential isotope labeling methods, uses two different mass tags ($^{12}C_9$- and $^{13}C_9$-isotope labels), which specifically label cysteine-containing peptides. Following their labeling, cysteine-containing peptides are affinity purified through the incorporated biotin moiety (Yi et al., 2005). Peptide pairs with 9 Da-mass difference are then detected by MS and their ion intensities are compared for relative quantitation. Similarly, Isotope Coded Protein Labeling (ICPL) (Schmidt et al., 2005) allows the differential isotope labeling of lysines and the differentially labeled peptides are analyzed by MS. Isobaric Tags for Relative and Absolute Quantitation (iTRAQ) also label lysines and have similar physico-chemical properties. However, iTRAQ tags have identical masses, thus relative abundance is not measured through the MS signal as with ICAT and ICPL but measured by MS/MS, determining the ion signal intensities of reporter ions of different mass that are generated by the MS/MS fragmentation process (Ross, 2006; Ross et al., 2004b). This method benefits from increased MS detection due to the contribution of all samples to the precursor ion signal. It was successfully applied to quantify the secreted proteome of Arabidopsis cells following the infection with the bacterial pathogen *Pseudomonas syringae* (Kaffarnik et al., 2009). Recently eight iTRAQ tags have become available allowing high multiplexing capacity (Ross, 2006; Ross et al., 2004b). Two-plex or six-plex Tandem Mass Tags or TMT are alternative isobaric tags that are similar to iTRAQ (Dayon et al., 2008). Many more stable isotope-labeling approaches are emerging such as RABA (Reductive Alkylation By Acetone, (Zhai et al., 2009)), ANIBAL (ANiline and Benzoic Acid Labeling of amino and carboxylic groups, (Boersema et al., 2009; Panchaud et al., 2008)) and stable isotope dimethyl labeling (Boersema et al., 2009). Many of these are quite attractive since they use low-cost reagents.

Rather than labeling peptides with chemical tags, isotope-labeling of peptides can also be performed by exchanging ^{16}O with ^{18}O during tryptic digestion when performed in heavy water ($H_2{}^{18}O$) (Yao et al., 2001). The 2 Da-mass shift is then detected via MS analysis.

Chemical labeling of peptides often modifies irreversibly specific amino acids. These modifications need to be considered when performing protein identification with search engines such as Mascot. Fortunately, most of the commonly used chemical labeling methods (e.g. ICAT, iTRAQ, ICPL, dimethylation, ^{18}O-exchange) are supported by Mascot.

However, despite of being a powerful approach for quantitative proteomics, chemical labeling does not overcome the variability between protein extracts (degradation, handling variability), as the labeling occurs after the extraction of proteins or even after tryptic digestion. In addition, costs of the labeling chemicals also need to be taken into consideration, as it can be a limiting factor for large-scale experiments.

19.1.2.3 Metabolic Labeling

Metabolic labeling, i.e. labeling proteins at the time of protein synthesis, by incorporation of stable isotopes such as ^{15}N is an alternative strategy to chemical labeling. Samples are typically pooled at the harvesting stage prior to most of the sample handling including protein extraction. Labeling at this early stage avoids any technical bias that may occur later in the sample preparation workflow – for instance due to differences in protein extraction, fractionation or proteolysis. One of the main metabolic labeling strategies is Stable Isotope Labeling with Amino acids in Cell culture (SILAC) (Mann, 2006), where cells are grown in the presence of an amino acid that is labeled with ^{13}C and/or ^{15}N. Quantitation is then typically performed by comparing the MS signal intensities of samples grown in medium containing unlabeled ^{12}C/^{14}N-arginine or one of the labeled arginines, i.e. ^{13}C/^{14}N-, ^{12}C/^{15}N- or ^{13}C/^{15}N-containing arginine. SILAC has been used to label proteins in Arabidopsis cell cultures, but because of the photosynthetic capacity of plants to fix carbon from atmospheric CO_2, complete labeling is typically prevented (Gruhler et al., 2005). Partial labeling renders quantitation more difficult, especially as the level of labeling must first be established empirically and the uptake of the label is dependent on the exposure to light. Moreover, the price of labeled arginine makes SILAC an expensive technique for the quantitative analysis of whole and mature Arabidopsis plants.

Alternative methods to SILAC have also been developed for metabolic labeling using ^{15}N or ^{13}C (reviewed by Heck and Krijgsveld (Heck and Krijgsveld, 2004)), and have been successfully applied to *E. coli* (Ross et al., 2004a), mammalian cells (Conrads et al., 2001) and yeast (Kolkman et al., 2006). In these specific cases, the media contained mixed organic and inorganic nitrogen sources estimated to contain 98–99% ^{15}N, which was supplied as ammonium and/or inorganic nitrate salt. This metabolic labeling approach was also successful for the quantitative analysis of whole organisms such as the nematode model *Caenorhabditis elegans* and the fruit fly *Drosophila melanogaster* that were fed on ^{15}N-labeled *E. coli* and yeast, respectively (Krijgsveld et al., 2003). For these organisms, which require a more complex feeding medium, it was still possible to have a near 100%-labeling efficiency. Similarly, whole rats were labeled through a protein-free diet supplemented with ^{15}N algal cells (Wu et al., 2004).

More recently, Arabidopsis cell cultures were labeled with ^{15}N in a traditional Murashige and Skoog salt medium containing sucrose and inorganic ^{15}N-nitrate as the sole nitrogen source for quantifying proteins and nitrogen-containing metabolites (Benschop et al., 2007; Engelsberger et al., 2006). Similarly, young seedlings were grown in this medium in shaking liquid cultures (Huttlin et al., 2007; Nelson et al., 2007). However, as the Murashige and Skoog medium is relatively rich, the plants have to be grown in vitro and in sterile conditions which is less convenient, especially when the plants have to be grown to maturity.

There is another metabolic labeling method that is inexpensive, robust and suitable for quantitative plant proteomics. Its basic principle has previously been employed for metabolic ^{15}N-labeling of whole hydroponically grown mature potato plants for the structural analysis of abundant tuber proteins such as patatin by nuclear magnetic resonance (NMR) spectroscopy (Ippel et al., 2004). This labeling method uses ^{15}N-salts in a hydroponic medium and is equally applicable to quantitative MS-based proteomic analysis of whole and mature plants, which has led to the introduction of Hydroponic Isotope Labeling of Entire Plants (HILEP) for relative quantitative proteomic analysis by mass spectrometry and is described and reviewed in this chapter.

19.1.2.4 Hydroponic Isotope Labeling of Entire Plants (HILEP)

HILEP in combination with MS-based proteomic quantitation was developed and first employed in a proof-of-principle study for quantifying the proteome of Arabidopsis leaves treated with hydrogen peroxide to mimic oxidative stress response (Palmblad et al, 2007, Bindschedler et al, 2008). Since hydroponic cultures are ideal for tightly controlling plant growth conditions, the use of HILEP provides optimal experimental conditions for the growth and labeling of whole and mature plants in an inexpensive and relatively simple medium. It enables the quantitative analysis at the molecular level of mature plants responding to environmental stimuli or diseases.

Hydroponics is plant cultivation in the absence of soil, using a nutrient solution instead. The term itself originates from the concatenation of two Greek words, meaning *water-based labour*. Most of the nutrients are provided as solutes in the hydroponic medium, which mainly consists of water and salts of potassium, phosphate, nitrate, ammonium, calcium chloride and magnesium sulfate. Plants can grow in the absence of organic nutrients as they can fix atmospheric carbon through photosynthesis. Hydroponic cultivation of plants is a very ancient technique established long before photosynthesis was discovered and understood. Indeed, it was probably the method of cultivation used in The Hanging Gardens of Babylon in 600 BC. It certainly had been used by the Egyptians around that time and by the Romans in ca. 200 BC. In the eleventh century, the Aztecs were using hydroponic cultivation on a regular basis, creating "floating gardens". The technology was brought back to Europe by the Spanish Conquistadores and since then has been scientifically investigated. Further development during and after World War II has led to the use of hydroponics as a modern cultivation technique for intensive and high yield farming (http://www.hydroponicsonline.com/lessons/History/lesson2-1history.htm; (Jones, 1982, 2005)). Recently, the near 100% complete metabolic labeling of entire and mature plants using hydroponics was first described in the production of 3.6 kg of labeled potato tubers for the 3D structural analysis of abundant proteins such as patatin and Kunitz-type protein inhibitors using NMR spectroscopy (Ippel et al., 2004).

As mentioned earlier plants are photosynthetic organisms and thus are able to fix carbon from atmospheric CO_2. Therefore, ^{13}C-labeling would be inappropriate if near 100% metabolic labeling is required. Stable isotope labeling with ^{15}N is more suitable, especially as ^{15}N-salts are cheaper. Near 100%-labeling of proteins with ^{15}N gives the advantage of having two distinct isotopic envelopes for the ^{14}N- and

^{15}N-labeled peptides, thus avoiding the requirement to use complex mathematical tools for estimating the ratio between the light (unlabeled) and heavy (labeled) peptides or to use techniques such as Subtle Modification of Isotope Ratio Proteomics or SMIRP (Whitelegge et al., 2004). The latest advances in quantitative plant proteomics including stable isotope labeling have recently been reviewed (Oeljeklaus et al., 2009; Thelen and Peck, 2007). It appears that the use of metabolic ^{15}N-labeling is becoming the method of choice for in vivo labeling of plant proteins for quantitative proteomics using MS (Bindschedler et al., 2008; Hebeler et al., 2008; Huttlin et al., 2007; Nelson et al., 2007; Palmblad et al., 2007a, 2008). Most of these techniques resemble HILEP or are variations of HILEP such as Stable Isotope Labeling for *In Planta* quantitative proteomics (SILIP), which was employed to label tomato plants grown in a hydroponic medium containing inorganic soil substrates such as river sand (Schaff et al., 2008). It was also shown that HILEP can be employed for multiplex quantitation of up to four different samples (Palmblad et al., 2008).

19.2 Procedure

19.2.1 Hydroponic Cultivation and the Generation of ^{15}N-Labelled Products

Optimal conditions for the metabolic ^{15}N-labeling of Arabidopsis plants using hydroponic cultivation were established in previous studies (Bindschedler et al., 2008; Palmblad et al., 2007a, 2008) based on a modified Hoagland's hydroponic medium (Hoagland, 1920), which was adapted using other published protocols (Huttlin et al., 2007; Huttner and Bar-Zvi, 2003; Noren et al., 2004). Protocol 1 (see Annex) details the hydroponic medium formulation as used in (Bindschedler et al., 2008). Protocol 2 (see Annex) describes the hydroponic cultivation of *Arabidopsis thaliana* from seed to mature plant. The general workflow of the experimental procedure is shown in Fig. 19.1a.

To facilitate the identification of ^{15}N-labeled peptides, it is preferable to label plants with the highest possible percentage of ^{15}N in the nitrogen-containing salts, as the maximum peak intensity of the isotopic envelope of peptides from plants grown in media with salts containing 90% ^{15}N is already shifted by 1 Da compared to plants grown in a 100% ^{15}N medium (Zhang et al., 2009). Salts containing 98–99% ^{15}N are widely commercially available at a modest cost compared to labeled amino acids as used in SILAC. As a typical example, Fig. 19.1b shows the MS spectrum of a ^{14}N/^{15}N-peptide pair and the MS/MS spectra of each peptide species. ^{15}N-salts were provided as 98+% ammonium and potassium nitrate. To produce 1 g of hydroponically grown ^{15}N-labeled leaves costs about £5 using the protocols described in this chapter. Arabidopsis plants exhibit a healthy appearance when grown under these conditions, and no phenotypic differences can be detected between hydroponically grown ^{14}N and ^{15}N plants (see Fig. 19.2).

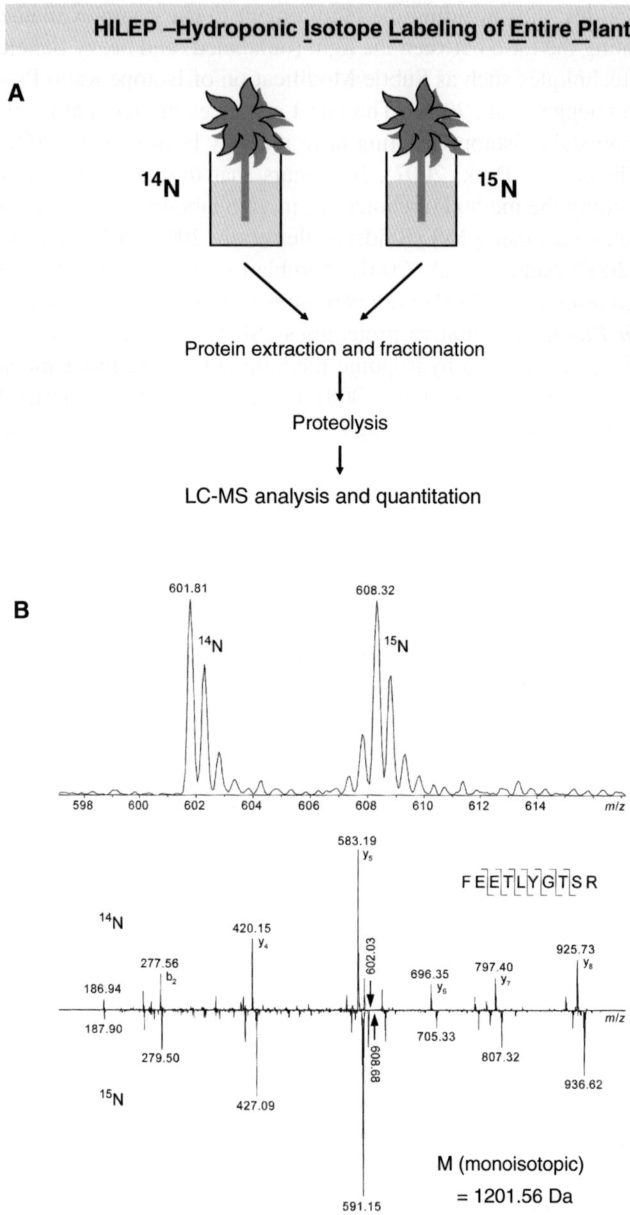

Fig. 19.1 General HILEP workflow (**a**) for the relative quantitation of plant proteomes. (**b**) MS and MS/MS spectra of the ^{14}N- and ^{15}N-labeled FEETLYGTSR peptide (monoisotopic mass of 1201.56 Da). In part reprinted from Phytochemistry, 69, Laurence V. Bindschedler, Magnus Palmblad and Rainer Cramer, Hydroponic isotope labelling of entire plants (HILEP) for quantitative plant proteomics; an oxidative stress case study, 1962–1972, Copyright (2008) ©, with permission from Elsevier

Fig. 19.2 Photographs of plants grown hydroponically. (**a**) Two weeks-old plants in 200 μl-pipette tips in the original pipette tip box. Plants were grown on 0.7% agar and half-strength hydroponic medium. The last third of the tip has to be cut to allow root growth. (**b**) Plants transferred to lunch boxes containing full-strength hydroponic medium. Five-weeks-old plants are shown. The medium is aerated with an aquarium pump. (**c**) and (**d**) Seven-weeks-old healthy-looking plants grown in ^{14}N-hydroponic medium (**c**) and ^{15}N-hydroponic medium (**d**). In part reprinted from Phytochemistry, 69, Laurence V. Bindschedler, Magnus Palmblad and Rainer Cramer, Hydroponic isotope labelling of entire plants (HILEP) for quantitative plant proteomics; an oxidative stress case study, 1962–1972, Copyright (2008) ©, with permission from Elsevier

19.2.2 Protein Sample Preparation

After harvesting the plant material an equal amount of fresh weight of the ^{14}N-sample and the ^{15}N-sample are typically pooled prior to further processing. This way, both samples are processed together, avoiding variability in sample handling, loss and degradation. The preparation workflow may then include any type of protein extraction, cell/protein fractionation (e.g. protein precipitation, intercellular washing fluid (IWF) protein extraction, membrane protein enrichment), enzymatic treatment, polypeptide purification/enrichment, analytical preparation and other sample processing steps (see protocols and methods in other chapters).

19.2.3 Data Analysis

Several bioinformatics tools have been developed for the relative quantitation of proteomes (for review see (Mueller et al., 2008) and find a list of free software at http://www.ms-utils.org/). Unlike quantitation with some chemical tags or with $^{18}O/^{16}O$-labeling during proteolysis, the mass difference between light (^{14}N) and heavy (^{15}N) peptides is dependent on the amino acid composition of the peptide. Even for peptides labeled with heavy/light arginine in SILAC experiments, the mass difference is usually fixed, since most proteolytic steps in proteomic analyses are performed with trypsin which cuts after arginine and lysine. Thus, early bioinformatics tools for MS-based proteomic quantitation using methods such as SILAC needed to be modified for peptide quantitation based on global ^{15}N-labeling (Palmblad et al., 2007a).

In references (Bindschedler et al., 2008; Palmblad et al., 2007a), an automated data analysis pipeline is described for the routine application of HILEP in an integrated analytical workflow using the freely available Trans-Proteomic Pipeline (TPP) software suite, which can be down-loaded from http://tools.proteomecenter.org/software.php (Keller et al., 2005). The main steps in this data analysis workflow were as follows:

1. LC-MS/MS raw data files were converted into mzXML files.
2. Peak list files compatible with the Mascot protein search engine and with TPP (e.g. mgf files) were generated.
3. ^{14}N- and ^{15}N-proteins were identified separately with Mascot, generating dat result files (http://www.matrixscience.com), which were converted to ^{14}N-pepXML or ^{15}N-pepXML files in the TPP suite.
4. In the TPP suite, the ^{14}N-pepXML files and their corresponding MS/MS raw datafiles (mzXML files) were then merged using the xINTERACT program. The whole process was repeated for the ^{15}N-pepXML files and their corresponding MS/MS raw data files (mzXML files).
5. Peptide ratios were calculated using the XPRESS algorithm (Han et al., 2001). PeptideProphet (Keller et al., 2002) was then used to show peptide ratios and to allow statistical analysis of protein abundance.
6. Protein identifications and relative protein abundances were then displayed in a single protXML file using the ProteinProphet viewer (Nesvizhskii et al., 2003) in TPP.
7. Results were extracted into a spreadsheet for further statistical analysis.

Fig. 19.3 Mass spectra of identified peptides from Arabidopsis plants grown hydroponically in ^{14}N- or ^{15}N-medium. Six-weeks-old plants were treated with 100 mM hydrogen peroxide. Treated (^{14}N) and control (^{15}N) plants were pooled together when harvested 24 h later and leaf proteins were extracted from the apoplast (see experimental details in Bindschedler et al, 2008). Examples for different expression ratios are shown: (**a**) DTDILAAFR (Rubisco large SU; AtCg00490), monoisotopic mass of 1020.5 Da. (**b**) ALDENLLASPEK (probable endo-chitinase; At2g43570), monoisotopic mass of 1298.6 Da. (**c**) SGLVNEAAIDK (beta-xylosidase; At5g64570), monoisotopic mass of 1116.7 Da

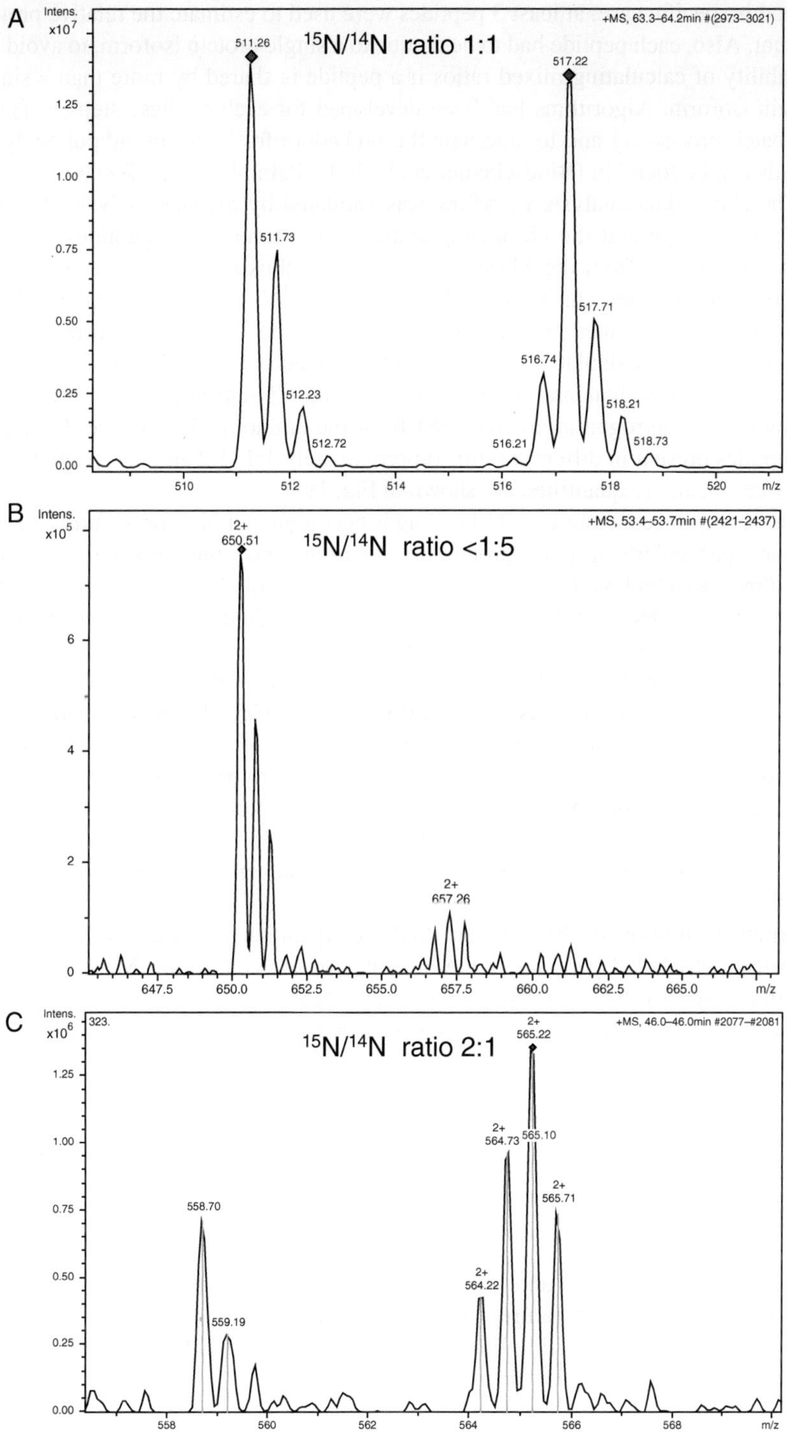

Fig. 19.3 (continued)

In this specific case, at least 3 peptides were used to estimate the relative protein amount. Also, each peptide had to be unique to a single protein isoform, to avoid the possibility of calculating mixed ratios if a peptide is shared by more than a single protein isoform. Algorithms had been developed for each of these steps to facilitate batch processing and to automate the procedure for high-throughput analysis. Details can be found in (Bindschedler et al., 2008; Palmblad et al., 2007a).

The above data analysis workflow was validated by manual analysis. The *m/z* value, the charge and the chromatographic elution time of the peptides of interest were retrieved from the Mascot search results, thus allowing the extraction of the corresponding ion chromatographic peak. The ^{14}N- and ^{15}N-isotope envelopes were manually examined and the total areas of both the ^{14}N- and the corresponding ^{15}N-peptide were calculated to estimate the abundance ratio. The mass difference between the most abundant peaks in the ^{14}N- and ^{15}N-envelope was correlated to the number of nitrogen atoms deduced from the amino acid sequence. Examples of peptides present in different ratios (approximately 1:1, 1:2 and 5:1, respectively) that were manually quantified are shown in Fig. 19.3.

However, since metabolic ^{15}N-labeling is becoming popular for quantitative proteomics, particularly in plant proteomics, there are now alternatives to the above workflow. Nonetheless, the number of ready-to-use tools for the quantitative analysis of ^{15}N-labeled proteins is still limited. The MSQuant tool and the XPRESS algorithm in TPP were originally developed for SILAC and ICAT, respectively, and need some supplementary programs to be applicable to global ^{15}N-labeling. The use of MSQuant is further restricted as it was only designed to be compatible with three MS instruments and for the use of Mascot ((Mueller et al., 2008), http://www.ms-utils.org). The RelEx quantitation software seems to be suitable for any type of labeling but is restricted to the analysis of Finnigan Xcalibur datafiles (MacCoss et al., 2003). On the other hand the TPP pipeline supports most types of MS/MS data as well as several search engines, such as X!Tandem, Phenyx, MASCOT and SEQUEST.

Another addition to the software tools for quantitative proteomic analysis of globally ^{15}N-labeled peptides is the quantitation toolbox within Mascot Distiller (Matrix Science). This toolbox is an integrated system that does not require programming skills, as all the processes from generating peak lists to peptide quantitation and statistics are integrated in the same software package for straightforward end-user applications.

19.3 Summary

HILEP is a cost-effective and robust method for accurate quantitative analysis of proteomes of whole and mature plants. It produces healthy-looking plants in a controlled environment, requiring little "gardening" time. HILEP is readily applicable to various plant systems and bioanalytical applications.

Annex

Protocol 1: Hydroponic Medium Formulation for Growing Arabidopsis thaliana Plants

Hydroponic medium formulation (as described in (Bindschedler et al., 2008)).

1. Prepare the following stock solutions and store them at 4°C:

Stock solutions	MW	Amount in 50 mL
1 M KNO_3	101.0/102.0[a]	5.05/5.1[a] g
1 M $CaCl_2$	147.0	7.35 g
1 M $MgSO_4 \cdot 7H_2O$	246.5	12.3 g
1 M KH_2PO_4	136.1	6.8 g
1 M $NH_4\ NO_3$	236.0/238.0[a]	11.8/11.9[a] g
70 mM Fe-EDTA	367.1	0.13 g

[a]For ^{15}N solution

2. Prepare the micronutrient stock solution and store at -20°C:

Micronutrients	MW	Amount in 100 mL
50 mM H_3BO_3	61.8	0.31 g
10 mM $MnSO_4 \cdot H_2O$	169.0	0.17 g
1.5 mM $ZnSO_4 \cdot 7H_2O$	287.5	0.043 g
1 mM $CuSO_4 \cdot 5H_2O$	249.7	0.025 g
0.58 mM Na_2MoO_4	241.9	0.014 g

3. To prepare 1 L of full-strength hydroponic solution, mix

Stock solutions	Volume (mL)
KNO_3	2.5
$CaCl_2$	1
$MgSO_4$	0.75
KH_2PO_4	0.5
$NH_4\ NO_3$	0.5
Fe-EDTA solution	1
Micronutrients solution	1

and make up to 1 L in distilled water (store at 4°C or sterilize if kept at room temperature).

Sterile, half-strength solution containing 0.7% agar is used to germinate seeds (for 10–15 days). Seedlings are then transplanted into full-strength solution.

Protocol 2: Growing Arabidopsis thaliana Plants for HILEP

Growing Arabidopsis plants in ^{14}N or ^{15}N hydroponic medium (Bindschedler et al., 2008).

1. Seed sterilization as described by Noren et al. (Noren et al., 2004).
   Place the required amount of seeds in a microcentrifuge tube. Wash these successively in:

   a. 70% Ethanol 5 min
   b. 0.5% SDS 15 min

      Rinse three times with sterile distilled water.
      At this stage the seeds can be stored in sterile water in the fridge for up to 1 week.
2. Autoclave a box of 200 μL-pipette tips (ensuring that the box has a transparent lid as it will be used to create a mini-greenhouse).
3. Make up a 0.7% agar solution containing half-strength hydroponic solution and autoclave it to ensure sterility.
4. Aseptically fill the sterile pipette tips with the molten agar solution (at ∼50–60°C). Initially, add only a few 10–20 μL-droplets and let them solidify, sealing the tips before filling them completely. Make sure there is a convex dome of agar as the gel will shrink when drying. Cover with the transparent lid and let the agar set.
5. Place one seed in the centre of each pipette tip cone using a 100–200 μL-pipettor under aseptic conditions. Close the pipette tip box and seal with parafilm™, vernalise the seeds by placing the mini-greenhouse at 4°C (fridge) for at least 48 h. Alternatively, the vernalisation can take place after the seeds are sterilized and stored in water (step 1) prior to adding the seeds to the agar.
6. Place the mini-greenhouse in a growth cabinet under a short day light regime, 10 h of light at 23°C and 14 h darkness at 20°C with 60% humidity.
7. After 10–15 days, before the roots reach the lower part and grow out of the tip, remove the parafilm™ seal. Cut the end of the tip at one third of its length and replace the lid without sealing. At this stage the growth conditions are non-sterile but keep the system as clean as possible.
8. To grow the plants hydroponically, paint 1 L-lunch boxes with black acrylic paint on the outside to avoid algae growth. An example of an ideal box size is 21 cm × 14 cm × 5 cm. The lids should have a rubber seal and four clamps for tight closure. Drill twelve holes (4×3) in the lid to hold 12 plantlets in each box. Wash the boxes thoroughly and sterilize the inner surface with ethanol or by exposure to UV light. Once prepared, fill the boxes with one liter of full-strength hydroponic solution.
9. After 12–15 days (4 leaves stage), transfer the tips that contain a healthy seedling into the holes of the perforated box lid. Plants are then placed back in the growth cabinet, and left to grow until they are at the age required. Top up the

hydroponic solution whenever necessary. Plants should have large healthy looking leaves with no necrosis or chlorosis. Roots should have formed a network in the hydroponic solution.

10. (Optional) Drill an additional hole in the lid to hold tubing for oxygenating the solution via an aquarium pump. A pump for a 45 L-aquarium can aerate four boxes. If an aquarium pump is used, the growth medium should be changed weekly as contaminants may grow quicker under these conditions. This was observed even when a 0.22 μm-filter was added between the pump and the solution.

Useful Websites

http://www.ionsource.com/functional_reviews/CompassXport/CompassXport.htm
http://tools.proteomecenter.org
http://www.matrixscience.com
http://www.ms-utils.org (http://www.ms-utils.org/pepXML2Excel.html)

References

Benschop, J.J., Mohammed, S., O'Flaherty, M., Heck, A.J., Slijper, M., and Menke, F.L. (2007). Quantitative phospho-proteomics of early elicitor signalling in Arabidopsis. Mol Cell Proteomics *6*, 1198–1214.

Bindschedler, L.V., Palmblad, M., and Cramer, R. (2008). Hydroponic isotope labelling of entire plants (HILEP) for quantitative plant proteomics; An oxidative stress case study. Phytochemistry *69*, 1962–1972.

Boersema, P.J., Raijmakers, R., Lemeer, S., Mohammed, S., and Heck, A.J. (2009). Multiplex peptide stable isotope dimethyl labeling for quantitative proteomics. Nat Protoc *4*, 484–494.

Conrads, T.P., Alving, K., Veenstra, T.D., Belov, M.E., Anderson, G.A., Anderson, D.J., Lipton, M.S., Pasa-Tolic, L., Udseth, H.R., Chrisler, W.B., *et al.* (2001). Quantitative analysis of bacterial and mammalian proteomes using a combination of cysteine affinity tags and 15 N-metabolic labeling. Anal Chem *73*, 2132–2139.

Dayon, L., Hainard, A., Licker, V., Turck, N., Kuhn, K., Hochstrasser, D.F., Burkhard, P.R., and Sanchez, J.C. (2008). Relative quantification of proteins in human cerebrospinal fluids by MS/MS using 6-plex isobaric tags. Anal Chem *80*, 2921–2931.

Engelsberger, W.R., Erban, A., Kopka, J., and Schulze, W.X. (2006). Metabolic labeling of plant cell cultures with K15NO3 as a tool for quantitative analysis of proteins and metabolites. Plant Methods *2*, 14.

Gruhler, A., Schulze, W.X., Matthiesen, R., Mann, M., and Jensen, O.N. (2005). Stable isotope labeling of Arabidopsis thaliana cells and quantitative proteomics by mass spectrometry. Mol Cell Proteomics *4*, 1697–1709.

Han, D.K., Eng, J., Zhou, H., and Aebersold, R. (2001). Quantitative profiling of differentiation-induced microsomal proteins using isotope-coded affinity tags and mass spectrometry. Nat Biotechnol *19*, 946–951.

Hebeler, R., Oeljeklaus, S., Reidegeld, K.A., Eisenacher, M., Stephan, C., Sitek, B., Stuhler, K., Meyer, H.E., Sturre, M.J., Dijkwel, P.P., *et al.* (2008). Study of early leaf senescence in Arabidopsis thaliana by quantitative proteomics using reciprocal 14N/15N labeling and difference gel electrophoresis. Mol Cell Proteomics *7*, 108–120.

Heck, A.J., and Krijgsveld, J. (2004). Mass spectrometry-based quantitative proteomics. Expert Rev Proteomics *1*, 317–326.

Hoagland, D. (1920). Optimum nutrient solutions for plants. Science *52*, 562–564.

Hughes, M.A., Silva, J.C., Geromanos, S.J., and Townsend, C.A. (2006). Quantitative proteomic analysis of drug-induced changes in mycobacteria. J Proteome Res *5*, 54–63.

Huttlin, E.L., Hegeman, A.D., Harms, A.C., and Sussman, M.R. (2007). Comparison of full versus partial metabolic labeling for quantitative proteomics analysis in Arabidopsis thaliana. Mol Cell Proteomics *6*, 860–881.

Huttner, D., and Bar-Zvi, D. (2003). An improved, simple, hydroponic method for growing Arabidopsis thaliana. Plant Mol Biol Rep *21*, 59–63.

Ippel, J.H., Pouvreau, L., Kroef, T., Gruppen, H., Versteeg, G., van den Putten, P., Struik, P.C., and van Mierlo, C.P. (2004). In vivo uniform (15)N-isotope labelling of plants: using the greenhouse for structural proteomics. Proteomics *4*, 226–234.

Ishihama, Y., Oda, Y., Tabata, T., Sato, T., Nagasu, T., Rappsilber, J., and Mann, M. (2005). Exponentially modified protein abundance index (emPAI) for estimation of absolute protein amount in proteomics by the number of sequenced peptides per protein. Mol Cell Proteomics *4*, 1265–1272.

Jones, J.B. (1982). Hydroponics – Its History and Use in Plant Nutrition Studies. J Plant Nutr *5*, 1003–1030.

Jones, J.B. (2005). Hydroponics. A Practical Guide for the Soilless Grower, 2nd edition, CRC Press, Boca Raton, London, New York, Washington, D.C.

Kaffarnik, F.A., Jones, A.M., Rathjen, J.P., and Peck, S.C. (2009). Effector proteins of the bacterial pathogen Pseudomonas syringae alter the extracellular proteome of the host plant, Arabidopsis thaliana. Mol Cell Proteomics *8*, 145–156.

Keller, A., Eng, J., Zhang, N., Li, X.J., and Aebersold, R. (2005). A uniform proteomics MS/MS analysis platform utilizing open XML file formats. Mol Syst Biol *1*, 2005 0017.

Keller, A., Nesvizhskii, A.I., Kolker, E., and Aebersold, R. (2002). Empirical statistical model to estimate the accuracy of peptide identifications made by MS/MS and database search. Anal Chem *74*, 5383–5392.

Kolkman, A., Daran-Lapujade, P., Fullaondo, A., Olsthoorn, M.M., Pronk, J.T., Slijper, M., and Heck, A.J. (2006). Proteome analysis of yeast response to various nutrient limitations. Mol Syst Biol *2*, 2006 0026.

Krijgsveld, J., Ketting, R.F., Mahmoudi, T., Johansen, J., Artal-Sanz, M., Verrijzer, C.P., Plasterk, R.H., and Heck, A.J. (2003). Metabolic labeling of C. elegans and D. melanogaster for quantitative proteomics. Nat Biotechnol *21*, 927–931.

MacCoss, M.J., Wu, C.C., Liu, H., Sadygov, R., and Yates, J.R., 3rd (2003). A correlation algorithm for the automated quantitative analysis of shotgun proteomics data. Anal Chem *75*, 6912–6921.

Mann, M. (2006). Functional and quantitative proteomics using SILAC. Nat Rev Mol Cell Biol *7*, 952–958.

Marouga, R., David, S., and Hawkins, E. (2005). The development of the DIGE system: 2D fluorescence difference gel analysis technology. Anal Bioanal Chem *382*, 669–678.

Mueller, L.N., Brusniak, M.Y., Mani, D.R., and Aebersold, R. (2008). An assessment of software solutions for the analysis of mass spectrometry based quantitative proteomics data. J Proteome Res *7*, 51–61.

Nelson, C.J., Huttlin, E.L., Hegeman, A.D., Harms, A.C., and Sussman, M.R. (2007). Implications of 15N-metabolic labeling for automated peptide identification in Arabidopsis thaliana. Proteomics *7*, 1279–1292.

Nesvizhskii, A.I., Keller, A., Kolker, E., and Aebersold, R. (2003). A statistical model for identifying proteins by tandem mass spectrometry. Anal Chem *75*, 4646–4658.

Noren, H., Svensson, P., and Andersson, B. (2004). A convenient and versatile hydroponic cultivation system for Arabidopsis thaliana. Physiologia Plantarum *121*, 343–348.

Oeljeklaus, S., Meyer, H.E., and Warscheid, B. (2009). Advancements in plant proteomics using quantitative mass spectrometry. J Proteomics *72*, 545–554.

Ong, S.E., and Mann, M. (2005). Mass spectrometry-based proteomics turns quantitative. Nat Chem Biol *1*, 252–262.

Palmblad, M., Bindschedler, L.V., and Cramer, R. (2007a). Quantitative proteomics using uniform (15)N-labeling, MASCOT, and the trans-proteomic pipeline. Proteomics *7*, 3462–3469.

Palmblad, M., Mills, D.J., and Bindschedler, L.V. (2008). Heat-shock response in Arabidopsis thaliana explored by multiplexed quantitative proteomics using differential metabolic labeling. J Proteome Res *7*, 780–785.

Palmblad, M., Mills, D.J., Bindschedler, L.V., and Cramer, R. (2007b). Chromatographic alignment of LC-MS and LC-MS/MS datasets by genetic algorithm feature extraction. J Am Soc Mass Spectrom *18*, 1835–1843.

Panchaud, A., Hansson, J., Affolter, M., Bel Rhlid, R., Piu, S., Moreillon, P., and Kussmann, M. (2008). ANIBAL, stable isotope-based quantitative proteomics by aniline and benzoic acid labeling of amino and carboxylic groups. Mol Cell Proteomics *7*, 800–812.

Ross, A., Kessler, W., Krumme, D., Menge, U., Wissing, J., van den Heuvel, J., and Flohe, L. (2004a). Optimised fermentation strategy for 13C/15N recombinant protein labelling in Escherichia coli for NMR-structure analysis. J Biotechnol *108*, 31–39.

Ross, P., Dey, S., Pillai, S., Daniels, S., Williamson, B., Guertin, S.., Minkoff, M., Chen, X., Purkayastha, B., Pappin, D (2006). Protein Quantitation Using a Novel 8-plex Set of Isobaric Peptide Labels. 54th ASMS Conference, Seattle, WA *ThP 62512*.

Ross, P.L., Huang, Y.N., Marchese, J.N., Williamson, B., Parker, K., Hattan, S., Khainovski, N., Pillai, S., Dey, S., Daniels, S., *et al.* (2004b). Multiplexed protein quantitation in Saccharomyces cerevisiae using amine-reactive isobaric tagging reagents. Mol Cell Proteomics *3*, 1154–1169.

Schaff, J.E., Mbeunkui, F., Blackburn, K., Bird, D.M., and Goshe, M.B. (2008). SILIP: a novel stable isotope labeling method for in planta quantitative proteomic analysis. Plant J *56*, 840–854.

Schmidt, A., Kellermann, J., and Lottspeich, F. (2005). A novel strategy for quantitative proteomics using isotope-coded protein labels. Proteomics *5*, 4–15.

Silva, J.C., Denny, R., Dorschel, C.A., Gorenstein, M., Kass, I.J., Li, G.Z., McKenna, T., Nold, M.J., Richardson, K., Young, P., *et al.* (2005). Quantitative proteomic analysis by accurate mass retention time pairs. Anal Chem *77*, 2187–2200.

Thelen, J.J., and Peck, S.C. (2007). Quantitative proteomics in plants: choices in abundance. Plant Cell *19*, 3339–3346.

Timms, J.F., and Cramer, R. (2008). Difference gel electrophoresis. Proteomics *8*, 4886–4897.

Unlu, M., Morgan, M.E., and Minden, J.S. (1997). Difference gel electrophoresis: A single gel method for detecting changes in protein extracts. Electrophoresis *18*, 2071–2077.

Wang, G., Wu, W.W., Zeng, W., Chou, C.L., and Shen, R.F. (2006). Label-free protein quantification using LC-coupled ion trap or FT mass spectrometry: Reproducibility, linearity, and application with complex proteomes. J Proteome Res *5*, 1214–1223.

Whitelegge, J.P., Katz, J.E., Pihakari, K.A., Hale, R., Aguilera, R., Gomez, S.M., Faull, K.F., Vavilin, D., and Vermaas, W. (2004). Subtle modification of isotope ratio proteomics; an integrated strategy for expression proteomics. Phytochemistry *65*, 1507–1515.

Wu, C.C., MacCoss, M.J., Howell, K.E., Matthews, D.E., and Yates, J.R., 3rd (2004). Metabolic labeling of mammalian organisms with stable isotopes for quantitative proteomic analysis. Anal Chem *76*, 4951–4959.

Yao, X., Freas, A., Ramirez, J., Demirev, P.A., and Fenselau, C. (2001). Proteolytic 18O labeling for comparative proteomics: model studies with two serotypes of adenovirus. Anal Chem *73*, 2836–2842.

Yi, E.C., Li, X.J., Cooke, K., Lee, H., Raught, B., Page, A., Aneliunas, V., Hieter, P., Goodlett, D.R., and Aebersold, R. (2005). Increased quantitative proteome coverage with (13)C/(12)C-based, acid-cleavable isotope-coded affinity tag reagent and modified data acquisition scheme. Proteomics *5*, 380–387.

Zhai, J., Liu, X., Huang, Z., and Zhu, H. (2009). RABA (reductive alkylation by acetone): a novel stable isotope labeling approach for quantitative proteomics. J Am Soc Mass Spectrom *20*, 1366–1377.

Zhang, Y., Webhofer, C., Reckow, S., Filiou, M.D., Maccarrone, G., and Turck, C.W. (2009). A MS data search method for improved 15N-labeled protein identification. Proteomics *9*, 4265–4270.

Chapter 20
Efficient Strategies for Analysis of Low Abundance Proteins in Plant Proteomics

Olga A. Koroleva and Laurence V. Bindschedler

Abstract Proteomic studies alongside with other post-genomic approaches are rapidly progressing in plant biology. This trend is reflected by the increased number of publications related to plant proteomics in the last several years. While proteomic studies span from the global characterisation of proteomes (the protein complements) in whole plants and organs to the more detailed characterisation of particular proteins, there is an increasing demand for methods allowing sophisticated (and preferably quantitative) analysis of a variety of cells, compartments and organelles, post-translational protein modifications, protein complexes and their dynamics. The common problem is the high dynamic range of the proteome, and many researchers are looking for the ways to improve the recovery of naturally low-abundant plant proteins. The aim of this chapter is to provide a review of a variety of approaches, rather than specific protocols, established for the isolation of specific fractions of proteins suitable for analysis by mass spectrometry. We focus on the sample preparation methodology from general to specific applications in plant proteomics and on advances made in the characterisation of global plant proteomes and plant-specific compartments. We also describe the advantages of metabolic labeling techniques suited for quantitative plant proteomic studies and discuss tandem affinity purification of plant protein complexes and perspectives of single-cell proteomics in plants.

Keywords Plant compartments · Plant proteomics · Plant sample preparation · Single cell proteomics · Tandem affinity purification

20.1 Introduction

Opportunities for systematic proteomic studies in plants have been opened by completing the sequencing of the genome of the first model plant *Arabidopsis thaliana* (Arabidopsis Genome Initiative, 2000), followed by the rice genome (International

---

O.A. Koroleva (✉)
School of Biological Sciences/BioCentre, University of Reading, Reading, Berkshire, RG6 6AS, UK; Current address: Pursuit Dynamics plc, Huntingdon, Cambridgeshire, PE29 6HB, UK
e-mail: olgakoroleva@pursuitdynamics.com

A.R. Ivanov, A.V. Lazarev (eds.), *Sample Preparation in Biological Mass Spectrometry*, DOI 10.1007/978-94-007-0828-0_20,

Rice Genome Sequencing Project, 2005) and, more recently, the nuclear and chloroplast genome of *Brachypodium distachyon* (Bortiri et al., 2008; Huo et al., 2008) and the draft genome sequence of an economically important crop cucumber, *Cucumis sativus* (Huang et al., 2009a). Several other genome projects are underway, and the TIGR Plant Transcript Assemblies database (http://plantta.jcvi.org/) already contains a collection of transcript sequences including full-length and partial cDNAs and expressed sequence tags (ESTs) from over 200 plant species (Childs et al., 2007). Functional and post-genomic approaches are rapidly developing in plant biology. These include general characterisation of proteomes in whole plants or organs, cell types or subcellular compartments and organelles, composition of protein complexes, post-translational protein modifications as well as quantitative proteomic studies. The most complete proteome map for the model plant *Arabidopsis thaliana* has recently been produced by Baerenfaller and co-authors (2008), who used different organs and developmental stages, together with undifferentiated cell cultures to generate a complex proteomic data set for the evaluation of Arabidopsis gene models (open reading frames, ORFs) and the proteome dynamics. This study showed the ability of proteomics to improve plant genome annotation by presenting evidence for more than fifty new gene models, including seven from intergenic regions. The power of a new proteogenomics approach to annotate previously incorrect gene models, over 10% of the sequenced genome, on the basis of identified peptides, was further demonstrated by Castellana and co-authors (2008). The proteomic strategies also have a great potential to study non-model plant species of economic importance (Carpentier et al., 2008). The success of proteomics studies strongly depends on the good quality of protein samples and thus requires development of robust and efficient workflows. All proteomic approaches require establishment of optimised protocols which allow rapid, reproducible and efficient extraction and fractionation of proteins or purification of protein complexes from plant material.

20.2 Specific Features of Plants Which Have Implications for Mass Spectrometry

20.2.1 Protein Content in Plant Tissues is Low

Compared to animal cells where the protein concentration is usually in the range of 20–30 mg protein/g of tissue, in the vegetative tissues of plants, the protein content is approximately 20 times lower, in the range of 1–2 mg protein/g of tissue (USDA National Nutrient Database, 2009). Several factors contribute to this, such as the large vacuoles which occupy up to 99% of the volume in some cells and which contain a very low amount of protein. In addition, cell walls contribute substantially to the total plant biomass, whilst having relatively low protein content. The cell walls represent yet another obstacle that further reduces the protein recovery from plant tissues. The extracellular matrix formed by cell walls can be a very rigid structure that needs harsh conditions to be broken up either mechanically or by enzymatic digestion to release the cell contents. In addition, plant tissues contain large amounts of secondary metabolites, especially phenolic compounds, known to interfere with protein extraction and electrophoresis separation, thus further reducing the efficiency of mass-spectrometry-based protein analysis. Moreover, in addition to having low protein levels, the plant cells have high amounts of the other major biopolymers such as polysaccharides and nucleic acids, which may also affect the protein yield during extraction.

20.2.2 Plant-Specific Compartments

Several subcellular compartments are common for all eukaryotes, and proteomic analysis showed many similarities between plants and animals in sets of proteins isolated from nuclei (Bae et al., 2003; Calikowski et al., 2003; Choudhary et al., 2009; Khan and Komatsu, 2004; Pandey et al., 2008), nucleoli (Koroleva et al., 2007; Pendle et al., 2005), mitochondria (Huang et al., 2009b; Millar et al., 2001) and endomembrane compartments (Dunkley et al., 2006; Eubel et al., 2008; Groen et al., 2008). However, plants have a number of plant-specific compartments which have to be taken into account as potential contributors to the total tissue proteome.

Proteomic analysis of subcellular components in Arabidopsis provided localisation information for ~2600 proteins, which have been integrated in the SUBA database (http://www.suba.bcs.uwa.edu.au). The SUBA database encompasses 10 distinct subcellular locations, over 6743 predicted non-redundant proteins corresponding to 51% of the Arabidopsis unique ESTs (Heazlewood et al., 2007). To promote understanding of organelle dynamics, including organelle function, biogenesis, differentiation, movement and interactions with other organelles, the Plant Organelles Database (PODB; http://podb.nibb.ac.jp/Organellome) was launched by Mano and co-authors (Mano et al., 2008).

20.2.2.1 Cell Walls, Apoplast and Associated Proteins

The cell walls and vacuoles are essential components of plant cells, since they play major roles in the growth, development and structure of the plant and are considered to be plant specific organelles in their own right. Though small vacuoles are present in animal cells, they are very different from the large vacuoles in many plant tissues such as epidermal cells. Cell walls are also present in bacteria and fungi, but the compositions of these cell walls are very different from the plant cell walls. Taken together, plant vacuoles and cell walls provide the mechanical rigidity required for the upright position of terrestrial plants, by combining the scaffolding and rigidity of the cell walls and the high turgor pressure resulting from the accumulation of solutes in the vacuoles. In addition, the cell walls not only act to protect plant cells from the surrounding biotic and abiotic environment, but they also influence and control the cell shape and size and thus the internal structure of all plant tissues. The cell wall is considered as a rigid extracellular matrix with continuity into the extracellular space or apoplast. It is difficult to strictly define the boundaries between these two compartments. Plasmodesmata are membranous structures which connect the cytoplasm of two adjacent plant cells through cell walls and they are involved in short distance signalling across the cell wall barrier. Isolation of plasmodesmata from purified cell walls allowed identifying plasmodesmata-associated proteins (Bayer et al., 2004; Simpson et al., 2009). Cell wall proteomics has greatly advanced the description of the proteins associated with this plant-specific compartment. Many of the cell wall proteins are involved in modification of cell wall components and structure, defence against pathogens, signalling and interactions with plasma membrane proteins at the cell surface, as reviewed by Jamet et al. (2008a). Over 800 of proteins belonging to the plant cell wall and apoplast from Arabidopsis and rice have been deposited into the WallProtDB database, www.polebio.scsv.ups-tlse.fr/WallProtDB/ (Albenne et al., 2009), including those identified in previous studies (Bayer et al., 2006; Cho et al., 2009; Irshad et al., 2008). The cell wall proteome is difficult to analyse since many proteins, such as extensins and arabinogalactan proteins are covalently or ionically bound to the cell walls. A common feature of cell wall hydroxyproline-rich glycoproteins is that they are heavily glycosilated and are resistant to heat and proteolysis. Loosely bound proteins can be desorbed from the cell walls using salts such as calcium chloride, sodium chloride or lithium chloride (Millar et al., 2009). Such proteins are often present in the medium of cell cultures (Chivasa et al., 2005), or can be extracted from the apoplast (Kwon et al., 2005). Specific extraction and deglycosylation methods using stringent conditions are required to release more tightly bound cell wall proteins (Jamet et al., 2008a,b).

20.2.2.2 Vacuoles and Vacuolar Proteomics

Vacuoles have multiple functions, such as storage compartments and provide the necessary turgor pressure for cell wall extension and cell elongation. Meristematic plant cells contain a large number of small vacuoles, whilst mature cells usually contain one large central vacuole, occupying up to 99% of the cell volume in epidermal cells. The vacuolar proteome in *Arabidopsis thaliana* was initially characterised for vacuoles isolated from suspension cell culture, when 163 proteins were identified by Shimaoka and co-authors (Shimaoka et al., 2004). Analysis of vacuolar fraction derived from rosette leaves protoplasts led to the identification of 402 proteins (Carter et al., 2004). It included a large proportion of membrane proteins, identified in a tonoplast-enriched fraction analyzed separately by one-dimensional (1-D) Sodium Dodecyl Sulfate (SDS) – Polyacrylamide Gel Electrophoresis (PAGE) followed by nano-scale liquid chromatography (nLC) and tandem mass spectrometry (MS/MS). However, the bulk of vacuolar protein content was represented by a myrosinase-associated single protein (Carter et al., 2004). A challenging goal of further characterisation of the components of the vacuolar membrane, which contains less than 1% of the total cellular proteins, was achieved by Schmidt and co-authors (2007), who purified vacuoles of closely related species cauliflower (*Brassica oleracea*) and used complementary extraction procedures such has saline and alkaline washing, acetone and chloroform/methanol extraction for maximal protein recovery. The vacuolar proteins were identified by analyzing the data against the Arabidopsis database, leading to identification of 102 cauliflower integral proteins and 214 peripheral proteins, including the vacuolar pyrophosphatase, which was found to be the most prominent protein (Schmidt et al., 2007).

20.2.2.3 Chloroplasts and Associated Proteins

Chloroplasts are organelles that are absent in animal cells. Chloroplasts are descendents of a cyanobacterial endosymbiont, however many chloroplast protein genes of endosymbiont origin are encoded by the chromosomal DNA located in the nucleus (Ishikawa et al., 2009). Their best known function is photosynthesis, however, they also perform assimilation of nitrogen and sulphur and synthesise many essential plant compounds, such as amino acids, nucleotides, lipids, plant hormones and secondary metabolites. The chloroplasts contain large amounts of proteins associated with the photosynthetic machinery. Rubisco (ribulose-bisphosphate carboxylase/oxygenase) is a super-abundant protein in plant leaf and other tissues. The total amount of Rubisco constitute about 30% of the total protein content in most plant leaves and therefore in whole tissue extracts can represent up to 40% of soluble proteins (McCabe et al., 2001; Parry et al., 2003). The presence of highly abundant proteins, such as Rubisco, is a problem for mass spectrometry (MS) analysis, since the dominance of abundant peptide ions will hinder the analysis of low abundance proteins due to the ion suppression phenomenon.

Sequential extractions with salt, detergent, and organic solvents, followed by multidimensional protein separation steps, led to identification of 154 proteins in the Arabidopsis peripheral and integral thylakoid membrane proteome (Friso et al., 2004). Proteome dynamics during plastid differentiation in rice has been quantitated using a 2D gel combined with a shotgun proteomics approach (Kleffmann et al., 2007). The most comprehensive cover of the chloroplast proteome (1325 proteins, including 916 nuclear-encoded) was recently achieved by performing 1D PAGE nLC MS/MS based analysis of ten independent chloroplast preparations from Arabidopsis thaliana (Zybailov et al., 2008).

20.2.2.4 Phloem and Xylem – and Associated Proteins

Phloem and xylem, the two major components of the vascular bundles, are the highways for transport of water, nutrients and metabolites in higher plants. They are also important for providing structural rigidity to the whole plant as well as playing a role in long-distance signalling and responses to biotic and abiotic stress. Their specific function as a long-distance transport system in vascular plants requires a range of proteins involved in maintenance of the structure, long-distance communication and defence against pathogens. While xylem sap exudates can be collected by using the root pressure method, pure phloem sap can be obtained by a unique method which depends on the assistance of phloem-feeding insects. The small size of the model plant Arabidopsis and subsequently the very small volumes of the phloem sap that can be collected using the aphid stylectomy method, as little as 10–300 pL of exudate per stylet (Deeken et al., 2002), renders such approach unsuitable for proteomic investigations. However other plant species produce a sufficient amount of phloem exudate for successful proteomic analysis. To collect sufficient amounts of phloem sap from barley plants, an optimized stylectomy set-up was applied requiring 30 plants and 600 aphids (*Rhopalosiphum padi*), yielding on average 10 μL of phloem sap within 6 h of sampling (Gaupels et al., 2008). The limited amounts of rice xylem and phloem samples were overcome by the use of nLC-MS/MS, leading to the successful identification of proteins and peptides, including signal transduction proteins, putative transcription factors, stress response factors and metabolic enzymes (Aki et al., 2008).

The Cucurbita species and pumpkin in particular, have always been the favourite model plants for phloem studies because of relative ease of obtaining large amounts of free-flowing phloem sap. Analysis of pumpkin phloem exudates by LC-MS/MS allowed the identification of 1,209 different consensi, of which 1,121 could be annotated using BLAST search analyses against Arabidopsis thaliana, rice, and poplar, demonstrating the high conservation of the phloem proteome in diverse species (Lin et al., 2009). Sets of phloem proteins that function in RNA binding, mRNA translation, ubiquitin-mediated proteolysis, and macromolecular and vesicle trafficking were found. The results indicate that protein synthesis and turnover, previously considered being absent in enucleate sieve elements, have a potential to occur within the angiosperm phloem translocation stream.

Brassica napus, a close relative of Arabidopsis thaliana, was chosen as a model plant for phloem research. Using separation by 1-D gel electrophoresis, high-resolution 2-D gel electrophoresis and MS/MS analysis, 140 phloem proteins were successfully identified (Giavalisco et al., 2006). Phloem filament protein PP1 and phloem lectin PP2 are the two major phloem proteins which occur in all dicots and can represent more than half of the total phloem proteins (Kehr and Rep, 2006). Acidification precipitation followed by neutralization can be used to remove these proteins (Kehr and Rep, 2006) and thus to improve detection of the rest of the phloem proteome.

Xylem exudates driven by root-pressure were collected from cut *Brassica napus* stems, and subjected to 2-DE, and MS/MS analysis, resulting in the identification of 69 proteins, including proteins potentially involved in defence-related reactions and cell wall metabolism (Kehr et al., 2005). Analysis of the xylem sap proteome in maize revealed proteins related to xylem differentiation, primary and secondary cell wall synthesis, defence mechanisms and programmed cell death (Alvarez et al., 2006). In comparison, nearly 6000 unique proteins were identified in the proteome of Populus developing xylem by application of shotgun tandem MS profiling (Kalluri et al., 2009).

20.3 Methods of Sample Preparation for Whole-Tissue Plant Proteomics

20.3.1 Protein Extraction

The protein extraction protocol used should be specifically tailored depending on the overall design of the analytical workflow, and the fractionation method employed.

When a total protein extract is required, extraction in denaturing conditions is advantageous, since less protein degradation occurs in comparison to native conditions. However, when subcellular fractionation is being carried out, or when a particular protein complex needs to be preserved intact for affinity purification or a study involving protein ligands (see Section 20.4), only extractions in native conditions are suitable. Prior to extraction, the plant material has to be homogenised. The most popular method includes freezing the plant material in liquid nitrogen and manual mechanical disruption of the plant tissues with a pestle and mortar, optionally in the presence of small amounts of quartz sand. However the manual technique is very time-consuming, non-standardised in terms of the final size of the particles and is limited by the amount of tissue processed (usually not more than 10 g in a single round). Automated tissue homogenizers are better suited for high-throughput sample preparation from large amounts of tissue. Cryogenic laboratory mills (for example, 6770 Freezer/Mill from Spex SertiPrep Group, UK) are more efficient (usually less than 2 min are required for grinding of one to four samples, 1–5 g each), providing standardised grinding protocols with the added advantage of maintaining sample temperature low. There are also instruments capable of homogenising fresh samples without the freezing step – such as high-pressure disruption using the HAIVA cell disruptor (Constant Systems Ltd, UK), which offers a range of flow rates, temperature control and continuous or "single-shot" mode of action. The HAIVA appliance is based on the principle of French Press, however a simple procedure of sample introduction, higher pressure range and high speed of sample processing under controlled, contained and repeatable conditions allows production of higher quality protein extracts from plant cell suspension cultures (O.A. Koroleva, unpublished).

20.3.1.1 Protein Extraction in Denaturing Conditions

A number of alternative protocols for total protein extraction from plant tissues are available, and the comparative advantages of typical procedures have been extensively discussed (Thiellement et al., 2006). Many of the economically important plant species contain high levels of interfering compounds and several specific protocols for purification of protein fractions have been developed, as recently reviewed by Carpentier and co-authors (2008). However, the majority of the protocols that have successfully been used with diverse plant tissues for total protein extraction in denaturing conditions are variations of just two key approaches: trichloroacetic acid (TCA)/acetone precipitation or phenol extraction coupled with ammonium acetate precipitation, both often complemented with a combination of detergents and reducing agents.

A typical TCA/acetone precipitation protocol involved protein precipitation from another extraction buffer or directly from frozen tissue powder, using 10% w/v TCA and 0.07% v/v 2-mercaptoethanol solution in cold acetone (optionally the co-precipitant sodium deoxycholate can be added for higher yield), followed by sequential centrifugation and washing steps with cold 0.07% v/v 2-mercaptoethanol in acetone. The TCA/acetone solution can efficiently remove contaminants such as salts, pigments and lipids and inhibits protease activity.

The widely used phenol extraction protocol consists of extraction of proteins in chosen extraction buffer, then removal of interfering substances using Tris-HCL saturated phenol, followed by precipitation by 0.1 M ammonium acetate in cold methanol. This method was reported to be more efficient in removing of interfering substances for 2-D gel applications than TCA/acetone.

A less common protocol is precipitation by chloroform/methanol (Wessel and Flugge, 1984), which is recommended for the quantitative precipitation of soluble as well as hydrophobic proteins from dilute solutions. Methanol, chloroform and water (4:1:3 v/v) are successively added to one volume of aqueous protein extract, mixed and after centrifugation the top aqueous layer is removed. Four volumes of methanol are added and the sample centrifuged again, the supernatant removed and the pellet washed twice in acetone/0.07% 2-mercaptoethanol, to remove lipids.

A mixture of chaotropes, detergents and reducing agents is necessary to resuspend and solubilise the pelleted proteins after precipitation. Typically, a combination of 7 M urea and 2 M thiourea as chaotropes, 2–4% CHAPS detergent, 20 mM of dithiothreitol (DTT) or 0.07% 2-mercaptoethanol as reducing agents and several plant protease inhibitors (or commercially available cocktails) is used (Thiellement et al., 2006). Prior to sensitive applications (such as direct analysis by Matrix-Assisted Laser Desorption Ionization (MALDI) MS), certain detergents and salts present in protein solutions can be removed using cation exchange chromatography tips such as OMIX SCX pipette tips (Varian).

Quite often a simplified and rapid total protein extraction from plant tissue in denaturing conditions (to avoid protein degradation) is required for general qualitative or quantitative proteome comparison. Commercially available 2D Clean up kit (GE Healthcare) is generally designed for removing contaminants such as detergents, salts, lipids, phenolics, and nucleic acids which interfere with isoelectrofocusing of extracted proteins. In order to speed up and simplify the extraction procedure, we precipitated proteins directly from frozen leaf tissue ground in liquid nitrogen, thus omitting the step of extracting proteins in an extraction buffer prior to the precipitation step. Then the purification protocol was followed as recommended by the manufacturer. This protein extraction protocol is very fast (less than 90 min) and suitable for either 2-D gel (illustrated on Fig. 20.1) or direct LC-MS/MS analysis.

20.3.1.2 Protein Extraction in Native Conditions

For applications such as the analysis of protein complexes (Section 20.5), native gel electrophoresis or cellular fractionation, proteins need to be extracted in native conditions. A "native" extraction solution typically contains 10–50 mM buffer such as Tris-HCL, HEPES-KOH, MES or sodium phosphate, at pH 6–8, either sodium chloride at 100–200 mM or sucrose/polyols at 200–400 mM as osmotic agents, a reducing agent such as 1–10 mM DTT or 2-mercaptoethanol to avoid protein oxidation, and 1–10 mM EDTA. A variety of detergents can be used, commonly one of the following: Tween 20, NP-40, CHAPS, Triton X-100, reviewed by Rabilloud

Fig. 20.1 Fluorescence 2-D Difference Gel Electrophoresis (DIGE) of Mago-RNAi mutant and wild-type Arabidopsis plants, using samples prepared by protein precipitation/ extraction from frozen powder with 2D Cleanup kit and labeled by CyDye DIGE Fluor minimal dyes (GE Healthcare)

and co-authors (2006). To minimise protein degradation due to endogenous proteases, plant specific protease inhibitor cocktails can be used, with the exception of specific protein purification protocols relying on proteolytic cleavage of an epitope tag (Section 20.5.2). Initial addition of insoluble polyvinylpolypyrrolidone (PVP) or sodium metabisulphite (1–5 mM) (0.5–1%) can help to remove phenolic compounds (Charmont et al., 2005). The plant material (usually stored in –80°C prior to sample extraction) has to be homogenized in liquid N_2 and extracted by five to ten volumes of buffer per weight equivalent for an efficient extraction, and kept on ice. To reduce the protein sample viscosity due to the presence of nucleic acids, samples can be treated with nucleases such as benzonase (Boschetti et al., 2009). Precipitation of the crude protein extract with ammonium sulfate has the advantage of concentrating the sample while keeping the proteins in a native and enzymatically active form and also partially removing nucleic acids and glycans; however the native protein complexes can not be preserved. The protein sample must be desalted after the ammonium sulfate precipitation for sensitive downstream applications.

20.3.2 Overcoming the High Dynamic Range of Protein Concentrations in the Leaf Proteome

The high dynamic range of protein concentrations in the plant proteome poses a challenge for 2D-gel separations and MS analysis due to ion signal suppression, preventing the detection of low-abundance proteins. Sample fractionation is crucial for successful identification of low-abundance proteins by MS, since the most abundant peptides tend to dominate the mass spectrum and mask any minor peaks. Fractionation approaches, ranging from electrophoretic or chromatographic separations to exploitation of physico-chemical properties of a specific subset of proteins, are available to deal with the issue of high dynamic range.

Sample pre-fractionation by immuno-affinity chromatography, using specific antibodies to deplete the most abundant proteins, is one of the ways to improve recovery of low-abundance proteins. This strategy has been recently applied to the leaf proteomes to deplete Rubisco using an anti-Rubisco immuno-affinity column (Cellar et al., 2008). Immuno-affinity columns are commercially available; however, this approach is limited by the availability of specific antibodies, thus only a limited number of proteins can be removed. Sequential precipitation of proteins with increasing concentrations of PEG is an alternative method to remove/deplete Rubisco (Widjaja et al., 2009; Xi et al., 2006). Other affinity–based approaches can be used to fractionate and/or specifically isolate proteins. For example, poly-U and polyA affinity binding columns were used to analyse low abundance nucleotide–binding proteins in Arabidopsis leaves (Ni et al., 2010; Xu et al., 2007).

Another strategy for depletion of highly abundant proteins exploits "equalisation" of the sample via interaction with a combinatorial hexapeptide ligand library, which is comprised of millions of different peptides coupled to beads or other chromatographic supports (ProteoMiner, Bio-Rad). Each specific protein can bind only to a certain type of peptide ligand and all peptides are present in a limiting amount in the combinatorial peptide ligand library (CPLL). Subsequently, when a total protein extract is passed over CPLL beads, the abundant proteins will quickly saturate their binding sites, thus a large proportion of the abundant proteins will be removed in the flow-through fraction, leading to depletion of the abundant proteins from the bound fraction. The low-abundant proteins will be retained on their specific ligands. A large amount of starting material is required to enrich sufficient amounts of low-abundance proteins to allow their detection in downstream applications. This "equalisation" strategy using CPLL was originally developed for blood serum samples (Sennels et al., 2007), and has since been successfully applied to several other animal systems. Recently, the CPLL approach has been tested on plant leaf proteome, with the aim of depleting Rubisco from total protein extracts. A traditional CPLL approach did not make a significant improvement, however the use of carboxylated hexapeptides in the CPLL led to a substantial increase in the proteome coverage (Boschetti et al., 2009). Moreover, the most recent developments have shown promising results (Fig. 20.2). In this new study, Rubisco was severely depleted when the plant protein extract was bound on the CPLL for depletion in mild acidic conditions rather than at a neutral pH (Fasoli et al., 2011).

To summarise this section, protein fractionation followed by depletion of highly abundant proteins and/or enrichment of the low abundance proteins, is a promising strategy to decrease the high dynamic range of protein concentrations. This allows improved resolution of low abundance proteins on 2D-gels, thus increasing the number of proteins detected and identified by MS.

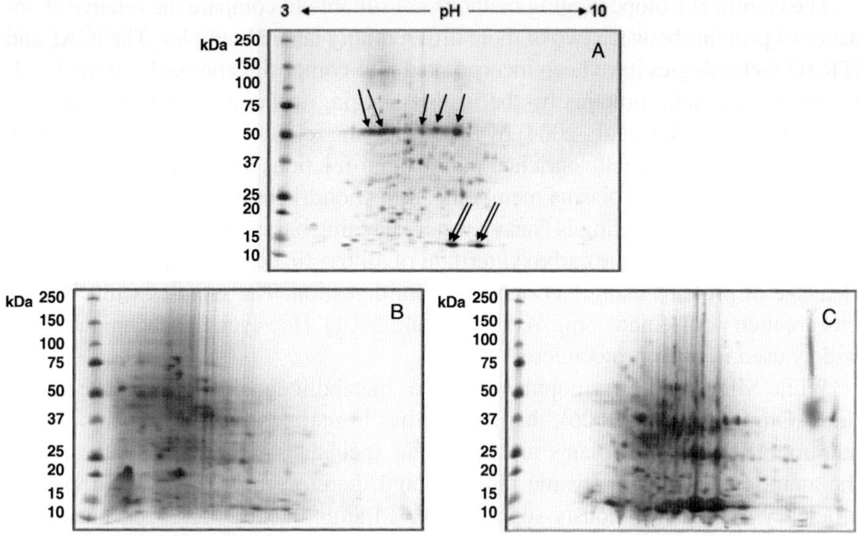

Fig. 20.2 Two-dimensional electrophoresis of protein extract from spinach leaves before (**a**) and after treatment with ProteoMiner at pH 4.0 (**b**) and 9.3 (**c**). All analyzed samples were first adsorbed onto the IPG pH 3–10 strips. 150 mg of total proteins in 150 μL volume were loaded in each experiment. Single arrows indicate isoforms of Rubisco 56 kDa subunit; double arrows indicate isoforms of Rubisco 14 kDa subunit. Staining with colloidal Coomassie blue. Reproduced with permission from article published in *J Proteomics*, Fasoli E, Farinazzo A, Sun CJ, Kravchuk AV, Guerrier L, Fortis F, Boschetti E, Righetti PG. (2010), Interaction between proteins and peptide libraries in proteome analysis: pH involvement for a larger capture of species, Copyright Elsevier (2011)

20.4 Labeling Techniques for Quantitative Proteomics in Plants

Mass spectrometry can be used for comparison and quantification of different samples (Mann, 1999; Ong and Mann, 2005). For the relative quantification studies, the protein samples can be differentially labelled and then mixed together so that the relative proportion of each peptide in the whole complex mixture originating from each sample can be identified after MS analysis of the mixture. The use of isotopes for labeling provides the most accurate method for protein identification and quantification in complex mixtures (Steen and Mann, 2004). Most of the general techniques available for quantitative proteomics (reviewed by Ong and Mann (2005)), are readily applicable to plant proteomics analysis, reviewed by Thelen and Peck (2007) and discussed elsewhere in this volume. The methods for the relative quantification studies can be classified into two main groups: chemical tags such as ^{18}O, ICAT, ICPL and iTRAQ (see Appendix); and metabolic labeling such as stable isotope labeling with amino acids in cell culture (SILAC) or inorganic ^{15}N incorporation-based amino acid labeling.

The chemical isotope tagging methods are suitable to compare the relative abundance of proteins between two or more differentially labeled samples. The ICAT and iTRAQ technologies have been incorporated into composite approach for the localization of organelle proteins by the isotope tagging technique (LOPIT) (Dunkley et al., 2006; Dunkley et al., 2004) by comparing the relative ratios of tagged protein peptides from of organelle enriched microsomal fractions allowing to discriminate ER, Golgi, vacuolar or plasma membrane, mitochondria or plastids.

Oxygen isotope labeling is "heavy water" labeling based on incorporation of one or two ^{18}O atoms into the carboxyl termini of all tryptic peptides during proteolytic cleavage of proteins using ^{18}O rich water for digestion. The samples with ^{16}O and ^{18}O are then pooled and compared (Yao et al., 2001). However, this technique is not widely used in plant proteomics.

While SILAC is a very popular choice of metabolic labeling in animal cell cultures (Ong and Mann, 2006), this method has limitations for use in plants. Being autotrophic organisms, plants are metabolic specialists which can synthesise all the amino acids from inorganic nitrogen, and therefore have lower incorporation efficiency of the exogenously supplied labeled amino acids. Moreover, plant cells can also very rapidly interconvert amino acids. As a result, the labeling efficiency achieved using exogenous amino acid feeding of Arabidopsis cell cultures has been found to be only 70–80% (Gruhler et al., 2005). Consequently a substantial and unpredictable portion of the non-labeled peptides from the heavy isotope-labeled sample will appear to belong to the unlabeled sample, thus requiring reciprocal labeling which complicates analysis.

Another group of methods is based on metabolic labeling. This approach takes advantage of plant metabolic capacities and in contrast to SILAC, metabolic labeling with ^{15}N has been shown to result in >98% incorporation in both intact plants (Ippel et al., 2004) and cell cultures (Engelsberger et al., 2006), while also being more cost effective. In this approach proteins are labelled in vivo. Samples can be pooled prior to protein extraction, thus eliminating the possibility of differential treatment during sample processing and allowing relative quantitation within the same mass spectrum, while reducing the need for extensive replicates. Plants are photosynthetic and thus do not need any organic compounds and can be grown hydroponically in a medium containing only inorganic salts. Hydroponic isotope labeling of entire plants (HILEP) has proven to be very cost efficient and robust method to completely label whole mature plants. Nearly 100% of ^{15}N labeling efficiency was achieved in Arabidopsis plants by growing them to the fully expended rosette stage in hydroponic media containing 2.5 mM ^{15}N potassium nitrate and 0.5 mM ^{15}N ammonium nitrate (Bindschedler et al., 2008; Palmblad et al., 2007). A similar quantitative proteomics method, SILIP (Stable isotope labeling method *in planta)*, was developed for labeling of growing tomato plants in sand in a greenhouse environment (Schaff et al., 2008). HILEP and SILIP can be used to investigate the proteomics of whole mature plants undergoing abiotic stress, senescence or plant pathogen interactions. The HILEP technology is described in length in another chapter of this book.

Another advantage of metabolic labeling is that it can be combined with chemical labeling. Such "double labeling" can provide additional confidence in experimental results for quantitative analysis of up– and down regulated proteins in mutants. As a proof-of-concept approach, we have tested the combination of metabolic ^{15}N labeling with 2-D Difference Gel Electrophoresis (DIGE) for comparative proteomic analysis. Metabolically labelled wild-type Arabidopsis plants and knock-down mutant of nuclear protein Mago (Koroleva et al., 2009) were grown as described (Bindschedler et al., 2008). Rosette leaves were collected from ^{14}N– grown mutant and ^{15}N –grown wild-type plants, frozen and ground to powder in liquid nitrogen. Protein extraction was performed directly from frozen powder. Aliquots of 60–70 mg were allocated into Eppendorf tubes, mixed with 300 μL of cold TCA-based precipitant from the 2D Cleanup kit (GE Healthcare) and extraction performed as per the manufacturers' protocol. The final protein pellet was air dried for 5 min and resuspended in 100 μL of sample buffer (7 M urea, 2 M thiourea, 4% CHAPS, 40 mM DTT) and pH adjusted to 8.5. The protein concentration was measured using the 2D Quant kit (GE Healthcare) and the total amounts of protein in the samples were estimated as 0.6–1 μg per pellet, approximately 1% of the initial plant material. Fifty microgram of protein was used for each labeling reaction with CyDye DIGE Fluor minimal dyes (GE Healthcare). The labelled samples were combined, separated on isoelectric focusing strips (pH 3–10, 18 cm) in the first dimension on Ettan IPGphor (GE Healthcare), and in the second dimension by SDS-PAGE (Ettan DALT , (GE Healthcare). The high quality protein preparation produced using this protocol is illustrated by the example DIGE image shown in Fig. 20.1. The analysis of the DIGE image (Fig. 20.1) using the DeCyder software (GE Healthcare) indicated that expression levels changed more than 2-fold for over 140 proteins in the mutant (O.A. Koroleva, unpublished data).

20.5 Affinity Purification of Plant Protein Complexes

Originally developed for yeast and mammalian protein affinity purification, protein tagging strategies are becoming standard tools for testing protein-protein interactions or isolation of protein complexes in plant molecular biology. However, the analysis of the composition of protein complexes still remains a significant challenge. Many interactions are transient in nature and components of these unstable complexes can be easily dissociated during affinity purification procedures. Reducing of the stringency of the purification can preserve these interactions, but will result in concomitant increase in non-specific contaminants. The common limitation of any affinity-tag purification approach is that the total protein yield is dependent on the level of expression of the tagged protein. As discussed above, the general protein content in plants is low, thus exacerbating the problem.

The success of the protein purification procedure depends on the protein topology and the location of the epitope tag in relation to the specific motifs and signal peptides. For optimum affinity purification efficiency, the epitope tag should be maximally exposed on the surface of the protein complex. The potential effect of the epitope tag on the conformation or the charge state of the tagged protein, which can diminish the stability of the protein complex, should be considered. It is very important to ensure that the expression of the tagged protein is high; otherwise neither protein purification protocol will provide enough protein for MS analysis. Moreover, the expression of the recombinant tagged protein will lead to a competition between the tagged protein and the endogenous target protein for binding to the interacting protein partners, unless a mutant plant lacking the native protein was used. Therefore, the strong constitutive 35S CaMV promoter rather than endogenous promoters can be a better choice for the enhanced gene expression of recombinant protein production in plants. In a comparison study conducted by Van Leene et al. (2007), all recombinant proteins accumulated at higher levels when expressed under control of the constitutive 35S promoter compared with the endogenous promoters. Thus, purifications of tagged proteins produced under control of a strong constitutive promoter provided the best coverage of protein complex detection (Van Leene et al., 2007).

20.5.1 Single Protein Tags

A range of tags are currently available, providing a choice of tagging and purification strategies in plant proteomics studies. Many tags have been successfully used to study protein-protein interaction in plants, including Protein A and G, GFP, CBP, GST, HA, His, MBP, c-Myc, FLAG, StrepII, Biotin and TBP (Kato et al., 2003; Koroleva et al., 2009; Koroleva et al., 2004; Peckham et al., 2006; Wilchek and Bayer, 1990; Witte et al., 2004; Zanetti et al., 2005; Zhong et al., 2003), explanation of abbreviations is given in Appendix and Table 20.1. Several tags used for purification of protein complexes in plants are listed in Table 20.1. Often a combination of epitope tags is required to study protein interactions. For example, a combination of His, GST and MBP tagged proteins was used to show that APETALA1 (AP1) and SEPALLATA3 (SEP3), both MADS box DNA-binding proteins, interacted with SEU protein (Sridhar et al., 2004; Sridhar et al., 2006). However not all tags are compatible with the MS/MS analysis of the purified protein complexes. Elution of Protein A or G from immunoglobulin G (IgG) columns requires low pH, which would destroy protein complexes. In addition, the presence of IgG-derived peptides in the eluted immunopurified protein fractions, may mask the detection of peptides from the target proteins. Therefore IgG-binding domains of protein A or protein G can not be used as single tags for MS proteomics, but only as the first-step purification tags, which undergo proteolytic cleavage before the second round of purification in tandem affinity purification protocols.

20.5.2 Tandem Affinity Purification (TAP-Tags)

The Tandem Affinity Purification (TAP) procedure, originally developed by Rigaut et al. (1999), is used to pool down target proteins (tagged by composite TAP-tags) and their associated protein complexes. The original TAP-tag consists of two IgG-binding domains of protein A, a specific protease cleavage site of the tobacco etch

Table 20.1 Affinity purification tags used for isolation and MS/MS analysis of plant protein complexes

Protein tag	Acronym of tag construct	Components	Binding agent	Number of purified (tagged) proteins	Number of ligands identified	References
Single epitope tags						
Hexahistidine	His_6	Six histidines	Ni^{2+} -NTA, Co^{2+} -CMA (Talon)	2	–	Witte et al. (2004)
Biotin	Biotin	Biotin from tomato	Streptavidin, Strep-tactin	1	86	Zhong et al. (2003)
Biotin carboxyl carrier protein domain of 3-methylcrotonyl CoA carboxylase from Arabidopsis	BCCD	Biotin carboxyl carrier protein domain	Streptavidin magnetic beads	1 (5)	10	Qi and Katagiri (2009)
Strep II	Strep II	Biotin-like epitope	Streptavidin magnetic beads	2	1	Witte et al. (2004)
Tandem affinity purification tags						
Improved N (C)–terminal Tandem Affinity Purification	NTAPi and CTAPi	Tandem Protein A domains; TEV protease site; calmodulin binding protein (CBP)	IgG-agarose; Calmodulin (CAM)-agarose	2	2	Rohila et al. (2004),
				39 (41)	over 100 (23 complexes)	Rohila et al. (2006),
				45 (88)	over 300 (13 complexes)	Rohila et al. (2009),
				2	1	Witte et al. (2004),
				1	11	Van Aken et al. (2007),
				1	5	Batelli et al. (2007),

Table 20.1 (continued)

Protein tag	Acronym of tag construct	Components	Binding agent	Number of purified (tagged) proteins	Number of ligands identified	References
				1 1 1 6	2 6 39 (incl. non specific) 38	Zhao et al. (2008a), Takahashi et al. (2008), Koroleva et al. (2007), Zhao et al. (2009)
Modified N/C TAPi with GFP reporter)	pKNTAP and pKCTAP	Two domains of the protein A (ZZ); TEV protease site; CBP	IgG-agarose; Calmodulin (CAM)-agarose	6	186 (43 confirmed)	Van Leene et al. (2007)
Alternative (C)–terminal Tandem Affinity Purification	CTAPa	Tandem Protein A; HRV protease site; His_6, c-Myc_9	IgG-agarose; Ni^{2+} -NTA, Talon resin	1 (30)	2	Rubio et al. (2005)
Protein A– FLAG -StrepII	SFZZ	Two domains of the protein A (ZZ); TEV protease site; 3xFLAG; Strep II	IgG Sepharose beads; Streptavidin beads	1	3	Van Leene et al. (2008)
Protein G and Streptavidin-binding peptide	GS-TAP	Two domains of protein G; two TEV cleavage sites; streptavidin-binding peptide.	IgG Sepharose beads; Streptavidin beads	2	19 (confirmed)	Van Leene et al. (2008)

Table 20.1 (continued)

Protein tag	Acronym of tag construct	Components	Binding agent	Number of purified (tagged) proteins	Number of ligands identified	References
Ubiquitin (Ub) – hexahistidine	6His-UBQ	Ubiquitin (Ub) monomers N-terminally tagged with hexahistidine	GST–UBA; Ni^{2+} -NTA	1	54	Saracco et al. (2009)
HA–PreScission–Biotin	HBP	HA epitope*; HRV protease site; Biotin carboxyl carrier protein domain of 3-methylcrotonyl CoA carboxylase from Arabidopsis	anti-HA antibody*; Streptavidin magnetic beads	1 (5)	10	Qi and Katagiri (2009)

* has not been used for purification

virus (TEV) protease and a calmodulin-binding peptide (CBP) (Rigaut et al., 1999). The technique was explored in several biological systems including plants, and several variations of the TAP-tag were developed (Table 20.1). The TAP strategy is usually based on two (or more) rounds of protein purification in native conditions, in order to improve the purification of protein complexes for subsequent proteomic analysis. Incorporation of a specific protease site between two or more tags allows removal of the distal tag used in the first round of purification (usually Protein A or G), thus eliminating the non-specific background associated with IgG fragments. Essentially, any combination of epitope tags which includes a proteolytic site between the tags, with proximal tag remaining to be attached to the protein for the second round of purification not requiring antibody-based purification, is suitable for proteomic analysis by mass spectrometry. So far, several principal TAP-tags and their modifications (Table 20.1) have been developed for plants (Qi and Katagiri, 2009; Rohila et al., 2004; Rohila et al., 2006; Rubio et al., 2005; Van Leene et al., 2007; Van Leene et al., 2008).

The original TAP-tag (Rigaut et al., 1999) has been adapted to improve its expression in plants by optimisation of the codon sequence, removing a cryptic splice site, a nuclear localisation signal, the polyadenylation sites as well as the AT– or GC-rich regions, and by the inclusion of castor bean catalase intron 1. This tag was named TAPi (Rohila et al., 2004; Rohila et al., 2006). The creation of GATEWAY-compatible binary vectors, NTAPi and CTAPi (for tagging of the N– or C– terminus of the protein respectively), allowed high-throughput cloning for expression of tagged proteins. This tagging strategy was applied to study interaction partners of 129 TAP-tagged rice protein kinases and in total, 84 rice kinases were purified, of those 36 were isolated as complexes with one or more interacting proteins (Rohila et al., 2006; Rohila et al., 2009). Modified N/C TAPi tags in combination with the GFP reporter protein in Arabidopsis were used with success to study cell cycle related protein complexes, allowing the identification of 28 new molecular associations and confirming 14 previously described interactions (Van Leene et al., 2007).

We used NTAPi purification strategy for isolation and proteomic analysis of proteins interacting with nuclear transport factor p15h2 (At1g27970), a small (14 kDa) protein. Using the modified protocol of NTAPi purification, the NTAPi-p15h2 bait protein was isolated along with a number of associated proteins (Koroleva et al., 2007). Since another nuclear transport factor, p15h3 (At1g27310), was present among the co-purified proteins, it is likely that these proteins form a stable heterodimer or higher order complex associated with nuclear pores (Koroleva et al., 2007). Other examples of successful TAPi-tag purification approach are the isolation of a Cf-9 protein complex (420 kDa) in tobacco by Rivas et al. (2002) and the cleavage and polyadenylation specificity factor complex in Arabidopsis (Zhao et al., 2009a). However, there are several drawbacks in using of NTAPi and CTAPi tags. Native proteins containing calmodulin binding domains are usually co-purified with the NTAPi or CTAPi tagged protein (Koroleva et al., 2007). In addition, the TEV protease has a very low activity below 16°C. Thus, proteolytic removal of the protein A tag has to be performed at 16°C, which may lead to proteolysis by endogenous proteases. Protease inhibitors cannot be added to the reaction buffer because they would inhibit the TEV protease itself, with the exception of TEV-compatible E-64 inhibitor, but the latter inhibits only cysteine proteases.

The GS-tag, originally developed for the isolation of mammalian protein complexes by Burckstummer, et al. (2006), has been reported to give a tenfold increase in bait recovery when compared to the traditional TAP-tag. The GS-tag, which combines two IgG-binding domains of protein G and a streptavidin-binding peptide, separated by two TEV cleavage sites, has been recently adapted for plants by Van Leene and co-authors (2008). This system has significant advantages over the original TAP-tag, including reduced background levels and higher complex yield. It is also expected that replacement of the TEV cleavage sites in the GS tag with the rhinovirus 3C protease cleavage site will provide better protein complex stability during purification (Van Leene et al., 2008).

An alternative tandem TAP-tag for the isolation of plant protein complexes, the pC-TAPa construct, is also a Gateway-compatible vector, allowing convenient recombination of ORFs from pre-existing Gateway entry clones. The pC-TAPa tag includes three epitope tags (two protein A binding domains, six histidines and nine myc epitope repeats) and a 3C HRV protease site (Rubio et al., 2005). The use of 3C HRV protease is advantageous due to its higher activity at lower temperatures compared to the TEV protease, however this is offset by its higher cost. Replacement the original CBP module prevented the non-specific co-purification of endogenous calmodulin-binding proteins. However only a single protein complex has been characterized with the CTAPa tag so far (Rubio et al., 2005).

A novel strategy for TAP-tag protein purification was recently suggested, using the biotin carboxyl carrier protein domain (BCCD) of 3-methylcrotonyl CoA carboxylase (MCCA) from Arabidopsis (Qi and Katagiri, 2009). This protein domain is readily biotinylated in plants and in Escherichia coli (Wang et al., 1994) and was used in development of the HA–PreScission–Biotin (HBP) tag. The BCCD domain of the HPB tag is biotinylated in vivo, and the tagged protein can be captured by streptavidin beads on the basis of biotin-straptavidin interaction. The high affinity of this interaction allowed efficient recovery of tagged proteins and stringent purification to reduce non-specific binding of other proteins (Qi and Katagiri, 2009; Zhong et al., 2003). The HPB tag was recently successfully applied to the challenging task of purification of low-abundance plasma-membrane protein complexes (Qi and Katagiri, 2009). However the authors preferred one-step only affinity purification based on the biotin–streptavidin interaction using magnetic streptavidin beads, because it produced much higher yields with adequate purity of the tagged protein compared to the two-step TAP procedure. The biotin affinity purification after microsomal fractionation and chemical cross-linking/solubilisation allowed successful identification of the bait protein and its 10 potential interactors (Qi and Katagiri, 2009). However, the proximal location of HA in the HPB tag does not allow proteolytic cleavage or the removal of antibody fragments during the second purification step, rendering the preparation sub-optimal for LC-MS/MS.

Another TAP method to isolate ubiquitin–protein conjugates from Arabidopsis that exploits a stable transgenic line expressing a synthetic poly-UBQ gene encoding ubiquitin (Ub) monomers N-terminally tagged with hexahistidine has recently been decribed (Saracco et al., 2009). Sequential enrichment by Ub-affinity and nickel chelate-affinity chromatography, and followed by two-dimensional liquid chromatography and MS, led to identification of 54 non-redundant targets, expressed by as many as 90 possible isoforms (Saracco et al., 2009).

Two step purification approaches using TAP-tags were developed for isolation of higher purity protein complexes when compared to the single-tag approaches. However, in many cases one-step affinity purification methods can be more successful, since the protocols are easier and faster, and result in a higher yield of tagged proteins. Purification via the TAP tag requires approximately 5–6 h due to the combination of two purification steps and an intermediate protease cleavage, compared to 1–2 h for a single round of affinity purification. Experimental data suggests that a single tag may be no less suitable and efficient than a TAP-tag. Three purification strategies based on the StrepII, His(6) and TAP-tags were tested for their ability to purify one membrane-anchored protein kinase, and one soluble protein, from leaf extracts of *Nicotiana benthamiana* (Witte et al., 2004). Transiently expressed StrepII-tagged membrane protein was purified in less than 60 min and was directly suitable for enzymatic or MS analyses, allowing the identification of its phosphorylation sites. In contrast, purification of the same protein via His tag yielded partially oxidised protein of low purity. Soluble protein SGT1b was isolated with equal success as either TAP-tagged or StrepII-tagged protein with similar yield and high purity, however purification of the StrepII-tagged protein was considerably easier and faster. Using either tagging strategy, the same co– purifying protein was identified, suggesting that both the StrepII and TAP tags are suitable for purification of protein complexes from plant material (Witte et al., 2004).

For one round of TAPa purification, 15–90 g of fresh weight Arabidopsis plant tissues were used (Rubio et al., 2005), while 50–60 g were needed for extraction of TAPi protein complexes in rice (Rohila et al., 2006). When transgenic Arabidopsis suspension culture was used as a source for protein extraction, the culture volume was scaled up to 10 L and cells in exponential phase of growth were harvested on a sintered glass filter (Van Leene et al., 2007). In another NTAPi –based TAP study, 100 g of the cultured cells were used for a typical TAP experiment (Zhao et al., 2009b). Frozen cell suspension material (15 g) typically provided 200–300 mg of total protein (Van Leene et al., 2007). Some recombinant proteins can be expressed at high levels in cell culture, thus in these cases smaller samples of plant material are sufficient to yield adequate amounts of TAP-tagged protein for purification. Transiently expressed nuclear export factor p15h2 with attached NTAPi tag, isolated following modified TAP protocol, led to a successful purification of the NTAPi-p15h2 bait protein, altogether with isolation of 39 associated proteins from 5 g of pelleted Arabidopsis culture cells (ca 50 mL) (Koroleva et al., 2007).

One of the ways to prevent protein loss between TAP and MS/MS analysis is to use TCA protein precipitation, as an alternative to separating proteins on a gel and cutting out bands with subsequent digestion, and tryptic digestion of the pellet in a minimal volume (such as using ½ of reaction volumes recommended in conventional digestion protocols, resulting in final digest volume of 20–30 μL). For successful identification of proteins interacting with nuclear transport factor p15h2 using TAP tag purification (Koroleva et al., 2007), TAP eluates were combined and cold 100% TCA was added to a final concentration of 20% and samples incubated overnight at 4°C. After centrifugation (15 min, 12,000 g in a benchtop centrifuge at 4°C), the supernatant was removed and the resulting pellet was washed twice by addition of 300 μL of ice-cold acetone and then dried in an open Eppendorf tube for 30 min at room temperature. This modification of the TAP procedure avoided losses of yield at the stage of gel separation, in-gel digestion and extraction of peptides from the gel, which can be 50% or more. It also minimized the time required for the MS analysis, as all isolated proteins were identified simultaneously during a single 1 h of nLC – MS/MS analysis.

20.6 Single-Cell Proteomics

The definition of proteomics is identification of the complete set of proteins expressed by a genome, cell, tissue or whole organism at a given time under defined conditions (Wilkins et al., 1997). Ideally, this data set should be quantitative and include information of the post-translational modifications of each protein. Tremendous improvements in the accuracy, sensitivity, speed and throughput of the MS instruments and development of powerful analytical software led to fast progress in proteomics research in the last few years. However MS-based analytical approaches to study the location and dynamics of all proteins at the single-cell level are not yet available. Applying proteomics to a single cell is a challenge. The volume of a typical single plant cell usually is in the range of 20 pL–1 nL (Koroleva et al., 2000; Koroleva et al., 1998) and consequently the expected total amount of protein is extremely small, not more than 1 ng (i.e., in the range of 1–100 fmol) per cell. The concentration of an individual protein will comprise only a very small fraction of this total amount. The current techniques allow researchers to tag and overexpress known proteins inside a cell, but due to the lack of protein amplification techniques (similar to PCR for DNA amplification) it is not possible to analyze a whole proteome or to determine differences in protein expression between individual cells. Only the most abundant peptides could be detected from small samples derived from individual cells. There are two major directions in single-cell proteomics: either identification of abundant proteins in specialised single cells ("true" single cell proteomics) or studying homogenous populations of a single cell type. Cell callus and cell cultures, or cells of the same type mechanically separated from the whole tissue, are expected to have the same identity and therefore a population of such cells can be pooled for the analysis of their proteome.

The overexpression of a gene of interest is the only available technique which could lead to an increase in spectral counts for respective peptide ions. This was a successful approach adopted by Kondo et al. (2006) for identification of Clavata (CLV3) peptide in Arabidopsis cells by MALDI imaging. The authors used Arabidopsis meristem calli of a transgenic cell line overexpressing the respective gene for the characterization of the structure of the mature peptide MCLV3 encoded by the CLV3 gene. MALDI-TOF (time– of– flight) MS was performed in situ and a specific ion was observed related to a peptide containing 12 amino acid residues from Arg70 to His81 in CLV3, in which two of three proline residues were modified to hydroxyproline (Ph or Hyp). The MCLV3 was the only peptide specifically detected in CLV3-overexpressing calli and, therefore, was concluded to be the only CLV3-derived peptide present in the tissue (Kondo et al., 2006).

Guard cells are a unique cell type situated on the surface of plant tissues which regulate plant gas exchange and water loss by controlling the size of the stomatal openings. They can be isolated as protoplasts by specific digestion of epidermal pieces with cell wall–degrading enzymes. Purified guard cells protoplasts were used for single-cell-type proteome analysis in order to determine their protein composition (Zhao et al., 2008). Moreover, differences in protein complement between purified guard cells and mesophyll cells have been recently addressed through a global comparative proteomics analysis (Zhu et al., 2009). Using iTRAQ tagging and two-dimensional LC-MS, 1458 non-redundant proteins were identified in both guard cells and mesophyll cells of *Brassica napus* leaves. Stringent statistical analysis indicated that a total of 74 proteins were enriched in guard cells (Zhu et al., 2009).

The proteome of isolated soybean root hair cells (separated from roots by filtering through a wire mesh) was defined by 2D-PAGE separation of the total protein extract combined with shotgun proteomics approach (Brechenmacher et al., 2009). In total, 1,492 different proteins were identified in soybean root hair proteome, of which 1,134 were identified by shotgun analysis and 527 by 2D-PAGE followed by MS. Only 169 proteins were identified by both approaches, which emphasises the advantage of using both methods (Brechenmacher et al., 2009). The soybean root hair proteome data provided useful insight into the metabolic activities of a single, differentiated plant cell type.

A compromise between single-cell sampling and a whole-tissue sample is the use of a laser capture microdissection (LCM) for isolation of closely positioned/related group of cells. It partially overcomes the problem of tissue heterogeneity and contamination. Due to a low protein content, the sample does not need any further fractionation prior to LC or MALDI –MS, however only the most abundant (or marker) proteins can be identified. The LCM technique has been combined with isotope-coded affinity tag (ICAT) technology and two-dimensional liquid chromatography to investigate the qualitative and quantitative proteomes of carcinoma cells (Li et al., 2004). A total of 644 proteins were qualitatively identified on dissected cells, and 261 proteins were unambiguously quantitated (Li et al., 2004). While there are many other examples of LCM technique coupled with liquid chromatography–tandem mass spectrometry for proteomic analysis in biomedical research, fewer reports are available for plant tissues. LCM has been used to generate tissue-specific protein profiles of isolated Arabidopsis vasculature samples (Schad et al., 2005). Total samples of 2 μg of protein were isolated from approximately 20,000 vascular bundle cells. Proteins were digested and peptides analysed by LC-MS/MS. As a result, 33 specific proteins were identified (Schad et al., 2005). In maize, LCM isolation of root pericycle cells and separation of the most abundant, soluble pericycle proteins by 2D gel electrophoresis followed by LC-MS/MS led to identification of 20 proteins (two of which were preferentially expressed in the pericycle), thus defining maize pericycle proteome (Dembinsky et al., 2007).

20.7 Concluding Remarks

The higher plants are elaborate biological systems, and this is reflected in the complexity and dynamics of their proteomes and metabolomes. In addition to the large variation in the protein composition between different cells and tissues discussed in this chapter, alternative splicing and posttranslational modifications further contribute to the complexity of the proteome. Therefore, combinations of approaches and methods as well as development of new protocols are continually required for the insight into the functional aspects of plant growth and development.

Acknowledgements The authors are grateful to Dr Katy Cooke (University of Reading) for critical reading of the manuscript.

Appendix: List of Abbreviations

BCCD	Biotin Carboxyl Carrier protein Domain
CBP	Calmodulin Binding Protein
CPLL	Combinatorial Peptide Ligand Library
DIGE	Difference Gel Electrophoresis
DTT	Dithiothreitol
EST	Expressed Sequence Tag
GFP	Green Fluorescent Protein
GST	Glutathione S-Transferase
HA	Haemagglutinin
HILEP	Hydroponic Isotope Labeling of Entire Plants
HIS	Histidine
HPB	HA–PreScission–Biotin
ICAT	Isotope Coded Affinity Tags
ICPL	Isotope Coded Protein Label
IgG	Immunoglobulin G
iTRAQ	Isobaric Tag for Relative and Absolute Quantification
LC	Liquid Chromatography
MALDI	Matrix-Assisted Laser Desorption Ionization
MBP	Maltose Binding Protein
MS	Mass Spectrometry
MS/MS	Tandem Mass Spectrometry
nLC	Nano Liquid Chromatography
ORF	Open Reading Frame
PAGE	PolyAcrylamide Gel Electrophoresis
PVP	Polyvinylpolypyrrolidone

35S CaMV	Promotor of Cauliflower mosaic virus 35S transcript
SDS	Sodium Dodecyl Sulfate
SILAC	Stable Isotope Labeling with Amino acids in Cell culture
SILIP	Stable Isotope Labeling in Planta
Strep	Streptavidin
TAP	Tandem Affinity Purification
TBP	TATA Binding Protein
TCA	Trichloroacetic Acid
TOF	Time-Of-Flight

References

Aki, T., Shigyo, M., Nakano, R., Yoneyama, T., and Yanagisawa, S. (2008). Nano scale proteomics revealed the presence of regulatory proteins including three FT-Like proteins in phloem and xylem saps from rice. Plant Cell Physiol *49*, 767–790.

Albenne, C., Canut, H., Boudart, G., Zhang, Y., San Clemente, H., Pont-Lezica, R., and Jamet, E. (2009). Plant cell wall proteomics: mass spectrometry data, a trove for research on protein structure/function relationships. Mol Plant *2*, 977–989.

Alvarez, S., Goodger, J.Q., Marsh, E.L., Chen, S., Asirvatham, V.S., and Schachtman, D.P. (2006). Characterization of the maize xylem sap proteome. J Proteome Res *5*, 963–972.

Arabidopsis Genome Initiative (2000). Analysis of the genome sequence of the flowering plant Arabidopsis thaliana. Nature *408*, 796–815.

Bae, M.S., Cho, E.J., Choi, E.Y., and Park, O.K. (2003). Analysis of the Arabidopsis nuclear proteome and its response to cold stress. Plant J *36*, 652–663.

Baerenfaller, K., Grossmann, J., Grobei, M.A., Hull, R., Hirsch-Hoffmann, M., Yalovsky, S., Zimmermann, P., Grossniklaus, U., Gruissem, W., and Baginsky, S. (2008). Genome-scale proteomics reveals Arabidopsis thaliana gene models and proteome dynamics. Science *320*, 938–941.

Bayer, E.M., Bottrill, A.R., Walshaw, J., Vigouroux, M., Naldrett, M.J., Thomas, C.L., and Maule, A.J. (2006). Arabidopsis cell wall proteome defined using multidimensional protein identification technology. Proteomics *6*, 301–311.

Bayer, E., Thomas, C.L., and Maule, A.J. (2004). Plasmodesmata in Arabidopsis thaliana suspension cells. Protoplasma *223*, 93–102.

Bindschedler, L.V., Palmblad, M., and Cramer, R. (2008). Hydroponic isotope labelling of entire plants (HILEP) for quantitative plant proteomics; An oxidative stress case study. Phytochemistry *69*, 1962–1972.

Bortiri, E., Coleman-Derr, D., Lazo, G.R., Anderson, O.D., and Gu, Y.Q. (2008). The complete chloroplast genome sequence of Brachypodium distachyon: Sequence comparison and phylogenetic analysis of eight grass plastomes. BMC Res Notes *1*, 61.

Boschetti, E., Bindschedler, L.V., Tang, C., Fasoli, E., and Righetti, P.G. (2009). Combinatorial peptide ligand libraries and plant proteomics: A winning strategy at a price. J Chromatogr A *1216*, 1215–1222.

Brechenmacher, L., Lee, J., Sachdev, S., Song, Z., Nguyen, T.H., Joshi, T., Oehrle, N., Libault, M., Mooney, B., Xu, D., *et al.* (2009). Establishment of a protein reference map for soybean root hair cells. Plant Physiol *149*, 670–682.

Burckstummer, T., Bennett, K.L., Preradovic, A., Schutze, G., Hantschel, O., Superti-Furga, G., and Bauch, A. (2006). An efficient tandem affinity purification procedure for interaction proteomics in mammalian cells. Nat Methods *3*, 1013–1019.

Calikowski, T.T., Meulia, T., and Meier, I. (2003). A proteomic study of the arabidopsis nuclear matrix. J Cell Biochem *90*, 361–378.

Carpentier, S.C., Panis, B., Vertommen, A., Swennen, R., Sergeant, K., Renaut, J., Laukens, K., Witters, E., Samyn, B., and Devreese, B. (2008). Proteome analysis of non-model plants: A challenging but powerful approach. Mass Spectrom Rev *27*, 354–377.

Carter, C., Pan, S., Zouhar, J., Avila, E.L., Girke, T., and Raikhel, N.V. (2004). The vegetative vacuole proteome of Arabidopsis thaliana reveals predicted and unexpected proteins. Plant Cell *16*, 3285–3303.

Castellana, N.E., Payne, S.H., Shen, Z., Stanke, M., Bafna, V., and Briggs, S.P. (2008). Discovery and revision of Arabidopsis genes by proteogenomics. Proc Natl Acad Sci USA *105*, 21034–21038.

Cellar, N.A., Kuppannan, K., Langhorst, M.L., Ni, W., Xu, P., and Young, S.A. (2008). Cross species applicability of abundant protein depletion columns for ribulose-1,5-bisphosphate carboxylase/oxygenase. J Chromatogr B Anal Technol Biomed Life Sci *861*, 29–39.

Charmont, S., Jamet, E., Pont-Lezica, R., and Canut, H. (2005). Proteomic analysis of secreted proteins from Arabidopsis thaliana seedlings: Improved recovery following removal of phenolic compounds. Phytochemistry *66*, 453–461.

Childs, K.L., Hamilton, J.P., Zhu, W., Ly, E., Cheung, F., Wu, H., Rabinowicz, P.D., Town, C.D., Buell, C.R., and Chan, A.P. (2007). The TIGR plant transcript assemblies database. Nucleic Acids Res *35*, D846–851.

Chivasa, S., Simon, W.J., Yu, X.L., Yalpani, N., and Slabas, A.R. (2005). Pathogen elicitor-induced changes in the maize extracellular matrix proteome. Proteomics *5*, 4894–4904.

Cho, W.K., Chen, X.Y., Chu, H., Rim, Y., Kim, S., Kim, S.T., Kim, S.W., Park, Z.Y., and Kim, J.Y. (2009). Proteomic analysis of the secretome of rice calli. Physiol Plant *135*, 331–341.

Choudhary, M.K., Basu, D., Datta, A., Chakraborty, N., and Chakraborty, S. (2009). Dehydration-responsive nuclear proteome of rice (Oryza sativa L.) illustrates protein network, novel regulators of cellular adaptation, and evolutionary perspective. Mol Cell Proteomics *8*, 1579–1598.

Deeken, R., Geiger, D., Fromm, J., Koroleva, O., Ache, P., Langenfeld-Heyser, R., Sauer, N., May, S.T., and Hedrich, R. (2002). Loss of the AKT2/3 potassium channel affects sugar loading into the phloem of Arabidopsis. Planta *216*, 334–344.

Dembinsky, D., Woll, K., Saleem, M., Liu, Y., Fu, Y., Borsuk, L.A., Lamkemeyer, T., Fladerer, C., Madlung, J., Barbazuk, B., *et al.* (2007). Transcriptomic and proteomic analyses of pericycle cells of the maize primary root. Plant Physiol *145*, 575–588.

Dunkley, T.P., Hester, S., Shadforth, I.P., Runions, J., Weimar, T., Hanton, S.L., Griffin, J.L., Bessant, C., Brandizzi, F., Hawes, C., *et al.* (2006). Mapping the Arabidopsis organelle proteome. Proc Natl Acad Sci USA *103*, 6518–6523.

Dunkley, T.P., Watson, R., Griffin, J.L., Dupree, P., and Lilley, K.S. (2004). Localization of organelle proteins by isotope tagging (LOPIT). Mol Cell Proteomics *3*, 1128–1134.

Eubel, H., Meyer, E.H., Taylor, N.L., Bussell, J.D., O'Toole, N., Heazlewood, J.L., Castleden, I., Small, I.D., Smith, S.M., and Millar, A.H. (2008). Novel proteins, putative membrane transporters, and an integrated metabolic network are revealed by quantitative proteomic analysis of Arabidopsis cell culture peroxisomes. Plant Physiol *148*, 1809–1829.

Fasoli, E., D'Amato, A., Kravchuk, A.V., Boschetti, E., Bachi, A., Righetti, P.G. (2011). Popeye strikes again: The deep proteome of spinach leaves. J Proteomics *1*, 74(1):127–36. Epub 2010 Nov 5.

Friso, G., Giacomelli, L., Ytterberg, A.J., Peltier, J.-B., Rudella, A., Sun, Q., and Wijk, K.J.v. (2004). In-depth analysis of the Thylakoid membrane proteome of Arabidopsis thaliana chloroplasts: New proteins, new functions, and a plastid proteome database. Plant Cell *16*, 478–499.

Gaupels, F., Buhtz, A., Knauer, T., Deshmukh, S., Waller, F., van Bel, A.J., Kogel, K.H., and Kehr, J. (2008). Adaptation of aphid stylectomy for analyses of proteins and mRNAs in barley phloem sap. J Exp Bot *59*, 3297–3306.

Giavalisco, P., Kapitza, K., Kolasa, A., Buhtz, A., and Kehr, J. (2006). Towards the proteome of Brassica napus phloem sap. Proteomics *6*, 896–909.

Groen, A.J., de Vries, S.C., and Lilley, K.S. (2008). A proteomics approach to membrane trafficking. Plant Physiol *147*, 1584–1589.

Heazlewood, J.L., Verboom, R.E., Tonti-Filippini, J., Small, I., and Millar, A.H. (2007). SUBA: The Arabidopsis subcellular database. Nucleic Acids Res *35*, D213–218.

Huang, S., Li, R., Zhang, Z., Li, L., Gu, X., Fan, W., Lucas, W.J., Wang, X., Xie, B., Ni, P., *et al.* (2009a). The genome of the cucumber, Cucumis sativus L. Nat Genet. Advance online publication.

Huang, S., Taylor, N.L., Narsai, R., Eubel, H., Whelan, J., and Millar, A.H. (2009b). Experimental analysis of the rice mitochondrial proteome, its biogenesis, and heterogeneity. Plant Physiol *149*, 719–734.

Huo, N., Lazo, G.R., Vogel, J.P., You, F.M., Ma, Y., Hayden, D.M., Coleman-Derr, D., Hill, T.A., Dvorak, J., Anderson, O.D., *et al.* (2008). The nuclear genome of Brachypodium distachyon: analysis of BAC end sequences. Funct Integr Genomics *8*, 135–147.

International Rice Genome Sequencing Project (2005). The map-based sequence of the rice genome. Nature *436*, 793–800.

Irshad, M., Canut, H., Borderies, G., Pont-Lezica, R., and Jamet, E. (2008). A new picture of cell wall protein dynamics in elongating cells of Arabidopsis thaliana: Confirmed actors and newcomers. BMC Plant Biol *8*, 94.

Ishikawa, M., Fujiwara, M., Sonoike, K., and Sato, N. (2009). Orthogenomics of photosynthetic organisms: bioinformatic and experimental analysis of chloroplast proteins of endosymbiont origin in Arabidopsis and their counterparts in Synechocystis. Plant Cell Physiol *50*, 773–788.

Jamet, E., Albenne, C., Boudart, G., Irshad, M., Canut, H., and Pont-Lezica, R. (2008a). Recent advances in plant cell wall proteomics. Proteomics *8*, 893–908.

Jamet, E., Boudart, G., Borderies, G., Charmont, S., Lafitte, C., Rossignol, M., Canut, H., and Pont-Lezica, R. (2008b). Isolation of plant cell wall proteins. Methods Mol Biol *425*, 187–201.

Kalluri, U.C., Hurst, G.B., Lankford, P.K., Ranjan, P., and Pelletier, D.A. (2009). Shotgun proteome profile of Populus developing xylem. Proteomics *9*, 4871–4880.

Kato, Y., Hazama, A., Yamagami, M., and Uozumi, N. (2003). Addition of a peptide tag at the C terminus of AtHKT1 inhibits its Na+ transport. Biosci Biotechnol Biochem *67*, 2291–2293.

Kehr, J., Buhtz, A., and Giavalisco, P. (2005). Analysis of xylem sap proteins from Brassica napus. BMC Plant Biol *5*, 11.

Kehr, J., and Rep, M. (2007). Protein extraction from xylem and phloem sap. Methods Mol Biol *355*, 27–35.

Khan, M.M., and Komatsu, S. (2004). Rice proteomics: Recent developments and analysis of nuclear proteins. Phytochemistry *65*, 1671–1681.

Kleffmann, T., von Zychlinski, A., Russenberger, D., Hirsch-Hoffmann, M., Gehrig, P., Gruissem, W., and Baginsky, S. (2007). Proteome dynamics during plastid differentiation in rice. Plant Physiol *143*, 912–923.

Kondo, T., Sawa, S., Kinoshita, A., Mizuno, S., Kakimoto, T., Fukuda, H., and Sakagami, Y. (2006). A plant peptide encoded by CLV3 identified by in situ MALDI-TOF MS analysis. Science *313*, 845–848.

Koroleva, O.A., Calder, G., Pendle, A.F., Kim, S.H., Lewandowska, D., Simpson, C.G., Jones, I.M., Brown, J.W., and Shaw, P.J. (2009). Dynamic behavior of Arabidopsis eIF4A-III, putative core protein of exon junction complex: fast relocation to nucleolus and splicing speckles under hypoxia. Plant Cell *21*, 1592–1606.

Koroleva, O.A., Davies, A., Deeken, R., Thorpe, M.R., Tomos, A.D., and Hedrich, R. (2000). Identification of a new glucosinolate-rich cell type in Arabidopsis flower stalk. Plant Physiol *124*, 599–608.

Koroleva, O.A., Farrar, J.F., Deri Tomos, A., and Pollock, C.J. (1998). Carbohydrates in individual cells of epidermis, mesophyll, and bundle sheath in barley leaves with changed export or photosynthetic rate. Plant Physiol *118*, 1525–1532.

Koroleva, O., McKeown, P., Pendle, A.F., and Shaw, P. (2007). Proteomic Analysis of the Plant Nucleolus. In Plant Proteomics, J. Šamaj, J. Thelen, eds. (Berlin, Heidelberg, New York, Springer), pp. 247–269.

Koroleva, O.A., Tomlinson, M., Parinyapong, P., Sakvarelidze, L., Leader, D., Shaw, P., and Doonan, J.H. (2004). CycD1, a putative G1 cyclin from Antirrhinum majus, accelerates the cell cycle in cultured tobacco BY-2 cells by enhancing both G1/S entry and progression through S and G2 phases. Plant Cell *16*, 2364–2379.

Kwon, H.K., Yokoyama, R., and Nishitani, K. (2005). A proteomic approach to apoplastic proteins involved in cell wall regeneration in protoplasts of Arabidopsis suspension-cultured cells. Plant Cell Physiol *46*, 843–857.

Li, C., Hong, Y., Tan, Y.X., Zhou, H., Ai, J.H., Li, S.J., Zhang, L., Xia, Q.C., Wu, J.R., Wang, H.Y., *et al.* (2004). Accurate qualitative and quantitative proteomic analysis of clinical hepatocellular carcinoma using laser capture microdissection coupled with isotope-coded affinity tag and two-dimensional liquid chromatography mass spectrometry. Mol Cell Proteomics *3*, 399–409.

Lin, M.K., Lee, Y.J., Lough, T.J., Phinney, B.S., and Lucas, W.J. (2009). Analysis of the pumpkin phloem proteome provides insights into angiosperm sieve tube function. Mol Cell Proteomics *8*, 343–356.

Mann, M. (1999). Quantitative proteomics? Nat Biotechnol *17*, 954–955.

Mano, S., Miwa, T., Nishikawa, S.-i., Mimura, T., and Nishimura, M. (2008). The plant organelles database (PODB): A collection of visualized plant organelles and protocols for plant organelle research. Nucl Acids Res *36*, D929–937.

McCabe, M.S., Garratt, L.C., Schepers, F., Jordi, W.J., Stoopen, G.M., Davelaar, E., van Rhijn, J.H., Power, J.B., and Davey, M.R. (2001). Effects of P(SAG12)-IPT gene expression on development and senescence in transgenic lettuce. Plant Physiol *127*, 505–516.

Millar, A.H., Sweetlove, L.J., Giege, P., and Leaver, C.J. (2001). Analysis of the Arabidopsis mitochondrial proteome. Plant Physiol *127*, 1711–1727.

Millar, D.J., Whitelegge, J.P., Bindschedler, L.V., Rayon, C., Boudet, A.M., Rossignol, M., Borderies, G., and Bolwell, G.P. (2009). The cell wall and secretory proteome of a tobacco cell line synthesising secondary wall. Proteomics *9*, 2355–2372.

Ni, R.J., Shen, Z., Yang, C.P., Wu, Y.D., Bi, Y.D., and Wang, B.C. (2010). Identification of low abundance polyA-binding proteins in Arabidopsis chloroplast using polyA-affinity column. Mol Biol Rep *37*, 637–641.

Ong, S.E., and Mann, M. (2005). Mass spectrometry-based proteomics turns quantitative. Nat Chem Biol *1*, 252–262.

Ong, S.E., and Mann, M. (2006). A practical recipe for stable isotope labeling by amino acids in cell culture (SILAC). Nat Protoc *1*, 2650–2660.

Palmblad, M., Bindschedler, L.V., and Cramer, R. (2007). Quantitative proteomics using uniform (15)N-labeling, MASCOT, and the trans-proteomic pipeline. Proteomics *7*, 3462–3469.

Pandey, A., Chakraborty, S., Datta, A., and Chakraborty, N. (2008). Proteomics approach to identify dehydration responsive nuclear proteins from chickpea (Cicer arietinum L.). Mol Cell Proteomics *7*, 88–107.

Parry, M.A., Andralojc, P.J., Mitchell, R.A., Madgwick, P.J., and Keys, A.J. (2003). Manipulation of rubisco: The amount, activity, function and regulation. J Exp Bot *54*, 1321–1333.

Peckham, G.D., Bugos, R.C., and Su, W.W. (2006). Purification of GFP fusion proteins from transgenic plant cell cultures. Protein Expr Purif *49*, 183–189.

Pendle, A.F., Clark, G.P., Boon, R., Lewandowska, D., Lam, Y.W., Andersen, J., Mann, M., Lamond, A.I., Brown, J.W., and Shaw, P.J. (2005). Proteomic analysis of the Arabidopsis nucleolus suggests novel nucleolar functions. Mol Biol Cell *16*, 260–269.

Qi, Y., and Katagiri, F. (2009). Purification of low-abundance Arabidopsis plasma-membrane protein complexes and identification of candidate components. Plant J *57*, 932–944.

Rigaut, G., Shevchenko, A., Rutz, B., Wilm, M., Mann, M., and Seraphin, B. (1999). A generic protein purification method for protein complex characterization and proteome exploration. Nat Biotechnol *17*, 1030–1032.

Rivas, S., Romeis, T., and Jones, J.D. (2002). The Cf-9 disease resistance protein is present in an approximately 420-kilodalton heteromultimeric membrane-associated complex at one molecule per complex. Plant Cell *14*, 689–702.

Rohila, J.S., Chen, M., Cerny, R., and Fromm, M.E. (2004). Improved tandem affinity purification tag and methods for isolation of protein heterocomplexes from plants. Plant J *38*, 172–181.

Rohila, J.S., Chen, M., Chen, S., Chen, J., Cerny, R., Dardick, C., Canlas, P., Xu, X., Gribskov, M., Kanrar, S., *et al.* (2006). Protein-protein interactions of tandem affinity purification-tagged protein kinases in rice. Plant J *46*, 1–13.

Rohila, J.S., Chen, M., Chen, S., Chen, J., Cerny, R.L., Dardick, C., Canlas, P., Fujii, H., Gribskov, M., Kanrar, S., *et al.* (2009). Protein-protein interactions of tandem affinity purified protein kinases from rice. PLoS One *4*, e6685.

Rubio, V., Shen, Y., Saijo, Y., Liu, Y., Gusmaroli, G., Dinesh-Kumar, S.P., and Deng, X.W. (2005). An alternative tandem affinity purification strategy applied to Arabidopsis protein complex isolation. Plant J *41*, 767–778.

Saracco, S.A., Hansson, M., Scalf, M., Walker, J.M., Smith, L.M., and Vierstra, R.D. (2009). Tandem affinity purification and mass spectrometric analysis of ubiquitylated proteins in Arabidopsis. Plant J *59*, 344–358.

Schad, M., Lipton, M.S., Giavalisco, P., Smith, R.D., and Kehr, J. (2005). Evaluation of two-dimensional electrophoresis and liquid chromatography--tandem mass spectrometry for tissue-specific protein profiling of laser-microdissected plant samples. Electrophoresis *26*, 2729–2738.

Schaff, J.E., Mbeunkui, F., Blackburn, K., Bird, D.M., and Goshe, M.B. (2008). SILIP: a novel stable isotope labeling method for in planta quantitative proteomic analysis. Plant J *56*, 840–854.

Schmidt, U.G., Endler, A., Schelbert, S., Brunner, A., Schnell, M., Neuhaus, H.E., Marty-Mazars, D., Marty, F., Baginsky, S., and Martinoia, E. (2007). Novel tonoplast transporters identified using a proteomic approach with vacuoles isolated from cauliflower buds. Plant Physiol *145*, 216–229.

Sennels, L., Salek, M., Lomas, L., Boschetti, E., Righetti, P.G., and Rappsilber, J. (2007). Proteomic analysis of human blood serum using peptide library beads. J Proteome Res *6*, 4055–4062.

Shimaoka, T., Ohnishi, M., Sazuka, T., Mitsuhashi, N., Hara-Nishimura, I., Shimazaki, K.-I., Maeshima, M., Yokota, A., Tomizawa, K.-I., and Mimura, T. (2004). Isolation of intact vacuoles and proteomic analysis of tonoplast from suspension-cultured cells of Arabidopsis thaliana. Plant Cell Physiol *45*, 672–683.

Simpson, C., Thomas, C., Findlay, K., Bayer, E., and Maule, A.J. (2009). An Arabidopsis GPI-anchor plasmodesmal neck protein with callose binding activity and potential to regulate cell-to-cell trafficking. Plant Cell *21*, 581–594.

Sridhar, V.V., Surendrarao, A., Gonzalez, D., Conlan, R.S., and Liu, Z. (2004). Transcriptional repression of target genes by LEUNIG and SEUSS, two interacting regulatory proteins for Arabidopsis flower development. Proc Natl Acad Sci USA *101*, 11494–11499.

Sridhar, V.V., Surendrarao, A., and Liu, Z. (2006). APETALA1 and SEPALLATA3 interact with SEUSS to mediate transcription repression during flower development. Development *133*, 3159–3166.

Steen, H., and Mann, M. (2004). The ABC's (and XYZ's) of peptide sequencing. Nat Rev Mol Cell Biol *5*, 699–711.

Thelen, J.J., and Peck, S.C. (2007). Quantitative proteomics in plants: Choices in abundance. Plant Cell *19*, 3339–3346.

Thiellement, H., Zivy, M., Damerval, C., and Mechin, V. (eds.) (2006). Plant Proteomics: Methods and Protocols. Series: Methods in Molecular Biology, Vol. 355 (Humana Press, Totowa, New Jersey, USA), 399 p.

Van Leene, J., Stals, H., Eeckhout, D., Persiau, G., Van De Slijke, E., Van Isterdael, G., De Clercq, A., Bonnet, E., Laukens, K., Remmerie, N., *et al.* (2007). A tandem affinity purification-based technology platform to study the cell cycle interactome in Arabidopsis thaliana. Mol Cell Proteomics *6*, 1226–1238.

Van Leene, J., Witters, E., Inze, D., and De Jaeger, G. (2008). Boosting tandem affinity purification of plant protein complexes. Trends Plant Sci *13*, 517–520.

Wang, X., Wurtele, E.S., Keller, G., McKean, A.L., and Nikolau, B.J. (1994). Molecular cloning of cDNAs and genes coding for beta-methylcrotonyl-CoA carboxylase of tomato. J Biol Chem *269*, 11760–11768.

Wessel, D., and Flugge, U.I. (1984). A method for the quantitative recovery of protein in dilute solution in the presence of detergents and lipids. Anal Biochem *138*, 141–143.

Widjaja, I., Naumann, K., Roth, U., Wolf, N., Mackey, D., Dangl, J.L., Scheel, D., and Lee, J. (2009). Combining subproteome enrichment and Rubisco depletion enables identification of low abundance proteins differentially regulated during plant defense. Proteomics *9*, 138–147.

Wilchek, M., and Bayer, E.A. (1990). Applications of avidin-biotin technology: Literature survey. Methods Enzymol *184*, 14–45.

Wilkins, M.R., Lindskog, I., Gasteiger, E., Bairoch, A., Sanchez, J.C., Hochstrasser, D.F., and Appel, R.D. (1997). Detailed peptide characterization using PEPTIDEMASS--a World-Wide-Web-accessible tool. Electrophoresis *18*, 403–408.

Witte, C.P., Noel, L.D., Gielbert, J., Parker, J.E., and Romeis, T. (2004). Rapid one-step protein purification from plant material using the eight-amino acid StrepII epitope. Plant Mol Biol *55*, 135–147.

Xi, J., Wang, X., Li, S., Zhou, X., Yue, L., Fan, J., and Hao, D. (2006). Polyethylene glycol fractionation improved detection of low-abundant proteins by two-dimensional electrophoresis analysis of plant proteome. Phytochemistry *67*, 2341–2348.

Xu, Y., Wang, B.C., and Zhu, Y.X. (2007). Identification of proteins expressed at extremely low level in Arabidopsis leaves. Biochem Biophys Res Commun *358*, 808–812.

Yao, X., Freas, A., Ramirez, J., Demirev, P.A., and Fenselau, C. (2001). Proteolytic 18O labeling for comparative proteomics: model studies with two serotypes of adenovirus. Anal Chem *73*, 2836–2842.

Zanetti, M.E., Chang, I.F., Gong, F., Galbraith, D.W., and Bailey-Serres, J. (2005). Immunopurification of polyribosomal complexes of Arabidopsis for global analysis of gene expression. Plant Physiol *138*, 624–635.

Zhao, H., Xing, D., and Li, Q.Q. (2009a). Unique features of plant cleavage and polyadenylation specificity factor revealed by proteomic studies. Plant Physiol *151*, 1546–1556.

Zhao, H., Xing, D., and Li, Q.Q. (2009b). Unique features of plant cleavage and polyadenylation specificity factor revealed by proteomic studies. Plant Physiol *151*, 1546–1556.

Zhao, Z., Zhang, W., Stanley, B.A., and Assmann, S.M. (2008). Functional proteomics of Arabidopsis thaliana guard cells uncovers new stomatal signaling pathways. Plant Cell *20*, 3210–3226.

Zhong, J., Haynes, P.A., Zhang, S., Yang, X., Andon, N.L., Eckert, D., Yates, J.R., 3rd, Wang, X., and Budworth, P. (2003). Development of a system for the study of protein-protein interactions in planta: Characterization of a TATA-box binding protein complex in Oryza sativa. J Proteome Res *2*, 514–522.

Zhu, M., Dai, S., McClung, S., Yan, X., and Chen, S. (2009). Functional differentiation of Brassica napus guard cells and mesophyll cells revealed by comparative proteomics. Mol Cell Proteomics *8*, 752–766.

Zybailov, B., Rutschow, H., Friso, G., Rudella, A., Emanuelsson, O., Sun, Q., and van Wijk, K.J. (2008). Sorting signals, N-terminal modifications and abundance of the chloroplast proteome. PLoS One *3*, e1994.

Chapter 21
Plant Plasma Membrane Proteomics: Challenges and Possibilities

Anders Laurell Blom Møller, Katja Witzel, Annelies Vertommen, Vibeke Barkholt, Birte Svensson, Sebastien Carpentier, Hans-Peter Mock, and Christine Finnie

Abstract Plasma membrane proteins are challenging to study due to their hydrophobicity and low abundance. Extraction of plasma membranes from plants is complicated by the rigid plant cell wall. The most commonly used method to enrich for plant plasma membranes is the aqueous-polymer two-phase system. Many gel-based separation techniques have been used in plasma membrane proteomics studies including one-dimensional SDS-PAGE, classical two-dimensional electrophoresis based on isoelectric focussing and SDS-PAGE, double SDS-PAGE, Blue Native-PAGE and 16-BAC-PAGE. Some of the difficulties associated with gel-based separation of hydrophobic proteins can be bypassed by the use of gel-free approaches which involve in-solution separation of proteins or liquid chromatography-based separation of peptides. Mass spectrometry is subsequently used for identification of proteins. An overview is presented of the most commonly used methods for analysis of plant plasma membrane proteomes, their strengths and weaknesses.

Keywords 2D-16-BAC/SDS electrophoresis · Aqueous polymer two-phase system · Double SDS-PAGE · Native PAGE · Solution-phase isoelectric focusing

21.1 Introduction

Plants, being static organisms, must perceive and react to changes in their environment in order to survive stresses such as pathogens, temperature, water and nutrient availability. The plasma membrane plays an important role in this process since almost all environmental changes are initially perceived by plasma membrane

C. Finnie (✉)
Enzyme and Protein Chemistry, Department of Systems Biology, Technical University of Denmark, Lyngby, Denmark
e-mail: csf@bio.dtu.dk

A.R. Ivanov, A.V. Lazarev (eds.), *Sample Preparation in Biological Mass Spectrometry*, DOI 10.1007/978-94-007-0828-0_21,

proteins with sensory and signaling roles. Other plasma membrane proteins are transporters responsible for uptake of nutrients and water by plant cells. An in-depth knowledge and characterization of plasma membrane proteomes in different tissues, developmental stages and environmental conditions is therefore essential for a good understanding of plant responses to environment. Such knowledge should contribute substantially to targeted breeding or gene technology approaches for development of crop plants for the future. This review will focus on the techniques available for isolation and analysis of plant plasma membrane proteomes.

21.2 Isolation of Plasma Membranes from Plant Tissues

Isolation of plant plasma membranes is complicated by the presence of a rigid plant cell wall, containing pectins, hemicellulose, and cellulose, which defines the size and shape of the cell and acts as a protective external matrix (Mohnen, 2008; Sarkar et al., 2009). The cell wall makes tissue disruption a challenging issue in many plant proteomics studies.

21.2.1 Homogenisation of Plant Tissue

Efficient homogenisation of the plant tissue is critical for reproducibility and is necessary to obtain a sufficient yield of cell membranes. The method of choice is likely to differ for different plant tissues and species. Ideally, the method should result in cell disruption while leaving organelles intact. Tissue homogenisation is often achieved using pestle and mortar in liquid nitrogen or directly in the extraction buffer (Komatsu et al., 2009), however it should also be considered that freezing and thawing of the tissue will affect membrane integrity with possible negative influence on yield and purity. Alternatively a blender can be used for homogenising the plant tissue directly in the extraction buffer (Santoni, 2007). Sucrose (Komatsu et al., 2009; Mitra et al., 2009) or glycerol (Mitra et al., 2007) are usually included in the extraction buffer as osmolytes to ensure intact membrane vesicles. This osmotic concentration will be maintained throughout the whole isolation process.

21.2.2 Isolation of Microsomes

After cell lysis, cell debris is removed by filtration through a microcloth. Subsequently, sequential centrifugations are used for initial fractionation of homogenates. Typically, an initial centrifugation at 26,000×*g* (25 min) is used to pellet remaining cell wall debris, nuclei, mitochondria and chloroplasts (Santoni,

2007) followed by 50,000 × *g* (30 min) to separate the microsomal from the cytosolic fraction (Larsson et al., 1987).

21.2.3 Methods for Enrichment of Plasma Membranes

Plasma membranes represent a small fraction of the total cellular membranes, thus methods are needed to enrich specifically for the plasma membrane. The most commonly used method is the aqueous-polymer two-phase system (Larsson et al., 1987). When the polymers (Polyethylene glycol (PEG) 3350 and Dextran T500) are mixed in water at a critical concentration two phases are formed. In a phase system with an overall composition of 6.5% (w/w) PEG 3350, 6.5% (w/w) Dextran T500 and 87% (w/w) water, an upper phase with 0.18, 9.13 and 90.69% and a lower phase with 18.21%, 2.05 and 79.74% (w/w) Dextran T500, PEG 3350 and water, respectively, is formed at 0°C (Albertsson and Tjerneld, 1994). Most of the plant plasma membrane vesicles are located in the PEG 3350-enriched upper phase whereas endomembranes partition in the Dextran T500-enriched lower phase. Since this is an enrichment rather than a purification procedure, the upper phase still contains some endomembranes and likewise, some plasma membranes will also be lost to the lower phase. Yield and purity of plasma membranes can be increased by re-extraction of the lower phases with a fresh upper phase, re-extraction of upper phases with a fresh lower phase and combining the resulting upper phases (Larsson et al., 1987). A fraction with higher purity can subsequently be obtained by using a second two-phase system in which the affinity ligand wheat germ agglutinin (WGA) is coupled to dextran. WGA binds glycoproteins associated with the plasma membrane, thus plasma membrane vesicles will be trapped in the dextran-enriched phase (Persson et al., 1991; Persson and Jergil, 1994).

Since the efficiency of the two-phase system depends on physical properties of the membranes, which may differ between species, tissues or growth conditions, the composition of the two-phase system should ideally be optimised for each sample type. This involves testing a series of polymer concentrations, typically between 6.0 and 6.6% (w/w) and salt (KCl) concentrations in the range of 0–5 mM (Larsson et al., 1987; Santoni, 2007). Partitioning is highly sensitive to small changes in the composition of the phase system, so it is advisable to prepare sufficient batches for several phase systems to ensure comparable results (Brooks and Norris-Jones, 1994). These batches can be stored at –20°C. Since Dextran T500 is hygroscopic, different batches may contain varying amounts of water, which should be determined prior to use (for method see Note 1 in Santoni (2007)) (Brooks and Norris-Jones, 1994; Santoni, 2007). Temperature control throughout the experiment is also critical since small fluctuations in temperature affect phase composition (Albertsson and Tjerneld, 1994; Brooks and Norris-Jones, 1994; Santoni, 2007; Schindler and Notwang, 2006).

Free-flow electrophoresis has been used as an alternative to two-phase partitioning for separation of plasma membranes and tonoplasts from microsomes from *Arabidopsis thaliana* (Bardy et al., 1998). Microsomes are injected into a stream of buffer across which a perpendicular electrical field enables separation of membrane vesicles according to their size:charge ratio. Free-flow electrophoresis enables separation of both tonoplast and the plasma membrane from the same homogenate as well as large scale purification of membrane proteins in milligram quantities (Bardy et al., 1998). Furthermore the isolated vesicles retain their functions allowing assay of protein activities. When free-flow electrophoresis is employed after the aqueous-polymer two-phase system, high purity fractions of plasma membranes can be obtained (Bardy et al., 1998).

Affinity purification of plasma membrane vesicles using magnetic bead-coupled antibodies recognising plasma-membrane-specific proteins has been commercially developed (Qiagen) enabling isolation of plasma membrane proteins from mammalian cells. Alternatively, one-step affinity-chromatography based on the ability of the lectin concanavalin A (ConA) to bind glycosylated membrane proteins has been used to isolate plasma membranes from rat liver, PC-3 and HeLa cells (Lee et al., 2008).

Affinity-based purification of plasma membranes is of course highly dependent on the ligand used, which must be specific for a protein or proteins known to be sufficiently abundant and exclusively present in the plasma membrane. So far, this approach has not been developed for plant systems.

21.3 Removal of Associated Membrane Proteins, Soluble Proteins and Lipids

In addition to integral membrane proteins, isolated membrane vesicles contain peripheral membrane proteins, non-specifically associated soluble proteins and soluble proteins which are trapped within the vesicles upon tissue homogenisation. In a proteomics study, trapped and non-specifically associated soluble proteins are undesirable; therefore strategies are needed for their removal. Unfortunately, it is often difficult to distinguish between these proteins and biologically relevant peripheral membrane proteins.

Membrane-associated proteins can be stripped from membranes by sonication or washes in high pH or high ionic strength buffers (Blonder et al., 2004; Marmagne et al., 2007; Mobius et al., 2005; Pedersen et al., 2003; Rolland et al., 2006; Santoni et al., 1999; Zhang et al., 2006). Soluble proteins trapped inside vesicles can be removed by additional washing steps in high pH buffer (above pH 11) or in low concentrations of the mild non-ionic detergent Brij, which opens the vesicles and produces predominantly cytoplasmic side-out vesicles (Alexandersson et al., 2004; Johansson et al., 1995).

For proteomic analysis removal of membrane lipids is preferable since they interfere with gel electrophoresis, trypsin digestion and subsequent mass spectrometric identification (Simões-Barbosa et al., 2000). Removal of lipids can be facilitated by protein precipitation in a mixture of methanol and chloroform (Wessel and Flügge, 1984).

21.3.1 Quality Control of Plasma Membrane Fractions

Since most methods for plasma membrane isolation are enrichment procedures, the resulting preparations are unlikely to be completely free of endomembranes or nonspecifically associated soluble proteins. Especially for proteomics studies, where the aim is often to identify the protein complement of a specific cellular compartment, it is important to determine the purity of the sample under analysis. This is usually estimated by an enzymatic assay (Komatsu et al., 2009) or immunological detection (Hynek et al., 2006, 2009) of proteins with known cellular locations. Most commonly in plant studies, the plasma membrane H^+-ATPase is used as a marker. In enzymatic assays, the ATPase activity of the plasma membrane H^+-ATPase is compared to total ATPase activity of the sample, by assaying in the presence of the inhibitors vanadate, azide and nitrate that inhibit the plasma membrane, vacuolar and mitochondrial ATPase activities, respectively (Carletti et al., 2008; Komatsu et al., 2009; Santoni et al., 1998).

21.4 Gel-Based Proteomics

Strategies for analysis of plasma membrane proteomes can be divided into gel-based and gel-free approaches. Both methods have advantages and drawbacks and will be discussed in the following sections.

21.4.1 Classical Two-Dimensional Gel Electrophoresis of Membrane Proteins

In the first dimension of classical 2D gel electrophoresis (2DE), proteins are separated along a pH gradient by isoelectric focussing during which they migrate to their isoelectric point (pI). The second dimension consists of a SDS-PAGE separation perpendicular to the first. While this approach allows high resolution separations of soluble protein extracts, it is highly problematic for hydrophobic membrane proteins. The isoelectric focussing step is incompatible with the presence of high amounts of charged compounds such as salts and ionic detergents, which means that strong detergents such as SDS cannot be used to solubilise hydrophobic proteins. Secondly, membrane proteins are highly prone to aggregation and precipitation at their isolelectric point, preventing their transfer to the second dimension SDS-PAGE gel. Despite these challenges, 2DE approaches for membrane proteins have been developed based on different detergents/surfactants and chaotrope combinations for membrane protein solubilisation in the first dimension (Luche et al., 2003; Santoni et al., 2000; Tastet et al., 2003). Alkaline washes (in urea-NaOH) of membrane vesicles in combination with a mixture of detergents (β-dodecyl-n-maltoside or ASB14) (Santoni, 2007), were used for a better recovery of integral membrane proteins by classical 2D gel electrophoresis. This strategy enabled identification of aquaporins and P-type ATPases from 2DE gels. Recently, plasma membrane proteins from soybean were compared using gel-free LC MS/MS and 2DE (Komatsu et al., 2009) showing that most of the identified proteins in the gel-based method were membrane-associated proteins while the gel-free method additionally resulted in identification of integral plasma membrane proteins.

21.4.2 Native PAGE

Blue native (BN)-PAGE is used for separation of protein complexes ranging from 10 to 10000 kDa, and is applicable both to membrane and soluble protein complexes. To facilitate isolation and solubilisation of membrane protein complexes prior to electrophoresis, anionic, cationic, zwitterionic or non-ionic detergents can be used in low concentrations. While anionic and cationic detergents are more efficient at preventing protein aggregation, zwitterionic or non-ionic detergents are less likely to denature the proteins. Generally, the non-ionic detergents digitonin and β-dodecyl-n-maltoside are used (Reisinger and Eichacker, 2006). The negatively charged dye Coomassie Blue G250 is added, which binds to hydrophobic regions of proteins and facilitates their migration towards the anode (Schägger and von Jagow, 1991). Since the proteins are in their native state, complexes are maintained during BN-PAGE electrophoresis. By applying a denaturing second dimension separation on SDS-PAGE, dissociation of complexes enables resolution of the individual components. BN-PAGE followed by either an SDS-PAGE (Lüthje et al., 2009; Wittig et al., 2006) or a second native-PAGE dimension (Juszczuk and Rychter, 2009; Wumaier et al., 2009) has proved to be a powerful method for 2DE membrane proteomics. After 2D-BN/BN-PAGE, further resolution can be achieved by applying a third dimension SDS-PAGE separation (Schägger and Pfeiffer, 2000) or even double SDS-PAGE (see Section 4.3) (Meyer et al., 2007; Rais et al., 2004; Wumair et al., 2009). Adding additional dimensions increases resolution, enabling separation of larger super-complexes that otherwise would be difficult to analyse due to their size and complexity. A drawback of these methods is often the amount of material required, typically in the order of 400 μg membrane protein, making them impractical for scarce samples.

BN-PAGE has been successfully used to study plant mitochondrial membrane complexes (Heinemeyer et al., 2009; Juszczuk and Rychter, 2009; Wittig et al., 2006), chloroplast membrane proteins (Reisinger and Eichacker, 2006), plasma membrane complexes (Kjell et al., 2004) and redox proteins in the plasma membrane (Lüthje et al., 2009). In clear native PAGE, Coomassie Blue G250 is omitted from the first dimension separation such that proteins migrate according to their native charge under the running conditions (pH 7.5) (Schägger et al., 1994).

Native gel running conditions have the advantage that structure and activity of proteins and protein complexes may be retained, enabling "in-gel" activity measurements, which have been done for mitochondrial enzyme complexes (Jung et al., 2000). Protein complexes may also be electroeluted from gels and reconstituted, which was done for further analysis of the chloroplast FoF1-ATP Synthase, CFo and CF1 (Neff and Dencher, 1999) Another example is the electroelution from a native gel of the ATP synthase from spinach thylakoid membranes, which was subsequently crystallised in two-dimensions for structural studies (Poetsch et al., 2000).

21.4.3 Chromatographic Separation of Proteins Combined with 1D-SDS-PAGE

Fractionation of proteins by liquid chromatography can be combined with a second dimension of SDS-PAGE separation. Reversed-phase (RP) chromatography with elution of proteins by 2-propanol has been used to fractionate plasma membrane proteins isolated from barley seed tissues, based on their hydrophobicity (Hynek et al., 2006, 2009). Reversed-phase batch chromatography on C4 resin was used to fractionate plasma membrane-enriched samples from barley seed aleurone layers. This approach enabled more hydrophilic proteins to be eluted at a lower concentration of 2-propanol before elution of more hydrophobic proteins in 80% (v/v) 2-propanol (Hynek et al., 2006). Following SDS-PAGE separation, bands were cut from the gel and analysed by LC-MS/MS, enabling identification of integral membrane proteins (Hynek et al., 2006). A similar approach based on spin columns was applied to plasma membrane enriched fractions from germinating barley embryos (Hynek et al., 2009) and barley leaves (Fig. 21.1). RP-HPLC fractionation (Fig. 21.1) with gradient elution by 2-propanol allowed a more reproducible fractionation of plasma membrane proteins when compared to batch chromatography with stepwise elution.

21.4.4 Double SDS-PAGE

Double SDS PAGE is an alternative method for separating membrane proteins. The strong anionic detergent SDS is one of the most powerful detergents to solubilize proteins. Hydrophobic proteins have a higher affinity for this detergent compared with soluble proteins. This higher affinity results in an anomalous migration pattern during SDS-PAGE and forms the basis for double SDS-PAGE (Akiyama and Ito, 1985). Since the anomalous migration of hydrophobic proteins also depends on the acrylamide concentration, a different acrylamide concentration is applied in both dimensions. For example, a large pored gel ($\leq$ 10% acrylamide) can be used for

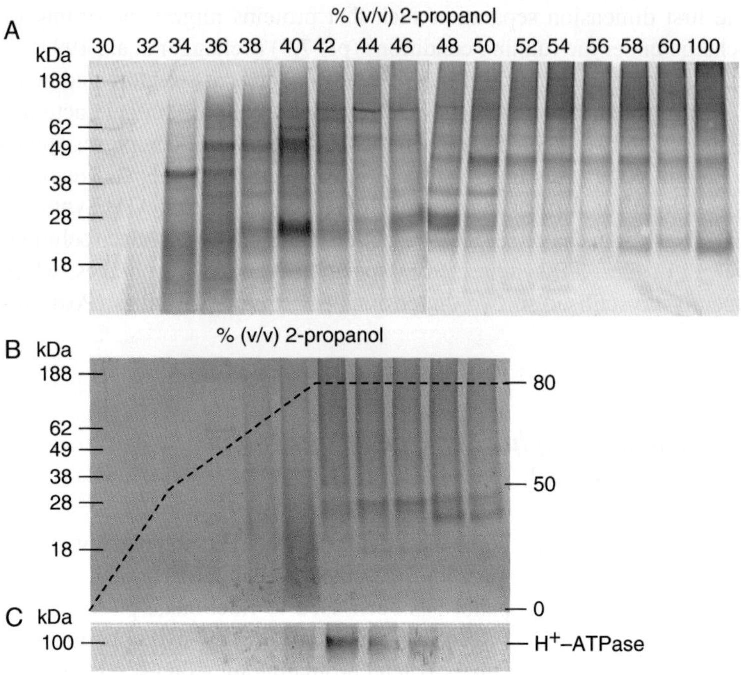

Fig. 21.1 Reversed-phase chromatography coupled with SDS-PAGE. A. Batch reversed-phase separation. Plasma membrane proteins (60 μg) from barley leaves were loaded onto a micro spin-column containing Nucleosil 300–5 C4 (Hynek et al., 2009) and eluted by a step-wise gradient of 2-propanol (30–100% [v/v]). B. Reversed-phase HPLC. Plasma membrane proteins (50 μg) from barley leaves were loaded onto a home-packed Nucleosil 300–5 C4 column (4.6 mm × 45 mm) and eluted with a gradient of 2-propanol. Eluted fractions were separated by SDS-PAGE and silver stained. C. Western blotting of the corresponding HPLC fractions from B, using an antibody raised against the plasma membrane H^+-ATPase

the first dimension SDS-PAGE. During this separation the excessive charge effect of the hydrophobic proteins predominates the molecular sieving effect (Beyreuther et al., 1980) and they migrate anomalously. Subsequently, a lane of the first dimension gel is excised and laid on top of a second dimension gel with narrow pores ($\geq$ 15% acrylamide). In this case, the smaller pore size of the gel is restrictive and the extra negative charge will not affect the migration pattern of membrane proteins. Membrane and soluble proteins migrate at the same speed and distance. The soluble proteins will form a diagonal line after the second dimension, whereas the aberrant migration of more hydrophobic proteins will cause them to lie off the diagonal.

The technique can be further optimized through addition of nonionic modifiers such as urea and glycerol to enhance the differential mobility effect (Rais et al., 2004; Williams et al., 2006). However, if identification of highly hydrophobic proteins is desired, urea should be avoided (Rais et al., 2004). Resolution, especially for low molecular weight proteins, can be further improved by using Tricine (Rais et al.,

2004; Schägger and von Jagow, 1987) or Bicine (Williams et al., 2006) SDS-PAGE instead of the Laemmli buffer system during the first dimension.

However, because SDS is utilized twice, the ability to separate individual proteins from each other remains limited (Fig. 21.2). The order of the resolving power is clearly double SDS-PAGE < alternative cationic/anionic detergent techniques (e.g. 16-BAC, see Section 21.4.5) << isoelectric focusing (IEF)-based techniques. This means that the "off-diagonal" spots still contain several hydrophobic proteins when complex samples are analysed, making double SDS-PAGE not ideal for quantitative membrane proteomics. Moreover, double SDS-PAGE is labour intensive.

Despite these disadvantages, double SDS-PAGE for plasma membrane proteomics provides results complementary to those obtained by other techniques and is suited to study highly hydrophobic proteins (Burré et al., 2006).

21.4.5 2D-16-BAC/SDS Electrophoresis

The 2D-16-BAC/SDS technique, based on the cationic detergent 16-benzyldimethyl-n-hexadecylammonium chloride (16-BAC) in the first dimension and the anionic detergent SDS in the second, has a higher resolving power than double SDS-PAGE, but still much less than classical 2DE separations since the migration of proteins in both dimensions is determined by the molecular weight (Fig. 21.3). The combination of 16-BAC and SDS was initially used to study the integral membrane proteome of carbonate-treated synaptic and clathrin-coated vesicles (Hartinger et al., 1996). Since the two detergents have different intrinsic protein binding properties, an improved separation is obtained compared to double SDS-PAGE (e.g. Braun et al., 2007). The original 16-BAC-PAGE was based on a potassium phosphate buffer system (Macfarlane, 1983). The resolving power of the system was later improved by optimizing the concentration of 16-BAC and implementing a methoxyacetic acid/ acetic acid buffer system (Kramer, 2006). A further improvement to 16-BAC/SDS-PAGE involved using thinner 16-BAC gels to increase the resolution (Wenge et al., 2008). Another cationic detergent, cetyl trimethyl ammonium bromide (CTAB) has also been combined with SDS-PAGE, for analysis of membrane proteins from T-cells (Helling et al., 2006).

The 16-BAC/SDS-PAGE system (Kramer, 2006) was used to separate membrane protein fractions from barley leaves (Fig. 21.3). When MALDI-TOF-TOF mass spectrometry was used to analyse protein spots, all identified proteins were abundant, membrane-associated proteins and no integral membrane proteins were identified. The use of LC-MS/MS might improve the number of protein identifications since due to the low resolution of the technique, as is the case with double SDS-PAGE, several proteins are expected to be present in a single spot. The use of liquid chromatography may also enable identification of lesser abundant, integral membrane proteins. LC-MS/MS analysis of 16-BAC/SDS-PAGE spots was used to identify 42 integral membrane proteins from yeast mithocondria (Zahedi et al., 2005) or 12 integral membrane proteins from rat synapses (Coughenour et al., 2004).

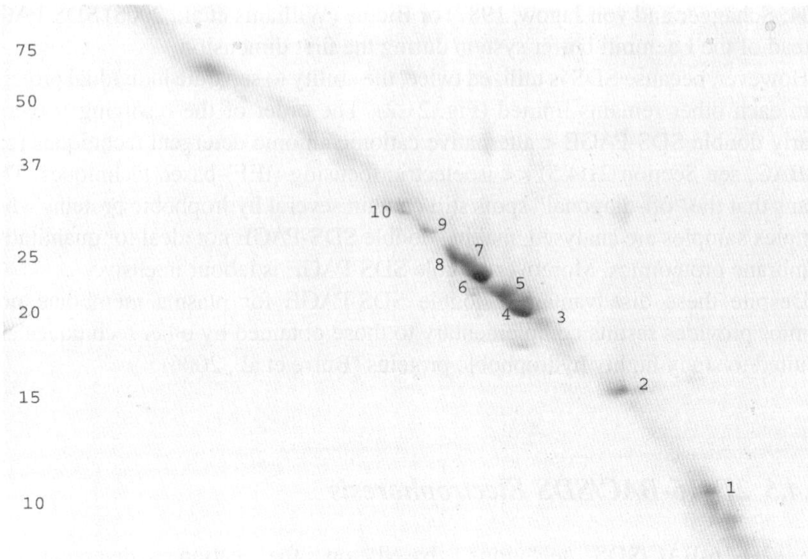

	Closest Homologue	TMD
1	PSAG (photosystem I subunit G) [*Arabidopsis thaliana*]	2
	PSAH-1 (photosystem I subunit H-1) [*Vitis vinifera*]	1
2	PSAL (photosystem I subunit L) [*Oryza sativa*]	2
3	cytochrome b(6) subunit of the cytochrome b6f complex [*Angelphytum bahiense*]	4
4	putative light-harvesting chlorophyll a/b binding protein [*Vitis vinifera*]	2
5	Lhcb6 [*Arabidopsis thaliana*]	2
	Lhca2 [*Oryza sativa*]	0
6	putative light-harvesting chlorophyll a/b binding protein [*Vitis vinifera*]	3
	Lhca3 [*Vitis vinifera*]	1
	Lhcb3 [*Vitis vinifera*]	2
7	Lhcb5 [*Vitis vinifera*]	2
8	Lhcb1.2 [*Vitis vinifera*]	2
	Lhcb5 [*Arabidopsis thaliana*]	2
9	porin, putative [*Oryza sativa*]	37 beta
	Lhcb4.2 [*Ricinus communis*]	2
10	porin, putative [*Oryza sativa*]	37 beta

Fig. 21.2 Double SDS-PAGE. Double SDS gel of proteins extracted in a mixture of chloroform (C) and methanol (M) (ratio C/M: 5/4) from an endomembrane fraction of banana leaves. The first dimension gel was a 10% acrylamide SDS gel, the second dimension gel contained 15% acrylamide. Molecular masses (kDa) of standard proteins are indicated on the left. The gel was Coomassie Blue G250 stained. The table shows the closest homologues of the proteins present in the numbered spots as well as the number of transmembrane domains (TMD) they contain, as predicted by the Aramemnon website (http://aramemnon.botanik.uni-koeln.de/). In case the TMD were β-sheets, this is mentioned, otherwise the number indicates the number of α-helices

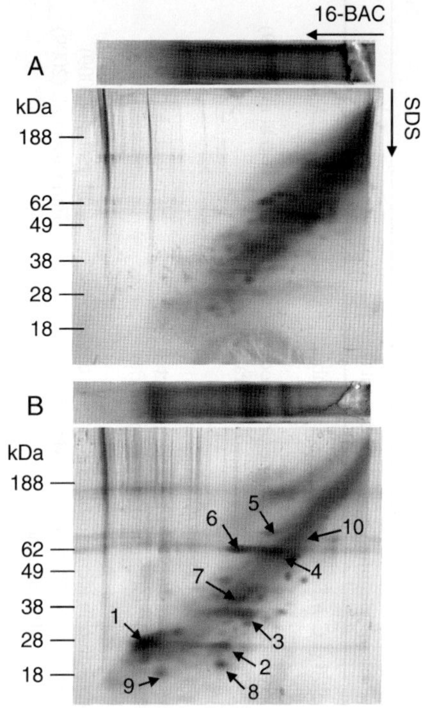

1. Chlorophyll a/b binding protein
2. Photosystem I reaction ceter subunit II
3. Chloroplastic ATP synthase gamma subunit
4. Rubisco large subunit
5. ATP synthase CF1 beta subunit
6. ATP synthase CF1 alpha subunit
7. Ferredoxin-NADP(H) oxidoreductase
8. Photosystem I reaction ceter subunit IV
9. Photosystem I reaction ceter subunit III
10. Vacuolar ATP synthase catalytic subunit A

Fig. 21.3 2D-16-BAC/SDS-PAGE. Plasma membrane and endomembrane fractions (10 μg from barley leaf) separated by 16-BAC-PAGE. 2D-16-BAC/SDS-PAGE in which the lanes from plasma (**a**) and endomembrane (**b**) containing 10 μg protein were placed onto the second dimension SDS-PAGE gels. The numbered protein spots were excised, digested by trypsin and analysed using MALDI-TOF/TOF mass spectrometry. Proteins identified from the spots indicated in B are listed

21.5 Comparison of Gel-Based Plant Plasma Membrane Proteome Studies

A comparison of the gel-based approaches taken so far for plant plasma membrane proteomics (Table 21.1) shows that for identification of integral membrane proteins, most success has been achieved using SDS-PAGE-based separation coupled to protein identification by LC-MS/MS, whereas classical 2DE (IEF/SDS-PAGE) based studies tend to result mainly in the identification of peripheral membrane proteins. In general, a second dimension separation by SDS-PAGE is preferred, since this both provides suitable conditions for solubilization of hydrophobic proteins and presents the proteins in a state suitable for enzyme digestion and subsequent mass spectrometry. Since no gel-based method developed so far enables high resolution

Table 21.1 Gel-based proteomic studies of plant plasma membranes

Plant	Tissue, conditions	Methods	Proteins identified	Comments	Reference
Poplar	Cambium/phloem layer, xylem, leaves	SDS-PAGE	956 (LC-MS/MS)	213 membrane proteins	Nilsson et al. (2010)
Soybean	Root, hypocotyl responding to flooding stress	IEF/SDS-PAGE or gel free LC-MS/MS	54 (Edman degradation, MALDI-TOF and LC-MS/MS) 90 (gel free LC-MS/MS)	8 predicted integral membrane proteins	Komatsu et al. (2009)
Barley	Germinating embryo	Reversed-phase chromatography, SDS-PAGE	61 (LC-MS/MS of 14 SDS-PAGE bands)	28 predicted integral membrane proteins	Hynek et al. (2009)
Cauliflower	Redox-related proteins	BN/SDS-PAGE	No protein identification	8 protein complexes	Lüthje et al. (2009)
Rice	Suspension cells, detergent-resistant fraction	SDS-PAGE	192 (LC-MS/MS of SDS-PAGE bands)	Many integral membrane proteins	Fujiwara et al. (2009)
Tobacco	Detergent-resistant fraction, changes in response to elicitors	SDS-PAGE	350 (LC-MS/MS of 20 SDS-PAGE bands)	Differential quantitative analysis by $^{14}N/^{15}N$ isotope labeling	Stanislas et al. (2009)
Rice	Root, cold-stressed	SDS-PAGE IEF/SDS-PAGE	78 (LC-MS/MS)	Integral membrane proteins	Hashimoto et al. (2009)
Rice	Root, responding to salt-stress	IEF/SDS-PAGE	18 (MALDI-TOF/TOF)	Membrane associated salt-stress responsive proteins	Cheng et al. (2009)
Maize	Root, responding to humic substances	IEF/SDS-PAGE	42 (LC-MS/MS)	Membrane associated proteins	Carletti et al. (2008)
Arabidopsis	Seedlings, responding to brassinosteroids	IEF/SDS-PAGE Difference gel electrophoresis (DIGE) comparative analysis	19 (LC-MS/MS)	Membrane associated proteins	Tang et al. (2008)

Table 21.1 (continued)

Plant	Tissue, conditions	Methods	Proteins identified	Comments	Reference
Rice	Root, responding to salt	IEF/SDS-PAGE	3 (MALDI-TOF-TOF)	Membrane associated proteins	Malakshah et al. (2007)
Rice	Suspension cells, responding to chitooligosaccharides	IEF/SDS-PAGE	8 (MALDI-TOF-TOF)	Membrane associated proteins	Chen et al. (2007a)
Rice	Suspension cells, expressing resistance gene *Xa21*	IEF/SDS-PAGE	8 (MALDI-TOF-TOF)	Membrane associated proteins	Chen et al. (2007b)
Duniella salina (alga)	Salt stress	BN/SDS-PAGE	55 (LC-MS/MS)	60% integral membrane or associated proteins	Katz et al. (2007)
Barley	Aleurone layers	Reversed-phase chromatography, SDS-PAGE	46 (LC-MS/MS of 10 SDS-PAGE bands)	36 predicted integral membrane proteins	Hynek et al. (2006)
Medicago truncatula	Root, detergent resistant fraction	SDS-PAGE	270 (LC-MS/MS of 14 SDS-PAGE bands)	Many integral membrane proteins	Lefebvre et al. (2007)
Arabidopsis	Cell culture	SDS-PAGE	446 (LC of 11 SDS-PAGE bands)	136 integral membrane proteins	Marmagne et al. (2007)
Medicago truncatula	Root, arbuscular mycorrhiza-related proteins	SDS-PAGE or 2D-LC	78 (LC/MS/MS of 12 SDS-PAGE bands)	44 integral membrane proteins	Valot et al. (2006)
Tobacco	Suspension cells, detergent resistant fraction	SDS-PAGE	145 (LC-MS/MS of 10 SDS-PAGE bands)	Many integral membrane proteins	Morel et al. (2006)
Arabidopsis	Callus, detergent resistant fraction	SDS-PAGE, IEF/SDS-PAGE, Difference gel electrophoresis (DIGE) analysis	(LC-MS/MS from SDS-PAGE bands and 2D spots)	Membrane associated proteins in 2D spots; integral membrane proteins in SDS-PAGE bands	Borner et al. (2005)

Table 21.1 (continued)

Plant	Tissue, conditions	Methods	Proteins identified	Comments	Reference
Spinach	Leaf	BN/SDS-PAGE	9 (ESI-QTOF)	Probably 3 integral membrane proteins	Kjell et al. (2004)
Rice	Rice seedlings and cultured suspension cells	IEF/SDS-PAGE	(MALDI-TOF)	Many integral membrane proteins	Tanaka et al. (2004)
Arabidopsis	Root	SDS-PAGE, IEF/SDS-PAGE	(MALDI-TOF and ESI-MS/MS)	A targeted study of aquaporins	Santoni et al. (2003)
Arabidopsis	Leaf	IEF/SDS-PAGE	36(MALDI-TOF and ESI-MS/MS)	5 integral membrane proteins identified	Santoni et al. (2000)
Arabidopsis	Callus	IEF/SDS-PAGE	18(MALDI-TOF and ESI-MS/MS)	Only membrane associated proteins were identified	Prime et al. (2000)
Arabidopsis	Root and leaf	IEF/SDS-PAGE	No protein identification	Proteins were identified in Santoni et al. 1998	Santoni et al. (1999)
Arabidopsis	Leaf	IEF/SDS-PAGE	(RP-HPLC and protein sequencing)	82 spots were microsequenced	Santoni et al. (1998)

separation of hydrophobic proteins, the method of choice for protein identification is LC-MS/MS which provides additional separation at the peptide level and thus enables identification of a greater number of proteins. For this reason, gel-free proteomic techniques offer distinct advantages for membrane proteins. However, a completely gel-free approach might be troublesome for the identification of proteins from poorly characterized species (Carpentier et al., 2008). Further issues pertaining to gel-free separation approaches are discussed below.

21.6 Gel-Free Separation Approaches

As discussed above, gel-based technologies are often limited for the analysis of plasma membrane proteins. Several approaches were developed in order to overcome the drawbacks posed by gel-based analysis of membrane proteins, including in-solution separation of proteins and LC-based separation of peptides. Some of the most recent examples are presented in the following.

21.6.1 Solution-Phase Separation of Proteins

Prefractionation of complex protein samples results in high resolution separation and facilitates the detection of low abundant proteins. Thereby, reduction of sample complexity improves subsequent analysis, either gel-based or gel-free. The microscale in-solution IEF method fractionates samples into well-defined pH pools using a series of chambers separated by large pore partition membrane discs of distinct isoelectric point ranges (Han and Speicher, 2008). The method has been applied to the analysis of the Plasmodium proteome (Nirmalan et al., 2007), combined with difference gel electrophoresis (DIGE) technology (Han et al., 2008) or the analysis of posttranslational modifications (Schulenberg and Patton, 2004) in animal tissues. Through prefractionation of a mouse brain protein extract a total number of 2673 polypeptides were identified on 2DE gels, including 255 hydrophobic structures with up to 7 predicted transmembrane domains (Myung and Lubec, 2006). A similar device was developed by Righetti et al. (2007) and applicability was shown for human and bacterial cells. Although published for some years now, no publication has yet described the implementation of the method for plant protein analysis and yet, the usefulness of in-solution phase IEF in plant membrane proteomics is obvious. Barley leaf plasma membranes were isolated by two-phase partitioning and subjected to solution-phase IEF. Subsequent separation of protein fractions on 2DE gels allowed for the identification of various proteins with biotin attachment domains and more than 5 predicted transmembrane domains (Witzel and Mock, unpublished results). As this method appears to be reproducible and reliable, reports proving the value for plant membrane proteomics are expected in the future.

Fractionation by free flow electrophoresis techniques is achieved completely in solution. This liquid fractionation device allows the separation of charged molecules and organelles in a matrix-free environment according to their electrophoretic mobility or isoelectric point depending on the applied electrophoretic conditions (Canut et al., 1999). In the past, this method was applied for the isolation of whole plasma membranes from tree (Toll et al., 1995) and wheat leaves (Egger et al., 1992), Arabidopsis (Bardy et al., 1998) and Vitis cell-suspension culture (Deswarte et al., 1994) as well as for the separation of tonoplast membranes from Arabidopsis (Barkla et al., 2007). So far, the fractionation of membrane proteins and peptides was demonstrated only in animal systems (Malmstrom et al., 2006; Weber et al., 2004), but in theory the technique should also be suitable for the separation of highly hydrophobic plant plasma membrane proteins.

21.6.2 Gel-Free Peptide Separation

One of the major technical advancements in the field of proteomics is development of gel-free mass spectrometry based methods for the separation, quantification and identification of proteins and peptides (Lambert et al., 2005). Several approaches were developed to separate complex protein digests by online or offline liquid chromatography, usually based on reversed-phase chromatography, coupled with matrix-assisted laser-desorption ionization (MALDI) or electrospray ionization (ESI) MS (Speers and Wu, 2007) (Fig. 21.4). LC-MALDI is typically performed offline, where LC eluates are deposited onto a MALDI target as discrete spots and MS and MS/MS experiments are performed afterwards (Wang et al., 2009). However, no reports on the LC-MALDI analysis of plant tissues have been published to date. More widely used in bottom-up approaches is LC-ESI MS. Dunkley et al. (2006) applied this method to localize organelle proteins by isotope tagging. Fractions from a density gradient centrifugation of membrane preparations

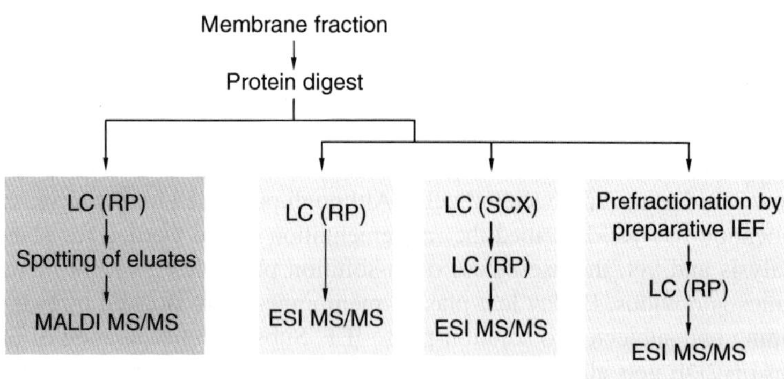

Fig. 21.4 Offline vs. online gel-free technologies for identification of membrane proteins. In offline modus, peptide mixtures are separated by LC and eluting fractions are deposited onto MALDI sample plates for MS and MS/MS experiments. Online separation involves RP-LC coupled directly to ESI-MS, SCX-LC and RP-LC (MudPIT) or prefractionation methods to reduce sample complexity

were obtained from Arabidopsis callus cultures. Proteins were digested and differentially labelled with stable isotope tags (iTRAQ) that give distinct low mass signature ions with a mass shift of 1 Da during MS/MS fragmentation, which are used for protein quantification (Ross et al., 2004). Subsequent prefractionation of tryptic peptides by strong cation exchange (SCX) chromatography and LC-ESI MS/MS led to the identification of 689 proteins, 92 of which could be assigned to the plasma membrane, including receptor kinase homologs, membrane transporters and glycosylphosphatidylinositol-anchored proteins. Additionally, the localization of a previously uncharacterized plasma membrane protein was confirmed by green fluorescent protein-fusion studies in Tobacco leaves. Nelson et al. (2006) applied a different stable isotope labelling strategy to map the Arabidopsis plasma membrane proteome and estimate the enrichment of plasma membrane proteins in the preparations. Here, trypsin cleavage was used to incorporate two ^{18}O atoms into the C-terminus of peptides from endomembrane preparations, while proteins from plasma membrane preparations were digested in normal water with the usual ^{16}O isotope. Samples were pooled, fractionated by SCX chromatography and analysed by LC-ESI MS/MS. The mass shift of 4 Da in the MS survey was used to differentiate between proteins originating from plasma membrane or endomembrane preparations. Overall, 70 proteins were found to be enriched in plasma membrane preparations by this approach. The issue of contaminations from co-enriched organelles was also addressed by comparing their own dataset with others', leading to the conclusion that about a quarter of the proteins identified in such studies can be regarded as contaminants. Mitra et al. (2009) developed an optimized protocol for the analysis of Arabidopsis plasma membrane proteins in order to improve peptide identification and reduce sample loss after the tryptic cleavage. A chloroform extraction was applied after protein digestion to eliminate phytochemical contaminants and residual proteolytic activity, followed by SCX chromatography and LC-ESI MS/MS. Among the 272 identified proteins were 77 membrane-specific transporters and 20 leucine-rich repeat receptor-like kinases. The great potential of LC-ESI MS/MS in analyzing hydrophobic proteins was recently demonstrated by Nagaraj et al. (2008) in an animal system. Membrane preparations from several mouse tissues were solubilised with various detergents (Triton X-100, 3-[(3-Cholamidopropyl)dimethylammonio]-1-propanesulfonate (CHAPS), SDS) and proteins were digested in-solution with endoproteinase Lys-C. Produced peptides were analyzed by LC-ESI MS/MS and resulted in the identification of 956 known or putative membrane proteins with about 18% annotated to the plasma membrane. Strategies for solubilisation and proteolytic digestion of integral membrane proteins for identification by mass spectrometry have been reviewed recently (Lu et al., 2008).

From a physiological viewpoint, the elucidation of posttranslational modifications is one of the most challenging objectives of membrane proteomics studies. Advances in detection techniques have enabled the large-scale identification of phosphorylation sites. Pioneering work in this regard was performed by Nühse et al. (2004, 2007). Plasma membranes from Arabidopsis cell suspension cultures were isolated, proteins digested and phosphorylated peptides enriched by metal-ion affinity chromatography (Nühse et al., 2004). Overall, more than 300 phosphorylation sites were identified, 50 of which were mapped on receptor-like kinases. The data was deposited in a database for plant phosphorylation sites (http://plantsp.genomics.purdue.edu/html). The iTRAQ labelling approach was then used to characterize phosphorylation/dephosphorylation events in the plasma membrane in response to pathogen attack (Nühse et al., 2007).

Multidimensional protein identification technology (MudPIT) enables the online separation of peptides based on both their charge and hydrophobicity using cation exchange chromatography coupled to RP chromatography and placed in-line with ESI MS/MS (Kline and Wu, 2009). The study of Lee et al. (2007) aimed firstly at comparing MudPIT with the classical 1D LC-ESI MS/MS for shotgun proteomics and secondly, at identifying transmembrane proteins from organelles of Arabidopsis leaves by MudPIT. A total of 2342 proteins were found in microsomes, nuclei, plastids and mitochondria and 328 proteins contained a predicted transmembrane domain. In a similar comparative approach, Morel et al. (2006) investigated detergent-resistant membrane (DRM) fractions in tobacco BY-2 cell culture, either by 1D LC-ESI MS/MS, MudPIT or gas phase fractionation (GPF) MS/MS. There is great interest in functional characterization of these microdomains since recent studies indicate their involvement in biotic interactions, developmental and signalling processes (Zappel and Panstruga, 2008). The most efficient method for protein identification from this plasma membrane subfraction proved to be 1D LC-ESI MS/MS with 145 hits, followed by MudPIT and GPF with 54 and 43 identified proteins, respectively (Morel et al., 2006). First conclusions were drawn on the physiological function of DRMs contrasting the major plasma membrane functions, e.g. transport, signalling and metabolic processes, with the role of DRM in stress response, cellular trafficking and cell wall metabolism. The applicability of MudPIT to the analysis of hydrophobic membrane proteins was recently demonstrated in a study on membrane fractions from mouse skeletal muscle cells (Elschenbroich et al., 2009). Authors aimed at comparing MudPIT with the recently introduced OFFGEL fractionator (Agilent Technologies) for the high resolution separation of peptides and revealed comparable results using both platforms with about 1400 identified proteins.

21.7 Concluding Remarks

Current gel-free concepts have a great impact on shotgun mapping approaches and allowed for the identification of a large number of putative plant plasma membrane proteins, including post-translational modifications. However, gel-based proteomic approaches in conjunction with in-solution prefractionation techniques represent a valuable alternative, especially for investigating protein complexes or protein isoforms that are insufficiently resolved by MS-based techniques. The characterization of plant plasma proteins remains a challenging task where gel-free and gel-based proteomics approaches are more likely to be complementary than exclusive.

Acknowledgements Anja Thoe Fuglsang (University of Copenhagen) is thanked for providing antibodies. Work in the authors' laboratories was supported by the Danish Research Council for Technology and Production Sciences, the Danish Natural Science Research Council, Centre for Advanced Food Studies (LMC), The German Research Foundation (DFG), The German Federal Ministry of Education and Research (BMBF), the Flanders research foundation (FWO Belgium) and the EU COST action FA0603 "Plant Proteomics in Europe"

References

Akiyama, Y., and Ito, K. (1985). The SecY membrane component of the bacterial protein export machinery: analysis by new electrophoretic methods for integral membrane proteins. EMBO J *4*, 3351–3356.

Albertsson, P.Å., and Tjerneld, F. (1994). Phase diagrams. Methods Enzymol *228*, 3–13.

Alexandersson, E., Saalbach, G., Larsson, C., and Kjellbom, P. (2004). Arabidopsis plasma membrane proteomics identifies components of transport, signal transduction and membrane trafficking. Plant Cell Physiol *45*, 1543–1556.

Bardy, N., Carrasco, A., Galaud, J.P., Pont-Lezica, R., and Canut, H. (1998). Free-flow electrophoresis for fractionation of Arabidopsis thaliana membranes. Electrophoresis *19*, 1145–1153.

Barkla, B.J., Vera-Estrella, R., and Pantoja, O. (2007). Enhanced separation of membranes during free flow zonal electrophoresis in plants. Anal Chem *79*, 5181–5187.

Beyreuther, K., Bieseler, B., Ehring, R., Griesser, H.W., Mieschendahl, M., Müller-Hill, B., and Triesch, I. (1980). Investigation of structure and function of lactose permease of Escherichia coli. Biochem Soc Trans *8*, 675–676.

Blonder, J., Terunuma, A., Conrads, T.P., Chan, K.C., Yee, C., Lucas, D.A., Schaefer, C.F., Yu, L.R., Issaq, H.J., Veenstra, T.D., and Vogel, J.C. (2004). A proteomic characterization of the plasma membrane of human epidermis by high-throughput mass spectrometry. J Invest Dermatol *123*, 691–699.

Borner, G.H., Sherrier, D.J., Weimar, T., Michaelson, L.V., Hawkins, N.D., Macaskill, A., Napier, J.A., Beale, M.H., Lilley, K.S., and Dupree, P. (2005). Analysis of detergent-resistant membranes in Arabidopsis. Evidence for plasma membrane lipid rafts. Plant Physiol *137*, 104–116.

Braun, R.J., Kinkl, N., Beer, M., and Ueffing, M. (2007). Two-dimensional electrophoresis of membrane proteins. Anal Bioanal Chem *389*, 1033–1045.

Brooks, D.E., and Norris-Jones, R. (1994). Preparation and analysis of 2-phase systems. Methods Enzymol *228*, 14–27.

Burré, J., Beckhaus, T., Schägger, H., Corvey, C., Hofmann, S., Karas, M., Zimmermann, H., and Volknandt, W. (2006). Analysis of the synaptic vesicle proteome using three gel-based protein separation techniques. Proteomics *6*, 6250–6262.

Canut, H., Bauer, J., and Weber, G. (1999). Separation of plant membranes by electromigration techniques. J Chromatogr B Biomed Sci Appl *722*, 121–139.

Carletti, P., Masi, A., Spolaore, B., Polverino De Laureto, P., De Zorzi, M., Turetta, L., Ferretti, M., and Nardi, S. (2008). Protein expression changes in maize roots in response to humic substances. J Chem Ecol *34*, 804–818.

Carpentier, S.C., Panis, B., Vertommen, A., Swennen, R., Sergeant, K., Renaut, J., Laukens, K., Witters, E., Samyn, B., and Devreese, B. (2008). Proteome analysis of non-model plants: A challenging but powerful approach. Mass Spectrom Rev *27*, 354–377.

Chen, F., Li, Q., and He, Z. (2007). Proteomic analysis of rice plasma membrane-associated proteins in response to Chitooligosaccharide elicitors. J Integr Plant Biol *49*, 863–870.

Chen, F., Yuan, Y., Li, Q., and He, Z. (2007). Proteomic analysis of rice plasma membrane reveals proteins involved in early defense response to bacterial blight. Proteomics *7*, 1529–1539.

Cheng, Y., Qi, Y., Zhu, Q., Chen, X., Wang, N., Zhao, X., Chen, H., Cui, X., Xu, L., and Zhang, W. (2009). New changes in the plasma-membrane-associated proteome of rice roots under salt stress. Proteomics *9*, 3100–3114.

Coughenour, H.D., Spaulding, R.S., and Thompson, C.M. (2004). The synaptic vesicle proteome: A comparative study in membrane protein identification. Proteomics *4*, 3141–3155.

Deswarte, C., Peyrebrune, S., Canut, H., Roustan, J.P., and Fallot, J. (1994). Purification of plasma-membrane vesicles from Vitis-vinifera cv Gamay suspension cells by free-flow electrophoresis. Vitis *33*, 99–100.

Dunkley, T.P.J., Hester, S., Shadforth, I.P., Runions, J., Weimar, T., Hanton, S.L., Griffin, J.L., Bessant, C., Brandizzi, F., Hawes, C., Watson, R.B., Dupree, P., and Lilley, K.S. (2006). Mapping the Arabidopsis organelle proteome. Proc Natl Acad Sci USA *103*, 6518–6523.

Egger, A., Morre, J., Sellden, G., and Sandelius, A.S. (1992). Isolation of plasma-membrane fractions from green wheat leaves by free-flow electrophoresis or 2-phase partition alone or in combination. Physiol Plant *84*, 121–133.

Elschenbroich, S., Ignatchenko, V., Sharma, P., Schmitt-Ulms, G., Gramolini, A.O., and Kislinger, T. (2009). Peptide separations by on-line MudPIT compared to isoelectric focusing in an Off-Gel format: Application to a membrane-enriched fraction from C2C12 mouse skeletal muscle cells. J Proteome Res *8*, 4860–4869.

Fujiwara, M., Hamada, S., Hiratsuka, M., Fukao, Y., Kawasaki, T., and Shimamoto, K. (2009). Proteome analysis of detergent-resistant membranes (DRMs) associated with OsRac1 mediated innate immunity in rice. Plant Cell Physiol *50*, 1191–200.

Han, M.J., Herlyn, M., Fisher, A.B., and Speicher, D.W. (2008). Microscale solution IEF combined with 2-D DIGE substantially enhances analysis depth of complex proteomes such as mammalian cell and tissue extracts. Electrophoresis *29*, 695–705.

Han, M.J., and Speicher, D.W. (2008). Microscale isoelectric focusing in solution: A method for comprehensive and quantitative proteome analysis using 1-D and 2-D DIGE combined with MicroSol IEF prefractionation. Methods Mol Biol *424*, 241–256.

Hartinger, J., Stenius, K., Högemann, D., and Jahn, R. (1996). 16-BAC/SDS-PAGE: A two-dimensional gel electrophoresis system suitable for the separation of integral membrane proteins. Anal Biochem *240*, 126–133.

Hashimoto, M., Toorchi, M., Matsushita, K., Iwasaki, Y., and Komatsu, S. (2009). Proteome analysis of rice root plasma membrane and detection of cold stress responsive proteins. Protein Pept Lett *16*, 685–697.

Heinemeyer, J., Scheibe, B., Schmitz, U.K., and Braun, H.P. (2009). Blue native DIGE as a tool for comparative analyses of protein complexes. J Proteomics *72*, 539–544.

Helling, S., Schmitt, E., Joppich, C., Schulenborg, T., Mullner, S., Felske-Muller, S., Wiebringhaus, T., Becker, G., Linsenmann, G., Sitek, B., Lutter, P., Meyer, H.E., and Marcus, K. (2006). 2-D differential membrane proteome analysis of scarce protein samples. Proteomics *6*, 4506–4513.

Hynek, R., Svensson, B., Jensen, O.N., Barkholt, V., and Finnie, C. (2009). The plasma membrane proteome of germinating barley embryos. Proteomics *9*, 3787–3794.

Hynek, R., Svensson, B., Jensen, O.N., Barkholt, V., and Finnie, C. (2006). Enrichment and identification of integral membrane proteins from barley aleurone layers by reversed-phase chromatography, SDS-PAGE, and LC-MS/MS. J Proteome Res *5*, 3105–3113.

Johansson, F., Olbe, M., Sommarin, M., and Larsson, C. (1995). Brij 58, a polyoxyethylene acyl ether, creates membrane vesicles of uniform sidedness. A new tool to obtain inside-out (cytoplasmic side-out) plasma membrane vesicles. Plant J *7*, 165–173.

Jung, C., Higgins, C.M., and Xu, Z. (2000). Measuring the quantity and activity of mitochondrial electron transport chain complexes in tissues of central nervous system using blue native polyacrylamide gel electrophoresis. Anal Biochem *286*, 214–223.

Juszczuk, I.M., and Rychter, A.M. (2009). BN-PAGE analysis of the respiratory chain complexes in mitochondria of cucumber MSC16 mutant. Plant Physiol Biochem *47*, 397–406.

Katz, A., Waridel, P., Shevchenko, A., and Pick, U. (2007). Salt-induced changes in the plasma membrane proteome of the halotolerant alga Dunaliella salina as revealed by blue native gel electrophoresis and nano-LC-MS/MS analysis. Mol Cell Proteomics *6*, 1459–1472.

Kjell, J., Rasmusson, A.G., Larsson, H., and Widell, S. (2004). Protein complexes of the plant plasma membrane resolved by Blue Native PAG. Physiol Plant *121*, 546–555.

Kline, K.G., and Wu, C.C. (2009). MudPIT analysis: Application to human heart tissue. Methods Mol Biol *528*, 281–293.

Komatsu, S., Wada, T., Abaléa, Y., Nouri, M.Z., Nanjo, Y., Nakayama, N., Shimamura, S., Yamamoto, R., Nakamura, T., and Furukawa, K. (2009). Analysis of plasma membrane proteome in soybean and application to flooding stress response. J Proteome Res *8*, 4487–4499.

Kramer, M.L. (2006). A new multiphasic buffer system for benzyldimethyl-n-hexadecylammonium chloride polyacrylamide gel electrophoresis of proteins providing efficient stacking. Electrophoresis *27*, 347–356.

Lambert, J.P., Ethier, M., Smith, J.C., and Figeys, D. (2005). Proteomics: From gel based to gel free. Anal Chem 77(12):3771–3787.

Larsson, C., Widell, S., and Kjellbom, P. (1987). Preparation of high-purity plasma membrane. Methods Enzymol *148*, 558–568

Lee, Y.C., Block, G., Chen, H., Folch-Puy, E., Foronjy, R., Jalili, R., Jendresen, C.B., Kimura, M., Kraft, E., Lindemose, S., Lu, J., McLain, T., Nutt, L., Ramon-Garcia, S., Smith, J., Spivak, A., Wang, M.L., Zanic, M., and Lin, S.H. (2008). One-step isolation of plasma membrane proteins using magnetic beads with immobilized concanavalin A. Protein Exp Purif *62*, 223–229.

Lee, J., Garrett, W.M., and Cooper, B. (2007). Shotgun proteomic analysis of Arabidopsis thaliana leaves. J Sep Sci *30*, 2225–2230.

Lefebvre, B., Furt, F., Hartmann, M.A., Michaelson, L.V., Carde, J.P., Sargueil-Boiron, F., Rossignol, M., Napier, J.A., Cullimore, J., Bessoule, J.J., and Mongrand, S. (2007). Characterization of lipid rafts from Medicago truncatula root plasma membranes: A proteomic study reveals the presence of a raft-associated redox system. Plant Physiol *144*, 402–418.

Lu, B., McClatchy, D.B., Kim, J.Y., and Yates, J.R. III (2008). Strategies for shotgun identification of integral membrane proteins by tandem mass spectrometry. Proteomics *8*, 3947–3955.

Luche, S., Santoni, V., and Rabilloud, T. (2003). Evaluation of nonionic and zwitterionic detergents as membrane protein solubilizers in two-dimensional electrophoresis. Proteomics *3*, 249–253.

Lüthje, S., Hopff, D., Schmitt, A., Meisrimler, C.N., and Menckhoff, L. (2009). Hunting for low abundant redox proteins in plant plasma membranes. J Proteomics *72*, 475–483.

Macfarlane, D.E. (1983). Use of benzyldimethyl-n-hexadecylammonium chloride ("16-BAC"), a cationic detergent, in an acidic polyacrylamide gel electrophoresis system to detect base labile protein methylation in intact cells. Anal Biochem *132*, 231–235.

Malakshah, S.N., Habibi Rezaei, M., Heidari, M., and Salekdeh, G.H. (2007). Proteomics reveals new salt responsive proteins associated with rice plasma membrane. Biosci Biotechnol Biochem *71*, 2144–2154.

Malmstrom, J., Lee, H., Nesvizhskii, A.I., Shteynberg, D., Mohanty, S., Brunner, E., Ye, M.L., Weber, G., Eckerskorn, C., and Aebersold, R. (2006). Optimized peptide separation and identification for mass spectrometry based proteomics via free-flow electrophoresis. J Proteome Res *5*, 2241–2249.

Marmagne, A., Ferro, M., Meinnel, T., Bruley, C., Kuhn, L., Garin, J., Barbier-Brygoo, H., and Ephritikhine, G. (2007). A high content in lipid-modified peripheral proteins and integral eceptor kinases features in the arabidopsis plasma membrane proteome. Mol Cell Proteomics *6*, 1980–1996.

Meyer, B., Wittig, I., Trifilieff, E., Karas, M., and Schägger, H. (2007). Identification of two proteins associated with mammalian ATP synthase. Mol Cell Proteomics *6*(10), 1690–1699.

Mitra, S.K., Gantt, J.A., Ruby, J.F., Clouse, S.D., Goshe, M.B. (2007). Membrane proteomic analysis of Arabidopsis thaliana using alternative solubilization techniques. J Proteome Res *6*, 1933–1950.

Mitra, S.K., Walters, B.T., Clouse, S.D., and Goshe, M.B. (2009). An efficient organic solvent based extraction method for the proteomic analysis of Arabidopsis plasma membranes. J Proteome Res *8*, 2752–2767.

Moebius, J., Zahedi, R.P., Lewandrowski, U., Berger, C., Walter, U., and Sickmann, A. (2005). The human platelet membrane proteome reveals several new potential membrane proteins. Mol Cell Proteomics *4*, 1754–1761.

Mohnen, D. (2008). Pectin structure and biosynthesis. Curr Opin Plant Biol *11*, 266–277.

Morel, J., Claverol, S., Mongrand, S., Furt, F., Fromentin, J., Bessoule, J.J., Blein, J.P., and Simon-Plas, F. (2006). Proteomics of plant detergent-resistant membranes. Mol Cell Proteomics *5*, 1396–1411.

Myung, J.K., and Lubec, G. (2006). Use of solution-IEF-fractionation leads to separation of 2673 mouse brain proteins including 255 hydrophobic structures. J Proteome Res *5*, 1267–1275.

Nagaraj, N., Lu, A.P., Mann, M., and Wisniewski, J.R. (2008). Detergent-based but gel-free method allows identification of several hundred membrane proteins in single LC-MS runs. J Proteome Res *7*, 5028–5032.

Neff, D., and Dencher, N.A. (1999). Purification of multisubunit membrane protein complexes: Isolation of chloroplast FoF1-ATP synthase, CFo and CF1 by blue native electrophoresis. Biochem Biophys Res Commun *259*, 569–575.

Nelson, C.J., Hegeman, A.D., Harms, A.C., and Sussman, M.R. (2006). A quantitative analysis of Arabidopsis plasma membrane using trypsin-catalyzed O-18 labeling. Mol Cell Proteomics *5*, 1382–1395.

Nilsson, R., Bernfur, K., Gustavsson, N., Bygdell, J., Wingsle, G., and Larsson, C. (2010). Proteomics of plasma membranes from poplar trees reveals tissue distribution of transporters, receptors and proteins in cell wall formation. Mol Cell Proteomics *9*, 368–387.

Nirmalan, N., Flett, F., Skinner, T., Hyde, J.E., and Sims, P.F.G. (2007). Microscale solution Isoelectric focusing as an effective strategy enabling containment of hemeoglobin-derived products for high-resolution gel-based analysis of the Plasmodium falciparum proteome. J Proteome Res *6*, 3780–3787.

Nühse, T.S., Bottrill, A.R., Jones, A.M.E., and Peck, S.C. (2007). Quantitative phosphoproteomic analysis of plasma membrane proteins reveals regulatory mechanisms of plant innate immune responses. Plant J *51*, 931–940.

Nühse, T.S., Stensballe, A, Jensen, O.N., and Peck, S.C. (2004). Phosphoproteomics of the Arabidopsis plasma membrane and a new phosphorylation site database. Plant Cell *16*, 2394–2405.

Pedersen, S.K., Harry, J.L., Sebastian, L, Baker, J., Traini, M.D., McCarthy, J.T., Manoharan, A., Wilkins, M.R., Gooley, A.A., Righetti, P.G., Packer, N.H., Williams, K.L., and Herbert, B.R. (2003). Unseen proteome: Mining below the tip of the iceberg to find low abundance and membrane proteins. J Proteome Res *2*, 303–311.

Persson, A., and Jergil, B. (1994). Rat liver plasma membranes. Methods Enzymol *228*, 489–496.

Persson, A., Johansson, B., Olsson, H., and Jergil, B. (1991). Purification of rat liver plasma membranes by wheat-germ-agglutinin affinity partitioning. Biochem J *273*, 173–177.

Poetsch, A., Neff, D., Seelert, H., Schägger, H., and Dencher, N.A. (2000). Dye removal, catalytic activity and 2D crystallization of chloroplast H^+-ATP synthase purified by blue native electrophoresis. Biochim Biophys Acta *1466*, 339–349.

Prime, T.A., Sherrier, D.J., Mahon, P, Packman, L.C., and Dupree, P. (2000). A proteomic analysis of organelles from Arabidopsis thaliana. Electrophoresis *21*, 3488–3499.

Rais, I., Karas, M., and Schägger, H. (2004). Two-dimensional electrophoresis for the isolation of integral membrane proteins and mass spectrometric identification. Proteomics *4*, 2567–2571.

Reisinger, V., and Eichacker, L.A. (2006). Analysis of membrane protein complexes by blue native PAGE. Proteomics *6*(Suppl 2), 6–15.

Righetti, P.G., Citterio, A., and Girault, H. (2007). Gel-free IEF in a membrane-sealed multicompartment cell for proteome prefractionation. Electrophoresis *28*, 1860–1866.

Rolland, N., Ferro, M., Ephritikhine, G., Marmagne, A., Ramus, C., Brugière, S., Salvi, D., Seigneurin-Berny, D., Bourguignon, J., Barbier-Brygoo, H., Joyard, J., and Garin, J. (2006). A versatile method for deciphering plant membrane proteomes. J Exp Bot *57*, 1579–1589.

Ross, P.L., Huang, Y.L.N., Marchese, J.N., Williamson, B., Parker, K., Hattan, S., Khainovski, N., Pillai, S., Dey, S., Daniels, S., Purkayastha, S., Juhasz, P., Martin, S., Bartlet-Jones, M., He, F., Jacobson, A., and Pappin, D.J. (2004). Multiplexed protein quantitation in Saccharomyces cerevisiae using amine-reactive isobaric tagging reagent. Mol Cell Proteomics *3*, 1154–1169.

Santoni, V. (2007). Plant plasma membrane protein extraction and solubilization for proteomic analysis. Methods Mol Biol *335*, 93–109.

Santoni, V., Doumas, P., Rouquié, D., Mansion, M., Rabilloud, T., and Rossignol, M. (1999). Large scale characterization of plant plasma membrane proteins. Biochimie *81*, 655–661.

Santoni, V., Kieffer, S., Desclaux, D., Masson, F., and Rabilloud, T. (2000). Membrane proteomics: Use of additive main effects with multiplicative interaction model to classify plasma membrane proteins according to their solubility and electrophoretic properties. Electrophoresis *21*, 3329–3344.

Santoni, V., Molloy, M., and Rabilloud, T. (2000). Membrane proteins and proteomics: Un amour impossible? Electrophoresis *21*, 1054–1070.

Santoni, V., Rouquié, D., Doumas, P., Mansion, M., Boutry, M., Degand, H., Dupree, P., Packman, L., Sherrier, J., Prime, T., Bauw, G., Posada, E., Rouzé, P., Dehais, P., Sahnoun, I., Barlier, I., and Rossignol, M. (1998). Use of a proteome strategy for tagging proteins present at the plasma membrane. Plant J *16*, 633–641.

Santoni, V., Vinh, J., Pflieger, D., Sommerer, N., and Maurel, C. (2003). A proteomic study reveals novel insights into the diversity of aquaporin forms expressed in the plasma membrane of plant roots. Biochem J *373*, 289–296.

Sarkar, P., Bosneaga, E., and Auer, M. (2009). Plant cell walls throughout evolution: towards a molecular understanding of their design principles. J Exp Bot *60*, 3615–3635.

Schägger, H., and Pfeiffer, K. (2000). Supercomplexes in the respiratory chains of yeast and mammalian mitochondria. EMBO J *19*, 1777–1783.

Schägger, H., Cramer, W.A., and von Jagow, G. (1994). Analysis of molecular masses and oligomeric states of protein complexes by blue native electrophoresis and isolation of membrane protein complexes by two-dimensional native electrophoresis. Anal Biochem *217*, 220–230.

Schägger, H., and von Jagow, G. (1987). Tricine-sodium dodecyl sulfate-polyacrylamide gel electrophoresis for the separation of proteins in the range from 1 to 100 kDa. Anal Biochem *166*, 368–379.

Schägger, H., and von Jagow, G. (1991). Blue native electrophoresis for isolation of membrane protein complexes in enzymatically active form. Anal Biochem *199*, 223–231.

Schindler, J., and Nothwang, H.G. (2006). Aqueous polymer two-phase systems: Effective tools for plasma membrane proteomics. Proteomics *6*, 5409–5417.

Schulenberg, B., and Patton, W.F. (2004). Combining microscale solution-phase isoelectric focusing with multiplexed Proteomics((R)) dye staining to analyze protein post-translational modifications. Electrophoresis *25*, 2539–2544.

Simões-Barbosa, A., Santana, J.M., and Teixeira, A.R. (2000). Solubilization of delipidated macrophage membrane proteins for analysis by two-dimensional electrophoresis. Electrophoresis *21*, 641–644.

Speers, A.E., and Wu, C.C. (2007). Proteomics of integral membrane proteins--theory and application. Chem Rev *107*, 3687–3714.

Stanislas, T., Bouyssie, D., Rossignol, M., Vesa, S., Fromentin, J., Morel, J., Pichereaux, C., Monsarrat, B., and Simon-Plas, F. (2009). Quantitative proteomics reveals a dynamic association of proteins to detergent-resistant membranes upon elicitor signaling in tobacco. Mol Cell Proteomics *8*, 2186–2198.

Tanaka, N., Fujita, M., Handa, H., Murayama, S., Uemura, M., Kawamura, Y., Mitsui, T., Mikami, S., Tozawa, Y., Yoshinaga, T., and Komatsu, S. (2004). Proteomics of the rice cell: Systematic identification of the protein populations in subcellular compartments. Mol Genet Genomics *271*, 566–576.

Tang, W., Deng, Z., Oses-Prieto, J.A., Suzuki, N., Zhu, S., Zhang, X., Burlingame, A.L., and Wang, Z.Y. (2008). Proteomics studies of brassinosteroid signal transduction using prefractionation and two-dimensional DIGE. Mol Cell Proteomics *7*, 728–738.

Tastet, C., Charmont, S., Chevallet, M., Luche, S., and Rabilloud, T. (2003). Structure-efficiency relationships of zwitterionic detergents as protein solubilizers in two-dimensional electrophoresis. Proteomics *3*, 111–121.

Toll, E., Castillo, F.J., Crespi, P., Crevecoeur, M., and Greppin, H. (1995). Purification of plasma membranes from leaves of conifer and deciduous tree species by phase partitioning and free-flow electrophoresis. Physiol Plant *95*, 399–408.

Valot, B., Negroni, L., Zivy, M., Gianinazzi, S., and Dumas-Gaudot, E. (2006). A mass spectrometric approach to identify arbuscular mycorrhiza-related proteins in root plasma membrane fractions. Proteomics *6*, 145–155.

Wang, N., Young, J.B., and Li, L. (2009). Liquid chromatography MALDI MS/MS for membrane proteome analysis. Methods Mol Biol *528*, 295–231.

Weber, G., Islinger, M., Weber, P., Eckerskorn, C., and Völkl, A. (2004). Efficient separation and analysis of peroxisomal membrane proteins using free-flow isoelectric focusing. Electrophoresis *25*, 1735–1747.

Wenge, B., Bönisch, H., Grabitzki, J., Lochnit, G., Schmitz, B., and Ahrend, M.H. (2008). Separation of membrane proteins by two-dimensional electrophoresis using cationic rehydrated strips. Electrophoresis *29*, 1511–1517.

Wessel, D., and Flügge, U.I. (1984). A method for the quantitative recovery of protein in dilute solution in the presence of detergents and lipids. Anal Biochem *138*, 141–143.

Williams, T.I., Combs, J.C., Thakur, A.P., Strobel, H.J., and Lynn, B.C. (2006). A novel Bicine runningbuffer system for doubled sodium dodecyl sulfate – polyacrylamide gel electrophoresis of membrane proteins. Electrophoresis *27*, 2984–2995.

Wittig, I., Braun, H.P., and Schägger, H. (2006). Blue native PAGE. Nat Protoc *1*, 418–428.

Wumaier, Z., Nübel, E., Wittig, I., and Schägger, H. (2009). Chapter 8: Two-dimensional native electrophoresis for fluorescent and functional assays of mitochondrial complexes. Methods Enzymol *456*, 153–168.

Zahedi, R.P., Meisinger, C., and Sickmann, A. (2005). Two-dimensional benzyldimethyl-n-hexadecylammonium chloride/SDS-PAGE for membrane proteomics. Proteomics *5*, 3581–3588.

Zhang, L.J., Wang, X.E., Peng, X., Wei, Y.J., Cao, R., Liu, Z., Xiong, J.X., Yin, X.F., Ping, C., and Liang, S. (2006). Proteomic analysis of low-abundant integral plasma membrane proteins based on gels. Cell Mol Life Sci *63*, 1790–804.

Zappel, N.F., and Panstruga, R. (2008). Heterogeneity and lateral compartmentalization of plant plasma membranes. Curr Opin Plant Biol *11*, 632–640.

Part VII
Affinity Interaction and Biochemical Enrichment Techniques

Chapter 22
Affinity and Chemical Enrichment for Mass Spectrometry-Based Proteomics Analyses

Guillaume O. Adelmant, Job D. Cardoza, Scott B. Ficarro, Timothy W. Sikorski, Yi Zhang, and Jarrod A. Marto

Abstract Collective improvements in mass spectrometry, separations, enrichment techniques, and related sample processing, along with integrated data analytic strategies, have advanced the objectives of proteomics studies from generation of peptide and protein catalogs to construction of dynamic networks that seek to correlate changes in protein expression and post-translational modification status to biological state. While researchers pursue the analysis of increasingly complex mixtures, the goal of compiling a comprehensive proteome sequence at an organism level remains beyond our reach. Nevertheless the use of proteomics techniques to characterize specific pathways, modifications, and protein interactions has provided a wealth of insight not otherwise available for biological studies. Continued progress in targeted analyses will yield sufficient temporal and spatial resolution in proteomics studies to support predictions of phenotype as a function of biological perturbation.

Keywords Tandem affinity purification · Immobilized template purification · Phosphorylation · Protein complexes · Phosphopeptide enrichment

22.1 Introduction

Biological phenotypes result from a multitude of interactions between proteins, nucleotides, and small molecule metabolites (Ideker et al., 2001). As a result, an ever-growing number of hypothesis-driven studies designed to characterize cellular response to stress, infection, disease, and other biological or environmental perturbations, now include mass spectrometry-based proteomics data (Gavin et al., 2002, 2006; Goh et al., 2007; Hakes et al., 2008; Kocher and Superti-Furga, 2007; Pawson and Warner, 2007). Despite technical improvements in mass spectrometry instrumentation, overall proteome coverage and sample throughput are low as

---

J.A. Marto (✉)
Department of Cancer Biology and Blais Proteomics Center, Dana-Farber Cancer Institute, Department of Biological Chemistry and Molecular Pharmacology, Harvard Medical School, Boston, MA 02115, USA
e-mail: jarrod_marto@dfci.harvard.edu

A.R. Ivanov, A.V. Lazarev (eds.), *Sample Preparation in Biological Mass Spectrometry*, DOI 10.1007/978-94-007-0828-0_22,

compared to analogous data from DNA and RNA sequencing efforts. Limited detection and dynamic range on current mass spectrometers represent the major obstacles to "complete" proteome sequence analysis. (Aebersold and Mann, 2003; Hille et al., 2001) As a result of these limitations, researchers often pursue experimental strategies designed to target protein sub-classes or specific post-translational modifications. (Adelmant and Marto, 2009; Dunn et al., 2009) This chapter will detail various methods to purify multi-component protein complexes and enrich phosphorylated peptides from biological extracts.

22.2 Analysis of Multi-component Complexes Based on Epitope-Tagged Protein Targets

22.2.1 General Comments

The purification of native or intact protein complexes followed by analysis of their constituent components by mass spectrometry is one of the most powerful strategies to discover the cellular function of proteins and more broadly to map protein interaction networks that control cellular processes. The purification of protein complexes from cells, tissues or model organisms is probably the most challenging step of this strategy: ideally purification will result in the isolation of specific, physiologically relevant binding partners of a given target protein, and yield sufficient quantities of material for robust, reliable mass spectrometry-based identification of all protein members. Affinity tags, added to the coding sequence of a protein target of interest facilitate the purification process by alleviating the need for protein-specific antibodies and provide an opportunity to achieve some level of standardization of the purification process. Typically, very high levels of specificity can be obtained through the use of two successive affinity purification steps, each targeting a different tag. As first described, the tandem affinity purification (TAP) procedure utilized a combination of protein-A and calmodulin-binding peptide (CBP) tags, respectively (Rigaut et al., 1999). This strategy has been used to map protein-protein interactions in yeast (Gavin et al., 2002) and mammalian cells (Bouwmeester et al., 2004). Despite subsequent improvements in their design (Burckstummer et al., 2006), this tag combination has proven to be relatively inefficient for use in mammalian cells. Recently the protein-A and calmodulin-binding tags have been supplanted by shorter, more efficient epitope tags (Gloeckner et al., 2007; Nakatani and Ogryzko, 2003). In this section, we will describe the typical analysis workflow, from cell culture and sub-cellular fractionation, tandem affinity purification procedure and quality control, up to analysis by mass spectrometry.

Stable expression of an affinity tagged protein can be achieved through various methods, including, (i) viral-based integration of the corresponding DNA sequence into the genome of a model cell line (HeLa cells or other mammalian cell line (Groisman et al., 2003)), (ii) homologous recombination in mice, *Saccharomyces cerevisiae* (Rigaut et al., 1999; Testa et al., 2003), and Saccharomyces pombe (Tasto et al., 2001), or (iii) transgenic expression in Drosophila melanogaster (Veraksa et al., 2005). Detailed protocols for these methods are beyond the scope of this report, and interested readers are encouraged to consult the references above for further information. In addition to the aforementioned strategies, it is worth noting that target proteins may also be purified following transient transfection of a suitable expression vector (Gloeckner et al., 2009), or, in cases where expression of the protein of interest is toxic, from a vector with an inducible promoter. The choice of an expression system (in tissue culture cells or in mice tissues) and a cell line will impact the purification process. Nonetheless, the generic protocol described here, initially optimized for tissue culture cells, has been successfully used for the tandem affinity purification of tagged proteins in a number of experimental contexts in our lab. Similarly, this protocol may be used for either FLAG-HA or FLAG-SII purifications, pending minor modifications (see below). Finally, note that the amount of starting material necessary for a successful purification can vary widely and is primarily dependent upon both the expression level of the target protein and its inherent biochemical behavior. As a general guideline for tissue culture cells, such as HeLa S3 or HEK293, the amount of starting material may range from a single 15 cm plate up to several high density spinner flasks.

Whenever feasible, cells should be fractionated into their sub-cellular compartments (nucleus, chromatin, and cytoplasm). This represents not only an easy, albeit crude, first fractionation step, but may also facilitate purification of compartment-specific complexes.

22.2.2 Sub-cellular Fractionation

The following protocol, adapted from the Roeder lab (Dignam et al., 1983) is particularly well-suited for sub-cellular fractionation of HeLa-S3 cells. For adherent cells, the differential digitonin fractionation method should be employed instead (Ramsby and Makowski, 1999).

22.2.2.1 Reagents and Materials

- Dounce homogenizers:
  - 7 mL (large). Kontes, Fisher Scientific cat#: 885301-0007
  - 2 mL (small). Kontes, Fisher Scientific cat#: 885301-0002
- Centrifuge tubes:
  - 50 mL conical tubes
  - 15 mL conical tubes
  - 1.5 mL micro-centrifuge tubes
- Micrococcal S7 nuclease: Roche Applied Sciences cat# 10107921001
- Protease inhibitors: Roche Applied Sciences cat# 11 873 580 001
- Phosphatase inhibitors: Sigma Cat# P2850
- Ribo-nuclease (RNase) inhibitors: Invitrogen cat# 100000840
- Phosphate Buffered Saline (PBS). Keep at 4°C

22.2.2.2 Buffers

- Hypotonic buffer. Keep at 4°C
  - Tris pH 7.3: 10 mM
  - KCl: 10 mM
  - $MgCl_2$: 1.5 mM
- Micrococcal nuclease (MNase) buffer:
  - 20 mM Tris-HCl, pH 7.5
  - 100 mM KCl
  - 2 mM $MgCl_2$
  - 1 mM $CaCl_2$
- FLAG IP and HA IP buffers: see Section 22.2.3.2

22.2.2.3 Protocol

Before Starting

- Chill all buffers on ice. Before use, add protease, and optionally, phosphatase and RNase inhibitors.
- Add 2-mercaptoethanol to the hypotonic buffer and keep on ice. 25 mL should be sufficient for $\approx 6 \times 150$ mm plates.
- Chill Dounce homogenizers on ice.
- Pre-cool all centrifuges, rotors and tubes (50, 15, and 1.5 mL).

Cell Collection

- Without removing the culture media, collect cells using a cell scraper and transfer to 50 mL conical tubes placed on ice.
- Centrifuge at 500 $\times$g for 3 min.
- Decant the culture media by pouring into a biological waste container.
- Pipette 10 mL of ice-cold PBS into all 50 mL conical tubes. Resuspend the pellet by pipetting up and down.
- Add cold PBS to bring the volume to 50 mL. Mix by inversion.
- Centrifuge at 500 $\times$g for 3 min.
- Remove PBS by pouring contents into a biological waste container.
- If several 50 mL conical tubes per cell line were used, pipette 10 mL of PBS in one of them, resuspend the cell pellet by pipetting up and down, transfer to the next tube, resuspend the second cell pellet and so on. Transfer the combined cell suspension to a 15 mL conical tube (you may combine up to three pellets in one 15 mL tube).
- Centrifuge at 500 $\times$g for 3 min.
- Carefully remove the supernatant. Record the "packed cell" volume.
- Resuspend the washed cell pellet using 5 $\times$ the "packed cell" volume of complete hypotonic buffer.

- Centrifuge at 800 ×g for 3 min.
- The pellet is very loose at this stage: remove the supernatant with a pipette. Record the "swollen cell" volume.

Note: This volume should be 50–80% larger than the "packed cell" volume, indicating appropriate cell swelling.

- Resuspend the swollen cell pellet with 1 × the "packed cell" volume of complete hypotonic buffer.

Note: the optimal volume of complete hypotonic buffer to use at this stage should be experimentally determined for each cell line, but typically falls between 0.5 and 3 time the "packed cell" volume.

- Incubate on ice for 10 min.
- Transfer a ≈ 5 μL aliquot of the swollen cell suspension to a glass slide, cover with a coverslip and set aside.
- Transfer the remaining cell suspension to a pre-cooled Dounce homogenizer.
- Insert the glass pestle ("loose") and slowly pestle up and down for 10 cycles (a cycle is 1 down and 1 up).
- Set the pestle aside (you may put it in the tube used for the last centrifugation, kept on ice). Transfer a ≈ 5 μL aliquot to a glass slide, cover with a coverslip.
- Examine the cell spreads before and after Dounce homogenization. The cells before disruption should have a swollen appearance. After disruption, nearly all cells should be broken open, the nuclei clearly isolated, and abundant debris should be visible between them. If more than ≈ 10% of cells remain intact, add 2 more up and down cycles and evaluate the extent of cell disruption again. Continue with a few more cycles if necessary but do not go beyond a total of 15 cycles (nuclei will begin to fragment).
- Distribute the content of the homogenizer between several 1.5 mL centrifuge tubes, labeled "N".
- Centrifuge at 1000 ×g for 10 min.
- Transfer the supernatant to a new 1.5 mL centrifuge tube labeled "C1".
- Resuspend the nuclear pellet "N" with 1 mL of complete hypotonic buffer.
- Centrifuge both the supernatant "C1" and the resuspended nuclear pellet "N" as above.
- Transfer the supernatant from the "C1" tube to a new 1.5 mL centrifuge tube labeled "C2". This is the crude cytoplasmic fraction. Further centrifugation at 100,000 ×g for 30 min will result in a purer, membrane-free S100 fraction.
- Remove the supernatant from the "N" tube, and discard. The pellet is the crude nuclear fraction. A further purification step on a sucrose gradient, while not necessary, will result in a purer, membrane-free nuclear fraction.

- Snap freeze these fractions, properly labeled, in liquid nitrogen and place at –80°C. Do not leave tubes in liquid nitrogen for more than a few minutes.

Note: For HeLa-S3 cells, the fractionation of 6 dense 15 cm plates (≈ $2e^8$ cells), should result in about 1–2 mL of cytoplasmic fraction and a nuclear pellet volume of about 300–500 μL.

Protein Extraction

Thaw nuclear pellets and/or the cytoplasmic fractions on ice.

Note: this may take up to 1 h for a 1.5 mL volume.

Cytoplasmic Fraction

- Add NaCl to a *final* concentration of 150 mM
- Add Tris, pH 7.5 to a *final* concentration of 50 mM
- Add NP40 to a *final* percentage of 0.05%
- For example, if the volume of the cytoplasmic fraction is 1.2 mL, add 45 μL of 5 M NaCl, 75 μL of 1 M Tris pH 7.5, 7.5 μL of 10% NP40. Add hypotonic buffer to the 1.5 mL graduation mark.
- Incubate end over end for 15 min at 4°C
- Pellet cell debris at 20,000 ×g for 10 min at 4°C.
- Transfer the supernatant to a new 1.5 mL micro-centrifuge tube.
- Proceed with the FLAG IP on the "fresh" extract. Do not freeze.

Note: This not an "extraction" per se. This step is intended to adjust the salt concentration and pH and add a small percentage of NP40 to increase the IP efficiency.

Nuclear Fraction

- Lyse the nuclear pellet using 3–4 volumes of FLAG IP buffer, pipetting up and down until all clumps have disappeared.
- Incubate end over end for 30 min at 4°C
- Pellet cell debris at 20,000 ×g for 10 min at 4°C.
- Transfer the supernatant to new 1.5 mL micro-centrifuge tubes.
- Proceed with the FLAG IP on the "fresh" extract. Do not freeze.
- The insoluble pellet may be further processed for crude chromatin preparation.

Chromatin fraction

- Starting with the insoluble pellet obtained after detergent lysis of purified nuclei or, as an alternative, from a pellet obtained after detergent lysis of whole cells: Disrupt and homogenize the insoluble pellet by pipetting up and down several times using ~ 5 pellet volume of MNase buffer.
- Re-pellet the washed (crude) chromatin pellet for 5 min at 10,000×*g*, 4°C.

- Thoroughly resuspend the pellet in one pellet volume of MNase buffer containing 3 units of MNase per μl.
- Incubate at room temperature (or at 4°C) with constant vortexing until complete solubilization of the chromatin. The solution should appear homogenous with no visible clumps. If necessary, add 3 more units of MNase, pipette the sample up and down with a small-bore pipet tip to break apart clumps and continue the incubation.
- Dilute the solubilized chromatin with an equal volume of a 2× concentrated immuno-precipitation buffer with low NP40 content (such as 2× concentrated HA-IP buffer).
- Incubate at 4°C for 10–30 min, vortexing.
- Pellet the insoluble material for 10 min at 20,000×g, 4°C.
- Proceed with the FLAG IP on the "fresh" extract. Do not freeze.

Note: If the bait is a DNA binding protein, nucleosomes may co-purify with the target protein, leading to the "contamination" with core histones and other abundant chromatin-associated factors. Furthermore, incomplete chromatin digestion generates DNA "ends". Factors that have a high affinity for this particular chromatin structure, such as the abundant Ku70/Ku80 heterodimer, will indirectly co-purify with the bait through a "DNA bridge". In general, the presence of abundant chromatin-associated proteins in the purified complex should always be regarded with suspicion.

22.2.2.4 Quality Controls

- The fractionation efficiency can be analyzed by western blot analysis using antibodies against common sub-cellular markers (GAPDH for the cytoplasm, RNA Polymerase II for the nucleus and any core histones for the chromatin). In this case, small aliquots (≈20 μL) of the crude cytoplasmic fraction (or S100) and of the nuclear fraction (after protein extraction, see below) should be saved during the fractionation procedure. Equivalent amount of protein from each fraction should then be analyzed by immuno-blotting. The chromatin fraction can also be analyzed by polyacrylamide gel electrophoresis and coomassie staining (histones are easily recognized by their migration around 15 kDa).
- The distribution of the bait between sub-cellular fractions should be determined as well. Note that it is common for a bona fide nuclear protein to leach in the cytoplasmic fraction during the fractionation process. Although it may be acceptable to proceed with the purification of a nuclear protein from a cytoplasmic fraction, this may lead to artificial association between the protein of interest and abundant cytoplasmic proteins.
- To evaluate the extent of chromatin digestion, take an aliquot (∼10%) of the chromatin before and after MNase digestion, extract with phenol/chloroform, precipitate the de-proteinated DNA, and run the resulting material on a 1.8% agarose gel. A single band around 150 bp indicates optimal digestion of the chromatin down to mono-nucleosomes.

22.2.3 Tandem Immuno-Affinity Purification

22.2.3.1 Reagents and Materials

- 0.2 mm flat gel-loading tips: Costar Cat# 4884
- Centrifuge filters: Millipore Cat# UFC30HV00 (these filters have low protein binding characteristics)
- FLAG-agarose conjugate: Sigma cat # A2220
- FLAG peptide: Sigma cat # F3290
- HA-agarose conjugate: Santa Cruz cat # sc-7392AC
- HA peptide: Covance cat # PEP-101P
- Gel for Silver Stain analysis: Invitrogen cat # NP0323BOX
- Benchmark protein ladder: Invitrogen cat # 10747-012
- FLAG antibody (HRP-labeled) for western blot analysis: Sigma cat# A8592
- Protease inhibitors: Roche Applied Sciences cat# 11 873 580 001
- Phosphatase inhibitors: Sigma Cat# P2850
- Ribo-nuclease (RNase) inhibitors: Invitrogen Cat# 100000840

22.2.3.2 Buffers

All buffers should be prepared from stock solutions the day of the experiment. Protease inhibitors must be added to all buffers just prior use. The addition of phosphatase inhibitors and ribo-nuclease inhibitors is optional. Note however, that these inhibitors may drastically affect the yield and specificity of the purified target protein complex. Most commercial ribo-nuclease inhibitors are protein-based. The final elution buffers should be prepared without these inhibitors.

Stock Solutions (to be diluted):

- NaCl: 5 M
- Tris pH 7.5: 1 M
- EDTA: 0.5 M
- NP40: 10%
- Glycerol: 50%

FLAG IP buffer:

- NaCl: 150 mM
- Tris pH 7.5: 50 mM
- EDTA: 1 mM
- NP40: 0.5%
- Glycerol: 10%

HA-IP buffer:

- NaCl: 150 mM
- Tris pH 7.5: 50 mM
- EDTA: 1 mM

- NP40: 0.05%
- Glycerol: 10%

FLAG peptide stock solution (5 mg/mL):

- Dissolve lyophilized peptide in 200 mM Tris pH 8.0

FLAG peptide elution buffer:

- Dilute FLAG peptide stock solution ten-fold in HA-IP buffer

HA peptide stock solution (4 mg/mL):

Dissolve lyophilized peptide in:

- NaCl: 150 mM
- Tris pH 7.5: 50 mM
- EDTA: 1 mM

HA peptide elution buffer:

- Dilute HA peptide stock solution ten-fold in HA-IP buffer

22.2.3.3 Protocol

Before Starting

Protein concentration and extract volume must be normalized across all samples within a set before the start of the purification process. This is particularly important when the purified protein complexes are to be analyzed by quantitative LC-MS/MS, either to calculate fold enrichment between control and bait IPs for single step purifications, or to measure changes in composition within a protein complex (see below).

- Place the FLAG-agarose vial on ice to bring its content to ≈4°C.
- Add protease and optionally, phosphatase and RNase inhibitors to pre-chilled buffers before use.
- Pre-cool all centrifuges, rotors and tubes (50, 15, and 1.5 mL).

FLAG Immuno-Precipitation

Immuno-Purification

- For each tube of IP, transfer 20 μL of the FLAG-agarose slurry to a 1.5 mL centrifuge tube.

Note: this amount of FLAG-agarose slurry is sufficient to immuno-precipitate a *very* abundant cellular protein tagged with the FLAG epitope from several milligrams of total protein from a cytoplasmic fraction or nuclear extract. While using more slurry may speed-up the kinetic of the capture, it will also result in the co-purification of higher amount of contaminants.

- Wash the slurry three times with 400 μL of FLAG-IP buffer.
- For each 20 μL of slurry initially transferred, re-suspend the washed beads using 40 μL of FLAG-IP buffer.
- Transfer 40 μL of the washed beads to each IP tube.
- Incubate for 4 h at 4°C, rotating the tubes end-over-end.

Washes

- Pellet the FLAG-agarose beads at 7000 ×g for 30 s.
- Transfer most of the supernatant to a new tube (or discard directly).

Note: Do not pipette the pelleted beads

- Using a flat 0.2 mm gel loading tip, pipette the remaining extract and transfer to above tube (or discard).

Note: Careful removal of extract at this stage minimizes carry-over of abundant cellular proteins

- Add 500 μL of FLAG-IP buffer.
- Vortex briefly.
- Pellet the FLAG-agarose beads at 7000 ×g for 30 s.
- Remove most of the supernatant. See note above.
- Repeat the wash cycle.
- Remove most of the supernatant.
- Repeat the wash cycle one last time using the *HA-IP* buffer.
- Remove most of the supernatant.
- Using a flat gel-loading tip, aspirate as much of the remaining wash buffer as possible.

Elution

- Add 20 μL of the *FLAG-peptide elution* buffer.
- Incubate on a vortex platform at 4°C for 30 min, making sure that the beads remain in suspension.
- Pellet the beads at 7000 ×g for 30 s.
- Wash a centrifuge filter with 200 μL HA-IP buffer. Centrifuge briefly and discard the flow-through.
- Using a flat 0.2 mm gel-loading tip, transfer the bead supernatant to the pre-washed centrifuge filter. Spin immediately.

Note: Do not transfer the beads at this stage. Keep the first elution on ice during the second elution.

- Add 20 μL of the *FLAG-peptide elution* buffer to the beads left in the tube.
- Incubate on a vortex platform at 4°C for 30 min.
- Pellet the beads at 7000 ×g for 30 s.
- Pipette up-and-down to re-suspend the bead pellet. Transfer the *slurry* to the centrifuge filter used previously, and spin immediately to collect both FLAG elutions in a single tube.

HA Immuno-Precipitation

Immuno-Purification

- For each IP, transfer 20 μL of the HA-agarose slurry to a 1.5 mL centrifuge tube.
- Wash the slurry three times with 400 μL of HA-IP buffer.
- For each 20 μL of slurry initially transferred, resuspend the washed beads using 20 μL of HA-IP buffer.
- Transfer 20 μL of the washed beads to each IP tube.
- Incubate O/N at 4°C on a vortex platform. Make sure that the beads remain in suspension.

Note: End-over end rotation does not work well with such small volumes.
The length of the HA-IP can be shortened to about 4 h without affecting its efficiency. However, an overnight incubation is usually more practical.

Washes

- Following the technique used for the FLAG-IP washes, but using only *200* μL of the *HA-IP buffer*, wash the beads 3 times.
- Using the flat gel-loading tip, aspirate as much of the remaining buffer as possible.

Elution

- Add 20 μL of the *HA-peptide elution* buffer.
- Incubate on a vortex platform at 4°C for 30 min, making sure that the beads remain in suspension.
- Pellet the beads at 7000 ×g for 30 s.
- Using a flat 0.2 mm gel-loading tip, transfer the supernatant to a centrifuge filter (pre-washed with 200 μL HA-IP buffer, see above), and spin immediately. *Do not transfer the beads at this stage*. Keep the first eluate on ice during the second elution.
- Add 20 μL of the *HA-peptide elution* buffer to the beads left in the tube.
- Incubate on a vortex platform at 4°C for 30 min.

- Pellet the beads at 7000 ×g for 30 s.
- Using a flat 0.2 mm gel-loading tip, transfer the supernatant to a centrifuge filter used previously, and spin immediately. *Do not transfer the beads at this stage.* Keep the combined eluates on ice during the third elution.
- Perform a third elution as indicated above.
- Pipette up-and-down to resuspend the bead pellet. Transfer the *slurry* to the centrifuge filter used previously, and spin immediately to collect the three HA eluates in a single tube.
- Freeze samples at –80°C.

Note: Two HA elutions steps may be sufficient for most protein complexes.

StrepTactin affinity purification can essentially be performed as the HA purification described above. The following modification should be made:

- StrepTactin buffer has the same composition as the HA buffer but is made with Tris pH 8.0.
- The elutions are performed with 2.5 mM desthio-biotin, diluted from a 5 × stock made in water.

22.2.3.4 Quality Controls

For the first few rounds of purifications, it is recommended to take samples for western blot analysis at several steps of the purification (this is only necessary for the tagged protein IP sample, not for the control IP). The percentages of each fraction indicated below are given as guidelines only. A higher or lower percentage may be required depending on the expression level of the tagged protein and its behavior during the purification.

Sample Collection

FLAG IP

- The "input (F)", between 0.1 and 0.5% of the extract volume.
- The "flow through (F)", taken after the FLAG IP after pelleting the beads. Same amount (in % of the fraction's volume) as the "input (F)".
- The "bound (F)" (or captured) material corresponds to an aliquot of the mixed bead *slurry* taken during the last wash cycle. Between 1 and 4% of the slurry.
- The "eluate (F)", taken from the combined two elutions. Same % as the "bound (F)".

Note: For silver stain analysis, save 5–10% of the combined FLAG elutions for *both* the control IP, and the tagged protein IP sample.

HA or StrepTactin IP

- The "input (HA/SII)" for this IP is the same as the "eluate (F)" from the FLAG IP, no need to take a sample.

- The “flow through (HA/SII)”, between 2 and 4% of the volume.
- The “bound (HA/SII)” material, between 2 and 4% of the mixed bead *slurry* volume.
- The “eluate (HA/SII)”, taken from the combined three elutions, between 2 and 4% of the total volume.

Note: For silver stain analysis, save between 10 and 20% of the HA/SII elution for *both* the control IP, and the tagged protein IP sample.

Gel Analysis

Aliquots saved at different stages of the purification may be run on regular PAGE. The anti-FLAG antibody or an antibody directed against the protein of interest may be used for western blot detection.

For silver stain analysis, pre-cast gradient gels are strongly recommended. These gels provide very clear background and minimal gel-to-gel variability. As a molecular weight marker for the silver stain, the Benchmark protein ladder (Invitrogen) provides very clear, well defined bands. Dilute 1 μL of the stock in 100 μL of 1× LDS loading buffer before each run. Load 10 μL per lane; discard any unused diluted MW marker. Home-made silver staining reagents or a variety of commercial kits can be used for the staining stage.

22.2.3.5 Anticipated Results

- Yield: Successful tandem affinity purifications are characterized by high capture efficiency and efficient elutions at both the FLAG and the HA IP stages. This can easily be estimated by western blot analysis of small aliquots taken at different stages of the purification. While the efficiency of the FLAG and StrepTactin affinity purification are typically very high for most proteins, the efficiency of HA affinity purification may be protein target dependent. In our experience the degree of protein-dependent variability is most pronounced at the elution step.
- Specificity: In addition to high yield, success in tandem affinity purifications is characterized by the absence of co-purifying proteins in the control IP. By contrast, several distinct bands, in addition to the tagged protein itself should be clearly visible on the silver stained gel.

22.2.4 Single-Step Immuno-Affinity Purification

While tandem affinity purification often provides both high yield and specificity with respect to binding partners of the target protein, there are scenarios in which a single-step affinity purification is preferred. For example, as mentioned above, the HA stage can be plagued by low yield, often at the elution step. Moreover, the additional time and sample manipulation required for the second affinity purification may lead to unacceptable losses of sub-stoichiometric or transient protein interactors. In these cases, a single step FLAG IP may be performed according to the first part of the protocol described above. However, the FLAG IP is notoriously "dirty" as many contaminant proteins co-purify with the tagged protein. Even in the case of abundant and "well-behaved" protein targets, it is virtually impossible to identify *bona fide* interactors from contaminants by simply comparing the lists of proteins detected in the control and the tagged protein IPs. Fortunately, the enrichment factor for each protein can be determined by relying on quantitative LC-MS/MS strategies (Pflieger et al., 2008). In addition, the chemical properties of amino acid side chains may be used to further simplify the analysis of FLAG-only IPs. Cysteine, for example, possesses a nucleophilic thiolate side chain that is very amenable to chemical modification under mildly basic conditions (pH 7.5–8.5). Moreover, the low frequency of cysteine occurrence within the proteome facilitates the identification of a maximum number of proteins based on a minimum number of peptides. In fact, Fenyö et al. (Fenyö et al., 1998) demonstrated that even with a liberal mass error of $\pm$ 0.5 Da, the number of possible protein assignments in *S. cerevisiae* based on peptides of mass 800–5000, that also contained at least one cysteine, was reduced by as much as ten-fold as compared to protein assignments based on all possible peptides (i.e., no cysteine constraint). This study led to a number of investigations over the last decade that described a variety of reversible (Liu et al., 2007; Palani et al., 2008; Raftery, 2008; Shiu et al., 2009) and irreversible (Chen et al., 2008; Clements et al., 2005; Dai et al., 2005; Giron et al., 2009; Gygi et al., 1999, 2002) thiol-reactive agents for cysteine peptide enrichment and, in the cases of ICAT (Gygi et al., 1999, 2002) and QCET (Liu et al., 2007), subsequent peptide-level protein quantitation.

The reduced sample complexity that results from enrichment of cysteine-containing peptides translates directly to more efficient use of LC-MS/MS duty cycle. In essence more unique protein IDs are expected from the same number of MS/MS scans for samples enriched for peptides that contain cysteine. These benefits also extend to quantitative analyses, particularly in the case of MS/MS-based labels such as iTRAQ (Ross et al., 2004) and TMT (Thompson et al., 2003). Here, reduced sample complexity minimizes precursor contamination during MS/MS, and hence improves the accuracy of peptide quantification. The protocol outlined below for cysteine capture may be used in a variety of contexts, including studies that require stable isotope labeling for quantitative analysis (see Section "Cysteine Peptide Enrichment").

22.2.5 *Processing of Immuno-Affinity Purified Proteins for Mass Spectrometry Analysis*

The overall sample processing workflow for component analysis of a purified protein complex by LC-MS/MS is similar to that used for typical proteomics experiments: trypsin digestion, peptide clean-up/desalting and concentration. However, the relatively low abundance of proteins in the purified sample and the presence of specific contaminants require some additional processing steps before LC-MS/MS analysis.

22.2.5.1 Reagents and Materials

Note: All working solutions are prepared fresh from stock reagents listed here on the day of use.

- SelfPack POROS 50R2 and 10R2 reversed-phase packing (Applied Biosystems)
- SelfPack POROS 20HS strong cation exchange packing (Applied Biosystems)
- RapiGest™ SF (Waters)
- DL-dithiothreitol/DTT (Fluka)
- Iodoacetamide, SigmaUltra/IAA (Sigma)
- Sequencing grade modified trypsin, 500 ng/μL (Promega)
- Trifluoroacetic acid/TFA (Sigma)
- UltimAR® Acetonitrile/MeCN (Mallinckrodt)
- Sep-Pak® tC18 μElution Plate with 10 mg resin per well (Waters)
- Triethylammonium bicarbonate buffer, 1.0 M, pH 8.5/TEAB (Sigma)
- Activated Thiol Sepharose 4B beads (GE Healthcare)
- 0.2 mm flat gel-loading tips (Costar)
- iTRAQ 4-plex or 8-plex (Sciex)
- TMT Label Reagent Set (Thermo Scientific)

22.2.5.2 Buffers

- Reversed Phase (RP) Buffer A1: 0.1% TFA in water
- RP Buffer B1: 0.1% TFA, 80% MeCN
- RP Buffer A2: 0.2 M acetic acid (HOAc) in water
- RP Buffer B2: 0.2 M HOAc, 70% MeCN
- Strong Cation eXchange (SCX) Buffer A: 0.1% Formic acid, 25% MeCN
- SCX Buffer B: 0.1% Formic acid, 25% MeCN, 300 mM KCl
- 20 mM Sodium Phosphate buffer, pH 8.0
- 0.5 M TEAB adjusted to pH 8.0

22.2.5.3 Column Construction

Reversed Phase Column

- Cut a 25 cm long segment of 200 μm I.D. × 360 μm O.D. fused silica capillary
- Put a frit at one end of the capillary (see Section "Fritting Procedure 1" for details)
- Prepare a 50% slurry of POROS 50R2 resin in 100 % MeCN
- Load a 6–10 cm length of the reversed-phase resin into the fritted capillary using a helium-pressurized loading vessel set at 700 psi.

Note: A magnetic stirring rod placed in the slurry and a magnetic stirrer placed underneath the pressurized loading vessel will help keep the resin suspended in the MeCN during loading.

Note: The binding capacity of a 10 cm × 200 μm column is ~ 40 μg of peptide. Always use a column with a binding capacity that matches the expected peptide amount. Lower capacity columns can be prepared using a 75 μm I.D. × 360 μm

O.D. fused silica capillary. In this case, POROS 10R2 reversed phase media should be used for packing.

- Rinse the column with 10 column volumes of RP Buffer A1 in the pressurized loading vessel at 700 psi

Note: RP columns can be kept indefinitely after preparation. If the column is not used immediately, dry the column in the pressurized loading vessel at 200 psi for 10 min. If desired, put a second frit at the opposite end of the resin bed after drying.

Reversed Phase Column Conditioning (Optional)

The following procedure is designed to saturate irreversible binding sites on the column, thereby minimizing sample loss. This only needs to be performed on a new column. Previously used columns may be reused if they have been rinsed with several reversed-phase gradients.

- Load 100 fmol of enolase protein digest with the pressurized loading vessel.
- Rinse the column with an HPLC reversed-phase gradient (Agilent 1100 system is suitable for this), maintaining a pressure below 70 bar, such as:
  - 100% RP Buffer A2 to 100% RP Buffer B2 in 10 min
  - 100% RP Buffer B2 for 1 min
  - 100% RP Buffer B2 to 100% RP Buffer A2 in 1 min
  - 100% RP Buffer A2 for 5 min

Strong Cation Exchange Column

- Prepare a fritted capillary column as described above
- Prepare a 50% slurry of POROS 20HS resin in 100% MeCN
- Load a 6–10 cm length of the reversed-phase resin into the fritted capillary using a helium-pressurized loading vessel set at 700 psi.

Note: The binding capacity of an SCX column is slightly larger than that of a RP column

- Rinse the column with 10 column volumes of SCX Buffer B followed by 10 column volumes of SCX Buffer A in the pressurized loading vessel at 700 psi

Note: SCX columns should be used immediately after preparation.

22.2.5.4 Protocol

Reduction, Denaturation, Alkylation and Digestion

- Thaw purified samples on ice.
- Add RapiGest SF (1% stock) to get a final concentration of 0.1%
- Add DTT (1 M stock) to get a final concentration of 10 mM. Vortex.

- Heat at 56°C for 30 min
- Cool samples on ice
- Add iodoacetamide (1 M stock) to a final concentration of 20 mM.

Note: Skip this alkylation step if cysteine enrichment or HA peptide removal need to be performed.

- Incubate in the dark for 30 min at room temperature
- Add trypsin at a ratio between 1:20 and 1:100 relative to the estimated amount of protein in the sample.

Note: For single step FLAG IP, the FLAG peptide used for the elution corresponds to ~ 16 μg of tryptic site.

- If necessary, adjust the pH to ~ 8 using 1 M Tris pH 8.0.
- Digest overnight at 37°C
- Cleave RapiGest by adding TFA (stock at 10%) to a final concentration of 0.5%
- Incubate for 30 min at 37°C
- Pellet the insoluble RapiGest by-products by centrifugation at 20,000 ×g for 20 min.

Reversed Phase Tryptic Peptide Desalting

Using a micro-capillary column

- Load tryptic digest onto a micro-capillary reversed phase column
- Wash with 5–100 μL RP Buffer A1
- Elute with 50 μL RP Buffer B1
- Dry sample under vacuum.
- Store sample at –80°C until ready to use.

Using a 96-well micro-elution tC18 μElution Plate

- Condition one well of a reversed-phase tC18 μElution Plate with 400 μL RP Buffer B1.
- Rinse with 2 × 400 μL RP Buffer A1
- Load sample, retaining flow-through
- Rinse with 2 × 400 μL RP Buffer A1
- Elute with 2 × 50 μL RP Buffer B1
- Dry eluate under vacuum

Note: This method may be used for sample containing between 25 and 250 μg of peptide.

HA Peptide Removal

A typical protein complex purified by tandem affinity contains ≈ 24 μg of HA peptide, an amount that would dramatically interfere with chromatographic separations and with mass spectrometry analysis. Fortunately, the HA peptide sequence used in our lab and described in this protocol contains an N-terminal cysteine residue, allowing simple removal on Activated Thiol Sepharose 4B beads.

Important: After protein denaturation and reduction, skip the alkylation reaction.

- Reconstitute the dried tryptic peptides with 40 μL of 20 mM Sodium phosphate buffer, pH 8.0.

Note: For efficient capture, the pH of the reconstituted peptide solution should be between 7.5 and 8.5.

- Add 20 μL of a 50% slurry of washed Activated Thiol Sepharose 4B.
- Incubate at room temperature on a vortex platform for 30 min.
- Pellet the beads and collect the supernatant using a 0.2 mm flat pipet tip.
- Briefly rinse the beads with 20 μL of 20 mM Sodium phosphate buffer, pH 8.0.
- Pellet the beads and combine with the first supernatant.
- Purify the HA-depleted tryptic peptides from the combined supernatants by reversed-phase chromatography.

Note: This step should reduce the amount of HA peptide by at least two orders of magnitude.

NP40 and Desthio-Biotin Removal

The final HA elution contains 0.05% of NP40. Due to its hydrophobicity and molecular heterogeneity, NP40 will be carried over throughout the sample processing and will contaminate the last part of an LC-MS/MS run. However, NP40 is non-ionic and can efficiently be removed by SCX (or SAX) chromatography. Similarly, desthio-biotin is uncharged in 0.1% formic acid and will not bind efficiently to SCX resin.

- Solubilize the dried peptides resulting from the last reversed-phase chromatography in SCX buffer A.
- Inject the solubilized peptides on a SCX micro-capillary column of appropriate binding capacity.
- Rinse the column with 100 μL of SCX buffer A
- Isocratically elute bound peptides using 20–50 μL SCX buffer B.
- Remove acetonitrile by vacuum centrifugation.

Cysteine Peptide Enrichment

The following procedure is typically used to analyze single-step FLAG IPs in our lab. It is also used, with or without labeling, to reduce the complexity of other peptide mixtures.

Important: After protein denaturation and reduction, skip the alkylation reaction.

Cysteine Capture

- Measure approximately 25 μL Activated Thiol Sepharose 4B beads, enough to yield ~ 50 μL of swollen beads
- Wash the beads three times with 1 mL of doubly-deionized water (beads will swell)

Note: Some of the beads can be as small as 45 μm and may difficult to pellet by centrifugation. Therefore, it is best to remove supernatant from the bead pellet using a 0.2 mm flat gel loading tip to limit bead loss.

- Wash the beads once with 1 mL 0.5 M TEAB pH 8.5
- Re-suspend the beads with 0.5 M TEAB pH 8.5 to obtain 25% slurry.
- Solubilize the dried peptides resulting from the last reversed-phase chromatography in 20 μL of 0.5 M TEAB pH 8.5.
- Add 50 μL of the bead slurry (equivalent to 12.5 μL packed beads) to the sample

Note: 1 μL of packed beads is sufficient to capture cysteine peptides derived from an initial protein amount ranging between 1 and 5 μg
Note: For efficient capture, the pH of the reconstituted peptide solution should be between 7.5 and 8.5.

- Incubate the solution for 1 h at 37°C on a vortex at approx. 1000 rpm

Note: Vortex frequency required to maintain beads in suspension will vary with solution volume, size of bead pellet and incubation temperature.

- Pellet the beads at 3300 ×g for 30 s.
- Collect the supernatant with a 0.2 mm flat gel loading tip and discard.
- Wash the bead pellet three times with 500 μL 0.5 M TEAB pH 8.5

Cysteine Peptide Labeling (optional)

As mentioned earlier, the distinction between contaminants and specific interactors in single-step affinity purification relies on the quantitative measurement of their relative abundance between the control and the bait IP. For MS/MS-based quantification (iTRAQ, TMT) however, the accuracy of quantitative measurements can be compromised for samples with a relatively high complexity, such as those resulting from single-step FLAG IPs. As described above, enrichment of cysteine-containing peptides can greatly reduce the complexity of FLAG-only IPs, improving the ability to discern relevant protein interactions from background contaminants.

- For each sample in a set, capture cysteine-containing peptides on Thiol-activated Sepharose beads as described above.
- After the last wash, carefully remove any excess of buffer.

- Add 20 μL of 0.5 M TEAB pH 8.5.
- Reconstitute each iTRAQ reagent with 80 μL of ethanol. Vortex briefly and collect liquid to the bottom of the tubes with a quick centrifugation.
- Add 80 μL of reconstituted iTRAQ to the appropriate bead slurry
- Vortex immediately.
- To keep beads in suspension during labeling, incubate on a vortexing platform for 30 min at room temperature.
- Quench any unreacted iTRAQ reagent by adding 20 μL of 20 mM Tris, pH 8.0 to each reaction. Vortex.
- Carefully transfer each bead slurry to a single new 1.5 mL tube.
- Pellet the combined beads at 3300 ×g for 15 s.
- Remove the supernatant.
- Wash the combined beads twice with 5–10 bead volumes of 0.5 M TEAB pH 8.5.
- After the last wash, carefully remove any excess of buffer.

Note: the volume of iTRAQ reagent in each tube is sufficient to label 100 μg of tryptic peptides. Adjust the volume according to the expected amount of cysteine-containing peptide immobilized on the Thiol-Activated Sepharose, keeping the percentage of ethanol during the reaction above 70%. iTRAQ reagents are very unstable in aqueous solutions. It is not recommended to store iTRAQ reagent once reconstituted.

Elution and Alkylation

- Add 50 μL 10 mM DTT/0.5 M TEAB pH 8.5
- Incubate on vortex for 30 min at 1400 rpm and 37°C
- Pellet the combined beads at 3300 ×g for 15 s.
- Collect the eluate with a 0.2 mm flat gel loading tip
- Repeat the elution step and combine the two eluates.
- Alkylate cysteine residues by adding iodoacetamide to a final concentration of 20 mM
- Incubate for 30 min in the dark at room temperature

Cysteine peptide cleanup

- Load enriched cysteine peptides onto a micro-capillary reversed phase column

Note: Cysteine-containing peptides constitute a small percentage (∼10%) of a sample's total mass; the use of micro-capillary column format for cleanup is recommended to minimize sample loss.

- Wash with 20 μL RP Buffer A1
- Elute with 30 μL RP Buffer B1

- Dry sample under vacuum
- Store sample at –80°C until ready to use

Note: A second cysteine enrichment may be performed after the first capture to ensure maximum recovery. The requirement for a second enrichment can be evaluated by analyzing 10% of the first and second eluates by LC-MS/MS. A single capture should yield 95–99% cysteine enrichment.

22.3 Analysis of Multi-component Complexes Based on Immobilized DNA Templates

The interactions between many proteins and their nucleic acid binding sites are a fundamental part of almost all nuclear processes. These dynamic interactions facilitate remodeling of nuclear architecture, alter the subnuclear localization of genomic regions and RNAs, and regulate the recruitment, activity, and processivity of the enzymes involved in nucleic acid metabolism. Proteins are recruited to nucleic acids through numerous mechanisms. Some interactions are DNA sequence specific, while other protein/nucleic acid interactions are mediated through epigenetic mechanisms such as DNA methylation, covalent histone modifications, and alterations in chromatin structure, which can affect cooperative binding with other protein complexes. The determination and characterization of protein/nucleic acid interactions is pivotal to our understanding of how the genome is organized, maintained, and expressed. Traditional methods for the identification of nucleic acid binding proteins often involve a combination of time consuming chromatographic-based steps, each requiring verification via functional assay, prohibiting the application of these techniques on a proteome-wide scale (Mittler et al., 2009).

The use of immobilized oligonucleotides to capture specific proteins from a cell extract provides a useful tool that is more amenable to proteomic level analysis. However, this approach is often plagued by high background levels which can complicate the identification of *bona fide* protein-protein interactions. Recent advances in mass spectrometry and related proteomics technologies have facilitated the robust analysis of protein complexes that specifically nucleate around an immobilized template.

This technique has been successfully applied to a variety of systems. Ranish and colleagues used this approach to identify proteins that associate with gene promoters in a TATA binding protein dependent manner (Ranish et al., 2003). In another study, previously unknown sequence specific transcription factors were identified with an immobilized DNA encoding a regulatory region of the muscle creatine kinase gene (Himeda et al., 2004, 2008). The approach has been used to identify novel factors that associate specifically with methylated CpG islands (Mittler et al., 2009). In addition, the proteomics technique has been utilized in the context of RNA templates as well, including mRNAs, tRNAs, and miRNAs (Butter et al., 2009; Heo et al., 2009). It is clear that the intersection of nucleic acid and protein biochemistry will benefit significantly from mass spectrometry-based assays directed at the analysis of proteins isolated via immobilized templates.

22.3.1 Design and Construction of Immobilized DNA Templates

22.3.1.1 Buffers

- 1 × Acetate Transcription Buffer
  - 150 mM Potassium Acetate, pH 7.6
  - 20 mM HEPES, pH 7.6
  - 1 mM EDTA
  - 5 mM MgOAc
- Blocking Buffer
  - 1× acetate transcription buffer (pH 7.6)
  - 60 mg/mL casein
  - 5 mg/mL poly vinyl pyrrolidone
  - 2.5 mM DTT

Adjust pH to 8.0 using KOH, store at –20°C in 1 mL aliquots

22.3.1.2 Template Design

Design of appropriate experimental and control templates is the first and perhaps most important step for a successful immobilized template assay. Typically these templates are biotinylated at one end, and bound to streptavidin coated beads.

- Chose an element that will provide selective recruitment of target proteins from an extract. This could be a specific sequence element (for example, a promoter sequence or an enhancer region), a specific structural conformation (RNA hairpin, thymidine dimer), or a particular modification (DNA methylation, histone acetylation).
- The final length of the template can vary. The most common streptavidin coated beads used in these experiments have maximum binding efficiencies for DNAs <1 kilobase, but much longer templates (>4 kb) will still bind and form functional templates.
- Include a linker region between biotin and the experimental element intended as the specific nucleation site for proteins. The linker minimizes steric hindrance of the beads and facilitates access to the target element; most experiments work well with linkers of ≈300 bp.
- The control and experimental templates should be identical except for the specific element. In particular, it is important to match the lengths of the respective templates, as the degree of non-specific binding tends to scale with template length. For example, the false positive rate may be excessive if the experimental template is longer than that for the control.

22.3.1.3 Template Construction

- PCR is the most common method used for template construction. Simply design primers to the 5' and 3' end of the nucleic acid you wish to create for your template, and include a biotin moiety at the 5' end of one of the primers (most DNA synthesis facilities can include a biotin molecule at the end of primers).
- The easiest way to generate distinct control and experimental templates is to encode the constructs on different plasmids and use the same primers for each PCR reaction.

Note: In some cases, for example a template designed to mimic DNA damage, it is not possible to include a biotinylated end via PCR. However, these templates can often be synthesized to include a biotin at one end. Biotin may be added to an end of a preexisting template as well. A restriction enzyme is used to cleave the template, forming a sticky end. Next, the Klenow fragment of DNA polymerase can be used to fill in the protracted end in the presence of a biotinylated dNTP to form a blunt ended, biotinylated template.

- The biotinylated template can be immobilized onto any streptavidin-coated bead. Both agarose and magnetic beads work well, although the background associated with the latter is often lower. The stability of the streptavidin-biotin interaction facilitates the use of high salt washes (1–2 M) that in-turn minimize nonspecific protein associations with the beads. Re-equilibration is readily accomplished with 3–4 washes of binding buffer.

Nucleic Acid Immobilization

- Wash Dynabeads M-280 Streptavidin (Dynal) 2 times in 400 μL Tris-EDTA (pH 7.5) + 1 M NaCl Note: all washes can be done by simply flicking the tube until beads are in suspension, followed by a 15 s centrifuge spin at 100 ×g and isolation of beads with a magnet.
- Resuspend beads in TE (pH 7.5) + 1 M NaCl and 0.1% NP-40 to a final concentration of 10 mg/mL and 0.003% NP-40.
- Add 8.6 ng biotinylated nucleic acid/μg beads to tube along with 5 M NaCl equal to ≈0.25× the volume of DNA added.
- Incubate for 30 min at room temperature while rotating (to make sure beads don't sink to bottom during incubation).
- Wash beads in 400 μL TE (pH 7.5) + 1 M NaCl.
- Resuspend beads in 1 μL Blocking Buffer per 1 μg of beads.
- Incubate 25 min at room temperature while rotating.
- Wash beads 3 times, with 400 μL Transcription Buffer.
- Resuspend beads in Transcription Buffer to a final concentration of 10 mg beads/mL. Use templates the same day – do not store. Keep on ice until ready to use.

22.3.2 Isolation and Purification of Proteins with Immobilized Templates

Details for incubation of extract with immobilized templates will be experiment-dependent and often will be determined empirically. For example, transcription factors are often found at very low copy number per cell, and therefore may require incubation of relatively large quantities of extract with low levels of DNA template in order to obtain protein yields suitable for MS/MS analysis. Similar to the immunoprecipitation experiments described above (Section 22.2.1), a physiological salt concentration (≈150 mM) typically works well, although factors that exhibit high affinity for consensus binding sites may tolerate higher salt concentrations during purification, which can often reduce background (see below). In addition, weak detergents such as NP-40 are often included during incubation, although the final concentration is empirically determined. Unlike immunoprecipitations, incubation of immobilized templates in cellular extract is often performed at room temperature for 45 min to 1 h.

A high background of proteins that bind promiscuously to nucleic acids is the most common hurdle encountered in proteomics assays based on purification by immobilized templates. Several techniques may be used to address this problem. First, isolation of nuclei prior to generation of lysate will minimize contamination with cytoplasmic proteins. Also, several washes, each with increasing stringency, can be used after the incubation step. Again, specific detergent concentrations may vary depending upon the proteins of interest. Transcription pre-initiation complexes, for example, form through the cooperative interactions between many proteins and DNA, and can easily be disrupted by detergent concentrations that exceed 0.05%. Blocking the beads with casein, BSA, or polyvinyl pyrrolidone prior to incubation of the template may reduce nonspecific interactions with the resin. Finally, addition of competitor nucleic acid (DNA/RNA) is an effective means of reducing background levels. Many of the most abundant contaminants in these assays are proteins that bind non-specifically to DNA or RNA. As a result, use of 2–10× of competitor DNA during incubation of the extract and templates, will reduce non-specific background. Nucleic acid competitors can also be included in the washes to help remove any non-specifically bound proteins from the immobilized template. Conditions for template elution must be matched to downstream sample processing steps. For example, SDS will efficiently elute specifically bound proteins and contaminants, providing high yield but also increased background. Moreover, additional steps will be required to remove SDS prior to mass spectrometry analysis. An effective alternative is to design the template with a single restriction site upstream of the specific nucleic acid element. After the wash steps, immobilized complexes bound to the adjacent DNA sequence are eluted from the beads by addition of an appropriate restriction enzyme. This method minimizes contamination that results from non-specific binding of contaminant proteins to the beads.

22.3.2.1 Buffers

- 50 mM Tris, pH 7.5, 250 mL
  - 1.5 g Tris in 250 mL H_2O
  - Adjust pH to 7.5 with HCl
  - Before use, add 4.6 mg DTT/mL
- YPD/S, 2.5 L at room temp. (YPD with 1 M sorbitol)
  - 25 g Yeast extract
  - 50 g Bactopeptone
  - 50 g Dextrose (glucose)
  - 455 g Sorbitol
  - H_2O to 2.5 L total
- YPD/S, 1.5 l (4 degrees)
  - 15 g Yeast extract
  - 30 g Bactopeptone
  - 30 g Dextrose
  - 273 g Sorbitol
  - Add water to 1.5 l
- 1 M Sorbitol, 1.5 L (4 degrees)
  - 273 g Sorbitol
  - H_2O to 1.5 l
- Buffer A
  - 18% Ficoll 400 (or polysucrose 400)
  - 10 mM Tris 7.5
  - 20 mM Potassium Acetate
  - 5 mM Magnesium Acetate
  - 1 mM EDTA
  - 0.5 mM Spermidine
  - 0.15 mM Spermine

Note: 3 mM DTT and protease inhibitors are added just prior to use
Note: The Ficoll takes many hours to dissolve and frequently is stirred overnight.

- Buffer B
  - 100 mM Tris acetate pH 7.9
  - 50 mM potassium acetate
  - 10 mM MgSO4
  - 20% glycerol
  - 2 mM EDTA

3 mM DTT and protease inhibitors added just prior to use

- Buffer C
  - 20 mM HEPES pH 7.6
  - 10 mM MgSO4
  - 1 mM EGTA
  - 20% glycerol

3 mM DTT and protease inhibitors are added just before use

- Dialysis Buffer
  - Buffer C + 75 mM Ammonium Sulfate
- 1× Transcription Wash Buffer
  - 1× transcription buffer (pH 7.6)
  - 0.05% NP-40
  - 2.5 mM DTT

Make up just before use

22.3.2.2 Nuclear Extract Preparation from Saccharomyces cerevisiae

This protocol was adapted from: (Ponticelli and Struhl, 1990; Ranish et al., 1999):

Cell Culture

- 4 l of *Saccharomyces cerevisiae* are grown in YPD (3% glucose) to A600 of 3–4.

Extract Preparation

- Harvest Cells in 1 liter bottles (2000 × g for 12 min). Pour off media.
- Resuspend cell pellets in 35 mL 50 mM Tris 7.5, 30 mM DTT. Usually this can be done by gently shaking the centrifuge bottles. Leave cells in 1 l bottles. Incubate at 30°C for 15 min, with gentle shaking. After incubation, combine all cells into 1 bottle, and pellet cells (2000 × g for 12 min)
- Carefully drain excess liquid and weigh cells. Expected yield is 20–35 g cells, this protocol works best in this range. Anything less than 18 g will give poor extracts. If cells are overgrown, Zymolyase will work poorly in spheroplasting cells. Scale the experiment according to the above guidelines.
- Resuspend in 20 mL YPD/S. Add 15 mL 2 M sorbitol. Make fresh stock of 1 mg/mL Zymolyase 100T. Add 15 mL of this stock of Zymolyase to resuspended cells. Incubate at 30° with occasional gentle mixing.

Note: Zymolyase 100T is reportedly contaminated with proteases (although less so as compared to 20T), so extra care is needed to wash spheroplasts. Zymolyase is not highly soluble.

- Check progress of spheroplasting every 15 min, as follows: Mix 4 μL of cells with 4 μL 1% SDS on a glass slide. Observe the number of cell ghosts under microscope. Incubate cells until about 80% spheroplasts are obtained. This can take anywhere from 15 min to 2.5 h. If cells are spheroplasting slowly after 1 h, an extra 5–10 mL of zymolyase can be added. After spheroplasting has reached about 80%, add 100 mL YPD/S (room temp) and pellet cells (2000 × g for 12 min).

- Resuspend cells in 250 mL YPD/S (room temp) and incubate at 30°C for 30 min to allow cells to recover. The resuspension of spheroplasts works best if a small volume (~50 mL) of YPD/S is first added and cells are resuspended using a baking spatula, although you never reach a homogenous mixture. Then add the remaining YPD/S.
- Pellet cells (2000 ×g for 12 min) and resuspend in 200 mL cold YPD/S (4°C). Resuspend as in the previous step. Keep everything cold from this point on. Cells can be kept on ice for an hour or so if other cells are still spheroplasting. Pellet cells (2000 ×g 12 min).
- Resupsend in another 200 mL cold YPD/S. These washes make sure that all the zymolyase and potential contaminant proteases are gone before you lyse the cell membrane.
- Pellet cells (2000 × g for 12 min) and resuspend in 250 mL cold 1 M sorbitol.
- Pellet cells (2000 × g for 12 min) and carefully drain sorbitol media. Resuspend in 100 mL Buffer A at 4°C containing DTT and protease inhibitors. Transfer to 250 mL beaker.
- Pass cell suspension 2–3× through homogenizer at 500 RPM at 4°C. Transfer homogenized cells to centrifuge tubes.

Note: If a homogenizer is not available, then dounce the spheroplasts 3×.

- Spin 3000 ×g for 8 min in SS34 rotor. Transfer supernatant to new centrifuge tubes. Do not worry about the slimy loose pellet that also transfers. Repeat.
- Spin supernatant 3000 × g for 5 min. Transfer supernatant to new centrifuge tubes. Repeat. It is likely that a significant amount of loose pellet will remain after the last spin; this should not adversely affect extract quality.
- Transfer supernatant to 50 mL centrifuge tubes and pellet crude nuclei. Spin 20,000 ×g for 30 min in SS34 rotor. Remove supernatant by dumping and discard. Drain pellets.
- Resuspend crude nuclear pellets (includes other cellular debris, but no cytoplasmic soluble proteins) with a small spatula in 10 mL Buffer B and transfer to 50 mL Falcon tube. The prep can be stopped at this point. Quick freeze and store resuspended nuclear pellets at –70°C overnight.
- Thaw nuclei on ice and measure volume. Add 3 M ammonium sulfate (pH 7.5) to 0.5 M final concentration (1/5 original volume of nuclei) and immediately mix and incubate on roller in cold room for 30 min. After 10 min, break up any lumps with a glass rod. This step lyses nuclei.
- Transfer to Ti70 thick walled ultracentrifuge tubes and spin at 130,000 ×g for 90 min at 4°C.
- Carefully remove supernatant with 5 mL pipette (and Pasteur pipette if necessary) being careful to avoid disturbing the pellet. Try to avoid aspirating white floaty

material as well. Transfer to 50 mL Falcon tube to get an approximate volume of supernatant. Then transfer to new ultracentrifuge tube.

- Add 0.35 g solid ammonium sulfate per 1 mL of supernatant and immediately incubate on cold room roller for 30 min. The ammonium sulfate can be added all at once if a number of preps are being done. However, it is best if ammonium sulfate is added slowly while mixing either with a glass rod or low vortex. Verify that the pH remain above 7. Adjust pH with 1 M NaOH if necessary.
- Spin in Ti70 at 15,500 ×g for 20 min at 4°C. Remove supernatant by dumping and re-spin pellets at 15,500 ×g for 5 min (may be unnecessary if the pellet is already well packed). Carefully remove all remaining supernatant with a Pasteur pipette.
- Resuspend pellets in Buffer C (with no Ammonium Sulfate) containing DTT and protease inhibitors. Depending on protein pellet size, resuspend in 0.5–3.5 mL buffer. This can be done with a wooden applicator or a blue pipette tip depending on the amount of protein.
- Dialyze nuclear extract vs Buffer C + 75 mM ammonium sulfate at 4°C. Dialyze *vs.* 500 mL buffer with 3 changes of buffer over 4.5 h total.
- Aliquot extract, flash freeze in liquid nitrogen and store at –70°C. Extracts should be 25–50 mg/mL in protein.

Anticipated results: The final concentration of the extract is usually in the range of 25–50 mg/mL of protein. The total yield normally is ~100–200 mg of protein. Extract quality/activity can be measured using standard in vitro transcription assays.

22.3.2.3 Immobilized Template Pull-Downs with Yeast Nuclear Extracts

Note: The following protocol, as first described in (Ranish et al., 1999), is suitable for isolation of transcription pre-initiation complexes. Minor adjustments will likely be required with different experimental systems

This assay is performed using 100 μL reactions.

- Set up the following reaction mix (1× Mix):
  - 20 μL 5× Transcription Buffer
  - 2.52 μL 0.1 M DTT
  - 0.5 μL 10% NP-40
  - 16.98 μL H_2O
  - 40 μL total volume

Note: Can also include protease inhibitors if working with labile proteins. Can include Hexokinase/Glucose to deplete endogenous nucleotides

- Make 40 μL aliquots of Mix.
- To each aliquot, add ddH20 to give a FINAL volume (after addition of extract/amanitin/competitor DNA/immobilized template) of 100 μL

- Add nuclear extract to each aliquot (usually 100–400 μg nuclear extract per reaction titrate each nuclear extract prep for optimal complex assembly).
- Add 1 μL alpha-amanitin (1 mg/mL stock)
- Incubate reactions for 10 min at room temp.
- Centrifuge 7500 ×g for 2 min at 4°C. Transfer supernatants to new tubes (add any recombinant proteins after this step).
- Add 1 μg HaeIII digested E. coli genomic DNA to each reaction.

Note: Titrate to decrease background but maintain complex assembly

- Add 96 ng of Gal4-VP16 per 5 μL immobilized template to immobilized template. Incubate 10 min at room temperature while rotating
- To each nuclear extract mix, add 5 μL of immobilized template (+ volume of activator).
- Incubate reactions 40 min at room temperature with occasional gentle vortexing to prevent beads from settling.
- Wash templates 3× 400 μL of 1× Transcription Wash Buffer.
- Resuspend beads in 40 μL of restriction enzyme buffer.
- Elute templates from the beads by digesting with 60 units of restriction enzyme for 30 min, 37°C.
- Transfer eluate to new tube, store at –20°C.

Anticipated results:

- Specificity: Successful immobilized template pull-downs are characterized by strong enrichment, as compared to the control, of a target protein captured on the experimental template. Enrichment efficiency can readily probed by immunoblot. Typically, analysis of approximately one-half the extract, prepared as described above, is adequate for immunoblot. The protocol may be scaled-up ≈5-fold for LC-MS/MS analysis.
- Yield: We typically find that the enriched proteins are almost always eluted with co-purifying contaminants. Although the immobilized template pull-downs yield several micrograms of protein, we estimate that the portion of proteins that are enriched on the experimental template constitute only a small fraction of the total, eluted protein mass.

22.3.3 Processing of Proteins Eluted from Immobilized DNA Templates for Mass Spectrometry Analysis

22.3.3.1 General Comments

Quantitative MS/MS analysis of complexes formed on immobilized templates often requires steps that are not typical necessary for immunoprecipitation assays. First, residual DNA must be removed as it can suppress the ionization efficiency of peptides during LC-MS analysis. In addition, DNA can interfere with amine-reactive stable isotope labels often used for relative quantification. DNA can be easily digested to NTPs with DNAaseI or Benzoase prior to tryptic digestion. We find that NTPs do not bind efficiently to reversed-phase resins, and hence are removed during peptide clean-up. DNA may also be removed through peptide clean-up on a cation exchange column.

Although the methods described above will reduce non-specific binding and background, protein samples isolated with immobilized templates will undoubtedly be highly complex, and require some method of fractionation prior to MS/MS analysis. For example, Himeda and colleagues analyzed 53 ion exchange chromatography fractions in conjunction with the ICAT quantitation strategy, and identified >900 proteins associated with a template encoding a muscle specific, gene enhancer region. (Himeda et al., 2004) While a description of applicable separation strategies is outside the scope of this report, it is clear that additional protein- and peptide-level fractionation will significantly improve mass spectrometry data in immobilized template assays.

22.3.3.2 Immobilized Template Eluate Preparation for ITRAQ Based LC-MS/MS Analysis

- Add 1 M ammonium bicarbonate to all samples to a final concentration of 50 mM
- Add Rapigest™ SF (Waters) to each sample to final concentration of 0.1%.
- Add 2 M DTT to each sample to final concentration of 10 mM.
- Incubate samples at 56°C for 1 h.
- Add 0.5 M iodoacetamide to each sample to a final concentration of 55 mM.
- Incubate samples at room temperature for 45 min in the dark.
- Add trypsin (protein:enzyme ratio of 20:1) to each sample and incubate overnight at 37°C.
- Cleave Rapigest reagent by addition of TFA to final concentration of 0.5% and then incubate samples at 37°C for 35 min
- Purify peptides from each sample with reversed phase clean up (either column or Sep-Pak microelution plate).
- Lyophilize and then resuspend in 30 μL 500 mM tri-ethyl ammonium bicarbonate.
- Label with ITRAQ reagents (as per manufacturers' protocol) for 1 h at room temperature.
- Neutralize excess reagent by bringing each sample to a final concentration of 20 mM Tris.
- Combine samples, and reduce to dryness in a centrifugal concentrator.
- Resuspend in 0.1% formic acid, 25% acetonitrile.
- Remove nucleic acid from sample with strong cation exchange chromatography.
- Use varying concentrations of KCl (20–200 mM) to fractionate peptides from the SCX column or bump elute with.1% formic acid, 25% acetonitrile, 500 mM KCl.
- Reduce concentration of acetonitrile 10× by centrifugal concentration. Resuspend sample in 0.1% formic acid
- Fractionate sample further and analyze with LC-MS/MS

22.3.3.3 Anticipated Results

The use of one (or more) stages of protein or peptide fractionation in addition to online LC (-MS) will yield in excess of 3000 peptides that map to 500–900 proteins. This number will be highly dependent on the system used and the experimental setup. Analysis of a small aliquot of the final template elution by SDS-PAGE and silver staining will provide a qualitative readout of sample complexity, and hence the degree of fractionation required.

22.4 Chemical- and Immunoaffinity-Based Strategies for Analysis of Phosphorylation

22.4.1 General Comments

Reversible modification of proteins by phosphorylation has evolved as an important mechanism for regulation of protein activity, stability, trafficking, as well as protein–protein interaction. In addition, phosphorylation mediates a majority of dynamic cellular processes. For example, receptor protein tyrosine kinases (RPTKs) and protein tyrosine kinases (PTKs) are perhaps the most extensively studied oncogenes. RPTKs internalize extra-cellular signals through a series of tyrosine phosphorylation events, both on RPTKs and downstream substrates, upon ligand-induced dimerization. Mutation or over-expression of RPTKs or PTKs often causes dysregulation of otherwise tightly controlled signaling networks, and can disrupt the critical balance between cell proliferation/differentiation and apoptosis. Tyrosine phosphorylation in particular controls many of the key functions of RPTKs, cytosolic PTKs and downstream signaling molecules. A global and dynamic understanding of the tyrosine phosphorylation events will reveal the key differences among normal and deregulated signaling networks, and lead to more intelligent and effective therapeutic approaches. These data and observations have catalyzed a plethora of efforts directed at the generation of comprehensive signaling maps in mammalian systems. It is hypothesized that such studies will improve our understanding of cellular physiology and also facilitate the development of small molecule therapeutics.

Recently mass spectrometry has emerged as the method of choice for rapid and sensitive identification of protein phosphorylation. However, phosphopeptides present unique challenges as compared to their non-modified counterparts. Most importantly, because the majority of protein phosphorylation is tightly regulated, the stoichiometry of modification on any one protein can be very low. As a result, many techniques have been introduced to facilitate efficient enrichment of phosphoproteins and – peptides from complex biological mixtures. Anecdotal evidence suggests that phosphopeptides are particularly susceptible to surface absorption; hence sample cleanup and handling should be kept to a minimum whenever possible.

22.4.2 Strong Cation Exchange (SCX) Chromatography

Strong cation exchange (SCX) chromatography has been used for enrichment of phosphorylated peptides (Beausoleil et al., 2004). This technique relies on the fact that phosphate groups bear a negative charge at pH ~2.7 which reduces phosphopeptide retention on the SCX resin relative to other unmodified peptides. Thus a significant fraction of phosphorylated peptides elute early in the salt gradient. However, some unmodified peptides co-elute, especially in later fractions, often necessitating further enrichment using other techniques (described below).

22.4.2.1 Equipment

- HPLC
- Column: 200 × 2.1 mm polysulfoethyl A (5 μm, 200 Å pores) from PolyLC, Columbia MD. A guard column of the same resin is recommended.
- Injector and Loop (1 mL)
- Optional: UV detector (monitor 214 nm)
- Optional: Fraction collector

22.4.2.2 Buffers

- solvent A = 5 mM KH_2PO_4 pH 2.7/25% acetonitrile
- solvent B = solvent A containing 350 mM KCl

Note: SCX solvents should be primed out of the system with water when the pump is not in use to help minimize salt corrosion of the pump.

22.4.2.3 Protocol

- Configure the LC system
  - Flow rate = 200 μL/min
  - Gradient: 0–15% B in 5 min, 15–60% B in 20 min, 60–100% B in 5 min
  - Before analysis of standards or enrichment of phosphopeptides, condition the column with a blank gradient
  - Before phosphopeptide enrichment, first verify SCX column performance by injection of ~1–10 nmol of suitable peptide standards (monitor absorbance at 214 nm). Peaks should be no more than ~1 min wide at base.

After analysis of standards, condition the column with a blank gradient before enrichment of phosphorylated peptides

- Desalt sample
  - Equilibrate a C18 well (Waters tC18 plate, 100 mg/well) with 1 mL of 80% MeCN/0.1% acetic acid, then with 1 mL of 0.1% acetic acid
  - Load the peptide mixture
  - Wash the resin with 2× 1 mL 0.1% acetic acid

- Elute with 500 μL 80% acetonitrile
- Dry in speedvac to ~50 μL
- Make up volume to ~250 μL with solvent A

- Fractionate sample
  - Place sample into loop (1 mL)
  - Wash sample onto column for 15 min (3× loop volume)
  - Take loop out of line
  - Start gradient and fraction collector

  Collect 1 min fractions

Note: In our experience, 2–3 of the 1 min fractions are highly enriched for phosphopeptides. Others may require additional enrichment with an orthogonal technique (see below).

22.4.3 Immobilized Metal Affinity Chromatography

Immobilized metal affinity chromatography (IMAC) is a separation strategy originally developed by Porath (Porath et al., 1975) that exploits the unique selectivity of metal ions towards various protein and peptide functional groups. This technique utilizes a chelating moiety (usually IDA or NTA, with three or four coordination sites, respectively) that is charged with a specific metal ion to impart the desired selectivity. After the discovery that iron (Fe^{3+}) had significant affinity for phosphate ions, Fe^{3+}-IDA-IMAC was used for enrichment of phosphoproteins (Andersson and Porath, 1986), and a few years later for isolation of phosphopeptides prior to mass spectrometric analysis (Nuwaysir and Stults, 1993). Early work coupling IMAC enrichment with MS-based phosphopeptide analysis utilized IDA charged with Fe^{3+} or Ga^{3+}, and demonstrated that phosphopeptide selectivity was higher with the latter (Posewitz and Tempst, 1999). However, in complex mixtures, the selectivity was typically <50%, spurring the development of methods designed to further improve selectivity. Ficarro et al. demonstrated that conversion of peptide carboxylic acid side chains to their corresponding methyl esters dramatically increased phosphopeptide yield and selectivity for Fe^{3+}-IDA resin (Ficarro et al., 2002). This approach was further refined (Moser and White, 2006; Ndassa et al., 2006), and then recently the use of Fe^{3+} or Ga^{3+} with the NTA chelator at acidic pH (0.1% TFA) demonstrated high selectivity phosphopeptide enrichment without methyl ester modification or the addition of chemical competitors (Ficarro et al., 2009; Tsai et al., 2008).

22.4.3.1 Methyl Ester Modification of Peptides

Reagents and Materials

- 0.5 dram glass vial
- Anhydrous acetyl chloride, or thionyl chloride
- Anhydrous methanol

Safety

Reaction should be performed in a fume hood while wearing appropriate safety gear including gloves, goggles, and lab coat. Acetyl and thionyl chlorides are toxic and corrosive and should be handled with care according to manufacturer's MSDS.

References

(Hunt et al., 1986; Moser and White, 2006; Ndassa et al., 2006).

Protocol

- Place a clean, dry micro stir bar in a 0.5 dram glass vial (for up to 2 mL reagent), and add 1 mL anhydrous methanol.
- With stirring, add 160 μL of anhydrous acetyl chloride or, alternatively, 40 μL of thionyl chloride.

Note: If the liquid bumps, add reagent more slowly.

- Cap the vial and let the reaction stir at RT for 5 min. After 5 min, reagent is ready for use.
- Add the appropriate amount of reagent to dry peptide (∼1 mL reagent/1 mg tryptic peptides)

Note: It may be necessary to sonicate (∼1–10 min) to suspend the peptide residue in the reagent.

- Let the mixture incubate for 2 h at RT.
- After 2 h, briefly spin down liquid in a microcentrifuge tube, and dry peptides in a centrifugal concentrator (no heat). Dry residue may be stored at –80°C.

Note: Cleanup peptides by reversed-phase chromatography before this procedure for best results.
Note: Minimize exposure to air/moisture during procedure.

22.4.3.2 Fe-IDA Enrichment of Methylated Phosphopeptides

Reagents and Materials

- 50% slurry of Ni-IDA-agarose beads (Novagen)
- Magnetic rack

Buffers

- 100 mM EDTA, pH 8.0
- 10 mM $FeCl_3$
- 1:1:1 methanol:acetonitrile:0.01% acetic acid
- 100 mM ammonium phosphate, pH 8.0

Protocol

- Pipette 20 μL of a 50% slurry of Ni-IDA-agarose beads (Novagen) into a fresh Sarstedt tube.

Note: 20 μL of the 50% slurry can be used to enrich phosphopeptides from approximately 800 μg of methylated tryptic peptides

- Using a magnetic rack, wash the beads 3 times with 400 μL water.
- After removing the supernatant, incubate the resin for 30 min with 400 μL 100 mM EDTA (end-over-end rotation)
- Discard the supernatant, and wash the beads 3 times with 400 μL water.
- Incubate beads 30 min with 600 μL 10 mM $FeCl_3$ (end-over-end rotation)
- Discard the supernatant, and wash the resin 3 times with 400 μL water, and finally with 400 μL 1:1:1 methanol:acetonitrile:0.01% acetic acid
- Reconstitute methyl ester modified peptides in 1:1:1 methanol:acetonitrile:0.01% acetic acid (~1 μg/μL) and add sample to the beads (after removing the supernatant)
- Incubate beads with sample 30 min to bind the phosphopeptides (end-over-end rotation)

Note: The 1:1:1 methanol:acetonitrile:0.01% acetic acid (~1 μg/μL) is critical to the enrichment. Lower pH buffers can disrupt phosphopeptide binding.
Note: Peptides should be at a concentration of no more than 1 μg/μL. Higher concentrations of peptide can lower the selectivity.
Note: Peptides should be methyl ester modified according to the procedure above.

- Remove the supernatant, and save for later analysis if necessary.

Note: Supernatant can be used to normalize for differences in labeling between different iTRAQ channels in a quantitative experiment.

- Wash the beads 3 times with 400 μL 1:1:1 methanol:acetonitrile:0.01% acetic acid (vortex mix each time).
- Remove the supernatant, and add 50 μL 100 mM ammonium phosphate, pH 8.0.
- Elute for 10 min
- Using the magnetic rack, carefully transfer the eluate to a new fresh 1.5 mL centrifuge tube.
- Add 50 μL water to the beads, vortex (residual phosphate buffer keeps the solution basic). Pool the supernatant with the previous elution.
- Dry samples by centrifugal concentration to an appropriate volume, and acidify. Transfer to 96-well plate (Sarstedt) and analyze by LC-MS.

22.4.3.3 Fe-NTA Enrichment of Phosphopeptides

Reagents and Materials

- 5% slurry of Ni-NTA-agarose beads (Qiagen)
- Magnetic rack

Buffers

- 100 mM EDTA, pH 8.0
- 10 mM $FeCl_3$
- 1:1:1 methanol:acetonitrile:0.01% acetic acid
- 80% acetonitrile/0.1% TFA
- 1:20 ammonia/water

Protocol

- Pipette 200 μL of a 5% slurry of Ni-NTA-agarose beads (Qiagen) into a fresh 1.5 mL centrifuge tube.

Note: 200 μL of the 5% slurry can be used to enrich phosphopeptides from approximately 800 μg of tryptic peptides

- Using a magnetic rack, remove the bead supernatant, and wash the beads 3 times with 400 μL water.
- After removing the supernatant, incubate the resin for 30 min with 400 μL 100 mM EDTA (end-over-end rotation)
- Discard the supernatant, and wash the beads 3 times with 400 μL water.
- Discard the supernatant, and incubate beads 30 min with 600 μL 10 mM $FeCl_3$ (end-over-end rotation)
- Discard the supernatant, and wash the resin 3 times with 400 μL water, and finally with 400 μL 1:1:1 methanol:acetonitrile:0.01% acetic acid
- Reconstitute peptides in 80% acetonitrile/0.1% TFA and add sample to the beads (after removing the supernatant)

Note: The 0.1% TFA is critical to the enrichment. Using 0.1% acetic acid or 0.1% formic acid significantly lowers the specificity.
Note: Peptides should be at a concentration of $\leq$1 μg/μL. Higher concentrations of peptide can lower the selectivity.
Note: Peptides should be desalted before adding to the beads as salts/buffers can significantly lower the selectivity. For example, neglecting to desalt peptides after iTRAQ labeling can result in very low selectivity.
Note: Desalted iTRAQ reactions can have a slight reddish tinge (from reagent by-products). We have observed that this does not affect the selectivity.

- Incubate beads with sample 30 min to bind the phosphopeptides (end-over-end rotation)
- Remove the supernatant, and save for later analysis if necessary.

Note: Supernatant can be used to normalize for differences in labeling between different iTRAQ channels in a quantitative experiment.

- Wash the beads 3 times with 400 μL 80% MeCN, 0.1% TFA (vortex mix each time).
- Remove the supernatant, and add 50 μL 1:20 ammonia:water, and vortex.
- Elute for 10 min
- Using the magnetic rack, carefully transfer the eluate to a new fresh 1.5 mL centrifuge tube.
- Add 50 μL water to the beads, vortex (residual ammonia keeps the solution basic). Pool the supernatant with the ammonia elution.
- Dry samples by centrifugal concentration to an appropriate volume, and acidify. Transfer to 96-well plate (Sarstedt) and analyze by LC-MS.

22.4.4 Metal Oxide Affinity Chromatography

Although it was reported as early as 1990 that titanium dioxide could selectively bind phosphate (Matsuda et al., 1990), it was not applied toward phosphopeptide enrichment for proteomics applications until 2003(Pinkse et al., 2004). The metal oxide performed similarly to Fe^{3+}-IDA-IMAC in that peptides containing multiple acidic groups were also enriched, and furthermore that methyl ester modification could negate this undesired interaction. A few years later, Larsen (Larsen et al., 2005) reported that the specificity toward phosphate groups could be significantly increased by addition of DHB to the load and rinse buffers; the authors speculated that the carboxylates effectively competed with acidic side chains of non-modified peptides to increase the specificity of phosphopeptide enrichment. Since this initial report, other compounds including lactic acid have been demonstrated to act in a similar fashion, avoiding potential interference by DHB during LC-MS/MS analysis (Sugiyama et al., 2007). Parallel work in several laboratories described additional metal oxides, including zirconium (Kweon and Hakansson, 2006), hafnium (Rivera et al., 2009), and niobium (Ficarro et al., 2008) for use as affinity resins in phosphopeptide studies. These data revealed biases in various metal oxides; for example, phosphopeptides identified after enrichment via niobium pentoxide exhibited a bias towards basic residues, while zirconium oxide displayed a bias for singly phosphorylated peptides (both as compared to enrichment by titanium dioxide). These results illustrate the potential for greater phosphoproteome coverage through the use of multiple metal oxides in parallel.

22.4.4.1 Titanium Dioxide Enrichment of Phosphopeptides from Complex Mixtures

Reagents

- Titanium dioxide (GL sciences)

Buffers

- Loading buffer: 6 mL MeCN, 1.5 mL water, 2.5 mL lactic acid, 500 μL TFA
- Washing buffer: 8 mL MeCN, 1.8 mL water, 200 μL TFA

22.5.4.1.3 Protocol

- Weigh 10–50 mg titanium dioxide in a 1.5 mL centrifuge tube.

Note: 10 mg has a capacity for at least 100 μg of total protein digest.

- Add a volume of 80% MeCN/0.1% TFA equal to 10× the weight in mg.
- Vortex the tube to resuspend the oxide particles, and transfer an amount of resin of appropriate capacity to a fresh tube
- Centrifuge at 13,900×*g* for 2 min to pellet the resin, and remove the supernatant
- Reconstitute the sample in 400 μL (or other appropriate amount) of loading buffer, add sample to the resin, and incubate for 30 min at room temperature with end-over-end rotation

Note: For best results, samples should be desalted (and concentrated if applicable; but ideally not to dryness), and diluted 5–10× in loading buffer

- Centrifuge at 13,900 ×g for 2 min to pellet the resin, and pipette the supernatant to a fresh tube for further analysis, if applicable
- Wash the resin twice with 400 μL (or other appropriate amount) of loading buffer, once with wash buffer, and once with 80% MeCN/0.1% TFA: add the solution, vortex, centrifuge as above, and discard the supernatant.
- Add 50 μL 1:20 ammonia: water (see Section 22.4.5.1 on carriers if appropriate), vortex, and elute for 5 min with shaking. Centrifuge at 13,900 ×g for 2 min to pellet the resin, and pipette the eluate to a fresh tube.
- Add 50 μL water to the resin, vortex, centrifuge at 13,900 ×g for 2 min to pellet the resin, and pool the supernatant with the ammonia elution.
- Dry the pooled elution fractions to an appropriate volume, acidify and analyze by LC-MS/MS.

22.4.5 Experimental Format for Phosphopeptide Enrichment

The format used for phosphopeptide enrichment is an important experimental parameter; given the penchant for phosphopeptide absorption to surfaces, there is a trade-off between yield and sample throughput. For analysis of even modest numbers of samples, the use of microcapillary columns and manual pressurized loading vessels quickly becomes impractical. Nonetheless analysis of samples from materials of limited supply (i.e. clinical materials, primary cells, etc.) will likely benefit from the additional effort associated with manual processing. However, when sample numbers preclude manual manipulation, the use of carriers and/or automated sample handling systems can be adjusted to minimize losses. The sections below discuss a variety of formats for phosphopeptide enrichment.

22.4.5.1 Batch Mode

This mode of enrichment is performed directly in microcentrifuge tubes, and is best suited to small numbers of samples (<10). Although the use of magnetic beads in this mode greatly facilitates washing steps by immobilizing the beads, non-magnetic resins may simply be centrifuged to remove the supernatant. To minimize sample losses, carriers can be added to the elution buffer. These carriers prevent not only general loss of peptides to surfaces, but also phosphopeptide-specific losses which are thought to occur on metal oxide surfaces encountered throughout the LC system and autosampler, or on trace metal ions immobilized on reversed-phase resins typically used to separate phosphopeptides prior to or online with mass spectrometry analysis. These interactions may be minimized by addition of EDTA or citrate to the elution buffer (Winter et al., 2008).

Buffers

Preparation of elution buffer with carriers

- 1 mL 1:20 ammonia:water, 30 μL 100 mM EDTA, 3.2 μL 8 M GuHCl, 5 pmol Glu Fibrinopeptide B, 5 pmol angiotensin I.

22.4.5.2 Microcapillary Column

Peptide losses can be particularly acute when processing small quantitites of sample (<1 pmol), even when carriers are used. This is especially relevant to phosphopeptide analysis of single protein digests because post-enrichment mixtures may contain only a few phosphopeptides at low abundance. To circumvent this hurdle, enrichment may be performed directly in fused silica capillaries. In this scenario, non-phosphorylated peptides in the sample actually serve as carriers until they depart in the column flow-through. Once bound to the column resin, phosphopeptides can be directly eluted in-line to a reversed-phase pre-column for LC-MS/MS analysis. Columns can be packed with IMAC-based materials or metal oxide resins, and fritted using a variety of techniques (sintering, commercial microfilters, or silicate solutions). Although this approach offers perhaps the best recovery of phosphorylated peptides, the added manual manipulations greatly restrict sample throughput.

Note: When eluting to a reversed-phase column, make sure the resin is compatible with the elution buffer pH. Many silica-based stationary phases are not stable at pH > 8.

Note: Phosphate-based elution buffers need to be rinsed thoroughly from the reversed-phase capillary column prior to mass spectrometry analysis to avoid formation of gas phase phosphate clusters.

Fritting Procedure 1: Soldering Iron, Lithium Silicate

Reagents and Materials

- Lithium Silicate solution (PQ Corporation, Valley Forge, PA)
- Tetramethyl ammonium silicate (Sigma-Aldrich, St. Louis, MO)
- Formamide
- 0.1% formic acid
- Soldering iron
- Fused silica capillary tubing

Protocol

- Pipette 100 μL Lithium Silicate solution (PQ Corporation, Valley Forge, PA) into a fresh tube, add 20 μL tetramethyl ammonium silicate (Sigma-Aldrich, St. Louis, MO), and vortex
- Add 10 μL formamide, and vortex
- Take a piece of 360 μm O.D. × 200 μm I.D. fused silica (20–30″ long), and dip it into the silicate solution to draw the solution about 1–2″ into the capillary
- Heat a soldering iron with a flat tip to 375°C. Place the fused silica against the iron flat away from the liquid solution.
- Move slowly towards the silicate solution, keeping good contact with the tip of the soldering iron. Pass the soldering iron over the liquid so that polymerization of a 1–2 mm frit occurs after a few seconds.
- When polymerized, wash remaining solution out with 0.1% formic acid.

Note: Careful positioning of the soldering iron is required to achieve the desired frit length.
Note: Make sure that heating is even and that the frit is not segmented

Fritting Procedure 2: Oven, Potassium Silicate

Reagents and Materials

- Oven
- Fused silica capillary
- Potassium Silicate solution (PQ Corporation)
- Formamide
- Microcentrifuge
- Microcapillary cutter
- 0.1% formic acid

Protocol

- Pipette 450 μL Potassium Silicate solution (PQ Corporation) into a fresh tube, add 88 μL of formamide (Sigma), vortex and centrifuge at 13,900 ×g for 30 s.

Pipette 250 μL of the solution from the top to a new fresh tube, discard the lower layer.

- Take a piece of 360 μm O.D. × 200 μm I.D. fused silica (20–30″ long), and dip it into the silicate solution to draw the solution about 1–2″ into the capillary
- Place capillaries in an oven for 1 h at 100°C.
- When polymerized, wash frits with 0.1% formic.
- Cut to desired length

Fritting Procedure 3: Torch Sintering

Reagents and Materials

- Licrosphere silica particles (5 μm, Merck)
- Fused silica capillary tubing
- Microtorch (Microflame, Inc.)

Protocol

- Take a piece of 360 μm O.D. × 200 μm I.D. fused silica (20–30″ long), and use a microtorch butane flame to remove ~1–2 mm of the polyimide from the end of the capillary. Use a kimwipe to remove charred material.
- Obtain a 0.5 dram vial filled with a ~5 mm layer of Lichrosphere silica particles (5 μm).
- Dip the exposed end of the capillary into the vial of silica particles ~10× without touching the bottom, and then ~10× hitting the bottom.
- Use the microtorch nitrous oxide flame to sinter the frit

Note: Frit must be heated sufficiently to sinter the silica; overheating will result in a fused capillary.
Note: Procedure 1 allows specific placement of the frit within the fused silica capillary. Procedure 3 tends to give the highest flow rates, however, the exposed glass is fragile and easily broken.

Fritting procedure 4: Commercial Filter Frit

Reagents and Materials

- PEEK in-line microfilter assembly (P/N M-520; Upchurch Scientific)

Protocol

- Place a piece of 360 μm O.D. × 200 μm I.D. fused silica into an UpChurch microfilter (P/N M520). The PEEK filter acts as a frit and retains particles

Procedure: Preparation of a Microcapillary NTA-Silica Column

Reagents and Materials

- Ni-NTA spin column from Qiagen (P/N 31014)
- Acetonitrile

Protocol

- Obtain a Ni-NTA spin column from Qiagen (P/N 31014). Remove the packing material, and place in a glass 0.5 dram vial.
- Add 1 mL acetonitrile, vortex, remove supernatant quickly in an effort to remove fines, but keep larger particles in the vial. Repeat ~10–15×. Add ~500 μL acetonitrile.

Note: The washing step is crucial, otherwise fines quickly clog the frit

- Add a stirbar into the slurry, and pack into a fritted capillary using a pressurized loading device (any of the fritting methods in Section 22.4.5.2 can be used).

22.4.5.3 Automated Platforms

Enrichment of phosphopeptides is a multi-step process that can be easily automated using existing robotic or LC platforms. For example, using a series of valves and LC pumps, phosphopeptide enrichment can be performed in-line, and eluted to a precolumn for LC-MS/MS analysis (Ficarro et al., 2005). In addition, the advent of metal oxide and IMAC-based magnetic beads enables automated enrichment of phosphopeptides in a 96-well format (Ficarro et al., 2009). Carriers can be used to minimize losses to surfaces that would otherwise occur as a result of off-line elution (see section above on batch mode enrichment for additional details about carriers).

Automated Fe^{3+}-NTA Enrichment of Phosphopeptides Using the KingFisher Magnetic Bead Processor

Reagents and Materials

- Kingfisher Magnetic Bead Processor (ThermoFisher Scientific)
- Kingfisher shallow 96-well plates and tip comb (ThermoFisher Scientific)
- 5% slurry of Ni-NTA-agarose beads (Qiagen)

Buffers

- 100 mM EDTA, pH 8.0
- 10 mM $FeCl_3$
- 1:1:1 methanol:acetonitrile:0.01% acetic acid
- 80% acetonitrile/0.1% TFA
- 1:1 acetonitrile/1:20 ammonia/water

Protocol

- Prepare a series of 8 96-well plates (KingFisher shallow 96-well plates, ThermoFisher Scientific, San Jose, CA): (1) beads plate; (2) beads wash plate; (3) sample plate; (4) wash plate 1; (5) wash plate 2; (6) wash plate 3; (7) elution plate; (8) tip plate

Note: Beads or buffers are added to each well for which sample exists in the sample plate (up to 96).

- Plate 1, Beads plate: Add Fe^{3+}-activated NTA beads (see Section 22.4.3.3 for Fe^{3+} activation of beads; 50 μL of the 5% suspension) in 200 μL of 1:1:1 acetonitrile:methanol:0.01% acetic acid.
- Plate 2, Beads Wash plate: 200 μL of 80% acetonitrile with 0.1% TFA.
- Plate 3, Sample Plate: Desalted peptides (up to 100 μg) in 200 μL 80% MeCN/0.1% TFA.
- Plates 4, 5, and 6, Wash plates 1–3: 200 μL 80% MeCN/0.1% TFA.
- Plate 7, Elution plate: 50 μL of 1:1 acetonitrile/1:20 ammonia:water, pre-washed with 200 μL of acetonitrile for 20 min.
- Plate 8, Tip Plate: Contains a tip comb

Note: Amount of beads, extract, and carrier can be adjusted as appropriate for the given application. For example, enrichment of phosphopeptides from silver stained bands can be accomplished using 20 μL of the 5% slurry of Qiagen beads, with a sample volume of 50 μL rather than 200 (Ficarro et al., 2009).

- Create a program for the KingFisher magnetic bead processor (ThermoFisher Scientific, San Jose, CA) that performs the following operations: Bead pick up from the beads plate (bind time 1 min, speed medium, 5 collections), bead wash in beads wash plate (wash time = 1 min, speed medium, 5 collections), phosphopeptide capture from sample plate (release time = 1 min, speed = slow; wash time = 30 min, speed = slow; collections = 5), 3 beads washes in wash plates 1–3 (release time = 1 min, speed = slow; wash time = 1 min, speed = slow; collections = 5), and elution of phosphopeptides into the elution plate (wash time = 1 min, speed = bottom very slow, 5 collections).
- Transfer elutions to an autosampler plate (96 well PCR plate, Sarstedt, Newton, NC), dry to 5–10 μL by vacuum centrifugation, acidify with 10% TFA, reconstitute to 30 μL with 0.1% TFA, and analyze by LC/MS.

22.5 Immunoaffinity Methods for Enrichment of Phosphotyrosine

22.5.1 Protein-Level Immunoaffinity Purification

Over the years, protein tyrosine phosphorylation was mainly studied via specific antibodies and western blot. However, it is difficult and time-consuming to develop high quality antibodies for large number of proteins. Using antibodies directed against phosphotyrosine, Blagoev et al. (Blagoev et al., 2004) showed that mass spectrometry could be used to identify and quantify a large number of tyrosine phosphorylated proteins isolated simultaneously from cells stimulated with epidermal growth factor for various time periods. However, protein-level purification can reduce the number of phosphotyrosine identifications due to an excess of non-phosphorylated peptides present in the mixture. In addition, the three dimensional structure of full length proteins may render the phosphotyrosine site inaccessible to the antibody, again limiting the number of phosphorylation sites identified.

22.5.2 Peptide-Level Immunoaffinity Purification

It is known that multiple tyrosine sites on RPTKs are phosphorylated upon stimulation and each can mediate interaction with a different protein partner, and hence activate distinct signaling pathways. Rush et al. (Rush et al., 2005) demonstrated that phosphotyrosine antibodies could be used to purify phosphotyrosine containing peptides directly from tryptic digests of cell lysates. Immunopurification at the peptide level provided a greater number of phosphotyrosine sites, due, in-part, to the reduced background of non-phosphorylated peptides. Shortly thereafter, Zhang et al. (Zhang et al., 2005) improved the specificity and yield of peptide-level immunoaffinity purification by incorporation of an additional IMAC-based enrichment step which was also compatible with the use of iTRAQ stable isotope tags (Ross et al., 2004) for relative quantification of signaling cascades. This approach facilitated the quantification of more than 100 tyrosine phosphorylation sites mediated through the epidermal growth factor receptor. Data acquired across multiple time points facilitated functional categorization via self-organized maps (Zhang et al., 2005). The protocol below is based on the original work of Zhang et al., and further optimized in our laboratory.

22.5.3 Protocol

22.5.3.1 Cell Lysis and Protein Digestion

- Lyse 10–20 million cells with 1 mL of 8 M urea supplemented with 1 mM activated Na_3VO_4.

Note: Cells should be rinsed with ice-cold PBS to remove extra media prior to lysis.

- Perform BCA (cat #: 23225, ThermoFisher Scientific), according to manufacturer's protocol, on a small aliquot (~20 μL) of each sample to quantify total protein.
- Reduce proteins with 10 mM DTT at 50°C for 1 h, then cool to room temperature and alkylate with 55 mM iodoaceteamide at room temperature for 1 h in the dark.

Note: Temperature during reduction should not exceed 60°C.

- Dilute lysates with 4 mL of 100 mM NH_4HCO_3 to a final urea concentration of ~1.6 M.
- Add trypsin at a 1:50 enzyme:protein ratio.
- Incubate samples at 37°C overnight on a rotator.
- Acidify digests with 10% TFA to a final pH 3–4, and centrifuge at 12,000 ×g to pellet debris.
- Cleanup peptides by reversed-phase (96-well microelution plate), and bump elute with 25% MeCN/0.1% HOAc.
- Lyophilize peptides in 100 μg aliquots, and store at –20°C until further use.

22.5.3.2 iTRAQ Labeling

- Resuspend each peptide aliquot (100 μg) with 30 μL of 0.5 M triethylammonium bicarbonate.

Note: One unit of iTRAQ reagent is sufficient to label ≤100 μg peptide; an excess of sample will result in incomplete labeling.

- Resuspend each iTRAQ reagent in ethanol following the manufacturer's protocol.
- Add iTRAQ reagents into samples according to an appropriate experimental design. For example: iTRAQ-114 → Sample 1, iTRAQ-115 → Sample 2, iTRAQ-116 → Sample 3, iTRAQ-117 → Sample 4.
- Briefly vortex each tube and then incubate at room temperature for 1 h.

Note: pH should be ≥ 7.5 to ensure complete labeling.

- Combine samples, dry in a centrifugal concentrator, and store at –80°C until needed.

22.5.3.3 Immunoprecipitation of Phosphotyrosine Peptides

Note: All steps were done at 4°C unless otherwise specified.

IP Buffer: 100 mM HEPES, 100 mM NaCl, pH 7.4 (Keep at 4°C).

- Phosphotyrosine antibodies (20 μL of pTyr-100 from Cell Signaling Technology and 20 μL of PT-66 from Sigma) were mixed in 200 μL of IP buffer.

Note: The use of multiple antibodies for peptide IP reduces bias and increases coverage.

- Resuspend protein G-agarose beads (Calbiochem) on a rotator. Transfer 20 μL of a 50% slurry into the antibody solution.
- Incubate for 8 h.

- Centrifuge agarose beads at 1000 ×g for 2 min.
- Rinse agarose beads with 400 μL of IP buffer for 5 min.
- Reconstitute iTRAQ labeled peptides in 400 μL of IP buffer.
- Adjust pH to ~7.4 with 1 N NaOH.

Note: Inaccurate pH will result in low capture efficiency and/or high nonspecific binding.

- Transfer peptide samples (4×100 μg) to 5 μL of rinsed protein G-agarose beads (with bound antibody) and incubate overnight on a rotator.
- Centrifuge agarose beads at 1000 ×g for 2 min. Transfer supernatant to a fresh tube.
- Rinse (3×) agarose beads with 400 μL of IP buffer for 5 min.
- Elute peptides with 60 μL of 100 mM glycine, 7% MeCN, pH 2.5 at room temperature for 15 min.
- Centrifuge agarose beads at 1000 ×g for 2 min.
- Transfer supernatant into a 600 μL microcentrifuge tube.

22.5.3.4 Fe(III)-NTA Enrichment of Phosphotyrosine Containing Peptides

Note: Fe(III)-NTA column needs to be freshly conditioned prior to use.

- Load eluted peptides onto a preconditioned Fe(III)-NTA column (200 μm ID×6 cm) at ≈2 μL/min. See Section 22.5.3.5 below.
- Rinse column with 80% MeCN, 0.1% TFA at 10 μL/min for 10 min.
- Rinse column with 0.01% acetic acid at 10 μL/min for 10 min.
- Elute peptides from Fe(III)-NTA column directly onto a reversed-phase precolumn (100 μm ID×4 cm long, POROS 10R2) with 50 μL of 100 mM $(NH_4)_2HPO_4$, 1 mM EDTA at 2 μL/min.

Note: Online elution to reversed-phase precolumn minimizes sample losses.

- Rinse reversed phase precolumn with 100 μL 0.1% acetic acid, and then place inline with HPLC effluent for LC-MS/MS analysis.

22.5.3.5 Fe(III)-NTA Column

Note: Column packing and conditioning are carried out on a pressurized sample loading vessel.

- Remove resin from a Ni(II)-NTA spin column (cat #: 31014, Qiagen)
- Rinse resins with MeCN, and then pack into a fritted 200 μm ID×6 cm fused silica tube.
- Rinse column with 100 mM EDTA at 10 μL/min for 10 min to remove Ni(II).
- Rinse column with 0.1% TFA at 10 μL/min for 10 min to remove EDTA.

- Rinse column with 100 mM $FeCl_3$ at 10 μL/min for 10 min
- Rinse column with 80% MeCN/0.1% TFA to remove extra $FeCl_3$. The column is now ready for phosphopeptide enrichment.

22.5.4 Anticipated Results

The authors could routinely identify and quantify 100–500 unique phosphotyrosine containing peptides from a single 2 h LC-MS/MS analysis. The exact number of peptides is highly dependent on the biological system and the instrument types. Generally, large numbers of peptides (>300) were observed from cancer cell lines with constitutively activated kinases, such as K562, and small numbers of peptides (~100) were observed from unstimulated stem cells. To estimate the level of tyrosine phosphorylation of a given sample, whole cell lysate of the sample, along with that of K562, can be probed with anti-phosphotyrosine antibodies, and compared by western blot.

References

Adelmant, G., and Marto, J.A. (2009). Protein complexes: The forest and the trees. Expert Rev Proteomics *6*, 5–10.

Aebersold, R., and Mann, M. (2003). Mass spectrometry-based proteomics. Nature *422*, 198–207.

Andersson, L., and Porath, J. (1986). Isolation of phosphoproteins by immobilized metal (Fe^{3+}) affinity chromatography. Anal Biochem *154*, 250–254.

Beausoleil, S.A., Jedrychowski, M., Schwartz, D., Elias, J.E., Villen, J., Li, J., Cohn, M.A., Cantley, L.C., and Gygi, S.P. (2004). Large-scale characterization of HeLa cell nuclear phosphoproteins. Proc Natl Acad Sci USA *101*, 12130–12135.

Blagoev, B., Ong, S.E., Kratchmarova, I., and Mann, M. (2004). Temporal analysis of phosphotyrosine-dependent signaling networks by quantitative proteomics. Nat Biotechnol *22*, 1139–1145.

Bouwmeester, T., Bauch, A., Ruffner, H., Angrand, P.O., Bergamini, G., Croughton, K., Cruciat, C., Eberhard, D., Gagneur, J., Ghidelli, S., *et al.* (2004). A physical and functional map of the human TNF-alpha/NF-kappa B signal transduction pathway. Nat Cell Biol *6*, 97–105.

Burckstummer, T., Bennett, K.L., Preradovic, A., Schutze, G., Hantschel, O., Superti-Furga, G., and Bauch, A. (2006). An efficient tandem affinity purification procedure for interaction proteomics in mammalian cells. Nat Methods *3*, 1013–1019.

Butter, F., Scheibe, M., Morl, M., and Mann, M. (2009). Unbiased RNA-protein interaction screen by quantitative proteomics. Proc Natl Acad Sci USA *106*, 10626–10631.

Chen, S.H., Hsu, J.L., and Lin, F.S. (2008). Fluorescein as a versatile tag for enhanced selectivity in analyzing cysteine-containing proteins/peptides using mass spectrometry. Anal Chem *80*, 5251–5259.

Clements, A., Johnston, M.V., Larsen, B.S., and McEwen, C.N. (2005). Fluorescence-based peptide labeling and fractionation strategies for analysis of cysteine-containing peptides. Anal Chem *77*, 4495–4502.

Dai, J., Wang, J., Zhang, Y., Lu, Z., Yang, B., Li, X., Cai, Y., and Qian, X. (2005). Enrichment and identification of cysteine-containing peptides from tryptic digests of performic oxidized proteins by strong cation exchange LC and MALDI-TOF/TOF MS. Anal Chem *77*, 7594–7604.

Dignam, J.D., Lebovitz, R.M., and Roeder, R.G. (1983). Accurate transcription initiation by RNA polymerase II in a soluble extract from isolated mammalian nuclei. Nucleic Acids Res *11*, 1475–1489.

Dunn, J.D., Reid, G.E., and Bruening, M.L. (2009). Techniques for phosphopeptide enrichment prior to analysis by mass spectrometry. Mass Spectrom Rev *28*.

Fenyö, D., Qin, J., and Chait, B.T. (1998). Protein identification using mass spectrometric information. Electrophoresis *19*, 998–1005.

Ficarro, S.B., Adelmant, G., Tomar, M.N., Zhang, Y., Cheng, V.J., and Marto, J.A. (2009). Magnetic bead processor for rapid evaluation and optimization of parameters for phosphopeptide enrichment. Anal Chem *81*, 4566–4575.

Ficarro, S.B., McCleland, M.L., Stukenberg, P.T., Burke, D.J., Ross, M.M., Shabanowitz, J., Hunt, D.F., and White, F.M. (2002). Phosphoproteome analysis by mass spectrometry and its application to Saccharomyces cerevisiae. Nat Biotechnol *20*, 301–305.

Ficarro, S.B., Parikh, J.R., Blank, N.C., and Marto, J.A. (2008). Niobium(V) oxide (Nb2O5): Application to phosphoproteomics. Anal Chem *80*, 4606–4613.

Ficarro, S.B., Salomon, A.R., Brill, L.M., Mason, D.E., Stettler-Gill, M., Brock, A., and Peters, E.C. (2005). Automated immobilized metal affinity chromatography/nano-liquid chromatography/electrospray ionization mass spectrometry platform for profiling protein phosphorylation sites. Rapid Commun Mass Spectrom *19*, 57–71.

Gavin, A.C., Aloy, P., Grandi, P., Krause, R., Boesche, M., Marzioch, M., Rau, C., Jensen, L.J., Bastuck, S., Dumpelfeld, B., *et al.* (2006). Proteome survey reveals modularity of the yeast cell machinery. Nature *440*, 631–636.

Gavin, A.C., Bosche, M., Krause, R., Grandi, P., Marzioch, M., Bauer, A., Schultz, J., Rick, J.M., Michon, A.M., Cruciat, C.M., *et al.* (2002). Functional organization of the yeast proteome by systematic analysis of protein complexes. Nature *415*, 141–147.

Giron, P., Dayon, L., Mihala, N., Sanchez, J.C., and Rose, K. (2009). Cysteine-reactive covalent capture tags for enrichment of cysteine-containing peptides. Rapid Commun Mass Spectrom *23*, 3377–3386.

Gloeckner, C.J., Boldt, K., Schumacher, A., Roepman, R., and Ueffing, M. (2007). A novel tandem affinity purification strategy for the efficient isolation and characterisation of native protein complexes. Proteomics *7*, 4228–4234.

Gloeckner, C.J., Boldt, K., and Ueffing, M. (2009). Strep/FLAG Tandem Affinity Purification (SF-TAP) to Study Protein Interactions. In Current Protocols in Protein Science, J.E. Coligan et al., eds. Chapter 19, Unit 19, 20, John Wiley & Sons, Inc.

Goh, K.I., Cusick, M.E., Valle, D., Childs, B., Vidal, M., and Barabasi, A.L. (2007). The human disease network. Proc Natl Acad Sci USA *104*, 8685–8690.

Groisman, R., Polanowska, J., Kuraoka, I., Sawada, J., Saijo, M., Drapkin, R., Kisselev, A.F., Tanaka, K., and Nakatani, Y. (2003). The ubiquitin ligase activity in the DDB2 and CSA complexes is differentially regulated by the COP9 signalosome in response to DNA damage. Cell *113*, 357–367.

Gygi, S.P., Rist, B., Griffin, T.J., Eng, J., and Aebersold, R. (2002). Proteome analysis of low-abundance proteins using multidimensional chromatography and isotope-coded affinity tags. J Proteome Res *1*, 47–54.

Gygi, S.P., Rist, B., Gerber, S.A., Turecek, F., Gelb, M.H., and Aebersold, R. (1999). Quantitative analysis of complex protein mixtures using isotope-coded affinity tags. Nat Biotechnol *17*, 994–999.

Hakes, L., Pinney, J.W., Robertson, D.L., and Lovell, S.C. (2008). Protein–protein interaction networks and biology – what's the connection? Nat Biotechnol *26*, 69–72.

Heo, I., Joo, C., Kim, Y.K., Ha, M., Yoon, M.J., Cho, J., Yeom, K.H., Han, J., and Kim, V.N. (2009). TUT4 in concert with Lin28 suppresses microRNA biogenesis through pre-microRNA uridylation. Cell *138*, 696–708.

Hille, J.M., Freed, A.L., and Watzig, H. (2001). Possibilities to improve automation, speed and precision of proteome analysis: A comparison of two-dimensional electrophoresis and alternatives. Electrophoresis *22*, 4035–4052.

Himeda, C.L., Ranish, J.A., Angello, J.C., Maire, P., Aebersold, R., and Hauschka, S.D. (2004). Quantitative proteomic identification of six4 as the trex-binding factor in the muscle creatine kinase enhancer. Mol Cell Biol *24*, 2132–2143.

Himeda, C.L., Ranish, J.A., and Hauschka, S.D. (2008). Quantitative proteomic identification of MAZ as a transcriptional regulator of muscle-specific genes in skeletal and cardiac myocytes. Mol Cell Biol *28*, 6521–6535.

Hunt, D.F., Yates, J.R., Shabanowitz, J., Winston, S., and Hauer, C.R. (1986). Protein sequencing by tandem mass spectrometry. Proc Natl Acad Sci USA *83*, 6233–6237.

Ideker, T., Thorsson, V., Ranish, J.A., Christmas, R., Buhler, J., Eng, J.K., Bumgarner, R., Goodlett, D.R., Aebersold, R., and Hood, L. (2001). Integrated genomic and proteomic analyses of a systematically perturbed metabolic network. Science *292*, 929–934.

Kocher, T., and Superti-Furga, G. (2007). Mass spectrometry-based functional proteomics: From molecular machines to protein networks. Nat Methods *4*, 807–815.

Kweon, H.K., and Hakansson, K. (2006). Selective zirconium dioxide-based enrichment of phosphorylated peptides for mass spectrometric analysis. Anal Chem *78*, 1743–1749.

Larsen, M.R., Thingholm, T.E., Jensen, O.N., Roepstorff, P., and Jorgensen, T.J. (2005). Highly selective enrichment of phosphorylated peptides from peptide mixtures using titanium dioxide microcolumns. Mol Cell Proteomics *4*, 873–886.

Liu, T., Qian, W.J., Camp, D.G., 2nd, and Smith, R.D. (2007). The use of a quantitative cysteinyl-peptide enrichment technology for high-throughput quantitative proteomics. Methods Mol Biol *359*, 107–124.

Matsuda, H., Nakamura, H., and Nakajima, T. (1990). New ceramic Titania: Selective adsorbent for organic phosphates. Anal Sci *6*, 911–912.

Mittler, G., Butter, F., and Mann, M. (2009). A SILAC-based DNA protein interaction screen that identifies candidate binding proteins to functional DNA elements. Genome Res *19*, 284–293.

Moser, K., and White, F.M. (2006). Phosphoproteomic analysis of rat liver by high capacity IMAC and LC-MS/MS. J Proteome Res *5*, 98–104.

Nakatani, Y., and Ogryzko, V. (2003). Immunoaffinity purification of mammalian protein complexes. Methods Enzymol *370*, 430–444.

Ndassa, Y.M., Orsi, C., Marto, J.A., Chen, S., and Ross, M.M. (2006). Improved immobilized metal affinity chromatography for large-scale phosphoproteomics applications. J Proteome Res *5*, 2789–2799

Nuwaysir, L.M., and Stults, J.T. (1993). Electrospray ionization mass spectrometry of phosphopeptides isolated by on-line immobilized metal-ion affinity chromatography. J Am Soc Mass Spectrom *4*, 662–669.

Palani, A., Lee, J.S., Huh, J., Kim, M., Lee, Y.J., Chang, J.H., Lee, K., and Lee, S.W. (2008). Selective enrichment of cysteine-containing peptides using SPDP-functionalized superparamagnetic Fe(3)O(4)@SiO(2) nanoparticles: Application to comprehensive proteomic profiling. J Proteome Res *7*, 3591–3596.

Pawson, T., and Warner, N. (2007). Oncogenic re-wiring of cellular signaling pathways. Oncogene *26*, 1268–1275.

Pflieger, D., Junger, M.A., Muller, M., Rinner, O., Lee, H., Gehrig, P.M., Gstaiger, M., and Aebersold, R. (2008). Quantitative proteomic analysis of protein complexes: Concurrent identification of interactors and their state of phosphorylation. Mol Cell Proteomics *7*, 326–346.

Pinkse, M.W., Uitto, P.M., Hilhorst, M.J., Ooms, B., and Heck, A.J. (2004). Selective isolation at the femtomole level of phosphopeptides from proteolytic digests using 2D-NanoLC-ESI-MS/MS and titanium oxide precolumns. Anal Chem *76*, 3935–3943.

Ponticelli, A.S., and Struhl, K. (1990). Analysis of Saccharomyces cerevisiae his3 transcription in vitro: Biochemical support for multiple mechanisms of transcription. Mol Cell Biol *10*, 2832–2839.

Porath, J., Carlsson, J.A.N., Olsson, I., and Belfrage, G. (1975). Metal chelate affinity chromatography, a new approach to protein fractionation. Nature *258*, 598–599.

Posewitz, M.C., and Tempst, P. (1999). Immobilized Gallium(III) affinity chromatography of phosphopeptides. Anal Chem *71*, 2883–2892.

Raftery, M.J. (2008). Enrichment by organomercurial agarose and identification of cys-containing peptides from yeast cell lysates. Anal Chem *80*, 3334–3341.

Ramsby, M.L., and Makowski, G.S. (1999). Differential detergent fractionation of eukaryotic cells. Analysis by two-dimensional gel electrophoresis. Methods Mol Biol *112*, 53–66.

Ranish, J.A., Yi, E.C., Leslie, D.M., Purvine, S.O., Goodlett, D.R., Eng, J., and Aebersold, R. (2003). The study of macromolecular complexes by quantitative proteomics. Nature Genet *33*, 349–355.

Ranish, J.A., Yudkovsky, N., and Hahn, S. (1999). Intermediates in formation and activity of the RNA polymerase II preinitiation complex: Holoenzyme recruitment and a postrecruitment role for the TATA box and TFIIB. Genes Dev *13*, 49–63.

Rigaut, G., Shevchenko, A., Rutz, B., Wilm, M., Mann, M., and Seraphin, B. (1999). A generic protein purification method for protein complex characterization and proteome exploration. Nat Biotechnol *17*, 1030–1032.

Rivera, J., Choi, Y.S., Vujcic, S., Wood, T., and Colon, L. (2009). Enrichment/isolation of phosphorylated peptides on hafnium oxide prior to mass spectrometric analysis. Analyst *134*, 31–33.

Ross, P.L., Huang, Y.N., Marchese, J.N., Williamson, B., Parker, K., Hattan, S., Khainovski, N., Pillai, S., Dey, S., Daniels, S., *et al.* (2004). Multiplexed protein quantitation in Saccharomyces cerevisiae using amine-reactive isobaric tagging reagents. Mol Cell Proteomics *3*, 1154–1169.

Rush, J., Moritz, A., Lee, K.A., Guo, A., Goss, V.L., Spek, E.J., Zhang, H., Zha, X.M., Polakiewicz, R.D., and Comb, M.J. (2005). Immunoaffinity profiling of tyrosine phosphorylation in cancer cells. Nat Biotechnol *23*, 94–101.

Shiu, H.Y., Chan, T.C., Ho, C.M., Liu, Y., Wong, M.K., and Che, C.M. (2009). Electron-deficient alkynes as cleavable reagents for the modification of cysteine-containing peptides in aqueous medium. Chemistry *15*, 3839–3850.

Sugiyama, N., Masuda, T., Shinoda, K., Nakamura, A., Tomita, M., and Ishihama, Y. (2007). Phosphopeptide enrichment by aliphatic hydroxy acid-modified metal oxide chromatography for nano-LC-MS/MS in proteomics applications. Mol Cell Proteomics *6*, 1103–1109.

Tasto, J.J., Carnahan, R.H., McDonald, W.H., and Gould, K.L. (2001). Vectors and gene targeting modules for tandem affinity purification in schizosaccharomyces pombe. Yeast *18*, 657–662.

Testa, G., Zhang, Y., Vintersten, K., Benes, V., Pijnappel, W.W., Chambers, I., Smith, A.J., Smith, A.G., and Stewart, A.F. (2003). Engineering the mouse genome with bacterial artificial chromosomes to create multipurpose alleles. Nat Biotechnol *21*, 443–447.

Thompson, A., Schafer, J., Kuhn, K., Kienle, S., Schwarz, J., Schmidt, G., Neumann, T., Johnstone, R., Mohammed, A.K., and Hamon, C. (2003). Tandem mass tags: A novel quantification strategy for comparative analysis of complex protein mixtures by MS/MS. Anal Chem *75*, 1895–1904.

Tsai, C.F., Wang, Y.T., Chen, Y.R., Lai, C.Y., Lin, P.Y., Pan, K.T., Chen, J.Y., Khoo, K.H., and Chen, Y.J. (2008). Immobilized metal affinity chromatography revisited: pH/acid control toward high selectivity in phosphoproteomics. J Proteome Res *7*, 4058–4069.

Veraksa, A., Bauer, A., and Artavanis-Tsakonas, S. (2005). Analyzing protein complexes in Drosophila with tandem affinity purification-mass spectrometry. Dev Dyn *232*, 827–834.

Winter, D., Seidler, J., Ziv, Y., Shiloh, Y., and Lehmann, W.D. (2008). Citrate boosts the performance of phosphopeptide analysis by UPLC-ESI-MS/MS. J Proteome Res *8*, 418–424.

Zhang, Y., Wolf-Yadlin, A., Ross, P.L., Pappin, D.J., Rush, J., Lauffenburger, D.A., and White, F.M. (2005). Time-resolved mass spectrometry of tyrosine phosphorylation sites in the epidermal growth factor receptor signaling network reveals dynamic modules. Mol Cell Proteomics *4*, 1240–1250.

Chapter 23
A Multidimensional Chromatography Strategy Using HILIC and IMAC for Quantitative Phosphoproteome Analysis

Dean E. McNulty, Michael J. Huddleston, and Roland S. Annan

Abstract Mass spectrometry-based protein phosphorylation analysis on a proteome-wide scale remains a formidable challenge, hampered by the complexity and dynamic range of protein expression on the global level and multi-site phosphorylation at substoichiometric ratios at the individual protein level. Reduction of sample complexity and enrichment of the phosphopeptide pool is a necessary prerequisite for global phosphoproteomics. Metal affinity chromatography and strong cation exchange chromatography, either alone or in tandem, have emerged as the most widely used chromatographic and enrichment strategies. However, each is not without shortcomings. The former suffers from selectivity over highly acidic peptides, while the latter is a low resolution technique that provides little in the way of enrichment for phosphorylated peptides. Here we describe a phosphopeptide fractionation scheme using hydrophilic interaction chromatography (HILIC) which both enriches the phosphopeptide pool and efficiently fractionates the remaining peptides. When HILIC is used in front of metal affinity chromatography, the selectivity of the metal affinity resin is improved to greater than 90%. The lack of significant numbers of nonphosphorylated peptides also allows for more efficient use of the mass spectrometer duty cycle in that the instrument spends nearly all of its time sequencing phosphopeptides.

Keywords Phosphorylation · Mass spectrometry · Hydrophilic Interaction chromatography · Proteomics · Phosphoproteome

23.1 Introduction

The analysis of the phosphoproteome by mass spectrometry is hampered by the enormous complexity of the proteome and the large dynamic range of protein expression (de Godoy et al., 2006). Many phosphorylated proteins are less

---

R.S. Annan (✉)
Proteomics and Biological Mass Spectrometry Laboratory, GlaxoSmithKline, Collegeville, PA 19426, USA
e-mail: roland.s.annan@gsk.com

A.R. Ivanov, A.V. Lazarev (eds.), *Sample Preparation in Biological Mass Spectrometry*, DOI 10.1007/978-94-007-0828-0_23,

abundant proteins involved in signaling and regulation of cellular function. Most phosphoproteins contain between 1 and 20 phosphorylation sites and these sites are largely occupied at substoichiometric levels. Many phosphorylation sites also occur in combination with other modifications and multiple phosphorylation sites can exist in a single peptide. All of these factors suggest that the phosphoproteome is likely to be every bit as complex as the general proteome, but with a dynamic range spanning an additional two orders of magnitude and that the phosphopeptides exist in the presence of a very large excess of nonphosphorylated peptides.

The most common strategy for enriching phosphopeptides in mixtures is the use of immobilized metal affinity chromatography (IMAC) (Nuwaysir and Stults, 1993) or titanium dioxide (TiO_2) (Pinkse et al., 2004) micro columns (Erdjument-Bromage et al., 1998). While phosphopeptides have a very high affinity for these materials, acidic peptides (containing aspartic acid and glutamic acid) and peptides containing histidine also have a high binding affinity. The conversion of peptide carboxylate groups to methyl ester derivatives has been shown to restore the selectivity of both IMAC (Ficarro et al., 2002) and TiO_2 (Pinkse et al., 2004). It has also been demonstrated that including dihydroxybenzoic acid or glycolic acid in the loading buffer improves the selectivity of phosphopeptide binding to TiO_2 (Larsen et al., 2005).

Strong cation exchange chromatography (SCX) was initially suggested as an alternative to metal chelating resins as a means of enriching phosphopeptides (Beausoleil et al., 2004). Although it appears highly reproducible, it does not provide a very high level of enrichment, and many phosphopeptides do not elute in the enriched fractions. Taking fractions throughout the entire SCX gradient, however, and enriching these via metal affinity chromatography results in a satisfactory strategy that yields typically 5,000–15,000 phosphopeptide identifications albeit not without using starting materials in the range 8–15 mg of whole cell lysate and extensive replicate analysis (Olsen et al., 2006; Dephoure et al., 2008).

The benefits of a 2D chromatography strategy for peptide-based expression proteomics are well established. Assuming the two separation modes are orthogonal, the total peak capacity of the system or its maximum ability to separate components in a mixture is the product of the individual peak capacity for each mode (Horvath et al., 2009). Reversed-phase chromatography (RP), the final separation in nearly all peptide-based proteomic strategies is a high resolution technique that can achieve optimized peak capacities of 1,000 (Gilar et al., 2004) though 300–400 is more typical. On the other hand, SCX (PolySulfoethyl Aspartamide) which is the most commonly utilized first dimension separation has a peak capacity of only 60–80 for tryptic peptides. The total peak capacity of a SCX-RP system is further compromised by the relatively low orthogonality of the two separation modes (Gilar et al., 2005). Replacing SCX with a high resolution chromatographic separation at the peptide level, which is truly orthogonal to RP, would dramatically improve the resolving power of 2D systems for proteomics. When coupled with an enrichment strategy like IMAC or TiO_2 this would make for a powerful strategy of the analysis of the phosphoproteome.

Hydrophilic interaction chromatography (HILIC) (Alpert, 1990) is a high resolution separation technique where retention increases with increasing polarity (hydrophilicity) of the peptide, opposite to the trends observed in RPLC (Yoshida and Okada, 1999). HILIC has been shown to have the highest degree of orthogonality to RPLC of all commonly used peptide separation modes (Gilar et al., 2005). This and the suitability of HILIC for the highly efficient separation of polar analytes prompted us to employ HILIC as part of a multi-dimensional separation strategy for phospho-proteomics. Here we show that HILIC as a first dimension separation in combination with IMAC and reversed-phase liquid chromatography-tandem mass spectrometry is an efficient, easy to employ strategy for quantitative phosphoproteomics.

23.2 Materials and Methods

23.2.1 Cell Culture and Lysis

The human breast carcinoma cell line BT474 was purchased from the American Type Culture Collection (ATCC). Cells were differentially labeled by SILAC, where one cell population was grown in medium with light (normal) L-arginine and L-lysine, while the other cell population was grown in heavy (isotopic) [U-^{13}C6, ^{15}N4]-L-Arginine and [U-^{13}C6]-L-Lysine (Invitrogen). A small molecule inhibitor of Her2 was synthesized by GlaxoSmithKline and prepared as a stock solution in DMSO. BT474 cells were treated either with 1 μM inhibitor (heavy) or vehicle (light) for 4 h after which they were lysed directly in the plates using 1 mL ice-cold 8 M urea/0.4 M NH_4HCO_3, containing Phosphatase Inhibitor Cocktails I + II (Sigma). One T150 flask was used for each condition. Cells were scraped, and sonicated prior to clarification by centrifugation. After determining the protein concentration for each condition, the light and heavy cells were mixed to give a 1:1 ratio. The pooled lysate was reduced with 4.5 mM DTT for 30 min at 37°C, and alkylated with 10 mM iodoacetamide for 30 min at room temperature in the dark. The sample was diluted 1:4 with water, and 250 μL of immobilized trypsin slurry was added (Pierce). The sample was digested at room temperature with rocking overnight. Digests were acidified to 1% TFA, desalted on a 1 g Sep-Pak cartridge (Waters), and lyophilized to dryness.

23.2.2 Hydrophilic-Interaction Chromatography (HILIC)

Preparative chromatographic separations were performed on a Beckman System Gold 126 HPLC (Beckman-Coulter) using a 4.6 × 250 mm TSKgel Amide-80 5 μm particle column (TOSOH Biosciences). 2 mg of pooled, desalted BT474 tryptic digest were loaded in 80% solvent B (98% acetonitrile with 0.1% TFA). Solvent A consists of 98% water with 0.1% TFA. The peptides were eluted with an inverse gradient that starts at 80% B, hold for 5 min, followed by 80% B to 60% B in

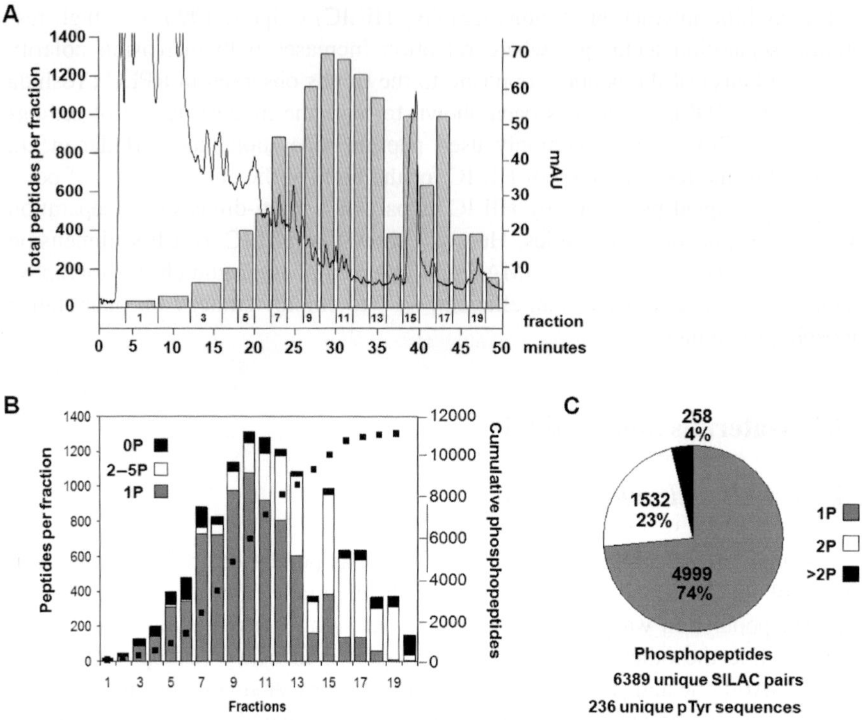

Fig. 23.1 Phosphoproteomic analysis of SILAC labeled B474 breast cancer cells. After pooling 1 mg of cell lysate from treated (light) and untreated (heavy) cells, the sample was digested with trypsin and the peptides fractionated by HILIC. Each fraction was enriched for phosphopeptides using IMAC and analyzed by reversed-phase LC-MS/MS. (**a**) UV-based elution profile of tryptic peptides from BT474 lysate. Tryptic peptides were separated using an optimized phosphopeptide gradient and pooled as indicated. A_{280} UV absorbance profile and gradient trace are shown. Superimposed on the UV trace is the total number of peptides (phosphorylated and nonphosphorylated) identified in each fraction after IMAC enrichment; (**b**) Percentage of nonphosphorylated, single and multiply phosphorylated peptides identified in each fraction. Summed cumulative total of phosphopeptides are indicated by the *dotted line*. Cumulative and per fraction peptide counts include redundant sequences. (**c**) Unique phosphopeptides from BT474 cells

40 min and finally 0%B in 5 min. Twenty 1 mL fractions were collected throughout the gradient, as shown in Fig. 23.1a, and used directly for IMAC as described below.

23.2.3 Immobilized Metal Affinity Chromatography (IMAC)

Twenty μL of 50% slurry (10 μL gel) of PHOS-Select Iron Affinity Gel (Sigma) was added directly to the HILIC fractions and vortexed rapidly for 30 min. The fractions were transferred into a 0.22 μm Nylon Spin-X Centrifuge Tube Filter (Corning) and centrifuged at 8,200×g for 30 s. The unbound material was discarded, and the gel was washed once with 500 μL of 250 mM acetic acid with 30%

acetonitrile and then again with 500 μL of water. The gel was eluted with 100 μL of 400 mM ammonium hydroxide with vortexing for 10 min. The eluant was collected and lyophilized to dryness.

23.2.4 Liquid Chromatography–Mass Spectrometry (LC-MS)

The IMAC enriched HILIC fractions were analyzed directly using an Agilent 1100 NanoLC equipped with an autosampler and an auxiliary binary pump. The LC was interfaced directly to a Thermo LTQ-Orbitrap XL mass spectrometer. Peptides were loaded onto a 200 μm × 5 mm PepSwift Monolithic PS-DVB trap column at 20 μL/min using the capillary pump. The peptides were eluted from the trap column and separated on a 100 μm × 5 cm analytical column of the same material using an acetonitrile:water gradient at 0.5 μL/min. Mobile phase A: 0.2 % formic acid; mobile phase B: acetonitrile with 0.2% formic acid; Load solution: 0.05% HFBA. The gradient elution program starts at 2%B (5 min hold), followed by 2–15%B in 60 min, 15–30%B in 30 min, and finally 30–95%B in 5 min with a 5 min hold.

The LTQ-Orbitrap XL uses an instrument method configured for data dependent tandem mass spectrometry (LC-MS/MS) acquisition. Full range MS spectra were acquired in profile mode from *m/z* 400 to 2,000 in the Orbitrap analyzer at a resolution of 30,000 (with no preview scan) followed by up to 10 MS/MS spectra acquired in the LTQ. Parameters for MS/MS acquisition included an isolation width of 3 *m/z*, normalized collision energy of 30% and all precursors must exceed a minimum threshold of 5,000. Precursor selection for MS/MS occurs only if a charge state is determined as 2+ or higher. Dynamic exclusion was enabled with a repeat count of 1, a repeat duration of 240 s and an exclude list size of 500. AGC target values and maximum ion injection times were set to 1E6 and 750 ms for the Orbitrap MS scans and 1E4 and 200 ms for the LTQ MS/MS scans. Both were set to use a single microscan. All Orbitrap MS data was acquired with the full-scan injection waveform on and calibrated using a single ion lock mass at *m/z* 445.

23.2.5 Peptide Identification and Quantitation

All data was processed by MaxQuant (Cox and Mann, 2008) using Mascot 2.2 and the IPI human database v3.52, both forward and reverse. Searches were performed for fully tryptic peptides containing up to one missed cleavage, fixed carboxyamidomethyl-cysteine modification and phosphorylation on serine, threonine and tyrosine set as a variable modification. Mass tolerances for the initial search are 7 ppm on the precursor and 0.5 Da on the fragment ions. Secondary filters employed by MaxQuant included correct SILAC label and number of arginine and lysine residues, both of which corresponded to the SILAC pair detected, and a combined Mascot plus PTM score of 28 (Olsen et al., 2006). MaxQuant filters were set to a peptide FDR of 0.01 and a peptide false positive rate (FPR) to 0.1% based on the PTM score. MaxQuant output was visualized using Spotfire.

23.3 Results

Because phosphopeptides exhibit increased retention on HILIC relative to non-phosphorylated peptides, we optimized the gradient with the intent to segregate that part of the digested lysate which does not contain phosphorylated peptides into the flow-thru while providing optimal resolution of the phosphopeptide containing fractions during the gradient elution. This resulted in a separation which provides significant enrichment and excellent fractionation of phosphopeptides (see Fig. 23.1a). Based on UV absorbance, more than 50% of the eluted peptides from SILAC labeled BT474 cells were contained in the first two fractions, yet after IMAC we identified a total of only 35 phosphorylated peptides in these fractions.

We reported previously that HILIC fractionation dramatically improves the selectivity of IMAC (McNulty and Annan, 2008). Furthermore, because the HILIC mobile phases contain only acetonitrile, water and acid, the IMAC resin can be added directly to the fractions without the need for desalting. As can be seen in Fig. 23.1a,b, in spite of the very large number of peptides suggested to be contained in these two fractions, only a very small number (31) of nonphosphorylated peptides were captured. The enrichment provided by the HIILIC itself is evidenced by the fact that 35 unique phosphopeptides were identified in the two fractions, with only four being found in fraction 1. The hydrophilicity-based separation on HILIC yields fractions where the phosphopeptides in each fraction compete favorably with the nonphosphorylated peptides, without the need for derivatization or additives. Figure 23.1a shows that as the overall peptide load decreases, during the development of the HILIC gradient, the number of peptides captured by IMAC increases and stays fairly high until near the end. These captured peptides are largely phosphorylated, with nonphosphorylated peptides constituting only a very small percentage of any given fraction (Fig. 23.1b). Overall, the IMAC capture resulted in >92% selectivity based on MS/MS identification of phosphopeptides vs. non-phosphopeptides (Fig. 23.1b). This is consistent with our previous work where we reported a similar selectivity for HeLa cells.

The increasing hydrophilicity of the peptides eluting late in the HILIC gradient is evident by the increasing number of multiply phosphorylated peptides identified from fraction 10 onward (Fig. 23.1b). In fact, beginning with fraction 14, the majority of peptides captured by IMAC in each sample were multiply phosphorylated. In our previous work we reported that 10% of the peptides recovered by the HILIC-IMAC strategy were multiply phosphorylated. In this work the multiply phosphorylated peptides account for 27% of the total (Fig. 23.1c). We attribute this increased percentage largely to the use of a high performance mass spectrometer in these studies. Multiply phosphorylated peptides tend to be larger and thus have an ambiguous charge state determination in a low resolution mass spectrometer. Multiply phosphorylated peptides are also prone to highly abundant neutral losses of H_3PO_4 yielding fewer structurally useful fragment ions. The LTQ-Orbitrap records MS spectra at a resolution of 60,000 and with the use of an internal lock mass, measures precursor ion masses at better than 1 ppm. While multiply phosphorylated peptides still produce analytically less useful fragment ion spectra in the LTQ, the combination of high resolution and accurate *m/z* determination in the Orbitrap, produce correct charge state assignments and ultimately highly accurate peptide masses. An accurate precursor ion mass is critical to confidently assigning sequences to peptides with poor fragment ion spectra.

Another change in the HILIC-IMAC LC-MS instrumentation we made prior to this work was substituting a 100 μm monolithic PS-DVB analytical column in place of the standard 75 μm packed C18 reversed phase column. We have found that this column produces lower levels of background, making it easier to detect and sequence low abundance peptides. In addition, larger peptides tend to chromatograph better on the monolithic column than the standard packed columns. Both of these attributes are likely to enhance the identification of multiply phosphorylated peptides as they tend to produce weaker signals and chromatography more poorly.

After removing redundancy we identified 6,389 unique phosphorylated SILAC pairs for which we were able to extract quantitative data on their response to inhibition of Her2 signaling. Interestingly, this included 236 unique phosphotyrosine sequences (Fig. 23.2a). Among those proteins expected to be involved in Her2

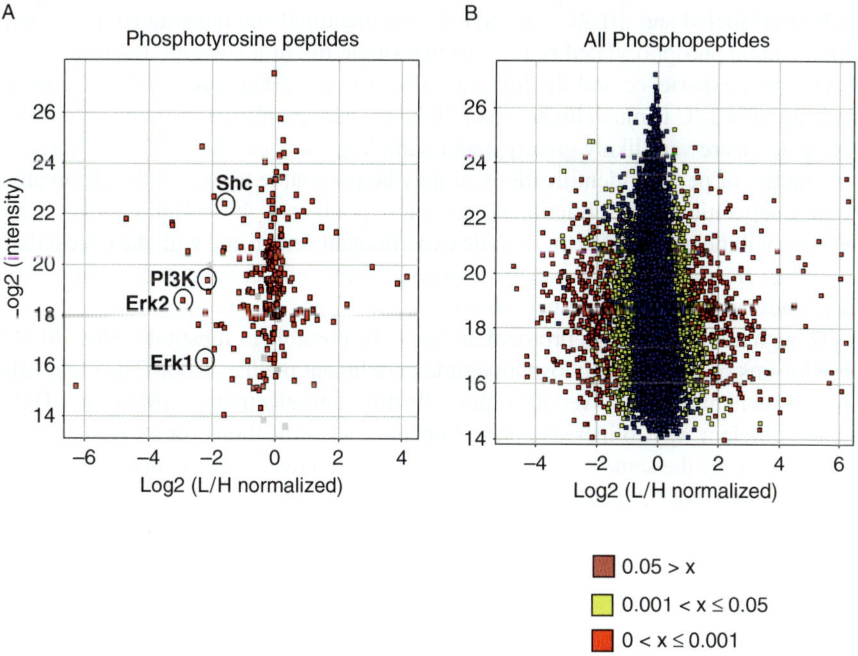

Fig. 23.2 Quantitative phosphoproteomic analysis of Her2 inhibition in B474 breast cancer cells. (**a**) More than 230 unique phosphotyrosine containing sequences were identified in the HILIC IMAC protocol. Their SILAC ratios are plotted against their intensity. Select members of the Her2 signaling pathway are indicated; (**b**) SILAC ratios for all 11,324 phosphopeptides identified by MaxQuant and Mascot are plotted against their intensity. Peptides are color coded as indicated according to their significance (*p* value). More than 250 phosphopeptides undergo a statistically significant change ($p > 0.001$)

signaling, we identified down-regulation of phosphotyrosine sites on Her2, Erk1 and 2, Shc, and PI3K. Overall, more than 250 individual phosphorylation sites undergo a statistically significant change ($p > 0.001$) in response to Her2 inhibition (Fig. 23.2b).

23.4 Conclusions

We have described a highly efficient, highly selective phosphoproteomic strategy that uses a combination of HILIC and IMAC to enrich phosphopeptides from protein lysates. The enriched fractions were analyzed using a high resolution monolithic RP separation column interfaced to a high performance LTQ-Orbitrap mass spectrometer. Database searching with Mascot showed that the combination of HILIC and IMAC resulted in a greater than 92% selectivity for phosphopeptides in this system. The HILIC chromatography efficiently enriched singly and multiply phosphorylated peptides into separate pools, eliminating the observed biased of IMAC toward multiply phosphorylated peptides (Thingholm et al., 2008). In the described system, a single IMAC protocol was used for all fractions. Using a single T150 flask each of unlabeled and SILAC labeled cells we identified and quantitated more than 6,000 unique phosphorylated sequences in a single run of the twenty fraction set.

Our own experience and the literature (Gilar et al., 2005) suggest that the peak capacity of HILIC is much higher than SCX. By taking only 20 fractions across the gradient, we are very likely greatly under-sampling the separation. Thus to increase the number of peptide identifications, it may be preferable to extend the chromatographic separation and/or fractionate into more pools (e.g. 40×1 min pools). This decision will be made based on specific experimental design and with the caveat that additional fractionation will require increased sample processing and significantly more dedicated mass spectrometer time.

Though not tested in this protocol, it should be possible to substitute other IMAC based resins or alternative phosphopeptide enrichment media such as TiO_2 and still obtain excellent results. Given the reportedly different selectivity of IMAC and TiO_2 a more efficient way to increase phosphopeptide identifications for a given sample may be to apply the same sample to the two resins, analyzing each once, rather than simply making multiple analyses of the same sample.

References

Alpert, A.J. (1990). Hydrophilic-interaction chromatography for the separation of peptides, nucleic acids and other polar compounds. J Chromatogr *499*, 177–196.

Beausoleil, S.A., Jedrychowski, M., Schwartz, D., Elias, J.E., Villen, J., Li, J., Cohn, M.A., Cantley, L.C., and Gygi, S.P. (2004). Large-scale characterization of HeLa cell nuclear phosphoproteins. Proc Natl Acad Sci USA *101*, 12130–12135.

Cox, J., and Mann, M. (2008). MaxQuant enables high peptide identification rates, individualized p.p.b.-range mass accuracies and proteome-wide protein quantification. Nat Biotechnol *26*, 1367–1372.

de Godoy, L.M., Olsen, J.V., de Souza, G.A., Li, G., Mortensen, P., and Mann, M. (2006). Status of complete proteome analysis by mass spectrometry: SILAC labeled yeast as a model system. Genome Biol *6*, R50.

Dephoure, N., Zhou, C., Villen, J., Beausoleil, S.A., Bakalarski, C.E., Elledge, S.J., and Gygi, S.P. (2008). A quantitative atlas of mitotic phosphorylation. PNAS *105*, 10762–10767.

Erdjument-Bromage, H., Lui, M., Lacomis, L., Grewal, A., Annan, R.S., McNulty, D.E., Carr, S.A., and Tempst, P. (1998). Examination of micro-tip reversed-phase liquid chromatographic extraction of peptide pools for mass spectrometric analysis. J Chromatogr A *826*, 167–181.

Ficarro, S.B., McCleland, M.L., Stukenberg, P.T., Burke, D.J., Ross, M.M., Shabanowitz, J., Hunt, D.F., and White, F.M. (2002). Phosphoproteome analysis by mass spectrometry and its application to Saccharomyces cerevisiae. Nat Biotechnol *20*, 301–305.

Gilar, M., Daly, A.E., Kele, M., Neue, U.D., and Gebler, J.C. (2004). Implications of column peak capacity on the separation of complex peptide mixtures in single- and two-dimensional high-performance liquid chromatography. J Chromatogr A *1061*, 183–192.

Gilar, M., Olivova, P., Daly, A.E., and Gebler, J.C. (2005). Orthogonality of separation in two-dimensional liquid chromatography. Anal Chem *77*, 6426–6434.

Horvath, K., Fairchild, J.N., and Guiochon, G. (2009). Generation and limitations of peak capacity in online two-dimensional liquid chromatography. Anal Chem *81*, 3879–3888.

Larsen, M.R., Thingholm, T.E., Jensen, O.N., Roepstorff, P., and Jorgensen, T.J. (2005). Highly selective enrichment of phosphorylated peptides from peptide mixtures using titanium dioxide microcolumns. Mol Cell Proteomics *4*, 873–886.

McNulty, D.E., and Annan, R.S. (2008). Hydrophilic-Interaction Chromatography reduces the complexity of the phosphoproteome and improves global phosphopeptide isolation and detection. Mol Cell Proteomics *7*, 971–980.

Nuwaysir, L.M., and Stults, J.T. (1993). Electrospray ionization mass spectrometry of phosphopeptides isolated by on-line immobilized metal-ion affinity chromatography. J Am Soc Mass Spectrom *4*, 662–667.

Olsen, J.V., Blagoev, B., Gnad, F., Macek, B., Kumar, C., Mortensen, P., and Mann, M. (2006). Global, in vivo, and site-specific phosphorylation dynamics in signaling networks. Cell *127*, 635–648.

Pinkse, M.W., Uitto, P.M., Hilhorst, M.J., Ooms, B., and Heck, A.J. (2004). Selective isolation at the femtomole level of phosphopeptides from proteolytic digests using 2D-NanoLC-ESI-MS/MS and titanium oxide precolumns. Anal Chem *76*(14), 3935–3943, PMID:15253627.

Thingholm, T.E., Jensen, O.N., Robinson, P.J., and Larsen, M.R. (2008). SIMAC – A phosphoproteomics strategy for the rapid separation of monophosphorylated from Multiply Phosphorylated Peptides. Mol Cell Proteomics *7*, 661–671.

Yoshida, T., and Okada, T. (1999). Peptide separation in normal-phase liquid chromatography: Study of selectivity and mobile phase effects on various columns. J Chromatogr A *840*, 1–9.

Chapter 24
Glycomic Analysis of Membrane-Associated Proteins

Diarmuid T. Kenny, Liaqat Ali, Samah Issa, and Niclas G. Karlsson

Abstract Analysis of the glycosylation of membrane proteins is important for understanding the interaction and signaling of a particular cell type. This chapter describes the pathway from the enrichment of membrane proteins from cells and tissues, and subsequent release and analysis of both *N*-linked and *O*-linked oligosaccharides from these enriched fractions. Negative ion LC-MS and LC-MS^2 using graphitized carbon as separation media has shown to be a very effective way to separate and characterize both *N*-linked and *O*-linked oligosaccharides, and in combination with efficient sample preparation provides a platform for glycomic discovery for both clinical samples and cell culture models for studying glycosylation. The chapter also describes how high molecular weight glycoproteins can be isolated from the membrane protein pool by SDS-agarose/polyacrylamide composite gel electrophoresis and individual glycoprotein bands excised and subjected to the glycomic analysis.

Keywords Glycomics · LC-MS · Oligosaccharides · Membrane proteins · Mucin

Abbreviations

IPA	2-propanol
APS	Ammoniumpersulfate solution
CHAPS	Cholamidopropyl)dimethylammonio]-1-propanesulfonate
DTT	Dithiothreitol
IAA	Iodoacetamide
MFGM	Milk fat globular membranes
PVP	Polyvinyl pyrolidone
PVDF	Polyvinylidene difluoride
RPM	Revolutions per minute
SDS-AgPAGE	Sodium dodecyl sulphate-agarose/polyacrylamide composite gel electrophoresis

N.G. Karlsson (✉)
Department of Medical Biochemistry, Göteborg University, 405 30 Gothenburg, Sweden
e-mail: niclas.karlsson@medkem.gu.se

A.R. Ivanov, A.V. Lazarev (eds.), *Sample Preparation in Biological Mass Spectrometry*, DOI 10.1007/978-94-007-0828-0_24,

THF	Tetrahydrofuran
TEMED	Tetramethylethylenediamine
Tris	Tris(hydroxymethyl)aminomethane

24.1 Introduction

With the increasing awareness of the glycosylation as a mediator of extracellular signaling (Gabius, 2008), and that cells are surrounded by a glycosylated layer (i.e. the glycocalyx) containing glycolipids, glycoproteins and proteoglycans (Reitsma et al., 2007), there is a need to understand the structural elements of the glycosylation mediating this signaling. There is also evidence that the expression of various oligosaccharide epitopes on the cell surface is not static, and the cellular signaling is altered because of the environment due to stimulation by cytokines and by microbes and pathogens (Lindén et al., 2008). Glycomics of the cell surface glycoproteins is an efficient approach to identify immunologically important oligosaccharide epitopes such as sialyl-Lewis x, blood group ABO antigens, sulfation and Lewis type. This would help to understand the interaction potential of the cell, and guide further investigation to identify specific glycoprotein expressing these oligosaccharide epitopes important for the interaction. Hence, glycomics is often the first step in a glycobiological research initiative, similar to as proteomics is for providing leads into the biology of protein regulation and interaction. With the increased sophistication in glycomics technology, the use of MALDI-MS and ESI-MS has revolutionized the field, and the current methodology development has greater focus on the area of sample preparation and oligosaccharide derivatization. By far, the most common approach for mammalian glycomic investigation is to utilize enzymatic (*N*-linked oligosaccharides) or chemical (*O*-linked oligosaccharides) approaches to release the oligosaccharides from a purified protein or a mixture of proteins. Oligosaccharides are then labelled on the reducing end (Mechref and Novotny, 2002) or subjected to a global derivatization (modification of all hydroxyls in the oligosaccharide chain) in order to improve the oligosaccharides MS-compatibility, or to improve separation (HPLC and CE) (North et al., 2009), or both. Instead, we have developed an LC-MS protocol where the derivatization is kept to a minimum (reduction), and graphitized carbon has shown to be capable of separating oligosaccharide based both on isomers and size. Utilizing the negative ion mode for detection it has been shown that the sensitivity for oligosaccharides is sub-picomole using capillary flow (Schulz et al., 2002), and down to femtomole for nanoflow LC-MS (Karlsson et al., 2004), providing a platform for biological analysis of both small scale cell culture (Holmén et al., 2004; Robinson et al., 2003) and clinical samples (Larsson et al., 2009; Schulz et al., 2007, 2005). With appropriate enrichment of cell membrane-associated proteins using either sodium carbonate extraction for cell culture and tissue (Karlsson et al., 2004; Robinson et al., 2003) or churning (milk fat globular membranes (MFGM)) (Wilson et al., 2008), we can extract information about the glycome by releasing and analyzing the glycosylation from the pool of membrane proteins. An additional successful approach is to further separate the membrane proteins by gel electrophoresis (Wilson et al., 2008), and with our interest in membrane bound mucins, we have developed a protocol for separating mucin type molecules using SDS-agarose/polyacrylamide composite gel electrophoresis (SDS-AgPAGE) capable of separating heavily glycosylated proteins up to several million Dalton. Combining this approach with the release of oligosaccharides from individual mucin bands has shown to be a very effective way to investigate the glycosylation of high molecular weight glycoproteins (Holmén et al., 2004; Larsson et al., 2009; Schulz et al., 2002, 2007, 2005; Wilson et al., 2008).

24.2 Methods

Preparation of total membrane-bound proteins and high molecular weight membrane bound glycoproteins for glycomic analysis of *N*-linked and *O*-linked oligosaccharides is performed according to Fig. 24.1. After isolation of membrane-bound proteins by precipitation, the precipitated fractions enriched with membrane-bound proteins were solubilized and blotted to PVDF (Polyvinylidene Difluoride) membranes by dot blotting, or by semi dry blotting following SDS-AgPAGE (high molecular mass glycoproteins, such as mucins). If not specified, all the reagents were from Sigma-Aldrich (St Louis, MO) of highest analytical grade and 18 MΩ/cm water is from a MilliQ system (Millipore, Billerica, MA).

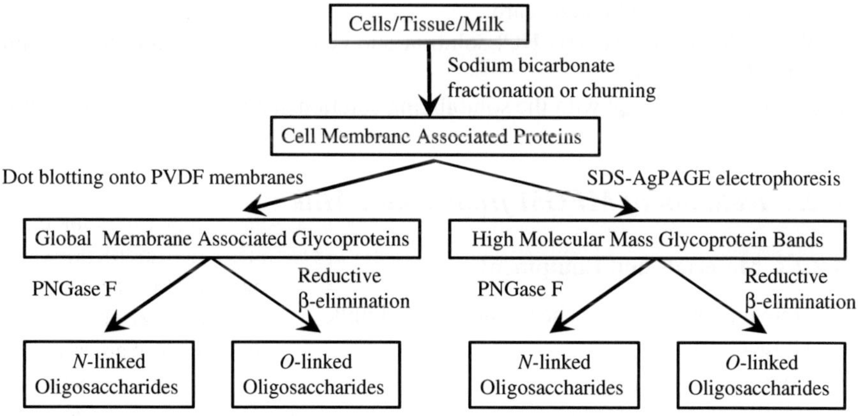

Fig. 24.1 The approach workflow for global and high molecular weight glycomic analysis of membrane-associated glycoproteins

24.2.1 Enrichment of Membrane Proteins from Mammalian Cells and Tissues

24.2.1.1 Materials and Equipment

Lyophilized mammalian cells or tissue.

Solvent A, soluble cytoplasmic and loosely bound membrane protein extraction reagent (100 mM sodium carbonate)

Solvent B, solubilizing solution (7.0 M urea 2.0 M thiourea, 40 mM tris(hydroxymethyl)aminomethane (Tris) base, 4% w/v 3-[(3-cholamidopropyl)dimethylammonio]-1-propanesulfonate (CHAPS)

Tris-HCl solution, 50 mM Tris base titrated to pH 7.3 with HCl

A sonicator with a sonication probe (Soniprep 150, MSE).

24.2.1.2 Protocol

1. Weigh out 160 mg of lyophilized cells or tissue and transfer to a 15 mL plastic tube.
2. Add 10 mL of pre-chilled solvent A
3. Sonicate the solution using a sonication probe at 40% maximum intensity for 15 s and place immediately on ice for 1 min. Repeat sonication a total of four times with repeated cooling on ice.
4. Post sonication, transfer the solution to a beaker. Add solvent A to a final volume of 100 mL and stir for 5 min at 100–400 RPM using a magnet bar and a magnetic stirrer.
5. Transfer solution to ultracentrifuge tubes. Centrifuge tubes at 5,000*g* for 10 min at 4°C to pellet cell debris.
6. Transfer the supernatant to new ultracentrifuge tubes. Centrifuge tubes at 115,000*g* for 75 min at 4°C.
7. Remove supernatant and discard.
8. Wash pellets with the Tris-HCl solution and centrifuge at 115,00*g* for 75 min (4°C). Repeat once.
9. Solubilize membrane with the solubilizing solution and prepare for dot blotting.

24.2.2 Isolation of MFGM from Whole Milk

24.2.2.1 Materials and Equipment

Human, bovine, pig or sheep unprocessed milk

Ultracentrifuge

Vortex

Heating block or incubator (55°C)

Bench top centrifuge

Lyophilizer.

24.2.2.2 Protocol

1. Ultracentrifuge milk samples at 13,000*g* for 30 min at 4°C. This process separates the skim milk fraction from the cream.
2. Vortex the cream for 2 × 10 min intervals and allow to rest in between for approximately 2–3 min.
3. Incubate the samples at 55°C for 5 min.

4. The fat membranes are isolated from the butter oil by centrifugation at 2,000*g* for 10 min at 25°C.
5. An equal volume of MilliQ water is added to the membranes and the samples are vortexed and incubated at 55°C for additional 5 min.
6. Centrifuge the samples at 2,000*g* for 10 min at 25°C.
7. The water layer is removed and the insoluble fat membranes are lyophilized.

24.2.3 Casting of High Molecular Weight Composite SDS-AgPAGE Gels for Separation of Glycoproteins

24.2.3.1 Materials and Equipment

Agarose (LE, Analytical Grade, Promega).
Acrylamide 40% solution, Acrylamide/Bis-acrylamide 37.5:1 (BioRad, Hecules, CA).
40 % Ammonium persulfate solution (APS) (Note 1).
Tetramethylethylenediamine (TEMED), (BioRad)
5× Tris-HCl buffer; 1.875 M, pH 8.1
50% Glycerol solution.
Casting equipment (BioRad Miniprotean or equivalent).
Peristaltic pump Pharmacia Biotech P-1 pump.
Glass plates (e.g. Miniprotean with 1 mm spacer and combs (BioRad))
Gradient mixer (15 mL), (Sigma-Aldrich), Part number Z34,039-1
Oven, 60°C.
Magnetic stirrer and bar for gradient mixer.
Lower gel solution (for Miniprotean 1 mm spacer, for other formats increase or decrease volumes accordingly): Agarose (0.08 g) is mixed with 5× Tris HCl buffer pH 8.1 (1.6 mL), 50% glycerol (1.6 mL) and MilliQ water (3.6 mL).
Upper gel solution (for Miniprotean 1 mm spacer, for other formats increase or decrease volumes accordingly): Agarose (0.04 g) is mixed with 5× Tris HCl buffer pH 8.1 (1.6 mL) and MilliQ water (6.4 mL).

24.2.3.2 Casting SDS-Ag/PAGE Gels

1. Mount the gel casting equipment (glass plates) and place in 60°C oven, together with the gradient mixer and the pump. Connect the pump to the gradient mixer with the supplied rubber tubing.
2. Prepare the lower and the upper gel solutions and heat in a microwave until the agarose has melted and then place immediately in the 60°C oven (Note 2).
3. Allow the lower gel solution to cool down to 60°C, and then add acrylamide (1.2 mL of 40% solution). Add the upper and lower gel solutions (4 mL of each for the MiniProtean system with 1 mm spacer) to each chamber of the gradient mixer, respectively. Add APS (2 μL) and TEMED (2 μL) to each of the two

chambers. The lower gel solution is stirred with a magnet bar and a magnetic stirrer.

4. Open the two valves of the gradient mixer and start the pumping. The gel must be cast as quickly as possible (within 2 min), loading the gradient from the top of the gel.
5. Put a comb (1 mm) into the top of the gel or pour isobutanol on the top of the gel (for preparative gel) and polymerize at room temperature for 1–2 h. The gels can be stored at + 4°C wrapped in a plastic film with wet paper for at least a week or wrapped in a plastic bag with the gel buffer and kept it in the fridge for months.

24.2.4 Sample Preparation and Running of SDS-AgPAGE

24.2.4.1 Materials and Equipment

Precast SDS-AgPAGE gel
Dithiothreitol (DTT)
Iodoacetamide (IAA)
2× Sample buffer; (0.75 M Tris-HCl pH 8.1, 2% SDS, 0.01% bromophenol blue, 60% glycerol).
Gel electrophoresis equipment (e.g BioRad Miniprotean).
Running buffer; 192 mM boric acid, 1 mM EDTA , 0.1% SDS, pH 7.6 adjusted with Tris base.

24.2.4.2 Protocol

1. Mix the sample in 2× Sample buffer and reduce it with 50 mM DTT for 30–40 min at 95°C.
2. Alkylate the sample with 125 mM IAA for 1 h.
3. Remove the comb carefully from the SDS-AgPAGE gel, place the gel in the electrophoresis equipment and add running buffer. The wells are washed 4–5 times carefully with running buffer prior to use.
4. Completely fill the inner container and fill the outer container 30–40% with the running buffer.
5. Load about 20–30 μL of sample in each well and close the lid.
6. Run the gel at 10–15 mA for 1 h and after that 25–30 mA until the dye front has migrated out of the gel.

24.2.5 Semidry Blotting of High Molecular Mass Glycoproteins to Membranes after SDS-AgPAGE

24.2.5.1 Materials and Equipment

SDS-AgPAGE with separated high molecular mass membrane glycoproteins as described above.
10% SDS solution

Methanol

10 × Cathode buffer Stock; 250 mM Tris base, 400 mM α-amino-n-hexanoic acid.

Cathode buffer solution (for 500 mL); add 50 mL of 10 × Cathode buffer stock and 0.5 mL of the 10% SDS solution. Add MilliQ water to a final volume of 500 mL.

10 × Anode buffer Stock; 250 mM Tris base

Anode buffer solution (200 mL); add 20 mL of 10 × Anode buffer Stock and 40 mL of methanol.

Add MilliQ water to a final volume of 200 mL.

10 × Ion Trap Stock; 3.0 M Tris base

Ion Trap solution (100 mL); Add 10 mL of 10 × Ion trap stock and fill up with water to 100 mL.

Blot papers (BioRad, catalog no. 1703965), (For Miniprotean, cut 6 pieces of blotting paper to a size of 133 mm × 78 mm. For other sizes, adjust accordingly).

Millipore Immobilon-P transfer membrane (For Miniprotean cut the membrane to a size of 134 mm × 80 mm)

Blotting apparatus (e.g. Trans-Blot SD semi-Dry Transfer cell (BioRad)).

24.2.5.2 Protocol

1. Soak the membrane piece for 2 min in methanol.
2. Soak 2 blot papers in Cathode buffer solution for 10 min.
3. Soak 2 blot papers and the wetted membrane from step 1 in Anode buffer solution for 10 min.
4. Soak 1 blot paper in Ion trap solution for 10 min.
5. When the gel finished running (the bromophenol blue dye migrated out of the gel), transfer the gel into MilliQ water and soak the gel for 2 min.
6. Replace water with Cathode buffer solution and incubate the gel for 5 min.
7. Place the blot paper soaked in Ion trap solution on a flat surface. Gently run a roller or a 50 mL plastic tube over the papers to remove air.
8. Place the two blot papers soaked in Anode buffer solution on top of the rolled blot papers (Note 3).
9. Place the blot membrane on the stack.
10. Place the gel on the stack. Make sure no air bubbles are trapped between the gel and the blot membrane.
11. Place the two blot papers soaked in Cathode buffer solution on top of the stack. Gently roll over the stack to remove air.
12. Transfer the stack to the blotting apparatus.
13. Close the cassette. Put some ice on the top to keep it cool. The blotting is performed at 3–5 mA/cm^2 for 40 min.
14. After blotting the bands are visualized with carbohydrate or protein stains like Alcian blue or Direct Blue. The membrane is dried and stored safely (Fig. 24.2).

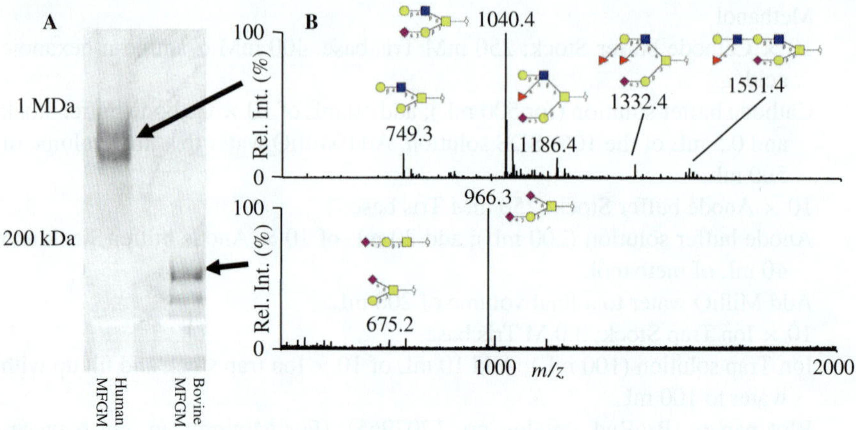

Fig. 24.2 High molecular weight glycoproteins (mucins) from milk fat globular membranes (MFGM). MFGM proteins was subjected to SDS-AgPAGE and blotted onto membranes and visualized by Alcian blue (**a**). Subsequent release of *O*-linked oligosaccharides and analysis by LC-MS showed that human MFGM protein glycosylation was more complex than bovine (Reprinted from Wilson et al. (2008) with permission from ACS Publication). *Cartoons* displayed were generated by Glycoworkbench (Ceroni et al., 2008) (**b**)

24.2.6 Dot Blotting of Membrane-Bound Glycoproteins

24.2.6.1 Materials and Equipment

Membrane proteins Dot-blot apparatus (BioRad).
Vacuum pump.
Blot papers (BioRad).
Millipore Immobilon-P transfer membrane.
Methanol.

24.2.6.2 Protocol

1. Clean and dry dot blot apparatus.
2. Place Gasket support plate in vacuum manifold.
3. Place sealing on top of gasket support plate.
4. Cut a PVDF membrane to 115 mm by 80 mm. Wet the membrane with Methanol for 2 min and soak in MilliQ water for 2 min. (always use forceps).
5. Soak a pre-cut filter paper in MilliQ water for two min.
6. Place the wet filter paper on the gasket support plate.
7. Place the wetted PVDF membrane on top of the filter paper.
8. Assemble the dot blot according to manufacturer's instructions.
9. Pipette 10 μL of prepared membrane samples into each well of the dot blot.
10. Add 50 μL of water to each well with sample to wash the sides of the wells.
11. Use a vacuum pump attached to the dot blot apparatus to pass the samples through the PVDF membrane

12. Disassemble the dot blot apparatus and remove the PVDF membrane.
13. Stain the PVDF membrane in an appropriate stain (Direct blue 71, BioRad) and destain (use a 40% Ethanol, 10% acetic acid destain solution for Direct Blue 71).

24.2.7 Staining of Blotted Membranes

24.2.7.1 Materials and Equipment

Alcian blue stain solution: 25% ethanol, 10% acetic acid, 0.125% (w/v) alcian blue.

Alcian blue destaining solution: methanol.

Direct blue 71(DB71) stain stock solution: 1% (w/v) DB71 (Bio-Rad) in MilliQ water.

DB71 destaining solution: 10% acetic acid, 40% ethanol in MilliQ water.

DB71 stain solution: 8 mL of DB71 stock solution in 100 mL of DB71 destaining solution (Note 4).

24.2.7.2 Protocol

1. Wet the blot with wetting methanol.
2. Discard methanol and add 20 mL of Alcian blue stain solution (acidic sugars) or DB71 stain solution (proteins).
3. Shake the blots for 20 min.
4. Pour off and keep Alcian blue or DB71 stain in a sealed container (stain can be reused up to five times).
5. Add 20 mL of destaining solution and shake for 5 × 2 min.
6. Dry the blot, either in air, or in vacuum. The background will diminish as the blot dries.
7. The stained blot can be imaged in visible, reflected light. If faintly stained bands are not visible, definition can be improved by wetting the blot with wetting solution before imaging.

24.2.8 Release and Clean-up of O-linked Oligosaccharides

24.2.8.1 Materials and Equipment

β-elimination solution; 50 mM NaOH, 0.5 M $NaBH_4$.

Methanol.

HCl.

Glacial acetic acid.

Desalting solution; 1% glacial acetic acid in methanol.

Oven set to 50°C.

Vacuum centrifuge.
Cation exchange beads AG 50WX8 (BioRad). Prepare cation exchange resin suspension by washing AG50WX8 resin 5 times in an equal volume of methanol, then 5 times in an equal volume of MilliQ water and store in an equal volume of methanol.
C18 Zip Tip (Millipore).
Table top centrifuge.

Prepare cation exchange columns in advance by adding 25 μL cation exchange suspension to ZipTips (1 ZipTip per sample) and place columns in 1.5 mL Eppendorf tubes, and centrifuge for 10 s. Wash columns with 2 × 60 μL methanol, and 2 × 60 μL 1.0 M HCl, centrifuge for 10 s for each wash. Remove the columns from the 1.5 mL Eppendorf tubes, discard the flow-through, and then return the columns to the Eppendorf tubes. Wash columns with 2 × 60 μL of water, centrifuging for 10 s after each addition. Remove the columns from the 1.5 mL Eppendorf tubes, discard the flow through. Place the columns in labelled new 0.5 mL eppendorf tubes.

24.2.8.2 Protocol

1. Cut out the relevant glycoprotein bands from the blots and place each in a microtiter plate well or alternatively a 0.5 mL eppendorf tube.
2. Wet each band with ~2 μL methanol depending on the size of the blot piece (2 μL for ~1–2 mm wide band).
3. Add 20 μL of β-elimination solution to each well (Note 5).
4. Cover and incubate overnight (16 h) at 50°C. (Note 6).
5. Neutralize solution by adding 1 μL glacial acetic acid to each well. Solutions should immediately fizz due to production of H_2 gas.
6. Pipette solutions from the neutralized samples to the corresponding cation exchange columns prepared in advance, centrifuge for 10 s.
7. Wash wells with 2 × 25 μL MilliQ water, load onto columns and centrifuge 10 s for each wash.
8. Elute residual glycans from the columns by washing each with 30 μL MilliQ H_2O and Centrifuge for 10 s.
9. Remove cation exchange columns from the Eppendorf tubes, and dry the eluted glycan solutions (remaining in the tubes) to completion in a vacuum centrifuge (45–60°C). White salt residue (borate) should be visible in each 0.5 mL Eppendorf tubes at this stage.
10. Discard used columns.
11. Remove borate residues by addition of 50 μL desalting solution and dry to completion in a SpeedVac (45–60°C).
12. Repeat the addition of desalting solution and subsequent evaporation until no further residues are observed (4–5 times).
13. Dissolve the released oligosaccharides in MilliQ water (usually 10–50 μL) for analysis with LC-MS (Fig. 24.3).

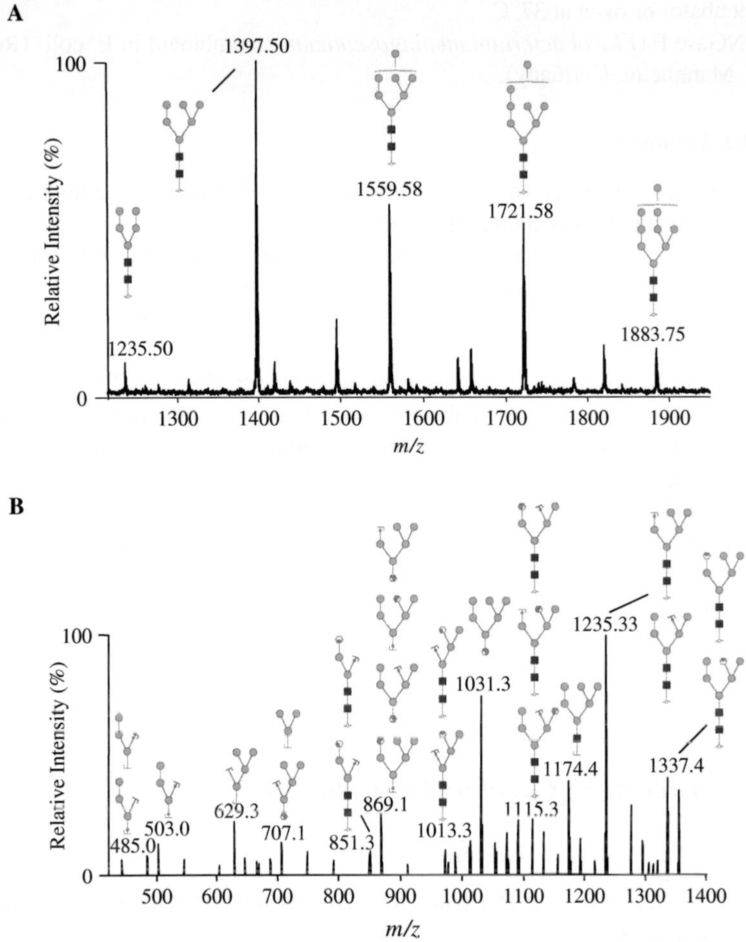

Fig. 24.3 Identification of high mannose type oligosaccharides from CHO-cell membranes using LC-MS and LC-MS^2. High mannose oligosaccharide alditols (RT 22–24 min) detected by LC-MS after PNGase F release and reduction (**a**), and MS^2-spectrum of the component with the [M–H]$^-$-ion of *m/z* 1721.6 (**b**). *Cartoons* displayed were generated by Glycoworkbench (Ceroni et al., 2008)

24.2.9 Release of N*-linked Oligosaccharides for LC-MS Analysis*

24.2.9.1 Materials and Equipments

Polyvinyl Pyrolidone (PVP) 40 Solution; 1 % (w/v) in 50% methanol.
Clean scalpel or blade.
96 well plate and adhesive cover tape.
Methanol.
Vacuum centrifuge.

Incubator or oven at 37°C.

PNGase F (*Flavobacterium meningosepticum* recombinant in E. coli) (Roche, Mannheim, Germany).

24.2.9.2 Protocol

1. Fill the wells of a microtiter plate with 100 μL of 1% (w/v) PVP solution in 50% (v/v) methanol to wet and block the membrane.
2. Cut proteins spots visualized by Direct Blue or Alcian blue from the PVDF membrane and place them in separate wells of the microtiter plate.
3. Shake the spots for 5 min the 1% PVP solution and wash three times for 5 min with water.
4. Add 5 μL of PNGaseF (0.5 U/ μL) and incubate for 15 min at 37°C, add 10 μL of water to each well and incubate overnight at 37°C. Add 100 μL of water to surrounding sample wells and seal the plate to create a humidified environment so as to reduce evaporation.
5. Sonicate the samples (in the 96 well plate) for approximately 5 min.
6. Transfer liquid into a new microtiter plate and add 20 μL of water to the blot.
7. Repeat step 5 and 6 and combine the liquids containing the released oligosaccharides removed from the blots.
8. Dry down the combined oligosaccharide solutions and dissolve in MilliQ water for analysis with LC-MS or continue with reduction using the protocol above for release of *O*-linked oligosaccharides (Fig. 24.4).

24.2.10 Packing of Capillary LC-Columns

24.2.10.1 Materials and Equipment

Fused silica capillary 250 μm i.d., 370 μm o.d. (Polymicro Technologies, Phoenix, AZ).

1/16 in. union (ZRU1.5, 0.25 mm bore, Vici AG, Schenkon, Switzerland).

Steel screen (0.5 or 1 μm pores Vici AG).

PEEK tubing 20–25 mm, 400 μm i.d., (Upchurch Scientific, Oak Harbor, WA)

Hypercarb (graphitized carbon particles) 5 μm particle size, (Thermo Scientific, Waltham, MA).

Vespel ferrule (370 μm i.d, Vici AG).

Tetrahydrofuran (THF).

2-propanol (IPA).

Ultrasonic bath.

Small magnetic bar (7 × 2 mm).

A slurry container made of a piece of 55 mm × 4.6 mm stainless steel column (volume around 1 mL with an external male 1/16 in connection in outer end filled by slurry).

Compressed air driven packing pump (Maximator, type M72, Zorge/Harz Schmidt Kranz Groupe, Germany).

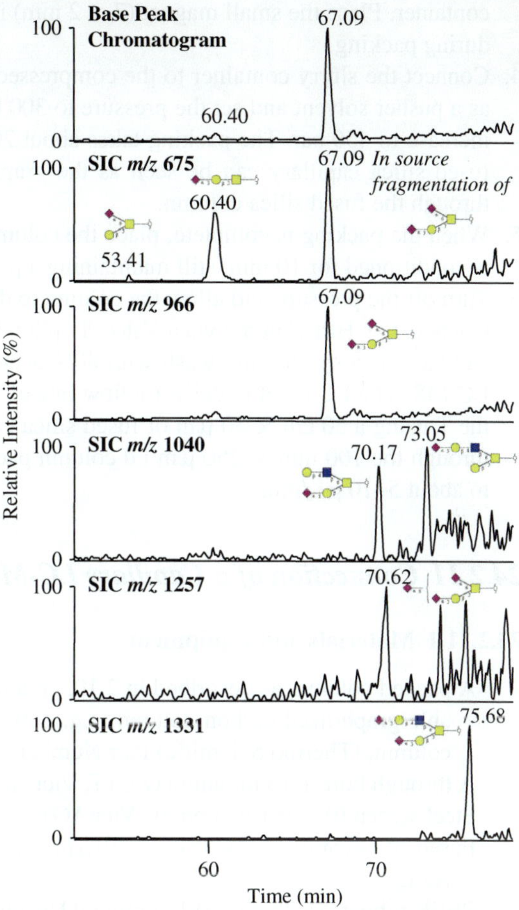

Fig. 24.4 LC-MS of *O*-linked oligosaccharide alditols from ovarian cancer tissue membrane-associated glycoproteins. Base peak chromatogram and Selective Ion Chromatogram (**SIC**) of components detected in the sample. *Cartoons* displayed were generated by Glycoworkbench (Ceroni et al., 2008)

Binary HPLC pump (Hewlett Packard 1100 series).
Micro volume tee-assembly 1/16" 15 mm bore T-split. (Valco-T, Vici AG, pr).
Acetonitrile.
Ammonium bicarbonate.

24.2.10.2 Protocol

1. Connect a fused silica capillary of 10–12 cm length (250 μm i.d, 370 μm o.d.) to the 1/16 in. union containing the steel screen using 20–25 mm of PEEK tubing as a sleeve.
2. Connect with a Vespel ferrule to the fused silica capillary to the outer end of the slurry container.
3. Prepare a slurry of about 15–20 mg of graphitized carbon in 1.2 mL THF/IPA (90:10), sonicate in an ultrasonic bath for 4 min, and transfer to the slurry

container. Place the small magnet (7 × 2 mm) in the slurry container for stirring during packing.

4. Connect the slurry container to the compressed air driven pump with methanol as a pusher solvent and set the pressure to 300 bar at the start and after 3–5 min increase to 450 bar. The packing takes about 20–25 min and the packing in the fused silica capillary can be seen as the graphite particles are clearly visible through the fused silica column.
5. When the packing is complete, place the column in the ultrasonic bath where it is conditioned for 10 min, still maintaining a pressure of 450 bar.
6. Turn off the pressure and allow the column to depressurize.
7. Connect to a HPLC pump via a Valco-T split with a 50 cm × 50 μm fused silica capillary as a splitter and wash with 80% acetonitrile before connecting to the LC-MS. The LC is set to deliver a flow rate of 250 μL/min and the flow split in the T using a 50 cm × 50 μm of fused silica capillary as restrictor gives a flow through the 100 mm × 250 μm i.d column packed with 5 μm graphite particle to about 5–10 μL/min.

24.2.11 Connection of a Capillary LC-MS Column to MS

24.2.11.1 Materials and Equipment

A packed column, as described in 2.10, or, alternatively, a commercially available graphitized carbon column (e.g. 100 mm × 3.0 mm 5 μm Hypercarb column, (Thermo Scientific) Part Number 35005-103030).

A through bore 1/16 in. union (ZU1T, Vici AG).

Steel screen (0.5 or 1 μm pores, Vici AG).

Fused silica capillary 30–35 cm × 50 μm i.d. (Polymicro technologies, Phoenix AZ).

PEEK tubing 20–25 mm (0.127 mm i.d Upchurch Scientific, Oak Harbor, WA).

Ferrule, finger tight (F-336NX, Fttg NT PEEK Headless) Upchurch Scientific).

Binary pump (Agilent 1100 Agilent technologies, Palo Alto, CA).

Ion trap mass Spectrometer (LTQ, Thermo Electron Corp, San Jose CA) equipped with a standard electrospray module.

Autosampler (HTC-PAL, CTC Analytics, Zwingen, Switzerland) with a 2 μL sample loop.

24.2.11.2 Protocol

1. Exchange a through-bore union with a steel screen to the inlet of the packet sleeved capillary column
2. Place the PEEK tubing as a transfer line from the injector directly to the top of the steel screen in the union and fix with a finger tight ferrule. Use a seal tight ferrule with a fitting NT PEEK to connect the transfer line to the injector.
3. Connect the column outlet transfer line (30–40 cm × 50 μm of a fused silica capillary tubing) with a ferrule to the ion source.

4. Set the pump to deliver a flow of 250 μL/min and the flow split in a T using a 50 cm × 50 μm fused silica capillary as restrictor. The flow rate through the 10 cm × 250 μm i.d column packed with 5 μm graphite particles should be about 5–10 μL/min.

24.2.12 Tuning of LC-MS for Oligosaccharide Analysis

24.2.12.1 Materials and Equipments

Maltopentaose solution (10 μg/mL) in 10 mM ammonium bicarbonate (40% acetonitrile: 60% water).
Ion trap mass spectrometer with standard electrospray interface, as described above.
250 μL syringe.
Syringe pump.
PEEK tubing and connections from the syringe to the electrospray source.

24.2.12.2 Protocol

1. Fill the syringe with the maltopentaose solution.
2. Connect the syringe to the PEEK tubing and Insert the syringe in the pump.
3. Pump at 4–5 μL/min and perform the tuning on the $[M–H]^-$-ion of *m/z* 827 in negative ion mode (usually the autotuning function on the instrument is sufficient, but the temperature and the gas flow may have to be optimized manually).

24.2.13 Analysis of Oligosaccharides using LC-MS

24.2.13.1 Materials and Equipment

Released oligosaccharides in water as described above.
Tuned mass spectrometer as described above.
Connected graphitized carbon capillary column.
Solvent A; 10 mM ammonium bicarbonate, pH 8.0.
Solvent B; 10 mM ammonium bicarbonate in 80% acetonitrile, pH 8.0.
Wash solution;10 mM ammonium bicarbonate), pH 8.0.

24.2.13.2 Protocol

1. Connect the graphitized carbon capillary column to the injector and to the ion source and start the LC system setting the pump to deliver a flow of 0.250 mL/min.

Table 24.1 Gradient for graphitized carbon LC-MS

No.	Time (min)	Flow (ml/min)	Solvent A	Solvent B
1	0	0.250	100	0
2	7	0.250	95	5
3	46	0.250	70	30
4	46.1	0.250	20	80
5	54	0.250	20	80
6	54.1	0.250	100	0
7	68	0.250	100	0
8	68	0.250	100	0
9	68	0.250	100	0

2. Let the pressure of the LC system stabilize and let the column equilibrate with 100% of solvent A for 20–30 min.
3. Set the scan range (*m/z* 380–2000) and the gradient according to Table 24.1.
4. Inject the released oligosaccharides dissolved into MilliQ water and start the gradient.

Notes

1. Always make fresh solution of APS because the free radical produced by the unstable APS is necessary for the polymerization and atmospheric oxygen is a free radical scavenger of these radicals.
2. Do not add acrylamide until the solvent has cooled to 60°C oven because it autopolymerizes quickly.
3. The rolling tool should be rolled over gently to remove the air between the filter papers and between membrane and gel, otherwise the transfer will not be efficient.
4. DB71 stain stock solution can be stored in a sealed container for up to 3 months at room temperature. DB71 stain solution can be stored in a sealed container for up to 3 weeks at room temperature.
5. Always prepare fresh NaOH/$NaBH_4$ solution on the day of use.
6. If any solution has evaporated and condensed on the adhesive coversheet of microtiter plates during overnight incubation, do not return this solution to the well (results in sugar polymer contamination).

References

Ceroni, A., Maass, K., Geyer, H., Geyer, R., Dell, A., and Haslam, S.M. (2008). GlycoWorkbench: A tool for the computer-assisted annotation of mass spectra of glycans. J Proteome Res *7*, 1650–1659.

Gabius, H.-J. (2008). Glycans: Bioactive signals decoded by lectins. Biochem Soc Trans *036*, 1491–1496.

Holmén, J.M., Karlsson, N.G., Abdullah, L.H., Randell, S.H., Sheehan, J.K., Hansson, G.C., and Davis, C.W. (2004). Mucins and their O-Glycans from human bronchial epithelial cell cultures. Am J Physiol Lung Cell Mol Physiol *287*, L824–834.

Karlsson, N.G., Wilson, N.L., Wirth, H.J., Dawes, P., Joshi, H., and Packer, N.H. (2004). Negative ion graphitised carbon nano-liquid chromatography/mass spectrometry increases sensitivity for glycoprotein oligosaccharide analysis. Rapid Commun Mass Spectrom *18*, 2282–2292.

Larsson, J.M., Karlsson, H., Sjovall, H., and Hansson, G.C. (2009). A complex, but uniform O-glycosylation of the human MUC2 mucin from colonic biopsies analyzed by nanoLC/MSn. Glycobiology *19*, 756–766.

Lindén, S.K., Sutton, P., Karlsson, N.G., Korolik, V., and McGuckin, M.A. (2008). Mucins in the mucosal barrier to infection. Mucosal Immunol *1*, 183–197.

Mechref, Y., and Novotny, M.V. (2002). Structural investigations of glycoconjugates at high sensitivity. Chem Rev *102*, 321–369.

North, S.J., Hitchen, P.G., Haslam, S.M., and Dell, A. (2009). Mass spectrometry in the analysis of N-linked and O-linked glycans. Curr Opin Struct Biol *19*, 498–506.

Reitsma, S., Slaaf, D.W., Vink, H., van Zandvoort, M.A., and oude Egbrink, M.G. (2007). The endothelial glycocalyx: Composition, functions, and visualization. Pflugers Arch *454*, 345–359.

Robinson, L.J., Karlsson, N.G., Weiss, A.S., and Packer, N.H. (2003). Proteomic analysis of the genetic premature aging disease Hutchinson Gilford progeria syndrome reveals differential protein expression and glycosylation. J Proteome Res *2*, 556–557.

Schulz, B.L., Packer, N.H., and Karlsson, N.G. (2002). Small-scale analysis of O-linked oligosaccharides from glycoproteins and mucins separated by gel electrophoresis. Anal Chem *74*, 6088–6097.

Schulz, B.L., Sloane, A.J., Robinson, L.J., Prasad, S.S., Lindner, R.A., Robinson, M., Bye, P.T., Nielson, D.W., Harry, J.L., Packer, N.H., et al. (2007). Glycosylation of sputum mucins is altered in cystic fibrosis patients. Glycobiology *17*, 698–712.

Schulz, B.L., Sloane, A.J., Robinson, L.J., Sebastian, L.T., Glanville, A.R., Song, Y., Verkman, A.S., Harry, J.L., Packer, N.H., and Karlsson, N.G. (2005). Mucin glycosylation changes in cystic fibrosis lung disease are not manifest in submucosal gland secretions. Biochem J *387*, 911–919.

Wilson, N.L., Robinson, L.J., Donnet, A., Bovetto, L., Packer, N.H., and Karlsson, N.G. (2008). Glycoproteomics of milk: Differences in sugar epitopes on human and bovine milk fat globule membranes. J Proteome Res *7*, 3687–3696.

Chapter 25
Multi-lectin Affinity Chromatography (M-LAC) Combined with Abundant Protein Depletion for Analysis of Human Plasma in Clinical Proteomics Applications

Marina Hincapie, Tatiana Plavina, and William S. Hancock

Abstract In this chapter we describe a robust sample preparation method for proteomic analysis of human plasma and discovery of candidate protein biomarkers. The method consists of depletion of the most abundant plasma proteins and multi-lectin affinity chromatography (M-LAC) to fractionate plasma into flow-through (unbound) and bound fractions, followed by nano-LC-MS/MS analysis of the digested proteins and label-free comparative quantitation. The performance of the method is monitored by several quality control checkpoints to assure reproducibility of the proteomic analysis. This approach reduces the complexity of plasma samples and significantly improves the overall dynamic range of protein detection, enabling detection of tissue leakage proteins present in plasma in ng/mL concentrations. In addition, the lectin-based fractionation of depleted plasma targets the plasma glycoproteome, permitting the identification of glycoproteins with disease-related differences in glycoform populations. The proteomics workflow was applied to the analysis of plasma from diseased patients affected with psoriasis and healthy donors. As an example of the effectiveness of the method we will present data from this biomarker study.

Keywords Clinical proteomics · Glycoproteins · Lectin affinity chromatography · Biomarker discovery · Sample preparation

25.1 Introduction

The challenges and limitations of using human plasma or serum for the identification of protein biomarkers have been highlighted in numerous publications (Anderson and Anderson, 2002). Despite the difficulties associated with the complexity of these fluids the quest for circulating markers with diagnostic or prognostic value to

---

M. Hincapie (✉)
The Barnett Institute of Chemical and Biological Sciences, Northeastern University, Boston, MA 02115, USA
e-mail: m.hincapie@neu.edu

A.R. Ivanov, A.V. Lazarev (eds.), *Sample Preparation in Biological Mass Spectrometry*, DOI 10.1007/978-94-007-0828-0_25,

disease states is still being actively pursued. This is due to the inherent advantages of serum or plasma including; availability, ease of collection and utility in clinical studies. The extensive dynamic range of protein concentrations, heterogeneity of protein populations due to post-translational modifications and the complexity of subsequent tryptic digests all limit the power of the LC/MS proteomic analysis in both sensitivity and reproducibility (Jacobs et al., 2005; Yates et al., 2006). To increase the sensitivity of protein detection numerous methods based on selective subtraction of high abundance proteins or enrichment of sub-proteomes have been applied to simplify the plasma proteome (Echan et al., 2005; Fang and Zhang, 2008; Liu et al., 2005; Martosella et al., 2005; Yocum et al., 2005).

As greater than 50% of circulatory proteins are glycosylated, the natural property of lectins as glycan-binding proteins has made them an obvious choice for studying the plasma or serum glycoproteome. Immobilized lectins, either as a single lectin (Dwek et al., 2001), series of lectins (Serial Lectin Affinity Chromatography (SLAC)) (Cummings and Kornfeld, 1982; Qiu and Regnier, 2005), or multi-lectin combinations (Yang and Hancock, 2004), have been used to reduce the complexity of plasma or serum samples prior to analysis by LC-MS and have been shown to be effective for identifying differences in glycoprotein abundance. Although lectins can display overlapping specificities their binding properties are sufficiently different to target specific populations of glycoprotein motifs and detect subtle changes in protein glycosylation. Because a wide variety of lectins are commercially available, the use of lectins in glycoproteomic research has spread. Additionally, changes in protein glycosylation have been associated with a variety of disease conditions such as cancer (Dennis et al., 1987; Kreunin et al., 2007) and immune disorders (Carlsson et al., 2008; Purcell et al., 2008). Therefore, while much of proteomics research has been focused on protein quantitative analysis there is a growing recognition of the significance of glycosylation (particularly in the case of circulating proteins) and alterations to the glycan structure as potential disease biomarker targets (Block et al., 2005; Dai et al., 2006).

Our laboratory has developed the Multi-Lectin Affinity Chromatography (M-LAC) technique to capture a significant (> 50%) portion of the plasma glycoproteome in a single step. The technology has been applied to protein biomarker discovery studies to evaluate changes in protein concentrations (Dayarathna et al., 2008; Plavina et al., 2008) and to detect disease-induced changes in protein glycosylation (Yang and Hancock, 2006). The M-LAC column is made up of a mixture of three immobilized lectins, either on an agarose or on a polymeric support and contains the following lectins: concavanalin A (ConA) which has selectivity for branched-mannosidic structures, a common glycan motif in hybrid and biantennary complex type *N*-Glycans (Becker et al., 1975); wheat germ agglutinin (WGA) which recognizes *N*-acetyl-glucosamine(GlcNAc) and also has affinity for sialic acid (Neu5Ac) containing structures (Bakry et al., 1991), and jacalin (JAC) lectin which has specificity toward galactosyl-1-3-*N*-Acetylgalactosamine (GalNAc) and for *O*-linked glycoproteins (Saulsbury, 1997). The ability of M-LAC to enrich glycoproteins is attributed to the combination of specificities of the three lectins, which were selected to target the major glycosylation motifs of plasma glycoproteins;

resulting in near comprehensive capture of the plasma glycoproteome. In a recent publication we reported up to 10 fold enhancement in binding affinities with the multiple lectin format compared to the corresponding individual lectins (ConA, WGA and JAC) (Ralin, 2008).

To further enhance the performance of the M-LAC technology with respect to reproducibility of the proteomic analysis, sample throughput and to minimize sample manipulation, a new generation of M-LAC has been developed. The same lectins as used in the original agarose-based M-LAC (ConA, WGA and JAC) have been covalently attached to a polystyrene-divinylbenzene support matrix coated with a cross-linked polyhydroxylated polymer (POROS). Due to the favorable pressure/flow characteristics of the media the column can be operated using high pressure liquid chromatography (HPLC) instrumentation (Kullolli et al., 2008) for automation of the overall sample processing prior to mass spectrometry (MS) analysis (Kullolli et al., 2010).

In this chapter we present the overall sample preparation strategy for analysis of human plasma in the context of biomarker discovery. Here we describe details of the sample preparation at the protein and peptide levels prior to the analysis by LC-ESI-MS/MS. We show the effectiveness of combining depletion of the most abundant proteins in plasma (albumin and IgG) with glycoprotein enrichment in identifying plasma proteins of medium (low μg/mL to medium/low ng/mL) abundance. Furthermore, we illustrate the power of this approach for clinical proteomic applications by presenting the results of a study in which plasma from psoriasis affected patients and healthy controls were profiled and several candidate disease protein biomarkers identified.

25.2 Experimental Methods

25.2.1 Collection and Storage of Plasma Samples

Sample collection, handling, storage and downstream processing are known to greatly influence the results of proteomic analysis and are crucial parameters for reliable analysis of biological samples. Samples should be collected in a uniform manner strictly following established standard operation procedures. For proteomic analysis plasma is commonly chosen over serum due to its shorter processing time between venipuncture and freezing, resulting in better sample integrity. In addition, plasma avoids the changes introduced by activation of the coagulation cascade and platelet-derived factors. To obtain plasma, blood is collected into tubes containing an anticoagulant, EDTA or citrate being the preferred anticoagulants for proteomics analysis (Omenn GS et al., 2005). Ideally, samples should be aliquoted upon sample collection, prior to freezing, to avoid unnecessary freeze/thaw cycles and stored at –70°C or below.

For the development of the proteomics platform and its application to the biomarker study described in this chapter blood was collected by venipuncture into Becton Dickinson (Franklin Lakes, NJ) Vacutainer citrate tubes and immediately centrifuged at 3,000×g for 10 min to separate blood cells from the plasma fluid. The separated plasma was aliquoted, frozen within 1 h from the time of collection and stored at –70°C.

25.2.2 Sample Preparation for Proteomic Analysis

Samples are quickly thawed by immersing the tube in cold water and centrifuged (3,000×g) to remove any particulates or clots and the supernatant is carefully removed. Each plasma sample is spiked with an internal standard for normalization during protein quantitation.

In this study bovine fetuin was added to plasma samples at a concentration of 250 μg/mL.

25.2.2.1 Depletion of Albumin and Immunoglobulins G (IgGs) Prior to M-LAC Fractionation

The following protocol is used for depletion of albumin and IgG, followed by fractionation on agarose-immobilized M-LAC. For depletion of albumin and IgG, individual POROS Protein G (0.2 mL) and POROS anti-HSA (2.0 mL) cartridges (Life Technologies, Carlsbad, CA) are connected with a short piece of PEEK tubing and used as per the manufacturer's instructions, with the exception of modification to the buffer system to maintain a uniform composition of buffers between the depletion of abundant proteins and M-LAC fractionation. To allow direct loading of depleted plasma onto the M-LAC column the same binding buffer (25 mM Tris, 0.5 M NaCl, 1 mM $MnCl_2$, 1 mM $CaCl_2$, 0.02% sodium azide, pH 7.4) is used for both affinity steps. Chromatography can be performed on any standard HPLC system but the automated multi-column method described below requires an HPLC with column switching capabilities. Prior to sample loading the cartridges are equilibrated with 5 column volumes (CV) of binding buffer at 2.5 mL/min. Plasma spiked with bovine fetuin (50 μL) is diluted 1:10 with binding buffer, filtered (0.45 μm spin filter) and injected onto the depletion cartridges at a flow rate of 1.0 mL/min. The flow rate is then increased to 2.5 mL/min for the remainder of the fractionation. The cartridges are washed with 5 CV of binding buffer and the first 4 mL of the flow-through, which contains the plasma depleted of albumin and IgG is collected. The bound albumin and IgGs are eluted with 7 CV of elution buffer (100 mM glycine, pH 2.5) and the eluate is collected. The cartridges are then regenerated with 4 CV of 0.25 M Tris, 1 M NaCl, pH 7.5 and re-equilibrated with 5 CV of binding buffer prior to processing of the next sample.

25.2.2.2 Preparation of Agarose M-LAC Columns for Operation Under Gravity Flow

M-LAC columns are prepared by packing agarose-immobilized lectins into an empty 5 mL plastic disposable Zeba column (Thermo Fisher, Waltham, MA). The empty column is clamped onto a laboratory ring stand and filled halfway with

M-LAC binding buffer. The three individual agarose-bound lectins are mixed in a 1:1:1 (v/v/v) ratio by adding 0.6 mL of a 50% slurry of each lectin (Vector Laboratories, Burlingame, CA) to an empty polypropylene tube and then are transferred to the prepared Zeba column. The buffer is allowed to pass through the column to pack the media. The resulting M-LAC column which is approximately 1 mL in volume, is immediately equilibrated with 5 mL of M-LAC binding buffer, capped at the top and the bottom, and stored in an upright position at 4°C for up to a week.

Preparation of polystyrene M-LAC columns for operation by HPLC: Lectins are reacted with the POROS-20 AL support, which is coated with aldehyde functional groups. The solid support is prepared by resuspending 500 mg of the POROS- 20 AL in 2 mL of 100 mM HEPES buffer, pH 8.2 to yield about 1.5 mL of wet beads. The beads are washed 3 times with 5 mL of HEPES buffer using centrifugation (4,000×*g* for 5 min) to sediment the beads; each time the supernatant is removed and discarded. After the last wash, 1.5 mL of 100 mM HEPES buffer, pH 8.2 is added to produce 3.0 mL of a 50% slurry. Each lectin solution is prepared in a 15 mL polypropylene tube as follows: 7 mg of ConA are dissolved in 0.5 mL of 100 mM HEPES, pH 8.2, 0.75 M sodium sulfate, 0.2 M methyl-alpha-D-mannopyranoside, 0.2 methyl-alpha-D-glucopyranoside, 1 mM $CaCl_2$ and 1 mM $MnCl_2$; 7 mg of WGA are dissolved in 0.5 mL of 100 mM HEPES, pH 8.2, 1 M sodium sulfate, 0.5 M *N*-acetyl glucosamine, and 7 mg of JAC are dissolved in 0.5 mL of 100 mM HEPES, pH 8.2, 1 M sodium sulfate, 0.1 M melibiose. One mL of the beads (50% slurry) is transferred to each of the three lectin solutions. Sodium cyanoborohydride is added to each tube to a final concentration of 20 mM, and the coupling reaction is allowed to proceed for 6 h at room temperature (RT) with constant rocking. At the end of the incubation the reaction is quenched with 4 mL of 0.5 M Tris-HCl, pH 7.4 and a second aliquot of sodium cyanoborohydride (20 mM final concentration) is added; incubation is continued at RT for an additional 2 h to block any remaining reactive sites. The beads are collected by centrifugation for 5 min at 4,000×*g*; the supernatant is removed and saved for protein concentration measurement. To determine the amount of protein immobilized to the POROS beads a protein concentration assay is performed on the pre-conjugated and post-conjugated lectin solutions. The difference in concentrations is the amount of protein that was conjugated to the beads. The beads are washed 4 times with M-LAC binding buffer (25 mM tris, 0.5 M sodium chloride, 1 mM $CaCl_2$, 1 mM $MnCl_2$ and 0.02% sodium azide, pH 7.4). Equal volumes of the conjugated lectins are combined into a single tube (1:1:1). The beads are then packed under high pressure into a PEEK column (4.6 mm × 50 mm) using a Self Pack device (Life Technologies, Carlsbad, CA) following the manufacturer's instructions.

25.2.2.3 Fractionation of Depleted Plasma Using M-LAC

Depleted plasma can be fractionated by M-LAC using either agarose-immobilized lectins run under gravity flow or on POROS-immobilized lectins for automated HPLC fractionation. The two methods are described below.

Manual fractionation of depleted plasma using M-LAC under gravity flow (Method I): Plasma depleted of albumin and IgG is further fractionated using M-LAC under gravity flow. The column is clamped on a ring stand; the bottom and top caps are removed, and 5 mL of binding buffer are passed through the column. The bottom of the column is capped and the depleted plasma (3.5 mL) is added to the M-LAC column. The top of the column is capped and the sample is incubated with M-LAC at room temperature for 30 min with continuous rocking. At the end of the incubation the column is re-mounted on the ring stand, the top and bottom are uncapped and the flow-through (unbound) fraction is collected. The lectins in the M-LAC column are further washed with 4 mL of M-LAC binding buffer; this wash is combined with the flow through fraction. An additional 5 mL of binding buffer is run through the column and the resulting wash is discarded. Captured glycoproteins are eluted with 4 mL of elution buffer. Two different methods of protein desorption have been evaluated and shown to be comparable (Kullolli et al., 2008). In one approach, glycoproteins are desorbed by competitive saccharide elution using a buffer that contains a cocktail of sugar inhibitors (25 mM Tris pH 7.4, 0.5 M NaCl, 0.2 M methyl-α-mannopyranoside, 0.2 M methyl-D-glucopyrannoside, 0.5 M *N*-acetyl-glucosamine, 0.1 M melibiose or 0.8 M galactose), while the second approach utilizes mild acidic conditions (100 mM acetic acid). When the fractionation is to be performed using the automated M-LAC workflow as described below, acetic acid is preferred due to its lower viscosity compared to the saccharide cocktail.

The two M-LAC fractions, unbound and bound, are solvent exchanged with 25 mM Tris, pH 7.4, and concentrated to approximately 50 μL using a 4 mL Amicon 5 kDa molecular weight cut off filter (Millipore, Bedford, MA). Each filter is washed with an additional 50 μL of buffer and this wash is combined with the concentrated fraction. The volume of each sample is adjusted to 150 μL with 25 mM Tris, pH 7.4. A 50 μL aliquot is denatured by diluting it 1:5 with trypsin digestion buffer (100 mM ammonium bicarbonate, 6 M guanidine, pH 8.0) for subsequent trypsin digestion (see below). A second aliquot of 50 μL is retained for QC purposes, and the remaining sample is frozen and stored at –70°C.

Fractionation of plasma using an automated HPLC platform (Method II): The method is a 3-column multidimensional chromatographic separation that is optimized for fractionation of 50 μL of human plasma. The configuration of the multidimensional fractionation is shown in Fig. 25.1. The depletion of albumin and IgG is coupled with M-LAC glycoprotein fractionation and on-line desalting using a reversed-phase (RP) "trap" column. The depletion columns are placed on valve position 1 (V1); 0.2 mL Protein G column (PG) and 2.0 mL anti-HSA column are connected with PEEK tubing; the PG column is placed in front of the albumin column. The M-LAC column (4.6 mm × 50 mm), is in valve position 2 (V2), and the POROS 50 R1 (4.6 mm × 50 mm) reversed-phased (RP) trap is positioned on valve 3 (V3). The particle size and characteristics of the POROS media utilized in all three columns allows sample fractionation at high flow rates with low backpressures, which is ideal for high throughput sample preparation. In addition, the POROS 50 R1 reversed-phase media has good binding and elution properties for proteins

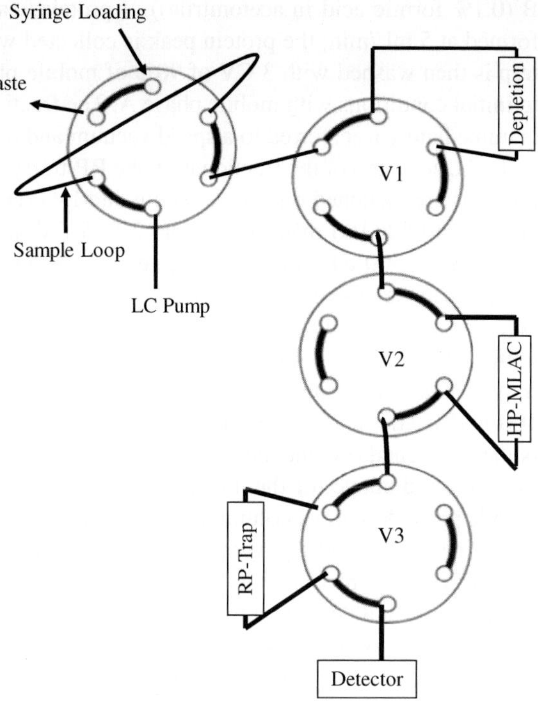

Fig. 25.1 Schematic representation of the on-line 3D-dimensional HPLC system. The sample fractionation workflow consists of a serial configuration of depletion columns containing anti-albumin antibody and protein A (V1) with in-line multi-lectin affinity chromatography (M-LAC) (V2) and in-line desalting using reversed-phase (V3). The flowpath is shown in the injection mode

resulting in high protein recoveries. Chromatography is monitored at 280 nm and is performed at room temperature. In our laboratory the method has been performed using a BioCAD Vision (Life Technologies, Carlsbad, CA) or a Prominence 2D HPLC (Shimadzu Scientific Instruments, Columbia MD). The columns operational procedure is executed as follows: prior to sample application the depletion columns and M-LAC are on-line and are equilibrated with 15 CV of binding buffer (25 mM Tris, 0.5 M NaCl, 1 mM $MnCl_2$, 1 mM $CaCl_2$ pH 7.4, 0.02% sodium azide). At the end of the equilibration step the affinity (depletion + M-LAC) columns are taken off line and the RP trap is placed on-line and is equilibrated with 10 CV of mobile phase A: (0.1% formic acid in 5% acetonitrile, 95% water) at 2.5 mL/min. Plasma (50 μL) is diluted 1:10 with binding buffer and injected into the sample loop in the load position; the sample loop is then switched to the inject position with all three affinity columns and the RP trap on-line. To allow sufficient time for protein binding to the respective affinity columns the loading is performed at a flow rate of 0.5 mL/min for 30 min. The flow rate is then increased to 2.5 mL/min. and the columns are washed with 4 CV of binding buffer. As sample is loaded any unretained proteins that flow through both the depletion and the M-LAC columns (unbound fraction) are concentrated on the RP trap. Following the washing step both depletion and M-LAC columns are taken-off line, and the RP trap is left on-line. The desalting (5 CV of buffer A), and step elution (5 CV of 70% mobile phase

B (0.1% formic acid in acetonitrile)) of proteins captured by the RP trap are performed at 5 mL/min; the protein peak is collected with a fraction collector. The RP trap is then washed with 3 CV of 90% of mobile phase B and then re-equilibrated to initial conditions with mobile phase A. The fraction containing unbound proteins is immediately transferred to a speed vacuum and taken to dryness.

In the next step of the fractionation the RP-trap is taken off-line and the depletion columns are switched on-line; albumin and IgG retained by the depletion columns are eluted with 100 mM glycine pH 2.5 (5 CV at 2.5 mL/min), and the proteins collected. A 500 μL aliquot is retained for QC purposes; the remaining abundant protein can be discarded or stored at –70°C for further analysis. The depletion columns are immediately neutralized and regenerated with 5 CV of neutralization buffer (0.25 M Tris, 1 M NaCl, pH 7.5) at 2.5 mL/min. Next, the M-LAC and RP trap are placed back on-line, and glycoproteins bound to the M-LAC are eluted with 100 mM acetic acid (5 CV at 5 mL/min) onto the RP trap. The M-LAC column is then switched off-line and glycoproteins are eluted from the RP trap and dried, as performed above for the unbound fraction. The RP column is equilibrated with mobile phase A and the column is switched off-line. The M-LAC is placed back on-line to neutralize and regenerate the column. The system is then ready for another cycle of plasma fractionation. The dried, unbound and bound protein fractions are reconstituted and denatured in 50 μL trypsin digestion buffer (100 mM ammonium bicarbonate, 6 M guanidine, pH 8.0). Ten microliters are removed for protein assay, the unbound fraction is diluted 1:5 and the bound fraction 1:10 with water before measuring the protein concentration. A 20 μL aliquot from each sample is transferred to another eppendorf tube and stored at –70°C. The remaining 20 μL are digested with trypsin as described below.

25.2.3 Protein Assay

The total protein concentration of all the fractions resulting from depletion and M-LAC fractionation is measured using the Coomassie Blue Plus assay (Pierce, Rockland, IL). Total protein recoveries are calculated and used as a QC checkpoint to evaluate the efficiency of the combined depletion and M-LAC steps.

25.2.4 In-Solution Trypsin Digestion of Protein Fractions

Denatured proteins are reduced with 5 mM dithiothreitol (DTT) for 30 min at 60°C and alkylated with 15 mM iodoacetamide for 30 min at room temperature in the dark. The reaction is then quenched by addition of second aliquot of 5 mM DTT. Samples are diluted 5 fold with 100 mM ammonium bicarbonate pH 8.0 to bring the guanidine-HCl concentration down to 1.2 M. Proteins are proteolysed with sequencing-grade trypsin (Promega, Madison, WI) at a 1:40 w/w ratio and samples are incubated for 16 h at 37°C. Digestion is terminated by acidification with 1% formic acid.

25.2.5 Peptide Extraction from Complex Protein Digest

Peptides resulting from protein digestion are desalted and separated from any non-digested or partially digested proteins using RP-HPLC. The desalting and peptide extraction is performed at 1.5 mL/min and is monitored at both 280 and 214 nm. The digest is injected onto a Discovery BIO C18 cartridge (3 μm, 4.6 mm × 30 mm; Sigma-Aldrich, St Louis, MO). The removal of salts is accomplished by washing the cartridge with 5CV of 0.1% formic acid. Peptides are step eluted with 30% acetonitrile, more hydrophobic peptides, large protein fragments or non-digested protein are step eluted with 90% acetonitrile. Peptides eluted with 30% organic solvent are concentrated under vacuum to a volume of approximately 20–30 μL using speed vacuum centrifugation and the volume of the samples is adjusted to 50 μL with 0.1% formic acid. Twenty-five μL of the solution containing the peptides is aliquoted into two different autosampler vials. One vial is stored at –70°C and the duplicate vial is used for nano LC-ESI-MS/MS peptide separation.

As a QC checkpoint the peak area at 214 nm of both 30% and 90% step elutions is integrated and the ratio of the peak areas is used to quantitatively assess the digestion efficiency. Using this approach approximately 75% of a complex protein digest results in good quality peptides for LC analysis.

25.2.6 Peptide Analysis by Nano-LC-ESI-MS/MS

The complex peptide mixture is separated using a 2D nano-HPLC system. In our laboratory the chromatographic peptide separation is performed using an Eksigent LC system (Dublin, CA) on a reversed-phase C18 capillary (175 mm × 0.075 mm i.d.) column, in-house packed with Magic C18 media, particle size 5 μm, pore size 300 Å (Michrom Bioresources, Auburn, CA) and directly connected to an electrospray (ESI) ion trap mass spectrometer. For the MS analysis of peptides by ESI, we use a linear ion trap (LTQ) instrument (Thermo Fisher, Waltham, MA). The electrospray conditions are: temperature of the ion transfer tube, 245°C; spray voltage, 2.0 kV; normalized collision energy, 35%. The mass spectrometer is operated in the data dependant mode to switch automatically between MS and MS/MS using MS acquisition software (Xcalibur 2.0, Thermo Fisher, Waltham, MA). Each MS full scan (mass range of *m/z* 400–1,600) is followed by MS/MS scans of the 7 most intense peaks.

Peptides (5 μL/injection) are first loaded at a higher flow rate (1 μL/min) and concentrated on a peptide Captrap cartridge (Michrom Bioresources, Auburn, CA). The peptides are then eluted onto the capillary column and chromatographed at 250 nL/min with a linear gradient from 2% solvent B to 40% solvent B over 120 min, then from 40% to 90% solvent B over 15 min and then isocratically at 90% solvent B for 15 min. The composition of solvent A is 0.1% (v/v) of formic acid in HPLC grade water and that of solvent B 0.1% (v/v) formic acid in acetonitrile.

25.2.7 Database Search and Protein Identification

Proteins are identified by searching the raw MS/MS spectra against the latest version of the human protein database Swiss-Prot using the search engine SEQUEST (Thermo Fisher, Waltham, MA). The database consists of normal and reversed protein sequences to facilitate estimation of the false positive rate (Elias and Gygi, 2007). Trypsin is specified as the digestion enzyme with up to two missed cleavages and carboxyamidomethylation is designated as a fixed modification of cysteine. Only peptides resulting from tryptic cleavages are searched. The SEQUEST results are filtered based on correlation score (X_{corr}) versus charge state, with values selected to obtain highly confident peptide identifications: Xcorr 1.9, 2.2, 3.75 for singly, doubly and triply charged peptide ions, respectively; all with dCn $\geq$ 0.1. Further statistical validation of MS/MS spectra to peptide sequence assignments is performed using the Peptide Prophet program (Keller et al., 2002); we use a cutoff at 0.95 peptide probability to eliminate low confidence protein identifications.

Relative abundance of peptides and corresponding proteins is assessed using label free quantitation that uses the spectral count method as a semi-quantitative approach to first compare differences in protein abundance and then select potential candidate proteins for peak area integration of the extracted ion chromatogram (EIC)-based quantitation.

For QC purposes, three unique nonhuman peptides of the spiked internal standard (bovine fetuin) are selected. The retention time and the peak area of the EIC for these is evaluated to assess the performance of the nano-HPLC separation and ESI-MS/MS, respectively.

25.3 Results and Discussion

25.3.1 Overview of Method Workflow

All the steps of sample preparation leading up to the analysis by LCMS, that is, protein separation, trypsin digestion and peptide separation could dramatically affect the quality of the data. The results of any clinical study can be easily confounded if changes unrelated to disease pathogenesis are introduced due to irreproducibility of any of the steps in the workflow. To reduce the overall variability in biomarker studies our laboratory has developed a robust and reproducible proteomics platform, the protocols for each step of the process are described in the Experimental Section and shown in Fig. 25.2. The development of the method and performance of all seven steps of the workflow including the QC metrics are presented in detail in the published work of Plavina et al. (2007). The workflow consists of the following steps: (1) spiking plasma samples with an internal standard (bovine fetuin is used in this study); (2) removal of albumin and IgG from plasma using HPLC; (3) M-LAC fractionation of the depleted plasma either manually (gravity flow) or automated (HPLC); (4) trypsin digestion of the unbound and bound fractions; (5) off-line sample clean up to remove salts/reagents from the trypsin reaction and to

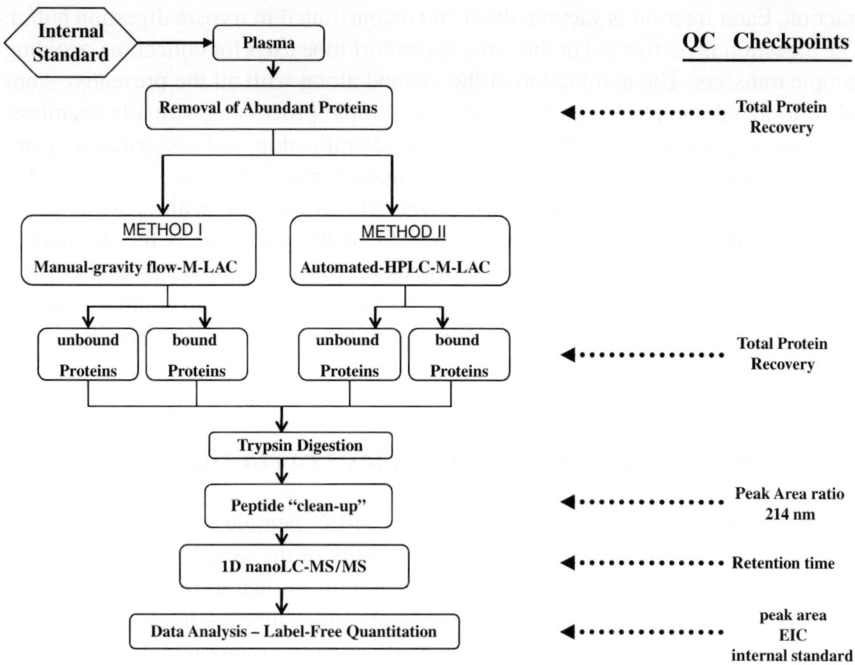

Fig. 25.2 Workflow for plasma analysis. The various steps for sample preparation at the protein and peptide level and analysis by LC-ESI-MS/MS are illustrated in the flowchart

extract the tryptic peptides from the complex digest; (6) nano-LC-ESI MS/MS analysis of peptides in a data-dependent mode; (7) label-free relative peptide and protein quantitation. This method has become a standard procedure in our laboratory for biomarker discovery studies.

25.3.2 An Automated On-Line 3D HPLC System for Protein Fractionation

Sequential, multi-step, manual protein fractionation is often less effective than a fully automated, hands-off process. One significant drawback of manual processes is that effluent from one column often needs to be desalted and/or concentrated before applying it to the next column leading to extensive sample manipulation, losses and less reproducible workflows. To circumvent these issues we developed an on-line-3D separation to automatically perform abundant protein depletion, sample enrichment by M-LAC and desalting/concentration by RP prior to trypsin digestion (Kullolli et al., 2010). The chromatographic procedure is accomplished with software-controlled column switching valves allowing for multi-dimensional chromatography. The fractionation of plasma results in only 2 fractions per sample for downstream LCMS/MS analysis: a flow-through (unbound) and a retained (bound) fraction. Each fraction is vacuum dried and reconstituted in trypsin digestion buffer. The digestion is performed in the same eppendorf tube used for collection, avoiding sample transfers. The automation of the method along with all the preventive steps taken to avoid sample losses has made our sample processing virtually seamless, resulting in good reproducibility of protein identification and comparative quantitation. Compared to the first generation agarose-based M-LAC, the automated platform gives an improved overall recovery (up from 75% with agarose media to >90% with POROS media). A key benefit of this platform is the throughput capabilities for sample processing. Fractionation of plasma or serum and column regeneration is accomplished in less than one hour per sample. In addition, analyzing only two fractions per sample permits replicate analysis and larger samples sets to be included in biomarker studies.

25.3.3 Peptide Extraction from Complex Protein Digest

The reproducibility and robustness of the nano-LC peptide separation and ESI-MS/MS analysis is greatly dependent on the quality of the samples introduced onto the capillary column. Incomplete tryptic digestion is known to frequently occur in complex protein mixtures and some proteins are inherently resistant to proteolysis (Klammer and MacCoss, 2006), negatively affecting the reproducibility of peptide separation and sequencing. Furthermore, introduction of non-digested or partially digested proteins could lead to irreversible protein binding or incomplete protein elution, which often results in sample build up and high column back pressure, consequently reducing the lifetime of the capillary column and separation efficiency. In order to address these issues we implemented a key step in the sample preparation prior to proteomic analysis by MS. It utilizes RP-HPLC to remove salts from the trypsin digestion mixture and to separate fully digested peptides from very hydrophobic peptides and any non-digested or partially digested proteins. This peptide extraction step dramatically improves the longevity of the capillary column; in our laboratory we routinely perform up to 200 runs on each capillary column without deterioration of column performance. This is a crucial consideration for large scale biomarker studies, where the samples must be analyzed under the same analytical and column conditions.

25.3.4 Assessment of Sample Preparation and Performance at the LC-ESI-MS/MS Level

The internal standard spiked into plasma prior to sample preparation is a good indicator of the analytical variability of the proteomic platform. As mentioned earlier (Experimental Section 25.2.7), we assess the reproducibility of the analysis by selecting three unique nonhuman peptides of the spiked internal standard (bovine fetuin) that are consistently detected by the mass spectrometer. In addition, peptides of different intensities are also used to evaluate the performance of the LC-ESI-MS/MS step. In our studies, the coefficient of variation (CVs) of peak areas of EIC used for peptide quantitation ranged from 1% for the high intensity peptide ions to 26% for the low intensity peptide ions. High variability for peptide quantitation of low intensity peptides is likely due to low signal-to-noise ratio as well as to interferences of higher abundance peptides. The observed variability is in good agreement with the results of a comprehensive study of label-free quantitation applied to plasma and cell extracts in which the reproducibility of relative quantitation for more than 90% of peptides was within 20% (Wang et al., 2006).

25.3.5 Depletion of Two Abundant Proteins in Combination with M-LAC Fractionation Improves the Dynamic Range of Protein Identifications

We have improved the dynamic range and number of proteins identified by combining depletion of abundant proteins with M-LAC, compared to fractionation by M-LAC alone. In a study where plasma depleted of albumin and IgG was further fractionated using M-LAC under gravity flow, the number of identified proteins (combined unbound and bound fraction, average of three runs) increased significantly: from 120 for M-LAC alone to 191 for the combined method (Plavina et al., 2007). A subset of these proteins is presented in Table 25.1. The reference concentrations of some of the identified proteins in serum or plasma were obtained from a literature search, for example, calgranulin A (8.4 ng/mL), neutrophil defensin 3 (42 ng/mL) and cartilage glycoprotein-39 (28 ng/mL) (Polanski and Anderson, 2006). Proteins of intermediate abundance, present in plasma at mid ng/mL to low μg/mL concentrations, such as L-selectin (reported 600 and 17 ng/mL in two different studies) and C-reactive protein (2 μg/mL) were consistently detected with 4–8 unique peptides, whereas the same proteins were detected with at most 2–3 unique peptides or not at all when only M-LAC was used during the sample preparation step. The results indicate that the method allows confident identification of some proteins present in plasma at low ng/mL concentrations. It is important to note that the selected proteins listed in Table 25.1 were either identified only when albumin (non-glycosylated) and IgG (glycosylated) were depleted prior to M-LAC, or identified with higher confidence if compared to using M-LAC alone.

Our earlier publications discussed M-LAC primarily as a glycoprotein enrichment technique to evaluate changes in glycoprotein abundance and to examine changes in protein glycosylation. Considering the improved depth in the analysis of the unbound fraction M-LAC is clearly a valuable protein fractionation tool if combined with depletion of abundant proteins. Thus the removal of albumin and IgGs coupled with M-LAC allows a more comprehensive analysis of the plasma proteome and good enrichment of the plasma glycoproteome.

It is also worthwhile to point out that M-LAC can be used in combination with other depletion methods. Several levels of depletion can be achieved with other commercially available multi antibody depletion columns that target the 6, 7, 12, 14

Table 25.1 List of selected medium and lower level proteins identified in plasma with M-LAC only, compared to depletion plus M-LAC

M-LAC fraction	Protein name	SwissProt accession no.	Reported normal plasma or serum concentration, ng/mL, (reference)	Mean # of unique peptides, $n = 2$[a]	
				M-LAC only	depletion + M-LAC
Bound (glycosylated)	Ankyrin 3	Q5VXD5		0.0	2.0
	ADAM 9	Q13443		0.0	1.5
	ADAMTS 18	Q8TE60		0.0	1.5
	Breast cancer type 2 susceptibility protein	P51587		0.0	2.5
	T-cadherin	P55290		0.0	2.0
	CD5 antigen-like	O43866		2.0	7.5
	CD44	P16070	223[b]	0.0	2.0
	Cholinesterase	P06276		3.5	10.5
	Cartilage glycoprotein-39	P36222	28[b]	0.0	1.5
	Colon carcinoma kinase-4	Q13308		0.0	2.0
	Inositol 1, 4, 5-triphosphate receptor type 2	Q14571		0.5	2.0
	Laminin B2	P11047		0.0	2.0
	L-selectin	P14151	601[b], 17[c]	2.0	4.5
	Mitogen-activated protein kinase 11	Q15759		0.0	2.0
	Monocyte differentiation antigen CD14	P08571		0.0	2.5
	Phosphatidylinositol-glycan-specific phospholipase D1	P80108		0.0	3.0
	Serine/threonine protein kinase Nek-3	P51956		0.0	1.0
	T-lymphocyte surface antigen Ly-9	Q9HBG7		0.0	2.5
	Von Willebrandt factor	P04275	112[b]	3.5	9.5
Unbound	Cadherin-related tumor suppressor homolog	Q14517		0.0	2.0
	Calgranulin A	P05109	8.4[b]	0.0	2.0
	Carbonic anhydrase 1	P00915		1.5	4.5
	Ciliary dynein heavy chain 5	Q8TE73		0.0	2.0

Table 25.1 (continued)

M-LAC fraction	Protein name	SwissProt accession no.	Reported normal plasma or serum concentration, ng/mL, (reference)	Mean # of unique peptides, $n = 2$[a]	
				M-LAC only	depletion + M-LAC
	C-reactive protein	P08185	2020[b]	1.0	6.5
	Cystatin C	P01034	320[b]	0.0	4.0
	Integrin alpha 5	P06756		0.0	2.0
	Low density lipoprotein receptor-related protein 1B	Q9NZR2		0.0	2.5
	Mast/stem cell growth factor receptor	P10721	333[b]	0.0	2.5
	Microtubule-actin crosslinking factor 1, isoform 4	Q96PK2		0.0	3.0
	Midasin	Q9NU22		0.0	3.5
	Nesprin 2	Q8NF91		0.0	2.5
	Neutrophil defensin 3	P59666	42[b]	0.0	2.0
	Platelet basic protein	P02775		0.0	3.5
	Profilin 1	P07737		0.0	3.0
	Protocadherin gamma A9	Q9Y5G9		0.0	2.0
	Proto-oncogene tyrosine-protein kinase Fes/Fps	P07332		0.0	1.5
	Proto-oncogene tyrosine-protein kinase FGR	P09769		0.0	1.5
	Ryanodyne receptor 1	P21817		0.0	2.5
	Ryanodyne receptor 2	Q92736		0.0	3.5
	Zinc finger protein 40 (MHC binding protein 1)	P15822		0.0	2.0

[a]Mean results based on duplicate analysis. If the protein was identified by ≥ two peptides in one of the duplicate runs, even the single peptide identification from the second run was considered and included in the calculations of the mean number of identified unique peptides. All identifications were done with the following filtering criteria: for peptides, Xcorr 1.9, 2.2, 3.75 for +1, +2 and +3 charged peptide ions, respectively; dCn ≥ 0.1; for proteins, ≥ 95% confidence of identification using ProteinProphet™ software.

[b, c]Reported concentrations from two different sources; [b, c] Haab et al. (2005) varied more than 30 fold.

or 20 most abundant plasma proteins. However, as pointed out earlier (see discussion in Section 25.2.2.1) the buffer systems suggested by the manufacturers need to be modified to assure compatibility with the M-LAC system. In some cases, as was observed for depletion of 12 abundant proteins using the IgY-based (SEPPRO) columns (Sigma Aldrich, St Louis, MO), adjustment of the buffer system showed no effect on the performance of the depletion column and was successfully implemented in the automated workflow (data not shown). On the other hand, in the case of the MARS column (Agilent, Palo Alto, CA) a reduced efficiency in protein depletion occurred when the manufacturer's recommended binding buffer was replaced with the M-LAC binding buffer. Therefore, depletion using the MARS and M-LAC steps needs to be performed independently and require a buffer exchange step between the two affinity columns.

25.3.6 Application of the Proteomics Method to Biomarker Discovery

We present selected data from a biomarker discovery study performed to identify candidate biomarkers of psoriasis. Twenty plasma samples from healthy individuals (control group) and 20 plasma samples from individuals with moderate to severe psoriasis (psoriasis group) were analyzed in this study. Samples from the two groups were carefully matched to represent the same distribution of age, gender and race to facilitate proper comparative analysis. Samples were analyzed in a random order to minimize biases of sample preparation. Plasma samples were fractionated and analyzed according to the workflow presented in Fig. 25.2 using the M-LAC gravity flow (method I). Analysis of 40 samples led to the confident identification of more than 600 proteins. Among these, eleven proteins were found to have significantly different concentrations between control and psoriasis plasma samples. One of the identified candidate markers, was galectin 3-binding protein (G3BP), we detected differences in concentration between the two sample sets, as determined from the mean peak area ratios of the EIC and later verified by ELISA measurements (see below). Galectin 3-binding protein (SwissProt accession no. Q08380) is an extracellular matrix and secreted protein that promotes integrin-mediated cell adhesion. Its normal concentration in human serum is in the low μg/mL range. The results of the proteomic analysis indicated that G3BP is significantly elevated in psoriasis plasma samples when compared to healthy controls (the ratio of median values for peak area 2.4, $p < 0.0001$, Fig. 25.3a). These data were independently confirmed using an ELISA method. As shown in Fig. 25.3b, the concentration of G3BP was found to be significantly higher in samples of psoriasis patients compared to those of control donors (the ratio of median values for concentrations 2.0, $p < 0.0001$). The median concentration of G3BP in normal plasma was 6.7 μg/mL, which is in good agreement with a previously reported value of 9 μg/mL of G3BP in serum. As illustrated in Fig. 25.3c, the mass spectrometry-based measurements and ELISA concentration values for G3BP were found to be in good correlation.

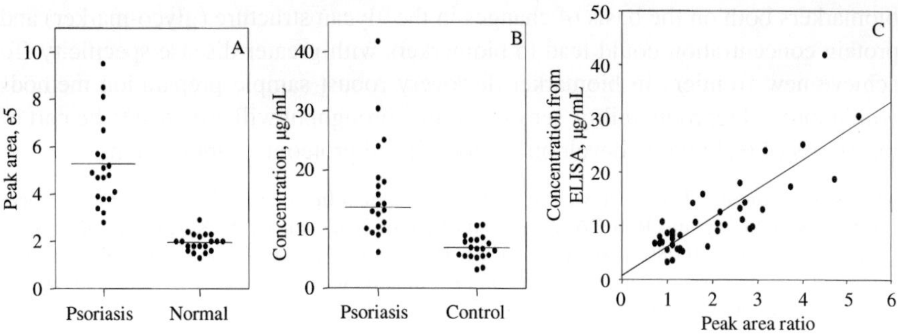

Fig. 25.3 Plasma concentrations of galectin 3 binding protein (G3BP) as assessed by (**a**) mass-spectrometry-based relative quantitation (peak area ratios of EIC) and (**b**) ELISA assay; (**c**) Correlation between mass spectrometry-based relative quantitation and ELISA measurements of G3BP. Horizontal lines show the medians

25.4 Conclusion

The method described herein that combines depletion of albumin and IgG with M-LAC fractionation for sample preparation, followed by nano LC-ESI-MS/MS analysis has been applied to plasma proteome studies demonstrating substantial depth of analysis. Using this workflow numerous tissue leakage proteins present in plasma at low ng/mL concentrations could be detected. Furthermore, using label-free quantitation, the method is able to accurately quantify changes in abundance of proteins present in plasma at low µg/mL concentrations as demonstrated by the example presented here for G3BP. The fully automated version of the method improved the reproducibility and protein recoveries making it a method of choice, especially for biomarker studies involving large sample sets. The moderate throughput of the method contrasts favorably with many methods commonly used for in-depth analysis of complex biological samples that separate each sample into numerous fractions, use more than one dimension of chromatographic separation and therefore, have a very low throughput.

An important aspect of the method is that the selective removal of most abundant proteins affords the opportunity to analyze both the unbound and bound fractions from M-LAC. This allows both a global survey of the plasma proteome and a more targeted approach to interrogate the plasma glycoproteome.

Finally, analysis of glycosylated plasma proteins is an emerging area aimed at exploiting disease-associated changes in the post-translational modification of glycans that are being pursued as a novel class of biomarkers (glyco-markers). The use of M-LAC permits detection of modified glycoform populations based on differential lectin binding. Specific glycoforms can be further isolated and characterized to investigate modifications due to aberrant glycosylation. While a detailed characterization of glycan structures is a major challenge, current developments in analytical tools and new methodologies are rapidly evolving; in the near future it will be feasible to carry out a more comprehensive analysis. The identification of candidate biomarkers both on the basis of changes in the glycan structure (glyco-marker) and protein concentration could lead to biomarkers with greater disease specificity. To achieve new frontiers in biomarker discovery robust sample preparation methods with improved reproducibility, sensitivity and throughput will continue to be part of the efforts of technology development for clinical proteomics applications.

Acknowledgements This work was supported by National Institutes of Health/National Cancer Institute (NIH/NCI) grants RO1 CA122591 and U01-CA128427 and by the Korean Research WCU grant R31-2008-000-10086-0. W. S. H and M .H. disclose that they have a financial interest in current efforts by Northeastern University and PeptiFarma to licence the M-LAC technology for biomarker discovery. Contribution Number 960 from the Barnett Institute.

References

Anderson, N.L., and Anderson, N.G. (2002). The human plasma proteome: History, character, and diagnostic prospects. Mol Cell Proteomics *1*, 845–867.

Bakry, N., Kamata, Y., and Simpson, L.L. (1991). Lectins from *Triticum vulgaris* and *Limax flavus* are universal antagonists of botulinum neurotoxin and tetanus toxin. J Pharmacol Exp Ther *258*, 830–836.

Becker, J.W., Reeke, G.N., Jr., Wang, J.L., Cunningham, B.A., and Edelman, G.M. (1975). The covalent and three-dimensional structure of concanavalin A. III. Structure of the monomer and its interactions with metals and saccharides. J Biol Chem *250*, 1513–1524.

Block, T.M., Comunale, M.A., Lowman, M., Steel, L.F., Romano, P.R., Fimmel, C., Tennant, B.C., London, W.T., Evans, A.A., Blumberg, B.S., *et al.* (2005). Use of targeted glycoproteomics to identify serum glycoproteins that correlate with liver cancer in woodchucks and humans. Proc Natl Acad Sci USA *102*, 779–784.

Carlsson, J., Gullstrand, C., Ludvigsson, J., Lundstrom, I., and Winquist, F. (2008). Detection of global glycosylation changes of serum proteins in type 1 diabetes using a lectin panel and multivariate data analysis. Talanta *76*, 333–337.

Cummings, R.D., and Kornfeld, S. (1982). Fractionation of asparagine-linked oligosaccharides by serial lectin-agarose affinity chromatography. A rapid, sensitive, and specific technique. J Biol Chem *257*, 11235–11240.

Dai, Z., Liu, Y.K., Cui, J.F., Shen, H.L., Chen, J., Sun, R.X., Zhang, Y., Zhou, X.W., Yang, P.Y., and Tang, Z.Y. (2006). Identification and analysis of altered alpha1,6-fucosylated glycoproteins associated with hepatocellular carcinoma metastasis. Proteomics *6*, 5857–5867.

Dayarathna, M.K., Hancock, W.S., and Hincapie, M. (2008). A two step fractionation approach for plasma proteomics using immunodepletion of abundant proteins and multi-lectin affinity chromatography: Application to the analysis of obesity, diabetes, and hypertension diseases. J Sep Sci *31*, 1156–1166.

Dennis, J.W., Laferte, S., Waghorne, C., Breitman, M.L., and Kerbel, R.S. (1987). Beta 1–6 branching of Asn-linked oligosaccharides is directly associated with metastasis. Science *236*, 582–585.

Dwek, M.V., Ross, H.A., Streets, A.J., Brooks, S.A., Adam, E., Titcomb, A., Woodside, J.V., Schumacher, U., and Leathem, A.J. (2001). Helix pomatia agglutinin lectin-binding oligosaccharides of aggressive breast cancer. Int J Cancer *95*, 79–85.

Echan, L.A., Tang, H.Y., Ali-Khan, N., Lee, K., and Speicher, D.W. (2005). Depletion of multiple high-abundance proteins improves protein profiling capacities of human serum and plasma. Proteomics *5*, 3292–3303.

Elias, J.E., and Gygi, S.P. (2007). Target-decoy search strategy for increased confidence in large-scale protein identifications by mass spectrometry. Nat Methods *4*, 207–214.

Fang, X., and Zhang, W.W. (2008). Affinity separation and enrichment methods in proteomic analysis. J Proteomics *71*, 284–303.

Haab, B.B., Geierstanger, B.H., Michailidis, G., Vitzthum, F., Forrester, S., Okon, R., Saviranta, P., Brinker, A., Sorette, M., Perlee, L., *et al.* (2005). Immunoassay and antibody microarray analysis of the HUPO Plasma Proteome Project reference specimens: Systematic variation between sample types and calibration of mass spectrometry data. Proteomics *5*, 3278–3291.

Jacobs, J.M., Adkins, J.N., Qian, W.J., Liu, T., Shen, Y., Camp, D.G., 2nd, and Smith, R.D. (2005). Utilizing human blood plasma for proteomic biomarker discovery. J Proteome Res *4*, 1073–1085.

Keller, A., Nesvizhskii, A.I., Kolker, E., and Aebersold, R. (2002). Empirical statistical model to estimate the accuracy of peptide identifications made by MS/MS and database search. Anal Chem *74*, 5383–5392.

Klammer, A.A., and MacCoss, M.J. (2006). Effects of modified digestion schemes on the identification of proteins from complex mixtures. J Proteome Res *5*, 695–700.

Kreunin, P., Zhao, J., Rosser, C., Urquidi, V., Lubman, D.M., and Goodison, S. (2007). Bladder cancer associated glycoprotein signatures revealed by urinary proteomic profiling. J Proteome Res *6*, 2631–2639.

Kullolli, M., Hancock, W.S., and Hincapie, M. (2008). Preparation of a high-performance multi-lectin affinity chromatography (HP-M-LAC) adsorbent for the analysis of human plasma glycoproteins. J Sep Sci *31*, 2733–2739.

Kullolli, M., Hancock, W.S., and Hincapie, M. (2010). Automated platform for fractionation of human plasma glycoproteome in clinical proteomics. Anal Chem *82*, 115–120.

Liu, T., Qian, W.J., Chen, W.N., Jacobs, J.M., Moore, R.J., Anderson, D.J., Gritsenko, M.A., Monroe, M.E., Thrall, B.D., Camp, D.G., 2nd, *et al.* (2005). Improved proteome coverage by using high efficiency cysteinyl peptide enrichment: The human mammary epithelial cell proteome. Proteomics *5*, 1263–1273.

Martosella, J., Zolotarjova, N., Liu, H., Nicol, G., and Boyes, B.E. (2005). Reversed-phase high-performance liquid chromatographic prefractionation of immunodepleted human serum proteins to enhance mass spectrometry identification of lower-abundant proteins. J Proteome Res *4*, 1522–1537.

Omenn, G.S., States, D.J., Adamski, M., Blackwell, T.W., Menon, R., Hermjakob, H., Apweiler, R., Haab, B.B., Simpson, R.J., Eddes, J.S., Kapp, E.A., Moritz, R.L., Chan, D.W., Rai, A.J., Admon, A., Aebersold, R., Eng, J., Hancock, W.S., Hefta, S.A., Meyer, H., Paik, Y.K., Yoo, J.S., Ping, P., Pounds, J., Adkins, J., Qian, X., Wang, R., Wasinger, V., Wu, C.Y., Zhao, X., Zeng, R., Archakov, A., Tsugita, A., Beer, I., Pandey, A., Pisano, M., Andrews, P., Tammen, H., Speicher, D.W., and Hanash, S.M. (2005). Overview of the HUPO Plasma Proteome Project: Results from the pilot phase with 35 collaborating laboratories and multiple analytical groups, generating a core dataset of 3020 proteins and a publicly-available database. Proteomics *5*, 3226–3245.

Plavina, T., Hincapie, M., Wakshull, E., Subramanyam, M., and Hancock, W.S. (2008). Increased plasma concentrations of cytoskeletal and Ca2+-binding proteins and their peptides in psoriasis patients. Clin Chem *54*, 1805–1814.

Plavina, T., Wakshull, E., Hancock, W.S., and Hincapie, M. (2007). Combination of abundant protein depletion and multi-lectin affinity chromatography (M-LAC) for plasma protein biomarker discovery. J Proteome Res *6*, 662–671.

Polanski, M.A., and Anderson, N.L (2006). A list of candidate cancer biomarkers for targeted proteomics. Biomarker Insights *2*, 1–48.

Purcell, A.W., van Driel, I.R., and Gleeson, P.A. (2008). Impact of glycans on T-cell tolerance to glycosylated self-antigens. Immunol Cell Biol *86*, 574–579.

Qiu, R., and Regnier, F.E. (2005). Comparative glycoproteomics of N-linked complex-type glycoforms containing sialic acid in human serum. Anal Chem *77*, 7225–7231.

Ralin, D.W., Dultz S.C., Silver, J.E., Travis, J.C., Kullolli, M., Hancock, W.S., Hincapie, M. (2008). Kinetic analysis of glycoprotein–lectin interactions by label-free internal reflection ellipsometry. Clin Proteomics *4*, 37–46.

Saulsbury, F.T. (1997). Alterations in the O-linked glycosylation of IgA1 in children with Henoch-Schonlein purpura. J Rheumatol *24*, 2246–2249.

Wang, G., Wu, W.W., Zeng, W., Chou, C.L., and Shen, R.F. (2006). Label-free protein quantification using LC-coupled ion trap or FT mass spectrometry: Reproducibility, linearity, and application with complex proteomes. J Proteome Res *5*, 1214–1223.

Yang, Z., and Hancock, W.S. (2004). Approach to the comprehensive analysis of glycoproteins isolated from human serum using a multi-lectin affinity column. J Chromatogr A *1053*, 79–88.

Yang, Z., Harris, L.E., Palmer-Toy, D.E., and Hancock, W.S. (2006). Multilectin affinity chromatography for characterization of multiple glycoprotein biomarker candidates in serum from breast cancer patients. Clin Chem *52*, 1897–1905.

Yates, J.R., Cociorva, D., Liao, L., and Zabrouskov, V. (2006). Performance of a linear ion trap-orbitrap hybrid for peptide analysis. Anal Chem *78*, 493–500.

Yocum, A.K., Yu, K., Oe, T., and Blair, I.A. (2005). Effect of immunoaffinity depletion of human serum during proteomic investigations. J Proteome Res *4*, 1722–1731.

Chapter 26
Analysis of Glycopeptides Using Porous Graphite Chromatography and LTQ Orbitrap XL ETD Hybrid MS

Terry Zhang, Rosa Viner, Zhiqi Hao, and Vlad Zabrouskov

Abstract In this study, several glycoproteins: bovine α1-acid glycoprotein, fetuin, and human α1-acid glycoprotein, were analyzed using nanoLC-MS/MS. The performance of three different stationary phases – C_8, C_{18}, and porous graphite – was systematically evaluated and optimized for glycopeptide separation. A porous graphite column was chosen as the optimum stationary phase for glycoprotein analysis. An LTQ Orbitrap XL hybrid mass spectrometer equipped with electron transfer dissociation (ETD) was used to acquire, within a single analysis, high-resolution, high-mass-accuracy full MS spectra, collision-induced dissociation (CID) MS/MS spectra for glycan composition analysis, and ETD MS/MS spectra for peptide structure information. As a result, characterization of glycopeptides was achieved within a single LC-MS/MS run.

Keywords ETD · Glycoprotein · N-glycosylation · LTQ Orbitrap XL ETD · Porous graphite

26.1 Introduction

Glycosylation of Asn, Ser or Thr is arguably the most common known post-translational modification (PTM), resulting in a multitude of highly heterogeneous protein isoforms (Morelle et al., 2006; Peterman and Mulholland, 2006). While the physicochemical differences among the glycosylated protein molecules are often minute, their characterization remains a great analytical challenge. To date, although there are reports of glycoproteome analyses based on electrophoresis techniques, liquid chromatography coupled to multi-stage mass spectrometry (LC-MS/MS) offers advantages in speed, sensitivity, and automation (Kaji et al., 2008). It remains the most powerful and versatile technique for elucidation of glycopeptide structure.

V. Zabrouskov (✉)
Thermo Fisher Scientific, San Jose, CA 95134, USA
e-mail: vlad.zabrouskov@thermofisher.com

A.R. Ivanov, A.V. Lazarev (eds.), *Sample Preparation in Biological Mass Spectrometry*, DOI 10.1007/978-94-007-0828-0_26,

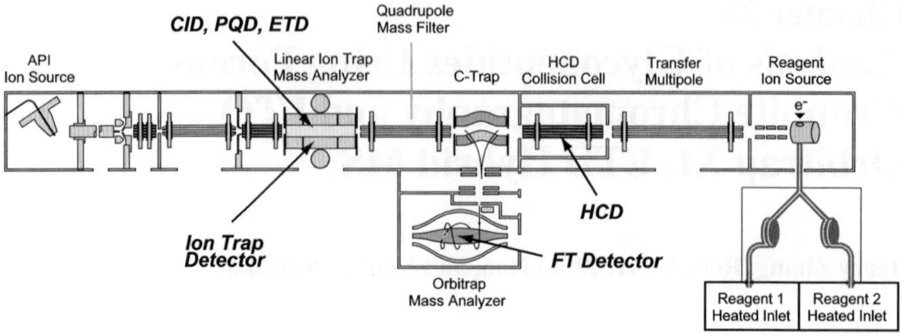

Fig. 26.1 A schematic diagram of an LTQ Orbitrap XL hybrid mass spectrometer equipped with ETD

LC-MS/MS analysis does, however, have its own challenges. In addition to the difficulties in capturing minor glycopeptides efficiently from samples containing large numbers of non-glycosylated peptides, the commonly used collision-induced dissociation has limitations for determining the modification site due to the labile nature of the glycan attachment to the peptide ion (Kaji et al., 2008). CID MS/MS predominantly generates fragment ions from cleavages of glycosidic bonds without breaking amide bonds (Morelle et al., 2006; Peterman and Mulholland, 2006). In contrast to CID, electron transfer dissociation (ETD) preserves labile PTMs while cleaving peptide bonds, making the identification of modification sites possible (Syka et al., 2004; Alley Jr et al., 2007). Since glycosylated proteins and the resulting peptides are generally very heterogeneous, their mass spectra are highly complex. Consequently, high-quality liquid chromatography, high mass resolution, and accurate mass measurements of ETD precursors and fragments are essential for glycopeptide analysis.

The LTQ Orbitrap XL mass spectrometer is a high-resolution, accurate-mass hybrid mass spectrometer that includes both linear ion trap and Orbitrap mass analyzers as shown in Fig. 26.1. It can be equipped with an ETD source (McAlister et al., 2007). Beyond the optional ETD, there are three fragmentation modes available on this instrument: collision-induced dissociation (CID), higher-energy collisional dissociation (HCD), and pulsed-Q dissociation (PQD). Since CID and ETD often provide complementary structural information for glycopeptides, coupling high-resolution, high-mass-accuracy full MS with alternating CID and ETD MS/MS makes comprehensive glycopeptide characterization more likely in single analysis (Fig. 26.2) (Syka et al., 2004; Alley Jr et al., 2007).

26.2 Experimental Conditions

This section describes instruments and methods employed for sample preparation, HPLC separation, and mass spectrometric analysis.

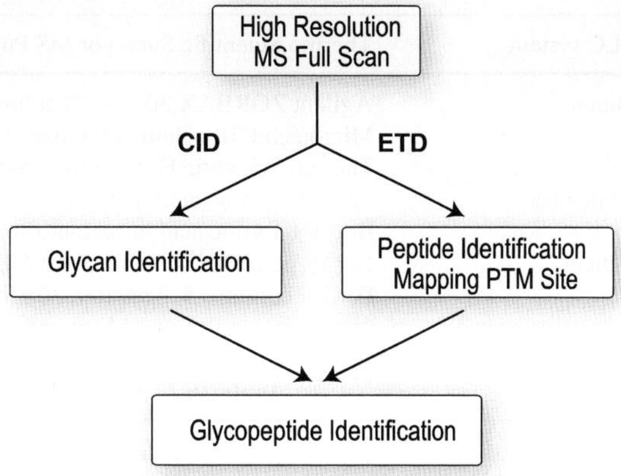

Fig. 26.2 A flow chart for glycopeptide analysis using an LTQ Orbitrap XL ETD hybrid mass spectrometer

26.2.1 Sample preparation

Glycoproteins purchased from Sigma were denatured in 0.1% SDS 50 mM Tris HCl buffer (pH 8.5), reduced with 5 mM (final concentration) DTT for 1 h at 60°C and alkylated with 25 mM (final concentration) iodoacetamide for 2 h in the dark at room temperature. The reduced and alkylated proteins were precipitated with acetone (1:6 (v:v) of protein mixture vs acetone) for 4 h at –20°C, protein pellets were recovered by centrifugation at 15 K rpm for 10 min on tabletop centrifuge, and air dried after acetone removal.

For digestion 50 uL of 25 mM ammonium bicarbonate buffer, pH 7.8 containing 10% AcN and 1 μg of trypsin (Thermo Scientific, Rockford, IL) was added to each protein pellet and mixture was incubated for 16 h (overnight) at 37°C. After digestion samples were concentrated up to 10–15 uL using Speed Vac and diluted 5–10 times with 5% FA and analyzed by nano LC/MS/MS.

26.2.2 HPLC Separation

HPLC instrumentation, columns, and separations condition are listed in Table 26.1.

26.2.3 MS Analysis

A Thermo Scientific LTQ XL linear ion trap mass spectrometer equipped with ETD was used to assess the ability of three different stationary phases to separate glycosylated peptides. A Thermo Scientific LTQ Obritrap XL hybrid mass spectrometer equipped with ETD was used to analyze the structures of the glycoproteins. Operating parameters for both instruments are listed in Table 26.2.

Table 26.1 LC Separation

HPLC system	Thermo Scientific Surveyor MS Pump with a flow splitter
Columns	Agilent ZORBAX 300SB C8 column (75 μm × 5 cm); Microtech C18 column (150 μm × 10 cm) Thermo Scientific Hypercarb porous graphite column (75 μm × 5 cm)
Mobile phases	A: 0.1% formic acid in water B: 0.1% formic acid in acetonitrile
Gradient	For Hypercarb™ column 5–50% B in 30 min For RP columns 5–35% B in 30 min

Table 26.2 MS Analysis

Mass spectrometer	Thermo Scientific LTQ XL™ linear ion trap mass spectrometer with ETD and nano-ESI source
Spray voltage	2 kV
Capillary temp	160°C
Capillary voltage	35 V
Tube lens	125 V
MSn target	1e4
Mass range	50–2000 *m/z* or 100–4000 *m/z*
Anion reagent	Fluoranthene
Anion reagent isolation	On
Anion target	2e5
Max anion injection time	50 ms
ETD reaction time	75–150 ms
Mass spectrometer	Thermo Scientific LTQ Orbitrap XL ETD hybrid
Mass range	400–2000 *m/z*, resolution 60,000–100,000 @ *m/z* 400
FT MS AGC target	5e5
FT MS/MS AGC target	1e5, 3 amu isolation width
MS/MS resolution	7,500 FWHM @ m/z 400, 3 microscans
Monoisotopic precursor pelection	Off
Reject charge states	1, 2 and unassigned
Exclusion mass tolerance	10 ppm
Max ion time FT MS	500 ms
Max ion time FT MS/MS	500 ms
MS/MS mass range	100–2000 *m/z*
Survey scan	Source CID at 65 V for *m/z* 204

26.2.4 Data Processing

Thermo Scientific Xtract software was used for deconvolution of multiply charged precursor and MS/MS spectra. Generated peak list of MH+ masses was imported into the GlycoMod tool from the Swiss-Prot website. The GlycoMod was used to assign possible oligosaccharide structures and compositions. The following search criteria were used: mass tolerance 5 ppm, Na or K adducts, trypsin digestion with 2 possible missed cleavage sites, protein sequence, HexNac, Hex, NeuAc and fucose as possible *N*-linked monosaccharide residues.

26.3 Results and Discussion

26.3.1 Choosing a Stationary Phase for Glycopeptide Analysis

Three stationary phases were evaluated for HPLC separation of glycosylated peptides. One pmol of a bovine α1-acid glycoprotein digest was injected onto the C_{18} column versus 500 fmol onto the porous graphite column. Figure 26.3 shows the extracted ion chromatograms of *m/z* 1706.3, 1138.5 and 867.3 belonging to 2+, 3+ and 4+ molecular ions of the bi-antennary glycopeptides $_{91}$CVY**N**CSFIK$_{99}$. The intensities of the 2+ and 3+ glycopeptide precursor ions from analysis with the porous graphite column were similar to the intensities from analysis with the C_{18} column despite 50% lower on-column load. This was likely due to the higher affinity of porous graphite for hydrophilic peptides. In addition, chromatography using a porous graphite stationary phase promoted formation of abundant higher-charge-state metal-adducted precursor ions, which improved detection limits for glycopeptides.

Figure 26.4 shows a high-resolution deconvoluted spectrum of these glycopeptides. At least one potassium adduct was observed for each glycoform. As demonstrated in Fig. 26.3, formation of a metal adduct is likely responsible for producing higher-charge-state ions. The $(M+K+3H)^{4+}$ precursor at *m/z* 867.3 was the dominant peak for glycopeptide 5 (Fig. 26.4) while $(M+3H)^{3+}$ precursor at *m/z* 1138.6 was the dominant signal for glycopeptide 2. Formation of higher-charge precursors can be explained by partial neutralization of sialic acid negative charges by metal cations (Newton et al., 2005; Medzhhradszky et al., 2007).

Bovine and human α1-acid glycoproteins contain five N-glycosylation sites with hybrid-type glycan structures (Snovida et al., 2006; Treuheit et al., 1992). Without any prior enrichment, four out of five glycopeptides were detected and identified using the porous graphite column compared to two out of five peptides on the C_{18} column and three out of five peptides on the C_8 column (Table 26.3). Only the largest and most hydrophobic peptides were not detected using any of the phases.

Similar results were obtained for a bovine fetuin digest. All bovine fetuin glycopeptides were detected using C_{18} (Peterman and Mulholland, 2006) or C_8 chromatography. However, they were mostly present as lower-charge species, resulting in poorer MS/MS ETD spectra provided insufficient information for unambiguous identification. Previously, Alley and co-workers (Alley Jr et al., 2007) reported that fetuin glycopeptides could not be observed after separation on a graphite HPLC Glycan Chip. The study presented here indicated that larger hydrophobic peptides elute much later in the gradient. Nevertheless, three out of four fetuin peptides were identified after separation on the porous graphite column. Only the largest O-glycosylated 246–306 peptide was not detected. Figure 26.5 shows an example of an ETD spectrum of 5+ precursor at *m/z* 1307 of bovine fetuin tri-antennary peptide $_{72}$RPTGEVYDIEIDTLETTCHVLDPTPLA**N**CSVR$_{103}$, which was successfully identified after separation on porous graphite column. After optimization, all nano LC-MS/MS analyses of glycopeptides were performed using a porous graphite column and 1 pmol of glycoprotein digest.

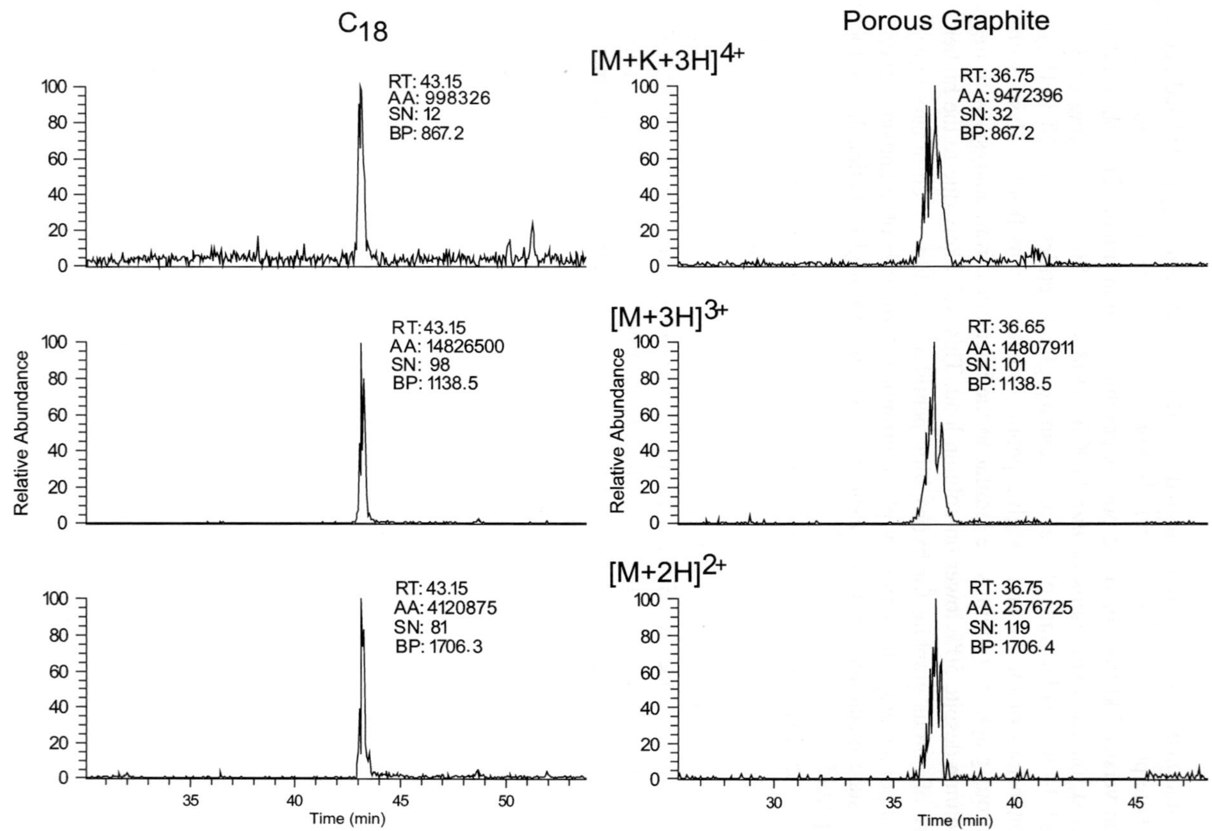

Fig. 26.3 A comparison of extracted ion chromatograms at different charge states from analysis of bovine α1-acid glycoprotein bi-antennary peptide $_{91}$CVY*N*CSFIK$_{99}$ using C_{18} and porous graphite columns. ***N*** corresponds to the modified residue

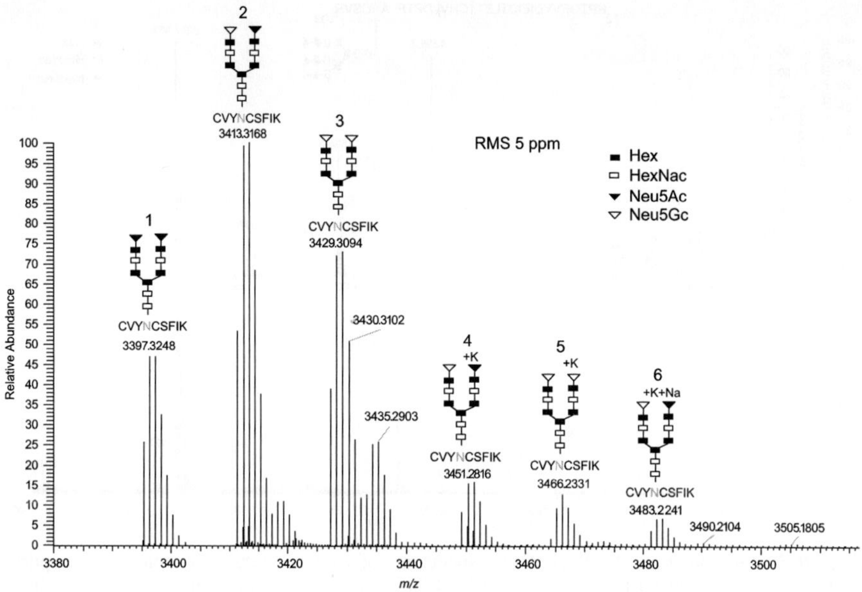

Fig. 26.4 A deconvoluted full MS spectrum of bovine α1-acid glycoprotein bi-antennary peptide $_{91}$CVY*N*CSFIK$_{99}$. *N* corresponds to the modified residue

Table 26.3 Bovine α1-acid glycoprotein glycopeptides detected by nano LC-MS/MS

	Type of LC column		
Peptides	Graphite	C18	C8
R$_{103}$QNGTLSK$_{109}$V	•		
R$_{53}$NPEYNK$_{58}$S	•		
K$_{91}$CVYNCSFIK$_{99}$I	•	•	•
R$_{128}$TFMLAASWNGTK$_{139}$N	•	•	•
$_{19}$QSPECANLMTVAPITNATMDLLSGK$_{43}$W			•

26.3.2 MS/MS Analysis of Glycopeptides

Following a strategy described by Peterman and Mulholland (2006), an LTQ Orbitrap XL^{TM} hybrid mass spectrometer equipped with ETD was used to identify the glycopeptides. Characteristic oxonium ions at *m/z* 204 and/or 366 generated by in-source CID were further fragmented by MS^2 "a pseudo MS^3 event" (Peterman and Mulholland, 2006). This allowed for highly sensitive and selective detection of the eluting glycosylated peptide ion(s) at a given retention time.

High-resolution accurate-mass full-scan MS spectra were acquired in the Orbitrap mass analyzer, as shown in Fig. 26.6, followed by data-dependent MS/MS

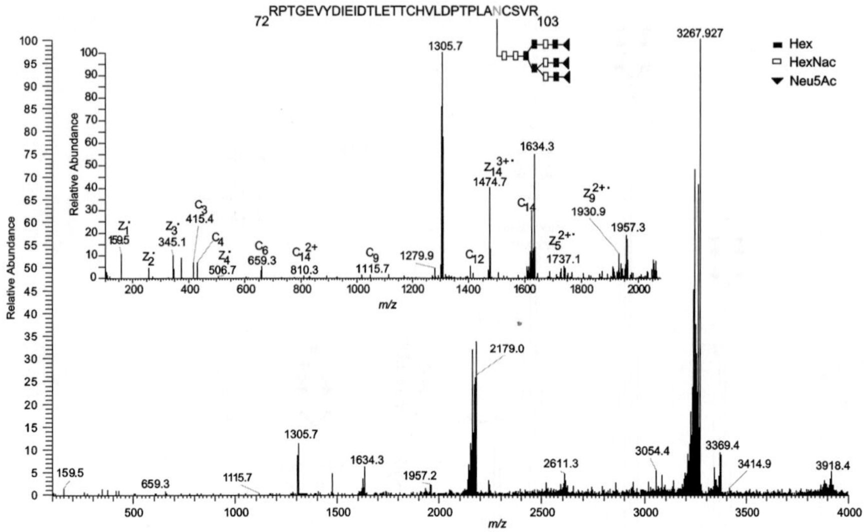

Fig. 26.5 An ion trap MS/MS ETD spectrum of bovine fetuin tri-antennary peptide $_{72}$RPTGEVYDIEIDTLETTCHVLDPTPLA*N*CSVR$_{103}$ (5+, *m/z* 1307). Insert: the magnified region for scan range 100–2000 *m/z*. ***N*** corresponds to the modified residue

alternating CID and ETD scans with fragments analyzed in either the ion trap or Orbitrap mass analyzers. The resulting CID spectra of the 3+ precursors (Fig. 26.6c) were used for glycan structure elucidation and the ETD spectra of the 4+ precursors (Fig. 26.6b) were used for glycopeptide identification as demonstrated in Figs. 26.7 and 26.8 for the bovine α1-acid glycoprotein peptide $_{103}$Q*N*GTLSK$_{109}$.

The Fig. 26.7 inset shows the CID spectrum of the precursor $[M+3H]^{3+}$ at *m/z* 990.395 measured in the Orbitrap mass analyzer. The oxonium fragment ion at *m/z* 366.139 confirmed this precursor as a glycopeptide. The spectrum was deconvoluted and glycan composition was assigned as shown in Fig. 26.7. This glycoform contains one each of N-acetylneuraminic (Neu5Ac) and N-glycolyeuraminic (Neu5Gc) acids at the glycan branch end as determined by the presence of 657/673 fragment pair. The product ion at *m/z* 950.481 corresponds to $[M\ HexNAc+H]^{+}$. All measured masses were within 5 ppm of their theoretical values which helped unambiguously assign the glycan portion as a bi-sialated glycopeptide structure.

Higher-charged metal-adducted precursor ions (4+) were further selected for ETD analysis. From the ETD MS/MS spectrum of the precursor $[M+K+O+3H]^{4+}$ as shown in Fig. 26.8, the glycosylation site was clearly identified at Asn 104 based on an almost complete series of ***c/z***$^{\bullet}$ ions. No significant loss of carbohydrate was detected and, as expected, several of the observed glycan-containing fragments retained potassium ions. The high resolution, high mass accuracy, and low chemical noise of the Orbitrap mass analyzer significantly benefited ETD analysis of glycopeptides with charge states 4+ and above, allowing straightforward deconvolution and interpretation of spectra.

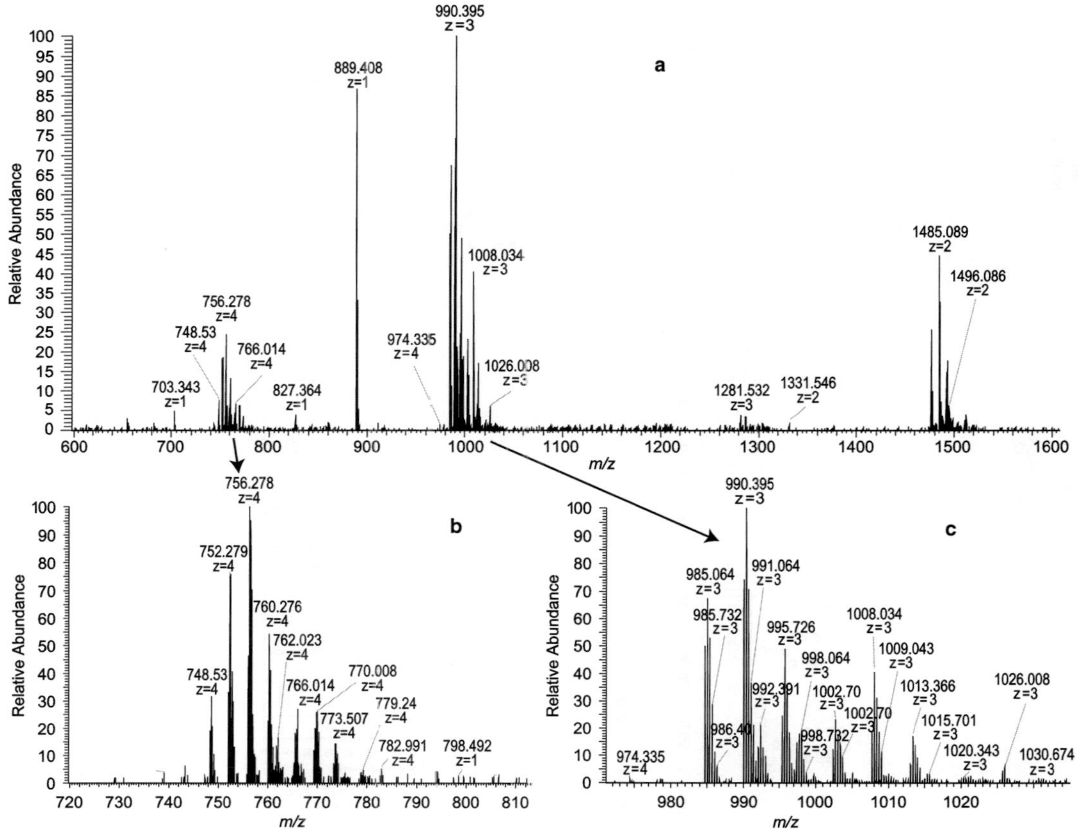

Fig. 26.6 (**a**) An MS spectrum; (**b, c**) spectra of charge states 4+, 3+ showing highly heterogeneous glycoforms; and (**d**) a deconvoluted full MS spectrum of bovine α1-acid glycoprotein bi-antennary peptide $_{103}$Q*N*GTLSK$_{109}$. *N* corresponds to the modified residue

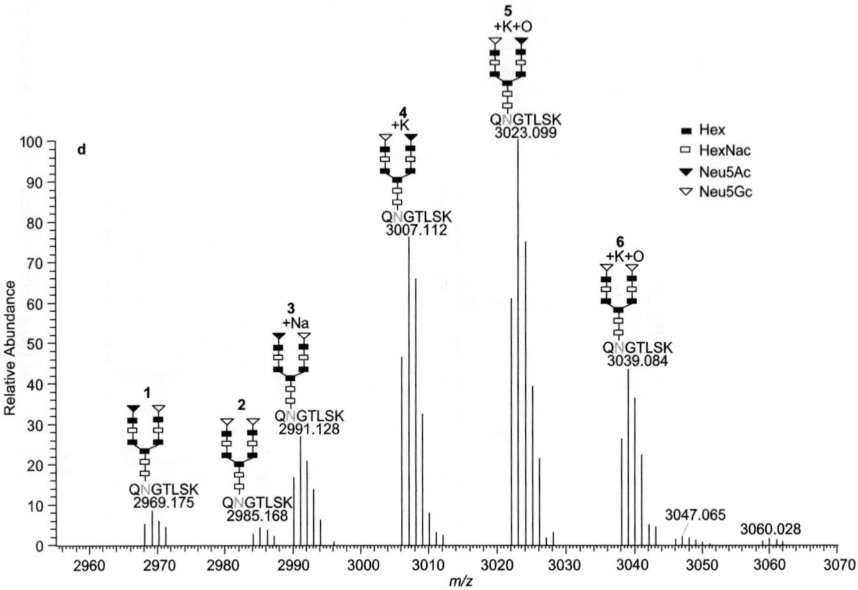

Fig. 26.6 (continued)

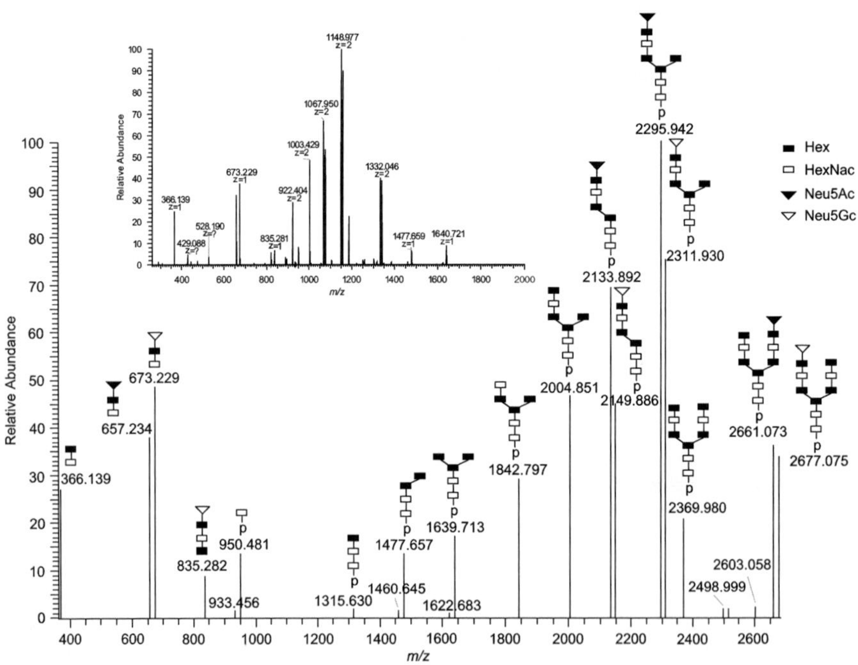

Fig. 26.7 A deconvoluted Orbitrap CID MS/MS spectrum of bovine α1-acid bi-antennary glycopeptide $_{103}$Q*N*GTLSK$_{109}$. The insert shows the original Orbitrap CID spectrum of the 3+ precursor at *m/z* 990.395 (Fig. 26.6d, structure 1). *N* corresponds to the modified residue

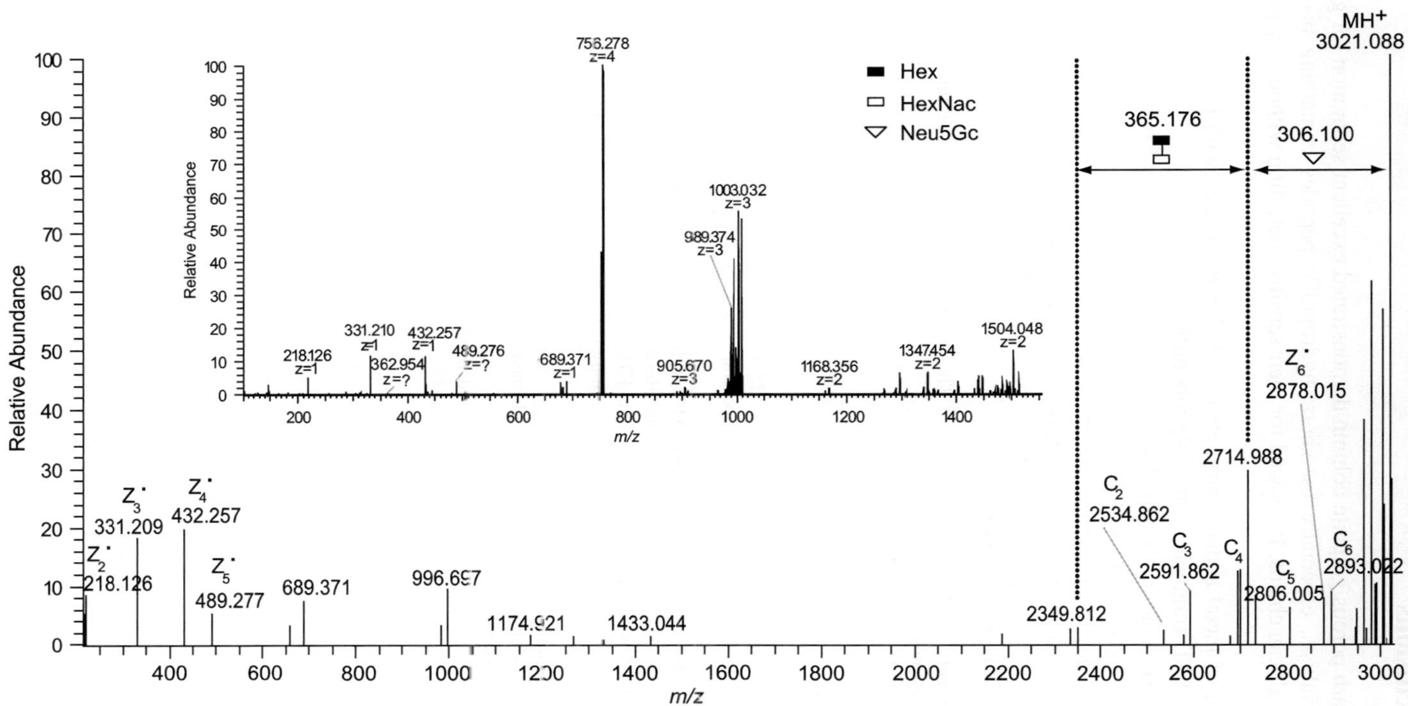

Fig. 26.8 A deconvoluted Orbitrap ETD MS/MS spectrum of bovine α1-acid bi-antennary glycopeptide $_{103}$Q*N*GTLSK$_{109}$. The insert shows the original Orbitrap ETD spectrum of [M+K+O+3H]$^{4+}$ at *m/z* 756.278 ion (Fig. 26.6d, structure 5). *N* corresponds to the modified residue

26.4 Conclusions

The Hypercarb porous graphite column demonstrated excellent separation for glycopeptide analysis, especially for short, hydrophilic peptides containing bi- or triantennary glycan chains. It allowed for their sensitive detection without any prior enrichment.

Formation of metal adducts promoted evolution of species with higher charge states, aiding ETD fragmentation of glycopeptides.

ETD preserved labile glycans, facilitating the identification of both the peptide of interest and the site of modification.

Combining peptide structural information obtained by ETD and the glycan composition information obtained by CID enabled confident identification and characterization of glycopeptides within a single LC-MS/MS analysis using an LTQ Orbitrap XL ETD mass spectrometer.

Acknowledgement We thank David Fisher at Thermo Fisher Scientific for editing and proof reading of this manuscript.

References

Alley Jr., W.R., Merchref, Y., and Novotny, M.V. (2007). Using Graphitized Carbon for Glycopeptides Separations Prior to Mass Spectral Detection. Proceedings of the 55th ASMS Conference (Indianapolis, IN).

Alley Jr., W.R., Merchref, Y., and Novotny, M.V. (2009). Characterization of glycopeptides by combining collision-induced dissociation and electron-transfer dissociation mass spectrometry data. Rapid Commun Mass Spectrom *23*, 161–170.

Kaji, H, and Isobe, T. (2008). Liquid chromatography/mass spectrometry (LC-MS)-based glycoproteomics technologies for cancer biomarker discovery. Clin Proteomics *4*, 14–24.

McAlister, G.C., Phanstiel, D., Good D.M., Berggren, W.T., and Coon, J.J. (2007). Implementation of electron-transfer dissociation on a hybrid linear ion trap-orbitrap mass spectrometer. Mol Cell Proteomics *6*, 1942–1951.

Medzihradszky, K.F., Guan, S., Maltby, D.A., and Burlingame, A.L. (2007). Sulfopeptide fragmentation in electron-capture and electron-transfer dissociation. J Am Soc Mass Spectro *18*(9), 1617–1624

Morelle, W., Canis, K., Chirat, F., Faid, V., and Michalski, J-C. (2006). The use of mass spectrometry for the proteomic analysis of glycosylation. Proteomics *6*, 3993–3915.

Newton, K.A., Amunugama, R., and McLuckey, S.A. (2005). Gas-phase ion/ion reactions of multiply protonated polypeptides with metal containing anions. J Phys Chem A *109*(16), 3608–3616.

Peterman, S.M., and Mulholland, J.J. (2006). A novel approach for identification and characterization of glycoproteins using a hybrid linear ion trap/FT-ICR mass spectrometer. J Am Soc Mass Spectrom *17*(2), 68–79.

Snovida, S.I., Chen, V.C., Krokhin, O., and Perreault, H. (2006). Isolation and identification of sialylated glycopeptides from bovine α1-acid glycoprotein by off-line capillary electrophoresis MALDI-TOF mass spectrometry. Anal Chem *78*, 6556–635

Syka J.E., Coon J.J., Schroder M.J., Shabanowitz J, and Hunt D.F. (2004). Peptide and protein sequence analysis by electron transfer dissociation mass spectrometry. Proc Natl Acad Sci USA *101*, 9528–9533.

Treuheit, M.J., Costello, C.E., and Halsall, H.B. (1992). Analysis of the five glycosylation sites of human α1-acid glycoprotein. Biochem J *283*, 105–112.

Part VIII
Methods for Quantitative Proteomics

Chapter 27
Stable Isotopic Labeling for Proteomics

Keith Ashman, María Isabel Ruppen Cañás, Jose L. Luque-Garcia, and Fernando García Martínez

Abstract Since the coining of the term Proteome, the field of Proteomics has developed rapidly and come to rely heavily on mass spectrometry, initially for protein identification but recently the use of stable isotopic labels for protein quantitation has grown in importance. This trend has been driven by improvements in the mass spectrometers and reagents but also by the need to understand the molecular dynamics of cells. Since proteins are important effector molecules in most biochemical processes, key questions to be answered are: how much is present and when. In this chapter we describe the various methods currently available to quantitate proteins based on stable isotope protein labeling and discuss their merits as well as some of the issues still to be addressed.

Keywords Proteomics · Quantitation · Stable isotope labeling · iTRAQ · SILAC · Fractionation · HPLC · Mass spectrometer · MALDI · Electrospray · Peptides · Amines · Protein · Derivatization · Biomarkers · Metabolic · Stoichiometry · Chromatography

Abbreviations

ACN	Acetonitrile
AQUA	Absolute quantification
CID	Collision induced disassociation
CILAT	Cleavable isobaric labeled affinity tag
ESI	Electrospray ionisation
iCAT	Isotopically coded affinity tags
iTRAQ	Isobaric tags for relative and absolute quantitation
HILIC	Hydrophilic interaction chromatography
HPLC	High performance liquid chromatography
ICPL	Isotope-coded *protein label*
ICPMS	Inductively coupled plasma mass spectrometry

K. Ashman (✉)
Centro Nacional de Investigaciones Oncológicas (CNIO), E-28029 Madrid, Spain
e-mail: kashman@cnio.es

A.R. Ivanov, A.V. Lazarev (eds.), *Sample Preparation in Biological Mass Spectrometry*, DOI 10.1007/978-94-007-0828-0_27,

IgG	Immunoglobulin
IPG	Immobilized pH gradient
IPG-IEF	Immobilized pH gradient isoelectric focusing
LC-MALDI MS/MS	Liquid chromatography matrix assisted laser desorption ionization tandem mass spectrometry
LC-MS	Liquid chromatography mass spectrometry
MALDI	Matrix assisted laser desorption ionisation
MMTS	Methyl methane-thiosulphonate
MS	Mass spectrometry
MS/MS	Tandem mass spectrometry
Nano RP LC	Nano reversed phase liquid chromatography
NHS	*N*-Hydroxysuccinamide
PBS	Phosphate buffered saline
QconCAT	Artificial proteins representing a quantification concatamer
qTOF MS/MS	Quadrupole time of flight tandem mass spectrometry
RP LC	Reversed phase liquid chromatography
SILAC	Stable isotope labeling by amino acids in cell culture
SCX	Strong cation exchange chromatography
SISCAPA	Stable isotope standards and capture by anti-peptide antibodies
TCA	Tri-chloroacetic acid
TCEP	Tris (2-carboxyethyl) phosphine
TEAB	Triethylammonium bicarbonate
TMT	Tandem Mass Tag Technology
TOF/TOF	Time of flight/time of flight

27.1 Introduction

Biological systems are dynamic. They contain many molecular species (DNA, RNA, proteins, carbohydrates, lipids. . . .) that interact with each other, resulting in complex series of physico-chemical, spatial and temporal changes. Despite tremendous advances in molecular biology, in the broadest sense of that term, it is still impossible to describe in molecular detail how even a virus or bacterium work. The flow of information from DNA ⇒ RNA ⇒ Protein ⇒ molecular complexes ⇒ organelles ⇒ cells ⇒ tissues ⇒ organisms ⇒ ecosystems is extremely difficult to follow. One thing is clear: the complexity of the molecular interaction networks increases rapidly along the information chain. The field of systems biology attempts to understand how these various layers combine to make the system more than merely the sum of its parts. One of the core problems facing systems biology is the translation of two dimensional, linear genomic information into multidimensional molecular interaction and function space.

This may be summarized as:

(1) The identification of the individual molecules, along with their structures i.e. who are the players.

(2) The analysis of molecular interactions i.e. who do they play with.
(3) The characterization both chemical and temporal of secondary modifications such as methylation, glycosylation, acetylation, meristylation, phosphorylation, ubiquitination... i.e. how do they effect each other.
(4) The stoichiometry and function of molecular complexes i.e. how many are involved and what do they do.
(5) The temporal and spatial stoichiometry of molecular interactions i.e. when and where is all this going on.

Another question might be: why does any of this take place, but that is one for the philosophers to ponder.

Since proteins play many crucial roles in cells, much effort has been expended in developing technologies for quantitative proteomics, to facilitate the analysis of dynamic biological processes, because knowing how the amounts and activities of proteins change over time provides important clues as to state of the system. Isotopic labeling of proteins, or other molecules for that matter (Boyer, 1954; De Leenheer and Thienpont, 1992; Lane et al., 2009; Schnolzer et al., 1996), is not a new technique but the development of sensitive, fast, highly accurate mass spectrometers has opened up the field of quantitative proteomics to the wide application of this method for protein quantitation. In the following pages we will review the current practice.

27.2 Chemical Labeling of Peptides with Stable Isotopes

Chemical tagging of peptides is a long established methodology, see review by (Leitner and Lindner, 2009). Stable isotope peptide labeling reagents are now finding wide use in quantitative proteomics. The ideal reagent would have many of the following characteristics:

(a) All peptides in a sample are labeled.
(b) The mass difference between the labeled isoforms is greater than the natural isotopic distribution (in the case of peptides at least 4 Da).
(c) The labeling is quantitative.
(d) The isotopic isoforms can be enriched.
(e) The sample treatment after labeling is minimal.
(f) The spectral interpretation is not affected by the presence of residual labeling reagent.
(g) The ionization efficiency is enhanced by labeling.
(h) The labeling does not affect chromatographic retention time.
(i) The labeling allows multiplexing of two or more samples.
(j) The method is applicable to all types of biological samples (e.g. plasma, cells, tissues...).

In the following sections the methods commonly in use to chemically label peptides with stable isotopes are described.

27.2.1 Heavy Oxygen $^{16}O_2/^{18}O_2$

The $^{16}O_2/^{18}O_2$ strategy for quantitative proteomics relies on the enzymic incorporation of heavy oxygen into the peptide bond of transpeptidation products and into the non-hydrolyzed substrate, catalyzed by serine proteases (Fenselau and Yao, 2009). Catalytic $^{18}O_2$ exchange has been widely described using trypsin but other enzymes such as Glu-C, chymotrypsin and Lys-C have been used with success (Antonov et al., 1981; Mirgorodskaya et al., 2000; Reynolds et al., 2002; Schnolzer et al., 1996). In $^{18}O_2$ labeling, two atoms of heavy oxygen, derived from heavy water $H_2{}^{18}O$, are introduced at the carboxy termini of each peptide generated as a protein is cleaved (note, the carboxy terminus of the protein is not labeled). The first ^{18}O atom is introduced upon the cleavage of the peptidic amide bond and the second ^{18}O atom introduction is catalyzed when the cleaved peptide is bound as an intermediate to the enzyme and this second bond is cleaved in presence of heavy water (Table 27.1). The resulting heavy peptides have a mass 2 ($1\times{}^{18}O$) or 4 ($2\times{}^{18}O$) Da higher than their light counterpart and the two molecular species can be resolved in medium-high resolution mass spectrometers (Stewart et al., 2001). Due to the characteristics of the exchange reaction the critical parameters affecting the labeling efficiency are:

(a) ^{16}Oxygen back-exchange through hydrolysis.
(b) Effect of pH.
(c) Nature of the peptides.
(d) The type of enzyme used. For example, immobilized enzyme allows the quick removal of the enzyme from the reaction, thereby reducing back hydrolysis, which can remove the incorporated label.
(e) Relative concentration of $^{16}O_2/^{18}O_2$ in the reaction/digestion buffer (water enriched >95% with ^{18}O will give good efficiency of labeling).
(f) Biased sample loss during sample handling.

One of the main drawbacks of this technique is that the exchange reaction is not always complete for all the peptides, resulting in a complex isotopic pattern due to the overlap of the natural envelopes of unlabeled, singly labeled and doubly labeled peptides. This effect requires the use of complex software applications to accurately calculate the ratio of incomplete labeled peptides (Ramos-Fernandez et al., 2007) To overcome this disadvantage, the labeling reaction has been decoupled from the enzymic digestion (Fenselau and Yao, 2009; Yao et al., 2003) since the optimal kinetics of the two reactions are slightly different. It was shown that peptides with C-terminal arginine have a higher binding constant, K_m to the enzyme than peptides with C-terminal lysine, but K_{cat} (turnover rate) values are similar producing a bias toward amino-R labeling. To reduce this effect the reaction time should be extended in order to have all the peptides labeled with the same intensity. Using this so-called decoupled labeling procedure, $^{16}O_2/^{18}O_2$ labeling has been shown to be comparable in accuracy to SILAC labeling for quantitative proteomics (Gevaert et al., 2008).

Table 27.1 The various isotopic labels available, their source, site of labeling and structure

Label	Supplier	Modified site	Structure
iTRAQ isobaric Tags for Relative and Absolute Quantitation (4 channels, 8 channels) (Ross et al., 2004)	Applied Biosystems (Life Technologies)	Free amino groups	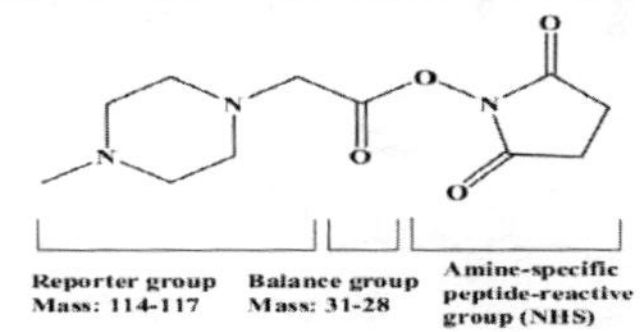
mTraq a non isobaric version of iTRAQ (Holzmann et al., 2009)	Applied Biosystems (Life Technologies)	Free amino groups	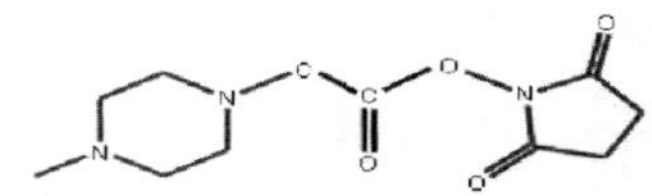
Silac Stable Isotopic Labeling using Amino Acids in Cell Culture, metabolic labeling of proteins (2 channels N15, C13). Typically Lysine or Arginine are used as the labeled amino acids (Ong et al., 2002)	Thermo Scientific, Pierce Dundee Cell products Cambridge Isotope laboratories	Metabolic Incorporation Lys and Arg	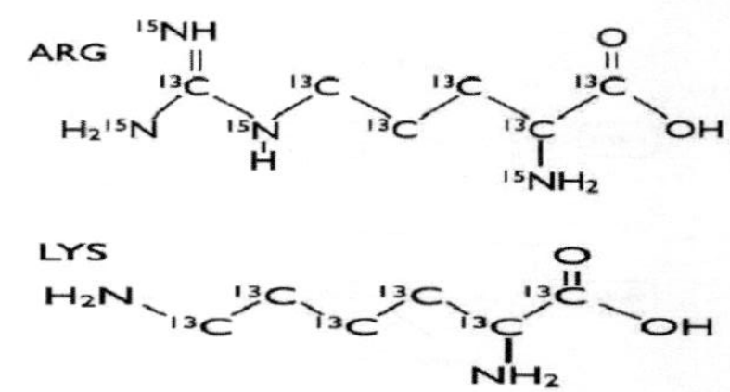

Table 27.1 (continued)

Label	Supplier	Modified site	Structure
^{18}O peptide based (2 channel) Mechanism of trypsin-catalyzed incorporation of ^{18}O into proteolytic fragments. One atom of ^{18}O is introduced by the reaction leading to cleavage of the peptide bond. A second ^{18}O atom is introduced by repeated binding/hydrolysis cycles with the proteolytic peptide fragment as a pseudosubstrate. (Ramos-Fernandez et al., 2007)	Cambridge Isotope Laboratories Marshall Isotopes Ltd.	Enzymic Incorporation Commonly at Lys and Arg via Trypsin	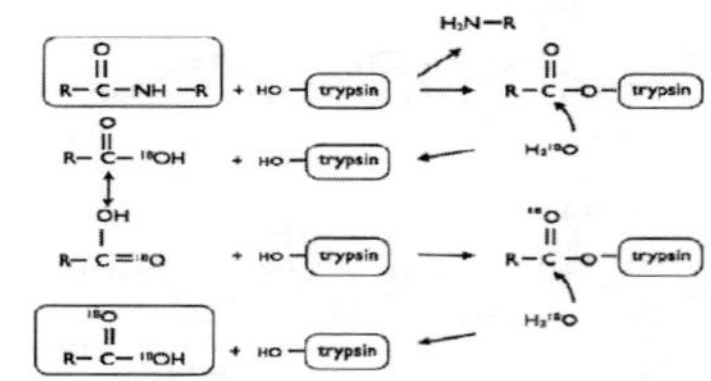
iCAT (isotopically coded affinity tag) (2 channel) (Gygi et al., 1999)	ABSciex	Thiol group	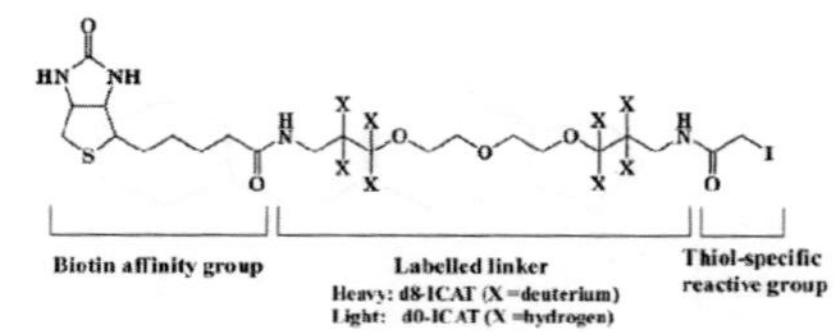
CILAT (Zeng and Li, 2009)	Not commercially available	Thiol group	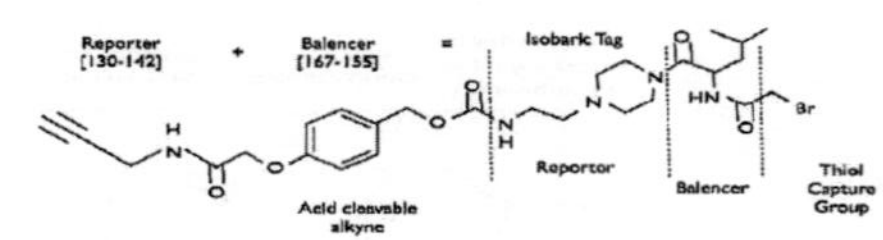
TMT Tandem Mass Tag Technology Peptide based on same approach as iTRAQ (2 or 6 channels) (Thompson et al., 2003)	Thermo Scientific, Pierce	Free amino groups	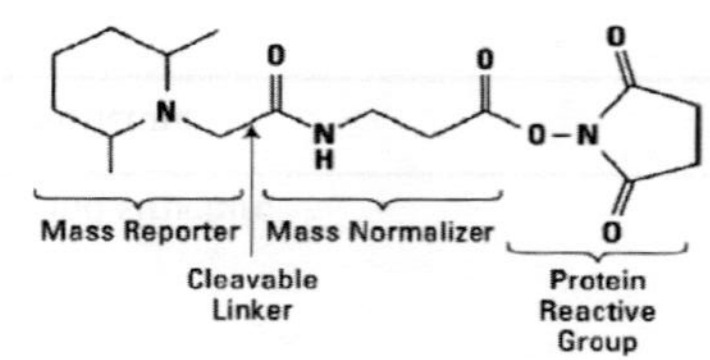

Table 27.1 (continued)

Label	Supplier	Modified site	Structure
ICPL Isotope-coded *protein label* (3 channels) (Schmidt et al., 2009)	Serva, Bruker	Free amino groups	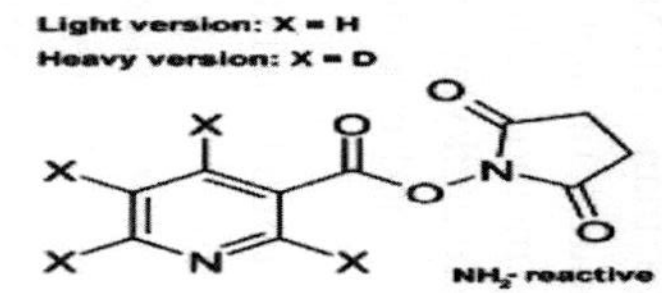
Reductive amination (Dimethyl labeling). This method uses a simple reagent, formaldehyde, to label free amino groups. (Hsu et al., 2003)	Aldrich, Fluka, Isotec.	Amino groups Except n-term Proline	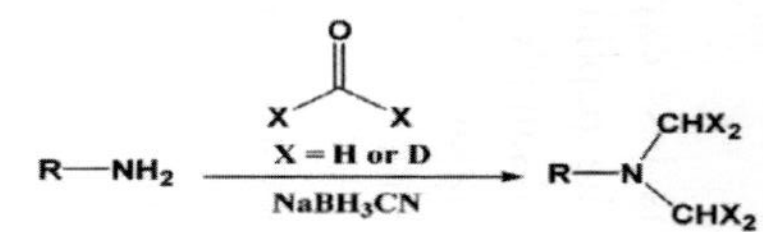
ICP-MS-Based strategies for protein quantification eg. MECAT(Ahrends et al., 2009; Bettmer et al., 2009)	The Proteome Factory www. proteomefactory.com	Thiol group	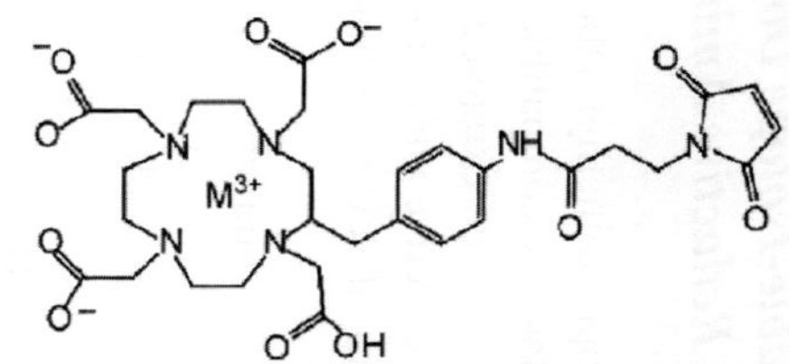 metal chelating spacer maleimide thiol label

27.2.2 *Stable-Isotope Dimethyl Labeling by Reductive Amination*

Stable isotopic dimethyl labeling, is less expensive than iTRAQ or SILAC, is extremely fast, is compatible with the peptide products of any protease and its applicability can be extended to most samples. This method relies upon the reaction called reductive amination (also known as *reductive* alkylation) and the use of combinations of several isotopomers of formaldehyde and cyanoborohydride (see Table 27.1).

In reductive amination all primary amines in a peptide mixture (e.g. tryptic digest) are converted, through the formation of an intermediate Schiff base and subsequent reduction with cyanoborohydride, into dimethylamines resulting in peptide triplets that differ in mass by a minimum of 4 Da (or 8 Da when cleaved after a lysine residue) (Hsu et al., 2003). This labeling strategy produces peaks differing by 28 mass units for each derivatized site relative to its non-derivatized counterpart and 4 mass units for each derivatized isotopic pair.

The suitability of this method has been demonstrated in several applications including phospho-proteome comparison in zebrafish embryos (Lemeer et al., 2008). Another advantage of this approach is that it can be automated by performing the reaction on line with LC-MS (Boersema et al., 2009). One drawback of this procedure, is possible shifts in HPLC retention times due to the increased of hydrophobicity of dimethylated peptides versus their non-labeled counterparts. To overcome this, it is important to completely label the peptides. It has been reported that partial resolution of light and heavy tags during LC separation is likely to occur for hydrogen-based isotopes, which may reduce the accuracy of quantitation (Zhang et al., 2002).

27.2.3 *iTRAQ: Isobaric Tags for Relative and Absolute Quantitation*

iTRAQ Reagents are a multiplexed set of four (Ross et al., 2004) or eight (Choe et al., 2007) isobaric reagents which are amine specific and yield labeled peptides which are identical in mass and hence appear as single peaks in MS scans, but which produce diagnostic, low-mass MS/MS signature ions allowing for quantitation of up to eight different samples simultaneously.

Table 27.1 shows the structure of the iTRAQ label. The complete molecule consists of a reporter group (based on *N*-methylpiperazine), a mass balance group (carbonyl) and a peptide-reactive group (NHS ester). The overall mass of reporter and balance components of the molecule are kept constant by using differential isotopic enrichment with ^{13}C, ^{15}N, and ^{18}O atoms. The number and position of enriched centers in the ring has no effect on chromatographic or MS behavior. The reporter group ranges in mass from *m*/*z* 114 to 117 (for the 8-plex iTRAQ reagents: 113–119, 121), while the balance group ranges in mass from 28 to 31 Da, such

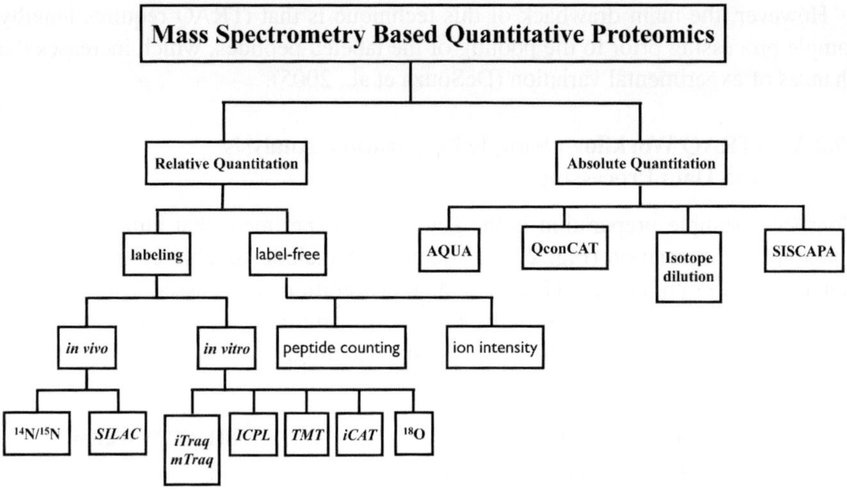

Fig. 27.1 Scheme showing the various methods of mass spectrometry based protein quantitation

that the combined mass remains constant (145 Da) for each of the four reagents. The reporter ions occur in a "silent region" of the typical peptide MS/MS spectrum, which results in reduced interference from background noise in the quantitative data. When reacted with a peptide, the tag forms an amide linkage to any peptide amine (N-terminal or ε-amino group of lysine). These amide linkages fragment in a similar fashion to backbone peptide bonds when subjected to CID. Following fragmentation of the tag amide bond, however, the balance (carbonyl) moiety is lost (neutral loss), while charge is retained by the reporter group fragment. A mixture of four identical peptides each labeled with one member of the multiplex set appears as a single, unresolved precursor ion in MS (identical m/z). This results in increased signal intensity and an increased probability of correct peptide identification, particularly for non-abundant proteins (Yan and Chen, 2005). Following CID, the four reporter group ions appear as distinct masses (114–117 Da). All other sequence-informative fragment ions (b-, y-, etc.) remain isobaric. The relative concentration of the peptides and the proteins in the samples is thus deduced from the relative intensities of the corresponding reporter ions. In contrast to ICAT and similar mass-difference labeling strategies, quantitation is thus performed at the MS/MS stage rather than in MS.

iTRAQ is often used to identify potential biomarkers from clinical specimens that could be used for the development of analytical methods for diagnosis and drug discovery. In these kinds of studies, the multiplexing ability afforded by the iTRAQ reagents allows the comparative proteome analysis of different biological samples, resulting in the identification of protein clusters and pathways that show consistent trends of differential expression. This has turned out to be particularly useful for following biological systems over multiple time points or, more generally, for comparing multiple treatments in the same experiment.

However, the main drawback of this technique is that iTRAQ requires lengthy sample processing prior to the pooling of the labeled peptides, which increases the chances of experimental variation (DeSouza et al., 2005).

27.2.3.1 iTRAQ Workflow: Sample Preparation, Analysis and Data Processing

Consistent sample preparation is the key to any experiment that aims to provide quantitative information (Fig. 27.2). The need to include suitable control samples cannot be over emphasized. iTRAQ sample preparation is discussed in some detail below, but many of the points raised apply to any quantitative proteomics experiment based on chemical or metabolic stable isotope labeling.

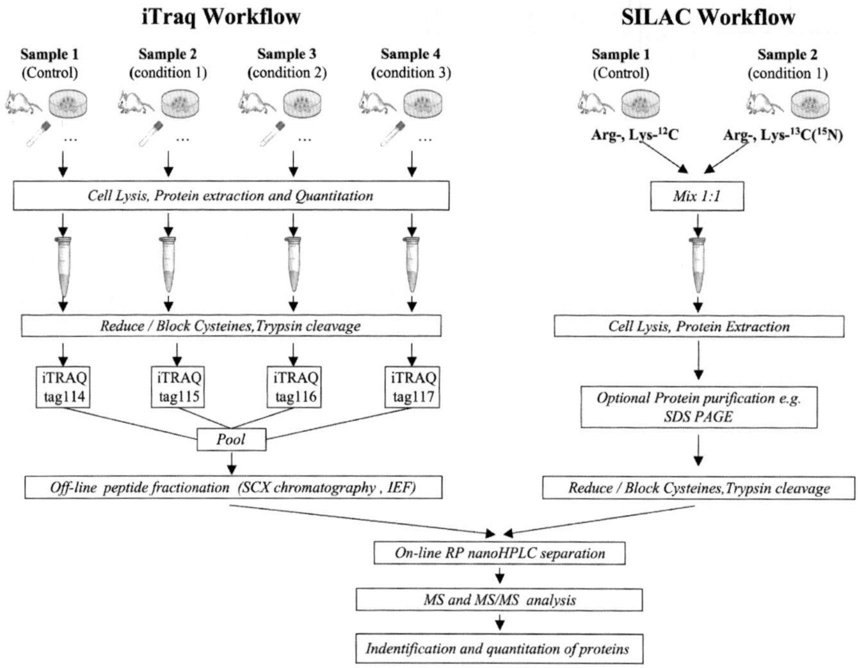

Fig. 27.2 The basic outlines of iTRAQ and SILAC labeling experimental strategies. For iTRAQ, proteins are extracted either from cells, tissue, growth medium, etc. The amount of protein is measured, the proteins are cleaved with trypsin, then labeled with an iTRAQ reagent (4 are shown but an 8-plex version is available). At this point the samples are pooled for further separation and analysis by MS/MS. For SILAC, cells are grown in either "light" (control) or "heavy" (condition) SILAC medium containing amino acids with stable isotopes (^{13}C, ^{15}N, or ^{2}H). After completion of labeling and differential treatment of SILAC cultures, cells or protein extracts are combined. Peptides are generated by incubation of proteins with trypsin and analyzed by mass spectrometry. Intensities of unlabeled and stable isotope labeled peptide pairs in the mass spectrum can be used to calculate relative abundances in light and heavy SILAC samples. These basic schemes can be altered by introducing additional steps during sample purification, e.g., affinity purification of selected proteins, chromatographic fractionation or gel electrophoresis

iTRAQ Sample Preparation and Labeling

Biological systems are chemically highly active and this activity not only continues during sample preparation, but also is likely to be changed by the process itself. For example, human cells produce well over 500 proteases (Lopez-Otin and Bond, 2008), many of which will be present and active in cell lysates unless they are inhibited. Disruption of cells also brings molecules into contact that may normally be cloistered from one another (Villanueva et al., 2009). The first step is to extract proteins from complex biological samples (human and mice tissues from different organs, cells, etc.) (Grassl et al., 2009). Different lysis buffers as well as triethylammonium bicarbonate (TEAB) or PBS with protease inhibitors are used for this purpose, followed by boiling, homogenization, sonication, etc. of the sample extract. Insoluble debris are pelleted by centrifugation and protein extracts can be cleaned-up overnight by acetone or TCA precipitation, collected by centrifugation, dried in air and resuspended in TEAB. When dealing with plasma samples, a preliminary depletion process (Bhat et al., 2005) for the most abundant proteins (in most cases albumin and IgG) must be carried out (Song et al., 2008).

To ensure complete labeling, it is necessary to measure the protein concentration before labeling. Total protein concentration is usually measured with the Coomassie (Bradford) or bicinchonic acid assays (Protein Assay Kit, Bio-Rad, Hercules, CA; Sigma, St. Louis, MO; Pierce Biotechnology, Rockford, IL.).

Equal amounts of protein (50–100 μg) from each sample are denatured with 2 % SDS, reduced with 50 mM tris (2-carboxyethyl) phosphine (TCEP) at 60ºC for 1 h and the cysteine residues are subsequently alkylated with 200 mM methyl methanethiosulphonate (MMTS) at room temperature for 15 min. Protein enzymic cleavage is typically carried out overnight with trypsin (enzyme/substrate 1:20, *w/w*) at 37°C. Each tryptic digest is then labelled according to manufacturer's instructions with one isobaric amine-reactive tag by adding ethanol (70% v/v final ethanol) and allowed to react for 1 hour at room temperature. The different reactions are then combined and evaporated to dryness. From this point on the reproducibility due to sample-specific variation is eliminated or much reduced.

Additionally, a protocol has been described for intact protein labeling using iTRAQ (Wiese et al., 2007). Labeling of intact proteins can be advantageous since it allows for further protein separation steps on the combined samples. However, trypsin does not cleave modified lysine residues, which leads to significantly longer peptides that generally are more difficult to identify by MS. Also, very high labeling efficiencies are required in case further protein separation is desired prior to MS analysis, since incomplete labeling impairs the resolving power achievable with, e.g., 1D and 2D gel electrophoresis.

iTRAQ Peptide Fractionation

Multidimensional fractionation is required to reduce the complexity of the peptide mixtures to be analyzed by MS/MS. This improves sequence coverage and the quantitative results. Therefore, the peptide mixtures obtained after labeling are often separated offline by strong cation exchange (SCX) chromatography or by isoelectric focusing. For SCX, peptides are loaded onto columns such as PolySulfoethyl A containing anionic polymer (poly-2-sulfethyl aspartamide) (PolyLC, Columbia, MD) at low pH (pH<3) and eluted with a linear gradients of increasing salt (KCl) concentrations. Fractions are collected at short intervals and later pooled together according to variations in peak intensity. Samples are then dried for subsequent RP nanoLC fractionation. Prior to this step, some authors prefer to desalt the eluates from the ion exchange step using tools such as Sep-Pak cartridges or ZipTips™ (Millipore, Milford, MA).

Immobilized pH gradient isoelectric focusing (IPG-IEF) has emerged as a highly promising alternative to strong-cation exchange fractionation as the first separation dimension in shot-gun proteomics (Cargile et al., 2005), especially for membrane proteome analysis (Chick et al., 2008; Eriksson et al., 2008). Depending on the pH range used in the focusing step, it is possible to reduce sample complexity, which increases the possibility of identifying low-abundance proteins. Narrow range IPG-strips for peptide focusing and application gels can be purchased from GE Healthcare Bio-Sciences (AB, Uppsala, Sweden). Other fractionation strategies make use of HILIC and RP chromatography (McNulty and Annan, 2008), or high and low pH RP combinations (Dowell et al., 2008).

Analysis by Nano Reversed Phase Liquid Chromatography (Nano RP LC) MS/MS

Each fraction previously collected is further separated by nano RP LC and peptides are eluted using a binary gradient of up to 170 min. Packed particle (e.g. C18 PepMap (75–100 μm ID × 150 mm) (Dionex-LC Packings, Amsterdam, The Netherlands)) or monolithic RP C18 columns (e.g. PS-DVB (200-μm-ID × 5 cm) analytical column (Dionex-LC Packings)) with flow rates ranging from 200 nL/min to 2.7 μL/min are typically used.

If LC MALDI MS/MS workflows are applied, then a nanoLC system equipped with a spotting device such as a Probot™ MALDI (Dionex-LC Packings) are needed. The column effluent is mixed directly in a 1:1–1:2 ratio with the MALDI matrix solution (2–7 mg/ml α-cyano-4-hydroxycinnamic acid in 70% ACN with ammonium citrate or ammonium phosphate) through a μ-Tee connection, before spotting onto MALDI target plates with a fixed speed of 5–20 s/well. The matrix usually contains an internal standard control for mass calibration (Fibrinopeptide B–Glufib-). Plates are then analysed on instruments such as an ABI 4700, 4800 or 5800 Analyzer equipped with TOF/TOF ion optics (Applied Biosystems, Framingham, MA; AB Sciex Concord, Ontario, Canada) or Ultraflex TOF/TOF (Bruker, Bremen, Germany).

For electrospray analysis, nanoLC systems are directly interfaced to hybrid ESI Quadrupole time-of-flight tandem mass spectrometers (QTOF MS/MS) such as Qstar (AB Sciex, Concord, Ontario Canada) or LTQ Linear Ion Trap (Thermo Fisher Scientific, Waltham, MA).

27.2.4 Isotope Dilution

Although the above methods allow the comparison of expressed proteins i.e. their relative quantitation, the absolute quantity of a targeted protein cannot be easily measured from a complex mixture. The AQUA (Absolute Quantification) strategy (Gerber et al., 2007; Gerber et al., 2003) is a technique that addresses this issue. This approach uses ^{13}C-labeled reference peptides and tandem MS to measure expression in terms of number of molecules per cell. Reference peptides, corresponding to specific, pre-defined proteolytic fragments of proteins of interest, are synthesized with an incorporated stable isotope. Samples containing the protein of interest are then subjected to complete digestion with trypsin, followed by the addition of the isotopically labeled peptide in a known amount. LC-MS is subsequently used to quantitate the naturally occurring peptide from the protein of interest by comparison with the level of the corresponding internal standard. This technique is made possible by the use of stable isotopes to generate peptide standards that are structurally identical but occur 2–10 Daltons higher in mass to charge. This enables the quantitation of peptides in a stoichiometric sense using the synthetic peptide as internal standard. A variation on the general principal of isotope dilution, termed SISCAPA (Anderson et al., 2004; Whiteaker et al., 2009) uses anti-peptide antibodies to capture peptides of interest, primarily from serum, along with a spiked in synthetic heavy standard, to improve the selectivity and sensitivity of the method. An interesting alternative to chemical peptide synthesis of the heavy peptides is called QconCAT (Rivers et al., 2007). Standard peptides are coded in a synthetic gene, from which an artificial protein is made. The synthetic gene is expressed in media containing heavy isotopes and after purification of the protein, cleavage with trypsin produces an equimolar mixture of all labeled standard peptides.

27.3 Metabolic Incorporation of Stable Isotopes

The premise behind this technique is simple. Cells (Beynon and Pratt, 2005) or even mice (Kruger et al., 2008) are fed either with amino acids containing natural isotopic patterns or with amino acids synthesized to contain heavy isotopes which leads to heavy isotope labeled proteins. When peptides are subsequently prepared for analysis by mass spectrometry it is possible by comparing the signals from the heavy and light peptides to measure the relative amounts of protein generated under different growth conditions.

27.3.1 Stable Isotope Labeling by Amino Acids in Cell Culture (SILAC)

Stable isotope labeling by amino acids in cell culture (SILAC) has become a commonly used metabolic labeling strategy that uses stable isotope-labeled amino acids added to the growth medium to encode cellular proteomes for quantitative analysis

(Ong et al., 2002). The general SILAC strategy (Fig. 27.2) is a straightforward approach in which one cell population is labeled with a light amino acid (control) and another cell population is labeled with its heavy counterpart (condition 1). The heavy amino acid can contain ^{2}H, ^{13}C or ^{15}N instead of ^{1}H, ^{12}C, or ^{14}N. Incorporation of the heavy amino acid into a peptide leads to a known mass shift compared to the peptide that contains the light version of the same amino acid (for example, 6 Da in the case of $^{13}C_6$-Lys), but no other chemical changes. After obtaining approximately 100% labeling (typically after 5–6 cell doublings), the cells from the two states are mixed and their proteomes extracted and analyzed by mass spectrometry (MS). Each peptide appears as a pair in the mass spectra, the peptide with a lower mass contains the light amino acid and originates from a protein belonging to the control, and the peptide with a higher mass contains the heavy amino acid and originates from the same protein but from a source subjected to different growth conditions. Because the light and heavy amino acids are chemically identical, except for their mass difference, the ratio of the peak intensities in the mass spectrometer directly yields the ratio of the proteins in the cell population corresponding to state A versus state B (Mann, 2006).

SILAC is characterized by two main aspects: 100% labeling efficiency can be achieved allowing the cells to grow in the culture media for 5–6 doublings of the cell population, and both samples (states A and B) can be combined at the very beginning of the analytical process (right after completion of the labeling). These two features make SILAC one of the most reliable methods for protein quantification developed so far as it avoids errors that may be introduced during the sample preparation procedures. This is especially important when using complex protein fractionation/purification methods, which are prone to introduce errors in the final results if carried out independently for the two samples.

SILAC medium can be prepared identically to the medium used in common cell cultures. However, other potential sources of unlabeled amino acids have to be previously dialyzed away since they can compete with the heavy amino acids, resulting in only partial protein labeling. When selecting a SILAC heavy amino acid to replace the natural light counterpart present in the common media, there are several criteria that should be taken into account (see Table 27.2):

(a) Although in principle, any amino acid could be used, ideally, the SILAC amino acid should be essential for survival of the cultured cells, and such is the case of lysine (Lys), methionine (Met) and leucine (Leu). Arginine (Arg) is not an essential amino acid but has been shown to be essential in many cultured cell lines (Ong et al., 2003; Ong and Mann, 2006). Moreover, it is also important to consider the enzyme used for protein digestion. Since trypsin is the most common proteolytic enzyme used and it cleaves every Arg and Lys residue, the use of labeled Arg and Lys will assure that all the peptides will contain at least one isotopic label; thus, making possible the quantitation of all the identified peptides.
(b) The mass shift observed between the light and the heavy peptides should be of at least 4 Da to avoid overlapping isotopic distributions.

Table 27.2 Possible stable amino acid precursors for metabolic labeling

Amino acid	Accessible atoms, representative stable isotope-labeled variants	Comments
Alanine C:3, N:1, *H:(3 + 1)*a*	U-[$^{13}C_3$]Ala, 3-[^{13}C]Ala, 1-[^{13}C]Ala, 2,3,3,3-[$^2H_4$]Ala	Transamination to pyruvate, and thence to lactate, which makes the alanine pool metabolically very mobile.
Arginine C:6, N:4, H:(6 + 1)	U-[$^{13}C_6$]Arg, Guanidino-[$^{15}N_2$]Arg, 2,3,3,4,4,5,5-[$^2H_7$]Arg	Susceptible to the action of arginase to form urea and ornithine, which might result in loss of guanidine label. Ornithine converted to glutamate semialdehyde, which is cyclized to pyrroline 5 carboxylate, a precursor of proline. Appearance of ^{13}C label through metabolism of Arg to Pro seen experimentally (Berger et al., 2002). Convenient in analysis of tryptic peptides (and all endopeptidase ArgC-generated peptides), as all Arg-terminated peptides will contain a single stable isotope label. Helps in *y* ion assignment of MS/MS spectra for peptide sequencing.
Aspartate, asparagine glutamate, glutamine,	Many variants	The acidic amino acid and their amides are so actively metabolized and play such a central role in amino acid deamination that label might be very rapidly lost or redistributed. If these amino acids are to be used it would be important to track the fate of the label in preliminary experiments.
Glycine C:2, N:1, H:(2)	U-[$^{13}C_2$]Gly, 2,2-[$^2H_4$]Gly	Maximal 2 amu separation of heavy and light causes difficulty in separation of natural [^{13}C] isotopomer envelopes.
Histidine C:6, N:3, H:(4 + 1)	U-[$^{13}C_6$]His, U-[$^{15}N_3$]His, 1,2,2-[$^2H_3$]His	Essential amino acid. Histidine is degraded to glutamate.
Isoleucine C:6, N:1, H:(9 + 1)	U-[$^{13}C_6$]Ile	Essential amino acid. Allows differentiation between isoleucine and leucine.
Leucine C:6, N:1, H:(9 + 1)	U-[$^{13}C_6$]Leu, 2,3,3,3,5,5,5,5′,5′,5′-[$^2H_{10}$]Leu	Essential amino acid. Rapidly transaminated, removing the a carbon deuteron (Pratt et al., 2002a, 2002b). Allows differentiation between Leu and Ile residues.

Table 27.2 (continued)

Amino acid	Accessible atoms, representative stable isotope-labeled variants	Comments
Lysine C:6, N:2, H: (8 + 1)	U-[$^{13}C_6$]Lys, 4,4,5,5 [$^2H_4$]Lys, 3,3,4,4,5,5,6,6 [$^2H_8$]Lys	Essential amino acid. Degraded to acetyl-CoA and could in principle recycle label to other amino acids . Convenient in analysis of tryptic peptides (and all LysC-generated peptides), as all Lys-terminated peptides will contain a single stable isotope label. Helps in *y* ion assignment of MS/MS spectra for peptide sequencing. Readily available in 4-atomic mass unit variants.
Methionine C:5, N:1, H:(7 + 1)	U-[$^{13}C_5$]Met, methyl-[$^2H_3$]Met	The *S*-methyl group can be a highly metabolically active methyl donor and label specific to this position could be very labile. Methionine is a precursor of cysteine.
Phenylalanine C:9, N:1, H:(7 + 1)	Ring-[$^{13}C_6$]Phe, U-[$^{13}C_9$]-Phe, Ring (2,3,4,5,6)-[$^2H_5$]Phe, a,b,b,2,3,4,5,6-[$^2H_7$]Phe	Essential amino acid. Phenylalanine is readily converted to tyrosine by phenylalanine hydroxylase.
Serine C:3, N:1, H:(3 + 1)	U-[$^{13}C_3$]-Ser, 2,3,3-[$^2H_3$]-Ser	Degraded to pyruvate. Synthesized from glycolytic pathway precursor. Maximal 3-atomic mass unit separation of heavy and light causes difficulty in separation of natural [^{13}C] isotopomer envelopes. Marks serine-containing peptides to aid studies on serine phosphorylation.
Tryptophan C:11, N:2, H:(7 + 1)	U-[$^{13}C_{11}$]-Trp	Given the scarcity of tryptophan in proteomes, there seems little justification for use of this amino acid.
Tyrosine C:9, N:1, H:(7 + 1)	Ring-[$^{13}C_6$]-Tyr, U-[$^{13}C_9$]-Tyr, b,b -[$^2H_2$]-Tyr, ring (2,3,5,6)-[^{2}H4]-Tyr	Degraded to acetyl-CoA. Can be used to label tyrosine-containing peptides to aid studies on tyrosine phosphorylation.
Valine C:5, N:1, H:(7 + 1)	U-[$^{13}C_5$]-Val, [^{15}N],U-[$^{13}C_5$]-Val, [$^2H_8$]-Val	Essential amino acid. Readily transaminated, removing the a carbon deuteron.

*The designation h(n+1) refers to the number of hydrogen atoms (n) other than those associated with the carboxyl and amino groups in addition to the a–carbon hydrogen (referred to by "+1").

(c) The characteristics of isotopic peptides in reversed phase chromatography have to be considered. SILAC amino acids containing deuterium (^{2}H) might be cheaper than those containing ^{13}C or ^{15}N. However, the effect of ^{2}H in the retention time is higher, what may cause the light and the heavy peptides to separate by HPLC, and that can considerably complicate the data analysis (Zhang et al., 2002).

(d) Metabolic interconversion of Arg to proline (Pro) in certain cell lines has to be accounted for. This conversion has been observed in several cell lines and it usually occurs when an excess of Arg is provided to the cells in the culture media. The resulting converted proline peptide divides the heavy peptide signal, causing inaccuracy when compared with the light peptide signal. This is of particular concern as it can effect up to half of all peptides in a SILAC experiment. Also, the reverse metabolic conversion of Pro to Arg can occur in Arg-deficient media. Several alternatives have been proposed to compensate for and limit this problem such as the titration of the Arg concentration for each new cell line by culturing the cells with varying amounts of Arg and analyzing the degree of incorporation by MS. Recently, it has been shown (Bendall et al., 2008) that the addition of as little as 200 mg/liter of Pro to the SILAC media can render the conversion of Arg to Pro completely undetectable without observable back conversion from the Pro supplement.

(e) It has been reported that certain cell lines grow slower in media containing SILAC amino acids. Although this phenomenon has been rarely observed, it can often be solved by increasing the concentration of dialyzed serum in the culture media (Iliuk et al., 2009; Imami et al., 2010).

(f) Another important factor that has to be considered when designing a SILAC experiment is the nature of the cells used. Culture of immortalized cell lines is relatively simple; however, some modifications need to be implemented when using primary cells. Primary culture usually refers to differentiated cells that are harvested from a living organism and maintained in culture. Spellman et al. studied the proteomes of neuronal molecular and cellular dynamics combining SILAC and primary neuronal cell culture (Spellman et al., 2008). To assure that SILAC could be applied to non-dividing neurons, which do not have the advantage of label dilution effects enjoyed by dividing cells, they tracked the levels of label incorporation for individual proteins and showed that the majority of proteins did indeed achieve significant levels of stable isotope labeling. Calculation of the percent of label incorporation in a parallel experiment enables correction of the final ratios obtained in the SILAC experiment (Zhang et al., 2009).

27.4 Data Processing and Analysis

Quantitative proteomics experiments generate large amounts of data, which can only be processed and evaluated using appropriate software. It is beyond the scope of this article to discuss software in detail and the reader is referred to several recent papers that discuss this topic in more detail (Brusniak et al., 2008; Cox et al., 2009; Mueller et al., 2008). There are numerous applications, both commercial and open source available (see Table 27.3), making it difficult to find an appropriate program for the analysis of data generated by a specific instrument. In addition to the limited applicability of a program to different MS platforms, practical factors such as file compatibility, computational requirements, user friendliness, data visualization, variations in the sample preparation protocols, etc., are critical factors that affect the choice of a data analysis program. A major advancement in file compatibility of LC-MS data was made through the development of generic MS file formats for proteomics data (Orchard et al., 2009) which provide the facility to import, archive, and process LC-MS data from most mass spectrometer types into instrument independent analysis platforms (Keller et al., 2005; Orchard et al., 2009). Another important consideration is the computation time required by the software to process the data. Proteomics studies often contain a large number of LC-MS measurements and require considerable computational resources to complete data analysis within a reasonable amount of time. Finally, complementary to the automated processing of large data sets, data visualization is important to assess the quality of the individual processing steps carried out by the software.

Prior to quantification proteins first have to be identified, using a search engine such as Mascot (Matrix Science, London, United Kingdom) (Perkins et al., 1999), Paragon^{TM} algorithm (Applied Biosystems, MDS Sciex) (Shilov et al., 2007; Tang et al., 2005), X!Tandem (GPM, Beavis Informatics Ltd.) (Craig and Beavis, 2003), OMSSA (http://pubchem.ncbi.nlm.nih.gov/omssa/) (Geer et al., 2004), Sequest (Thermo Fisher Scientific, Waltham, MA) (Gatlin et al., 2000)... (for further information the reader is referred to (Deutsch et al., 2008; Nesvizhskii et al., 2007)).

After identification the quantitative data have to be generated using a suitable application (see Table 27.3). In order to identify clusters of proteins showing consistent trends of differential expression, the data can be subjected to *K*-means cluster analysis using the tools in the statistical software SPSS (SPSS, Chicago, Illinois) or alternative statistical approaches.

Gene Ontology classification of the identified proteins can be carried out using tools like Protein Center (Proxeon Bioinformatics, Odense, Denmark) or Celera Discovery System (Celera Genomics, Rockville, MD) (Kerlavage et al., 2002) These tools can be used to organize the proteins into major cellular components, molecular functions, biological processes, etc. Additionally, software packages such as Ingenuity Pathways Analysis software (http://www.ingenuity.com), Ingenuity Systems®, Redwood City, CA, USA) or Cytoscape software (Shannon et al., 2003) can be used to overlay the quantified proteins onto molecular networks developed from information contained in the literature, based on known regulatory relationships such as protein interactions, modifications, regulation of expression, etc. These networks can also be associated to biological functions and/or diseases.

Table 27.3 Software packages available for processing datasets generated by Mass Spectrometry from stable isotope labeled proteomics samples

Software	Operating systems	Tested data types	Input format	Compatible labels	Availability
Multi-Q (Lin et al., 2006)	Windows Web version,	Mascot/Sequest results	mzXML	iTRAQ	http://ms.iis.sinica.edu.tw/Multi-Q
iTracker (Griffin et al., 2007)	Linux, OSX, Windows	via TPP	.mgf, .dat	iTRAQ	http://www.cranfield.ac.uk/health/researchareas/bioinformatics/page3201.jsp
Libra (Keller et al., 2005)	Linux, OSX, Windows	via TPP	mzXML	iTRAQ	http://tools.proteomecenter.org/wiki/
ProQuant	Windows	QStar, Qtrap	raw file	iTRAQ	http://www.appliedbiosystems.com
ProteinPilot	Windows	QStar, Qtrap, Maldi-Tof/Tof	raw file	iTRAQ	http://www.appliedbiosystems.com
XPRESS	Linux, OSX, Windows	LTQ, OrbiTrap, Qtof, FT-LTQ	mzXML	ICAT, ICPL, SILAC	http://tools.proteomecenter.org/wiki
ASAPRatio (Li et al., 2003)	Linux, OSX, Windows	LTQ, OrbiTrap, Qtof, FT-LTQ	mzXML	^{2}H, ICAT, SILAC, ICPL	http://tools.proteomecenter.org/wiki/
PeakPicker	Windows	Maldi-Tof/Tof	raw file	ICPL	http://www.appliedbiosystems.com
WARP-LC	Windows	Qtof	raw file	ICPL	http://www.bdal.com/
ZoomQuant (Halligan et al., 2005)	Linux, OSX, Windows	LTQ	raw file	^{18}O	http://proteomics.mcw.edu/zoomquant/
STEM (Shinkawa et al., 2005)	Windows	Mascot results	.pkl file	^{18}O	http://www.sci.metro-u.ac.jp/proteomicslab/STEMDLP-0.html

Table 27.3 (continued)

Software	Operating systems	Tested data types	Input format	Compatible labels	Availability
MSQuant (Mortensen et al., 2009; Schulze and Mann, 2004)	Windows	QStar, Qtof, FT-LTQ	raw file, .wiff .dat	SILAC, ICAT, ICPL, ^{18}O, ^{15}N, mTraq, iTRAQ	http://msquant.alwaysdata.net/
MaxQuant (Cox et al., 2009)	Windows	QStar, Qtof, FT-LTQ	raw file, .wiff .dat	SILAC	http://www.maxquant.org/
Mascot Distiller	Windows, Linux	Qtof,	wiff .dat	^{18}O, AQUA, ICPL, iTRAQ, SILAC.	http://www.matrixscience.com
ProteinScape	Windows	Ion trap Maldi-Tof/Tof	raw file	AQUA, ICAT, iTRAQ, ICPL, SILAC+	http://www.bdal.com/
CPAS	Linux, OSX, Windows	QStar,Qtof, FT-LTQ	mzXML	^{18}O, SILAC, ICAT	http://proteomics.fhcrc.org/CPL/home.html
CENSUS (Park et al., 2008)	Linux, OSX, Windows	FT-LTQ	mzXML MS1/MS2	^{15}N, ^{18}O, SILAC, iTRAQ,	http://fields.scripps.edu/census/index.php
msInspect (Bellew et al., 2006)	Linux, OSX,	QStar, Qtof, FT-LTQ	mzXML	^{18}O	http://proteomics.fhcrc.org/CPL/msinspect/index.html
Spectrum Mill	Windows	QStar, Qtof, FT-LTQ	.RAW, .wiff, .pkl	^{15}N, ^{18}O, SILAC, iTRAQ	http://www.chem.agilent.com

27.5 Conclusions

From the forgoing discussion it should be apparent that undertaking a quantitative proteomics by stable isotope labeling experiment requires bringing together a combination of biology (the question and samples), chemistry (the labeling and separation), physics (mass spectrometry) and bioinformatics (the data analysis). It is only by close collaboration between these disciplines that the best results will be obtained, and our understanding of biology moved forward.

References

Ahrends, R., Pieper, S., Neumann, B., Scheler, C., and Linscheid, M.W. (2009). Metal-coded affinity tag labeling: A demonstration of analytical robustness and suitability for biological applications. Anal Chem *81*, 2176–2184.

Anderson, N.L., Anderson, N.G., Haines, L.R., Hardie, D.B., Olafson, R.W., and Pearson, T.W. (2004). Mass spectrometric quantitation of peptides and proteins using stable isotope standards and capture by anti-peptide Antibodies (SISCAPA). J Proteome Res *3*, 235–244.

Antonov, V.K., Ginodman, L.M., Rumsh, L.D., Kapitannikov, Y.V., Barshevskaya, T.N., Yavashev, L.P., Gurova, A.G., and Volkova, L.I. (1981). Studies on the mechanisms of action of proteolytic enzymes using heavy oxygen exchange. Eur J Biochem *117*, 195–200.

Bellew, M., Coram, M., Fitzgibbon, M., Igra, M., Randolph, T., Wang, P., May, D., Eng, J., Fang, R., Lin, C., *et al.* (2006). A suite of algorithms for the comprehensive analysis of complex protein mixtures using high-resolution LC-MS. Bioinformatics *22*, 1902–1909.

Bendall, S.C., Hughes, C., Stewart, M.H., Doble, B., Bhatia, M., and Lajoie, G.A. (2008). Prevention of amino acid conversion in SILAC experiments with embryonic stem cells. Mol Cell Proteomics *7*, 1587–1597.

Berger, S.J., Lee, S.W., Anderson, G.A., Pasa-Tolic, L., Tolic, N., Shen, Y., Zhao, R., and Smith, R.D. (2002). High-throughput global peptide proteomic analysis by combining stable isotope amino acid labeling and data-dependent multiplexed-MS/MS. Anal Chem *74*, 4994–5000.

Bettmer, J., Montes Bayon, M., Encinar, J.R., Fernandez Sanchez, M.L., Fernandez de la Campa Mdel, R., and Sanz Medel, A. (2009). The emerging role of ICP-MS in proteomic analysis. J Proteomics *72*, 989–1005.

Beynon, R.J., and Pratt, J.M. (2005). Metabolic labeling of proteins for proteomics. Mol Cell Proteomics *4*, 857–872.

Bhat, V.B., Choi, M.H., Wishnok, J.S., and Tannenbaum, S.R. (2005). Comparative plasma proteome analysis of lymphoma-bearing SJL mice. J Proteome Res *4*, 1814–1825.

Boersema, P.J., Raijmakers, R., Lemeer, S., Mohammed, S., and Heck, A.J. (2009). Multiplex peptide stable isotope dimethyl labeling for quantitative proteomics. Nat Protoc *4*, 484–494.

Boyer, P.D. (1954). Spectrophotometric study of the reaction of protein sulfhydryl groups with organic mercurials. J Am Chem Soc *76*, 4331–4337.

Brusniak, M.Y., Bodenmiller, B., Campbell, D., Cooke, K., Eddes, J., Garbutt, A., Lau, H., Letarte, S., Mueller, L.N., Sharma, V., *et al.* (2008). Corra: Computational framework and tools for LC-MS discovery and targeted mass spectrometry-based proteomics. BMC Bioinformatics *9*, 542.

Cargile, B.J., Sevinsky, J.R., Essader, A.S., Stephenson, J.L., Jr., and Bundy, J.L. (2005). Immobilized pH gradient isoelectric focusing as a first-dimension separation in shotgun proteomics. J Biomol Tech *16*, 181–189.

Chick, J.M., Haynes, P.A., Molloy, M.P., Bjellqvist, B., Baker, M.S., and Len, A.C. (2008). Characterization of the rat liver membrane proteome using peptide immobilized pH gradient isoelectric focusing. J Proteome Res *7*, 1036–1045.

Choe, L., D'Ascenzo, M., Relkin, N.R., Pappin, D., Ross, P., Williamson, B., Guertin, S., Pribil, P., and Lee, K.H. (2007). 8-plex quantitation of changes in cerebrospinal fluid protein expression in subjects undergoing intravenous immunoglobulin treatment for Alzheimer's disease. Proteomics *7*, 3651–3660.

Cox, J., Matic, I., Hilger, M., Nagaraj, N., Selbach, M., Olsen, J.V., and Mann, M. (2009). A practical guide to the MaxQuant computational platform for SILAC-based quantitative proteomics. Nat Protoc *4*, 698–705.

Craig, R., and Beavis, R.C. (2003). A method for reducing the time required to match protein sequences with tandem mass spectra. Rapid Commun Mass Spectrom *17*, 2310–2316.

De Leenheer, A.P., and Thienpont, L. M. (1992). Applications of isotope dilution-mass spectrometry in clinical chemistry, pharmacokinetics, and toxicology. Mass Spectrum Rev *11*, 249–307.

DeSouza, L., Diehl, G., Rodrigues, M.J., Guo, J., Romaschin, A.D., Colgan, T.J., and Siu, K.W. (2005). Search for cancer markers from endometrial tissues using differentially labeled tags iTRAQ and cICAT with multidimensional liquid chromatography and tandem mass spectrometry. J Proteome Res *4*, 377–386.

Deutsch, E.W., Lam, H., and Aebersold, R. (2008). Data analysis and bioinformatics tools for tandem mass spectrometry in proteomics. Physiol Genomics *33*, 18–25.

Dowell, J.A., Frost, D.C., Zhang, J., and Li, L. (2008). Comparison of two-dimensional fractionation techniques for shotgun proteomics. Anal Chem *80*, 6715–6723.

Eriksson, H., Lengqvist, J., Hedlund, J., Uhlen, K., Orre, L.M., Bjellqvist, B., Persson, B., Lehtio, J., and Jakobsson, P.J. (2008). Quantitative membrane proteomics applying narrow range peptide isoelectric focusing for studies of small cell lung cancer resistance mechanisms. Proteomics *8*, 3008–3018.

Fenselau, C., and Yao, X. (2009). $18O_2$-labeling in quantitative proteomic strategies: A status report. J Proteome Res *8*, 2140–2143.

Gatlin, C.L., Eng, J.K., Cross, S.T., Detter, J.C., and Yates, J.R., 3rd (2000). Automated identification of amino acid sequence variations in proteins by HPLC/microspray tandem mass spectrometry. Anal Chem *72*, 757–763.

Geer, L.Y., Markey, S.P., Kowalak, J.A., Wagner, L., Xu, M., Maynard, D.M., Yang, X., Shi, W., and Bryant, S.H. (2004). Open mass spectrometry search algorithm. J Proteome Res *3*, 958–964.

Gerber, S.A., Kettenbach, A.N., Rush, J., and Gygi, S.P. (2007). The absolute quantification strategy: Application to phosphorylation profiling of human separase serine 1126. Methods Mol Biol *359*, 71–86.

Gerber, S.A., Rush, J., Stemman, O., Kirschner, M.W., and Gygi, S.P. (2003). Absolute quantification of proteins and phosphoproteins from cell lysates by tandem MS. Proc Natl Acad Sci USA *100*, 6940–6945.

Gevaert, K., Impens, F., Ghesquiere, B., Van Damme, P., Lambrechts, A., and Vandekerckhove, J. (2008). Stable isotopic labeling in proteomics. Proteomics *8*, 4873–4885.

Grassl, J., Westbrook, J.A., Robinson, A., Boren, M., Dunn, M.J., and Clyne, R.K. (2009). Preserving the yeast proteome from sample degradation. Proteomics *9*, 4616–4626.

Griffin, T.J., Xie, H., Bandhakavi, S., Popko, J., Mohan, A., Carlis, J.V., and Higgins, L. (2007). iTRAQ reagent-based quantitative proteomic analysis on a linear ion trap mass spectrometer. J Proteome Res *6*, 4200–4209.

Gygi, S.P., Rist, B., Gerber, S.A., Turecek, F., Gelb, M.H., and Aebersold, R. (1999). Quantitative analysis of complex protein mixtures using isotope-coded affinity tags. Nat Biotechnol *17*, 994–999.

Halligan, B.D., Slyper, R.Y., Twigger, S.N., Hicks, W., Olivier, M., and Greene, A.S. (2005). ZoomQuant: An application for the quantitation of stable isotope labeled peptides. J Am Soc Mass Spectrom *16*, 302–306.

Holzmann, J., Pichler, P., Madalinski, M., Kurzbauer, R., and Mechtler, K. (2009). Stoichiometry determination of the MP1-p14 complex using a novel and cost-efficient method to produce an equimolar mixture of standard peptides. Anal Chem *81*(24), 10254–10261.

Hsu, J.L., Huang, S.Y., Chow, N.H., and Chen, S.H. (2003). Stable-isotope dimethyl labeling for quantitative proteomics. Anal Chem *75*, 6843–6852.

Iliuk, A., Galan, J., and Tao, W.A. (2009). Playing tag with quantitative proteomics. Anal Bioanal Chem *393*, 503–513.

Imami, K., Sugiyama, N., Tomita, M., and Ishihama, Y. (2010). Quantitative proteome and phosphoproteome analyses of cultured cells based on SILAC labeling without requirement of serum dialysis. Mol Biosyst *6*, 594–602.

Keller, A., Eng, J., Zhang, N., Li, X.J., and Aebersold, R. (2005). A uniform proteomics MS/MS analysis platform utilizing open XML file formats. Mol Syst Biol *1*, 2005 0017.

Kerlavage, A., Bonazzi, V., di Tommaso, M., Lawrence, C., Li, P., Mayberry, F., Mural, R., Nodell, M., Yandell, M., Zhang, J., *et al.* (2002). The Celera discovery system. Nucleic Acids Res *30*, 129–136.

Kruger, M., Moser, M., Ussar, S., Thievessen, I., Luber, C.A., Forner, F., Schmidt, S., Zanivan, S., Fassler, R., and Mann, M. (2008). SILAC mouse for quantitative proteomics uncovers kindlin-3 as an essential factor for red blood cell function. Cell *134*, 353–364.

Lane, A.N., Fan, T.W., Higashi, R.M., Tan, J., Bousamra, M., and Miller, D.M. (2009). Prospects for clinical cancer metabolomics using stable isotope tracers. Exp Mol Pathol *86*, 165–173.

Leitner, A., and Lindner, W. (2009). Chemical tagging strategies for mass spectrometry-based phospho-proteomics. Methods Mol Biol *527*, 229–243, x.

Lemeer, S., Jopling, C., Gouw, J., Mohammed, S., Heck, A.J., Slijper, M., and den Hertog, J. (2008). Comparative phosphoproteomics of zebrafish Fyn/Yes morpholino knockdown embryos. Mol Cell Proteomics *7*, 2176–2187.

Li, X.J., Zhang, H., Ranish, J.A., and Aebersold, R. (2003). Automated statistical analysis of protein abundance ratios from data generated by stable-isotope dilution and tandem mass spectrometry. Anal Chem *75*, 6648–6657.

Lin, W.T., Hung, W.N., Yian, Y.H., Wu, K.P., Han, C.L., Chen, Y.R., Chen, Y.J., Sung, T.Y., and Hsu, W.L. (2006). Multi-Q: A fully automated tool for multiplexed protein quantitation. J Proteome Res *5*, 2328–2338.

Lopez-Otin, C., and Bond, J.S. (2008). Proteases: Multifunctional enzymes in life and disease. J Biol Chem *283*, 30433–30437.

Mann, M. (2006). Functional and quantitative proteomics using SILAC. Nat Rev Mol Cell Biol *7*, 952–958.

McNulty, D.E., and Annan, R.S. (2008). Hydrophilic interaction chromatography reduces the complexity of the phosphoproteome and improves global phosphopeptide isolation and detection. Mol Cell Proteomics *7*, 971–980.

Mirgorodskaya, O.A., Kozmin, Y.P., Titov, M.I., Korner, R., Sonksen, C.P., and Roepstorff, P. (2000). Quantitation of peptides and proteins by matrix-assisted laser desorption/ionization mass spectrometry using (18)O-labeled internal standards. Rapid Commun Mass Spectrom *14*, 1226–1232.

Mortensen, P., Gouw, J.W., Olsen, J.V., Ong, S.E., Rigbolt, K.T., Bunkenborg, J., Cox, J., Foster, L.J., Heck, A.J., Blagoev, B., *et al.* (2009). MSQuant, an open source platform for mass spectrometry-based quantitative proteomics. J Proteome Res *9*(1), 393–403.

Mueller, L.N., Brusniak, M.Y., Mani, D.R., and Aebersold, R. (2008). An assessment of software solutions for the analysis of mass spectrometry based quantitative proteomics data. J Proteome Res *7*, 51–61.

Nesvizhskii, A.I., Vitek, O., and Aebersold, R. (2007). Analysis and validation of proteomic data generated by tandem mass spectrometry. Nat Methods *4*, 787–797.

Ong, S.E., Blagoev, B., Kratchmarova, I., Kristensen, D.B., Steen, H., Pandey, A., and Mann, M. (2002). Stable isotope labeling by amino acids in cell culture, SILAC, as a simple and accurate approach to expression proteomics. Mol Cell Proteomics *1*, 376–386.

Ong, S.E., Kratchmarova, I., and Mann, M. (2003). Properties of 13C-substituted arginine in stable isotope labeling by amino acids in cell culture (SILAC). J Proteome Res *2*, 173–181.

Ong, S.E., and Mann, M. (2006). A practical recipe for stable isotope labeling by amino acids in cell culture (SILAC). Nat Protoc *1*, 2650–2660.

Orchard, S., Hoogland, C., Bairoch, A., Eisenacher, M., Kraus, H.J., and Binz, P.A. (2009). Managing the data explosion. A report on the HUPO-PSI Workshop. August 2008, Amsterdam, The Netherlands. Proteomics *9*, 499–501.

Park, S.K., Venable, J.D., Xu, T., and Yates, J.R., 3rd (2008). A quantitative analysis software tool for mass spectrometry-based proteomics. Nat Methods *5*, 319–322.

Perkins, D.N., Pappin, D.J., Creasy, D.M., and Cottrell, J.S. (1999). Probability-based protein identification by searching sequence databases using mass spectrometry data. Electrophoresis *20*, 3551–3567.

Pratt, J.M., Petty, J., Riba-Garcia, I., Robertson, D.H., Gaskell, S.J., Oliver, S.G., and Beynon, R.J. (2002a). Dynamics of protein turnover, a missing dimension in proteomics. Mol Cell Proteomics *1*, 579–591.

Pratt, J.M., Robertson, D.H., Gaskell, S.J., Riba-Garcia, I., Hubbard, S.J., Sidhu, K., Oliver, S.G., Butler, P., Hayes, A., Petty, J., *et al.* (2002b). Stable isotope labelling in vivo as an aid to protein identification in peptide mass fingerprinting. Proteomics *2*, 157–163.

Ramos-Fernandez, A., Lopez-Ferrer, D., and Vazquez, J. (2007). Improved method for differential expression proteomics using trypsin-catalyzed 18O labeling with a correction for labeling efficiency. Mol Cell Proteomics *6*, 1274–1286.

Reynolds, K.J., Yao, X., and Fenselau, C. (2002). Proteolytic 18O labeling for comparative proteomics: Evaluation of endoprotease Glu-C as the catalytic agent. J Proteome Res *1*, 27–33.

Rivers, J., Simpson, D.M., Robertson, D.H., Gaskell, S.J., and Beynon, R.J. (2007). Absolute multiplexed quantitative analysis of protein expression during muscle development using QconCAT. Mol Cell Proteomics *6*, 1416–1427.

Ross, P.L., Huang, Y.N., Marchese, J.N., Williamson, B., Parker, K., Hattan, S., Khainovski, N., Pillai, S., Dey, S., Daniels, S., *et al.* (2004). Multiplexed protein quantitation in Saccharomyces cerevisiae using amine-reactive isobaric tagging reagents. Mol Cell Proteomics *3*, 1154–1169.

Schmidt, A., Bisle, B., and Kislinger, T. (2009). Quantitative peptide and protein profiling by mass spectrometry. Methods Mol Biol *492*, 21–38.

Schnolzer, M., Jedrzejewski, P., and Lehmann, W.D. (1996). Protease-catalyzed incorporation of 18O into peptide fragments and its application for protein sequencing by electrospray and matrix-assisted laser desorption/ionization mass spectrometry. Electrophoresis *17*, 945–953.

Schulze, W.X., and Mann, M. (2004). A novel proteomic screen for peptide-protein interactions. J Biol Chem *279*, 10756–10764.

Shannon, P., Markiel, A., Ozier, O., Baliga, N.S., Wang, J.T., Ramage, D., Amin, N., Schwikowski, B., and Ideker, T. (2003). Cytoscape: A software environment for integrated models of biomolecular interaction networks. Genome Res *13*, 2498–2504.

Shilov, I.V., Seymour, S.L., Patel, A.A., Loboda, A., Tang, W.H., Keating, S.P., Hunter, C.L., Nuwaysir, L.M., and Schaeffer, D.A. (2007). The Paragon Algorithm, a next generation search engine that uses sequence temperature values and feature probabilities to identify peptides from tandem mass spectra. Mol Cell Proteomics *6*, 1638–1655.

Shinkawa, T., Taoka, M., Yamauchi, Y., Ichimura, T., Kaji, H., Takahashi, N., and Isobe, T. (2005). STEM: A software tool for large-scale proteomic data analyses. J Proteome Res *4*, 1826–1831.

Song, X., Bandow, J., Sherman, J., Baker, J.D., Brown, P.W., McDowell, M.T., and Molloy, M.P. (2008). iTRAQ experimental design for plasma biomarker discovery. J Proteome Res *7*, 2952–2958.

Spellman, D.S., Deinhardt, K., Darie, C.C., Chao, M.V., and Neubert, T.A. (2008). Stable isotopic labeling by amino acids in cultured primary neurons: Application to brain-derived neurotrophic factor-dependent phosphotyrosine-associated signaling. Mol Cell Proteomics *7*, 1067–1076.

Stewart, II, Thomson, T., and Figeys, D. (2001). 18O labeling: A tool for proteomics. Rapid Commun Mass Spectrom *15*, 2456–2465.

Tang, W.H., Halpern, B.R., Shilov, I.V., Seymour, S.L., Keating, S.P., Loboda, A., Patel, A.A., Schaeffer, D.A., and Nuwaysir, L.M. (2005). Discovering known and unanticipated protein modifications using MS/MS database searching. Anal Chem *77*, 3931–3946.

Thompson, A., Schafer, J., Kuhn, K., Kienle, S., Schwarz, J., Schmidt, G., Neumann, T., Johnstone, R., Mohammed, A.K., and Hamon, C. (2003). Tandem mass tags: A novel quantification strategy for comparative analysis of complex protein mixtures by MS/MS. Anal Chem *75*, 1895–1904.

Villanueva, J., Nazarian, A., Lawlor, K., and Tempst, P. (2009). Monitoring peptidase activities in complex proteomes by MALDI-TOF mass spectrometry. Nat Protoc *4*, 1167–1183.

Whiteaker, J.R., Zhao, L., Anderson, L., and Paulovich, A.G. (2009). An automated and multiplexed method for high throughput peptide immunoaffinity enrichment and multiple reaction monitoring mass spectrometry-based quantification of protein biomarkers. Mol Cell Proteomics *9*(1), 184–196.

Wiese, S., Reidegeld, K.A., Meyer, H.E., and Warscheid, B. (2007). Protein labeling by iTRAQ: A new tool for quantitative mass spectrometry in proteome research. Proteomics *7*, 340–350.

Yan, W., and Chen, S.S. (2005). Mass spectrometry-based quantitative proteomic profiling. Brief Funct Genomic Proteomic *4*, 27–38.

Yao, X., Afonso, C., and Fenselau, C. (2003). Dissection of proteolytic 18O labeling: Endoprotease-catalyzed 16O-to-18O exchange of truncated peptide substrates. J Proteome Res *2*, 147–152.

Zeng, D., and Li, S. (2009). Improved CILAT reagents for quantitative proteomics. Bioorg Med Chem Lett *19*, 2059–2061.

Zhang, G., Fenyo, D., and Neubert, T.A. (2009). Evaluation of the variation in sample preparation for comparative proteomics using stable isotope labeling by amino acids in cell culture. J Proteome Res *8*, 1285–1292.

Zhang, R., Sioma, C.S., Thompson, R.A., Xiong, L., and Regnier, F.E. (2002). Controlling deuterium isotope effects in comparative proteomics. Anal Chem *74*, 3662–3669.

Chapter 28
Standard Operating Procedures and Protocols for the Preparation and Analysis of Plasma Samples Using the iTRAQ Methodology

Leanne B. Ohlund, Darryl B. Hardie, Monica H. Elliott, Alexander G. Camenzind, Derek S. Smith, Jennifer D. Reid, Gabriela V. Cohen Freue, Axel P. Bergman, Mayu Sasaki, Lisa Robertson, Robert F. Balshaw, Raymond T. Ng, Alice Mui, Bruce M. McManus, Paul A. Keown, W. Robert McMaster, Carol E. Parker, and Christoph H. Borchers

Abstract The Applied Biosystems iTRAQ technique is based on derivatization of peptide samples with mass-balanced tags. The samples are then combined, and quantitation is based on reporter ion abundances in the MS/MS spectrum. The iTRAQ technique, with its capability for the analysis of either 4 or 8 samples in a single LC-MS/MS analysis, is one of the most powerful techniques available today for "shotgun" quantitative proteomics. However, it is not an easy technique, and involves multiple – and critical – digestion, separation, and derivatization steps. Our laboratory has spent several years developing over 40 standard operating procedures (SOPs) for the generation of reproducible iTRAQ data. These protocols are designed for generating and analyzing iTRAQ data on the Applied Biosystems 4800 MALDI-TOF/TOF instrument and the Applied Biosystems QSTAR ESI-MS/MS, which are two of the most commonly-used instruments for iTRAQ analysis. We have also included sections on the Applied Biosystems ProteinPilot software which is used for analyzing the data and obtaining protein expression ratios. Standard operating procedures are obviously designed to be used with these specific instruments. For laboratories with other instruments, we have tried to include sufficient annotation and descriptions of the procedures so that these protocols can be adapted for use on other equipment. Likewise, many of these protocols are not specific to iTRAQ, and can be used for sample preparation and analysis for other assays as well. For more complete and detailed protocols, the reader is referred to the on-line supplementary information on our website at http://www.proteincentre.com/itraq-book-chapter-supplemental-information.

Keywords iTRAQ · LC-ESI-MS/MS · LC-MALDI · Protein quantification · SCX

C.H. Borchers (✉)
Department of Biochemistry and Microbiology, University of Victoria, Victoria, BC, V8W 3P6; University of Victoria-Genome BC Proteomics Centre, #3101-4464 Markham Street, Vancouver Island Technology Park, Victoria, BC, V8Z 7X8, Canada
e-mail: christoph@proteincentre.com

A.R. Ivanov, A.V. Lazarev (eds.), *Sample Preparation in Biological Mass Spectrometry*, DOI 10.1007/978-94-007-0828-0_28,

28.1 Introduction

Most quantitative proteomics approaches are based on the quantitation of peptides as surrogates for their parent proteins. In other words, proteins are digested into peptides (usually using proteolytic enzymes), and the concentration of protein present in the original sample is assumed to be proportional to the concentration of a peptide from that protein.

The iTRAQ technique was first described by Ross et al. (2004) and is now being commercialized by Applied Biosystems (now AB Sciex). iTRAQ (isobaric tags for relative and absolute quantification) is a chemical labeling technique that allows the simultaneous relative quantitation of up to 4 or 8 samples in a single mass spectrometry experiment. This is done by chemically labeling the amino-termini of the peptides with a series of 4 (4-plex) or 8 (8-plex) different labels or "tags". These differentially labeled samples are then mixed together and analyzed in the same mass spectrometric analysis.

The iTRAQ labels are *isobaric* tags, all of which shift the molecular weight of the peptide by the same amount in the MS spectrum (Fig. 28.1a). This approach involves a 3-part tag: the reporter region which shows the differences between the tags, a balance group so that the sum of the masses of the balance and reporter ion portions of the tag always stay the same, and the group which reacts with the free amino groups – the N-terminus and lysines in the peptide. The isobaric nature of the tag means that even when the samples are combined, only one mass needs to be selected for MS/MS. In the MS/MS spectrum, the y-ions from the different labeled versions of the peptides coincide, and these are the ions used to identify the protein. Only the low-mass reporter ions in the MS/MS spectrum are different, and their relative abundances reflect the abundance of the peptide in the different samples. Reporter ions are detected at *m/z* 114, 115, 116, 117 in a 4-plex experiment; *m/z* 113, 114, 115, 116, 117, 118, 119, and 121 in an 8-plex experiment (Fig. 28.1b).

The iTRAQ approach is currently one of the most popular techniques for relative quantitation, and has been used for many studies on biomarker identification, as well as other studies done to determine molecular pathways. However, recent evidence indicates that this technique may actually "compress" the true protein expression ratios (DeSouza et al., 2007, 2009). The maximum expression ratios that can be observed in discovery experiments using iTRAQ seem to be only $\sim$3 to 5, while larger differences are observed in the validation experiments using mTRAQ, DIGE, or isotopically-labeled standard peptides. This most likely affects the other mass-balanced labeling methods as well. Nevertheless, the consensus is that the trends are valid even if the expression ratios may not be exact. Another problem is that, even though *m/z* 120 has been "skipped" because the phenylalanine immonium ion is at *m/z* 120, in some spectra high levels of phenylalanine can "spill over" and the ^{13}C isotope peak can affect the *m/z* 121 abundance (Ow et al., 2009). In addition, during iTRAQ-based quantitation – and effectively all other relative quantitation techniques – the assumption is made that the majority of the protein concentrations do not change. In iTRAQ, a protein that doesn't change is used to "normalize" the expression ratios. For this reason, iTRAQ is best used in those discovery biomarker

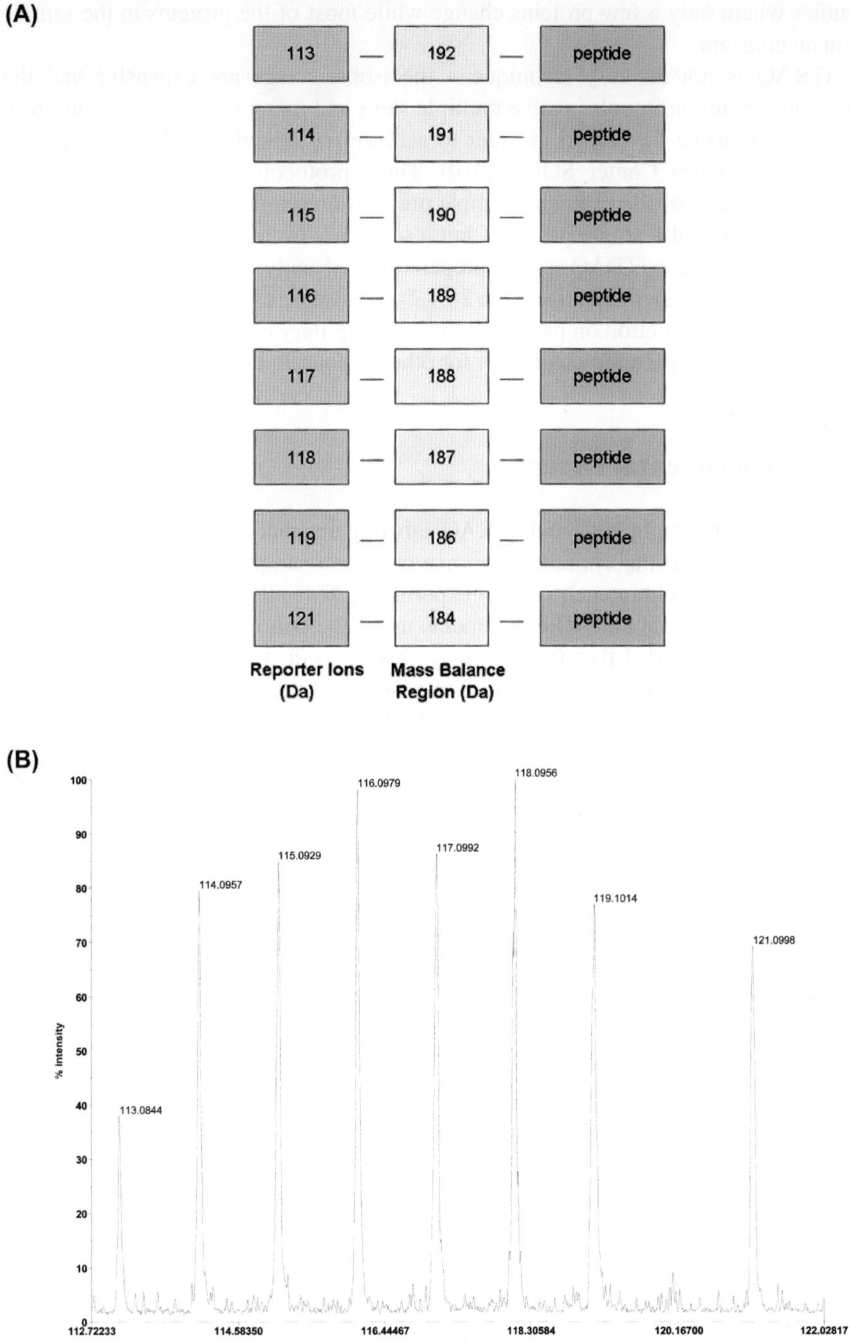

Fig. 28.1 (**a**) Schematic representation of the mass-balanced iTRAQ 8-plex reagent, (**b**) an MS/MS spectrum showing the reporter region of an iTRAQ-labeled peptide. Reprinted from Elliott et al. (2009) with permission

studies where only a few proteins change while most of the proteins in the sample remain constant.

iTRAQ is not an easy technique – the isobaric tags are expensive and the isobaric-tagging protocols involve multiple steps and many standard operating procedures had to be developed in order to achieve reproducible and reliable results (UVic_Proteomics_Center_SOPs, 2010). These protocols are given in this book chapter. Because of the extensive sample preparation required for iTRAQ, what we present here is not a single protocol, but a series of seventeen protocols, not all of which are specific to iTRAQ sample preparation and analysis. Thus, for example the section on Plasma Depletion (Section 28.4.2) can also be used prior to 2D-gel analysis. Likewise, the section on packing nanoscale capillary columns (Section 28.4.10) can be used to prepare these columns for other nanoscale LC separations.

28.2 Experimental Design

While there are eight different iTRAQ labeling reagents available, experiments can be designed to incorporate more than eight test samples by using a reference (control) sample within each iTRAQ experiment. This allows cross-comparison of results from multiple iTRAQ experiments using the common reference sample as a reference standard. Likewise, you may choose to do four samples in duplicate rather than eight different samples. Some researchers advocate performing each experiment in quintuplicate to obtain statistically significant results.

There are several different types of replicate experiments that can be performed: from technical replicates (the same iTRAQ-labeled sample analyzed multiple times), to experimental replicates (same sample labeled with iTRAQ multiple times), and biological replicates (individual samples taken from the same treatment group and processed and iTRAQ labeled independently) (Fig. 28.2). The variation of iTRAQ quantitation between replicates and different replicate types was investigated in a study on yeast and bacteria (Gan et al., 2007). The authors found that

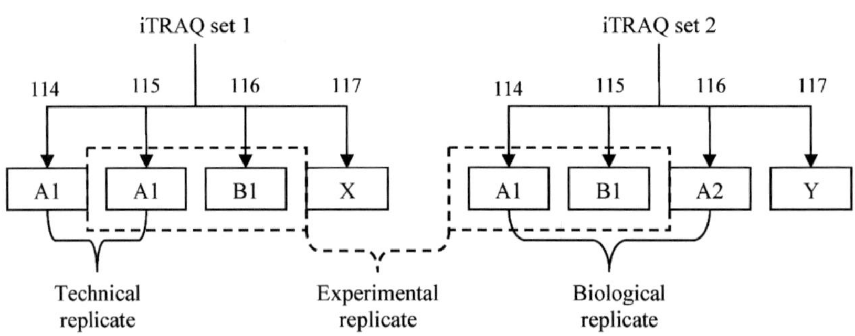

Fig. 28.2 Different types of variation encountered in iTRAQ experiments. (Reprinted from Gan et al. (2007) © 2007 by American Chemistry Society. Reprinted with permission of American Chemical Society via the Copyright Clearance Center)

biological variation was the most significant source of error (±25%). The technical variation between different labels within the same iTRAQ-labeled sample was 11%, with only 0.2% due to the MS portion of the assay, while the experimental variation between iTRAQ labeling experiments on *different* labeling experiments from the *same* biological sample was ~23%). Not surprisingly, they also found a strong correlation between the percent sequence coverage and the percent coefficient of variation (% CV) of the experimental replicates (i.e., the higher the allowable % CV, the higher the % coverage). In this study, allowing ±50% CV resulted in ~88% coverage in protein quantitation between biological replicates. Replicate analyses of the *same* iTRAQ-labeled sample (a technical replicate) gave 95% quantitation coverage with ±30% CV. Of course, lower-abundance proteins are the most difficult to detect and quantitate, so a "trade-off" has to be made between confidence level of quantitation and protein coverage. This is a decision that one has to make when the data is analyzed. The % CV at all of these stages affects the "believability" of your iTRAQ results – how much of an observed difference it takes before you can be certain that the difference is real.

28.2.1 LC-MALDI-MS/MS or LC-ESI-MS/MS?

A quadrupole and a TOF (time-of flight) are two types of mass analyzers. MS/MS (tandem mass spectrometry), which is needed for experiments using iTRAQ, can be done on an instrument with two mass analyzers, separated by a collision cell (Parker et al., 2005). The Applied Biosystems (AB) 4800 has two TOF analyzers, and is usually referred to as a MALDI-TOF/TOF; the AB QSTAR has a quadrupole, an rf-only quadrupole that functions as a collision cell, and a TOF, and is technically a QqTOF instrument, but is often referred to as a QTOF.

The decision on whether to use LC-MALDI-MS/MS or LC-ESI-MS/MS will, or course, depend on what instruments and software are available. In general, the reporter ion abundances are lower on ESI instruments than in MALDI (Wiese et al., 2007), and it should be kept in mind that not all ESI instruments give good reporter ions sensitivities, depending on the instrument design and the collision energies used. However, if the instrument can be tuned to give reasonable reporter ion abundances, LC-ESI-MS/MS can be used in place of LC-MALDI-MS/MS.

MALDI instruments, particularly the AB 4800 and the AB 5800, have been extensively used for iTRAQ. Because of the large numbers of spots analyzed, matrix buildup is a common problem, requiring frequent source cleaning. For this reason, the newer MALDI 5800 instrument includes a "self-cleaning" source. Electrospray instruments that have been successfully used for iTRAQ include the AB QSTAR and the AB 4000 Q-TRAP. Based on our experience, the Micromass Q-Tof API-US gives very low reporter ion abundances. The "basic" Thermo Fisher Orbitrap also gives very low reporter ion abundances, but the detection sensitivity is somewhat improved by their HCD (high collision energy dissociation) software option, and is *very* much improved by the external C-trap option. For a comparative study of

ESI and MALDI on iTRAQ-labeled samples, the reader is referred to Kuzyk et al. (2009).

The instructions given in Sections 28.4.11, 28.4.12, and 28.4.13 are for use with LC-MALDI and the AB 4800 MALDI-TOF/TOF instrument. The instructions given in Sections 28.4.14, 28.4.15, and 28.4.16 are for use with the AB QSTAR Pulsar i QqTOF LC-ESI-MS/MS instrument. This is not a simple procedure, but is a combination of multiple protocols spanning several weeks of work, and depends on the choice of ESI or MALDI-MS/MS. For a single 8-plex sample set, sample collection and depletion stage takes approximately 4 days. After this, it takes 7–10 workdays from BCA protein quantitation through to data analysis. It is, however, possible to "multitask", especially for the sample preparation stages. This results in an average time of 1–2 weeks per 8-plex (see Fig. 28.3). All of these time estimates, however, assume that you have at least 3 full-time technicians to devote to this process, and that there are no delays due to instrument downtime or problems with instrument validation.

28.3 Materials and Equipment

28.3.1 Sample Collection

28.3.1.1 Blood and Biopsy Collection

1. 6.0 mL-EDTA blood collection tubes (VWR, CABD367863) or P-100 plasma collection tubes (Becton Dickinson)

28.3.1.2 Secondary Containers

1. 1.2 mL self standing barcoded cryogenic vials (VWR, 73521-052)
2. Tough-Spots^{TM} blue dot labels (RPI Corp., 247131-B)

28.3.1.3 Processing and Storage

1. 3.0 mL transfer pipettes
2. 12 × 75 mm polystyrene tubes no caps
3. Nitrile or latex examination gloves
4. Ice bath
5. Cold block
6. Refrigerated centrifuge capable of spinning down standard blood collection tubes)
7. –80°C freezer
8. –20°C freezer

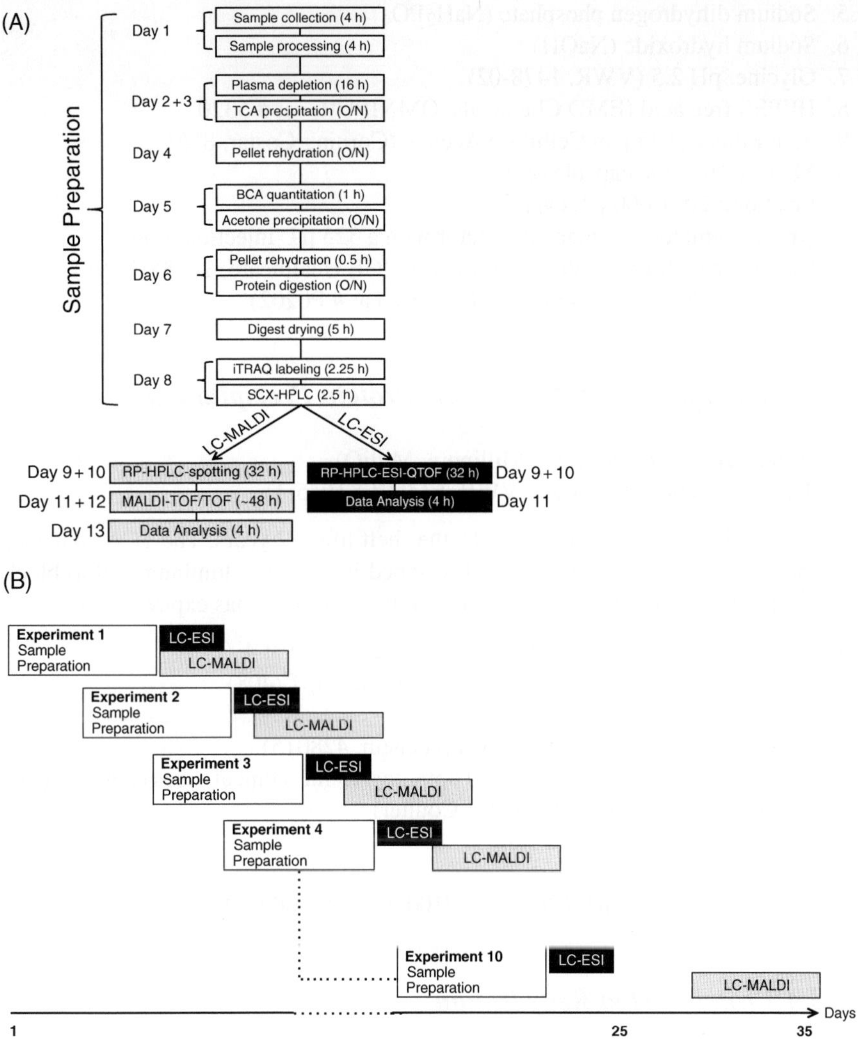

Fig. 28.3 Timeline for iTRAQ experiments. (**a**) The timeframe for one 8-plex iTRAQ experiment takes from start to finish (in days) and how much time each of the steps in the procedure takes – the difference in time between ESI and MALDI results from de-coupling the LC separation from the MS step; (**b**) To save time, experiments can be "multi-tasked" or overlapped instead of performing them in series

28.3.2 Plasma Depletion

1. Custom depletion column (Sigma IgY-14 LC5/Seppro Mix 14)
2. Water, deionized (DI) (Millipore, MilliQ)
3. Amersham/GE Healthcare; ÄKTA Prime Liquid Chromatography system
4. 10x PBS buffer (EMD Chemicals, OMNIPUR, cat # 6506)

5. Sodium dihydrogen phosphate (NaH_2PO_4)
6. Sodium hydroxide (NaOH)
7. Glycine, pH 2.5 (VWR, 1478-02)
8. HEPES free acid (EMD Chemicals, OMNIPUR, cat # 5320
9. Spin Filters: 0.45 μm Cellulose Acetate (Corning Costar, 8163)
10. Micro-centrifuge capable of 9000×*g*
11. Pipettors: 20–1,000 μL capacity
12. An autosampler or a manual injector with a 325 μL injection loop
13. Computer with Prime View 5.0 software (GE Healthcare, 11-0003-59)
14. 0.01% Azide cleaning solution (Teknova, cat # P0202)

28.3.3 Trichloroacetic Acid (TCA) Protein Precipitation

1. Water, DI (deionized water, Millipore, MilliQ)
2. Trichloroacetic acid (TCA) (VWR, EMD TX1045-1)

   NOTE: Store TCA solution at 4°C; the shelf life is 2 years. The TCA container must be made of inert plastic and wrapped in cloth or aluminum foil to block out light. If particulates have formed in the solution, it has expired.

3. Acetone (ACS grade or better) that has been stored at 4°C.
4. Triethylammonium bicarbonate (TEAB) (Sigma, T7408)
5. Urea, ACS grade
6. Sodium dodecyl sulfate (SDS) (Calbiochem, 428015)
7. Refrigerated centrifuge capable of spinning 15 mL conical tubes at 3000×*g* for 60 min (Allegra × 12R, Beckman Coulter)
8. Glass Pasteur pipettes and bulb
9. Refrigerator, at 4°C
10. Pipettor 200–1,000 μL (Pipetman P100, GF-F123602, Mandel)

28.3.4 Protein Pellet Rehydration

1. Water, deionized (MilliQ, Millipore)
2. Urea, ACS grade
3. Triethylammonium bicarbonate (TEAB) (Sigma, T7408)
4. Sodium dodecyl sulfate (SDS) (Calbiochem, 428015)

   CAUTION: SDS is harmful if inhaled, use appropriate precautions such as mask and gloves.

5. 2 mL 0.45 μm spin filters tubes (Genway, SF-100)
6. Pipettors 200–1,000 μL (Pipetman P100, GF- F123602, Mandel)
7. Pipettors 50–200 μL (Pipetman P200, GF- F123601, Mandel)
8. Pipette tips 1000 μL (Axygen, T-1000-CL, VWR CA22234-086)
9. 1.5 mL Eppendorf tubes (VWR, 20170-038)

28.3.5 BCA (Bicinchoninic Acid) Protein Quantitation

1. 0.5 M TEAB (part of AB iTRAQ kit, AB 4352156)
2. Water, LC-MS grade (Fisher, W6-4)
3. Costar flat bottom 96-well plate (Fisher, 07-200-39)
4. Plastic adhesive cover (Fisher 5-500-32)
5. Ascent Software (ThermoLabsystems, v2.6)
6. Multiskan Ex Plate Reader (ThermoLabsystems, 51118170)
7. BCA Reagent A (Sigma-Aldrich, B9643)
8. BCA Reagent B (Sigma-Aldrich, C2287)
9. Bovine Serum Albumin (BSA) protein standard (Sigma-Aldrich, P0914)
10. 1.5 mL microcentrifuge tubes (VWR, 20170-378)
11. 14 mL polypropylene tubes (Fisher, 14-959-11B)
12. 10 mL serological pipette (Fisher, 13-678-12E)

28.3.6 Acetone Protein Precipitation

1. Cold acetone, HPLC grade (Fisher, A949-4)
2. 1.5 mL microcentrifuge tubes (VWR 20170-378)
3. SDS, electrophoresis grade (Fisher, BP166-100)

28.3.7 Protein Digestion

1. iTRAQ dissolution buffer, 0.5 M TEAB (AB, 4352136)
2. Centrifuge (Mikro 2, Hettich International)
3. TCEP, 50 mM (part of AB iTRAQ kit, 4326685)
4. Cysteine blocking reagent (part of AB iTRAQ kit, 4326685)
5. Promega trypsin, modified porcine, sequencing grade (Fisher PR V5111)
6. Lab-Line incubator (Taylor Scientific, 19-1230-03)
7. Isotemp Dry Bath (Fisher, Model 145)

28.3.8 Peptide iTRAQ Labeling (4-plex)

1. Centrifuge (Eppendorf, 5415D)
2. iTRAQ dissolution buffer, 0.5 M (AB 4352137)
3. Ethanol (part of AB iTRAQ kit, AB 4352156)
4. iTRAQ Reagent 114 (AB 4352137)
5. iTRAQ Reagent 115 (AB 4352153)
6. iTRAQ Reagent 116 (AB 4352154)
7. iTRAQ Reagent 117 (AB 4352155)

   NOTE: Use the corresponding reagents from the 8-plex kit, if desired (AB 4390811)

8. HPLC load buffer (Buffer A) (10 mM potassium phosphate, 25% acetonitrile at pH 2.5)
9. 14 mL polypropylene tube (Fisher, 14-959-11B)
10. pH strips (pH range 2–9, Fisher, M95783)
11. 1.5 mL microcentrifuge tubes (VWR, 20170-378)
12. Speedvac (Thermo Electron Co., SPD1010-115)
13. Acetonitrile (ACN), HPLC grade (Fisher, A998-4)
14. Potassium phosphate (Sigma, P-0662)

28.3.9 Strong Cation Exchange (SCX) – First-Dimension Separation

1. Applied Biosystems Vision Workstation (model 440)
2. Centrifuge (Eppendorf 5415D)
3. 14 mL polypropylene tubes (Fisher, 14-959-11B)
4. 1.5 mL microcentrifuge tubes (VWR, 20170-378)
5. SCX Load Buffer (Buffer A) (10 mM KH_2PO_4, 25% ACN)
6. SCX Buffer B (10 mM KH_2PO_4, 0.5 M KCl, 25% ACN, pH 2.75)
7. SCX Buffer C (10 mM KH_2PO_4, 1 M KCl, 25% ACN, pH 2.75)
8. α-Endorphin (human) (Sigma, E6136)
9. Angiotensin I (human) (Sigma, A9650)
10. Angiotensin II (human) (Sigma, A9525)
11. Phosphoric acid, 85% (Fisher, A242-1)
12. pH strips, pH range 2–9 (Fisher, M95783)
13. 2.5 mL Hamilton gas-tight syringe (Fisher, 13-684-108)
14. Methanol, HPLC grade (Fisher, A452-4)
15. Water, LC-MS grade (Fisher W6-4)
16. Acetonitrile, LC-MS grade (Fisher, A955-4)
17. 2.0 mL polypropylene tube (Rose Scientific, NC-508-GRD)
18. 2.0 mL polypropylene tube caps (Rose Scientific, 506)
19. Speedvac (Themo Electron Co., SPD1010-115)
20. SCX HPLC column Polysulfoethyl A, 100 × 4.6 mm, 5 μm 300 Å (PolyLC Inc, 104SE0503)

28.3.10 Preparation of Analytical Reversed-Phase C18 Capillary Column

1. High pressure cell (Mass Evolution Inc., 1 1000 110)
2. Magic C18AQ, 5 μm, 100 Å packing material (Chromatographic Specialties, MBR9996610000)
3. IntegraFrit Capillary (New Objective, IF360-75-50-N-S)
4. Methanol, HPLC grade (Fisher A452-4)
5. Water, LC-MS grade (Fisher, W6-4)
6. Analytical Balance (Sartorius, CP64)

7. Helium Gas (He: Air Gas UHP300C001) or Argon Gas(Ar: Air Gas UHP300C)
8. 2.0 mL polypropylene tubes (Rose Scientific, NC-508-GRD)
9. 2.0 mL polypropylene tube caps (Rose Scientific, 506)
10. Magnetic stir plate (Corning, PC-353)

28.3.11 Preparation of Recrystallized MALDI Matrix and MALDI Matrix Solution

28.3.11.1 For Matrix Recrystallization

1. α-cyano-4-hydroxycinnamic acid (CHCA) (Sigma, C2020-10G)
2. Ethanol, 99% (Fisher, A995-4)
3. Water, LC-MS grade (Fisher, W6-4)
4. Glass bottle, 500 mL (VWR, 8900-238)
5. Magnetic/hot plate (VWR, 12365-382)
6. Stir bar (Fisher, 14-513-60)
7. Whatman #1 filter paper, 5.5 cm (Fisher, 09-805B)
8. Buchner funnel (Fisher, 10-356C)
9. Side-arm Ehrlenmeyer flask, 1 L (Fisher, 10-181F)
10. Erlenmeyer flask, 2 L (Fisher, FB-500-2000)
11. Freeze dryer/lyophilizer (Thermo Electron Corp., modulyod-115)
12. Brown glass vial, 4 mL (Fisher, 02-912-361)

28.3.11.2 For Matrix Solution

1. Acetonitrile, LC-MS grade (Fisher, A955-4)
2. Water, LC-MS grade (Fisher, W6-4)
3. Trifluoroacetic acid (Thermo Scientific/Pierce, 28903)
4. CHCA, recrystallized (see Section 28.4.11.1)
5. Ammonium citrate dibasic, ACS grade (Sigma-Aldrich, A-8170)
6. 5 mL glass 1.5 dram vial (VWR, 66011-063)
7. Analytical balance (Sartorius, CP64)
8. Microcentrifuge (Eppendorf, 5415D)

28.3.12 Reversed-Phase Nano-HPLC Separation of the SCX Fractions

1. Eksigent NanoLC-1Dplus HPLC (Eksigent Technologies)
2. Eksigent AS1 autosampler (Spark Holland/model 920)
3. Eksigent Software (Eksigent Technologies, v2.08)
4. Probot Micro Fraction Collector (LC Packings /Dionex)
5. µCarrier Software (LCPackings/Dionex, v2.0 build 1620)
6. C18 pepMap trap column (LC Packings/Dionex, 160454)
7. Magic C18 analytical column (see Section 28.4.10)

8. Opti-TOF LC-MALDI plate inserts (AB, 1018469)
9. Water, LC-MS grade (Fisher, W6-4)
10. Autosampler vials (Life Science, 30300P-1232)
11. Snap cap lids (Life Science, 5250S-11)
12. Methanol, HPLC grade (Fisher, A452-4)
13. Acetonitrile, LC-MS grade (Fisher, A955-4)
14. Trifluoroacetic acid (TFA) (Thermo Scientific/Pierce, 28903)
15. MALDI matrix solution (see Section 28.4.11.2)
16. 4700 mass standards kit (AB, 4333604)
17. BSA digest (Michrom Bioresources, PTD/00001/1516)

28.3.13 MALDI TOF/TOF Analysis

1. 4800 MALDI TOF/TOF Analyzer (AB, 4383680)
2. 4000 Series Explorer Software (AB, 1019077)
3. Magnetic sample plate holder (AB, 1016492)

28.3.14 ESI-QTOF Calibration

1. API QSTAR Pulsar I, or newer model (AB/Sciex)
2. Proxeon nanospray capillaries (Proxeon, ES380)
3. Proxeon capillary holder (Proxeon, ES286)
4. Mini-Centrifuge (Fisher, 05-090-100)
5. Sex pheromone inhibitor (SPI) iPD1 calibration standard (Bachem, H-9985)
6. Cesium iodide (CsI) calibration standard (Sigma-Aldrich, C-8643)
7. Water, LC-MS grade (Fisher, W6-4)

28.3.15 ESI-QTOF Validation

1. LC Packings Famos autosampler (LC Packings/Dionex)
2. LC Packings Switchos (LC Packings/Dionex)
3. LC Packings UltiMate (LC Packings/Dionex)
4. Water, LC-MS grade (Fisher, W6-4)
5. Autosampler vials (Life Science, 30300P-1232)
6. Snap-cap lids (Life Science, 5250S-11)
7. Methanol, HPLC grade (Fisher, A452-4)
8. Acetonitrile, LC-MS grade (Fisher, A9554)
9. BSA digest (Michrom Bioresources, PTD/00001/15)
10. Zorbax 300SB-C18 Guard Column 5 mm × 0.3 mm, 5 μm particle size (Agilent, 5065-9913)
11. Magic C18AQ analytical column (see Section 28.4.10)
12. 10 μm emitter tip (New Objective, FS360-20-10-N-20-C12)
13. Cheminert nanovolume-tee (Valco–VICI, H-C-NTFPK)
14. Cheminert nanovolume union (Valco–VICI, C-NEU.5FPK)

28.3.16 ESI-QTOF Analysis

1. Water, LC-MS grade (Fisher, W6-4)
2. Acetonitrile, LC-MS grade (Fisher, A9554)
3. Formic acid (FA) (Fluka, 56302-50ML-F)

28.4 Procedures

28.4.1 Sample Collection

Plasma contains proteases that can digest plasma proteins. These proteases should be inhibited by EDTA or protease inhibitors as soon as possible – through the use of special collection tubes, this can even be done during the plasma collection process. Two types of blood collection tubes (Vacutainers) are now recommended for Plasma collection: the BD EDTA-containing tubes, and the BD-P100 tubes which are "pre-loaded" with protease inhibitors. The use of serum is NOT recommended for proteomics studies, because of the inherent lack of reproducibility of the formation of serum from plasma.

Plasma samples should be collected in Becton Dickinson (BD) Lavender EDTA tubes or in BD P100 tubes (Omenn et al., 2005), using the manufacturer's new protocol (Aguilar-Mahecha et al., 2011).

28.4.1.1 Using Best Practices for Blood Collection

1. The phlebotomist will collect study blood immediately following clinical blood collection using BD P-100 or purple top (EDTA) tubes
2. Allow at least 10 s for a complete blood draw to take place. Ensure that the blood has stopped flowing into each tube before removing the tube from the holder.
3. Invert the tubes gently ten times to ensure proper mixing.
4. Use the clinical labels to label every tube.
5. Immediately place all tubes in an ice bath.
6. Samples must be processed within 2 h.

28.4.1.2 EDTA or P-100 Tube Processing

NOTE: Samples have to be processed (aliquoted and spun down) before freezing. After processing, the samples should be stored at –80°C to prevent degradation and proteolysis.

1. Spin the EDTA or P-100 tubes in a refrigerated centrifuge as per standard clinical laboratory procedure.
2. Verify that the supplied vials match the patient study ID. Set up the vials for aliquoting in a cold block or an ice bath.
3. After centrifugation, place the samples in an ice bath until the plasma and buffy coat are removed from the cells.

NOTE: The aliquots MUST be kept cold prior to freezing.

4. Using a transfer pipette, remove the plasma from the EDTA tube into a 12 × 75 mm plastic tube, and mix by gently drawing the plasma up and down in the tube several times.
5. Once mixed, draw the entire amount of plasma up into the pipette and add 5 drops to each 1.2 mL cryogenic vial labeled with a blue dot (P). If there is plasma left over, distribute evenly between the aliquots. If there is not enough plasma for 5 drops per vial, a smaller number of vials should be aliquoted.
6. Cap all aliquots. Ensure that the vials are correctly labeled, and put them into a labeled biohazard bag.
7. Freeze immediately at –80°C.

28.4.1.3 Specimen Storage

1. Immediately after collection and processing, samples are stored at –80°C. Any movement of specimens should be done at –80°C or below.

   NOTE: The shelf-life of processed plasma samples has not been determined, but the shelf-life of plasma at –80°C is 6 months, so it would be best to keep samples for <6 months.

28.4.2 Plasma Depletion

Plasma is an extremely challenging matrix for proteomics studies. Over 90% of the protein content of plasma is made up of a few high abundance proteins (Anderson and Anderson, 2002). Protein concentrations range from >40 mg/mL (albumin) to low pg/mL levels, with 22 proteins making up > 99% of the blood proteome (Simpson et al., 2008). In order to "dig deeper" into the proteome, it is often necessary to deplete the plasma of high-abundance proteins. Usually, antibodies are used to do this, and several manufacturers have produced "kits" for removing various numbers of high-abundance proteins. Among these are the Sigma's Seppro IgY12 or Seppro IgY14, as described below. Caution: when you remove the target protein, you may also be removing proteins that bind to it, as well as non-specific binders to the antibody. (*NOTE*: We have also used the Agilent MARS depletion system. In our experience, the HPLC-based version of this system seems to give better results than the cartridge form.)

The protocol below is for the use of the Seppro IgY14 column, which is designed to remove the 14 most abundant proteins (Sigma-Aldrich, 2008). This procedure describes the method used to remove 90 to 95% of the total protein content found in human plasma. The plasma proteins are depleted using a micro bead column with avian (chicken) antibody (IgY)-antigen and specialized buffers for sample loading, washing and eluting. These 14 proteins are: Albumin, IgG, α1-antitrypsin, IgA, IgM, transferrin, haptaglobin. α-1-acid glycoprotein, α-2-macroglobin, HDL-apolipoprotein AI, HDL-apolipoprotein AII, fibrinogen, apolipoprotein E, and complement C3. This protocol is written for an Amersham (GE Healthcare) ÄKTA

Prime low pressure Liquid Chromatography system, with Amersham's Prime View software (GE_Healthcare, 2002). If you have a different HPLC system, this protocol will need to be adapted for your particular instrument.

28.4.2.1 Solvent Preparation

1. Dilution buffer (100 mM PBS, pH 7.6 at 4°C)
    1. Dilute 100 mL of 10x commercial PBS buffer in ~900 mL DI H_2O.
    2. Dilution buffer may also be prepared by dissolving 1.2 g of NaH_2PO_4 in ~900 mL of DI H_2O. Adjust pH to 7.55 at room temperature with NaOH (at 4°C the pH changes to 7.6) and top up with DI water to 1000 mL.
2. Stripping Buffer (100 mM Glycine, pH 2.5)
    1. 10x Stripping buffer is provided by the column manufacturer. Dilute buffer 10 times in DI H_2O.
    2. Stripping buffer may also be prepared as follows: dissolve 7.51 g glycine in ~900 mL DI H_2O and titrate pH to 2.5 with HCl. Bring to 1000 mL with DI H_2O.
3. Neutralizing Buffer (100 mM HEPES, pH 8.0)
    1. Dissolve 2.38 g of HEPES free acid in ~900 mL DI H_2O and titrate pH to 7.77 with NaOH. Bring to 1000 mL with DI H_2O.

28.4.2.2 Instrument Set-up Procedure

1. Turn on the PC operating the LC system, then the ÄKTA prime LC system, and initiate the instrument by clicking on the ÄKTA prime icon.
2. Set up buffers on the ÄKTA Prime system in the following order:

Channel A	Dilution Buffer
Channel B	Stripping Buffer
Position 2	Neutralizing buffer
Position 3	Distilled water

3. Ensure that all lines are purged with their respective buffers at a flow rate of 5 mL/min for 5 min per line.

   CAUTION: Remove the column when purging lines.
4. Set up the LC time according to Table 28.1.
   The total buffer volume consumed per run = 107.5 mL
5. Attach the column and equilibrate it in Dilution Buffer (Channel A) for 10 min at a flow rate of 1.0 mL/min. Ensure that the UV signal is stable.

Table 28.1 Set-up table for LC gradient and fraction collection for sample depletion

Run parameters								
Instrument program number				1		Fraction size	12 mL	
Cycle	Total time (min)	Time (min)	Fraction collection	% Dilution	% Stripping	% Neutralizing	Flow rate (mL/min)	Valve position
Equilibration	0.0–15	15	OFF	100	0	0	2.0	Load
Injection	15.1–19.9	5	OFF	100	0	0	0.3	Inject
Elution 1	20.0–40.0	20	ON	100	0	0	0.3	Inject
Elution 2	40.1–45.0	5	ON	100	0	0	2.0	Inject
Stripping	45.1–55.0	10	OFF	0	100	0	2.0	Inject
Neutralizing	55.1–65.0	10	OFF	0	0	100	2.0	Inject
Equilibration	65.1–75.0	10	OFF	100	0	0	2.0	Inject

28.4.2.3 HPLC Depletion

NOTE: A detailed protocol is provided by the manufacturer of the column. The 3 buffers used are provided by the manufacturer as 10x stock solutions. If your HPLC does not have the capacity for 3 solvents, then switch buffer 3 in and out at the appropriate times in the gradient.

An example of a depletion chromatogram is shown in Fig. 28.4.

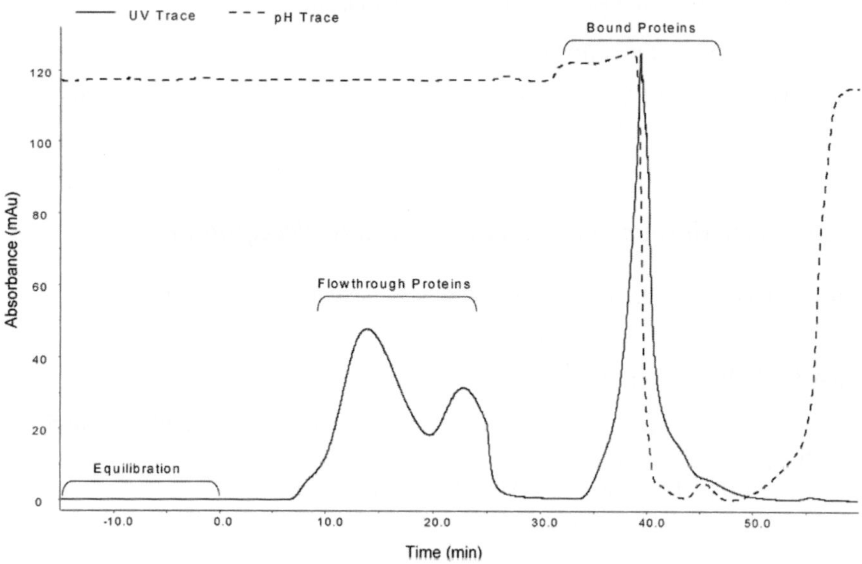

Fig. 28.4 Standard depletion profile. Flowthrough proteins are retained for further analysis; fraction collection can be adjusted to account for minor retention time drift ensuring that the full fraction is collected. Bound proteins consist of the 14 proteins retained by the column, which are removed by stripping with glycine buffer

1. Sample Preparation
   1. Dilute 130 μL of plasma with 520 μL of 1x Dilution Buffer (represents a 5x dilution)
   2. Transfer the plasma solution into a 0.45 μm spin filter and spin for 1 min at 9000×*g*
2. Sample Depletion
   1. Start the LC program and, using a sterile syringe, inject the sample into the sample loop. Leave the syringe attached to the Luer lock fitting during the run.
   2. Use the following gradient (see Table 28.1, maximum column pressure = 100 psi):

Wash for 20 min, with 100% 1 × dilution buffer
At 23 min, switch to 100% 1 × stripping buffer
At 38 min, switch to 100% 1 × neutralization buffer
At 46 min, re-equilibrate the column with 100% dilution buffer
At 55 min, the run is complete.

3. Once the fractions have been collected, proceed to Section 28.4.3 for the TCA precipitation procedure.
4. Store the column at 4°C in 0.02% sodium azide.

   Never freeze the column

5. Wash the ÄKTA Prime system monthly with 0.01% sodium azide solution.
6. Make Stripping Buffer weekly

28.4.3 Trichloroacetic acid (TCA) Protein Precipitation

This protocol describes the concentration and clean up of the samples.

28.4.3.1 Solvent Preparation

1. Urea solution: Measure about 2 mL of ACS urea reagent and dilute with 5 mL DI H_2O (~4 M solution).
2. Dilute 1.0 M TEAB to 100 mM with distilled water.
3. Mix 5 g of SDS in 100 mL DI H_2O to make 5% SDS.

28.4.3.2 Sample Preparation

1. Add TCA to the flowthrough fraction so that the final TCA concentration is 10%. The flowthrough fraction should be approximately 12.6 mL, so adding 1.4 mL of TCA would be needed to make a 10% solution.

   NOTE: Wash the pipettor with water immediately after using TCA (TCA is extremely corrosive).

2. Keep the flowthrough-plus-TCA solution at 4°C for 2–16 h to precipitate the proteins.
3. Set the centrifuge at 4°C, 3000×g for 60 min, and spin the flowthrough fractions.
4. After the spin has completed, a white pellet should be observed at the bottom of the conical tube. Occasionally a pellet may form that is not visible.
5. Using a glass Pasteur pipette gently decant off the supernatant buffer. Leave about 1 mm of buffer over the pellet.
6. Gently add ~3 mL of chilled acetone using a new Pasteur pipette.

7. Repeat steps (5) and (6) two more times, so that the pellet has been washed a total of three times.
8. On the last acetone wash remove the acetone, but make sure not to let the pellet dry. Rehydrate the pellet (Section 28.4.4) within 3–5 min after removing the acetone.

28.4.4 Protein Pellet Rehydration

1. Rehydration buffer must be prepared fresh and has a one-day expiration date. Combine the following, scaling up proportionally for the number of samples needing to be rehydrated: 1000 μL urea solution, plus 1000 μL TEAB 100 mM, plus 200 μL 5% SDS (for a total volume of 2200 μL). The total volume of rehydration buffer needed for each sample is **300 μL**.
2. Store the samples at 4°C overnight.
3. If the pellet is still visible, gently vortex the sample until the solution is clear.
4. Transfer to 2 mL spin filter tubes.
5. Spin samples at 9000×*g* for 1 min and transfer to 1.5 mL microfuge tubes.
6. Store at –20°C.

28.4.5 BCA (Bicinchoninic Acid) Protein Quantitation

In the iTRAQ protocol, 4 or 8 samples (4-plex or 8-plex iTRAQ reagent) are compared in a single MS analysis. For best results, the same amount of each label should be used, in order to give a 1:1:1:1 (or 1:1:1:1:1:1:1:1) ratio for proteins – and their corresponding reporter ions – that are NOT differentially expressed. This means that the same TOTAL amount of *protein* should be digested for each sample analyzed, and each iTRAQ label.

In this protocol for measuring protein concentration, based on the Sigma BCA Kit (Sigma-Aldrich, 2005) a dilution series of the standard and the samples is plated in duplicate in a flat-bottom 96 well plate, and a calibration curve is calculated from this dilution series. The protein concentrations in the samples are read from this calibration curve. This procedure, of course, can be done manually or with any UV spectrophotometer that measures absorbance at 570 nm. The procedure below is for a Thermo Multiskan 96-well plate reader with Ascent software.

NOTE: Keep samples on ice throughout this procedure. Use a tissue to blot excess sample and reagents from pipette tips. Samples must be acetone precipitated immediately after BCA quantitation. Do not put in the freezer for another day. A freeze-thaw cycle of the protein samples may affect the total amount of protein (proteins may precipitate), and then the concentrations determined by BCA would no longer apply.

28.4.5.1 Preparation of a Dilution Series in a 96-Well Plate

1. Prepare the following solutions:
   1. Plate Diluent: 1 mL water.
   2. BSA Standard Solution: 0.2 mg/mL BSA
2. Dilute samples:

   NOTE: Add 5–20 μL of sample (the volume depends on how concentrated you think the sample is – add less if you know that the sample is really concentrated) to a 1.5 mL labeled tube. Record this volume.

   1. Bring all samples to a final volume of 200 μL with DI H_2O.
   2. Vortex to mix, and centrifuge for ∼10 s.

   NOTE: Be sure to include a *BLANK* in the quantification.

3. Turn plate around so that well A1 is at the bottom left. Pipette 25 μL, water into all the #11 and #12 wells that are being used. Then, 20 μL into #9 and #10 wells. Keep repeating this pattern (in duplicate) decreasing the volume by 5 μL each time until you arrive at wells #1 and #2, which don't receive any plate diluent.

   NOTE: Steps 3 and 4 are more efficient with an electronic multichannel pipette.

4. Turn plate around so that well A1 is at the top left. Pipette 25 μL of BSA into wells A1 and A2. Then add 20 μL into wells A3 and A4. Keep decreasing the volume by 5 μL each time in duplicate until you arrive at wells #11 and #12 which don't receive any BSA.
5. Repeat these steps for each sample. Sample 1 will go across Row B and Sample 2 will go across Row C and so on (see Table 28.2).

28.4.5.2 Measurement of Optical Density

NOTE: If you have different equipment in the laboratory, use the analogous methods for your own equipment.

1. Turn on Multiskan Ex Plate Reader then open Ascent software. Wait for confirmation that the reader is connected to the software before proceeding. If the reader does not connect, try restarting the computer.
2. Prepare BCA reagent by doing a 1/50 dilution of reagent B (blue) into reagent A (clear). If preparing 15 mL, this would be 300 μL reagent B into 14.7 mL of reagent A. Invert and pour into weighing boat.
3. Using a multichannel pipette, dispense 200 μL of BCA Reagent into each well starting with column 1 and ending with column 12.
4. Set Ascent program to shake for 10 s at 300 rpm.
5. Cover plate with plastic adhesive and incubate for 15 min at 60°C.

Table 28.2 Preparation of a dilution series in a 96-well plate

Sample		1	2	3	4	5	6	7	8	9	10	11	12
BSA	**A**	25 µL BSA 0 µL H_2O	25 µL BSA 0 µL H_2O	20 µL BSA 5 µL H_2O	20 µL BSA 5 µL H_2O	15 µL BSA 10 µL H_2O	15 µL BSA 10 µL H_2O	10 µL BSA 15 µL H_2O	10 µL BSA 15 µL H_2O	5 µL BSA 20 µL H_2O	5 µL BSA 20 µL H_2O	0 µL BSA 25 µL H_2O	0 µL BSA 25 µL H_2O
1	**B**	25 µL Sample 1 0 µL H_2O	25 µL Sample 1 0 µL H_2O	20 µL Sample 1 5 µL H_2O	20 µL Sample 1 5 µL H_2O	15 µL Sample 1 10 µL H_2O	15 µL Sample 1 10 µL H_2O	10 µL Sample 1 15 µL H_2O	10 µL Sample 1 15 µL H_2O	5 µL Sample 1 20 µL H_2O	5 µL Sample 1 20 µL H_2O	0 µL Sample 1 25 µL H_2O	0 µL Sample 1 25 µL H_2O
2	**C**	25 µL Sample 2 0 µL H_2O	25 µL Sample 2 0 µL H_2O	20 µL Sample 2 5 µL H_2O	20 µL Sample 2 5 µL H_2O	15 µL Sample 2 10 µL H_2O	15 µL Sample 2 10 µL H_2O	10 µL Sample 2 15 µL H_2O	10 µL Sample 2 15 µL H_2O	5 µL Sample 2 20 µL H_2O	5 µL Sample 2 20 µL H_2O	0 µL Sample 2 25 µL H_2O	0 µL Sample 2 25 µL H_2O
3	**D**	25 µL Sample 3 0 µL H_2O	25 µL Sample 3 0 µL H_2O	20 µL Sample 3 5 µL H_2O	20 µL Sample 3 5 µL H_2O	15 µL Sample 3 10 µL H_2O	15 µL Sample 3 10 µL H_2O	10 µL Sample 3 15 µL H_2O	10 µL Sample3 15 µL H_2O	5 µL Sample 3 20 µL H_2O	5 µL Sample 3 20 µL H_2O	0 µL Sample 3 25 µL H_2O	0 µL Sample 3 25 µL H_2O
4, etc.	**E**	25 µL Sample 4 0 µL H_2O	25 µL Sample 4 0 µL H_2O	20 µL Sample 4 5 µL H_2O	20 µL Sample 4 5 µL H_2O	15 µL Sample 4 10 µL H_2O	15 µL Sample 4 10 µL H_2O	10 µL Sample 4 15 µL H_2O	10 µL Sample 4 15 µL H_2O	5 µL Sample 4 20 µL H_2O	5 µL Sample 4 20 µL H_2O	0 µL Sample 4 25 µL H_2O	0 µL Sample 4 25 µL H_2O

NOTE: There is no point in incubating longer – the signal will not improve.

6. Remove plate from incubation, and cool to room temperature in the dark for 5–10 min, or until the plate is no longer hot to touch.
7. *Carefully* remove plastic adhesive cover and place plate on plate reader.
8. Set Ascent to shake for 10 s at 300 rpm, then measure in continuous mode with a 570 nm filter. Measure OD of each well in the plate.
9. Open *Excel Quant Sheet* (*available as a download at* http://www.proteincentre.com/itraq-book-chapter-supplemental-information)
10. Copy OD results from Ascent into *Excel Quant Sheet.*
11. Enter sample names corresponding to OD data in the spreadsheet.
12. Enter volume of sample that was originally diluted to 200 μL (see Section 28.4.5.1, Step 2) under the *Volume Precipitated* column for each sample. This value should be 5–20 μL.
13. Look at the range values on the spreadsheet: If the range is ≤0.15 then the quantitation can be used. If not, look at the raw data that was pasted in from Ascent and delete any duplicates that do not follow the general trend. If you cannot tell which value to delete, or if there are no outliers, then the quantitation must be redone.
14. Observe the R^2 value of the plotted curve, if it is not at least 0.99 then the quantitation must be redone.
15. Make sure that at least three data points from each sample fall within the linear standard curve range (will be highlighted yellow).
16. Formulas embedded in the spreadsheet will automatically calculate the average concentration of each sample as well as the volume of sample containing the 100 μg needed for the iTRAQ analysis.Save a copy of this file.
17. Acetone protein precipitation (Section 28.4.6) must be performed on the same day as the BCA protein quantitation.

28.4.6 Acetone Protein Precipitation

This step in the procedure is a 2-day clean-up step involving precipitation of the proteins from the plasma, using acetone. This removes small molecules, peptides, and salts. We do an acetone precipitation on the volume of sample that contains 100 μg of protein, as determined by the preceding BCA protein quantitation.

NOTE: If you add too much protein, you won't get complete labeling. Too much or too little protein will also affect the desired 1:1:1:1 ratio of the reporter ions for "non-differentially-expressed" proteins.

28.4.6.1 Acetone Precipitation – Day 1

1. Centrifuge sample for 2 min.
2. Aliquot 10 volumes of ice cold acetone per sample to each labeled microcentrifuge tube.

3. Add 100 μg of sample (as calculated by the quant spreadsheet) to the acetone.
4. Invert tubes 3 times to mix and precipitate overnight at –20°C.
5. Store original samples in the –20°C freezer.

28.4.6.2 Acetone Precipitation – Day 2

NOTE: Check that tabletop centrifuge is at 4°C before starting.

1. Centrifuge precipitated samples at 4°C for 10 min at 13,000 rpm.
2. Without disturbing the protein pellets, use a pipette to remove the acetone and centrifuge 2 min at maximum speed.
3. Use a pipette to remove any remaining acetone and immediately add 30 μL iTRAQ Dissolution Buffer (0.5 M TEAB) and 3 μL 2% SDS, to each sample. Use a new pipette tip each time. Vortex and centrifuge.

   NOTE: Do not allow acetone pellet to dry completely.

4. Store samples at 4°C for a minimum of 4 h to a maximum of overnight, until they are resolubilized.

28.4.7 Protein Digestion

Most quantitative proteomics procedures are based on the quantitation of peptides, not proteins – this is often called "bottom-up" proteomics, because protein identification and protein quantitation are done on protein digests. Mass spectrometric analysis is typically more sensitive for peptides than for proteins, and – because mass accuracy is a function of mass – the analysis of peptides gives more accurate peptide and fragment-ion mass information than does the analysis of intact proteins. Also, because of internal energy considerations, it is difficult to obtain fragmentation (i.e., sequence information) from large peptides (> *m/z* 2500) or proteins, using conventional instruments.

In quantitative "bottom-up" proteomics, one is making the assumption that the concentration of a protein can be determined from the concentration of one of its proteolytic peptides –one mole of protein in the original sample will generate in other one mole of a peptide in the digest. Making this assumption, however, means that the digestion step is extremely critical, and must be efficient and – most important – reproducible. Improving the digestion is still an active area of research which we and other laboratories are trying to improve (Proc et al., 2010).

Since iTRAQ is based on low-mass fragment ions, usually from tryptic peptides, a tryptic digestion protocol is given here. However, the buffer used in this protocol is not the "standard" ammonium bicarbonate normally usually used for tryptic digestion. The reason for this is the amine-reactivity of the iTRAQ reagent. In order to avoid primary amines, the buffer contained in the iTRAQ kit is TEAB, triethylammonium bicarbonate. The digestion protocol given here uses TCEP, a reducing agent, to reduce the disulfide bridges, allowing the proteins to "open up" for more efficient digestion. A cysteine blocking agent, MMTS (methyl methanethiosulfonate), is used to prevent these disulfides from re-forming S-S-linked dipeptides.

1. Remove TCEP from freezer and thaw. 50 mM TCEP is stored at –20°C.
2. Remove samples from 4°C refrigerator, where they were stored after resolubilization (Section 28.4.6).
3. Vortex and centrifuge for ∼10 s.
4. Add 2 μL, 50 mM TCEP to each sample.
5. Vortex and centrifuge ∼10 s.
6. Reduce for 1 h at 60°C.
7. Cool samples to room temperature, and then centrifuge for ∼10 s.
8. Add 1 μL Cysteine Blocking Reagent.
9. Vortex to mix and centrifuge ∼10 s.
10. Incubate at room temperature for 10 min.
11. Prepare 1 mL of 100 mM iTRAQ Dissolution Buffer from stock.
12. Add 250 μL of 100 mM Dissolution Buffer to each of 2 bottles of trypsin (20 μg in each).

    NOTE: 2 or 4 bottles are needed for a 4- or 8-plex experiment respectively. The trypsin is pooled to make sure the amount added to each sample is equal. This also helps if some of the trypsin is stuck on the lid. There is more than enough trypsin to digest all of the samples even if one of the vials were to be empty.

13. Mix trypsin solution with a pipette. Pool trypsin into one bottle.
14. Add 120 μL of trypsin solution to each sample.
15. Vortex and centrifuge for ∼10 s.
16. Digest overnight at 37°C.

28.4.8 Peptide iTRAQ Labeling (4-plex or 8-plex)

NOTE: These are adapted from the instructions for the 4-plex kit and 8-plex kits (Applied_Biosystems, 2004b,c; 2007).

1. Centrifuge digested samples for 2 min at 13,000 rpm. Speedvac each sample for 55 min, then allow to air dry until no liquid remains.
2. Add 30 μL of iTRAQ Dissolution Buffer to dried samples.
3. Vortex and centrifuge for ∼10 s.
4. Bring ethanol/isopropanol and iTRAQ Reagents to room temperature.
5. Centrifuge the solutions containing the labeling reagents for ∼10 s.
6. For a 4-plex experiment, add 70 μL ethanol to each iTRAQ Reagent. For an 8-plex experiment, add 50 μL of isopropanol.
7. Vortex for 1 min to mix well.
8. Centrifuge for ∼10 s.
9. Add all of one label to one sample and vortex immediately. Repeat this step, adding the other labels to the rest of the samples.

NOTE: Be sure to record which label is added to which sample.

10. Centrifuge all samples for ~10 s.
11. Label for 1 h for 4-plex, or 2 h for 8-plex, at room temperature.
12. Add 1 mL SCX Buffer A to each sample for 4-plex, 400 μL to each sample for 8-plex, vortex to mix. Proceed to Strong Cation Exchange HPLC Protocol (Section 28.4.9), or store samples at –20°C.

28.4.9 Strong Cation Exchange (SCX) – First-Dimension Separation

Strong cation exchange chromatography is often used as the first dimension in a multidimensional (i.e., a MudPIT (Wolters et al., 2001)-type) separation of the iTRAQ-labeled peptides. This is done with a larger ID column than the one that will be used in the second-dimension separation. Although SCX-HPLC is does not have as high a chromatographic resolution as reversed-phase HPLC it does provide an orthogonal fractionation of the labeled peptides. Usually, 90 fractions are collected, but only 10 to 20 of these fractions – the ones with the highest UV absorbance – are analyzed by mass spectrometry. Fractions with weak absorbance can be set aside, because they typically do not contain many peptides. An example of an SCX separation is shown in Fig. 28.5.

28.4.9.1 Vision Column Equilibration Procedure

1. Prepare SCX buffers using LC-MS grade water and acetonitrile:
   SCX Load Buffer A (10 mM KH_2PO_4, 25% ACN, pH 2.75)
   SCX Buffer B (10 mM KH_2PO_4, 0.5 M KCl, 25% ACN, pH 2.75)
   SCX Buffer C (10 mM KH_2PO_4, 1 M KCl, 25% ACN, pH 2.75)

   NOTE: When acetonitrile and water mix they outgas. Do not tighten the lid while making the buffers.

2. Filter all buffers with a 0.45 μm filter to remove any particulates.
3. Prime lines with new buffers.

28.4.9.2 HPLC System Validation

1. Prepare test mix by adding 12.5 μL of 0.25 μg/μL α-endorphin solution; 1.5 μL of a 2.17 μg/μL angiotensin I solution; and 2.0 μL of a 1.67 μg/μL angiotensin II to 5 mL SCX buffer A in a 14 mL polypropylene tube.

   NOTE: Standards can be frozen in aliquots at –20°C. Make 15 μL aliquots for α-endorphin and 5 μL aliquots for angiotensin I and II. Do not freeze-thaw standards.

2. Add 5 μL of 7 M phosphoric acid to the 5 mL of standard text mix from Step 1.
3. Vortex gently, keeping liquid out of lid.
4. Check to make sure that the pH is below 3 with a pH strip.

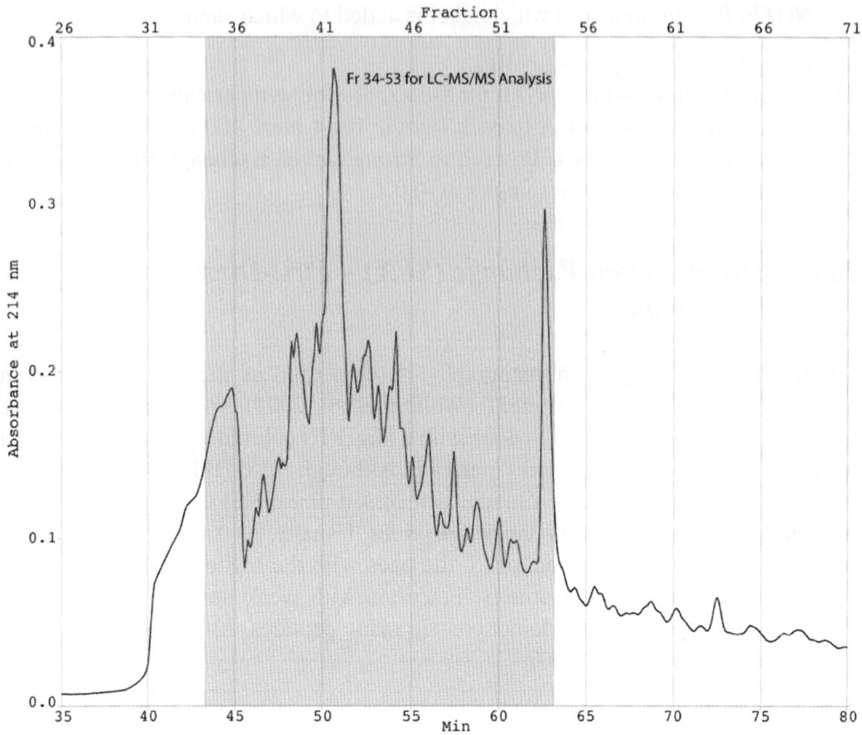

Fig. 28.5 A representative SCX chromatogram from separation of iTRAQ-labeled peptides from a 4-plex experiment, showing fraction collection times. Fractions 34–53 were chosen for RP-HPLC-MALDI analysis based on UV absorbance

5. Go to Vision and Clear Strip Chart recorder.
6. Run "*Standards*" method.
   The HPLC will switch from position B to A. The method will begins with 2 min SCX buffer A at 0.5 mL/min.
7. Inject your standards into the 5 mL loop.
8. After 2 min, the HPLC will switch to position B and the standards will be in line with the column.

   NOTE: Fractions are not collected during this method.

9. The method continues with the following steps:

   100% A to 100% B for 15 min.
   100% B for 5 min.
   100% B to 100% C for 15 min.
   100% C for 5 min
   100% A for 10 min
10. After method has finished, open the chromatogram.

11. With this HPLC set-up (column, flow rates, gradient, etc.), the validation test is considered a success if the following criteria are met:

Peak	Retention time (min)	Absorbance
α-endorphin	24 ± 5	≥ 0.1 A_{214}
Angiotensin II	29 ± 5	≥ 0.1 A_{214}
Angiotensin I	32 ± 5	≥ 0.1 A_{214}

If they didn't pass, then the standards will need to be re-run.

NOTE: This is the validation step for new columns as well as new buffers.

28.4.9.3 Running the Blank

1. In Vision software, select *Config* in Control panel, then Edit Config.
2. Clear Strip Chart recorder
3. Click *File* in Control Panel, then *Run Method.*
4. Select *blankiTRAQ.met.* Click *OK*

   The "BlankiTRAQ" method used for running blanks is as follows:

   Flow rate 0.5 mL/min;

   The gradient used is:

   0–15 min, linear gradient from 100% A to 100% B
   15–20 min, 100% B
   20–35 min, linear gradient from 100% B to 100% C
   35–40 min, 100% C
   40–50 min, 100% A

5. Click *Run Method with System Configuration.*
6. Enter file name following the example, for example, "Blank20100901"
7. Click **OK**.
8. *Clear strip chart recorder.*
9. Run SCX buffer A at 1 mL/min.
10. Wait until peak levels off on Strip Chart Recorder, then select *zero* in *Control Panel.*

28.4.9.4 Sample Preparation

1. Bring samples to room temperature.
2. Vortex to mix, and centrifuge for ~10 s.
3. Pool all 4 or 8 individual iTRAQ-labeled samples in a 14 mL polypropylene tube.
4. Wash each tube with 50 μL buffer A.
5. Vortex each tube for 10 s and centrifuge for 30 s.
6. Add buffer A wash from each tube to the14 mL tube containing pooled samples.
7. Bring total volume in 14 mL tube to 5 mL with buffer A.

8. Add 25 μL, 7 M phosphoric acid to pooled samples until the pH is between 2.5 and 3.0. Vortex gently to avoid getting any sample into lid.
9. Check pH by pipetting 5–10 μL of sample onto the acid portion of a pH strip. Do not dip pH strip into the sample.
10. Centrifuge the tube for 2 min at 6000×*g*.
11. Carefully remove the tube from centrifuge and transfer the supernatant to a new 14-mL polypropylene tube. Avoid disturbing the pellet at bottom of tube, if there is one.

28.4.9.5 Vision Workstation Set Up

1. Fill the fraction collector tube rack with 90 2.0-mL polypropylene tubes. Place rack into the AFC 2000 fraction collector on the Vision Workstation.
2. Rinse Hamilton syringe with water, and then with methanol, and then again with water.
3. Run SCX Buffer A buffer over the HPLC for 25 min – no run method required.
4. Zero the UV detector on the control panel.
5. Click stop on the control panel.
6. Clear Fraction collector
7. Clear Strip Chart recorder

28.4.9.6 SCX HPLC on Vision Workstation

1. Select File in *Control Panel* and *Run Method.*
2. Select the *iTRAQ SCX* method.
3. Click *Run Method with System Configuration.*
4. Enter sample into File Name field ending with the file extension ".bio".

   NOTE: Do not click *OK*.

5. Fill Hamilton syringe with sample.
6. Click *OK* on Vision software.
7. Wait until loop switches from position B to A.
8. *The iTRAQ SCX method contains the following steps*:

   Wash with SCX Buffer A for 2 min at 0.5 mL/min while the loop is off-line.
   NOTE: This is to ensure that there is no salt present in the line or on the column.
   Manually inject 5 mL of sample into a 5 mL loop.
   0–2 min Wash for 2 min, then make injection at 0.5 mL/min.
   2–22 min Collect the effluent in 1 mL fractions. This is just the flowthrough.
   22–102 min Begin linear gradient from 5 to 100% B at 0.5 mL/min. Collect the effluent in 0.5 mL fractions.
   102–112 min Hold at 100% B; stop fraction collection.
   112–118 min Linear gradient 100% B to 100% C.
   118–128 min Hold at 100% C at 1 mL/min.
   128–148 min Run a linear gradient from 100% C to 100% A at 1 mL/min

NOTE: The iTRAQ SCX method (Section 28.4.9.6) and the blank method (Section 28.4.9.3) are alternated to provide a wash between each sample.

28.4.9.7 SCX Chromatogram and Drying Fractions

1. When SCX HPLC has finished, examine the chromatogram.
2. Adjust *X*-axis min and max, so that all fractions are visible.
3. Save and print current graph.
4. From the SCX Chromatogram, determine which fractions have a UV absorbance above the baseline and label the corresponding tubes with their fraction number.
5. Speedvac fractions for 90 min or until volume is less than 150 μL.
6. Store fractions at –20°C until ready for MS analysis.

28.4.10 Preparation of Analytical Reversed-Phase C18 Nano LC Column

This section describes the preparation of a capillary LC column that will be used for the LC-MALDI spotting and ESI-LC MS/MS. If a commercial column is used, skip this section.

28.4.10.1 C18 Slurry Preparation

1. Weigh out 5 mg of Magic C18AQ 5 μm 100 Å packing material and place in a 2.0 mL polypropylene tube.
2. Add 500 μL of 100% methanol.
3. Add a magnetic bean stirrer, cap the tube and flick the tube to mix.

28.4.10.2 Packing the Column

1. Place the high pressure cell on top of a magnetic stir plate.
2. Remove the cap of the tube containing the C18 slurry and, using forceps, place the tube inside the pressure cell.
3. Turn on the stirring and verify that mixing is occurring by using a flashlight to see inside the tube.
4. Secure the lid of the pressure cell by tightening the set screws. Insert an empty IntegraFrit column, frit end up, and carefully slide the column down into the cell until contact is made with the bottom of the Vision tube. Raise the column about half a centimeter so the end of the column is off the base of the Vision tube and the magnetic bean stirrer, allowing the slurry to enter into the bottom of the empty column.
5. Tighten the nut to secure the column. Finger tight is adequate.
6. Open the valve of the helium (argon can also be used) gas cylinder and set the regulator to 1000 psi.

7. Turn the valve on the pressure cell to the pack position. A droplet forming at the frit and a visual inspection of C18 beads migrating up the column verifies the column is packing correctly.
8. Use a flashlight and a ruler to measure column length.
9. Once the column reaches 14 cm turn the gas tank regulator and the magnetic stir plate off. Wait ~20 min for the pressure on the packed column to reach zero and verify the column is 15 cm in length.
10. Turn the valve on the pressure cell to the vent position and loosen the nut holding the column. Remove the column from the pressure cell.
11. If not using the column immediately, push both ends of the column through 2 holes made in the cap of a 2.0 mL polypropylene tube filled with 50% MeOH for storage.
12. Label the column with packing material and date.

    NOTE: For analysis by LC-MALDI-MS/MS, proceed to Section 28.4.11; for analysis by LC-ESI-MS/MS, skip to Section 28.4.14.

28.4.11 Preparation of Recrystallized MALDI Matrix and MALDI Matrix Solution

The matrix (as in "*matrix*-assisted laser desorption ionization") co-crystallizes with the sample as the solution dries and assists in the transfer of energy to the sample during the ionization process, although the exact mechanism of ion formation is still not understood (Alves et al., 2006; Chang et al., 2007). The most commonly-used matrix for peptides is α-cyano-4-hydroxycinnamic acid (CHCA).

Some CHCA batches from vendors are of lower quality and may have a darker brownish color instead of a light, bright yellow hue. These impurities lead to decreased sensitivity and poor spectra. CHCA matrix is also available commercially in highly-purified form. This "MS-grade" matrix is acceptable for use straight out of the bottle, but it often comes in limited quantities and at a higher price. We have found that it is possible to use lower-grade CHCA if it is purified first. Recrystallization improves the quality of the CHCA and, in turn, the sensitivity and reproducibility of sample acquisition. The recrystallization procedure is outlined in the first part of this section.

28.4.11.1 CHCA Matrix Recrystallization Procedure

Day 1

1. Place 1 L distilled water in a 4°C cooler an hour before starting.
2. Pour 300 mL of ethanol into a 500 mL bottle and add a clean stir bar.
3. In a fume hood, warm alcohol on hot plate to 35°C (this takes about 20 min) with medium stirring.
4. Weigh out 5 g of CHCA and add to the warm ethanol.
5. Continue to heat and increase the stirring rpm to 400–600.

6. Stir until CHCA is completely dissolved and then remove from heat.
7. Vacuum filter the solution into a clean 1-L side-arm Erlenmeyer flask (it helps to moisten the filter paper with a little water to get it to stick to the Buchner funnel). Impurities in the matrix will remain on the filter paper, which can be discarded.
8. Transfer filtrate to a clean 2 L flask and add 2 volumes of cold distilled water. A yellow precipitate will form immediately.
9. Store flask in 4°C cooler for 24 h. Wrap in foil to keep the solution in the dark because CHCA is light-sensitive.

Day 2

10. Set up the vacuum filtration system again.
11. Remove flask from the cooler and vacuum filter the solution.
12. Wash the CHCA residue with 100 mL cold water.
13. Scrape the residue off the filter paper and into a clean 4 mL brown glass vial.
14. Completely dehydrate the CHCA by lyophilizing overnight.
15. Store recrystallized CHCA in the dark at 4°C.

28.4.11.2 MALDI Matrix Solution Procedure

NOTE: Matrix solution must be made fresh before use and tubes kept wrapped in aluminum foil as it is light sensitive.

4 mL of matrix solution is required for each complete iTRAQ experiment, including a BSA validation procedure. ~1.5 mL is used for each automated Probot spotting of matrix of 2 plates.

1. Weigh out 12 mg of recrystallized CHCA and 7.2 mg of ammonium citrate and add to the glass vial.
2. Add 2.8 mL of a acetonitrile and 1.2 mL of DIH_2O to the vial.
3. Carefully add 4 μL of TFA to the vial.
4. Vortex well to mix. Yellow crystals should go into solution.
5. Wrap in aluminum foil to protect from light. Store at room temperature up to 3 days.

28.4.12 Reversed-Phase Nano-HPLC Separation of the SCX Fractions

The SCX procedure (Section 28.4.9) uses a high-salt buffer, so before proceeding to the next step (reversed-phase separation), the samples are usually desalted. In these protocols, desalting is done with a long "wash" while the peptides are bound to the trapping column.

In this section, we describe the procedure for reversed-phase high performance liquid chromatography (HPLC) separation of the iTRAQ labeled SCX fractions, and direct spotting of the eluent onto a 4800 MALDI target plate using an Eksigent HPLC (Eksigent, 2004, 2005) and a Dionex Probot Microfraction collector (Dionex/LC_Packings, 2002).

28.4.12.1 Install C18 Columns on the HPLC

1. Make a new Magic C18 column according to Section 28.4.10.
2. Connect the Magic C18 column to the HPLC and flush column with 5% Solvent A at 1000 nL/min for ~15 min in order to re-pack any of the C18 resin that may have decompressed after the high-pressure packing.
   Solvent A – 2% ACN, 0.1% TFA
   Solvent B – 98% ACN, 0.1% TFA
3. Install a new C18 PepMap trapping column and flush with 100% Solvent A at 20 μL/min for ~15 min.
4. Change the buffer composition and flow rate on the nano pump to 5% B and 300 nL/min, which are the method start conditions, and let flow for ~10 min to equilibrate the analytical column. Check for a small droplet forming at the end of the Magic C18 column, which indicates that the system does not have any major leaks.
5. Connect the analytical column to the spotter.

   NOTE: Most spotters have the option to tee-in matrix solution for continuous matrix addition while sample is spotted on the target. We have found that spotting the matrix and the sample together can lead to complications such as matrix crystallization in the tee (especially for overnight HPLC runs) which causes blockages. Uneven distribution of matrix leading to inconsistent spot size and placement was also a problem for the subsequent automated mass spectrometry acquisition. For these reasons, we now apply the matrix *after* the HPLC run, provided that the size of the sample droplet spotted is at least 100 nL. Using the matrix over-spotting method, we have observed increased peptide signal compared to continuous matrix addition (Hardie et al., 2006).

28.4.12.2 Validate HPLC and MALDI-TOF Systems

It is necessary to evaluate the quality of the chromatography and mass spectrometry before analyzing the iTRAQ SCX fractions. There are many standard peptide mixtures that can be used to test column performance. We suggest using a BSA tryptic digest standard (100 fmol) to validate and optimize the HPLC-MALDI system (See on-line supplemental material).

1. Check that peptides elute in the sample spots (by generating a base peak chromatogram to see where peptides elute; and adjusting the gradient to center the elution profile)
2. Pick a peak to monitor retention time, intensity, and ion score (this is a good indication of when the instrument source is contaminated and needs to be cleaned to improve sensitivity)
3. With new columns, it may take more than one injection of BSA to achieve best results because of possible irreversible adsorption of BSA peptides to the packing material and to achieve proper equilibration of the columns.
4. A 1-h HPLC method is sufficient to separate BSA peptides. See Table 28.3 for suggested HPLC gradient.

Table 28.3 Suggested reversed-phase HPLC gradients for 1- and 2-h methods

Time (min)	1-h method %	
0	5	
10	5	
15	10	
45	60	
48	80	
50	5	5
60	5	15
100	–	40
105	–	75
110	–	5
120	–	5

5. See Section 28.4.13 for suggested MALDI-TOF/TOF analysis.
6. MALDI-TOF/TOF BSA data can be analyzed using Mascot (www.matrixscience.com) or equivalent database search engine.
7. For the Michrom BSA digest the following Mascot search parameters are recommended:

   Database: Uniprot-Swissprot
   Taxomony: all entries
   Enzyme: Trypsin
   Missed cleavages: 1
   Fixed modification: Carboxymethyl (C)
   Variable modifications: Deamidated (NQ), Oxidation (M)
   Peptide tolerance: 0.25 Da
   MS/MS tolerance: 0.15 Da
   Peptide charge: 1+, monoisotopic
   Quantitation: none
   Data format: Mascot generic
   Instrument: MALDI-TOF-TOF

8. The BSA validation is considered successful if the following specifications are met:

   Number of peptides identified >25
   BSA percent sequence coverage >35%
   Total ion score >1000
   ppm error <30

   NOTE: You should get numbers significantly higher than these if all is running smoothly, usually 35 peptides, 50% sequence coverage, 1800 score, 10–20 ppm.

28.4.12.3 Prepare SCX Fractions for HPLC

1. Decide, based on the UV absorbance observed on the SCX chromatogram, which SCX fractions will be further separated by either a 1- or 2-h HPLC gradient. This choice depends on sample complexity as indicated by the UV absorbance

the SCX chromatogram, and also on experience. Consider performing an initial "test" iTRAQ experiment where any fraction showing UV absorbance above the baseline is analyzed in order to determine typical peptide elution patterns. Sample complexity can be charted as a simple histogram showing the number of precursors detected by MS per fraction, or, after data analysis, the number of peptides identified.

2. Bring each SCX fraction to 150 μL with HPLC Solvent A (2% ACN, 0.1% TFA) and transfer to labeled autosampler vials.

28.4.12.4 HPLC Set-Up

1. Create HPLC methods. The following parameters are suggested:
   Capillary flow rate (Ch 1) – 20 μL/min
   Nano flow rate (Ch 2) – 300 nL/min
   30 μL injection volume (1/5 SCX fraction)
   15 min trap column washing
   10 min equilibration of columns at end of method
   Gradients for 1- and 2-h separations are shown in Table 28.3.

28.4.12.5 Spotting Set-Up

1. Place clean 384-well target plates in the tray of the spotter. Alternatively, unetched targets can also be used if markings are present to align the target, but modifications to the target templates in the software may be required. Wipe plates with HPLC-grade methanol to remove any smudges or dust.

   NOTE: Stainless steel target plates can be infinitely reused if cleaned according to the manufacturer's instructions. Keep away from concentrated acids and throw out a plate if accidentally dropped on the floor or bench because this may affect mass calibration across the plate due to surface warping.

2. Perform a target plate alignment procedure.

   NOTE: For contact-dependant spotters such as the Probot, the Z-axis alignment is critical for spot deposition of a small volume (<300 nL). For consistent results, we have found it best to align the Z-axis so that the tip of the column makes contact with the target.

   NOTE: Be very careful – the column frit is fragile and the column will be ruined if it breaks.

3. In μCarrier 2.0, set up a spotting method with following parameters:
   60 s fraction collection time
   Wait for contact start signal
   Collection on 48 spots (2 rows on 384-well plate) for a 1-h LC method
   Collection on 96 spots (4 rows on 384-well plate) for a 2-h LC method

   (See on-line supplementary material for application file creation).

4. Start the spotting application. The Probot will wait for a signal from the autosampler at the specified time. Once spotting starts, fractions are collected every minute. Watch the first row of spots to ensure that the Probot is making contact with the plate and evenly spotting. To fill two 384-well plates, HPLC spotting will take 16 h.

28.4.12.6 Automated Matrix Over-Spotting

1. Check that the plates from the day before spotted properly.
2. Make a fresh batch of MALDI Matrix (see Section 28.4.11)
3. Prime the syringe delivering the matrix solution to the spotter tee. It may be necessary to plug one end of the tee to direct matrix through the spotter needle.
4. Align target plates for accurate spotting.
5. Set up a spotting method to add 700 nL of matrix solution to each sample spot.

28.4.12.7 Calibrant Spotting

NOTE: This procedure is performed once per target. Calibrant spots can be re-used for instrument calibration for analysis of subsequent LC runs.

1. Prepare 4700 Cal Mix according to dilution instructions from Applied Biosystems, in a 0.5 μL microcentrifuge tube; use fresh matrix as diluent.
2. Spot 1 μL of Cal Mix/matrix solution on MALDI plate calibration spots; air dry at room temperature.
3. Store spotted plates in their cases at room temperature until ready for 4800 MS analysis.

28.4.13 MALDI-TOF/TOF Analysis

This section describes the operation of the Applied Biosystems 4800. These procedures were adapted from the following reference materials: (Applied_Biosystems/MDS_SCIEX, 2001a, 2005).

1. Use the LC-MALDI spot set template to create a spot set on the 4800, or use existing spot set for that target. (See on-line supplementary material for spot set template design)
2. Use compressed nitrogen gas to blow any dust or fibers off the target and load the plate into the instrument.
3. Perform plate alignment.
4. Acquire spectrum on a calibrant spot in MS Reflector mode

   NOTE: The instrument high voltage should be on for a minimum of 20 min before calibrating for best mass accuracy.

5. Adjust laser intensity setting and other tuning parameters until the following resolution and signal to noise (S/N) specifications are met:

Peptide name	*m/z*	Resolution (*R*)
des-Arg1-Bradykinin	904.4681	>10,000
Angiotensin I	1296.6853	>12,000
Glu1-Fibrinopeptide B	1570.6774	>13,000
ACTH (1–17)	2093.0867	>14,000
ACTH (18–39)	2465.1989	>13,000
ACTH (7–38)	3657.9294	>10,000

All peaks must have a signal to noise (S/N) ratios >100.
See Fig. 28.6a and b for examples of the spectra.

6. Update internal calibration values.
7. Acquire a new MS spectrum without internal calibration peak matching and inspect mass accuracy of *m/z* values. All masses must be within a 25 ppm error as below:

Peptide name	*m/z*	Δ Mass
desArg1-Bradykinin	904.4681	+/− 0.023
Angiotensin I	1296.6853	+/− 0.032
Glu1-Fibrinopeptide B	1570.6774	+/− 0.039
ACTH (1–17)	2093.0867	+/− 0.052
ACTH (18–39)	2465.1989	+/− 0.062
ACTH (7–38)	3657.9294	+/− 0.091

8. If resolution, S/N, and mass accuracy specifications are met, continue with a plate model calibration to obtain optimal calibration across the plate.
9. Acquire an MS spectrum on a sample spot and adjust laser setting and other parameters for optimal signal of peptides. Note the sample spot number and *m/z* values of a few peaks.
10. Open an MS/MS method and set the precursor to Angiotensin I (*m/z* 1296.68). Acquire an MS/MS spectrum.
11. Ensure that the spectrum contains the following peaks with corresponding resolution and S/N:

Mass	Resolution	S/N
110.072	>1000	>100
513.283		
784.411	>3000	>100
1165.591		
1181.658	>4000	>100
1183.601	>4000	>100

12. Acquire MS/MS spectra of peptides noted in step 9.

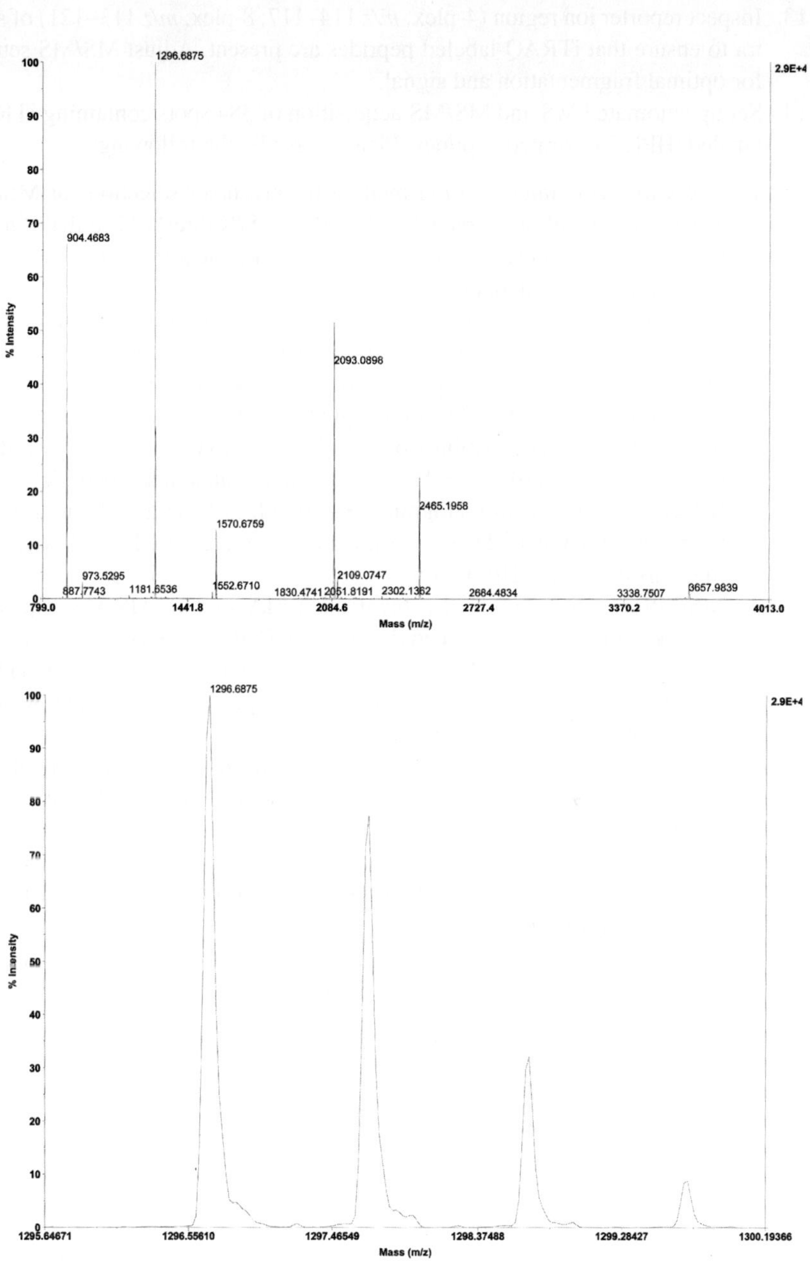

Fig. 28.6 (**a**) A representative MALDI-MS spectrum of the AB calibration standards (the 4700 mass standards kit, see item 16 in Section 28.3.12). (**b**) An expanded region of the same mass spectrum showing the mass resolution

13. Inspect reporter ion region (4-plex: *m/z* 114–117; 8-plex: *m/z* 113–121) of spectra to ensure that iTRAQ-labeled peptides are present. Adjust MS/MS settings for optimal fragmentation and signal.
14. Set up automated MS and MS/MS acquisition of 384 spots containing iTRAQ-labeled, HPLC separated peptides. Please consider the following:

    1. Use a job-wide interpretation method for precursor selection for MS/MS. The software will pick precursors based on a S/N threshold so that if a peptide elutes over more than one spot, MS/MS is acquired from the spot in which it is most abundant.
    2. A minimum S/N filter of 30–50 is suggested along with exclusion of precursors within 200 resolution. In our experience, the optimum number of precursors to acquire per spot is around 30, which is before sample consumption significantly affects the signal. For the same reason, it is best to do the precursor acquisition from weakest to strongest signals, in order to obtain the best possible MS/MS spectra on low-abundance peptides.
    3. Set the Minimum Chromatogram Peak Width to 1, with 200 ppm fraction-to-fraction tolerance. These recommendations are based on 60 s spotting intervals during the HPLC run.
    4. MS/MS fragmentation is possible using a 1 kV or 2 kV MS/MS acquisition method with collision-induced dissociation (CID) engaged or not. We use a 2 kV method with CID for detection of low-mass fragments in the iTRAQ reporter ion region (4-plex: *m/z* 114–117; 8-plex: *m/z* 113–121), which is important for accurate quantitation.
    5. The number of laser shots per MS/MS spectrum can be adjusted automatically by the software based on S/N thresholds of fragment peaks, or the number of laser shots can be left at a set number. Adjust these settings on iTRAQ labeled peptides to build up the signal from the reporter ions for quantitation. We use 1250 shots per spectrum (25 sub spectra) for all MS/MS acquisitions.
    6. An iTRAQ experiment is typically spotted on more than one plate. If using an instrument with multiple plate loading capabilities, make sure to queue up the calibration and acquisition of samples on the second plate.

15. Proceed to Section 28.4.17 for data analysis.

28.4.14 ESI-QTOF Calibration

The information in this section was adapted from the following manuals for the QSTAR: (Applied_Biosystems, 2004a; Applied_Biosystems/MDS_SCIEX, 2001a,b). Please see Proxeon_Biosystems (2002) for additional details on the operation of the electrospray source, and (Dionex/LC_Packings, 2001a, b, 2003) for additional details on the operation of the on-line Dionex/LC Packings nanoLC system.

28.4.14.1 Preparation of Calibration Solutions

1. Prepare a calibration solution of 30 pmol/μL SPI iPD1 and 15 pmol/μL CsI. This working calibration solution is stable for 2 weeks.
2. Use forceps and capillary cutter to shorten a NanoES capillary.
3. Place the shortened NanoES capillary in the upper part of the capillary holder and tighten the screw to hold it in place.
4. Pipette 4 μL of the iPD1 peptide/ CsI working calibration solution into the NanoES capillary. Gel-loading pipette tips are recommended.
5. Centrifuge the capillary holder containing the capillary for 15 s to spin the calibration solution to the tip of the capillary.
6. Loosen the screw and use forceps to remove the capillary. Hold to light to see level of liquid at bottom of capillary, re-centrifuge if necessary.
7. Place the NanoES capillary in the NanoES source head fitted on the QSTAR Pulsar i MS.

28.4.14.2 Initiating Nano-Spray for TOF-MS Calibration

1. In manual tuning mode, enter an ion source gas 1 (GS1) value of 2.
2. With the high voltage deactivated, slowly move the tip into the orifice plate and scratch the tip with a side-to-side or up-and-down motion to break open the capillary tip.
3. Move the nanospray tip back and down so that the spray is off-axis relative to the orifice at the 4 o'clock position, 8–10 mm away from the curtain plate. If you see liquid spraying you have opened the tip too much! If this happens transfer the sample to a new tip and start again.
4. Set the ionspray voltage to 1100–1400 V, CAD gas setting to 3, accumulation time 1 s and TOF masses 110–1100 amu and update the Q1 transmission window and TOF extraction parameters. Accumulate 10 scans.
5. For the QSTAR Pulsar i, the resolution (FWHM) should be at minimum 4000 and 8000 for the Cs^+ (*m/z* 132.9054) and SPI iPD1 (*m/z* 829.5398 peaks, respectively, with peak intensities of ~10,000 counts.
6. If this is the case, re-calibrate the TOF and set as instrument default.

28.4.15 ESI-QTOF and HPLC Validation

In this section, we describe the procedure for validating the reversed-phase chromatography of the Dionex/LC Packings HPLC system using BSA standards and detection by mass spectrometry. This chapter was adapted from the following manuals from Dionex/LC packings: (Dionex/LC_Packings, 2001a, b, 2003), and the Applied Biosystems QS Analyst QS Getting Started Manual (Applied_Biosystems/MDS_SCIEX, 2001a).

28.4.15.1 Installation of Analytical and Trapping C18 Columns on the HPLC

1. Prepare RP-HPLC solvents as follows:

   Solvent A: acetonitrile:water (2:98 (v/v)) with 0.1% formic acid
   Solvent B: acetonitrile:water (98:2 (v/v)) with 0.1% formic acid
2. Refer to Section 28.4.10 for preparing a Magic C18 analytical column.
3. Remove used C18 columns.
4. Replace the Zorbax trapping column by inserting in the column holder and tightening the fittings on the connecting tubing from the valve.

   Do not over-tighten!
5. The Magic C18 analytical column needs a nanovolume tee at the non fritted end to apply high voltage. A ~2.5 cm piece of PEEK tubing, a Valco-VICI Cheminert tee and platinum wire are needed to build this fitting.
6. Thread the capillary through the PEEK sleeve until protruding and trim ~1 cm. Push the capillary back until it is flush with the end of the sleeve and then insert in the Cheminert nano union attached to the connecting tubing coming from the HPLC. Finger-tighten the fitting.
7. Check for leaks by flowing Solvent A for about 10 min.
8. Gently insert the frit end of the column into the PEEK sleeve of the nanospray head. Ensure that the column tip is pushed all the way into the fitting before finger-tightening the union.
9. Replace the emitter tip at the other end of the Cheminert nano union only if it has been bumped or rubbed – otherwise it is fine to reuse it until it clogs or does not produce a stable spray.
10. Slide the nanospray head in the holder and tighten in place.
11. Attach the high voltage lead to the nanovolume tee electrode (see Fig. 28.7).

    NOTE: We have found that applying the high voltage pre-column rather than post-column helps to prevent bubbles, and results in fewer spray/signal drop-outs at the emitter tip.
12. Now adjust the emitter tip into spray position, which is approximately the 4 o'clock position relative to the orifice and about 5–8 mm away from the curtain plate. *Never spray directly into the orifice!*
13. Start the pumps and ensure that there is flow to the emitter tip. Observe the nanospray plume with a red or green laser pointer.

28.4.15.2 Validation of LC-MS/MS Using a BSA Standard

1. Prepare 20 fmol/μL BSA Standard Solution using HPLC-Solvent A (2% ACN, 0.1% FA) in a labeled autosampler vial. Vortex and flick to move contents to bottom and place in autosampler tray.
2. Equilibrate the LC-MS system using method start conditions. In this case, Analyst 1.1 software will control both the MS and HPLC.

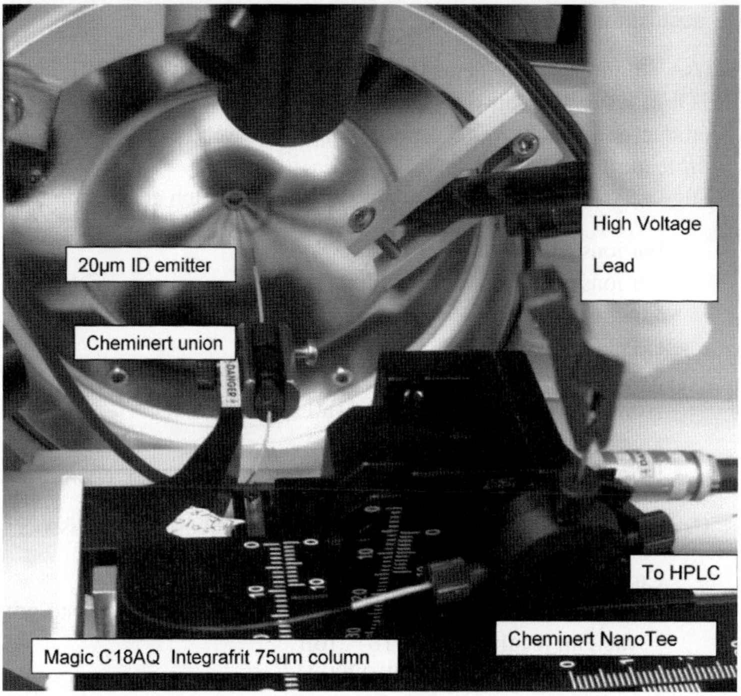

Fig. 28.7 Connecting the capillary column to the QSTAR source. Note the high voltage is applied *before* the analytical C18 column at the nanovolume tee electrode to minimize nanospray issues such as bubbles and dropped spray

3. Check on nanospray stability in manual tuning mode.
4. Adjust the position of the nanospray head for optimal signal.
5. For QSTAR Pulsar i, about 15,000–20,000 counts should be observed for the TIC baseline. About 200 counts should be observed for background contaminants (such as polysiloxane peaks at *m/z* 445.1 and 429.1) in the TOF MS scan. Noise should be minimal. Look for spray stability – if spitting or droplet formation is observed a new emitter tip should be installed or tip voltage increased 50 V until a stable spray is observed using a laser pointer.
6. Build an acquisition method with the following HPLC parameters:

   Capillary pump injection and trapping flow-rate:
   30 μL/min; Solvent A; Trapping time: 5–10 min
   Equilibration time at end of gradient: 5 min
   Analytical pump flow rate: 300 nL/min;
   Gradient: T = 0 min, 5% B; T = 5 min, 5% B; T = 30 min, 28% B; T = 41 min, 43% B; T = 43 min, 80% B; T = 45 min, 80% B; T = 47 min, 0% B; T = 60 min, 0% B

7. MS acquisition method parameters:

    1. Use an information dependent acquisition method consisting of a 1 s TOFMS survey scan over the mass range 400–1200 amu and two 2.5-s product-ion scans over the mass range 100–1500 amu with the *Enhance All* feature enabled.
    2. Suggested information-dependent acquisition switch criteria:

        For ions greater than: *m/z* 400.000
        For ions smaller than: *m/z* 1200.000
        Charge state: 2 to 5
        Exceeds: 20 counts
        Switch after: 1 spectra
        Exclude former target ions for 180 s
        ions tolerance: 100.000 ppm
        Ignore peaks within: 6.0 amu window

    3. Use the following Analyst 1.1 settings for collision-induced dissociation:

        Curtain gas is set at 25
        Nitrogen as the CAD gas and software value is adjusted to produce a vacuum reading of 4×10^{-5} Torr
        Ionization tip voltage 1900–2000 V

8. Queue up an injection of 100 fmol BSA using the 1 h method created in the previous steps.
9. Start the sample queue.
10. Open the data file while it is acquiring to monitor data quality.
11. At the end of the run the TIC should look similar to Fig. 28.8.

    NOTE: It may take more than one injection of 100 fmol of BSA to achieve good results.

28.4.15.3 Searching BSA Data with Mascot

1. To perform a Mascot search, run the current Mascot script for Analyst (download from Matrix Science, http://www.matrixscience.com).
2. The Mascot search parameters will open in a new browser window. Enter your name and e-mail address and select the following parameters**:** Database: Uniprot-Swissprot; Taxonomy: Mammalia; Fixed modifications: Carboxymethyl(C); Variable modifications: Deamidation (NQ), Oxidation (M); Peptide tolerance; 0.3 Da; MS/MS tolerance: 0.15 Da; Peptide charge: 2+ and 3+; Mascot generic data format, and ESI-QUAD-TOF instrument.
3. Start search. It should only take a few minutes.
4. The top hit should be bovine serum albumin (BSA) from the species *Bos taurus*. The score should be >1000, the number of peptides >35, and matching sequence coverage of the BSA protein should be >45%.

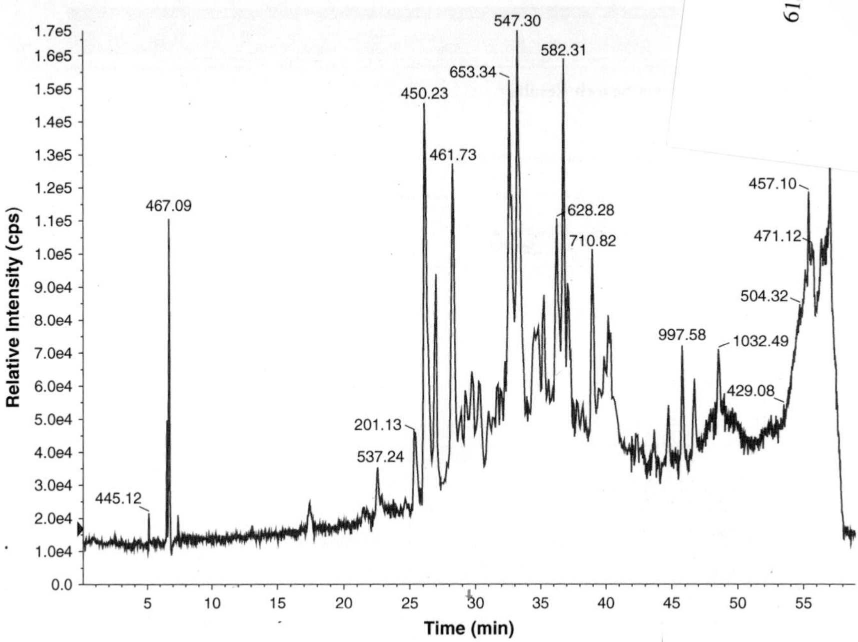

Fig. 28.8 TIC trace from the QSTAR for a tryptic digest of BSA

5. Inspect the retention time and ion score for the 582 ion to monitor any HPLC gradient drift and poor MS/MS fragmentation. The results should look like Fig. 28.9a
6. Generate an extracted ion chromatogram (XIC) for the 582.5 and 653.8 ions to inspect signal intensity and chromatographic peak shape. The results should look like Fig. 28.9b.

28.4.16 ESI-QTOF Analysis

Instead of off-line LC spotting, followed by MALDI-MS/MS analysis of the plates (Sections 28.4.11–28.4.13), on-line LC-ESI-MS/MS analysis can be performed on the SCX-separated fractions.

The LC analysis is performed using an integrated Famos autosampler (Dionex/LC_Packings, 2001a), Switchos II switching pump (Dionex/LC_Packings, 2001b), and UltiMate micro pump system (Dionex/LC_Packings, 2003) with a hybrid quadrupole-TOF mass spectrometer (QSTAR Pulsar i) equipped with a nano-electrospray ionization source and fitted with a 10 μm fused-silica emitter tip. This section on operation of the Q-STAR was adapted from the following instruction manuals: (Applied_Biosystems, 2004a; Applied_Biosystems/MDS_SCIEX, 2001a).

(A)

(B)

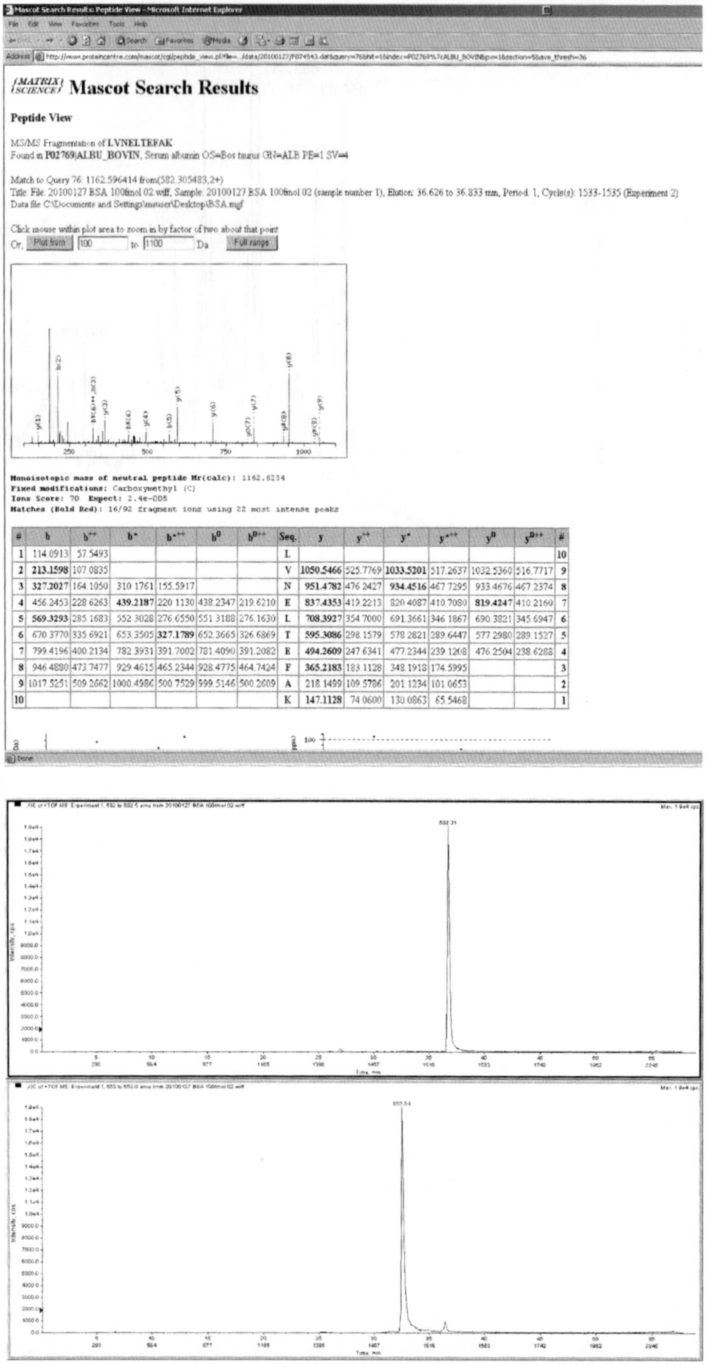

#	b	b++	b*	b*++	b0	b0++	Seq.	y	y++	y*	y*++	y0	y0++	#
1	114.0913	57.5493					L							10
2	**213.1598**	107.0835					V	**1050.5466**	525.7769	**1033.5201**	517.2637	1032.5360	516.7717	9
3	**327.2027**	164.1050	310.1761	155.5917			N	**951.4782**	476.2427	**934.4516**	467.7295	933.4676	467.2374	8
4	456.2453	228.6263	**439.2187**	220.1130	438.2347	219.6210	E	**837.4353**	419.2213	820.4087	410.7080	**819.4247**	410.2160	7
5	**569.3293**	285.1683	552.3028	276.6550	551.3188	276.1630	L	**708.3927**	354.7000	691.3661	346.1867	690.3821	345.6947	6
6	670.3770	335.6921	653.3505	**327.1789**	652.3665	326.6869	T	**595.3086**	298.1579	578.2821	289.6447	577.2980	289.1527	5
7	799.4196	400.2134	782.3931	391.7002	781.4090	391.2082	E	**494.2609**	247.6341	477.2344	239.1208	476.2504	238.6288	4
8	946.4880	473.7477	929.4615	465.2344	928.4775	464.7424	F	**365.2183**	183.1128	348.1918	174.5995			3
9	1017.5251	509.2662	1000.4986	500.7529	999.5146	500.2609	A	218.1499	109.5786	201.1234	101.0653			2
10							K	**147.1128**	74.0600	130.0863	65.5468			1

Fig. 28.9 Mascot results for a typical BSA evaluation run

28.4.16.1 Gradient Setup

1. Capillary pump injection and trapping flow-rate: 40 μL/min. Analytical gradient flow rate (for a 65-min method): 300 nL/min; T = 0 min, 0% B; T = 5 min, 0% B; T = 5.1 min, 5% B; T = 30 min, 28% B; T = 41 min, 43% B; T = 43 min, 80% B; T = 45 min, 80% B; T = 47 min, 0% B; T = 55 min, 0% B.
   Solvent A Equilibration time: 5 min and switched back in-line 5 min before end of gradient

   NOTE: If the observed A_{215} is greater than 0.1 for any fraction collected during the SCX fractionation, a 135 min gradient (5–80% solvent B) is used to compensate for the higher concentration of peptides in that fraction.

2. Analytical gradient flow rate (for a 135-min method): 300 nL/min; T = 0 min, 0% B; T = 5 min, 0% B; T = 5.1 min, 5% B; T = 70 min, 20% B; T = 100 min, 40% B; T = 112 min, 80% B; T = 115 min, 80% B; T = 116 min, 0 % B; T = 135 min, 0% B. Solvent A Equilibration time: 5 min and switched back in-line 5 min before end of gradient

28.4.16.2 QSTAR Acquisition Parameters

1. Scan setup
   1. An information-dependent acquisition method consisting of a 1-s TOF/MS survey scan of mass range 400–1600 amu and two 2.5 s product ion scans of mass range 100–1500 amu with the enhance all feature enabled is used.
   2. The two most intense peaks over 20 counts, with charge state 2–5 are selected for fragmentation and a 6 amu window is used to prevent the peaks from the same isotopic cluster from being fragmented again.
   3. Once an ion has been selected for MS/MS fragmentation it is put into an exclude list for 180 s. Analyst 1.1 settings for information dependent acquisition are as follows: Switch Criteria – For ions greater than: *m/z* 400.000; For ions smaller than: *m/z* 1600.000; With charge state: 2 to 5; Which exceeds: 20 counts; Switch after: 1 spectra; Exclude former target ions: For: 180 s; Ions Tolerance: 100.000 ppm; Ignore peaks within: 6.0 amu window
   4. Use the following Analyst 1.1 settings for collision-induced dissociation: Curtain gas is set at 25, Nitrogen is plumbed as the CAD gas and software value is adjusted to produce a vacuum reading of 4×10^{-5} Torr; Ionization tip voltage 1900–2000 V.

28.4.16.3 Sample Analysis

1. Bring SCX fractions (from Section 28.4.9) to a final volume of 150 μL with 2% acetonitrile/0.1% formic acid (FA) in water.

   NOTE: For ESI analysis, use FA as ion-pairing reagent, not TFA.

2. Inject 27 μL (~1/5 of the total volume).

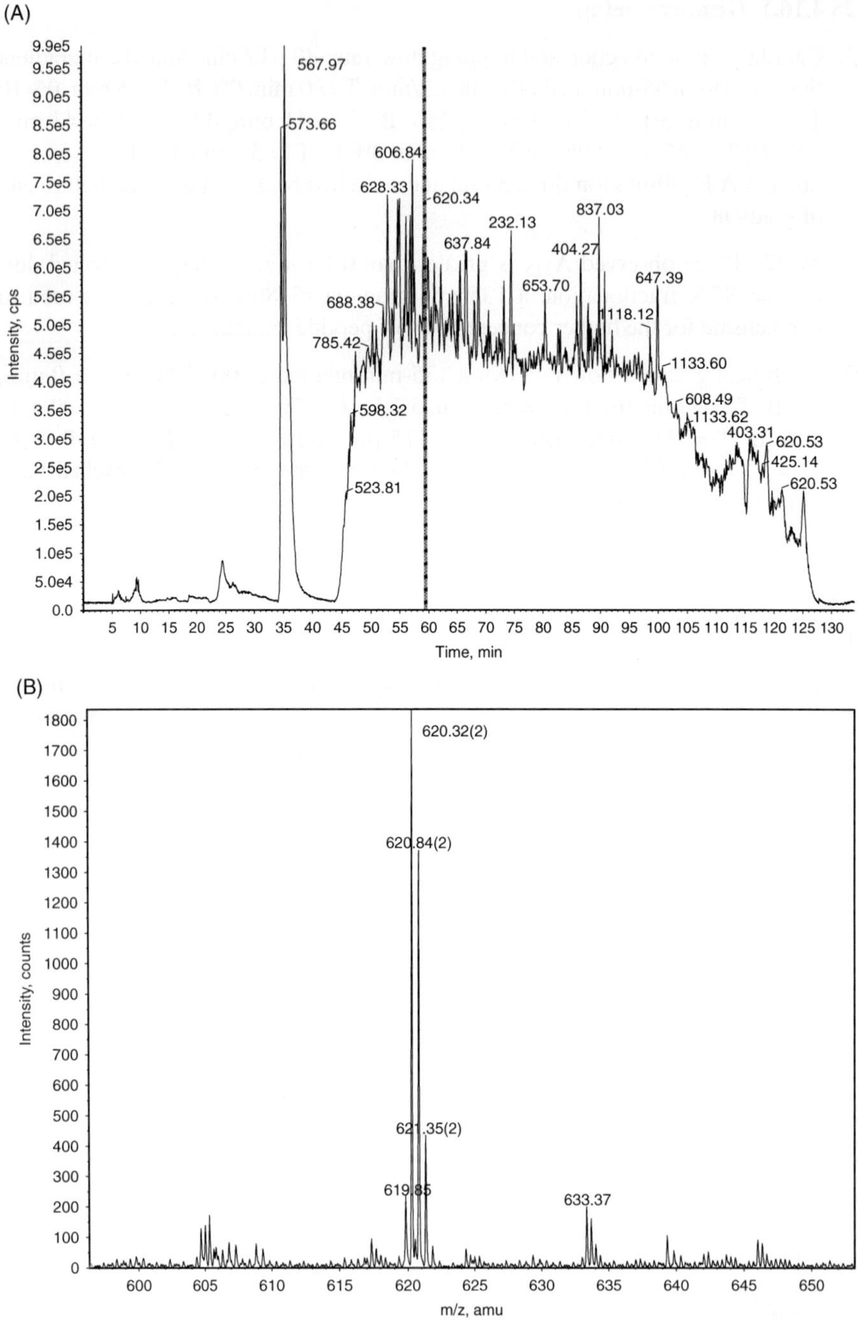

Fig. 28.10 (**a**) Representative Total Ion Chromatogram for an 8-plex 135 min analysis; (**b**) Representative TOF-MS spectrum for an 8-plex 135 min analysis; (**c**) Representative MS/MS spectrum for an 8-plex 135 min analysis; (**d**) Representative MS/MS spectrum showing 8-plex reporter ion series from 135 min analysis

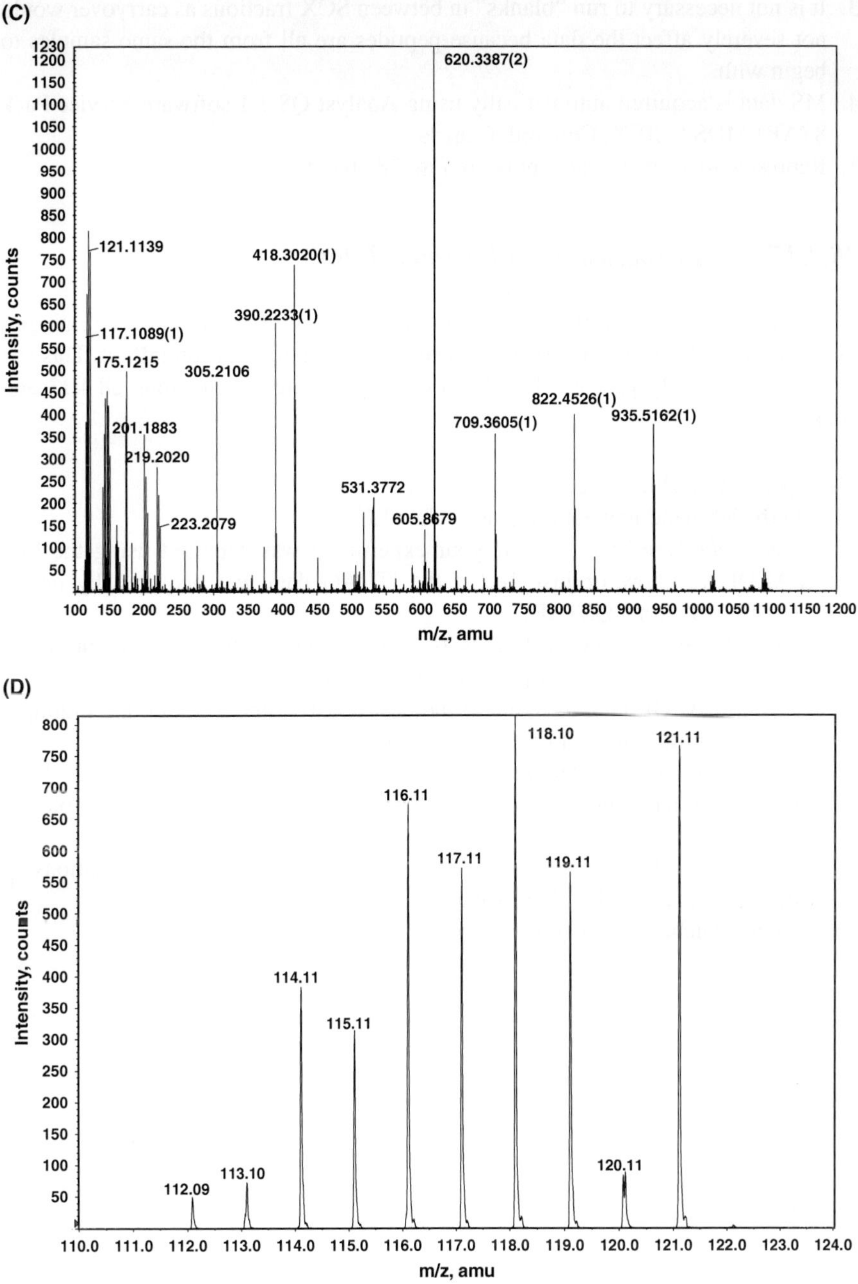

Fig. 28.10 (continued)

3. It is not necessary to run "blanks" in between SCX fractions as carryover would not severely affect the data because peptides are all from the same samples to begin with.
4. MS data is acquired automatically using Analyst QS 1.1 software Service Pack 8 (ABI MDS SCIEX, Concord, Canada).
5. Representative spectra are shown in Fig. 28.10a–d.

28.4.17 Data Processing with ProteinPilot

AB Sciex provides a link from their website http://www.absciex.com/mk/get/SOFTWARE_PROTEINPILOT to a downloadable version of ProteinPilot. This can be used to remotely process the MS/MS data generated in Sections 28.4.13 and 28.4.16.

1. Open ProteinPilot v3.0.
2. On the left hand navigation panel, select LC.
3. Select *Add TOF/TOF Data* if your experiment was acquired using the LC-MALDI workflow or just *Add* if the iTRAQ data was acquired by online LC-MS/MS with a QSTAR.
4. Select the appropriate .wiff files from the iTRAQ experiment if analyzing QSTAR data. Hold shift key to select multiple files.
5. For LC-MALDI data, navigate to the desired experiment folder and highlight the spot set files one at a time and click *Add.*
6. In the main window select *Paragon Method.*
7. Set *Paragon* parameters for iTRAQ analysis which include selecting 4-plex or 8-plex analysis and checking the quantitation box.
8. Set cysteine alkylation to MMTS, and include any other modifications of interest, such as methionine oxidation.
9. Select a database and begin search.

Acknowledgements The authors would like to thank the organizations for funding work relevant to these methods including the PROOF Centre of Excellence, Genome Canada, Novartis, IBM, Genome BC, University of British Columbia VP Research, St. Paul's Hospital Foundation, BC Transplant, and the James Hogg iCAPTURE Centre. We would also like to thank Drs. Steven Parnell and Oscar Alzate for critical reading of this manuscript.

References

Aguilar-Mahecha, A., Buchanan, M., Kuzyk, M.A., Borchers, C.H., and Basik, M. (2011). Comparison of blood collection tubes and processing protocols for plasma proteomics studies. Manuscript in preparation.

Alves, S., Fournier, F., Afonso, C., Wind, F., and Tabet, J.-C. (2006). Gas-phase ionization/desolvation processes and their effect on protein charge state distribution under matrix-assisted laser desorption/ionization conditions. Eur J Mass Spectrom *12*, 369–383.

Anderson, N.L., and Anderson, N.G. (2002). The human plasma proteome: History, character, and diagnostic prospects. Mol Cell Proteomics *1*, 845–867.

Applied_Biosystems (2004a). Analyst QS/BioAnalyst Tutorial.

Applied_Biosystems (2004b). iTRAQ^{TM} Reagents Amine-Modifying Labeling Reagents for Multiplexed Relative and Absolute Protein Quantification Chemistry Reference Guide Part Number 4351918, Rev. A.

Applied_Biosystems (2004c). iTRAQ^{TM} Reagents Amine-Modifying Labeling Reagents for Multiplexed Relative and Absolute Protein Quantitation Protocol, Rev. C. http://\penalty-\@Mwwwabsciexcom/LITERATURE/cms_041463pdf.

Applied_Biosystems (2007). iTRAQ^{TM} Reagents-8plex Amine-Modifying Labeling Reagents for Multiplexed Relative and Absolute Protein Quantitation Protocol Guide Part Number 4375249, Rev. B.

Applied_Biosystems/MDS_SCIEX (2001a). Analyst QS Getting Started Manual Part Number 1001933 B August 2001.

Applied_Biosystems/MDS_SCIEX (2001b). API QSTAR Pulsar I Hardware Manual, Part Number 1001553 B April 2001.

Applied_Biosystems/MDS_SCIEX (2005). 4800 MALDI TOF/TOF Analyzer Getting Started Guide Part Number 4352078 Rev. C April 2005.

Chang, W.C., Huang, L.C.L., Wang, Y.-S., Peng, W.-P., Chang, H.C., Hsu, N.Y., Yang, W.B., and Chen, C.H. (2007). Matrix-assisted laser desorption/ionization (MALDI) mechanism revisited. Anal Chim Acta *582*, 1–9.

DeSouza, L.V., Grigull, J., Ghanny, S., Dube, V., Romaschin, A.D., Colgan, T.J., and Siu, K.W.M. (2007). Endometrial carcinoma biomarker discovery and verification using differentially tagged clinical samples with multidimensional liquid chromatography and tandem mass spectrometry. Mol Cell Proteomics *6*, 1170–1182.

DeSouza, L.V., Romaschin, A.D., Colgan, T.J., and Siu, K.W.M. (2009). Absolute quantification of potential cancer markers in clinical tissue homogenates using multiple reaction monitoring on a hybrid triple quadrupole/linear ion trap tandem mass spectrometer. Anal Chem *81*, 3462–3470.

Dionex/LC_Packings (2001a). Famos Autosampler for Capillary- and Nano HPLC User's Manual. Part Number 160557.

Dionex/LC_Packings (2001b). Switchos Advanced Microcolumn Switching Device User's Manual. Part Number 162013.

Dionex/LC_Packings (2002). Micro Fraction Collector User's Manual. Part Number 161403.

Dionex/LC_Packings (2003). UltiMate Capillary and Nano HPLC System User's Manual. Part Number 160534.

Eksigent (2004). Eksigent NanoLC User Manual is from 2004.

Eksigent (2005). Eksigent Nano1D Software User Guide.

Elliott, M., Smith, D., Kuzyk, M., Parker, C.E., and Borchers, C.H. (2009). Recent trends in quantitative proteomics. J Mass Spectrom *44*, 1637–1660.

Gan, C.S., Chong, P.K., Pham, T.K., and Wright, P.C. (2007). Technical, experimental, and biological variations in isobaric tags for relative and absolute quantitation (iTRAQ). J Proteome Res *6*, 821–827.

GE_Healthcare (2002). Äkta Prime User's Manual; catalogue number: 18-1135-24, Edition AE.

Hardie, D., Jakubowski, P., Jackson, A., and Ohlund, L. (2006). Investigation of LC-MALDI Matrix Spotting for Improved TOF/TOF Analysis by Combined Spotting and Acquisition Strategies. Presented at the 54th ASMS Conference on Mass Spectrometry and Allied Topics, Seattle, WA, May 28–June 1, 2006.

Kuzyk, M.A., Ohlund, L.B., Elliott, M.H., Smith, D., Qian, H., Delaney, A., Hunter, C.L., and Borchers, C.H. (2009). A comparison of MS/MS-based, stable-isotope-labeled, quantitation performance on ESI-quadrupole TOF and MALDI-TOF/TOF mass spectrometers. Proteomics *9*, 3328–3340.

Omenn, G.S., States, D.J., Adamski, M., Blackwell, T.W., Menon, R., Hermjakob, H., Apweiler, R., Haab, B.B., Simpson, R.J., Eddes, J.S., *et al.* (2005). Overview of the HUPO Plasma

Proteome Project: Results from the pilot phase with 35 collaborating laboratories and multiple analytical groups, generating a core dataset of 3020 proteins and a publicly-available database. Proteomics *5*, 3226–3245.

Ow, S.Y., Salim, M., Noirel, J., Evans, C., Rehman, I., and Wright, P.C. (2009). iTRAQ underestimation in simple and complex mixtures: "The Good, the Bad and the Ugly". J Proteome Res *8*, 5347–5355.

Parker, C.E., Warren, M.R., Loiselle, D.R., Dicheva, N.N., Scarlett, C.O., and Borchers, C.H. (2005). Identification of components of protein complexes. Methods Mol Biol *301*, 117–151.

Proc, J.L., Kuzyk, M.A., Hardie, D.B., Yang, J., Smith, D.S., Jackson, A.M., Parker, C.E., and Borchers, C.H. (2010). A quantitative study of the effects of chaotropic agents, surfactants, and solvents on the digestion efficiency of human plasma proteins by trypsin. J Proteome Res *9*, 5422–5437.

Proxeon_Biosystems (2002). Nano-Electrospray Installation and Operations Manual.

Ross, P.L., Huang, Y.N., Marchese, J.N., Williamson, B., Parker, K., Hattan, S., Khainovski, N., Pillai, S., Dey, S., Daniels, S., *et al.* (2004). Multiplexed protein quantitation in Saccharomyces cerevisiae using amine-reactive isobaric tagging reagents. Mol Cell Proteomics *3*, 1154–1169.

Sigma-Aldrich (2005). BCA Protein Assay Kit Technical Bulletin. http://wwwsigmaaldrichcom/etc/medialib/docs/Sigma/Bulletin/qpbcabulPar0001Filetmp/qpbcabulpdf.

Sigma-Aldrich (2008). Seppro_column_user's_guide. http://wwwsigmaaldrichcom/etc/medialib/docs/Sigma/Bulletin/sep030bulPar0001Filetmp/sep030bulpdf.

Simpson, R.J., Bernhard, O.K., Greening, D.W., and Moritz, R.L. (2008). Proteomics-driven cancer biomarker discovery: Looking to the future. Curr Opin Chem Biol *12*, 72–77.

UVic_Proteomics_Center_SOPs (2010). http://www.proteincentre.com/services/standard-operating-procedures.

Wiese, S., Reidegeld, K.A., Meyer, H.E., and Warscheid, B. (2007). Protein labeling by iTRAQ: A new tool for quantitative mass spectrometry in proteome research. Proteomics *7*, 340–350.

Wolters, D.A., Washburn, M.P., and Yates, J.R., III (2001). An automated multidimensional protein identification technology for shotgun proteomics. Anal Chem *73*, 5683–5690.

Part IX
Sample Preparation in Functional Proteomics

Chapter 29
Methods for the Isolation of Phosphoproteins and Phosphopeptides for Mass Spectrometry Analysis: Toward Increased Functional Phosphoproteomics

Tapasree Goswami and Bryan A. Ballif

Abstract Reversible phosphorylation is arguably the best understood, if not the most important mechanism a cell employs to dynamically regulate protein (and thereby cellular) activity. In recent years the field of proteomics has acquired an impressive tool kit with which to study protein phosphorylation on a more global scale. These tools build upon classical techniques and incorporate novel methodologies, all propelled by rapid advances in mass spectrometers that function with increased speed, sensitivity and resolution. The depth and breadth of phosphoproteomic analyses are becoming increasingly comprehensive from both cumulative and single-study perspectives. Phosphoproteomics is now poised to become increasingly functional as experimental design incorporates the amassed wealth of current biological understanding with novel technologies and quantitative mass spectrometry. In this chapter, three important phosphoproteomic sample preparation methodologies are described: classical phospho-dependent affinity chromatography, tandem strong cation exchange-immobilized metal affinity chromatography (SCX-IMAC), and immunoaffinity chromatography. Each method is discussed with an eye toward designing experiments that maximize phospho-dependent functional understanding.

Keywords Functional proteomics · Phosphoproteomics · Quantitative mass spectrometry · Phospho-dependent Protein:Protein interactions · SCX-IMAC · Immunoaffinity

Abbreviations

AQUA	Absolute Quantification
ApoER2	ApoE Receptor 2
ATP	Adenosine Triphosphate
ATM	Ataxia-Telangiectasia Mutated

B.A. Ballif (✉)
Department of Biology and Vermont Genetics Network Proteomics Facility, University of Vermont, Burlington, VT 05405, USA
e-mail: bballif@uvm.edu

A.R. Ivanov, A.V. Lazarev (eds.), *Sample Preparation in Biological Mass Spectrometry*, DOI 10.1007/978-94-007-0828-0_29,

ATR	Ataxia-Telangiectasia Mutated and Rad3-Related
BRCT	Breast Cancer 1 C-terminal
CST	Cell Signaling Technology
Crk	CT10-Regulator of Kinase
CrkL	CT10-Regulator of Kinase-Like
Dab1	Disabled-1
FHA	Forkhead-Associated
GST	Glutathione-S Transferase
IMAC	Immobilized Metal Affinity Chromatography
LC	Liquid Chromatography
MH2	Mother Against Decapentaplegic Homology 2
MS	Mass Spectrometry
$PhPO_4$	Phenyl Phosphate
PI 3-K	Phosphoinositide 3-Kinase
PLC	Phospholipase C
PTB	Phosphotyrosine-Binding
pY-Dab1	Phosphotyrosyl-Disabled-1
PTM	Posttranslational Modification
RTK	Receptor Tyrosine Kinase
RSK	90 Kilodalton Ribosomal S6 Kinase
Sf9	*Spodoptera frugiperda*-9
SFK	Src Family Tyrosine Kinase
SDS-PAGE	Sodium Dodecyl Dulfate Polyacrylamide Gel Electrophoresis
SH2	Src Homology 2
SCX	Strong Cation Exchange Chromatography
TCA	Trichloroacetic Acid
VLDLR	Very Low Density Lipoprotein Receptor

29.1 Introduction

Eukaryotes devote between 2 and 3% of their genomes to reversible protein phosphorylation, encoding kinases and phosphatases whose critical regulatory roles modulate a myriad of cellular processes (Forrest et al., 2006; Hunter and Plowman, 1997; Johnson and Hunter, 2005; Manning et al., 2002a,b). Phosphoproteomics is a sub-discipline of proteomics that aims to identify and characterize phosphoproteins, or phosphorylation sites, frequently in a comprehensive way from cell or tissue extracts. Phosphoproteomics grew out of classical protein biochemistry and mass spectrometry-based analytical chemistry and now bridges all manner of chemical and biological disciplines. Like explorers that toil from a mountain base for a more expansive view at the summit, scientists developing and employing phosphoproteomic technology began with the challenging goal of generating a descriptive, yet "global" survey of phosphorylation sites in a given cell type or tissue (Lemeer and Heck, 2009; Macek et al., 2009; Nita-Lazar et al., 2008; Rogers and Foster, 2009).

As phosphoproteomic methodology has advanced, and as mass spectrometers have emerged with higher sensitivity, speed and resolution, increasingly comprehensive analyses have been conducted, some reporting the identification of more than ten thousand phosphorylation sites in a single study (Swaney et al., 2009; Zhai et al., 2008). If each protein from the roughly 25,000 protein-encoding genes in humans had only 4 possible phosphorylation sites, then 100,000 phosphorylation sites would be expected in a global analysis. Currently, the PhosphoSitePlus Database (www.phosphosite.org) reports a total of 96,978 identified, non-redundant mammalian phosphorylation sites on 13,374 proteins. We already appear to have an impressive view, even if we are not yet at the proverbial summit. Indeed, such exploratory cataloging has been invaluable to biologists and biochemists who can now "look-up" phosphorylation sites on their proteins of interest, examine the motif in which they lie, and begin to generate and test hypotheses as to how and to what effect such phosphorylation may be influencing their proteins of interest.

Given the important methodological progress over the last decade, phosphoproteomics is now prepared to become increasingly "functional," as researchers are now able to design experiments that not only ask at which sites a particular protein is phosphorylated, but also temporal and spatial questions of a protein's phosphorylation profile. Indeed phosphoproteomics is emerging as a primary force in systems biology given large cohorts of proteins can be analyzed in a single sample. This is especially true when phosphoproteomics is coupled to quantitative mass spectrometry approaches (de la Fuente van Bentem et al., 2008; Dengjel et al., 2009; Matsuoka et al., 2007; Nita-Lazar et al., 2008). Whether asking a very targeted question or stepping back from the trees to see the forest, functionality in proteomics, as in any biological discipline, comes through appropriate experimental design and careful evaluation of the data. In this chapter both targeted and large-scale phosphoproteomic analyses are discussed. Insights are given into experimental design and data analysis so as to augment the "functional" side of functional phosphoproteomics.

29.2 Classical Phosphoproteomics and the Enrichment of Phosphorylated Proteins or Proteins that Interact with Phosphorylated Proteins

Since the pioneering work of Krebs and Fisher (Krebs and Fischer, 1956) in the 1950s, scientists have appreciated that phosphorylation can alter a protein's structure and function in part by virtue of local structural adjustments imparted by phosphate's negative charge at typical cellular pH. Functional outcomes of protein phosphorylation include changes in enzymatic activity, subcellular localization, stability and Protein:Protein interactions (Pawson and Scott, 2005). Many proteins have structural features that specifically interact with a linear sequence of amino acids within the context of a phosphorylated amino acid. These include binding pockets for phosphoserine or phosphothreonine in 14-3-3, FHA, MH2, WD40, BRCT, WW, FF and POLO-Box domains as well as binding pockets for phosphotyrosine in Src Homology 2 (SH2) and Phosphotyrosine-binding (PTB) domains (Mohammad and Yaffe, 2009; Pawson and Kofler, 2009).

Classically a phosphoproteomic study asks a very targeted question involving phospho-dependent events such as: At what residue does kinase X phosphorylate protein Y? or What proteins interact with protein Y when Y is phosphorylated at residue Z? Although classical phosphoproteomics may enjoy less fan-fare than large-scale, high-throughput phosphoproteomics, each approach can be elegant if built on the core framework of experimental science: an important question and a well-designed experiment with proper controls. The more the question has supporting genetic, cell biological, biochemical, evolutionary or medical context, the riper the proteomic experiment is for deriving functional understanding.

29.2.1 Case Study I: Methods for the Identification of Phospho-dependent Protein: Protein Interactions

This section provides an example of a classical phosphoproteomic study that ultimately led to increased functional understanding of the Reelin/Disabled-1 (Dab1) signaling pathway. Its description provides an introduction to salient issues involved in classical phosphoproteomics.

29.2.1.1 Choosing the Question and Formulating the Hypothesis

In the last few years of the twentieth century, a number of biochemical and genetic studies (Howell et al., 1997a,b; Rice et al., 1998; Sheldon et al., 1997; Ware et al., 1997) were conducted that began to delineate the signal transduction pathway downstream of Reelin, a secreted protein that plays essential roles in the positioning of neurons during the development of the vertebrate central nervous system. Reelin was found to bind to the lipoprotein receptors, Very Low Density Lipoprotein Receptor (VLDLR) and ApoE Receptor 2 (ApoER2) (Hiesberger et al., 1999). Furthermore, it was shown that primary forebrain cultures stimulated with Reelin lead to tyrosine phosphorylation on the intracellular adapter protein Disabled-1 (Dab1), which interacts constitutively with the intracellular domains of VLDLR and ApoER2. The importance of Dab1 tyrosine phosphorylation in Reelin signaling is profound, as mice harboring a knock-in allele of *dab1* encoding for tyrosine to phenylalanine substitutions, such that its protein product cannot be phosphorylated on tyrosine, phenocopies loss of Dab1 or loss of Reelin (specifically referred to as a Reeler phenotype) (Howell et al., 2000). This led to the general hypothesis that Dab1 phosphorylated on tyrosine binds to embryonic brain proteins that are critical to transducing the Reelin signal.

29.2.1.2 Designing the Experiment with Appropriate Controls

A number of methodologies exists to biochemically purify a given protein's binding partners and have been reviewed previously (Phizicky and Fields, 1995). The two most common methods in use today include multiple variations of affinity chromatography (the focus of this section) and co-immunoprecipitation (co-IP). For phospho-dependent interactions these methodologies apply generally with some modifications (see below). In this section the words "bait" and "prey" are used to indicate the protein "bait" for which interacting protein partners "prey" are sought.

In the context of the phosphotyrosyl-Dab1 (pY-Dab1) example, an appropriate functional proteomics experiment would be designed to identify proteins that bind to pY-Dab1 but not to Dab1 that is not phosphorylated on tyrosine. Yet, before proceeding directly to experimental design, it is wise to be guided by as much information from previous experiments as possible. Notwithstanding, there can arrive a state where small apparent advances in design may bring diminishing returns and not be an efficient use of time. This is particularly true given that proteomic experiments are often discovery-based, serving as platforms for further experimentation or validation. Nevertheless, a careful, well-designed experiment should yield the most robust and cleanest results. One additional piece of data might then be considered in our example: mice homozygous for a *dab1* knock-in allele that truncates the carboxyl-terminal half of the protein appear wildtype (Herrick and Cooper, 2002), suggesting that the amino-terminal half of Dab1can fold appropriately and, at least with two gene copies present, is sufficient to effectuate normal development. Therefore an appropriate bait protein in this case would be the amino-terminal half of Dab1 phosphorylated on the critical tyrosine residues (Y185, Y198, Y220 and Y232). The most appropriate negative control bait would be the same but with the four tyrosine residues mutated to phenylalanine to render them incapable of being phosphorylated. It is also frequently desirable to incorporate amino- and carboxyl-terminal tags to permit double affinity purification when possible and when the protein's function or structure isn't compromised by the tags. Double affinity purification can dramatically reduce background and limit the affinity chromatography procedure to full-length bait molecules. An often overlooked aspect to functional proteomics is the design of a method that targets interacting prey proteins that are present in the same cells (even better the same sub-cellular locale) and at the same time/cell state as the bait protein. Thus, toward identifying proteins likely to be more functionally relevant interactors with pY-Dab1 one would do well to first examine cytosolic proteins from embryonic brain at a time when Reelin signaling is highly active. Given the heterogeneity of most tissue extracts, however, one must realize that not every cell in a developing brain is executing canonical Reelin signal transduction. Again one has to balance the efforts that would be required to generate an extract from very specific, perhaps low abundance cell types with the time required and available for the validation phase of a proteomics project. Functional proteomics might not, therefore, argue that one begin a search for Reelin signaling components in the adult liver, but at the same time, it might not argue that one must begin the experiment with exclusively Reelin-responsive cell types. In our example an embryonic day 16.5 murine embryonic forebrain and midbrain extract was ultimately chosen, a time when Reelin signaling is highly active, if even in a subset of cells in these tissues.

One final aspect of the bait purification will be discussed – preparing a phosphorylated bait protein. In our example Sf9 cells were used to produce the amino terminal half of Dab1 with an amino-terminal His_6 tag and a carboxyl-terminal GST tag. Prior to lysis the Sf9 cells were treated with sodium vanadate and hydrogen peroxide to inhibit tyrosine phosphatases and thereby boost tyrosine phosphorylation based on the activity of endogenous tyrosine kinases. This of course will not work for every experiment as all desired kinase activities may not be present or induced in Sf9 cells. Alternatives can include the incubation of the purified bait protein with the relevant kinase in an in vitro kinase reaction. Whatever method is chosen, high stoichiometry of phosphorylation is desirable.

29.2.1.3 Choosing Appropriate Lysis and Binding Buffers

In phospho-dependent Protein:Protein interactions as with any biochemical Protein:Protein interaction study, one must balance the strength of a given lysis and binding buffer's capacity to solubilize prey proteins with the buffer's capacity to facilitate or preserve the desired molecular interaction. Solubilization, particularly of hydrophobic membrane proteins, may require harsh detergents and salts that might disrupt only but the strongest of Protein:Protein interactions. Should such harsh conditions be required to solubilize a subset of the cell's proteome, detergents and salts could be diluted toward concentrations more appropriate for Protein:Protein interactions provided that the target prey proteins don't fall out of solution at the lower detergent and salt concentrations. In our example above of pY-Dab1-interacting proteins, a lysis buffer was chosen that was known to preserve phospho-dependent and -independent Protein:Protein interactions and to solubilize at least some transmembrane receptors. Biochemical assays of growth factor-stimulated Phosphoinositide 3-Kinase (PI 3-K) activity have historically been conducted by using anti-phosphotyrosine antibodies to immunoprecipitate the activated receptor tyrosine kinases (RTK)–or other tyrosine phosphorylated proteins in the cell–to which the SH2 domain of the p85 regulatory subunit of PI 3-K is recruited. The immune complex is then subjected to an in vitro lipid kinase reaction, typically using ^{32}P-γ-ATP as a tracer, and lipid products are resolved using thin layer chromatography. The striking feature of this purification as regards Protein:Protein interactions is that it requires two phospho-dependent interactions (the interaction of the antibody with the tyrosine phosphorylated growth factor receptor and the interaction of the p85 SH2 domain with the RTK) and one non-phospho-dependent interaction (the interaction of the p85 regulatory subunit with the active p110 catalytic subunit of PI 3-K). However, this buffer is only one of many that have been used to biochemically purify phospho-dependent Protein:Protein interactions (Table 29.1).

29.2.1.4 Preparation of Purified Phospho-Dependent Complexes for MS analysis

Once phospho-dependent protein complexes are achieved using either affinity chromatography or co-IP it is often wise to separate the complexes so as to maximize the opportunity of identifying prey proteins. Both affinity chromatography and IP require significant amounts of bait protein or antibody respectively, both of which can be problematic in MS/MS analyses. If, for example, such a complex were digested with trypsin directly, the resulting mixture would have bait or antibody peptides dramatically overrepresented. If such a mixture were loaded directly onto a reversed-phase LC-MS/MS platform for shotgun sequencing, then the mass

Table 29.1 Major constituents of lysis/binding buffers successfully used for phospho-dependent protein: protein interactions

Reference	Ballif	Benzinger	van der Geer	Jin	Meek	Yang	Elia
Phospho-interaction domain	SH2, 14-3-3	14-3-3	PTB	14-3-3	14-3-3	SH2	POLO-Box
Detergent	1% NP-40	1% Triton X-100	1% Triton X-100	0.5% NP-40	None- (Hypotonic)	1% NP-40, 0.25% DOC	0.5% NP-40
Buffer	25 mM Tris-HCl, pH 7.4	10 mM Tris-HCl pH 8.0	50 mM HEPES pH 7.5	50 mM HEPES pH 7.5	50 mM Tris-HCl, pH 7.5	50 mM Tris-HCl, pH 7.5	25 mM Tris-HCl, pH 7.5
Salt	137 mM NaCl	150 mM NaCl	150 mM NaCl	150 mM NaCl	150 mM NaCl	150 mM NaCl	125 mM NaCl
*Other	10% glycerol		10% glycerol	10% glycerol			

Indicated in the table are the phospho–interaction domains, the last name of the first author from which the information was taken, and the major buffer constituents and concentrations used for lysis and/or binding assays. *Not included in the table are common reagents such as protease inhibitors, phosphatase inhibitors, reducing agents and chelating reagents. References are as follows: Ballif et al. (2004, 2006), Benzinger et al. (2005), van der Geer et al. (1996), Jin et al. (2004), Meek et al. (2004), Yang et al. (2009), Elia et al. (2003)

spectrometer would spend an inordinate amount of time isolating and fragmenting peptide ions originating from the bait or antibody at the expense of not fragmenting (and therefore not identifying) ions of less relative abundance (peptides from potential prey proteins). If, however, the peptide mixture were subjected to multidimensional chromatography prior to MS (e.g. MudPIT analysis (Wolters et al., 2001)) or if proteins in the complex were first separated by SDS-PAGE prior to tryptic digestion the negative consequences of overrepresented bait could be reduced. In either case, one could further reduce the interference of the bait by selectively eluting the prey from the bait such that the bait remains immobilized on the solid phase matrix. Selectively eluting prey proteins typically involves using either high salt concentrations (to disrupt the electrostatic interactions between the phosphate and the phospho–protein interaction domain), or using a specific competing phosphopeptide. After either elution procedure, the salt or the peptide would need to be removed prior to MS analysis. This can be achieved by either dialysis or more typically precipitation and concentration with acetone or trichloro-acetic acid. The resulting pellet can then be subjected to tryptic digestion directly or protein bands can be digested in-gel following SDS-PAGE.

29.2.1.5 Executing the Initial Experiment

In the pY-Dab1 example individual cultures of Sf9 cells were infected with one of three constructed expression viruses: (1) His_6-GST (2) His_6-Dab1(1–252, Y4F)-GST or (3) His_6-Dab1(1–252)-GST. 1–252 denotes the specific amino acid residues in Dab1, and Y4F corresponds to the tyrosine to phenylalanine substitutions in Dab1 as indicated above. Prior to lysis, all cultures were treated with sodium vanadate and hydrogen peroxide to inhibit tyrosine phosphatases as discussed above. After lysis, expressed proteins were enriched on a nickel resin, eluted with imidazole, and then purified again using glutathione sepharose as described (Ballif et al., 2004). Approximately 60 mg of protein extract from embryonic day 16.5 murine forebrains and midbrains was prepared and subjected to a series of chromatography steps as shown in Fig. 29.1a, which included critical pre-clearing/subtraction steps to enrich for pY-Dab1-interacting proteins. After extensive washing a series of elution steps was tested to determine appropriate conditions to elute pY-Dab1-interacting proteins without eluting the GST-fusion proteins from the glutathione sepharose. 25 mM phenyl phosphate ($PhPO_4$) was first used to compete with phosphotyrosine, followed by increasing concentrations of $MgCl_2$ for general chaotropic disruption. Eluted proteins were subjected to TCA precipitation followed by SDS-PAGE and silver staining (Fig. 29.1b). These initial results showed that, to one degree or another, each elution condition released a few proteins consistently and specifically from the column containing pY-Dab1. These protein bands were excised and subjected to in-gel tryptic digestion and LC-MS/MS analysis. The names for the identified proteins are indicated in Fig. 29.1b. In general, the $MgCl_2$ elutions gave very clean results and thus this experiment was repeated except that more starting material was used (150 mg of embryonic brain extract) and only two elutions were performed step-wise (200 mM and 1 M $MgCl_2$). Eluates from individual columns

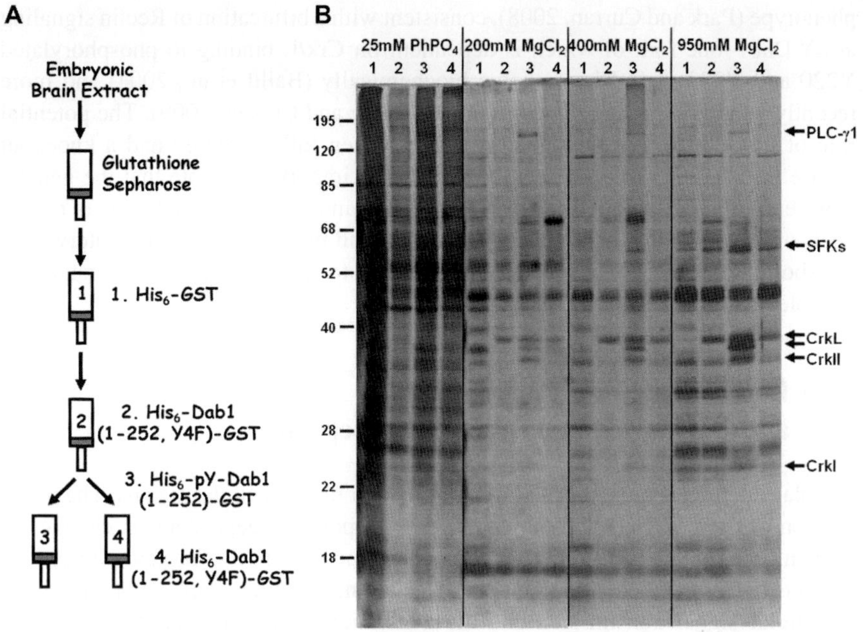

Fig. 29.1 Experimental workflow and results of affinity chromatography to purify pY-Dab1-interacting proteins. **(a)** Stepwise affinity chromatography scheme with numbered columns harboring approximately 300 μg of the indicated fusion proteins purified from a baculoviral/Sf9 cell expression system. **(b)** Stepwise elutions with either phenyl phosphate ($PhPO_4$) or $MgCl_2$ were performed as indicated on each column. Eluates were individually precipitated with TCA and then subjected to SDS-PAGE and silver staining. Molecular weights in kDa are indicated at left and proteins subsequently identified by LC-MS/MS are indicated at right. All samples were loaded on one 10% polyacrylamide gel pieced together electronically to straighten lanes and remove interspaced molecular weight markers

were combined and TCA precipitated. The results of this second experiment were very clean, showed robust contrast, and have been published elsewhere (Ballif et al., 2004). As indicated, the major proteins identified in this analysis were the Crk and CrkL (Crk/L) family of adaptor proteins, the Src family of tyrosine kinases (SFKs) and Phospholipase C-γ1 (PLC- γ1).

29.2.1.6 Assessing the Functional Relevance of the Identified pY-Dab1-Interacting Proteins

In addition to strong biochemical data supporting a role for SFKs and Crk/L in Reelin signaling (Arnaud et al., 2003; Ballif et al., 2004, 2003), there is now also strong genetic support. Experiments conducted subsequent to the proteomic analysis described above include the generation of a $Src^{-/-}$, $Fyn^{-/-}$ compound knockout mouse that exhibited a Reeler-like phenotype (Kuo et al., 2005), and the generation of a $Crk^{-/-}$, $CrkL^{-/-}$ brain conditional knockout mouse that showed a partial Reeler phenotype (Park and Curran, 2008), consistent with a bifurcation of Reelin signaling at pY-Dab1–one fork of which is dependent on Crk/L binding to phosphorylated Y220 and Y232 of Dab1, as shown biochemically (Ballif et al., 2004) and more recently genetically using knock-in alleles (Feng and Cooper, 2009). The potential role of PLC- $\gamma 1$ in Reelin signaling has not been fully explored and a knockout mouse for PLC- $\gamma 1$ has not been reported. It is important to note that the genetic work exploring the relationship between Crk/L and Dab1 followed 5 years on the heels of the proteomic and biochemical work, and thus emphasizes the interwoven collaboration between biochemistry and genetics toward a functional understanding of molecular mechanisms.

29.3 Large-Scale Phosphoproteomics and the Enrichment of Phosphopeptides

Cellular function is often dictated by extracellular or intracellular signals. The transduction of these signals and the appropriate response is dependent on changes in protein function or their properties. It has become evident that protein phosphorylation is one of the most important posttranslational modifications (PTMs) involved in cellular signal transmission and the generation of specific cellular responses. The chemical transfer of a phosphate group from ATP to (most commonly) serine, threonine or tyrosine residues is rapid, reversible, can be highly specific, and is a mechanism that has been conserved through evolution. Phosphorylation of proteins by protein kinases is a dynamic process by which the function and localization of thousands of proteins are regulated within cells.

In its simplest and canonical form, a signal transduction pathway is initiated by one signal, progresses through a series of subsequent Protein:Protein interactions, and finally results in specific gene regulation. However, many signaling cascades can act in concert and can regulate non-transcriptional cellular activities. Almost at each step of a signaling cascade, particularly those involving the activation of a kinase, there is an opportunity for cross-talk between signaling pathways. The activation of each kinase might lead to many proteins becoming phosphorylated, and each of these phosphorylated substrates might participate in multiple signaling pathways. A single signal can thus ultimately influence multiple signal transduction cassettes operating within a cell and hence cellular function is largely dependent on the complete signaling network operating within the cell.

Although the proteome and phosphoproteome may both be general indicators of the metabolic, developmental, growth, or differentiation status of a cell, the phosphoproteome is more specifically an indicator of the sum of the dynamic kinase- and phosphatase-dependant signaling events in the cell at any given time. Identification of the most inclusive phosphoproteome of cells or organ systems is thus a prerequisite for a quantitative description and prediction of the net signaling function within that cell or organ. Changes in the phosphoproteomes of cells after various stimuli, after pharmacological challenge, or in tissues at different stages of development or disease will therefore help to elucidate crucial pathways involved in manifesting the biological change. It is clear, therefore, that robust methodology to achieve comprehensive and quantitative phosphoproteomic descriptions must be established to have functional understanding.

Perhaps the biggest challenge in identifying phosphorylation events is the low stoichiometry at which proteins are often phosphorylated. Additionally, most phosphorylation events are transient in nature – a signal-dependent phosphorylation event must be dephosphorylated by a phosphatase to "reset" the system for repeated signaling. Hence, phosphopeptides are much more rarely identified in the presence of large excesses of unphosphorylated peptides. It is clear that conventional, "shotgun" proteomic analysis is therefore not suitable for the identification of low abundance phosphorylation events. To this end, strategies to enrich phosphopeptides from a pool of phosphorylated and unphosphorylated peptides have been developed.

For typical mass-spectrometric analyses proteins, and thus phosphoproteins, are first digested with trypsin as the derivation of sequence information is more efficient with peptide fragments. While the complexity of the sample increases many fold after tryptic digestion, the increased efficiency of phosphorylation site identification from peptides outweighs (in most cases) the complexity challenge. Hence most protocols of enrichment start with trypsin-digested protein extracts.

Some of the commonly used approaches for the enrichment of phosphoproteins/phosphopeptides involve formation of chemical derivatives; β-elimination or phosphoramidate adduct formation; immunoprecipitation using phospho-motif specific antibodies (see Section 29.3.2 below); solution-charge-dependant enrichment using strong cation exchange chromatography (SCX) or Electrostatic Repulsion-Hydrophilic Interaction Chromatography (ERLIC); Lewis acid-base interaction with TiO_2; and affinity purification via complex formation between chelated metal ions and phosphate groups (immobilized metal ion affinity chromatography, IMAC) (Thingholm et al., 2009). TiO_2 appears to have a greater affinity for phosphorylated peptides compared to IMAC columns containing Fe, and thus elution of multiply phosphorylated species from TiO_2 columns may be hindered giving an apparent preference of TiO_2 for singly-phosphorylated peptides (Thingholm et al., 2009). Below we discuss the use of different lengths of Fe-based IMAC columns used in parallel to more efficiently capture singly- and doubly-phosphorylated species. Indeed, a number of combination approaches are commonly used and, as expected, combination strategies often provide more comprehensive phosphopeptide enrichment than individual methods. Combination methods include: SIMAC (sequential elution from IMAC) in which IMAC columns are eluted using acidic or alkaline buffers and the acid eluate and IMAC flowthrough are further purified using TiO_2; and SCX-IMAC in which SCX fractions are further enriched using IMAC (Thingholm et al., 2009).

While methodological advances in phosphopeptide enrichment are moving us closer towards comprehensive and quantitative identification of phosphoproteomes, the major stumbling blocks in harnessing the power of mass spectrometry are in the bioinformatic analysis of the data and functional validation of the results. Ways of combining the latter more seamlessly will be important in the future. Optimization of phosphopeptide enrichment methods, however, will increase the comprehensive scope of phosphoproteomic analyses and thereby enable an increased understanding of phosphoproteomic mass-spectrometry data in terms of functionality.

29.3.1 Case Study II: Methods for Enhancing the Large-Scale Enrichment and Identification of Phosphopeptides Using SCX-IMAC

Given that it is of paramount importance that phosphpeptides are enriched efficiency prior to large-scale phosphopeptide identification, we have compared the efficiencies of several different strategies. These include SCX alone, SDS-PAGE fractionation followed by SCX-IMAC (Fe-based), and SCX-IMAC. Of these methods we identified the largest numbers of phosphopeptides from the sequential use of SCX followed by IMAC. However, as Fe-based IMAC has been considered less efficient than other methods such as TiO_2 at identifying singly-phosphorylated species, we sought to further optimize this procedure for a single SCX-IMAC platform that would more efficiently retrieve singly-phosphorylated species. The series of experimental steps involved in this optimization are discussed below and we hope will prove useful for others in their design of phoshopeptide enrichment methods. We also point readers to consult an in-depth methods description of the SCX-IMAC approach largely as we have used it (Villen and Gygi, 2008). We therefore proceed directly to the consideration of optimization of phosphopeptide enrichment using Fe-based IMAC.

29.3.1.1 Optimization of IMAC-based Phosphopeptide Enrichment

While IMAC has been used and discussed by many researchers, the goal of this study was to determine conditions for increased recovery of phosphopeptides. During the optimization process we addressed the following considerations: (1) Volume of IMAC resin (2) Batch binding versus column binding, and (3) Extent of IMAC resin washing prior to elution. To monitor the success of the optimization process we used internal ^{13}C-, ^{15}N-labeled absolute quantification (AQUA) peptide standards (Gerber et al., 2007; Gerber et al., 2003) to measure the recovery and specificity of IMAC enrichment under different conditions. "Heavy" and "light" AQUA peptides consisted of two pairs of mammalian-specific peptides, one pair was singly phosphorylated and the other was doubly phosphorylated. The heavy forms of the internal standards were introduced into a desalted SCX fraction resuspended in 100 μL of IMAC binding buffer (250 mM actetic acid, 30% acetonitrile) containing approximately 0.2 mg of tryptic peptides from yeast whole cell extract. After an IMAC enrichment was performed using a given set of conditions, the IMAC eluate was recovered, an equal amount of the light internal standard was added to the eluate. The eluate was then desalted and subjected to LC-MS/MS and percent recovery was established as a ratio of the heavy to light AQUA standards. Thus, the singly- and doubly- phosphorylated tracer peptides would enable the evaluation of the efficiency of a particular condition with respect to its propensity to enrich singly- and doubly-phosphorylated peptides. In this study desalted and dried fraction 9 of the 14 SCX fractions collected from the SCX chromatography was resuspended in 400 μL of IMAC binding buffer containing the "heavy" AQUA standards. This sample was then divided it into 4 aliquots. Fraction 9 in our runs was one of the four consecutive fractions identified from previous SCX-IMAC experiments as rich in phosphophospho-peptides. In these studies the IMAC resin used was Phos-Select (Sigma-Aldrich, St. Louis, MO). Each 100 μL peptide aliquot was then used for one of the following four optimization conditions: (1) 10 μL (50% slurry) of IMAC resin, batch binding, (2) 30 μL resin, batch binding, (3) 10 μL resin, column binding, and (4) 30 μL resin, column binding. Each of these resin volumes and phosphopeptide binding conditions was further repeated using various numbers of washes and multiple elution steps to determine whether these variables influenced phosphopeptide recovery/enrichment.

For batch binding, the resin was equilibrated with IMAC binding buffer and then shaken vigorously on a vortex mixer with one of the 100 μL peptide aliquots for 1 h. After binding the supernatant was removed and the resin was washed with IMAC binding buffer. Phosphopeptides were eluted by vortexing the resin in 70 μL of IMAC elution buffer (50 mM Tris-NH_3, pH 10.0). This alkaline pH elution was repeated two more times and each time the supernatant was collected into tubes containing 20 μL of 10% formic acid for neutralization. Columns for IMAC enrichment were prepared in gel-loading tips (see Fig. 29.2f). The capillary end of each tip was pinched to create a constriction that allows liquid to flow through but retains the IMAC resin. IMAC columns of either 10 or 30 μL of resin were set up in this manner. Since the constriction in the capillary severely restricts thc flow, positive pressure must be applied during column binding, wash and elution steps. To enable this a 15 mL syringe attached with the middle piece of a micropipette tip as an improvised adapter was inserted into the wide bore end of the gel loading tip and the syringe plunger was pushed into the barrel of the syringe. 100 μL (one of the four peptide aliquots) was applied to each IMAC column. The flowthrough was recirculated once more over the column to increase phosphopeptide recovery. The columns were then washed with IMAC binding buffer to varying extents. Phosphopeptides were eluted in 70 μL of IMAC elution buffer and eluates were collected in tubes containing 20 μL of 10% formic acid. Two elutions were performed per column and then combined. IMAC eluates were then spiked with the "light" AQUA peptides, desalted using StAGE tips (Rappsilber et al., 2003), dried and processed for LC-MS/MS using a linear ion trap-orbitrap hybrid mass spectrometer (LTQ-Orbitrap; ThermoElectron) as described previously (Ballif et al., 2008).

The data obtained from one such optimization experiment is shown in Fig. 29.2. 30 μL of IMAC resin purified more phosphopeptides than 10 μL (29.2b), although the specificity was somewhat reduced using 30 μL of resin (29.2c). In general the column method led to increased phosphopeptide enrichment (29.2b and c). Moreover, while the doubly-phosphorylated AQUA standard was enriched with about the same efficiency in all methods tested, recovery of the singly-phosphorylated AQUA standard was less efficient overall, but was higher when 30 μL of resin was used compared to 10 μL and was highest when purified using a

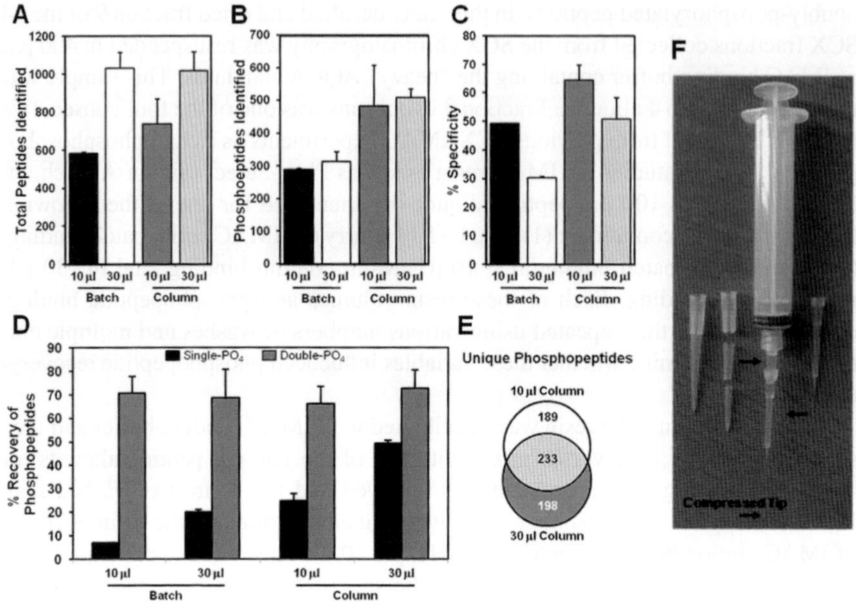

Fig. 29.2 Optimization of IMAC method for phosphopeptide enrichment. The total number of peptides (**a**), and phosphopeptides (**b**) enriched and the specificity (**c**) of enrichment using 10 or 30 μL IMAC resin by batch or column method, as indicated shown as bar graphs (*n*=3). (**d**) The recovery efficiency (percentage of total added) of singly- (Single-PO4) or doubly- (Double-PO4) phosphorylated reference standard using 10 or 30 μL IMAC resin by batch or column method, as indicated, are shown as bar graphs (*n*=3). Error bars are standard errors of the mean. (**e**) Number of unique phosphopeptides enriched by 10 μL (189) or 30 μL (198) IMAC columns alone or common to both (233) shown as a Venn diagram. (**f**) Photograph showing the construction of IMAC column purification apparatus

column binding method. Interestingly, while there was a large amount of similiarity in the unique phosphopeptides enriched by either the 10 or the 30 μL column, about 40% of total phosphopeptides enriched by the 10 μL column were unique to the 10 μL column and thus not identified from the 30 μL column. Hence a combination of the phosphopeptides enriched by both 10 and 30 μL resin amounts would allow deeper coverage. We also found that eluting the column twice, increased the total phophopeptides identified and washing the column four times after binding reduced the extent of non-phosphorylated peptides in the eluates without loss of phosphopeptides (data not shown) and thus improved the specificity.

29.3.2 Case Study III: Immunoaffinity Enrichment of Motif-Specific Phosphopeptides

29.3.2.1 Background and Relevance

Multiple avenues of research have been interfacing for many years toward the generation of tools enabling a more clear understanding of the function (particularly the identification of target substrates) of protein kinases. These include genetic ablation, mRNA knockdown, in vitro biochemical assays, ex vivo cell biological assays, and the generation of specific pharmacological inhibitors. Empirically through these lines of experimentation, as well as confirming structural studies, it was shown that the structural constraints within the catalytic cleft of a kinase led to a preferred substrate target sequence. In these studies the target sequence was shown to be a linear sequence of amino acids surrounding the phosphate acceptor residue (Songyang and Cantley, 1995; Songyang et al., 1996).

As more experimental data accumulated, bioinformatic tools arose such as Scansite (Obenauer et al., 2003), that can be used for in silico prediction of a kinase's substrates. These tools come with limitations, as noted by their authors, including the inherent challenge that the structural features of the catalytic cleft of many kinases are sufficiently similar so as to preclude unambiguous assignment of a given substrate to a given kinase. Substrates predicted using such programs are therefore most appropriately classified as potential targets of kinase families. In addition to the emergence of these bioinformatics tools critical phospho-specific antibody tools were developed. Phospho-specific antibodies began appearing in the late 1980's with generic antibodies such as the anti-phosphotyrosine monoclonal antibody, 4G10 (Druker et al., 1989). Subsequently many phospho-specific antibodies were generated, designed to recognize a known phosphorylated amino acid within the context of surrounding residues *in cis* (e.g. phospho(S/T)Q or phosphoTP). These antibodies became instrumental in aiding researchers to uncover novel kinase substrates or specific target residues within a known substrate. Initially these anibodies would be used in one of two ways: (a) immuno-purification by protein IP or (b) immuno-tracking by western blotting and subsequent ablation by site-directed mutagenesis. In the immuno-purification method, proteins would be immunoprecipitated from a cell extract using a phosphomotif-specific antibody. The cell extract would be derived from untreated cells for the control IP (for example), the experimental IP would be performed on extracts from cells exposed to a stimulus known to activate a given kinase. Proteins (identified by MS) that immunoprecipitated specifically upon stimulation were candidate substrates of the activated kinase(s) of interest. An excellent example of this was seen when the generic phospho(S/T)Q antibody was used following irradiation of cells leading to DNA damage and the activation of the DNA damage response kinases, ATM and ATR. ATM and ATR are biased to phosphorylate substrates on serine or threonine when followed by a glutamine. IPs of proteins using these antibodies following DNA damage ultimately resulted in the identification of novel ATM/ATR substrates (Cortez et al., 2004). In the second approach using immuno-tracking, a protein would be immunoprecipitated not with the phospho-specific antibody, but with an antibody directed at recognizing the protein independent of its state of modification. The immune complex would then subjected to western blotting with the phosphomotif-specific antibody. An increase in reactivity following a given stimulus is indicative that a kinase that phosphorylates that motif is being activated. Typically, the immunoprecipitated protein is then subjected to MS analysis to determine the presence of phosphorylation sites lying within such motifs. Validation of the phosphorylation event is typically followed by site-directed mutagenesis of the phosphorylated residue to one that cannot be phosphorylated and thereby the immuno-reactivity with the phosphomotif antibody is abolished. These studies are often further validated using pharmacological inhibitors or mRNA knock-down (Anjum et al., 2005; Ballif et al., 2005; Roux et al., 2004).

Although both of these methods continue to be important, an important breakthrough was made by John Rush and colleagues at Cell Signaling Technology when the IP step was attempted at the level of tryptic peptides, rather than at the protein level (Rush et al., 2005). The major advantage to IP from a pool of peptides is that once the peptides in the immune complex are subjected to MS analysis the identity of the protein as well as the specific amino acid in the peptide that is phosphorylated can be determined. Given the advances in quantitative MS this immunoaffinity peptide purification has great promise for functional phosphoproteomics. Indeed, large-scale experiments have recently been conducted using this combined methodology (Boersema et al., 2009; Luo et al., 2008; Matsuoka et al., 2007; Rikova et al., 2007). Given the example of identifying novel ATM/ATR substrates listed above, it is worth mentioning that the peptide IP approach was used with phospho(S/T)Q antibodies in combination with SILAC (Stable Isotope Labeling with Amino acids in Cell culture (Ong et al., 2002)) to identify more than 700 phospho(S/T)Q sites whose relative abundance increased more than four-fold following DNA damage by ionizing radiation (Matsuoka et al., 2007). Furthermore, this same study made use not only of generically-designed phosphomotif-specific antibodies, but also antibodies that were raised to recognize specific phospho(S/T)Q sites in specific proteins. More than 60 of these antibodies were used and they recognized a number of peptides harboring linear contextual sequences similar to the sequences to which the antibodies were raised (Matsuoka et al., 2007).

Methods for conducting phosphopeptide-IPs have been outlined extensively elsewhere (Ballif et al., 2008; Rush et al., 2005), but here we discuss these methodologies *vis-à-vis* motifs that can be less amenable to standard proteomic protocols. Specifically we address phospho-specific motifs that contain arginine or lysine amino-terminal to the site of phosphorylation. An example of such a motif is the canonical target motif for basophilic kinases such as Akt or RSK isoforms. These, as do many other, kinases prefer to phosphorylate a serine or threonine residue three amino acids downstream of an arginine, or less-preferred lysine ((R/K)XX(S/T), where X is any amino acid). These kinases have an even higher preference for the target sequence when there is an additional arginine (or lysine) residue in the -5 position ((R/K)X(R/K)XX(S/T)). It is obvious that this poses a challenge for standard proteomic protocols that rely on trypsin, given that a digest to completion with trypsin would typically remove the arginine and lysine residues amino-terminal to the phosphorylated residue, and thus obliterate the epitope recognized by the cognate phosphomotif-specific antibody.

29.3.2.2 Phosphopeptide IPs, Proteases and Phospho-specific Antibodies Recognizing Basic Residues Amino-terminal to the Phosphorylation Site

In separate experiments we digested proteins from HEK 293 cells with trypsin and chymotrypsin. Each digest was independently desalted and subjected to phosphopeptide IP with an antibody recognizing the canonical Akt/RSK target motif. The methods used to conduct these experiments were published previously (Ballif et al.,

Table 29.2 Results summary of anti-phosphomotif ((R/K)X(R/K)XX(S/T)#) peptide IPs

Enzyme	Micrograms HEK 293 protein digested	Total peptides identified	Total phospho-peptides identified	Total unique[a] phospho-peptides
Chymotrypsin	10	287	57	40
Trypsin	30	438	215	120

The indicated amounts of HEK 293 cell protein extract was digested as indicated and subjected to phosphomotif peptide IPs. Peptides were eluted from immune complexes using 0.15% trifluoroacetic acid. Desalted peptides were subjected to reversed-phase LC-MS/MS analysis in a linear ion trap-orbitrap hybrid mass spectrometer (LTQ-orbitrap, Thermo Electron) essentially as described previously (Ballif et al., 2008). Raw data were searched against the forward and reverse IPI human concatenated databases and subjected to Sequest searches with no enzyme specificity for chymotryptic digests and tryptic specificity for tryptic digests. Data were filtered to a less than 1% false discovery rate as described (Elias and Gygi, 2007). [a]Unique indicates peptides that had unique phosphorylation sites as some (primarily chymotryptic) peptides were seen in duplicate having slightly truncated or extended sequences depending on the sites of proteolysis

2008; Rush et al., 2005) with additional details provided in the legend to Table 29.2. Although these experiments were not performed head-to-head, the goal of their presentation here is to give the reader a sense of their success as well as to provide experimental examples for discussion.

In the chymotryptic digest approximately 10 mg of protein was used, whereas 30 mg of protein was used in the tryptic digestion. Approximately 30 μg of phosphomotif-specific antibody (CST #9611) was used for each IP. The results of the peptide immunoprecipitations are presented generally in Tables 29.2 and 29.3 and the complete list of unique phosphopeptides is given in Table 29.4. As shown in

Table 29.3 Categorized motif results of anti-phosphomotif peptide IPs

Peptides	RXRXXS#
99	XXRXXS#
21	nXXS#
7	RXXXXS#
5	nXXXXS#
4	nXXXS#
3	nXS#
1	RXXXS#
1	RXS#
19	Misc.

Motifs generated from the 160 identified phosphopeptides described in Table 29.2 and shown in Table 29.4. For ease of alignment, arginine (R) is representative of arginine or lysine, and phosphorylated serine (S#) is representative of phosphorylated serine, threonine or an occasional tyrosine. An amino-terminal residue is indicated by nX. Any amino acid is indicated by "X". The 19 phosphopeptides in the miscellaneous category had no discernable amino-terminal pattern although in 10 of the 19 the site of phosphorylation lay with 5 amino acids of a carboxyl-terminal arginine or lysine residue. The "Peptides" column indicates the number of peptides in each category

Table 29.4 Peptides identified in anti-phosphomotif ((R/K)X(R/K)XX(S/T)#) peptide IPs

Gene sym.	Peptide sequence–(R/K)XX(S/T)#	R	z	XCorr	dCn2	E	D	A1	P	A2	P	A3	P
AAK1	R.RILS#DVTHSAVFGVPASK.S	0	3	2.655	0.283	T		4	Y?	–	–	–	–
ACOT2	R.YRADT#LGELDLER.A	1	2	2.554	0.325	T	1	47	N	–	–	–	–
ACTBL2	M.KIKIIAPPERKYS#VW.I	9	3	3.098	0.307	C		6	N?	–	–	–	–
AKAP13	R.RAET#FGGFDSHQMNASK.G	3	3	2.688	0.295	T	4	57	N	–	–	–	–
AKAP13	V.SLPRRAET#FGGFDSHQM.N	3	2	2.167	0.251	C		79	N	–	–	–	–
AKT1S1	F.GDLPRPRLNT#SDFQ.K	1	2	1.940	0.231	C	1	19	Y	–	–	–	–
ANLN	K.RLLT#SITTK.S	2	2	2.212	0.271	T		0	N?	–	–	–	–
AP1GBP1	F.RDRS#NTLNEKPALPVIRDKYKDLTGEVEENERY.A	1	5	3.654	0.197	C		16	Y	–	–	–	–
APC	L.VRQS#TFIKEAPS@PT.L	0	2	2.096	0.213	C		8	N?	15	N	–	–
ARHGEF12	R.YRTAS@QGPQTDSVIQNSENIK.A	1	3	3.252	0.305	T		11	Y	–	–	–	–
ARHGEF18	R.RAET#FAGYDCTNSPTK.N	2	2	2.761	0.489	T		65	N	–	–	–	–
BRCA1	R.KRRPT#SGLHPEDFIK.K	4	3	2.443	0.271	T		25	Y	–	–	–	–
C10orf47	R.SRSFT#LDDESLK.Y	2	2	3.073	0.274	T		37	Y	–	–	–	–
C1orf164	R.IRATT#LLEGGFR.G	4	2	1.643	0.300	T		18	N	–	–	–	–
C22orf19	R.RRPT#LGVQLDDK.R	0	3	3.388	0.116	T	3	>99	Y	–	–	–	–
C9orf140	R.RRHT#IASGVDCGLLK.Q	0	3	2.773	0.265	T		53	N	–	–	–	–
CACNB2	R.GLS#RQET#FDSETQESR.D	10	2	2.448	0.385	T		75	N	36	Y	–	–
CACNB2	L.SRQET#FDSETQESRDSAYVEPKEDY.S	10	3	2.457	0.294	C		5	Y?	–	–	–	–
CAMK2D	M.HRQET#VDCLKKF.N	6	2	1.813	0.215	C	1	>99	Y	–	–	–	–
CCT2	F.GSRVRVDST#AKVAEIEHAEKEKMKEKVERIL.K	0	4	3.165	0.135	C		0	Y?	–	–	–	–
CDC2	F.GIPIRVYT#HEVVTLW.Y	0	2	2.348	0.262	C		12	Y	–	–	–	–
CDC2L5	R.LYSSEESRPYT@NK.V	2	2	2.228	0.268	T	1	16	Y	–	–	–	–
CEP170	F.VQSSGRIRQPS#VDLTDDDQTSSVPHSAISDIMSSDQETY.S	4	4	4.456	0.379	C	3	56	Y	–	–	–	–
CRKRS	Y.NSEESRPYT#NKVITLW.Y	0	2	2.800	0.294	C		26	Y	–	–	–	–
CTDP1	R.KRQPS@M*SETMPLYTLCK.E	1	3	2.643	0.215	T	2	11	Y	–	–	–	–
CYTSA	R.SRKGS#SGNASEVSVACLTER.I	1	3	3.802	0.414	T		0	Y?	–	–	–	–
DARS	R.VRDLT#IQKADEVVWVR.A	0	3	3.084	0.331	T	1	>99	N	–	–	–	–
DCBLD1	L.RAHT#FSAQSGY.R	1	2	2.483	0.244	C		29	N	–	–	–	–
DENND4C	H.ALERRSS#LPLDHGSPAQENPESEKS#SPAVSRSKTF.T	3	5	3.721	0.126	C		9	Y?	2	Y?	–	–
DSC1	R.LGEESIRGHT#LIKN.-	1	3	3.203	0.334	T		11	N	–	–	–	–
EHBP1L1	R.LRRPS#VNGEPGSVPPPR.A	3	3	3.373	0.281	T	2	80	N	–	–	–	–
EPB41L4B	R.RHS#T#FKASNPVIAAQLCSK.T	3	3	5.313	0.381	T	2	33	N	31	N	–	–
EPHA2	R.VSIRLPS#TSGSEGVPFR.T	1	3	2.841	0.353	T		0	Y?	–	–	–	–

Table 29.4 (continued)

Gene sym.	Peptide sequence–(R/K)XX(S/T)#	R	z	XCorr	dCn2	E	D	A1	P	A2	P	A3	P
EXOC2	R.GSS#FQSGRDDT#WR.Y	1	2	3.018	0.468	T		0	Y?	72	N	–	–
EZH2	N.NSSRPST@PTIN.V	1	2	2.351	0.266	C		0	Y?	–	–	–	–
FAM44A	K.RSLT#VSDDAESSEPER.K	1	2	3.198	0.311	T		16	Y	–	–	–	–
FRYL	R.RSNT#LDIMDGR.I	3	2	1.955	0.329	T	4	43	Y	–	–	–	–
GARNL1	I.VRQKT#VDIDDAQILPR.S	3	2	3.058	0.293	C	4	>99	Y	–	–	–	–
GPATC2	K.RRPS#SNLNNNVR.G	1	3	2.343	0.310	T		19	N	–	–	–	–
GSK3A	R.ARTSS#FAEPGGGGGGGGGPGGSASGPGGTGGGK.A	0	3	6.024	0.541	T		0	Y?	–	–	–	–
GTF2F1	R.LRLDT#GPQSLS#GKS@TPQPPSGK.T	0	3	2.867	0.291	T	2	13	Y	24	Y	13	Y
GTF3C1	L.GVVRCPRVRKNSS@TDQGSDEEGSLQKEQESAM.D	3	4	2.772	0.149	C		3	Y?	–	–	–	–
HBXAP	R.LHRIET#DEEES#CDNAHGDANQPAR.D	2	3	3.878	0.557	T	1	>99	Y	0	Y?	–	–
HSPC148	K.YRQTT#QDAPEEVR.N	0	2	3.060	0.332	T		0	Y?	–	–	–	–
HSPH1	K.VRVNT#HGIFTISTASMVEK.V	2	3	3.574	0.351	T	2	54	N	–	–	–	–
IRS4	R.RCGT#LGAQPDGEPAALAAAAAAEPPFYK.D	0	3	3.605	0.223	T	2	91	N	–	–	–	–
KIAA1522	R.ERRS#TVLGLPQHVQK.E	2	3	3.357	0.363	T		0	Y?	–	–	–	–
KIAA1804	K.KGCT#WGPNSIQMK.D	1	2	3.053	0.391	T	1	39	Y	–	–	–	–
KIF13B	R.RSAT@LSGS#ATNLASLTAALAK.A	1	3	3.683	0.209	T		0	N?	0	N?	–	–
KIF13B	K.GRWES#QQDVSQTTVSR.G	1	3	3.906	0.330	T	1	98	Y	–	–	–	–
LUZP1	Y.TQRSST@DFSELEQPRSCLF.E	2	3	3.353	0.095	C		5	N?	–	–	–	–
LYN	R.TIYVRDPT#SNK.Q	0	2	1.956	0.291	T		29	Y	–	–	–	–
MARK3	R.RTAT#YNGPPASPSLSHEATPLSQTR.S	6	4	3.024	0.143	T	1	4	Y?	–	–	–	–
MELK	R.LRLSS@FSCGQASATPFTDIK.S	0	3	3.115	0.234	T	1	6	Y?	–	–	–	–
MLLT4	L.GQMRTQS@LNPAPF.S	5	2	1.861	0.211	C		6	N?	–	–	–	–
MRE11A	R.GRADT#GLETSTR.S	0	2	1.996	0.435	T	1	77	N	–	–	–	–
NDRG1	R.SRS@HT#SEGAHLDITPNSGAAGNSAGPK.S	2	3	3.779	0.210	T	2	22	Y	15	Y	–	-
OSTM1	K.RLKS#STS#FANIQENSN.-	0	2	3.599	0.438	T		20	Y	18	Y	–	–
PDAP1	R.ARQYT#SPEEIDAQLQAEK.Q	0	2	3.334	0.404	T	1	16	N	–	–	–	–
PDCD4	R.RLRKNS@SRDSGRGDS#VSDSGSDALRSGL.T	2	4	3.012	0.217	C	–	7	Y?	11	Y	–	–
PEBP1	K.NRPTS#ISWDGLDSGK.L	0	2	3.433	0.339	T	6	24	Y	–	–	–	–
PEBP1	L.TPTQVKNRPTS#ISWDGLDSGKLY.T	0	3	3.406	0.229	C	–	12	Y	–	–	–	–
PHACTR4	K.RKDT#LAMK.L	2	2	2.101	0.200	T	–	>99	N	–	–	–	–
PKN1	F.GLCKEGMGYGDRTST#F.C	4	3	3.028	0.165	C	1	28	Y	–	–	–	–
PLK1	F.GLATKVEYDGERKKT#LCGTPNYIAPEVL.S	1	3	3.253	0.231	C	–	20	Y	–	–	–	–
PPP1CA	K.YGQFSGLNPGGRPIT#PPR.N	2	3	2.894	0.347	T	9	56	Y	–	–	–	–

Table 29.4 (continued)

Gene sym.	Peptide sequence–(R/K)XX(S/T)#	R	z	XCorr	dCn2	E	D	A1	P	A2	P	A3	P
PPP1CC	K.KKPNATRPVT#PPRGM.I	0	3	2.95	0.271	C	–	8	N?	–	–	–	–
PRKACA	K.GRTWT#LCGTPEYLAPEIILSK.G	14	2	3.614	0.440	T	3	10	Y?	–	–	–	–
PRKCD	F.GESRAST#FCGTPDYIAPEIL.Q	1	2	4.045	0.357	C	3	38	Y	–	–	–	–
RAP1GA1	K.RSFT#FGAYGGVDK.S	0	2	2.524	0.321	T	–	10	N?	–	–	–	–
RBM14	F.GNSTGGFDGQARQPT#PPFF.G	1	2	3.291	0.233	C	2	>99	Y	–	–	–	–
REPS1	R.RQSS@SYDDPWKITDEQR.Q	3	3	3.189	0.277	T	–	0	Y?	–	–	–	–
RICTOR	R.IRTLT#EPSVDFNHSDDFTPISTVQK.T	1	3	3.129	0.345	T	–	5	Y?	–	–	–	–
RICTOR	L.SSESKTSNRRIRTLT@EPSVDFNHSDDFTPISTVQKTL.Q	1	5	2.979	0.121	C	2	6	N?	–	–	–	–
RPL27	K.RNQS#FCPTVNLDK.L	2	2	2.123	0.224	T	1	26	N	–	–	–	–
SHKBP1	R.RSNT#M*PPNLGNAGLLGR.M	1	3	3.049	0.331	T	–	8	N?	–	–	–	–
SLC19A1	R.GRCET#SASELER.M	3	2	1.861	0.332	T	–	19	N	–	–	–	–
SNX1	K.RRFS#DFLGLYEK.L	6	3	2.565	0.282	T	3	>99	Y	–	–	–	–
SNX2	K.RRFS#DFLGLHSK.L	0	3	2.801	0.223	T	–	90	Y	–	–	–	–
SPAG9	K.RSST#LSQLPGDK.S	5	2	2.653	0.302	T	–	16	Y	–	–	–	–
SR-A1	Y.RQRS#PS#PAPAPAPAAAAGPPTRKKS.R	0	4	3.666	0.245	C	–	>99	Y	>99	Y	–	–
STX6	Y.GRLDRELQRANS#HFIEEQQAQQQL.I	0	4	3.170	0.173	C	–	>99	Y	–	–	–	–
SVIL	M.NARYQT#QPVTLGEVEQVQ.S	3	2	3.639	0.428	C	–	37	Y	–	–	–	–
TBC1D1	R.RRANT#LSHFPIECQEPPQPAR.G	0	4	2.696	0.058	T	2	9	Y?	–	–	–	–
TBC1D4	R.RAHT#FSHPPSSTK.R	1	2	2.162	0.405	T	–	37	Y	–	–	–	–
TENC1	Y.NTTEPAVRWDS#YENF.N	3	2	3.155	0.441	C	–	16	Y?	–	–	–	–
TIAM1	R.RAKT#TQDVNAGEGSEFADSGIEGATTDTDLLSR.R	1	3	4.905	0.562	T	–	10	N?	–	–	–	–
TMPO	K.GPPDFSSDEEREPT#PVLGSGAAAAGR.S	2	3	3.658	0.383	T	–	39	Y	–	–	–	–
TNS3	L.GGRLRKLS#LGQYDNDAGGQLPF.S	1	3	3.315	0.278	C	–	42	Y	–	–	–	–
TPD52L2	K.SFEDRVGT#IK.S	6	2	2.332	0.375	T	1	>99	Y	–	–	–	–
TPX2	R.IRMPT#KEDEEEDEPVVIK.A	1	3	3.155	0.378	T	–	>99	N	–	–	–	–
TSC2	L.VHPPSHSKAPAQTPAEPTPGYEVGQRKRLIS@SVEDFTEFV.-	3	5	3.354	0.230	C	–	0	Y?	–	–	–	–
UBAP2L	R.RYPS#SISSS#PQK.D	2	2	2.129	0.340	T	–	13	Y	13	Y	–	–
USP27X	R.RRIT#SSFTIGLR.G	0	3	2.192	0.305	T	2	37	N	–	–	–	–
VAPA	R.KVAHSDKPGS#TSTASFR.D	2	3	2.936	0.246	T	3	0	Y?	–	–	–	–
WASF1	K.RHPS#TLPVISDAR.S	0	2	2.799	0.316	T	3	26	N	–	–	–	–
WDR59	L.VVSHSRYPS#FTSSGSCSSMSDPGLNTGGW.N	2	3	3.701	0.444	C	–	11	N?	–	–	–	–
ZNF511	R.HRIPS#TICFGQGAAR.G	0	2	2.244	0.378	T	1	16	N	–	–	–	–

Table 29.4 (continued)

Gene sym.	Peptide sequence–(R/K)XX(S/T)#	R	z	XCorr	dCn2	E	D	A1	P	A2	P	A3	P
ZNF609	R.VRTNS#MGSATGPLPGTK.V	0	2	1.964	0.354	T	–	21	Y	–	–	–	–
Gene Sym.	*Peptide Sequence–RXS#*	*R*	*z*	*XCorr*	*dCn2*	*E*	*D*	*A1*	*P*	*A2*	*P*	*A3*	*P*
PKN3	F.GLCKEGIGFGDRTS#TF.C	0	2	2.350	0.215	C	–	0	Y?	–	–	–	–
Gene Sym.	*Peptide Sequence–nXX(S/T)#*	*R*	*z*	*XCorr*	*dCn2*	*E*	*D*	*A1*	*P*	*A2*	*P*	*A3*	*P*
ARCN1	R.RNT#LEWCLPVIDAK.N	1	2	3.685	0.474	T	1	>99	N	–	–	–	–
BAHD1	R.RRT#NGWVPVGAACEK.A	0	3	2.992	0.253	T	1	>99	N	–	–	–	–
MARK1	R.RNT#YVCER.T	2	2	2.712	0.275	T	–	32	N	–	–	–	–
MARK3	R.RNT#YVCSER.T	6	2	2.772	0.328	T	–	38	Y	–	–	–	–
NUMA1	R.RRT#TQIINITMTK.K	2	3	3.443	0.112	T	2	13	Y	–	–	–	–
RHBDF1	R.RGT#ADWFGVSK.D	1	2	2.438	0.322	T	–	58	N	–	–	–	–
SRRM1	R.RQS#PSPST#RPIR.R	1	2	2.808	0.344	T	–	31	Y	9	Y?	–	–
AKT1S1	R.LNT#SDFQK.L	1	2	2.661	0.328	T	–	26	Y	–	–	–	–
APBA2BP	R.SST#WSPGSSDTGR.S	2	2	2.206	0.314	T	–	0	N?	–	–	–	–
C20orf74	R.SFS#LSWR.S	0	2	1.924	0.282	T	–	31	Y	–	–	–	–
CDK3	R.TYT#HEVVTLWYR.A	1	2	3.696	0.467	T	2	13	Y	–	–	–	–
EIF4G1	R.SFS#KEVEER.S	5	2	2.197	0.282	T	–	12	Y	–	–	–	–
FAM62A	R.LGT#QTFCSR.V	1	2	2.070	0.286	T	–	52	N	–	–	–	–
HSP90AB1	K.SLT#NDWEDHLAVK.H	4	2	2.323	0.253	T	–	0	N?	–	–	–	–
KIAA1804	K.GCT#WGPNSIQM*K.D	1	2	2.435	0.269	T	1	37	Y	–	–	–	–
PRKACA	K.GRT#WTLCGTPEYLAPEIILSK.G	14	3	3.315	0.228	T	1	9	Y?	–	–	–	–
PRKACA	R.TWT@LCGTPEYLAPEIILSK.G	14	2	3.483	0.406	T	3	9	Y?	–	–	–	–
PRKG1	K.TWT#FCGTPEYVAPEIILNK.G	2	2	3.443	0.466	T	1	0	Y?	–	–	–	–
UBAP2L	R.TAT#EEWGTEDWNEDLSETK.I	5	2	4.888	0.565	T	–	0	N?	–	–	–	–
USP25	R.TPT#EVWR.D	2	2	2.508	0.307	T	1	52	Y	–	–	–	–
USP7	R.SRY@T#YLEK.A	2	2	2.020	0.309	T	–	9	N?	–	–	–	–
Gene Sym.	*Peptide Sequence–RXXXT#*	*R*	*z*	*XCorr*	*dCn2*	*E*	*D*	*A1*	*P*	*A2*	*P*	*A3*	*P*
AP1M1	K.LETGAPRPPAT#VT#NAVSWR.S	0	3	1.959	0.196	T	2	23	Y	8	Y?	–	–
Gene Sym.	*Peptide Sequence nXXXT#*	*R*	*z*	*XCorr*	*dCn2*	*E*	*D*	*A1*	*P*	*A2*	*P*	*A3*	*P*
ATP6V0A1	K.HLGT#LNFGGIR.V	1	2	2.102	0.305	T	–	>99	Y	–	–	–	–
PAK4	K.LAAGRPFNT@YPR.A	0	3	2.324	0.269	T	–	10	Y?	–	–	–	–
RAB13	K.TITT#AYYR.G	9	2	2.005	0.281	T	–	18	N	–	–	–	–
SLC12A7	K.VQMT#WTR.E	1	2	2.304	0.244	T	–	14	Y	–	–	–	–

Table 29.4 (continued)

Gene sym.	Peptide sequence–(R/K)XX(S/T)#	R	z	XCorr	dCn2	E	D	A1	P	A2	P	A3	P
Gene Sym.	*Peptide Sequence–nX(T/S)#*	*R*	*z*	*XCorr*	*dCn2*	*E*	*D*	*A1*	*P*	*A2*	*P*	*A3*	*P*
CACNA2D4	K.TS#ALLWLLLLGTSLSPAWGQAK.I	2	3	2.288	0.202	T	–	12	N	–	–	–	–
DDX19B	R.KT#ASWLAAELSK.E	7	2	2.194	0.223	T	–	19	N	–	–	–	–
STEAP3	R.ES#TIKFTLPTDHALAEK.T	1	3	3.030	0.278	T	–	0	Y?	–	–	–	–
Gene Sym.	*Peptide Sequence nXXXX(S/T)#*	*R*	*z*	*XCorr*	*dCn2*	*E*	*D*	*A1*	*P*	*A2*	*P*	*A3*	*P*
CDK3	K.IGEGT#YGVVYK.A	4	2	2.474	0.220	T	1	16	Y	–	–	–	–
UBTF	R.M*VLCS#QQWK.L	2	2	2.342	0.262	T	–	>99	N	–	–	–	–
Unknown	R.MLLVS#AGCSS#SSPNTTPPLR.Q	0	4	2.371	0.094	T	–	13	N	0	N?	–	–
KPNA2	R.NNQGT#VNWSVDDIVK.G	1	2	2.796	0.208	T	–	55	N	–	–	–	–
PLRG1	K.HSVHT#LVFR.S	1	2	2.486	0.323	T	–	46	N	–	–	–	–
Gene Sym.	*Peptide Sequence–(E/R/K)XXXX(S/T)#*	*R*	*z*	*XCorr*	*dCn2*	*E*	*D*	*A1*	*P*	*A2*	*P*	*A3*	*P*
TLK2	R.HGASFT#EQWTDGYAFQNLIK.Q	2	3	3.064	0.293	T	–	8	N?	–	–	–	–
MAPK3	R.IADPEHDHTGFLT#EY#VATR.W	2	3	3.177	0.313	T	1	37	Y	16	Y	–	–
MGC2803	R.KGGPGS@TLSFVGK.R	0	2	2.649	0.362	T	1	0	N?	–	–	–	–
SETD1A	R.RKTVSFS@AIEVVPAPEPPPATPPQAK.F	1	3	3.554	0.187	T	–	3	N?	–	–	–	–
RFAM101B	L.RRFLSS#VELEAAELPGSDDLSDEC.-	0	3	2.811	0.218	C	–	8	N?	–	–	–	–
SMN2	F.RRGTGQS#DDS#DIWDDTALIKAY.D	3	3	3.031	0.220	C	–	8	Y?	24	Y	–	–
C14orf173	L.ERRSSWY#VDASDVLTTEDPQCPQPL.E	1	3	4.323	0.371	C	–	5	Y?	–	–	–	
Gene Sym.	*Peptide Sequence–Miscellaneous*	*R*	*z*	*XCorr*	*dCn2*	*E*	*D*	*A1*	*P*	*A2*	*P*	*A3*	*P*
ATE1	K.KEEPQELLQS#CDFVGEK.L	2	3	2.981	0.243	T	–	>99	Y	–	–	–	–
C10orf81	K.SPPSQLFS#SVTSWKKR.F	1	3	2.934	0.058	T	–	8	N?	–	–	–	–
DOCK7	R.LPNTYPNSSSPGPGGLGGSVHY#ATM*AR.S	2	3	2.798	0.176	T	–	31	N	–	–	–	–
UBN2	K.YGGFYINTGT#LQFR.Q	0	2	3.547	0.426	T	2	25	N	–	–	–	–
TMEM16F	R.SSAFGT#LNWFTK.V	1	2	3.256	0.436	T	2	64	N	–	–	–	–
UBN1	K.YGGFYINSGT#LQFR.Q	0	2	3.700	0.406	T	–	17	N	–	–	–	–
SCAMP4	K.AQTEWNTGT#WR.N	1	2	2.800	0.360	T	–	13	N	–	–	–	–
SLC7A3	K.TVDLDPGT#LYVHSV.-	0	2	4.077	0.512	T	–	31	N	–	–	–	–
FAM60A	K.AAAEKPEEQGPEPLPIS#TQEW.-	1	2	3.109	0.224	T	1	13	Y	–	–	–	–

Table 29.4 (continued)

Gene sym.	Peptide sequence–(R/K)XX(S/T)#	R	z	XCorr	dCn2	E	D	A1	P	A2	P	A3	P
FN1	R.DLEVVAATPT@SLLISWDAPAVTVR.Y	12	4	2.440	0.155	T	–	13	N	–	–	–	–
KIAA0776	R.SVFMSS@TTSASGTGRKR.T	1	3	1.935	0.257	T	–	0	N?	–	–	–	–
MAPK14	R.HTDDEMT#GYVATR.W	3	2	3.550	0.538	T	1	31	Y	–	–	–	–
OSBPL11	K.GSLPSGTT#IEWLEPK.I	0	2	2.599	0.257	T	–	17	N	–	–	–	–
PCYT2	R.QIDSGSNLTT@DLIVQR.I	0	2	2.922	0.274	T	–	0	N?	–	–	–	–
PRKG1	K.TWTFCGT@PEYVAPEIILNK.G	2	3	3.989	0.338	T	–	2	N?	–	–	–	–
PUS7	K.NQTQLNTT#WLR.-	0	2	2.760	0.322	T	–	6	?	–	–	–	-
UBXD2	Y.RLRTQDDGEDENNT#WNGNSTQQM.-	0	3	3.476	0.319	C	–	35	N	–	–	–	–
PRKCZ	G.PGDTTSTFCGT@PNYIAPEIL.R	2	2	3.395	0.251	C	–	7	N?	–	–	–	–
PRKCA	H.MMDGVTT#RTF.C	0	2	2.470	0.303	C	–	12	Y	–	–	–	–

Phosphopeptides are sorted by category of motif (see Table 29.3) and then by Gene Symbol. "#" follows phosphorylated residues. "@" follows phosphorylated residues when adjusted by Ascore (Beausoleil et al., 2006) analysis. "R" is the redundancy in the IPI database. "z" is the charge of the peptide ion. XCorr and dCn2 values (dCn2 compares top hit with the next hit that is a different amino acid sequence) are indicated. "E" is the enzyme used for digestion ("T" or "C" for trypsin and chymotrypsin respectively). ' D" is the number of duplicate peptides identified. A1-A3 are the Ascore values for each phosphorylation site proceeding amino- to carboxyl-terminal. "P" indicates PhosphoSitePlus and "Y" and "N" denote presence or absence in the PhosphoSitePlus database respectively. "?" denotes that the Ascore was less than 11 and therefore the identification is ambiguous in Table 29.2, 272 total phosphopeptides were identified between the two IPs, representing 160 unique phosphopeptides and 175 unique phosphorylation sites. Only four phosphorylation sites were common between the two IPs. The difference in total numbers of phosphopeptides identified between the two preparations (see Table 29.2) could be due simply to the difference in the total amount of starting material or perhaps a technical detail in washing that led to a higher percentage of phosphopeptides identified in the tryptic digest relative to the chymotrypic digest (49 and 20% respectively). It is also very likely that the chymotryptic digest created greater sample complexity than the tryptic digest owing to less specificity found in chymotrypsin activity. It is well-known that trypsin has much higher specificity cutting carboxyl-terminal to lysine and arginine (Olsen et al., 2004) than chymotrypsin does for its preferred cutting carboxyl-terminal to the bulky aromatic residues tyrosine, phenylalanine and tryptophan (Berezin et al., 1970). Indeed it is common to see chymotryptic activity carboxyl-terminal to methionines, leucines, asparagines and glutamines or other amino acids. This effectively lowers the molar representation of any one peptide by distributing stretches of the protein into multiple peptide species. This may result in some peptides having less chance of meeting minimum ion count requirements to be chosen for MS/MS analysis.

Given the fidelity of trypsin, it could be asked why we detected any peptides at all in the phosphopeptide IP using tryptic peptides. For the vast majority of phosphopeptides identified from the tryptic digest, this can be explained by ragged-ends, where two or more arginine or lysine residues are clustered near the amino-terminus and thus an arginine or lysine was left after digestion. Additionally, trypsin is inhibited by arginine and lysines followed by proline residues and these were also observed in our dataset. As trypsin can also be inhibited by post-translational modifications proximal to arginine or lysine, this could be playing a role toward preserving at least one amino-terminal arginine. One perhaps surprising possibility enabling the identification of phosphopeptides by this motif-specific antibody, however, is that the positively charged amino-terminus of the peptide may substitute to a degree for a basic residue recognized by the phosphomotif-specific antibody. In our dataset, 99 of the 160 phosphopeptides had (R/K)XX(S/T)# motifs (where # denotes phosphorylation) due to ragged ends, but there were 21 peptides that had nXX(S/T)# motifs (where n denotes the amino terminus of the first X). Although 8 of the 21 have an arginine as one of the X residues, most do not and suggest that indeed the basic amino terminus is facilitating (or at least not hindering) antibody recognition. Table 29.3 categorizes the types of motifs that could be identified within the dataset.

29.3.2.3 Phosphorylation Site Assignments and Novelty of Findings

It is well appreciated by the authors that phosphorylation site assignment following MS/MS analysis of a peptide with more than one amino acid capable of being phosphorylated isn't perfect using standard MS/MS search algorithms such as Sequest. To build a confidence score around a Sequest assignment, without requiring manual validation, we used the Ascore phosphorylation analysis program developed by Beausoleil et al. (2006). All data reported herein used the phosphorylation sites and scores as reported by the Ascore program, which made relatively few adjustments to assignments made by Sequest. However, although the Ascore does make an assignment, the assignment that is made comes with a confidence score to assist in determining if the site is correctly localized. In the data presented here for motif analysis we used the Ascore assignment even when the score wasn't sufficiently high to localize the site without ambiguity. The rationale is that any ambiguity is most likely located between neighboring residues and thus the assigned residue is likely to be correct or "proximally correct". Nevertheless, all Ascores are reported in Table 29.4 and scores of 11 and above were conservatively taken as "confidently localized". In total, 117 of the 175 identified phosphorylation sites had Ascores of 11 or above. Of these sites, 48 were novel, or at least had not yet been curated and added to the PhosphoSitePlus database.

29.3.2.4 Future Directions for Phosphomotif-Specific IPs

Given the great power of phosphomotif-specific antibodies as a way to selectively enrich for a subset of phosphosubstrates from very complicated mixtures, their use has great promise. Negotiating the methodology with motifs containing arginine or lysine residues amino-terminal to the site of phosphorylation can be managed, to a degree, as shown herein. However, fully tryptic and chymotryptic digests will only facilitate the recovery of a relatively small fraction of the desired phosphopeptides. The need for a high-specificity, low-cost, complementary enzyme to trypsin in the field of proteomics looms large. It is the hope that protein engineering, protease methodology or protease purification will ultimately lead to a wide-array of such enzymes. In the meantime, we discuss the following suggestions in addition to chymotryptic and tryptic digests as shown above: (a) Limited tryptic digestion ($\sim$1.5 h) can generate peptides that may still have the arginines and lysines needed for ideal antibody recognition. The drawback will be an increase in total sample requirement, and perhaps the loss of some peptides as they become too large; (b) Use of Lys-C, Asp-N or Glu-C. Lys-C is an enzyme highly specific to cleaving the peptide bond carboxyl-terminal to lysine. Employing Lys-C would retain arginines and increase the coverage of desired phosphopeptides. The drawbacks, however, are again the average size of peptides increases making some less amenable to LC-MS/MS analysis. Furthermore, Lys-C is roughly forty times the price of MS-grade trypsin. Asp-N cuts amino-terminal to aspartic acid residues and in highest purity can be rather specific but roughly costs 100 times that of MS-grade trypsin. Glu-C cuts carboxyl-terminal to glutamic acid and can also be rather specific in its highest purity. MS-grade Glu-C is roughly four times the price of MS-grade trypsin. Exploration of these alternative enzymes will be an important avenue for large-scale phoshopoproteomic analyses as even the use of trypsin for non-phosphoproteomic analyses typically only provides roughly 50% coverage of the proteome.

29.4 Conclusion

In this chapter we have discussed the use of a classical phosphoproteomic approach, guided by genetics, to tease out functionally-important phospho-dependent Protein:Protein interactions. We have also discussed a simple method to nearly double phosphopeptide identifications in large-scale SCX-IMAC studies. Finally, we have discussed immunoaffinity phosphoproteomics toward identifying phosphopeptides that fall within target motifs of kinase classes or families. Each of these strategies, when carefully designed, can lead toward greater functional understanding particularly as they integrate biological knowledge with quantitative mass spectrometry, studies of dynamic systems, and insightful bioinformatics.

Acknowledgements We thank Lionel Arnaud who was a critical collaborator in the Cooper lab during the development and execution of the methodology to identify pY-Dab1-interacting proteins. We thank Jon Cooper (Fred Hutchinson Cancer Research Center) for support and discussions as well as additional members from his laboratory, particularly Abir Mukherjee, Tara Herrick and Priscilla Kronstad-O'Brien. We thank Steve Gygi (Harvard Medical School) for support and discussions, and members of his laboratory for the same, in particular, Scott Gerber, Wilhelm Haas and Sean Beausoleil. Judit Villén and Xue Li from the Gygi lab played an important role in the initial characterization and development of the SCX-IMAC methods presented here. We thank Steve Elledge (Harvard Medical School) and two members from his laboratory, Shuhei Matsuoka and Chunshui Zhou. Shuhei performed the anti-Akt phospho-substrate peptide IP from tryptic peptides and Chunshui gave helpful insights during the development of the SCX-IMAC methods and provided the tryptic yeast peptides. We thank John Blenis (Harvard Medical School) and two members of his lab, Philippe Roux and Jessie Hanrahan who assisted in the preparation of the cells for the Anti-Akt phoshpho-substrate peptide IP from chymotryptic peptides. We thank John Rush (Cell Signaling Technology) for supplying the AQUA reference peptides. The authors were supported primarily by the following grants during the work presented here and during the writing of this manuscript: NIH 5T32CA09657; NIH HG00041; NSF IOS 1021795; Vermont Genetics Network and NIH/NCRR P20RR16462.

References

Anjum, R., Roux, P.P., Ballif, B.A., Gygi, S.P., and Blenis, J. (2005). The tumor suppressor DAP kinase is a target of RSK-mediated survival signaling. Curr Biol *15*, 1762–1767.

Arnaud, L., Ballif, B.A., Forster, E., and Cooper, J.A. (2003). Fyn tyrosine kinase is a critical regulator of disabled-1 during brain development. Curr Biol *13*, 9–17.

Ballif, B.A., Arnaud, L., Arthur, W.T., Guris, D., Imamoto, A., and Cooper, J.A. (2004). Activation of a Dab1/CrkL/C3G/Rap1 pathway in reelin-stimulated neurons. Curr Biol *14*, 606–610.

Ballif, B.A., Arnaud, L., and Cooper, J.A. (2003). Tyrosine phosphorylation of disabled-1 is essential for reelin-stimulated activation of akt and Src family kinases. Brain Res Mol Brain Res *117*, 152–159.

Ballif, B.A., Cao, Z., Schwartz, D., Carraway, K.L., 3rd., and Gygi, S.P. (2006). Identification of 14-3-3epsilon substrates from embryonic murine brain. J Proteome Res *5*, 2372–2379.

Ballif, B.A., Carey, G.R., Sunyaev, S.R., and Gygi, S.P. (2008). Large-scale identification and evolution indexing of tyrosine phosphorylation sites from murine brain. J Proteome Res *7*, 311–318.

Ballif, B.A., Roux, P.P., Gerber, S.A., MacKeigan, J.P., Blenis, J., and Gygi, S.P. (2005). Quantitative phosphorylation profiling of the ERK/p90 ribosomal S6 kinase-signaling cassette and its targets, the tuberous sclerosis tumor suppressors. Proc Natl Acad Sci USA *102*, 667–672.

Beausoleil, S.A., Villen, J., Gerber, S.A., Rush, J., and Gygi, S.P. (2006). A probability-based approach for high-throughput protein phosphorylation analysis and site localization. Nat Biotechnol *24*, 1285–1292.

Benzinger, A., Muster, N., Koch, H.B., Yates, J.R., 3rd., and Hermeking, H. (2005). Targeted proteomic analysis of 14-3-3 sigma, a p53 effector commonly silenced in cancer. Mol Cell Proteomics *4*, 785–795.

Berezin, I.V., Levashov, A.V., and Martinek, K. (1970). On the modes of interaction between competitive inhibitors and the alpha-chymotrypsin active centre. FEBS Lett *7*, 20–22.

Boersema, P.J., Foong, L.Y., Ding, V.M., Lemeer, S., van Breukelen, B., Philp, R., Boekhorst, J., Snel, B., den Hertog, J., Choo, A.B., *et al.* (2009). In depth qualitative and quantitative profiling of tyrosine phosphorylation using a combination of phosphopeptide immuno-affinity purification and stable isotope dimethyl labeling. Mol Cell Proteomics *9*, 84–99.

Cortez, D., Glick, G., and Elledge, S.J. (2004). Minichromosome maintenance proteins are direct targets of the ATM and ATR checkpoint kinases. Proc Natl Acad Sci USA *101*, 10078–10083.

de la Fuente van Bentem, S., Mentzen, W.I., de la Fuente, A., and Hirt, H. (2008). Towards functional phosphoproteomics by mapping differential phosphorylation events in signaling networks. Proteomics *8*, 4453–4465.

Dengjel, J., Kratchmarova, I., and Blagoev, B. (2009). Receptor tyrosine kinase signaling: A view from quantitative proteomics. Mol Biosyst *5*, 1112–1121.

Druker, B.J., Mamon, H.J., and Roberts, T.M. (1989). Oncogenes, growth factors, and signal transduction. N Engl J Med *321*, 1383–1391.

Elia, A.E., Rellos, P., Haire, L.F., Chao, J.W., Ivins, F.J., Hoepker, K., Mohammad, D., Cantley, L.C., Smerdon, S.J., and Yaffe, M.B. (2003). The molecular basis for phosphodependent substrate targeting and regulation of Plks by the Polo-box domain. Cell *115*, 83–95.

Elias, J.E., and Gygi, S.P. (2007). Target-decoy search strategy for increased confidence in large-scale protein identifications by mass spectrometry. Nat Methods *4*, 207–214.

Feng, L., and Cooper, J.A. (2009). Dual functions of Dab1 during brain development. Mol Cell Biol *29*, 324–332.

Forrest, A.R., Taylor, D.F., Fink, J.L., Gongora, M.M., Flegg, C., Teasdale, R.D., Suzuki, H., Kanamori, M., Kai, C., Hayashizaki, Y., *et al.* (2006). PhosphoregDB: The tissue and sub-cellular distribution of mammalian protein kinases and phosphatases. BMC Bioinformatics *7*, 82.

Gerber, S.A., Kettenbach, A.N., Rush, J., and Gygi, S.P. (2007). The absolute quantification strategy: Application to phosphorylation profiling of human separase serine 1126. Methods Mol Biol *359*, 71–86.

Gerber, S.A., Rush, J., Stemman, O., Kirschner, M.W., and Gygi, S.P. (2003). Absolute quantification of proteins and phosphoproteins from cell lysates by tandem MS. Proc Natl Acad Sci USA *100*, 6940–6945.

Herrick, T.M., and Cooper, J.A. (2002). A hypomorphic allele of dab1 reveals regional differences in reelin-Dab1 signaling during brain development. Development *129*, 787–796.

Hiesberger, T., Trommsdorff, M., Howell, B.W., Goffinet, A., Mumby, M.C., Cooper, J.A., and Herz, J. (1999). Direct binding of reelin to VLDL receptor and ApoE receptor 2 induces tyrosine phosphorylation of disabled-1 and modulates tau phosphorylation. Neuron *24*, 481–489.

Howell, B.W., Gertler, F.B., and Cooper, J.A. (1997a). Mouse disabled (mDab1): A src binding protein implicated in neuronal development. EMBO J *16*, 121–132.

Howell, B.W., Hawkes, R., Soriano, P., and Cooper, J.A. (1997b). Neuronal position in the developing brain is regulated by mouse disabled-1. Nature *389*, 733–737.

Howell, B.W., Herrick, T.M., Hildebrand, J.D., Zhang, Y., and Cooper, J.A. (2000). Dab1 tyrosine phosphorylation sites relay positional signals during mouse brain development. Curr Biol *10*, 877–885.

Hunter, T., and Plowman, G.D. (1997). The protein kinases of budding yeast: Six score and more. Trends Biochem Sci *22*, 18–22.

Jin, J., Smith, F.D., Stark, C., Wells, C.D., Fawcett, J.P., Kulkarni, S., Metalnikov, P., O'Donnell, P., Taylor, P., Taylor, L., *et al.* (2004). Proteomic, functional, and domain-based analysis of in vivo 14-3-3 binding proteins involved in cytoskeletal regulation and cellular organization. Curr Biol *14*, 1436–1450.

Johnson, S.A., and Hunter, T. (2005). Kinomics: Methods for deciphering the kinome. Nat Methods *2*, 17–25.

Krebs, E.G., and Fischer, E.H. (1956). The phosphorylase b to a converting enzyme of rabbit skeletal muscle. Biochim Biophys Acta *20*, 150–157.

Kuo, G., Arnaud, L., Kronstad-O'Brien, P., and Cooper, J.A. (2005). Absence of fyn and src causes a reeler-like phenotype. J Neurosci *25*, 8578–8586.

Lemeer, S., and Heck, A.J. (2009). The phosphoproteomics data explosion. Curr Opin Chem Biol *13*, 414–420.

Luo, W., Slebos, R.J., Hill, S., Li, M., Brabek, J., Amanchy, R., Chaerkady, R., Pandey, A., Ham, A.J., and Hanks, S.K. (2008). Global impact of oncogenic src on a phosphotyrosine proteome. J Proteome Res *7*, 3447–3460.

Macek, B., Mann, M., and Olsen, J.V. (2009). Global and site-specific quantitative phosphoproteomics: Principles and applications. Annu Rev Pharmacol Toxicol *49*, 199–221.

Manning, G., Plowman, G.D., Hunter, T., and Sudarsanam, S. (2002a). Evolution of protein kinase signaling from yeast to man. Trends Biochem Sci *27*, 514–520.

Manning, G., Whyte, D.B., Martinez, R., Hunter, T., and Sudarsanam, S. (2002b). The protein kinase complement of the human genome. Science *298*, 1912–1934.

Matsuoka, S., Ballif, B.A., Smogorzewska, A., McDonald, E.R., 3rd., Hurov, K.E., Luo, J., Bakalarski, C.E., Zhao, Z., Solimini, N., Lerenthal, Y., *et al.* (2007). ATM and ATR substrate analysis reveals extensive protein networks responsive to DNA damage. Science *316*, 1160–1166.

Meek, S.E., Lane, W.S., and Piwnica-Worms, H. (2004). Comprehensive proteomic analysis of interphase and mitotic 14-3-3-binding proteins. J Biol Chem *279*, 32046–32054.

Mohammad, D.H., and Yaffe, M.B. (2009). 14-3-3 proteins, FHA domains and BRCT domains in the DNA damage response. DNA Repair (Amst) *8*, 1009–1017.

Nita-Lazar, A., Saito-Benz, H., and White, F.M. (2008). Quantitative phosphoproteomics by mass spectrometry: Past, present, and future. Proteomics *8*, 4433–4443.

Obenauer, J.C., Cantley, L.C., and Yaffe, M.B. (2003). Scansite 2.0: Proteome-wide prediction of cell signaling interactions using short sequence motifs. Nucleic Acids Res *31*, 3635–3641.

Olsen, J.V., Ong, S.E., and Mann, M. (2004). Trypsin cleaves exclusively C-terminal to arginine and lysine residues. Mol Cell Proteomics *3*, 608–614.

Ong, S.E., Blagoev, B., Kratchmarova, I., Kristensen, D.B., Steen, H., Pandey, A., and Mann, M. (2002). Stable isotope labeling by amino acids in cell culture, SILAC, as a simple and accurate approach to expression proteomics. Mol Cell Proteomics *1*, 376–386.

Park, T.J., and Curran, T. (2008). Crk and crk-like play essential overlapping roles downstream of disabled-1 in the reelin pathway. J Neurosci *28*, 13551–13562.

Pawson, T., and Kofler, M. (2009). Kinome signaling through regulated protein-protein interactions in normal and cancer cells. Curr Opin Cell Biol *21*, 147–153.

Pawson, T., and Scott, J.D. (2005). Protein phosphorylation in signaling–50 years and counting. Trends Biochem Sci *30*, 286–290.

Phizicky, E.M., and Fields, S. (1995). Protein-protein interactions: Methods for detection and analysis. Microbiol Rev *59*, 94–123.

Rappsilber, J., Ishihama, Y., and Mann, M. (2003). Stop and go extraction tips for matrix-assisted laser desorption/ionization, nanoelectrospray, and LC/MS sample pretreatment in proteomics. Anal Chem *75*, 663–670.

Rice, D.S., Sheldon, M., D'Arcangelo, G., Nakajima, K., Goldowitz, D., and Curran, T. (1998). Disabled-1 acts downstream of reelin in a signaling pathway that controls laminar organization in the mammalian brain. Development *125*, 3719–3729.

Rikova, K., Guo, A., Zeng, Q., Possemato, A., Yu, J., Haack, H., Nardone, J., Lee, K., Reeves, C., Li, Y., *et al.* (2007). Global survey of phosphotyrosine signaling identifies oncogenic kinases in lung cancer. Cell *131*, 1190–1203.

Rogers, L.D., and Foster, L.J. (2009). Phosphoproteomics–finally fulfilling the promise? Mol Biosyst *5*, 1122–1129.

Roux, P.P., Ballif, B.A., Anjum, R., Gygi, S.P., and Blenis, J. (2004). Tumor-promoting phorbol esters and activated ras inactivate the tuberous sclerosis tumor suppressor complex via p90 ribosomal S6 kinase. Proc Natl Acad Sci USA *101*, 13489–13494.

Rush, J., Moritz, A., Lee, K.A., Guo, A., Goss, V.L., Spek, E.J., Zhang, H., Zha, X.M., Polakiewicz, R.D., and Comb, M.J. (2005). Immunoaffinity profiling of tyrosine phosphorylation in cancer cells. Nat Biotechnol *23*, 94–101.

Sheldon, M., Rice, D.S., D'Arcangelo, G., Yoneshima, H., Nakajima, K., Mikoshiba, K., Howell, B.W., Cooper, J.A., Goldowitz, D., and Curran, T. (1997). Scrambler and yotari disrupt the disabled gene and produce a reeler-like phenotype in mice. Nature *389*, 730–733.

Songyang, Z., and Cantley, L.C. (1995). SH2 domain specificity determination using oriented phosphopeptide library. Methods Enzymol *254*, 523–535.

Songyang, Z., Lu, K.P., Kwon, Y.T., Tsai, L.H., Filhol, O., Cochet, C., Brickey, D.A., Soderling, T.R., Bartleson, C., Graves, D.J., *et al.* (1996). A structural basis for substrate specificities of protein ser/thr kinases: Primary sequence preference of casein kinases I and II, NIMA, phosphorylase kinase, calmodulin-dependent kinase II, CDK5, and erk1. Mol Cell Biol *16*, 6486–6493.

Swaney, D.L., Wenger, C.D., Thomson, J.A., and Coon, J.J. (2009). Human embryonic stem cell phosphoproteome revealed by electron transfer dissociation tandem mass spectrometry. Proc Natl Acad Sci USA *106*, 995–1000.

Thingholm, T.E., Jensen, O.N., and Larsen, M.R. (2009). Analytical strategies for phosphoproteomics. Proteomics *9*, 1451–1468.

van der Geer, P., Wiley, S., Gish, G.D., and Pawson, T. (1996). The shc adaptor protein is highly phosphorylated at conserved, twin tyrosine residues (Y239/240) that mediate protein-protein interactions. Curr Biol *6*, 1435–1444.

Villen, J., and Gygi, S.P. (2008). The SCX/IMAC enrichment approach for global phosphorylation analysis by mass spectrometry. Nat Protoc *3*, 1630–1638.

Ware, M.L., Fox, J.W., Gonzalez, J.L., Davis, N.M., Lambert de Rouvroit, C., Russo, C.J., Chua, S.C., Jr., Goffinet, A.M., and Walsh, C.A. (1997). Aberrant splicing of a mouse disabled homolog, mdab1, in the scrambler mouse. Neuron *19*, 239–249.

Wolters, D.A., Washburn, M.P., and Yates, J.R., 3rd. (2001). An automated multidimensional protein identification technology for shotgun proteomics. Anal Chem *73*, 5683–5690.

Yang, G., Li, Q., Ren, S., Lu, X., Fang, L., Zhou, W., Zhang, F., Xu, F., Zhang, Z., Zeng, R., *et al.* (2009). Proteomic, functional and motif-based analysis of C-terminal src kinase-interacting proteins. Proteomics *9*, 4944–4961.

Zhai, B., Villen, J., Beausoleil, S.A., Mintseris, J., and Gygi, S.P. (2008). Phosphoproteome analysis of drosophila melanogaster embryos. J Proteome Res *7*, 1675–1682.

Chapter 30
Adipose Tissue Lysis and Protein Extraction Followed by MS-based Proteomic Profiling Reveals Constituents of Oxidative Stress in Obesity

Emily A. Freeman, Vera Gross, Ilyana Romanovsky, Alexander V. Lazarev, and Alexander R. Ivanov

Abstract This chapter discusses the use of alternating hydrostatic pressure (Pressure Cycling Technology, or PCT) and optimized solvents for detergent-free disruption of white adipose tissue samples for proteomic analyses. This technique can increase efficiency of protein recovery from mouse fat tissue specimens as compared to conventional homogenization and dissolution techniques. Protein extracts generated using PCT were analyzed by SDS-PAGE, 2D-electrophoresis, and nanoflow HPLC coupled to high resolution, high mass accuracy tandem mass spectrometry. As a result, a novel pressure cycling-assisted liquid-liquid extraction and fractionation method has been developed to achieve nearly complete tissue dissolution and fractionation of lipids and proteins into distinct liquid phases. This method allows for efficient extraction of highly hydrophobic proteins typically underrepresented in the extracts obtained using conventional methods. A pilot analysis of several genetically distinct model mouse lines and diets has revealed trends in protein expression which may be linked to obesity and metabolic syndrome progression, or also serve as potential drug targets. Proteomic analysis using 2D-gels and 1D-gel LC MS/MS revealed greater representation of proteins involved in lipid, glucose, and energy metabolism, as well as acute phase response and oxidative stress response signaling pathways linked to obesity. The pilot results of comparative proteomic analyses of adipose tissue harvested from animals of different genotypes and exposed to different diets were analyzed further using gene ontology and functional pathway analysis bioinformatics tools. Lipid fractions obtained from the tissue samples using the newly developed sample preparation and fractionation platform were collected for future lipid mass spectrometry-based profiling studies.

Keywords Adipose tissue · MS-based proteomics · Pathway analysis · Pressure cycling · Protein profiling · Tissue lysis

---

A.R. Ivanov (✉)
HSPH Proteomics Resource, Department of Genetics and Complex Diseases, Harvard School of Public Health, Boston, MA 02115, USA
e-mail: aivanov@hsph.harvard.edu

A.R. Ivanov, A.V. Lazarev (eds.), *Sample Preparation in Biological Mass Spectrometry*, DOI 10.1007/978-94-007-0828-0_30,

30.1 Introduction

The era of "omics"-based studies has introduced massive advancements in the quality and quantity of data that can be collected and analyzed from vastly different sample types. The rule that "the end-stage results are only as good as the sample preparation" is still very much applicable, and this is one of the principle reasons adipose tissue is among the lesser-mapped mammalian tissue proteomes. Adipose is a connective tissue composed primarily of fat cells known as adipocytes or lipocytes, with smaller percentages of other cells such as fibroblasts and macrophages (Cinti, 2005). Adipose serves as a massive store for various energy types, as well as an essential component of the endocrine system, producing hormones such as leptin, adiponection, and TNFα (Kershaw and Flier, 2004). There are two distinct types of adipose known as white adipose tissue (WAT) and brown adipose tissue (BAT); WAT is generally assumed to be more closely correlated to metabolic diseases such as type 2 Diabetes and metabolic syndrome due to the fact that it is the most prevalent fat tissue type in the obese adult human body (Gnacinska et al., 2009). The presence of large amounts of WAT is also a hallmark for many metabolic syndromes such as diabetes and hyperlipidaemia. Though it is intuitive that the only methods to change the total fat mass of the human body would be through changes in adipocyte number and size, it has been shown that the overall number of adipocytes in a human body is generally a constant that is established in childhood (Spalding et al., 2008); this suggests that the majority of changes adipose cells undergo in transition to and maintenance of the state of obesity concerns their biochemical properties and constituents, such as the proteome, and has little to do with changes purely in population numbers. Metabolic and physiological disorders such as type 2 diabetes, obesity, and cancer are directly correlated to energy metabolism, lipid biosynthesis and secretion of signaling proteins in adipose tissue (Furukawa et al., 2004; Trayhurn, 2005; Trayhurn and Wood, 2004, 2005), thus studies focusing on micro and macro changes of the lipid and protein profiles of afflicted individuals may provide critical information for targeted therapies and drug treatments. In addition, prior research has already investigated the associations between obesity and metabolic diseases that are closely linked to increases in oxidative stress and related inflammatory and stress response pathways (Curtis et al., 2010; Furukawa et al., 2004).

Tissues such as adipose, brain, and blood that contain a higher lipid content present challenges in isolating their protein constituents from the lipid portion due to protein–lipid interactions and the hydrophobic characteristics of lipid-associated proteins (Lanne et al., 2001). Methods used for protein isolation from adipose and similar tissues often involve the use of harsh chemicals and laborious manual techniques, both of which have great potential to negatively impact the protein sample, reduce reproducibility, and consume great amounts of time.

Conventional detergent-based tissue lysis and protein solubilization methods applied to tissues like adipose tend to produce highly variable results, especially with respect to functionally relevant hydrophobic proteins from mitochondria, ER, plasma membrane and lipid droplets. Abundant sample-derived lipids tend to sequester detergents into micelles, thus interfering with protein extraction. This resulting hindrance has made adipose one of the less-extensively mapped proteomes of the mammalian body, despite its pivotal role in metabolic processes. Here we describe an improved extraction method that utilizes PCT and specialized organic solvents for the dissolution of tissues, as well as efficient partitioning and recovery of both proteins and lipids from a single sample. Gel electrophoresis, liquid chromatography and high mass accuracy/resolution mass spectrometry have been used to identify and quantitatively analyze proteins derived from mouse lines with distinct genetic and dietary conditions. The proteomic data thus acquired have enabled subsequent pathway analysis leading to preliminary observations of molecular mechanisms involved in obesity and associated metabolic conditions, particularly activation of oxidative stress response pathways.

30.2 Materials and Methods

30.2.1 WAT Sample Collection and Lysis

Approximately equal aliquots (100±15 mg) of white adipose tissue (WAT) from the abdominal fat pads of wild type (WT) and obese leptin-deficient ob/ob mice (Jackson labs) fed either a normal diet or a high fat diet were harvested (Fig. 30.1). Samples were extracted with the ProteoSolve-SB kit (Pressure BioSciences, South Easton, MA) as has been described previously (Gross et al., 2008). Simultaneous sample homogenization and fractionation was carried out in FT 500 PULSE Tubes (Pressure BioSciences) subjected to alternating hydrostatic pressure generated in the Barocycler NEP-3229 (Pressure BioSciences) for 20 cycles at room temperature. Each cycle consisted of 20 s at 241 MPa followed by 20 s at atmospheric pressure.

30.2.2 WAT Lysate Fractionation

The adipose tissue samples were fractionated using a combination of centrifugation and sodium dodecyl sulfate polyacrylamide gel electrophoresis (SDS-PAGE). Following the pressure cycling, the samples were centrifuged for 15 min at 12,000×*g* to promote complete separation of liquid phases. Unless otherwise noted, the bottom liquid layer (polar fraction) from each tube was collected, while the remaining top layer was stored for subsequent analysis of lipid fractions. Solvent was evaporated by centrifugation under vacuum to a final volume of approximately 5–10 μL. The samples were then reconstituted in either 2x Laemmli SDS-PAGE buffer or in 2D sample buffer (7 M urea, 2 M thiourea, 4% CHAPS), both supplemented with protease inhibitor cocktail, to provide compatibility with the desired downstream analysis methods.

For 2D-gel analysis of samples from animals with varying genotypes, each sample was run in triplicate, at the minimum. Samples were reduced and alkylated prior to isoelectric focusing (IEF), which was performed on immobilized pH gradient

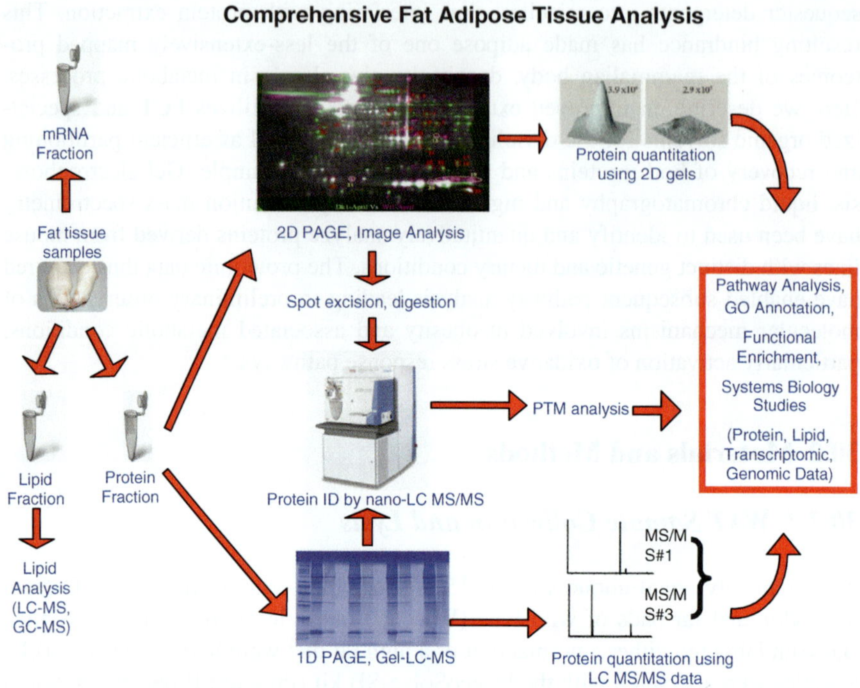

Fig. 30.1 Overview of experimental design. Adipose tissues were harvested and fractionated using pressure cycling technology and the ProteoSolve-SB kit. The protein fractions were analyzed either by 2D-gel comparative analysis coupled with LC-MS, or by GeLC-MS

strips (linear pH gradient 3–10, Bio-Rad) for 100,000 Volt-hours at 10,000 V on an IsoelectrIQ2 integrated IEF instrument (Proteome Systems Woburn, MA). Pre-cast 4–12% gels (Bio-Rad) were used for 1D SDS-PAGE analysis of samples corresponding to variations in diet. Three mice were used for each dietary condition, and each individual sample was run in triplicate and reduced/alkylated ahead of SDS-PAGE. Gels were stained with either ProteomIQ Blue colloidal Coomassie stain (Proteome Systems) or SYPRO Ruby (Invitrogen), scanned, and analyzed with Phoretix software (Nonlinear Dynamics Ltd, UK) to identify statistically significant differentially extracted proteins.

Gel spots from the 2D-gels corresponding to proteins demonstrating differential abundance were excised and processed using a conventional in-gel trypsin digestion protocol. Thirteen bands were excised from each 1D gel lane for GeLC-MS/MS profiling analysis. For all in-gel digestions, samples were reduced with 100 mM DTT and alkylated with 50 mM IAA to prevent reversible oxidation of thiols. Gel pieces were washed and dehydrated in 100% acetonitrile before rehydration in 200 μL of 0.5 ng/μL trypsin solution (sequencing-grade modified porcine trypsin from Promega) in 25 mM ammonium bicarbonate (pH 8.6) with 2 mM calcium chloride. Samples were digested while shaking overnight at 37°C.

Protein digests (5–10 μL) were separated on a C18 solid phase extraction trapping column (300 μm i.d. × 5 mm) (Dionex, Sunnyvale, CA) connected on-line to a 100-μm i.d. × 12 cm nano-LC reversed-phase in-house packed fused silica column (PicoFrit, pulled tip of 8 μm i.d.; New Objective, Woburn MA); the stationary phase: Magic C18AQ, 3 μm, 100 Å (Michrom Bioresources, Auburn, CA) using a linear gradient of acetonitrile in 0.1% formic acid. The eluate was supplied to either an LTQ Orbitrap or an LCQ Deca XP Plus mass spectrometer (Thermo Fisher Scientific, San Jose, CA) by nanoelectrospray. The MS data .raw files acquired by the mass spectrometers using the Xcalibur software (version 2.0.6; Thermo Electron, CA) were copied to the Sorcerer IDA2 search engine (version 3.5 RC2; Sage-N Research, Thermo Electron, CA) and submitted for database searches using the SEQUEST-Sorcerer algorithm (version 4.0.4). Methionine, histidine, and tryptophane oxidation [+15.994915 atomic mass units (amu)] and cysteine alkylation [+57.021464 amu with iodoacetamide derivative] were set as differential modifications. No static modifications or differential posttranslational modifications were employed. A peptide mass tolerance equal to 30 ppm and a fragment ion mass tolerance equal to 0.8 amu were used in all searches. The monoisotopic mass type, semitryptic peptide termini, and up to 2 missed cleavages were included in all searches. The database search results were filtered and validated using PeptideProphet and ProteinProphet software platforms of the Trans-Proteomic Pipeline (TPP, version 4.0.2, Institute for Systems Biology, WA). The false positive rate (FPR) was calculated as the number of peptide matches from a "reverse" database divided by the total number of "forward" protein matches, and expressed as a percentage. The filtering was performed to bring the FPR to < 1% for all datasets.

Protein pathway analysis was done using two separate software suites; Metacore (GeneGo Inc, St.Joseph, MI) and Ingenuity Pathway Analysis (IPA) (Ingenuity Systems, Redwood City, CA). Data uploaded into the platforms used identified proteins' Uniprot accession numbers as well as quantitative values from analyses using 2DGE images or LC-MS quantification. For analysis of samples from 2D gels, relative spot intensity/volume values were normalized to the WT samples. Spectral counting data for LC-MS/MS profiling of 1D-gel samples was extracted from Peptide Prophet output files.

30.3 Results and Discussion

This study focused on the preparation and analysis of the less-extensively mapped proteome of white adipose tissue using several alternative analytical techniques for comparative proteomic profiling. The adipose tissue harvested from mice of different genotypes and diets was fractionated by PCT and organic solvent-based chemistry using the ProteoSolve-SB kit (Fig. 30.1). This protocol achieves extensive separation of the protein-, lipid-, and nucleic acid-enriched fractions of the adipose tissue, although only the protein portion was analyzed for the current study.

The experiments utilized the ob/ob mouse model, which has a deficiency in leptin, a hormone critical for fatty-acid and metabolic homeostasis (Andreev et al.,

2007). Accordingly, these animals exhibit obesity and hyperinsulinemia due to overly active and abundant pancreatic islets. A thorough review of the use and findings from this mouse line has been published (Lindstrom, 2007). The ob/ob animal model was selected for these experiments because it is an excellent model for elucidating molecular mechanisms involved in type 2 diabetes, obesity, and metabolic stress factors, and provides an ample supply of adipose tissue. Studies were designed to compare the ob/ob mouse tissues to those of the WT strain in the C57BL/6 J background. Additionally, to study the interplay of diet, metabolic stress and disease on tissue proteomes, we compared the adipose of WT mice fed a regular diet (RD) and WT mice fed a high-fat diet (HFD).

Images of adipose tissue protein extracts on 1D and 2D gels were examined for protein expression changes using image analysis software prior to LC-MS/MS protein identification. Gel spots or bands of putative differentially expressed proteins were trypsin digested for profiling with MS, and relative expression values were compared between different diets and genotypes. Resulting protein identification data were submitted to IPA and Metacore bioinformatics pathway analysis suites to elucidate changes in activation of protein pathways, molecular networks, and enrichment of gene ontology (GO) terms.

As has already been demonstrated (Freeman et al., 2009; Gross et al., 2008), the protein extracts prepared from WAT using pressure cycling-assisted lysis require minimal clean-up and are compatible with 1D and 2D PAGE, as well as liquid chromatography coupled with tandem mass spectrometry. This novel tissue lysis and protein extraction technique allows for improved recovery of highly hydrophobic and complex-forming proteins when compared to conventional techniques that use chaotropic agents and detergents at atmospheric pressure, as was previously demonstrated (Freeman and Ivanov, 2010). Our initial analysis focused on the differences in proteomic profiles of genetically obese (ob/ob) mice compared to the wild type. We used 2DGE to observe the most pronounced changes in protein profiles by using comparative image analysis software, providing quantification of protein spot abundances and image alignment. A minimum of 3 technical replicate gels for each mouse model condition were included in the study. All pairs of gel images were compared to improve statistical significance of detected protein expression changes, as demonstrated in the example in Fig. 30.2a. Using protein identification lists generated from LC-MS/MS data for excised gel spots, the data were initially analyzed

Fig. 30.2 (**a**) An example of 2D-gel images of adipose samples from different mouse genotypes and comparative image analysis. Gel images were overlaid to identify differential spot expression. Each pair of gel images was merged and aligned prior to differential quantitative image analysis; the image of overlaid gels shows mutual expression in spots which are *white*, while unique proteins are identified using the individual *gel colors* (examples are *circled* in *yellow*). Quantitative data resulting from comparative analysis of three gel pairs was used to elucidate changes in WAT between each two genotypes or diets. (**b**) An example of one of the highest coverage protein networks identified from the 2DGE experiments. The network demonstrates high recovery of proteins corresponding to metabolic disease and the lipid metabolism pathway. Identified proteins are shown in color

A)

B)

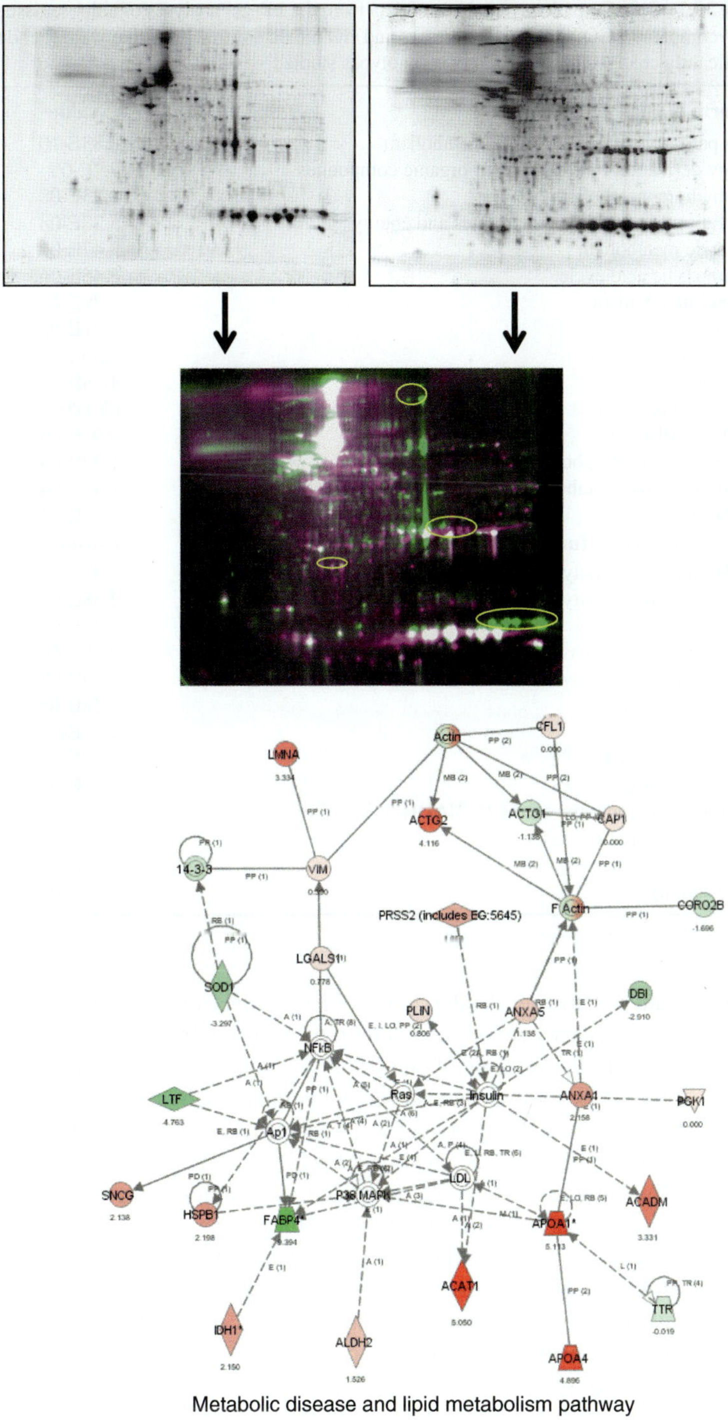

Fig. 30.2 (continued)

Table 30.1 GeneGo pathway analysis of data from the 2D-gel analyses of adipose tissue (WT and ob/ob compared). GO biological processes and GO molecular functions that were found to be most significantly changed between the two genotypes studied

Go term biological processes	***P*-value**
Main pathways of carbohydrate metabolism	3.83E-10
Energy derivation by oxidation of organic compounds	7.7E-09
Cabohydrate metabolism	2.95E-08
Generation of precursor metabolites and energy	5.28E-08
Aldehyde metabolism	3.47E-06
Tricarboxylic acid cycle	4.46E-06
Alcohol metabolism	1.29E-05
Glycolysis	3.71E-05
Glucose metabolism	4.36E-05
Metabolism	4.74E-05
Glucose catabolism	6E-05
NADP metabolism	6.94E-05
Carboxylic acid metabolism	1.39E-04
Monosaccharide metabolism	1.45E-04
Gluconeogenesis	2.51E-04
Go term molecular functions	***P*-value**
Oxidoreductase activity	1.89E-07
Oxidoreductase activity, acting on the CH-OH group	2.58E-07
NAD binding	6.43E-06
Isocitrate dehydrogenase (NADP+) activity	9.36E-06
Oxidoreductase activity, acting on NADH or NADPH	8.86E-05
Catalytic activity	2.25E-04
Hydro-lyase activity	2.22E-03
Coenzyme binding	4.08E-03
Cofactor binding	4.77E-03
Oxidoreductase activity, acting on NADH or NADPH	4.89E-03
Electron transporter activity	6.17E-03
Electron carrier activity	1.13E-02
Lyase activity	1.71E-02

using Ingenuity Pathway Analysis and GeneGo Metacore (GeneGo Inc.) software suites for preliminary network and protein pathway analysis. These initial data sets resulting from the comparison of ob/ob and WT samples revealed the most dramatic protein differences to be related to carbohydrate, aldehyde, alcohol and glucose metabolism as well as gluconeogenesis (*p*-values of 3.85E-10, 3.47E-06, 1.29E-05, 4.36E-05, and 2.51E-04, respectively) (Table 30.1). Similarly, the same data showed molecular functions corresponding to oxidation/reduction, enzyme catabolism, and electron transfer pathways (Table 30.1). In Fig. 30.2b we have shown an example of a protein pathway relating to lipid metabolism which was well-covered in terms of protein identifications. Of particular interest was the prevalence of oxidative stress pathways among the highest scoring networks, suggesting a strong connection between obesity, metabolic syndromes or diseases, and the cell's natural response to reactive oxygen species (ROS) and coping mechanisms. We also observed excellent coverage of plasma-membrane and nuclear proteins associated with some of the

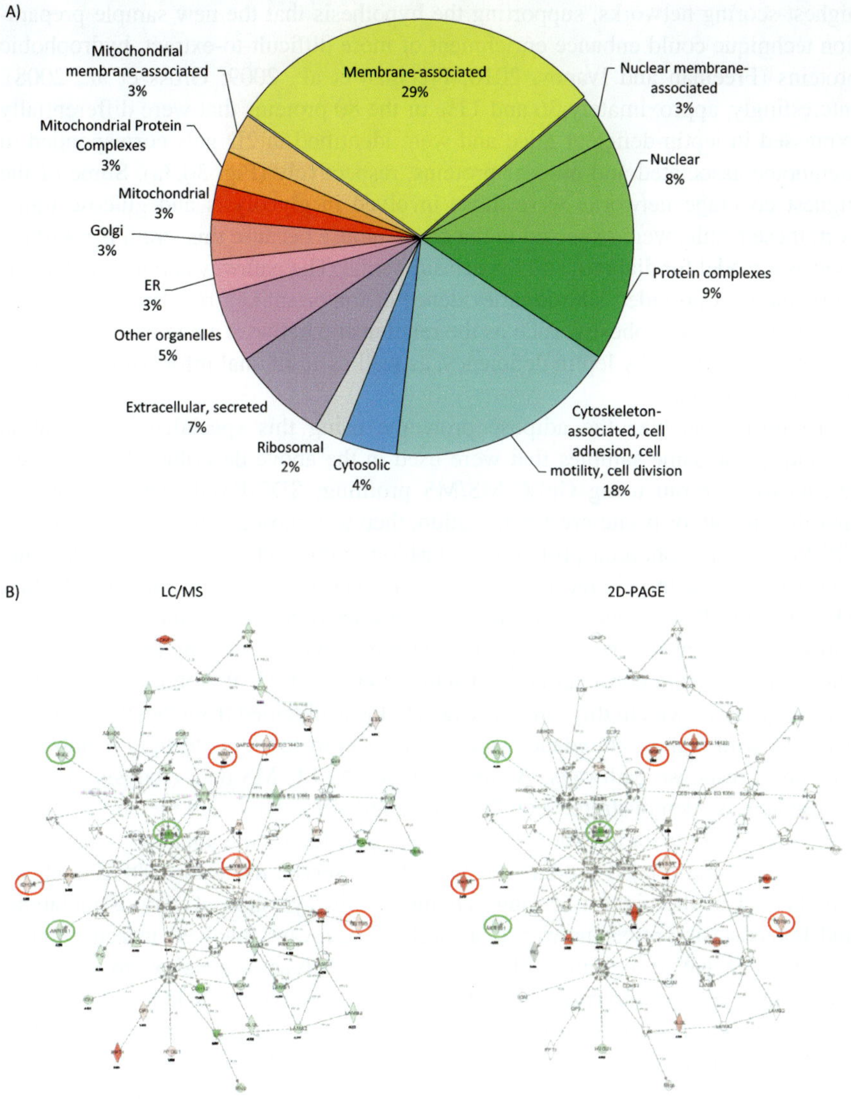

Fig. 30.3 **(a)** All protein identifications made using the 2DGE experiments were sorted into generalized subcellular localization GO categories and colored accordingly. The predominance of expression differences was observed in membrane-associated proteins (29%, *yellow*). **(b)** Comparison of GeLC-MS and 2DGE data for ob/ob and WT mice. An ob/ob to WT ratio of protein abundance was used for differential analysis. Ratios higher than 1.5 are indicated in shades of red and the ratios lower than 1.5 in shades of green. The protein network represents the acute phase response signaling pathway

highest-scoring networks, supporting the hypothesis that the new sample preparation technique could enhance enrichment of more difficult-to-extract, hydrophobic proteins (Freeman and Ivanov, 2010; Freeman et al., 2009; Gross et al., 2008). Interestingly, approximately 36 and 11% of the 80 proteins that were differentially expressed in leptin-deficient mice and were identified in 2D gels corresponded to membrane-associated and nuclear proteins, respectively (Fig. 30.3a). Some of the highest coverage networks were those involved in glycolysis and glucose transport; these results were expected in the ob/ob mouse because this strain is routinely used as a model for diabetes and hyperinsulinemia. The pathway and molecular network analyses provided additional evidence of important key regulatory and stress response features of obesity, such as the relationship between acute oxidative stress and obesity induced by leptin deficiency, as well as additional information on other proteome networks.

To further analyze the adipose proteome using this specialized preparation technique, the same samples that were used in the above described 2DGE experiments were re-run using GeLC-MS/MS profiling; SDS-PAGE was used as the first-dimension for lysate pre-fractionation, then was subsequently analyzed in LC-MS/MS. In this "shotgun proteomics" fashion, entire gel lanes were divided into sections, excised, in-gel digested with trypsin, and submitted to on-line LC-MS/MS. The resulting lists of accession numbers were examined using the same analysis software suites, as described above, with spectral counts reflecting the relative protein quantities from chromatograms (Schmidt et al., 2007; Washburn et al., 2002). The analyses showed highly similar results to those obtained from the 2D-gel image analysis, with highest abundance value changes corresponding to lipid metabolism and stress response pathways. As an example, the LC-MS data analysis of ob/ob mouse adipose demonstrated excellent coverage of the fatty-acid metabolism pathway (over 60%) (data not shown), and was supported by the abundant amounts of Acetyl-CoA and related proteins, as was also observed using the 2DGE analysis. Another set of the highest scoring canonical pathways showed various metabolic and homeostasis protein networks, as well as pathways involved in acute oxidative stress (negative logarithm of p-values): triacylglycerol metabolism = 4, lipid metabolism = 22, carbohydrate metabolism = 21, response to oxidative stress = 6. Comparing the results from quantitative analyses of 2D-gel images and GeLC-MS/MS confirmed that the two methods were predominately in agreement, showing similar patterns of protein and network expression changes between the ob/ob and WT adipose samples despite vastly different proteomic coverage provided by these two approaches (approximately 80 differentially regulated proteins for 2D PAGE and hundreds of proteins for GeLC-MS/MS). In comparing the output of the two methods, differentially abundant gel spots derived from 2D-gels provided 30–40 IDs each, that were based upon quantitative differences observed in 2D-gel images; in contrast, a single GeLC experiment could provide over 600 individual quantifiable protein IDs. It is important to note that a portion of the total difference in number of identifications between the two methods is owed to the experimenters' discretion as to how many 2DGE spots to analyze using LC-MS/MS. Interestingly, despite the quantified differences, both approaches were able to detect major changes in WAT proteomes induced by obesity. The IPA platform revealed pathways influenced by the state of obesity, such as the acute phase response pathway shown in Fig. 30.3b; the comparison of results from GelC and 2D-gel images analyses shows similar patterns of up- and down-regulation for associated indentified proteins. The few minor exceptions are likely due to the sensitivity of the LC-MS analysis, differential recovery and sensitivity of the two analytical techniques, as well as ambiguities in protein identification and quantitation for some 2D-gel spots.

In the second study, the role of diet and its relationship to obesity and stress states was examined in wild type mice fed a regular diet (RD) or a high-fat diet (HFD). Abdominal fat pad samples were collected and processed as described above. Three animals were assigned to each diet (RD or HFD), and samples from these animals were run in triplicate gel lanes for a total of 9 replicate samples per diet.

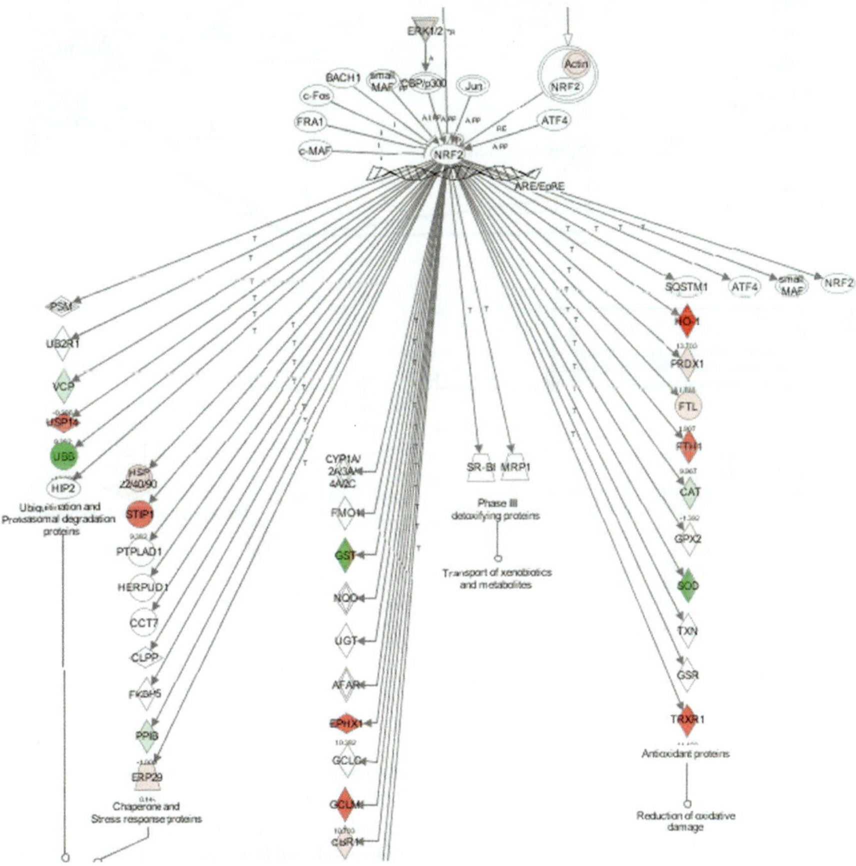

NRF-2 Oxidative stress response pathway

Fig. 30.4 The presented GeLC-MS data demonstrate quantitative protein changes in an oxidative stress response network in an obese animal induced by high fat diet when compared to the WT animal fed a regular diet

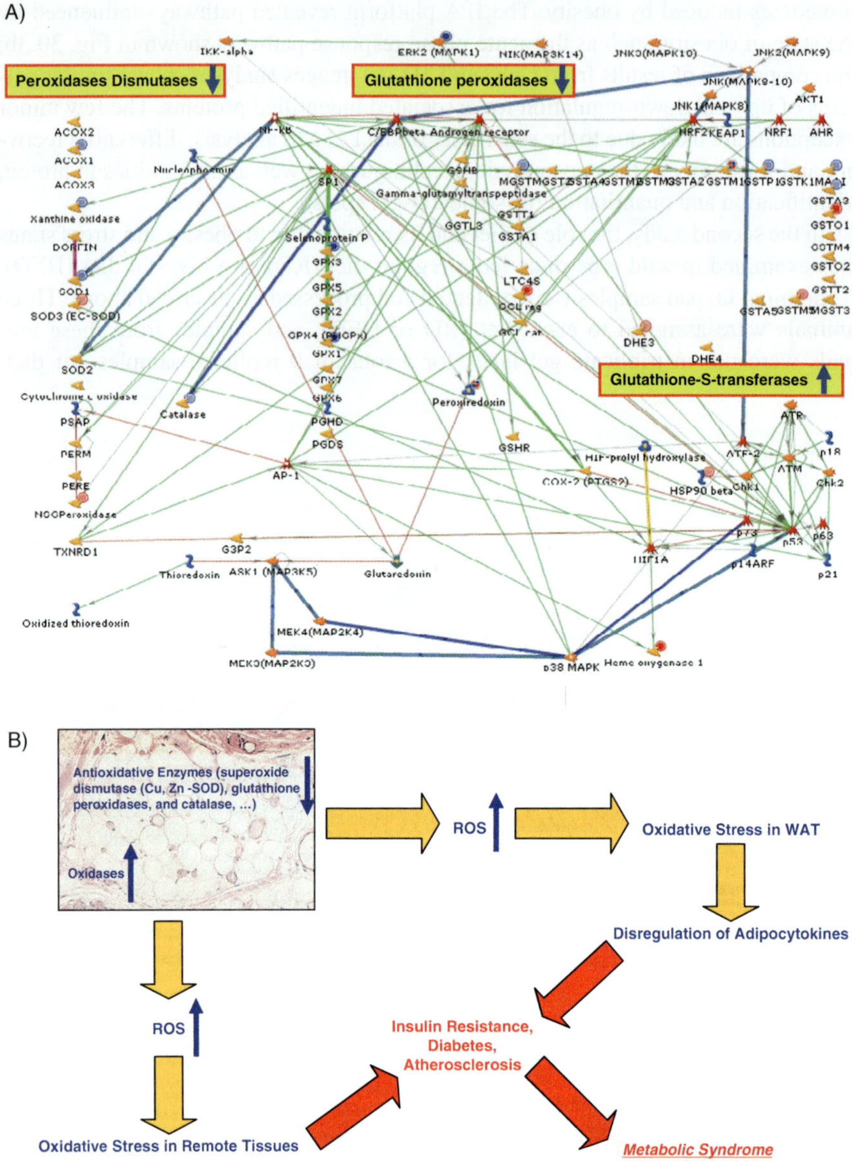

Fig. 30.5 (**a**) An example of one of the most dramatic expression changes in a protein network corresponding to a change in diet for the WT mouse. The high fat diet mouse (HFD) is shown here normalized to the results from the WT mouse with regular diet (RD). *Red circles* above protein symbols indicate an increase in expression for the HFD model, and *blue circles* indicate a decrease in relation to the RD model. Corresponding enzyme families are highlighted in the text boxes, along with their relative expression changes. (**b**) The working model of molecular mechanisms involved in obesity-induced oxidative stress in *white* adipose tissue. In brief, the model depicts how the state of obesity influences expression changes in proteins involved in oxidase and reductase enzymatic cascades. This leads to the increase of reactive oxidative species locally in adipose tissue and in remote tissues. The resulting disregulation of WAT secretory functions triggers manifestation of various implications of metabolic syndrome

In the animals consuming the HFD, the overall percentage of body fat increased as expected, and subsequent proteomic profile analyses also indicated heightened metabolic stress and related stress response proteins, including markers of oxidative stress (Fig. 30.4). The analyses indicate up- and down-regulation of multiple protein classes critical to the oxidative stress response and metabolism of ROS (Fig. 30.5a). These findings support a currently proposed model of the role of oxidative stress in obesity (Curtis et al., 2010) (Fig. 30.5b); in essence, the model first identifies adipose as a tissue responsible for down-regulating antioxidative enzymes such as peroxidases and dismutases, while at the same time increasing oxidase enzymes like glutathione-S-transferase. These fluctuations in enzymatic expression may lead to increased levels of ROS circulating throughout the body and adipose tissue, promoting a heightened state of oxidative stress and stress-response pathways in the fat and surrounding tissues. We observed these changing levels of associated proteins (Fig. 30.5a), and believe this provides support to the concept that high fat diet and obesity conditions are likely to lead to more extreme and chronic metabolic conditions such as type 2 diabetes, insulin resistance, etc (Fig. 30.5b).

30.4 Conclusions

Overall, we have illustrated a novel protocol for efficient extraction and fractionation of lipid-, protein-, and DNA/RNA-enriched fractions from white adipose tissue harvested from animal models of varying physiological and genetic conditions. Two complementary analytical approaches, 2DGE and GeLC-MS, were used to compare changes in the adipose proteomes of mouse models of obesity. Results obtained by both methods revealed very similar general patterns of changes in protein networks related to metabolic diseases and processes, supporting the pivotal role for both innate and environmental factors in syndromes like type 2 diabetes and hyperlipidemia. Specifically, in each of the data sets we observed activation of oxidative stress response pathways, encouraging additional research into the roles of obesity and diet in the onset and progression of metabolic diseases and disorders and their intimate relationship with cellular stress pathways. The results from the current study give a broad overview of the processes occurring in these tissues and provide strong support for the popular theory that altered expression of proteins related to oxidative stress is intimately linked to obesity and the eventual onset of more serious associated conditions. The pilot data reported here have demonstrated that effective and reproducible methods of sample preparation are crucial in order to provide an excellent foundation for downstream analysis techniques that produce consistent, reliable, and meaningful results. Further proteomic experiments on a larger scale will not only contribute to a better understanding of the physiological and biochemical pathways involved in metabolic conditions, but will also serve as a gateway to the discovery of new drug targets and therapeutics focusing on obesity and metabolic disorders.

Acknowledgements We would like to thank Pressure Biosciences and Drs. Haiming Cao and Gökhan Hotamisligil for providing pilot fat adipose tissue samples. We are thankful for the financial support of the Department of Genetics and Complex Diseases at the Harvard School of Public Health.

References

Andreev, V.P., Dwivedi, R.C., Paz-Filho, G., Krokhin, O.V., Wong, M.L., Wilkins, J.A., and Licinio, J. (2007). Dynamics of plasma proteome during leptin-replacement therapy in genetically based leptin deficiency. Pharmacogenomics J *7*, 666–685.

Cinti, S. (2005). The adipose organ. Prostaglandins Leukot Essent Fatty Acids *73*, 9–15.

Curtis, J.M., Grimsrud, P.A., Wright, W.S., Xu, X., Foncea, R.E., Graham, D.W., Brestoff, J.R., Wiczer, B.M., Ilkayeva, O., Cianflone, K., *et al.* (2010). Downregulation of adipose glutathione S-transferase A4 leads to increased protein carbonylation, oxidative stress, and mitochondrial dysfunction. Diabetes *59*, 1132–1142.

Freeman, E., and Ivanov, A.R. (2010). Proteomics Under Pressure: Development of Essential Sample Preparation Techniques in Proteomics Using Ultra-high Hydrostatic Pressure (Boston, MA, Harvard School of Public Health).

Freeman, E., Margolin, Y., and Ivanov, A.R. (2009). Searching for efficient and high-throughput alternatives for essential sample preparation techniques in mass-spectrometry-based functional proteomics. In ABRF (Memphis, TN).

Furukawa, S., Fujita, T., Shimabukuro, M., Iwaki, M., Yamada, Y., Nakajima, Y., Nakayama, O., Makishima, M., Matsuda, M., and Shimomura, I. (2004). Increased oxidative stress in obesity and its impact on metabolic syndrome. J Clin Invest *114*, 1752–1761.

Gnacinska, M., Malgorzewicz, S., Stojek, M., Lysiak-Szydlowska, W., and Sworczak, K. (2009). Role of adipokines in complications related to obesity: A review. Adv Med Sci *54*, 150–157.

Gross, V., Carlson, G., Kwan, A.T., Smejkal, G., Freeman, E., Ivanov, A.R., and Lazarev, A. (2008). Tissue fractionation by hydrostatic pressure cycling technology: The unified sample preparation technique for systems biology studies. J Biomol Tech *19*, 189–199.

Kershaw, E.E., and Flier, J.S. (2004). Adipose tissue as an endocrine organ. J Clin Endocrinol Metab *89*, 2548–2556.

Lanne, B., Potthast, F., Hoglund, A., Brockenhuus von Lowenhielm, H., Nystrom, A.C., Nilsson, F., and Dahllof, B. (2001). Thiourea enhances mapping of the proteome from murine white adipose tissue. Proteomics *1*, 819–828.

Lindstrom, P. (2007). The physiology of obese-hyperglycemic mice [ob/ob mice]. Sci World J *7*, 666–685.

Schmidt, M.W., Houseman, A., Ivanov, A.R., and Wolf, D.A. (2007). Comparative proteomic and transcriptomic profiling of the fission yeast schizosaccharomyces pombe. Mol Syst Biol *3*, 79.

Spalding, K.L., Arner, E., Westermark, P.O., Bernard, S., Buchholz, B.A., Bergmann, O., Blomqvist, L., Hoffstedt, J., Naslund, E., Britton, T., *et al.* (2008). Dynamics of fat cell turnover in humans. Nature *453*, 783–787.

Trayhurn, P. (2005). Endocrine and signalling role of adipose tissue: New perspectives on fat. Acta Physiol Scand *184*, 285–293.

Trayhurn, P., and Wood, I.S. (2004). Adipokines: Inflammation and the pleiotropic role of white adipose tissue. Br J Nutr *92*, 347–355.

Trayhurn, P., and Wood, I.S. (2005). Signalling role of adipose tissue: Adipokines and inflammation in obesity. Biochem Soc Trans *33*, 1078–1081.

Washburn, M.P., Ulaszek, R., Deciu, C., Schieltz, D.M., and Yates, J.R., 3rd. (2002). Analysis of quantitative proteomic data generated via multidimensional protein identification technology. Anal Chem *74*, 1650–1657.

Chapter 31
Protocols for LC-MS/MS-Based Quantitative Analysis of Proteolytic Substrates from Complex Mixtures

Mari Enoksson, Miklós Békés, Laurence M. Brill, and Khatereh Motamedchaboki

Abstract The identification of protease substrates from complex mixtures is a challenging task, especially when studying proteolysis in vivo. Even if the substrates of a signaling pathway are identified, only detection of the cleavage sites will shed light on the identity of the protease whose activity has resulted in that cleavage. Identifying cleavage sites on a proteome-wide scale allows the determination of cleavage profiles (proteloytic signatures) as well as protease substrates. Here, we present two complimentary strategies to identify proteolytic cleavage sites from complex samples. Both methods rely on the specific labeling of N-terminal amines at the protein level. These amines correspond to native N-termini of proteins and the N-termini of cleavage fragments generated by a proteolytic event. Labeling at the protein level allows the enrichment of "native" and experimentally induced N-termini by various selection methods after tryptic digestion of the sample. N-terminally labeled peptides are then identified by tandem mass-spectrometry using an LTQ Orbitrap XL. The protocols presented here are widely applicable to practically any biological system driven by proteases, in vitro as well as in vivo. They allow for the identification of protease substrates, detection of the cleavage sites and the determination of proteolytic signatures.

Keywords Proteolysis · Quantitative proteomics · N-terminomics · high resolution LC-MS/MS

31.1 Introduction

Proteolysis, the enzymatic hydrolysis of a peptide bond by a protease on a target protein substrate, is involved in many different biological pathways. More than 2% of the human genome encodes proteases (Quesada et al., 2009). This irreversible

---

K. Motamedchaboki (✉)
NCI Cancer Center Proteomics Facility, Sanford|Burnham Medical Research Institute, La Jolla, CA 92037, USA
e-mail: motamed@sanfordburnham.org

A.R. Ivanov, A.V. Lazarev (eds.), *Sample Preparation in Biological Mass Spectrometry*, DOI 10.1007/978-94-007-0828-0_31,

change of the target protein can either cause its activation (gain-of-function) or inactivation (loss-of-function). To achieve loss-of-function, a majority of the total target protein pool needs to be cleaved, whereas only a small portion of gain-of-function substrate cleavage may be necessary to initiate or abrogate a signaling pathway.

Many protease pathways interact with each other. For example, one specific protease may cleave several different substrates, and one specific substrate may be cleaved by different proteases. Therefore, studying proteolytic pathways in vivo can be a challenging task, especially if the target substrates are low abundance proteins prior to proteolytic processing. The development of different proteomics methods have been very useful for the protease field, and there is a growing body of literature describing gel-based methods (Bredemeyer et al., 2004; Brockstedt et al., 1998; Dix et al., 2008) and gel-free formats (Dean and Overall, 2007; Enoksson et al., 2007; Gevaert et al., 2003; Impens et al., 2008; Kleifeld et al., 2010; Mahrus et al., 2008; McDonald et al., 2005; Timmer et al., 2007; Van Damme et al., 2005; Van Damme et al., 2009) to study both protease cleavage specificity and protease-driven biology. Most of the gel-free procedures rely on enrichment of N-terminal peptides from the sample. Proteolytic activity results in a new N-terminus at the cleavage site. This new N-terminus, along with the original protein N-terminus, can be specifically labeled by amine reactive reagents after chemically blocking other reactive groups in the protein prior to the labeling step. After digestion of the proteins with trypsin or another specific protease during sample preparation, the labeled N-terminal peptides are selected, enriched and subsequently analyzed by high mass accuracy, high resolution LC-MS/MS. Mapping of the identified N-terminal peptide on the protein sequence provides both substrate identification and proteolytic cleavage site information. The main advantage of these "N-terminomics" approaches is reduction of sample complexity, which enables detection of lower abundance substrates.

31.2 Detecting Protolytic Substrates by N-Terminus Specific Labeling

31.2.1 Sample Derivatization Enabling Specific Labeling of Protein N-Termini

Here, we present two proteomics protocols employing the same basic chemical derivatization procedure but using different types of amine-reactive labels (see Fig. 31.1) (Enoksson et al., 2007; Timmer et al., 2007). Specific labeling of protein and cleavage product N-termini is achieved by blocking lysine and cysteine side chains prior to the N-terminal-labeling step. First, the protein sample is rapidly denatured and reduced by boiling the protein sample in 6 M guanidine and 10 mM DTT. Cysteine side chains are alkylated, and the ε-amines on the lysine side chains are converted to homoarginine by *O*-methylisourea. Guanidination of lysine ε-amines has been extensively used for years (Kimmel, 1967), but the results can vary depending on how the experiment is carried out. *O*-methylisourea can react with both

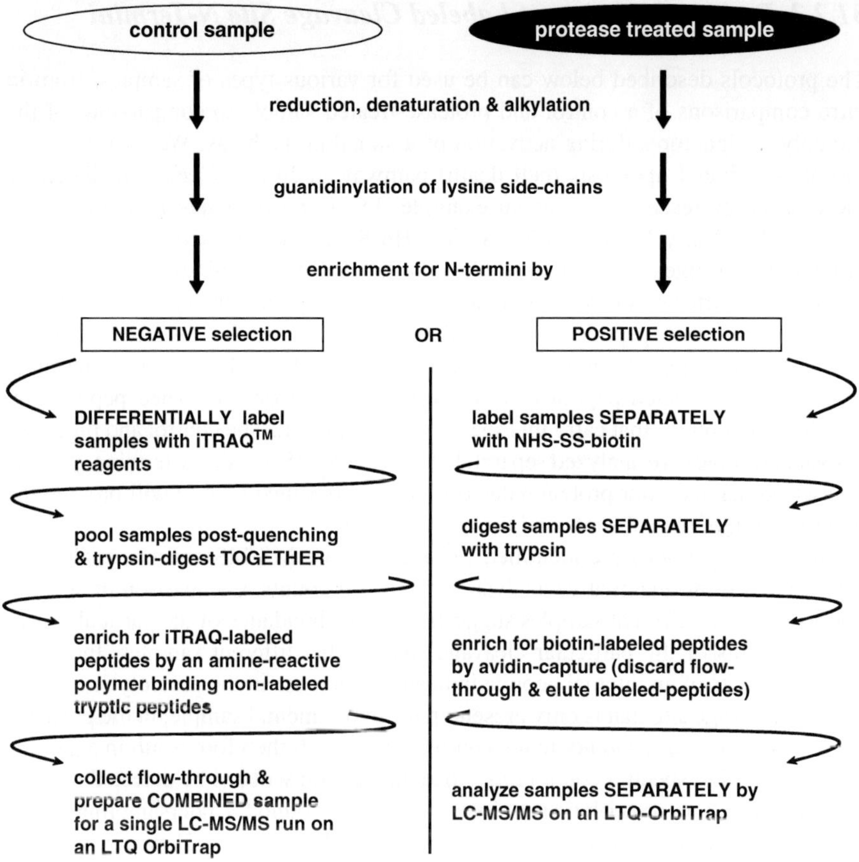

Fig. 31.1 N-terminomics Workflow. Schematic representation of sample preparation to identify proteolytic cleavage sites by N-terminomics. Briefly, experimental (protease-treated/protease-activated) and control samples are simultaneously prepared to specifically label N-termini. The procedure allows for "negative" and "positive" selection techniques. Both result in enrichment of labeled N-termini corresponding to "housekeeping" cleavage events and to experimentally induced neo-N-termini (see text for details) as a consequence of proteolytic activity. Finally, high accuracy mass-spectrometry and database searching identify the peptides corresponding to the labeled N-termini

the lysine side chains and N-termini. Therefore, it is important to follow the protocol below to get a specific guanidination reaction with only lysine side chains (Timmer et al., 2007). Peptides with arginine at the C-terminus tend to be more readily detected by MS/MS than those with lysine at the C-terminus. Thus, this reaction not only facilitates specific N-terminal labeling, but also enhances detection of the peptides (Beardsley et al., 2000; Brancia et al., 2000). After desalting the sample, the protein N-termini – including N-termini generated by proteolysis in vivo or experimentally – are specifically labeled with amine-reactive labels for subsequent analysis by an optimized LC-MS/MS method for peptide/protein identification and quantification. See Section 31.2.4 for experimental details.

31.2.2 Positive Selection of Labeled Cleavage Site N-Termini

The protocols described below can be used for various types of samples, from in vitro comparisons of a control and protease-treated sample to comparisons of the proteolytic signatures during activation of a signaling pathway. We have used the receptor-mediated apoptosis (cell death) pathway in Jurkat T cells, mediated by the caspase protease family, as an example. In the *positive selection procedure*, a cleavable affinity label (EZ-link Sulfo-NHS-SS-Biotin) is used (see Fig. 31.1). After separate digestion of the samples, the labeled N-termini are enriched using avidin resin. After washing away unlabeled tryptic peptides, the *positively selected* labeled N-terminal peptides are eluted by reduction of the disulfide linker in the label. This step leaves a mass addition of +87.99828 Da on the N-terminal residue. Thus, labeled peptides can be differentiated from unlabeled peptides by filtering for peptides that carry this specific mass addition. The control and protease-activated samples are analyzed separately by LC-MS/MS and the N-terminal peptide datasets generated after protein database searches (see text below) will provide the identification of the substrate and the cleavage site.

Individual peptides are identified by one or more MS/MS spectra, enabling a relative and semi-quantitative analysis *via* spectral counts. Comparison of spectral counts between different samples suggests relative abundance of the particular peptide (and thus of the particular cleavage site) in the different samples. Increased spectral counts of an internal peptide indicate an increase in the proteolytic event in vivo. A cleavage site that is only present in the experimental sample, in the presence of an active protease, and not in the control sample will therefore result in a peptide with a very high spectral count ratio, since the peptide will not be detected at all in the sample lacking proteolytic activity (zero spectral counts). Undersampling can be an issue in MS/MS analyses, especially with complex samples. However, we have found that the positive selection procedure results in a relatively simple mixture. In tests of gradients of 60, 120 and 180 min, the number of peptide identifications was almost as high in 60 min gradients as in the longer gradients. Thus, it seems possible that sensitivity could be a bigger challenge, in spite of the sensitive instrumentation. Since only a few (or even one) peptide per protein are detected using this method, our MS/MS method includes a dynamic exclusion repeat count of two, rather than one, to attempt to improve quantitative estimates from spectral counts. However, we have not tested instrument methods with differing repeat counts. In addition, every sample should be analyzed at least three times (three technical replicates) to improve the reliability of comparisons between samples in an experiment. Also, at least three experimental (biological) replicates should be performed. The recommended minimal threshold for spectral counts depends on the experimental set-up. As a rule of thumb, single hits are excluded. Figure 31.2 is an example of a comparison of spectral counts between peptides identified under apoptotic and non-apoptotic conditions. Increased spectral counts of peptides found in the apoptotic sample over the control indicate cleavage events as a result of apoptosis. For example, a ratio of 1.0 of spectral counts of apoptotic sample vs. total counts means this N-terminus is only detected in the apoptotic sample. As expected, most of the N-terminal peptides

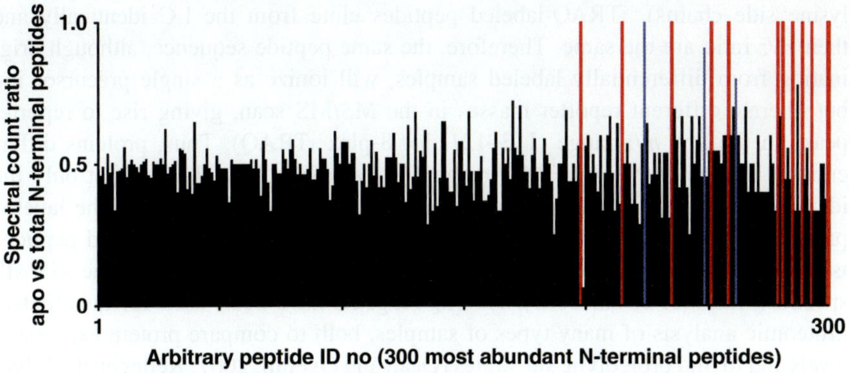

Fig. 31.2 Example data-set of a "positive-selection" N-terminomics. Jurkat T cells were untreated (control) or treated with FasL for induction of receptor-mediated apoptosis (apo). The 300 most abundant N-terminal peptides (highest spectral counts) are presented in the figure and results are shown as the spectral count ratio of apoptotic samples (apo)/total spectral counts. *Red bars*: P1=Asp cleavages apo/total >0.75, *Black/blue bars*: Other cleavages (P1≠Asp) apo/total >0.75. Note the similar abundance in the most abundant peptides (ratio apo/total ≈ 0.5), while D cleavages (presumably caspase substrates) are very over-represented in apoptotic samples

with a spectral count ratio in apo/all peptides >0.75 were cleavages after Asp, and thus presumably were caspase substrates. However, there were examples of other types of substrates that were present in apoptotic samples. Many of these substrates were also detected by the negative selection procedure (see Section 31.2.3). Several variants of this labeling procedure have been published recently. They all rely on the basic principle of selective identification of N-termini that have been cleaved in experimental samples by a variation of this positive selection methodology. An engineered variant of subtiligase has been used for the specific labeling of the N-termini (Mahrus et al., 2008). However, enzymatic preferences for labeling of certain N-terminal residues cannot be excluded. Also, a method very similar to the positive selection strategy, but using phenyl isothiocyanate (PITC) as the lysine side chain blocking reagent, was recently reported (Xu et al., 2009). PITC labels all amino-groups, and the protein sample was treated with trifluroacetic acid (TFA) to make the N-termini available for NHS-SS-Biotin labeling. TFA treatment leads to removal of both the blocking group and the N-terminal (P1') residue, and the peptide was thus shortened by one amino acid.

31.2.3 Negative Selection of Labeled Cleavage Site N-Termini

A potentially more precise quantitative approach is the *negative selection procedure*, using a modified iTRAQ (isobaric tagging for relative and absolute quantification) method (Ross et al., 2004). Each of the iTRAQ reagents contains a unique reporter group that differs in mass from the other reporters, and a balancer group containing an amine-reactive group, which labels free amines (protein/peptide N-termini and lysine side chains). iTRAQ-labeled peptides elute from the LC identically and their *m/z* ratio are the same. Therefore, the same peptide sequence, although originating from differentially labeled samples, will ionize as a single precursor ion but liberate different reporter masses in the MS/MS scan, giving rise to reporter peaks in the low *m/z* range (113–121, for 8-plex iTRAQ). Thus, proteins differentially labeled with one iTRAQ reagent in each sample will allow not only the identification of the labeled peptide but also relative quantification of the labeled protein between different samples. Relative quantification of the labeled peptides is achieved by analyzing the ratio of the reporter ion peak heights in the MS/MS spectra (Phanstiel et al., 2009). iTRAQ reagents have been used for quantitative proteomic analysis of many types of samples, both to compare protein expression levels and to find proteolytic substrates (Dean and Overall, 2007; Keller et al., 2008). In the manufacturer's protocol, protein samples are denatured, reduced, alkylated and digested prior to iTRAQ labeling. N-termini (of all peptides) and lysine side chains are available for labeling. The labeling approach presented here differs from the manufacturer's protocol. Similar to the positive N-terminomics selection procedure (see Section 31.2.2), this protocol is focused on detecting proteolytic substrates (Enoksson et al., 2007). Although iTRAQ lacks the affinity tag required for positive selection, the isobaric labels on the N-termini prevent binding to the amine-reactive resin, resulting in localization of labeled peptides in the flow-through fraction, and thus giving "negative selection" of protease cleavage products. This is in contrast to unlabeled peptides, derived from sample preparation, and lacking an N-terminus resulting from the "experimental" protease. For example, the extent of a proteolytic cleavage event in vivo can be followed over time by taking samples at different time points, labeling whole proteins in the different samples with iTRAQ reagents carrying different reporters, and analyzing peptides derived from these samples simultaneously *via* LC-MS/MS. By following the same initial chemical derivatization protocol as in the positive selection procedure, prior to iTRAQ labeling, only the N-termini in the sample can be specifically labeled with the iTRAQ reagent when lysine side chains were previously blocked by guanidination. In brief, the samples are denatured, reduced, alkylated and lysine side chains are blocked as described in Section 31.2.1 (see Fig. 31.1). The N-termini in separate protein samples are subsequently labeled with isobaric iTRAQ labels containing different reporters. The differentially labeled samples are pooled and digested together. This is in contrast to the more typical iTRAQ protocol, where the digestion occurs prior to labeling and the samples thus are digested separately, which may result in the introduction of quantitative variability. Proteomics sample reproducibility may in fact be affected if digestion occurs separately since the enzyme used for digestion may cleave with slightly different efficiencies in different samples (Picotti et al., 2007), and differential sample losses, due to sample handling, could occur. Since only a fraction of all peptides in the samples are labeled with the iTRAQ reagent, we have developed an enrichment procedure that increases the probability of detecting the peptides of interest. The pooled peptide mixture sample is incubated with amine-reactive resin. This resin will bind tryptic (unlabeled) peptides, while peptides containing an iTRAQ label at the N-terminus will fail to bind the resin and thus remain in the flow-through fraction, thus being enriched (separated) from unlabeled peptides. This *negative selection* step thus allows for enrichment of the iTRAQ labeled N-termini. Following LC-MS/MS, the raw data is searched against a database of choice to identify the peptides and the peak heights of the iTRAQ reporters are extracted to yield quantitative iTRAQ ratios. A 1:1 ratio of two different iTRAQ reporters means that the labeled peptide was present in both samples in approximately equal amounts. This is deduced because the intensity of both iTRAQ reporters was equal, shown by a peak height ratio of 1:1. In contrast, a ratio of 10:1 means that one of the samples had ten times the amount of that peptide, indicating that the increased amount of the peptide is the result of that specific experimental condition. An example of iTRAQ ratios obtained from an experiment comparing apoptotic vs. non-apoptotic cells, prepared by "negative-selection" N-terminomics, is shown in Fig. 31.3.

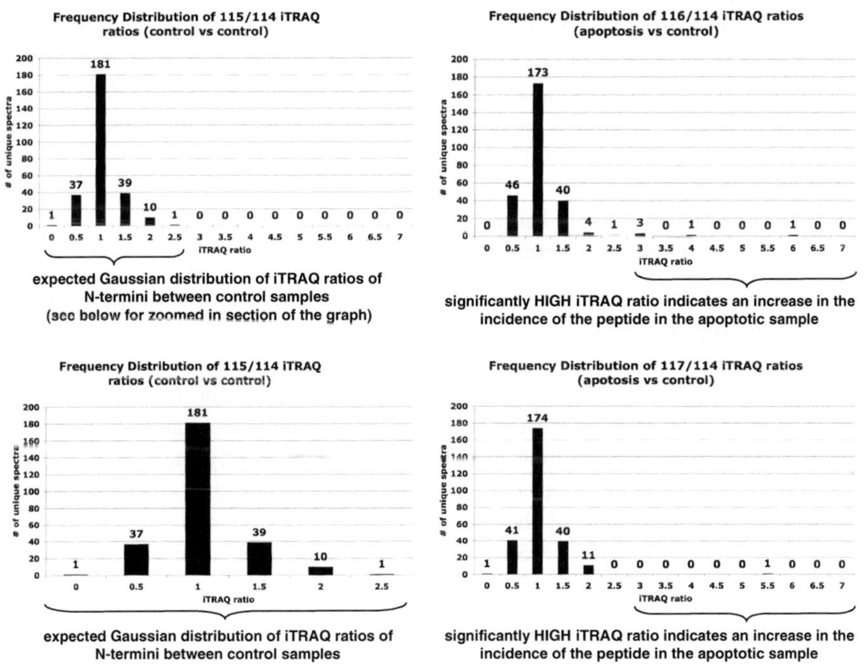

Fig. 31.3 Example data-set of a "negative selection" N-terminomics experiment. Apoptotic and control samples were prepared and analyzed by N-terminomics using iTRAQ-labeling on N-termini and negative selection, resulting in enrichment of iTRAQ-labeled N-termini. Control samples were labeled by iTRAQ 114 and 115, while apoptotic samples were labeled by iTRAQ 116 and 117. Control vs. control ratios (115/114) were expected to have a Gaussian distribution centered around 1, indicating the presence of the given peptides in equimolar amounts in the two samples. Peptides with significantly higher (apoptotic vs. control) iTRAQ ratios (116/114 or 117/114) were over-represented in the two apoptotic samples, which were labeled with the iTRAQ reagent containing the reporter with a mass of 116 or 117 Da, indicating that they resulted from the caspase-proteolytic cascade initiated during apoptosis

31.2.4 LC-MS/MS Analysis

31.2.4.1 Analysis of Samples Derived from the Positive Selection Procedure

Reversed-phase (RP) HPLC (Michrom Bioresources, Inc., Auburn, CA) is used to separate peptide mixtures, and is coupled directly to an LTQ Orbitrap XL mass spectrometer (Thermo Fisher Scientific, San Jose, CA). The LC-MS/MS instrumentation was recently described in detail (Brill et al., 2009). Briefly, solvent A is 0.1% formic acid, and solvent B is 100.0% acetonitrile. The RP gradient is 2.0% solvent B to 5.0% B from 0.0 to 2.0 min, 5.0% B to 35.0% B from 2.1 to 60.0 min, 80.0% B from 60.1 to 66.0 min and 2.0% B from 66.1 to 76.0 min. The analytical column is a 0.2 × 150 mm packed with Magic C18 AQ resin, 3 μm particles, 200 Å pore size (Michrom), flow rate of 2.0 μl/min coupled to an ADVANCE ionization source (Michrom). Monoisotopic precursor selection and charge state screening are enabled in the MS/MS method. The MS/MS method is a top four data-dependent method in which precursor ions, *m*/*z* from 300–2000 are scanned in the Orbitrap at a resolution of 60,000 in the profile mode, and MS/MS scans, using collision-induced dissociation (CID) of the four most abundant precursors are performed in the linear ion trap. CID occurs under standard conditions in the linear ion trap at 35% collision energy. Dynamic exclusion is enabled for 60 s repeat duration and 60 s exclusion duration with a repeat count of 2.

31.2.4.2 Analysis of iTRAQ Labeled Peptide Samples from Negative Selection

Typically, iTRAQ-labeled samples have been analyzed by time-of-flight (TOF) mass-spectrometers, such as a q-TOF or MALDI-TOF-TOF. However recent advances in hybrid ion trap-Orbitrap mass spectrometers, notably the inclusion of the higher energy collision-induced dissociation (HCD) cell in the "back end" of the instrument, have expanded the repertoire of instruments capable of performing iTRAQ analysis. In addition, the results are improved by the mass accuracy and resolution of the Orbitrap mass analyzer, and the Orbitrap is easily capable of scanning the *m*/*z* range of the reporter ions. The RP gradient is the same as that described above, but the MS/MS method is a top six, data dependent method. Precursor ions are scanned from *m*/*z* 300–2000 in the Orbitrap at a resolution of 30,000. The three most abundant precursor ions are fragmented in the HCD cell at 50% collision energy, and the resulting fragment ions, with a first *m*/*z* ratio of 100, are scanned in the Orbitrap at a resolution of 15,000. A threshold of 1e5 for MS/MS is specified, and dynamic exclusion enabled as described above. Subsequently, the same three precursor ions are fragmented by CID and scanned in the linear ion trap. Dynamic exclusion parameters are the same as described above. Applying both dissociation techniques on the same precursor ion provides effective peptide identification by CID and also allows quantification of iTRAQ reporter ions with high resolution and mass accuracy, as well as low quantitative variability (typically within ±10%). In our experience, relatively few peptide IDs are obtained from the peptides fragmented by HCD, because high collision energies are required to fully liberate reporter ions from precursors for optimal quantitative accuracy. These high energies often result in relatively sparse b- and y-ion series. However, extensive b- and y-ion series are frequently obtained via CID.

31.2.4.3 Peptide Identification and Database Searching

Peptides are identified from MS/MS spectra using Sorcerer™-SEQUEST® (SageN Research, San Jose, CA). Searches against the appropriate protein database, with a precursor mass tolerance of 10.0 ppm, specify static carbamidomethylation of Cys residues and differential guanidination of lysine residues, differential iTRAQ or NHS-SS-Biotin (+87.99828 Da) labeling of N-termini and lysine residues, differential oxidation of Met residues, and semi-enzyme (e.g. semi-tryptic) specificity. In addition, to match the same precursor ion with a high confidence peptide ID to the one giving rise to the iTRAQ reporter ratios, we use Proteome Discoverer 1.0 (Thermo Fisher Scientific). Although this software offers extensive proteomics analysis capabilities, only iTRAQ data analysis is discussed in this chapter.

Proteome Discoverer 1.0 utilizes raw data files from the LTQ Orbitrap XL and allows database searches using either Mascot or SEQUEST search algorithms. The SEQUEST-derived peptide ID, using MS/MS scans in the CID mode, and iTRAQ quantification using MS/MS scans derived from HCD data are utilized. With Proteome Discoverer, one can automatically match peptide IDs with iTRAQ reporter ratios based on fragmentation of the same precursor ion, and automatically calculate the iTRAQ ratios for a given spectrum.

31.2.5 Data Interpretation

The hypothesis and experimental design will indicate data interpretation methodology. Databases and web-based tools that can provide some useful information are, for example, protein databases such as SwissProt/Uniprot (http://www.uniprot.org/), which contain information about cleavage of propeptides, signal peptides or mitochondrial localization peptides, while other more protease-directed databases, such as CutDB (http://cutdb.burnham.org/) (Igarashi et al., 2007), PoPS (Boyd et al., 2005) or the caspase-oriented database Casbah (http://bioinf.gen.tcd.ie/casbah/) (Luthi and Martin, 2007), can be useful when comparing results to previously published data. Determining the cleavage site preference of a specific protease or defining the proteolytic signature can be facilitated by calculating the prevalence of the residues surrounding the cut sites. This can be facilitated by fairly simple web-based tools that calculate the sequence pattern around the cleavage sites, for example web logo (http://weblogo.berkeley.edu/) or icelogo (http://code.google.com/p/icelogo/) (Colaert et al., 2009). The active site of a protease is comprised of a cleft lined with residues that define the protease specificity. These residues thus make key contributions to the specificity pocket of the protease. The substrate peptide chain residues fit into these pockets. The numbering of the cleavage site residues originates at the scissile bond. Residues toward the N-terminus of the scissile bond are "unprimed" (P1, P2, P3 etc.) but sites toward the C-terminus are "primed" (P1', P2', P3' etc.) (Schechter and Berger, 1967). Web logo and/or icelogo comparisons thus reveal the prevalence of different residues at P4, P3, P2 etc. The advantage with icelogo is that the user defines the reference set to which the experimental data should be compared (Colaert et al., 2009). This probability-based method may thus reflect the data more accurately.

The basic chemical derivatization procedure has been validated using the positive selection N-terminomics approach (Timmer et al., 2007). We have used this method on several biological models, from bacteria and yeast to cell lines, tissue and plasma samples (Table 31.1) (Enoksson et al., 2007; Timmer et al., 2007; Timmer et al.,

Table 31.1 N-terminomics at work

Applications of N-terminomics to date

Biological Phenomena	Technique used	Publication	Significance
initiator methionine (Met) removal (in vivo)	N-terminomics (positive selection, labeling N-termini with a biotin-tag)	Timmer et al. (2007)	Establishment of a widely applicable technique (specific labeling of N-termini) to profile cleavage sites in vivo, highlighting, among many, the specificity of Methionine aminopeptidease in *E. coli*
Apoptosis (cell-free, cytochrome *c* induced)	N-terminomics (iTRAQ-labeling of N-termini)	Enoksson et al. (2007)	Utilizing iTRAQ-labeling at the protein level to identify proteolytic cleavage sites from a complex sample during cell-free apoptosis
Determinants of protease substrates	N-terminomics (positive selection, labeling N-termini with a biotin-tag)	Timmer et al. (2009)	Using N-terminomics to identify and validate protease substrates to determine the kinetic and structural constraints of proteolysis

The future use of N-terminomics

Biological Phenomena	Potential significance
Coagulation pathway	Identification of physiologically relevant "bystander cleavages" during the blood coagulation cascade as orchestrated by extra-cellular serine proteases and/or aminopeptidases of the blood
Inflammation	Shedding light on the extent of proteolytic signaling during inflammation, both intra-cellularly (as orchestrated by caspase-1) and extra-cellularly (through the action of MMPs)
Invasion of tumor cells	Determining physiologically important substrate profiles (and thus the protease being responsible) during malignant tumor formation

The table summarizes the versatility of N-terminomics through highlighting its applicability in a wide variety of biologically relevant systems significantly controlled by proteolysis

2009). The method is sensitive and detection of low copy substrates is possible. In fact, the dynamic range of substrates identified in *E. coli* samples spans over 10^3, detecting as few as 200 protein copies per cell (Timmer et al., 2009). Using cell-free apoptosis as a model, we also demonstrated that substrates that cannot be detected by Western Blotting can be detected by these methods (Enoksson et al., 2007). Importantly, these methods do not aim to find every proteolytic substrate in the sample. This is very difficult, if not practically impossible. As with other proteomics methods, identification of a peptide depends on the fact that the peptide can be detected by LC-MS/MS. For example, an N-terminal peptide that is too long (> approximately 30 residues) will be difficult to detect with the instrumentation we use (but instrumentation equipped with electron transfer dissociation may be more amenable to analysis of longer peptides). Thus, it is important to keep in mind that these methods can miss detection of certain substrates. However, in some cases using a different enzyme to generate peptides, such as Glu-C or chymotrypsin, could increase the number of peptides and proteins identified.

31.3 Materials

31.3.1 Sample Derivatization Prior to Labeling (Positive and Negative Selection)

Guanidine hydrochloride, Acros Scientific, #120230025
Dithiothreitol (DTT), Biovectra, #1365
Iodoacetamide (IAA), Sigma, I1149
O-methylisourea, Acros organics, #233030250
HEPES, Fisher Scientific, BP 310-1
Urea, Mallinckrodt Baker, #8642-20
AG 501-X8, Bio-Rad, #142-6424
Bio-Rad protein assay kit, Bio-Rad, #500-0006
PD-10 columns, GE Healthcare, #170850-01
Ammonium Bicarbonate (NH_4HCO_3), Acros Organics, #393210010
Sodium Chloride (NaCl), Fisher Scientific, BP 358-212

31.3.2 Labeling

EZ Link Sulfo-NHS-SS-Biotin, Pierce/Thermo Fisher, #21331
iTRAQ reagent Multiplex Kit, Applied Biosystems, #4352135
Hydroxylamine, Sigma, H2391
Glycine, Fisher Scientific, BP381-500
Sequencing grade modified trypsin, Promega, #V511
High capacity immobilized Neutravidin, Pierce/Thermo Fisher, #29204
Disposable gravity columns, Clontech, #635606
Tris(2-carboxyethyl)phosphine hydrochloride (TCEP), Invitrogen, #T2556
Affi-Gel 10/15 Gel, Bio-Rad, #153-6098

31.3.3 Sample Clean-Up

Sep-Pak C18 cartridges, Waters, WAT066905
Protein low binding tubes, Sarstedt, #72.690.600
5,5-Dithiobis(2-nitrobenzoic acid) (DTNB), Sigma, D-1830
Acetonitrile (ACN), Alfa Aesar, #22927
N,N-Dimethylformamide, anhydrous (DMF), Sigma, #227056
Trifluoroacetic acid, TFA, Sigma T-6508

31.4 Sample Preparation

31.4.1 Sample Derivatization Prior to Labeling

1. Denaturation and reduction of the sample: Add final concentrations of 6 M Guanidine-HCl and 10 mM DTT to the sample and boil it for 10 min. It is important to perform this step immediately after harvesting the sample in order to avoid post-lysis proteolysis (see Note 1).
2. Alkylation of Cys side chains: Cool to room temperature and add IAA to 30 mM final concentration. Incubate in the dark for 30 min at RT.
3. Guanidination of lysine side chains: Adjust pH to ~10.3 using NaOH. Add *O*-methylisourea to the sample to 0.5 M final concentration. Immediately adjust pH to ~10.3 using NaOH and incubate sample overnight at 4°C.
4. Desalting step: Buffer exchange into 8 M fresh urea buffer (8 M urea, 50 mM HEPES pH 7.8) using (for example) PD10-columns, according to the manufacturer's protocol. De-ionize and filter the urea with AG 501-X8 resin and a 0.45 μM pore size filter before use (see Note 2). Add the sample (maximum volume of 1.8 mL to minimize small molecule carry-over) and collect the resin flow-through in small fractions, usually 0.3–0.5 mL/fraction, in order to minimize sample volume and to avoid small molecule carry-over (see Note 3). Measure absorbance at 280 nm (A_{280}) of the different fractions, pool the fractions containing protein based on A_{280} and measure the final protein concentration using a protein assay, for example the Bio-Rad Protein Assay. Equalize protein concentrations for the different samples by adding 8 M urea, 50 mM HEPES pH 7.8.
5. Continue with positive selection (see Section 31.4.2) or negative selection procedure (see Section 31.4.3).

31.4.2 Positive Selection Procedure Using NHS-SS-Biotin

1. N-terminal labeling: After initial chemical derivatization (Section 31.4.1), incubate the sample with 2 mM EZ link Sulfo-NHS-SS-Biotin for 1.0 h at 37°C.
2. Quench un-reacted label: Add 50 mM glycine/50 mM NH_4HCO_3 pH 7.8 for 30 min at 37°C.

3. Add 40 mM hydroxylamine for 20 min at room temperature to prevent possible side chain labeling (possible low incidence of labeling of serine, threonine and histidine) (see Note 4).
4. Desalting step: Buffer exchange the sample into 6 M Urea, 150 mM NaCl, 150 mM NH_4HCO_3 pH 7.8 using PD-10 columns as described under Section 31.4.1. Collect the sample in small fractions to ensure exclusion of free biotin (which elutes from the PD-10 column at later elution times; see Note 3). Measure A_{280} of the different fractions, pool the fractions containing protein and dilute the sample 1:3 with dH_2O.
5. Add sequencing-grade modified trypsin and incubate overnight at 37°C. Optimal ratio (w/w) of trypsin:protein in the sample is 1:50.
6. Boil samples for 5.0 min, centrifuge at 10,000×*g* and save the supernatant.
7. Enrichment of labeled peptides: Wash high-capacity neutravidin resin in high salt buffer (20 mM NH_4HCO_3 pH 7.8, 1.0 M NaCl) followed by low salt buffer (20 mM NH_4HCO_3 pH 7.8). Dilute resin in 20 mM NH_4HCO_3 to 50% slurry and apply to sample. Incubate for 1.0 h at RT. The amount of resin to be used ultimately depends on the sample and the amount of free N-termini. In general, 0.5 mL resin is sufficient for protein derived from 50–100 million Jurkat cells. It is however recommended to test if a higher resin volume is necessary for a new type of sample.
8. Wash extensively – with minimum ten resin volumes – of low salt, then high salt, then low salt NH_4HCO_3 buffer.
9. Elution of labeled N-terminal peptides: Add 25 mM TCEP in 20 mM NH_4HCO_3 for 30 min at RT. Repeat the elution step and pool the eluates. Wash the resin in 20 mM NH_4HCO_3 and pool with the eluates to maximize peptide recovery.
10. Desalt the peptide sample using Sep-Pak C18 cartridges according to the manufacturer's protocol. Elute the sample in 50–80% ACN, 0.1% TFA.
11. Quantify the peptide samples according to the manufacturer's protocol using DTNB assay (Ellman's reagent).
12. Dry the peptide sample in a speedvac and analyze by LC-MS/MS (see Section 31.4.4).

31.4.3 Negative Selection Procedure Using iTRAQ Labeling

1. N-terminal labeling: After basic chemical derivatization (Section 31.4.1), add iTRAQ reagent (containing reagents liberating the *m/z* 114, 115, 116 or 117 reporter, when using the 4-plex iTRAQ kit) dissolved in 70 μL of ethanol to an individual sample (50–200 μg protein per sample). Incubate 1.0 h at 37°C (see Note 5).
2. Pool differentially labeled samples from step one, and evaporate the ethanol (do not dry to completion, decrease the volume by ca. 70–80 μl) in a speed vac.
3. Dilute the mixed iTRAQ sample with 50 mM HEPES, pH 7.8 to result in a concentration of 2 M urea. Add sequencing-grade modified trypsin (optimal ratio

(w/w) of trypsin:protein in the sample is 1:50) and incubate overnight at 37°C. Boil samples for 5.0 min, centrifuge at 10,000×*g* for 10.0 min and save the supernatant.

4. Enrichment of labeled peptides: Wash amine-reactive resin (Affi-gel 10 and Affi-Gel 15) with cold deionized H_2O immediately before use. Add 50 mM HEPES, pH 7.8 to achieve a 50% slurry and add to the peptide mixture. Incubate at 4°C for a minimum of 4.0 h.
5. Collect flow-through fraction which is unbound to the resin. This contains the iTRAQ labeled N-terminal peptides. Wash resin with 50 mM HEPES, pH 7.8 and pool the flow-through fractions (see Note 6).
6. Desalt the peptide sample using Sep-Pak cartridges according to the manufacturer's protocol. Elute the sample in 50–80% ACN, 0.1% TFA.
7. Dry the peptides in a speed vac and analyze by LC-MS/MS (see Section 31.4.4)

31.4.4 LC-MS/MS Analysis

1. Re-suspend the final peptide pellet in a suitable volume (10–50 μl) of 2.0% ACN, 0.1% TFA.
2. Dilute the sample 10-fold in 2.0% ACN, 0.1% TFA.
3. Inject a trial sample (ca. 1.0%) of diluted peptides onto the peptide captrap (Michrom), for a RP separation with a 15 min gradient (the same as described above with the exception of the gradient length). This step is to determine how much sample to inject in step 4.
4. Inject the optimized quantity of peptides to obtain abundant, but not excessive signal in the mass spectrometer (we strive for peak total ion chromatogram signal in the mid-e8 range). A RP gradient of 60 min is optimal, in our experience. Use the top four, data-dependent method for peptides from the positive selection procedure, and the top six, data-dependent method for peptides from the negative selection procedure.
5. LC-MS/MS method parameters are described above.

31.4.5 Database Search and Data De-convolution

1. Search the raw data using standard search algorithms (such as SEQUEST or Mascot) using the parameters described above. We recommend use of a false discovery rate (FDR) of 0.02 or less. To obtain the FDR, we use ProteinProphet (Institute for Systems Biology, Trans-Proteomic Pipeline). Alternatively, searches may also be performed against a decoy database using standard methods (Elias and Gygi, 2007) to estimate the FDR.
2. Curate the list of identified peptides by categorizing native N-termini and non-target protein cleavage events such as removal of initiator Met, signal peptides, propeptides etc., and identify cleavage events unique to the experimental condition under investigation.

a. In the case of "positive-selection" samples; extract spectral counts for each unique N-terminus and compare spectral count ratios between or among different experimental conditions.
b. In the case of "negative selection" samples, match peptide identifications achieved by CID to the iTRAQ-reporter ratios generated by HCD scans (software such as Proteome Discoverer has built-in functions to match CID to HCD spectra based on the *m/z* of the precursor ion as described above). Note that some peptides are also identified on the basis of the HCD spectra, but we have identified far more peptides on the basis of the CID spectra.

3. Compare the peptides that demonstrate unique cleavages, induced by the experimental condition, to existing protease cleavage databases.

Notes

1. Avoid all reagents that contain primary amines (such as Tris buffers) during sample preparation. These reagents will react with the N-terminal labels and will thus prevent efficient labeling.
2. Deionization of urea is necessary in order to eliminate free amines that otherwise can react with the label.
3. The PD-10 desalting steps are very important in order to avoid carryover of small molecules. For example DTT from step one will reduce the disulfide linker if it is carried over to the labeling step, carry over of small amine-containing molecules will react with the label and excess label will saturate the avidin resin.
4. NHS-SS-Biotin may react with serine, threonine and histidine to a very low extent (less than 7% of all peptides (Timmer et al., 2007)). This side chain labeling can be prevented by hydroxylamine treatment. Since the data sets should be filtered for peptides containing an N-terminal mass addition, this potentially only constitutes a problem when there is side-chain labeling of the P1' residue after digestion.
5. Sample volume during labeling should be minimized. If the protein is dilute, protein precipitation is recommended using, for example, TCA/acetone precipitation. The samples can be re suspended either in iTRAQ Dissolution Buffer, or in fresh 100 mM NH_4HCO_3 buffer, pH 8.0 containing 8.0 M urea. Note that the protein concentration of the samples needs to be normalized for total protein amount before iTRAQ labeling.
6. It is recommended to check the amine-reactive resin efficiency by measuring the protein content of the sample before and after incubation with the resin using a standard protein assay, for example the Bio-Rad protein assay kit. Approximately 80–90% of the sample should remain on the amine-reactive resin.

Acknowledgements We thank John Timmer, Eric Wildfang and Wenhong Zhu for assistance with development of some of the methods, Guy Salvesen for support and key discussions, and the NCI Cancer Center at the Sanford|Burnham Medical Research Institute for support.

References

Beardsley, R.L., Karty, J.A., and Reilly, J.P. (2000). Enhancing the intensities of lysine-terminated tryptic peptide ions in matrix-assisted laser desorption/ionization mass spectrometry. Rapid Commun Mass Spectrom *14*, 2147–2153.

Boyd, S.E., Pike, R.N., Rudy, G.B., Whisstock, J.C., and Garcia de la Banda, M. (2005). PoPS: A computational tool for modeling and predicting protease specificity. J Bioinform Comput Biol *3*, 551–585.

Brancia, F.L., Oliver, S.G., and Gaskell, S.J. (2000). Improved matrix-assisted laser desorption/ionization mass spectrometric analysis of tryptic hydrolysates of proteins following guanidination of lysine-containing peptides. Rapid Commun Mass Spectrom *14*, 2070–2073.

Bredemeyer, A.J., Lewis, R.M., Malone, J.P., Davis, A.E., Gross, J., Townsend, R.R., and Ley, T.J. (2004). A proteomic approach for the discovery of protease substrates. Proc Natl Acad Sci USA *101*, 11785–11790.

Brill, L.M., Motamedchaboki, K., Wu, S., and Wolf, D.A. (2009). Comprehensive proteomic analysis of Schizosaccharomyces pombe by two-dimensional HPLC-tandem mass spectrometry. Methods *48*, 311–319.

Brockstedt, E., Rickers, A., Kostka, S., Laubersheimer, A., Dorken, B., Wittmann-Liebold, B., Bommert, K., and Otto, A. (1998). Identification of apoptosis-associated proteins in a human Burkitt lymphoma cell line. Cleavage of heterogeneous nuclear ribonucleoprotein A1 by caspase 3. J Biol Chem *273*, 28057–28064.

Colaert, N., Helsens, K., Martens, L., Vandekerckhove, J., and Gevaert, K. (2009). Improved visualization of protein consensus sequences by iceLogo. Nat Methods *6*, 786–787.

Dean, R.A., and Overall, C.M. (2007). Proteomics discovery of metalloproteinase substrates in the cellular context by iTRAQ labeling reveals a diverse MMP-2 substrate degradome. Mol Cell Proteomics *6*, 611–623.

Dix, M.M., Simon, G.M., and Cravatt, B.F. (2008). Global mapping of the topography and magnitude of proteolytic events in apoptosis. Cell *134*, 679–691.

Elias, J.E., and Gygi, S.P. (2007). Target-decoy search strategy for increased confidence in large-scale protein identifications by mass spectrometry. Nat Methods *4*, 207–214.

Enoksson, M., Li, J., Ivancic, M.M., Timmer, J.C., Wildfang, E., Eroshkin, A., Salvesen, G.S., and Tao, W.A. (2007). Identification of proteolytic cleavage sites by quantitative proteomics. J Proteome Res *6*, 2850–2858.

Gevaert, K., Goethals, M., Martens, L., Van Damme, J., Staes, A., Thomas, G.R., and Vandekerckhove, J. (2003). Exploring proteomes and analyzing protein processing by mass spectrometric identification of sorted N-terminal peptides. Nat Biotechnol *21*, 566–569.

Igarashi, Y., Eroshkin, A., Gramatikova, S., Gramatikoff, K., Zhang, Y., Smith, J.W., Osterman, A.L., and Godzik, A. (2007). CutDB: A proteolytic event database. Nucleic Acids Res *35*, D546–549.

Impens, F., Van Damme, P., Demol, H., Van Damme, J., Vandekerckhove, J., and Gevaert, K. (2008). Mechanistic insight into taxol-induced cell death. Oncogene *27*, 4580–4591.

Keller, M., Ruegg, A., Werner, S., and Beer, H.D. (2008). Active caspase-1 is a regulator of unconventional protein secretion. Cell *132*, 818–831.

Kimmel, J.R. (1967). Guanidination of proteins. Methods Enzymol *11*, 584–589.

Kleifeld, O., Doucet, A., auf dem Keller, U., Prudova, A., Schilling, O., Kainthan, R.K., Starr, A.E., Foster, L.J., Kizhakkedathu, J.N., and Overall, C.M. (2010). Isotopic labeling of terminal amines in complex samples identifies protein N-termini and protease cleavage products. Nat Biotechnol *28*, 281–288.

Luthi, A.U., and Martin, S.J. (2007). The CASBAH: A searchable database of caspase substrates. Cell Death Differ *14*, 641–650.

Mahrus, S., Trinidad, J.C., Barkan, D.T., Sali, A., Burlingame, A.L., and Wells, J.A. (2008). Global sequencing of proteolytic cleavage sites in apoptosis by specific labeling of protein N termini. Cell *134*, 866–876.

McDonald, L., Robertson, D.H., Hurst, J.L., and Beynon, R.J. (2005). Positional proteomics: Selective recovery and analysis of N-terminal proteolytic peptides. Nat Methods *2*, 955–957.

Phanstiel, D., Unwin, R., McAlister, G.C., and Coon, J.J. (2009). Peptide quantification using 8-plex isobaric tags and electron transfer dissociation tandem mass spectrometry. Anal Chem *81*, 1693–1698.

Picotti, P., Aebersold, R., and Domon, B. (2007). The implications of proteolytic background for shotgun proteomics. Mol Cell Proteomics *6*, 1589–1598.

Quesada, V., Ordonez, G.R., Sanchez, L.M., Puente, X.S., and Lopez-Otin, C. (2009). The Degradome database: Mammalian proteases and diseases of proteolysis. Nucleic Acids Res *37*, D239–243.

Ross, P.L., Huang, Y.N., Marchese, J.N., Williamson, B., Parker, K., Hattan, S., Khainovski, N., Pillai, S., Dey, S., Daniels, S., *et al.* (2004). Multiplexed protein quantitation in Saccharomyces cerevisiae using amine-reactive isobaric tagging reagents. Mol Cell Proteomics *3*, 1154–1169.

Schechter, I., and Berger, A. (1967). On the size of the active site in proteases. I. Papain. Biochem Biophys Res Commun *27*, 157–162.

Timmer, J.C., Enoksson, M., Wildfang, E., Zhu, W., Igarashi, Y., Denault, J.B., Ma, Y., Dummitt, B., Chang, Y.H., Mast, A.E., *et al.* (2007). Profiling constitutive proteolytic events in vivo. Biochem J *407*, 41–48.

Timmer, J.C., Zhu, W., Pop, C., Regan, T., Snipas, S.J., Eroshkin, A.M., Riedl, S.J., and Salvesen, G.S. (2009). Structural and kinetic determinants of protease substrates. Nat Struct Mol Biol *16*, 1101–1108.

Van Damme, P., Martens, L., Van Damme, J., Hugelier, K., Staes, A., Vandekerckhove, J., and Gevaert, K. (2005). Caspase-specific and nonspecific in vivo protein processing during Fas-induced apoptosis. Nat Methods *2*, 771–777.

Van Damme, P., Maurer-Stroh, S., Plasman, K., Van Durme, J., Colaert, N., Timmerman, E., De Bock, P.J., Goethals, M., Rousseau, F., Schymkowitz, J., *et al.* (2009). Analysis of protein processing by N-terminal proteomics reveals novel species-specific substrate determinants of granzyme B orthologs. Mol Cell Proteomics *8*, 258–272.

Xu, G., Shin, S.B., and Jaffrey, S.R. (2009). Global profiling of protease cleavage sites by chemoselective labeling of protein N-termini. Proc Natl Acad Sci USA *106*, 19310–19315.

Part X
Characterization of Membrane Proteins

Chapter 32
Mass Spectrometry-Based Proteomics Analyses of Integral Membrane Proteins

Eric Bonneil, Sylvain Brunet, Michel Jaquinod, Joseph P.M. Hui, Anik Forest, and Pierre Thibault

Abstract The present work compares two proteomics approaches for the analysis of integral membrane proteins (IMPs). Membrane proteins from human U937 monocytes were enriched in IMP following alkaline solvent extraction to solubilize peripheral membrane proteins. Cell extracts of IMPs were separated either by gel electrophoresis followed by in-gel tryptic digestion and liquid chromatography-tandem mass spectrometry (1D-gel-LC-MS/MS) or digested directly with trypsin to release peptides for subsequent analyses by two-dimensional liquid chromatography-tandem mass spectrometry (2D-LC-MS/MS). Both approaches yielded reproducible results with 95% of the matched ions showing less than ±1 min variation in retention time and less than 30% variation in intensity for all sample replicates examined. We also observed a similar proportion (~30%) of identified IMPs using both proteomics approaches. However, the gel-free 2D-LC-MS/MS technique led to a higher number of identified proteins (up to 583 for monocyte sample) and an increase in sequence coverage for transmembrane proteins corresponding to 20 μg of total cell extract. The application of label-free quantitative proteomics is also demonstrated for the identification of differentially abundant IMPs from human U937 monocytic cells exposed to phorbol ester. Protein expression profiling experiments enabled the identification of differentially abundant IMPs including cell growth antigen (CD98), the iron trafficking protein (sideroflexin) and LAMP2, a protein which shuttles between the lysosome and the plasma membrane.

Keywords Two dimensional liquid chromatography · SDS PAGE · Membrane proteins · Sequence coverage · Protein expression

---

P. Thibault (✉)
Caprion Proteomics, Montréal, QC, Canada; Present address: Institute for Research in Immunology and Cancer, Université de Montréal, Montréal, QC, Canada
e-mail: pierre.thibault@umontreal.ca

A.R. Ivanov, A.V. Lazarev (eds.), *Sample Preparation in Biological Mass Spectrometry*, DOI 10.1007/978-94-007-0828-0_32,

32.1 Introduction

Membrane proteins are encoded by 20–30% of sequenced genomes and perform a wide range of cellular functions including membrane potential stabilization, signal transduction and cell volume regulation (Alberts, 2002). They are usually divided into two categories: "integral membrane proteins" (IMPs) anchored through the lipid bilayer by at least one transmembrane domain and "peripheral membrane proteins" that bind to the surface of the membranes through acyl chains (e.g. isoprenoid, myristoyl or glycosylphosphatidyl inositol tails) anchored in the bilayer or through noncovalent interactions with IMPs, phospholipids or other membrane-associated proteins. IMPs are involved in almost all aspects of cell biology including lipid synthesis, cell adhesion, and the exchange of matter, energy and information between the cell surface and the cytosol. Consequently, integral membrane proteins, in particular those spanning across the plasma membrane, are of great diagnostic and therapeutic values.

Peripheral membrane proteins can be conveniently solubilized in alkaline solvents (e.g. 0.1 M carbonate, pH 11) whereas lipid bilayer and IMPs require detergents (e.g. NP40, triton X-100, etc.) for proper solubilization, IMPs are amphiphatic, and are composed of hydrophilic and hydrophobic regions. They are difficult to solubilize in aqueous buffers, focused poorly in the first dimension (isoelectric focusing) of a 2D gel electrophoresis, tend to aggregate during the gel separation and their hydrophobic tryptic peptides are poorly extracted from the gel (Bunai et al., 2005). Those limitations have seriously compromised the usefulness of 2D-gel electrophoresis for the analysis of transmembrane proteins (Rabilloud, 2009; Santoni et al., 2000).

On the other hand, sodium dodecyl sulfate polyacrylamide gel electrophoresis (SDS-PAGE) combined with liquid chromatography-tandem mass spectrometry (1D gel-LC-MS/MS) and two-dimensional liquid-chromatography coupled to tandem mass spectrometry (2D LC-MS/MS) have shown greater promises for the analysis of membrane proteins (Fountoulakis et al., 2004; Heintz et al., 2009; Jiang et al., 2007; Mitra et al., 2009a, b, Natera et al., 2008; Romijn and Yates, 2008; Schirle et al., 2003; Speers and Wu, 2007). Both techniques provide significant advantages for the analysis of membrane proteins. SDS can improve the solubility of membrane proteins prior to their separation and analysis by 1D gel-LC-MS/MS, whereas the digestion of membrane extracts followed by 2D LC-MS/MS can release hydrophilic peptides from hydrophobic proteins irrespective of their isoelectric points (pI) or molecular weight, an ideal situation for lower abundance proteins like IMPs (Cordwell, 2006). In the present work, we compared the analytical merits of 1D gel-LC-MS/MS and 2D-LC-MS/MS approaches in terms of reproducibility, number of identified IMPs, sequence coverage, and application to protein expression profiling.

32.2 Materials and Methods

32.2.1 Cell Culture

U937 (human monocyte like, histiocytic lymphoma cells, form ATCC, Manassas, VA) were grown in RPMI-1640 formulation (Wisent, St Bruno, Canada), supplemented with 10% fetal calf serum (Wisent, St Bruno, Canada), 1% of Pen-Strep (Gibco-BRL, Grand Island, NY) at 37°C in a 5% CO_2 atmosphere. For cell differentiation experiments, freshly washed cells were seeded in 150 cm^2 flask at a density of 10^6 cells/mL in 100 mL of culture medium. Concentrated phorbol 12-myristate 13-acetate (PMA) stock solution (Sigma, St Louis, MO) in dimethyl-sulfoxide, DMSO (Sigma, St Louis, MO) was added directly into the flask to a final concentration of 25 μM. Cells were left in presence of the PMA for 48 h before the extraction of membrane proteins.

32.2.2 Extraction and Isolation of Membrane Proteins

After 48 h following differentiation of macrophages, cells (monocytes and macrophages) were washed in phosphate buffer saline, PBS (Wisent, St Bruno, Canada) before being gently detached (macrophages) or re-suspended (monocytes) into ice cold homogenization buffer (10 mM Tris-HCL (Sigma, St Louis, MO), 250 mM sucrose (Sigma, St Louis, MO), at pH 7.4 supplemented with 1X complete protease inhibitor tablet (Roche, Nutley, NJ). Cells were placed in a cavitation device (Parr Instrument Co, Molline, IL) and subjected to nitrogen gas pressurization at 1000 PSI for 15 min on ice followed by rapid pressure release. Cavitated cells were then collected through the exhaust valve of the chamber and centrifuged at 400×g, 4°C for 10 min to pellet down unbroken cells and nuclei. The post nuclear supernatant was then collected and further centrifuged at 100,000×g, 4°C for 60 min to pellet membrane proteins. The pellet was re-suspended in homogenization buffer (10 mM Tris-HCL, 250 mM sucrose, at pH 7.4). Protein concentration was measured using BCA protein assay kit reagents (Pierce Biotechnology Inc., Rockford, IL) and adjusted to 5 mg/ml with homogenization buffer before being aliquoted in small volumes (100 μL) and snap frozen in liquid nitrogen. Alternately, the membrane protein pellets were directly suspended in Tris-HCl 50 mM, sucrose 100 mM, $MgCl_2$ 5 mM pH 7.2 (Sigma, St Louis, MO). 10 units of Omnicleave DNAse (Epicentre, Madison, MI) were added to each sample and digestion of remaining DNA was carried out at 25°C for 1 h. Volumes were reduced to 50 μL in a Speed-Vac to which was added 900 μL of 100 mM sodium carbonate pH 11 (Sigma, St Louis, MO) before incubation at 4°C for 90 min under gentle mixing. Samples were centrifuged at 55,000 rpm (100,000×g) at 4°C for 90 min. Pellets were retrieved and reconstituted in 100 μL of 100 mM ammonium bicarbonate (Sigma, St Louis, MO) pH 8.0 and 900 μL of 0.55 M NaCl (Sigma, St Louis, MO). Samples were divided in two aliquots. One sample was separated on 1D-gel, bands were digested and analyzed by LC-MS. The other aliquot was digested in solution and analyzed by 2D-LC-MS.

32.2.3 In-Gel Digestion and LC Separation

Samples were diluted in the Laemmli/ammonium bicarbonate buffer solution (2:1) and boiled at 100°C for 5 min. After cooling at room temperature, the samples were loaded on a precast 12% SDS-PAGE (Biorad, Hercules, CA). Protein bands were reduced with 10 mM dithiothreitol (Sigma, St Louis, MO) in 50 mM ammonium bicarbonate (Sigma, St Louis, MO) at 56°C for 1 h and alkylated with 55 mM iodoacetamide (Sigma, St Louis, MO) in 50 mM ammonium bicarbonate at room temperature in the dark for 1 h. 24 bands were excised and dehydrated with acetonitrile, ACN (Fisher Scientific, Whitby, ON). All digestions were performed overnight using modified porcine trypsin (Promega, Madison, WI) in 50 mM ammonium bicarbonate, 2 mM $CaCl_2$ (Sigma, St Louis, MO), 0.1 M urea (Sigma, St Louis, MO), 10% ACN pH 8.1. Supernatants were collected and in-gel peptides were extracted with 90% ACN, 0.5 M urea. Peptide samples were dried in a Speed-Vac (Thermo Savant, Millford, MA), redissolved in 0.2% formic acid (EM Science, Mississauga, ON) 10% ACN and injected onto a 300-μm × 5-mm Waters C_{18} precolumn (Waters, Millford, MA). HPLC was carried out on a Waters Cap-LC system. Mobile phases were 0.2% formic acid (solvent A) and ACN with 0.2% formic acid (solvent B). Peptides were eluted and separated on an in-house 150-μm × 10-cm home-made C_{18} column (Jupiter 5 μm, 300 Å, Phenomenex, Torrance, CA) using a gradient of 10–60% (solvent B) in 56 min.

32.2.4 In-Solution Digestion and 2D-LC Separation

Membrane samples were dried and resuspended in 50 mM ammonium carbonate pH 8.1/methanol (50/50). Trypsin was added (enz/prot: 1/25) and samples were incubated overnight at 37°C. Samples were evaporated to dryness and reconstituted in formic acid 0.2%, ACN 5% at a concentration of 0.5 μg/μL. The 2D-LC system consists of a strong cation exchange (SCX) column 5 mm × 360-μm (polyLC, Columbia, MD), a 300-μm × 5-mm Waters C_{18} precolumn and a 150-μm ×10-cm in-house C_{18} analytical column (Jupiter 5 μm, 300 Å, Phenomenex, Torrance, CA) coupled to a CapLC nanoLC (Waters, Millford, MA). Peptides were eluted from the SCX column using sequential fractions of 0, 55, 65, 80, 100, 150, 200, 300, 400, 500, 750 and 1 M ammonium acetate, pH 3.5. After adsorption on the C_{18} precolumn, they were separated on the analytical column using a gradient of 5–60% (solvent B) in 56 min.

32.2.5 Mass Spectrometry

Electrospray mass spectra were recorded on a quadrupole time-of-flight mass spectrometer Q-TOF Ultima (Waters, Millford, MA). Calibration of the instrument was made by infusion of the peptide Glu-Fib B (Sigma, St Louis, MO) solution at concentration of 83 fmol/μL. To generate the highest quality MS/MS spectra, we used the following data-dependant acquisition parameters: survey scan (MS only) range *m/z* 400–1500, 1 s scan time, 1–3 precursor ions selected based on charge state (+2, +3, and +4). For each MS/MS scan, collisional activation of selected precursors was obtained using argon as a target gas, the *m/z* range was extended to *m/z* 50–2000, scan times used ranged from 1 to 4 s (signal dependent), and a charge state-dependent collision energy profile was used.

32.2.6 Creating Exclusion Lists

A script was developed to generate exclusion lists of all precursor ions (*m/z* and RT values from "auto.txt" file of MassLynx) from each LC-MS/MS analysis. This exclusion list was generated for all selected ions of previous MS/MS experiments and accumulated precursor ions of each respective salt fraction or gel band. An exclusion window of $\pm m/z$ 0.8 centered on the major isotopes with a time window of ± 1.5 min were used in all experiments. This iterative process was repeated at least three to five times, depending on fraction abundance and resulted in the generation of exclusion lists containing thousands of ions.

32.2.7 Peptide Detection and Clustering

A script was developed to convert the raw data files from the MassLynx software into text files representing all *m/z*, intensity and time values above a user-defined intensity threshold, typically 15 counts. These text files were then processed using in-house software enabling data reduction, peptide detection and alignment of peptide ion maps (*m/z*, retention time, intensity) for all samples (Kearney and Thibault, 2003). A list of unique peptide ions was obtained by clustering ions based on their respective charge, *m/z* and time within user-specified tolerances (typically $\pm$ 0.1 *m/z* and ± 0.15 min).

32.2.8 Database Searching

Database searches were performed against a non-redundant NCBI database using Mascot (Matrix Science, London, UK) selecting human species. Database searches allowed for a variable modification of Cys with iodoacetamide (in-gel digestion), variable oxidation of methionine, variable deamidation (Asp and Gln) and trypsin digest, up to two missed cleavages. Parent ion and fragment ion mass tolerances were both set at ± 0.1 Da. Peptide identification with a minimum score of 26 were selected and corresponded to a false discovery rate of <2%. The number of transmembrane domains was determined for each protein with the TMHMM software 2.0 (Center for Biological Sequence Analysis, Lyngby, Denmark).

32.2.9 Gene Ontology (GO) Terms

Uniprot accession numbers were retrieved from their corresponding NCBI accession number using the Uniprot website (www.uniprot.org). Approximately 30% of the accession numbers could not be retrieved from the website. Consequently, the Uniprot accession numbers were retrieved manually from the NCBI web page for each protein when possible. The website Babelomics v.3.1 (www.fatigo.org) was used to generate the GO term list for each analysis. When several GO terms were attributed to a protein, the GO term redundancy was removed by first prioritizing the GO term GO:0016021 integral to membrane and secondly the GO term GO:0016020 membrane. Each GO term category corresponds to a single GO term except for "membrane" which includes GO:0016020 (membrane), GO:0005792 (microsomal membrane) and GO:0005905 (coated pits) and "others" which contains all the remaining GO terms.

32.3 Results and Discussion

32.3.1 Sample Preparation

Preparation of an enriched membrane fraction prior to mass spectrometry analysis is essential to reduce sample complexity of the U937 cell extract. Sub-cellular fractionation by sucrose gradient ultracentrifugation has been favored over that of colloidal silica coating (Rahbar and Fenselau, 2004; Rahbar and Fenselau, 2005) or aqueous-polymer two-phase partitioning (Schindler and Nothwang, 2006; Schindler and Nothwang, 2009). Colloidal silica requires optimization of bead binding and lysis protocols that can vary significantly according to the cell types examined. The aqueous-polymer two-phase partitioning uses polyethylene glycol (PEG). PEG can introduce undesired contaminants that can significantly compromise the chromatographic adsorption and the sensitivity of the LC-MS/MS system.

IMPs were enriched using high pH and high salt extraction that convert sealed vesicles into flat membrane sheets, thereby solubilizing peripheral membrane proteins while IMPs can be reconstituted in detergent buffers (Fischer et al., 2006; Pasini et al., 2006). To reduce sample loss, lipids were removed by ultracentrifugation at 4°C for 90 min instead of using methanol/chloroform precipitation. Samples were divided in two aliquots. One aliquot was separated by SDS-PAGE and the entire gel was cut in 20 bands each subsequently digested in an ammonium bicarbonate/urea buffer prior to LC-MS analysis. In contrast to the 1D-gel-LC-MS approach, detergents cannot be used in the sample preparation for 2D-LC since anionic detergents like SDS create ionic pair with the positively charged peptides and impair their adsorption on the SCX column. The SCX column and C_{18} pre-column are on-line during sample loading to maximize the capture on the C_{18} pre-column of peptides not retained on the SCX column. Consequently, any negatively charged hydrophobic detergent could bind to the C_{18} pre-column preventing proper peptide adsorption. Previous experiments have shown that detergents bind quite strongly to C_{18} and require extensive washes for removal to reduce the interference of SDS on electrospray ionization (Vissers et al., 1996). MS compatible detergents have been introduced a few years ago (Norris et al., 2003; Yu et al., 2004). They have a labile ketal functional group between the hydrophilic head group and the hydrophobic tail that can be cleaved upon hydrolysis. The hydrophobic functionality is not soluble in water. We reasoned that some of the hydrophobic peptides might be solubilized in the detergent hydrophobic tail-rich phase precluding them from being analyzed together with other more hydrophilic peptides. Aqueous-organic solvent systems (ACN-water, methanol-water) are an alternative to detergents for facilitating protein digestion. Unlike most detergents, organic-aqueous solutions can be easily removed by evaporation, reducing sample handling and potential loss. We performed several tests by varying the methanol/water ratio and found that buffers containing methanol/water 1/1 provided higher digestion yields (data not shown).

32.3.2 Comparison of Analytical Approaches for the Analysis of Membrane Protein Extracts

In a preliminary experiment we analyzed a 5 μg membrane protein extract from U937 monocyte cells using 1D gel-LC-MS/MS and 2D LC-MS/MS. Peptide elution from the SCX column was optimized to 12 fractions whereas the 1D-gel was cut into 24 bands to give the higher number of identifications in both cases (data not shown). An outline of the procedure and results obtained for both approaches is shown in Fig. 32.1. Although the number of assigned MS/MS spectra is similar between these analyses (576 and 623), the number of identified proteins is

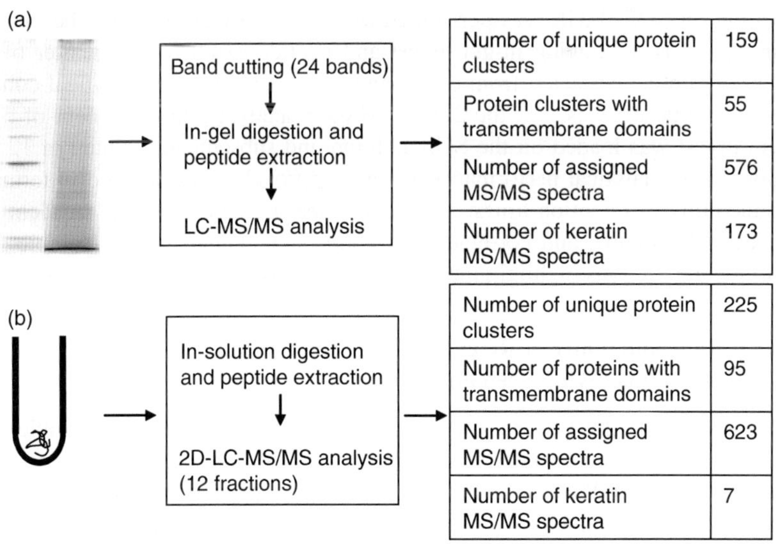

Fig. 32.1 Analysis of U937 human monocyte membrane extract using (**a**) SDS-PAGE protein separation (5 μg), band excision, in-gel trypsin digestion followed by LC-MS/MS analysis of each band and (**b**) in-solution tryptic digestion of 5 mg of membrane with separation of tryptic peptides by 2D-LC-MS/MS. The number of proteins identified by each approach is shown on the corresponding tables

significantly higher in 2D-LC-MS/MS (225 proteins) compared to 1D-gel-LC-MS/MS (159 proteins). The number of identified integral membrane proteins (IMPs) with transmembrane domains is correspondingly more significant in the 2D-LC-MS/MS experiment (95 proteins) compared to 1D-gel-LC-MS/MS (55 proteins). Amongst IMPs uniquely identified in the 2D-LC-MS/MS approach were glycoproteins such as CD98 or Niemann-Pick C1 protein that are glycosylated and are not typically well resolved on SDS-PAGE gels. Also, the in-gel digestion can lead to additional contaminants such as human keratins that can further complicate the LC-MS/MS analyses. This was clearly evidenced when comparing the number of keratin MS/MS spectra identified from both approaches (7 of 623 identified peptides for the 2D-LC-MS/MS vs 159 of 576 identified peptides for the 1D-gel-LC-MS/MS). The fully automated process of the 2D-LC-MS/MS analysis reduces the possibility of keratin contamination. Comparison of the results obtained for 5 μg of monocyte membrane proteins suggests that 2D-LC-MS/MS approach yields a higher number of IMPs for a relatively small amount of protein extract.

32.3.3 MS-Based Proteomics Approaches for the Analysis of Differentiated Cell Model Systems

In order to improve the number of identified IMPs from membrane extracts, subsequent experiments were performed using 20 μg of monocyte and macrophage membrane extracts. The increased amounts of protein extracts required optimization of separation procedures to address the higher sample complexity. Protein separation on SDS-PAGE was performed on a 10 × 0.1 cm gel and the gel was cut in 24 bands. Following in-gel digest and peptide extraction, a quarter of the amount of peptide extracted from each band was loaded on the analytical column for LC-MS/MS analysis. For the 2D-LC-MS/MS analysis, an equivalent of 5 μg protein digest was loaded on the SCX column and subsequently eluted in 12 salt fractions. The analytical performance of each approach was evaluated in terms of reproducibility of retention times, abundance measurements, number of identified peptides and proteins, and correlation of abundance changes across sample sets. The following sections highlight the relevant figures of merit for each approach.

32.3.3.1 Reproducibility of Retention Times and Peptide Abundance Measurements

Both methods (i.e. 2D-LC-MS and 1D-gel-LC-MS) have to be reproducible with respect to retention time to correlate peptide ions across sample replicates and to enhance protein identification in iterative exclusion experiments (Section 32.3.3.2). In addition, the reproducibility of peptide abundance is also important to obtain statistically relevant differences across sample replicates and conditions. The reproducibility of peptide retention times and intensities is presented in Fig. 32.2 for representative analyses from a 2D-LC salt fraction and from the LC-MS analysis of a 1D-gel band. Relative standard deviation for matched peptide retention times correlation between replicate runs is excellent (Average RSD of 1% for 2D LC and

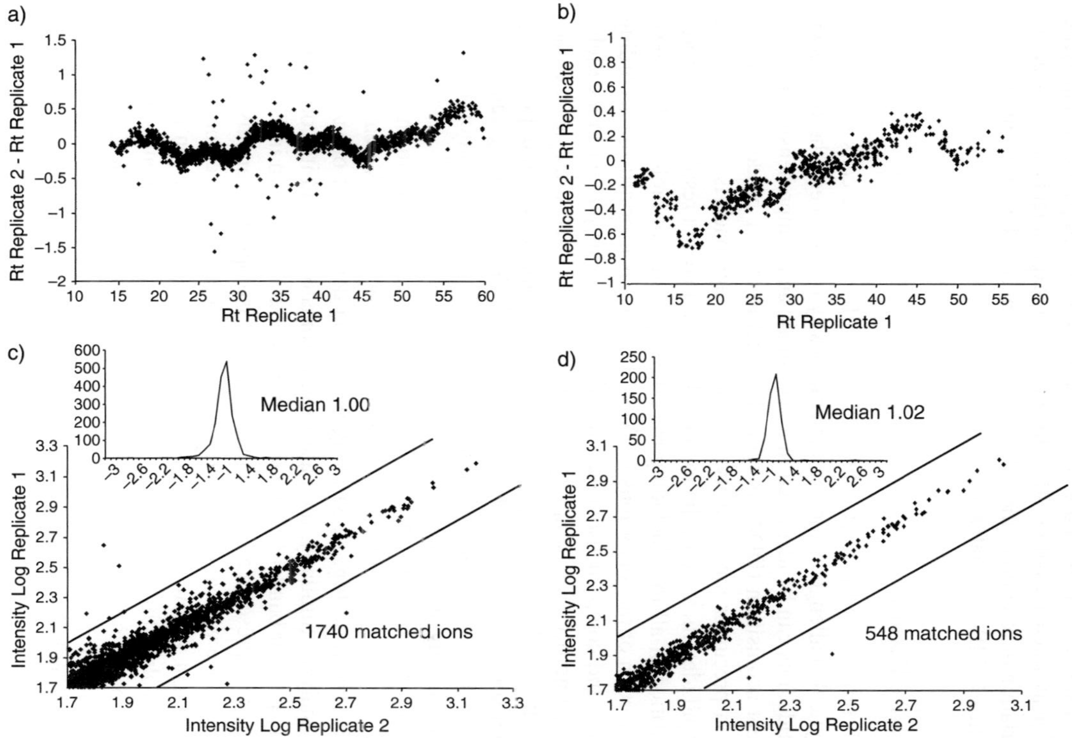

Fig. 32.2 Reproducibility of retention time and peak intensity. Variation of retention time for matched peptides obtained from (**a**) 50 mM ammonium acetate salt fraction of 2D-LC-MS/MS analysis and (**b**) in-gel digest of band 9 from 1D-gel-LC-MS/MS. Log-log plots of peak intensities for two replicates of (**c**) 50 mM ammonium acetate salt fraction from 2D-LC-MS/MS and of (**d**) band 12 from 1D-gel-LC-MS/MS of U937 control monocytes membrane extract

0.58% for 1D-gel) for both methods with a Rt variation of less than ±1 min for more than 95% of the peptides. Figures 32.2c,d show log-log plots for peak intensities of matched peptides for these two approaches. As evidenced from the scatter plot of the 1D-gel-LC-MS analyses, more than 95% of the matched ions showed less than 20% of peak intensity variation for consecutives analyses. The on-line 2D-LC-MS showed slightly larger values with 95% of the matched peptides displaying a variation of peak intensity below 30%. In subsequent experiments we considered maximum variation of retention times of ±1.5 min when correlating retention times across different LC-MS/MS analyses.

32.3.3.2 Iterative Exclusions

The study of complex cell extracts obtained from differentiated cells represent a sizeable analytical challenge as the sample complexity often exceed the frequency at which peptide ions can be identified in a systematic and comprehensive manner. Data-dependent acquisition (DDA) is typically used to trigger product ion acquisition based on precursor ion intensity. Although this approach is convenient for obtaining a rapid identification of the ion population, it may not result in a comprehensive acquisition of all ions including trace-level peptides of biological interest. Undirected selection of precursor ions for MS-MS analysis is prioritized according to ion intensity, and comprehensive identification of proteins present in a cellular extract requires improved ion sampling strategies. One of these strategies includes the use of iterative exclusion to collect peptide ions without resampling previously acquired precursor ions. The threshold for MS to MS/MS switching can be set according to the distribution of ion intensities with increasing acquisition time for precursor of decreasing abundance.

A typical base peak 2D-LC-MS chromatogram corresponding to the 50 mM ammonium acetate fraction of the membrane extract from the PMA-treated monocyte sample is shown on Fig. 32.3a along with a ion density plot of *m/z* vs time for the same analysis. Similarly, the base peak chromatogram and its corresponding ion density plot for band 12 of the same protein extract is shown in Fig. 32.3b. The ion density plot presents a peptide map where all isotopic ions are displayed in the *m/z* vs time plane with a logarithmic intensity scale showing a color transition from black to bright yellow for progressively more intense peptide ion intensities. Peptide detection algorithms have been developed to identify the monoisotopic ion of each peptide based on their corresponding isotopic ion distribution. This complex LC-MS data set can then be reduced to a peptide map listing all ions according to their mass (*m/z* and charge), retention time and abundance values (Kearney and Thibault, 2003). For example, the peptide maps for the LC-MS analyses shown in Figs. 32.3a,b contain 1962 and 739 multiply charged peptide ions, respectively. The number of peptide ions detected in a given peptide map run often exceeds the total number of MS/MS that can be acquired on a Q-TOF instrument. A usual duty cycle consists of the acquisition of an MS survey scan of 1 s followed by the MS/MS acquisition of the three most abundant multiply charged precursors. Each MS/MS scan is typically acquired for 2–3 s to obtain good quality MS/MS spectra resulting

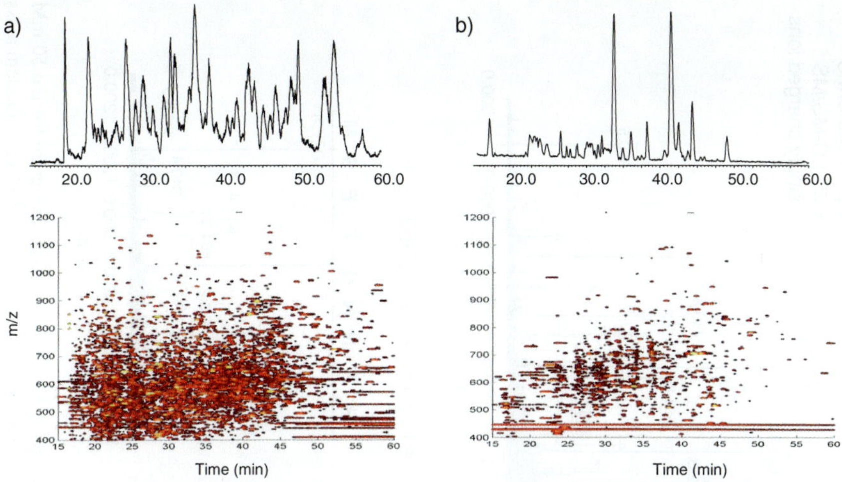

Charge	# peptides
2	1834
3	124
4	4
Total	1962

Charge	# peptides
2	539
3	148
4	52
Total	739

Fig. 32.3 Elution profiles of peptides detected in 2D-LC-MS and 1D-gel-LC-MS analysis of membrane proteins from U937 human monocytes stimulated with PMA. (**a**) A 2D-LC-MS base peak chromatogram and a ion density plot (*m/z* vs time) for the 50 mM ammonium acetate fraction. A total of 5 μg of membrane digest was loaded on the SCX column. Peptide elution was obtained using 12 salt fractions of increasing ionic strength. (**b**) A base peak chromatogram and a ion density plot for band 12. A total of 20 μg were loaded on the gel and 24 bands were excised, in-gel trypsin digested prior to 1D-gel-LC-MS analysis. For each band, peptide extracts were reconstituted in 32 μL, and 7.5 μL was injected for each analysis

in a maximum number of 1028 MS/MS spectra for an elution window of 40 min. To circumvent this under-sampling problem and to maximize the number of acquired MS/MS spectra, we performed iterative exclusion runs to prevent the reacquisition of ions sequenced by MS/MS in preceding analyses (Bendall et al., 2009; Hui et al., 2003). Unless otherwise indicated, a total of three iterative exclusion runs were performed for each sample.

For 2D-LC-MS/MS analyses a total of 5 μg of tryptic digest was loaded on the SCX precolumn and 12 salt fractions were injected for each sample replicate (n=3). For 1D-band analysis by LC-MS/MS (1D-gel-LC-MS/MS), a quarter of the band digest volume was injected and the peptides were separated and sequenced by tandem mass spectrometry. An exclusion list was automatically generated after each LC-MS/MS run according to the script described in Section 32.2.6. An example of the application of iterative exclusion to enhance peptide identification is shown in Fig. 32.4 for the 50 mM ammonium acetate fraction of the 2D-LC-MS.MS analyses of a membrane extract from PMA treated monocytes. The complexity of the peptide distribution can be visualized from the base peak chromatogram (Fig. 32.4a). For

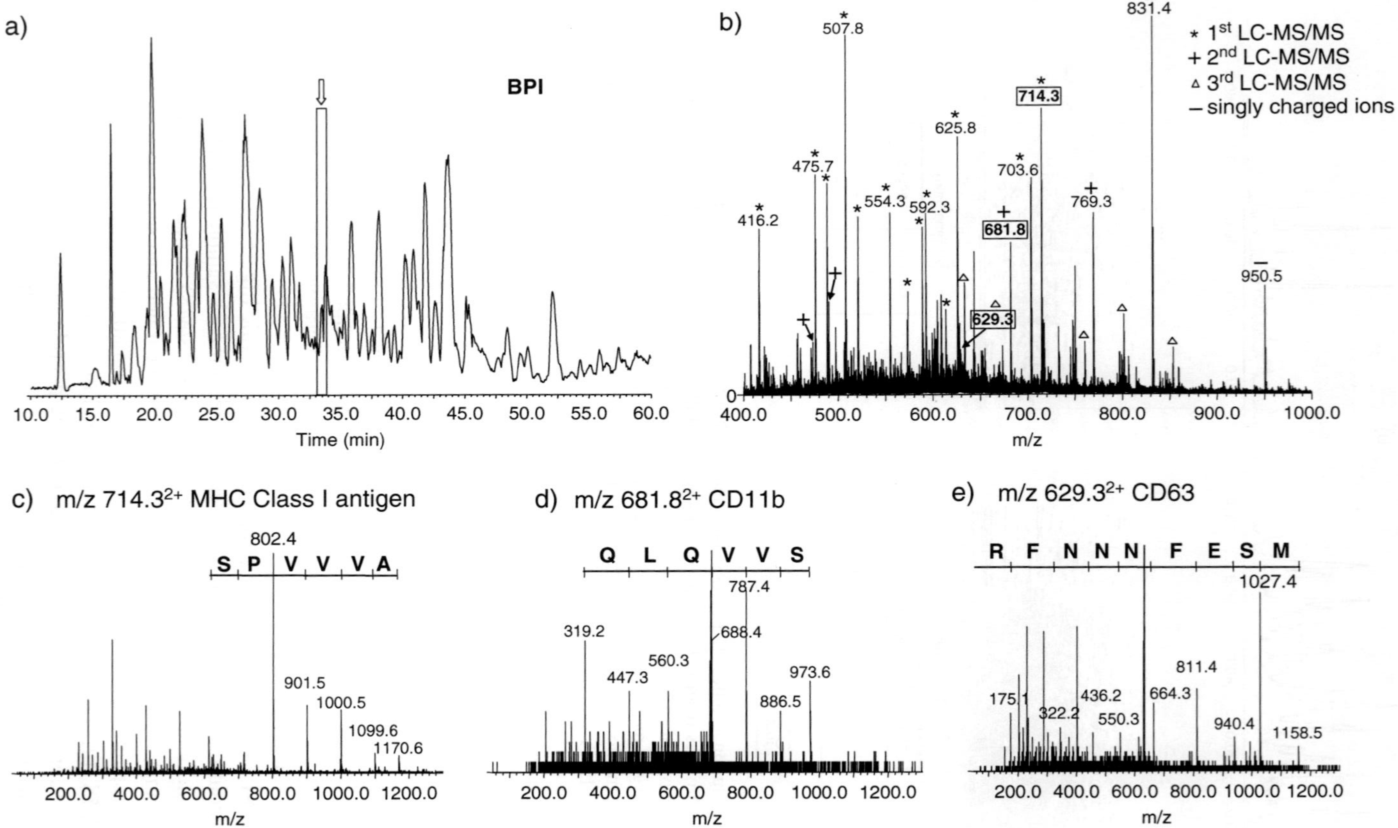

Fig. 32.4 Enhancement of peptide identification using 2D-LC-MS/MS analysis with iterative exclusion. (**a**) A base peak ion chromatogram for the 50 mM ammonium acetate salt fraction of the membrane extract from PMA-treated monocytes. (**b**) An integrated mass spectrum across 33 and 34 min. Tandem mass spectra of doubly-protonated precursor ions (**c**) *m/z* 714.3 corresponding to WAAVVVPSGEEQR of MHC class I antigen, (**d**) *m/z* 681.8 corresponding to GFGQSVVQLQGSR from CD11b and (**e**) *m/z* 629.3 corresponding to VMSEFNNNFR from CD63

convenience, the average mass spectrum of a 1.0 min time segment for peptide eluting at 33 min is shown in Fig. 32.4b. As observed in Fig. 32.4b, the most abundant ions were selected for MS/MS sequencing during the first LC-MS/MS run and enabled the identification of abundant trans-membrane proteins such as the MHC class I antigen (Fig. 32.4c). Subsequent iterative exclusion runs facilitated the identification of co-eluting peptides from lower abundance proteins such as CD11b (Fig. 32.4d) and CD63 (Fig. 32.4e). The overall efficiency of iterative exclusion to enhance peptide identification is presented in Fig. 32.5 for the number of identified peptides for both the 1D-gel-LC-MS/MS and 2D-LC-MS/MS approaches. Figures 32.5a,b show the progression in the number of unique peptides identified for each 2D-LC-MS/MS runs obtained for the analyses of control and PMA-treated U937 monocyte cells, respectively. The use of iterative exclusion enabled the targeted acquisition of new peptide ions in replicate runs. For example, we observed a gain of 37 and 10% in the acquisition of new peptide ions in the second and third 2D-LC-MS/MS runs of the control cells (Fig. 32.5a). The enhancement in peptide identification was also correlated with an increase in the number of identified proteins from 385 to 583 between the first and third runs. A similar gain in the number of newly identified peptides was also observed for the 1D-gel-LC-MS/MS approach where we obtained an overall increase of 40 and 8% in the second and third runs for all bands of the control monocyte cells (Fig. 32.5b). Generally, the proportional increase in newly identified peptides decreased with the number of exclusion runs varying from 56, 31 and 13% for the first, second and third runs, respectively. Consistent with previous reports, we observed diminishing returns in the cumulative

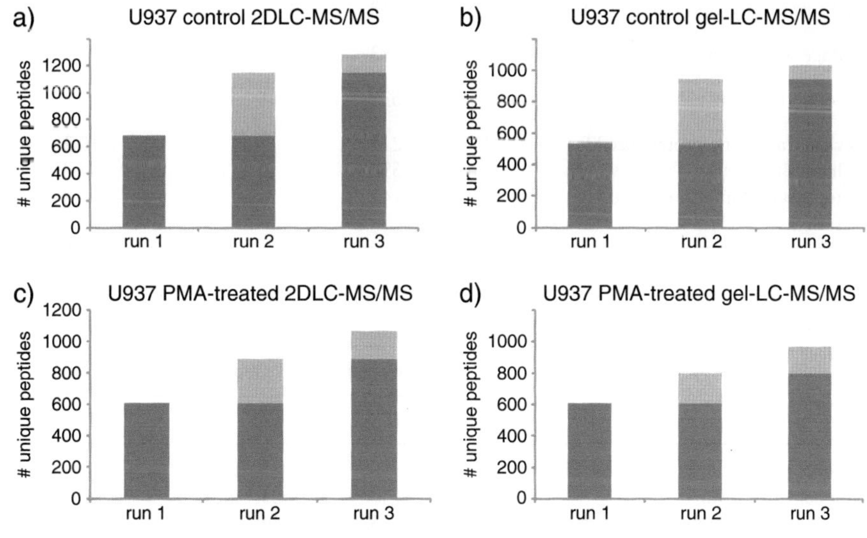

Fig. 32.5 Overall distribution of identified peptides using iterative exclusion with 2D-LC-MS/MS (**a** and **b**) and 1D-gel-LC-MS/MS (**c** and **d**). Bar charts show the increasing number of identified peptides (*black*: cumulative number of identified peptides; *grey*: unique peptides identified in a specified run). Membrane proteins from control monocytes (**a** and **c**) and PMA-treated monocytes (**b** and **d**)

number of unique peptides beyond the third exclusion run (Bendall et al., 2009; Hui et al., 2003). The use of iterative exclusion not only facilitated the identification of lower abundance proteins, but also enhanced sequence coverage (Table 32.1). For example, the sequence coverage of Ribophorin I and solute carrier 25 increased from 11% to ~30%. For convenience, the number of MS/MS acquired and assigned for all peptides and proteins identified by the two approaches is summarized in Table 32.2. As observed in Table 32.2, the numbers of multiply charged peptide ions is comparable in all experiments while approximately one third of the acquired MS/MS (with a peptide signature) were matched to their cognate proteins following a database search. However, only 6–8% of all acquired MS/MS spectra resulted in a unique assignment when accounting for redundant acquisition and spectra of low quality. Interestingly, the proportion of unique assigned MS/MS spectra was generally lower for the 1D-gel-LC-MS/MS (~10%) compared to the 2D-LC-MS/MS (~12%) approach. This variation was partly attributed to the larger band broadening

Table 32.1 Increase of sequence coverage with iterative 2D LC-MS/MS analyses

Proteins	Accession #	# AAs	Run 1 (%)	Run 2 (%)	Run 3 (%)
Solute carrier family 25 member 4	4502099	298	10	27	49.6
Na^+/K^+ ATPase	21361181	1023	5.6	9.6	15.3
MHC Class I antigen	14029717	273	16	16	23.8
ATP synthase H+ transporting mitochondrial F1 complex	32189394	529	15.1	22.3	26.9
Cell surface glycoprotein CD11b	72063	1153	5.3	9.3	11.2
Voltage dependent anion channel 1	4507879	283	0	12.4	14.5
Ribophorin I	4506675	631	8.4	13.6	17
Vesicle trafficking protein sec22b	4759086	215	4.6	4.6	10.2
Calnexin	10716563	592	17.6	33.2	33.2
Mitochondrial import receptor Tom22	9910382	142	0	0	33.8

Table 32.2 Comparison of 1D-gel separation followed by 1D-gel-LC-MS/MS and 2D-LC-MS/MS for the analysis of 20 μg of membrane extracts

	Control monocytes		PMA-treated monocytes	
	1D-gel	2D-LC	1D-gel	2D-LC
Number of total multiply charged ions	11869	12913	14248	13999
Number of MS/MS	14269	15504	14943	19009
Number of unique multiply charged ions	8390	10914	7716	11681
Number of MS/MS with a peptide signature	11318	10648	8203	9595
Number of assigned MS/MS	3446	4162	2210	3504
Number of unique assigned peptides	1034	1283	967	1066
Number of unique protein clusters	386	583	397	455
Number of peptides/protein	2.7	2.2	2.4	2.3

observed in the SDS-PAGE experiment where protein microheterogeneity (e.g. isoforms and glycoforms) (Kuster et al., 2001) can give rise to a broader spread of tryptic peptides than ion exchange chromatography. These results are also in agreement with those shown in Fig. 32.1 where a higher number of identified proteins were observed for the 2D-LC-MS/MS approach. Overall, the use of iterative exclusion for either MS-based proteomics approaches enabled to maximize protein identification while minimizing sample consumption.

32.3.3.3 Distribution of Membrane and Membrane-Associated Proteins

The comparison of the numbers of proteins and associated Gene Ontology (GO) terms obtained for both 1D-Gel-LC-MS/MS and 2D-LC-MS/MS is shown in Fig. 32.6. For each sample, more than 50% were membrane proteins with more

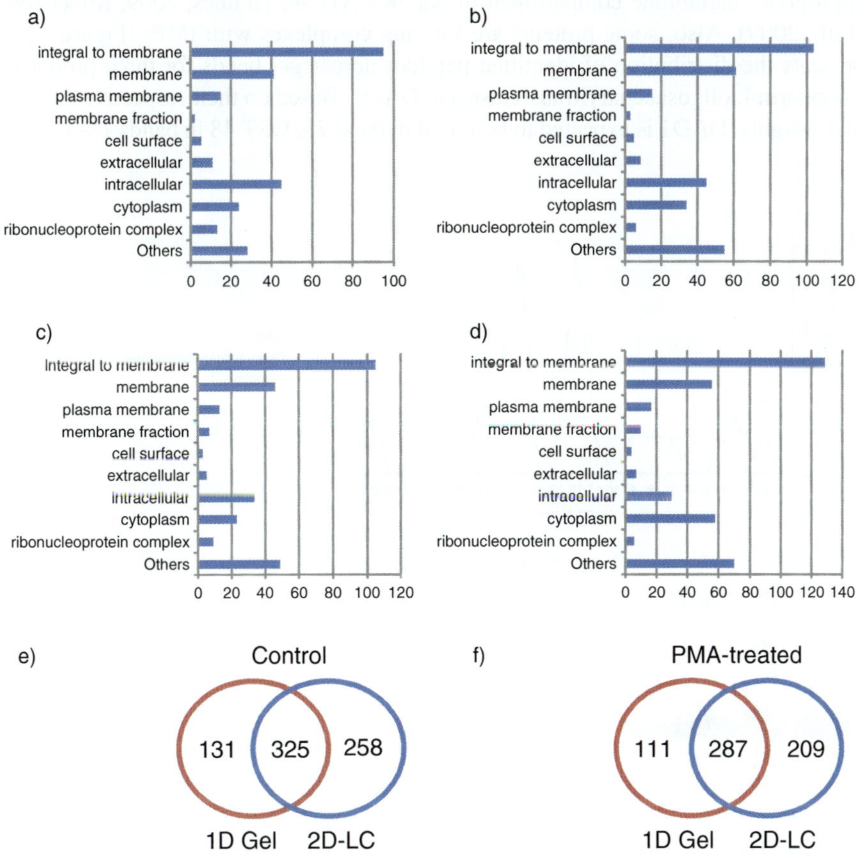

Fig. 32.6 Gene ontology (GO) terms of protein distribution for membrane protein extracts analyzed by 2D-LC-MS/MS and 1D-gel-LC-MS/MS. Bar charts representing the number of proteins identified for each GO term for control (**a** and **b**) and PMA-treated (**c** and **d**) cells analyzed by 1D-gel-LC-MS/MS (**a** and **c**) and 2D-LC-MS/MS (**b** and **d**). Venn diagrams comparing the number of identified proteins for control (**e**) and PMA-treated (**f**) monocyte cells

than 30% corresponding to IMPs. The relative protein distribution is comparable for both approaches, but shows some differences for the two sample types (a vs c and b vs d). Interestingly, the number of intracellular proteins representing approximately 10% of all identified proteins is increased from 35 to 44 when comparing control to PMA-treated monocytes cells. Among these are Rab proteins including Rab 10b and Rab 11 that were only identified following PMA treatment. The main membrane-associated proteins are ribosomal proteins involved in protein translocation across membranes (Kramer et al., 2009), Rab and Ras—related proteins involved in the process that underlie the targeting and fusion of transport vesicles with their appropriate acceptor membranes (Omerovic and Prior, 2009; Wickner and Schekman, 2008) and cytoskeletal proteins. While cytoskeletal proteins are directly linked to the membrane of the organelles to give the cell its shape, specific proteins of that class are known to interact with transmembrane proteins. For example, actin and spectrin identified in this study are involved in the activation and targeting to the appropriate membrane compartment of Na^+/K^+ ATPase (Baines, 2009; Robertson et al., 2009). Also, some proteins are forming complexes with IMPs. Figure 32.7 presents the distribution of identified peptides across gel bands for three proteins: ribophorin I, oligosaccharyltransferase and DAD1. Based on their respective molecular weights, DAD1 is expected to be found in band 23, OST 48 in bands 15–17 and

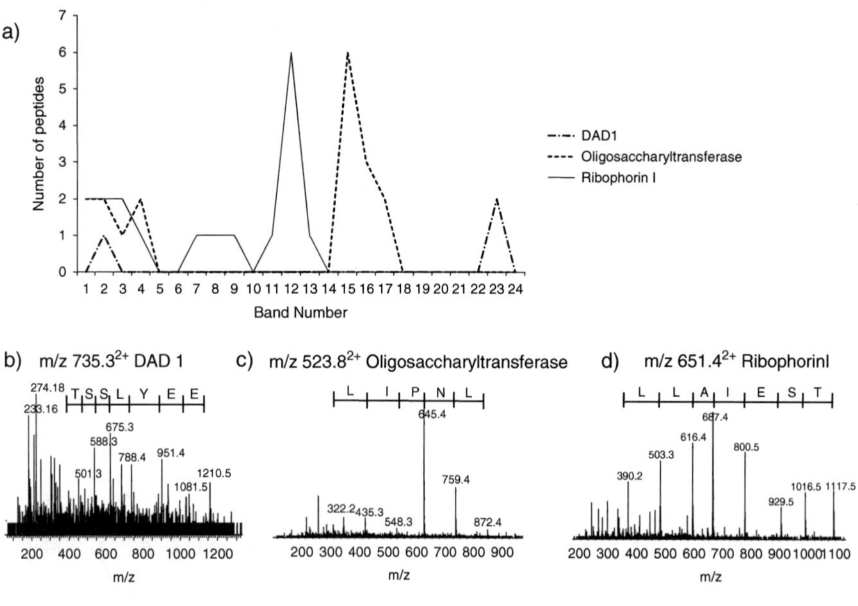

Fig. 32.7 An abundance profile of DAD1, oligosaccharyltransferase and ribophorin according to the gel band number of membrane extract from control U937 human monocytes. (**a**) Reconstructed electropherograms showing the number of peptides identified for each proteins Tandem mass spectra obtained in band 2 for precursors at (**b**) *m/z* 735.3 corresponding to FLEEYLSSTPQR from DAD1 (**c**) *m/z* 523.8 corresponding to SSLNPILFR from oligosaccharyltransferase and (**d**) *m/z* 651.4 corresponding to ALTSEIALLQSR from ribophorin I

ribophorin I in bands 11–13. Ribophorin is glycosylated at amino-acid position 299 and could migrate at a higher molecular weight position (bands 7–9). However, the distribution of these proteins according to band position (Fig. 32.7a) revealed interesting profiles. For example, all three proteins were found in band 2 as confirmed by the three MS/MS spectra (Figs. 32.7b–d) suggesting their presence as a protein complex of a larger molecular weight. Indeed, these proteins are known to form a complex catalyzing the N-linked glycosylation process of protein in the endoplasmic reticulum (Kelleher and Gilmore, 2006). These three proteins were also identified by 2D-LC-MS/MS although the protein complex was digested upon trypsin proteolysis.

Both MS-based proteomics approaches provided comparable results in terms of membrane protein distribution where a sizable proportion of identified proteins overlapped as shown in the Venn diagram of Figs. 32.6e–f. For example, 58 and 71% of proteins identified by 1D-gel-LC-MS/MS were identified by 2D-LC-MS/MS leaving 131 and 111 proteins identified by 1D-gel-LC-MS/MS only in control and PMA-treated monocytes, respectively. In contrast, 258 and 209 proteins were identified by 2D-LC-MS/MS only in the control and PMA-treated monocyte samples, respectively. A comparison of sequence coverage obtained for both approaches is presented in Table 32.3 for several transmembrane proteins. Clearly, a higher number of identified peptides was obtained when using in-solution digestion including proteins comprising several transmembrane domains like solute carrier family 2 member 3, oligosaccharyl-transferase STT3, and nicotinamide nucleotide

Table 32.3 Comparison of the number of identified peptides for transmembrane proteins between 1D-gel-LC-MS/MS and 2D-LC-MS/MS

		Control monocytes		PMA-treated monocytes	
Proteins with transmembrane domains	Number of TMDs	1D-gel	2D-LC	1D-gel	2D-LC
Oligosaccharyl transferase STT3 subunit	13	1	3	0	3
Nicotinamide nucleotide transhydrogenase	12	0	1	0	3
Niemann-Pick disease C1 type	12	3	3	5	5
Solute carrier family 2, member 3	11	2	4	3	4
Potential phospholipid-transporting ATPase VB	9	0	0	0	1
Vacuolar proton translocating ATPase	8	0	0	3	0
Na^+/K^+ ATPase alpha	10	10	12	7	11
CD97 antigen	7	0	0	0	2
70-kd peroxisomal membrane protein	5	0	2	0	0
CD63 antigen	4	1	2	2	2
Ribophorin II	4	2	3	1	0
Sideroflexin	3	6	3	4	5
ADP/ATP carrier protein	3	10	12	10	9
EMMPRIN	2	1	3	3	2
4F2 heavy chain antigen (CD98)	1	7	10	7	8
Calnexin	1	13	16	3	11

transhydrogenase. It is noteworthy that the abundant proteins Na^+/K^+ ATPase baring ten transmembrane domains was identified by 11 peptides using 2D-LC-MS/MS compared to only 7 peptides for the 1D-gel-LC-MS/MS strategy. The ability to perform in solution digestion in an organic solvent may greatly enhance the digestion of IMPs.

Furthermore, these results suggest that the 2D-LC-MS/MS approach yielded a higher number of identified membrane proteins possibly due to the higher digestion yield and peptide recovery obtainable with in solution digestion.

32.3.3.4 Quantitative Proteomics Using a Label-Free Approach

The U937 monocytes undergo cellular differentiation to macrophage upon exposure to PMA. This differentiation process is playing an important role in inflammation conditions, autoimmune diseases and in the formation of atherosclerotic plaques (Finstad et al., 1998). Quantitative profiling based on the incorporation of heavy atoms at the protein level (ICAT) or at the peptide level (iTRAQ, SILAC) has been implemented for membrane protein analysis (Bisle et al., 2006; Guo et al., 2007; McClatchy et al., 2007). One of the limitations of chemical derivatization approaches is that their relatively limited dynamic range which may not be sufficient to cover expression levels spreading over a few orders of magnitude (Chen and Pramanik, 2009).

Label-free quantitative proteomics (Bridges et al., 2007; Brusniak et al., 2008; Kearney and Thibault, 2003; Mueller et al., 2008; Old et al., 2005; Ono et al., 2006) is relatively simple and inexpensive to implement and minimizes side reactions that can be observed with isotope labeling approaches. However, regular and rigorous quality control must be performed to ensure reproducible chromatography and mass spectrometry conditions. In this study, we performed quantitative proteomics using a label-free approach and compared the changes in protein abundance of monocytes upon PMA treatment. An example of this is shown in Fig. 32.8 for CD97 and CD71, two plasma membrane proteins that were down-regulated upon PMA treatment. For convenience, Table 32.4 compares the changes of abundance of randomly selected membrane proteins. Intensity measurements were summed for peptides eluting in more than one salt fraction or being extracted from more than one band. For example, Lamp-2 and CD98 were up-regulated by a factor of 2 and 3 upon PMA treatment. The Na^+/K^+ ATPase plasma membrane protein did not show any meaningful change in abundance whereas sideroflexin was down-regulated two-fold following PMA incubation. It is noteworthy that both methods showed very comparative results for fold change measurements with less than 18% when comparing different peptide intensity ratios.

Earlier reports on CD98 and Lamp-2 proteins also concurred with the change of protein abundance observed in this study. In human peripheral blood mononuclear cells, both Lamp-1 and 2 exhibit a steep increase in their cell surface expression under phorbol ester differentiation (Kannan et al., 1996). Similarly, Sahraoui et al. reported an increased expression of 4F2 antigen (CD98) on acute lymphoblastic leukemia T cells upon stimulation by PMA treatment (Sahraoui et al., 1992). The 4F2 cell surface expression was also reported to be up-regulated when monocytes

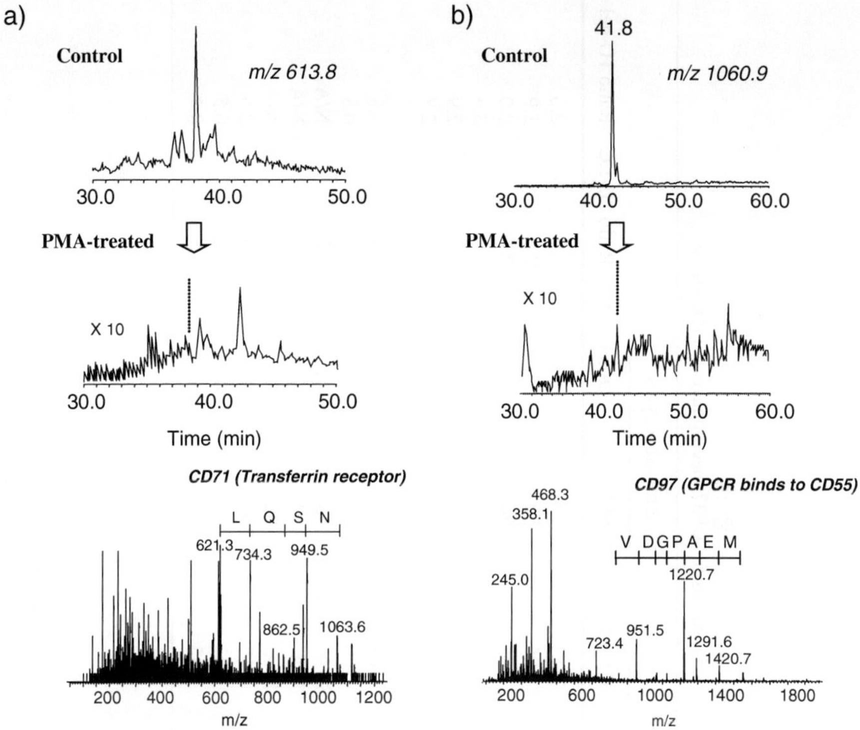

Fig. 32.8 Quantitative proteomics (label-free) analysis of membrane proteins from control and PMA-treated U937 human monocytes using 2D-LC-MS/MS (100 mM ammonium acetate fraction). (**a**) A reconstructed ion chromatogram for the doubly-protonated ion *m/z* 613.8 in control and PMA-treated samples along with the MS/MS spectrum corresponding to peptide YNSQLLSFVR from CD71 transferrin receptor. (**b**) A reconstructed ion chromatogram for the doubly-protonated ion *m/z* 1060.9 in control and PMA-treated samples along with the MS/MS spectrum corresponding to peptide LVDELMEAPGDVEALAPPVR from CD97

mature to dentritic cells under granulocyte-macrophage colony-stimulating factor (GM-CSF) treatment (Woodhead et al., 1998).

Phorbol esters are known to induce expression of interleukin-2 which in turn up-regulates Na^+/K^+ ATPase (Marakhova et al., 2005). Very little is known about sideroflexin although it plays a role in intracellular iron trafficking to mitochondria (Zheng et al., 2003). An earlier report by Sahraoui et al. indicated that the transferrin receptor CD71, another key protein in iron metabolism, was down-regulated under PMA stimulation. Both proteins might work coercively in regulating biological process relating to iron mobilization (Sahraoui et al., 1992).

32.4 Conclusions

The 2D-LC-MS/MS approach has become an increasingly popular alternative to gel electrophoresis for large-scale proteomic analyses. In the present work, we compared this approach to gel electrophoresis (SDS-PAGE) combined with

Table 32.4 Comparative profiling expression for selected proteins for U937 macrophage and monocyte

			PMA-treated		Control monocytes			
Proteins	Peptide *m/z*	Peptide sequence	2D-LC	1D-gel	2D-LC	1D-gel	Ratio 2D-LC	Ratio 1D-gel
LAMP-2	494.3	IPLNDLFR	3490	2213	1819	1091	1.9	2.0
	551.3	YLDFVFAVK	117	96	50	51	2.3	1.9
	656.9	GILTVDELLAIR	843	105	354	53	2.4	2.0
CD98	686.4	LLTSFLPAQLLR	386	72	115	21	3.3	3.4
	753.4	GQSEDPGSLLSLFR	842	127	311	44	2.7	2.9
	626.8	EDFDSLLQSAK	423	128	136	46	3.1	2.8
Na^+/K^+ ATPase	915.4	GVGIISEGNETVEDIAAR	598	46	464	53	1.3	0.8
	618.9	LNIPVSQVNPR	220	108	218	142	1.0	0.8
	616.3	AVAGDASESALLK	477	200	420	252	1.3	0.8
CD71	522.6	VEYHFLSPYVSPK	78	N/A	168	N/A	0.5	N/A
	602.8	LLNENSYVPR	212	N/A	350	N/A	0.6	N/A
	781.4	LAVDEEENADNNTK	47	N/A	102	N/A	0.5	N/A
Sideroflexin-1	790.4	SSMSVTSLEAELQAK	408	112	762	208	0.5	0.5
	635.8	VGIPVTDENGNR	524	227	933	486	0.6	0.5
	750.9	NILLTNEQLESAR	389	335	820	499	0.5	0.7

LC-MS/MS for the analysis of membrane proteins. The challenge lay in the intrinsic nature of this class of hydrophobic and often glycosylated proteins. These proteins are generally poorly resolved and their distribution can spread over several consecutive bands due to microheterogeneity associated to N- and O-linked glycosylation. In this study, we analyzed membranes proteins from both control and PMA-treated monocyte cells using a relatively small amount of protein extracts (5–20 μg). For the 20 μg analysis, 583 proteins were identified in the 2D-LC-MS/MS analysis of monocyte membrane extract compared to 456 in the 1D-gel-LC-MS/MS approach. Both MS-based proteomics strategies enabled the successful identification of more than 30% of integral membrane proteins (IMPs). A higher number of IMPs were identified in control and PMA-treated monocytes cells using 2D-LC-MS/MS. This approach also led to higher sequence coverage compared to the 1D-gel-LC-MS/MS. For example, the abundant Na^+/K^+ ATPase comprising ten transmembrane domains was identified by three peptides in 1D-gel-LC-MS/MS compared to eight distinct peptides using the 2D-LC-MS/MS approach. These observations suggest several limitations of the SDS-PAGE approach when limited amount of proteins are manipulated. Aside from the time consuming manipulation and keratin contamination that SDS-PAGE is subjected to, the sensitivity of this approach is also limited by the lower digestion yield and peptide recovery from in-gel trypsin proteolysis. However, one of the significant advantages of the SDS-PAGE approach is the separation of intact proteins enabling the confinement of all relevant peptides in the same protein band thus facilitating protein assignment including the identification of protein isoforms and variants. Finally, quantitative proteomics (label-free) can be achieved with equal success using 1D-gel-LC-MS/MS and 2D-LC-MS/MS and provided comparable expression profiling measurements for the cell model system examined.

References

Alberts, B. (2002). Molecular Biology of the Cell (New York, NY, Garland Science).

Baines, A.J. (2009). Evolution of spectrin function in cytoskeletal and membrane networks. Biochem Soc Trans *37*, 796–803.

Bendall, S.C., Hughes, C., Campbell, J.L., Stewart, M.H., Pittock, P., Liu, S., Bonneil, E., Thibault, P., Bhatia, M., and Lajoie, G.A. (2009). An enhanced mass spectrometry approach reveals human embryonic stem cell growth factors in culture. Mol Cell Proteomics *8*, 421–432.

Bisle, B., Schmidt, A., Scheibe, B., Klein, C., Tebbe, A., Kellerman, J., Siedler, F., Pfeiffer, F., Lottspich, F., and Oesterhelt, D. (2006). Quantitative profiling of the membrane proteome in a halophilic archaeon. Mol Cell Proteomics *5*, 1543–1558.

Bridges, S.M., Magee, G.B., Wang, N., Williams, W.P., Burgess, S.C., and Nanduri, B. (2007). ProtQuant: A tool for the label-free quantification of MudPIT proteomics data. BMC Bioinformatics *8*(Suppl 7), S24.

Brusniak, M.Y., Bodenmiller, B., Campbell, D., Cooke, K., Eddes, J., Garbutt, A., Lau, H., Letarte, S., Mueller, L.N., Sharma, V., Vitek, O., Zhang, N., Aebersold, R., and Watts, J.D. (2008). Corra: Computational framework and tools for LC-MS discovery and targeted mass spectrometry-based proteomics. BMC Bioinformatics *9*, 542.

Bunai, K., Nozaki, M., Kakeshita, H., Nemoto, T., and Yamane, K. (2005). Quantitation of de novo localized (15).N-labeled lipoproteins and membrane proteins having one and two transmembrane segments in a Bacillus subtilis secA temperature-sensitive mutant using 2D-PAGE and MALDI-TOF MS. J Proteome Res *4*, 826–836.

Chen, G., and Pramanik, B.N. (2009). Application of LC/MS to proteomics studies: Current status and future prospects. Drug Discov Today *14*, 465–471.

Cordwell, S.J. (2006). Technologies for bacterial surface proteomics. Curr Opin Microbiol *9*, 320–329.

Finstad, H.S., Drevon, C.A., Kulseth, M.A., Synstad, A.V., Knudsen, E., and Kolset, S.O. (1998). Cell proliferation, apoptosis and accumulation of lipid droplets in U937-1 cells incubated with eicosapentaenoic acid. Biochem J *336*(Pt 2), 451–459.

Fischer, F., Wolters, D., Rogner, M., and Poetsch, A. (2006). Toward the complete membrane proteome: High coverage of integral membrane proteins through transmembrane peptide detection. Mol Cell Proteomics *5*, 444–453.

Fountoulakis, M., Tsangaris, G., Oh, J.E., Maris, A., and Lubec, G. (2004). Protein profile of the HeLa cell line. J Chromatogr A *1038*, 247–265.

Guo, Y., Singleton, P.A., Rowshan, A., Gucek, M., Cole, R.N., Graham, D.R., Van Eyk, J.E., and Garcia, J.G. (2007). Quantitative proteomics analysis of human endothelial cell membrane rafts: Evidence of MARCKS and MRP regulation in the sphingosine 1-phosphate-induced barrier enhancement. Mol Cell Proteomics *6*, 689–696.

Heintz, D., Gallien, S., Wischgoll, S., Ullmann, A.K., Schaeffer, C., Kretzschmar, A.K., Van Dorsselaer, A., and Boll, M. (2009). Differential membrane proteome analysis reveals novel proteins involved in the degradation of aromatic compounds in Geobacter metallireducens. Mol Cell Proteomics *8*, 2159–2169.

Hui, J.P.M., Tessier, S., Butler, H., Badger, J., Kearney, P., Carrier, A., and Thibault, P. (2003). Iterative Exclusion and Directed MS-MS from Trace Level Protein Digests of Differentiated Cell Extracts. In American Society for Mass Spectrometry (ed.). 53rd ASMS Conference on Mass Spectrometry and Allied Topics (Nashville, TN).

Jiang, X., Feng, S., Tian, R., Han, G., Jiang, X., Ye, M., and Zou, H. (2007). Automation of nanoflow liquid chromatography-tandem mass spectrometry for proteome analysis by using a strong cation exchange trap column. Proteomics *7*, 528–539.

Kannan, K., Stewart, R.M., Bounds, W., Carlsson, S.R., Fukuda, M., Betzing, K.W., and Holcombe, R.F. (1996). Lysosome-associated membrane proteins h-LAMP1 (CD107a) and h-LAMP2 (CD107b) are activation-dependent cell surface glycoproteins in human peripheral blood mononuclear cells which mediate cell adhesion to vascular endothelium. Cell Immunol *171*, 10–19.

Kearney, P., and Thibault, P. (2003). Bioinformatics meets proteomics–bridging the gap between mass spectrometry data analysis and cell biology. J Bioinform Comput Biol *1*, 183–200.

Kelleher, D.J., and Gilmore, R. (2006). An evolving view of the eukaryotic oligosaccharyltransferase. Glycobiology *16*, 47R–62R.

Kramer, G., Boehringer, D., Ban, N., and Bukau, B. (2009). The ribosome as a platform for co-translational processing, folding and targeting of newly synthesized proteins. Nat Struct Mol Biol *16*, 589–597.

Kuster, B., Krogh, T.N., Mortz, E., and Harvey, D.J. (2001). Glycosylation analysis of gel-separated proteins. Proteomics *1*, 350–361.

Marakhova, I., Karitskaya, I., Aksenov, N., Zenin, V., and Vinogradova, T. (2005). Interleukin-2-dependent regulation of Na/K pump in human lymphocytes. FEBS Lett, *579*, 2773–2780.

Mcclatchy, D.B., Dong, M.Q., Wu, C.C., Venable, J.D., and Yates, J.R., 3rd (2007). 15 N metabolic labeling of mammalian tissue with slow protein turnover. J Proteome Res *6*, 2005–2010.

Mitra, S.K., Clouse, S.D., and Goshe, M.B. (2009a). Enrichment and preparation of plasma membrane proteins from Arabidopsis thaliana for global proteomic analysis using liquid chromatography-tandem mass spectrometry. Methods Mol Biol *564*, 341–355.

Mitra, S.K., Walters, B.T., Clouse, S.D., and Goshe, M.B. (2009b). An efficient organic solvent based extraction method for the proteomic analysis of Arabidopsis plasma membranes. J Proteome Res *8*, 2752–2767.

Mueller, L.N., Brusniak, M.Y., Mani, D.R., and Aebersold, R. (2008). An assessment of software solutions for the analysis of mass spectrometry based quantitative proteomics data. J Proteome Res *7*, 51–61.

Natera, S.H., Ford, K.L., Cassin, A.M., Patterson, J.H., Newbigin, E.J., and Bacic, A. (2008). Analysis of the Oryza sativa plasma membrane proteome using combined protein and peptide fractionation approaches in conjunction with mass spectrometry. J Proteome Res *7*, 1159–1187.

Norris, J.L., Porter, N.A., and Caprioli, R.M. (2003). Mass spectrometry of intracellular and membrane proteins using cleavable detergents. Anal Chem 75, 6642–6647.

Old, W.M., Meyer-Arendt, K., Aveline-Wolf, L., Pierce, K.G., Mendoza, A., Sevinsky, J.R., Resing, K.A., and Ahn, N.G. (2005). Comparison of label-free methods for quantifying human proteins by shotgun proteomics. Mol Cell Proteomics *4*, 1487–1502.

Omerovic, J., and Prior, I.A. (2009). Compartmentalized signalling: Ras proteins and signalling nanoclusters. FEBS J *276*, 1817–1825.

Ono, M., Shitashige, M., Honda, K., Isobe, T., Kuwabara, H., Matsuzuki, H., Hirohashi, S., and Yamada, T. (2006). Label-free quantitative proteomics using large peptide data sets generated by nanoflow liquid chromatography and mass spectrometry. Mol Cell Proteomics *5*, 1338–1347.

Pasini, E.M., Kirkegaard, M., Mortensen, P., Lutz, H.U., Thomas, A.W., and Mann, M. (2006). In-depth analysis of the membrane and cytosolic proteome of red blood cells. Blood *108*, 791–801.

Rabilloud, T. (2009). Membrane proteins and proteomics: Love is possible, but so difficult. Electrophoresis, *30*(Suppl 1), S174–S180.

Rahbar, A.M., and Fenselau, C. (2004). Integration of Jacobson's pellicle method into proteomic strategies for plasma membrane proteins. J Proteome Res *3*, 1267–1277.

Rahbar, A.M., and Fenselau, C. (2005). Unbiased examination of changes in plasma membrane proteins in drug resistant cancer cells. J Proteome Res *4*, 2148–2153.

Robertson, A.S., Smythe, E., and Ayscough, K.R. (2009). Functions of actin in endocytosis. Cell Mol Life Sci *66*, 2049–2065.

Romijn, E.P., and Yates, J.R., 3rd (2008). Analysis of organelles by on-line two-dimensional liquid chromatography-tandem mass spectrometry. Methods Mol Biol *432*, 1–16.

Sahraoui, Y., Allouche, M., Perrakis, M., Clemenceau, C., Jasmin, C., and Georgoulias, V. (1992). Phorbol myristate acetate-induced expression of high-affinity interleukin 2 receptors and production of interleukin 2 by human acute lymphoblastic leukemia T cells. Leukemia *6*, 295–303.

Santoni, V., Molloy, M., and Rabilloud, T. (2000). Membrane proteins and proteomics: Un amour impossible? Electrophoresis *21*, 1054–1070.

Schindler, J., and Nothwang, H.G. (2006). Aqueous polymer two-phase systems: Effective tools for plasma membrane proteomics. Proteomics *6*, 5409–5417.

Schindler, J., and Nothwang, H.G. (2009). Isolation of plasma membranes from the nervous system by countercurrent distribution in aqueous polymer two-phase systems. Methods Mol Biol *564*, 335–340.

Schirle, M., Heurtier, M.A., and Kuster, B. (2003). Profiling core proteomes of human cell lines by one-dimensional PAGE and liquid chromatography-tandem mass spectrometry. Mol Cell Proteomics 2, 1297–1305.

Speers, A.E., and Wu, C.C. (2007). Proteomics of integral membrane proteins—Theory and application. Chem Rev *107*, 3687–3714.

Vissers, J.P., Chervet, J.P., and Salzmann, J.P. (1996). Sodium dodecyl sulphate removal from tryptic digest samples for on-line capillary liquid chromatography/electrospray mass spectrometry. J Mass Spectrom *31*, 1021–1027.

Wickner, W., and Schekman, R. (2008). Membrane fusion. Nat Struct Mol Biol *15*, 658–664.

Woodhead, V.E., Binks, M.H., Chain, B.M., and Katz, D.R. (1998). From sentinel to messenger: An extended phenotypic analysis of the monocyte to dendritic cell transition. Immunology *94*, 552–559.

Yu, Y.Q., Gilar, M., and Gebler, J.C. (2004). A complete peptide mapping of membrane proteins: A novel surfactant aiding the enzymatic digestion of bacteriorhodopsin. Rapid Commun Mass Spectrom *18*, 711–715.

Zheng, H., Ji, C., Zou, X., Wu, M., Jin, Z., Yin, G., Li, J., Feng, C., Cheng, H., Gu, S., Xie, Y., and Mao, Y. (2003). Molecular cloning and characterization of a novel human putative transmembrane protein homologous to mouse sideroflexin associated with sideroblastic anemia. DNA Seq *14*, 369–373.

Chapter 33
Matrix-Assisted Laser Desorption/Ionization-Mass Spectrometry of Hydrophobic Proteins in Mixtures

Rachel R. Ogorzalek Loo

Abstract MALDI-MS sample/matrix preparations tailored to display the hydrophobic protein constituents of mixtures are described. Sinapinic, ferulic, DHB, 4-hydroxybenzylidene malononitrile, and 2-mercaptobenzothiazole matrices perform well in these formulations, effective in revealing lipid-acylated proteins and the abundant, water-insoluble ATPase proteolipid from *Escherichia coli* whole cell mixtures. Among the cocktails examined, 50% H_2O/33% 2-propanol/17% formic acid is preferred, primarily for its slow rate of serine/threonine formylation, an artifactual modification induced by many formic acid-containing solutions. The rates of protein formylation measured for different formic-acid containing cocktails will guide solvent selection for other hydrophobic protein applications, such as extracting intact proteins from polyacrylamide gels or solubilizing membrane proteins for ESI-MS. Additional strategies for hydrophobic protein analysis, such as detergent addition and liquid-liquid extraction are also discussed briefly.

Keywords Ion suppression · Lipoproteins · Membrane proteins, Mixture analysis · Whole cell analysis

33.1 Introduction

MALDI-MS excels in many applications that examine the intact proteins comprising complex mixtures, including the profiling of whole bacterial cells (Fenselau and Demirev, 2002; Lay, 2002; Ogorzalek Loo et al., 2002) and MALDI-MS imaging. (Chaurand et al., 2002) A frequent limitation of these methods, however, has been the restricted set of proteins detected. Analytical challenges often manifest as the invisibility of an expected protein, known to be highly abundant in a sample and readily detected when presented as an isolated species, but absent

R.R. Ogorzalek Loo (✉)
Department of Biological Chemistry, Molecular Biology Institute, UCLA-DOE Joint Institute for Genomics & Proteomics, University of California-Los Angeles, Los Angeles, CA 90095, USA
e-mail: rloo@mednet.ucla.edu

A.R. Ivanov, A.V. Lazarev (eds.), *Sample Preparation in Biological Mass Spectrometry*, DOI 10.1007/978-94-007-0828-0_33,

when dispersed in a complex mixture. For example, MALDI mass spectra from *E. coli* cells are dominated by ions from ribosomal and other basic proteins, yet exclude or under-represent many of the most abundant proteins in the cell, including Braun's lipoprotein, acyl carrier protein, outer membrane proteins, and ATP synthase. Switching matrices often fails to change the list of detected masses significantly, although changes to cultivation conditions or growth phase can alter spectra dramatically. The proteins displayed or ionized are thought to be dictated by ion suppression effects, solubility, protein abundance, basicity, and other factors.

Methods for MALDI analysis of membrane proteins and hydrophobic polypeptides have been reported, (Allmaier et al., 1995; Barnidge et al., 1997; Green-Church and Limbach, 1998; Rosinke et al., 1995; Schaller, 2000; Schey et al., 1992) sometimes employing solvents or detergents similar to those successful for ESI-MS. (le Maire et al., 1993; Ogorzalek Loo et al., 1994; Schaller et al., 1997, 1993) Most efforts to improve protein detection have focused on isolated or over-expressed proteins and protein complexes, i.e., samples with few proteins. Here we describe solvents, additives, and sample/matrix deposition techniques tailored to enhance the detection of hydrophobic proteins within cellular mixtures, radically changing the spectra obtained.

Formic acid (FA), an excellent protein solvent, figures prominently in our formulations. Unfortunately, this carboxylic acid can esterify (formylate) protein side chains, leading to undesirable protein heterogeneity. Thus, we compare the extent to which several solvent and matrix formulations covalently modify proteins, aiding in the selection of compositions least likely to artifactually impose heterogeneity. This information will assist both MALDI and ESI-MS analyses of unfractionated protein mixtures, purified polypeptides, and intact proteins extracted from polyacrylamide gel slices. (Cohen, 2006; Cohen and Chait, 1997; Ogorzalek Loo, 2004; Ogorzalek Loo et al., 1997)

33.2 Experimental

Sinapinic acid, ferulic acid, 2-mercaptobenzothiazole (2-MBT), 2,5-dihydroxybenzoic acid (DHB), and 3-indole acrylic acid were obtained from Sigma-Aldrich and evaluated as matrices; however, indole acrylic acid was unstable in high concentrations of formic acid. The matrix 4-hydroxybenzylidene malononitrile was obtained from Lancaster Synthesis (Pelham, NH). "Standard matrix" was prepared from sinapinic acid (saturated) in 33% CH_3CN/67% H_2O/0.1% trifluoroacetic acid (TFA) (v/v/v). A higher organic concentration (50% CH_3CN/50% H_2O/0.1% TFA (v/v/v)) was used to formulate saturated ferulic acid, 2-MBT, and 4-hydroxybenzylidene malononitrile. DHB was prepared in the latter solvent at concentrations of 20 or 80 mg/mL or as a saturated solution. "Formic acid-based matrix" or FA-CAN, used to obtain many of the spectra illustrated here, was prepared as a saturated solution of the UV-absorbing solid in 60% formic acid/40% CH_3CN.

Lyophilized *Escherichia coli* K12 cells (Sigma-Aldrich) and *Methanosarcina acetivorans* C2A cell pellets were examined. The methanogenic archae were cultivated anaerobically as single cells in a 100 mM methanol, 200 mM NaCl-based medium at 37°C (Sowers and Gunsalus, 1988). Pelleted cells were stored at –80°C, until resuspension for immediate analysis. Dried *E. coli* cells, estimated to contain 50% protein by weight, were resuspended in the desired solvent to approximately 2.5 mg dry cells/mL, while *M. acetivorans* cells were resuspended in volumes appropriate to yield protein concentrations of ~1 mg/mL. Water was employed as the resuspension solvent for "standard matrices", and 1 μL of H_2O-suspended cells was mixed with 9 μL of standard matrix in a micro-centrifuge tube for sample deposition. With formic acid-based conditions, cells were resuspended and vortexed in FA (60% formic acid/40% CH_3CN (v/v)) immediately prior to deposition and matrix application, in order to maintain high formic acid concentrations in the sample and matrix solutions, avoiding premature protein precipitation.

Attention to spotting technique was important in obtaining spectra from FA formulations. Notably, successful methods yield larger matrix crystals. Because FA matrix spreads freely on the probe surface, only 0.2 μL of sample was deposited by quickly touching the pipet tip to the probe. Immediately thereafter, 0.2 μL of FA matrix (ferulic, 2-MBT, or DHB) was deposited atop the droplet, without additional mixing. (Sample spots allowed to dry prior to matrix addition did not reveal proteins as high in mass as did those that remained wet.) Because crystals did not form reliably with ferulic acid, appearing only 50% of the time, sample spots in that matrix were prepared in duplicate. For sinapinic acid FA matrix, crystal "seeding" prior to spotting of sample and matrix was essential to obtain spectra reliably. Thus, crystals were first seeded by spotting 0.2 μL of the FA sinapinic acid matrix onto the MALDI probe. After the matrix dried, 0.2 μL sample and 0.2 μL matrix were spotted without mixing as per the above protocol. Figure 33.1 illustrates matrix deposits obtained from the recommended procedures. Analyses were performed within an hour of solvent addition, displaying little, if any, formylation or oxidation. However, matrix-spotted samples maintained overnight (even under vacuum) displayed significant formylation, as described previously. (Ehring et al., 1997) These formic-acid based matrix solutions performed well for at least a week, stored in the dark at room temperature.

Reactivities of five formic acid-containing solutions were evaluated by dissolving lyophilized proteins in them and subsequently depositing the mixtures with matrix on a sample stage. Sample and matrix for the five solutions were formulated as follows: *Kim cocktail:* The sample was dissolved in 25 μL of 70% CH_3CN/30% formic acid; a saturated solution of matrix was prepared by dissolving sinapinic acid in 70% CH_3CN/30% H_2O or 70% CH_3CN/29.9% H_2O/0.1% TFA. (Kim et al., 2001) *Cohen cocktail (FWI):* The sample was dissolved in 25 μL of 50% H_2O/33% 2-propanol/17% formic acid, while a saturated solution of matrix was prepared by dissolving α-cyano-4-hydroxycinnamic acid in the same solvent. (Cohen and Chait, 1997; Cohen et al., 1996) *Feick cocktail (FAPH):* The sample was dissolved in 25 μL of 50% formic acid/25% CH_3CN/15% 2-propanol/10% H_2O. (Ehring et al., 1997; Feick and Shiozawa, 1990) Matrices were prepared by dissolving

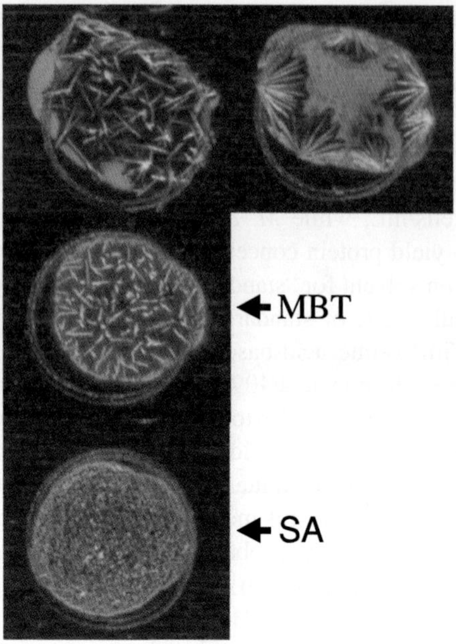

Fig. 33.1 *E. coli* cells suspended in 2:1 formic acid:CH_3CN spotted onto the sample plate with mercaptobenzothiazole (MBT), ferulic acid, and sinapinic acid (SA) matrix, as described. From Ogorzalek Loo and Loo (2007) Reproduced with permission

either α-cyano-4-hydroxycinnamic acid (saturated) in 1:1 CH_3CN:0.2% TFA (v/v) or sinapinic acid (saturated) in 1:2 CH_3CN:0.15% TFA (v/v). *Cadene cocktail:* The sample was dissolved in 25 μL of 50% formic acid/33% 2-propanol/17% water (v/v/v). An aliquot of sample solubilized in the Cadene cocktail was spotted followed by an aliquot of matrix (saturated) in FAPH. However, this cocktail is typically applied as a thin-film. (Cadene and Chait, 2000; Fenyo et al., 2007) *3:2 Formic Acid/Acetonitrile (FA-CAN):* As described above, protein samples were dissolved in 60% formic acid/40% CH_3CN and saturated matrix solutions were prepared from the same composition. Sample and matrix deposition were performed as described above, with spots illustrated in Fig. 33.1.

Aqueous stock solutions of 1–2% w/v perfluorooctanoic acid (pentadecafluorooctanoic acid or PFOA) were incubated at 37°C for 15–30 min. to dissolve the detergent completely. When employed, PFOA was added to both sample and matrix. D-Sorbitol was optionally added to 1.5% (w/v) of matrix solution.

TFA (sequencing grade) and formic acid (98–100%, extra pure) were obtained from Pierce (Rockford, IL) and Aldrich (Milwaukee, WI), respectively. PFOA was acquired from TCI America (Portland, OR), while HPLC-grade water, isopropanol, and acetonitrile were purchased from EM Science (Darmstadt, Germany). D-Sorbitol was obtained from Acros Organics (Morris Planes, New Jersey).

MALDI mass spectra were obtained on an Applied Biosystems Voyager DE-STR time-of-flight (TOF) mass spectrometer (Framingham, MA) operated in linear mode with 337 nm irradiation and positive ion detection.

Formic acid is dangerously caustic to skin. Proper eye and skin protection are required when handling concentrated formic acid solutions. Contact with 2-MBT should be avoided, to prevent sensitization to this allergenic compound. Because the rubber additive is widely distributed in everyday products, allergies to 2-MBT are problematic.

33.3 Results and Discussion

33.3.1 How Does Formic Acid Modify Proteins?

The utility of formic acid for protein dissolution is well known. It has supported membrane protein purifications and analyses, has solubilized refractory protein aggregates from Huntington's, Alzheimer's, and Parkinson's patients, and has extracted *intact* proteins from polyacrylamide gels. Despite this utility, formic acid also presents problems for analysts: O-formylation of serine and threonine residues and cleavage of acid-labile bonds, particularly Asp-Pro. The former problem, esterification of serines and threonines, can be reduced by careful solvent selection and by reducing temperature (Bollhagen et al., 1995; Heukeshoven and Dernick, 1982; Kienhuis et al., 1959; Narita, 1959; Sheumack and Burley, 1988).

Some previous reports listed formyl addition to amino groups as a potential modification, but little experimental support exists. The original suggestion apparently surfaced from a report on failed Edman sequence analyses that were ascribed to blocked N-termini, thought to arise from incubation in pyridine formate buffer (Levy et al., 1981; Shively et al., 1982). However, other studies found no evidence for N-terminal or lysine acylation after formic acid solubilization, (Kienhuis et al., 1959; Sheumack and Burley, 1988; Tarr and Crabb, 1983; Vanfleteren, 1989) although formylation has been documented from more severe performic acid oxidation conditions. (Dai et al., 2005) It is possible that the blocked N-termini cited in the Edman sequencing work instead reflected O–>N acyl shifts on N-terminal serine or threonine residues, experienced during subsequent exposure to *high* pH, *i.e.*, during phenyl isothiocyanate (PITC) coupling. Acyl groups migrate reversibly between an N-terminal residue's β-hydroxy and α-amino functional groups, with N–>O acyl shifts induced by acid and O–>N acyl shifts induced by base (Iwai and Ando, 1967).

Similar cautions have been raised regarding protein deamidation caused by formic acid, although negligible deamidation was observed in sequence-analyzed polypeptides following exposure (Feick and Shiozawa, 1990; Vanfleteren, 1989). Succinimide intermediate-driven deamidation actually occurs at *neutral and high pH*, catalyzed by base, heat, and ionic strength, and retarded by organic solvents (Daniel et al., 1996; Geiger and Clarke, 1987; Johnson et al., 1989). Asn deamidation under acidic (pH 1-2) conditions occurs slowly and via a different intermediate, from water attack on a protonated carbonyl (Daniel et al., 1996). When hydrolyzing proteins in a 70°C vapor of aqueous pentafluoropropionic acid, Gobom, et al. (1999) observed asparagine and glutamine deamidation *only* when the water content exceeded 50%. All of these observations suggest that deamidation is of less concern than previously feared.

Formic acid has been linked to tryptophan and methionine oxidation, with extent of modification scaling to length of exposure. Cohen, et al. (1999) attributed the reaction to trace peroxides, prescribing either aging of formic acid solutions for > 2 weeks or spiking with 10 mM methionine as scavenger.

Some bond cleavage is unavoidable when using concentrated formic acid, with cleavages N- or C- terminal to aspartic acid, and especially between Asp-Pro residues the most common. Its protein-dependent impact on analyses has been discussed previously, (Ogorzalek Loo and Loo, 2007) but will be lessened by preparing samples quickly (Sheumack and Burley, 1988) and at low temperature (Bollhagen et al., 1995). At high acid concentrations, limited water molecules decrease the hydrolysis rate, but concomitantly increase the extent of formylation.

33.3.2 Analyzing Complex Mixtures

Mixture spectra obtained from standard and formic acid-based sinapinic acid matrices reveal similar numbers of ions from *E. coli* cells: 94 singly and 20 doubly-charged ions from standard versus 119 and 20 for formic acid (Figs. 33.2 and 33.3). Nevertheless, the spectra differ dramatically, with no more than 10 protein ions in common. Several lipoproteins, apparent from their heterogeneous lipid acylation (Fig. 33.2b inset), were observed from the formic acid preparation. Although apparent in low complexity samples, (Ogorzalek Loo et al., 2001, 2002), hydrophobic lipoproteins such as Braun's lipoprotein (*lpp*, UniProt P69776) are generally suppressed in whole cell mixture spectra. Their absence is striking, when one considers that an *E. coli* cell contains 500,000 free molecules of *lpp*, the major outer membrane lipoprotein (Inouye et al., 1976). This dependence on low sample complexity hints at the interplay between a protein's solubility, the solubility of other components, and how those two factors influence matrix incorporation.

Similarly dominant in formic acid-based solutions is an 8282 +/– 4 Da ion, absent or very weak under standard conditions, consistent with *atpL* (P68699), the ATP synthase C chain (Fig. 33.2). Referred to as the "ATPase proteolipid" due to its unusual solubility profile, this very abundant *E. coli* protein is insoluble in water and was first purified from chloroform/methanol mixtures. The complete list of *E. coli* proteins ascribed to these whole cell profiles supplements a previous publication (Ogorzalek Loo and Loo, 2007).

Similar strategies to address complex mixture analyses have been applied by Walker, et al. (2005) who found that MALDI-MS of 5–100-kDa proteins desorbed from dried immobilized pH gradient gels benefits when 2:3:3:2 formic acid: water: acetonitrile: isopropanol is included in the sinapinic acid matrix. Schey and colleagues (Thibault et al., 2008) profiled mammalian lens and retina proteins by spotting small volumes of 7:3 formic acid:hexafluoroisopropanol onto the tissues, followed by sinapinic or DHB in acetonitrile/water/0.1% TFA. To enhance the yield of multiply charged ions for MALDI-TOF-TOF analysis, Liu and Schey (2008) deposited an ultra-thin-layer of matrix with 1:3:2 formic acid:water:CH_3CN spotting solution.

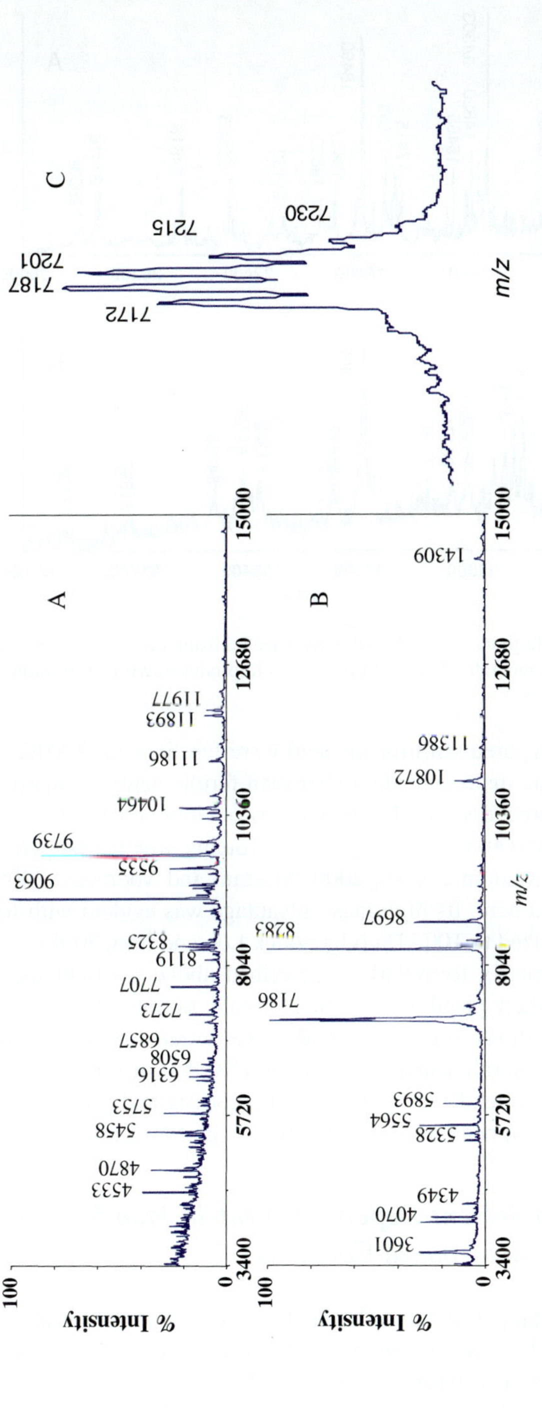

Fig. 33.2 MALDI mass spectra obtained from lyophilized *E. coli* cells desorbed from sinapinic acid using **(a)** standard, and **(b)** formic acid-based matrix conditions. Inset: expansion of *m/z* 7000–7500 revealing heterogereity characteristic of lipid acylation in *lpp* (P69776). From Ogorzalek Loo and Loo (2007) Reproduced with permission

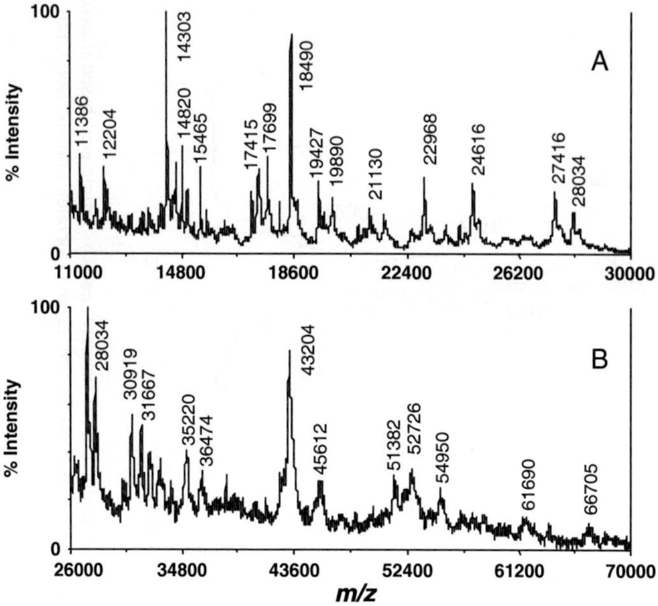

Fig. 33.3 Medium and high *m/z* range *E. coli* ions detected from a formic acid-based preparation of sinapinic acid. From Ogorzalek Loo and Loo (2007) Reproduced with permission

Other matrices prepared with formic acid were ferulic acid, 2-MBT, DHB, and 4-hydroxybenzylidene malononitrile. Other than ferulic acid, the matrices yielded spectra similar to those discussed, but with no obvious advantages. Ferulic acid surpasses sinapinic acid at detecting high mass proteins, whether prepared with conventional solvents, (Madonna et al., 2000; Meetani and Voorhees, 2005; Nilsson, 1999) or as described here. Its high mass advantage was evident with formic acid, displaying proteins sized to 100 kDa (Ogorzalek Loo and Loo, 2007).

We rely on the matrix formulations described above to supplement standard matrix conditions when profiling complex protein samples by MALDI-MS. In transferring these methods to tryptic peptides, we revealed previously unobserved species, but with an overall sensitivity perhaps 1/10 that of standard preparations. Clearly our methods extend the range of both peptides and proteins accessed by MALDI profiling, but appear most suited to profiling protein mixtures.

33.3.3 Time- and Solvent-Dependent Protein Modification in Formic Acid-Based Cocktails

It is best to analyze samples incubated in high concentrations of formic acid shortly after acid addition, because the number of formate esters increases with time. Reactivities of formic acid-based cocktails ("FA", "Kim", "Cohen", "Feick", and "Cadene" described in Section 33.2) were evaluated by dissolving proteins in them

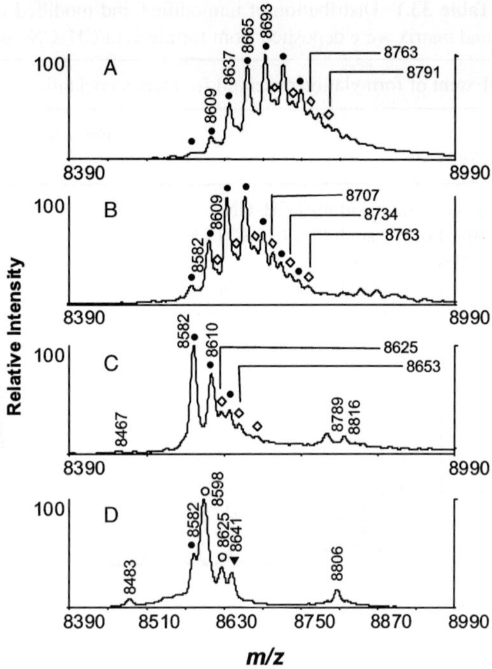

Fig. 33.4 MALDI mass spectra acquired from ubiquitin deposited 1 day after solubilization in (**a**) 60% formic acid/40% CH_3CN, (**b**) Feick cocktail: 50% formic acid/25% CH_3CN/15% 2-propanol/10% H_2O, (**c**) Cadene cocktail: 50% formic acid/33% 2-propanol/17% H_2O, and (**d**) Cohen cocktail: 50% H_2O/33% 2-propanol/17% formic acid. Symbols signify peaks corresponding to multiply formylated forms of ubiquitin with: ◊ no oxidation (+0 Da), • single oxidation (+16 Da), o double oxidation (+32 Da), and ▼ triple oxidation (+48 Da). From Ogorzalek Loo and Loo (2007) Reproduced with permission

and immediately dispensing spots with matrix onto the sample stage. Mass analyses were performed at several time points. Solubilized proteins were also incubated at room temperature for 21 h, after which sample and matrix were deposited and analyzed.

Formylation continues even *after* sample/matrix crystallization and even on sample targets within the mass spectrometer (Ehring et al., 1997; Ogorzalek Loo and Loo, 2007); *e.g.,* ubiquitin displayed no formylation when analyzed immediately after deposition in 3:2 FA-CAN, yet 1.5 h later the same spot displayed protein ~18% formylated (+28 Da), and 14% oxidized and formylated (+42 Da) (Ogorzalek Loo and Loo, 2007). Similar amounts of unmodified protein were yielded by the Kim and Feick cocktails 1.5 h after spotting, but with *reproducibly more +42 Da than +28 Da-modifications*. No formylation was observed 1.5 h after matrix deposition with the Cohen cocktail (Table 33.1).

The extent of formylation was independent of matrix, depending only on protein, solvent composition, and incubation time. Ubiquitin, deposited 1 day after solubilization in 3:2 FA, revealed proteins formylated (+28) from 1 to 9 times, and partially oxidized (Fig. 33.4a). Slightly less formylation was displayed by the Feick composition (Fig. 33.4b), while the Cadene cocktail, equal to Feick in formic acid, but absent CH_3CN, showed much less formylation (Fig. 33.4c). Major peaks from the Cohen mix correspond to singly and doubly oxidized ubiquitin, with little formylation (Fig. 33.4d).

Table 33.1 Distribution of unmodified and modified ubiquitin observed 1.5 h. after the sample and matrix were deposited from formic acid/CH_3CN-containing solvents

Extent of formylation observed for matrix cocktails				
Cocktail	Unmodified %	Oxidized (+16)	Formylated (+28) %	Both (+42) %
60% Formic acid:40%CH_3CN	75	–	20	15
50% Formic acid:25% CH_3CN:15% IPA:10% H_2O	75	–	20	20
30% Formic acid:70% CH_3CN	70	–	10	20

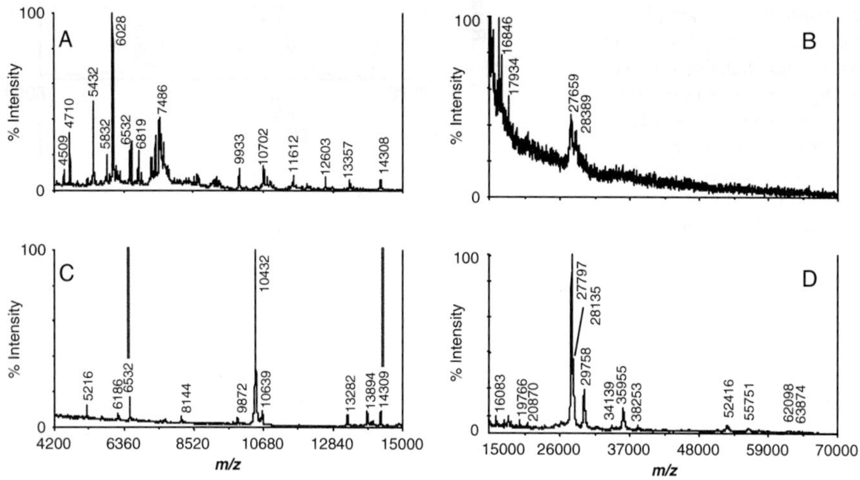

Fig. 33.5 (**a**) Low *m/z* range MALDI-MS of *Methanosarcina acetivorans* cell pellets suspended in H_2O and mixed 1:4 with "standard" matrix (i.e., sinapinic acid (saturated) and 0.1% (v/v) TFA in 1:2 CH_3CN:H_2O, and (**b**) High *m/z* range, prepared as in "a". (**c**) Low *m/z* range MALDI-MS obtained by mixing H_2O-suspended cell pellets 1:1 with PFOA-containing matrix (i.e., sinapinic acid (saturated) and 1 % PFOA in 1:2 CH_3CN:H_2O), and (**d**) High *m/z* range, prepared as in "c". Only 2 peaks are common to both spectra. From Ogorzalek Loo and Loo (2007) supplemental material. Reproduced with permission

The propensity for protein formylation among the cocktails examined is: Cohen<Cadene<Kim<Feick<3:2 FA, with avoidance of CH_3CN and/or reduced formic acid content key to reducing this side reaction. Clearly, the small equilibrium constant of acid-catalyzed Fischer esterifications magnifies the importance of the formic acid:hydroxyl ratio. Given that esterifications can be driven to completion by removing water, it is to be expected that a solubilization cocktail with only formic acid and acetonitrile would promote formylation. Finally, scavenger inclusion (*e.g.,* methionine, tryptophan, phenol, thioglycolic acid, thiodiglycol, or 2-mercaptoethanol) is recommended to reduce oxidation.

33.3.4 Detergent Addition

Detergents (surfactants) are often essential additives to maintain protein solubility during isolation, and purification. Their use in MALDI-MS of complex protein mixtures and isolated proteins has been highlighted previously (Bornsen, 2000; Rosinke et al., 1995; Schaller, 2000; Schey et al., 1992). The non-ionic saccharide *n*-octyl-β-D-glucoside has been especially useful, combining a high critical micelle concentration (CMC) with inherently low spectral background. It has been used successfully at concentrations both above and below its ~20 mM CMC. Striking differences in the distribution of proteins detected from cellular mixtures have also been noted after inclusion of Triton X-100 or TWEEN 80 at concentrations exceeding their CMC (Bornsen, 2000; van Adrichem et al., 1998). Perfluorooctanoic acid (PFOA), when used at concentrations approaching its 9.7 mM CMC (0.4% w/v), also yields informative, yet remarkably different spectra from those of standard preparations. Figure 33.5 demonstrates that *M. acetivorans* cells examined with and without PFOA-containing matrix share only 2 major peaks, with the PFOA preparation benefiting high *m/z* detection. A consideration in employing high detergent concentrations is that matrix crystals can be sparsely deposited on the sample stage, taxing the analyst. These conditions, while far from optimal for standard, water-soluble analytes, do facilitate measurements of hydrophobic components.

33.3.5 Saccharide Matrix Additives

Saccharide co-matrices, particularly fucose, have been employed to reduce fragmentation and enhance reproducibility, signal persistence and intensity (Distler and Allison, 2001; Gluckmann and Karas, 1999; Gusev et al., 1995; Koester et al., 1992; Peterson et al., 2002; Shahgholi et al., 2001). We found sorbitol to be a beneficial co-matrix, enhancing medium and higher *m/z* performance in some cases, while generally promoting homogeneous crystallization (Ogorzalek Loo and Loo, 2007). Its inclusion (1.0–1.5% w/v) was helpful when analyzing samples low in concentration. The previously employed fucose and fructose monosaccharides, an aldose and ketose, respectively, are reducing sugars that can potentially react with amines to form Schiff bases. In contrast, sorbitol, an unreactive monosaccharide alditol, is less likely to covalently modify proteins.

Uncharged saccharides' compatibility with protein analyses by MALDI-MS presents a second opportunity in sample preparation. For washing cell pellets, 250 mM sucrose can substitute for PBS (phosphate buffered saline) to provide a salt-free means to maintain osmotic strength.

33.3.6 In-Situ Liquid-Liquid Extraction

An in-situ liquid–liquid extraction (LLE) method has been described and applied to simple peptide and protein mixtures from 3 to 30 kDa. (Kjellstroem and Jensen,

2003) In this unique method, hydrophobic solutes within an aqueous droplet (5% formic acid, 0.1% TFA, or 100% H_2O) diffuse into immiscible ethyl acetate or ethyl acetate/matrix solution. Basically, an aqueous peptide or protein solution is spotted, after which an aliquot of an ethyl acetate organic phase is placed on top of the aqueous drop. The droplets assemble into a central aqueous phase surrounded by a distinct organic phase. In ~30 S the organic phase evaporates, after which the aqueous droplet is transferred to a new position on the sample stage. Analysis can be performed from both the aqueous and organic extracts, as desired. Bacteriorhodopsin and a myristoylated peptide were among the species successfully analyzed with this rapid method.

33.4 Conclusions

Methods enabling different ensembles of proteins to be detected from mixtures by MALDI-MS have been discussed. Formic acid-based solutions are particularly effective for visualizing hydrophobic proteins, and appropriate solvent compositions can reduce or even eliminate artifactual O-formylation. Detergent addition is also useful, but can complicate matrix deposition when employed at the near-CMC concentrations essential to detecting the most hydrophobic species. Liquid-liquid extraction, particularly applied multiple times, is a rapid method to enhance spectra for hydrophobic species. Sorbitol addition to sinapinic acid matrix promotes homogeneous crystallization and, in some cases, enhances high mass ion detection. Sorbitol addition benefited the analysis of dilute samples, although little benefit was observed when adequately concentrated protein solutions were examined. Because alditol saccharides cannot react with amines to form Schiff bases, sorbitol may be preferable to fucose or fructose as a matrix additive.

Acknowledgments The author thanks Robert P. Gunsalus, Unmi Kim, and Kristy Sandler of the University of California-Los Angeles Department of Microbiology, Immunology & Molecular Genetics for the *Methanosarcina acetivorans* cultures employed in these studies and Philip C. Andrews (University of Michigan) for inspiration and for his appreciation of formic acid. Support was provided by the U.S. Department of Energy through the UCLA-DOE Laboratory of Genomics & Proteomics (DEFC03-87 ER60615), by the David Geffen School of Medicine at UCLA, and by the UCLA Molecular Biology Institute. The UCLA Mass Spectrometry and Proteomics Technology Center was established and equipped with a generous gift from the W.M. Keck Foundation.

References

Allmaier, G., Schaffer, C., Messner, P., Rapp, U., and Mayer-Posner, F.J. (1995). Accurate determination of the molecular weight of the major surface layer protein isolated from *Chlostridium thermosaccharolyticum* by time-of-flight mass spectrometry. J Bacteriol *177*, 1402–1404.

Barnidge, D.R., Dratz, E.A., Sunner, J., and Jesaitis, A.J. (1997). Identification of transmembrane tryptic peptides of rhodopsin using matrix-assisted laser desorption/ionization time-of-flight mass spectrometry. Protein Sci *6*, 816–824.

Bollhagen, R., Schmiedberger, M., and Grell, E. (1995). High-performance liquid chromatographic purification of extremely hydrophobic peptides: Transmembrane segments. J Chromatogr A *711*, 181–186.

Bornsen, K.O. (2000). Influence of Salts, Buffers, Detergents, Solvents, and Matrices on MALDI-MS Protein Analysis in Complex Mixtures. In Protein and Peptide Analysis: New Mass Spectrometric Applications, J.R. Chapman, ed. (Totowa, NJ, Humana Press), pp. 387–404.

Cadene, M., and Chait, B.T. (2000). A Robust, detergent-friendly method for mass spectrometric analysis of integral membrane proteins. Anal Chem *72*, 5655–5658.

Chaurand, P., Schwartz, S.A., and Caprioli, R.M. (2002). Imaging mass spectrometry: A new tool to investigate the spatial organization of peptides and proteins in mammalian tissue sections. Curr Opinion Chem Biol *6*, 676–681.

Cohen, S.L. (2006). Ozone in Ambient air as a source of adventitious oxidation. A mass spectrometric study. Anal Chem *78*, 4352–4362.

Cohen, S.L., and Chait, B.T. (1997). Mass Spectrometry of whole proteins eluted from sodium dodecyl sulfate-polyacrylamide gel electrophoresis gels. Anal Biochem *247*, 257–267.

Cohen, S.L., Halaas, J.L., Friedman, J.M., Chait, B.T., Bennet, L., Chang, D., Hecht, R., and Collins, F. (1996). Human leptin characterization. Nature *382*, 589.

Cohen, S.L., Ward, G., and Tsai, P.-K. (1999). MALDI-MS Characterization of Human Papillomavirus Protein. Paper presented at: Proceedings of the 47th ASMS Conference on Mass Spectrometry and Allied Topics (Dallas, TX, American Society for Mass Spectrometry).

Dai, J., Zhang, Y., Wang, J., Li, X., Lu, Z., Cai, Y., and Qian, X. (2005). Identification of degradation products formed during performic oxidation of peptides and proteins by high-performance liquid chromatography with matrix-assisted laser desorption/ionization and tandem mass spectrometry. Rapid Comm Mass Spectrom *19*, 1130–1138.

Daniel, R.M., Dines, M., and Petach, H.H. (1996). The denaturation and degradation of stable enzymes at high temperatures. Biochem J *317*, 1–11.

Distler, A.M., and Allison, J. (2001). Improved MALDI-MS analysis of oligonucleotides through the use of fucose as a matrix additive. Anal Chem *73*, 5000–5003.

Ehring, H., Stromberg, S., Tjernberg, A., and Noren, B. (1997). Matrix-Assisted laser desorption/ionization mass spectrometry of proteins extracted directly from sodium dodecyl sulfate-polyacrylamide gels. Rapid Comm Mass Spectrom *11*, 1867–1873.

Feick, R.G., and Shiozawa, J.A. (1990). A high-yield method for the isolation of hydrophobic proteins and peptides from polyacrylamide gels for protein sequencing. Anal Biochem *187*, 205–211.

Fenselau, C., and Demirev, P.A. (2002). Characterization of intact microorganisms by MALDI mass spectrometry. Mass Spectrom Rev *20*, 157–171.

Fenyo, D., Wang, Q., DeGrasse, J.A., Padovan, J.C., Cadene, M., and Chait, B.T. (2007). MALDI sample preparation: The Ultra thin layer method. JoVE *3*, e192.

Geiger, T., and Clarke, S. (1987). Deamidation, isomerization, and racemization at asparaginyl and aspartyl residues in peptides. Succinimide-linked reactions that contribute to protein degradation. J Biol Chem *262*, 785–794.

Gluckmann, M., and Karas, M. (1999). The initial ion velocity and dependence on matrix, analyte, and preparation method in ultraviolet matrix-assisted laser desorption/ionization. J Mass Spectrom *34*, 467–477.

Gobom, J., Mirgorodskaya, E., Nordhoff, E., Hojrup, P., and Roepstorff, P. (1999). Use of vapor-phase acid hydrolysis for mass spectrometric peptide mapping and protein identification. Anal Chem *71*, 919–927.

Green-Church, K.B., and Limbach, P.A. (1998). Matrix-assisted laser desorption/ionization mass spectrometry of hydrophobic peptides. Anal Chem *70*, 5322–5325.

Gusev, A.I., Wilkinson, W.R., Proctor, A., and Hercules, D.M. (1995). Improvement of signal reproducibility and matrix/comatrix effects in MALDI analysis. Anal Chem *67*, 1034–1041.

Heukeshoven, J., and Dernick, R. (1982). Reversed-phase high-performance liquid chromatography of virus proteins and other large hydrophobic proteins in formic acid containing solvents. J Chromatogr *252*, 241–254.

Inouye, S., Takeishi, K., Lee, N., DeMartini, M., Hirashima, A., and Inouye, M. (1976). Lipoprotein from the outer membrane of Escherichia coli: Purification, paracrystallization, and some properties of its free form. J Bacteriol *127*, 555–563.

Iwai, K., and Ando, T. (1967). N–>O acyl rearrangement. Methods Enzymol *11*, 263–282.

Johnson, B.A., Shirokawa, J.M., Hancock, W.S., Spellman, M.W., Basa, L.J., and Aswad, D.W. (1989). Formation of isoaspartate at two distinct sites during in vitro aging of human growth hormone. J Biol Chem *264*, 14262–14271.

Kienhuis, H., Blasse, G., and Matze, J. (1959). Action of anhydrous formic acid on peptides and proteins. Nature *184*, 2015–2016.

Kim, Y.J., Freas, A., and Fenselau, C. (2001). Analysis of viral glycoproteins by MALDI-TOF mass spectrometry. Anal Chem *73*, 1544–1548.

Kjellstroem, S., and Jensen, O.N. (2003). In situ liquid-liquid extraction as a sample preparation method for matrix-assisted laser desorption/ionization MS analysis of polypeptide mixtures. Anal Chem *75,* 2362–2369.

Koester, C., Castoro, J.A., and Wilkins, C.L. (1992). High-resolution matrix-assisted laser desorption/ionization of biomolecules by fourier transform mass spectrometry. J Am Chem Soc *114*, 7572–7574.

Lay, J.O., Jr. (2002). MALDI-TOF mass spectrometry of bacteria. Mass Spectrom Rev *20*, 172–194.

le Maire, M., Deschamps, S., Moller, J., Le Caer, J.P., and Rossier, J. (1993). Electrospray ionization mass spectrometry on hydrophobic peptides electroeluted from sodium dodecyl sulfate-polyacrylamide gel electrophoresis. Application to the topology of the sarcoplasmic reticulum Ca^{2+} ATPase. Anal Biochem *214*, 50–57.

Levy, W.P., Rubinstein, M., Shively, J., Del Valle, U., Lai, C.-Y., Moschera, J., Brink, L., Gerber, L., Stein, S., and Pestka, S. (1981). Amino acid sequence of a human leukocyte interferon. Proc Natl Acad Sci USA *78*, 6186–6190.

Liu, Z., and Schey, K.L. (2008). Fragmentation of multiply-charged intact protein ions using MALDI TOF-TOF mass spectrometry. J Am Soc Mass Spectrom *19*, 231–238.

Madonna, A.J., Basile, F., Ferrer, I., Meetani, M.A., Rees, J.C., and Voorhees, K.J. (2000). On-probe sample pretreatment for the detection of proteins above 15 kda from whole cell bacteria by matrix-assisted laser desorption/ionization time-of-flight mass spectrometry. Rapid Comm Mass Spectrom *14*, 2220–2229.

Meetani, M.A., and Voorhees, K.J. (2005). MALDI mass spectrometry analysis of high molecular weight proteins from whole bacterial cells: pretreatment of samples with surfactants. J Am Soc Mass Spectrom *16*, 1422–1426.

Narita, K. (1959). Reaction of anhydrous formic acid with proteins. J Am Chem Soc *81*, 1751–1756.

Nilsson, C.L. (1999). Fingerprinting of *Helicobacter pylori* strains by matrix-assisted laser desorption/ionization mass spectrometric analysis. Rapid Comm Mass Spectrom *13*, 1067–1071.

Ogorzalek Loo, R.R. (2004). Gel Electrophoresis. In The Encyclopedia of Mass Spectrometry: Applications in Biochemistry, Biology, and Medicine, M.L. Gross, and. R. Caprioli, eds. (New York, NY, Elsevier), pp. 36–47.

Ogorzalek Loo, R.R., Cavalcoli, J.D., VanBogelen, R.A., Mitchell, C., Loo, J.A., Moldover, B., and Andrews, P.C. (2001). Virtual 2-D Gel electrophoresis: visualization and analysis of the *E. coli* proteome by mass spectrometry. Anal Chem *73*, 4063–4070.

Ogorzalek Loo, R.R., Dales, N., and Andrews, P.C. (1994). Surfactant effects on protein structure examined by electrospray ionization mass spectrometry. Protein Sci *3*, 1975–1983.

Ogorzalek Loo, R.R., Du, P., Loo, J.A., and Holler, T. (2002). In-vivo labeling: A glimpse of the dynamic proteome and additional constraints for protein identification. J Am Soc Mass Spectrom *13*, 804–812.

Ogorzalek Loo, R.R., and Loo, J.A. (2007). Matrix-assisted laser desorption/ionization-mass spectrometry of hydrophobic proteins in mixtures using formic acid, perfluorooctanoic acid, and sorbitol. Anal Chem *79*, 1115–1125.

Ogorzalek Loo, R.R., Mitchell, C., Stevenson, T.I., Martin, S.A., Hines, W., Juhasz, P., Patterson, D., Peltier, J., Loo, J.A., and Andrews, P.C. (1997). Sensitivity and mass accuracy for proteins analyzed directly from polyacrylamide gels: Implications for proteome mapping. Electrophoresis *18*, 382–390.

Peterson, J., Allikmaa, V., Subbi, J., Pehk, T., and Lopp, M. (2002). Structural deviations in poly(amidoamine) dendrimers: A MALDI-TOF MS analysis. Eur Polymer J *39*, 33–42.

Rosinke, B., Strupat, K., Hillenkamp, F., Rosenbusch, J., Dencher, N., Krueger, U., and Galla, H.-J. (1995). Matrix-Assisted laser desorption/ionization mass spectrometry (MALDI-MS) of membrane proteins and non-covalent complexes. J Mass Spectrom *30*, 1462–1468.

Schaller, J. (2000). Analysis of Hydrophobic Proteins and Peptides by Mass Spectrometry. In Protein and Peptide Analysis: New Mass Spectrometric Applications, J.R. Chapman, ed. (Totowa, NJ, Humana Press), pp. 425–437.

Schaller, J., Pellascio, B.C., and Schlunegger, U.P. (1997). Analysis of hydrophobic proteins and peptides by electrospray ionization mass spectrometry. Rapid Comm Mass Spectrom *11*, 418–426.

Schey, K.L., Papac, D.I., Knapp, D.R., and Crouch, R.K. (1992). Matrix-assisted laser desorption mass spectrometry of rhodopsin and bacteriorhodopsin. Biophys J *63*, 1240–1243.

Schindler, P.A., Van Dorsselaer, A., and Falick, A.M. (1993). Analysis of hydrophobic proteins and peptides by electrospray ionization mass spectrometry. Anal Biochem *213*, 256–263.

Shahgholi, M., Garcia, B.A., Chiu, N.H.L., Heaney, P.J., and Tang, K. (2001). Sugar additives for MALDI matrices improve signal allowing the smallest nucleotide change (A:T) in a DNA sequence to be resolved. Nucleic Acids Res *29*, e91/91–e91/10.

Sheumack, D.D., and Burley, R.W. (1988). Separation of lipid-free egg yolk proteins by high-pressure liquid chromatography using solvents containing formic acid. Anal Biochem *174*, 548–551.

Shively, J.E., Hawke, D., and Jones, B.N. (1982). Microsequence analysis of peptides and proteins. iii. artifacts and the effects of impurities on analysis. Anal Biochem *120*, 312–322.

Sowers, K.R., and Gunsalus, R.P. (1988). Adaptation for growth at various saline concentrations by the Archaebacterium *Methanosarcina thermophila*. J Bacteriol *170*, 998–1002.

Tarr, G.E., and Crabb, J.W. (1983). Reverse-phase high-performance liquid chromatography of hydrophobic proteins and fragments thereof. Anal Biochem *131*, 99–107.

Thibault, D.B., Gillam, C.J., Grey, A.C., Han, J., and Schey, K.L. (2008). MALDI tissue profiling of integral membrane proteins from ocular tissues. J Am Soc Mass Spectrom *19*, 814–822.

van Adrichem, J.H.M., Boernsen, K.O., Conzelmann, H., Gass, M.A.S., Eppenberger, H., Kresbach, G.M., Ehrat, M., and Leist, C.H. (1998). Investigation of protein patterns in mammalian cells and culture supernatants by matrix-assisted laser desorption/ionization mass spectrometry. Anal Chem *70*, 923–930.

Vanfleteren, J.R. (1989). Sequential sodium dodecyl sulfate-polyacrylamide gel electrophoresis and reversed-phase chromatography of unfolded proteins. Anal Biochem, 385–390.

Walker, A.K., Rymar, G., and Andrews, P.C. (2005). Virtual 2-D gel electrophoresis by MALDI mass spectrometry. Proteomic Protoc Handb, 417–429.

Part XI
Methods for MALDI MS Enabled Biomedical Applications

Chapter 34
MALDI MS-Based Biomarker Profiling of Blood Samples

Ali Tiss, Celia J. Smith, and Rainer Cramer

Abstract Differential MS analysis of blood samples from diseased and control subjects is increasingly being employed in the hunt for biomarkers that can detect disease at an early stage. For diagnostic tests, in particular for population screening, robust protocols are required that can offer high-throughput analysis, ideally at high mass spectrometric sensitivity. To achieve this, blood samples need to be collected, prepared and analyzed in a standardized manner that minimizes potential bias. Simple purification methods combined with MALDI MS profiling have so far been championed for providing the best approach. In this chapter, we describe an adapted and validated protocol based on a simple and fast solid-phase extraction technique using ZipTips®. This protocol facilitates the purification of potential blood biomarkers in a few steps for mass spectral biomarker pattern diagnostics using MALDI. It is suitable for use in an automated high-throughput and potentially clinical environment and has the advantage of only requiring a few microlitres of blood plasma or serum. The presented protocol has been tested over several years in our laboratory and found to be more reproducible and suitable for plasma and serum profiling than similar methodologies based on magnetic bead purification.

Keywords Biomarkers · MALDI · Mass spectrometry · Plasma/serum profiling · ZipTip® technology

Abbreviations

ACN	Acetonitrile
CHCA	α-Cyano-4-Hydroxycinnamic Acid
LMW	Low Molecular Weight
MALDI	Matrix-Assisted Laser Desorption/Ionization
PCS	Polypeptide Calibrant Standard
QC	Quality Control

R. Cramer (✉)
The BioCentre, University of Reading, RG6 6AS, Reading, UK;
Department of Chemistry, Harborne Building, University of Reading, RG6 6AS, Reading, UK
e-mail: r.k.cramer@reading.ac.uk

A.R. Ivanov, A.V. Lazarev (eds.), *Sample Preparation in Biological Mass Spectrometry*, DOI 10.1007/978-94-007-0828-0_34,

SST	System Suitability Test
TFA	Trifluoroacetic Acid
TOF	Time-Of-Flight

34.1 Introduction

The ultimate goal of biomarker research is to establish straightforward, cost-effective, minimally invasive clinical tests with high sensitivity and specificity for early disease detection. Profiling blood and blood-derived products based on mass spectrometry (MS) is becoming a powerful tool to mine the physiologically informative molecules contained in this biofluid (Petricoin et al., 2002; Rosenblatt et al., 2004; Villanueva et al., 2004). By its nature and availability, blood has the potential to be an ideal medium for elucidating the current state of health. During the last few years many studies, using matrix-assisted laser desorption/ionization (MALDI) time-of-flight (TOF) MS coupled with bioinformatics tools, have been reported as being capable of distinguishing between the polypeptide patterns of healthy individuals and patients diagnosed with various diseases (de Noo et al., 2006; Petricoin et al., 2002; Villanueva et al., 2006; Ye et al., 2007). However, none of the biomarkers identified in these studies have so far been progressed towards their clinical use (Diamandis, 2006; Omenn, 2006).

One of the major obstacles for the direct discovery of new biomarkers in blood is due to its complexity, which includes the presence of salts, lipids, and other biological matrix compounds and the high dynamic range of proteins, making it extremely difficult to detect and analyze disease-specific biomarkers by MS. For example, when untreated plasma or serum is analyzed on a standard MALDI target plate, it does not produce any useful spectra. Thus, to overcome these obstacles there is an obvious need for improved methods to facilitate MALDI MS analysis of blood without losing its diagnostic power. Any such method needs to be employable across many different laboratories and to incorporate high standards and quality control throughout the entire analytical workflow including sample collection, preparation, MS analysis, and data processing and mining (Check, 2004; Coombes et al., 2005; Diamandis, 2004).

As a consequence, detailed studies have been reported with the aim of standardizing protocols and reducing systematic biases. Indeed, it has been shown that many non-biological variations can be introduced during blood sample collection, preparation, and storage, adding to possible deficiencies in study design and inappropriate data analysis approaches (Tiss et al., 2007; Villanueva et al., 2005, 2004; West-Norager et al., 2007). Some of the resulting standardized protocols are now used with more confidence in discovering and validating new biomarkers with the hope of reducing the gap between the discovery of MS-based candidate biomarkers and their clinical application.

The protocol described in this chapter has been developed and refined over a period of more than 5 years, in particular for the analysis of blood serum. It has emerged from extensive tests that have been undertaken in order to establish a

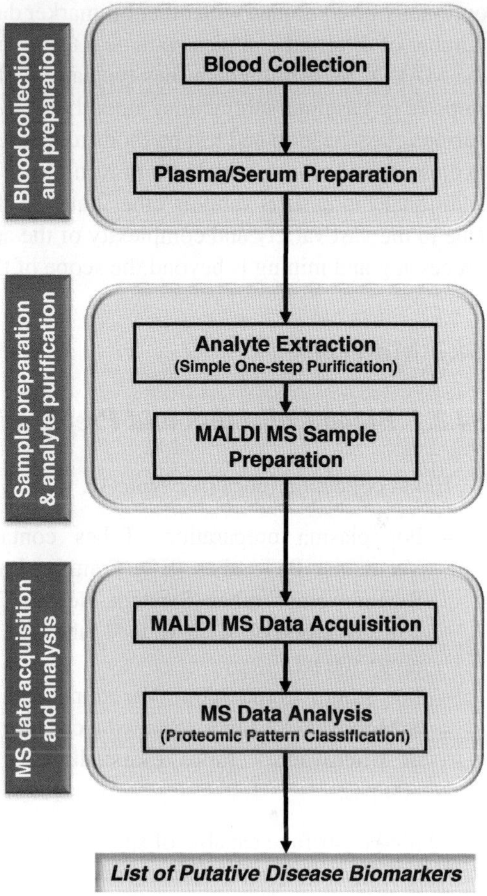

Fig. 34.1 Workflow showing the major steps required for MALDI MS profiling and biomarker discovery from blood

high-throughput protocol that provides both precise and accurate measurements. The protocol is based on MALDI-TOF MS profiling of blood samples that are purified employing commercial ZipTip® technology. As such it is a simple and robust method for providing mass spectral patterns that can be used for sample classification and, thus, potentially for clinical diagnostics and early disease detection.

The overall workflow (see Fig. 34.1) is composed of three main steps which are (i) sample collection, (ii) sample preparation and analyte purification, and (iii) MS data acquisition and analysis. It is suitable for both manual and automated sample preparations and analyses (Tiss et al., 2007). The automated sample preparation procedures have been developed in our laboratory based on an easy-to-operate and cost-effective robotic platform, a CyBi®-disk robot (CyBio AG, Jena, Germany) equipped with a 96-piston head and modified for the use of ZipTips®. The time needed for the robotic preparation of up to 90 samples is only about 30 min. The emphasis of this protocol is on the collection of high-quality data, which can then optionally be combined with other biomarker data or metadata and processed using a number of different complex statistical techniques and bioinformatics. In principle, the MALDI MS profile data files typically undergo a series of data pre-processing steps (e.g. baseline subtraction, smoothing, alignment) before being subjected to various classification and test methods to identify biomarkers (that are differentially expressed in disease), which are in turn tested either alone or in combination with additional biomarkers (and/or other data) to determine their diagnostic accuracy. Due to the vast variety and complexity of the approaches possible in this field, data processing and mining is beyond the scope of this protocol.

34.2 Materials

34.2.1 Blood Collection and Preparation (in Clinical Lab)

1. Blood collection tubes (see Note 1):
   - For plasma preparation: Tubes containing an anti-coagulant, such as Vacutainer PST tubes (BD, Franklin Lakes, NJ, USA, cat. no. 367960) or S-Monovette® tubes (Sarstedt, Newton, NC, USA, cat. no. 02.1065) for up to 7 mL of blood or Microvette® tubes (Sarstedt, cat. no. 20.1309) for volumes up to 300 μL.
   - For serum preparation: Vacutainer SST tubes (BD, cat. no. 367988) or S-Monovette® tubes (Sarstedt, cat. no. 02.1063) for up to 7 mL of blood or Microvette® tubes (Sarstedt, cat. no. 20.1308) for volumes up to 300 μL.
2. Chilled centrifuge capable of spinning blood collection tubes at 13,000 × *g*.
3. Cryovials (e.g. cryostraws) and cryoboxes for sample storage (Fisher Scientific, Loughborough, UK, cat. no. FB71035).
4. Natural microcentrifuge tubes: 0.2 and 1.5 mL (Bioquote, York, UK, cat. nos. BA4300F and 16130).

34.2.2 Sample Pre-processing and Sub-sampling (in Clinical or MS Lab)

1. Standard plasma or serum (see Note 2).
2. Cryoboxes for sample storage (Fisher Scientific, cat. no. FB71035).
3. Natural microcentrifuge tubes: 0.2 and 1.5 mL (Bioquote, cat. nos. BA4300F and 16130).
4. 96-well microtitre plates with lids (Fisher Scientific, cat. no. TUL-962-060A).
5. Low-profile Thermostrips (ABgene, Epson, UK, cat. no. AB0773).
6. Virkon® disinfection solution (Fisher Scientific, cat. no. HYG-205-030 J). See Note 3 for solution preparation.

34.2.3 Analyte Extraction, Purification and Analysis (in MS Lab)

1. Acidification solution (2% TFA, pH $\approx$ 0.5).
2. Equilibration and washing solution (0.1% TFA, pH $\approx$ 2).
3. Elution solution (50% ACN/0.1% TFA).
4. α-cyano-4-hydroxycinnamic acid (CHCA) (Bruker Daltonics, Bremen, Germany, cat. no. 201344).
5. Standard plasma or serum (same as in the previous section).
6. Protein and peptide MS calibration standards (Bruker Daltonics, cat. nos. 206355 and 222570).
7. C18 ZipTips® (Millipore, supplied by Fisher Scientific, cat. no. FDR-597-040G).
8. Natural microcentrifuge tubes: 0.2 and 1.5 mL (Bioquote, cat. nos. BA4300F and 16130).
9. 96-well microtitre plates (Fisher Scientific, cat. no. TUL-962-060A).
10. 30 mL-Teflon® bottle (Fisher Scientific, cat. no. BTK-260-010 V).
11. pH 0-6 indicator sticks (Fisher Scientific, cat. no. FB33005).
12. Isofreeze PCR rack (Alpha-Laboratories, Eastleigh, UK, cat. no. LW5990Y).
13. AnchorChip-600 384-spot MALDI target plates (Bruker Daltonics, cat. no. 209513).

34.3 Methods

Users of this protocol should have read and understood all relevant Risk Assessments and Health and Safety rules and regulations and SOPs for the handling of blood and blood-derived samples before starting any work.

The various procedures detailed in the following protocol can be grouped into four independent steps:

(i) Phlebotomy and plasma/serum generation and separation which usually occur in clinical laboratories or health care centres.
(ii) Sub-sampling the plasma/serum samples which can be performed either immediately after blood collection at the collection site or later in the MS laboratory. At the end of this step small, ready-to-analyse sample volumes are stored in microtitre plates.
(iii) Purification of analytes (mainly polypeptides) from plasma/serum: This part of the protocol is critical and involves binding the analytes to a chromatographic phase followed by a washing step at which non-bound compounds are removed. The analytes are then eluted off the chromatographic support before being mixed with the matrix and spotted onto a MALDI target plate. This step is carried out in the MS laboratory (see Fig. 34.2).

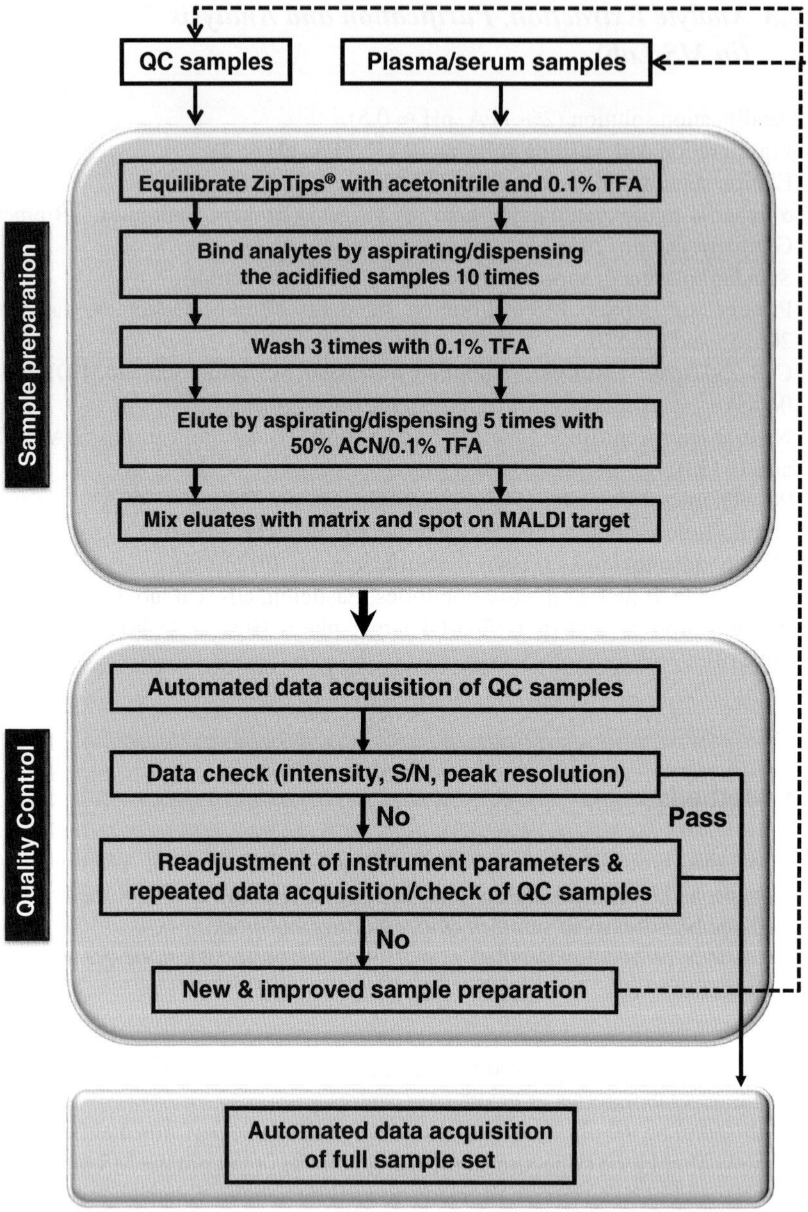

Fig. 34.2 Workflow diagram showing the various steps of plasma/serum sample preparation using ZipTips® and quality control before MS data acquisition

(iv) MS data acquisition and analysis: This last step consists of collecting MALDI mass spectra under strict QC procedures from the samples prepared above (see Fig. 34.2). The obtained data are then analyzed using appropriate software and bioinformatics tools.

34.3.1 Collecting Blood Samples and Preparing Plasma/Serum

1. Collect blood in the appropriate tube (either for plasma or serum) and label it with the corresponding sample details. Gently mix the blood by inverting the tube 10 times (see Note 4). The following steps are suitable for blood volumes between tens of microlitres and tens of millilitres. For other amounts of blood, this protocol (including consumables and equipment) might need to be adjusted.
2. *(a)* For plasma preparation:

   – Centrifuge the blood sample at 2,000×*g* for 5 min at 4°C.
   – Decant the plasma (top layer) to appropriately sized and labelled cryovials, taking care not to disturb the middle whitish layer (mononuclear cells and platelets) or the bottom layer (red blood cells).

   (b) For serum preparation:

   – Allow blood to clot for 1 h at room temperature keeping the tube in the upright position (see Note 5).
   – Spin tubes at 1,400×*g* for 10 min at room temperature.
   – Transfer the serum (upper phase) to appropriately sized and labelled cryovials. If resources (time, staff and sample storage) allow, as many aliquots of the appropriate volume as possible should be made in order to minimize freeze-thaw cycles.

3. Immediately store all samples at –80°C.
4. Transport samples to the MS laboratory on dry ice. On arrival they must be immediately stored at –80°C (see Note 6).

34.3.2 Pre-processing and Sub-sampling of Plasma/Serum Samples

This protocol assumes that plasma or serum samples will be sub-sampled into microtitre plates before the analyte extraction procedure. The following steps detail this procedure (see Note 7).

1. Remove the sample tubes from the freezer, and allow the samples to thaw on an ice-bath (see Note 8).
2. Randomize samples before any further steps are taken to avoid any systematic error due to the sample order or history.
3. Transfer the samples from cryostraws to microcentrifuge tubes (this step is only necessary when cryostraws are used such as for long-term, large-scale sample storage):

   – Remove the first straw from the freezer and scan the barcode on the straw or, if no bar codes are used, write down the sample details.
   – Using scissors, cut the straw just below the closure at the top.

- Invert the straw and place the cut end in a 1.5 mL-tube. Now cut the other end of the straw and dispose of the cut-off pieces in the Virkon® solution. Wash the scissor blades with ethanol.
- Lift the straw slightly to allow the thawed serum to drain into the tube. Use the pipettor with a tip attached to blow air through the straw from the top down, in order to expel the last few drops of plasma/serum into the tube. Place the empty straw into the Virkon® solution. Use the pipettor with the same tip still attached to mix the content of the tube, ensuring that the whole sample is thoroughly thawed and homogeneous.

4. Sub-sampling from microcentrifuge tubes to microtitre plates:
   - Transfer 5 μL of plasma/serum to the corresponding well in one or more microtitre plates as required, leaving on each microtitre plate 6 wells vacant for the use of control samples as detailed later (see Note 9). Record the position of each sample on a microtitre plate plan (see Note 10).
   - From the remaining volume sub-sample as many aliquots as required into 200 μL-tubes or thermostrips (these can be later used for single point analysis).
   - Pipette 5 μL-aliquots of the standard plasma/serum sample (thawed and homogenized) to 3 of the 6 empty wells on the microtitre plates containing samples and record their position on the microtitre plate plans. This standard sample will be used as initial control for the whole run. The 3 remaining empty wells are later used for post-preparation control samples (see SST solution, Section 3.3.1).
5. Cover each microtitre plate with a cover lid and secure the lid in place with a rubber band. Put each plate inside a small clear plastic bag, and place the bags together with a copy of the plate plan in a labelled sample box.
6. Immediately freeze the samples and store them at –80°C until required.

34.3.3 Extracting Analytes from Plasma/Serum Samples

This protocol describes the manual extraction of mainly polypeptidic analytes from blood plasma/serum which is suitable if the number of samples is just a few. It requires the samples to be in a microtitre plate format suitable for the use of an eight-channel pipettor. However, this protocol can also be applied using single tubes but needs to be adjusted accordingly. For large sample sets and high throughput extraction a liquid handling robot is recommended.

34.3.3.1 Preparation of Solutions (see Note 11)

1. Matrix solution (0.5 mg/mL CHCA in ethanol:acetone; 2:1): Add 1 mL of acetone to an 1.5 mL-tube containing 3 mg of CHCA and vortex until the CHCA is dissolved. Add 4 mL ethanol and 1 mL acetone to a 30 mL-Teflon® bottle,

Table 34.1 Polypeptide calibrant list used for external calibration

Polypeptide	Ion type	Average mass (m/z)
Peptide calibration standard II mixture *(Bruker Daltonics, #222570)*		
Bradykinin Fragment 1-7	$[M+H]^+$	757.858
Angiotensin II	$[M+H]^+$	1047.189
Angiotensin I	$[M+H]^+$	1297.486
Substance P	$[M+H]^+$	1348.642
Bombesin	$[M+H]^+$	1620.860
Renin Substrate Tetradecapeptide Porcine	$[M+H]^+$	1760.026
ACTH clip 1-17	$[M+H]^+$	2094.427
ACTH clip 18-39	$[M+H]^+$	2466.681
Somatostatin 28	$[M+H]^+$	3149.574
Protein calibration standard I mixture *(Bruker Daltonics, #206355)*		
Insulin	$[M+H]^+$	5734.56
Ubiquitin I	$[M+2H]^{2+}$	4283.45
Cytochrome C	$[M+2H]^{2+}$	6181.05
Ubiquitin I	$[M+H]^+$	8565.89

then add the CHCA solution to give a total volume of 6 mL. Mix the solution and keep at 4°C in the dark until needed.

2. Polypeptide Calibrant Standard (PCS) solution: Add 125 μL of the 0.1% TFA solution to a vial of peptide calibration standard (see Materials for details) and vortex thoroughly to solubilize the peptides. Do the same with a second vial for the protein calibration standard (see Materials for details). Transfer 25 μL from the first vial to a 1.5 mL-tube and add all the contents of the second vial plus 600 μL of 0.1% TFA. Mix and aliquot at 4 μL. Store the aliquots at –20°C (see Table 34.1 for a list of PCS components).
3. System Suitability Test (SST) solution: A 96-well plate containing 5 μL of the standard plasma/serum in each well is prepared in advance using the steps detailed in Section 3.3.2. (steps 1–4.; see below). Sample eluates are pooled, mixed and then aliquotted at 4 μL. Aliquots are stored at –80 °C until required (see Note 12).

34.3.3.2 Sample Clean-Up and MALDI Sample Preparation

1. Preparing the samples

   - Remove the aliquotted samples from the freezer and thaw them on an Isofreeze PCR rack keeping their temperature at no more than 4°C.
   - Immediately after thawing, add 5 μL of 2% TFA to each sample well and aspirate/dispense several times to ensure good mixing.

2. Preparing the ZipTips® (see Note 13).

   - Load 8 ZipTips® onto a multi-channel pipettor.
   - Aspirate 10 μL of ACN onto each ZipTip® and dispense it to waste. Repeat twice.
   - Aspirate 10 μl of 0.1 % TFA onto each ZipTip® and dispense it to waste. Repeat twice.

3. Binding and washing the analytes

   - Slowly aspirate and dispense the sample solutions back into the same well. Repeat 9 times.
   - Aspirate 10 μL of 0.1 % TFA onto each ZipTip® and dispense it to waste (or collect if the washing fractions are to be analysed). Repeat twice.

4. Eluting the samples

   - Elute the analytes bound to each ZipTip® with 7 μL of 50% ACN/0.1% TFA by aspirating/dispensing the eluent 5 times through the ZipTips® into the wells of a new microtitre plate and keep it on the Isofreeze PCR rack.

5. Mixing the eluates with the matrix

   - Transfer 18 μL of matrix solution into the wells of a new microtitre plate.
   - Add 2 μL of each eluate to the 18 μL of matrix solution. Aspirate/dispense at least 5 times to mix.
   - Immediately spot 0.8 μL of the mixtures onto the anchors of the MALDI target. Spot each eluate 4 times, leaving 12 random anchors vacant for SST samples (see Notes 12 and 14).

6. MALDI calibrant and control sample preparation

   - Transfer 27 μL of the matrix solution to two 200 μL-tubes and close the lids tightly. Take a 4 μL-aliquot of each of the SST and PCS solutions from the freezer and let them thaw at room temperature.
   - Vortex the PCS solution and add a 3 μL-aliquot of it to one of the tubes containing matrix solution, and vortex/mix thoroughly. Spot 0.8 μL of this mixture onto each of 10 random calibrant positions on the MALDI target plate. These spots are used for external calibration. The amount of each polypeptide spotted is ca. 10 fmol for peptides and ca. 50 fmol for proteins.
   - Vortex the SST solution and add a 3 μL-aliquot of it to the other tube containing matrix solution, and vortex/mix thoroughly. Spot 0.8 μL of this mixture onto their allocated positions (12 per MALDI target plate, see Notes 9 and 12). These spots will be used to assess the performance of the MALDI sample preparation and MS analysis (see details below in Section 3.4.2).
   - Wait until all MALDI sample spots are completely dry (see Note 14).

34.4 MALDI MS Data Acquisition and Analysis

34.4.1 Data Acquisition

This protocol describes a typical workflow employing an Ultraflex II MALDI-TOF/TOF instrument (Bruker Daltonics) in the linear positive ion mode controlled by the FlexControl software v.3.0 with samples spotted on AnchorChip target plates having 384 anchors of 600 μm diameter.

1. Open the FlexControl software and load optimized FlexControl and AutoXecute methods (see Table 34.2 and Note 15). These settings must be optimized for each individual instrument.
2. Insert the MALDI target plate into the mass spectrometer. The plate should be analyzed within 2 h of MALDI sample spotting.
3. Calibrate the mass spectrometer using a single calibrant sample spot containing the PCS mixture (see Note 16). Data should be collected automatically by clicking on the tab "Run method on current spot".
4. Collect spectra from an SST sample spot to check that shots are accumulated and the AutoXecute evaluation criteria are met (see Note 17 and Fig. 34.2). If this is not the case, re-adjust the laser power accordingly and collect spectra from another SST spot to fulfil these conditions. Data should be collected automatically by clicking on the Tab "Run method on current spot". A typical mass spectrum obtained from a pooled human serum sample is shown in Fig. 34.3.
5. Load the correct autoXsequence file (containing the names and positions of each sample on the target as well as the file name and path where the spectra will be saved on the hard disk). This file can be generated by using the spreadsheet from the sample plate plan initially prepared.

Table 34.2 Ultraflex instrument settings optimized for the analysis of LMW plasma/serum analytes, using FlexControl software v3.0

Parameters	Values
Laser source	Nitrogen (337 nm)
Digitizer	2 GHz
Mass range for data acquisition	700–10,000 Da
Acquisition mode	Linear, positive ions
Ion acceleration	25 kV
Delayed extraction potential difference	1.6 kV
Lens potential	6 kV
Gating strength for matrix suppression	High (for ions <400 Da)
Delayed extraction	100 ns
Detector gain	7.5
Sample rate	0.5 GS/s
Data points per spectrum	32,163

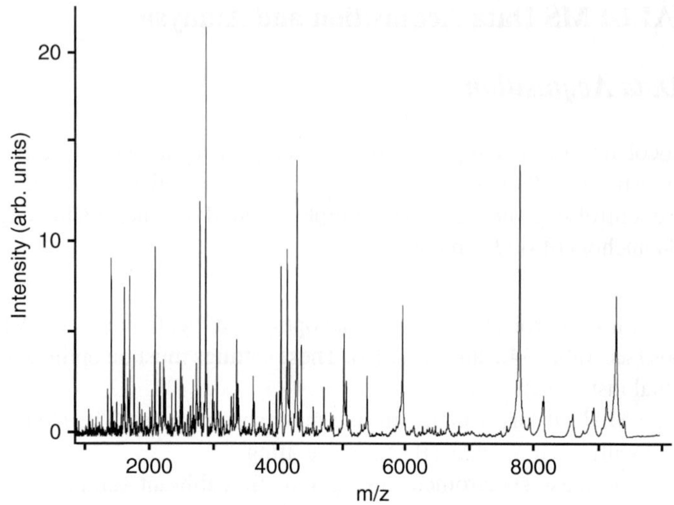

Fig. 34.3 MALDI-TOF MS profile of a pooled human blood serum sample obtained from purification of a 5 μL-aliquot using C18-ZipTips®. The mass spectrum was generated by ClinProTools software (V2.1) after baseline-subtraction, smoothing and normalization

6. Start the automatic run by clicking on the tab "Start automatic run". Data collection for a full 384-spot plate (spotted from a full 96-well microtitre plate) takes about 6 h.

34.4.2 Data Quality Control and Filtering/Processing

In the following steps FlexAnalysis 3.0 and ClinProTools 2.1 software (Bruker Daltonics) are used.

1. Within the FlexAnalysis program, open the folder containing all the acquired spectra by choosing the "Process" menu, then selecting "Batch check" and clicking on "Folder". Within the next window click on "Browse" and select the batch of sample data folders. Click on "Start Batch Check" and set the expected number of shots to 1,000. By clicking "OK" this will generate a report detailing the spectra with less than 1,000 shots which should then be excluded from further processing.
2. Within the ClinProTools program, load the spectra files of the SST samples by selecting "Open Model Generation Class" under the "File" menu. View the spectra in the pseudo-gel view.
3. Load the spectra of the standard plasma/serum samples, which will be depicted in the pseudo-gel view alongside the spectra of the SST samples. The two classes should give virtually identical spectra, in which case the overall preparation of the batch of samples is deemed acceptable. If the spectra are markedly different

or two or more spectra per sample have been excluded further investigation will be needed to determine the cause.

4. Load the spectra of each individual sample as a class into ClinProTools, and inspect them visually to determine any obvious differences within that class. If differences are found it may be necessary to exclude the entire data set for this sample on the basis of poor quality (see Note 18).

34.4.3 Preliminary Data Analysis

Curated spectra sets can be analysed using ClinProTools or exported from FlexAnalysis as tab-separated 2-column text files that contain the mass list with the corresponding intensities for use with other software packages for more detailed analysis of the data. For instance, ClinProTools can be used to determine peaks that are significantly different (in intensity or area) between 2 or more classes of spectra, by using the average spectra and clustering feature. It is also possible to use this program to generate other classification models (for example, Genetic Algorithms, Support Vector Machine) based on lists of discriminatory peaks. However, this higher level of data analysis and data mining is beyond the scope of this protocol.

Notes

1. Recommendations vary for the best anti-coagulant and whether to use plasma or serum (Omenn et al., 2005). For plasma analysis, this protocol has been developed using small mouse blood volumes collected with Microvette® tubes (Sarstedt, Newton, NC, USA, cat. no. 20.1309) containing heparin as anti-coagulant. For serum analysis, this protocol has been developed using human serum obtained from blood collected with Vacutainer SST tubes (BD, USA, Franklin Lakes, NJ, cat. no. 367988).
2. For plasma analysis, this protocol has been developed with commercial standard mice plasma from C57BL6 mice (Patricell, Nottingham, UK, cat. no. IMS-C57BL6-N-12). For serum analysis, this protocol has been developed with commercial standard human serum (Sigma-Aldrich, Dorset, UK, cat. no. S7023).
3. Virkon® solution: Add 10 g of Virkon® powder to 1 L of water. Mix thoroughly until completely dissolved, resulting in a pink coloured solution. Replace the solution when the pink colour has gone (about 1 week).
4. When collecting blood samples, a proper phlebotomy technique should be applied. Ideally, the same amount of blood is collected for each sample. Haemolysis should be avoided at this step as well as any other (subsequent) steps.
5. Blood should be allowed to clot for exactly 1 h, otherwise differences in the mass spectral profiles might arise due to different clotting times of the samples.
6. Avoid freeze-thaw cycles at all steps because they induce proteolysis (in the absence of anti-proteases) and the precipitation of peptides/proteins.
7. All sub-sampling must be carried out in an adequate safety cabinet. Equipment and consumables used for the purpose of sub-sampling should be housed inside the cabinet and should be labelled with biohazard labels (tubes, pipette tips, solutions, etc.). Before starting work, ensure that there are two 1 L Virkon® solutions available that are less than 1 week old and still pink. One of the solutions should be placed in a wide-mouthed screw cap jar for disposal of plastic consumables, and the other should be in a screw cap bottle for dealing with accidental spills.

8. The amount of time that individual samples are outside of the -80°C freezer must be kept to a minimum. All sample vials, tubes and plates must be kept on wet ice or Isofreeze PCR racks during the entire sub-sampling procedure. Standard plasma/serum samples should be initially aliquotted at 200–500 μL into 500-μL tubes and stored at –80 °C until required.
9. The sample volume available determines the number of replicates that can be prepared as 5 μL is needed for each preparation. All samples with a total volume of less than 5 μL should be excluded. It is preferable to prepare each sample at least 3 times. However, if this is not possible then for each individual sample as many replicates as possible should be made. Each 96-well microtitre plate is used for up to 90 sample aliquots to be run manually or on a liquid handling robot. The remaining 6 wells are vacant for the use of control samples as detailed later. Briefly, 3 vacant wells will be filled with standard plasma/serum samples that will provide controls for the entire sample preparation workflow. The other 3 vacant wells will remain empty in case the microtitre plate is used in an automated workflow. From these wells no samples will be prepared or spotted on the respective MALDI target plate positions, which then can be used for the preparation of SST samples (see Note 12).
10. In an automated workflow for large sample sets, the sub-sampling procedure can be carried out in two steps. First from tubes into parent microtitre plates, then from parent to child (replicate) microtitre plates. For example, in our lab we use a liquid handling CyBi®-Disk robot (CyBio AG) equipped with a 96-piston head for 25 μL-tips and 10 microtitre plate positions, specifically set-up for this task.
11. The matrix solution should be freshly prepared on the day of the experiment. The solutions of 2% TFA, 0.1% TFA and 50% ACN/0.1% TFA should be replaced weekly.
12. SST solutions/samples are used as controls to check both the performance of the mass spectrometer and the quality of the MALDI sample preparation.
13. Ensure that ZipTips® do not dry out or aspirate air during the entire procedure.
14. Ensure that the relative humidity in the lab is between 35 and 60% during MALDI sample preparation (incl. crystallization). Very low or very high humidity dramatically affects the crystallization and thus, the MALDI MS ion signal.
15. Data acquisition should be automated. In this protocol, an automated workflow is provided for an Ultraflex II TOF/TOF instrument using the instrument's "AutoXecute" function. Each spectrum is the sum of 1,000 single laser shots randomized over 10 positions within the same spot (100 shots/position) at a laser frequency of 25 Hz. Before each acquisition cycle of 100 laser shots, the new position on the target spot is pre-irradiated with 10 laser shots at 10% higher laser power to improve spectral quality. Evaluation parameters are set so that only spectra (of 100 shots) containing at least one peak with a resolving power of greater than 300 and a signal-to-noise ratio of more than 10 in the m/z range of 700–10,000 are accumulated.
16. The calibration can be checked on a second PCS spot and repeated if necessary.
17. Prior to analyzing an entire MALDI target, an SST sample spot should be used for adjusting the laser power and as a general system check. Using this control sample, the resolution and intensity of five major peaks across the mass range 1,800–8,200 Da are typically checked against previously obtained data. If necessary, the laser energy is adjusted to keep the intensity and resolution of these peaks within a range of ± 2 standard deviations. In addition, the overall performance of the mass spectrometer should be thoroughly checked every 2–3 weeks using peptide standards.
18. If the number of replicate spectra per sample falls below 3 as a result of spectra exclusion, it may be necessary to repeat the sample preparation, if this is possible.

References

Check, E. (2004). Proteomics and cancer: Running before we can walk? Nature *429*, 496–497.

Coombes, K.R., Morris, J.S., Hu, J., Edmonson, S.R., and Baggerly, K.A. (2005). Serum proteomics profiling–a young technology begins to mature. Nat Biotechnol *23*, 291–292.

de Noo, M.E., Deelder, A., van der Werff, M., Ozalp, A., Mertens, B., and Tollenaar, R. (2006). MALDI-TOF serum protein profiling for the detection of breast cancer. Onkologie *29*, 501–506.

Diamandis, E.P. (2004). Mass spectrometry as a diagnostic and a cancer biomarker discovery tool: Opportunities and potential limitations. Mol Cell Proteomics *3*, 367–378.

Diamandis, E.P. (2006). Validation of breast cancer biomarkers identified by mass spectrometry. Clin Chem *52*, 771–772; author reply 772.

Omenn, G.S. (2006). Strategies for plasma proteomic profiling of cancers. Proteomics *6*, 5662–5673.

Omenn, G.S., States, D.J., Adamski, M., Blackwell, T.W., Menon, R., Hermjakob, H., Apweiler, R., Haab, B.B., Simpson, R.J., Eddes, J.S., *et al.* (2005). Overview of the HUPO plasma proteome project: Results from the pilot phase with 35 collaborating laboratories and multiple analytical groups, generating a core dataset of 3020 proteins and a publicly-available database. Proteomics *5*, 3226–3245.

Petricoin, E.F., Ardekani, A.M., Hitt, B.A., Levine, P.J., Fusaro, V.A., Steinberg, S.M., Mills, G.B., Simone, C., Fishman, D.A., Kohn, E.C., *et al.* (2002). Use of proteomic patterns in serum to identify ovarian cancer. Lancet *359*, 572–577.

Rosenblatt, K.P., Bryant-Greenwood, P., Killian, J.K., Mehta, A., Geho, D., Espina, V., Petricoin, E.E., and Liotta, L.A. (2004). Serum proteomics in cancer diagnosis and management. Annu RevMed *55*, 97–112.

Tiss, A., Smith, C., Camuzeaux, S., Kabir, M., Gayther, S., Menon, U., Waterfield, M., Timms, J., Jacobs, I., and Cramer, R. (2007). Serum peptide profiling using MALDI mass spectrometry: Avoiding the pitfalls of coated magnetic beads using well-established ziptip technology. Proteomics *7*, 77–89.

Villanueva, J., Philip, J., Chaparro, C.A., Li, Y., Toledo-Crow, R., DeNoyer, L., Fleisher, M., Robbins, R.J., and Tempst, P. (2005). Correcting common errors in identifying cancer-specific serum peptide signatures. J Proteome Res *4*, 1060–1072.

Villanueva, J., Philip, J., Entenberg, D., Chaparro, C.A., Tanwar, M.K., Holland, E.C., and Tempst, P. (2004). Serum peptide profiling by magnetic particle-assisted, automated sample processing and MALDI-TOF mass spectrometry. Anal Chem *76*, 1560–1570.

Villanueva, J., Shaffer, D.R., Philip, J., Chaparro, C.A., Erdjument-Bromage, H., Olshen, A.B., Fleisher, M., Lilja, H., Brogi, E., Boyd, J., *et al.* (2006). Differential exoprotease activities confer tumor-specific serum peptidome patterns. J Clin Invest *116*, 271–284.

West-Norager, M., Kelstrup, C.D., Schou, C., Hogdall, E.V., Hogdall, C.K., and Heegaard, N.H. (2007). Unravelling in vitro variables of major importance for the outcome of mass spectrometry-based serum proteomics. J Chromatogr B Anal Technol Biomed Life Sci *847*, 30–37.

Ye, B., Gagnon, A., and Mok, S.C. (2007). Recent technical strategies to identify diagnostic biomarkers for ovarian cancer. Expert Rev Proteomic *4*, 121–131.

Chapter 35
Sample Preparation in Biological Analysis by Atmospheric Pressure Matrix Assisted Laser/Desorption Ionization (AP-MALDI) Mass Spectrometry

Appavu K. Sundaram, Berk Oktem, Jane Razumovskaya, Shelley N. Jackson, Amina S. Woods, and Vladimir M. Doroshenko

Abstract Rapid detection and identification of biological molecules is of utmost importance in a number of applications including clinical diagnosis and detection of foodborne pathogens as well as environmental contaminants. Mass spectrometry affords a high degree of accuracy (specificity) and sensitivity without a need for extensive sample purification. Atmospheric-pressure matrix-assisted laser desorption/ionization (AP-MALDI) mass spectrometry has become a useful technique in a number of important applications, including identification of proteins, and structural analysis of oligosaccharides, phosphopeptides, and lipids. Sample preparation methods for AP-MALDI mass spectrometry are relatively simple, rapid and can be easily automated for high throughput sample analysis. AP-MALDI MS/MS analysis has been successfully utilized to identify the presence of different types of bioagents such as *Escherichia coli (E. coli)*, spores of several *Bacilli*, *Saccharomyces cerevisiae* (yeast), ovalbumin, and *Clostridium botulinum* Neurotoxin Type A in environmental samples. AP-MALDI imaging mass spectrometry is an emerging technique for direct analysis of biological tissue sections that can be useful for profiling spatial distribution of drugs, peptides and proteins in tissue sections of plants, animals, and humans.

Keywords AP-MALDI MS · Proteomics · Tandem mass spectrometry · Tissue imaging · Tryptic digestion

35.1 Introduction

Rapid and accurate detection and identification of biological molecules is of utmost importance in a number of applications such as clinical diagnosis, detection of pathogens in contaminated food and detection of biowarfare agents in environmental

---

A.K. Sundaram (✉)
Science and Engineering Services Inc., Columbia, MD 21046, USA
e-mail: asundaram@apmaldi.com

A.R. Ivanov, A.V. Lazarev (eds.), *Sample Preparation in Biological Mass Spectrometry*, DOI 10.1007/978-94-007-0828-0_35,

samples (Demirev and Fenselau, 2008; Dover et al., 2009). Analysis and detection of species specific biomarkers (peptides, proteins, nucleic acids, lipids or carbohydrates) are widely used for identification of microbial pathogens by immunoassays, PCR, and mass spectrometry (Demirev and Fenselau, 2008; Edwards et al., 2006; Hang et al., 2008; Hindson et al., 2005; Lim et al., 2005; Rowe-Taitt et al., 2000; Sapsford et al., 2004; Stachowiak et al., 2007; Warscheid and Fenselau, 2004; Yao et al., 2002a; Gudlavalleti et al., 2008). Specificity and sensitivity of the detection technology depend heavily on the sample preparation prior to analysis by the respective method. While PCR can be very sensitive and specific, it requires extensive purification of the sample's DNA for analysis. Similarly, immunoassay also requires extensive sample purification to reduce non-specific binding of structurally similar molecules.

Mass spectrometry yields a high degree of accuracy (specificity) and sensitivity without a need for extensive sample purification. Development of soft ionization techniques such as electrospray ionization mass spectrometry (ESI MS) and matrix assisted laser desorption/ionization mass spectrometry (MALDI MS) has significantly improved characterization and quantification of thermally labile biomolecules such as proteins and peptides (Demirev and Fenselau, 2008; Doroshenko and Cotter, 1996; Fenselau and Demirev, 2001; Whiteaker et al., 2004). MALDI MS and ESI MS in combination with high-performance liquid chromatography (HPLC) are widely used in biological research, especially in the pharmaceutical industry.

Mass spectrometry has greatly enhanced proteomics research, especially microbial proteomics. Top down proteomics, in which there is no need to enzymatically digest the proteins prior to analysis by MS, has been successfully used to analyze *Bacillus* spores for species identification. Bottom-up proteomics on the other hand involves proteolytic digestion of the proteins, followed by MS and MS/MS analysis of the resulting peptides. Fragmentation patterns of each peptide provide peptide's sequence and are analyzed by bioinformatics tools which search protein databases to identify the proteins. Traditional sample preparation for MS-based proteomic analysis involves expensive and time consuming techniques such as 2D electrophoresis and high-performance liquid chromatography for purification of analytes from complex mixtures.

Atmospheric-pressure matrix-assisted laser desorption/ionization (AP-MALDI) mass spectrometry has become a useful technique in a number of important applications, including the identification of proteins, the structural analysis of oligosaccharides, lipids and the study of phosphopeptides (Doroshenko et al., 2002; Laiko et al., 2002; Madonna et al., 2003; Pribil et al., 2005; Seggern et al., 2003; Sundaram et al., 2008b; Tan et al., 2004). With the tandem MS (MS/MS) capabilities offered by ion trap mass spectrometers, AP-MALDI MS/MS has become a powerful tool for confirmation of protein identities. Sample preparation for AP-MALDI is relatively simple and rapid without the added step of loading the sample into vacuum. Recently developed pulsed dynamic focusing (PDF) technique was shown to significantly improved the sensitivity of AP-MALDI in ion trap mass spectrometry. AP-MALDI MS/MS analysis has been successfully utilized to identify the presence of bioagents such as *Escherichia coli (E. coli)*, spores of several *Bacilli*, *Saccharomyces cerevisiae* (yeast), ovalbumin, and *Clostridium botulinum* Neurotoxin Type A (Oktem et al., 2007; Sundaram et al., 2008a; Taranenko et al., 2005a,b).

MALDI imaging mass spectrometry is an emerging technique for direct analysis of biological tissue sections (Caprioli et al., 1997). MALDI imaging can be used to profile spatial distribution of drugs, lipids, peptides, and proteins in tissue sections of plants, animals and humans (Jackson et al., 2007; Wang et al., 2008; Liu et al., 2007; Maddalo et al., 2005; Reyzer et al., 2003). Although AP-MALDI MS imaging has previously been reported for the cases where infrared lasers were employed, the imaging was limited to application of tissues with high water content, such as plant tissues (Li et al., 2007). Here we report AP-MALDI imaging using a high repetition rate UV laser. It offers important benefits as it is a softer ionization technique compared with vacuum MALDI and it can be used for thicker tissue slices without drying. AP-MALDI also enables the use of volatile matrices such as DHA (2, 6-dihydroxyacetophenone). AP-MALDI source coupled with ion-trap, QTOF or FT-ICR instruments provides MS/MS possibility for AP-MALDI tissue imaging.

35.2 Materials and Methods

35.2.1 Reagents

Trifluoroacetic acid (TFA), LC MS grade acetonitrile, ammonium hydroxide, LCMS grade water, 1-octadecanethiol, absolute ethanol, ammonium bicarbonate, α-cyano-4-hydroxy-cinnamic acid (CHCA), and 2, 5 dihydroxy benzoic acid (DHB) were purchased from Sigma-Aldrich Chemical Company (St. Louis, MO, USA). Trypsin immobilized on polystyrene beads was purchased from Applied Biosystems (Foster City, CA, USA).

35.2.2 AP-MALDI MS

All mass spectrometry experiments were carried out on a Thermo Finnigan (San Jose, CA, USA) LCQ Deca XP ion trap mass spectrometer integrated with an AP-MALDI source with pulsed dynamic focusing (MassTech Inc., Columbia, MD, USA) using positive ionization mode (Doroshenko et al., 2002; Laiko et al., 2000). A high repetition rate (200 Hz) all-solid-state Nd:YAG laser ($\lambda = 355$ nm) was used. Laser pulse duration was approximately 5 ns, and the laser beam was focused to approximately 500 μm size spot. During analysis, the laser energy was attenuated to about 60 μJ/pulse for MS/MS data collection and to about 12 μJ/pulse for MS data collection. Sample spots and laser pulse were observed on a monitor using a CCD camera. Experiments were carried out in a repetitive laser shot mode (frequency 30 Hz). Operating conditions for all AP-MALDI MS and AP-MALDI MS/MS experiments were as follows: LCQ Deca XP automatic gain control (AGC) was turned off, ion injection time was 220 ms, and the temperature of the capillary was held at 280°C. Spectra were acquired over a 1 min collection time and then averaged.

For tissue imaging experiments the following set up was used. AP-MALDI source equipped with a 200 Hz Nd:YAG laser at 355 nm wavelength was coupled with LCQ Classic (Thermo Finnigan, San Jose CA). The commercial software for the AP-MALDI source includes constant speed raster motion option which provides scanning of target surface synchronously with the mass spectral data acquisition. Separate software was used to convert the recorded set of mass spectra to 2D-MS image. Laser spot size was a major concern for this application. Commercial AP-MALDI source comes with a standard 400 μm diameter optical fiber. For imaging we used a smaller diameter optical fiber (200 μm) for better beam quality. This was a compromise of four sets of data acquired with 50,100, 200 and 400 μm fibers (data not shown).

35.2.3 Microbial Sample Preparation

Bacillus cereus 14579, MS2 bacteriophage virus along with the host *E. coli* cells, and E. coli O157:H7 cells were purchased from ATCC (Manassas, VA). Spores of *B. cereus* were grown on new sporulation medium (Hang et al., 2008; Perdue et al., 2003) as needed. *E. coli* cells were grown in lysogeny broth (LB medium). MS2 bacteriophage was propagated on *E. coli* cells growing in LB medium.

AP-MALDI MS based identification of microorganisms utilizes the "bottom-up" proteomics approach and the on-probe sample preparation protocol on a C18 coated MALDI target plate is illustrated in Scheme 35.1 (Sundaram et al., 2006). In general, the microbial sample was first treated with an extraction reagent that selectively extracts only a few proteins from any particular microorganism. Extracted protein was then digested using trypsin and the tryptic peptides were rapidly cleaned by washing with LC MS grade water to remove the salts, detergents and other contaminants, while the peptides remained bound to the C18 groups on the target. Clean tryptic peptides were then co-crystallized with CHCA and analyzed by AP-MALDI MS and MS/MS. AP-MALDI MS/MS data was then further analyzed by bioinformatics (using MASCOT search engine, Matrix Science, London, UK) to search the publicly available NCBInr protein database. The MASCOT search result obtained

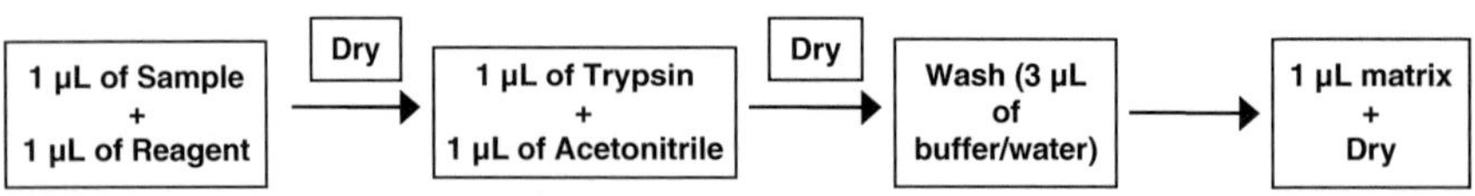

Agent specific extraction reagents:

Spores: 10% trifluoro acetic acid (TFA)
Virus: 14% ammonium hydroxide solution
Toxins: None

Scheme 35.1 On-probe sample processing protocol

after such an analysis provided the amino acid sequence (based on the fragment pattern) of the peptide analyzed, a list of proteins present in the database that contain that particular peptide sequence, and the organisms which express those proteins. Therefore, identification of an organism was made by analyzing the AP-MALDI MS/MS data for the unique peptide biomarkers of a species (peptides that are present only in that microorganism and not in others).

35.2.3.1 Bacterial Spores

Fenselau and co-workers have previously reported methods for rapid extraction of small acid soluble proteins (SASPs) from various *Bacillus* spores by treating with 10% trifluoroacetic acid and analyzing them by mass spectrometry (Hathout et al., 1999; Warscheid and Fenselau, 2003; Warscheid et al., 2003). We have modified this extraction protocol to suit the rapid analysis of microbial samples by AP-MALDI MS. To 1 μL of bacterial spores on a MALDI target plate an equal volume of 50% acetonitrile in 1% trifluoroacetic acid was added and the sample was allowed to dry out. As the sample dries out, SASPs were extracted out from the spores. The extracted SASPs were then digested by adding 1 μL of trypsin and incubating at 50°C until the sample dried out. At this stage the tryptic peptides were purified by the on-probe (C18 coated target plate) sample clean up method by washing with water. The purified tryptic peptides were then co-crystallized with CHCA and then analyzed by AP-MALDI MS and MS/MS.

35.2.3.2 Viruses

Detection and identification of viral particles by AP-MALDI MS was achieved by analyzing the coat proteins of the viruses (Thomas et al., 1998; Yao et al., 2002a,b). Coat proteins or capsid proteins of virus particles were extracted by treating them with an equal volume of 14% ammonium hydroxide solution, followed by digestion using trypsin. Tryptic peptides were then cleaned by the on-probe C-18 cleanup method, co-crystallized with CHCA and then analyzed by AP-MALDI mass spectrometry. AP-MALDI MS/MS data collected for the tryptic peptides were then used for searching the publicly available protein databases to identify the extracted coat protein, and hence, the virus that synthesizes this protein.

35.2.3.3 Toxins

Biological toxins (mostly proteinaceous in nature) were directly treated with trypsin and the resulting tryptic peptides were cleaned further by the on-probe clean up method as described above for the spores and viruses, co-crystallized with CHCA and then analyzed by AP-MALDI MS and MS/MS.

35.2.3.4 Vegetative Bacteria

Sample preparation of bacterial cells for analysis by AP-MALDI MS included treatment with equal volume of 1% trifluoroacetic acid in 50% acetonitrile to extract membrane bound proteins such as acyl carrier proteins, outer membrane proteins, and heat shock and cold shock proteins. Extraction of such proteins from vegetative cells occurs as the sample spot dries out. After the sample spot completely dried out, trypsin was added to digest the extracted proteins. After digestion the peptides were purified by the on-probe C18 clean up method, followed by co-crystallization with CHCA and AP-MALDI MS and MS/MS analysis.

35.2.4 Tissue Sample for Imaging

Rat brain samples were provided by Dr. Amina Woods. Rat brain tissue sections were 18 μm thick. The sections were cut using a cryostat (Leica Microsystems CM3050S). No optimal cutting temperature (OCT) compound was used as it has been reported to interfere with MALDI signal. Tissue sections were prepared with and without washing by 70% ethanol. A pneumatic nebulizer was used to aerosolize 2,6-dihydroxyacetophenone (20 mg/ml in 50% Ethanol) solution (Jackson et al., 2005a). Spray distance is critical as the droplets should not dry before they hit the tissue surface. Similar solutions of CHCA and 2,5-dihydroxy benzoic acid (DHB) were tried as well.

35.3 Results and Discussion

35.3.1 Analysis of Spores

Extraction of small acid soluble spore proteins (SASPs) from various *Bacillus* spores by treating with 10% trifluoroacetic acid followed by analysis using mass spectrometry has been reported by Fenselau and co-workers. SASPs are present only in spores and are consumed by enzymatic hydrolysis during the spore germination process. Therefore, SASPs serve as reliable targets for detecting the presence of spores. We have since modified this extraction protocol, as described in the experimental section, to suit the rapid analysis of microbial samples by AP-MALDI MS. Figure 35.1 shows the AP-MALDI MS spectrum recorded for the trypsin digested SASPs from *B. cereus* ATCC 14579 spores. Figure 35.2 shows the AP-MALDI MS/MS spectrum of the peptide ion with *m/z* value of 1534 from Fig. 35.1. Similar AP-MALDI MS/MS data is collected for all the ions of interest from Fig. 35.1 and the collected AP-MALDI MS/MS data are used for searching the NCBInr protein database to obtain the amino acid sequence of the peptides and hence the corresponding protein and the organism identification. Although majority of these peptides are common to many members of the *Bacillus* group, we can identify the unique peptide biomarker that is present only in *B. cereus* ATCC 14579 spores. From the database search analysis the peptide ion with *m/z* value of 1534 (with amino acid sequence: LVSLAEQQLGGYQK) was identified as the unique biomarker for *B. cereus* 14579 spores (see Table 35.1). Similarly, other bacterial spores can be subjected to this protocol to extract and digest the spore specific proteins, and to identify the biomarker peptides for spores of different bacteria.

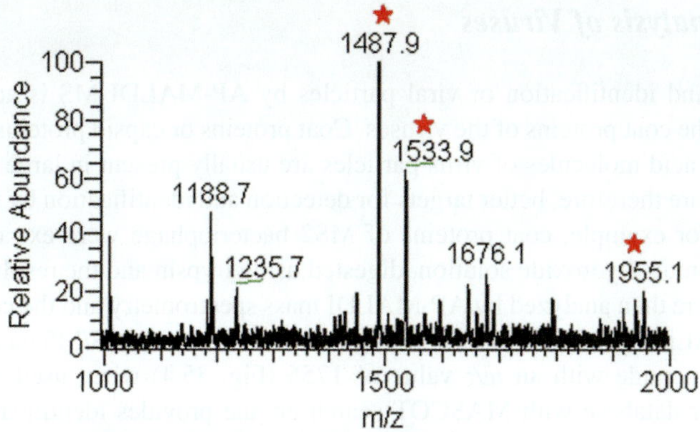

Fig. 35.1 An AP-MALDI MS spectrum of tryptically digested SASPs from *B. cereus*14579 spores. Peptides from SASPs are indicated with asterisks

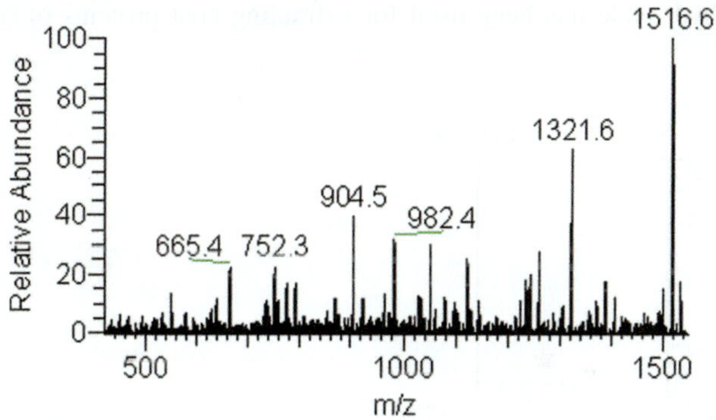

Fig. 35.2 An AP-MALDI MS/MS spectrum of the peptide ion with *m/z* = 1534 shown in Fig. 35.1

Table 35.1 Peptide biomarkers utilized in AP-MALDI MS/MS based detection of microbial pathogens and toxins

$[M+H]^+$ Observed	Amino acid sequence	Protein ID	Organism
1534	LVSLAEQQLGGYQK	SASP (gi|30022721)	*B. cereus* 14579
1755	VATQTVGGVELPVAAWR	Coat Protein (gi|15083)	MS2 bacteriophage
1972	DTFTNPEEGDLNPPPEAK	Botulinum Neurotoxin type A precursor (gi|399133)	*C. botulinum*
1715	ITTVQAAIDYINGHQA	Acyl carrier protein (gi|16129057)	*E. coli* K12
1655	LGYPITDDLDIYTR	Outer membrane protein A precursor (gi|71159605)	*E. coli* K12
1572	TFLSPWISNIHEK	Protective antigen (B. anthracis)	*B. anthracis*

35.3.2 Analysis of Viruses

Detection and identification of viral particles by AP-MALDI MS is achieved by analyzing the coat proteins of the viruses. Coat proteins or capsid proteins enclosing the nucleic acid molecules of virus particles are usually present in large number of copies and are therefore, better targets for detection and identification by mass spectrometry. For example, coat proteins of MS2 bacteriophage were extracted using 14% ammonium hydroxide solution, digested using trypsin and the resultant tryptic peptides were then analyzed by AP-MALDI mass spectrometry and the corresponding AP-MALDI spectrum is shown in Fig. 35.3. AP-MALDI MS/MS data collected for tryptic peptide with an *m/z* value of 1755 (Fig. 35.4) when used for searching NCBInr database with MASCOT search engine provides identification of the marker peptide (with amino acid sequence of VATQTVGGVELPVAAWR) for the MS2 group viruses and is therefore used as the biomarker peptide for detection and identification of MS2 bacteriophage. Therefore, treatment of viruses with 14% ammonium hydroxide has been used for extracting coat proteins of other viruses

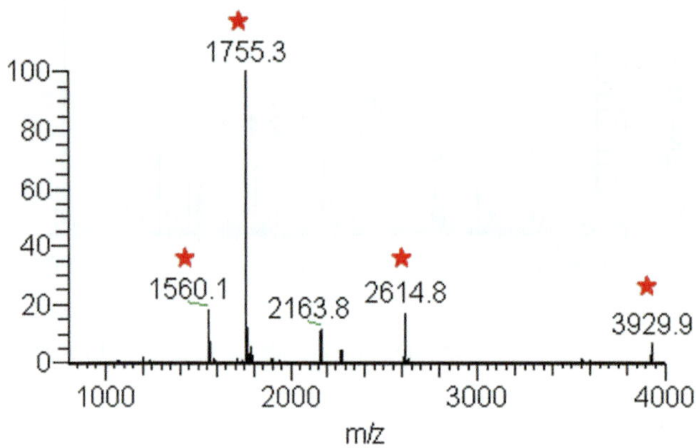

Fig. 35.3 An AP-MALDI MS spectrum of tryptically digested coat protein from MS2 bacteriophage. Peptides from the coat protein are indicated with asterisks

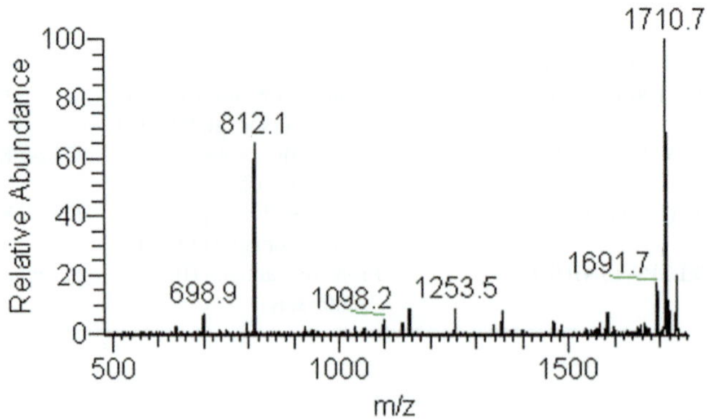

Fig. 35.4 An AP-MALDI MS/MS spectrum of the peptide ion with $m/z = 1755$ from Fig. 35.3

such as human adenovirus type 5, and then digested with trypsin before analysis by AP-MALDI MS and identification by a database-driven bioinformatics approach.

35.3.3 Analysis of Toxins

Toxins of biological origin are mostly proteinaceous in nature, with or without some posttranslational modifications in their structure. These type of toxins include botulinum neurotoxins, ricin, staphylococcus enterotoxin B etc. Micro-sequencing of the tryptic peptides of such toxins and bioinformatics analysis can be used effectively to detect and identify their presence in a sample. Toxin sample can be directly treated with trypsin, followed by rapid purification by the on-probe sample clean-up method and then analyzed by AP-MALDI MS and MS/MS. An AP-MALDI MS spectrum of Botulinum neurotoxin A is shown in Fig. 35.5 and the AP-MALDI MS/MS spectrum of the biomarker peptide ion with *m/z* value of 1972, is shown in Fig. 35.6. MASCOT database search using this data identified the presence of

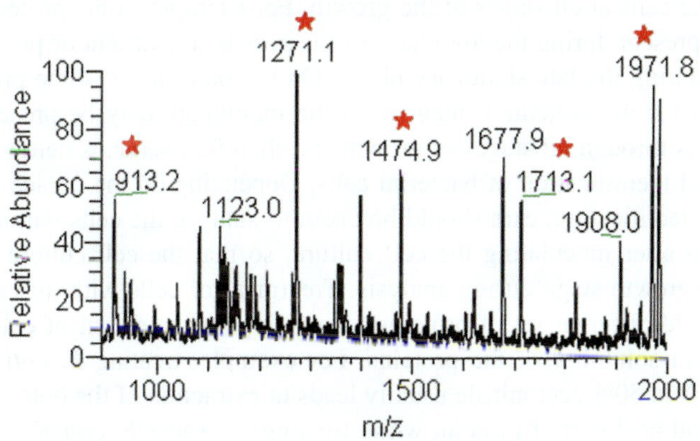

Fig. 35.5 An AP-MALDI MS spectrum of tryptically digested Botulinum neurotoxin type A. Peptides from the neurotoxin are indicated with asterisks

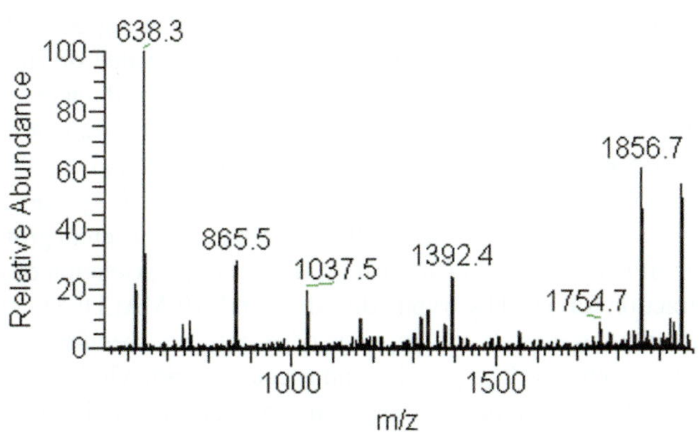

Fig. 35.6 An AP-MALDI MS/MS spectrum of the peptide ion with *m/z* = 1972 shown in Fig. 35.5

botulinum neurotoxin type A. Amino acid sequence of the identified peptide with a *m/z* value of 1972 was found to be DTFTNPEEGDLNPPPEAK, botulinum neurotoxin type A from *Clostridium botulinum*. Similar analysis of other toxins such as Ricin A chain, protective antigen from *B. anthracis* (Table 35.1), by subjecting them to tryptic digestion, followed by AP-MALDI MS/MS analysis and database search have enabled us to detect their presence in environmental samples and identify them using database searching bioinformatics approach.

35.3.4 Bacterial Cells

Sample preparation of bacterial cells for analysis by AP-MALDI MS includes treatment with 1% trifluoroacetic acid in 50% acetonitrile to extract membrane bound proteins such as acyl carrier proteins, outer membrane proteins, and heat shock and cold shock proteins. It should be noted that unlike in spores, a number of different proteins can be extracted from the cells due to their active metabolic status. Owing to their dynamic nature, care should be taken to target the proteins that are present in the cells at all stages of the growth. For example, some proteins may be abundantly present during the log phase of the growth and absent or present in low abundance during the late stationary phase. On the other hand, some proteins that are essential for the structural integrity of the membrane may be present in constant amounts through all stages of growth and therefore serve as better targets for detection and identification of bacterial cells. Depending on the choice of a target protein for identification, care should be taken to analyze the cells within a certain time interval after inoculating the cell culture, so that the cell culture should be in the same growth stage during analysis. Treatment of cells with different chemicals will yield different set of proteins, depending on the extent of cell lysis and pH dependant solubility of the proteins. For example, treating *E. coli* K12 cells with 1% TFA in 50% acetonitrile usually leads to extraction of the outer membrane protein A and acyl carrier protein, while treating the same *E. coli* K12 cells with 50 mM tris-carbonate buffer (pH 9) yields predominantly periplasmic protein. After extracting the required specific proteins, trypsin is then added to digest these proteins. An AP-MALDI MS spectrum of digested protein extracts from *E. coli* cells after treatment with 1% TFA in 50% acetonitrile is shown in Fig. 35.7. AP-MALDI MS/MS data were obtained for the peptide ion with *m/z* value of 1655 from the above figure is shown in Fig. 35.8. Database search performed against an NCBInr database using the MASCOT search engine revealed that this peptide ion with an *m/z* value of 1655 was unique for Gram negative enterobactors including *E. hervicola, E. coli, Salmonella* species, *Yersinia* species and *Shigella* species that belong to the family *Enterobacteriaceae*. Within the enterobactors, different proteins can be targeted for their detection and identification. For example, isolation of H-antigens by acidic extraction followed by tryptic digestion and AP-MALDI MS and MS/MS analysis enabled us to identify the peptides of H-antigens specific for *Salmonella* species. Similarly, other surface proteins and species specific virulence proteins can be isolated for different members of the group to achieve species level identification

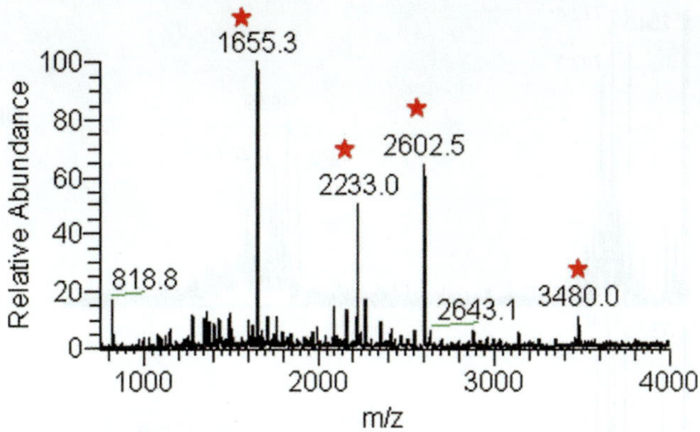

Fig. 35.7 An AP-MALDI MS spectrum of tryptically digested *E. coli* proteins. Tryptic peptides of *E. coli* related proteins are indicated with asterisks

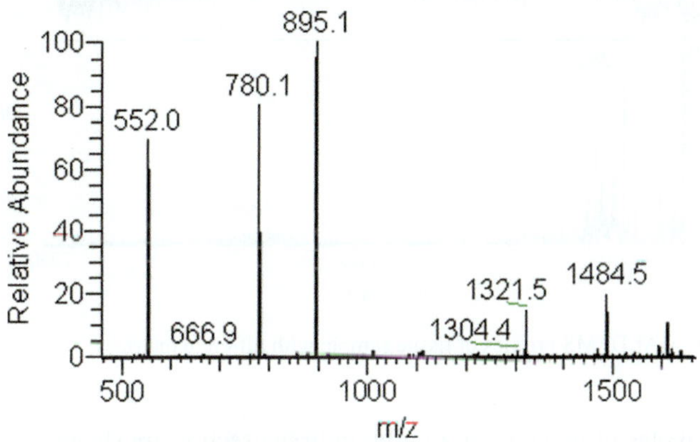

Fig. 35.8 An AP-MALDI MS/MS spectrum of the peptide ion with *m/z* = 1655 shown in Fig. 35.7

(peptide with a mass of 1715 Da from acyl carrier protein specific for *E. coli*, peptide with a mass of 1844 Da specific for *Yersinia pestis* CO92, etc.).

35.3.5 Tissue Imaging

The detection sensitivity is the primary concern for MS imaging. It is influenced by both the ionization efficiency of analytes and the localization of matrix crystals. Figure 35.9 shows MS profiles of similar tissue sections with different matrices. It is apparent from this data that the use of DHA provided better spectral quality than CHCA and DHB. Preliminary profiling is necessary prior to any imaging of the tissue section. Despite the high repetition rate of laser scanning typical imaging experiment- still requires a long time. Figure 35.10 shows a narrow range of the MS profile that is most relevant for the targeted tissue.

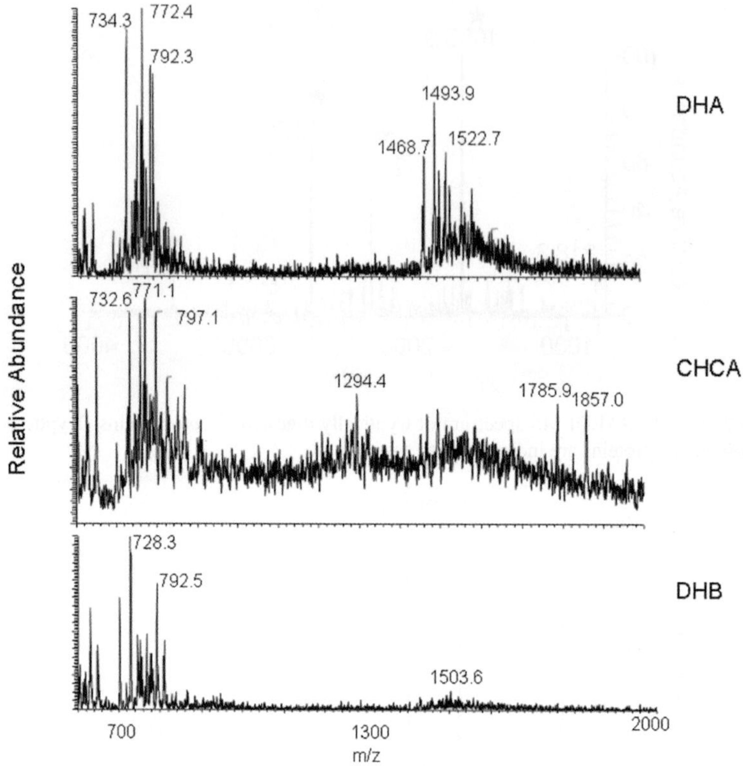

Fig. 35.9 AP-MALDI MS profiles of tissue sample with different matrices

Two examples of imaging of a rat brain tissue section are shown on Fig. 35.11. Both images were obtained from the same sample and data set where different mass units were visualized for each image. The ions observed at 734–760 are observed by Jackson et.al. as Phosphatidylcholine (PC) 32:0 and PC 34:1 (Jackson et al., 2005b).

35.4 Conclusions

In this chapter we have described the sample preparation methods for various applications of AP-MALDI MS. Biological samples are complex and contain a large number of molecules with different physical and chemical properties. These diverse molecules in the sample can potentially interfere with the MS analysis at various stages of sample preparations. Sample preparation methods for AP-MALDI MS analysis of biological samples described here are simple, cost-effective and can be easily automated. Detection and identification of microbial pathogens using AP-MALDI MS/MS-based proteomics approach is rapid, reproducible, and does not require any special species specific reagents or isotopic labels. Simple and high-speed identification of pathogens is of importance not only in case of bioterrorism

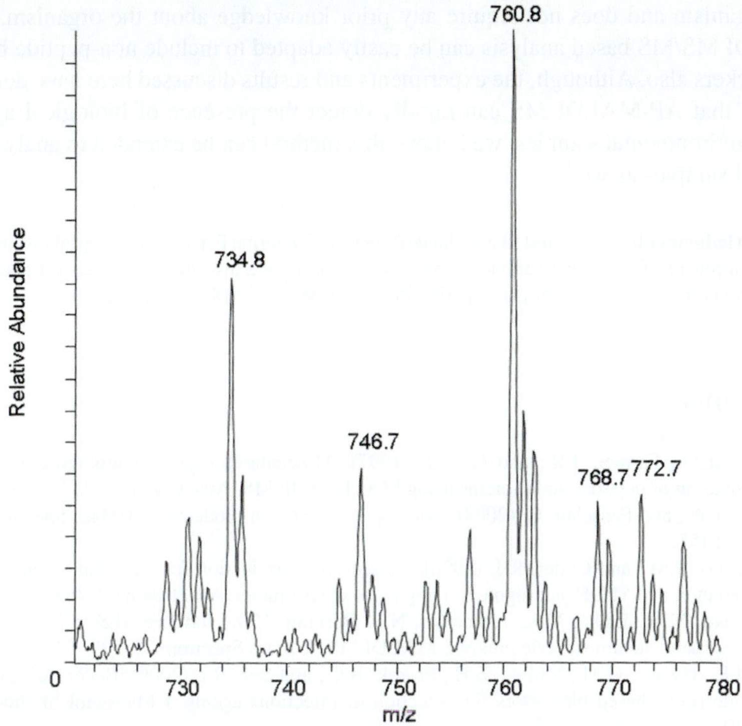

Fig. 35.10 Rat brain AP-MALDI MS profile

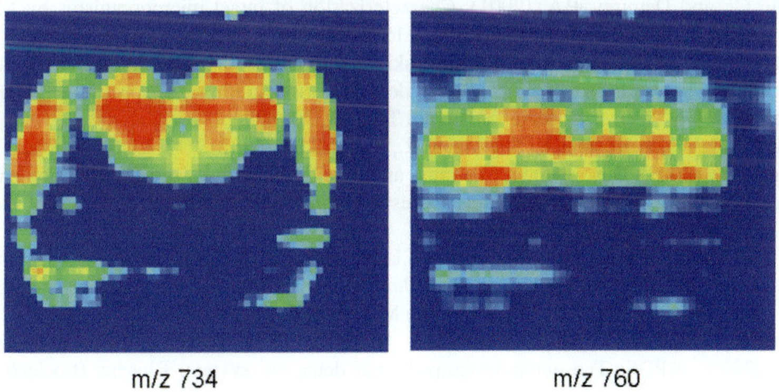

Fig. 35.11 AP MALDI MS images of rat brain at *m/z* 734 and *m/z* 760 (see text for details)

related incidents, but also in analysis of contaminated food and water consumption to prevent foodborne illnesses. AP-MALDI MS/MS-based proteomic analysis can also detect the presence of newly engineered or mutated pathogens, since different biomarker peptides can be identified by slight modification of the protocol. The AP-MALDI MS assay is based on the analysis of proteins/peptides expressed by the organism and does not require any prior knowledge about the organism. AP-MALDI MS/MS based analysis can be easily adapted to include non-peptide based biomarkers also. Although, the experiments and results discussed here have demonstrated that AP-MALDI MS can rapidly detect the presence of biological agents from environmental samples, we believe this method can be extended to analysis of clinical samples as well.

Acknowledgements We would like to thank Professor Catherine Fenselau for helpful discussions on development of microbial sample preparation methods. We also thank Dr. Victor Laiko, and Dr. Seshu Gudlavalleti for their participation in the AP-MALDI MS experiments.

References

Caprioli, R.M., Farmer, T.B., and Gile, J. (1997). Molecular imaging of biological samples: localization of peptides and proteins using MALDI-TOF MS. Anal Chem *69*, 4751–4760.

Demirev, P.A., and Fenselau, C. (2008). Mass spectrometry in biodefense. J Mass Spectrom *43*, 1441–1457.

Doroshenko, V.M., and Cotter, R.J. (1996). Pulsed gas introduction for increasing peptide CID efficiency n a MALDI/quadrupole ion trap mass spectrometer. Anal Chem *68*, 463–472.

Doroshenko, V.M., Laiko, V.V., Taranenko, N.I., Berkout, V.D., and Lee, H.S. (2002). Recent developments in atmospheric pressure MALDI MS. J. Mass Spectrom *221*, 39–58.

Dover, J.E., Hwang, G.M., Mullen, E.H., Prorok, B.C., and Suh, S.J. (2009). Recent advances in peptide probe-based biosensors for detection of infectious agents. J Microbiol Methods *78*, 10–19.

Edwards, K.A., Clancy, H.A., and Baeumner, A.J. (2006). *Bacillus anthracis*: toxicology, epidemiology and current rapid-detection methods. Anal Bioanal Chem *384*, 73–84.

Fenselau, C., and Demirev, P.A. (2001). Characterization of intact microorganisms by MALDI mass spectrometry. Mass Spectrom Rev *20*, 157–171.

Gudlavalleti, S.K., Sundaram, A.K., Razumovski, J., and Doroshenko, V. (2008). Application of atmospheric pressure matrix-assisted laser desorption/ionization mass spectrometry for rapid identification of *Neisseria* species. J Biomol Tech *19*, 200–204.

Hang, J., Sundaram, A.K., Zhu, P., Shelton, D.R., Karns, J.S., Martin, P.A., Li, S., Amstutz, P., and Tang, C.M. (2008). Development of a rapid and sensitive immunoassay for detection and subsequent recovery of *Bacillus anthracis* spores in environmental samples. J Microbiol Methods *73*, 242–246.

Hathout, Y., Demirev, P.A., Ho, Y.P., Bundy, J.L., Ryzhov, V., Sapp, L., Stutler, J., Jackman, J., and Fenselau, C. (1999). Identification of *Bacillus* spores by matrix-assisted laser desorption ionization-mass spectrometry. Appl Environ Microbiol *65*, 4313–4319.

Hindson, B.J., Makarewicz, A.J., Setlur, U.S., Henderer, B.D., Mcbride, M.T., and Dzenitis, J.M. (2005). APDS: The autonomous pathogen detection system. Biosens Bioelectron *20*, 1925–1931.

Jackson, S.N., Ugarov, M., Egan, T., Post, J.D., Langlais, D., Albert Schultz, J., and Woods, A.S. (2007). MALDI-ion mobility-TOFMS imaging of lipids in rat brain tissue. J Mass Spectrom *42*, 1093–1098.

Jackson, S.N., Wang, H.Y., and Woods, A.S. (2005a). In situ structural characterization of phosphatidylcholines in brain tissue using MALDI-MS/MS. J Am Soc Mass Spectrom, *16*, 2052–2056.

Jackson, S.N., Wang, H.Y., Woods, A.S., Ugarov, M., Egan, T., and Schultz, J.A. (2005b). Direct tissue analysis of phospholipids in rat brain using MALDI-TOFMS and MALDI-ion mobility-TOFMS. J Am Soc Mass Spectrom *16*, 133–138.

Laiko, V.V., Moyer, S.C., and Cotter, R.J. (2000). Atmospheric pressure MALDI/ion trap mass spectrometry. Anal Chem *72*, 5239–5243.

Laiko, V.V., Taranenko, N.I., Berkout, V.D., Musselman, B.D., and Doroshenko, V.M. (2002). Atmospheric pressure laser desorption/ionization on porous silicon. Rapid Commun Mass Spectrom *16*, 1737–1742.

Li, Y., Shrestha, B., and Vertes, A. (2007). Atmospheric pressure molecular imaging by infrared MALDI mass spectrometry. Anal Chem *79*, 523–532.

Lim, D.V., Simpson, J.M., Kearns, E.A., and Kramer, M.F. (2005). Current and developing technologies for monitoring agents of bioterrorism and biowarfare. Clin Microbiol Rev *18*, 583–607.

Liu, Q., Guo, Z., and He, L. (2007). Mass spectrometry imaging of small molecules using desorption/ionization on silicon. Anal Chem *79*, 3535–3541.

Maddalo, G., Petrucci, F., Iezzi, M., Pannellini, T., Del Boccio, P., Ciavardelli, D., Biroccio, A., Forli, F., Di Ilio, C., Ballone, E., Urbani, A., and Federici, G. (2005). Analytical assessment of MALDI-TOF Imaging Mass Spectrometry on thin histological samples. An insight in proteome investigation. Clin Chim Acta *357*, 210–218.

Madonna, A.J., Voorhees, K.J., Taranenko, N.I., Laiko, V.V., and Doroshenko, V.M. (2003). Detection of cyclic lipopeptide biomarkers from *Bacillus* species using atmospheric pressure matrix-assisted laser desorption/ionization mass spectrometry. Anal Chem *75*, 1628–1637.

Oktem, B., Sundaram, A.K., Shanbhag, S., Murphy, C., and Doroshenko, V.M. (2007). High Throughput Sample Preparation for Atmospheric Pressure MALDI-MS for Rapid Detection and Identification of Microorganisms. Proceedings of 55th American Society of Mass Spectrometry Conference. (Indianapolis, IN).

Perdue, M.L., Karns, J., Higgins, J., and Van Kessel, J.A. (2003). Detection and fate of *Bacillus anthracis* (Sterne) vegetative cells and spores added to bulk tank milk. J Food Prot *66*, 2349–2354.

Pribil, P.A., Patton, E., Black, G., Doroshenko, V., and Fenselau, C. (2005). Rapid characterization of *Bacillus* spores targeting species-unique peptides produced with an atmospheric pressure matrix-assisted laser desorption/ionization source. J Mass Spectrom *40*, 464–474.

Reyzer, M.L., Hsieh, Y., Ng, K., Korfmacher, W.A., and Caprioli, R.M. (2003). Direct analysis of drug candidates in tissue by matrix-assisted laser desorption/ionization mass spectrometry. J Mass Spectrom *38*, 1081–1092.

Rowe-Taitt, C.A., Golden, J.P., Feldstein, M.J., Cras, J.J., Hoffman, K.E., and Ligler, F.S. (2000). Array biosensor for detection of biohazards. Biosens Bioelectron *14*, 785–794.

Sapsford, K.E., Rasooly, A., Taitt, C.R., and Ligler, F.S. (2004). Detection of *Campylobacter* and *Shigella* species in food samples using an array biosensor. Anal Chem *76*, 433–440.

Seggern, C.E., Moyer, S.C., and Cotter, R. (2003). Liquid infrared atmospheric pressure matrix-assisted laser desorption/ionization ion trap mass spectrometer of sialylated carbohydrates. J Anal Chem *75*, 3212–3218.

Stachowiak, J.C., Shugard, E.E., Mosier, B.P., Renzi, R.F., Caton, P.F., Ferko, S.M., Van De Vreugde, J.L., Yee, D.D., Haroldsen, B.L., and Vandernoot, V.A. (2007). Autonomous microfluidic sample preparation system for protein profile-based detection of aerosolized bacterial cells and spores. Anal Chem *79*, 5763–5770.

Sundaram, A., Gudlavalleti, S., Oktem, B., Razumovskaya, J., Gamage, C., Serino, R., and Doroshenko, V. (2008a). Atmospheric Pressure MALDI-MS/MS Based Multiplexed, High Throughput Assay for Rapid Detection and Identification of Pathogens. 6th Annual ASM Biodefense and Emerging Diseases Research Meeting (Baltimore, MD).

Sundaram, A.K., Black, G.E., Taranenko, N.I., and Doroshenko, V.M. (2006). Rapid and Efficient Sample Clean up on Hydrophobic Probe Surface for MALDI Analysis (Seattle, WA, American Society of Mass Spectrometry).

Sundaram, A.K., Gudlavalleti, S.K., Oktem, B., Razumovskaya, J., Gamage, C., Serino, R.M., and Doroshenko, V.M. (2008b). Atmospheric Pressure MALDI-MS/MS Based High Throughput

Automated Multiplexed Assay System for Rapid Detection and Identification of Bioagents. Proceedings of 56th American Society of Mas Spectrometry Conference (Denver, CO).

Tan, P.V., Taranenko, N.I., Laiko, V.V., Yakshin, M.A., Prasad, C.R., and Doroshenko, V.M. (2004). Mass spectrometry of N-linked oligosaccharides using atmospheric pressure infrared laser ionization from solution. J Mass Spectrom *39*, 913–921.

Taranenko, N.I., Black, G.E., Serino, R.M., and Doroshenko, V.M. (2005a). Effect of Environmental Conditions on Identification of *Clostridium botulinum* Neurotoxin Type A Using AP-MALDI Mass Spectromotry. Proceedings of 53rd American Society of Mass Spectrometry Conference (San Antonio, TX).

Taranenko, N.I., Black, G.E., Serino, R.M., and Doroshenko, V.M. (2005b) Rapid Identification of Environmental Bacteria Using Atmospheric Pressure MALDI Ion Trap Mass Spectrometry. Proceedings of 53rd American Society of Mass Spectrometry Conference (San Antonio, TX).

Thomas, J.J., Falk, B., Fenselau, C., Jackman, J., and Ezzell, J. (1998). Viral characterization by direct analysis of capsid proteins. Anal Chem *70*, 3863–3867.

Wang, H.Y., Jackson, S.N., Post, J., and Woods, A.S. (2008). A minimalist approach to MALDI imaging of glycerophospholipids and sphingolipids in rat brain sections. Int J Mass Spectrom *278*, 143–149.

Warscheid, B., and Fenselau, C. (2003). Characterization of *Bacillus* spore species and their mixtures using postsource decay with a curved-field reflectron. Anal Chem *75*, 5618–5627.

Warscheid, B., and Fenselau, C. (2004). A targeted proteomics approach to the rapid identification of bacterial cell mixtures by matrix-assisted laser desorption/ionization mass spectrometry. Proteomics *4*, 2877–2892.

Warscheid, B., Jackson, K., Sutton, C., and Fenselau, C. (2003). MALDI analysis of *Bacilli* in spore mixtures by applying a quadrupole ion trap time-of-flight tandem mass spectrometer. Anal Chem *75*, 5608–5617.

Whiteaker, J.R., Warscheid, B., Pribil, P., Hathout, Y., and Fenselau, C. (2004). Complete sequences of small acid-soluble proteins from *Bacillus globigii*. J Mass Spectrom *39*, 1113–1121.

Yao, Z.P., Afonso, C., and Fenselau, C. (2002a). Rapid microorganism identification with on-slide proteolytic digestion followed by matrix-assisted laser desorption/ionization tandem mass spectrometry and database searching. Rapid Commun Mass Spectrom *16*, 1953–1956.

Yao, Z.P., Demirev, P.A., and Fenselau, C. (2002b). Mass spectrometry-based proteolytic mapping for rapid virus identification. Anal Chem *74*, 2529–2534.

Chapter 36
MALDI Imaging Mass Spectrometry for Investigating the Brain

Isabelle Fournier, Céline Mériaux, Maxence Wisztorski, Randeep Rakwal, and Michel Salzet

Abstract Dynamic properties of the nervous system can now be investigated through mass spectrometry technologies. Generally, the application of these powerful techniques requires the destruction of the specimen under study/examination, but recent technological advances have made it possible to directly analyze tissue sections and perform 2-D or 3-D molecular ions mapping. We review the increasing application of matrix-assisted laser desorption/ionization (MALDI) imaging to the analysis of molecular distributions of proteins and peptides in nervous tissues of both invertebrates and vertebrates, focusing in particular on recent studies of neurodegenerative diseases, and early efforts to implement assays of neuronal development.

Keywords Invertebrates · MALDI imaging · Neurodegenerative diseases · Peptides · Vertebrates

36.1 Introduction – From Lenses to MALDI Imaging

The oldest lens made of polished glass has been dated back to 700 years BC. This lens was discovered at Ninive, the last Assyrian establishment in Kurdistan. Ancient Roman writings have mentioned some enlarging tools like balls full of water or emerald sharpened like a concave lens. From ancient times human curiosity has forced us to observe the infinitely small. The first documented scientific usage of lenses appeared in the XVIIth century; before that all lenses were used to correct vision. The origin of the first microscope is yet still difficult to ascertain due to the fact that several types of microscopes were built during the same period. However, the best quality observations have been performed by Antoni Van Leewenhoeck and Robert Hooke. In 1625, Robert Hooke published in the journal Micrographia,

M. Salzet (✉)
MALDI Imaging Team, Laboratoire de Spectrométrie de Masse Biologique Fondamentale et Appliquée (FABMS), EA 4550, Université Nord de France, F-59655, Villeneuve d'Ascq Cedex, France
e-mail: michel.salzet@univ-lille1.fr

A.R. Ivanov, A.V. Lazarev (eds.), *Sample Preparation in Biological Mass Spectrometry*, DOI 10.1007/978-94-007-0828-0_36,

drawings of observations of plant cells, fungi and lice. The magnification was about 160×. Using his microscope, Robert Hooke also successfully demonstrated the circulation of the nettle poison in the finger. At the same time, Franscesco Stelluti published a book describing his observations using microscopes, while Marcello Malpighi was the first biologist to research the liver machinery and embryology using the microscope. Van Leewenhoeck was the first to describe bacteria living in the mouth, that he named the "levende dierkens", or living small animals. In 1650, the magnification of the microscope was increased due to biconvex lens but chromatic aberrations for example, surrounded the image, limiting the useable field of view. The dispersion of the light by the lens provoked these aberrations. In the XVIIIth century, John Dollond corrected this problem by adding a negative eyeball comprised of two plan-convex lenses. Since 1870, a standard model in Europe was designed with a tube almost 160 mm long and optics developed by Hartnack. Thereafter, Carl Zeiss with Ernst Abbe were the first who have created the wave theory of microscopic imaging. Combining his own efforts with the work of his predecessors, Abbe developed a mathematical treatment of the lens concept. Concurrently, with the development of novel types of glasses, the resolving power of microscopes was pushed to the limit of what was physically possible. In 1890, the images were more refined due to the development of coating techniques, and the field of view had become larger. Since this time, the basic microscope structure has not changed. To achieve higher resolution, it would be necessary to wait until the development of the electron microscope, introduced in 1930, with a magnification of 800,000×. The major advancements were made in the optics, the mechanics and most critically, the lighting. Since its invention, the microscope magnification ranges from 200 to 1500× on average, meaning that most of the observations are done at sizes ranging between 200 and 400 times smaller than the nominal size of the observed object. Just before the Second World War, two technical advances in optics were introduced, the phase contrast (invented by Zernike) and the interference contrast; this opened a door to the observation of almost transparent objects, like the living cells. The most recent developments include the microscopes with probes allowing for the detection of electric field with tunnel effect, magnetic force or atomic force. The limits of magnification are now about 1 billion times, permitting the observation of the surface of objects at the atomic scale.

In parallel with the microscope, immunohistochemical techniques permitting localization of antigens in tissues, cells, organisms, bacteria, virus etc., were introduced. Immunohistochemistry was born at the end of 1930. In 1940, Coons and collaborators developed an antibody labeled with a fluorescent tracer, the fluorescein (Coons et al., 1941). Development of novel fixation procedures (ethanol, methanol, acetone, picric acids and paraformaldehyde) (Gabe, 1964, 1967, 1969, 1972; Petit and Sahli, 1975; Steinbach, 1977; Laboux, 2004) were then introduced as well as tissue embedding in paraffin and procedures for antigen retrieval in immunocytochemistry studies (Curran, 1980; Beckstead, 1994; Ino, 2004). Currently, multiplex immunocytochemistry using quantum dots (Brock and Jovin, 2003; Jovin, 2003) as well as multiple antisera is on its way (Furuya et al., 2004). Therefore, in summary, tissue imaging with microscope allows us to localize several molecules of interests at a cellular or sub-cellular resolution. The breakdown limit is almost 50 different antigens detected simultaneously. The major current limitations of this technology are the tissue itself and the methods of its preservation. All developments made during the last seven decades on tissue preparation are becoming a foundation for matrix-assisted laser desorption/ionization (MALDI) mass spectrometry (MS) imaging (MALDI-MSI). Right now, the question should be: Can we go back to this background knowledge to improve the technology and what MALDI-MSI will bring in this field of tissue imaging?

Mass spectrometry-based protein analysis utilizes the mass-to-charge (*m/z*) ratio of ionized intact proteins, peptides or their fragments. A typical mass spectrometer consists of an ionization source, a mass analyzer and a detector for counting the number of analytes at each *m/z* ratio. With new emerging MS technologies, it is now possible to demonstrate that direct tissue analysis is feasible using MALDI sources. A major advantage of direct tissue analysis by MALDI is in elimination of time-consuming extraction, purification or separation steps, which have the potential for analyte losses and producing artifacts. After its introduction by B. Spengler in 1994 at the ASMS annual conference through the concept of Ion Imaging and Confocal Microscopy with a new Scanning UV-Laser Microprobe (Spengler, 1994), direct analyses using MALDI of tissue sections was developed (Li et al., 1994a; Dreisewerd et al., 1997; Jimenez et al., 1998; Chaurand et al., 1999; Fournier et al., 2003). These various studies performed by several groups demonstrated that acquisition of cellular expression profiles, while maintaining the cellular and molecular integrity, was feasible and with automation. The ability to reconstruct complex spectral data with imaging software could result in multiplexed imaging maps of selected bio-molecules within tissue sections (Caprioli et al., 1997; Stoeckli et al., 1999, 2001). Thus, direct tissue analysis obtained from tissue sections is convertible into 2-D maps, a method now known as MALDI-MSI, also call MALDI imaging. MALDI imaging combines the power of MS, namely exquisite sensitivity and unequivocal structural information, within an intact and unaltered morphological context.

The procedure has been developed as follows (Fig. 36.1): tissue sections from a fresh organ or a biopsy are placed on conductive glass slides (like Nickel or Indium-Tin-Oxide (ITO) coated microscope slides) and treated with organic solvents (chloroform, acetone and xylene) to remove lipids (Lemaire et al., 2006a). Tissue staining can be performed just like in classical histochemistry for classification of cells (Chaurand et al., 2004). Several dyes have been tested, of which some were found to be compatible with MALDI processes e.g., hematoxilin-eosin-safran (Lemaire et al., 2006b), cresyl violet (Chaurand et al., 2004) or some coumaryl dyes for lipids like Nile Blue, Sudan Black or Oil Red O (Arafah et al., 2009). Sections are then covered with matrices i.e., solid ionic matrices (Lemaire et al., 2006c) or liquid ionic matrices (Meriaux et al., 2010) followed by gold deposition in order to prevent charge effects (Wisztorski et al., 2006) before being introduced in the MALDI-TOF (time-of-flight) mass spectrometer for analysis. Next, a laser, typically a UV laser, is used to acquire a mass spectrum in each point of the surface area. Automated analysis of the entire tissue slice is obtained by acquiring mass spectra every 50–200 μm, providing representative information of each *m/z* ratio detected.

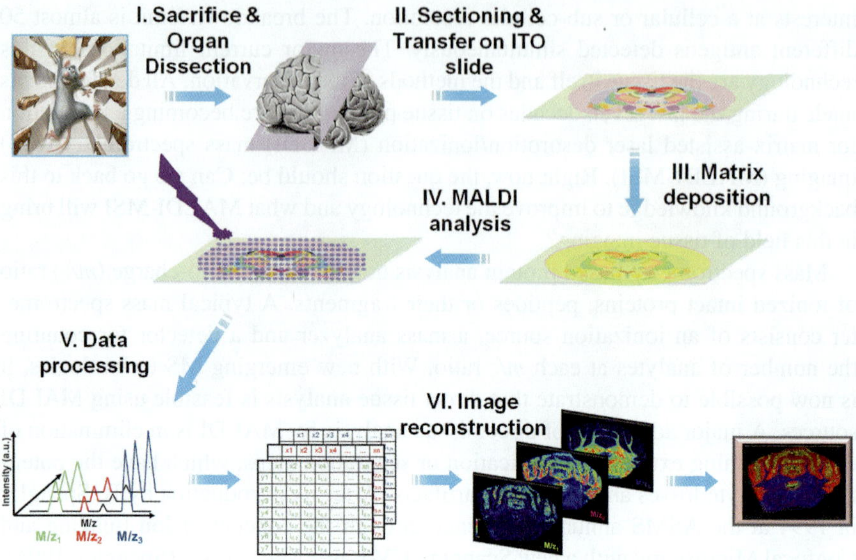

Fig. 36.1 Schematic representation of MALDI imaging procedures. Tissue sections from fresh organ or biopsy are transferred onto an ITO slide (Indium Tin Oxide). Sections are then covered with a specific matrix, depending on the nature of the bio-molecule under study. For peptides/proteins, very intense signals are obtained with α-cyano-4-hydroxycinnamic acid (α-CHCA) as a matrix. Once the sections are covered with α-CHCA (or another matrix), the target is introduced in the MALDI-TOF instrument for analysis. Next, the MALDI laser scans through a set of pre-selected locations (10–50 μm apart) in the tissue and mass spectra representative of the peptides/proteins (or lipids) present at each location are collected. Automated data collection takes 2–6 hours, depending on the number of points assayed, and followed by data analysis and image reconstruction using a software developed for MALDI-MSI

Each collected spectrum represents an average of several laser shots in order to obtain a statistically significant representation of the analyzed area. MALDI-MSI in its most current form is a point-to-point analysis. Analysis is typically obtained within 2–6 h and images are reconstructed using specific software (Stoeckli et al., 1999; Jardin-Mathe et al., 2008).

The image resolution is dependent on the size of image pixels and the number of pixels in the area studied. The number of pixels is limited by the precision of the X-Y stage controlling the sample plate. The X-Y stages offer very high precision that ensures a good reproducibility within an accuracy of 1–2 μm. However, it is also limited by data processing capabilities. Indeed, each pixel added to a sequence corresponds to a full mass spectrum of data that contains thousands of values for intensity versus *m/z*. The size of the data will thus depend on the mass analyzer used for the MALDI imaging experiments. Huge amounts of data are generated for an imaging sequence no matter what type of an instrument is used. For example, an image with a raster of 100 × 100 μm^2 of a whole rat brain section (roughly 1.5–2 cm by 1–1.5 cm) can consume more than 10 Gb. Present computers supplied with mass spectrometers are not well adapted to treat such enormous amounts of data, therefore, specialized systems must be used.

The challenge for improving MALDI-MSI resolution is to decrease the size of each pixel i.e., decrease the area irradiated by the laser. For a typical MALDI-MS analysis the size of the irradiated area is not of great concern and does not need to be minimized. Thus, most existing systems, whether home-made or commercially available instruments, generally are provided with a laser focused to achieve an irradiated area of roughly 100–150 μm in diameter. This produces images with a resolution R of 10,000 pixels/cm^2. In the field of MS, this works sufficiently well but for biologists this resolution is insufficient as this corresponds to an irradiation area of at least 5 × 5 cells at once. A single cell resolution for MALDI imaging would require a laser beam diameter of at least 20–25 μm. It is important to note that the number of pixels N is proportional to R^2. Thus, to obtain the maximum information, an increase by a factor of 2 in resolution will lead to a 4× increase in the number of pixels and data to be processed. The only direct way to decrease the size of the irradiated area is to better focus the laser beam. While this may be a challenge, it may well be achievable by laser physicists. A very high resolution system was tested by Spengler and colleagues for MALDI-MSI (Spengler and Hubert, 2002). Using a very specific setup of lenses that were added outside and inside the instrument on the laser beam path, a minimum resolution of ~0.5 μm of the pixel size was theoretically possible. Experimentally, the resolution was very close to this value for classic conditions of laser fluence and a homogeneous laser beam; however it can be larger under specific conditions. This resolution is far below the size of an average cell and would allow for sub-cellular imaging and the relatively fine observation of organelles. While such fine resolution is technically possible, actually obtaining data for peptides and proteins from tissue sections at such resolution is another matter. The first consideration is that decreasing the pixel size will also result in an important decrease in the number of molecules available for analysis. This means that the amount of proteins available on such a small surface area will probably be far below realistic detection threshold of the mass spectrometer. Moreover, several studies have confirmed that a drastic decrease in the ion yield is observed while decreasing the size of the irradiated area (Dreisewerd et al., 1995). Practically, a size of 50 μm is reasonable, but ion yields are well below detectable limit. MALDI is well known for its poor ability to produce ions. Only a small fraction of the molecules ejected from the solid surface will be present as ions for mass analysis. Regardless of the fact that our knowledge of MALDI processes has considerably improved over the past decade, we have not yet been able to increase the ion yield (Dreisewerd, 2003). Thus, MALDI-MSI resolution (Table 36.1) must only progress along with the fundamental progress in this field. Currently, dedicated systems for MALDI-MSI present better focusing systems of the laser beam capable of reaching down to the 10 μm pixel size and should at least provide the 50–75 μm pixel size while maintaining good analytical capacities (Holle et al., 2006). Recently Chaurand and collaborators have designed a new system that allows obtaining protein images at a resolution of 10 μm (Chaurand et al., 2007). The experimental setup combines a very carefully designed system of focusing lenses and an iris aperture to finely control the laser beam size with an ion source geometry that uses co-axial illumination of the sample to reduce radial distributions of ejected molecules and ions. By this approach increased intake of ions for MS analysis can be achieved. Such

Table 36.1 Comparison between MALDI imaging and histological techniques

	MALDI imaging	Histochemistry and immunocytochemistry
Spatial resolution	30 μm	100 nm to 10 μm
Primary structure	Yes	No
Nature of detected biomolecules	Lipids Neurotransmittors Peptides Proteins Drugs	Lipids Neurotransmittors Peptides Proteins mRNA Drugs
Sensitivity		Femtomole to picomole
Amount of biomolecules detected in one analysis	Between several hundred to thousands	50
Time of the analysis	30 min–6 h	1 day

systems are in the developmental stages, and not commercially available. Other systems that attempt to increase image resolution using alternative methods are however easily applicable to commercialized instruments. One approach, proposed by Sweedler and collaborators, relies on performing an overlap of pixels during the data acquisition (Jurchen et al., 2005). If the sample is irradiated sufficiently, so that all matrix material is removed from a spot, and if the sample is moved by only a fraction of the irradiated area diameter e.g., half of the diameter in a simple case, then the consecutive spot can only originate ions from the remaining half of its area that was not previously irradiated. This very simple method allows for increasing spatial resolution down to approximately 25 μm and can be used on all existing instruments. This approach, however, results in a considerable increase in the time of data acquisition. Other methods designed to be compatible with all instruments are currently being investigated. Wisztorski and colleagues have proposed using mask systems (Wisztorski et al., 2007). Again, a very complex system of laser focusing is not necessarily required, if covering the tissue with a mask presenting a network of apertures of a defined size. The laser beam is simply cut off when getting through the mask system and the irradiated area can be very well controlled. These systems allow for reduction in pixel size down to 30 μm while maintaining a high enough ion yield for analysis of peptides and proteins. Mask-based systems are of great interest as they lead to an increase in the ion yield when specific aperture shapes are used.

36.2 Investigation of the Invertebrates' Nervous System by MALDI-MSI

The earliest peptide profiling experiments on invertebrate nervous systems using MALDI-MS were carried out on mollusks, first on the gastropod *Lymnaea stagnalis* (Li et al., 1994a, 2000a; Jimenez et al., 1994; Jimenez and Burlingame, 1998) and later on several cephalopods (Li et al., 2000b). The experimental strategy of these studies was to compare peptide mass spectra patterns obtained from different parts of the nervous system, e.g., neuronal somata vs neurohemal organ axon terminals (Li et al., 1994a, b; Jimenez et al., 1994). This approach resulted in the detection of novel peptides, in addition to peptides previously identified by conventional molecular biological and peptide chemistry methods. In this manner, complex peptide processing and expression patterns could be discovered that were not detected using conventional methods.

Such a strategy, combining peptide fingerprinting of single neurons by MALDI, molecular cloning, peptide chemistry and electrospray ionization ESI-MS, has been developed to study the intricate processing pattern of a preprohormone expressed in identified neurons. In *L. stagnalis*, some experiments were conducted on neuroendocrine light yellow cells (LYCs) or caudodorsal cells. The LYCs are known to express a precursor, named prepro-LYCP (LYCPs, light yellow cell peptides). Prediction of its processing into three peptides, LYCP I, II, and III, at conventional dibasic processing sites flanking the peptide domains on the precursor, were confirmed by MS. However, MALDI analysis of individual LYCs revealed trimmed variant peptides derived from LYCP I and II. The variants were much more abundant than the intact peptides, indicating that LYCP I and II serve as intermediates in a peptide-processing sequence (Li et al., 1994a). Furthermore, MALDI also allowed detection of co-localization of novel peptides with the LYCPs (Li et al., 1994c). Caudodorsal cells of *Lymnaea* are known to initiate and coordinate ovulation and egg mass production and associated behaviour through the release of a complex set of peptides that are derived from the caudodorsal cell hormone-I (CDCH-I) precursor. Fingerprinting of peptides in the commissure by MALDI demonstrated the presence of all sequenced peptides and, in addition, could identify two other peptides derived from pro-CDCH-1, the beta 1- and beta 3-peptides (Li et al., 1994b). Recently, Sweedler and colleagues, studying the bee *Apis mellifera* genome, showed that 200 neuropeptides can be predicted, of which 100 were confirmed by MS. Moreover, this study opens the door of future molecular studies with the identification of 36 genes, 33 of which have not been previously reported (Hummon et al., 2006b). Using ToF-secondary ion mass spectrometry (ToF-SIMS) and MALDI-MS sample preparation methods, molecular ion maps with a high spatial resolution of cholesterol and the neuropeptide APGWamide were obtained (Altelaar et al., 2005, 2007; McDonnell et al., 2005). APGWamide was predominantly localized in the cluster of neurons that regulate male copulation behaviour of *Lymnaea*, which is in line with its biological activity (De Lange and van Minnen, 1998). Clearly, MALDI peptide profiling gives access to the most complete peptide representation in specific areas, and differential analysis of several distinct areas with MSI yields a representative map of all biomolecules present at one point in time.

In another recent report, MSI of neuropeptides in crustacean neuronal tissues (pericardial organ (PO) and brain) was used to reveal that two RFamide-family peptides and a truncated orcokinin peptide present a distinct localization from those of other members of their respective families. Over 30 previously sequenced neuropeptides were identified based on MS measurements. The MSI study at the organ-level elucidated the spatial relationships between multiple neuropeptide isoforms of the same family as well as the relative distribution of neuropeptide families (DeKeyser et al., 2007).

In insects, due to the recent progress in genome sequencing and on-going annotation(s), MALDI is now more extensively used. In the honeybee, over 200 neuropeptides have been discovered (Hummon et al., 2006b). In addition to the MS methods, MALDI-MSI has been performed allowing the localization of all these neuropeptides (Hummon et al., 2006a) or venom peptides (Francese et al., 2009). Similar studies have been performed in *Tribolium casteneum* (Amare and Sweedler, 2007) or *Anopheles gambiae* (Dani et al., 2008). In *Schistocerca gregaria*, orcokinins, a family of myotropic neuropeptides, have been identified using such a strategy (Hofer et al., 2005).

Our team has recently begun a series of peptidomic analyses followed by MALDI-MSI experiments of embryonic and adult medicinal leeches. One of our goals was to get information about the regeneration processes in adults by suspecting the re-expression of embryonic factors in the course of the regeneration process. For this purpose, we first tried to obtain molecular maps from whole mounted, opened embryos at different stages of development. These reference maps will provide us information of when and where specific proteins and peptides are first expressed, and whether such expression is stable or variable in time and space. Peptidomic analysis of whole embryos at different stages of developments (Table 36.2) shows differences between early (6, 8) and later (12a, 12b) embryonic stages. The peptidomic pattern is completely different with respect to smaller peptides in early stages compared to later stages. However, it is interesting to note that the peptide pattern is more similar between adult leech cord in course of regeneration and the early embryonic stages (Table 36.2). The common peptides that have been identified – glycoprotein CD6, FGF, plectin 1, netrin, cathepsin C, NADH oxidase and NADH dehydrogenase (Table 36.3) – are known to be involved in regeneration in vertebrates (for review see, International Review in Neuroscience, 2009). A 3-D map acquired using MALDI-MSI at the level of the adult ganglia for a peptide of *m/z* 2475, (GTRTMERSVRTSSQYASGGPMPN) corresponding to an N-terminal fragment of a novel intermediate filament protein *Hm*IF4 (Bruand et al., 2010), shows a differential distribution at the anterior part than at the posterior one (Fig. 36.2). In Fig. 36.3, where an anaesthetized adult leech was opened and pinned flat, after which transverse frozen sections were used for MALDI-MSI. The histological image (Fig. 36.3a) shows a part of the body wall, extending from the dorsal midline on the left past the ventral midline, with the central nervous system (CNS) as indicated. Distributions of ions in the indicated peaks of the spectrum in panel (c) are shown in panel (b) of Fig. 36.3. Interesting differences in localization are illustrated by these images. The ion *m/z* 2571 appears to be localized exclusively in ventral area, but absent from the nerve cord. The ion *m/z* 2982 appears to be distributed in a lateral region, including contiguous parts of both dorsal and ventral areas, and no presence in the area within and around the nervous system. Three ions present in the dorsal part of the body wall, *m/z* 3041, *m/z* 3056 and *m/z* 3555, were found to be differentially distributed within this region. The ion *m/z* 5130 was strongly

Table 36.2 Peptides identified at different leech embryonic stages (E12: 12 days, E8: 8 days, E6: 6 days) compared to the adult regeneration peptidome during the course of regeneration (6 h). Whole embryos and adult nerve cords 6 h after crushing were subjected to peptide extraction (Salzet et al., 1995) before separation by nanoLC and identification by ESI-MS

E12a	E12b	E6	E8	Adult regeneration (6 h)
		m/z 1242.47		
		***m/z* 1264.12**		***m/z* 1264.57**
		***m/z* 1490.65**		***m/z* 1490.66**
				***m/z* 1490.69**
		m/z 1520.61	***m/z* 1529.02**	***m/z* 1528.87**
		m/z 1541.12		***m/z* 1536.65**
		m/z 1560.67	*m/z* 1556.85	
		***m/z* 1586.20**	***m/z* 1583.43**	***m/z* 1582.87**
		m/z 1597.02	***m/z* 1604.27**	***m/z* 1607.84**
			***m/z* 1753.09**	***m/z* 1751.77**
		m/z 2040.91	*m/z* 2043.79	***m/z* 2099.06**
		m/z 2107.71	*m/z* 2104.39	***m/z* 2116.18**
		m/z 2148.65	*m/z* 2153.29	
		m/z 2188.48	*m/z* 2190.91	
		m/z 2221.52	*m/z* 2219.51	
m/z 2474.11	*m/z* 2474.11		*m/z* 2476.61	***m/z* 2471.25**
m/z 2524.69			*m/z* 2523.65	
m/z 3160.55	*m/z* 3160.81			
m/z 3512.25	*m/z* 3507.07	*m/z* 3514.16	*m/z* 3502.23	
m/z 3523.98	*m/z* 3527.53			
m/z 3540.39	*m/z* 3539.84			
m/z 4378.86	*m/z* 4377.95	*m/z* 4383.43	*m/z* 4375.86	
m/z 5431.75	*m/z* 5433.79			
m/z 5573.46	*m/z* 5568.99			
m/z 8428.45	*m/z* 8423.80			
m/z 10868.9	*m/z* 10866.00			

The values in bold means that these peptides have been identified. (These identified peptides are presented in Table 36.3)

localized in the nervous system and nearby, in a manner almost complementary to the ion *m/z* 2982. Last among these examples, the distribution of the ion *m/z* 3507 was the same as that observed in the embryo (see below), i.e., present exclusively within the nervous system.

Figure 36.4 shows examples from several embryos at the E12 stage of development, opened and mounted either with the inside of the embryo or the epidermis facing the laser beam. In one case, the endodermal layer was removed during dissection. The embryos were dried on a slide and a histological image was recorded prior to covering the tissues with a layer of matrix. Molecular images of different *m/z* values were reconstructed after data acquisition for each embryo. MSI images, superimposed on the histological images, are shown for ions at *m/z* 3507 and *m/z* 6443 (Fig. 36.4). For example, for the two embryos imaged from the internal surface, the ion *m/z* 3507 was exclusively localized in the nervous system, in ganglia, interganglionic connective nerves and lateral roots, proximal to the ganglia.

Table 36.3 Identified peptides common to both embryos and the adult nerve cord in regeneration process by nanoLC ion-trap in the MS/MS mode and interrogation of an EST database

Parent ion mass	Annotation	MQ-score	F-score	EST identification
m/z 1264.57	G.ALQSGQSMDEL.Q	0.302	2.06	Plectin 1 fragment
m/z 1490.66	S.FISTSEPQEQYY.L	0.581	−3.31	NADH oxydase
m/z 1490.69	D.FLSSCLQDHGNVTA.Y	0.998	1.73	FGF fragment
m/z 1528.87	K.LDIMFSLTQCFLV.L	−0.178	3.92	NADH deshy-drogenase
m/z 1536.65	LINNNNNNHNNNNQ.Q	0.521	4.93	Myosine kinase I
m/z 1582.87	N.IFSSKGSLSFLLGID.R	1.271	2.59	Ferroxidonin oxidoreduc-tase
m/z 1607.84	Q.NNKNDYNKTTRKL.T	1.242	3.54	Cathepsin C
m/z 1751.77	T.HQPSHGYCGEKAIPND.S	−0.044	1.52	Glycoprtein CD6
m/z 2099.06	Y.NALDTLVPAFTSEAPTGWLP.A	1.105	−2.19	Fatty acid desaturase 2
m/z 2116.18	S.SQVSRTRLLLWTNHHIR.T	0.560	4.34	Endonuclease-reverse transcriptase
m/z 2471.25	K.RKKSCMNSSKKTGIFSKFYTM.N	0.409	0.18	Serine/Threonine protein kinase

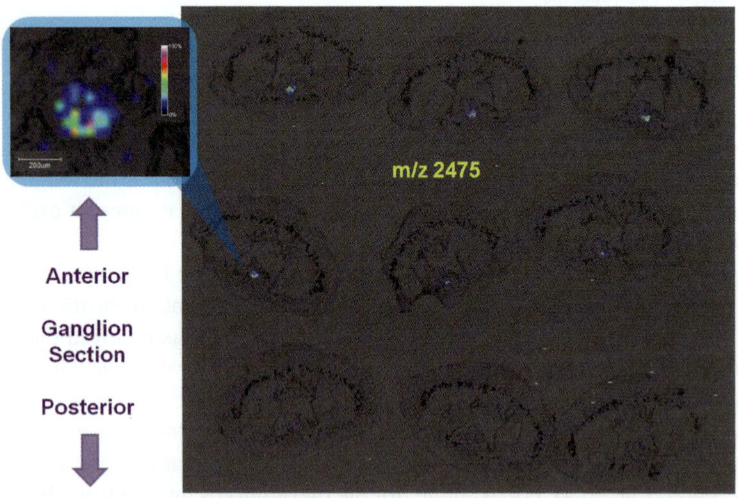

Fig. 36.2 Molecular images of a *Hm*IF4 fragment (m/z 2475) in serial sections of a leech ganglion showing a differential separation between anterior and posteriorparts (Meriaux et al., 2010)

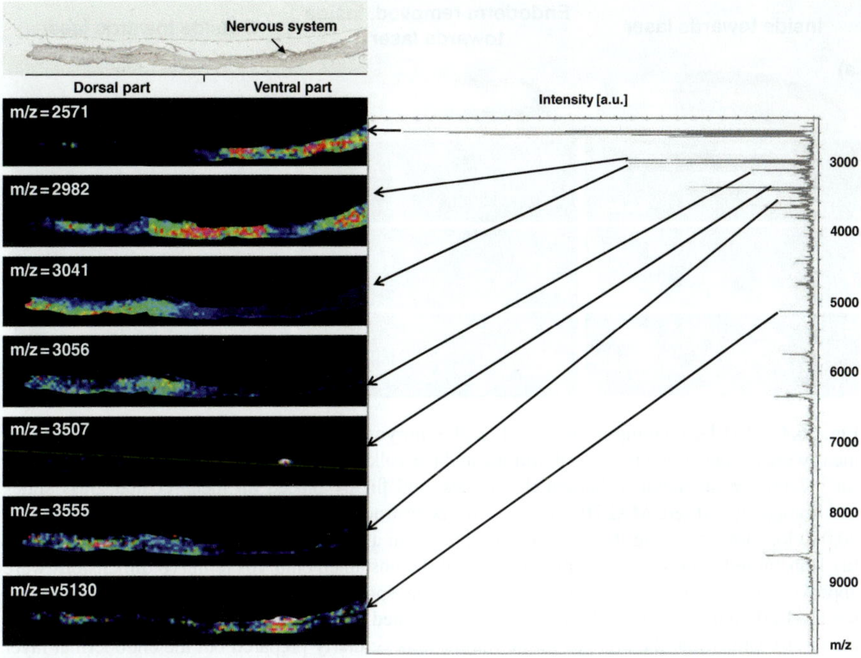

Fig. 36.3 MALDI imaging of a transverse section of the body wall of an adult leech. The specimen was opened along the dorsal midline, pinned flat, exposed for ~2 min to methanol to harden the tissues, frozen and sectioned (12 μm thickness) transversely. Sections were mounted on a metal-coated glass slide and imaged prior to being sprayed with α-CHCA matrix for MALDI analysis. **(a)** A low magnification, transmitted light digital image of a part of a transverse section, extending from the dorsal midline on the *left* (*dorsal* part), through the ventral midline (nervous system) and including the ventral region on the other side (ventral part). The tissue dimensions were approximately 0.5 mm × 14 mm. **(b)** MALDI images of the partial transverse section shown in **(a)**, reconstructed from mass spectra recorded in a raster of 2,000 points, 100 μm apart. Seven *m/z* values are shown, illustrating some of the different molecular distributions observed. **(c)** An example of a mass spectrum obtained at one location, indicating the peaks corresponding to the *m/z* values of the MALDI images shown in **(b)**. Relative abundances are colour coded, with *red-white* being high and *blue-black low*. With permission from the Journal of Developmental Neurobiology (Wisztorski et al., 2008)

In contrast, the ion *m/z* 6443 was not detected in the nerve cord but in nephridial structures, probably around the nephridiopores. The composite image illustrates very well this difference of localization between the two molecules. In comparison, for the embryo positioned with the epidermis towards the laser beam (right side in Fig. 36.4), no significant signal was observed for the *m/z* 3507 ion, confirming its localization within the CNS, while a reduced but detectable signal for the *m/z* 6443 ion was observed at the nephridiopores, which travel through the body wall and open to the outside in the epidermis. This peptide corresponds to antimicrobial peptides from the leech brain we recently characterized (Schikorski et al., 2008).

All these results demonstrate that MALDI-MSI allows us to obtain molecular maps of neuropeptides or neurohormones, reflecting the physiological state of the animal.

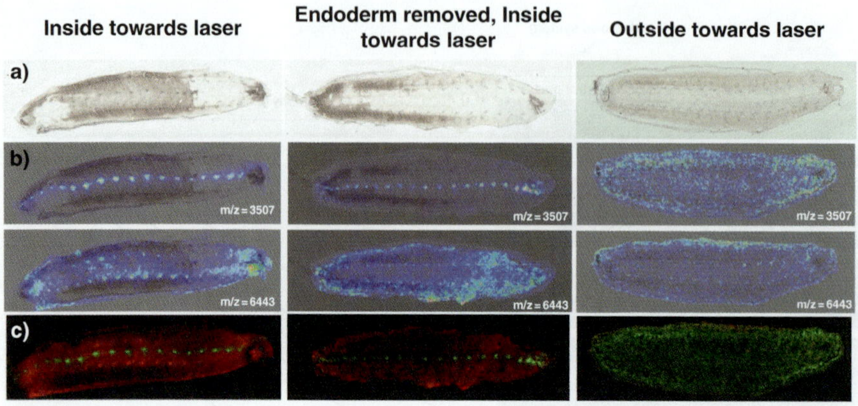

Fig. 36.4 MALDI imaging of dissected leech embryos in whole mount. Three stage E12 specimens were opened along the dorsal midline and the yolks were removed, then pinned flat, exposed for ~2 min to methanol to harden the tissues, and finally placed on metal-coated glass slides and immediately dried. MALDI images were reconstructed from the mass spectra collected from 20,000 locations covering the embryos completely in a rectangular raster of points 60 μm apart. **(a)** Transmitted light low-power micrographs of the unstained embryos (embryo dimensions were approximately 2.5 mm × 10 mm) taken prior to the spraying of α-CHCA matrix on the embryos for MALDI analysis. The embryo at left was mounted so the internal surface (endoderm) faced the laser beam, while the embryo in the middle was similarly prepared but the endodermal layer (~10–20 μm thick) was removed. The embryo at right was mounted with the epidermis towards the laser source. Anterior is to the right in all cases. **(b)** Reconstructed MALDI images showing the distributions of ions with *m/z* values of 3508 (*above*) and 6420 (*below*), for the three embryos. The *m/z* 3507 ions are maximally present in the segmental ganglia and interganglionic connective nerves, in the two animals analyzed with the inner surface towards the laser. It is essentially absent in the third animal, as might be expected since the CNS is close to the inner surface, indicating that the depth range into the tissue of the laser beam is less than ~100 μm. By contrast, the *m/z* 6420 ion is more widely distributed but with maxima associated with the nephridiopores. Signal associated with the nephridiopores is visible in the specimen on the right column as well, presumably because these structures extend through the body wall and open to the outside. **(c)** Superposition of the images shown in **(b)**, but here with abundance proportional to intensity in *green* (m/z 3508) and *red* (*m/z* 6420). With permission from the Journal of Developmental Neurobiology (Wisztorski et al., 2008)

36.3 Applications to Vertebrates' Central Nervous System

The rat brain was the first biological model used in MALDI-MSI studies because the organ's anatomy has been extensively characterized in the past. These studies resulted in several reported molecular maps of different neuropeptides (Caprioli et al., 1997; Langstrom et al., 2007; Wisztorski et al., 2008) as the one we recently performed on the spinal cord (Fig. 36.5). The spinal cord is organized into discrete, anatomically defined areas that include motor and sensory networks composed of chemically diverse cells. The MSI data presented here reveal the spatial distribution of multiple neuropeptides obtained within individual 10 μm sections of the rat spinal cord (Fig. 36.5). These MSI analyses revealed new insights into the

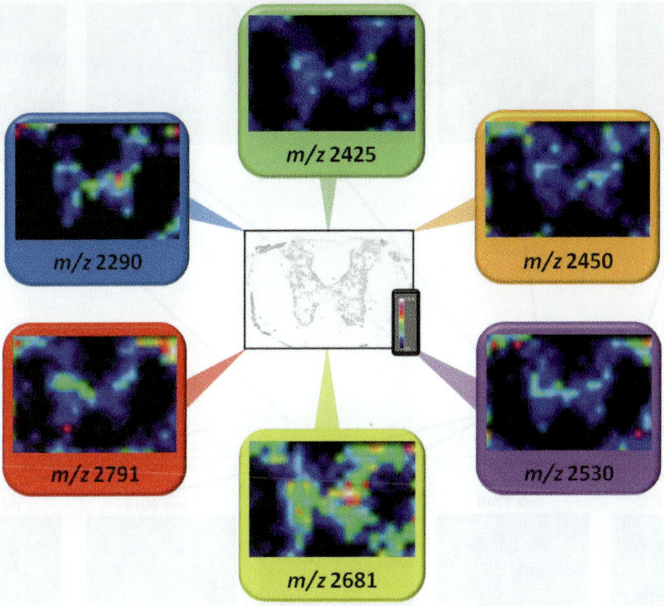

Fig. 36.5 Molecular images of peptides in the rat spinal cord (Meriaux et al., 2010)

chemical architecture of the spinal cord and set the stage for future imaging studies of the chemical changes induced by pain, anesthesia and drug tolerance. Similarly, it may be possible to obtain peptide maps of implanted embryonic mouse brain (Fig. 36.6) or lipid maps of gangliosides, which are particularly abundant in the central nervous system (CNS) and thought to play important role in memory, neuritogenesis, synaptic transmission and other neural functions (Sugiura et al., 2008). In these experiments, the C18-species were found widely distributed throughout the frontal brain whereas the C20-species selectively localized along the entorhinal-hippocampus projections, especially in the molecular layer (ML) of the dentate gyrus (DG). This points out to a specific localization of gangliosides in the brain. Taking the above consideration into account, it is important to develop MSI for pathological conditions like the neurodegenerative disease.

36.4 Applications to Brain Diseases

Neurodegenerative diseases induce proteome changes in the brain that can be investigated using neuroimaging techniques. For example, Parkinson's disease (PD) is the second most common neurological disease after Alzheimer's disease. PD is characterized by a selective degeneration of dopaminergic neurons in the *substantia nigra pars compacta* and by cytoplasmic inclusions (Lewy Bodies), where specific proteins like the α-synuclein are stored (Beal and Hantraye, 2001). The first MALDI

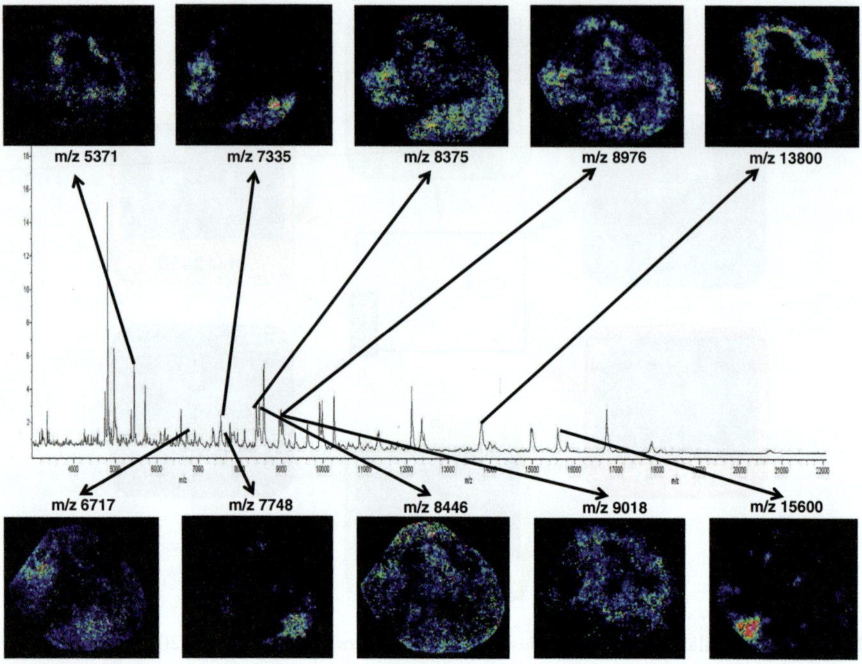

Fig. 36.6 Molecular images of mouse embryos implanted with stem cells (Meriaux et al., 2010)

tissue profiling studies on 6-hydroxydopamine (6-OHDA) Parkinson rat model have been performed by the Per Andrén group (Pierson et al., 2004). Several differences were found in the dopamine-depleted side of the rat brain when compared to the corresponding intact side, in calmodulin, cytochrome c and cytochrome c oxidase, for example, implicating denervation of dopamine neurons via the regulation of ubiquitin pathways, at least in a classical animal model of PD (Pierson et al., 2005). This study also emphasizes the utility of molecular profiling with MALDI-MSI, because it has the capacity to distinguish between metabolic fragments, conjugated proteins and posttranslational modifications (Pierson et al., 2005).

Recently, we examined the rat brain tissue using MALDI-MSI profiling combining the use of automatic spotting of a MALDI matrix with in situ enzymatic digestion of the FFPE (formalin fixed paraffin embedded) tissue from 6-OHDA unilaterally treated animals (Lemaire et al., 2006b) followed by nanoLC/MS-MS analysis (Table 36.1). These analyses confirmed that ubiquitin, trans-elongation factor 1, hexokinase and neurofilament M were down-regulated, as previously shown in both human and animal model tissues, whereas peroxidoredoxin 6, F1 ATPase and α-enolase were up-regulated (Stauber et al., 2008), which is in line with previous studies performed using classical proteomics or genomics (DNA microarray) approaches.

In addition, we identified three novel putative biomarkers, trans-elongation factor 1 (eEF1) and the collapsin response mediator proteins, (CRMP-1 and -2) using protein libraries (Stauber et al., 2008). Our observation of increases in CRMP-1 and CRMP-2 is in agreement with previously reported molecular profiling data (Barzilai et al., 2000). We speculate that CRMP factors might be good biomarker candidates for neurodegenerative diseases like PD or AD. This hypothesis can be tested by comparing CRMP-2 mRNA expression profiles and MALDI images of trypsin-digested tissue sections in controls and 6-OHDA treated animals. This comparison may lead to an observation that the peptide fragments are specifically localized in *Corpus callosum* in 6-OHDA treated animal, whereas the distribution is completely different in control animals (Stauber et al., 2008). This specific localization would be in line with the ones of neurodegenerative diseases (Stauber et al., 2008).

Similarly, Stoeckli's group has applied this new technology to study amyloid beta peptide distribution in brain sections from mice (Stoeckli et al., 2002, 2006). They demonstrated that the Aβ-(1–40) is by far the most abundant amyloid peptide. Three main regions can be distinguished: two very intense areas are located in the parietal and the occipital cortical lobes and the third one close to the lower part of the Sylvian fissure, i.e. in the hippocampal region. The normalized distributions of Aβ-(1–40) and Aβ-(1–42) show that they are the most abundant amyloid peptides. MSI gives access to the levels of known targets but also allows the mapping of the different targets with great accuracy, which is not possible when whole-brain extracts are analyzed (Stoeckli et al., 2002, 2006). These and other results mentioned above establish the great potential of MALDI-MSI as a new tool for the study of the molecular mechanisms responsible for neurodegenerative diseases.

MALDI-MSI is also a very appropriate tool for assaying the distribution of pharmaceuticals in rat brain tissue slices, which is critical information for new drug development. In fact, some studies of clozapine (Hsieh et al., 2006) and risperidone (O'Brien et al., 2006) have recently been performed by MALDI-MSI. The acquired data confirm that chronic risperidone treatment, which is accompanied by a behavioral phenotype of extrapyramidal origin, produces alterations in the striatal protein profile, possibly subsequent to blockade of the dopaminergic systems. These results suggest that possible mechanisms involved in antipsychotic drugs-induced extrapyramidal symptoms include metabolic dysfunction and oxidative stress (O'Brien et al., 2006).

36.5 Concluding Remarks

After 10 years since its invention, MALDI imaging has undergone many developments and has achieved a certain level of maturity, allowing it to be now used in many domains like proteomics, metabolomics, biomedical and pharmaceutical fields. In neurosciences, MALDI-MSI has become an essential tool for the detection, identification and characterization of the molecular components of biological processes, such as those responsible for the dynamic properties of the nervous system. This technique is still a young technology that is continuously evolving. Further developments are still needed, and we will be seeing these in the coming years. For example, improvements in resolution will be sought, since the minimum area that can be examined is currently in the range of several cells. Other limitations include the detectable mass range, which is typically between *m/z* 400 and 30 000. The lower limit is due to the use of matrices that masks the analysis below this *m/z* range. Various protocols and detection technologies are being tested to overcome the upper *m/z* range limitation, such as new tissue treatments, mass analyzers, or matrices (Franck et al., 2010).

Acknowledgements Supported by grants from Centre National de la Recherche Scientifique (CNRS), Ministère de L'Education Nationale, de L'Enseignement Supérieur et de la Recherche, Agence Nationale de la Recherche (ANR to IF) and Genoscope (to MS).

References

Altelaar, A.F., Luxembourg, S.L., McDonnell, L.A., Piersma, S.R., and Heeren, R.M. (2007). Imaging mass spectrometry at cellular length scales. Nat Protoc *2*, 1185–1196.

Altelaar, A.F., van Minnen, J., Jimenez, C.R., Heeren, R.M., and Piersma, S.R. (2005). Direct molecular imaging of *Lymnaea stagnalis* nervous tissue at subcellular spatial resolution by mass spectrometry. Anal Chem *77*, 735–741.

Amare, A., and Sweedler, J.V. (2007). Neuropeptide precursors in *Tribolium castaneum.* Peptides *28*, 1282–1291.

Arafah, K., Wisztorski, M., Croix, D., Fournier, I., and Salzet, M. (2009). Dye Assisted Laser Desorption Ionisation (DALDI): Introduction to a New Field of Investigation in Lipid Mass Spectrometry Imaging. 18th International Mass Spectrometry Conference (Bremen), Poster PMM-180.

Barzilai, A., Zilkha-Falb, R., Daily, D., Stern, N., Offen, D., Ziv, I., Melamed, E., and Shirvan, A. (2000). The molecular mechanism of dopamine-induced apoptosis: Identification and characterization of genes that mediate dopamine toxicity. J Neural Transm Suppl *60*, 59–76.

Beal, M.F., and Hantraye, P. (2001). Novel therapies in the search for a cure for Huntington's disease. Proc Natl Acad Sci USA *98*, 3–4.

Beckstead, J.H. (1994). A simple technique for preservation of fixation-sensitive antigens in paraffin-embedded tissues. J Histochem Cytochem *42*, 1127–1134.

Brock, R., and Jovin, T.M. (2003). Quantitative image analysis of cellular protein translocation induced by magnetic microspheres: Application to the EGF receptor. Cytometry A *52*, 1–11.

Bruand, J., Sistla, S., Mériaux, C., Dorrestein, P.C., Gaasterland, T., Ghassemian,M., Wisztorski, M., Fournier, I., Salzet, M., Macagno, E., and Bafna, V. (2011). Automated querying and identification of novel peptides using MALDI mass spectrometric imaging. J Proteome Res *10*(4), 1915–1928. Epub 2011 Mar 15. PubMed PMID:21332220.

Caprioli, R.M., Farmer, T.B., and Gile, J. (1997). Molecular imaging of biological samples: Localization of peptides and proteins using MALDI-TOF MS. Anal Chem *69*, 4751–4760.

Chaurand, P., Stoeckli, M., and Caprioli, R.M. (1999). Direct profiling of proteins in biological tissue sections by MALDI mass spectrometry. Anal Chem *71*, 5263–5270.

Chaurand, P., Schwartz, S.A., Billheimer, D., Xu, B.J., Crecelius, A., and Caprioli, R.M. (2004). Integrating histology and imaging mass spectrometry. Anal Chem *76*, 1145–1155.

Chaurand, P., Schriver, K.E., and Caprioli, R.M. (2007). Instrument design and characterization for high resolution MALDI-MS imaging of tissue sections. J Mass Spectrom *42*, 476–489.

Coons, A.H., Creech, H.J., and Jones, R.N. (1941). Immunological properties of an antibody containing a fluorescent group. Proc Soc Exp Biol *47*, 200–202.

Curran, R.C., and Gregory, J. (1980). Effects of fixation and processing on immunohistochemical demonstration of immunoglobulin in paraffin sections of tonsil and bone marrow. J Clin Pathol *33*, 1047–1057.

Dani, F.R., Francese, S., Mastrobuoni, G., Felicioli, A., Caputo, B., Simard, F., Pieraccini, G., Moneti, G., Coluzzi, M., della Torre, A., and Turillazzi, S. (2008). Exploring proteins in *Anopheles gambiae* male and female antennae through MALDI mass spectrometry profiling. PLoS One *3*, e2822.

DeKeyser, S.S., Kutz-Naber, K.K., Schmidt, J.J., Barrett-Wilt, G.A., and Li, L. (2007). Imaging mass spectrometry of neuropeptides in decapod crustacean neuronal tissues. J Proteome Res *6*, 1782–1791.

De Lange, R.P., and van Minnen, J. (1998). Localization of the neuropeptide APGWamide in gastropod molluscs by in situ hybridization and immunocytochemistry. Gen Comp Endocrinol *109*, 166–174.

Dreisewerd, K. (2003). The desorption process in MALDI. Chem Rev *103*, 395–426.

Dreisewerd, K., Kingstonb, R., Geraertsa, W.P.M., and Lia, K.W. (1997). Direct mass spectrometric peptide profiling and sequencing of nervous tissues to identify peptides involved in male copulatory behavior in *Lymnaea stagnalis.* Int J Mass Spectrom *169*, 291–299.

Dreisewerd, K., Schurenberg, M., Karas, M., and Hillenkamp, F. (1995). Influence of the laser intensity and spot size on the desorption of molecules and ions in matrix-assisted laser desorption/ionization with a uniform beam profile. Int J Mass Spectrom Ion Processes *141*, 127–148.

Fournier, I., Day, R., and Salzet, M. (2003). Direct analysis of neuropeptides by in situ MALDI-TOF mass spectrometry in the rat brain. Neuro Endocrinol Lett *24*, 9–14.

Francese, S., Lambardi, D., Mastrobuoni, G., la Marca, G., Moneti, G., and Turillazzi, S. (2009). Detection of honeybee venom in envenomed tissues by direct MALDI MSI. J Am Soc Mass Spectrom *20*, 112–123.

Franck, J., Longuespee, R., Wisztorski, M., Van Remoortere, A., Van Zeijl, R., Deelder, A., Mcdonnell, L., Salzet, M., and Fournier, I. (2010). MALDI mass spectrometry imaging of proteins exceeding 30 000 Da. Med Sci Monit *16*, BR293–299.

Furuya, T., Ikemoto, K., Kawauchi, S., Oga, A., Tsunoda, S., Hirano, T., and Sasaki, K. (2004). A novel technology allowing immunohistochemical staining of a tissue section with 50 different antibodies in a single experiment. J Histochem Cytochem *52*, 205–210.

Gabe, M. (1964). Histochemistry of secretion products. Biol Med (Paris) *53*, 641–674.

Gabe, M. (1969). Histological data on the endocrine pancreas of *Protopterus annectens* Owen. Arch Anat Microsc Morphol Exp *58*, 21–40.

Gabe, M. (1972). Relation between the abundance of glycol radicals and some tinctorial affinities of the product of hypothalamic neurosecretion. C R Acad Sci Hebd Seances Acad Sci D *274*, 549–551.

Gabe, M., and Saint Girons, H. (1967). Histological data on the tegument and cephalic epidermoid glands of Lepidosauria. Acta Anat (Basel) *67*, 571–594.

Hofer, S., Dircksen, H., Tollback, P., and Homberg, U. (2005). Novel insect orcokinins: Characterization and neuronal distribution in the brains of selected dicondylian insects. J Comp Neurol *490*, 57–71.

Holle, A., Haase, A., Kayser, M., and Hohndorf, J. (2006). Optimizing UV laser focus profiles for improved MALDI performance. J Mass Spectrom *41*, 705–716.

Hsieh, Y., Casale, R., Fukuda, E., Chen, J., Knemeyer, I., Wingate, J., Morrison, R., and Korfmacher, W. (2006). Matrix-assisted laser desorption/ionization imaging mass spectrometry for direct measurement of clozapine in rat brain tissue. Rapid Commun Mass Spectrom *20*, 965–972.

Hummon, A.B., Amare, A., and Sweedler, J.V. (2006a). Discovering new invertebrate neuropeptides using mass spectrometry. Mass Spectrom Rev *25*, 77–98.

Hummon, A.B., Richmond, T.A., Verleyen, P., Baggerman, G., Huybrechts, J., Ewing, M.A., Vierstraete, E., Rodriguez-Zas, S.L., Schoofs, L., Robinson, G.E., and Sweedler, J.V. (2006b). From the genome to the proteome: Uncovering peptides in the Apis brain. Science *314*, 647–649.

Ino, H. (2004). Application of antigen retrieval by heating for double-label fluorescent immunohistochemistry with identical species-derived primary antibodies. J Histochem Cytochem *52*, 1209–1217.

Jardin-Mathe, O., Bonnel, D., Franck, J., Wisztorski, M., Macagno, E., Fournier, I., and Salzet, M. (2008). MITICS (MALDI imaging team imaging computing system): A new open source mass spectrometry imaging software. J Proteomics *71*, 332–345.

Jimenez, C.R., and Burlingame, A.L. (1998). Ultramicroanalysis of peptide profiles in biological samples using MALDI mass spectrometry. Exp Nephrol *6*, 421–428.

Jimenez, C.R., Li, K.W., Dreisewerd, K., Spijker, S., Kingston, R., Bateman, R.H., Burlingame, A.L., Smit, A.B., van Minnen, J., and Geraerts, W.P. (1998). Direct mass spectrometric peptide profiling and sequencing of single neurons reveals differential peptide patterns in a small neuronal network. Biochemistry *37*, 2070–2076.

Jimenez, C.R., van Veelen, P.A., Li, K.W., Wildering, W.C., Geraerts, W.P., Tjaden, U.R., and van der Greef, J. (1994). Neuropeptide expression and processing as revealed by direct matrix-assisted laser desorption ionization mass spectrometry of single neurons. J Neurochem *62*, 404–407.

Jovin, T.M. (2003). Quantum dots finally come of age. Nat Biotechnol *21*, 32–33.

Jurchen, J.C., Rubakhin, S.S., and Sweedler, J.V. (2005). MALDI-MS imaging of features smaller than the size of the laser beam. J Am Soc Mass Spectrom *16*, 1654–1659.

Laboux, O., Dion, N., Arana-Chavez, V., Ste-Marie, L.G., and Nanci, A. (2004). Microwave irradiation of ethanol-fixed bone improves preservation, reduces processing time, and allows both light and electron microscopy on the same sample. J Histochem Cytochem *52*, 1267–1275.

Langstrom, B., Andren, P.E., Lindhe, O., Svedberg, M., and Hall, H. (2007). In vitro imaging techniques in neurodegenerative diseases. Mol Imaging Biol *9*, 161–175.

Lemaire, R., Wisztorski, M., Desmons, A., Tabet, J.C., Day, R., Salzet, M., and Fournier, I. (2006a). MALDI-MS direct tissue analysis of proteins: Improving signal sensitivity using organic treatments. Anal Chem *78*, 7145–153.

Lemaire, R., Desmons, A., Ducroy, P., Tabet, J.C., Salzet, M., and Fournier, I. (2006b). Direct Analysis and MALDI Imaging on Formalin Fixed Paraffin Embedded Tissue (FFPE): Application to Parkinson Disease. Proceedings of 54rd ASMS Conference on Mass Spectrometry (Seattle, WA).

Lemaire, R., Tabet, J.C., Ducoroy, P., Hendra, J.B., Salzet, M., and Fournier, I. (2006c). Solid ionic matrixes for direct tissue analysis and MALDI imaging. Anal Chem *78*, 809–819.

Li, L., Garden, R.W., and Sweedler, J.V. (2000a). Single-cell MALDI: A new tool for direct peptide profiling. Trends Biotechnol *18*, 151–160.

Li, K.W., Hoek, R.M., Smith, F., Jimenez, C.R., van der Schors, R.C., van Veelen, P.A., Chen, S., van der Greef, J., Parish, D.C., Benjamin, P.R., and Geraerts, W.P.M. (1994a). Direct peptide profiling by mass spectrometry of single identified neurons reveals complex neuropeptide-processing pattern. J Biol Chem *269*, 30288–30292.

Li, K.W., Jimenez, C.R., Van Veelen, P.A., and Geraerts, W.P. (1994c). Processing and targeting of a molluscan egg-laying peptide prohormone as revealed by mass spectrometric peptide fingerprinting and peptide sequencing. Endocrinology *134*, 1812–1819.

Li, L., Romanova, E.V., Rubakhin, S.S., Alexeeva, V., Weiss, K.R., Vilim, F.S., and Sweedler, J.V. (2000b). Peptide profiling of cells with multiple gene products: Combining immunochemistry and MALDI mass spectrometry with on-plate microextraction. Anal Chem *72*, 3867–3874.

Li, K.W., van Golen, F.A., van Minnen, J., van Veelen, P.A., van der Greef, J., and Geraerts, W.P. (1994b). Structural identification, neuronal synthesis, and role in male copulation of myomodulin-A of Lymnaea: A study involving direct peptide profiling of nervous tissue by mass spectrometry. Brain Res Mol Brain Res *25*, 355–358.

McDonnell, L.A., Piersma, S.R., Maarten Altelaar, A.F., Mize, T.H., Luxembourg, S.L., Verhaert, P.D., van Minnen, J., and Heeren, R.M. (2005). Subcellular imaging mass spectrometry of brain tissue. J Mass Spectrom *40*, 160–168.

Meriaux, C., Franck, J., Wisztorski, M., Salzet, M., and Fournier, I. (2010). Liquid ionic matrixes for MALDI mass spectrometry imaging of lipids. J Proteomics *73*(6), 1204–1218. Epub 2010 Feb 24. PubMed PMID:20188221.

O'Brien, E., Dedova, I., Duffy, L., Cordwell, S., Karl, T., and Matsumoto, I. (2006). Effects of chronic risperidone treatment on the striatal protein profiles in rats. Brain Res *1113*, 24–32.

Petit, J., and Sahli, F. (1975). Cytochemical and electron-microscopic study of the paraoesophageal bodies and related nerves in *Schizophyllum sabulosum* (L.), Diplopoda Julidae. Cell Tissue Res *162*, 367–375.

Pierson, J., Norris, J.L., Aerni, H.R., Svenningsson, P., Caprioli, R.M., and Andren, P.E. (2004). Molecular profiling of experimental Parkinson's disease: Direct analysis of peptides and proteins on brain tissue sections by MALDI mass spectrometry. J Proteome Res *3*, 289–295.

Pierson, J., Svenningsson, P., Caprioli, R.M., and Andren, P.E. (2005). Increased levels of ubiquitin in the 6-OHDA-lesioned striatum of rats. J Proteome Res *4*, 223–226.

Salzet, M., Bulet, P., Wattez, C., Verger-Bocquet, M., and Malecha, J. (1995). Structural characterization of a diuretic peptide from the central nervous system of the leech Erpobdella octoculata. Angiotensin II amide. J Biol Chem *270*, 1575–1582.

Schikorski, D., Cuvillier-Hot, V., Leippe, M., Boidin-Wichlacz, C., Slomianny, C., Macagno, E., Salzet, M., and Tasiemski, A. (2008). Microbial challenge promotes the regenerative process of the injured central nervous system of the medicinal leech by inducing the synthesis of antimicrobial peptides in neurons and microglia. J Immunol *181*, 1083–1095.

Spengler, B. (1994). Ion Imaging and Confocal Microscopy with a New Scanning UV- Laser Microprobe. 42nd Annual Conference on Mass Spectrometry and Allied Topics (Chicago, IL).

Spengler, B., and Hubert, M. (2002). Scanning microprobe matrix-assisted laser desorption ionization (SMALDI) mass spectrometry: Instrumentation for sub-micrometer resolved LDI and MALDI surface analysis. J Am Soc Mass Spectrom *13*, 735–748.

Stauber, J., Lemaire, R., Franck, J., Bonnel, D., Croix, D., Day, R., Wisztorski, M., Fournier, I., and Salzet, M. (2008). MALDI imaging of formalin-fixed paraffin-embedded tissues: Application to model animals of Parkinson disease for biomarker hunting. J Proteome Res *7*, 969–978.

Steinbach, P. (1977). Granular cells in the connective tissue of Helix pomatia L. (gastropoda, pulmonata). Histochemistry, ultrastructure, and results of polyacrylamide electrophoretic investigations. Cell Tissue Res *181*, 91–103.

Stoeckli, M., Farmer, T.B., and Caprioli, R.M. (1999). Automated mass spectrometry imaging with a matrix-assisted laser desorption ionization time-of-flight instrument. J Am Soc Mass Spectrom *10*, 67–71.

Stoeckli, M., Chaurand, P., Hallahan, D.E., and Caprioli, R.M. (2001). Imaging mass spectrometry: A new technology for the analysis of protein expression in mammalian tissues. Nat Med *7*, 493–496.

Stoeckli, M., Knochenmuss, R., McCombie, G., Mueller, D., Rohner, T., Staab, D., and Wiederhold, K.H. (2006). MALDI MS imaging of amyloid. Methods Enzymol *412*, 94–106.

Stoeckli, M., Staab, D., Staufenbiel, M., Wiederhold, K.H., and Signor, L. (2002). Molecular imaging of amyloid beta peptides in mouse brain sections using mass spectrometry. Anal Biochem *311*, 33–39.

Sugiura, Y., Shimma, S., Konishi, Y., Yamada, M.K., and Setou, M. (2008). Imaging mass spectrometry technology and application on ganglioside study; visualization of age-dependent accumulation of C20-ganglioside molecular species in the mouse hippocampus. PLoS One *3*, e3232.

Wisztorski, M., Brunet, L., Dreiserwer, K., Hillenkamp, F., Berkenkamp, S., Salzet, M., and Fournier, I. (2006). Effect of Metals Coating for UV MALDI-a-TOF Mass Spectrometry Imaging (MALDI MSI) and Direct Tissue Analysis in UV/IR MALDI-o-TOF Mass Spectrometry. Proceedings of 54rd ASMS conference on Mass Spectrometry (Seattle, WA).

Wisztorski, M., Croix, D., Macagno, E., Fournier, I., and Salzet, M. (2008). Molecular MALDI imaging: An emerging technology for neuroscience studies. Dev Neurobiol *68*, 845–858.

Wisztorski, M., Thomy, V. Verplanck, N., Stauber, J., Camart, J.C., Salzet, M., and Fournier, I. (2007). Use of Masks in MALDI-MSI: An Easy Tool for Increasing Spatial Resolution of Images by Decreasing Irradiated Area. Proceedings of 55rd ASMS conference on Mass Spectrometry (Indianapolis, IN).

Chapter 37
UV MALDI for DNA Analysis and the Developments in Sample Preparation Methods

Igor P. Smirnov

Abstract Matrix Assisted Laser Desorption-Ionization Mass Spectrometry (MALDI MS) is a powerful technique for the analysis of short DNA fragments and products of their enzymatic conversions, with majority of it's practical applications being developed in fields of organic chemistry of oligonucleotides and typing of genetic polymorphisms. While capable of providing unique structural information and demonstrating high speed of data acquisition, results of MALDI analysis strongly depends on the quality and the type of sample preparation method used and the MALDI matrices applied for laser desorption-ionization. This review is focused mainly on these two areas followed by brief description of the most recent developments in the DNA genotyping by MALDI MS.

Keywords Oligonucleotides · DNA · MALDI matrices · Sample preparation · Genotyping

37.1 Introduction

In the last two decades MALDI MS entered the field of biochemistry as a new revolutionary tool, allowing direct measurement of masses of biomolecules. It has been adopted for the analysis of all classes of biomolecules: nucleic acids (Perkel, 2008; Tost and Gut, 2006), protein and peptides (Ashcroft, 2003), lipids (Griesser et al., 2004; Schiller et al., 2007), carbohydrates (Dell and Morris, 2001; Harvey, 2001), and small molecules (Cohen and Gusev, 2002). In DNA research MALDI MS is playing a vital role across the entire field: from quality control of synthetic oligonucleotide primers and antisense drugs (Capaldi and Scozzari, 2008) to high-throughput genotyping applications (Pusch et al., 2002; Sauer, 2009; Tost, 2008; Tost and Gut, 2004) and screening and diagnostic research (Pusch and Kostrzewa,

I.P. Smirnov (✉)
Institute of Physical-Chemical Medicine, Moscow, Russian Federation; A.N. Belozersky Institute of Physico-Chemical Biology, Moscow State University, Moscow, Russian Federation
e-mail: smirnov_i@hotmail.com

A.R. Ivanov, A.V. Lazarev (eds.), *Sample Preparation in Biological Mass Spectrometry*, DOI 10.1007/978-94-007-0828-0_37,

2005). MALDI MS has several desirable features. It measures mass to charge ratio of a biomolecule directly and do not require fluorescent, or radioisotope labeling, and this reduces cost of analysis and removes extra steps associated with labeling and sample preparation. Since the matrix-assisted desorption/ionization process generates mostly singly charged ions (Karas et al., 2000), it makes possible analysis of complex mixtures without a necessity of prior separation of their components. The speed of analysis is very high, typically requiring only several seconds per sample for an analyte exhibiting a signal readily distinguishable from the background. Results of MALDI MS analysis are strictly depends on the choice of matrix composition and applied strategies for sample isolation and concentration steps. Hereafter the recent progress in this area will be outlined as well as necessary background information.

37.2 MALDI Matrices and Co-matrices

37.2.1 Matrices and Ammonium Salt Additives

Currently there is no comprehensive theory, which allows for an a priory design of the best MALDI matrix to use for a given class of analytes. Although, ionization mechanisms in MALDI and internal excitation process are subjects of significant research efforts (Christian et al., 2001; Hillenkamp et al., 2009; Karas and Krueger, 2003; Kinet et al., 2009; Kong et al., 2001), but are still not well understood. Therefore, the search for "an ideal" matrix belongs to the realm of experimentation with only few general guidelines (Horneffer et al., 2006; Strupat et al., 1991). A prospective MALDI matrix should meet the following initial requirements:

- Both a matrix and an analyte must be soluble in the same volatile solvent systems (it should be possible to evaporate the solvent under reasonable conditions). The matrix should crystallize with the analyte molecules incorporated homogeneously within the crystals.
- The matrix must adsorb light at the wavelength emitted by the MALDI laser. Upon laser irradiation, a suitable molecular plasma must form that allows for efficient analyte desorption and ionization.
- The matrix used for the desorption/ionization step (in other words, energy transfer) should cause minimum fragmentation of analyte molecules.

The matrices used in the UV MALDI are typically small organic compounds containing at least one aromatic ring and have UV adsorption at the MALDI laser's wavelength. The most common lasers used in commercial instruments are 337 nm (nitrogen gas laser), 355 and 266 nm (3-d and 4-th harmonic of Nd:YAG laser). It seems though, that a strong extinction coefficient (molar absorptivity) at the laser wavelength is not a prerequisite for a good matrix (Wu et al., 1993). Extinction coefficients are being measured in a solution at room temperature and not in the gas phase with the density of a solid material at several hundred degrees Celsius, as occurs during the laser pulse, where the properties of matrix molecules may be somewhat different. Most of the compounds experimentally determined as good UV-matrices contain a carboxyl or carbonyl functional group as well as OH-, NO_2- or NH_2-groups in ortho-and para-positions of aromatic rings.

Many matrices are used predominantly for only a particular class of biological molecules (or "analytes"). Co-matrices, defined as any chemicals added to the matrix, which improve the observed MALDI performance, also depend upon the analyte and matrix selected. In general, both the matrix and co-matrix chosen in a MALDI MS experiment strongly depend on the type of the analyte being studied. The optimal matrix composition that provides the best performance, for example, for DNA or synthetic polymer analysis, may generate very low quality spectra for proteins or no spectra at all, and vice versa. Besides, fine tuning may be necessary to achieve the best performance for a particular assay.

There are many good matrices available for peptides and proteins: α-cyano-4-hydroxycinnamic acid (CHCA), sinapinic acid, 2,5-dihydroxybenzoic acid (2,5-DHB), super DHB, ferulic acid and many others. Unfortunately, their performance for DNA analysis is rather poor (Hathaway, 1994; Koomen et al., 2000; Weng and Chen, 2004). Although for peptide-mimicking DNA analogs, such as peptide-nucleic acids (PNAs) (Bauer et al., 2004) and charge-tagged DNA (Sauer et al., 2006), MALDI MS analysis can be more efficiently done with peptide matrices.

The MALDI matrices most often used for DNA analysis are limited to just few compounds with 3-hydroxypicolinic acid (**I**) (3-HPA) being most universal and applied almost exclusively for DNA fragments larger that 15–20 bases in size.

The structures of the most commonly used matrices in DNA analysis are shown in the Fig. 37.1

3- Hydroxypicolinic acid (**I**, see Fig. 37.1) as a DNA matrix was first reported by Becker's group (Wu et al., 1993, 1994). It was found to work well in both negative and positive ion modes, unlike other matrices for which the negative ion mode is more preferential. Lubman's group successfully observed a 622 bp fragment using 3-HPA and nitrocellulose as a matrix support (Liu et al., 1995a). Generally, with 3-HPA it is possible to obtain high resolution spectra (> 1000 measured at full width at half maximum (FWHM)) for up to 60–70-mer oligonucleotides. Larger oligonucleotides exhibit lower peak resolution (as well as sensitivity) falling precipitously to the range of 50–20 FWHM. The sensitivity of MS detection for mixed-base 25-mer oligonucleotide is typically in the low femtomole range (5–20 fmol/μl) with a common dry droplet sample deposition protocol using the 3-HPA matrix. Generally, MALDI sensitivity of DNA analysis is dropping very quickly with the increase of the DNA length. For each additional 25 bases of DNA length it falls approximately 10-fold.

Other matrices employed for DNA analysis can exhibit better performance than 3-HPA in some aspects. For example, the detection of the longest fragment of DNA in UV MALDI was so far achieved using p-nitrophenol (**XI**) (Lin et al., 1999), the best sensitivity for short DNA oligomers – using 6-aza-2-thiothymine (ATT)/spermine) (**II**) (Asara and Allison, 1999), the highest tolerance to alkali metal contaminations – using 3,4-diaminobenzophenone (**XII**) (3,4-DBP) (Fu et al.,

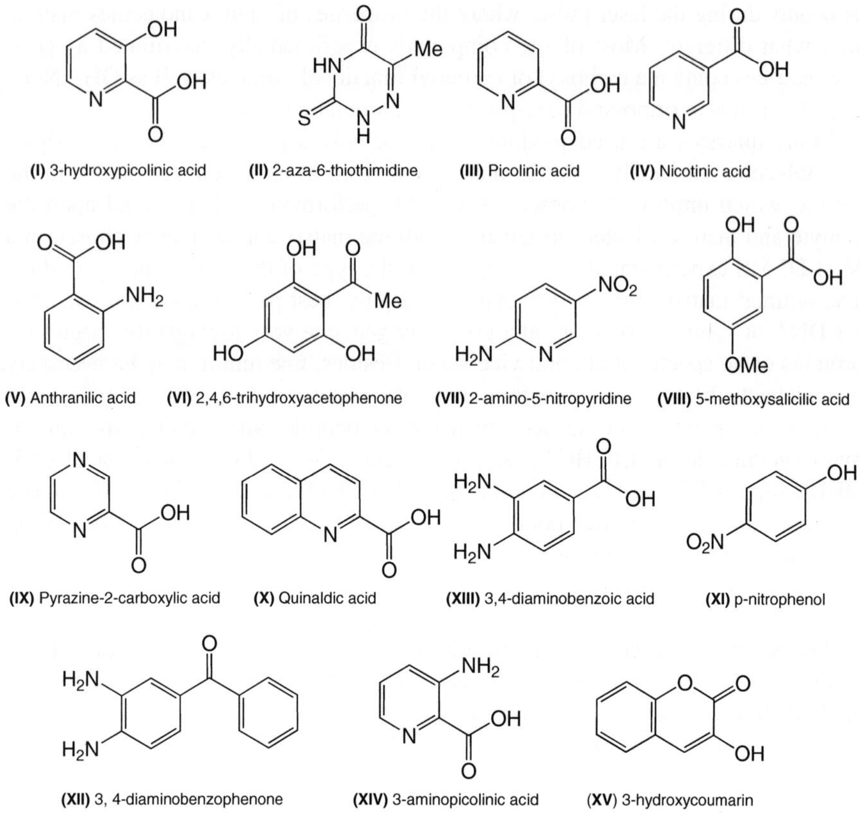

Fig. 37.1 The matrices used for DNA UV MALDI analysis. 3-hydroxypicolinic acid, (**I**); 2-aza-6-thiothimidine, (**II**); picolinic acid, (**III**); nicotinic acid, (**IV**); anthranilic acid, (**V**); 2,4,6-trihydroxyacetophenone, (**VI**); 2-amino-5-nitropyridine, (**VII**); 5-methoxysalicilic acid, (**VIII**); pyrazine-2-carboxylic acid, (**IX**); quinaldic acid, (**X**); p-nitrophenol, (**XI**); 3, 4-diaminobenzophenone, (**XII**); 3, 4-diaminobenzoic acid, (**XIII**); 3-aminopicolinic acid, (**XIV**); 3-hydroxycoumarin, (**XV**)

2006). However, these alternative matrices have drawbacks that hinder their broader application (such as significant fragmentation and "apurinization" observed for ATT and 3,4-DBP, or the requirement for special cooled sample stage in case of p-nitrophenol). Nevertheless, their implementation may be very beneficial for specific assays.

For MALDI MS analysis it is necessary to convert a DNA sample into the ammonium form or substituted alkylammonium form that would provide a single molecular ion upon desorption. If alkali cations are present on the polyphosphate DNA backbone, then not a single but a group of DNA molecular ions will be registered by a mass spectrometer. The observed difference in masses of these ions will depend on how many and what kind of alkali ions (usually xNa^+/yK^+) are attached to the DNA phosphate backbone. These extraneous peaks can significantly complicate data interpretation and reduce both sensitivity and resolution. In the most extreme cases such problems could render MALDI analysis impossible.

One way to reduce alkali adduct formation is an addition of ammonium salts to a MALDI matrix (Pieles et al., 1993). 3-HPA is a good chelator (Sun et al., 2003) each molecule of which is capable of coordinating two alkali metal ions (Popovic et al., 2007). Solutions of this matrix are particularly prone to extracting contaminating metal ions from surrounding environment such as dust particles or unclean container surfaces. This makes ammonium salt additives especially important. Besides, addition of ammonium salts (as co-matrices) appears to improve the quality of mass spectra possibly by influencing the dynamics of ion formation. It was shown that such co-matrices may limit proton exchange that is implicated in fragmentation (Christian et al., 2001; Simmons and Limbach, 1997, 1998). Currently the most commonly used co-matrix for 3-HPA is dibasic ammonium citrate (Zhu et al., 1996a). Koomen et al. (2000) reported that an addition to the 3-HPA/ammonium citrate solution 0.5% aqueous formic acid produced a better signal intensity. A variety of other ammonium salts have been tested (Li et al., 1998) and ammonium fluoride was found to be a good co-matrix as well. Interestingly, the presence of organic amines have negative effect on the performance of 3-HPA (Simmons and Limbach, 1997).

Most UV MALDI DNA matrices other than 3-HPA produce substantial "apurinized" DNA spectral peaks and lower sensitivity for oligonucleotides longer than 15–20 bases. However, they may have other advantages, depending on the matrix: such as providing more homogeneous surface with better spot-to-spot signal reproducibility, better alkali salt tolerance, higher sensitivity for short oligomers, or more intense post-source decay (PSD) fragmentation of short oligonucleotides and their cross-linked products with peptides, etc.

The most common alternatives for 3-HPA are several substituted acetophenones, most importantly 2,4,6-trihydroxyacetophenone (Pieles et al., 1993) (**VI**) (2,4,6-THAP) with diammonium citrate, as co-matrix. It is being used occasionally in a mixture with 2,3,4-THAP (Zhu et al., 1996a), or 2,6-dihydroxyacetophenone (Terrier et al., 2007). The optimization of matrix compositions based on THAP was described in several studies (Koomen et al., 2000; Lavanant and Lange, 2002; Pieles et al., 1993; Terrier et al., 2007; Zhu et al., 1996a). A saturated solution of 2,4,6-THAP in acetonitrile and 0.2 M triammonium citrate allowed to obtain spectra of dT_{10} (10 pmol/μl) in 0.1 M NaCl without producing significant intensities of sodium adducts (Lavanant and Lange, 2002). It was also applied for PSD sequence analysis of synthetic oligonucleotides up to 35-mer (Berhane and Limbach, 2003b; Lavanant and Lange, 2002; Zhu et al., 1997) and oligonucleotide-peptide complexes (Kuhn-Holsken et al., 2005). Interestingly, the fragmentation of oligonucleotides in THAP matrix was found to be strongly dependent on the matrix-to-analyte ratio (Christian et al., 2001). The THAP- based matrix was used also for PNA detection (Butler et al., 1996) and for analysis of metal-DNA complexes (with europium (Bourin et al., 1997b) and terbium (Bourin et al., 1997a)). THAP- based matrices used mostly for oligonucleotides shorter than 25 bases, much longer DNA can be analyzed only at the expense of significant "apurinization". It crystallizes uniformly and shows good signal reproducibility across the sample surface. It is compatible with high concentration of ammonium salt co-matrices, that helps to tolerate higher levels of alkali metals contaminations. Interestingly, 2,4,6-THAP (as well as 2,5-DHB) was found more thermostable compared to sinapinic acid and CHCA (Tarzi et al., 2009)

Another matrix frequently used for DNA analysis is 6-aza-2-thiothimidine (**II**) (ATT) (Garcia et al., 2002; Lecchi et al., 1995). It has a potential for the analysis of non-covalent complexes of short oligonucleotides (Ohara et al., 2006; Terrier et al., 2007), peptides (see very detailed comparative work of Zhel at el. (Zehl and Allmaier, 2003)) and DNA duplexes (Kirpekar et al., 1999). Like 2,4,6-THAP, ATT-based matrices are much less sensitive to alkali contaminations (compare to 3-HPA). It was possible to detect oligomerization products of activated nucleotides directly from the montmorillonite, mineral lithium-rich support, using ATT- and THAP- based matrices without any isolation/desalting steps (Zagorevskii et al., 2006), which was not possible using 3-HPA. 6-Aza-2-thiothimidine exhibits good solubility in acetone and acetonitrile and rapid evaporation of the solvent resulted in a very homogeneous crystal formation on the MALDI target surface (Garcia et al., 2002).

The main problem with using ATT (as well as THAP) is that it causes a significant fragmentation of oligonucleotides, and this diminishes sensitivity of MS analysis. Generally, the ability of a matrix to cause DNA fragmentation exhibits the following trend (Schulz et al., 2006): CHCA>>DHB>SA=THAP>ATT>3-HPA.

Other matrices are less commonly used for DNA analysis. 2-Amino-5-nitropyridine (**VII**) has been described (Fitzgerald et al., 1993) with ammonium halides (Cheng and Chan, 1996; Zhu et al., 1996b) and other salts (Li et al., 1998) as co-matrices. Ammonium fluoride was reported to provide excellent results (Cheng and Chan, 1996; Li et al., 1998). Nicotinic acid (3-pyridine carboxylic acid) (**IV**) (Beavis et al., 1989; Hettich and Buchanan, 1991) was used mostly with anthranilic acid as co-matrix. Picolinic acid (**III**) can be used by itself (Tang et al., 1994b) and as a co-matrix with 3-HPA (Kim et al., 2001; Liu et al., 1995a, b; Tang et al., 1994a; Zhu et al., 1996a). Anthranilic acid (2-aminobenzoic acid) has been characterized either as an undoped matrix (Nordhoff et al., 1992; Schieltz et al., 1992) (**V**) or with addition of co-matrices including ammonium citrate (Zhu et al., 1997), ammonium fluoride (was reported useful for PSD) (Chan et al., 2002), spermine (Asara and Allison, 1999), nicotinic acid (**IV**) (20%) (Stemmler et al., 1994, 1995) or nicotinic acid/ammonium citrate (Zhang and Gross, 2000). Anthranilic acid-based matrix was applied for the localization of an abasic site in modified oligodeoxynucleotides by MALDI PSD analysis (Zhang and Gross, 2002). There have also been reports of using of 3-aminopicolinic acid (Taranenko et al., 1994) (**XIV**), p-nitrophenol (**XI**) (Hunter et al., 1997; Lin et al., 1999; Williams and Fenselau, 1998), and frozen water (Hunter et al., 1997; Schieltz et al., 1992). 5-Methoxysalicilic acid (**VIII**) (Distler and Allison, 2001a) was applied for short oligonucleotide analysis as well.

The main progress on the MALDI matrices used in DNA research was achieved in 1990s. Among the developments of this decade were the introduction by Chen et al. (Chen and Chen, 2003) of 3,4- and 3,5-diaminobenzoic acid (**XIII**) doped into a sol-gel surface covering a MALDI target. This work was an extension of the group's prior research using 2,5-DHB (Lin and Chen, 2002) as the matrix. Good-quality mass spectra were obtained without extra desalting steps, and addition of 0.1% SDS to the analyte sample did not result in deterioration of the quality of the spectra. The largest detectable size for oligonucleotides was a 72-mer, and a detection limit of a 24-mer oligonucleotide was in the range of 20 fmol. Fu et al. (2006) introduced 3,4-diaminobenzophenone (**XII**) as a potentially useful matrix for oligonucleotide analysis, when high tolerance to the alkali salts is required. The mass spectrum of 23-mer (500 fmol/μl) was obtained in 0.4 M of NaCl (Fu et al., 2006; Xu et al., 2006) without observing significant sodium adduct formation. Quinaldic acid (**X**) (Song, 2003) was also reported as a possible matrix. Recently Zhu tested 2-pyrazinecarboxylic acid as a new co-matrix for 3-HPA (Zhou et al., 2004) (**IX**) with promising results. A 3-hydroxycoumarin (**XV**) matrix also was found to be potentially useful (Zhang et al., 2006).

37.2.2 Organic Amine Additives

Besides inorganic ammonium salts, organic amines have been studied as possible co-matrices. Although spermine was originally found to interfere with the 3-HPA matrix (Shaler et al., 1996) leading to the deterioration of spectra quality, other matrices have shown better results. Triethylamine and N-methylimidazole were tested with 2,4,6-THAP and ATT showing improvement in resolution and salt tolerance (Simmons and Limbach, 1997, 1998). It was previously reported that these organic bases helped to improve spectral quality in ESI MS of oligonucleotides (Greig and Griffey, 1995; Muddiman et al., 1996). Later spermine additives were applied to ATT and anthranilic acid/nicotinic acid mixtures by Asara et al. (Asara and Allison, 1999) and spermine/spermidine additives to 2,4,6-THAP by Vandell et al. (Vandell and Limbach, 1999). In both cases the presence of spermine (spermidine) suppress the formation of oligonucleotide alkali adducts more efficiently, than addition of diammonium citrate. Successful results were shown by Distler and Allison (Distler and Allison, 2001a) with the mixture of 5-MSA and spermine. Better sensitivity for 15-mer mixed base oligonucleotide was obtained when compared to 3-HPA/ammonium citrate formulations.

37.2.3 Sugars

Improved spot-to-spot uniformity of signal intensity and resolution of molecular ion peaks has been achieved by using sugar additives either by themselves or in combination with other additives (Distler and Allison, 2001b; Shahgholi et al., 2001). The intent was to introduce neutral none-UV adsorbing additives (Koster et al., 1992), which during co-desorption with matrix would provide (by itself and by their relatively cold decomposition products, CO_2 and H_2O) better conditions for the collisional cooling of excited analyte ions within a MALDI plume. Such additives were later refined for protein analysis (Gusev et al., 1995), using their secondary, unexpected, effect – the impact on a morphology of matrix crystals. As it is turned out, sugar additives to common matrices such as ferulic acid, 2,5-DHB and 5-MSA lead to formation of more homogeneous spots with better signal reproducibility across the sample surface and, therefore, making quantitative analysis more feasible. It also was found that D-ribose additive had diminishing effect on the 3-HPA matrix (Gusev et al., 1995). Therefore, sugar additives are not universal and do require optimization for each matrix and for every analyte class. For short oligonucleotide analysis using ATT and spermine the best results were obtained with fucose (Distler and Allison, 2001b). An extensive body of work on optimization of 3-HPA matrix composition was performed by Shahgholy et al. (2001) where fucose again provided the best quality spectra. However, these additives were studied using pure oligonucleotides and initial conclusions may not have been universally applicable. In many applications that involve real biological samples, positive effects form sugar additives may be less evident. In the author's experience, the effect of fucose as well as other sugar co-matrices may be less pronounced with samples containing complex mixture of salts, traces of surfactants and other contaminants, which is a common case in the analysis of enzymatic reaction products (in mini-sequencing, SNP typing, etc.) when no further purification from reaction media was employed prior to MALDI MS analysis.

37.2.4 Liquid Matrices

Another means of mitigating spot-to-spot signal intensity variation is employing so called liquid matrices introduced by Armstrong et al. (2001). These matrices are actually salts of the most common MALDI matrices with various organic bases, such as triethylamine, pyridine etc. (Tholey, 2009). In oligonucleotide analysis all attempts to produce useful 3-HPA salts were so far unsuccessful (Carda-Broch et al., 2003) and the best results to date were reported for methylimidazolium salt of CHCA. Due to excellent signal reproducibility it was possible to perform a quantitative analysis of short oligonucleotides (9-mers) mixture within the concentration range from 2 to 50 μM (Li and Gross, 2004). Original matrices CHCA and DHB are known to be too "hot" for DNA analysis causing extensive fragmentation, which seems to be the case for their salts as well. Ionic liquids were recently applied for PSD analysis of mixed base 6-mers on MALDI FTMS using aniline and dimethylaniline salts of CHCA (Jones et al., 2005).

37.2.5 Sample/Matrix Deposition Techniques

Sample/matrix deposition techniques in DNA MALDI MS analysis are mostly based on dried droplet method (Karas and Hillenkamp, 1988) where both analyte and matrix are mixed together in solution, spotted on the sample plate and allowed to be air dried. This usually takes several minutes depending upon solvent. The drying speed can be increased by placing a freshly spotted sample plate onto warm surface or into vacuum chamber. In the later case a more homogeneous surface is typically formed. Usually this step is preceded by treatment of oligonucleotide or oligonucleotide/matrix mixture with cation exchange beads in the $NH4^+$ -form (Nordhoff et al., 1992) (such as Dowex AG50W-X8 (Bio-Rad) or, SpectroCLEAN (Sequenome)). Less contaminated samples having low intrinsic concentrations of alkali metals or more concentrated samples can be cleaned up using cation exchange beads in H+-form (Langley et al., 1999).

Double layer sample preparation, which was developed for peptide analysis (Dai et al., 1999; Gobom et al., 2001), is also used for oligonucleotides, although the order of the deposition steps is reversed. First, aqueous DNA samples are dispensed and allowed to air dry on the MALDI target. Then the matrix dissolved in volatile organic solvents is applied on top of the DNA residue to form a thin homogeneous film. For this purpose Garcia et al. (2002) used 6-aza-2-thiothymine matrix dissolved in an ethanol/acetonitrile solvent blend containing ammonium citrate. This manner of sample preparation generates much better shot-to-shot and sample-to-sample reproducibility and essentially eliminates the need to search for "hot" spots. A similar approach was reported by Koomen et al. (2000) using CHCA, ferulic acid and 2,4,6-THAP for analysis of short oligonucleotides. Liu et al. modified this approach (Liu et al., 2003) using an AnchorChip sample stage. In this method diluted oligonucleotide solutions are first dispensed and allowed to partially dry and shrink to a small spot on the target. Subsequently, a small volume (0.1 μl) of saturated 3-hydroxypicolinic acid (3-HPA) solution is added. Samples prepared in this method were more homogenous and provide better signal reproducibility across the crystallized spot. It was possible to routinely detect oligonucleotides at a concentration level of 10 fmol/μl.

An interesting variation of the electrospray deposition approach (Erb and Owens, 2008; Hanton et al., 2004; Hensel et al., 1997; Wei et al., 2004) was reported by Molin et al. (2008), in which solutions of oligonucleotides with 3-HPA matrix were sprayed on to the MALDI target through a stainless steel sieve (38 μm, 450 mesh). That resulted in the formation of uniformly distributed microcrystals, providing excellent sensitivity and signal reproducibility. This technique has potential for automatic deposition system, but prior to practical application it has to be adapted to handle small (<10 μl) volumes of oligonucleotide/matrix solutions.

The MALDI signal variation from a sample can be significantly reduced if the sample spot size is comparable to the laser spot dimensions (Gobom et al., 1999; Vorm et al., 1994). Piezo-electric pipettors are commonly used to deposit such tiny amount of solutions. This deposition technique is compatible with the demands of automated high speed throughput genomic and proteomic applications, aiming to acquire spectra from thousands of samples per day. Spotting of only few nanolitres of concentrated sample solution onto a MALDI target is possible and low attomole sensitivity for small peptides was recently reported (Nordhoff et al., 2007). Applications of this technique to genotyping have been described (Tang, 2007; Tang et al., 1999) as well. A new sample target plate for DNA analysis made of SiO_2 with 30 nm Pt dots was reported recently for which a highly reproducible results were achieved (Honda et al., 2005). The microspotting approach has some disadvantages, though. First, there is a requirement to concentrate the analyte in just a few nanolitres of a pure solvent or matrix solution and, second, it is necessary to use expensive and complicated nano-volume liquid handling systems.

In this respect, more suitable for the regular laboratory are AnchorChips (Nordhoff et al., 2003; Schuerenberg et al., 2000) introduced by Bruker and comprised of MALDI plate coated with very hydrophobic Teflon-like polymer patterned with small hydrophilic spots or "anchors". Spots are arranged in the same pattern as a 384-well microtiter plate and are the same in shape and size (sample plates with four different sizes are available – 200, 400, 600, 800 μm in diameter). When matrix/analyte solution is deposited on to an anchor spot, it will crystallize upon evaporation inside the anchor, since it is surrounded by hydrophobic solvent-repelling surface. Therefore, it is possible to concentrate an analyte on such spot, moreover, the automated data acquisition becomes easier and more robust, since sample position and spot size are predetermined by the sample support. Such pre-structured supports simplify the liquid handling interface between mass spectrometer and analytical separation systems, such as HPLC (Perlman et al., 2007) and CE (Johnson et al., 2001).

37.2.6 MALDI Target Surface Coating and Special Supports

37.2.6.1 Nitrocellulose and Nafion

More homogeneous matrix crystallization and better spot-to-spot reproducibility was achieved on the regular stainless steel MALDI targets by implementing various types of coatings. Nitrocellulose was historically first one to be used for on-probe sample immobilization of proteins followed by MALDI MS (Mock et al., 1992). Later it was applied by the Lubman group as on-probe support for the analysis of PCR products (Liu et al., 1995a, b). Nafion also has been used as a coating (Bai et al., 1994a, b; Kim et al., 2001), improving matrix crystal homogeneity, and providing better salt tolerance levels due to its cation exchange properties. Additionally, it's slightly more hydrophobic surface results in improved liquid handling properties. Combined polymer/Nafion two-layer coating was reported by Kim et al. (2001). Initially, a MALDI target was coated by water-soluble polymers such as linear polyacrylamide, linear poly(ethylene oxide), or methyl cellulose. After evaporation of a solvent, the second layer of coating, Nafion was applied. Such supports provided more compact sample spots, smaller matrix crystals and made it easier to find "good spots".

37.2.6.2 Paraffin Wax Coating

Other avenues of making the MALDI target more hydrophobic were explored as well (Owen et al., 2003; Rechthaler et al., 2007). Paraffin wax coating (Hung et al., 1998; Wang et al., 2008), were also used. Recently Wang et al. (2008) demonstrated improvement in sensitivity for diluted peptides solutions using Paraffin M film, which was directly attached to the MALDI sample plate. Several groups had investigated poly(tetrafluoroethylene) (Hung et al., 1999) coating for the same purpose (Yuan and Desiderio, 2002).

37.2.6.3 Sol-Gel

Thin films of sol-gel that incorporated MALDI matrix as a dopant were prepared on Paraffin film as intermittent support, which then was attached on MALDI pate. It led to improvement of spot-to-spot reproducibility with 2,5-DHB matrix (Lin and Chen, 2002) for peptides and for short oligonucleotides (Chen and Chen, 2003). Later the same group applied 2-hydroxymethyl-15-crown-5- and 2-hydroxymethyl-18-crown-6-ethers as dopants during the sol–gel polymerization process to generate special substrates for MALDI sample deposition (Weng and Chen, 2004). Such substrate has been shown to function as a desalting material. The quality of oligonucleotide mass spectra can be dramatically improved this way without altering the method of sample preparation used for MALDI MS analysis. The sensitivity level of 20 fmol/μl for a 36-mer was demonstrated. Paraffin wax with immobilized matrix (2,5-DHB) was also recently applied for peptide analysis (Wei et al., 2009).

37.3 Sample Concentration/Desalting Methods

37.3.1 Cation Exchange Beads

When an oligonucleotide sample is not significantly contaminated and has reasonably high concentration (above 1 pmol/μl), treatment with cation exchange beads in ammonium form (Nordhoff et al., 1992) is usually sufficient (reports of using H+ form have been published occasionally (Langley et al., 1999)). Examples include synthetic oligonucleotides, many SNP reaction mixtures (Storm et al., 2003), which started with good quality PCR products and so on. In other cases, when the presence of non-volatile buffers may significantly compromise sensitivity and accuracy of MS, purification is necessary. If the concentration of the analyte permits, either microdialysis chamber can be used, or just a membrane floating on the surface of deionised water (Liu et al., 1996; Ragas et al., 2000). However, dialysis approaches cause sample dilution and are prohibitive for high-throughput applications as well.

37.3.2 Biotinylated Probes

Streptavidin (avidin)-coated support (beads or plates) designed to capture biotinylated DNA probes are widely used for MALDI-based assays (Bai et al., 2004; Chiu et al., 2000; Edwards et al., 2001; Hahner et al., 1999; Jurinke et al., 1997, 1998; Kim et al., 2002, 2004; Mengel-Jorgensen et al., 2004; Misra et al., 2007; Misra and Kim, 2009; Olejnik et al., 1996). Biotin is attached either to the primer 5'-end, or to the dideoxy- terminators. The main drawback is the cost of strepavidin (avidin)-coated solid supports (polymeric beads, magnetic beads, or microtiter plates). Recently a device was described based on avidin-coated beads, that allows for parallel processing of multiple samples in combination with a multichannel pipette (Misra and Kim, 2009). The authors reused the beads at least five times by implementing a simple regeneration protocol, which reduces the cost of SNP genotyping.

37.3.3 Reversed-Phase Supports

37.3.3.1 Particle Based Supports

For multiple samples and/or for samples with insufficient analyte concentration the reversed-phase (RP) microcolumn purification scheme proves to be very effective. For analysis of DNA enzymatic assays (SNP genotyping applications) the lab-made RP microcolumn with POROS 50 R1 was originally applied (Haff and Smirnov, 1997). Parallel purification processing feasibility was demonstrated using a 96-well filter plate with Waters Oasis HLB sorbent in peptide and oligonucleotide analysis, as well as for genotyping applications (Gilar et al., 2001; Hong et al., 2008; Sauer et al., 2004). Commercially available C-4, C-8 and C-18 ZipTips (Millipore Corp.) may be applied as well, when the number of samples is not overwhelming, although their binding capacity for oligonucleotides is several times lower compared to that of lab-made tips. The new MALDI Mass-PREP PROtarget plates (Waters Corp.) are now available. This solid-phase extraction (SPE) device uses a C18 stationary phase to bind, wash, and elute simultaneously 96 samples directly onto the plate. Recently the lab-made C18- magnetite beads were used together with microwave irradiation to absorb effectively oligonucleotides from aqueous solutions. After several washing steps immobilized oligonucleotides were eluted from the beads by the MALDI matrix solution (2,4,6-THAP/diammonium citrate) in 70% acetonitrile (Chen and Chen, 2007) followed by MS analysis. The presence of 6 M NaCl in the initial solution of mixed base 24-mer oligonucleotide did not degrade the resulting MALDI mass spectra. Oligonucleotides isolation by capture via adsorptive Genopure (Bruker) magnetic beads (Thongnoppakhun et al., 2009) is also frequently used.

37.3.3.2 Monolithic Supports

A new highly efficient type of the RP media, termed monolithic phases, has recently entered the area of bioanalytical separations (Ivanov, 2006; Josic and Clifton, 2007). A new type of MALDI sample plates using monolithic-based supports was introduced by Hattan and Vestal (Hattan and Vestal, 2008). Such plate contains collimated hole structures filled with monolithic reversed-phase. It allows one to capture and concentrate various analytes while simultaneously acting as a sink for carrier solvents. After the deposition is completed, the adsorbed samples are simultaneously eluted of each hole using a MALDI matrix/water/organic solution mixture via a special adapter. To date these plates have been used as an interface between HPLC separations and MALDI MS, but they have a potential for high-throughput genotyping assays to serve as an efficient active pre-structured support for parallel concentration and desalting SNP reaction products followed by MALDI analysis on the same plate.

37.3.4 DNA-Binding Polymers

Cationic and anionic polymers were successfully applied for isolation of biomolecules directly on MALDI plates (Kepper et al., 2006; Koenig, 2008; Smirnov et al., 2001; Xu et al., 2003). MALDI plate inserts for DNA isolation with attached to them DNA-binding polymers, such as polyethyleneimine, polyvinyl- pyrrolidone, ethoxylated polyethyleneimine (Smirnov et al., 2001) and later PAMAM dendrimer (Kepper et al., 2006) were used for SNP analysis. The binding between DNA and such coatings is strong enough to sustain washing, and, as a result, desalting and concentration can be performed in a single step. After DNA has been immobilized on the surface and the supernatant solution has been removed, a subsequent addition of a MALDI matrix releases analyte material from the surface of the polymeric coating, which then co-crystallizes with matrix. The mass spectrometric analysis is then performed directly from this support. This method of isolation/desalting improves sensitivity for various DNA fragments and tolerance to various buffer components, such as alkali metals and surfactants, making it attractive for high-throughput sample preparation and analysis of oligonucleotide mixtures by MALDI MS.

The demand for high throughput also pushes forward miniaturization as a set of the most technically compatible with the high speed of analysis, parallelization, and cost-effective liquid handling formats.

An informative review on this topic was published by Tu and Gross (Tu and Gross, 2009). Significant research efforts are dedicated to the development of various microfluidic devices, which sometimes called Micro Total Analysis Systems (μTAS). Additionally, the development of devices which employ 2-D structures with self assembling monolayers (Herzer et al., 2009) and nanoparticle-based 3-D structure functionalized microfabricated supports for DNA and RNA analysis (Berhane and Limbach, 2003a) are also underway.

37.4 Practical Aspects of Sample Preparation

Many aspects of DNA sample preparation are covered in reviews by Nordhoff (Nordhoff, 1996; Nordhoff et al., 2003) and Sauer (Sauer, 2007; Sauer et al., 2006).

For successful MALDI analysis it is absolutely critical to follow "MALDI hygiene rules" – use the powder-free gloves (latex or nitrile) and always thoroughly clean MALDI targets. When alkali adducts are still observable in spite of all measures taken, it may be a good idea to wash plasticware with deionized water prior to use. That includes all tubes for matrices, co-matrices and samples. In addition, pipette tips should be treated the same way. A common practice in our lab is to soak each new pipette tip in DI water by aspirating and dispensing DI water with a pipettor prior to aliquoting an actual sample or a matrix solution using that tip. This simple practice is greatly reducing the risk of very annoying complications in spectra caused by sodium and potassium adducts, so familiar to everybody in a MALDI MS laboratory.

When designing a workflow for MALDI MS, it should be taken into account that most frequently used oligonucleotides (standards, PCR and SNP primers, etc.) may be subjected to multiple freeze/thaw cycles potentially leading to degradation of analytes (Davis et al., 2000). The degree and speed of such degradation depends on the concentration, base composition, the length and to some degree the sequence of an oligonucleotide. Slow freeze/thaw cycles, that are common for a typical laboratory, were found to be most harmful. After 40 h and just 8 freeze/thaw cycles there was only 55% of a dA pentamer left intact (Davis et al., 2000). The robustness of strands is base-dependent in the order: T-mer > A-mer > C-mer > G-mer. To minimize a number of freeze/thaw cycles, it is recommended to divide the most common oligonucleotide stock solutions into aliquots before placing them in a freezer.

A pure 3-HPA stock solution is typically stored in presence of cation exchange beads in H^+ form. Without any additives it is stable in the dark at room temperature for a long time (up to several months). A matrix with all the additives is recommended to be prepared daily. In our laboratory we typically use a 10:1 mixture of the following solutions: 3-HPA (35 g/L) and dibasic ammonium citrate (0.2 M) in DI water by mixing them together in a properly cleaned 1.5 mL Eppendorf tube. For certain assays we found useful an addition of 0.1% trifluoracetic acid (TFA).

37.5 On Some Aspects of MALDI MS-Based Genotyping Application

MALDI MS applications for genotyping have been discussed in multiple reviews. The suggested reading would be the excellent review by Tost and Gut (2002) and other reviews providing more detailed coverage on the different aspects of genotyping research. Among those are publications by Gut (2004), Tost and Gut (2004, 2005, 2006), Sauer (2006, 2009), Oberacher (2008), van den Boom and Hillenkamp (2005), van den Boom and Berkenkamp (2007), Banoub et al. (2005), Push and Kostrzewa (2005) and others.

One approach that allows improved sensitivity and alkali metal tolerance involves reducing the length of analyzed portion of DNA probe. This can be accomplished by using cleavable tags containing a cleavable linker within an oligonucleotide probe and a certain tag, e.g. an easily ionizable small organic molecule providing increased sensitivity of mass spectrometric detection.

In most developed to date GOOD assay (Sauer et al., 2002; Sauer and Gut, 2003; Sauer et al., 2000a, b; Tost et al., 2004), the tag is comprised of a short part (typically 3 thiolated bases) of the original oligonucleotide probe. This thio-modified probe hybridizes one base upstream to the polymorphic site of the DNA template. After the extension of the probe by DNA polymerase with dideoxy terminators (which can be quaternised), the unmodified segment of the extended probe is hydrolized by exonuclease II (5'–>3'). The remaining short thio-modified part is alkylated with methyl iodide neutralizing the negative charge on the phosphorothioate groups. The MALDI detection sensitivity of these modified short oligonucleotide fragments is one to two orders of magnitude higher compare to regular phosphorodiester fragments of a similar size. This assay does not require any purification of the final product and uses α-cyano-4-hydroxycinammic acid and its methyl ester as MALDI matrices (Berlin and Gut, 1999; Gut et al., 1997).

There are several other research projects that are pursuing a similar goal by implementing photo-cleavable tags (Hahner et al., 1999; Olejnik et al., 1998, 1999; Wenzel et al., 2003), including the modified GOOD assay (Sauer et al., 2003), RNAse cleavable tags (Mauger et al., 2006, 2007; Mengel-Jorgensen et al., 2005; Wachter et al., 2008), and small organic cleavable tags (Birikh et al., 2008; Hammond et al., 2007).

In the RNAase cleavable tags approach one RNA nucleotide replaces a regular 2'-deoxynucleotide at a certain position (closer to 3' end) within the oligonucleotide probe during oligonucleotide synthesis (Mengel-Jorgensen et al., 2005; Wachter et al., 2008). Such modified primer is still complementary to the opposite strand, so annealing conditions are unchanged. Upon the extension of the probe, the products are isolated from the reaction mixture onto magnetic streptavidin coated beads by making use the biotin group attached to dideoxy terminators. The ribo-link is the cleaved with a mixture of RNases (RNase A, RNase T1 and RNase I) and this releases a shorter 3' portion of the probe into the solution, following MALDI MS analysis of the supernatant. The sensitivity of detection increases because the released probe segment is relatively short. This approach was successfully applied to conducting a 50-plex genotyping assay for the purpose of human identification (Wachter et al., 2008).

A promising approach to genotyping was recently demonstrated by Mauger et al. (2006, 2007), who used a new type of a chimeric DNA polymerase, capable of incorporating ribonucleotides into a growing DNA chain, followed by alkali hydrolysis of enzymatic reaction products and MALDI MS analysis of short oligonucleotide fragments.

An alternative strategy would be to use photo-cleavable linkers (Hahner et al., 1999; Hammond et al., 2007; Olejnik et al., 1998, 1999; Sauer et al., 2003; Wenzel et al., 2003), which can be cleaved shortly prior or directly during the MALDI desorption/ionization process. This technique has several advantages, since the cleavage step does not use wet chemicals and, therefore, no additional sample processing steps are necessary. It also provides short oligonucleotide cleavage products similar to the RNAse-based assay that results in the improved sensitivity as well as better tolerance toward alkali salts.

In another approach reported recently, the small organic tags (Birikh et al., 2008, 2009; Ustinov et al., 2008) containing derivatives of dimethoxytrityl, were used as cleavable groups subjects for MS detection. PCR products were obtained with one strand containing cleavable dimethoxytrityl type labels of various masses and the second strand containing a thio group at the 5'end. These PCR products were immobilized on a gold-coated slide via thio-groups and subsequently analyzed by a mass spectrometer (Birikh et al., 2008, 2009). No MALDI matrix is needed. This approach utilizes a single surface for affinity purification of extended probes followed by laser desorption/ionization MS (LDA MS) of the mass tag cleaved off the probe during a laser pulse. Hammond et al. (2007) reported an oligonucleotide single base extension assay implementing oligonucleotide probes labeled at the 5'-end with photocleavable, quaternised and brominated peptidic mass tags. These probes were extended by a mixture of the four dideoxynucleotides where one of them is biotinylated. The reaction was carried out four times, each time replacing one of the four ddNTPs with a corresponding biotinylated version. Biotinylated extension products were captured on streptavidin-coated beads and the captured products were directly analyzed by LDI-TOF MS after the washing step. The charged peptide moieties were cleaved from the beads by laser pulses and registered by a mass spectrometer without any matrix assistance. The genotype was determined by the presence of a molecular ion of the corresponding brominated mass marker.

The appeal of the small organic tags approach is that it does not require any matrix for ionization and all post enzymatic steps – immobilization, washing, and analysis can be done on the same surface in highly automated and reproducible manner.

37.6 Conclusions

UV MALDI MS is a very powerful instrumental approach for DNA analysis, and particularly for genotyping applications. Many promising developments have been achieved recently in the areas of new MALDI matrices and additives, monolithic supports, MALDI plate surface modifications involving affinity or adsorptive coatings, and particularly in development of the novel cleavable mass tags based upon RNA and small molecules. Laser desorption MS methods continue to play a key role in genotyping applications due to their compatibility with robotic high throughput interfaces and low cost per sample analysis.

References

Armstrong, D.W., Zhang, L.-K., He, L., and Gross, M.L. (2001). Ionic liquids as matrixes for matrix-assisted laser desorption/ionization mass spectrometry. Anal Chem *73*, 3679–3686.

Asara, J.M., and Allison, J. (1999). Enhanced detection of oligonucleotides in UV MALDI MS using the tetraamine spermine as a matrix additive. Anal Chem *71*, 2866–2870.

Ashcroft, A.E. (2003). Protein and peptide identification: The role of mass spectrometry in proteomics. Nat Prod Rep *20*, 202–215.

Bai, X., Kim, S., Li, Z., Turro, N.J., and Ju, J. (2004). Design and synthesis of a photocleavable biotinylated nucleotide for DNA analysis by mass spectrometry. Nucleic Acids Res *32*, 535–541.

Bai, J., Liu, Y.-H., Cain, T.C., and Lubman, D.M. (1994a). Matrix-assisted laser desorption/ionization using an active perfluorosulfonated ionomer film substrate. Anal Chem *66*, 3423–3430.

Bai, J., Liu, Y.-H., Lubman, D.M., and Siemieniak, D. (1994b). Matrix-assisted laser-desorption ionization mass-spectrometry of restriction enzyme-digested plasmid DNA using an active nafion substrate. Rapid Commun Mass Spectrom *8*, 687–691.

Banoub, J.H., Newton, R.P., Esmans, E., Ewing, D.F., and Mackenzie, G. (2005). Recent developments in mass spectrometry for the characterization of nucleosides, nucleotides, oligonucleotides, and nucleic acids. Chem Rev *105*, 1869–1915.

Bauer, O., Guerasimova, A., Sauer, S., Thamm, S., Steinfath, M., Herwig, R., Janitz, M., Lehrach, H., and Radelof, U. (2004). Multiplexed hybridizations of positively charge-tagged peptide nucleic acids detected by matrix-assisted laser desorption/ionization time-of-flight mass spectrometry. Rapid Commun Mass Spectrom *18*, 1821–1829.

Beavis, R.C., Chait, B.T., and Fales, H.M. (1989). Cinnamic acid derivatives as matrices for ultraviolet laser desorption mass spectrometry of proteins. Rapid Commun Mass Spectrom *3*, 432–435.

Berhane, B.T., and Limbach, P.A. (2003a). Functional microfabricated sample targets for matrix-assisted laser desorption/ionization mass spectrometry analysis of ribonucleic acids. Anal Chem *75*, 1997–2003.

Berhane, B.T., and Limbach, P.A. (2003b). Stable isotope labeling for matrix-assisted laser desorption/ionization mass spectrometry and post-source decay analysis of ribonucleic acids. J Mass Spectrom *38*, 872–878.

Berlin, K., and Gut, I.G. (1999). Analysis of negatively "charge tagged" DNA by matrix-assisted laser desorption/ionization time-of-flight mass spectrometry. Rapid Commun Mass Spectrom *13*, 1739–1743.

Birikh, K.R., Bernard, P.L., Shmanai, V.V., Malakhov, A.D., Shchepinov, M.S., and Korshun, V.A. (2009). SNP detection using trityl mass tags. Methods Mol Biol *578*, 345–361.

Birikh, K.R., Korshun, V.A., Bernad, P.L., Malakhov, A.D., Milner, N., Khan, S., Southern, E.M., and Shchepinov, M.S. (2008). Novel mass tags for single nucleotide polymorphism detection. Anal Chem *80*, 2342–2350.

Bourin, S., McStay, D., Lin, P.K.T., Duncan, G., and Lomax, J. (1997a). The Application of Matrix Assisted Laser Desorption Time of Flight Mass Spectrometry to the Study of DNA/Terbium Interactions. Sensors and Their Applications VIII. Proceedings of the Conference on Sensors and Their Applications, 8th, Glasgow, Sept 7–10, 1997, pp. 15–20.

Bourin, S., McStay, D., Lin, P.K.T., Duncan, G., and Lomax, J. (1997b). Study of DNA/Europium ion interaction by matrix assisted laser desorption time of flight mass spectrometry. Proc SPIE-Int Soc Opt Eng *2985*, 112–119.

Butler, J., Jiangbaucom, P., Huang, M., Belgrader, P., and Girard, J. (1996). Peptide nucleic-acid characterization by MALDI-TOF mass-spectrometry. Anal Chem *68*, 3283–3287.

Capaldi, D.C., and Scozzari, A.N. (2008). Manufacturing and analytical processes for 2′-O-(2-methoxy-ethyl)-modified oligonucleotides. Antisense Drug Technol (2nd Ed), *38*, 401–434.

Carda-Broch, S., Berthod, A., and Armstrong, D.W. (2003). Ionic matrices for matrix-assisted laser desorption/ionization time-of-flight detection of DNA oligomers. Rapid Commun Mass Spectrom *17*, 553–560.

Chan, T.W.D., Fung, Y.M.E., and Li, Y.C.L. (2002). A study of fast and metastable dissociations of adenine-thymine binary-base oligonucleotides by using positive-ion MALDI-TOF mass spectrometry. J Am Soc Mass Spectrom *13*, 1052–1064.

Chen, W.-Y., and Chen, Y.-C. (2003). Reducing the alkali cation adductions of oligonucleotides using sol-gel-assisted laser desorption/ionization mass spectrometry. Anal Chem *75*, 4223–4228.

Chen, W.-Y., and Chen, Y.-C. (2007). MALDI MS analysis of oligonucleotides: desalting by functional magnetite beads using microwave-assisted extraction. Anal Chem *79*, 8061–8066.

Cheng, S.-w., and Chan, T.W.D. (1996). Use of ammonium halides as co-matrices for matrix-assisted laser desorption/ionization studies of oligonucleotides. Rapid Commun Mass Spectrom *10*, 907–910.

Chiu, N.H.L., Tang, K., Yip, P., Braun, A., Koster, H., and Cantor, C.R. (2000). Mass spectrometry of single-stranded restriction fragments captured by an undigested complementary sequence. Nucleic Acids Res *28*, e31, ii–iv.

Christian, N.P., Reilly, J.P., Mokler, V.R., Wincott, F.E., and Ellington, A.D. (2001). Elucidation of the initial step of oligonucleotide fragmentation in matrix-assisted laser desorption/ionization using modified nucleic acids. J Am Soc Mass Spectrom *12*, 744–753.

Cohen, L.H., and Gusev, A.I. (2002). Small molecule analysis by MALDI mass spectrometry. Anal Bioanal Chem *373*, 571–586.

Dai, Y., Whittal, R.M., and Li, L. (1999). Two-layer sample preparation: A method for MALDI-MS analysis of complex peptide and protein mixtures. Anal Chem *71*, 1087–1091.

Davis, D.L., O'Brien, E.P., and Bentzley, C.M. (2000). Analysis of the degradation of oligonucleotide strands during the freezing/thawing processes using MALDI-MS. Anal Chem *72*, 5092–5096.

Dell, A., and Morris, H.R. (2001). Glycoprotein structure determination by mass spectrometry. Science *291*, 2351–2356.

Distler, A.M., and Allison, J. (2001a). 5-methoxysalicylic acid and spermine: A new matrix for the matrix-assisted laser desorption/ionization mass spectrometry analysis of oligonucleotides. J Am Soc Mass Spectrom *12*, 456–462.

Distler, A.M., and Allison, J. (2001b). Improved MALDI-MS analysis of oligonucleotides through the use of fucose as a matrix additive. Anal Chem *73*, 5000–5003.

Edwards, J.R., Itagaki, Y., and Ju, J. (2001). DNA sequencing using biotinylated dideoxynucleotides and mass spectrometry. Nucleic Acids Res *29*, e104/101–e104/106.

Erb, W.J., and Owens, K.G. (2008). Development of a dual-spray electrospray deposition system for matrix-assisted laser desorption/ionization time-of-flight mass spectrometry. Rapid Commun Mass Spectrom *22*, 1168–1174.

Fitzgerald, M.C., Parr, G.R., and Smith, L.M. (1993). Basic matrices for the matrix-assisted laser-desorption ionization mass-spectrometry of proteins and oligonucleotides. Anal Chem *65*, 3204–3211.

Fu, Y., Xu, S., Pan, C., Ye, M., Zou, H., and Guo, B. (2006). A matrix of 3,4-diaminobenzophenone for the analysis of oligonucleotides by matrix-assisted laser desorption/ionization time-of-flight mass spectrometry. Nuceic Acids Res *34*, e94.

Garcia, B.A., Heaney, P.J., and Tang, K. (2002). Improvement of the MALDI-TOF analysis of DNA with thin-layer matrix preparation. Anal Chem *74*, 2083–2091.

Gilar, M., Belenky, A., and Wang, B.H. (2001). High-throughput biopolymer desalting by solid-phase extraction prior to mass spectrometric analysis. J Chromatogr A *921*, 3–13.

Gobom, J., Nordhoff, E., Mirgorodskaya, E., Ekman, R., and Roepstorff, P. (1999). Sample purification and preparation technique based on nano-scale reversed-phase columns for the sensitive analysis of complex peptide mixtures by matrix-assisted laser desorption/ionization mass spectrometry. J Mass Spectrom *34*, 105–116.

Gobom, J., Schuerenberg, M., Mueller, M., Theiss, D., Lehrach, H., and Nordhoff, E. (2001). alpha -Cyano-4-hydroxycinnamic acid affinity sample preparation. A protocol for MALDI-MS peptide analysis in proteomics. Anal Chem *73*, 434–438.

Greig, M., and Griffey, R.H. (1995). Utility of organic bases for improved electrospray mass spectrometry of oligonucleotides. Rapid Commun Mass Spectrom *9*, 97–102.

Griesser, H.J., Kingshott, P., McArthur, S.L., McLean, K.M., Kinsel, G.R., and Timmons, R.B. (2004). Surface-MALDI mass spectrometry in biomaterials research. Biomaterials *25*, 4861–4875.

Gusev, A.I., Wilkinson, W.R., Proctor, A., and Hercules, D.M. (1995). Improvement of signal reproducibility and matrix/comatrix effects in MALDI analysis. Anal Chem *67*, 1034–1041.

Gut, I.G. (2004). DNA analysis of MALDI-TOF mass spectrometry. Hum Mutat *23*, 437–441.

Gut, I.G., Jeffery, W.A., Pappin, D.J.C., and Beck, S. (1997). Analysis of DNA by "charge tagging" and matrix-assisted laser desorption/ionization mass spectrometry. Rapid Commun Mass Spectrom *11*, 43–50.

Haff, L.A., and Smirnov, I.P. (1997). Single-nucleotide polymorphism identification assays using a thermostable DNA polymerase and delayed extraction MALDI-TOF mass spectrometry. Genome Res *7*, 378–388.

Hahner, S., Olejnik, J., Ludemann, H.-C., Krzymanska-Olejnik, E., Hillenkamp, F., and Rothschild, K.J. (1999). Matrix-assisted laser desorption/ionization mass spectrometry of DNA using photocleavable biotin. Biomol Eng *16*, 127–133.

Hammond, N., Koumi, P., Langley, G.J., Lowe, A., and Brown, T. (2007). Rapid mass spectrometric identification of human genomic polymorphisms using multiplexed photocleavable mass-tagged probes and solid phase capture. Org Biomol Chem *5*, 1878–1885.

Hanton, S.D., Hyder, I.Z., Stets, J.R., Owens, K.G., Blair, W.R., Guttman, C.M., and Giuseppetti, A.A. (2004). Investigations of electrospray sample deposition for polymer MALDI mass spectrometry. J Am Soc Mass Spectrom *15*, 168–179.

Harvey, D.J. (2001). Identification of protein-bound carbohydrates by mass spectrometry. Proteomics *1*, 311–328.

Hathaway, G.M. (1994). Characterization of modified and normal deoxyoligonucleotides by MALDI, time-of-flight mass spectrometry. Biotechniques *17*, 150–155.

Hattan, S.J., and Vestal, M.L. (2008). Novel three-dimensional MALDI plate for interfacing high-capacity LC separations with MALDI-TOF. Anal Chem *80*, 9115–9123.

Hensel, R.R., King, R.C., and Owens, K.G. (1997). Electrospray sample preparation for improved quantitation in matrix-assisted laser desorption/ionization time-of-flight mass spectrometry. Rapid Commun Mass Spectrom *11*, 1785–1793.

Herzer, N., Eckardt, R., Hoeppener, S., and Schubert, U.S. (2009). Sample target substrates with reduced spot size for MALDI-TOF mass spectrometry based on patterned self-assembled monolayers. Adv Func Mater *19*, 2777–2781.

Hettich, R., and Buchanan, M. (1991). Structural characterization of normal and modified oligonucleotides by matrix-assisted laser desorption Fourier-transform mass-spectrometry. J Am Soc Mass Spectrom *2*, 402–412.

Hillenkamp, F., Waefler, E., Jecklin, M.C., and Zenobi, R. (2009). Positive and negative analyte ion yield in matrix-assisted laser desorption/ionization revisited. Int J Mass Spectrom *285*, 114–119.

Honda, A., Sonobe, H., Ogata, A., and Suzuki, K. (2005). Improved method of the MALDI-TOF analysis of DNA with nanodot sample target plate. Chem Commun, 5340–5342.

Hong, S.P., Shin, S.-K., Lee, E.H., Kim, E.O., Ji, S.I., Chung, H.J., Park, S.N., Yoo, W., Folk, W.R., and Kim, S.-O. (2008). High-resolution human papillomavirus genotyping by MALDI-TOF mass spectrometry. Nat Protoc *3*, 1476–1484, S1476/1471.

Horneffer, V., Glueckmann, M., Krueger, R., Karas, M., Strupat, K., and Hillenkamp, F. (2006). Matrix-analyte-interaction in MALDI-MS: Pellet and nano-electrospray preparations. Int J Mass Spectrom *249/250*, 426–432.

Hung, K.C., Ding, H., and Guo, B. (1999). Use of poly(tetrafluoroethylene)s as a sample support for the MALDI-TOF analysis of DNA and proteins. Anal Chem *71*, 518–521.

Hung, K.C., Rashidzadeh, H., Wang, Y., and Guo, B. (1998). Use of Paraffin wax film in MALDI-TOF analysis of DNA. Anal Chem *70*, 3088–3093.

Hunter, J.M., Lin, H., and Becker, C.H. (1997). Cryogenic frozen solution matrixes for analysis of DNA by time-of-flight mass spectrometry. Anal Chem *68*, 3608–3612.

Ivanov, A.R. (2006). Polymeric monolithic capillary columns in proteomics. In Separation Methods in Proteomics, G.B. Smejkal, and A. Lazarev, eds. (New York, NY, Taylor & Francis-CRC), pp. 419–443.

Johnson, T., Bergquist, J., Ekman, R., Nordhoff, E., Schurenberg, M., Kloppel, K.D., Muller, M., Lehrach, H., and Gobom, J. (2001). A CE-MALDI interface based on the use of prestructured sample supports. Anal Chem *73*, 1670–1675.

Jones, J.J., Batoy, S.M.A.B., Wilkins, C.L., Liyanage, R., and Lay, J.O. (2005). Ionic liquid matrix-induced metastable decay of peptides and oligonucleotides and stabilization of phospholipids in MALDI FTMS analyses. J Am Soc Mass Spectrom *16*, 2000–2008.

Josic, D., and Clifton, J.G. (2007). Use of monolithic supports in proteomics technology. J Chromatogr, A *1144*, 2–13.

Jurinke, C., van den Boom, D., Collazo, V., Luechow, A., Jacob, A., and Koester, H. (1997). Recovery of nucleic acids from immobilized biotin-streptavidin complexes using ammonium hydroxide and applications in MALDI-TOF mass spectrometry. Anal Chem *69*, 904–910.

Jurinke, C., vandenBoom, D., and Koster, H. (1998). Asymmetric polymerase chain reaction improves streptavidin-biotin based purification of polymerase chain reaction products prior to matrix-assisted laser desorption/ionization time-of-flight mass spectrometric analysis. Rapid Commun Mass Spectrom *12*, 50–52.

Karas, M., Gluckmann, M., and Schafer, J. (2000). Ionization in matrix-assisted laser desorption/ ionization: Singly charged molecular ions are the lucky survivors. J Mass Spectrom *35*, 1–12.

Karas, M., and Hillenkamp, F. (1988). Laser desorption ionization of proteins with molecular masses exceeding 10000 daltons. Anal Chem *60*, 2299–2301.

Karas, M., and Krueger, R. (2003). Ion formation in MALDI: The cluster ionization mechanism. Chem Rev *103*, 427–439.

Kepper, P., Richard, R.T., Dahl, A., Lehrach, H., and Sauer, S. (2006). Matrix-assisted laser desorption/ionization mass spectrometric analysis of DNA on microarrays. Clin Chem *52*, 1303–1310.

Kim, S., Edwards John, R., Deng, L., Chung, W., and Ju, J. (2002). Solid phase capturable dideoxynucleotides for multiplex genotyping using mass spectrometry. Nuceic Acids Res *30*, e85.

Kim, Y., Hurst, G.B., Doktycz, M.J., and Buchanan, M.V. (2001). Improving spot homogeneity by using polymer substrates in matrix-assisted laser desorption/ionization mass spectrometry of oligonucleotides. Anal Chem *73*, 2617–2624.

Kim, S., Ulz, M.E., Nguyen, T., Li, C.-M., Sato, T., Tycko, B., and Ju, J. (2004). Thirtyfold multiplex genotyping of the p53 gene using solid phase capturable dideoxynucleotides and mass spectrometry. Genomics *83*, 924–931.

Kinet, C., Gabelica, V., Balbeur, D., and De Pauw, E. (2009). Electron detachment dissociation (EDD) pathways in oligonucleotides. Int J Mass Spectrom *283*, 206–213.

Kirpekar, F., Berkenkamp, S., and Hillenkamp, F. (1999). Detection of double-stranded DNA by IR-and UV-MALDI mass spectrometry. Anal Chem *71*, 2334–2339.

Koenig, S. (2008). Target coatings and desorption surfaces in biomolecular MALDI-MS. Proteomics *8*, 706–714.

Kong, Y., Zhu, Y., and Zhang, J.-Y. (2001). Ionization mechanism of oligonucleotides in matrix-assisted laser desorption/ionization time-of-flight mass spectrometry. Rapid Commun Mass Spectrom *15*, 57–64.

Koomen, J.M., Russell, W.K., Hettick, J.M., and Russell, D.H. (2000). Improvement of resolution, mass accuracy, and reproducibility in reflectedmode DE-MALDI-TOF analysis of DNA using cast evaporation-overlayer sample preparations. Anal Chem *72*, 3860–3866.

Koster, C., Castoro, J.A., and Wilkins, C.L. (1992). High-resolution matrix-assisted laser desorption ionization of biomolecules by Fourier-transform mass-spectrometry. J Am Chem Soc *114*, 7572–7574.

Kuhn-Holsken, E., Lenz, C., Sander, B., Luhrmann, R., and Urlaub, H. (2005). Complete MALDI-ToF MS analysis of cross-linked peptide-RNA oligonucleotides derived from nonlabeled UV-irradiated ribonucleoprotein particles. RNA *11*, 1915–1930.

Langley, G.J., Herniman, J.M., Davies, N.L., and Brown, T. (1999). Simplified sample preparation for the analysis of oligonucleotides by matrix-assisted laser desorption/ionisation time-of-flight mass spectrometry. Rapid Commun Mass Spectrom *13*, 1717–1723.

Lavanant, H., and Lange, C. (2002). Sodium-tolerant matrix for matrix-assisted laser desorption/ionization mass spectrometry and post-source decay of oligonucleotides. Rapid Commun Mass Spectrom *16*, 1928–1933.

Lecchi, P., Le, H., and Pannell, L. (1995). 6-Aza-a-thiothymine – a matrix for MALDI spectra of oligonucleotides. Nucleic Acids Res *23*, 1276–1277.

Li, Y.C.L., Cheng, S.-W., and Chan, T.W.D. (1998). Evaluation of ammonium salts as co-matrices for matrix-assisted laser desorption-ionization mass spectrometry of oligonucleotides. Rapid Commun Mass Spectrom *12*, 993–998.

Li, Y.L., and Gross, M.L. (2004). Ionic-liquid matrices for quantitative analysis by MALDI-TOF mass spectrometry. J Am Soc Mass Spectrom *15*, 1833–1837.

Lin, H., Hunter, J.M., and Becker, C.H. (1999). Laser desorption of DNA oligomers larger than one kilobase from cooled 4-nitrophenol. Rapid Commun Mass Spectrom *13*, 2335–2340.

Lin, Y.-S., and Chen, Y.-C. (2002). Laser desorption/ionization time-of-flight mass spectrometry on sol-gel-derived 2,5-dihydroxybenzoic acid film. Anal Chem *74*, 5793–5798.

Liu, C., Wu, Q., Harms, A., and Smith, R. (1996). On line microdialysis sample cleanup for electrospray-ionization mass-spectrometry of nucleic-acid samples. Anal Chem *68*, 3295–3299.

Liu, Y.H., Bai, J., Liang, X., Lubman, D.M., and Venta, P.J. (1995a). Use of a nitrocellulose film substrate in matrix-assisted laser desorption/ionization mass spectrometry for DNA mapping and screening. Anal Chem *67*, 3482–3490.

Liu, Y.-H., Bai, J., Zhu, Y., Liang, X., Siemieniak, D., Venta, P.J., and Lubman, D.M. (1995b). Rapid screening of genetic polymorphisms using buccal cell-DNA with detection by matrix-assisted laser-desorption ionization mass-spectrometry. Rapid Commun Mass Spectrom *9*, 735–743.

Liu, Y., Sun, X., and Guo, B. (2003). Matrix-assisted laser desorption/ionization time-of-flight analysis of low-concentration oligonucleotides and mini-sequencing products. Rapid Commun Mass Spectrom *17*, 2354–2360.

Mauger, F., Bauer, K., Calloway, C.D., Semhoun, J., Nishimoto, T., Myers, T.W., Gelfand, D.H., and Gut, I.G. (2007). DNA sequencing by MALDI-TOF MS using alkali cleavage of RNA/DNA chimeras. Nucleic Acids Res *35*, e62/61–e62/11.

Mauger, F., Jaunay, O., Chamblain, V., Reichert, F., Bauer, K., Gut, I.G., and Gelfand, D.H. (2006). SNP genotyping using alkali cleavage of RNA/DNA chimeras and MALDI time-of-flight mass spectrometry. Nucleic Acids Res *34*, e18/11–e18/18.

Mengel-Jorgensen, J., Sanchez, J.J., Borsting, C., Kirpekar, F., and Morling, N. (2004). MALDI-TOF mass spectrometric detection of multiplex single base extended primers. A study of 17 Y-chromosome single-nucleotide polymorphisms. Anal Chem *76*, 6039–6045.

Mengel-Jorgensen, J., Sanchez, J.J., Borsting, C., Kirpekar, F., and Morling, N. (2005). Typing of multiple single-nucleotide polymorphisms using ribonuclease cleavage of DNA/RNA chimeric single-base extension primers and detection by MALDI-TOF mass spectrometry. Anal Chem *77*, 5229–5235.

Misra, A., Hong, J.-Y., and Kim, S. (2007). Multiplex genotyping of cytochrome P450 single-nucleotide polymorphisms by use of MALDI-TOF mass spectrometry. Clin Chem *53*, 933–939.

Misra, A., and Kim, S. (2009). Microbead device for isolating biotinylated oligonucleotides for use in mass spectrometric analysis. Anal Biochem *384*, 96–100.

Mock, K.K., Sutton, C.W., and Cottrell, J.S. (1992). Sample immobilization protocols for matrix-assisted laser-desorption mass spectrometry. Rapid Commun Mass Spectrom *6*, 233–238.

Molin, L., Cristoni, S., Crotti, S., Bernardi, L.R., Seraglia, R., and Traldi, P. (2008). Sieve-based device for MALDI sample preparation. I. Influence of sample deposition conditions in oligonucleotide analysis to achieve significant increases in both sensitivity and resolution. J Mass Spectrom *43*, 1512–1520.

Muddiman, D.C., Cheng, X., Udseth, H.R., and Smith, R.D. (1996). Charge-state reduction with improved signal intensity of oligonucleotides in electrospray ionization mass spectrometry. J Am Soc Mass Spectrom *7*, 697–706.

Nordhoff, E. (1996). Matrix-assisted laser desorption/ionization mass spectrometry as a new method for the characterization of nucleic acids. TrAC, Trends Anal Chem *15*, 240–250.

Nordhoff, E., Ingendoh, A., Cramer, R., Overberg, A., Stahl, B., Karas, M., Hillenkamp, F., and Crain, P. (1992). Matrix-assisted laser desorption ionization mass-spectrometry of nucleic-acids with wavelengths in the ultraviolet and infrared. Rapid Commun Mass Spectrom *6*, 771–776.

Nordhoff, E., Lehrach, H., and Gobom, J. (2007). Exploring the limits and losses in MALDI sample preparation of attomole amounts of peptide mixtures. Int J Mass Spectrom *268*, 139–146.

Nordhoff, E., Schurenberg, M., Thiele, G., Lubbert, C., Kloeppel, K.-D., Theiss, D., Lehrach, H., and Gobom, J. (2003). Sample preparation protocols for MALDI-MS of peptides and oligonucleotides using prestructured sample supports. Int J Mass Spectrom *226*, 163–180.

Oberacher, H. (2008). On the use of different mass spectrometric techniques for characterization of sequence variability in genomic DNA. Anal Bioanal Chem *391*, 135–149.

Ohara, K., Smietana, M., and Vasseur, J.-J. (2006). Characterization of specific noncovalent complexes between guanidinium derivatives and single-stranded DNA by MALDI. J Am Soc Mass Spectrom *17*, 283–291.

Olejnik, J., Hahner, S., Ludemann, H., KrzymanskaOlejnik, E., Hillenkamp, F., and Rothschild, K. (1998). Photorelease and MALDI analysis of oligonucleotides bound to solid surfaces through photocleavable biotin. ASMS *97*, Abstract, A296.

Olejnik, J., Krzymanska-Olejnik, E., and Rothschild, K.J. (1996). Photocleavable biotin phosphoramidite for 5′-end-labeling, affinity purification and phosphorylation of synthetic oligonucleotides. Nucleic Acids Res *24*, 361–366.

Olejnik, J., Ludemann, H.-C., Krzymanska-Olejnik, E., Berkenkamp, S., Hillenkamp, F., and Rothschild, K.J. (1999). Photocleavable peptide-DNA conjugates: Synthesis and applications to DNA analysis using MALDI-MS. Nucleic Acids Res *27*, 4626–4631.

Owen, S.J., Meier, F.S., Brombacher, S., and Volmer, D.A. (2003). Increasing sensitivity and decreasing spot size using an inexpensive, removable hydrophobic coating for matrix-assisted laser desorption/ionisation plates. Rapid Commun Mass Spectrom *17*, 2439–2449.

Perkel, J. (2008). SNP genotyping: Six technologies that keyed a revolution. Nat Methods *5*, 447–453.

Perlman, D.H., Huang, H., Dauly, C., Costello, C.E., and McComb, M.E. (2007). Coupling of protein HPLC to MALDI-TOF MS using an on-target device for fraction collection, concentration, digestion, desalting, and matrix/analyte cocrystallization. Anal Chem *79*, 2058–2066.

Pieles, U., Zurcher, W., Schar, M., and Moser, H.E. (1993). Matrix-assisted laser-desorption ionization time-of-flight mass-spectrometry – a powerful tool for the mass and sequence-analysis of natural and modified oligonucleotides. Nuceic Acids Res *21*, 3191–3196.

Popovic, Z., Matkovic-Calogovic, D., Popovic, J., Vickovic, I., Vinkovic, M., and Vikic-Topic, D. (2007). Coordination modes of 3-hydroxypicolinic acid: Synthesis and structural characterization of polymeric mercury(II) complexes. Polyhedron *26*, 1045–1052.

Pusch, W., and Kostrzewa, M. (2005). Application of MALDI-TOF mass spectrometry in screening and diagnostic research. Curr Pharm Des *11*, 2577–2591.

Pusch, W., Wurmbach, J.-H., Thiele, H., and Kostrzewa, M. (2002). MALDI-TOF mass spectrometry-based SNP genotyping. Pharmacogenomics *3*, 537–548.

Ragas, J.A., Simmons, T.A., and Limbach, P.A. (2000). A comparative study on methods of optimal sample preparation for the analysis of oligonucleotides by matrix-assisted laser desorption/ionization mass spectrometry. Analyst *125*, 575–581.

Rechthaler, J., Rizzi, A., and Allmaier, G. (2007). A one-way hydrophobic surface foil as sample support for MALDI and off-line CZE/MALDI mass spectrometry: An alternative for low and high molecular mass compounds. Int J Mass Spectrom *268*, 131–138.

Sauer, S. (2006). Analysis of DNA variation by MALDI mass spectrometry: Recent developments and perspectives. Recent Dev Nucleic Acids Res 2, 1–14.

Sauer, S. (2007). The essence of DNA sample preparation for MALDI mass spectrometry. J Biochem Biophys Methods *70*, 311–318.

Sauer, S. (2009). DNA polymorphisms: Tools for detection. Wiley Encyclopedia Chem Biol *1*, 552–563.

Sauer, S., Gelfand David, H., Boussicault, F., Bauer, K., Reichert, F., and Gut Ivo, G. (2002). Facile method for automated genotyping of single nucleotide polymorphisms by mass spectrometry. Nucleic Acids Res *30*, e22.

Sauer, S., and Gut, I.G. (2003). Extension of the GOOD assay for genotyping single nucleotide polymorphisms by matrix-assisted laser desorption/ionization mass spectrometry. Rapid Commun Mass Spectrom *17*, 1265–1272.

Sauer, S., Kepper, P., Smyra, A., Dahl, A., Ferse, F.-T., Lehrach, H., and Reinhardt, R. (2004). Automated solid-phase extraction for purification of single nucleotide polymorphism genotyping products prior to matrix-assisted laser desorption/ionisation time-of-flight mass spectrometric analysis. J Chromatogr A *1049*, 9–16.

Sauer, S., Lechner, D., Berlin, K., Lehrach, H., Escary, J.-L., Fox, N., and Gut, I.G. (2000a). A novel procedure for efficient genotyping of single nucleotide polymorphisms. Nucleic Acids Res *28*, e13, ii–viii.

Sauer, S., Lechner, D., Berlin, K., Plancon, C., Heuermann, A., Lehrach, H., and Gut, I.G. (2000b). Full flexibility genotyping of single nucleotide polymorphisms by the GOOD assay. Nuceic Acids Res *28*, e100/101–e100/106.

Sauer, S., Lehrach, H., and Reinhardt, R. (2003). MALDI mass spectrometry analysis of single nucleotide polymorphisms by photocleavage and charge-tagging. Nuceic Acids Res *31*, e63/61–e63/10.

Sauer, S., Reinhardt, R., Lehrach, H., and Gut, I.G. (2006). Single-nucleotide polymorphisms: Analysis by mass spectrometry. Nat Protoc *1*, 1761–1771.

Schieltz, D., Chou, C., Luo, C., Thomas, R., and Williams, P. (1992). Mass-spectrometry of DNA mixtures by laser ablation from frozen aqueous-solution. Rapid Commun Mass Spectrom *6*, 631–636.

Schiller, J., Suss, R., Fuchs, B., Muller, M., Zschornig, O., and Arnold, K. (2007). MALDI-TOF MS in lipidomics. Front Biosci *12*, 2568–2579.

Schuerenberg, M., Luebbert, C., Eickhoff, H., Kalkum, M., Lehrach, H., and Nordhoff, E. (2000). Prestructured MALDI-MS Sample Supports. Anal Chem *72*, 3436–3442.

Schulz, E., Karas, M., Rosu, F., and Gabelica, V. (2006). Influence of the matrix on analyte fragmentation in atmospheric pressure MALDI. J Am Soc Mass Spectrom *17*, 1005–1013.

Shahgholi, M., Garcia, B.A., Chiu, N.H.L., Heaney, P.J., and Tang, K. (2001). Sugar additives for MALDI matrices improve signal allowing the smallest nucleotide change (A.T) in a DNA sequence to be resolved. Nucleic Acids Res *29*, e91/91–e91/10.

Shaler, T.A., Wickham, J.N., Sannes, K.A., Wu, K.J., and Becker, C.H. (1996). Effect of impurities on the matrix-assisted laser-desorption mass-spectra of single-stranded oligodeoxynucleotides. Anal Chem *68*, 576–579.

Simmons, T.A., and Limbach, P.A. (1997). The use of a co-matrix for improved analysis of oligonucleotides by matrix-assisted laser desorption/ionization time-of-flight mass spectrometry. Rapid Commun Mass Spectrom *11*, 567–572.

Simmons, T.A., and Limbach, P.A. (1998). Influence of co-matrix proton affinity on oligonucleotide ion stability inmatrix-assisted laser desorption/ionization time-of-flight mass spectrometry. J Am Soc Mass Spectrom *9*, 668–675.

Smirnov, I.P., Hall, L.R., Ross, P.L., and Haff, L.A. (2001). Application of DNA-binding polymers for preparation of DNA for analysis by matrix-assisted laser desorption/ionization mass spectrometry. Rapid Commun Mass Spectrom *15*, 1427–1432.

Song, F. (2003). Quinaldic acid as a new matrix for matrix-assisted laser desorption/ionization of nucleic acids. Rapid Commun Mass Spectrom *17*, 1802–1807.

Stemmler, E.A., Buchanan, M.V., Hurst, G.B., and Hettich, R.L. (1994). The structural characterization of polycyclic aromatic hydrocarbon dihydrodiol epoxide DNA adducts using matrix-assisted laser desorption/ionization Fourier transform mass spectrometry. Anal Chem *66*, 1274–1285.

Stemmler, E.A., Buchanan, M.V., Hurst, G.B., and Hettich, R.L. (1995). Analysis of modified oligonucleotides by matrix-assisted laser desorption/ionization Fourier-transform mass-spectrometry. Anal Chem *67*, 2924–2930.

Storm, N., Darnhofer-Patel, B., van den Boom, D., and Rodi, C.P. (2003). MALDI-TOF mass spectrometry-based SNP genotyping. Methods Mol Biol *212*, 241–262.

Strupat, K., Karas, M., and Hillenkamp, F. (1991). 2,5-Dihydroxybenzoic acid: A new matrix for laser desorption – ionization mass spectrometry. Int J Mass Spectrom Ion Proc *111*, 89–102.

Sun, C., Zheng, X., and Jin, L. (2003). Supramolecular formation via hydrogen bonding in the Zn(II), Mn(II) and Cu(II) complexes with 3-hydroxypicolinic acid. J Mol Struct *646*, 201–210.

Tang, K. (2007). Chip-based genotyping by mass spectrometry. Integr Biochips DNA Anal, 117–127.

Tang, K., Fu, D., Julien, D., Braun, A., Cantor, C., and Koster, H. (1999). Chip-based genotyping by mass spectrometry. Proc Natl Acad Sci U S A *96*, 10016–10020.

Tang, K., Taranenko, N.I., Allman, S.L., Chang, L.Y., and Chen, C.H. (1994a). Detection of 500-nucleotide DNA by laser-desorption mass-spectrometry. Rapid Commun Mass Spectrom *8*, 727–730.

Tang, K., Taranenko, N.I., Allman, S.L., Chen, C.H., Chang, L.Y., and Jacobson, K.B. (1994b). Picolinic-acid as a matrix for laser mass-spectrometry of nucleic-acids and proteins. Rapid Commun Mass Spectrom *8*, 673–677.

Taranenko, N.I., Tang, K., Allman, S.L., Ch'ang, L.Y., and Chen, C.H. (1994). 3-Aminopicolinic aid as a matrix for laser-desorption mass-spectrometry of biopolymers. Rapid Commun Mass Spectrom *8*, 1001–1006.

Tarzi, O.I., Nonami, H., and Erra-Balsells, R. (2009). The effect of temperature on the stability of compounds used as UV-MALDI-MS matrix: 2,5-dihydroxybenzoic acid, 2,4,6-trihydroxyacetophenone, alpha -cyano-4-hydroxycinnamic acid, 3,5-dimethoxy-4-hydroxycinnamic acid, nor-harmane and harmane. J Mass Spectrom *44*, 260–277.

Terrier, P., Tortajada, J., Zin, G., and Buchmann, W. (2007). Noncovalent Complexes Between DNA and Basic Polypeptides or Polyamines by MALDI-TOF. J Am Soc Mass Spectrom *18*, 1977–1989.

Tholey, A. (2009). Ionic liquids and mass spectrometry. In Ionic Liquids in Chemical Analysis, M. Koel, ed. (New York, NY, Taylor & Francis-CRC), pp. 371–395.

Thongnoppakhun, W., Jiemsup, S., Yongkiettrakul, S., Kanjanakorn, C., Limwongse, C., Wilairat, P., Vanasant, A., Rungroj, N., and Yenchitsomanus, P.-t. (2009). Simple, efficient, and cost-effective multiplex genotyping with matrix assisted laser desorption/ionization time-of-flight mass spectrometry of hemoglobin beta gene mutations. J Mol Diagn *11*, 334–346.

Tost, J. (2008). Methods for the genome-wide and gene-specific analysis of DNA methylation levels and patterns. In Epigenetics, J. Tost, ed. (Norfolk, Caister Academic), pp. 63–103.

Tost, J., and Gut, I.G. (2002). Genotyping single nucleotide polymorphisms by mass spectrometry. Mass Spectrom Rev *21*, 388–418.

Tost, J., and Gut, I.G. (2004). Genotyping single nucleotide polymorphisms by MALDI mass spectrometry. Adv Mass Spectrom *16*, 123–143.

Tost, J., and Gut, I.G. (2005). Genotyping single nucleotide polymorphisms by MALDI mass spectrometry in clinical applications. Clin Biochem *38*, 335–350.

Tost, J., and Gut, I.G. (2006). DNA analysis by mass spectrometry – past, present and future. J Mass Spectrom *41*, 981–995.

Tost, J., Kucharzak, R., Lechner, D., and Gut, I.G. (2004). The GOOD assay: a purification-free assay for genotyping by MALDI mass spectrometry. In PCR Technology: Current Innovations, T. Weissensteiner, H.G. Griffin, and A. Griffin, eds. (New York, NY, CRC Press), pp. 121–130.

Tu, T., and Gross, M.L. (2009). Miniaturizing sample spots for matrix-assisted laser desorption/ionization mass spectrometry. TrAC, Trends Anal Chem *28*, 833–841.

Ustinov, A.V., Shmanai, V.V., Patel, K., Stepanova, I.A., Prokhorenko, I.A., Astakhova, I.V., Malakhov, A.D., Skorobogatyi, M.V., Bernad, P.L., Khan, S., *et al.* (2008). Reactive trityl derivatives: Stabilised carbocation mass-tags for life sciences applications. Org Biomol Chem *6*, 4593–4608.

van den Boom, D., and Berkenkamp, S. (2007). MALDI-MS of Nucleic Acids and Practical Implementations in Genomics and Genetics. In MALDI MS: A Practical Guide to Instrumentation, Methods and Applications, F. Hillenkamp, and J. Peter-Katalinic, eds. (Weinheim, Wiley-VCH), pp. 131–179.

van den Boom, D., and Hillenkamp, F. (2005). Analysis of nucleic acids by mass spectrometry. In Analytical Techniques in DNA Sequencing, B.K. Nunnally, ed. (New York, NY, Taylor & Francis-CRC), pp. 85–105.

Vandell, V.E., and Limbach, P.A. (1999). Polyamine co-matrices for matrix-assisted laser desorption/ionization mass spectrometry of oligonucleotides. Rapid Commun Mass Spectrom *13*, 2014–2021.

Vorm, O., Roepstorff, P., and Mann, M. (1994). Improved resolution and very high sensitivity in MALDI TOF of matrix surfaces made by fast evaporation. Anal Chem *66*, 3281–3287.

Wachter, A., Mengel-From, J., Bɪɪrsting, C., and Morling, N. (2008). A 50 SNP-multiplex mass spectrometry assay for human identification. Forensic Sci Int Genet Suppl Ser *1*, 487–489.

Wang, J., Chen, R., Ma, M., and Li, L. (2008). MALDI MS sample preparation by using paraffin wax film: Systematic study and application for peptide analysis. Anal Chem *80*, 491–500.

Wei, H., Nolkrantz, K., Powell, D.H., Woods, J.H., Ko, M.C., and Kennedy, R.T. (2004). Electrospray sample deposition for matrix-assisted laser desorption/ionization (MALDI) and atmospheric pressure MALDI mass spectrometry with attomole detection limits. Rapid Commun Mass Spectrom *18*, 1193–1200.

Wei, Y., Mei, Y., Xu, Z., Wang, C., Guo, Y., Du, Y., and Zhang, W. (2009). A novel MALDI matrix for analyzing peptides and proteins: Paraffin wax immobilized matrix. Chinese J Chem *27*, 105–110.

Weng, M.-F., and Chen, Y.-C. (2004). Using sol-gel/crown ether hybrid materials as desalting substrates for matrix-assisted laser desorption/ionization analysis of oligonucleotides. Rapid Commun Mass Spectrom *18*, 1421–1428.

Wenzel, T., Elssner, T., Fahr, K., Bimmler, J., Richter, S., Thomas, I., and Kostrzewa, M. (2003). Genosnip: SNP genotyping by MALDI-TOF MS using photocleavable oligonucleotides. Nucleosides Nucleotides Nucleic Acids *22*, 1579–1581.

Williams, T., and Fenselau, C. (1998). p-nitroaniline/glycerol: A binary liquid matrix for matrix-assisted laser desorption/ionization analysis. Eur Mass Spectrom *4*, 379–383.

Wu, K.J., Shaler, T.A., and Becker, C.H. (1994). Time-of-flight mass-spectrometry of underivatized single-stranded-DNA oligomers by matrix-assisted laser-desorption. Anal Chem *66*, 1637–1645.

Wu, K.J., Steding, A., and Becker, C.H. (1993). Matrix-assisted laser desorption time-of-flight mass-spectrometry of oligonucleotides using 3-hydroxypicolinic acid as an ultraviolet-sensitive matrix. Rapid Commun Mass Spectrom *7*, 142–146.

Xu, S., Ye, M., Xu, D., Li, X., Pan, C., and Zou, H. (2006). Matrix with high salt tolerance for the analysis of peptide and protein samples by desorption/ionization time-of-flight mass spectrometry. Anal Chem *78*, 2593–2599.

Xu, Y., Bruening, M.L., and Watson, J.T. (2003). Non-specific, on-probe cleanup methods for MALDI-MS samples. Mass Spectrom Rev *22*, 429–440.

Yuan, X., and Desiderio, D.M. (2002). Protein identification with Teflon as matrix-assisted laser desorption/ionization sample support. J Mass Spectrom *37*, 512–524.

Zagorevskii, D.V., Aldersley, M.F., and Ferris, J.P. (2006). MALDI analysis of oligonucleotides directly from montmorillonite. J Am Soc Mass Spectrom *17*, 1265–1270.

Zehl, M., and Allmaier, G. (2003). Investigation of sample preparation and instrumental parameters in the matrix-assisted laser desorption/ionization time-of-flight mass spectrometry of noncovalent peptide/peptide complexes. Rapid Commun Mass Spectrom *17*, 1931–1940.

Zhang, L., and Gross, M. (2000). Matrix-assisted laser desorption/ionization mass spectrometry methods for oligodeoxynucleotides: Improvements in matrix, detection limits, quantification, and sequencing. J Am Soc Mass Spectrom *11*, 854–865.

Zhang, L.-K., and Gross, M.L. (2002). Location of abasic sites in oligodeoxynucleotides by tandem mass spectrometry and by a chemical cleavage initiated by an unusual reaction of the ODN with MALDI matrix. J Am Soc Mass Spectrom *13*, 1418–1426.

Zhang, Z., Zhou, L., Zhao, S., Deng, H., and Deng, Q. (2006). 3-Hydroxycoumarin as a new matrix for matrix-assisted laser desorption/ionization time-of-flight mass spectrometry of DNA. J Am Soc Mass Spectrom *17*, 1665–1668.

Zhou, L., Deng, H., Deng, Q., and Zhao, S. (2004). A mixed matrix of 3-hydroxypicolinic acid and pyrazinecarboxylic acid for matrix-assisted laser desorption/ionization time-of-flight mass spectrometry of oligodeoxynucleotides. Rapid Commun Mass Spectrom *18*, 787–794.

Zhu, Y.F., Chung, C.N., Taranenko, N.I., Allman, S.L., Martin, S.A., Haff, L., and Chen, C.H. (1996a). The study of 2,3,4-trihydroxyacetophenone and 2,4,6-trihydroxyacetophenoneas matrices for DNA detection in matrix-assisted laser-desorption ionization time-of-flight mass-spectrometry. Rapid Commun Mass Spectrom *10*, 383–388.

Zhu, Y.F., Taranenko, N.I., Allman, S.L., Martin, S.A., Haff, L., and Chen, C.H. (1996b). The effect of ammonium salt and matrix in the detection of DNA by matrix-assisted laser desorption/ionization time-of-flight mass-spectrometry. Rapid Commun Mass Spectrom *10*, 1591–1596.

Zhu, Y.F., Taranenko, N.I., Allman, S.L., Taranenko, N.V., Martin, S.A., Haff, L.A., and Chen, C.H. (1997). Oligonucleotide sequencing by fragmentation in matrix-assisted laser desorption/ionization time-of-flight mass spectrometry. Rapid Commun Mass Spectrom *11*, 897–903.

Part XII
Clinical Research and Applications

Chapter 38
Sample Preparation Techniques for Cancer Proteomics

Paul Dowling, Martin Clynes, and Paula Meleady

Abstract Cancer is an extremely complex and heterogeneous disease that reflects the genetic as well as protein changes within a cell. Understanding the molecular and cellular mechanisms that contribute to tumour formation, disease progression, and cancer cell migration and metastasis has been a major challenge in cancer research for many years. As a result, there are currently many effective therapies for the early detection and diagnosis of cancer; however cancer still remains a major cause of death and disease. Monitoring the protein expression pattern in clinical samples by proteomics technologies offers opportunities to discover potentially new biomarkers for early detection and diagnosis of cancer. In this chapter we will describe protocols for the preparation of samples for proteomic analysis from in vitro cancer cell line models, specifically on the preparation of proteins from whole cell lysates and proteins secreted into the culture media (i.e. conditioned media). We will also introduce protocols for the proteomic analysis of in vivo clinical samples from cancer patients using formalin-fixed paraffin-embedded (FFPE) tissue and biological fluids such as saliva and serum.

Keywords Cancer proteomics · Cancer cell lines · Conditioned media · Formalin-fixed paraffin-embedded tissue · Salivary proteomics · Immunodepletion

Abbreviations

ATCC	American Type Culture Collection
CM	Conditioned Medium
DIGE	Difference Gel Electrophoresis
ECACC	European Collection of Cell Cultures
FBS	Foetal Bovine Serum
FFPE	Formalin-Fixed Paraffin-Embedded
HMEC	Human mammary epithelial cells.

P. Meleady (✉)
National Institute for Cellular Biotechnology, Dublin City University, Dublin 9, Ireland
e-mail: paula.meleady@dcu.ie

A.R. Ivanov, A.V. Lazarev (eds.), *Sample Preparation in Biological Mass Spectrometry*, DOI 10.1007/978-94-007-0828-0_38,

PBS	Phosphate Buffered Saline
SF	Serum-Free
SILAC	Stable Isotope Labelling of Amino Acids in Culture

38.1 Introduction

Although many effective therapies exist for the early detection and diagnosis, cancer remains a major cause of death and disease. Cancer is an extremely complex disease that reflects the genetic as well as protein changes within a cell. Understanding the molecular and cellular mechanisms that contribute to tumour formation, disease progression, and cancer cell migration and metastasis has been a major challenge in cancer research for many years. Monitoring the protein expression pattern in clinical samples by proteomics technologies offers opportunities to discover potentially new biomarkers for early detection and diagnosis of cancer (Hanash, 2003). Proteomic technologies have been used for protein expression profiling of cancer cell lines (Planque et al., 2009; Keenan et al., 2009) and various types of clinical biological samples from cancer patients including tissue (Bouchal et al., 2009), serum (Dowling et al., 2007), plasma (Pernemalm et al., 2009), urine (Orenes-Piñero et al., 2007), and saliva (Dowling et al., 2008).

The first protein cancer marker, carcinoembryonic antigen (CEA), was identified in 1965 for the detection of colorectal cancer in patient serum (Gold and Freedman, 1965). Other biomarkers discovered in the 1970s and 1980s include prostate-specific antigen (PSA) for prostate cancer, CA-19 for colorectal and pancreatic cancer, CA-15-3 for breast cancer and CA-125 for ovarian cancer. However, not all biomarkers are effective in all clinical situations. For example, PSA is well established in clinical practice, but approximately one third of patients with an elevated PSA level often undergo unnecessary medical procedures because they do not have a malignant form of prostate cancer (Harris and Lohr, 2002). Many types of cancer such as lung carcinoma and melanoma do not have any significant biomarkers available to screen at the early stage of disease. Identification of new tumour biomarkers with predictive value is needed to allow early detection and treatment of cancer.

There have been many advances in the use of proteomic methods to study cancer cell biology over the last number of years (reviewed Qian et al., 2006; Mauri and Scigelova, 2009). However many challenges still remain including the dynamic range of proteins present in biological samples such as serum where potential biomarkers of interest may be present at very low concentrations thus requiring depletion of large serum proteins, and also the sensitivity and reproducibility of the LC and MS instrumentation.

In this chapter we will introduce protocols for the preparation of samples for proteomic analysis from both in vitro cancer cell line models and from in vivo clinical material from cancer patients.

38.2 Materials

38.2.1 Reagents

- Tissue culture flasks (T-75, T-175) – many suppliers (e.g. Sigma, Nunc, Costar, etc.)
- Cell culture medium (DMEM, RPMI, etc.) – many suppliers (e.g. Invitrogen, Sigma, Thermo Fisher Scientific, etc.)
- Foetal Bovine Serum (FBS) – many suppliers (e.g. Sigma, Lonza, etc.)
- 2.5% Trypsin solution (10×) (for cell culture) – Invitrogen/Gibco brand
- 21 gauge needles
- pH paper
- Bradford protein assay reagent (Biorad 500-0205)
- 5,000 Da molecular weight cut-off concentrators (Millipore, UFC 900524).
- Sequencing grade trypsin (Promega)
- Slide-A-Lyzer MINI Dialysis Unit, 2 kDa MWCO (Thermo Fisher Scientific)
- Protein Desalting Spin Columns (Thermo Fisher Scientific)
- Immunodepletion cartridges – many available including:
  - Sigma: (The ProteoPrep® 20 Plasma Immunodepletion Technology)
  - Beckman Coulter: (ProteomeLab IgY-12 High Capacity Primate-Optimized Proteome Partitioning Kits)
  - Agilent: (Multiple Affinity Removal Spin Cartridge-Hu 14)
- 0.22 μm spin filters (Thermo Fisher Scientific)
- Complete, Mini; Protease Inhibitor Cocktail Tablets containing PMSF, AEBSF, TLCK, Bestatin, pepstatin A, EDTA and E-64 (Roche)
- Rapigest (Waters)

38.2.2 Chemicals

High quality chemicals should be used for proteomic/mass spectrometry experimentation

- Sucrose, Tris, Urea, Thiourea, CHAPS, Magnesium Acetate, Ammonium Bicarbonate, SDS, Iodoacetamide, Dithiothreitol (DTT), trichloroacetic acid (TCA)

38.2.3 Solvents

High quality LC-MS grade solvents should be used where possible for proteomic/mass spectrometry experimentation

- Acetic acid, Ethanol, Xylene, Acetone, Acetonitrile, Trifluoroacetic acid (TFA)

38.2.4 Buffers/Solutions

- Phosphate-buffered saline (Dulbecco A) (PBSA). Prepare by adding 8 g of sodium chloride, 0.2 g of potassium chloride, 1.15 g of disodium hydrogen phosphate and 0.2 g of potassium dihydrogen phosphate to 1 L of distilled water. Sterilise by autoclaving. More simply, add one PBSA tablet (e.g. Oxoid #BR0014G and also many other suppliers) to 100 mL of distilled water, allow to dissolve, and sterilise by autoclaving.
- 0.25% Trypsin-EDTA solution for cell culture: Prepare by addition of 50 mL of 2.5% Trypsin (10×) stock solution and 10 mL of sterile 1% EDTA to 440 mL of sterile PBSA solution. Aliquot and store at –20°C until use.
- 20 mM Tris-HCl pH 9, containing 2% SDS.
- 50 mM ammonium bicarbonate.
- 10 mM DTT.
- 50 mM iodoacetamide in 50 mM ammonium bicarbonate.
- Trypsin solution for protein digestion: 20 μg of sequencing-grade trypsin in 200 μL of 50 mM ammonium bicarbonate buffer.
- 100% trichloroacetic acid (TCA): Add 1 g of TCA to 700 μL LC-MS water.
- 200 mM DTT.
- 100 mM iodoacetamide.
- Stopping solution: 4% acetonitrile and 0.2% TFA.

38.3 Protocols

38.3.1 Sample Preparation from Cancer Cell Line Models

Cancer cell lines have been extensively used to investigate how biological processes such as proliferation, apoptosis and migration become deregulated during the progression of cancer. The reason for this lies in the fact that biological fluids and tissue represent a high degree of complexity and variability, and limited numbers and amounts of samples are available for study. Cell line models are easy to maintain for longer times and are reproducible in terms of sample generation. However it is also important to note that cell lines are prone to genotypic and phenotypic drifts over time in culture (Osborne et al., 1987).

38.3.1.1 Sample Preparation for Whole Cell Lysates Prepared from Cultured Cell Lines

Cell lysates from cell lines provide an unlimited source of homogenous, self-replicating material which is free of contaminating stromal cells, and often easily cultured in simple standard media (Liotta and Petricoin, 2000). Investigations using cell lysates provide an overall view of cellular alterations in cells in response to specific conditions and therefore have been considered valuable research material for identification of biomarkers and drug targets. Cell lysates from cancer cell lines models have been used to investigate mechanisms of drug resistance in lung cancer cell lines (Keenan et al., 2009), cellular invasion in pancreatic cancer cell lines (Walsh et al., 2009) and changes associated with ErbB-2-over expression in breast cancer cell lines (Gharbi et al., 2002). Recently the systematic comparison of the membrane proteome of two closely related murine teratocarcinoma cell lines (F9B9 and F9DR), of which only one (F9DR) is capable of forming liver metastases in vivo was carried out using LC-MS based analyses (Roesli et al., 2009). Cancer cell lines have been also used to investigate alterations in the phosphoproteome during cancer progression. Because phosphorylated cytoskeletal proteins and chaperones mediate cell motility and apoptotic resistance, phosphoproteomic changes in the MCF10AT model of breast cancer progression were studied using a combination of phosphotyrosyl affinity enrichment, iTRAQ technology, and LC-MS/MS (Chen et al., 2007). A number of tyrosine kinases, phosphatases, and other signalling proteins including SLC4A7 were detected to undergo differential phosphorylation during disease progression.

In this protocol we describe a simple procedure for the preparation of whole cell lysates from cancer cell lines for proteomic analysis using 2D-DIGE. This method has been found to work very well in our laboratory for a wide range of cancer cell lines from many tissue types (e.g. lung, breast, melanoma, pancreas). The method can be easily adapted for use in gel-free, LC-MS/MS approaches.

Protocol

1. Before any work is commenced using cancer cell lines, it is extremely important to ensure your cells are healthy and free of microbial or fungal contamination. It is also very important to regularly check that your cancer cell line cultures are free of *Mycoplasma* infection (see Note 1).
2. Cells are seeded into T-75 cm^2 flasks at 5×10^5 cells per flask (see Note 2). Ensure that you set up enough biological replicate flasks in order to obtain statistically significant results.
3. Harvest exponentially growing cells by passaging the cells using 0.25% trypsin-EDTA solution to detach cells from the surface of the flasks, i.e. after the cultures have reached 60–70% confluency. 4 mL of 0.25% trypsin-EDTA solution is sufficient for a T-75 cm^2 flask (2 mL for T-25 cm^2 flask and 7–8 ml for T-175 cm^2 flask). Leave the cells to detach at 37°C for approximately 2–10 min (the length of time for detachment will be cell-line specific). Add an equal volume of culture medium containing FBS in order to inhibit the trypsin activity and transfer the cell suspension to a sterile 30 mL container. Centrifuge the cell suspension at $250\times g$ for 5 min at room temperature. (See Note 3).
4. Resuspend cell pellets in cold PBSA (at 4°C) and centrifuge at $250\times g$ for 5 min at room temperature. Take an aliquot for a cell count. Repeat this washing step to remove residual medium (including FBS proteins) and trypsin solution from the cell pellet.
5. Wash the cell pellets a final time with sucrose buffer (10 mM sucrose, 100 mM Tris, pH 8.0) to reduce the salt content from the PBSA wash solution in the final cell pellet.

6. Resuspend the cell pellet with 0.5 mL of lysis buffer such as 2D-DIGE lysis buffer (7 M urea, 2 M thiourea, 4% CHAPS, 30 mM Tris, 5 mM magnesium acetate, pH 8.5). We have found that this 0.5 mL volume of lysis buffer is sufficient for optimal lysis of a 60–70% confluent T-75 cm^2 flask with a final cell count of approximately $4–5\times10^6$ cells per flask. However this may need to be optimised for individual cancer cell lines.
7. Homogenise the lysate by gently passing it though a 21-gauge needle 5 times up and down using a 1 mL syringe. Avoid frothing the sample as this can cause protein denaturation.
8. Incubate the lysate for 1 h at room temperature with gentle shaking on a rocking platform.
9. After incubation, centrifuge the lysates using a refrigerated microcentrifuge at $3500\times g$ for 15 min at 4°C. Transfer the middle layer of the supernatant containing extracted protein to a fresh chilled Eppendorf tube. If the sample will be used for 2D-DIGE analysis check the pH of the sample by spotting 3 μL onto a pH indicator strip to ensure that it lies between pH 8.0 and 9.0.
10. A small aliquot of sample is used for protein estimation. Divide the sample into smaller aliquots to reduce freeze-thaw steps and store at –80°C until use.
11. For protein quantification we use a Bradford assay (Bio-Rad, 500-0205) that is compatible with urea and thiourea in the sample. 2 mg/mL of bovine serum albumin (BSA) solution is used as a known standard. Prepare dilutions of BSA stock for 0.125, 0.25, 0.5 and 1.0 mg/mL to generate a protein standard curve. Add 240 μL/well of Bradford protein assay reagent to the wells of a 96-well plate. Add 10 μL of protein standard dilution or sample (diluted 1:10) to the relevant wells of the 96-well plate. All samples are assayed in triplicate. Incubate the samples for 5 min with gentle shaking on a rocking platform to ensure even mixing of sample with protein assay reagent. Read the absorbance of each standard and sample at 595 nm using a plate reader. The concentration of the protein samples is determined from the plot of the absorbance at 595 nm versus the concentration of the protein standard.

38.3.1.2 Sample Preparation of Secreted Proteins from Conditioned Media

Secreted proteins can act locally and systemically in the body and play important roles in the regulation of cell physiology (growth and apoptosis, etc) and pathophysiology (invasion and motility). Therefore, the secretome reflects the functionality of a cell in a given environment (Chen et al., 2008). Biological fluids, such as serum, may contain the molecules of interest (biomarker or drug target) in too low concentrations which cannot be easily measured or purified, unless specific immunological reagents and highly sensitive ELISA methods are used (Hathout, 2007). Moreover, the secretome in a tumour microenvironment contains secreted proteins such as cytokines, chemokines, growth factors and proteases as well as the extracellular matrix, constituted by proteins, receptors and adhesion molecules, and this makes biological fluids a complex mixture to investigate and may mask potential biomarkers (Kulasingam and Diamandis, 2008). Recently the secretomes of cancer cell lines derived from the lung (Planque et al., 2009), prostate (Sardana et al., 2008), and ovary (Gunawardana et al., 2009) have been investigated using a two dimensional LC-MS/MS based approach. Secreted proteins from cancer cell lines have been analysed using other proteomic approaches such as SILAC and 2D-DIGE to investigate malignancy in breast cancer (Liang et al., 2009) and invasion in pancreatic cancer (Walsh et al., 2008).

In this protocol we describe a simple procedure for the preparation of secreted proteins from the conditioned media from cancer cell lines for proteomic analysis using 2D-DIGE. The method can be easily adapted for use in gel-free, LC-MS/MS approaches. Figure 38.1 shows an example of a differentially expressed protein identified by LC-MS as Maspin (SerpinB5) following a 2D-DIGE proteomic experiment comparing the secreted proteomes of two non-tumourigenic breast epithelial cell lines (MCF10A and HMEC) with two breast cancer cell lines (MCF7 and BT-20). Maspin is known to be a secreted protein and has tumour suppressor properties (Toillon et al., 2007); this shows the effectiveness of the protocol described below for the analysis of secreted proteins from cancer cell lines.

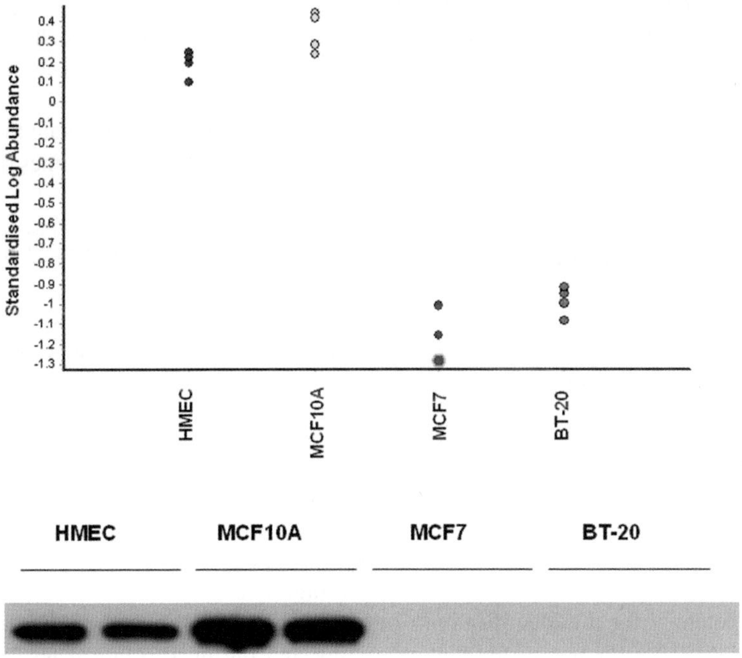

Fig. 38.1 Analysis of secreted proteins in the conditioned media (CM) of human breast epithelial cell lines, HMEC and MCF10A, compared to human breast cancer cell lines, MCF7 and BT-20, using the sample preparation method described in Section 38.3.1.2. *Upper panel*, Decyder 6.5 output from analysis of the 2D-DIGE gel images (four biological replicate samples for each cell line) comparing the standardised log abundance of the differentially regulated protein gel spot, subsequently identified by LC-MS as Maspin, from the four cell lines. *Lower panel*, western blot analysis to confirm the down-regulated expression of Maspin in the CM of MCF7 and BT-20 cells compared to HMEC and MCF10A cells

Protocol

1. Seed cells at 3×10^6 cells per flask in T-175 cm^2 flasks and allow the cells to grow until they are 50–60% confluent. Ensure that you set up enough biological replicate flasks in order to obtain statistically significant results.
2. In order to reduce the amount of serum proteins from the FBS present in the culture medium that may interfere with the proteomic analysis of the sample, a number of wash steps and incubation steps with serum-free (SF) basal medium are incorporated into the method (for most cell lines, the FBS concentration in the culture medium is 5–10%). Wash the cells three times with 10 mL of SF-basal media and then incubate the cells with 15 mL of SF basal media for 60 min at 37°C in an incubator.
3. After this time, wash the cells twice with SF basal media.
4. Add 15 mL of SF basal media to the cells and incubate for 72 h at 37°C.
5. After such time, collect the conditioned media (CM) from the cells by gently pouring or pipetting off the CM media into a 30 mL sterile container.
6. Centrifuge the CM for 15 min at $250 \times g$ at 4°C to remove cells and other insoluble material and store at –80°C until use.
7. At the time of analysis concentrate the CM using 5,000 Da molecular weight cut-off concentrators (Millipore, UFC 900524). A volume of 15 mL is concentrated to 500 μL by centrifuging at $1000 \times g$ at 4°C.
8. Prior to analysis by proteomic methods such as 2D-DIGE, precipitate the proteins using acetone precipitation. Add 1 volume of sample to 4 volumes of cold acetone (1:4 ratio) and incubate overnight at 4°C.
9. The following day centrifuge the sample at $1000 \times g$ for 20 min at 4°C.
10. Reconstitute the sample in the relevant lysis buffer (e.g. 2D-DIGE lysis buffer, see Section 38.3.1.1) and process as required
11. Use a small aliquot of sample for protein estimation (see Section 38.3.1.1) and divide the sample into smaller aliquots to reduce freeze-thaw cycles and store at –80°C until use.

38.3.2 Sample Preparation from Formalin-Fixed Paraffin-Embedded (FFPE) Cancer Tissue

Targeted proteomics research, based on the enrichment of cancer-relevant proteins from isolated cell populations selected from high-quality tissue specimens, offers great potential for the identification of diagnostic, prognostic, and predictive biological markers for use in the clinical setting and during preclinical testing and clinical trials, as well as for the discovery and validation of new protein drug targets. Formalin-fixed and paraffin-embedded (FFPE) tissue collections, with attached clinical and outcome information, are invaluable resources for conducting retrospective protein biomarker investigations and performing translational studies of cancer and other diseases (Sprung et al., 2009; Balgley et al., 2009). Recently it has been reported that quantification of proteins using iTRAQ^{TM} reagents can be applied to conduct large scale quantitative proteomics analysis of the vast archives of formalin fixed tissues (Jain et al., 2008). Jain and co-workers extracted proteins in the form of tryptic peptides utilizing a Liquid Tissue MS Protein Partitioning Kit (Expression Pathology). While the composition of the Liquid Tissue buffer is not available, other groups have successfully developed methods for extracting protein from FFPE tissue. For example, it was demonstrated that a lysis buffer containing 6 M guanidine hydrochloride incubated with FFPE tissue at high temperature led to the highest protein yield and the largest number of proteins identified, when compared with other methods (Jiang et al., 2007). Recently a method using filter-aided sample preparation (FASP) together with samples solubilised in high concentrations of SDS for FFPE analysis has been described (Ostasiewicz et al., 2010). Our protocol is similar in so far as we use high concentrations of SDS (2%) to solubilise proteins prior to dialysis.

Protocol

1. The first part of the protocol involves de-paraffinization of FFPE slides. FFPE slides are obtained from any source, for example tumour/normal paired tissue. Combine up to 2 sections, each with a thickness of about 10 μm and area of up to about 100 mm^2, in one preparation. Before de-paraffinization, slides are kept at room temperature for 1 h.
2. Immerse the slide in xylene for 10 min. Repeat this step using new xylene solution and immerse for 10 min.
3. Immerse the slide in 100% ethanol for 5 min, followed by another immersion in 85% ethanol for 1 min, followed by yet another immersion in 70% ethanol for 1 min.
4. Immerse the slide in high purity water for 1 min. Remove excess water from the slide, ensuring the sections are not allowed to dry out.
5. Transfer the sections to a clean Eppendorf tube using a razor blade (see Note 4).
6. To extract protein from the tissue sections add 50 μL of extraction buffer (20 mM Tris-HCl pH 9, containing 2% SDS) to the de-paraffinized FFPE tissues sections, vortex for 5 s and then incubate on a heating block at 100°C for 20 min.
7. Centrifuge the sample for 1 min at 1000×g, then vortex the sample for 5 seconds and incubate at 60°C for 2 h.
8. Following this incubation step, centrifuge the sample at 4°C for 10 min at 16,000×g. Transfer the supernatant (soluble fraction) to a new 1.5 mL Eppendorf tube.
9. Quantify the protein yield as previously described (see Section 38.3.1.1: Bradford assay). Using this protocol, the expected protein yield is about 10–30 μg of protein.
10. To prepare the samples for protein digestion ahead of LC-MS analysis the samples need to be dialyzed. Place the proteins collected in the supernatant in a dialysis unit (Thermo Fisher Scientific) and allow the sample to dialyze overnight at 4°C against 100 mM Tris-HCl at pH 8. After this time add urea to the supernatant to give a final concentration of 8 M.

11. Add 1 μL of reducing solution (10 mM DTT) and mix the sample by gentle vortex. Reduce the protein mixture for 1 h at room temperature.
12. Add 1 μL of alkylating solution (50 mM iodoacetamide in 50 mM ammonium bicarbonate) and mix the sample by gentle vortex. Alkylate the protein mixture for 1 h at room temperature.
13. Add 1 μL of reducing solution (10 mM DTT) to consume any unreacted iodoacetamide. Mix the sample by gentle vortex and allow the reaction to stand at room temperature for 1 h.
14. Reduce the urea concentration by diluting the reaction mixture 10-fold with 50 mM ammonium bicarbonate solution. Mix the solution by gentle vortex. This dilution reduces the urea concentration to approximately 0.6 M, a concentration at which the trypsin retains its activity.
15. Add 1 μg of sequencing-grade trypsin solution/25 μg protein. Mix the sample by gentle vortex and carry out the digestion overnight at 37°C.
16. Stop the reaction and adjust the pH of the solution to <6 by adding concentrated acetic acid as needed. Desalt the tryptic digests using Protein Desalting Spin Columns, lyophilize to dryness using a SpeedVac, and then store at –80°C until analysis.

38.3.3 Sample Preparation for Plasma/Serum Specimens Collected from Cancer Patients

There are numerous biological fluids, tissues, and cells that can be utilized to identify and quantify candidate biomarkers. Human plasma/serum is by far the most commonly used bio-fluid in biomarker discovery studies, but it also presents many difficult hurdles that most be overcome for successful analysis to be performed. Many biomarkers with clinical potential as diagnostics have concentrations in the low nanogram–picogram per milliliter range in blood, five to seven orders of magnitude lower that abundant proteins such as albumin. In this protocol, we describe the relatively new technology, immunodepletion that specifically removes highly abundant proteins, allowing researchers to analyse more low abundant proteins that are more likely to be of interest from a candidate biomarker perspective (Ernoult et al., 2010; Tonack et al., 2009). We have recently used this protocol, followed by 2D-DIGE analysis and subsequent MS, to identify a panel of proteins found to be differentially expressed between lung cancer and normal serum samples. Many of the proteins of interest identified were of low abundance, indicating how useful this protocol is to enrich for potential low level biomarkers (Dowling et al., 2007).

Protocol

1. The first step involves diluting human serum/plasma with the equilibration buffer (Buffer A) and filtering the diluted samples through a 0.22 μm spin filter to prevent clogging of spin cartridge frits by removing un-dissolved material. (Buffer A/Buffer B: These buffers are normally supplied with the

immunodepletion cartridge. Typically A is an equilibration buffer and B is an elution buffer that allows the high abundant proteins to be stripped off the capturing antibody).

2. Next, pipette the diluted sample on top of the immunodepletion spin cartridge, allow to incubate for 1–30 min (depending on the cartridge being used) and centrifuge slowly, typically for 1–2 min at 100×*g* (or lowest possible setting on centrifuge). The flow-through fraction is collected at this stage and placed in a fresh tube.
3. Wash the spin cartridge twice with Buffer A and allow to incubate for 5 min before centrifugation for 2–3 min at 100×*g* (or lowest possible setting on centrifuge). Collect this flow-through fraction and pool with the initial fraction from step 2. Label this the 'low-abundant proteins'.
4. At this stage the immuno-captured high abundant proteins are eluted using Buffer B, which is typically an acidic buffer which strips the proteins from the capturing antibody. Collect the flow-through fraction and place into a fresh tube. Label this the 'high-abundant proteins'.
5. The final stage involves re-equilibrating the spin cartridge with Buffer A by pushing it through the resin bed. This results in a re-equilibrated cartridge that can be used immediately to immunodeplete more plasma/serum or stored for future work.
6. For TCA/acetone precipitation of the proteins add a sufficient volume of 100% TCA to the 'low-abundant fraction' to reach a final TCA concentration of 10% (see Notes 5 and 6). The sample should turn milky white if proteins are present. Vortex and incubate on ice for 30 min.
7. Centrifuge the sample at 4°C for 30 min at 16,000×*g* until the supernatant is clear and there is a white pellet of protein at the bottom of the tube. Discard the supernatant.
8. To the pellet add 1 mL of –20°C acetone. Vortex the sample and incubate overnight at –20°C. Centrifuge at 4°C for 10 min at 16,000×*g*. Discard the supernatant.
9. To the pellet, add a fresh aliquot of –20°C acetone, vortex and incubate at –20°C for 30 min and centrifuge as above. Discard the supernatant and repeat. After the final centrifugation, discard the acetone fraction and dry the pellet completely.
10. At this stage the protein pellet can be directly digested with trypsin or dissolved in a buffer of your choice. The resulting peptides can be directly analysed by LC-MS or labeled with peptide tags that will allow for quantitation.

38.3.4 Sample Preparation from Saliva Collected from Head and Neck Cancer Patients

Saliva samples from certain disease patients store a large range of proteins that may be associated with the disease phenotype. Successful discovery and validation of these salivary proteins/peptides may result in their use as biomarkers or as part of a panel of biomarkers, leading to the use of this non-invasive bio-fluid for detecting and monitoring of particular diseases, including cancer. A saliva-based diagnostic approach would be extremely advantageous because of its relatively non-invasive approach. With the developments in proteomic technologies the identification of such diagnostic markers for early detection and disease recurrence employing saliva-based diagnostic tests are certainly attainable (Rao et al., 2009; Hu et al., 2008). Recently, we identified differentially expressed proteins in saliva from Head and Neck Squamous Cell Carcinoma (HNSCC) patients compared to a control group using the saliva proteome enrichment protocol described below, followed by 2D-DIGE and mass spectrometry (Dowling et al., 2008). The proximity of saliva to the developing tumour is undoubtedly a major factor in facilitating detection of these proteins, including beta fibrin and S100 calcium binding protein. These proteins are known to be involved in tumour progression, metastasis and angiogenesis. Figure 38.2 shows an example of a differentially regulated protein found in saliva from HNSCC patients compared to controls using the sample preparation method described below for saliva proteins.

Protocol

1. Collect unstimulated saliva (5 mL) from patients into a sterile container after their mouths are rinsed thoroughly with water.
2. Immediately snap freeze the saliva in liquid nitrogen and store at −80°C until analysis. Add 10 μL (100×) of protease inhibitor cocktail (Roche) to the saliva before storage to quench the high level of protease activity present in saliva.
3. Centrifuge the saliva sample (5 mL) at 4°C for 30 min at 10,000×g to discard cellular debris and un-dissolved material. (See Note 7).
4. Transfer the supernatants to 5 kDa membrane concentrators (Ultrafree 5 K cut-off, Millipore) and centrifuge at 4°C for 1 h at 5000×g to reduce the volumes to approximately 1 mL.
5. Quantify the protein yield as previously described (see Section 38.3.1.1: Bradford assay). Place approximately 1 mg of protein from the saliva supernatant into a fresh Eppendorf.
6. Adjust the saliva supernatant to contain 80% acetone by the addition of 4 volumes of HPLC-grade acetone to 1 volume of saliva supernatant. Incubate at –20°C overnight.
7. Centrifuge at 4°C for 20 min at 16,000×g and discard the supernatant.
8. Wash the pellet with two aliquots of 100% acetone at –20°C and then allow the protein pellet to briefly air-dry.
9. Re-suspend the protein pellet in 100 μL of trypsin digestion buffer (50 mM ammonium bicarbonate + 0.1% Rapigest) using a combination of vortexing and sonication.
10. Reduce the disulfides by adding 4 μL of 200 mM DTT and incubate at 37°C for 1 h.
11. Alkylate the solution by addition of 4 μL of 100 mM iodoacetamide and incubate for 15 min at room temperature in the dark.

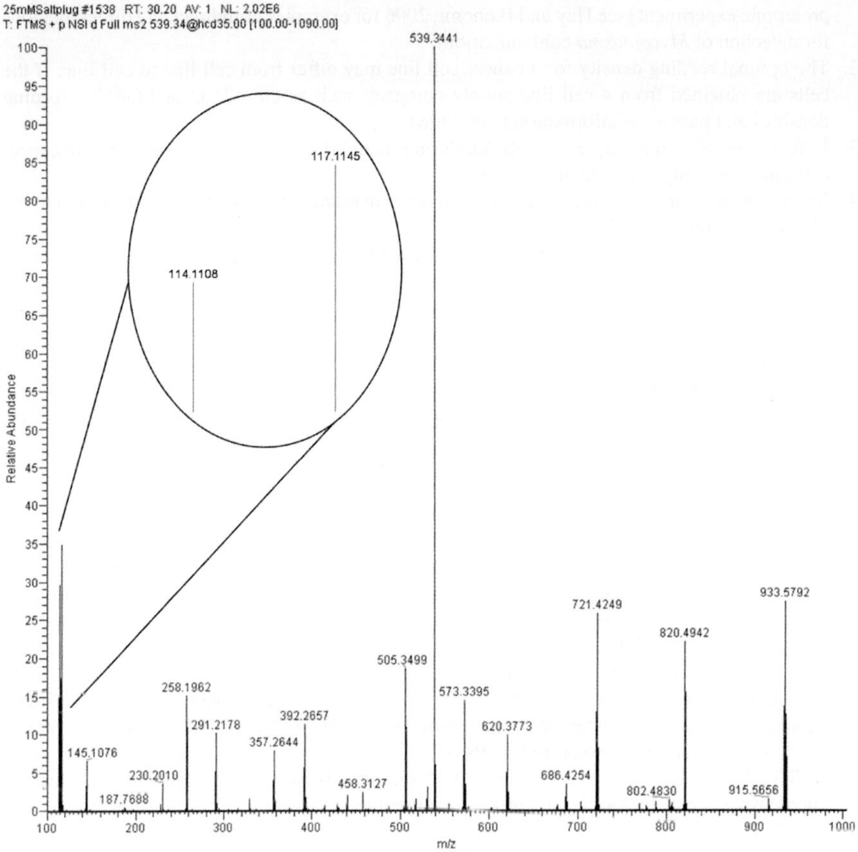

Fig. 38.2 Analysis of saliva proteins from Head and Neck Squamous cell carcinoma (HNSCC) patients compared to a control group using the sample preparation method described in Section 38.3.4. Proteins were digested with trypsin and subsequently labelled with iTRAQ reagents. This example shows a protein of interest, the MS/MS data used for identification and the differences in the 114:117 reporter ion ratio used for quantitation

12. Digest the saliva proteins using the trypsin solution at 37°C overnight (1 μg of sequencing-grade trypsin/25 μg protein).
13. Stop the reaction by acidifying the solution using the stop solution (4% acetonitrile and 0.2% TFA). Desalt the tryptic digests using Protein Desalting Spin Columns (Thermo Fisher Scientific), lyophilize to dryness using a SpeedVac, and then store at –80°C until analysis.

Notes

1. *Mycoplasma* contamination of cell lines has been found to alter cellular metabolism and growth, alter nucleic acid metabolism, etc. and thus it is essential that cell lines are regularly checked for *Mycoplasma* contamination, especially before commencement of an expensive

proteomic experiment (see Hay and Ikonomi, 2006 for examples of indirect and direct methods for detection of *Mycoplasma* contamination).

2. The optimal seeding density for a cancer cell line may differ from cell line to cell line. If the cells are obtained from a cell line supply company such as the ATCC or ECACC, seeding densities and passaging information is provided.
3. Refer to the following chapter for full details on subculturing adherent and non-adherent cancer cell lines (Meleady and O'Connor, 2006).
4. Extraction steps are performed in siliconized tubes in order to prevent protein adsorption to the plastic and limit protein loss.
5. The TCA used must be obtained as fresh as possible. TCA degrades with time to chloroform which will result in incomplete precipitation.
6. The acetone used must be HPLC grade. Poor quality acetone will result in poor recovery of proteins.
7. Strategies exist for the removal of high abundant salivary proteins, such as alpha amylase, that hamper the detection of low abundant proteins appearing in the disease state. Such strategies include the removal of alpha amylase by affinity adsorption to potato starch (Deutsch et al., 2008).

References

Balgley, B.M., Guo, T., Zhao, K., Fang, X., Tavassoli, F.A., and Lee, C.S. (2009). Evaluation of archival time on shotgun proteomics of formalin-fixed and paraffin-embedded tissues. J Proteome Res *8*, 917–925.

Bouchal, P., Roumeliotis, T., Hrstka, R., Nenutil, R., Vojtesek, B., and Garbis, S.D. (2009). Biomarker discovery in low-grade breast cancer using isobaric stable tags and two-dimensional liquid chromatography-tandem mass spectrometry (iTRAQ-2DLC-MS/MS) based quantitative proteomic analysis. J Proteome Res *8*, 362–373.

Chen, S.T., Pan, T.L., Juan, H.F., Chen, T.Y., Lin, Y.S., Huang, C.M. (2008). Breast tumour microenvironment: Proteomics highlights the treatments targeting secretome. J Proteome Res *7*, 1379–1387.

Chen, Y., Choong, L.Y., Lin, Q., Philp, R., Wong, C.H., Ang, B.K., Tan, Y.L., Loh, M.C., Hew, C.L., Shah, N., Druker, B.J., Chong, P.K., and Lim, Y.P. (2007). Differential expression of novel tyrosine kinase substrates during breast cancer development. Mol Cell Proteomics *6*, 2072–2087.

Deutsch, O., Fleissig, Y., Zaks, B., Krief, G., Aframian, D.J., and Palmon, A. (2008). An approach to remove alpha amylase for proteomic analysis of low abundance biomarkers in human saliva. Electrophoresis *29*, 4150–4157.

Dowling, P., O'Driscoll, L., Meleady, P., Henry, M., Roy, S., Ballot, J., Moriarty, M., Crown, J., and Clynes, M. (2007). Two-dimensional difference gel electrophoresis of the lung squamous cell carcinoma versus normal sera demonstrates consistent alterations in the levels of 10 specific proteins. Electrophoresis *28*, 4302–4310.

Dowling, P., Wormald, R., Meleady, P., Henry, M., Curran, A., and Clynes, M. (2008). Analysis of the saliva proteome from patients with head and neck squamous cell carcinoma reveals differences in abundance levels of proteins associated with tumour progression and metastasis. J Proteomics *71*, 168–175.

Ernoult, E., Bourreau, A., Gamelin, E., and Guette, C. (2010). A proteomic approach for plasma biomarker discovery with iTRAQ labelling and OFFGEL fractionation. J Biomed Biotechnol *2010*, Article ID 927917, 1–8.

Gharbi, S., Gaffney, P., Yang, A., Zvelebil, M.J., Cramer, R., Waterfield, M.D., and Timms, J.F. (2002). Evaluation of two-dimensional differential gel electrophoresis for proteomic expression analysis of a model breast cancer cell system. Mol Cell Proteomic *1*, 91–98.

Gold, P., and Freedman, S.O. (1965). Demonstration of tumour-specific antigens in human colonic carcinomata by immunological tolerance and adsorption techniques. J Exp Med *121*, 439–462.

Gunawardana, C.G., Kuk, C., Smith, C.R., Batruch, I., Soosaipillai, A., and Diamandis, E.P. (2009). Comprehensive analysis of conditioned media from ovarian cancer cell lines identifies novel candidate markers of epithelial ovarian cancer. J Proteome Res *8*, 4705–4713.

Hanash, S. (2003). Disease proteomics. Nature *422*, 226–232.

Harris, R., and Lohr, K.N. (2002). Screening for prostate cancer: An update of the evidence for the US Preventative Service Task Force. Ann Intern Med *137*, 917–929.

Hathout, Y. (2007). Approaches to the study of the cell secretome. Expert Rev Proteomic *4*, 239–248.

Hay, R.J., and Ikonomi, P. (2006). Detection of Microbial and Viral Contaminants in Cell Lines. In Cell Biology, A Laboratory Handbook (3rd Ed. Vol. 1), J. Celis, eds. (Elsevier), pp. 49–65.

Hu, S., Arellano, M., Boontheung, P., Wang, J., Zhou, H., Jiang, J., Elashoff, D., Wei, R., Loo, J.A., and Wong, D.T. (2008). Salivary proteomics for oral cancer biomarker discovery. Clin Cancer Res *14*, 6246–6252.

Jain, M.R., Liu, T., Hu, J., Darfler, M., Fitzhugh, V., Rinaggio, J., and Li, H. (2008). Quantitative proteomic analysis of formalin fixed paraffin embedded oral HPV lesions from HIV patients. Open Proteomic J *1*, 40–45.

Jiang, X., Jiang, X., Feng, S., Tian, R., Ye, M., and Zou, H. (2007). Development of efficient protein extraction methods for shotgun proteome analysis of formalin-fixed tissues. J Proteome Res *6*, 1038–1047.

Keenan, J., Murphy, L., Henry, M., Meleady, P., and Clynes, M. (2009). Proteomic analysis of multidrug-resistance mechanisms in adriamycin-resistant variants of DLKP, a squamous lung cancer cell line. Proteomics *9*, 1556–1566.

Kulasingam, V., and Diamandis, E.P. (2008). Tissue culture-based breast cancer biomarker discovery platform. Int J Cancer *123*, 2007–2012.

Liang, X., Huuskonen, J., Hajivandi, M., Manzanedo, R., Predki, P., Amshey, J.R., and Pope, R.M. (2009). Identification and quantification of proteins differentially secreted by a pair of normal and malignant breast-cancer cell lines. Proteomics *9*, 182–193.

Liotta, L., and Petricoin, E. (2000). Molecular profiling of human cancer. Nature Rev Gen *1*, 48–56.

Mauri, P., and Scigelova, M. (2009). Multidimensional protein identification technology for clinical proteomic analysis. Clin Chem Lab Med *47*, 636–646.

Meleady, P., and O′Connor, R. (2006). General Procedures for Cell Culture. In Cell Biology, A Laboratory Handbook (3rd Ed., Vol. 1), J. Celis, eds. (Elsevier), pp. 13–20.

Orenes-Piñero, E., Cortón, M., González-Peramato, P, Algaba, F., Casal, I., Serrano, A., and Sánchez-Carbayo, M. (2007). Searching urinary tumor markers for bladder cancer using a two-dimensional differential gel electrophoresis (2D-DIGE) approach. J Proteome Res *6*, 4440–4448.

Osborne, C.K., Hobs, K., and Trent, J.M. (1987). Biological differences among MCF-7 human breast cancer cell lines from different laboratories. Breast Cancer Res Treat *9*, 111–121.

Ostasiewicz, P., Zielinska, D.F., Mann, M., and Wiśniewski, J.R. (2010). Proteome, phosphoproteome, and N-glycoproteome are quantitatively preserved in formalin-fixed paraffin-embedded tissue and analyzable by high-resolution mass spectrometry. J Proteome Res *9*, 3688–3700.

Pernemalm, M., De Petris, L., Eriksson, H., Brandén, E., Koyi, H., Kanter, L., Lewensohn, R., and Lehtiö, J. (2009). Use of narrow-range peptide IEF to improve detection of lung adenocarcinoma markers in plasma and pleural effusion. Proteomics *9*, 3414–3424.

Planque, C., Kulasingan, V., Smith, C.R., Reckamp, K., Goodglick, L., and Diamandis, E.P. (2009). Identification of five candidate lung cancer biomarkers by proteomic analysis of conditioned media of four lung cancer cell lines. Mol Cell Proteomic *8*, 2746–2758.

Qian, W.J, Jacobs, J.M., Liu, T., Camp, D.G., 2nd, and Smith, R.D. (2006). Advances and challenges in liquid chromatography-mass spectrometry-based proteomics profiling for clinical applications. Mol Cell Proteomic *5*, 1727–1744.

Rao, P.V., Reddy, A.P., Lu, X., Dasari, S., Krishnaprasad, A., Biggs, E., Roberts, C.T., and Nagalla, S.R. (2009). Proteomic identification of salivary biomarkers of type-2 diabetes. J Proteome Res *8*, 239–245.

Roesli, C., Borgia, B., Schliemann, C., Gunthert, M., Wunderli-Allenspach, H., Giavazzi, R., and Neri, D. (2009). Comparative analysis of the membrane proteome of closely related metastatic and nonmetastatic tumor cells. Cancer Res *69*, 5406–5414.

Sardana, G., Jung, K., Stephan, C., and Diamandis, E.P. (2008). Proteomic analysis of conditioned media from the PC3, LNCaP, and 22Rv1 prostate cancer cell lines: Discovery and validation of candidate prostate cancer biomarkers. J Proteome Res *7*, 3329–3338.

Sprung, R.W., Jr, Brock, J.W., Tanksley, J.P., Li, M., Washington, M.K., Slebos, R.J., and Liebler, D.C. (2009). Equivalence of protein inventories obtained from formalin-fixed paraffin-embedded and frozen tissue in multidimensional liquid chromatography-tandem mass spectrometry shotgun proteomic analysis. Mol Cell Proteomic *8*, 1988–1998.

Toillon, R.A., Lagadec, C., Page, A., Chopin, V., Sautière, P.E., Ricort, J.M., Lemoine, J., Zhang, M., Hondermarck, H., and Le Bourhis, X. (2007). Proteomics demonstration that normal breast epithelial cells can induce apoptosis of breast cancer cells through insulin-like growth factor-binding protein-3 and maspin. Mol Cell Proteomic *6*, 1239–1247.

Tonack, S., Aspinall-O'Dea, M., Jenkins, R.E., Elliot, V., Murray, S., Lane, C.S., Kitteringham, N.R., Neoptolemos, J.P., and Costello, E. (2009). A technically detailed and pragmatic protocol for quantitative serum proteomics using iTRAQ. J Proteomic *73*, 352–356.

Walsh, N., Dowling, P., O'Donovan, N., Henry, M., Meleady, P., and Clynes, M. (2008). Aldehyde dehydrogenase 1A1 and gelsolin identified as novel invasion-modulating factors in conditioned medium of pancreatic cancer cells. J Proteomic *71*, 561–571.

Walsh, N., O'Donovan, N., Kennedy, S., Henry, M., Meleady, P., Clynes, M., and Dowling, P. (2009). Identification of pancreatic cancer invasion-related proteins by proteomic analysis. Proteome Sci *7*, 3.

Chapter 39
Sample Preparation of Primary Astrocyte Cellular and Released Proteins for 2-D Gel Electrophoresis and Protein Identification by Mass Spectrometry

Melissa A. Sondej, Philip Doran, Joseph A. Loo, and Ina-Beate Wanner

Abstract The neurotrauma field needs more accurate and sensitive tools to diagnose and evaluate damage to the brain or spinal cord. A controlled, in vitro mechanical trauma model to study protein changes during astrocyte activation after injury is used as a means to discover neurotrauma biomarkers. This model offers several advantages over analyzing cerebrospinal fluid (CSF) or microdialysates including increased sample amount, higher reproducibility between biological samples, enrichment of undegraded proteins due to fewer extracellular matrix proteases in the in vitro versus in vivo system and the ability to relate trauma changes to one cell type. This protocol describes an approach for preparing intact protein samples from conditioned medium (CM) and whole cell lysates of murine cortical astrocytes. The proteins from these samples are separated using 2-D gel electrophoresis, and the protein spot intensities are measured. Proteins showing statistically significant trauma induced changes are then identified using tandem mass spectrometry. This protocol contains optimization steps to enrich proteins of interest, to remove substances that interfere with isoelectric focusing (IEF) and to improve IEF conditions. With this procedure, the 2-D gels show high resolution spot separation and allow for high confidence protein identifications from 107 released and 22 whole cell lysates protein spots so far. Some proteins were identified from multiple spots, suggesting post-translational modification had occurred as a result of the mechanical trauma.

Keywords Astrocytes · Biomarker · Neurotrauma · Tandem mass spectrometry · Two-dimensional gel electrophoresis

---

M.A. Sondej (✉)
Department of Chemistry and Biochemistry, University of California, Box 951569/MSB 1424, Los Angeles, CA 90095, USA
e-mail: masondej@ucla.edu

A.R. Ivanov, A.V. Lazarev (eds.), *Sample Preparation in Biological Mass Spectrometry*, DOI 10.1007/978-94-007-0828-0_39,

39.1 Introduction

There is currently no clinically useful surrogate biomarker for traumatic injury of the central nervous system (CNS) (Hergenroeder et al., 2008; Kochanek et al., 2008). Analysis of trauma patients' CSF or microdialysates is compounded by (1) the dominance of serum-derived abundant proteins (albumin, immunglobulins) due to a breach in the blood brain barrier, (2) a high activity of extracellular matrix proteases and (3) the presence of abundant cytosolic proteins indicating cell damage and death (Hanrieder et al., 2009; Liu et al., 2006; Lakshmanan et al., 2010; Sjödin et al., 2010). Here we describe a new in vitro injury model combined with a sensitive, quantitative proteomic approach targeted to identify protein changes induced by mechanical trauma in astrocytes, the 'body-guards' of the brain.

Astrocytes are considered the most robust and abundant cells in the CNS (Allen and Barres, 2009). These cells control metabolite flow, recycle neurotransmitters and produce extracellular matrix around neurons. After CNS injury, astrocytes readily and actively respond in a process called reactive astrogliosis. This process describes well-known neuro-pathological changes of astrocytes becoming enlarged (hypertrophic), accumulating and proliferating around the injury site and forming the glial scar, a boundary around the injury core (Montgomery, 1994; Reier and Houle, 1988; Sofroniew, 2005). Reactive astrocytes increase amounts of the glial fibrillary acidic protein (GFAP) and related cytoskeletal proteins, and are known to secrete signaling molecules, inflammatory factors and neurotrophins into the extracellular space (Eddleston and Mucke, 1993; Ridet et al., 1997; Sandvig et al., 2004). Currently, there is no comprehensive profile of trauma-initiated protein changes in reactive astrocytes and their secretion products. Also the molecular mechanism of astrogliosis is not well understood.

Astrocytes can be directly activated by mechanical trauma, as they express mechanosensitive channels (Bowman et al., 1992; Di et al., 2000; Ostrow and Sachs, 2005). We have developed an in vitro injury model using mechanical trauma on purified differentiated astrocytes cultured on deformable membranes (Ellis et al., 1995; Neary et al., 2003; Wanner et al., 2008). An abrupt pressure pulse sends a brief blast wave deforming the membrane and stretching the cells throughout the culture in a reproducible and controlled manner. This blast pulse initiates dynamic morphological changes in astrocytes (de Vellis et al., 2009; Wanner et al., 2008). Activated in this manner, astrocytes increase production of key gliosis and glial scar proteins and become inhibitory for axonal regeneration of nerve cells (Wanner et al., 2008).

This protocol combines the high resolution separation of two-dimensional gel electrophoresis with the sensitivity of liquid chromatography-tandem mass spectrometry (LC-MS/MS) to determine trauma-initiated protein changes in reactive astrocytes and their released proteins. Two-dimensional electrophoresis, first introduced by O'Farrell in 1975, still gives the highest resolution for the separation of intact proteins (Friedman et al., 2009; O'Farrell, 1975). This technique traditionally uses denaturing isoelectric focusing (IEF) to separate proteins based on their isoelectric point (p*I*) which is followed by the second dimension, electrophoresis on a sodium dodecyl sulfate (SDS) polyacrylamide slab gel (PAGE) to separate proteins based on their molecular mass. Therefore, each location on the gel has a unique pH/molecular mass coordinate which results in a characteristic spot profile. This allows the separation of not only different proteins, but also isoforms of the same proteins including post-translational modifications such as truncation, phosphorylation and sialylation.

Diligent sample preparation is critical for producing consistent separation of intact proteins to achieve the reproducibility necessary for measuring stimulus-specific and statistically significant changes. Thus, a number of steps in the sample preparation were optimized resulting in consistently well focused 2-D gel spots (Fig. 39.1). Both samples types, the condition medium (CM) and the whole cell lysates, had a number of IEF interfering substances. Therefore, a number of precipitation methods were compared in order to remove these substances. The best results were obtained with trichloroacetic acid (TCA) precipitation followed by an acetone wash. However, precipitation alone was insufficient for sharply focused spots from CM. Therefore a series of different sample preparation techniques were applied to the CM sample prior to performing precipitation. The methodology that gave the highest quality 2-D gels included first concentrating the CM with a centrifugal concentrator, then desalting the sample using a desalting column and finally removing the residual albumin, that remained after a stepwise withdrawal prior to the trauma, with an albumin depletion column. For IEF, active rehydration, loading the sample onto the IPG strip using a low voltage, was superior over passive rehydration, using absorption. Further improvements were seen when a low voltage of 4,000 V was applied to the final isoelectric focusing step and the focusing time of this step was extended to 85,000 Vh.

39.2 Materials

39.2.1 Equipment and Software

1. Cell Injury Pressure Controller II (CIC II, Custom Design and Fabrication, Virginia Commonwealth University, VA) with a Nitrogen tank (Air Liquide Healthcare, Lake Mary, FL)
2. Centrifuge Beckman model J-6B, rotor JS-4.2 (Brea, CA), Sorvall model RC5C plus, rotor SA600 and Heraeus, Biofuge Fresco (Thermo Fisher Scientific, San Jose, CA)
3. PROTEAN IEF Cell, 11 cm Focusing Tray, Criterion Dodeca Cell and FX Pro Plus Scanner (Bio-Rad, Hercules, CA)
4. ProPicII spot cutter (DIGLAB Genomic Solution, Holliston, MA)
5. Progenesis SameSpots v3.3 2-D gel analysis software (Nonlinear Dynamics, Newcastle upon Tyne, UK)
6. LTQ-XL Orbitrap Mass Spectrometer fitted with a NanoSpray ion source (Thermo Fisher Scientific, San Jose, CA) and equipped with an Eksigent NanoLC-2D HPLC (Ekisgent, Dublin, CA)
7. MASCOT Daemon database search engine (Matrix Science, London, UK)

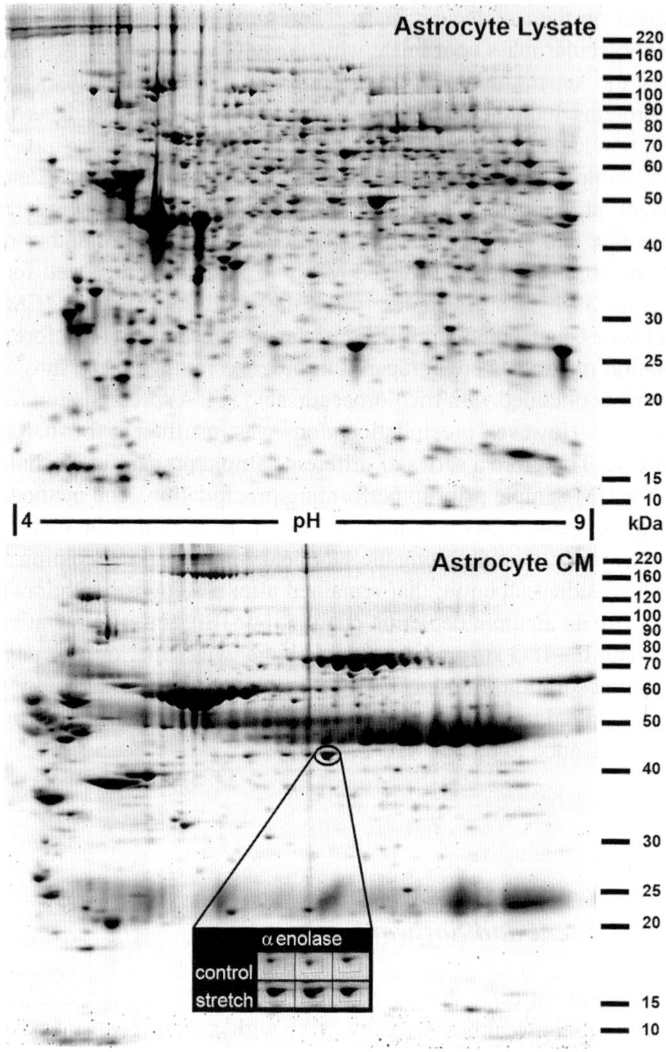

Fig. 39.1 2-D gel electrophoresis spot profile of astrocytic proteins. Shown are two representative 2-D gels with protein profiles of astrocyte whole cell lysates (lysates, *top*) and conditioned medium (CM, *bottom*) samples. Mouse astrocyte cells were grown on stretch plates and matured in vitro. The CM was collected and the cells lysed 5 h after mechanical stretching. After removal of interfering substances, the samples were subjected to isoelectric focusing on a 11 cm, pH 3–10 nonlinear IPG strip, which was then resolved in the second dimension on a 10.5–14% gradient SDS-PAGE. The resulting gels were stained overnight with SYPRO Ruby and destained for 3 days. Discrete protein spots are distributed over the entire molecular weight range (10–220 kDa) and show a wide p*I* spectrum (pH 4–9). The *encircled spot* in the astrocyte CM 2-D gel has been identified as α-enolase (observed molecular weight ~50 kDa, observed p*I* ~6.2). The *insert* shows the spot intensities of three experimental replicates of control versus stretched astrocytes. Stretching increased the average normalized volume 2.8-fold ($P < 0.004$) and was confirmed in two additional independent biological replicates

39.2.2 Supplies

39.2.2.1 Mechanical Trauma and Astrocyte Culture

1. DMEM/F12, DMEM/F12 with HEPES and sodium bicarbonate (Invitrogen, Carlsbad, CA)
2. Silastic membrane 6-well culture plates, (Bioflex, collagen I, Flexcell Intl., Hillsborough, NC)
3. Fetal bovine serum (Atlanta Biologicals, Lawrenceville, GA, USA)
4. Donor equine serum (Hyclone, Logan, UT)

39.2.2.2 Released Protein Preparation: Concentrating the Proteins in the Conditioned Medium, Desalting and Albumin Depletion

1. Dithiothreitol (DTT) (EMD, San Diego, CA)
2. Complete mini protease inhibitor cocktail (Roche, Indianapolis, IN)
3. Tris (Sigma-Aldrich, St Louis, MO)
4. Vivaspin 20 centrifugal concentrator, 3,000 Da molecular weight cut off (MWCO) (Sartorius Stedium Biotech, Bohemia, NY)
5. Sephadex G-25 prepacked PD-10 desalting spin columns with tube adapters (GE Healthcare, Piscataway, NJ)
6. Swellgel albumin removal kit (Thermo Scientific Pierce, Rockford, IL)

39.2.2.3 Soluble Cellular Protein Preparation: Astrocyte Lysis

1. Sodium pyrophosphate decahydrate and cell scrapers (blade length 1.8 cm, handle 25 cm) (Thermo Fisher Scientific, Pittsburgh, PA)
2. *p*-Nitrophenyl phosphate, nonidet P-40 (NP-40), sodium chloride and EDTA (Sigma-Aldrich, St. Louis, MO)
3. Oak Ridge centrifuge tubes, polypropylene copolymer (Nalgene, Rochester, NY)
4. 3 mL syringe with a 27½ gauge needle (BD Biosciences, Franklin Lakes, NY)
5. Easy pressure syringe filter holders (25 mm) (Pall Corporation, Port Washington, NY)

39.2.2.4 Protein Precipitation and Resuspension for IEF

1. Acetone, trichloroacetic acid (TCA), thiourea and bromophenol blue (Sigma-Aldrich, St. Louis, MO)
2. Urea, Omipur 99.5% grade (EMD, San Diego, CA)
3. 3-((3-cholamidopropyl)dimethylammonio)-1-propanesulfonic acid (CHAPS) (Thermo Scientific Pierce, Rockford, IL)
4. Soft glass stirring rods (catalog number 11-380A) (Thermo Fisher Scientific, Pittsburgh, PA)

39.2.2.5 Protein Concentration Measurement

1. Bio-Rad protein assay standard II (Bio-Rad, Hercules, CA)
2. Pierce 660 nm protein assay and Ionic Detergent Compatibility Reagent (Thermo Scientific Pierce, Rockford, IL)

39.2.2.6 First Dimensional Separation: Isoelectric Focusing

1. 11 cm, pH 3–10 nonlinear (NL) ReadyStrip immobilized pH gradient (IPG) strips, electrode wicks, disposable 11 cm rehydration/equilibration tray with lid, mineral oil, forceps and Bio-Lyte 3/10 ampholyte (Bio-Rad, Hercules, CA)
2. Filter paper/cellulose chromatography paper 35 × 43 cm (Whatman, Piscataway, NJ)

39.2.2.7 Second Dimensional Separation: SDS-PAGE Electrophoresis

1. 10.5–14% gradient precast Criterion Tris-HCl gel and low melt agarose (Bio-Rad, Hercules, CA)
2. Benchmark unstained protein ladder (Invitrogen, Carlsbad, CA)
3. α- iodoacetamide and glycerol, OmiPur 99% grade (EMD, San Diego, CA)
4. SDS and glycine (Thermo Fisher Scientific, Pittsburgh, PA)

39.2.2.8 SYPRO Ruby Staining, Imaging and 2-D Gel Analysis

1. Glacial acetic acid and methanol, HPLC grade (Thermo Fisher Scientific, Pittsburgh, PA)
2. Blot Box, extra large size (Denville, Metuchen, NJ)
3. SYPRO Ruby stain (Bio-Rad, Hercules, CA)

39.2.2.9 Mass Spectrometry

1. Acetonitrile, OmniSolv LC/MS grade (ACN) (EMD, San Diego, CA)
2. Formic acid, 99+% in ampules (Thermo Scientific Pierce, Rockford, IL)
3. Sequencing grade modified trypsin (Promega, Madison, WI)
4. Geloader pipet tips (Eppendorf, Hauppauge, NY)
5. LC/MS grade water (J. T. Baker, Phillipsburg, NJ)
6. Nano Trap column (3.5 cm, C18 100 μm) (CVC MicroTech, Fontana, CA)
7. Pre-packed C18 analytical column (Pico-Frit Biobasic, 75 μm × 10 cm, C18, 5 μm, 300 Å) (New Objective, Woburn, MA)
8. Ammonium bicarbonate and CalMix (Caffeine, the peptide MRFA, SDS and Sodium Taurocholate) (Sigma-Aldrich, St. Louis, MO)

39.2.3 Reagents and Buffers

Technique	Reagent name	Reagent composition	Preparation and storage conditions
Astrocyte Lysis	Cell Rinse Buffer	20 mM Tris, 5 mM EDTA, pH 7.4 and 150 mM NaCl	Filter sterilize and store at 4°C until use
	10X Lysis Buffer Stock	200 mM Tris, 50 mM EDTA, pH 7.4 and 1.5 M NaCl	Filter sterilize and store at 4°C until use
	Lysis Buffer	20 mM Tris, 5 mM EDTA, pH 7.4, 150 mM NaCl, 1% NP-40, 5 mM DTT, 10 mM sodium pyrophosphate, 30 mM *p*-nitrophenyl phosphate and complete protease inhibitors (1 mini tablet/10 ml)	Dilute 10X Lysis Buffer Stock. Filter sterilize. Prepare fresh prior to use. Pre-chill to 4°C
Desalting & albumin depletion	Tris Wash Buffer	10 mM Tris, pH 7.5	Pre-chill to 4°C
Protein precipitation	TCA Solution	100% w/v TCA	Store at 4°C in a brown bottle
	Acetone Solution	100% acetone	Store at −20°C in a brown bottle
Resuspension for IEF	Rehydration Solution	7 M urea, 2 M thiourea, 4% CHAPS, 0.2% Bio-Lytes pH 3–10, 0.0002% bromophenol blue and 0.4% DTT	Add DTT just prior to use
IPG equilibration	Equilibration buffer Buffer I	50 mM Tris-HCl, pH 8.8, 7 M urea, 20% glycerol, 2% SDS and 2% DTT	Add DTT just prior to use
	Equilibration buffer Buffer II	50 mM Tris-HCl, pH 8.8, 7 M urea, 20% glycerol, 2% SDS and 2.5% iodoacetamide	Add iodoacetamide just prior to use
SDS-PAGE electrophoresis	1X and 2X SDS-PAGE Running Buffer	1X = 25 mM Tris, pH 8.3, 192 mM glycine and 0.1% SDS	Make a 10X stock and dilute to 1X and 2X
	Agarose Overlay Solution	0.5% low melt agarose and 0.008% bromophenol blue in 1X SDS-PAGE Running Buffer	
SYPRO Ruby staining	Fixer/Destaining solution	10% methanol with 7% acetic acid	
HPLC	LC Buffer A	2% ACN with 0.1% formic acid	
	LC Buffer B	100% ACN with 0.1% formic acid	

39.3 Methods

39.3.1 Mechanical Trauma and Astrocyte Culture Adaptation for 2-D Gel Analysis

We use highly differentiated, postmitotic astrocytes that express proteins known to be present in astrocytes of mature brain. Primary astrocytes are isolated from four day old, neonatal mice cortices as previously described (de Vellis et al., 2009; Wanner et al., 2008). In vitro maturation of astrocytes is achieved by long-term culture in the presence of adult derived equine serum (Wanner, in press). To best study trauma-specific changes and to enrich for astrocytic proteins, withdrawal of serum from the medium (DMEM/F12) is necessary. This is done stepwise over the course of one week to avoid the serum removal being a stimulus. Also one week prior to stretching and harvesting, the HEPES-buffered DMEM/F12 medium is replaced with DMEM/F12 containing 29 mM sodium bicarbonate. HEPES is an amphoteric chemical that focuses within the pH gradient interfering with IEF (Smith, 2009). This substitution decreased the vertical streaking near the acidic end of the gels, as previously described by Westermeier (2002).

To quantify protein changes of mechanical trauma, it is important that the injury is widespread, reproducible and cellular material is not removed. Hence, we use a pressure controller (CIC II) to apply a defined abrupt nitrogen blast wave onto astrocytes grown on deformable collagenated silastic membranes (Bioflex plates, well 962 μm^2). Mouse astrocytes are stretched and exposed to a 50 ms overpressure of 2.6–2.8 psi. After 30 min, 5 and 24 h from the trauma, CM and whole cell lysates are harvested from 12 stretched and 12 unstretched cultures as described in Sections 39.3.2 and 39.3.3. The typical yield from 12 wells is approximately 6–8 $\times$ 10^6 astrocytes.

39.3.2 Released Protein Preparation: Concentrating Proteins in the Conditioned Medium, Desalting and Albumin Depletion, Desalting and Albumin Depletion

Astrocyte CM is centrifuged to remove insoluble material and then concentrated to 0.10–0.05 of the original volume using a 3,000 MWCO centrifugal concentrator. The concentrated sample is then desalted using a Sephadex G-25 desalting column to remove IEF interfering substances. The desalted CM is subjected to albumin depletion using spin-columns to enrich for astrocyte-specific proteins. Sample preparation of 24 mL CM from 12 astrocyte cultures typically yields approximately 0.6 mg protein after precipitation. This amount is sufficient for performing the protein assay and triplicate 2-D gels.

1. Transfer the CM from 12 wells (24 mL) into a 50 mL conical tube containing 0.02 g DTT and protease inhibitor (2.5 tablets, includes 1 mM EDTA). Gently

shake until the protease inhibitor tablet is dissolved. Keep samples on ice until all are collected.

2. Centrifuge for 5 min at 600×*g*, 4°C to remove any insoluble material. Transfer the supernatant without disturbing the pellet into a pre-chilled Vivaspin concentrator.
3. Concentrate the CM samples by centrifuging at 8,000×*g*, 4°C. After 60 min, check the volume of the sample and continue centrifuging it until the volume is reduced to $\leq$ 1.8 mL.
4. Desalting Column Preparation: Place the desalting column into a 50 mL falcon tube using a column adapter. According to the manufacturer's instructions, equilibrate a Sephadex G-25 desalting column with 5 × 5 mL Tris Wash Buffer, centrifuge for 2 min at 1,000×*g*, 4°C to remove any residual buffer and then transfer the column into a new 50 mL tube.
5. Upon removing the concentrator from the centrifuge, mix the concentrated CM sample by pipetting up and down. Transfer it to the top of a washed desalting column.
6. Centrifuge the column for 2 min at 1,000×*g*, 4°C.
7. Albumin Depletion Column Preparation: Hydrate the swellgel disks with MilliQ equivalent water according to the manufacturer's instructions.
8. Transfer 500 μL of the desalted eluate onto an albumin removal column, mix by gently vortexing and incubate on ice for 2 min to allow the albumin to bind to the resin. Centrifuge for 1 min at 12,000×*g*, 4°C.
9. Transfer the albumin depleted eluate into a new tube. Repeat step 8 for the remaining desalted CM sample (3–4X), using the same albumin removal column and combining all the eluates.
10. After the entire CM sample has been processed, wash the resin by applying 150 μL of Tris Wash Buffer, mix gently and incubate on ice for 1 min. Centrifuge for 1 min at 12,000×*g*, 4°C, and transfer the wash to the tube containing the eluates from the same sample.
11. Determine the final volume for each sample and proceed with Section 39.3.4.

39.3.3 Cellular Protein Preparation: Astrocyte Lysis

Whole cell lysates are obtained by homogenizing the cells in a solution containing 1% NP-40 detergent, DTT and protease and phosphatase inhibitors. To minimize protein oxidation, avoid creating bubbles during homogenization. The yield after precipitation from 12 wells of cultures containing approximately 6–8 × 10^6 astrocytes is 1.0–1.6 mg protein.

1. After the CM has been collected, briefly rinse cells from one well with 2 mL Cell Rinse Buffer. Remove the Cell Rinse Buffer from each well and add 200 μL Lysis Buffer. With a cell scraper, remove cells off the culture bottom. Tilt the plate on an angle and transfer the cells to a pre-labeled, round-bottom tube resting on ice. Twelve wells are combined into one tube.

2. Rinse the wells from one plate with an additional 600 μL Lysis Buffer, using the same 600 μL to rinse all 6 wells. Add the rinse to the tube with the lysate.
3. Homogenize cells by pipetting up and down 3X with P1000 and then slowly, pass the sample 5X through a 3 mL syringe with a 27 ½ gauge needle. Try to avoid creating bubbles and keep the samples chilled. Total cell lysis time from scraping to homogenization is about 30 min/sample.
4. Centrifuge for 10 min at 600×*g*, 4°C to remove any insoluble material.
5. Determine the volume of the supernatant. If the volume is less than 1.8 mL, transfer the supernatant into 2 mL microcentrifuge tube. If the volume is larger, transfer it to polypropylene Oak Ridge tubes. Proceed with Section 39.3.4.

39.3.4 Protein Precipitation and Resuspension in Rehydration Solution

Nonprotein-based ions, such as salts, nucleic acid and detergents must be removed from the sample or they will interfere with the isoelectric focusing causing horizontal streaking and empty regions. Ideally, proteins should be the only ions present in the sample. One method of removing interfering substances is protein precipitation which selectively separates the proteins from the contaminants and enables concentration of dilute samples. The disadvantage of precipitation is that 50% or more of the original protein amount can be lost and not all proteins can be readily resuspended after precipitation. The method described below uses TCA to precipitate the proteins followed by water then acetone washes to remove the residual TCA. The pellet is then resuspended in IEF Rehydration Solution.

1. Add 1/10 of the volume of the TCA Solution to the released and soluble protein samples. Mix by inverting several times. Incubate on ice for 15 min.
2. Centrifuge for 10 min at 17,400×*g*, 4°C.
3. Gently discard the supernatant. Centrifuge briefly again to bring any remaining liquid to the bottom of the tube.
4. Remove the remaining supernatant, making sure no visible liquid is left behind.
5. Add 100 μL of MilliQ equivalent water to each microcentrifuge tube or use 1 mL for each large centrifuge tube. Disperse the pellet using a glass rod or metal spatula. Note the pellet will not actually dissolve in the water.
6. Add 1 mL of Acetone Solution to each microcentrifuge tube or use 10 mL for each large centrifuge tube.
7. Vortex each tube for 30 s.
8. Incubate the tubes at –20°C. Over the course of the first hour of precipitation, temporally remove the tubes from the freezer three times and vortex them for 30 s. Continue incubation at –20°C overnight. (The overnight incubation can be skipped, but the yields will be lower.)

9. Centrifuge for 10 min at 17,400×g, 4°C. Remove the supernatant and then spin briefly a second time to remove any residual supernatant.
10. Air dry the pellet for 5 min, RT. Do not over dry the pellet or it will be difficult to resuspend it later.
11. Resuspend the pellet in 500–900 μL of Rehydration Solution by pipetting. Do not vortex because it could potentially oxidize the proteins (Westermeier, 2006 and 2010 (personal communication)).
12. Leave the sample resting on the bench top for 10–15 min. Pipette the sample up and down a second time to further break up the pellet. Freeze the samples at –80°C. The freezing helps the cell lysates to get into solution.

39.3.5 Protein Concentration Measurement

The Pierce 660 nm protein assay has a linear range of detection from 25 μg/mL up to 2 mg/mL and is compatible with the urea, thiourea, CHAPS and DTT concentrations used in the Rehydration Solution.

1. For the protein standard curve, prepare a series of 0 (blank), 0.125, 0.25, 0.5, 0.75, 1 and 2 mg/mL BSA in Rehydration Solution. Aliquot the BSA series into 35 μL and store at –80°C until use.
2. Thaw sample. Resuspend the sample by gently pipetting up and down.
3. To remove any unsolibilized material, centrifuge for 5 min at 12,000×g, 4°C. Transfer the supernatant to another 1.5 mL microcentrifuge tube.
4. Thaw BSA standard curve aliquots. Measure the protein concentration in triplicates for 35 μL of each sample according to the instruction for the Pierce 660 nm Protein Assay. The Ionic Detergent Compatibility Reagent needs to be added to the Protein Assay Reagent.
5. Aliquot your sample so that you only have to freeze-thaw once. Store at –80°C until use.

39.3.6 First Dimensional Separation: Isoelectric Focusing (IEF)

Proteins are amphoteric and depending on the pH of their surroundings, they can have a positive, a negative or zero net charge (Berkelman and Stenstedt, 1998). The first dimensional separation uses IEF to separate proteins based on their p*I*, the specific pH where the net charges of all the amino acids and modifications is zero. The immobilized pH gradient (IPG) strips come in a number different pH ranges, including a broad range pH 3–11 nonlinear (NL) IPG strip. The protocol described below uses active rehydration loading of the sample, applying a low voltage to the IPG strips as the sample is being loaded. This rehydration method is beneficial for some samples since it drives the salts to the electrodes, removes the proteases from the proteins and helps large proteins to enter the gel (Friedman et al., 2009).

39.3.6.1 IEF Sample Preparation

1. Dilute the sample to 0.5 μg/μL with Rehydration Solution and prepare 225 μL per IPG strip.
2. Centrifuge the diluted samples for 5 min at 16,000×g, 4°C to pellet any unsolubilized material.

39.3.6.2 Active Rehydration and Isoelectric Focusing

(1) Equilibrate the 11 cm pH 3–10 NL IPG strips to RT by laying them in the rehydration tray lid.
(2) Pipet 200 μL (100 μg) of the sample onto the bottom edge of the IEF tray. Each sample is prepared in triplicates.
(3) Use forceps to peel back the cover sheet from the IPG strip and place the acidic side (+) of the strip, gel side down, onto the sample, then slowly lower the basic side (-) until it also lays onto the sample. Be careful not to get the sample on the top side of the strip. Remove any air bubbles trapped beneath the strip.
(4) Cover the focusing tray with a disposable rehydration tray lid, and leave it on the bench top for ≈ 1 h.
(5) Slowly overlay each strip with ≈1.8 mL mineral oil also fill any remaining channels with oil (this is only necessary if the tray is old as discussed in Section 39.5 step 8) and then place the focusing tray lid on top of the IEF tray.
(6) Perform active rehydration overnight (15–16 h) using the Rehydration Program, selecting Active at 50 V and the default cell temperature of 20°C.
(7) The next day, place a wick using forceps at both ends of another, clean IEF tray. The wicks should touch the ends of the channel. Pipette 10 μL of MilliQ equivalent water onto the wicks.
(8) Stop the active rehydration program and remove an IPG strip from the tray using forceps.
(9) Place a piece of dry filter paper onto the bench top and then transfer the strip with gel side up onto that filter paper. The plastic side of the IPG strip needs to touch the filter paper. With a kimwipe wetted with MilliQ equivalent water, blot the excess mineral oil from the gel side of the strip. Repeat the blotting (see Section 39.5 step 9).
(10) Transfer the IPG strip, gel side down, to a channel of the tray prepared in step 7 with the + end of the strip at the + end of the tray.
(11) Slowly overlay each strip with ≈ 1.8 mL mineral oil, including empty channels (see Section 39.5 step 8). Make sure the wicks and strip make a good contact by using forceps to gently press down on both ends of the strip until air bubbles are no longer visible rising from the wick.
(12) Place the lid onto the focusing tray and then put the tray into the IEF Cell.
(13) Run the following isoelectric focusing program using a cell temperature of 20°C and a maximum current of 30 μA/strip: 100 V (linear) for 4 h, 250 V (linear) for 2 h, 4,000 V (linear) for 5 h and 4,000 V (rapid) for 85,000 VHr.

(14) After ≈ 8 h, replace the electrode wicks with new pre-wetted ones. The total IEF runs for ≈ 32 h.
(15) Proceed directly with the Section 39.3.7.1 step 2 or transfer the IPG strips into a disposable rehydration tray with the gel side facing up, wrap the tray with plastic wrap in order to prevent frost from entering the tray and store at –80°C until proceeding with the second dimension.

39.3.7 Second Dimensional Separation: SDS-PAGE Electrophoresis

During this step, the IPG strips are first equilibrated to the components of the SDS-PAGE Running Buffer. The equilibration buffer also includes urea and glycerol to diminish electroendosmotic effects that inhibit proteins from transfer into the SDS-PAGE gel (Sanchez et al., 1997) and DTT to reduce the disulfide bonds. The second equilibration buffer contains iodoacetamide instead of DTT to alkylate the sulfhydryl groups of the cysteine residues with carbamidomethyl groups in order to prevent streaking during the second dimension (Friedman et al., 2009). Subsequently, the IPG strips are placed on top of a SDS-PAGE gel. The separation is performed at high pH to minimize protein aggregation, and a low electrophoresis voltage is used to increase the sharpness of the spots.

39.3.7.1 IPG Equilibration

1. If appropriate, remove the IPG strips from the freezer and allow them to equilibrate to RT. Make sure the strips look clear and not opaque otherwise you will destroy the strips (10–15 min). Do not thaw the strips longer than 15–20 min since the diffusion of the proteins can reduce the sharpness of the protein spots.
2. Remove one of the IPG strips from the tray and blot the excess mineral oil 1X as described in step 9 of Section 39.3.6.2.
3. Transfer the strip into a channel within the disposable rehydration tray, gel side facing up.
4. Add 4 mL of Equilibration Buffer I to each channel containing a strip and allow the IPG strips to gently shake, completely submerged for 20 min on a rocking platform. The rehydration tray should be placed at a 20° angle in order to get better mixing.
5. After the 20 min incubation, carefully discard Buffer I from the rehydration tray by slowly pouring the liquid from the square side (not the inclined side) of the tray. Turn the tray upside down and gently tap it against a large kimwipe several times to remove excess buffer.
6. Add 4 mL of Equilibration Buffer II to each strip and repeat step 5.

39.3.7.2 SDS-PAGE Electrophoresis

1. Spot 3 μL of the Protein Ladder onto small pieces of filter paper (2 × 6 mm). Allow them to air dry at RT. They can be stored at –20°C in a 1.5 mL microcentrifuge tube until use.

2. Pre-chill the buffer in the Dodeca Criterion apparatus by turning on the circulating bath and setting it to 10°C. This will keep the Criterion buffer tank at 19°C during the electrophoresis.
3. Prepare the precast 10.5–14% gradient Criterion gels by removing them from the packaging and rinsing both sides of the plates with ddH_2O. Remove the combs and rinse the wells by squirting MilliQ equivalent water. Gently dry the wells from each gel by sticking a piece of filter paper in the well to absorb all the access water. Avoid tearing or pushing down the top of the gel.
4. Prior to applying the strips to the top of the SDS-PAGE gel, fill a 50 mL graduated cylinder with 1X SDS-PAGE Running Buffer and remove any bubbles from the surface using a Pasteur pipette.
5. Ten minutes before use, melt the Agarose Overlay Solution in the microwave.
6. After Section 39.3.7.1 step 6 is completed, pipet the Agarose Overlay Solution into the wells of one of the precast gels, avoiding introduction of air bubbles.
7. Remove an IPG strip from the tray and dip briefly (5 s) into the 50 mL graduate cylinder with the SDS-PAGE Running Buffer. When taking the strip out from the cylinder, drag the strip's plastic side against the wall of the cylinder to remove any bubbles.
8. Lay the strip with plastic side of the IPG strip, against the backside plate. The strip should be positioned with the (+) end against the well for molecular weight standards. Push the strip plastic backing (not the gel matrix) to work the strip to the bottom of the well.
9. Remove any air bubbles between the strip and the top of the SDS-PAGE gel by gently pushing on the plastic backing of the strip. If that doesn't work, remove the strip, break any bubbles on the Agarose Overlay or move them to the far edge of the cassette, and then place the strip back into the well.
10. Place the strip with the Protein Ladder into the small well.
11. Keep the gels vertical for a minimum of 10 min so that the agarose solidifies.
12. Slide the gel into the Criterion Dodeca Cell and fill the upper buffer chamber 7/8 full with 2X SDS-PAGE Running Buffer.
13. Electrophorese the gels at a constant 40 V for 10 min followed by a constant 120 V until the bromophenol blue front reached the bottom of the gel (2–2.5 h).

39.3.8 SYPRO Ruby Staining, Imaging and 2-D Gel Analysis

This protocol uses SYPRO Ruby stain to visualize the proteins. SYPRO Ruby is a fluorescent endpoint stain that has a detection limit of 1–10 ng protein, stains a broad range of proteins and has a dynamic range over 3 orders of magnitude. This stain has two excitation peaks (∼280 nm and ∼450 nm), allowing it to be visualized with a blue-light transilluminator, fluorescent laser imager and a UV light source such as the ones found on automated spot cutters. Spot intensity comparisons across all gels and conditions require keeping the staining, fixing and wash steps exactly the same. Finally, a 2-D gel analysis software such as Progenesis SameSpots is used to detect a spot and align it across all corresponding images. Protein level changes upon treatment are determined by background-corrected spot intensity measurements.

1. Gently break open the Criterion cassette and remove the IPG strip with the gel knife in order to prevent the strip from cracking the gel.
2. Place the gel in a Blot Box with MilliQ equivalent water, gently shake and then dump off the water. Repeat the rinse a second time.
3. Fill the Blot Box with 110 mL Fixer/Destain Solution. Fix the gel with shaking for a minimum of 1 h.
4. Rinse the gel in MilliQ equivalent water.
5. Dilute fresh SYPRO Ruby Stain 50:50 with previously used SYPRO Ruby stain to reduce the background and speckling. Using the packaging tray from the precast gel as the staining tray, stain the gel overnight with shaking in 85 mL of diluted SYPRO Ruby stain. Wrap the tray in aluminum foil to protect the fluorescent dye from light.
6. Transfer the gel into a Blot Box containing MilliQ equivalent water. Rinse the gel twice for 10 min with MilliQ equivalent water.
7. Destain the gel in 110 mL Fixer/Destain Solution for $\approx$ 7 h on a rocking platform.
8. Rinse the gel twice in MilliQ equivalent water just prior to imaging.
9. Scan the gel at a 100 μm resolution using a 532 nm excitation laser and a 555 nm LP emission filter of a fluorescence laser scanner. Reduce the photomultiplier tube setting if there any saturated pixels (except areas that will be excluded from the analysis). Save the image as a 16 bit tif image and use the image unedited by any other program for quantitative 2-D gel analysis.
10. To reduce the background further, replace the Fixer/Destain Solution with another 110 mL and continue destaining the gel overnight with shaking. Faint spots may destain. Re-scan the gel after rinsing it twice in water.
11. Analyze the spots on the gel images for quantitative difference using a 2-D gel analysis software such as the Progenesis SameSpots. Relative protein changes of spots showing a least a 1.8-fold difference with a probability of $P < 0.05$ are defined as statistically significant (multiple comparison ANOVA).

39.3.9 Mass Spectrometry

Spots which showed statistically significant differences between mechanically traumatized and control astrocyte cultures are excised from the gel. The proteins inside the gel plugs are digested with trypsin and the peptides are then eluted and dried with a SpeedVac. The resuspended peptides are separated based on their hydrophobicity on a reversed-phase high performance liquid chromatography (RP-HPLC) column. These separated peptides are directly injected into the mass spectrometer and fragmented to derive sequence information and protein identification. The mass spectrometer used in this example is a hybrid linear ion trap-Orbitrap

(LTQ-Orbitrap) which has high resolution (a maximum of 100,000 full width at half its maximum height) and mass accuracy (<5 ppm).

39.3.9.1 In-Gel Tryptic Digestion and Liquid Chromatography Separation

1. Excise any spots showing interesting changes between the mechanically traumatized and control astrocyte cultures. In this example, a robotic spot cutter was used.
2. Perform in-gel tryptic digestion on gel plugs as described by Shevchenko et al. (Shevchenko et al., 1996). All reagents are made with LC/MS grade water and in borosilicate glass scintillation vials. The digestion was performed in a laminar flow to minimize keratin contamination using dedicated equipment, supplies and reagents for keratin-free work.
3. Reconstitute the dried down peptides by adding 20 μL of LC Buffer A, vortexing them for 30 s and sonicating the tubes in a sonicator water bath for 15 min.
4. Centrifuge the samples for 10 min at 17,400×g to pellet any solid material. If any pellet material is visible, transfer the supernatant to a new tube using a very fine gel loading pipette tip. This is important, since even minute amounts of solid material will plug the tubing and the column from the nano HPLC system.
5. Load 10 μL (≈ 100–200 fmol) of the reconstituted peptides onto a C18 Nano Trap pre-column at a flow rate of 5 μL/min. Wash the column for 5 min with 2% acetonitrile containing 0.1% formic acid.
6. Separate the peptides are on a pre-packed C18 analytical RP-HPLC column using a flow rate of 300 nl/min and the following gradient: 1–37% LC Buffer B for 30 min, 37–85% LC Buffer B for 20 min and maintained at 85% LC Buffer B for 5 min. Re-equilibrate the column with 99% LC Buffer A for 15 min at the end of each run.

39.3.9.2 MS/MS Analysis

1. Using a Pico Tip (Silica Tip) emitter from New Objective (i.d. 30 μm), set the ESI voltage to 1.8 kV.
2. Acquire the MS/MS spectra in "data-dependent" mode (Full MS scan followed by MS/MS of the most intense peptide ions). This enables the acquisition of mass spectra in a non-redundant manner so that as many peptides as possible are surveyed from a complex mixture. Helium was employed as collision gas. Collision energies for maximum fragmentation were automatically calculated using empirical parameters based on the charge and mass-to-charge ratio of the peptide.
3. Set up an experimental method during which one full range MS scan (300–2,000 m/z) is acquired in the Orbitrap ($R = 100{,}000$) followed by the acquisition of seven data-dependent MS/MS scans in the ion-trap.
4. Calibrate the m/z scale of the Orbitrap weekly using CalMix, following the instructions provided by the manufacturer.

39.3.10 Peptide Sequence Identification Using MASCOT

The spectra are searched against the annotated genome using database search engine such as MASCOT to identify the protein present within a particular spot. It is important to update the sequence database regularly to increase the ability to characterize the protein of interest. It is not unusual for multiple proteins to be found in one spot.

1. Using MASCOT Daemon client, the raw data file from the LTQ-Orbitrap is searched against the International Protein Index (IPI) mouse database (version 3.57, 56,091 entries). The following search parameters are used: Enzyme: trypsin; Allow up to 2 missed cleavages; Variable modifications: carbamidomethylation (+57 Da), methionine oxidation (+16 Da); Peptide tolerance of ±10 ppm; and MS/MS tolerance of ±0.5 Da.
2. The following criteria are used for the identification of a protein within a 2-D gel spot. The identified protein must have a Mascot Score with a greater than 95% confidence and a minimum of two unique peptides with an ion score greater than 34. The peptide assignments for all the significant identification were also manually validated.

39.4 Results

With this targeted proteomic approach, we were able to determine a set of specific protein changes induced by mechanical trauma in astrocytes. Spot patterns of triplicate, SYPRO Ruby stained 2-D gels (Fig. 39.1) from stretched and control mouse astrocyte samples were compared. From the CM gels showing over 500 repeatedly detected spots, 107 significantly different spots (see step 11 of Section 39.3.8) were selected and so far 22 darkly stained spots were selected from the whole cell lysate gels. Most of the spots from the CM had up to 4-fold increase in protein levels upon trauma while a few decreased up to 2-fold (average P = 0.004, multiple comparison ANOVA). Selected spots were in-gel digested with trypsin, and the proteins were identified from these spots using LC-MS/MS. The majority of the identified spots contained only one protein while some of the spots had two or more proteins identified. With this protocol, we routinely achieved ion scores of greater than 34 with a confidence of over 95%. Multiple MS/MS queries matched to the protein identified with the sequence coverage ranging from 20 to 65% and the Standard Mascot Score ranging from 192 to 1,179 with the majority falling into the 400–600 range. Shown in Fig. 39.2 is an example of a typical spectrum obtained using the LTQ-XL Orbitrap mass spectrometer.

Trauma regulated proteins in CM include murine serum proteins, CSF proteins (Hanrieder et al., 2009; Liu et al., 2006; Sjödin et al., 2010), proteins known to be secreted by astrocytes (Keene et al., 2009) and typical astrocyte marker proteins (Wanner et al., manuscript in preparation). Comparing the proteins identified from CM to that of whole cell lysates indicate that the mechanical trauma also led to the uncontrolled release of proteins from transiently leaky cell membranes

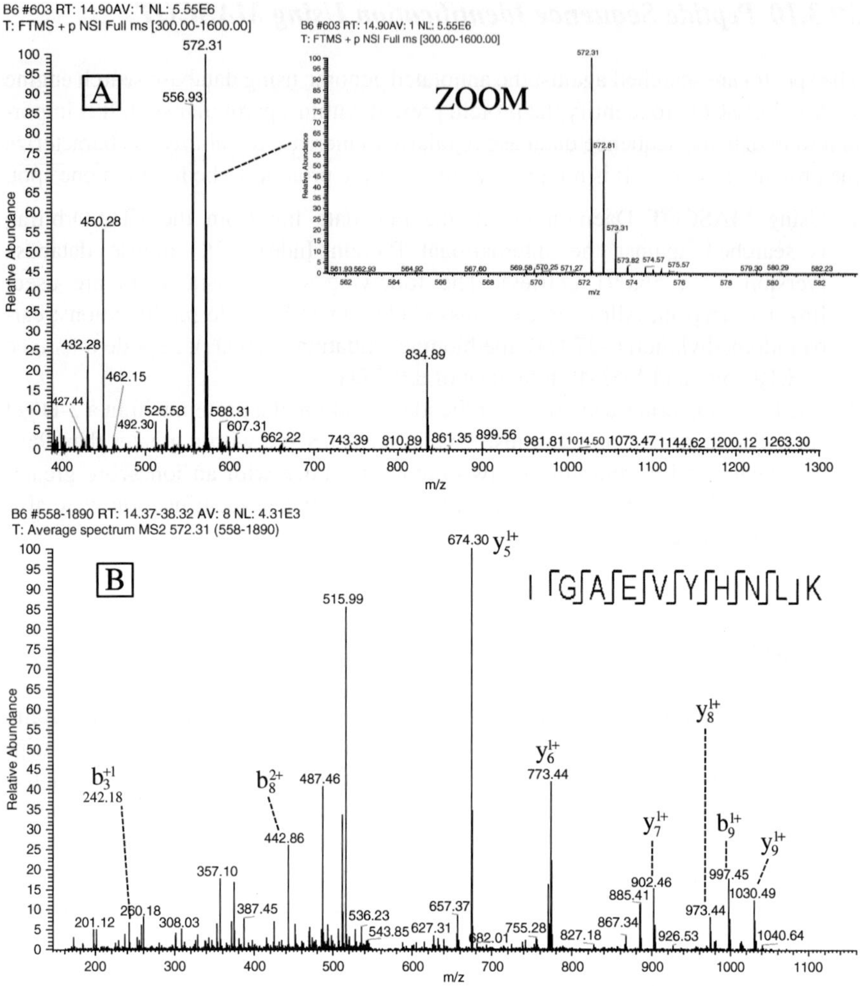

Fig. 39.2 LC-MS/MS analysis using the LTQ-XL Orbitrap mass spectrometer and the Mascot search engine. (**a**) Full scan MS1 mass spectrum from the Orbitrap showing the peptide precursor ion (572.31 *m/z*) (**b**) MS/MS fragment spectrum matched peptide IGAEVYHNLK (*m/z* 572.31). The MS/MS spectrum was searched against the IPI mouse database using Mascot. The insert shows the identified peptide sequence from α-enolase

(Amruthesh et al., 1993; Lamb et al., 1997) or dying cells (Ellis et al., 1995). The most intense spots of the whole cell lysate showed an increase level when stretched and were identified as metabolic proteins, reflecting the increased metabolic demand and de novo protein synthesis after injuring cultured astrocytes. This proteomic approach provides reproducible quantitative data and protein identification with high confidence. The identification of many trauma-regulated astrocytic proteins using an in vitro injury model is potentially useful for future neurotrauma biomarker development.

39.5 Notes

1. To obtain reproducible results from 2-D gel electrophoresis, everything needs to be done exactly the same way. This includes culture growth conditions and treatments, harvest times, reagents and volumes as well as the IEF and SDS-PAGE gel conditions. In order to achieve comparable spot intensities for the 2-D gel analysis, the same fixing, staining and destaining times and volumes should be utilized, and the same scanning settings applied to all gels.
2. Always use the highest grade of chemicals. All reagents for the sample preparation and 2-D gel should be made with MilliQ equivalent water.
3. During all steps, use powder-free nitrile gloves (no aloe). To minimize keratin and other protein contamination, wear a lab coat, do everything on as clean of surface as possible, use disposable lab supplies whenever possible and use a brand new staining tray dedicated to proteomic work.
4. Perform all sample preparation steps on ice, use only pre-chilled reagents (4°C), add proteases and phosphatase inhibitors and avoid vortexing samples to minimize proteolysis and oxidation since such protein modifications will appear on the 2-D gels as separate spots.
5. During centrifugation, be sure to position the tube in the same way each time so that the pellet is always in the same place. Always place the pipette tip away from the pellet to remove the supernatant.
6. Label the basic edge of the IPG strip with either a black or blue industrial strength sharpie. Other colors and brands will bleed off.
7. During the 1 h sample loading of the IPG strip, use the lid from the disposable rehydration tray to cover the IEF tray with the strips since the lid of the focusing tray puts pressure on the strips and may cause unequal loading.
8. As the IEF tray begins to age, samples can bleed into the neighboring empty channels along the electrode. If this happens, fill all wells of the IEF tray with mineral oil to prevent this from happening.
9. After active rehydration but before the IEF is started, the strip is blotted and transferred into a clean tray to remove any sample which was not absorbed by the strip. This step reduces streaking since the solution remaining outside of the gel can cause a parallel current where the proteins will not be focused (Castle et al.).
10. Pre-stained molecular weight protein standards are not visualized with SYPRO Ruby stain so use unstained standards.
11. SYPRO Ruby is a light sensitive fluorescence dye. Therefore, minimize light exposure after the staining and all subsequent steps by wrapping the staining container in aluminum foil or use a black Blot Box. Careful that the aluminum foil does not come in direct contact with the Fixer/Destain Solution or it will corrode onto your gel. The gels can be fixed, stained and destained on either a rocking platform or orbital shaker. However, best results are obtained with a rocking platform.

Acknowledgements We kindly acknowledge Dr. Puneet Souda for his help with the LC-MS/MS work and Reiner Westermeier for his expertise and in-depth discussions on running 2-D gels. The 2-D gel and LC-MS/MS work was performed in the W. M. Keck Proteomic Facility at the UCLA Molecular Instrumentation Center which was established with a grant from the W. M. Keck Foundation. This project is supported by the Craig Neilsen Foundation (for IBW, grant # 82776).

References

Allen, N.J., and Barres, B.A. (2009). Neuroscience: Glia – more than just brain glue. Nature *457*, 675–677.

Amruthesh, S.C., Boerschel, M.F., McKinney, J.S., Willoughby, K.A., and Ellis, E.F. (1993). Metabolism of arachidonic acid to epoxyeicosatrienoic acids, hydroxyeicosatetraenoic acids, and prostaglandins in cultured rat hippocampal astrocytes. J Neurochem *61*, 150–159.

Berkelman, T., and Stenstedt, T. (1998). 2-D Electrophoresis Using Immobilized pH Gradients: Principles and Methods, Edition AB. In Handbooks from Amersham Biosciences (Piscataway, NJ).

Bowman, C.L., Ding, J.P., Sachs, F., and Sokabe, M. (1992). Mechanotransducing ion channels in astrocytes. Brain Res *584*, 272–286.

Castle, L., Dale, E., Habers, A., Sadownick, B., Strong, W., Whitman, C., and Zhu, M. First Dimension Separation Methods. In 2-D Electrophoresis for Proteomics: A Methods and Product Manual, D. Garfin, and L. Heerdt, eds. Bulletin 2651 US/EG Rev B, (Hercules, CA, Bio-Rad Laboratories).

de Vellis, J., Ghiani, C., Wanner, I., and Cole, R. (2009). Preparation of Normal and Reactive Astrocyte Cultures. In Protocols for Neural Cell Culture, L. Doering, ed. (Totowa, NJ, Humana Press Inc. Springer), pp. 193–216.

Di, X., Goforth, P.B., Bullock, R., Ellis, E., and Satin, L. (2000). Mechanical injury alters volume activated ion channels in cortical astrocytes. Acta Neurochir Suppl *76*, 379–383.

Eddleston, M., and Mucke, L. (1993). Molecular profile of reactive astrocytes – implications for their role in neurologic disease. Neuroscience *54*, 15–36.

Ellis, E.F., McKinney, J.S., Willoughby, K.A., Liang, S., and Povlishock, J.T. (1995). A new model for rapid stretch-induced injury of cells in culture: Characterization of the model using astrocytes. J Neurotrauma *12*, 325–339.

Friedman, D.B., Hoving, S., and Westermeier, R. (2009). Isoelectric focusing and two-dimensional gel electrophoresis. Methods Enzymol *463*, 515–540.

Hanrieder, J., Wetterhall, M., Enblad, P., Hillered, L., and Bergquist, J. (2009). Temporally resolved differential proteomic analysis of human ventricular CSF for monitoring traumatic brain injury biomarker candidates. J Neurosci Methods *177*, 469–478.

Hergenroeder, G.W., Redell, J.B., Moore, A.N., and Dash, P.K. (2008). Biomarkers in the clinical diagnosis and management of traumatic brain injury. Mol Diagn Ther *12*, 345–358.

Keene, S.D., Greco, T.M., Parastatidis, I., Lee, S.H., Hughes, E.G., Balice-Gordon, R.J., Speicher, and D W, Ischiropoulos, H. (2009). Mass spectrometric and computational analysis of cytokine-induced alterations in the astrocyte secretome. Proteomics *9*, 768–782.

Kochanek, P.M., Berger, R.P., Bayir, H., Wagner, A.K., Jenkins, L.W., and Clark, R.S. (2008). Biomarkers of primary and evolving damage in traumatic and ischemic brain injury: Diagnosis, prognosis, probing mechanisms, and therapeutic decision making. Curr Opin Crit Care *14*, 135–141.

Lakshmanan, R., Loo, J.A., Drake, T., Leblanc, J., Ytterberg, A.J., McArthur, D.L., Etchepare, M., and Vespa, P.M. (2010). Metabolic crisis after traumatic brain injury is associated with a novel microdialysis proteome. Neurocrit Care *12*, 324–836.

Lamb, R.G., Harper, C.C., McKinney, J.S., Rzigalinski, B.A., and Ellis, E.F. (1997). Alterations in phosphatidylcholine metabolism of stretch-injured cultured rat astrocytes. J Neurochem *68*, 1904–1910.

Liu, T., Qian, W.J., Gritsenko, M.A., Xiao, W., Moldawer, L.L., Kaushal, A., Monroe, M.E., Varnum, S.M., Moore, R.J., Purvine, S.O., *et al.* (2006). High dynamic range characterization of the trauma patient plasma proteome. Mol Cell Proteomic *5*, 1899–1913.

Montgomery, D.L. (1994). Astrocytes: Form, functions, and roles in disease. Vet Pathol *31*, 145–167.

Neary, J.T., Kang, Y., Willoughby, K.A., and Ellis, E.F. (2003). Activation of extracellular signal-regulated kinase by stretch-induced injury in astrocytes involves extracellular ATP and P2 purinergic receptors. J Neurosci *23,* 2348–2356.

O'Farrell, P.H. (1975). High resolution two-dimensional electrophoresis of proteins. J Biol Chem *250*, 4007–4021.

Ostrow, L.W., and Sachs, F. (2005). Mechanosensation and endothelin in astrocytes – hypothetical roles in CNS pathophysiology. Brain Res Brain Res Rev *48*, 488–508.

Reier, P.J., and Houle, J.D. (1988). The glial scar: Its bearing on axonal elongation and transplantation approaches to CNS repair. Adv Neurol *47*, 87–138.

Ridet, J.L., Malhotra, S.K., Privat, A., and Gage, F.H. (1997). Reactive astrocytes: Cellular and molecular cues to biological function. Trends Neurosci *20*, 570–577.

Sanchez, J.C., Rouge, V., Pisteur, M., Ravier, F., Tonella, L., Moosmayer, M., Wilkins, M.R., and Hochstrasser, D.F. (1997). Improved and simplified in-gel sample application using reswelling of dry immobilized pH gradients. Electrophoresis *18*, 324–327.

Sandvig, A., Berry, M., Barrett, L.B., Butt, A., and Logan, A. (2004). Myelin-, reactive glia-, and scar-derived CNS axon growth inhibitors: Expression, receptor signaling, and correlation with axon regeneration. Glia *46*, 225–251.

Shevchenko, A., Wilm, M., Vorm, O., and Mann, M. (1996). Mass spectrometric sequencing of proteins silver-stained polyacrylamide gels. Anal Chem *68*, 850–858.

Sjödin, M.O., Bergquist, J., and Wetterhall, M. (2010). Mining ventricular cerebrospinal fluid from patients with traumatic brain injury using hexapeptide ligand libraries to search for trauma biomarkers. J Chromatogr B Anal Technol Biomed Life Sci *878*, 2003–2012.

Smith, R. (2009). Two-Dimensional Electrophoresis: An Overview. In Methods in Molecular Biology, D. Sheehan, and R. Tyther, eds. (Totowa, NJ, Humana Press), pp. 3–17.

Sofroniew, M.V. (2005). Reactive astrocytes in neural repair and protection. Neuroscientist *11*, 400–407.

Wanner, I.B., Deik, A., Torres, M., Rosendahl, A., Neary, J.T., Lemmon, V.P., and Bixby, J.L. (2008). A new in vitro model of the glial scar inhibits axon growth. Glia *56*(15), 1691–1709. PubMed PMID: 18618667.

Westermeier, R., and Naven, T. (2002). Trouble Shooting. In Proteomics in Practice. (Weinheim, Germany, Wiley-VCH GmbH), pp. 283–293.

Westermeier, R. (2006). Critical Aspects for Sample Preparation in Proteome Analysis. GE Healthcare Seminar at Human Proteome Organisation Fifth Annual World Congress (Long Beach, CA).

Part XIII
Sample Preparation Techniques in Metabolomics and Drug Discovery

Chapter 40
Mass Spectrometry-Based Metabolomics. Sample Preparation, Data Analysis, and Related Analytical Approaches

Oliver A.H. Jones, Lee D. Roberts, and Mahon L. Maguire

Abstract Metabolomics refers to the study of small-molecule metabolites (e.g. fats, sugars, nuclear acids) within biological samples such as cells, tissues, biofluids or even whole organisms. It has been found to be applicable to a wide range of fields, including the study of gene function, toxicology, plant sciences, environmental analysis, cancer, clinical diagnostics, nutrition and the discrimination of organism genotypes to name but a few. The approach combines high-throughput sample analysis with computer-assisted multivariate pattern-recognition techniques. A major challenge in metabolomics is to address the extremely diverse and complex nature of the subject matter and both past and future progress in the field depends in large part on the use and evolution of analytical techniques and instrumentation, especially mass spectrometry. This chapter therefore focuses on outlining and discussing current thinking behind the most commonly used analytical methodologies as well as associated multivariate data processing techniques.

Keywords Chromatography · metabolomics/metabonomics · Mass spectrometry · Multivariate statistics

40.1 Introduction

For many years the central molecular biological dogma has been that information flows directionally from genomic DNA through mRNA transcripts, to protein. However, each tier of organization is now known to depend on the other. In addition, the environment has a vital impact on gene expression and concentrations of mRNA transcripts, proteins, and even metabolites can all affect the genome (Fig. 40.1). This has lead to an explosion of research in the omic sciences in recent years. This

O.A.H. Jones (✉)
School of Engineering and Computing Sciences, University of Durham, Christopherson Building, Durham DH1 3LE, UK
e-mail: o.jones@gmail.com

A.R. Ivanov, A.V. Lazarev (eds.), *Sample Preparation in Biological Mass Spectrometry*, DOI 10.1007/978-94-007-0828-0_40,

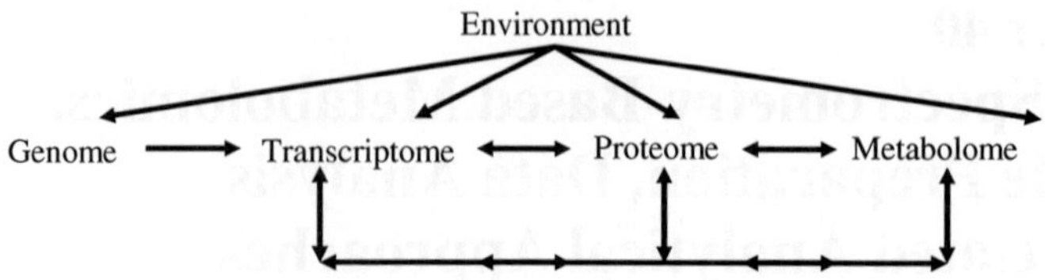

Fig. 40.1 Biological organisation of the '-omes

chapter will focus on the application of mass spectrometry to one of the newest; metabolomics.

Metabolomics may be defined as the analysis of thousands of the small molecules (metabolites), such as sugars, organic acids, amino acids, and nucleotides that are the products of cellular metabolism. The full metabolite complement (metabolome) of a cell, tissue, or indeed organism, can be used to give a "snapshot" of the underlying biochemistry. As in related techniques, such as genomics and proteomics, the metabolome is context dependent and will change in response to external factors such as disease or exposure to a toxin (Viant, 2006). The advantages of studying metabolites include cost-effectiveness and rapid measurements. This enables large numbers of samples to be processed quickly, thereby providing a high-throughput analytical tool with the potential for high statistical power. In addition, unlike many genes and proteins, metabolites are often conserved across species and so the detection methods and equipment used in one organism can usually be applied to another without the need for recalibration. This versatility means that metabolomics can be applied to studies in a wide variety of disciplines, including (but not limited to); drug toxicity and gene function (Nicholson et al., 2002), nutrition (Whitfield et al., 2004), microbiology (Bundy et al., 2005), cancer research (Denkert et al., 2006), pharmacology (Lindon et al., 2006) plant sciences (Schauer and Fernie, 2006) and environmental studies (Jones et al., 2008). A summary of the workflow for standard metabolomics based studies is shown in Fig. 40.2.

The concept of metabolic analysis is not new but metabolomics as a distinct field dates from around the mid-nineties (Robosky et al., 2002). One aspect of the technique which distinguishes it from previous metabolically based studies is the attempt to measure all metabolites simultaneously, often termed a "global" approach. When

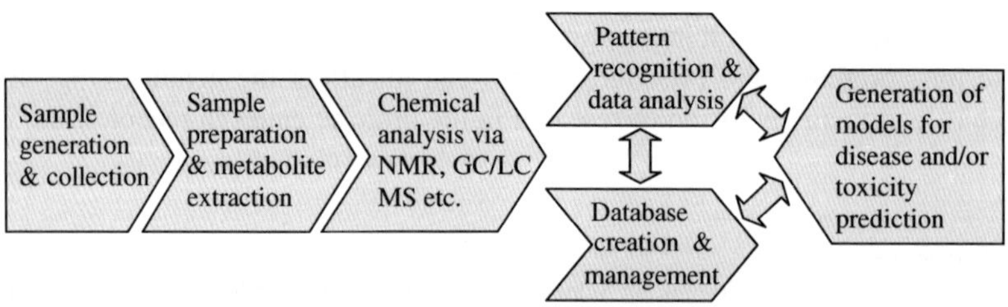

Fig. 40.2 Flow chart of the workflow for a standard metabolomics-based study (after Griffin (2004))

coupled with pattern recognition techniques (see Section 40.5) metabolomics makes a powerful investigative tool, with great potential for studying a wide range of issues; for example the biochemical effects of disease, or as a screening method for potential pharmacological agents. Indeed, it has proven to be highly sensitive for this type of analysis since metabolic perturbations often present much earlier than either tissue accumulation of toxins or induced histopathological changes (Griffin et al., 2000).

Somewhat confusingly, several terms have been coined to describe the process of combining global analytical tools and pattern recognition analysis to define the metabolic status of a tissue or organism. The term "metabolome" first emerged in 1998 in papers by Oliver (2002) and Tweeddale et al. (1998). The definitions were very similar with Oliver defining the metabolome as "the complete set of metabolites, the low molecular weight intermediates [for an organism and its organelles]" and Tweeddale defining it as "the total complement of metabolites in a cell".

The terms metabolomics and metabonomics are also both used fairly interchangeably to describe the use of analytical chemistry methods, coupled with statistics, to analyse the changes in the metabolome of a system, such as a biofluid, cell culture or tissue sample, to a perturbation. The two terms appeared in the literature within a year of each other (Fiehn et al., 2000; Nicholson et al., 1999) and both continue to be used, although metabolomics is more common. The terms, metabolic fingerprinting, metabolic foot printing and metabolic profiling are all in common use, frequently interchangeably, despite subtle differences in their definitions, thus making terminology complicated (and slightly perplexing) even for those working in the field. In this chapter, the term metabolomics will be used and taken to mean the combination of analytical tools with pattern recognition processes used to define a metabolic phenotype (metabotype) via a global approach (Oliver, 2002). This places the definition alongside terms such as genomics and proteomics, which represent the complete genetic profile and the complete protein expression in a cell, tissue or organism respectively.

Metabolomics is recognized as an important facet of the emerging science of systems biology. However, complete metabolome characterization has not been realized due to challenges in sample preparation, analytical limitations and data interpretation. This chapter will attempt to discuss each of these areas and outline current thinking behind the most commonly used methodologies as well as associated data processing techniques.

40.2 Sample Preparation and Metabolite Extraction

Metabolomics can be performed on virtually any biological material from cell cultures to organs to whole organisms (such as *Caenorhabditis elegans* or *Daphnia magna)* or even bacteria (Wu et al., 2010). Whatever the sample type, the aim is generally to measure as many metabolites as possible in order to increase the coverage of biochemical pathways, and give as complete a representation of the changes occurring in a system as a response to an external stimulus as possible. This is a huge challenge due to the extremely diverse and complex nature of the subject matter. Metabolites may range in concentration to the order $\sim 10^9$, have mass ranges of the order of $\sim$1500 amu and polarity ranges of $\sim 10^{20}$ (Atherton et al., 2006). Unsurprisingly therefore, no one approach can provide universal coverage of the metabolome and multiple techniques and preparation strategies are necessary for complete analysis. The most commonly used methods are nuclear magnetic resonance (NMR) spectroscopy and mass spectrometry (MS). The latter is the focus of this work. However, specific platforms are not a prerequisite for metabolomic analysis. Any technique capable of generating comprehensive metabolite measurements can potentially be used, including fourier transform- infrared and ion cyclotron spectroscopy (Aharoni et al., 2002; Timmins et al., 1998), high-performance (or high-pressure) liquid chromatography (HPLC) (Pham-Tuan et al., 2003), thin layer chromatography (Roberts et al., 2008), Raman[0] spectroscopy (Ellis and Goodacre, 2006), and capillary electrophoresis (Soga et al., 2003).

The first step for any metabolite study is the rapid termination of enzyme activity in the sample. This is important in order to achieve an accurate measure of the metabolome, since enzymatic activity will continue after the sample is taken, altering its metabolic content. Typically, samples are flash frozen in liquid N_2 to halt metabolism, after which they should be kept below $-20°C$ ($< -70°C$ if possible) in order to avoid any resumption of enzymatic activity. Enzymes can then be precipitated by treatment with acid or cold mixtures of organic solvents such as methanol, ethanol, acetone, or acetonitrile (and various mixtures thereof) (Pears et al., 2005).

The efficient extraction of metabolites from samples usually also requires the break up of tissues and cells. There are various methods for achieving this, the most common being grinding in a liquid N_2-cooled mortar and pestle, or by an electric tissue homogenizer directly in the extraction solvent. The latter is relatively straightforward but may not be practical for small amounts of tissue. Most samples can be ground down in a mortar and pestle but care must be taken to avoid partially thawing the tissues whilst grinding. The addition of dry ice or liquid N_2 to the mortar may help in this regard. In general, once the sample has been homogenized the extraction of metabolites is achieved by the application of cold solvents of varying polarities; the samples are usually also sonicated as part of this procedure. Solvents in common use include (but are not limited to): perchloric acid, acetonitrile-water, methanol-water, and methanol-chloroform-water, each solvent has advantages and disadvantages due to the large variety of physical and chemical properties of metabolites (Stentiford et al., 2005).

There are several published studies that compare the metabolic fingerprints of plants, micro-organisms, or biofluids extracted using different methodologies (Kaiser et al., 2009; Le Belle et al., 2002; Wu et al., 2010). Organic solvent extractions are generally quick and easy to perform and produce reasonable results. However, there is no one ideal way to simultaneously extract all metabolite types with high efficiency. For instance, whilst perchloric acid has been widely applied to precipitate proteins and extract hydrophilic metabolites in the past, this method has

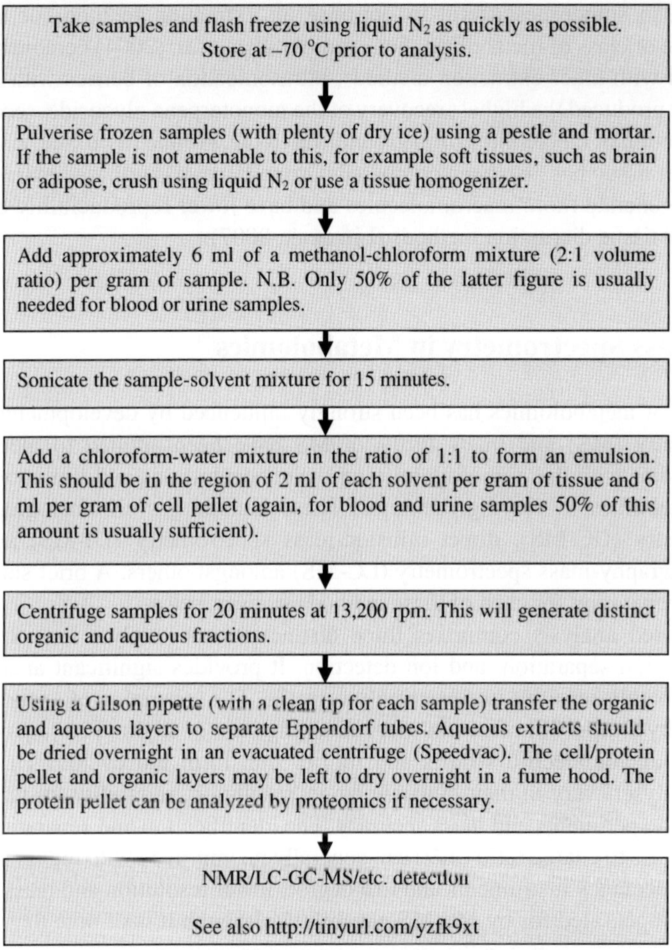

Fig. 40.3 Methanol–chloroform–water extraction method (LeBelle et al., 2002)

been shown to produce poor reproducibility between replicates (Lin et al., 2007). In contrast, acetonitrile-based extractions have been shown to have high reproducibility but poorer fractionation. Therefore, when one considers the yield and reproducibility of the technique, for both hydrophilic and hydrophobic metabolites, the methanol-chloroform-water extraction as outlined in Le Belle et al. (2002) has become the general method of choice (Lin et al., 2007). This extraction method is outlined in a simplified format in Fig. 40.3.

Methanol-chloroform-water extraction yields two fractions, both of which can be analyzed by mass spectrometry. The polar fraction will contain such compounds as amino acids, some lipids, sugars, choline metabolites and some aromatic compounds. The non-polar fraction will contain lipids and lipid soluble metabolites.

It is also sometimes useful to optimise an extraction method for a specific metabolite or class of metabolites. For example, Kim et al. (2004) demonstrated that compared with other extraction methods, ultrasonication of berries with methanol and water produced the highest recovery of the monoterpene glycoside, secologanin. The extraction of metabolites from ground wet tissue, ground dry tissue, and homogenized wet tissue has also been compared. Ground, wet tissue-based extraction was found to generate more macromolecules and have lower reproducibility compared with other tissue disruption methods (Lin et al., 2007).

40.3 Mass Spectrometry in Metabolomics

The field of metabolomics has been strongly influenced by developments in mass spectrometry (MS) and many forms of MS have been applied to metabolomic based studies with the aim of identifying the concentrations and fluxes of endogenous metabolites in varying scenarios. These include gas chromatography-mass spectrometry (GC-MS), direct infusion-mass spectrometry (DI-MS) and liquid chromatography-mass spectrometry (LC-MS) amongst others. A brief summary of sample preparation for each of these methods is given below.

MS based analysis comprises three distinct events: analyte ionisation, mass dependant ion separation, and ion detection. It provides significant analyte separation, is highly sensitive and extremely versatile. The separation of analytes can be achieved by a variety of mass analysers including time-of-flight (TOF), quadrupoles, magnetic and/or electric sectors and fourier transform ion cyclotron resonance. These analysers vary in their dynamic range, resolution, suitability for tandem-MS (MS/MS) experiments and mass accuracy. The identification and separation of analytes by mass to charge ratio (m/z) can generally be improved with higher resolution and mass accuracy instruments and to a degree, lower resolution and mass accuracy can be compensated for by MS/MS capability (although it is of note that there is a distinct correlation between the mass accuracy of an instrument and its relative cost). A detailed explanation of the principles and practises of these techniques is beyond the scope of this chapter (the interested reader is instead referred to McMaster and McMaster (2005)).

40.3.1 Gas Chromatography Mass Spectrometry

GC-MS is a combined analytical system. Compounds are first separated by GC, with eluting compounds then being detected via MS, traditionally using electron impact (EI) ionisation. The main advantages of GC-MS are (a) the increased sensitivity mentioned above and (b) the fact that compound identification is greatly facilitated by the availability of extensive, easily searchable databases of molecular fragmentation patterns. It has also proven to be a very robust and reproducible metabolomic technique. Some modern instruments also allow "two dimensional" chromatography

(Shellie et al., 2005). Here a short polar column is used in combination with the main analytical column, increasing the resolution of the device. The use of time-of-flight and/or ion trap mass spectrometers has also increased sensitivity compared to quadrupole based instruments. For these reasons GC-MS is often seen as the "gold standard" in metabolomics.

The disadvantages of GC-MS are that it is only useful for volatile, thermally stable compounds or those that can be rendered via various chemical reactions (which significantly increase the processing and analysis time per sample). The derivatisation of aqueous soluble metabolites has been conducted using a variety of methods; however a two step protocol using methoxylamine hydrochloride followed by N-methyl-N-(trimethylsilyl)-trifluoroacetamide (MSTFA) is most prevalently used in metabolomic analysis (Gullberg et al., 2004). The initial methoxymation converts the keto group of monosaccharides to an oxime; limiting the monosaccharide to the syn- and anti- isomers and this reduces the tautomeric forms of monosaccharides from 5 to 2, reducing the complexity of the resulting chromatograms considerably. The subsequent silylation reaction using MSTFA diminishes the intrinsic dipole-dipole interactions of the analyte molecules by substituting the active hydrogens in hydroxyl (OH), amine (NH) and thiol (SH) groups with a silyl group, thereby increasing the volatility and thermal stability of the metabolites.

The most frequent derivatization of organic phase metabolites is hydrolysis of complex lipids, to release esterified fatty acids, followed by the methylation of the fatty acids to form fatty acid methyl esters (FAMEs). A range of methods have been described in the literature to perform methylation of fatty acids. Methylation increases the volatility of the fatty acids whilst masking their polar functional carboxyl group (Morrison and Smith, 1964).

The similarity of molecular fragmentation patterns from structural isomers (such as those of sugar diastereomers) can still make compound identification difficult. In such cases accurate assignment must be resolved by using retention time indices of standards (such as those available at the Max Planck Institute of Molecular Plant Physiology online database at http://gmd.mpimp-golm.mpg.de/ in conjunction with the spectral analysis.

40.3.2 Liquid Chromatography Mass Spectrometry

LC-MS is another combined system, similar in principle to GC-MS but using a liquid mobile phase rather than gas. In common with GC-MS, it is more sensitive than NMR and has the added advantage that the need for sample volatility is eliminated, there is therefore no need for high analysis temperatures or derivatization reactions (although the latter can help to improve chromatographic resolution). This makes LC-MS a potentially global approach (in terms of the technique being capable of distinguishing-identifying a broad spectrum of metabolites and metabolite classes). Recent innovations such as ultra performance liquid chromatography (UPLC) have also shortened the chromatographic run time considerably (Cai et al., 2009; de Villiers et al., 2006).

40.3.3 Direct Infusion – Mass Spectrometry

DI-MS, in which the analyte mixture is directly infused into the mass spectrometer, also requires minimal sample preparation but suffers similar disadvantages to LC-MS. Factors for both include the fact that several different ionisation techniques are currently in use and these differ from those used in GC-MS. Typically electrospray ionisation (ESI) rather than electron impact is used for LC and DI-MS, but several other ionization techniques such as atmospheric pressure chemical ionisation (APCI) and atmospheric pressure photoionisation (APPI) are also in common use. As a result, spectral libraries to assist in sample identification for LC-MS lipidomic and metabolomic data are limited, partly because the exact results of ESI-MS are affected by the instrument type, ion source, ion source potentials, mobile phases and other factors affecting fragmentation patterns. However, a number of resources are available and these provide pertinent reference material and structural information of MS/MS fragmentation. Examples include LIPID MAPS (www.lipidmaps.org), The Lipid Library (http://www.lipidlibrary.co.uk/), the Cyber Lipid Centre (http://www.cyberlipid.org/), LIPIDAT (http://www.lipidat.ul.ie/) and LipidBank (http://lipidbank.jp/).

Additional problems include ion suppression (matrix effects) and the reproducibility of results (fragmentation patterns between instruments and laboratories can vary significantly). These issues make LC and DI-MS based metabolomics particularly demanding, although there is considerable potential for targeted analysis, especially for large and/or very polar compounds, such as hormones and lipids, which are not easily analysed by GC-MS (Roberts et al., 2008).

40.4 NMR in Metabolomics

40.4.1 Introduction

While mass spectrometry is increasingly used in metabolomic experiments, NMR spectroscopy has also proven to be a popular tool, and indeed, for many years was the main analytical method in use in the field. Again a detailed explanation of the principles and practises of this technique is beyond the scope of this chapter (the interested reader is instead referred to Keeler (2005)). However, it is useful to briefly compare the basic principles of both techniques here.

NMR has a number of advantages. For instance, unlike some forms of mass spectrometry, it does not require chemical modification of the sample in order to aid detection. The sample can instead be used almost "as is" greatly easing sample preparation. The non-destructive nature of NMR coupled with the lack of chemical derivatization also allows the sample to be reused for further analyses (e.g. mass spectrometry) at a later point in time. Recent improvements in NMR probes also allow the analysis of very small samples, for example mouse cerebral spinal fluid, which may only be present in quantities as low as $\sim$1–3 μl (Griffin et al.,

2002). Other NMR based approaches allow the quantification of metabolite concentrations in intact tissue, either in vivo or ex vivo which cannot be done by mass spectrometry to the same extent (although this may change in the near future (Ishida et al., 2003; Li et al., 2007). Magnetic resonance spectroscopy (MRS) also allows the non-invasive assessment of metabolite concentrations directly in vivo within a specific localised region and this has previously lead to the generation of proposed mechanisms of disease pathology (Griffiths et al., 2002).

40.4.2 Sample Preparation for NMR

There are a few factors to be considered when preparing a sample for study by NMR. The sample should be homogeneous with no sediment and it should be, as much as it is possible, free from large macromolecules, such as proteins, as these will give rise to distortions of the spectral baseline, making quantification more difficult. A reference compound must also be added to the sample; TMS (tetramethylsilane) or TSP (sodium 3-trimethyl-silyl-[2,2,3,3,d_4]-propionate) are commonly used as these have a defined frequency and give rise to a single peak at 0 ppm (which is then used to frequency correct the spectrum) which does not overlap with peaks from the vast majority of biologically interesting compounds. Modern NMR spectrometers also require the addition of deuterated solvents to the sample, typically in the form of D_2O or $CDCl_3$. This signal serves both to allow calibration of the strength of the applied magnetic field and also to provide a "lock signal" which is used to "shim" the sample (Keeler, 2005). This is performed to eliminate any magnetic field inhomogeneities which would otherwise cause the peaks in the spectrum to become broader, increasing overlap with neighbouring peaks thereby making metabolite identification more difficult. Using a deuterated solvent also has another benefit. Any solvent will also give rise to resonances in the NMR spectrum. As the solvent is typically the most abundant compound in the sample and peak area is proportional to the number of nuclei present in the sample, the solvent peak will be the largest in the spectrum by several orders of magnitude. This can lead to difficulties in the quantification of other, smaller peaks, since said smaller peaks will be squashed into the bottom range of the receiver along with the noise. The solvent peak can also overlap peaks arising from metabolites of interest, making quantification error prone and difficult. Solvent suppression modules in the NMR pulse sequence are being used to minimize the solvent signal, but this can be at the cost of distorting the spectral baseline near the solvent peak and suppressing resonances close to that of the solvent.

Deuterated solvents do not give rise to a ^{1}H signal; the use of 100% deuterated solvents thus obviates the need for excessive solvent suppression (along with the associated drawbacks). The sample may also require to be buffered to control the pH. It is best to avoid buffers that give rise to NMR signals that overlap those of metabolites of interest. For instance, phosphate buffer is commonly used for ^{1}H NMR samples, as it does not give rise to any ^{1}H resonances. Similarly, for ^{31}P spectra a buffer that does not contain phosphorus nuclei is preferred.

Preparation of samples for NMR analysis is also typically very simple. For low complexity samples it is sufficient to reconstitute in D_2O or 10% D_2O in water. For example, urine can be centrifuged to sediment any solids and the supernatant collected and D_2O and TSP added. Both of the fractions of methanol-chloroform extractions can be analysed by NMR. The polar fraction can be lyophilized and resuspended in phosphate buffer (~240 mM) in either 10 or 100% D_2O with the addition of TSP. In contrast the non-polar fraction should be resuspended in $CDCl_3$. It is advisable to use 100% $CDCl_3$ in preference to 10% $CDCl_3$ in $CHCl_3$ as the ^1H $CHCl_3$ resonances overlap with some aldehyde signals making identification and quantification difficult in these cases. Perchloric acid extracts can be treated similarly to the polar fraction of methanol-chloroform extraction (i.e. resuspension in deuterated buffer with the addition of TSP).

Nevertheless, whilst NMR is a highly reproducible, robust, non-destructive technique, which allows the simultaneous measurement of complex mixtures of many kinds of small molecule metabolites, its main disadvantage is that it lacks the sensitivity offered by mass spectrometry-based techniques, thus sample size/concentration can become an issue. While strategies are being developed to improve the sensitivity of NMR spectrometers, including the use of stronger magnetic fields, cryoprobes (which can increase the signal to noise ratio 3–4 fold by reducing thermal noise) and through the introduction of hyphenated techniques such as LC-NMR, it is unlikely NMR will offer better sensitivity than MS in the near future.

40.5 Multivariate Statistical Techniques

40.5.1 Introduction

As metabolomic analysis evolves, the number of metabolites measured and the coverage of the metabolome will continue to increase. The result is the generation of extensive amounts of raw data, containing the concentrations of a large array of metabolites, which must be contained in intricate data tables. Mining these vast resources for statistically relevant variation in metabolism requires potent statistical techniques, proficient in data reduction and simplification (Trygg et al., 2006).

Standard univariate techniques (such as Student's *t*-test) do not consider multiple correlations within the dataset, where the variation in one variable is related to the variation of one or more co-variables (Wold et al., 1984). In addition, due to the volume of data produced and the richness of spectral information, it is usually impractical (though not impossible) to assess the metabolic effect of stressors by univariate methods. Although there may be hundreds, if not thousands of variables within metabolomic data, there are certainly not hundreds of independent events taking place in the biological system under test. Important information is therefore more likely to be found within the patterns of correlation between variables, as opposed to within individual signals (Weljie et al., 2007).

By measuring the changes that occur (or do not occur) across many compounds and metabolic pathways, a much richer picture of the overall effects of a disease or drug-related perturbation on the metabolic network is obtained, than if the concentrations of only one or two directly affected compounds are measured. In such metabolomic analyses, the use of multivariate statistics coupled with sophisticated pattern recognition techniques have proven to be of value, since they consider all of the variables in a dataset simultaneously. A basic tenant of these techniques is to calculate a smaller number of latent variables. Latent variables are linear combinations of correlated variables which account for the same amount of variation present in the larger dataset, whilst reducing the dimensionality and minimising loss of information. Popular software used for this type of analysis includes SIMCA-P, (Umetrics, Umeå, Sweden), Pirouette (InfoMetrix, Woodinville, USA) Matlab (The MathWorks, Natick, USA) and R (http://www.r-project.org/) (Jones and Cheung, 2007).

Multivariate analysis has a number of advantages over univariate techniques in that it concurrently analyses all variables, taking into account their co-linear nature and it is robust in the face of missing data and noise. Thus multivariate analyses can minimize the risk of generating ambiguous results through the simultaneous assessment and correlation of all measured metabolites highlighting concerted patterns of change across the samples rather than through the isolated analysis of individual metabolites as is the case for univariate methods.

The primary multivariate statistical techniques used in metabolomics may be described as either unsupervised or supervised. Unsupervised techniques, such as principle components analysis (PCA), form the basis for multivariate data analysis (Purohit et al., 2004). They model the intrinsic variation within the dataset; these techniques are termed unsupervised as they do not take class membership into account. The starting point for these analyses is a data matrix with N rows (observations) and K columns (variables). PCA transforms the variables in the dataset into a smaller number of new latent variables called principal components (PCs). These PCs are combinations of the initial variables that describe the axes of greatest variation within the data. The components are mutually orthogonal (uncorrelated) and are calculated in order of decreasing contribution to the total variance of the original dataset (Eriksson et al., 2001). Most of the information in the dataset is described by the first few PCs, hence the reduction in the dimensionality of the data. Observations are assigned scores according to their projection in PC space. When displayed graphically on a score plot, samples with similar scores, and hence similarly correlated metabolic changes, will cluster together and away from groups with different scores and dissimilar metabolic changes.

To understand what each PC represents in relation to the original measurements, one must examine the loadings for the PCA model which can also be displayed graphically. The loadings for a PC describe the magnitude and direction of the contribution each variable makes to that PC; a large loading value for a metabolite would indicate that changes in concentration of that metabolite contribute strongly to the variation along that PC, whereas a small value would mean that a metabolite has little or nothing to do with variation along that PC (Eriksson et al., 2001).

Supervised techniques, such as Partial Least Squares – Discrimination Analysis (PLS-DA), use prior knowledge of class membership (e.g. control and disease groups), or regression trends to maximise separation between groups, or correlate data matrices by searching for changes in variables, which are correlated to class membership. Since they specifically look for variation associated with group membership, supervised techniques can be used to examine class separation which would otherwise be spread across three or more PCs of a PCA model. However, care must be taken to ensure data is not over fitted; supervised techniques therefore require some form of cross validation to ensure the statistical validity of results (Salek et al., 2007).

40.5.2 Validation of Multivariate Models

Validation of multivariate models requires examination of the model dimensionality. This is calculated by the degree of variation described by a model, and by the accuracy with which a model can predict the X data.

For both PLS-DA and PCA the variation in the model given by a component is known as R^2 which has a value between 0 and 1. The term R^2 is defined as R^2=1–RSS/$SXX_{tot.corr.}$, where RSS is the residual sum of squares of the data from the component, and $SXX_{tot.corr.}$ is the total variation in the X matrix post mean centering. If the R^2 was 1 that would indicate the total variation in the data matrix had been explained by the model (Eriksson et al., 1999).

Validation of PLS models generates two R^2 values defined as R^2X and R^2Y, describing the variation the model explains in the X matrix and the Y matrix. By increasing the components in a model, the R^2 can be increased, though this risks over-fitting the model, making it statistically insignificant; this can be avoided by cross validating the model.

The term Q^2, defined as Q^2=1–PRESS/$SXX_{tot.corr.}$, where PRESS is the sum of the squared differences between the predicted and real values (predictive residual sum of squares) is used as an indicator of the predictive nature of the model. For every successive component built for a model, the cross validation is calculated up to the point where generation of new components no longer improves the model's predictive power. Significantly predictive models are considered to have Q^2 values greater than 0.4 (Eriksson et al., 1999).

Outliers are readily identified in multivariate analysis; significantly deviant observations will be obvious in scores plots as the plots contain an ellipse which indicates the 95% confidence interval, established using a multivariate generalisation of the Student's *t*-test, known a Hotelling's T^2. It is of note that while some form of ellipse is typically plotted by most software packages, strictly speaking, a scores plot displays only the scores and not necessarily any form of confidence interval.Outliers can also be identified using the distance to model in X space defined as DModX; this equals the residual standard deviation of an observation. When DModX is larger than the D_{crit} (>95%) then the observation is a moderate outlier (Eriksson et al., 2001).

Much information may still be lost even in a rigorous statistical analysis. For this reason multivariate statistics used in conjunction with targeted univariate methods are most likely to yield the best and most complete results (Jones and Cheung, 2007). In addition, computer-based, supervised schemes are currently under development for application in metabolomics studies and this is where non linear techniques such as neural networks and genetic algorithms (which use machine learning based methods for the classification of samples) raise their, somewhat complicated, heads. (Goodacre, 2005; Kell, 2005).

40.6 Conclusions

Metabolomics is a rapidly developing field that has already made significant impacts in many different areas but technique does have potential drawbacks, of which it is important to be aware. These include its sensitivity to external influences, and the fact that since metabolomics is a static technique, it cannot generate data on dynamic processes, such as the flux rates of specific metabolic pathways. This is a problem common to many "omic" sciences, but can potentially be addressed by the use of isotope labelling studies (Birkemeyer et al., 2005; Roberts et al., 2009). However, providing studies are carefully set up and monitored, and the results interpreted with care, metabolomics holds great potential for the mass spectrometrist working in a wide range of fields.

References

Aharoni, A., Ric de vos, C.H., Verhoeven, H.A., Maliepaard, C.A., Kruppa, G., Bino, R., and Goodenowe, D.B. (2002). Nontargeted metabolome analysis by use of fourier transform ion cyclotron mass spectrometry. OMICS J Integr Biol *6*, 217–234.

Atherton, H., Bailey, N., Zhang, W., Taylor, J., Major, H., Shockcor, J., Clarke, K., and Griffin, J. (2006). A combined ^{1}H-NMR spectroscopy- and mass spectrometry-based metabolomic study of the PPAR alpha null mutant mouse defines profound systemic changes in metabolism linked to the metabolic syndrome. Physiol Genomic *27*, 178–186.

Birkemeyer, C., Luedemann, A., Wagner, C., Erban, A., and Kopka, J. (2005). Metabolome analysis: the potential of in vivo labeling with stable isotopes for metabolite profiling. Trends Biotechnol *23*, 28–33.

Bundy, J.G., Willey, T.L., Castell, R.S., Ellar, D.J., and Brindle, K.M. (2005). Discrimination of pathogenic clinical isolates and laboratory strains of *Bacillus cereus* by NMR-based metabolomic profiling. FEMS Lett *242*, 127–136.

Cai, S.-S., Syage, J.A., Hanold, K.A., and Balogh, M.P. (2009). Ultra performance liquid chromatography–atmospheric pressure photoionization-tandem mass spectrometry for high-sensitivity and high-throughput analysis of U.S. Environmental Protection Agency 16 Priority Pollutants Polynuclear Aromatic Hydrocarbons. Anal Chem *81*, 2123–2128.

de Villiers, A., Lestremau, F., Szucs, R., Gélébart, S., David, F., and Sandra, P. (2006). Evaluation of ultra performance liquid chromatography: Part I. Possibilities and limitations. J Chromatogr A *1127*, 60–69.

Denkert, C., Budczies, J., Kind, T., Weichert, W., Tablack, P., Sehouli, J., Niesporek, S., Konsgen, D., Dietel, M., and Fiehn, O. (2006). Mass spectrometry-based metabolic profiling reveals different metabolite patterns in invasive ovarian carcinomas and ovarian borderline tumors. Cancer Res *66*, 10795–10804.

Ellis, D.I., and Goodacre, R. (2006). Metabolic fingerprinting in disease diagnosis: biomedical applications of infrared and Raman spectroscopy. Analyst *131*, 875–885.

Eriksson, L., Johansson, E., Kettaneh-Wold, N., and Wold, S. (1999). Introduction to Multi- and Megavariate Data Analysis Using Projection Methods (PCA and PLS) (Umeå, Sweden, Umetrics), pp. 69–111.

Eriksson, L., Johansson, E., Kettaneh-Wold, N., and Wold, S. (2001). Multi- and Megavariate Data Analysis: Principles and Applications (Vol. 1) (Umeå, Sweden, Umetrics).

Fiehn, O., Kopka, J., Dörmann, P., Altmann, T., Trethewey, R.N., and Willmitzer, L. (2000). Metabolite profiling for plant functional genomics. Nat Biotechnol *18*, 1157–1161.

Goodacre, R. (2005). Making sense of the metabolome using evolutionary computation: Seeing the wood with the trees. J Exp Bot *56*, 245–254.

Griffin, J.L. (2004). The potential of metabonomics in drug safety and toxicology. Drug Discov Today Technol *1*, 285–293.

Griffin, J.L., Nicholls, A.W., Keun, H.C., Mortishire-Smith, R.J., Nicholson, J.K., and Kuehn, T. (2002). Metabolic profiling of rodent biological fluids via ^{1}H NMR spectroscopy using a 1 mm microlitre probe. Analyst *127*, 582–584.

Griffin, J.L., Walker, L.A., Troke, J., Osborn, D., Shore, R.F., and Nicholson, J.K. (2000). The initial pathogenesis of cadmium induced renal toxicity. FEBS Lett *478*, 147–150.

Griffiths, J.R., McSheehy, P.M.J., Robinson, S.P., Troy, H., Chung, Y.-L., Leek, R.D., Williams, K.J., Stratford, I.J., Harris, A.L., and Stubbs, M. (2002). Metabolic changes detected by *in vivo* magnetic resonance studies of HEPA-1 wild-type tumors and tumors deficient in hypoxia-inducible factor-1B (HIF-1B): Evidence of an anabolic role for the HIF-1 Pathway. Cancer Res *62*, 688–695.

Gullberg, J., Jonsson, P., Nordstrom, A., Sjostrom, M., and Moritz, T. (2004). Design of experiments: An efficient strategy to identify factors influencing extraction and derivatization of *Arabidopsis thaliana* samples in metabolomic studies with gas chromatography/mass spectrometry. Anal Biochem *331*, 283–295.

Ishida, Y., Nakanishi, O., Hirao, S., Tsuge, S., Urabe, J., Sekino, T., Nakanishi, M., Kimoto, T., and Ohtani, H. (2003). Direct analysis of lipids in single zooplankter individuals by matrix-assisted laser desorption/ionization mass spectrometry. Anal Chem *75*, 4514–4518.

Jones, O.A.H., and Cheung, V.L. (2007). An introduction to metabolomics and its potential application in veterinary science. Comp Med *57*, 436–442.

Jones, O.A.H., Griffin, J.L., Dondero, F., and Viarengo, A. (2008). Metabolic profiling of *Mytilus galloprovincialis* and its potential applications for pollution assessment. Mar Ecol Prog Ser *369*, 169–179.

Kaiser, K.A., Barding, G.A. Jr., and Larive, C.K. (2009). A comparison of metabolite extraction strategies for ^{1}H-NMR-based metabolic profiling using mature leaf tissue from the model plant *Arabidopsis thaliana*. Mag Res Chem *47*, S147–S156.

Keeler, J. (2005). Understanding NMR Spectroscopy (1st Ed) (Chichester, UK, Wiley).

Kell, D.B. (2005). Metabolomics, machine learning and modelling: towards an understanding of the language of cells. Biochem Soc Trans *33*, 520–524.

Kim, H.K., Choi, Y.H., Luijendijk, T.J.C., Rocha, R.A.V., and Verpoorte, R. (2004). Comparison of extraction methods for secologanin and the quantitative analysis of secologanin from *Symphoricarpos albus* ^{1}H-NMR. Phytochem Anal *15*, 257–261.

Le Belle, J., Harris, N., Williams, S., and Bhakoo, K. (2002). A comparison of cell and tissue extraction techniques using high-resolution ^{1}H-NMR spectroscopy. NMR Biomed *15*, 37–44.

Li, Y., Shrestha, B., and Vertes, A. (2007). Atmospheric pressure infrared MALDI Iimaging mass spectrometry for plant metabolomics. Anal Chem *80*, 407–420.

Lin, C.-Y., Wu, H., Tjeerdema, R.S., and Viant, M.R. (2007). Evaluation of metabolite extraction strategies from tissue samples using NMR metabolomics. Metabolomics *3*, 55–67.

Lindon, J.C., Holmes, E., and Nicholson, J.K. (2006). Metabonomics techniques and applications to pharmaceutical research and development. Pharmacol Res *23*, 1075–1088.

McMaster, M., and McMaster, C. (2005). LC/MS: A Practical User's Guide (Chichester, UK, Wiley).

Morrison, W.R., and Smith, L.M. (1964). Preparation of fatty acid methyl esters and dimethylacetals from lipids with boron fluoride-methanol. J Lipid Res *5*, 600–608.

Nicholson, J.K., Connelly, J., Lindon, J.C., and Holmes, E. (2002). Metabonomics: a platform for studying drug toxicity and gene function. Nat Rev Drug Discov *1*, 153–161.

Nicholson, J.K., Lindon, J.C., and Holmes, E. (1999). 'Metabonomics': understanding the metabolic responses of living systems to pathophysiological stimuli via multivariate statistical analysis of biological NMR spectroscopic data. Xenobiotica *29*, 1181–1189.

Oliver, S.G. (2002). Functional genomics: lessons from yeast. Phil Trans R Soc B *357*, 17–23.

Pears, M.R., Cooper, J.D., Mitchison, H.M., Mortishire-Smith, R.J., Pearce, D.A., and Griffin, J.L. (2005). High resolution ^{1}H NMR-based metabolomics indicates a neurotransmitter cycling deficit in cerebral tissue from a mouse model of Batten disease. J Biol Chem *280*, 42508–42514.

Pham-Tuan, H., Kaskavelis, L., Daykin, C.A., and Janssen, H.-G. (2003). Method development in high-performance liquid chromatography for high-throughput profiling and metabonomic studies of biofluid samples. J Chromatogr B *789*, 283–301.

Purohit, P.V., Rocke, D.M., Viant, M.R., and Woodruff, D.L. (2004). Discrimination models using variance-stabilizing transformation of metabolomic NMR data. OMICS J Integr Biol *8*, 118–130.

Roberts, L.D., McCombie, G., Titman, C.M., and Griffin, J.L. (2008). A matter of fat: an introduction to lipidomic profiling methods. J Chromatogr B *871*, 174–181.

Roberts, L.D., Virtue, S., Vidal-Puig, A., Nicholls, A.W., and Griffin, J.L. (2009). Metabolic phenotyping of a model of adipocyte differentiation. Physiol Genomics *39*, 109–119.

Robosky, L.C., Robertson, D.G., Baker, J.D., Rane, S., and Reily, M.D. (2002). *In vivo* toxicity screening programs using metabonomics. Combi Chem High Throughput Screening *5*, 651–662.

Salek, R.M., Maguire, M.L., Bentley, E., Rubtsov, D.V., Hough, T., Cheeseman, M., Nunez, D.J., Sweatman, B.C., Haselden, J.N., Cox, R., *et al.* (2007). A metabolomic comparison of urinary changes in type 2 diabetes in mouse, rat and man. Physiol Genomics *29*, 99–108.

Schauer, N., and Fernie, A.R. (2006). Plant metabolomics: towards biological function and mechanism. Trend Plant Sci *11*, 508–516.

Shellie, R.A., Welthagen, W., Zrostlikova, J., Spranger, J., Ristow, M., Fiehn, O., and Zimmermann, R. (2005). Statistical methods for comparing comprehensive two-dimensional gas chromatography-time-of-flight mass spectrometry results: Metabolomic analysis of mouse tissue extracts. J Chromatogr A *1086*, 83–90.

Soga, T., Ohashi, Y., Ueno, Y., Naraoka, H., Tomita, M., and Nishioka, T. (2003). Quantitative metabolome analysis using capillary electrophoresis mass spectrometry. J Proteome Res *2*, 488–494.

Stentiford, G.D., Viant, M.R., Ward, D.G., Johnson, P.J., Martin, A., Wenbin, W., Cooper, H.J., Lyons, B.P., and Feist, S.W. (2005). Liver tumors in wild flatfish: A histopathological, proteomic, and metabolomic study. OMICS J Integr Biol *9*, 281–299.

Timmins, É.M., Howell, S.A., Alsberg, B.K., Noble, W.C., and Goodacre, R. (1998). Rapid differentiation of closely related *Candida* species and strains by pyrolysis mass spectrometry and fourier transform infrared spectroscopy. J Clinical Microbiol *36*, 367–374.

Trygg, J., Holmes, E., and Lundstedt, T. (2006). Chemometrics in metabonomics. J Proteome Res *6*, 469–479.

Tweeddale, H., Notley-McRobb, L., and Ferenci, T. (1998). Effect of slow growth on metabolism of Escherichia coli, as revealed by global metabolite pool ("metabolome") analysis. J Bacteriol *180*, 5109–5116.

Viant, M.R. (2006). Revealing the metabolome of animal tissues using ^{1}H nuclear magnetic resonance spectroscopy. Methods Mol Biol *358*, 229–246.

Weljie, A.M., Dowlatabadi, R., Miller, B.J., Vogel, H.J., and Jirik, F.R. (2007). An inflammatory arthritis-associated metabolite biomarker pattern revealed by ^{1}H NMR spectroscopy. J Proteome Res *6*, 3456–3464.

Whitfield, P.D., German, A.J., and Noble, P.J. (2004). Metabolomics: an emerging post-genomic tool for nutrition. Br J Nutr *92*, 549–555.

Wold, S., Albano, C., Dunn, W.J., Edlund, U., Esbensen, K., Geladi, P., Hellberg, S., Johansson, E., Lindberg, W., and Sjöström, M. (1984). Multivariate Data Analysis in Chemistry (Holland, D. Reidel Publishing Company).

Wu, X.-H., Yu, H.-L., Ba, Z.-Y., Chen, J.-Y., Sun, H.-G., and Han, B.-Z. (2010). Sampling methods for NMR-based metabolomics of Staphylococcus aureus. Biotechnol J *5*(1), 75–84.

Chapter 41
Targeted Quantitative Analysis of Jasmonic Acid (JA) and Its Amino Acid Conjugates in Plant Using HPLC-Electrospray Ionization-Tandem Mass Spectrometry (ESI-LC-MS/MS)

Shigeru Tamogami, Ganesh Kumar Agrawal, and Randeep Rakwal

Abstract Jasmonic acid (JA) and its amino acid conjugates are critical signaling components involved in various aspects of plant growth, development and defense. Their constitutive and induced (due to a variety of environmental factors) levels vary from plant to plant and also from tissue to tissue within the same plant. To better understand the JA-mediated signaling and metabolic pathways, absolute quantification of JA and its conjugate(s) is essential in plants/tissues/organs during development and abiotic/biotic stresses. Here, we present an experimental protocol for extraction of JA and conjugates as well for absolute quantification applicable to both monocotyledonous (such as rice, maize) and dicotyledonous (such as *Arabidopsis*, Ox Knee) plants using high-pressure liquid chromatography-electrospray ionization-tandem mass spectrometry, which is still technically challenging. Our decade of experience has been blended in this protocol. In this chapter, we provide examples of a dicot plant and the absolute quantification of JA and JA-isoleucine in leaves.

Keywords Absolute quantification · Jasmonic acid · Jasmonoyl isoleucine · LC-MS/MS · Metabolites · Plant

41.1 Introduction

In nature, plants have evolved sophisticated defense mechanisms to protect themselves from a variety of environmental stresses (abiotic and biotic stresses). Jasmonic acid (JA) and its amino acid conjugates, such as JA-isoleucine (JA-Ile), are known mediators of signaling pathways leading to activation of plant defense responses (Reymond and Farmer, 1998; Agrawal et al., 2004; Wasternack, 2007; Smith et al., 2009). Jasmonic acid is biosynthesized from linolenic acid

---

S. Tamogami (✉)
Laboratory of Biologically Active Compounds, Department of Biological Production, Akita Prefectural University, Akita 010-0195, Japan
e-mail: tamo_chem@akita-pu.ac.jp

A.R. Ivanov, A.V. Lazarev (eds.), *Sample Preparation in Biological Mass Spectrometry*, DOI 10.1007/978-94-007-0828-0_41,

by sequential enzyme reactions of 13-lipoxygenase, allene oxide synthase, allene oxide cyclase, 12-oxo-phytodienoic acid reductase and three β-oxidation steps (Vick and Zimmerman, 1984; Hamberg and Gardner, 1992). Traditionally, gas chromatography-mass spectrometry (GC-MS) is used to determine JA level in plants. However, the use of GC-MS involves derivatization of the organic acids and long chromatographic separation at elevated temperatures that can cause thermal decomposition of the analytes, making accurate quantification of small amounts of JA by GC-MS rather difficult. Previously, high-pressure liquid chromatography-turboion spray-tandem mass spectrometry (LC-TIS-MS/MS; Tamogami and Kodama, 1998) and capillary LC electrospray MS/MS (Wilbert et al., 1998) techniques have also been applied to measure JA amino acid conjugates and JA/methyl jasmonate (MeJA). The later methods are non-destructive and do not require derivatization or elevated temperatures, and therefore, are powerful tools for the rapid and accurate measurement of small metabolites, such as JA and its amino acid conjugates. Our group mainly used LC-MS/MS and developed a rapid method for jasmonate quantification in small rice (model monocot crop plant) leaves sample (Tamogami and Kodama, 1998; Rakwal et al., 2002).

Recently, we brought further refinement to out protocol and have demonstrated a tight correlation between endogenous JA (and JA-Ile) production due to airborne MeJA and emission of volatile organic compounds (VOCs) in *Achyranthes bidentata*, a wild-plant species (Tamogami et al., 2008). This result along with a study published the same year (Wu et al., 2008) also clearly demonstrated that the plant-perceived MeJA is converted into JA and its amino acid conjugates. It should be noted that previous pioneering studies in 1990 (Farmer and Ryan, 1990) and 2001 (Preston et al., 2001; Kessler and Baldwin, 2001) had only suggested that MeJA is a potential airborne signal in nature (reviewed in Wasternack, 2007; Hail and Ton, 2008). Therefore, as the jasmonates are key signaling compounds of utmost biological importance (Chico et al., 2008; Katsir et al., 2008; Staswick, 2008; Fonseca et al., 2009), their precise and absolute quantification (Tamogami et al., 2008) is a necessity rather than choice.

Taken together, quantification of JA and JA-Ile in an absolute manner in time and space will definitely further our understanding on the complex signaling and metabolic networks mediated by JA in plants, during plant growth and protection against various abiotic and biotic stresses. In this chapter, we describe the experimental protocol, routinely used in our laboratories, for absolute quantification of both exogenous and endogenous jasmonates (JA and JA-Ile) in plant leaves along with our long-term experience in this area.

41.2 Materials

41.2.1 Plant Material

Seeds of *A. bidentata* (Ox Knee) were collected at Akita in Japan. Detailed conditions for plant materials (Tamogami et al., 2008) are as follows: Seeds were germinated and grown in a growth chamber (Biotron LPH-220S, Nippon Medical & Chemical Instruments, Osaka, Japan) maintained at 23°C for 12 h light and 12 h dark cycle with light intensity of 50 μmol/s/m^2, temperature of 23°C and relative humidity controlled at 60%. Plants were grown for 6 weeks and the shoots with fully opened four leaves were used as experimental material.

41.2.2 Jasmonate Preparation

(±)-MeJA was purchased from Sigma-Aldrich (St. Louis, MO). Standard conjugate compounds, such as (–)-JA-Ile, (–)-JA-leucine (Leu), were prepared using (±)-JA and amino acids as described previously (Kramell et al., 1988). (±)-Dehydrojasmonic acid (used as an internal standard) was prepared according to a previously reported method for MeJA synthesis (Büchi and Egger, 1971). (±)-$^2H_2$MeJA, namely methyl [9-^{2}H, 10-^{2}H]-jasmonate (double-deuterated at two olefinic positions) was prepared from (±)-methyl dehydrojasmonate in the presence of Lindlar catalyst following the same method (Büchi and Egger, 1971) with deuterium gas $^2H_2$ generated from CH_3O^2H (Sigma-Aldrich, St. Louis, MO, USA) and sodium metal. Atomic content of deuterium in the $^2H_2$JA was determined to be 99.5% by comparing the peak areas of ions at *m*/*z* 211 ($^2H_2$JA) and 209 (JA) by MS analysis. $^2H_2$JA was converted into conjugates with Ile or Leu (Fig. 41.1).

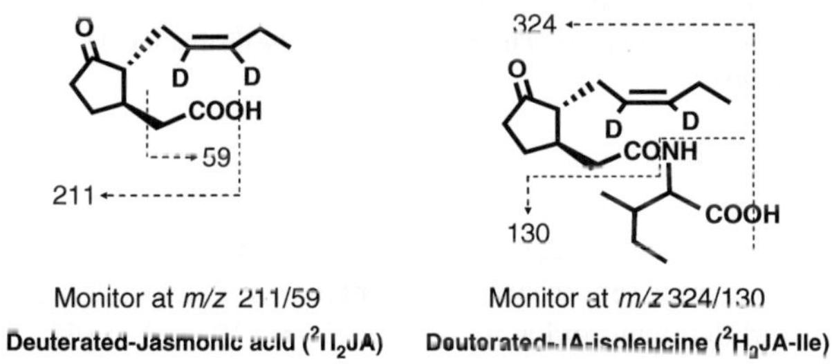

Fig. 41.1 Chemical structure of deuterated JA and its isoleucine conjugate (JA-Ile) with their precursor and product ions analyzed in the negative mode. D = deuterium

41.3 Sample Preparation

41.3.1 MeJA Application and Jasmonate Extraction

Plant materials (shoot with fully opened four leaves, approximately 4 g) and a paper disk containing $^2H_2$MeJA (2 mg) were enclosed in a 1 L glass container for 24 h. The experimental scheme is depicted in Fig. 41.2. Plant leaves (approximately 1 g) harvested at the desired time point(s) were used for jasmonate extraction with 60 mL of acetone (HPLC-grade, Wako, Tokyo, Japan) by grinding gently in a mortar and

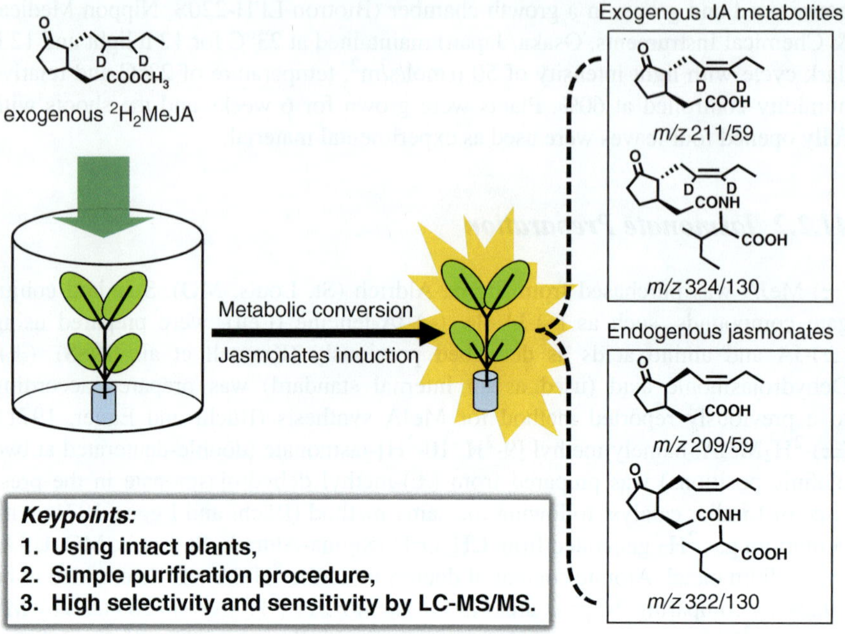

Fig. 41.2 *In planta* extraction and absolute quantification of jasmonates. Obtained JA metabolites and induced jasmonates were analyzed by MS/MS at the same time. Exogenous and endogenous jasmonates differ with two mass units, which is enough for the distinction. D = deuterium. Details are mentioned in the text

pestle. The acetone extract was passed through cotton (sterile, medical-grade) filter and concentrated in centrifugal vacuum rotary evaporator (Eyela, Tokyo, Japan). The concentrated extract (approximately 2 mL) was resuspended in HPLC-grade water (50 mL) and extracted twice with chloroform (20 mL) at pH 3. Organic layers were collected and concentrated under vacuum. The concentrated residue was resuspended with 2 mL of 80% aqueous methanol (MeOH; HPLC-grade, Wako) and passed through SepPak C18 light cartridges (Waters, Milford, MA, USA). The fraction thus obtained was concentrated to 200 μL using nitrogen gas. The sample was either immediately used for LC-MS/MS analysis or stored at 4°C for up to a week.

41.3.2 HPLC and Tandem MS Conditions for Jasmonate Measurement

Two microliters of the concentrated sample was injected into the ESI-LC-MS/MS system [HPLC, Agilent 1100 with *SCIEX* API-2000 LC-MS/MS; Column, Atlantis dC$_{18}$, 3 μm, 4.2 mm × 15 cm (Waters)]. Retention times of ^{2}H$_2$JA and JA were 6.03 min, and those of ^{2}H$_2$JA-Ile/Leu and JA-IIe/Leu were 6.16 min as per the

HPLC analysis and at a flow rate of 200 μL/min with aqueous 80% MeOH (v/v). The JA-Ile and JA-Leu were separated at a flow-rate of 200 μL/min with 50% aqueous MeOH (v/v); retention times of $^2H_2$JA-Ile and JA-Ile were 29.58 min and those of $^2H_2$JA-Leu and JA-Leu were 31.01 min. If JA-Ile/Leu separation is not required, aqueous 80% MeOH (v/v) can be used as an HPLC mobile phase to obtain higher sensitivity. Multiple reaction monitoring mode (MRM mode) was used to monitor ions. Synthetic standards were used to determine monitoring ion pairs by ESI-LC-MS/MS with some modifications to our previous method (Tamogami and Kodama, 1998), which are as follows: JA = *m/z* 209/59; 9,10-dehydrojasmonic acid (as an internal standard) = *m/z* 207/59; $^2H_2$JA = *m/z* 211/59; JA-Ile/Leu = *m/z* 322/130; and $^2H_2$JA-Ile/Leu = *m/z* 324/130. These ion pairs were monitored simultaneously. Absolute amounts were calculated with reference to synthetic standards. As jasmonate has carboxylic group, they (JA and amino acid conjugates) were analyzed in the negative ionization mode. Parent and the corresponding product ions are shown in Fig. 41.1. An important intermediate of JA biosynthesis, 12-oxo-phytodienoic acid (12-oxo-PDA) can also be analyzed at *m/z* 291/165 in the negative mode (Rakwal et al., 2002). A typical LC-MS/MS chromatogram of the leaf extract sample is shown in Fig. 41.3. Four kinds of jasmonates and internal standard (orange; *m/z* 207/59) were analyzed at the same time with the MRM mode.

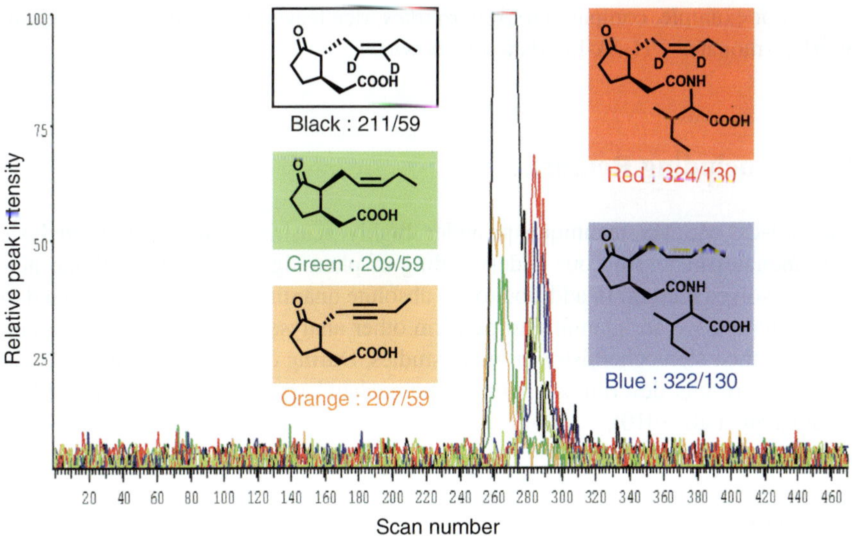

Fig. 41.3 A typical LC-MS/MS chromatogram of the leaf extract sample. The sample was prepared from a leaf extract exposed to airborne $^2H_2$MeJA for 24 h. The *colors of the chromatograms* correspond to the *colors of the compounds* shown in *inserts* and their MRM ion pairs. JA-Ile and JA-Leu (*m/z* 322/130) were eluted at the same retention time as 80% aqueous MeOH was used as the HPLC mobile phase. For simplicity, only JA-Ile structure is shown in the figure. Peak with black chromatogram ($^2H_2$JA, *m/z* 211/59) was saturated by size enlargement

41.4 Advantages of Using ESI-LC-MS/MS and Intact Plants in the Analyses of Jasmonates

Advantages of performing tandem MS (MS/MS) analysis are the high selectivity and sensitivity. High selectivity enables sample preparation with minimum purification steps and ensures that loss of very low abundance analytes is avoided. High sensitivity enables the detection of very small amounts of jasmonate metabolites in the sample. A previous study has shown that exogenous (airborne) MeJA applied to plants was converted into JA-Ile through JA inducing of de novo synthesized endogenous JA and JA-Ile (Tamogami et al., 2008). To determine in vivo the impact of airborne MeJA in perceived plant and its responses, a $^2H_2$MeJA treatment was used to distinguish the changes in the metabolites derived from $^2H_2$MeJA and synthesized de novo. This experimental strategy when combined with the protocol described here revealed for the first time that airborne $^2H_2$MeJA is incorporated into *Achyranthes* leaves and is undergoing metabolic conversion to $^2H_2$JA, $^2H_2$JA-Ile and $^2H_2$JA-Leu, which elicits the emission of VOCs and, importantly, the production of endogenous jasmonates JA-Ile/Leu and JA further amplifying VOC emissions (Tamogami et al., 2008). The experimental setup used is illustrated in Fig. 41.2. For such experiment, one needs to carefully select the plant. Based on our experience working with diverse plant species, various plants accumulate different level of jasmonates in response to stress and so the physiological meaning of jasmonates accumulation may vary form plant to plant and from condition to condition. For example, compared to *Achyranthes*, rice leaves accumulates low amount of $^2H_2$jasmonates ($^2H_2$JA-Ile; data not shown).

41.5 Concluding Remarks

The ESI-LC-MS/MS technique provides highly sensitive and selective analysis of jasmonates increasing our understanding of plant signaling and metabolic networks involved therein. In addition to the absolute quantitative analysis described in present report, we are planning to perform other analyses of jasmonates including resolving their stereochemistry in future studies. During revision of this manuscript, our recent results describing *cis-/trans*-stereochemistry of JA-Ile were published (Tamogami et al., 2010).

References

Agrawal, G.K., Tamogami, S., Han, O., Iwahashi, H., and Rakwal, R. (2004). Rice octadecanoid pathway. Biochem Biophys Res Commun *317*, 1–15.

Büchi, G., and Egger, B. (1971). A new synthesis of cyclopentenones. Methyl jasmonate and jasmine. J Org Chem *36*, 2021–2023.

Chico, J.M., Chini, A., Fonseca, S., and Solano, R. (2008). JAZ repressors set the rhythm in jasmonate signaling. Curr Opin Plant Biol *11*, 486–494

Farmer, E.E., and Ryan, C.A. (1990). Interplant communication: Airborne methyl jasmonate induces synthesis of proteinase inhibitors in plant leaves. Proc Natl Acad Sci USA *87*, 7713–7716.

Fonseca, S., Chini, A., Hamberg, M., Adie, B., Porzel, A., Kramell, R., Miersch, O., Wasternack, C., and Solano, R. (2009). (+)-7-iso-Jasmonoyl-L-isoleucine is the endogenous bioactive jasmonate. Nat Chem Biol *5*, 344–350.

Hail, M., and Ton, J. (2008). Long-distance signalling in plant defense. Trends Plant Sci *13*, 264–272.

Hamberg, M., and Gardner, H.W. (1992). Oxylipin pathway to jasmonates: Biochemistry and biological significance. Biochim Biophys Acta *1165*, 1–18.

Katsir, L., Chung, H., Koo, A., and Howe, G. (2008). Jasmonate signaling: A conserved mechanism of hormone sensing. Curr Opin Plant Biol *11*, 428–435.

Kessler, A., and Baldwin, I. (2001). Defense function of herbivore-induced plant volatile emissions in nature. Science *291*, 2141–2144.

Kramell, R., Schmidt, J., Schneider, G., Sembdner, G., and Schreiber, K. (1988). Synthesis of N-(jasmonoyl) amino acid conjugates. Tetrahedron *44*, 5791–5807.

Preston, C., Laue, G., and Baldwin, I. (2001). Methyl jasmonate is blowing in the wind, but can it act as a plant-plant airborne signal? Biochem System Eco *29*, 1007–1023.

Rakwal, R., Tamogami, S., Agrawal, G., and Iwahashi, H. (2002). Octadecanoid signaling component "burst" in rice (*Oryza sativa* L.) seedling leaves upon wounding by cut and treatment with fungal elicitor chitosan. Biochem Biophys Res Commun *295*, 1041–1045.

Reymond, P., and Farmer, E.E. (1998). Jasmonate and salicylate as global signals for defense gene expression. Curr Opin Plant Biol *1*, 404–411.

Smith, J.L., De Moraes, C.M., and Mescher, M.C. (2009). Jasmonate- and salicylate mediated plant defense responses to insect herbivores, pathogens and parasitic plants. Pest Manag Sci *65*, 497–503.

Staswick, P. (2008). JAZing up jasmonate signaling. Trends Plant Sci *13*, 66–71.

Tamogami, S., Agrawal, G.K., and Rakwal, R. (2010). An in planta technique for cis-/trans-stereochemical analysis of jasmonoyl isoleucine. J Plant Physiol *167*, 933–937.

Tamogami, S., and Kodama, O. (1998). Quantification of amino acid conjugates of jasmonic acid in rice leaves by high-performance liquid chromatography-turboionspray tandem mass spectrometry. J Chromatogr A *822*, 310–315.

Tamogami, S., Rakwal, R., and Agrawal, G.K. (2008). Interplant communication: Airborne methyl jasmonate is essentially converted into JA and JA-Ile activating jasmonate signaling pathway and VOCs emission. Biochem Biophys Res Commun *376*, 723–727.

Vick, B.A., and Zimmerman, D.C. (1984). Biosynthesis of jasmonic acid by several plant species. Plant Physiol *75*, 458–461.

Wasternack, C. (2007). Jasmonates: An update on biosynthesis, signal transduction and action in plant stress responses, growth and development. Ann Bot *100*, 681–697.

Wilbert, S.M., Ericsson, L.H., and Gordon, M.P. (1998). Quantification of jasmonic acid, methyl jasmonate, and salicylic acid in plants by capillary liquid chromatography electrospray tandem mass spectrometry. Anal Biochem *257*, 186–194.

Wu, J. Wang, L., and Baldwin, I. (2008). Methyl jasmonate-elicited herbivore resistance: Does MeJA function as a signal without being hydrolyzed to JA? Planta *227*, 1161–1168.

Chapter 42
Quantitative Analysis of Lipid Peroxidation Products Using Mass Spectrometry

Yasukazu Yoshida and Mototada Shichiri

Abstract It is important to assess the oxidative injury in vivo accurately and inclusively, as the oxidative stress induced by various oxidants in a random and destructive fashion is considered to play an important role in the pathophysiology of a number of human disorders. A method for the measurement of lipid peroxidation products in vivo, such as hydroxyoctadecadienoic acids (HODE) and 7-hydroxycholesterol (OHCh) determined by GC-MS and LC-MS/MS, is presented in this chapter.

Keywords 7-Hydroxycholesterol (7-OHCh) · Hydroxyeicosatetraenoic acid (HETE) · Hydroxyoctadecadienoic acid (HODE) · Lipid peroxidation · Mass spectrometry (MS)

42.1 Introduction

Lipid peroxidation products have received much attention as indicators of oxidative stress, since lipids are the most susceptible to oxidation in vivo. Thiobarbituric acid reactive substances (TBARS), malondialdehyde, lipid hydroperoxides and short-chain alkanes have been traditionally assessed. In this decade, F_2-isoprostanes and isofurans, which consist of a series of chemically stable prostaglandin F_2-like compounds formed by a mechanism independent of the cyclooxygenase pathway, have been widely assessed as oxidative stress markers in vivo (Morrow et al., 1990, 1992; Mori et al., 1999; Fessel et al., 2002; Basu, 2004). These compounds were detected and quantified in many biological samples such as plasma, saliva and urine by two major methods, mass-spectrometry and immunoassay. Many researchers reported the relevance of F_2-isoprostanes as stress markers with pathogenesis, for example, urinary 8-iso-prostaglandin $F_{2\alpha}$. On the other hand, hydroxyoctadecadienoic acid

---

Y. Yoshida (✉)
Health Research Institute (HRI), National Institute of Advanced Industrial Science and Technology (AIST), Ikeda, Osaka 563-8577, Japan
e-mail: yoshida-ya@aist.go.jp

A.R. Ivanov, A.V. Lazarev (eds.), *Sample Preparation in Biological Mass Spectrometry*, DOI 10.1007/978-94-007-0828-0_42,

(HODE) (Jira et al., 1998; Inouye et al., 1999; Cho et al., 2003) derived from linoleic acid and hydroxy- or 7-keto-cholesterol (Dzeletovic et al., 1995; Brown and Jessup, 1999; Brown et al., 1997; Ishii et al., 2002) have been paid attention to recently. There are few reports available that describe their detections in vivo, for example, formation of 9-hydroxy linoleic acid in diabetic erythrocyte membranes (Inouye et al., 1999) and hydroxy fatty acid in atherosclerotic patients (Jira et al., 1998).

One of the greatest needs in this area is the development of biomarkers for reliable measurement of oxidative stress in humans. Such biomarkers should be effective for monitoring healthy state and also for assessing the beneficial effects of antioxidants contained in foods, beverages, drugs and supplements. It may be used also to examine if oxidative stress is in fact involved in diseases and whether antioxidants can prevent, lessen or ameliorate the disease state. The oxidation products of linoleates have been measured with high performance liquid chromatography (HPLC) using the UV absorption at 230–235 nm due to conjugated diene. However, there is a limitation from the view point of sensitivity in this method. Recent advancements in mass spectrometry have demonstrated that coordination ion-spray mass spectrometry (CIS-MS) and electrospray ionization (ESI) or matrix-assisted laser desorption and ionization time-of-flight (MALDI-TOF) mass spectrometry can be used as powerful tools for the detection and identification of complex mixtures of lipid oxidation products (Schiller et al., 2004). The oxidation products of linoleates are now measured by gas chromatography-mass spectrometry (GC-MS) or high performance liquid chromatography-tandem mass spectrometry (LC-MS/MS) with high accuracy and sensitivity. We provide a short protocol for the quantitative analysis of hydroxyoctadecadienoic acid and 7-hydroxycholesterol by using mass spectrometry.

42.2 Materials

8-Iso-prostaglandin $F_{2\alpha}$ (8-iso-$PGF_{2\alpha}$), 8-iso-prostaglandin $F_{2\alpha}$-D_4 (8-iso-$PGF_{2\alpha}$-D_4), 5-hydroxyeicosa-6*E*, 8*Z*, 11*Z*, 14*Z*-tetraenoic acid (5-HETE), 12-hydroxyeicosa-5*Z*, 8*Z*, 10*E*, 14*Z*-tetraenoic acid (12-HETE), 15-hydroxyeicosa-5*Z*, 8*Z*, 11*Z*, 13*E*-tetraenoic acid (15-HETE), 13-hydroxy-9*Z*, 11*E*-octadecadienoic acid (13-(*Z*,*E*)-HODE), 9-hydroxy-10(*E*), 12(*Z*)-octadecadienoic acid (9-(*E*,*Z*)-HODE) and 13*S*-hydroxy-10*E*, 12*Z*-octadecadienoic-9,10,12,13-D_4 acid (13-HODE-D_4) were obtained from Cayman Chemical Company (MI, USA). 9-Hydroxy-10*E*,12*E*-octadecadienoic acid (9-(*E*,*E*)-HODE), 13-hydroxy-9*E*,11*E*-octadecadienoic acid (13-(*E*,*E*)-HODE), 10-hydroxy-8*E*, 12*Z*-octadecadienoic acid (10-(*Z*,*E*)-HODE) and 12-hydroxy-9*Z*, 13*E*-octadecadienoic acid (12-(*Z*,*E*)-HODE) were purchased from Larodan Fine Chemicals AB (Malmo, Sweden). 7α-Hydroxycholesterol (7α-OHCh) and 7β-hydroxycholesterol (7β-OHCh) were obtained from Steraloid Inc. (RI, USA) and their isotopes 7α-hydroxycholesterol-25, 26, 26, 26, 27, 27, 27-D_7 (7α-OHCh- D_7) and 7β-hydroxycholesterol-25, 26, 26, 26, 27, 27, 27-D_7 (7β-OHCh- D_7) were obtained from Medical Isotopes Inc. (NH, USA).

42.3 Sample Preparation

42.3.1 Plasma and Erythrocytes

Blood was obtained from human and animals in ethylenediaminetetraacetic acid (EDTA)-containing tubes (Terumo Corporation, Tokyo, Japan). Erythrocyte and plasma were separated by centrifugation (3,000 rpm for 10 min). Erythrocytes were washed twice with a four-fold volume of saline in order to remove plasma and white blood cells and then adjusted to a hematocrit value (HV) of 40% with saline. The erythrocytes (HV ca. 40%) were extracted with a four-fold volume of methanol containing 100 μM 2,6-di-*tert*-butyl-4-methylphenol (BHT) by vortexing and centrifugation (20,400×*g* at 4°C for 10 min) to obtain the hemolysate. The samples were stored at –80°C until the analyses.

42.3.2 Tissues

Tissue samples (for example, brain, liver and skin) were frozen in liquid nitrogen immediately after harvesting. The liver and brain were homogenized (Polytron PT-3100; Kinematica AG, Lucerne, Switzerland) in saline (liver or brain:saline = 1:3, w/w), and aliquots were diluted with saline. They were then subjected to the analyses.

42.4 Sample Processing

Tissue samples were homogenized with four times weight of saline with pestles and tubes (Bel-Art Products, NJ, USA) for 2 min. The homogenized sample (100 μL), plasma (200 μL) and hemolysate (200 μL) were mixed with 400, 300 and 500 μL saline, respectively. Subsequently, 500 μL methanol containing internal standards 8-iso-$PGF_{2\alpha}$-D_4 (50 ng), 13-HODE-D_4 (50 ng), 7α OHCh-D_7 (18 ng), 7β-OHCh-D_7 (19 ng) and 16-hydroxyhexadecanoic acid (70 ng) was added to the samples. The mixture was extracted with chloroform and ethyl acetate (chloroform/ethyl acetate = 4/1, v/v, 5 mL). The sample was mixed using a vortex mixer for 1 min and centrifuged at 1,500×*g* for 5 min at 4°C. The chloroform and ethyl acetate layer was concentrated to around 1 mL after the removal of the water layer and divided equally into two portions.

42.5 GC-MS Method for tHODE Measurement

Internal standards, 8-iso-$PGF_{2\alpha}$-D_4 and 9-HODE-D_4, and methanol were added to the plasma and the extracts from erythrocytes. The sample was centrifuged (3,000×*g* at 4°C for 10 min) and the supernatant was diluted with a four-fold volume of water and acidified (pH 3) using 2 N HCl. The acidified sample was centrifuged (3,000×*g* at 4°C for 10 min) and the supernatant was subjected to solid phase extraction.

42.5.1 Solid Phase Extraction

C_{18} cartridges (Waters Associates, Milford, MA) were preconditioned with aliquots of 2 mL methanol and 2 mL water (pH 3) with hydrogen chloride. Each sample was loaded to the cartridge at a flow rate of 1 mL/min and the cartridge was washed with 10 mL of water (pH 3) and 10 mL of acetonitrile/water (15/85, by volume). An elution was performed with 4 mL of hexane/ethyl acetate/isopropyl alcohol (30/65/5, by volume) at a flow rate of 1 mL/min. The eluate was then applied at a flow rate of 1 mL/min to the NH_2 cartridge preconditioned with hexane (5 mL). The cartridge was washed sequentially with 5 mL of hexane/ethyl acetate (30/70, by volume) and 5 mL of acetonitrile, and elution was performed with 5 mL of ethyl acetate/methanol/acetic acid (10/85/5, by volume) at a flow rate of 1 mL/min.

42.5.2 GC-MS Analysis

The solution was evaporated under the nitrogen flow and the silylating agent *N,O*-bis(trimethylsilyl)trifluoroacetamide (BSTFA) was added to the dried residue. The solution was vigorously mixed with a vortex mixer for 1 min and incubated for 60 min at 60°C in order to obtain the trimethylsilyl esters and ethers. The eluent obtained was diluted with acetone and then an aliquot of this sample was injected into a gas chromatograph (GC 6890 N; Agilent Technologies Co. Ltd., Palo Alto, CA, USA) equipped with a quadrupole mass spectrometer (5973 Network; Agilent Technologies) and a fused-silica capillary column (HP-5MS, 5% phenyl methyl siloxane, 30 m × 0.25 mm; Agilent Technologies). Helium was used as the carrier gas at a flow rate of 1.2 mL/min. The temperature of the column was programmed from 60 to 280°C at a rate of 10°C/min. The injector temperature was set at 250°C and the temperatures of transfer lines to the mass detector and ion source were 250 and 230°C, respectively. Electron energy of EI mass spectrometry was set at 70 eV. HODE and 8-iso-$PGF_{2\alpha}$ were identified on the basis of their retention times and mass patterns ($m/z = 440$, 369, and 225 for HODE, and 571 and 481 for 8-iso-$PGF_{2\alpha}$) and precursor ions at 440 and 481 were selected for the quantification of HODE and 8-iso-$PGF_{2\alpha}$, respectively, using the internal standards, 9-HODE-D_4 ($m/z = 444$) and 8-iso-$PGF_{2\alpha}$-D_4 ($m/z = 485$). These precursor ions were the most sensitive among the detected ions. With this method, 9-(*E,Z*)- and 13-(*Z,E*)-HODE, 9-(*E,E*)-HODE, 13-(*E,E*)-HODE and 8-iso-$PGF_{2\alpha}$ were measured simultaneously and isolated on the GC-MS chromatograph.

42.6 Analysis of 8-Iso-$PGF_{2\alpha}$, HETE, and HODE by LC-MS/MS

The divided chloroform and ethyl acetate solution was evaporated to dryness under the nitrogen flow. The derivatized sample was reconstituted with methanol and water (methanol/water = 70/30, v/v, 250 μL), and a portion of the sample (10 μL) was subjected to the LC-MS/MS analysis.

42.6.1 LC Conditions

LC was carried out on an ODS column (Hypersil Gold, 1.9 μm, 100 × 2.1 mm, Thermo Fisher Scientific, CA, USA) in a column oven (LC column casket CS-300B, Chromato Science Co. Ltd., Osaka, Japan) set at 30°C. The LC apparatus consisted of an autosampler (automatic sampling system AS-100, Bio-Rad Laboratories Co. Ltd., Tokyo, Japan), three pumps (Gilson models 306 and 307, Gilson Co. Inc., OH, USA), and a dynamic mixer (811D, Gilson). The eluent condition was a gradient comprising solvent A, 2 mM ammonium acetate in water (pH 6), and solvent B, methanol and acetonitrile (methanol/acetonitrile = 5/95), at a flow rate of 0.2 mL/min. The initial composition of the gradient was 80% A and 20% B. It was held for 2 min and then the composition was changed to 50% A and 50% B after 45 min.

42.6.2 MS Conditions

MS was carried out using a Thermo Finnigan TSQ Quantum Discovery Max, a triple-quadrupole mass spectrometer (Thermo Fisher Scientific, CA, USA) fitted with an electrospray ionization (ESI) source. ESI was carried out at a needle voltage of 4.2 kV. Nitrogen was used as the sheath gas (17 psi) and auxiliary gas (12 units). The capillary was heated to 280°C and the mass spectrometer was optimized to obtain the maximum sensitivity. A specific precursor-to-product-ion transition was carried out by selected reaction monitoring (SRM) after collision-induced dissociation in the negative mode. Argon was used as the collision gas and the collision cell pressure was 1.5 mTorr. The precursor, product ions, and collision energy were determined after the optimization of MS/MS as follows: m/z = 353.5 and 192.6–193.6 at 29 eV for 8-iso-$PGF_{2\alpha}$, m/z = 357.0 and 196.5–197.5 at 29 eV for 8-iso-$PGF_{2\alpha}$-D4, m/z = 319.0 and 114.5–115.5 at 10 eV for 5-HETE, m/z = 319.3 and 162.8–163.8 at 13 eV for 12-HETE, m/z = 319.3 and 202.5–203.5 at 10 eV for 15-HETE, m/z = 295.0 and 194.6–195.6 at 21 eV for both 13-(*Z*,*E*)-HODE and 13-(*E*,*E*)-HODE, m/z = 295.0 and 170.5–171.5 at 24 eV for both 9-(*E*,*Z*)-HODE and 9-(*E*,*E*)-HODE, m/z = 295.0 and 182.6–183.6 at 22 eV for both 10-(*Z*,*E*)-HODE and 12-(*Z*,*E*)-HODE and m/z = 299.0 and 197.6–198.6 at 26 eV for 13-HODE-D4. These data are summarized in Scheme 42.1.

42.7 Analysis of 7-OHCh, Cholesterol and Linoleic Acid by GC-MS

The other portion of the chloroform and ethyl acetate solution was also evaporated to dryness under nitrogen. A silylating agent (BSTFA, 30 μL) was added to the dried

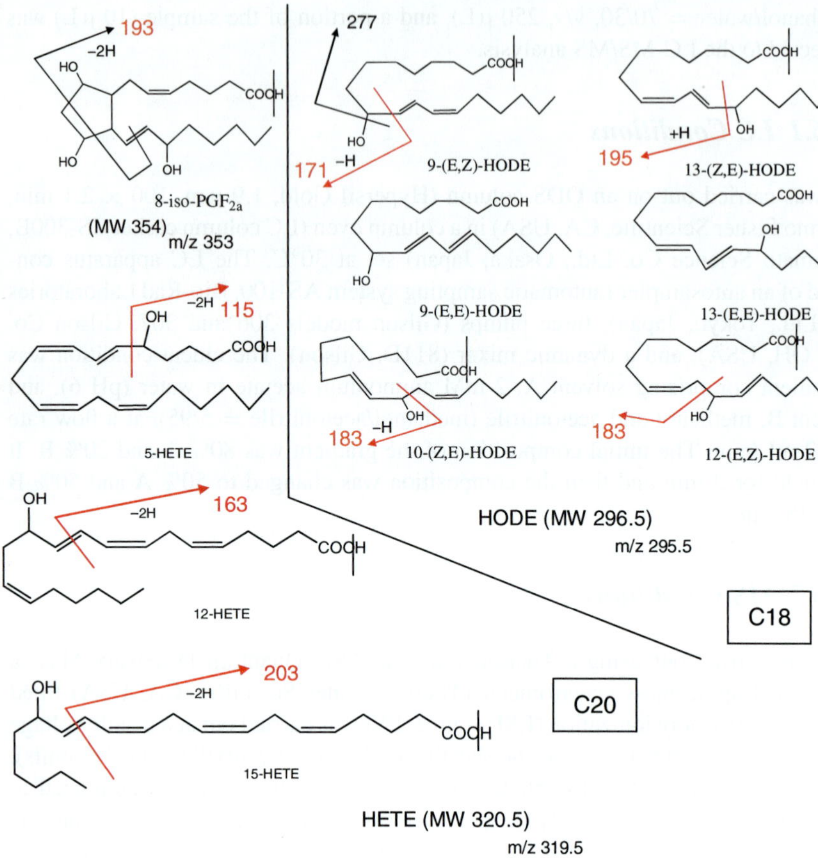

Scheme 42.1 The chemical structures with *m/z* of precursor and product ions of 8-iso-PGF$_{2\alpha}$, HODE and HETE

residue. The solution was vigorously mixed by vortexing for 0.5 min and incubated for 60 min at 60°C to obtain trimethylsilyl esters and ethers. An aliquot of this sample was injected into a gas chromatograph (GC 6890 N, Agilent Technologies) that was equipped with a quadrupole mass spectrometer (5973 Network, Agilent Technologies). A fused-silica capillary column (HP-5MS, 5% phenyl methyl siloxane, 30 m × 0.25 mm, Agilent Technologies) was used. Helium was used as the carrier gas at a flow rate of 1.2 mL/min. Temperature programming was carried out from 60 to 280°C at 10°C/min. The injector temperature was set at 250°C, and the temperatures of the transfer line to the mass detector and ion source are 250 and 230°C, respectively. Electron energy was set at 70 eV. 7α-OHCh, 7β-OHCh, 24S-OHCh, Ch and 18:2 were identified on the basis of their retention times and mass patterns; ions having *m/z* = 456 for 7α-OHCh and 7β-OHCh, 413 for 24S-OHCh, 458 for Ch, and 337 for linoleic acid were selected for quantification. 7α-OHCh, 7β-OHCh, 24S-OHCh, and Ch were identified quantitatively by using

7α-OHCh-D_7 and 7β-OHCh-D_7 as internal standards, and 18:2 was quantified by using 16-hydroxyhexadecanoic acid (HHDE) as an internal standard.

42.8 Concluding Remarks

The application of lipid oxidation products as biomarkers for diagnosis of disease progression, evaluation of therapies and health examination has been the focus of intensive studies. The usefulness and accuracy of lipid oxidation products as biomarkers should be validated in the future studies. There are obvious limitations of using the approach described herein for general healthcare practice, related to availability of sophisticated and expensive instrumentation such as LC-MS/MS. Therefore, the development of simpler and more convenient detection and quantification systems such as enzyme-linked immunosorbent assays will be needed for global use of the lipid oxidation product biomarkers. Our group is working towards this goal.

References

Basu, S. (2004). Isoprostanes: Novel bioactive products of lipid peroxidation. Free Rad Res *38*, 105–122.

Brown, A.J., and Jessup, W. (1999). Oxysterols and atherosclerosis. Atherosclerosis *142*, 1–28.

Brown, A.J., Leong, S., Dean, R.T., and Jessup, W. (1997). 7-Hydroperoxycholesterol and its products in oxidized low density lipoprotein and human atherosclerotic plaque. J Lipid Res *38*, 1730–1745.

Cho, H., Gallaher, D.D., and Csallany, A.S. (2003). Nonradiometric HPLC measurement of 13(S)-hydroxyoctadecadienoic acid from rat tissues. Anal Biochem 318, 47–51.

Dzeletovic, S., Breuer, O., Lund, E., and Diczfalusy, U. (1995). Determination of cholesterol oxidation products in human plasma by isotope-dilution mass-spectrometry. Anal Biochem *225*, 73–80.

Fessel, J.P., Porter, N.A., Moore, K.P., Sheller, J.R., and Roberts, L.J. (2002). Discovery of lipid peroxidation products formed in vivo with a substituted tetrahydrofuran ring (isofurans) that are favored by increased oxygen tension. Proc Natl Acad Sci USA *99*, 16713–16718.

Inouye, M., Mio, T., and Sumino, K. (1999). Formation of 9-hydroxy linoleic acid as a product of phospholipid peroxidation in diabetic erythrocyte membranes. Biochim Biophys Acta *1438*, 204–212.

Ishii, K., Adachi, J., Tomita, M., Kurosaka, M., and Ueno, Y. (2002). Oxysterols as indices of oxidative stress in man after paraquat ingestion. Free Radic Res *36*,163–168.

Jira, W., Spiteller, G., Carson, W., and Schramm, A. (1998). Strong increase in hydroxy fatty acids derived from linoleic acid in human low density lipoproteins of atherosclerotic patients. Chem Phys Lipids *91*, 1–11.

Mori, T.A., Croft, K.D., Puddey, I.B., and Beilin, L.J. (1999). An improved method for the measurement of urinary and plasma F2-isoprostanes using gas chromatography-mass spectrometry. Anal Biochem *268*, 117–125.

Morrow, J.D., Awad, J.A., Boss, H.J., Blair, I.A., and Roberts, J., II. (1992). Non-cyclooxygenase-derived prostanoids (F_2-isoprostanes) are formed in situ on phospholipids. Proc Natl Acad Sci USA *89*, 10721–10725.

Morrow, J.D., Harris, T.M., and Roberts, J., II. (1990). Noncyclooxygenase oxidative formation of a series of novel prostaglandins: Analytical ramifications for measurement of eicosanoids. Anal Biochem *184*, 1–10.

Schiller, J., Süβß, R., Arnhold, J., Fuchs, B., Leβig, J., Müller, M., Petkovic, M., Spalteholz, H., Zschörnig, O., and Arnold, K. (2004). Matrix-assisted laser desorption and ionization time-of-flight (MALDI-TOF) mass spectrometry in lipid and phospholipid research. Prog Lipid Res *43*, 449–488.

Chapter 43
Sample Preparation for Drug Metabolism Studies

Natalia Penner, Biplab Das, Caroline Woodward, and Chandra Prakash

Abstract Determination of metabolic stability and metabolite profiles of the new chemical entities (NCEs) plays a key role in the drug discovery and development process. At an early stage of drug discovery, information on the metabolic fate of the NCEs can direct medicinal chemists to synthesize metabolically more stable analogs by blocking sites of metabolism, and potentially creating NCEs with superior pharmacology and safety profiles. In the development stage, the focus is shifted to characterization of human metabolites, which enables the judicious selection of animal species used for safety evaluation and ensure that the selected animal species are exposed to all major metabolites formed in humans. The traditional method for metabolite characterization, HPLC-MS, requires extensive sample preparation prior to analysis. A correctly selected processing method results in cleaner samples, less background, and easier data interpretation. This chapter covers various aspects of sample preparation for studying metabolism of the drug candidates. An overview of the sample preparation approaches for both in vitro and in vivo studies including metabolic stability studies and screening for reactive metabolites is presented. Aspects of planning and conducting biotransformation in vivo studies starting from a sample collection, pooling, extracting drug-derived material are discussed. In addition, recommendations on how to investigate sample stability and to prevent degradation of metabolites are given.

Keywords Chemical derivatization for structural elucidation · Extraction of metabolites · Metabolite profiling · Sample preparation · Stabilization of samples

N. Penner (✉)
Department of Drug Metabolism and Pharmacokinetics, Biogen Idec, Cambridge, MA 02142, USA
e-mail: natasha.penner@biogenidec.com

A.R. Ivanov, A.V. Lazarev (eds.), *Sample Preparation in Biological Mass Spectrometry*, DOI 10.1007/978-94-007-0828-0_43,

43.1 Introduction

Characterization of metabolites of a drug candidate is a crucial part of the drug-discovery and development process. It is well established that the metabolic fate of an NCE plays a major role in defining its potential as a drug candidate. Considerable resources are invested into metabolite screening of new chemical entities (NCE) earlier in the drug discovery stage to minimize the risk of failure for drug metabolism and pharmacokinetics (DMPK) reasons. High performance liquid chromatography coupled with mass spectrometry (HPLC-MS) is the key analytical technique for metabolite characterization and quantitative analysis of drugs in modern drug discovery and development due to increased popularity and availability of highly sensitive mass spectrometers in the last decade.

A fast pace of instrumentation-related development and introduction of newest analytical technologies to everyday work does not change a simple fact that many biological samples cannot be analyzed by a mass spectrometer directly or cannot be injected onto an HPLC-MS system without pre-treatment. Samples routinely used for metabolite identification, such as plasma, require lengthy and often not straightforward sample processing to make them suitable for analysis. It is well known that significant time is spent on development of a satisfactory sample preparation method for the analysis of drug metabolites. Complexity of matrix may result in considerable signal suppression in mass spectrometry which makes metabolite detection and metabolite structural elucidation problematic. A correctly selected processing method results in cleaner samples, less background and easier data interpretation.

While dealing with samples for biotransformation studies, there are two main challenges:

- low concentration of drug-derived material in biological samples and
- presence of high levels of endogenous components such as proteins, lipids, bile acids, and components of the dosing formulation.

The combination of these two factors, which is common for biotransformation work, makes metabolite identification studies challenging and time-consuming. Therefore, three objectives of samples preparation for biotransformation studies are as follows: to bring level of drug-derived material to a level where analytical instruments can reliably detect it (pre-concentration), to remove endogenous interferences, and last but not least, to obtain a sample which is representative of a profile of drug-derived material in a biological sample of interest. The last point means that stability of a drug and its metabolites should be evaluated and thoroughly considered before starting sample processing. Precautions should be taken when working with unstable analytes.

When mass spectrometry alone is not sufficient, multiple analytical and wet chemistry techniques, such as chemical derivatization and NMR, should be applied to provide specific structural information. These techniques will be covered in this chapter.

43.2 Metabolite Identification in Discovery and Development

Metabolite identification in discovery and development is generally carried out using similar instrumentation and approaches (analytical strategies), although the questions to be answered are substantially different. Discovery work is focused on identification of metabolic soft spots and screening for the reactive metabolites to identify the best drug candidate. Safety of a drug in humans is a major concern during development. Regulatory requirements and a substantially different number of compounds going through discovery and development influence methods and conditions used for metabolite profiling work.

43.2.1 Preparation of In Vitro Samples

Metabolite identification in early discovery is performed in a high-throughput screening fashion using short generic methods usually without radioisotope labeling. Little attention is paid to customizing analysis methods; therefore, sample preparation methods are generally simple. In early discovery, in vitro systems, such as hepatic subcellular fractions or hepatocytes, are used to determine the metabolic fate of an NCE. The information from these studies can assist medicinal chemists in rationalizing synthetic approaches to create metabolically stable molecules. It also could lead to discovery of more potent compounds with a better safety profile. Sample preparation activities in vitro can be divided into two main categories:

- characterization of a metabolic fate of NCE in liver fractions (microsomes, S9, cytosol), hepatocytes, and with recombinant enzymes and
- evaluation of bioactivation potential by screening for the formation of reactive metabolites/intermediates.

43.2.1.1 Incubations with Hepatic Subcellular Fractions and Hepatocytes

A common strategy for an initial metabolism study is an incubation of a drug candidate with liver microsomes fortified with NADPH to assess phase I reactions. Next step is the evaluation of phase II reactions using hepatocytes or liver microsomes/S9/cytosol with appropriate cofactors as summarized in Table 43.1. Liver microsomes of different species including human, rat, dog, mouse, cynomolgus monkey and rabbit are available from a number of vendors (BD Gentest, Xenotech, In Vitro Technology and ADMET Technology), with protein content of approximately 20 mg/ml. Human intestinal microsomes are also available from BD Gentest. A typical microsomal or S9 incubation mixture of 300 μl in volume consists of the following components: 1 mg/ml of microsomal protein or S9 fraction, 10 μM test compound, 1 mM NADPH, 3.3 mM $MgCl_2$, and 100 mM potassium phosphate, pH 7.4. Final amount of DMSO in the incubation is below 0.1% and does not inhibit the activity of drug-metabolizing enzymes. All components listed above

Table 43.1 Common in vitro reactions for metabolism studies

Materials	Cofactors	Modifications
Liver microsomes	NADPH	Phase I reactions (oxidation, reduction, oxidative hydrolysis)
Liver microsomes, cytosol, S9	NADPH, UDPGA, PAPS, acetyl-coenzyme A, NAD, GSH	Phase I and phase II reactions, glutathione conjugation, glucuronidation, sulfation
Hepatocytes (fresh or cryopreserved)	No cofactors	Phase I and phase II reactions

NADPH – nicotinamide adenine dinucleotide phosphate; UDPGA – uridine diphosphate glucuronic acid; PAPS – 3′-phosphoadenosine-5′-phosphosulfate; NAD – nicotinamide adenine dinucleotide; GSH – glutathion

except NADPH are mixed in a vial (or 96-well plate), and reactions are then initiated by adding NADPH. Reactions are carried out at 37°C in a shaking incubator block (450 rpm). Incubation time depends upon the metabolic stability of a drug. To capture primary and secondary metabolites, reactions are terminated at two time points, for example at 15 and 45 min, with 300 μl of ice cold acetonitrile containing 0.1% formic acid. Control incubations are performed without NADPH. After reactions are completed, samples are centrifuged for 10 min at 10,000×*g*, and supernatants are used for LC-MS analysis.

For glucuronidation reactions, liver microsomes are treated with alamethicin, a pore-forming peptide, to get access for UDP-glucuronosyltransferases (UGTs) inside the membrane. UGTs are a superfamily of membrane-bound enzymes which active site is localized inside endoplasmic reticulum. Alamethicin allows free diffusion of a substrate, a cofactor, and products without affecting the gross membrane structure and intrinsic enzyme catalytic activity (Fisher et al., 2000).

Liver microsomes (1 mg/ml) in potassium phosphate, pH 7.4 (100 mM) are first incubated with alamethicin (25 μg/ml) for 10 min on ice, and then UDPGA (3 mM), a test compound (10 μM), $MgCl_2$ (3.3 mM), are added to a reaction mixture. The reaction mixture is further incubated at 37°C in a shaking incubator block (450 rpm) for 45 min. The reaction is terminated by addition of ice cold acetonitrile. Incubation buffer systems are available from BD Gentest: Soln A (cat# 451300) containing UDPGA and Soln B (cat#451320) containing alamethicin.

Cytosolic liver fractions mostly contain non-CYP enzymes, which may play a critical role in drug metabolism. These enzymes include alcohol and aldehyde dehydrogenases, epoxide hydrolases, sulfotransferases, glutathione transferases, *N*-acetyl transferases and methyl transferases (Brandon et al., 2003). Liver cytosolic fractions (20 mg/ml) are commercially available from the vendors such as BD Gentest and Xenotech. The content of a reaction mixture for incubation with cytosol depends on the type of an enzymatic reaction to be explored. For a typical sulfation reaction, a mixture would consist of 1 mg/ml cytosolic protein, 10 μM test compound, 1 mM PAPS (adenosine 3′-phosphate 5′-phosphosulfate) and 100 mM potassium phosphate, pH 7.4. Incubation conditions and sample processing are similar to the liver microsomes incubation procedure.

43.2.1.2 In Vitro Trapping Strategies for Reactive Metabolites

In vitro screening for reactive metabolites is important in early stages of drug discovery process. Reactive intermediates readily react with macromolecules such as proteins and/or nucleic acids by nucleophilic addition or substitution in biological systems. Identifying reactive sites on a compound not only provides guidance on chemical modification but also offers insight on the bioactivation pathway of a chemical series.

Trapping reagents forming conjugates that can be easily detected by LC-MS are widely used in the in vitro experiments (Prakash et al., 2008). Glutathione, glutathione ethyl ester, cysteine and *N*-acetylcysteine contain a free sulfhydryl group, a soft nucleophile, and therefore, readily react with a broad range of reactive soft electrophiles like Michael acceptors, epoxides, arene oxides, nitrenium ions and alkyl halides. For hard electrophiles like iminium ions ($-C{=}N^{+}R_1R_2$), cyanide (potassium cyanide, KCN) is effective for trapping. Trapping agents used for various functional groups are summarized in Table 43.2.

For a common trapping experiment, liver microsomes at 1 mg protein/ml are suspended in 1 ml of phosphate buffer (100 mM, pH 7.4) containing $MgCl_2$ (3 mM), GSH (5 mM) or *N*-acetylcysteine (5 mM) or $KCN/KC^{13}N^{15}$ (1 mM). The substrate is added to the final incubation mixture ensuring that the organic content of the reaction is below 0.2% (v/v). NADPH (1.2 mM) is added to start the reaction, and the mixture is incubated at 37°C for 60 min. The reaction is stopped by the addition of 1 volume of cold acetonitrile, and the reaction mixture is centrifuged at 14,000 rpm for 10 min. The supernatant is then injected for LC-MS/MS analysis. Control samples contain the liver microsomes and the substrates but lack either NADPH or a trapping reagent. Ticlopidine is a convenient positive control as during metabolic activation it forms both hard and soft electrophiles; therefore, it is capable of reacting with both glutathione and cyanide. Two glutathione conjugates and two cyano conjugates are observed in the positive control reaction (Fig. 43.1).

Table 43.2 Common trapping agents used for assessment of bioactivation

Trapping agents	Functional group
Glutathione (mono ester), mercaptoethanol, cysteine	Quinones, enones
Cyanide	Iminium ions
Semicarbazide	Aldehydes
Methoxylamine	
Lysine	Imides, aryl halides
Lysine + cysteine	Furan, epoxide
TEMPO (tetramethylpiperidin-*N*-oxyl)	Free radical trapping

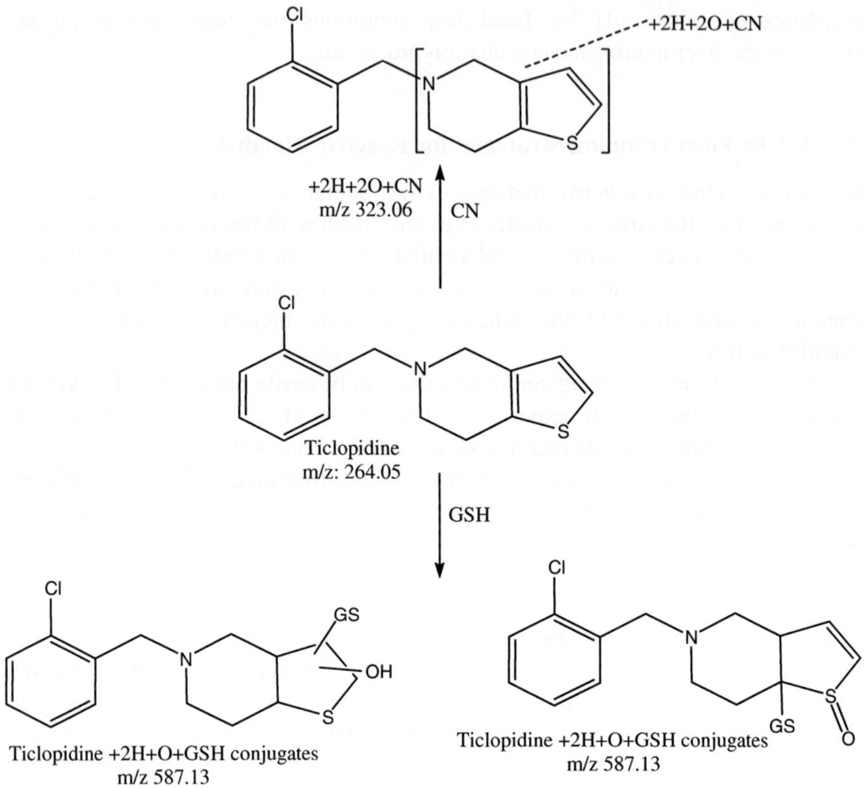

Fig. 43.1 Formation of GSH and cyano conjugates of ticlopidine

43.2.2 In Vivo Sample Collection

In development, a focus of drug metabolism studies is shifted to evaluation of a metabolic profile of a drug candidate in vivo (humans and animal species used for safety assessment). Specifics of in vivo studies requires multi-step sample preparation process which includes the following steps:

careful planning in the life part of a study;
well-thought sample collection procedure;
evaluation of the need for sample stabilization;
acceptable methods for extraction, reconstitution and analysis.

In the development stage a radiolabeled drug is usually available, and metabolite profiling is performed with long HPLC separation methods. HPLC–MS is coupled with in-line radioactive detector for the simultaneous acquisition of radiochromatographic profile, which is used for quantitative analysis, while LC-MS data are used

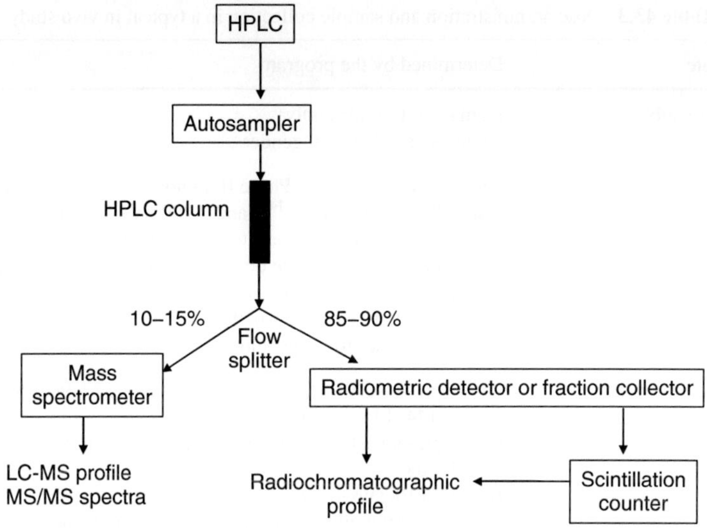

Fig. 43.2 System setup for metabolite profiling of radiolabeled samples

for structural elucidation. There have been a number of publications and reviews on the use of mass spectrometry for metabolite identification in both drug discovery and drug development setting (Gelhaus and Blair, 2009; Hop and Prakash, 2005; Kamel and Prakash, 2006; Ma et al., 2006; Prakash et al., 2007). Metabolic pathways of drug candidates provide guidance for the selection of animal species for toxicology studies in order to ensure that the selected animal species are exposed to all major metabolites formed in humans (Baillie et al., 2002).

A general system set-up for biotransformation analysis of radioactive samples is shown in Fig. 43.2. The effluent from an HPLC pump after a UV detector (optional, but useful) is split to direct approximately 10–20% of the flow into a mass spectrometer, while the majority of the flow goes into a flow scintillation analyzer.

The primary objectives of the radiolabeled studies are to determine the routes of elimination of a given compound in a given species, to characterize the metabolic profile and to identify circulating and excretory metabolites (Penner et al., 2009). A typical dose administration and sample collection outline is presented in Table 43.3. Duration of the in-life phase of a mass balance/biotransformation study is usually 168 h. In most cases this time is sufficient to collect ~90% of administered radioactivity. A complete plasma concentration–time profile is collected for up to five half-lives of the parent compound to characterize total radioactivity and unchanged drug in plasma or serum as well as major metabolites. Plasma collection for metabolite profiling depends on the pooling approach used and will be discussed in the next section. Blood is collected into tubes containing an anti-coagulant such as heparin or K_2EDTA.

Table 43.3 Dose administration and sample collection in a typical in vivo study

Dose route	Determined by the program
Number of subjects	Human: 4–6 (males only) Animals: 2–4 for each gender
Dose	Human: dose similar to Phase II, radioactivity as low as feasible, usually ~100 μCi for ^{14}C and ~200 μCi for ^{3}H Animal species: dose is selected based on toxicology studies (low or mid dose from single dose or 28-day studies), specific activity of the dose varies depending on specific activity of radiolabeled material, usually 100–150 μCi/kg for rodents, 20–25 μCi/kg for large animals
Sampling time points (h post-dose)	Urine: pre-dose, 0–12, 12–24, 24–48, 48–72, 72–96, 96–120, 120–144, 144–168 h, Feces: pre-dose, 0–24, 24–48, 48–72, 72–96, 96–120, 120–144, 144–168 h Blood for PK: up to 5 $t_{1/2}$ Blood for metabolite profiling: time points and volumes depend on a pooling strategy

43.2.3 Pooling of Biological Samples

Pooling is mixing of original samples to obtain a single representative sample. Pooling can be conducted across time points for one subject, across subjects for one time point, or both. To obtain a pooled excreta sample, individual samples should be combined in the amounts relative to their total volume/weight collected. An example of a pooling scheme for excreta is shown in Table 43.4. Before pooling, mass balance data should be evaluated for each subject separately to decide aliquots from which time points must be added to the pooled sample. For metabolite profiling of urine, bile, and feces, samples collected at different time intervals can be pooled to account for >90% of the radioactivity excreted in a given matrix. Ninety percent criteria may be lowered to prevent excessive sample dilution as for subject G1-06-F, where 24-48 h sample was excluded from the pool. Data for columns 1–4 are taken from a mass balance Debra report or similar, percent of each sample for pooling is defined by the analyst. Column 6 defines actual amounts of individual samples for executing a given pooling scheme. Information in columns 9 and 10 is needed for planning further experiments, for example, for calculating amount of sample required for injection into an HPLC-MS system to obtain a reasonable profile.

For plasma, several pooling approaches can be used depending on the amount of available sample and on the objective of the study (Penner et al., 2009). To obtain a pooled plasma sample for a single time point, equal volumes of each individual sample are mixed; to obtain a pooled plasma sample reflecting the area under the curve (AUC) of the parent drug and metabolites, volumes of individual samples

Table 43.4 A typical pooling scheme for excreta

1	2	3	4	5	6	7	8	9	10
Subject #	Timepoint, h	Total sample weight, g	Mean sample DPM/g	Percent of sample pooled	Weight pooled subject/time point, g	Total radioactivity of pooled sample, DPM	Total volume of pooled sample, g	Pooled sample DPM/g	Pooled sample ng-equiv/g
G1-01-M	6–24	270.0	113,077	5	13.5	1651655	24.7	66896	15185
	24–48	223.8	11,181	5	11.2				
G1-02-M	6–24	343.1	70,008	5	17.2	1357668	29.9	45430	10313
	24–48	254.6	12,308	5	12.7				
G1-03-M	0–6	22.0	683,334	5	1.1	1086794	18.0	60512	13736
	6–24	163.2	34,118	5	8.2				
	24–48	174.0	6520	5	8.7				
G1-04-F	6–24	101.9	175,277	5	5.1	965293	12.2	78896	17909
	24–48	142.8	10,120	5	7.1				
G1-05-F	0–6	48.5	548,915	5	2.4	1613352	8.0	200791	45579
	6–24	112.2	50,309	5	5.6				
G1-06-F	0–6	102.6	138,720	5	5.1	972649	25.1	38682	8781
	6–24	400.3	13,041	5	20.0				
	24–48	700.8	2252	5	0.0				
Data entry from mass balance part of a study				Defined by analyst	Sample amount to be pooled	Sum(Col4xCol6)	Sum(Col6)	Col7/Col8	Column 8 × conversion factor

Table 43.5 Example of volume calculation for AUC pooling of plasma

Time-point, h t_n	Volume to be pooled is proportional to Δt_n	Total sample volume to be pooled, µl V_n
0	1	40
1	2	80
2	3	120
4	4	160
6	4	160
8	6	240
12	16	640
24	12	480

Calculations are as follows:
$\Delta t_0 = t_1 - t_0$, $\Delta t_1 = t_2 - t_0 \ldots \Delta t_{n-1} = t_n - t_{n-2}$, $\Delta t_n = t_n - t_{n-1}$;
$V_n = k\Delta t_n$, where k is a proportionality constant, in this case $k = 40$.

should be calculated using the approach of Hamilton et al. (1981) and Hop et al. (1998) as shown in Table 43.5.

43.2.4 Sample Stabilization Techniques

Ensuring stability of samples for biotransformation analysis is a difficult task. Each sample for metabolite profiling is, in fact, a "black box" as there is no information about what is formed before the sample is analyzed. Therefore, it is important to take reasonable precautions when preparing a sample for metabolite identification in order to obtain a representative profile and prevent degradation of unstable metabolites.

There are known metabolic pathways leading to formation of the unstable metabolites. If a parent molecule contains a carboxylic acid group, it is likely that acylglucuronide conjugate would be formed. Acylglucuronides are unstable at neural and basic pH which are common for biological samples, and can degrade with a formation of the parent drug, thus, introducing a potentially large error in evaluation of a pharmacokinetic (PK) and metabolic profiles.

Instability of a drug and metabolites can also be a desired property, as in case of pro-drugs, which are designed to be unstable in biological conditions. After administration, pro-drugs form an active metabolite responsible for the biological activity. Rate of conversion of pro-drug to an active component can vary widely between species; therefore, it is important to capture an actual profile to model pharmacological and toxicological properties of pro-drugs intended for human use. Common classes of pro-drugs include esters converted by esterases (carboxylesterase (CE), butyrylcholinesterase (BChE) and paraoxonase (PON)), phosphates converted by phosphatases. Common approaches for stabilization of unstable metabolites are listed in Table 43.6. In addition, antioxidants such as ascorbic acid, sodium bisulfite, EDTA and mercaptoethanol could be added for the analytes that are chemically oxidized in biological fluids.

Table 43.6 Common approaches for sample stabilization

Metabolite/type of pro-drug	Method	Reference
Acyl-glucuronides	Acidification 4% v/v 2 M acetic acid in water and 1% v/v 0.5 M ascorbic acid 5% phosphoric acid to pH 1–2 Citric acid	Sparidans et al. (2008)
Esters enalapril–enalaprilat valaciclovir–aciclovir	Esterase inhibitors such as 0.1 mM Diisopropylfluorophosphate, and 0.1 mM eserine, sodium fluoride and phenylmethylsulfonyl fluoride	Berry et al. (2009)
Phosphates Psilocybin-psilocin	1% v/v phosphatase inhibitor such as Calbiochem Phosphatase Inhibitor Cocktail added to whole blood before separating plasma	
Amidases/hydrolases	Sodium fluoride and phenylmethylsulfonyl fluoride	

43.2.5 Extraction Methods

After pooling, samples are either directly injected onto HPLC-MS after centrifugation, or purified by further extraction. Metabolite detection in complex mixtures and biological samples is complicated by the presence of high level of endogenous material. Sensitivity and reliability of detection can be greatly improved by using sample preparation techniques.

Before any study sample is processed, method development is recommended to ensure that extraction procedures provide a good recovery of the drug-derived material. This approach can be conveniently implemented for radiolabeled samples. Initially, a radiolabeled parent compound can be spiked into a blank matrix, the spiked sample is then processed by an intended extraction method to measure the recovery of radioactivity. Another approach is to use a metabolite-rich sample, such as bile or urine collected from animals or in vitro preparation from microsomes or hepatocytes, for spiking into a blank matrix. In the latter case, an extraction recovery can also be calculated for metabolites.

Precipitation of proteins from a biological matrix by adding an organic solvent is the easiest and probably the most widely used procedure for extracting drugs and drug-related material from plasma, bile, and fecal homogenate. An organic solvent, usually acetonitrile, is added to the sample in a ratio of ~3:1, the mixture is well mixed and then centrifuged at ~3500–4000 rpm for 10–15 min. After supernatant is removed, the treatment can be repeated to achieve better recovery. For selected compounds, placing a mixture into a freezer overnight helps to increase extraction recovery. Different solvent mixtures can be used to achieve maximum recovery of drug-derived material. The recommended solvents are acetonitrile and methanol, addition of a small amount of DMSO can improve recovery.

A disadvantage of protein precipitation is that the resulting sample still contains high amount of an endogenous material, which makes the analysis difficult. Other well-established extraction methodologies, such as solid-phase extraction (SPE) and liquid–liquid extraction (LLE), can be applied to biological samples during metabolite identification studies and usually result in cleaner extract.

SPE is applicable to liquid samples such as plasma and urine. SPE is the most suitable technique for extracting drug-derived material/radioactivity from large volumes of a sample. A variety of SPE cartridges filled with media of various functionalities are offered by vendors. Reversed-phase or mixed-mode cartridges are the most widely used in metabolite identification studies since the majority of drugs and their metabolites are hydrophobic with weak acidic or basic properties. Before extraction, such cartridge is conditioned with methanol followed by water or appropriate aqueous solution (Jonsson et al., 2008). Sample is loaded onto a conditioned cartridge, washed with aqueous solution, and drug-derived material is then eluted with an organic solvent, usually methanol, acetonitrile or their mixture. Size of a cartridge is determined by a volume of the sample for extraction and estimated amount of a drug-derived material. Urine can be loaded onto an SPE cartridge directly; plasma can be diluted with water (1:1) to decrease viscosity.

An improved sample extraction procedure combining protein precipitation and SPE was used in (Prasad and Singh, 2009). After precipitation with acetonitrile, samples were centrifuged at 9000*g* for 10 min, and then a supernatant was separated and placed into a –20°C freezer. An organic layer was decanted from the aqueous frozen layer; the latter was thawed and loaded onto an SPE cartridge. The drug-derived material was eluted with acetonitrile, which was combined with the acetonitrile layer from protein precipitation. An additional step of freeze-liquid separation resulted in a cleaner extract and improved recovery of polar analytes.

Liquid–liquid extraction is another extraction procedure which is useful in biotransformation studies. LLE is based on the transfer of analyte(s) from one liquid phase to another, based on solubility. Liquid phases should be immiscible. Since one phase is usually water-based, the other one is organic such as ester (e.g. ethylacetate (Jerdi Mallorie et al., 2004)) or a mixture of dichloromethane/alcohol such as propanol (Jerdi Mallorie et al., 2004). Typical organic solvents used are hexane, chloroform, ethylacetate, or methyl *t*-butyl ether. After extraction, separation of two phases can be achieved by placing a sample into a freezer to freeze an aqueous part. Organic phase then can be separated – simply decanted – into a separate tube and dried.

A disadvantage of more thorough sample clean-up is the increased analysis time which is not acceptable in drug discovery. On-line approaches to sample preparation are an attractive option. Turbulent flow LC (TF-LC) – an accepted technique for quantitative bioanalysis (Herman, 2002, 2005; Herman and di Bussolo, 2009; Lim et al., 2001) – results in increased throughput and reduced ion suppression as a result of the on-line removal of endogenous proteinaceous material from plasma. This technique was used for the on-line cleanup of both plasma and bile samples for metabolite identification (Verdirame et al.; Lim et al., 2001; Herman and di Bussolo, 2009). Herman et al. employed a multiple injection approach for low-level metabolite characterization (Herman and di Bussolo, 2009). Other on-line sample preparation approaches, such as restricted access media (RAM) columns (Veuthey et al., 2004; Santos-Neto et al., 2007; Yamamoto et al., 2009), SPE (Chitneni et al., 2008; Dumasia et al., 2004; Gao et al., 2007; Stoob et al., 2005), and liquid-phase membrane extraction (Kuuranne et al., 2003; Schuegerl, 2005) were also used for metabolite identification. The application of such on-line extraction strategies is complicated by the risk of variable recovery of drug-related components during the extraction process, which is difficult to assess and difficult to monitor without a radiolabeled material. Differences in extraction recovery can lead to potentially significant errors in the assessment of major metabolic pathways. However, on-line approaches are still attractive due to their high throughput and improved metabolite detection.

43.2.6 Drying and Reconstitution

After extraction is complete, a supernatant is dried under stream of nitrogen or in a Speedvac; and a residue is then re-dissolved in solvent(s) compatible with an intended analytical method. Since the majority of separation is performed in the reversed-phase mode, the residue is usually dissolved in a mixture of mobile phases used for separation (various proportions of acetonitrile and a buffer). It is not recommended to use more then 50% of acetonitrile and/or methanol in a final sample injected onto a reversed-phase column, since it can greatly decrease a quality of separation, especially when a large volume of sample is injected. For compounds with low solubility it is beneficial not to dry extracted samples completely, or reconstitution may result in low recovery. Adding DMSO to a final sample (up to 50%) can help solubilization of a dried extract.

A sample reconstitution procedure can in some cases significantly affect quality of chromatographic separation, and therefore, sensitivity of an analytical method. A radiochromatographic profile of a fecal sample extracted with protein precipitation is shown in Fig. 43.3a. Clearly, the shape of chromatographic peaks is not satisfactory for a reliable data analysis. Analysis of LC-MS profile showed that chromatographic peaks of metabolites were split, which resulted in a loss of sensitivity and high background. There are multiple reasons for such peak splitting including:

- excessive organic solvent content in the injected sample and
- the presence of multiple forms of the same compound, e.g. protonated and deprotonated base/acid or complexes with endogenous compounds/metals, which exist in equilibrium and can be separated chromatographically.

It is hard to predict whether this problem would take place a priori, but a solution is usually simple: to decrease the organic content in the sample, to adjust pH, or to re-extract sample using an alternative method such as LLE to remove interference. In the example shown in Fig. 43.3, the extract A was mixed with 0.1% formic acid (1:2), centrifuged and re-injected again (Fig. 43.3b). This treatment removed peak

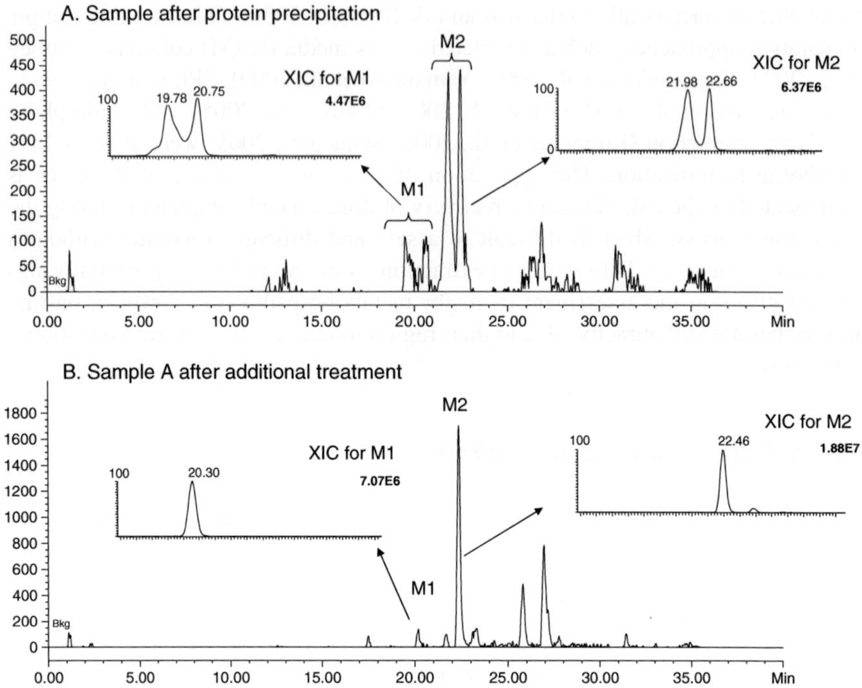

Fig. 43.3 Effect of extraction and reconstitution on chromatographic separation and sensitivity

splitting and significantly improved separation. This example clearly demonstrates that the sample reconstitution procedure plays an important role in obtaining high quality data.

43.2.7 Evaluation of Stability of the Drug and Metabolites

Evaluation of stability of the drug and its metabolites is an inherent part of a biotransformation study. While information about stability of a parent compound is part of any bioanalytical method development, stability of metabolites remains unknown. When a radiolabeled compound becomes available, it is strongly advised to test stability of a drug-derived material in various biological matrices under conditions typically encountered during metabolite identification work. Stability assessment should include at least short-term exposure to room temperature, autosampler, and freeze-thaw stability. Evaluation of long-term stability (3 months) can be also added to a study protocol.

To evaluate stability, radiochromatographic profiles of the same sample before and after stability test are compared (Fig. 43.4). No noticeable differences between radiochromatograms indicate little or no degradation of a drug-derived material.

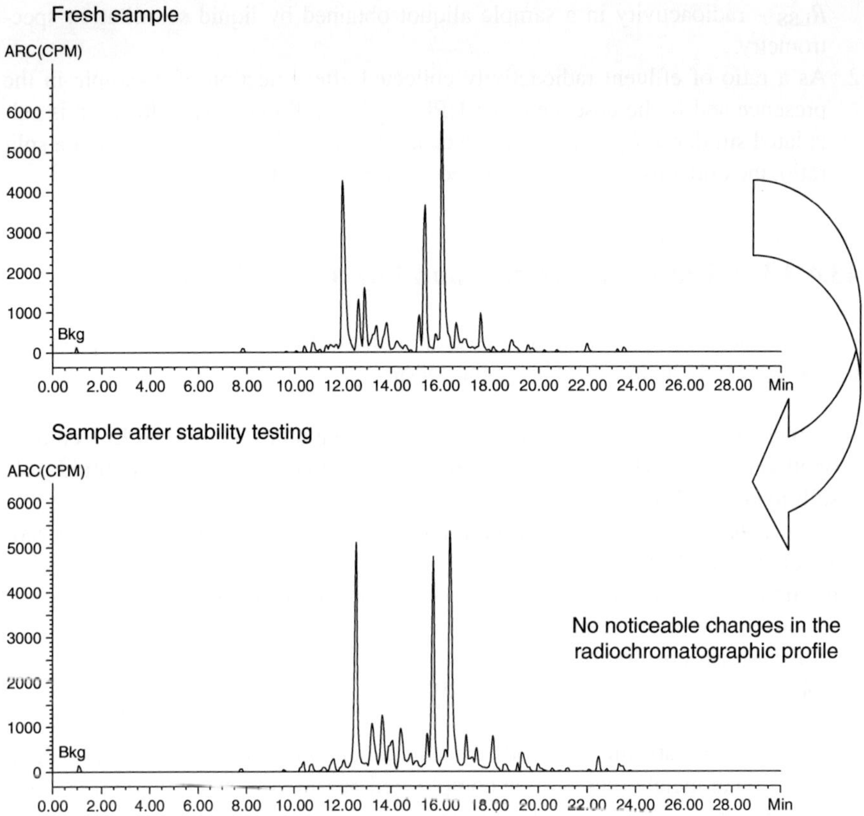

Fig. 43.4 Stability testing of biological samples for biotransformation

43.3 Column Recovery Calculations

Column recovery for the parent compound and drug-derived material should be calculated at a minimum for one metabolite-rich matrix such as bile, or for each matrix, to ensure that a drug-derived material is quantitatively eluted from an HPLC column. Column recovery can be calculated using two approaches:

1. As a ratio of effluent radioactivity collected after injection of sample ($R_{\text{post-column}}$) to radioactivity of the sample-spiked effluent from a blank sample (R_{spiked}).

$$R_{\text{post-column}} = \frac{\text{Volume of collected effluent} \times R_{\text{LSS}}}{\text{Volume counted} \times \text{injected volume} \times \text{split ratio}}$$

$$R_{\text{spiked}} = \frac{\text{Volume of collected effluent} \times R_{\text{LSS}}}{\text{Volume counted} \times \text{volume spiked}}$$

$$\text{Column recovery, \%} = \frac{R_{\text{post-column}}}{R_{\text{spiked}}} \times 100\%$$

R_{LSS} – radioactivity in a sample aliquot obtained by liquid scintillation spectrometry.

2. As a ratio of effluent radioactivity collected after injection of a sample in the presence and in the absence of an HPLC column. Radioactivity for both is calculated similar to $R_{post\text{-}column}$ in the example above. To avoid variation in a split ratio, the outlet to MS can be blocked (split ratio = 1).

43.4 Chemical Derivatization and Enzymatic Hydrolysis

Chemical derivatization, although largely under-utilized for metabolite profiling, can be extremely helpful to achieve the following:

- to increase retention time of hydrophilic metabolites on an HPLC column (methylation, acetylation, dansylation; derivatization with 1,2-dimethylimidazole sulfonyl chloride (DMSIC));
- to increase sensitivity of detection in a mass spectrometer (dansylation; derivatization with DMSIC);
- to improve stability of unstable metabolites (Liu and Hop, 2005);
- to confirm the presence of functional groups;
- to distinguish position of a modification such as oxidation, glucuronidation etc. (Johnson et al., 2003; Prakash et al., 2007a).

Common derivatization methods are summarized in Table 43.7. Structural elucidation of polar metabolites is often complicated because of low retention time, high mass spectrometric background, and low sensitivity. Chemical derivatization, such as dansylation, is useful for increasing the molecular weight and hydrophobicity of small polar metabolites to obtain better chromatographic separation. An increase in mass and an addition of a dimethylamino group can also improve the signal-to-noise ration of a metabolite in LC-MS (Prakash et al., 2007b).

A formation of unstable or reactive metabolites is encountered often especially in the discovery stage, when the structure is not yet optimized. The approaches to trapping unstable reactive metabolites with GSH, cyanide, etc. were described in Section 2.1.2. The use of other reagents, such as methoxylamine for the aldehyde group (Goldszer et al., 1981; Liu and Hop, 2005), ethyl chloroformate for stabilizing aminophenols (Idowu et al., 1995; Liu and Hop, 2005), acrylate for trapping thiols (Jemal and Hawthorne, 1997b; Jemal and Hawthorne, 1997a; Liu and Hop, 2005), hexafluoroacetylacetone for reacting with amidine group (Prakash and Cui, 1997), were also described in the literature for elucidation of unstable metabolites of the drugs in development. Chemical derivatization combined with LC-MS analysis is an extremely useful approach for structural elucidation of metabolites. Simple chemical reactions can be used to distinguish regioisomers, isobaric compounds, or to confirm presence of a functional group. A number of reviews are available in the literature (Liu and Hop, 2005; Lunn et al., 1998; Prakash et al., 2007; Quirke et al., 1994).

Table 43.7 Common derivatization reactions for metabolite identification (Quirke et al., 1994; Lunn et al., 1998; Liu and Hop, 2005)

Derivatization	Reaction	Short procedure
Prior to conducting derivatization, the sample should be dried under a stream of N_2 After a reaction is complete, samples usually should be dried and reconstituted in an appropriate mixture of solvents		
Acetylation	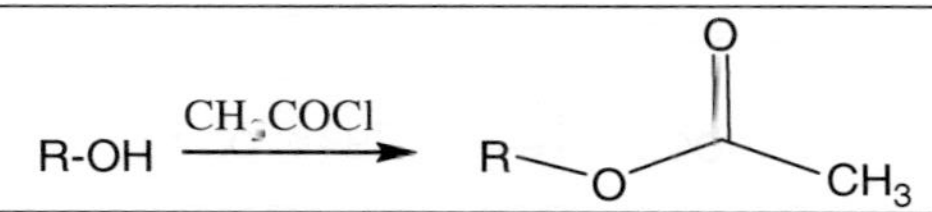	To a dried sample, add 200 µl of acetonitrile, vortex, and add 50 µl of acetyl chloride, incubate for 1 h, dry under N_2 (Schaefer et al., 1992)
	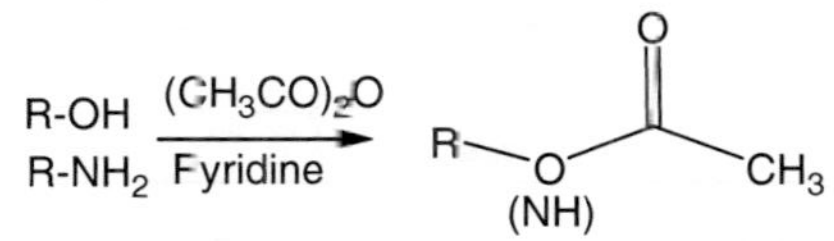	1. To a dried sample, add 200 µl of pyridine, incubate for 10 min at 37°C, then add 50 µl of acetic anhydride and stir for 1–2 h, dry under N_2 2. To approximately 10 µg of a metabolite, add 50 µl of pyridine–acetic anhydride (1:1), react for 15 min, dry (Schaefer et al., 1992)
	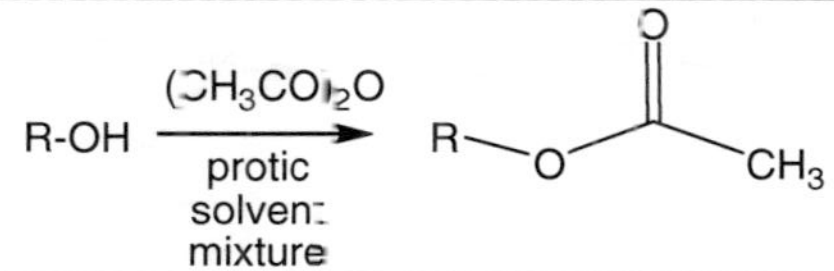	To approximately 10 µg of a metabolite, add 50 µl of water–ethanol–diisopropylethylamine (50:50:1), add 1 µl of acetic anhydride–acetonitrile (1:9), react for 15 min, dry (Schaefer et al., 1992)
	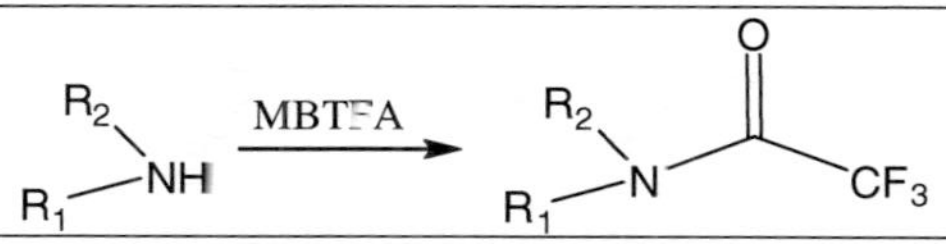	Dissolve a dry sample in acetonitrile, add 10 µl of MBTFA and incubate for 1 h, dry under N_2 (Schaefer et al., 1992; Shin et al., 1998) *N*-Methyl,bis-trifluoroacetamide (MBTFA)
Methylation/esterification	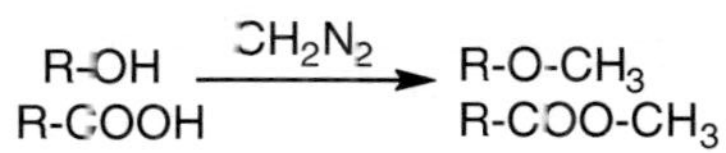	Dissolve a dried sample in 100 µl of methanol, cool to 0°C, bubble the solution with diazomethane produced in another flask from Diazald as described in (Lombardi, 1990). Use extreme caution! (Johnson et al., 2003; Kondo et al., 1996) Evaporate the solvent under N_2.

Table 43.7 (continued)

	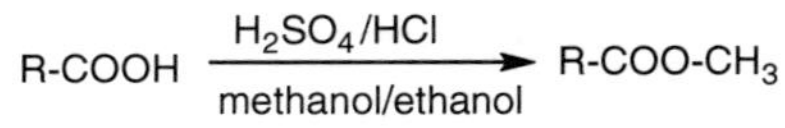	1. Add a solution of 10% sulfuric acid in methanol to the sample, incubate for 30 min at room temperature; extract the sample to remove the acid, evaporate the solvent under N_2. 2. Add 200 µl of a solution of ethanol saturated with HCl to the sample, incubate for 2 h at room temperature; extract the sample to remove the acid, evaporate the solvent under N_2.
N-Oxide formation	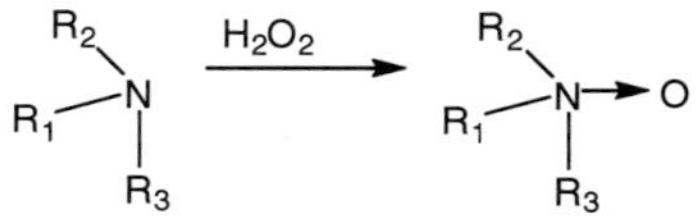	Add 100 µl of 3% (v/v) H_2O_2 to a dry sample; stirring incubate the mixture overnight, extract the sample or evaporate the solvent. This procedure works best for tertiary amines. Primary and secondary amines may give mixtures of products.
N-Oxide/sulfoxide reduction	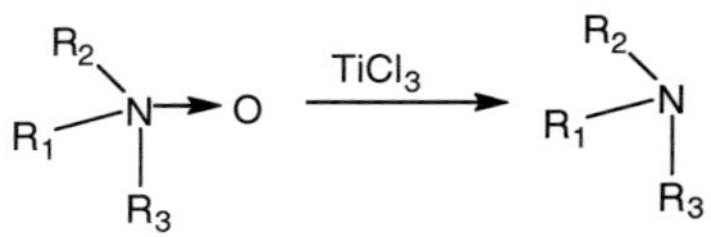	Reconstitute sample in 100 µl of mobile phase, add $TiCl_3$, 10% solution in 20% hydrochloric acid) in 10 µl increments until the color of the mixture remains a constant purple, incubate for 1–4 h at 0–5°C to reduce the possibility of sample degradation, neutralize acidity, extract, or dilute sample prior to LC-MS analysis.
Trimethylsilylation of R-OH and Ph-OH with *N*-trimethylsilylimidazole (Tri-Sil-Z)	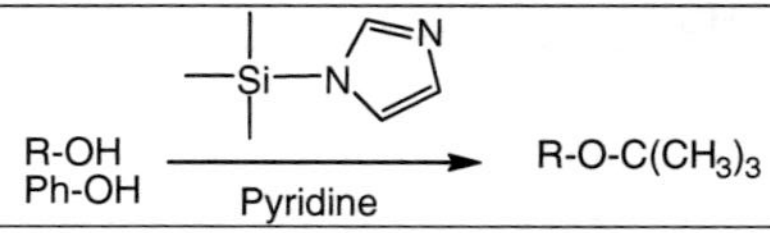	Dissolve a dry sample in 50 µl of acetonitrile, add 50 µl of Tri-Sil-Z in dry pyridine (1:4), incubate for 30 min at 50°C, dry under N_2 (Watanabe et al., 2003)
Amide hydrolysis	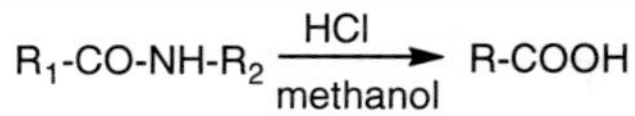	Add MeOH/HCl and incubate at 70°C for 30 min, dry. Due to the harshness of this procedure, multiple products may form because of unintended degradation of the metabolite
Oxidation of alcohols	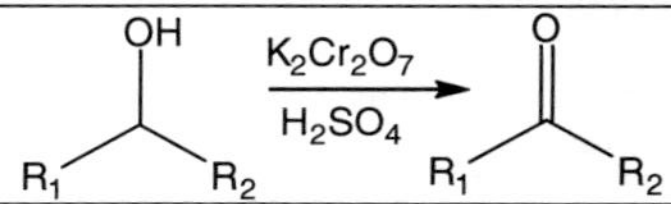	

Table 43.7 (continued)

Reduction of ketones, sulfoxides	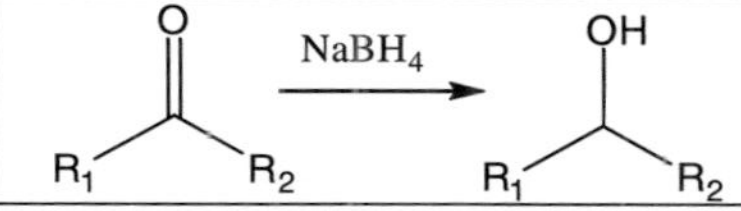	Add a pinch of sodium borohydride to 400 µl of a sample, mix well, incubate the mixture at the room temperature overnight, extract the sample or evaporate the solvent. (Elipe et al., 2003)
Chemical hydrolysis of glucuronides	R-O-gluc —HCl→ R-OH R-COO-gluc —NaOH→ R-COOH	1. Add 1 M HCl to the sample, incubate for 1 h at 50°C, extract the sample to remove the acid. 2. Add 1 N NaOH to the sample, incubate for 1 h at 50°C, extract the sample to remove the base.
Enzymatic hydrolysis of glucuronides/sulfate	R-O-gluc, R-O-sulf —Enzymatic hydrolysis→ R-OH	Add 0.5 ml of 0.1 M sodium acetate buffer pH 5.0 to 0.5 ml of a sample to adjust pH to 5, treat with 2500 units of β-glucuronidase/sulfatase (from Helix Pomatia, type H-1 with sulfatase activity); incubate the mixture in a shaking water bath at 37°C for 12 h, add 4 volumes of acetonitrile to precipitate proteins, centrifuge, and collect the supernatant, evaporate the solvent under N_2 and reconstitute in mobile phase.
Dansylation	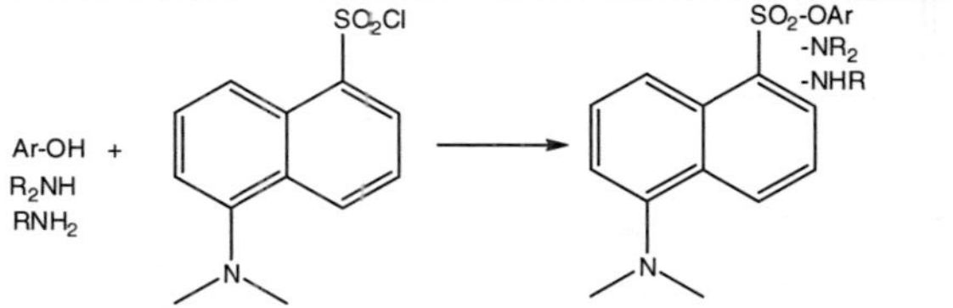	Add 100 µl of 100 mM $NaHCO_3$ solution (pH 8.5) to the residue, add 50 µl of dansyl chloride solution (1 mg/ml in acetone), incubate for 10 min at room temperature, evaporate the solvent (Dalvie and O'Donnell, 1998).
Guanidine/amidine group recognition	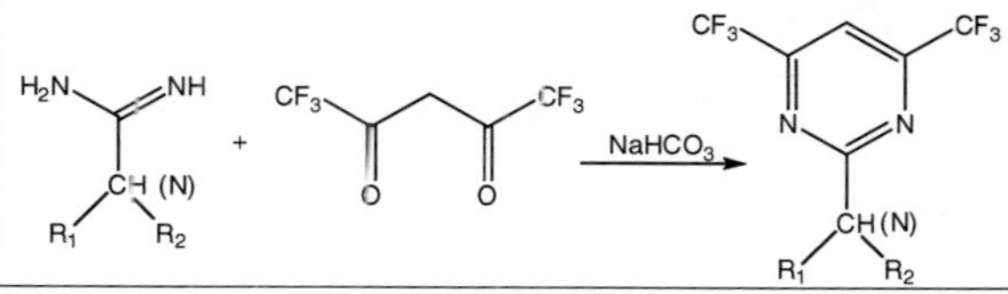	Add saturated $NaHCO_3$ to the sample, add 50 µl of hexafluoroacetyl-acetone in 500 µl toluene, incubate for 2 h at 100°C, dry (Malcolm and Marten, 1976).

Table 43.7 (continued)

Derivatization of phenols and amines with 1,2-dimethylimidazole sulfonyl chloride (DMSIC)	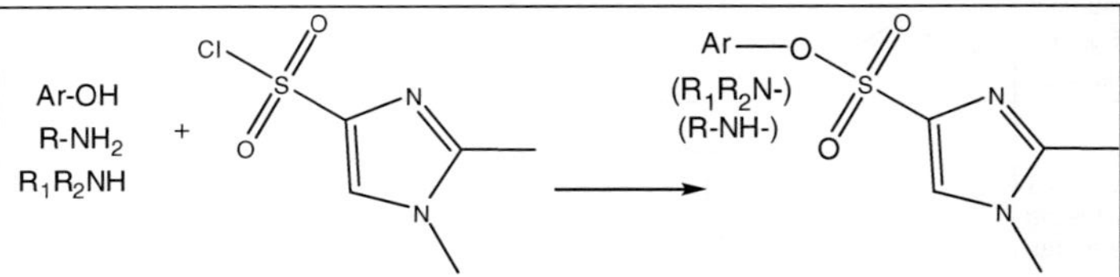	Salomonsson et al. (2008,2009)
Derivatization of thiols with acrylate		Add methylacrylate in 0.01 M phosphate buffer to the sample in phosphate buffer, sonicate with intermittent mixing for 10 min (Jemal and Hawthorne, 1997b; Jemal and Hawthorne, 1997a)
Derivatization of aldehydes with methoxylamine	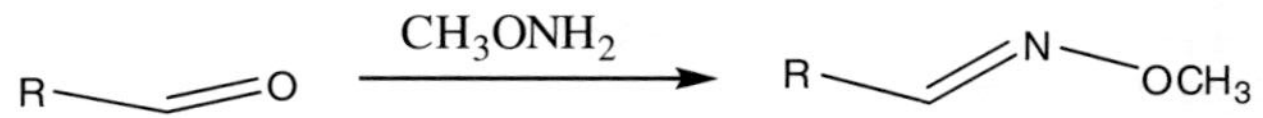	Goldszer et al. (1981)

Chemical derivatization can be successfully applied to distinguish the positions of glucuronidation (Schaefer et al., 1992; Kondo et al., 1996; Liu and Hop, 2005). This approach is based on selective acetylation of aromatic or aliphatic alcohols, or certain amines, by acetic anhydride in aqueous (protic) or non-aqueous (aprotic) solvents. Nucleophilic groups, such as amines and alcohols/phenols, are readily acetylated in non-aqueous solutions by acetic anhydride in the presence of a base such as pyridine. In aqueous solutions, the more nucleophilic amines are acetylated as well, but not the aliphatic hydroxyls. The groups which are blocked by glucuronides are not accessible to acetylation. Another approach is to use methylation by diazomethane (Kondo et al., 1996). The same approach can be used for distinguishing sites of hydroxylation.

The third approach, i.e. modification with DMISC which selectively reacts with phenols and amines similar to dansylchloride – was previously reported (Salomonsson et al., 2008, 2009). The latter was also successfully utilized for improving ionization of estrogens. Table 43.8 summarizes the use of derivatization for distinguishing isomers.

Table 43.8 Use of derivatization techniques for distinguishing sites of oxidation and glucuronidation (Liu and Hop, 2005; Schaefer et al., 1992)

Functional group	Acetylation in non-aqueous solution in the presence of pyridine	Acetylation in aqueous solution
Acetylation (Schaefer et al., 1992; Liu and Hop, 2005)		
Aliphatic alcohols		
R-OH	Yes	No
Ar-OH	Yes	Yes
Aliphatic amines		
$R\text{-}NH_2$, R_1R_2NH, $R_1R_2R_3N$, $Ar\text{-}NH_2$	Yes	Yes
$Ar_2\text{-}NH$	No	No
Methylation with diazomethane (Johnson et al., 2003)		
Aliphatic alcohols		
R-OH	No	
Ar-OH	Yes	
Aliphatic amines		
$R\text{-}NH_2$, R_1R_2NH, $R_1R_2R_3N$, $Ar\text{-}NH_2$	No	
Derivatization with DMSIC (Salomonsson et al., 2008, 2009)		
Aliphatic alcohols		
R-OH	No	
Ar-OH	Yes	
Aliphatic amines		
$R\text{-}NH_2$, R_1R_2NH, $R_1R_2R_3N$, $Ar\text{-}NH_2$	Yes	

R, aliphatic group; Ar, aromatic group

43.5 Sample Preparation for NMR

Since 1950s, nuclear magnetic resonance (NMR) has been used for the characterization of synthetic compounds. NMR allows the unambiguous determination of a compound structure by generating its unique ^{1}H and/or ^{13}C resonances characteristic of various functional groups and intramolecular environments. To ensure successful analysis, all samples must be processed by some preliminary method prior to submission for LC-NMR. The goal of sample preparation is to

- reduce background biological matrix;
- concentrate the metabolite of interest; and
- remove closely eluting peaks, which are disproportionately large.

It is not necessary to recover the metabolite quantitatively. A recovery should be sufficient to obtain the minimum sample amount for reliable structural elucidation. Routinely used for LC-MS, precipitation of proteins with an organic solvent is not a satisfactory sample preparation for an LC-NMR analysis. Each sample preparation procedure should be customized according to the source of the sample, the complexity of the matrix and the chemistry of the metabolite. SPE is recommended as the most general and universal method. In general, sample preparation for NMR should include at least one of the following techniques:

- Liquid–liquid extraction (a hexane wash is recommended to remove lipids from a metabolite extract)
- SPE (normal or reversed-phase using appropriate size cartridge)
- Preparative HPLC (reversed-phase, normal phase, or ion exchange)

Other techniques such as the use of ion-exchangers for bile salt removal or the use of affinity resins for specific moieties (e.g. glucuronides) isolation may also be useful in certain situations. The goal of a sample clean-up procedure is to balance purity with a sample loss; therefore, iterative preparative HPLC purifications should be avoided. Use of methanol should be avoided to limit interference in the NMR spectrometer.

43.6 Summary

Metabolism is a key determinant governing both pharmacokinetics and a clinical response of the majority of marketed drugs, and a great deal of effort is now directed at assessing metabolic stability and profiles in the early stages of drug development. In recent years, furthermore, increased attention is being paid by both pharmaceutical industries and regulatory agencies to the role of metabolites as potential mediators of the toxicological findings of new chemical entities, and on how best to deal with the issues, raised by metabolites during safety assessment programs for drug development. In the last three decades there has been an explosion in the development of robust, user friendly, and highly sensitive mass spectrometers, and a variety of chromatographic or sample preparation methodologies for the detection and characterization of metabolites in complex mixtures from both in vitro and in vivo studies. In addition, advances have also been made in software approaches, such as Metabolynx, Lightsight, Metworks or Intellixtract from various vendors, for processing data. Development will undoubtedly continue in all these areas in the future. Future applications of such techniques may enable a much more robust and appropriate assessment of the safety risk associated with drug candidates and their metabolic pathways.

Conducting a successful biotransformation study can be challenging; a low recovery of drug-derived material often has little room for improvement. Therefore, development of new, effective and efficient sample processing methods remains of high interest for drug metabolism scientists and analytical chemists. HPLC-MS/MS will continue to play a pivotal role in the identification of drug metabolites due to its high sensitivity and selectivity. The best combination for accomplishing rapid, accurate metabolite identification may require robotic systems for sample preparation, chromatography to minimize the suppression of ionization of drug-related material and in silico software to predict and find possible metabolites. These tools can be combined with LC-MS to determine exact mass measurements (accurate mass) for sample analysis, LC-MS-NMR, and chemical derivatization for further metabolite structure confirmation and elucidation.

References

Baillie, T.A., Cayen, M.N., Fouda, H., Gerson, R.J., Green, J.D., Grossman, S.J., Klunk, L.J., LeBlanc, B., Perkins, D.G., and Shipley, L.A. (2002). Drug metabolites in safety testing. Toxicol Appl Pharmacol *182*, 188–196.

Berry, L.M., Wollenberg, L., and Zhao, Z. (2009). Esterase activities in the blood, liver and intestine of several preclinical species and humans. Drug Metab Lett *3*, 70–77.

Brandon, E.F.A., Raap, C.D., Meijerman, I., Beijnen, J.H., and Schellens, J.H.M. (2003). An update on *in vitro* test methods in human hepatic drug biotransformation research: Pros and cons. Toxicol Appl Pharmacol *189*, 233–246.

Chitneni, S.K., Serdons, K., Evens, N., Fonge, H., Celen, S., Deroose, C.M., Debyser, Z., Mortelmans, L., Verbruggen, A.M., and Bormans, G.M. (2008). Efficient purification and metabolite analysis of radiotracers using high-performance liquid chromatography and on-line solid-phase extraction. J Chromatogr, A *1189*, 323–331.

Dalvie, D.K., and O'Donnell, J.P. (1998). Characterization of polar urinary metabolites by ion-spray tandem mass spectrometry following dansylation. Rapid Commun Mass Spectrom *12*, 419–422.

Dumasia, M.C., Morelli, I., and Teale, P. (2004). Detection of eltenac in the horse: identification of phase I metabolites in urine by capillary gas chromatography-mass spectrometry and the determination of excretion profile by liquid chromatography-mass spectrometry. Chromatographia *59*, S115–S121.

Gao, L., Cheng, X., Zhang, J., and Burns, D.J. (2007). A generic fast solid-phase extraction high-performance liquid chromatography/mass spectrometry method for high-throughput drug discovery. Rapid Commun Mass Spectrom *21*, 3497–3504.

Gelhaus, S.L., and Blair, I.A. (2009). LC-MS analysis in drug metabolism studies. Drugs Pharm Sci *186*, 355–372.

Goldszer, F., Tindell, G.L., Walle, U.K., and Walle, T. (1981). Chemical trapping of labile aldehyde intermediates in the metabolism of propranolol and oxprenolol. Res Commun Chem Pathol Pharmacol *34*, 193–205.

Hamilton, R.A., Garnett, W.R., and Kline, B.J. (1981). Determination of mean valproic acid serum level by assay of a single pooled sample. Clin Pharmacol Ther *29*, 408–413.

Herman, J.L. (2002). Generic method for on-line extraction of drug substances in the presence of biological matrices using turbulent flow chromatography. Rapid Commun Mass Spectrom *16*, 421–426.

Herman, J.L. (2005). The use of turbulent flow chromatography and the isocratic focusing effect to achieve on-line cleanup and concentration of neat biological samples for low-level metabolite analysis. Rapid Commun Mass Spectrom *19*, 696–700.

Herman, J.L., and di Bussolo, J.M. (2009). Turbulent-flow LC-MS: Applications for accelerating pharmacokinetic profiling and metabolite identification. Mass Spectrom Drug Metab Pharmacokinet, 311–340.

Hop, C.E.C.A., and Prakash, C. (2005). Metabolite identification by LC-MS: Applications in drug discovery and development. Prog Pharm Biomed Anal *6*, 123–158.

Hop, C.E.C.A., Wang, Z., Chen, Q., and Kwei, G. (1998). Plasma-Pooling Methods To Increase Throughput for *in Vivo* Pharmacokinetic Screening. J Pharm Sci *87*, 901–903.

Idowu, O.R., Peggins, J.O., Brewer, T.G., and Kelley, C. (1995). Metabolism of a candidate 8-aminoquinoline antimalarial agent, WR 238605, by rat liver microsomes. Drug Metab Dispos *23*, 1–17.

Jerdi Mallorie, C., Daali, Y., Oestreicher Mitsuko, K., Cherkaoui, S., and Dayer, P. (2004). A simplified analytical method for a phenotyping cocktail of major CYP450 biotransformation routes. J Pharm Biomed Anal *35*, 1203–1212.

Johnson, K., Shah, A., Jaw-Tsai, S., Baxter, J., and Prakash, C. (2003). Metabolism, pharmacokinetics, and excretion of a highly selective N-methyl-D-aspartate receptor antagonist, traxoprodil, in human cytochrome P450 2D6 extensive and poor metabolizers. Drug Metab Dispos *31*, 76–87.

Jonsson, G., Stokke Tone, U., Cavcic, A., Jorgensen Kare, B., and Beyer, J. (2008). Characterization of alkylphenol metabolites in fish bile by enzymatic treatment and HPLC-fluorescence analysis. Chemosphere *71*, 1392–1400.

Kamel, A., and Prakash, C. (2006). High performance liquid chromatography/atmospheric pressure ionization/tandem mass spectrometry (HPLC/API/MS/MS) in drug metabolism and toxicology. Curr Drug Metab *7*, 837–852.

Kondo, T., Yoshida, K., Yoshimura, Y., Motohashi, M., and Tanayama, S. (1996). Characterization of conjugated metabolites of a new angiotensin II receptor antagonist, candesartan cilexetil, in rats by liquid chromatograph/electrospray tandem mass spectrometry following chemical derivatization. J Mass Spectrom *31*, 873–878.

Kuuranne, T., Kotiaho, T., Pedersen-Bjergaard, S., Rasmussen, K.E., Leinonen, A., Westwood, S., and Kostiainen, R. (2003). Feasibility of a liquid-phase microextraction sample clean-up and liquid chromatographic/mass spectrometric screening method for selected anabolic steroid glucuronides in biological samples. J Mass Spectrom *38*, 16–26.

Lim, H.K., Chan, K.W., Sisenwine, S., and Scatina, J.A. (2001). Simultaneous Screen for Microsomal Stability and Metabolite Profile by Direct Injection Turbulent-Laminar Flow LC-LC and Automated Tandem Mass Spectrometry. Anal Chem *73*, 2140–2146.

Liu, D.Q., and Hop, C.E.C.A. (2005). Strategies for characterization of drug metabolites using liquid chromatography-tandem mass spectrometry in conjunction with chemical derivatization and on-line H/D exchange approaches. J Pharm Biomed Anal *37*, 1–18.

Lombardi, P. (1990). A rapid, safe and convenient procedure for the preparation and use of diazomethane. Chemistry & Industry

Lunn, G., Hellwig, L., and Cecchini, A. (1998). Handbook of Derivatization Reactions of HPLC (New York, NY, Wiley).

Ma, S., Chowdhury, S.K., and Alton, K.B. (2006). Application of mass spectrometry for metabolite identification. Curr Drug Metab *7*, 503–523.

Malcolm, S.L., and Marten, T.R. (1976). Determination of debrisoquin and its 4-hydroxy metabolite in plasma by gas chromatography/mass spectrometry. Anal Chem *48*, 807–809.

Prakash, C., and Cui, D. (1997). Metabolism and excretion of a new antianxiety drug candidate, CP-93,393, in cynomolgus monkeys. Identification of the novel pyrimidine ring cleaved metabolites. Drug Metab Dispos *25*, 1395–1406.

Prakash, C., Shaffer, C.L., and Nedderman, A. (2007). Analytical strategies for identifying drug metabolites. Mass Spectrom Rev *26*, 340–369.

Prakash, C., Sharma, R., Gleave, M., and Nedderman, A. (2008). *In vitro* screening techniques for reactive metabolites for minimizing bioactivation potential in drug discovery. Curr Drug Metab *9*, 952–964.

Prasad, B., and Singh, S. (2009). *In vitro* and *in vivo* investigation of metabolic fate of rifampicin using an optimized sample preparation approach and modern tools of liquid chromatography-mass spectrometry. J Pharm Biomed Anal *50*, 475–490.

Quirke, J.M.E., Adams, C.L., and Van Berkel, G.J. (1994). Chemical derivatization for electrospray ionization mass spectrometry. 1. Alkyl halides, alcohols, phenols, thiols, and amines. Anal Chem *66*, 1302–1315.

Schaefer, W.H., Goalwin, A., Dixon, F., Hwang, B., Killmer, L., and Kuo, G. (1992). Structural determination of glucuronide conjugates and a carbamoyl glucuronide conjugate of carvedilol: use of acetylation reactions as an aid to determine positions of glucuronidation. Biol Mass Spectrom *21*, 179–188.

Schuegerl, K. (2005). Extraction of primary and secondary metabolites. Adv Biochem Eng/Biotechnol *92*, 1–48.

Shin, H.-S., Park, B.-B., Choi, S.N., Oh, J.J., Hong, C.P., and Ryu, H. (1998). Identification of new urinary metabolites of famprofazone in humans. J Anal Toxicol *22*, 55–60.

Sparidans, R.W., Lagas, J.S., Schinkel, A.H., Schellens, J.H.M., and Beijnen, J.H. (2008). Liquid chromatography-tandem mass spectrometric assay for diclofenac and three primary metabolites in mouse plasma. J Chromatogr, B: Anal Technol Biomed Life Sci *872*, 77–82.

Stoob, K., Singer, H.P., Goetz, C.W., Ruff, M., and Mueller, S.R. (2005). Fully automated online solid phase extraction coupled directly to liquid chromatography-tandem mass spectrometry. J Chromatogr A *1097*, 138–147.

Watanabe, Y., Nakajima, M., Ohashi, N., Kume, T., and Yokoi, T. (2003). Glucuronidation of etoposide in human liver microsomes is specifically catalyzed by UDP-glucuronosyltransferase 1A1. Drug Metab Dispos *31*, 589–595.

Part XIV
Sample Preparation in Analysis of Exotic and Limited Availability Specimens

Chapter 44
Development of Micro-scale Sample Preparation and Prefractionation Methods in LC-MS-Based Proteomic Studies

Lan Dai, Chen Li, and David M. Lubman

Abstract One of the most significant difficulties in studying clinical specimens such as cancer stem cells by mass spectrometry based proteomics approaches arises from the small sample size which limits the use of conventional proteome profiling procedures. The ideal strategy to avoid sample loss is to entirely diminish sample preparation, but current mass spectrometry techniques do not allow direct identification/quantification of proteins present in complex biological matrices. In this chapter, we describe a workflow that combines a single-tube cell lysis procedure using PPS silent surfactant (3-[3-(1,1-bisalkyloxyethyl)pyridin-1-yl]propane-1-sulfonate) and a capillary isoelectric focusing (CIEF) step to specifically analyze a small number of cells. This workflow can be directly coupled to reversed-phase liquid chromatography (RPLC) followed by nanoflow rate electrospray ionization mass spectrometry (nanoESI- MS) to study the global proteome expression pattern with the proteins from 10^4 pancreatic cancer tissue stem cells (equal to <1 μg of protein material) which is 100 times less than the number of cells needed in a typical proteome profiling experiment.

Keywords Micro-scale · Sample preparation · MS-compatible surfactants · Capillary isoelectric focusing (CIEF)

44.1 Introduction

44.1.1 Significance of Developing and Optimizing a Sample Preparation Method Suited to Analyze Samples with Small Quantities

Proper sample preparation for MS-based analysis is a critical step in proteomics studies because the quality and reproducibility of sample extraction and preparation

---

D.M. Lubman (✉)
Department of Surgery, Medical School, University of Michigan Medical Center, Ann Arbor, MI 48109-0650, USA
e-mail: dmlubman@umich.edu

A.R. Ivanov, A.V. Lazarev (eds.), *Sample Preparation in Biological Mass Spectrometry*, DOI 10.1007/978-94-007-0828-0_44,

for downstream analysis significantly impacts the separation and identification results of LC-MS. It is also a time-consuming and labor-consuming step. Given the complexity of proteomes, there is no one standard method for preparing protein samples for MS analysis. Although a recent publication in *Nature Methods* (Wiśniewski et al., 2009) reported an unbiased sample preparation method for proteome analysis, to achieve the best sensitivity, sample treatment still must be chosen based on both the nature of the sample as well as the analytical techniques that will follow (Ahmed, 2009). There is a growing interest in analyzing the change of protein expression from presorted clinical samples such as cancer stem cells, micro-biopsies and laser capture microdissection (LCM). The difficulty of analyzing such samples arises from their small sample sizes, which require special sample processing to minimize sample loss in order to enable quantitative proteomic measurements (Wang et al., 2005). This is particularly true in comparative proteomics where we are usually most interested in the change of low-abundance proteins between experimental and control samples. Therefore, it is becoming increasingly important to develop and optimize the sample preparation methods with the ability to handle small quantities of sample, minimize sample loss, and work compatibly with downstream analytical techniques.

44.1.2 Improvements Targeted at Cell Lysis, Protein Digestion and Prefractionation

A typical shotgun proteomic strategy includes four major steps: (1) cell lysis, (2) protein digestion, (3) protein prefractionation, (4) follow-up LC-MS analysis. While the improvements in LC-MS analysis are beyond the scope of this chapter, here we focus on the improvements at each of the first three steps and use online nanoRPLC coupled with ESI-MS for peptide detection. The combination of a nanoRP HPLC and nanoESI source mass spectrometry is commonly employed in proteomic studies due to its high throughput and compatibility with various prefractionation techniques. Different sample preparation methods may be considered if any other type of mass spectrometry is used.

44.1.2.1 Prefractionation for Bottom-Up Proteomics

There are two prefractionation strategies for bottom-up proteomics. One includes one or two-dimensional gel or liquid-based separation of the proteins from cell lysate with electrophoresis or liquid chromatography followed by trypsin digestion and nanoLC-ESI-MS (Yan et al., 2003). This strategy is able to resolve proteins with PTMs and sometimes provide information on the protein mass and p*I*. Alternatively, the cell lysate is directly digested and prefractionated with electrophoresis or chromatography separation before nanoLC-ESI-MS (Washburn et al., 2001). For small amounts of sample (under 1 μg), the concern for the first strategy is that each protein fraction contains even less protein than the original amount, so that the efficiency of trypsin digestion of these protein fractions may be significantly lower than direct digestion of the cell lysate (Strader et al., 2006). In addition, separating intact proteins at the micro scale level still presents a challenging technical issue due to the size of intact proteins (Lee et al., 2001). Capillary electrophoresis (CE) is the only capillary-based method compatible with mass spectrometry for prefractionation of intact proteins with high molecular weight (Zhou and Johnston, 2004). However, precipitation of proteins in the CE environment may still occur, especially in capillary isoelectric focusing. Direct digestion of cell lysate followed by prefractionation of the resulting peptides is thus a preferable method for analysis of small amounts of sample. The disadvantage of peptide level prefractionation is the distribution of tryptic peptides from high-abundance proteins into multiple fractions, where ionization of peptides from low-abundance protein may be suppressed (Dowell et al., 2008).

44.1.2.2 Cell Lysis: PPS as a Preferred MS Compatible Surfactant

Cell lysis, including cell membrane disruption and protein solubilization, is the first step of sample preparation. For limited amounts of sample, a high yield of protein extraction is critical for the final detection and identification of low-abundance proteins. In the case of a pancreatic tissue stem cell study, the target cells (about <1% of total cells in the tissue are stem cells) are harvested from mice pancreas, labeled with specific surface markers, and presorted with a flow cytometer. Limited by the scale of the work, this process usually results in only 10^4 cells for analysis.

There are various methods of cell disruption and the homogenization methods used for proteomics study which can be divided into mechanical, ultrasonic pressure, freeze–thaw, osmotic, and detergent lysis (Bodzon-Kulakowska et al., 2007). Compared to other types of methods, ultrasonic pressure homogenization does not introduce impurities to the sample and causes minimal changes to the properties of the proteins. This mild method of cell disruption is best suited for the homogenization of presorted cells in water solution or small pieces of soft tissue and has thus been selected for our experiment.

To improve the solubility of the proteins from the disrupted cells, numerous cell lysis buffers and protocols have been developed which are targeted at different cell types and different protein categories. A typical cell lysis buffer contains chaotropes (e.g. urea), detergents (e.g. CHAPS/SDS), reducing agents (DTT) and protease inhibitors. However, these must be removed from the solution or diluted to a certain concentration before the digestion of proteins, since they may reduce the activity of the enzyme. Recently, the use of acid-cleavable surfactant and organic solvent have been proposed as a single-tube lysis protocol which may improve cell membrane disruption, protein solubilization and trypsin digestion without the addition of extra chemicals that may cause sample loss during extensive clean-up steps (Chen et al., 2007).

PPS Silent® Surfactant was initially designed, synthesized, and characterized to help solubilization of hydrophobic proteins for direct MALDI-MS analysis (Norris et al., 2003). The hydrophilic and hydrophobic chains of this surfactant are linked with an acid-cleavable bond. Following protein solubilization, breakage of this link eliminates the presence of surfactant-associated influence in the solution (shown in Fig. 44.1), such as foaming and aggregation. The lower panel of Fig. 44.1 shows the MALDI spectra of PPS before and after treatment with acid. Before the cleavage, the characteristic parent ion 430.59 ($[M+H]^+$) and 860.22 ($[2\ M+H]^+$) are

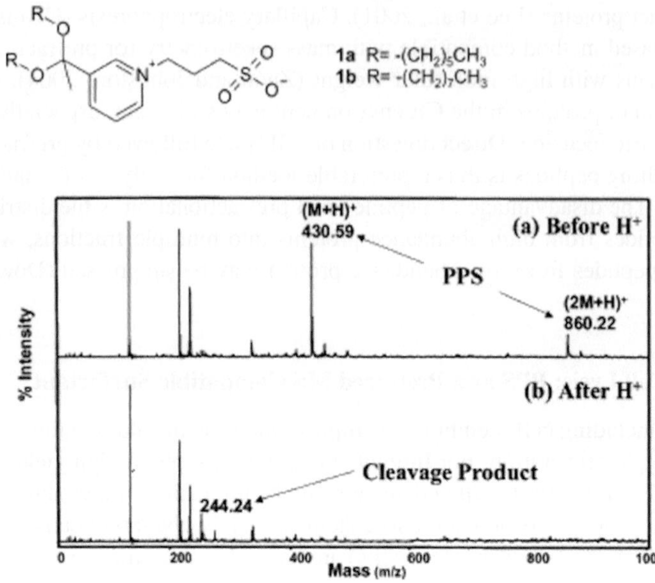

Fig. 44.1 *Upper panel* shows the chemical structure of PPS. *Lower panel* shows the MALDI mass spectra of PPS before and after acid cleavage. This figure is from Norris et al. (2003) with permission

dominant in the range where m/z is greater than 400. These two peaks have completely disappeared after the cleavage and the cleavage products are detected by acquiring a mass spectrum that contains low molecular weight fragments. Upon its commercial introduction as PPS Silent Surfactant for cell disruption and protein extraction by Protein Discovery, it was also found to increase the number of proteins identified by ESI/MS/MS in shotgun proteomic experiments (Chen et al., 2007). PPS Silent Surfactant is unique among special surfactants for protein solubilization in that it produces no oily film or cloudy pellet in the protein sample solution which may cause proteins losses. The degradation of the reagent through acid-cleavage circumvents the major limitation of conventional protein solubilization protocols where detergent suppresses the ionization of low abundance species. Considering the advantages of PPS surfactant, it is used with a single-tube protein extraction protocol in this study.

44.1.2.3 Prefractionation: CIEF as a Preferred Preparative Liquid Electrophoretic Method

A preparative or first dimension liquid separation is frequently used before reversed-phase HPLC-MS to maximize the resolution of the enormous number of peptides from the digestion of a cell lysate, thus the overlapping of the peptides can be reduced to optimize the performance of the identification. The available preparative liquid separations include ion-exchange, reversed-phase, normal-phase, size-exclusive chromatography and multiple types of electrophoretic separation

(Sandra et al., 2009). The chromatography-based separations often present undesired retention of peptides with strong interaction with the chromatographic resin and resulting poor recovery rate. In contrast, electrophoresis is usually performed in an open capillary, where peptides are separated based on migration under the influence of an electric field rather than based on retention. The inner wall of the capillaries used for electrophoresis is often coated with inert material to reduce the absorption of the analytes and the surface area available for interaction with analyte molecules is substantially lower when compared to chromatographic columns due to the absence of packing materials. Therefore, electrophoresis provides a much higher sample recovery rate than chromatographic methods. Capillary isoelectric focusing (CIEF) has emerged as a useful tool for protein/peptide prefractionation because of its high resolution and orthogonal separation mechanism versus RP-HPLC (Chen et al., 2003a,b). This p*I* based separation provides an optimal resolution of 0.01 pH unit, which indicates a peak capacity of 700 in a pH range from 3 to 10. CIEF focusing is conducted through establishing a linear pH gradient in the capillary with a series of carrier ampholytes. The peptides are positioned along the pH gradient according to their isoelectric points and concentrated in narrow bands. Once the focusing is complete, the concentrated peptide bands, maintained by an electric field, are hydrodynamically transferred into samples loops or 96-well plates for LC-MS injection. The capillary used in CIEF is usually neutrally coated with hydropropyl cellulose to prevent electroosmotic flow (EOF) and absorption of analytes. Thus, sample loss can be reduced to a minimal level, which is essential in handling small quantities of peptides. The carrier ampholytes for the creation of the pH gradient present in each fraction need to be removed by a nanotrap column filled with C18 material before the RPLC column due to their competition with peptides for electrospray ionization. Another electrophoretic first dimensional separation for shotgun proteomic study utilizes capillary isotachophoresis/capillary zone electrophoresis (CITP/CZE) (Guo et al., 2006), which separates and enriches the peptides depending on their abundance. Low-abundance peptides are better enriched than high-abundance ones. CITP/CZE is more robust and provides improved proteome coverage compared to CIEF. However, in comparative experiments, using CITP/CZE may introduce bias to the measurement of the peptides due to the abundance dependent enrichment; hence CIEF is preferred in our study.

44.2 Material and Methods

44.2.1 Biological Sample

About 10^4 pancreatic cancer stem cells were obtained from mice xenografts after sorting by flow cytometry and gently washed three times with cold PBS (pH 7.4) by repetitive pipetting, followed each time by centrifugation at $1000\times g$ for 5 min at 4°C. After the third wash, excessive PBS was gently removed after cell pellets precipitated at the bottom of the tube.

44.2.2 *Single-Tube Protein Extraction*

PPS (Protein Discovery, Knoxville, TN) powder was dissolved in 50 mM ammonia bicarbonate (pH 7.8) and was added in each tube to a final concentration of 0.2% (w/v). Approximately 100 μl of the cell suspension was then vortexed and incubated at 60°C for 10 min, followed by sonication in an ice-water bath for 2 h.

44.2.3 *Trypsin Digestion, Deactivation and PPS Cleavage*

An aliquot of 5 mM DTT was added and the mixture was incubated at 60°C for 30 min. After cooling, 5 mM iodoacetamide was added and the mixture was placed in the dark at room temperature for 30 min in order to allow the carboxymethylation reaction of cysteine residues. Fifty millimolar ammonia bicarbonate (pH 7.8) was then added at a dilution ratio of 1:5. The pH was re-checked by spotting 2 μl of sample on a pH strip to ensure a desirable basic condition (pH 7.8). Then L-1-tosylamido-2-phenyletyl chloromethylketone modified sequencing-grade porcine trypsin (Promega, Madison, WI) was added at an enzyme: substrate ratio of 1:50 (w/w). The mixture was incubated at 37°C in a water bath with agitation for 18 h. Formic acid (FA) was then added to a final concentration of 2% to quench the proteolysis. Following termination, the acidified mixture was placed in a 37°C water bath for 4 additional hours to facilitate the hydrolysis and cleavage of PPS.

44.2.4 *Sample Clean-Up*

Salts present in the tryptic digests often cause interruption of the CIEF separation due to high electric current. An acidified tryptic peptide mixture was desalted by a peptide micro-trap (Michrom, Auburn, CA) and eluted with 98% acetonitrile (ACN) and 0.3% FA, followed by complete drying using a SpeedVac concentrator (Labconco, Kansas City) and stored in a –80°C freezer for future use. All chemicals were purchased from Sigma unless otherwise specified.

Note, a high electrical current carried by the salt load will interfere with electrophoretic separation and reduces the resolution. Theoretically salts from previous sample preparation steps need to be removed if present in concentrations >100 mM. Applying a desalting step prior to CIEF is necessary in this experiment; otherwise a high concentration of salt would be present after drying the sample down.

44.2.5 *Fractionation by CIEF*

44.2.5.1 Coating of the Capillary

A 70 cm CIEF (100 μm i.d. 365 μm i.d.) capillary was coated with hydropropyl cellulose to eliminate EOF and absorption of analytes onto the capillary wall with a modified procedure from a previous publication (Chen et al., 2003a,b). The long capillary was initially filled with 5–10% hydroxypropyl cellulose solution. This polymer is difficult to dissolve and the solubilization process may take up to 12 h. The capillary was then slowly purged using nitrogen at 120 psi to remove the solution. To immobilize the polymer layer onto the capillary inner wall, the capillary was heated from 60 to 220°C at a rate of 5°C/min for 20 min, maintained at 220°C for another 10 min in a GC oven, then followed by cooling to room temperature. A well-coated capillary should have a lifetime up to a hundred analyses if stored properly. Loss of capillary coating can occur after prolonged use and can be observed by the increase of EOF, recognized through the delayed appearance of peaks during the focusing step, early migration of peaks during mobilization, and a loss of resolution. Gradual adsorption of peptides/proteins to the capillary wall can also cause deterioration of capillary performance in CIEF, which can be minimized by rinsing the capillary with diluted acid (e.g. 10 mM phosphoric acid) and water (Landers, 1996).

44.2.5.2 Configuration of CIEF

CIEF was performed on a Beckman PA800 instrument with off-line sample collection settings as shown in Fig. 44.2. The capillary was initially filled with sample gel which was made by mixing the CIEF gel (Beckman) with 2% ampholytes 3–10 (Beckman). Whenever a new capillary is used for the first time, the injection time needs to be tuned to make sure the capillary is filled with the sufficient sample volume but not excessively. Approximately 1 μg tryptic peptides was then reconstituted by cIEF sample gel and vortexed for 30 min to allow maximum protein recovery. 1 mM Sodium hydroxide solution (pH 11) and 0.1 M phosphate acid solution (pH 2) were employed as catholyte and anolyte, respectively. One end of the loaded capillary was immersed in the anolyte, while the other end of the capillary was contained inside of a coaxial metal tubing through which the catholyte was provided as a sheath flow through. The flow rate was controlled by a syringe pump at 5 μl/min, and was adjusted to ensure that a proper droplet formed at the outlet end to carry the fractionated peptides into individual wells in the 96-well sample plate. The outlet of the capillary was attached to a fraction collector so that each droplet could be deposited into each well in an automated fashion. Isoelectric focusing was performed at 21 kV (300 V/cm) over the entire capillary. The current decreased continuously as the peptides were focused and the process was considered complete after the current no longer changed. The focused bands of peptides were sequentially mobilized slowly under pressure towards the cathode and delivered as droplets with catholyte sheath flow into individual wells on a sample plate. The fractions were collected with a modified Beckman HPLC sample collector. Focusing and mobilization steps took approximately 100 min. This process was then repeated three times where each CIEF run utilized less than 300 ng of peptide material. Around 30 fractions were collected for each run and 3 technical replicates of prefractionated peptides were generated for follow-up LC-MS analysis.

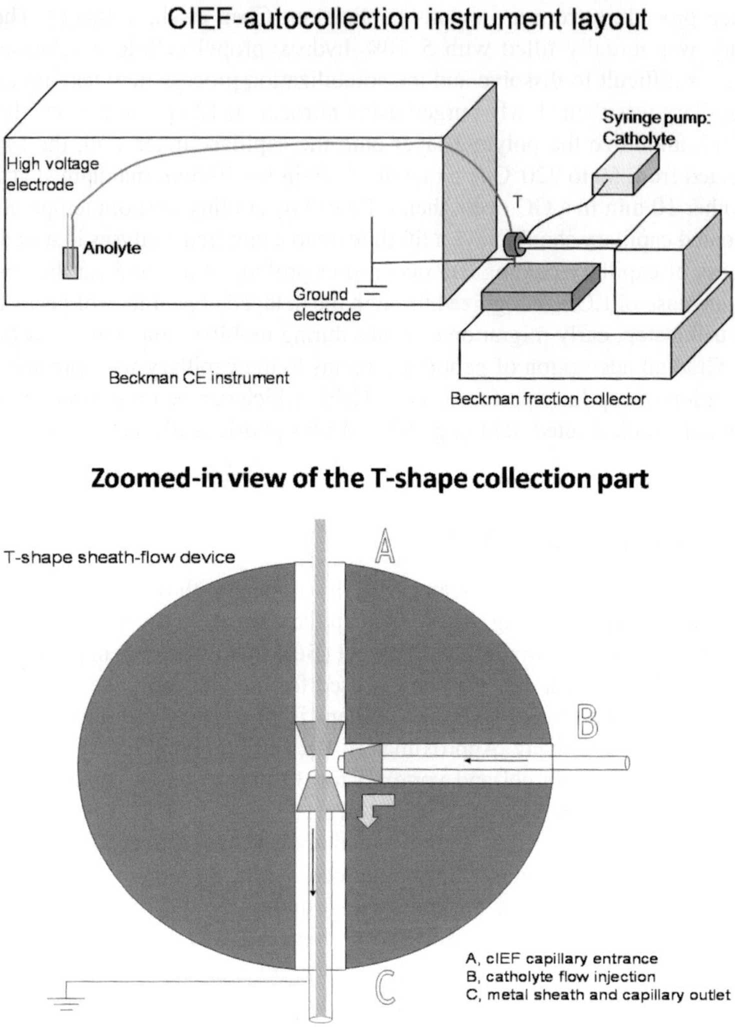

Fig. 44.2 *Upper panel* shows the overall layout of the offline CIEF collection system; *lower panel* shows the zoomed-in configuration of the fraction collection part. This figure is from Dai et al. (2009). Copyright Wiley-VCH Verlag GmbH & Co. KGaA. Reproduced with permission

44.3 Results and Discussion

An overview of the workflow used in these experiments is summarized in Fig. 44.3. This PPS-based protein extraction and digestion procedure requires 2 days of processing time. Our data shows that 10^4 of pancreatic cancer stem cells provide a sufficient amount of peptide sample for three replicate runs of CIEF followed by nanoHPLC-MS/MS. For comparison purposes, the PPS-based cell lysis and trypsin

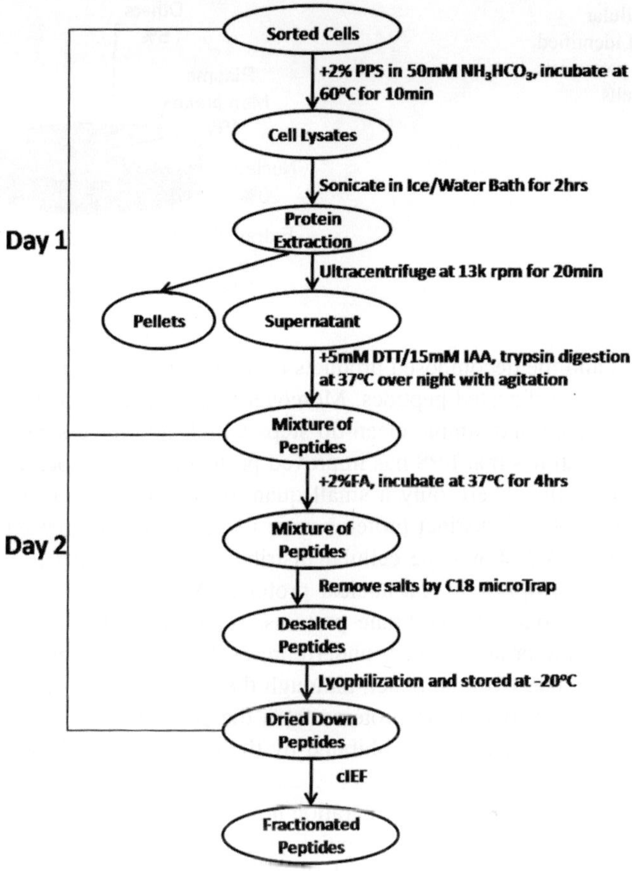

Fig. 44.3 Summary of the workflow

digestion method using 10^4 human pancreatic carcinoma cells (Mia PaCa cell line) in parallel with a commonly employed urea-based lysis and digestion protocol (7.5 M urea, 2.5 M thiourea, 12.5 v/v glycerol, 62.5 M Tris-HCl, 2.5% (w/v) *n*-octylglucoside (*n*-OG) and 1% v/v protease inhibitor cocktail). After digestion, the peptide mixtures prepared with these two methods were separately subjected to ESI-MS/MS analysis without prefractionation. The number of proteins identified with the PPS-based method was fivefold higher than proteins identified with the urea-based method.

Chen et al. (2007) reported a comprehensive study of the trypsin digestion efficiency of 5 μg of protein by MS compatible detergents in mixed organic-aqueous systems. The number of peptides/proteins identified by using PPS and RapigestSF (Waters) in Tris buffer ranked at the top among a variety of MS compatible detergents, and are a big improvement compared to the use of urea. This is mainly because protease is also subject to the denaturation effects of urea at the working

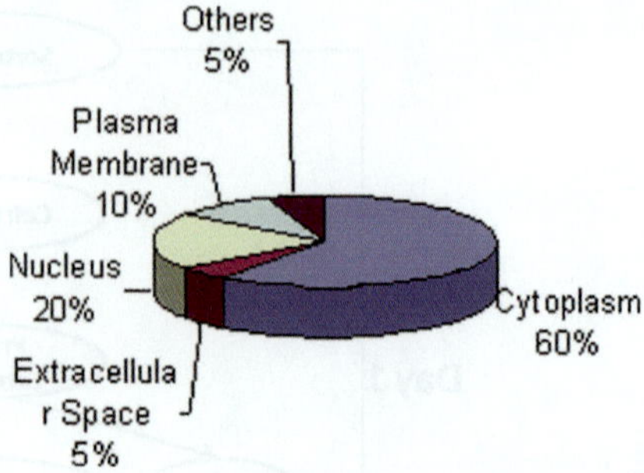

Fig. 44.4 Cellular distribution of identified proteins from pancreatic cancer stem cells

concentration and the decomposed products of urea under heat can block digestion sites or modify the digested peptides. Moreover, the acid-labile nature of the surfactant allows rapid and simple clean-up steps which minimize sample loss. Our experiment also shows that PPS has improved performance in shotgun proteomics over urea, especially where only a small quantity of protein is available. In our example, a total of 922 distinct proteins were identified from three replicate runs of CIEF. Figure 44.4 shows the cellular distribution of identified proteins where the majority are cytoplasmic and nuclear proteins. Around 10% of the total identified proteins are plasma membrane proteins, suggesting PPS has the ability to extract hydrophobic proteins. Other underrepresented membrane proteins may precipitate and associate with the pellet, although the pellet was not seen in our case. Resuspension of the membrane proteins from the pellet would be an ideal solution; however, the procedure is prohibited by the extremely small quantity of the sample.

Using CIEF as the preparative separation not only maximizes the sample recovery rate but also increases the potential peak capacity due to its orthogonal separation mechanism compared to RP-HPLC. Figure 44.5 shows a distribution plot

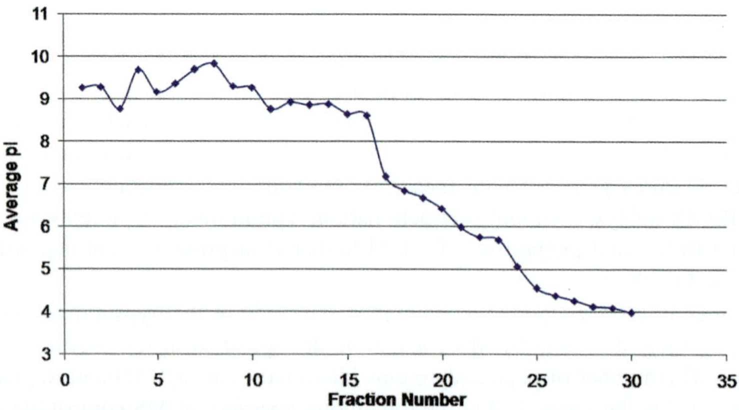

Fig. 44.5 p*I* distribution plot. Fraction number shown in the *X*-axis is plotted against the average of peptides' p*I* value within each fraction shown in the *Y*-axis

of the averaged theoretical isoelectric point of the peptides identified in the CIEF fractions using less than 300 ng of peptides. This pattern presents a non-perfect linearity which indicates that the peptides are separated based on their p*I*s. Peptides with p*I* in the region between 4 and 8 shows improved linearity implicating a higher resolution than the p*I* region from 8 to 10. Since CIEF is performed in an open capillary, resolution of the fractions collected at the end of capillary is very sensitive to back pressure caused by the on-line injection device coupled to nanoLC-MS. Off-line collection is preferable in maintaining the sharpness of the peptide bands with the use of a sheath-flow device at the collecting end of the capillary providing electric grounding and catholyte solution which also carries the eluted fractions into a 96-well plate (Dai et al., 2009). Compared to SCX, decreased smearing of peptides is observed with CIEF due to significant concentration of analytes (Chen et al., 2003a,b), thus it allows dividing the whole separation into more fractions with a low risk of further dilution of individual peptides. Typically 30 fractions were taken from CIEF for LC-MS based on consideration of total instrument time required (Balgley et al., 2008).

References

Ahmed, F.E. (2009). Sample preparation and fractionation for proteome analysis and cancer biomarker discovery by mass spectrometry. J Sep Sci *32*, 771–798.

Balgley, B.M., Wang, W.J., Song, T., Fang, X.P., Yang, L., and Lee, C.S. (2008). Evaluation of confidence and reproducibility in quantitative proteomics performed by a capillary isoelectric focusing-based proteomic platform coupled with a spectral counting approach. Electrophoresis *29*, 3047–3054.

Bodzon-Kulakowska, A., Bierczynska-Krzysik, A., Dylag, T., Drabik, A., Suder, P., Noga, M., Jarzebinska, J., and Silberring, J. (2007). Methods for samples preparation in proteomic research. J Chromatogr B *849*, 1–31.

Chen, E.I., Cociorva, D., Norris, J.L., and Yates, J.R. (2007). Optimization of mass spectrometry compatible surfactants for shotgun proteomics. J Proteome Res *6*, 2529–2538.

Chen, J.Z., Balgley, B.M., DeVoe, D.L., and Lee, C.S. (2003a). Capillary isoelectric focusing-based multidimensional concentration/separation platform for proteome analysis. Anal Chem *75*, 3145–3152.

Chen, J.Z., Gao, J., and Lee, C.S. (2003b). Dynamic enhancements of sample loading and analyte. J Proteome Res *2*, 249–254.

Dai, L., Li, C., Shedden, K.A., Misek, D.E., and Lubman, D.M. (2009). Comparative proteomic study of two closely related ovarian endometrioid adenocarcinoma cell lines using CIEF fractionation and pathway analysis. Electrophoresis *30*, 1119–1131.

Dowell, J.A., Frost, D.C., Zhang, J., and Li, L.J. (2008). Comparison of two-dimensional fractionation techniques for shotgun proteomics. Anal Chem *80*, 6715–6723.

Guo, T., Lee, C.S., Wang, W.J., DeVoe, D.L., and Balgley, B.M. (2006). Capillary separations enabling tissue proteomics-based biomarker discovery. Electrophoresis *27*, 3523–3532.

Landers, J.P. (1996). Handbook of Capillary Electrophoresis (2nd Ed, Chapter 4) (Boca Raton, FL, CRC Press), pp. 101–138.

Lee, S.W., Berger, S.J., Martinovic, S., Pasa-Tolic, L., Anderson, G.A., Shen, Y.F., Zhao, R., and Smith, R.D. (2001). Direct mass spectrometric analysis of intact proteins of the yeast large ribosomal subunit using capillary LC/FTICR. PNAS *99*, 5942–5947.

Norris, J.L., Porter, N.A., and Caprioli, R.M. (2003). Mass spectrometry of intracellular and membrane proteins using cleavable detergents. Anal Chem *75*, 6642–6647.

Sandra, K., Moshir, M., D'hondt, F., Tuytten, R., Verleysen, K., Kas, K., Francois, I., and Sandra, P. (2009). Highly efficient peptide separations in proteomics. Part 2: Bi- and multidimensional liquid-based separation techniques. J Chromatogr B *877*, 1019–1039.

Strader, M.B., Tabb, D.L., Hervey, W.J., Pan, C.L., and Hurst, G.B. (2006). Efficient and specific trypsin digestion of microgram to nanogram quantities of proteins in organic-aqueous solvent systems. Anal Chem *78*, 125–134.

Wang, H.X., Qian, W.J., Mottaz, H.M., Clauss, T.R.W., Anderson, D.J., Moore, R.J., Camp, D.G., Khan, A.H., Sforza, D.M., Pallavicini, M., Smith, D.J., and Smith, R.D. (2005). Development and evaluation of a micro- and nanoscale proteomic sample preparation method. J Proteome Res *4*, 2397–2403.

Washburn, M.P., Wolters, D., and Yates, J.R. (2001). Large-scale analysis of the yeast proteome by multidimensional protein identification technology. Nat Biotechnol *19*, 242–247.

Wiśniewski, J.R., Nagaraj, N., and Mann, M. (2009). Universal sample preparation method for proteome analysis. Nat Methods *6*, 359–362.

Yan, F., Subramanian, B., Nakeff, A., Barder, T.J., Parus, S.J., and Lubman, D.M. (2003). A comparison of drug-treated and untreated HCT-116 human colon adenocarcinoma cells using a 2-D liquid separation mapping method based upon chromatofocusing PI fractionation. Anal Chem *75*, 2299–2308.

Zhou, F., and Johnston, M.V. (2004). Protein characterization by on-line capillary isoelectric focusing, reversed-phase liquid chromatography, and mass spectrometry. Anal Chem *76*, 2734–2740.

Chapter 45
Revisiting Jurassic Park: The Isolation of Proteins from Amber Encapsulated Organisms Millions of Years Old

Gary B. Smejkal, George O. Poinar, Jr., Pier Giorgio Righetti, and Feixia Chu

Abstract Dominican Republic amber from the Oligo-Miocene epoch, 20–40 million years ago, was interrogated for residual protein. Tandem mass spectrometric analysis of trypsin digests of proteins from two silver-stained bands excised from SDS-PAGE led to the identification of 84 peptides from 19 *Saccharomyces* proteins. All peptides were identified from one high molecular weight gel band, suggesting a high degree of cross-linking of these proteins. This study reports the first experimental data on the identification of prehistoric proteins from amber.

Keywords Amber · *Hymenaea protera* · Oligo-Miocene epoch phenylacylthiazolium bromide · Mass spectrometry-based proteomics

45.1 Introduction

The mass extinction of the dinosaurs marking the Cretaceous-Tertiary boundary pales in magnitude compared to other lesser known mass extinction events such as the Permian-Triassic. Five major extinction events, and several smaller ones over the past 540 million years, have resulted in the extinction of 99% of all of the species that ever lived on earth. This is furthered by nearly two orders of magnitude, since by even the most conservative estimates, less than a tenth of all species presently inhabiting the earth are known, while the vast majority is still undiscovered (Wilson, 2007). Hence, it would appear that the field of biology is based entirely on what has been learned from fewer than 0.1% of all past and present species.

G.B. Smejkal (✉)
Harvard Medical School, Boston, MA 02115, USA
e-mail: gary_smejkal@hms.harvard.edu

F. Chu (✉)
Department of Molecular, Cellular & Biomedical Sciences, University of New Hampshire, Durham, NH, USA
e-mail: fvy2@unh.edu

A.R. Ivanov, A.V. Lazarev (eds.), *Sample Preparation in Biological Mass Spectrometry*, DOI 10.1007/978-94-007-0828-0_45,

In 1994 came the first reports of the isolation of DNA from a Late Cretaceous dinosaur bone exhumed from bituminous strata (Woodward et al., 1994). The following year, the cloning and sequencing of six putative dinosaur DNA fragments derived from a Cretaceous dinosaur egg fossil found in Xixia Basin, China (Yin et al., 1996), was later dismissed as the recovered sequences were found to be more closely related to fungi rather than to reptiles or birds (Wang et al., 1997), suggesting contamination and further fueling the growing consensus that dinosaurian DNA could not survive for such geological time spans.

Pawlicki (1995) first observed osteocytes in bone from an 80 million year old dinosaur *Tarbosaurus bataar*, in which approximately 10% of the cells stained with ethidium bromide suggested the presence of DNA, but not confirming its source. Using histochemical stains, the presence of carbohydrates and lipids was also observed (Pawlicki, 1977a,b). Later, Schweitzer et al. (2005, 2006) observed osteocytes and other "soft tissue replicates" preserved in demineralized *Triceratops horridus* and *Tyrannosaurus rex* bone, but failed to show evidence of residual DNA. Wick et al. (2001) showed the alleged soft tissues found preserved in a 50 million year old bat were due to the replacement of the original soft tissues by lawns of bacteria. Kaye et al. (2008) demonstrated that microbial biofilms form "endocasts" in which three-dimensional structure is preserved with microscopic detail, but in which the original organic material has been totally replaced by minerals. Raff et al. (2008) corroborated the finding, and identified DNA from soft tissue replicates as belonging to several bacterial taxa.

Perhaps the most promising circumstance enabling the preservation of DNA or proteins over millions of years comes from amber, the fossilized resin of several tree types, including leguminous trees. The unfossilized resins are comprised largely of terpenoids, labdanoids, and phenolics which rapidly dehydrate the included specimen, a prerequisite for preservation, as well as possess anti-bacterial, anti-fungal, and anti-inflammatory properties that intervene with usual decomposition. When the specimen was completely engulfed in the resinous flow (having to occur within seconds), it resulted in unprecedented preservation later observed in amber fossils. In one spectacular specimen, a soldier beetle is caught in the act of spraying its chemical defence on an attacking predator, the chemical plumes still visible in this freeze frame of the Middle Cretaceous biosphere from 100 million years ago (Poinar et al., 2007). Delicate structures and remarkable prehistoric scenarios are frequently preserved in amber (Fig. 45.1). Some spectacular examples include a 110 million year old spider web complete with visible droplets of adhesive and several adhered insect prey (Penalver et al., 2006), a pollinating bee with its mesoscutellum still intercalated with an orchid pollinarium (Ramírez et al., 2007), and two copulating fireflies juxtaposed for eternity (Poinar and Poinar, 1999).

Michael Crichton's novel *Jurassic Park* proposed the recovery of dinosaur DNA from the alimentary tracts of hematophagous insects preserved for millions of years in amber (Crichton, 1990). Though Crichton's work was purely fictional, amber inclusions have shown remarkable preservation of organisms at the tissue and cellular levels, and reptilian blood cells have been identified in partially digested blood meals from parasitic insects encapsulated in Cretaceous amber (Poinar and Poinar, 2004; Poinar, 2008). In fact, one theory is that dinosaur populations may have

Fig. 45.1 Preservation of fine structure in amber, including individual ommatidia of the compound eye of a formicine identified in Oligo-Miocene amber 20–30 million years old. Magnification bar equals 0.6 mm. Ants were relatively scarce during the explosive radiations of angiosperms and pollinating insects during the Early Cretaceous, representing less than 1% of all insect species at the time. Today, over 12,000 species are known, collectively comprising 15–25% of the terrestrial animal biomass

already been endangered by insect-borne diseases prior to the Chicxulub impact that triggered the mass extinctions that ended the Cretaceous Era (Poinar and Poinar, 2008).

Transmission electron microscopy has revealed that subcellular components such as nuclei, endoplasmic reticulum, ribosomes, and mitochondria were still intact in 40 million year old insects preserved in amber (Poinar and Hess, 1982). At the molecular level, DNA has been isolated from the 30 million year old fossil termite *Mastotermes electrodominicus* (DeSalle et al., 1992). Clones were reportedly chimeric in structure, with half of the clone identical to the termite sequence and the other half identical to a contaminating sequence (DeSalle et al., 1993). Contaminating sequences could have derived from microbial flora associated with the primary organism, and in support of that hypothesis, *Bacillus* DNA from the abdominal tissues of the extinct stingless bee *Proplebeia dominicana* was isolated and amplified (Cano et al., 1994). The oldest fossil DNA amplified and sequenced were fragments isolated from a 120 to 135 million year old amber-encapsulated nemonychid weevil (Cano et al., 1993).

From stringent DNA–DNA hybridization studies, *Staphylococcus* isolated from Eocene amber revealed similarities of only 38% to contemporary species, insinuating a previously undiscovered species (Lambert et al., 1998). The phylogenic significance of such analyses was exemplified using DNA sequences from Dominican amber to show the extinct leguminous tree *Hymenaea protera* was most closely related to the extant African species *H. verrucosa* (Poinar et al., 1993).

That DNA might survive geological time spans, and even then, under the most extraordinary circumstances, was further evidenced by Cano and Borucki (1995) who revived and cultured bacterial spores from 25 to 40 million year old amber inclusions. The oldest viable spores were reportedly isolated from a 250 million year old salt crystal (Vreeland et al., 2000, 2006; Nickle et al., 2002).

Termed "paleoproteomics" by Ostrom (Buckley et al., 2008), the search for biochemical sequences of phylogenic significance has shifted to proteins, which are thought to better endure geological time spans than DNA (Smejkal et al., 2009). As early as 1954, Abelson reported amino acids recovered from fossils hundreds of millions of years old, and based on his experiments, he proposed some amino acids might be stable for billions of years (Abelson, 1954, 1956). That original organic material might be preserved over such time spans was exemplified in one report of porphyrins and phycobilins from Precambrian fossils 1.7 to 2.6 billion years old (Kolesnikov and Egorov, 1977).

Later, Gurley et al. (1991) recovered amino acids from vertebra of a Late Jurassic sauropod *Seismosaurus*. Embery et al. (2000, 2003) reported non-collagenous proteins in the compact bone of a 120 million year old *Iguanodon bernissartensis*. Immunochemical analyses by Muyzer et al. (1992) and Schweitzer et al. (1999, 2007) implied that epitopes could be preserved over millions of years.

Recently, Asara et al. (2007) identified peptides with sequence homology to avian collagen from *T. rex*, further strengthening theories of an avian-theropod lineage and providing direct evidence of proteins surviving from Late Cretaceous skeletal elements. Mass spectrometry has also revealed peptides in bone exudates of an 80 million year old hadrosaur *Brachylophosaurus canadensis* (Schweitzer et al. 2009).

This chapter describes the isolation of high molecular mass protein aggregates from 30 to 40 million year old Dominican amber. Tandem mass spectrometry (MS/MS) identified several proteins with sequence homology to *Saccharomyces cerevisiae* proteins. While mass spectra do not unambiguously confirm the source of these peptides, the high degree of protein cross-linking suggests that they are not of contemporary origin.

45.2 Materials and Methods

45.2.1 Amber Specimens

Specimens were obtained from mines in the Cordillera Septentrional of the Dominican Republic. Dominican amber was produced by the leguminous tree, *Hymenaea protera* (Poinar, 1991; Poinar and Poinar, 1999). Dating of Dominican amber is still controversial with the latest proposed age of 15–20 million years based on foraminifera (Iturralde-Vincent and MacPhee, 1996) and the earliest as 30–45 million years based on coccoliths (Schlee, 1990). A range of ages for Dominican amber may be likely since the amber fossils are associated with turbiditic sandstones of the Upper Eocene to Lower Miocene (Draper et al., 1994). Dominican amber is secondarily deposited in sedimentary rocks, which makes a definite age determination difficult (Poinar and Mastalerz, 2000).

45.2.2 Sample Preparation

All procedures were performed under sterile conditions in a laminar flow hood. Amber pieces containing *Hymenaea protera* leaves and inflorescence (Fig. 45.2) were first scrubbed in 5% SDS with a dental brush, then heated to 90°C in 2% SDS and copiously rinsed in water. The amber was rinsed with 100% ethanol just prior to fracture. The amber was fractured with a sterile ceramic pestle, usually leaving plant material exposed on the faces of several amber fragments. Plant material was collected with a dental pick. Unavoidably, some amber remained associated with plant material such that the triturate resembled a salt and pepper mix

Plant triturates were extracted in 125 mM Tris-HCl pH 6.8 containing 2% SDS, 5 mM tributylphosphine, 20 mM aminoethylbenzene sulfonyl fluoride, 10 mM EDTA and 25 mM phenylacylthiazolium bromide (PTB; Prime Organics, Woburn, MA, USA). The samples were extracted for 100× 100 s in PULSE Tubes at 35,000 psi using a Barocycler NEP 3229 (Pressure Biosciences, South Easton, MA, USA). Each PULSE Tube was coupled to the insert of a sterile Ultrafree CL centrifugal filter (Millipore Corporation, Danvers, MA, USA) and evacuated centrifugally at 1000 RCF for 1 min.

Filtrates were applied directly to 8–16% total monomer concentration (%T) linear polyacrylamide gradient gels overlaid with a 4% T stacking gel (BioRad,

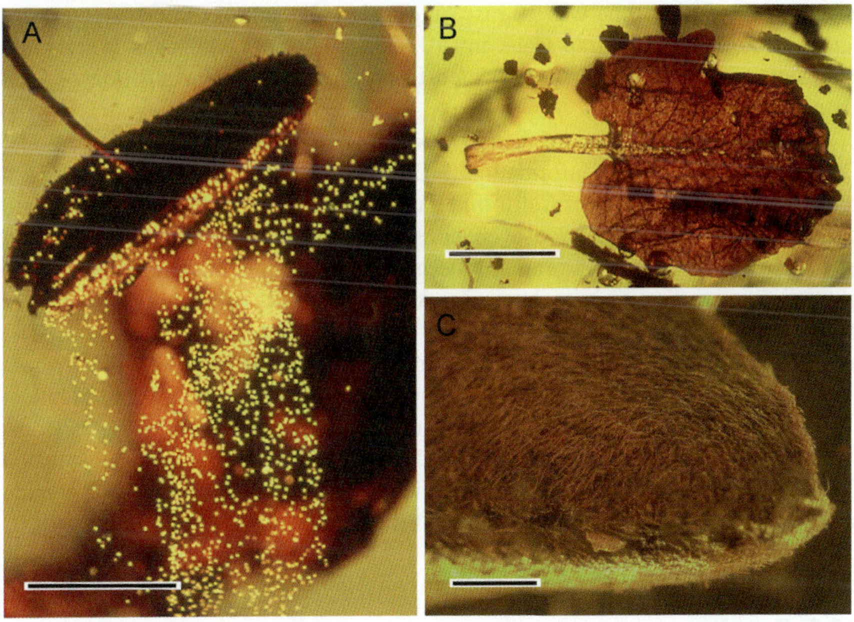

Fig. 45.2 Elements of *Hymenaea protera* preserved in Dominican Republic amber 20–40 million years old. **A**. Anther, filament, and dispersing pollen grains. Magnification bar equals 0.5 mm. **B**. Petal and claw. Bar equals 4 mm. **C**. High magnification of the tip of a stipule showing the remarkable preservation of filamentous trichomes. Bar equals 0.3 mm

Hercules, CA, USA). The gels were stained using the mass spectrometry-compatible SilverQuest Silver Stain Kit (Invitrogen, Carlsbad, CA, USA).

45.2.3 Mass Spectrometry

In-gel digestion of excised gel bands was performed as previously described (Chu et al., 2010). Bands were treated with DTT and iodoacetamide. For digestion, 100 ng of porcine, side chain-protected trypsin (Promega, Madison, WI, USA) was used for each gel band, and digestions were carried out at 37 °C for 4 h. Peptides were extracted from gel pieces with 50 μL of 50% acetonitrile, 2% acetic acid three times, and the extraction solution was dried down to approximately 10 μL.

Digests were analyzed by LC-MS/MS. Chromatographic separation was carried out on a 75 μm × 15 cm reversed-phase capillary column at a flow rate of 330 nL/min. Solvents (A: 0.1% formic acid in H_2O; B: 0.1% formic acid in acetonitrile) were delivered by a Thermo Surveyor MS Pump through a splitter, running a 5–35% solvent B gradient in 60 min. The HPLC eluent was introduced directly to the nanospray ion source of an LTQ Orbitrap XL mass spectrometer (Thermo Fisher Scientific Corporation, Waltham, MA, USA). LC-MS/MS data were acquired in a data dependent acquisition mode, cycling between one precursor ion scan measured in the Orbitrap and six CID scans measured in the LTQ. Typical performance characteristics for precursor scans were 30,000 resolution with 10 ppm mass measurement accuracy. Activation time was 30 ms. Automatic gain control (AGC) targets were set to 2,000,000 for Orbitrap scans and 10,000 for LTQ MS/MS scans. A dynamic exclusion of 60 s duration was selected to prevent repetitive acquisition of high abundance components.

The centroided peak lists of the CID spectra were searched against the National Center for Biotechnology Information (NCBI) protein database using Batch-Tag, a program in the University of California San Francisco Protein Prospector software package. Common modifications were considered, including carbamidomethylation of cysteine, acetylation at a protein N-terminus, methionine oxidation, N-terminal glutamine conversion to pyroglutamate. A mass tolerance window of 15 ppm for precursor ions and 0.7 Da for fragment ions was set for database search. The threshold for positive protein identification was at least one peptide with a Protein Prospector peptide score ≥22 and a peptide expectation value ≤0.01. Identical threshold for protein identification did not report any match from the randomized portion of the database when searched against the NCBI random concatenated database. Proteins with two or less identified peptides (Protein Prospector peptide score ≥15) were further inspected and confirmed manually.

45.3 Results

From mass spectra of trypsin digests, 84 peptides with sequence homology to 19 *S. cerevisiae* proteins were identified in amber isolates (Fig. 45.3, Table 45.1). The yeast is likely to be associated with plants and insects embedded in the amber.

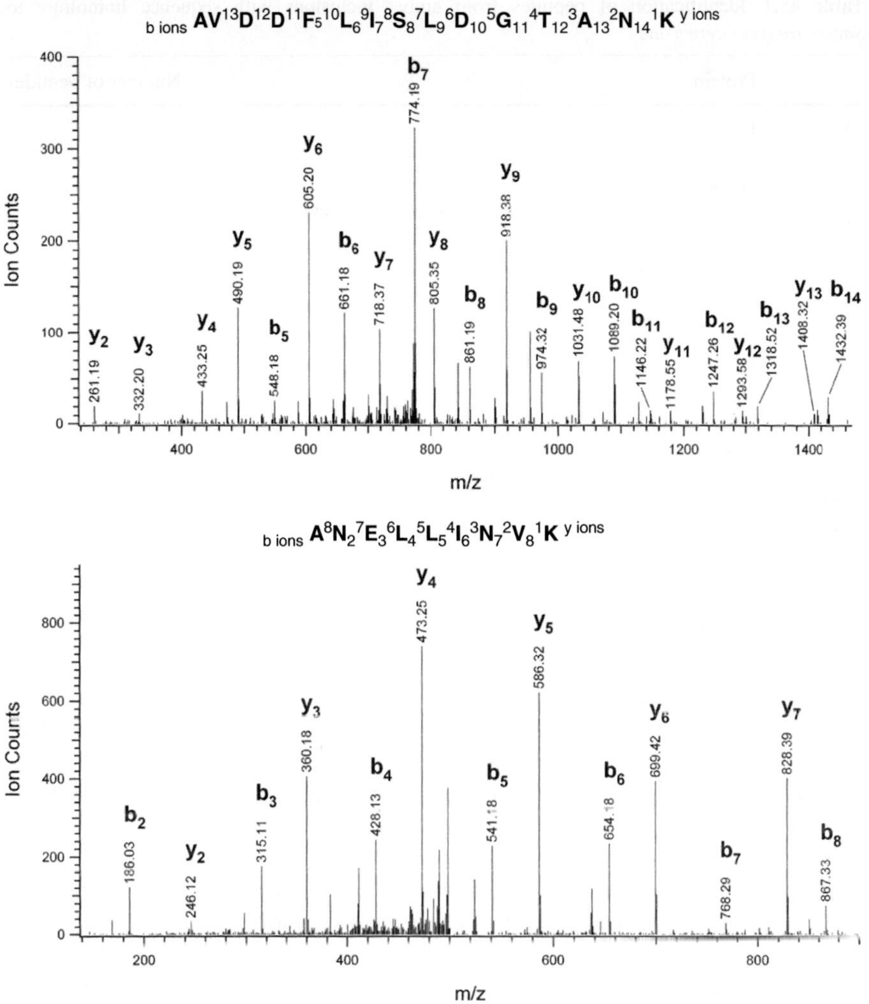

Fig. 45.3 LTQ-Orbitrap mass spectra identifying peptides of sequence homology to *Saccharomyces* enolase I (*top*) and alcohol dehydrogenase (*bottom*) isolated from 30 to 40 million year old amber inclusions

Experimental procedures for protein extraction, gel electrophoresis and mass spectrometric analysis were tightly controlled with blank samples. The fact that *S. cerevisiae* proteins were identified only in amber samples eliminates the possibility of contemporary contamination. More importantly, all these proteins were identified from a band at the interface of stacking and resolving gel, suggesting an extreme high degree of cross-linking for these proteins during amber formation. Mass spectrometric analysis of another gel band of amber sample did not lead to the identification of any known proteins. The high degree of cross-linking for the identified proteins provides evidence for the Oligo-Miocene origin of the

Table 45.1 Identification of peptides from amber inclusions with sequence homology to *Saccharomyces cerevisiae*

	Protein	Number of peptides
1	Enolase 1	11
2	Enolase 2	10
3	Alcohol dehydrogenase	9
4	Glyceraldehyde-3-phosphate dehydrogenase, isozyme 1	8
5	Phosphoglycerate kinase	7
6	Translation elongation factor EF-1 alpha	7
7	Pyruvate kinase	5
8	Fructose 1,6-bisphosphate aldolase	3
9	Mutase, phosphoglycerate	3
10	Glyceraldehyde-3-phosphate dehydrogenase, isozyme 3	3
11	Heat shock protein YJM789	3
12	Ribosomal L12B	2
13	Triosephosphate isomerase	2
14	Methionine and cysteine synthase	2
15	Ssa1p	2
16	Mitochondrial aldehyde dehydrogenase	2
17	Actin	2
18	40S ribosomal protein S18	2
19	Pyruvate decarboxylase	1
		84

proteins. Exclusion of these proteins from 4% polyacrylamide gels indicated molecular masses of several million Daltons, and failure of the aggregates to penetrate these gels proved to be an effective means for concentrating trace proteins from paleontological samples while concomitantly removing interfering substances such as SDS prior to trypsin digestion and LC-MS/MS.

Identifying multiple proteins of *S. cerevisiae* origin was initially surprising. Therefore, peptide sequences of top five proteins were further interrogated to verify the species assignment. For example, enolase 1 sequences from *S. cerevisiae* and *Hevea brasiliensis* (Para rubber tree, a member of the Fabaceae family) were compared. Although 7 of the 11 identified peptides hit conserved regions with high sequence similarity, none of the peptides has identical sequence between these two proteins. In fact, Blast NCBI protein database on the other four peptides in divergent regions only returned the *S. cerevisiae* protein. Collectively, these eleven peptides identified by tandem mass spectrometric analysis are likely of genuine *S. cerevisiae* origin, rather than due to the absence of *H. protera* genome sequence in the database.

45.4 Discussion

That yeast proteins would be preferentially isolated from amber is not an unexpected result, giving its predominance in the amber forest. Yeasts, defined as fungi that have at least a stage in their life cycle occurring as single cells that reproduce by budding or fission, are of common occurrence on flowers, leaves, wood and bark (Kreger-van Rij, 1973). These rapidly growing saprophytes, many of which are anaerobic fermentation, also occur in tree exudates and end up in fossilized resin. The earliest yeasts in amber are unicellular, budding cells described as *Peronosporites pythius* in Early Cretaceous Israeli amber dated at 130–135 million years old (Ting and Nissenbaum, 1986). Yeast-like fungi in the genera *Anthomycetes*, *Melanosphaerites*, *Arachnomycelium* and *Cladosporium* were also reported in 40–50 million year old Baltic amber (Grüss, 1931). Yeast colonies in Dominican amber consisted of reproducing cells that showed evidence of budding, compact cell clusters, hyphal protuberances and conjugation (Rikkinen and Poinar, 2002).

Indirect evidence of microorganisms in the alimentary tracts of fossil insects is revealed by the production of gases that exit from the rectum as flatus (Poinar, 2010). Wood-boring beetles, especially ambrosia beetles of the family Platypodidae, are notorious carriers of yeasts (Berryman, 1989: Baker, 1963). These beetles are common in Dominican amber and flatulence from them is considered to have been derived, at least in part, from yeast cells that continue to metabolize after the beetle has been asphyxiated in the resin (Fig. 45.4B).

Platypodids also carry yeast cells on the outer surface of their bodies (Baker, 1963) and in some instances the yeast has colonized areas in the amber adjacent

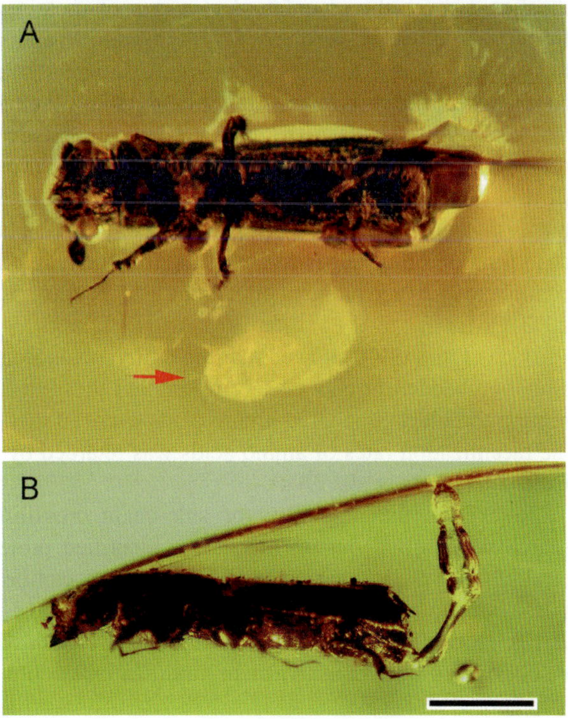

Fig. 45.4 Ambrosia beetles of the family Platypodidae commonly found in Dominican Republic amber. **A**. Yeast colony (*arrow*) probably utilizing the sugar component in the original resin as a source of nutrition. **B**. The release of gas produced by yeast and other microorganisms that continue to metabolize in the alimentary tract after the beetle has been asphyxiated in the resin. Magnification bar equals 1 mm

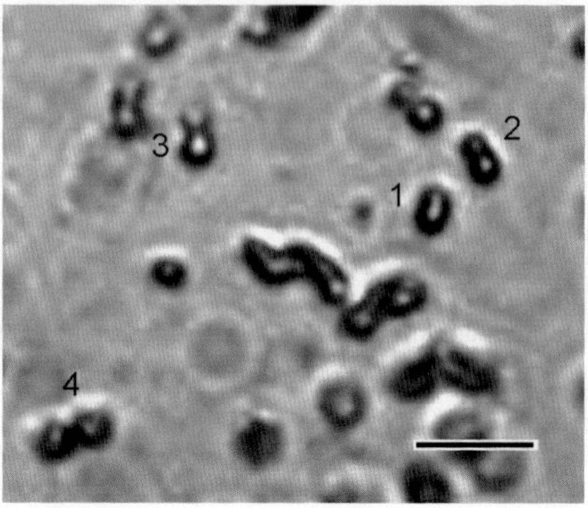

Fig. 45.5 Light microscopy showing budding yeast cells in Dominican Republic amber 20–30 million years old. Stages from single cell (*1*) to where the progeny cell is nearly equal in size to its parent (*4*) and intermediary stages (*2,3*) are shown. Magnification bar equals 6 μm. Yeast was identified from the colony associated with the platypodid in Fig. 45.4A

to the beetle (Fig. 45.4A), probably utilizing the sugar component in the original resin as a source of nutrition. Such yeast colonies comprise vegetative reproducing budding cells (Fig. 45.5). Thus, yeast in amber can be derived not only from wind-blown cells, but also from other plant and insect inclusions that have become entombed.

Over the course of millions of years, there is endless opportunity for the modification or even the complete destruction of proteins. Protein residues recovered from the calciferous matrices of oyster, mollusk, and scallop shells dating back to the Cretaceous and Jurassic periods have shown a gradual decrease in amino acid content with increasing age (Matter et al., 1969; Weiner et al., 1976; Akiyama and Wyckoff, 1970). However, for living organisms that are rapidly and completely consumed in terpenous resins, rapid dehydration would curtail indigenous proteolysis as well as the non-enzymatic hydrolysis of proteins. Because the amber environment is water-free, the hydrolysis of peptide bonds is not favored. For example, the desulfuration of cysteines to generate labile dehydro alanine residues that are susceptible to cleavage at the peptide bond is a reaction requiring water (Herbert et al., 2003).

Bada et al. (1999) showed that the amino acids from 40 to 100 million year old amber-embedded insects closely resembled those in contemporary insects, representing the oldest unaltered amino acids ever reported. Compared to other fossils, the racemization rates of amino acids was retarded by a factor of 10,000 in amber inclusions (Bada et al., 1994).

To explore the possibility of side-chain cross-linking, a mass modification search was carried out on the list of identified proteins as described (Chalkley et al., 2008). Side-chain cross-linked peptides can be considered as regular peptides, but having a big modification with unknown mass addition (Chu et al., 2010). However, no cross-linked peptides were identified thus far. It remains to be discovered what other covalent modifications could be introduced during amber formation.

Acknowledgements This work was supported by grants from Fondazione Cariplo (Milano) and PRIN 2008 (PGR and GBS), a research development grant from COLSA at UNH and the NH Agricultural Experiment Station (FC). GBS is an Affiliate Assistant Professor at the University of New Hampshire. GBS received additional support from the Novel Clinical and Translational Methodologies Program of Harvard Catalyst, The Harvard Clinical and Translational Science Center (NIH Award #UL1 RR 025758) and financial contributions from Harvard University and its affiliated academic health care centers.

References

Abelson, P.H. (1954). Organic constituents of fossils. Carnegie Inst Wash Yrbk *53*, 94–101.

Abelson, P.H. (1956). Paleobiochemistry. Sci Am *195*, 83–92.

Akiyama, M., and Wyckoff, R.W. (1970). The total amino acid content of fossil pecten shells. Proc Natl Acad Sci USA *67*, 1097–1100.

Asara, J.M., Schweitzer, M.H., Freimark, L.M., Phillips, M., and Cantley, L.C. (2007). Protein sequences from mastodon and Tyrannosaurus rex revealed by mass spectrometry. Science *316*, 280–285.

Bada, J.L., Wang, X.S., Poinar, H.N., Paabo, S., and Poinar, G.O. (1994). Amino acid racemization in amber-entombed insects: Implications for DNA preservation. Geochim Cosmochim Acta *58*, 3131–3135.

Bada, J.L., Wang, X.S., and Hamilton, H. (1999). Preservation of key biomolecules in the fossil record: current knowledge and future challenges. Philos Trans R Soc Lond B Biol Sci *354*, 77–86.

Baker, J.M. (1963). Ambrosia Beetles and Their Fungi with Particular Reference to Platypodus cylindrus. In Symbiotic Associations, P.S. Nutman, and B. Morse, eds. (Cambridge , MA, Cambridge University Press), pp. 232–265.

Berryman, A.A. (1989). Adaptive Pathways in Scolytid-Fungus Associations. In Insect-Fungus Interactions, N. Wilding, N.M. Collins, P.M. Hammond, and J.F. Webber, eds. (New York, NY, Academic), pp. 145–159.

Buckley, M., Walker, A., Ho, S.Y.W., Yang, Y., Smith, S., Ashton, P., Oates, J.T., Cappellini, E., Koon, H., Penkman, K., Elsworth, B., Ashford, D., Solazzo, C., Andrews, P., Strahler, J., Shapiro, B., Ostrom, P., Gandhi, H., Miller, W., Raney, B., Zylber, M.I., Gilbert, M.T.P., Prigodich, R.V., Ryan, M., Rijsdijk, K.F., Janoo, A., and Collins, M.J. (2008). Weighing the mass spectrometric evidence for authentic Tyrannosaurus rex collagen. Science *319*, 5859.

Cano, R.J., and Borucki, M.K. (1995). Revival and identification of bacterial spores in 25 to 40 million year old Dominican amber. Science *268*, 1060–1064.

Cano, R.J., Borucki, M.K., Higby-Schweitzer, M., Poinar, H.N., Poinar, G.O., Jr, and Pollard, K.J. (1994). Bacillus DNA in fossil bees: An ancient symbiosis? Appl Environ Microbiol *60*, 2164–2167.

Cano, R.J., Poinar, H.N., Pieniazek, N.J., Acra, A., and Poinar, G.O., Jr. (1993). Amplification and sequencing of DNA from a 120–135-million-year-old weevil. Nature *363*, 536–538.

Chalkley, R.J., Baker, P.R., Medzihradszky, K.F., Lynn, A.J., and Burlingame, A.L. (2008). In-depth analysis of tandem mass spectrometry data from disparate instrument types. Mol Cell Proteomic *7*, 2386–2398.

Chu, F., Baker, P., Chalkley, R.J., and Burlingame, A.L. (2010). Finding chimeras: a bioinformatic strategy for identification of cross-linked peptides. Mol Cell Proteomic *9*, 25–31.

Crichton, M. (1990). Jurassic Park (New York, NY, Knopf Publishing Group).

DeSalle, R., Barcia, M., and Wray, C. (1993). PCR jumping in clones of 30 million year old DNA fragments from amber preserved termites (*Mastotermes electrodominicus*). Experientia *49*, 906–909.

DeSalle, R., Gatesy, J., Wheeler, W., and Grimaldi, D. (1992). DNA sequences from a fossil termite in Oligo-Miocene amber and their phylogenetic implications. Science *257*, 1933–1936.

Draper, G., Mann, P., and Lewis, J.F. (1994). Hispaniola. In Caribbean Geology: An Introduction, S. Donovan, and T.A. Jackson, eds. (Kingston, Jamaica, The University of the West Indies Publishers Association), pp. 129–150.

Embery, G., Milner, A., Waddington, R.J., Hall, R.C., Langley, M.S., and Milan, A.M. (2000). The isolation and detection of non-collagenous proteins from the compact bone of the dinosaur Iguanodon. Connect Tissue Res *41*, 249–259.

Embery, G., Milner, A.C., Waddington, R.J., Hall, R.C., Langley, M.S., and Milan, A.M. (2003). Identification of proteinaceous material in the bone of the dinosaur Iguanodon. Connect Tissue Res *44*, 41–46.

Grüss, J. (1931). Die Urform des *Anthomyces* Reukaufii und andere Einschlüsse in den Bernstein durch Insekten verschleppt. Wochensch Brauerei *48*, 63–68.

Gurley, L.R., Valdez, J.G., Spall, W.D., Smith, B.F., and Gillette, D.D. (1991). Proteins in the fossil bone of the dinosaur, Seismosaurus. J Protein Chem *10*, 75–90.

Herbert, B.R., Hopwood, F., Oxley, D., McCarthy, J., Laver, M., Grinyer, J., Goodall, A., Williams, K., Castagna, A., and Righetti, P.G. (2003). Beta-elimination: an unexpected artifact in proteomic analysis. Proteomics *3*, 826–831.

Iturralde-Vincent, M.A., and MacPhee, R.D.E. (1996). Age and paleogeographic origin of Dominican amber. Science *273*, 1850–1852.

Kaye, T.G., Gaugler, G., and Sawlowicz, Z. (2008). Dinosaurian soft tissues interpreted as bacterial biofilms. PLoS One *7*, 2808.

Kolesnikov, M.P., and Egorov, I.A. (1977). Porphyrins and phycobilins in Precambrian rocks. Orig Life *8*, 383–390.

Kreger-van Rij, N.J.W. (1973). Endomycetales, Basidiomycetous yeasts, and related fungi. In The Fungi, An advanced Treatise (Vol. IVA), G.C. Ainsworth, F.K. Sparrow, A.S. Sussman, eds. (New York, NY, Academic), pp. 11–32.

Lambert, L.H., Cox, T., Mitchell, K., Rossello-Mora, R.A., Del Cueto, C., Dodge, D.E., Orkand, P., and Cano, R.J. (1998). Staphylococcus succinus sp. nov., isolated from Dominican amber. Int J Syst Bacteriol *48*, 511–518.

Matter, P., Davidson, F.D., and Wyckoff, R.W. (1969). The composition of fossil oyster shell proteins. Proc Natl Acad Sci USA *64*, 970–972.

Muyzer, G., Sandberg, P.A., Knapen, M.H.J., Vermeer, C., Collins, M., and Westbroek, P. (1992). Preservation of the bone protein osteocalcin in dinosaurs. Geology *20*, 871–874.

Nickle, D.C., Learn, G.H., Rain, M.W., Mullins, J.I., and Mittler, J.E. (2002). Curiously modern DNA for a "250 million-year-old" bacterium. J Mol Evol *54*, 134–137.

Pawlicki, R. (1995). Histochemical demonstration of DNA in osteocytes from dinosaur bones. Folia Histochem Cytobiol *33*, 183–186.

Pawlicki, R. (1977a). Histochemical reactions for mucopolysaccharides in the dinosaur bone. Acta Histochem *58*, 75–78.

Pawlicki, R. (1977b). Topochemical localization of lipids in dinosaur bone by means of Sudan B black. Acta Histochem *59*, 40–46.

Penalver, E., Grimaldi, D.A., and Delclos, X. (2006). Early Cretaceous spider web with its prey. Science *312*, 1761.

Poinar, G.O., Jr. (1991). Hymenaea protera sp. n. (Leguminosae: Caesalpinoideae) from Dominican amber has African affinities. Experientia *47*, 1075–1082.

Poinar, G.O., Jr. (2008). *Leptoconops nosopheris* sp. n. (Diptera: Ceratopogonidae) and *Paleotrypanosoma burmanicus* gen. n., sp. n. (Kinetoplastida: Trypanosomatidae), a biting midge-trypanosome vector association from the Early Cretaceous. Mem Inst Oswaldo Cruz *103*, 468–471.

Poinar, G.O., Jr. (2010). Fossil Flatus: Indirect Evidence of Intestinal Microbes. In Fossil Behavior Compendium, A.J. Boucot, and G.O. Poinar, Jr., eds. (Boca Raton, FL, CRC Press), pp. 22–25.

Poinar, G.O., Jr, and Hess, R. (1982). Ultrastructure of a 40 million year old insect tissue. Science *215*, 1241–1242.

Poinar, G.O., Jr., Marshall, C.J., and Buckley, R. (2007). One hundred million years of chemical warfare by insects. J Chem Ecol *33*, 1663–1669.

Poinar, G.O., Jr., and Mastalerz, M. (2000). Taphonomy of fossilized resins: determining the biostratinomy of amber. Acta Geol Hispan *35*, 171–182.

Poinar, G.O., Jr., and Poinar, R. (1999). The Amber Forest. Princeton University Press, Princeton, NJ. pp. 129–188.

Poinar, G.O., Jr., and Poinar, R. (2004). Evidence of vector-borne disease of Early Cretaceous reptiles. Vector Borne Zoon Dis *4*, 281–284.

Poinar, G.O., Jr., and Poinar, R. (2008). What bugged the dinosaurs? (Princeton, NJ, Princeton University Press).

Poinar, H., Poinar, G.O., Jr., and Cane, R.J. (1993). DNA from an extinct plant. Nature *363*, 677.

Raff, E.C., Schollaerta, K.L., Nelson, D.E., Donoghue, P.C.J., Thomas, C.W., Turner, F.R., Steina, B.D., Dong, X., Bengtson, S., Huldtgren, T., Stampanoni, M., Chongyu, Y., and Raff, R.A. (2008). Embryo fossilization is a biological process mediated by microbial biofilms. Proc Natl Acad Sci USA *105*, 19360–19365.

Ramírez, S.R., Gravendeel, B., Singer, R.B., Marshall, C.R., and Pierce, N.E. (2007). Dating the origin of the Orchidaceae from a fossil orchid with its pollinator. Nature *448*, 1042–1045.

Rikkinen, J., and Poinar, G.O., Jr. (2002). Yeast-like fungi in Dominican amber. Karstenia *42*, 29–32.

Schlee, D. (1990). Das Bernstein-Kabinett. Stuttg Beitr Naturkunde *28*, 100.

Schweitzer, M.H., Suo, Z., Avci, R., Asara, J.M., Allen, M.A., Arce, F.T., and Horner, J.R. (2007). Analyses of soft tissue from Tyrannosaurus rex suggest the presence of protein. Science *316*, 277–280.

Schweitzer, M.H., Watt, J.A., Avci, R., Knapp, L., Chiappe, L., Norell, M., and Marshall, M. (1999). Beta-keratin specific immunological reactivity in feather-like structures of the Cretaceous alvarezarid, *Shuvuuia deserti*. J Exp Zool *285*, 146–157.

Schweitzer, M.H., Wittmeyer, J.L., and Horner, J.R. (2006). Soft tissue vessels and cellular preservation in vertebrate skeletal elements from the Cretaceous to the present. Proc R Soc B *274*, 183–197.

Schweitzer, M.H., Wittmeyer, J.L., Horner, J.R., and Toporski, J.K. (2005). Soft tissue vessels and cellular preservation in Tyrannosaurus rex. Science *307*, 1952–1955.

Schweitzer, M.H., Zheng, W., Organ, C.L., Avci, R., Suo, Z., Freimark, L.M., Lebleu,VS, Duncan, M.B., Vander Heiden, M.G., Neveu, M.G., Lane, W.S., Cottrell, J.S., Horner, J.R., Cantley, L.C., Kalluri, R., and Asara, J.M. (2009). Biomolecular characterization and protein sequences of the Campanian Hadrosaur B. Canadensis. Science *324*, 626–631.

Smejkal, G.B., Poinar, G.O., and Righetti, P.G. (2009). Will amber inclusions provide the first glimpse of a Mesozoic proteome? Exp Rev Proteomic *6*, 1–4.

Ting, W.S., and Nissenbaum, A. (1986). Fungi in Lower Cretaceous amber from Israel. In Special publication, Exploration and Development Research Center (Miaoli, Taiwan, Chinese Petroleum Corporation), pp. 1–27.

Vreeland, R.H., Rosenzweig, W.D., and Powers, D.W. (2000). Isolation of a 250 million-year-old halotolerant bacterium from a primary salt crystal. Nature *407*, 897–900.

Vreeland, R.H., Rosenzweig, W.D., and Ventosa, A. (2006). Fatty acid and DNA analyses of Permian bacteria isolated from ancient salt crystals reveal differences with their modern relatives. Extremophiles *10*, 71.

Wang, H.L., Yan, Z.Y., and Jin, D.Y. (1997). Reanalysis of published DNA sequence amplified from Cretaceous dinosaur egg fossil. Mol Biol Evol *14*, 589–591.

Weiner, S., Lowenstam, H.A., and Hood, L. (1976). Characterization of 80-million-year-old mollusk shell proteins. Proc Natl Acad Sci USA *73*, 2541–2545.

Wick, G., Kalischnig, G., Maurer, H., Mayerl, C., and Müller, P.U. (2001). Really old-palaeoimmunology: immunohistochemical analysis of extracellular matrix proteins in historic and prehistoric material. Exp Gerontol *36*, 1565–1579.

Wilson, E.O. (2007). That's Life. New York Times, September 6, 2007.
Woodward, S.R., Weyand, N.J., and Bunnell, M. (1994). DNA sequence from Cretaceous period bone fragments. Science *661*, 229–232.
Yin, Z., Chen, H., Wang, Z., Zhang, Z., Zou, Y., Fang, X., and Wang, H. (1996). Sequence analysis of the cytochrome b gene fragment in a dinosaur egg. Yi Chuan Xue Bao *23*, 190–195.

Part XV
Sample Treatment in Biodefense, Forensics and Infectious Diseases

Chapter 46
Inactivation and Extraction of Bacterial Spores for Systems Biological Analysis

Bradford S. Powell and Robert J. Cybulski

Abstract Bacterial spores resist lysis and extraction by common methods of physical breakage or chemical dissolution. This hinders sample preparation for systems biology analysis. Sample heterogeneity caused by non-synchronized development, intrinsic variation in structure, and spurious germination triggered by handling all increase the complexity of sample content, which complicates detailed molecular characterization of spores. Mass spectrometry of spores is further constrained by requirements for sample pre-treatment due to the problematic nature of their cross-linked layers and abundant ionic content. Finally, safety is a principal concern in handling pathogenic specimens such as *Bacillus anthracis*, which requires dependable means for complete decontamination before transfer and analysis outside of a biocontainment environment. Traditional methods of spore inactivation are necessarily harsh and typically render sample material unsuitable for detailed spectral analysis. This chapter summarizes the variety and complexity of spores as sample material for study, reviews methods of spore decontamination and lysis, and then introduces two recently developed approaches to extract spore proteins for mass spectral analysis. One method employs ultra-high pressure cycling to achieve complete inactivation with extraction in a single tube, useful for whole analysis of the spore proteome. The other employs focused microwave energy to create exposure-controlled partial extracts of spore components as a probe of variations in averaged spore structure between preparation lots. Both methods are shown to decontaminate *B. anthracis* specimens while leaving proteins intact for mass spectrometry, including measurement of biomolecular markers intended for qualitative and quantitative comparisons between spore preparations. These enabling technologies are suitable for incorporation into standardized approaches for basic science, diagnostics, product development, or forensics purposes.

Keywords Anthrax decontamination · Barocyler · Focused microwave energy · PCT · Spore proteome · Spore sample preparation · Ultra high pressure cycling

B.S. Powell (✉)
Division of Bacteriology, US Army Medical Research Institute of Infectious Diseases, Fort Detrick, MD 21702, USA
e-mail: bpowell@mris.com

A.R. Ivanov, A.V. Lazarev (eds.), *Sample Preparation in Biological Mass Spectrometry*, DOI 10.1007/978-94-007-0828-0_46,

Abbreviations

Abs	absorbance
BME	2-mercaptoethanol
BclA	*Bacillus* collagen-like antigen
cfu	colony-forming unit
C	core
CX	cortex
DPA	dipicolinic acid
DTT	dithiothreitol
ELISA	enzyme-linked immunosorbent assay
em	electromagnetic
EX	exosporium
FACS	fluorescence-activated cell sorting
FMT	focused microwave technology
HN	Hair-like nap, comprised of glycosylated BclA
IM	inner membrane
LC	liquid chromatography
MHz	10^6 Hertz
MS	mass spectrometry
MS/MS	tandem mass spectrometry
μm	micrometer, 10^{-6} m
nm	nanometer, 10^{-9} m
OM	outer membrane
PCT	pressure cycling technology
PPE	personal protective equipment
SASP	small, acid-soluble protein
SC	spore coat
TCEP	*tris*(2-carboxyethyl)phosphine
UDS	spore extraction buffer containing 8 M urea in 0.005 M, 0.05 M dithioerythritol, 1% SDS, cyclohexyl aminomethane sulfonic acid (CHES), pH 9.8
UV	ultraviolet
W	Watt

46.1 Introduction

Bacterial spores are natural wonders but make problematic samples for study. Their importance is indisputable in aspects of medicine, industry, agriculture and commerce. Spores, particularly endospores, are compact and practically inert yet retain the capacity for rapid resumption of vegetative growth in a suitable environment, sometimes even after extended exposure to heat, UV irradiation, desiccation, or harsh chemicals. These remarkable properties greatly encumber their physical and chemical disintegration and molecular recovery for mass spectrometry (MS), which is increasingly being used during basic and applied analytical research. Improved methods for spore sample preparation will aid proteomics and other systems biology studies, including medical biological products testing and the development of diagnostic assays for biohazards analysis and forensics. Beyond resistance to lysis, the complex composition of spores further complicates efforts to study their molecular constituents. Because of these difficulties, the study of spore biology has historically focused on structural inspection of spore morphology and on genetic and biochemical investigations of the processes of spore genesis, called sporulation, and their vegetative outgrowth, called germination. Recent surveys using SDS-PAGE (Cote et al., 2009; Moody et al., 2010; Severson et al., 2009) and MS (Callahan et al., 2009; Cheng et al., 2008; Daubenspeck et al., 2004; Dickinson et al., 2004; Kuwana et al., 2002; Lai et al., 2003; Liu et al., 2004; Mukhopadhyay et al., 2009; Ryzhov et al., 2000; Stump et al., 2005; Waller et al., 2005) have begun to build knowledge of endospore protein content. However, proteomic study of purified spores for detailed comparisons between strains or production lots are constrained by sample preparation. The intractability and variability of spores impacts not only their preparation and content but collection and analysis of data as well. Also impacting their preparation and analysis as experimental samples is the fact that spores are not uniform, whether inspected within one purified batch or among several preparations. Long-observed variation in individual spores regarding their extent of formation, outgrowth, biological activity, refractile quality or spore coat fine architecture may be due to heterogeneity of spore ultra-structure or content, though direct causality has yet to be demonstrated. This non-homogeneity of spores, considered by some as genetic stochastic fluctuation or partial penetrance (Veening et al., 2008), may underlie variability of individual spores to resist extraction and could underlie observed non-alignment of early spore proteomes reported by different investigators (Huang et al., 2004; Jagtap et al., 2006; Liu et al., 2004). Differences in sample preparation methodology including inactivation, dissolution and extraction may also contribute to non-alignment of findings between laboratories. After lysis, pre-analytical spore sample work-up is important to achieve consistent ion analysis and to prevent detector interference or instrument fouling due to their polymeric, cross linked protective layers, abundant core metals and salts, and other simple ionic compounds. With regard to all of these issues, a harmonized effort toward standardization of methods for spore sample preparation would facilitate comparisons of findings within and among laboratories to promote a greater understanding of spore biology. Standardized methods, which must include robust means to safely and completely inactivate infectious or lethal spore specimens without destroying analytes, are needed to support necessary sample collection, treatment and handling in the discovery and development of improved medical treatments against anthrax (Friedlander and Little, 2009), botulism (Cheng et al., 2008; Kalb et al., 2009) and other spore-borne diseases. The retention of analyte molecular structure is a critical concern for the inactivation, recovery and clean-up of sample for MS. Unfortunately, current effective and safety-approved methods such as autoclaving, paraformaldehyde or γ-irradiation (DeQueiroz and Day, 2008) are unsuitable because they chemically damage the biomolecules intended for analysis. Commonly practiced methods of spore decontamination in preparation for MS will be discussed with comparison to two newly developed approaches that employ either focused microwave technology (FMT) or ultra-high pressure cycling technology (PCT).

Recent attention has been given toward developing improved vaccines and new medical interventions against anthrax whose testing of efficacy in animal models of disease requires reproducible lots of viable spores (Friedlander and Little, 2009). The exploration of new sample decontamination and extraction methods described here was pursued as part of an effort to better understand long observed variability in biological activity between spores produced by different strains of *Bacillus anthracis* (Auerbach and Wright, 1955; Fellows et al., 2001; Little and Knudson, 1986). The PCT and FMT methods are relatively simple, safe and inexpensive compared to many conventional approaches, and are suitable for subsequent systems biology, unlike popular detergent-reducing agent extractions. In addition, these methods are readily applicable for the treatment and extraction of molecular analytes from samples of other bacterial spore types. Each represents an entry into the pursuit for a simplified, robust, safe and more universal approach to spore inactivation and extraction that will help expand the understanding of unique and unifying aspects of spore biology.

46.1.1 Spores as Diverse Sample Material for Study

The ability of spores to withstand prolonged physical and chemical onslaught while retaining the sensitivity and capability to infect and grow has motivated intensive study of spore biology since their first published description 130 years ago (Sternbach, 2003). The bacterial spore has recently received wider interest as experimental and unknown samples for detailed molecular analysis or unequivocal detection, particularly *B. anthracis* (Kuwana et al., 2002; Lai et al., 2003; Liu et al., 2004), the etiological cause of anthrax and a CDC Category A biological threat agent (Dixon et al., 1999). Sample preparation represents a significant concern and frequent bottleneck throughout research in the life sciences. This barrier is especially true for the MS of spores, which is further hindered by the hazardous nature of some specimens. Except for the identification of a few biomarker assays of whole native spores (Dickinson et al., 2004; Ryzhov et al., 2000), a wider application of MS toward compositional analysis depends first on the development of improved, and perhaps unified, methods for spore sample inactivation, extraction, and preparation. Their compact, complex and durable architecture is a formidable obstacle to molecular dissection for detailed investigation. Put simply, spores survive extreme environments because they are dehydrated, mineralized (largely with calcium and magnesium) and genetically adapted with core and coat components of structural and enzymatic resistance (Belliveau et al., 1992; Coleman et al., 2007; Gerhardt and Marquis, 1990; Klobutcher et al., 2006). They are inherently difficult to deconstruct. Finally, the dangerous nature of some spore samples elicits multiple physical and procedural hurdles to assure safety and security for worker, environment, and public.

Despite these challenges, a few recent reports have presented the first proteome profiling of *Bacillus* spores (Kuwana et al., 2002; Lai et al., 2003; Liu et al., 2004). However, while scientific understanding of spore biology has advanced without extensive application of MS, future discoveries in spore composition, ultrastructure and component function await alternative approaches in processing and handling which enable the application of systems biology approaches.

While proteomic analysis of spores has been largely hindered by previously-mentioned challenges, much about spores has already been learned through genetic and microscopic studies. This knowledge base helps guide sample preparation and data interpretation for MS. The number and diversity of spores and organisms that sporulate is too large to be covered in this review, but the study of each would benefit from the advancement of sample preparation methods for MS and systems biological analysis. The common soil bacterium *Bacillus subtilis* continues to serve as the best characterized model system for discussion and comparative investigations of spore biology (Errington, 1993, 2003, 2010; Piggot, 2002), but a brief overview here will indicate the diversity of bacterial spores awaiting mass spectral analysis. While fungal and plant spores arise in association with sexual or asexual reproduction, most bacterial sporulation is a route to survival, disbursement and infection. The spore is the infectious life form of many etiologic microbes important to public and agricultural health. Bacterial spores, $\sim$1 μm in the length, contain germplasm of DNA, RNA, enzymes, structural and storage proteins and peptides, lipids, carbohydrates, metabolites, and ions, all protected within a multilayered, sometimes ornamented coat of high physico-chemical durability (Aronson and Fitz-James, 1976; Driks, 1999; Henriques and Moran, 2007). Bacterial sporulation is generally understood to generate one endospore from the death of one sporangium or mother cell, as detailed below for the archetypical endospore. While the gram-positive genera of *Bacillus* and *Clostridium* are the only clades known to form endospores, there are many more taxonomic groups of bacteria that create various resilient, resting stage spore-like forms in their life cycles and whose study by MS will also benefit from improved methods for sample preparation. Some other notable gram-positive endospore formers are *Brevibacillus* (Logan et al., 2002), *Frankia* (Baker and Torrey, 1980), *Paenibacillus* (Collins et al., 1994), *Sporolactobacillus* (Hess et al., 1979), and *Sporosarcina* (Mazanec et al., 1965). Some spore-formers which stain gram-negative though still related to *Bacillus* and *Clostridium* includes species of *Desulfitobacterium*, *Heliobacterium* (Bryantseva et al., 2000; Ormerod et al., 1996) and Haloalkaliphilic bacteria (GenomesOnline, 2009; Rieu-Lesme et al., 1995; Zhilina et al., 2001). *Sporomusa* (Kuhner et al., 1997), *Propionispora* (Biebl et al., 2000), and *Acetonema* (Kane and Breznak, 1991) are more distant gram-negative staining and sporulating taxa within the *Sporomusa*-group of the Clostridia class. Though exospores and cysts are not endospores, their advanced molecular study also benefits from improved methods for sample treatment and processing. In contrast to endospores, exospores are partitioned from multi-genomic cells on aerial mycelia of branching bacteria, such as the well known antibiotic producers *Streptomyces* (Lyons and Pridham, 1971), and *Thermoactinomycete* (Lacey, 1971), a causative agent of Farmer's lung disease. Many *Actinomycetes* show an unusual feature in that their spore-like forms are motile and contain flagella preformed in sporangia (Vesselinova and Ensign, 1996). Several species of *Clostridia* were shown long ago to be ornamented with appendages comprising filaments, ribbons or sheets, with fine hirsute or capped structure and attached in various polar or lateral arrangements to the spore body, coat, or surface (Rode and Slepecky, 1971). Finally, some gram-negative bacteria also create durable, resting, infectious forms akin to spores. Although they are not nearly as hearty as endospores, these durable forms include the small cell variant (SCV) of *Coxiella* (McCaul and Williams, 1981), the elementary bodies (EB) of *Chlamydia* (Constable, 1959; Miyashita et al., 1993; Popov et al., 1991), *Ehrlichia* and *Anaplasma* (Hildebrandt et al., 1973), the enigmatic spheroid bodies of Spirochetes with electron-dense inclusions (*Trepomena*, *Leptospira*, and *Borrelia*) (Ovcinnkov and Delektorskij, 1972) plus the microcysts of *Azotobaceter* (Sadoff, 1973), *Spirillum* (Williams and Rittenberg, 1956), *Sporocytophaga* (Sly et al., 1999) and the gliding bacteria like *Myxococcus* (Adye and Powelson, 1961). These various compact, resistant, sometimes ornately decorated life-stage forms are each subjects of passionate scientific inquiry and growing interest whose deeper understanding requires more effective means of sample preparation, including effective methods for mass spectral analysis. There exists a wealth of diversity in spore morphology, ultrastructure, and biological activity which invites development of more universal means to extract molecular constituents for analysis and comparison of findings among fields of inquiry. The remaining discussion of this review focuses on true endospores, which are difficult and dangerous samples for treatment and processing for MS.

46.1.2 Spore Structure

The bacterial spore coat is a principal impediment to sample inactivation, extraction, and analysis. As a sample for study, a batch of spores is neither pure nor uniform due to its biology and structure. Attributes of both the core and the coat confer to *Bacillus* endospores the capacity for metabolically inert survival for decades or more under conditions of starvation, desiccation, debilitating temperature extremes, physical shear, or chemical attack, and a sensitive capacity to detect and respond to favorable growth environments (Christopher et al., 1997; Vreeland et al., 2000). Upon binding specific signal molecules (nutrient germinants), the spore transforms into a vegetative cell through a series of events known collectively as germination. To accomplish these dual traits of inert resistance and sensitive response, bacteria efficiently package genetic material deep within a multilayered protective structure capable of repelling numerous harsh stressors while maintaining membrane-bound sensors that are responsive to minute signals (Aronson and Fitz-James, 1976; Holt and Leadbetter, 1969). The center or core of the spore ("C" in Fig. 46.1) contains genomic and plasmid DNA associated with numerous protective small acid-soluble proteins (Setlow, 1995). The core structure is surrounded by two membranes ("IM" and "OM", Fig. 46.1), interspaced by a thick, specialized peptidoglycan layer known as the cortex ("CX", Fig. 46.1) (Popham, 2002). The inner membrane contains

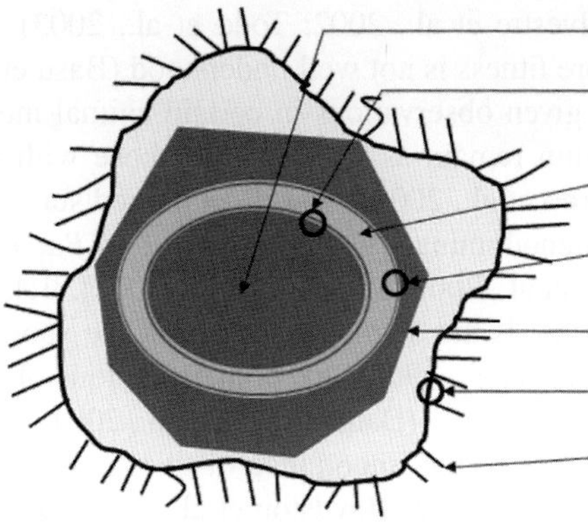

Fig. 46.1 Shown is a stylized diagram of canonical spore architecture. Spore core (C), inner spore membrane (IM), spore cortex (CX), outer spore membrane (OM), spore coat (SC), exosporium basal layer (EX), and the *Bacillus* Collagen-Like Antigen, BclA, which forms the hair-like nap (HN) on the spore surface. These spore layers and their structural components are differentially sensitive to chemical and physical treatment, but, as described, are generally difficult to extract

receptors which initiate germination under the proper conditions, while the cortex contains lytic enzymes that respond to early germination and aid irreversible degradation of spore structure (Giorno et al., 2007, 2009; Moir et al., 2002; Setlow, 2003). The outer membrane is the initiation site for assembly of the spore coat ("SC", Fig. 46.1), which contributes substantially to the chemically resistant qualities of the spore (Cho et al., 2006; Driks, 1999; Henriques and Moran, 2007; Laaberki and Dworkin, 2008; Setlow, 2006). The spore coat is a layered, interconnected shield consisting primarily of protein, though ~6% of the solubilized material of the spore coat is carbohydrate, and at least two proteins are glycosylated (Pandey and Aronson, 1979). These layers, with their characteristic bumps and ridges, allow changes in spore volume while providing mechanical strength and biochemical resistance (Driks, 1999; Giorno et al., 2009; Henriques and Moran, 2007; Nicholson et al., 2000). Deciphering the content, location, molecular interactions, and functional mechanisms of spore components are topics of intrigue or ongoing intensive investigation for several different bacteria.

Five phenotypic categories of outer spore structure have been suggested, including smooth, warty, spiny, hairy, and knobby (Lyons and Pridham, 1971). Differences in these phenotypes can be attributed to differences in spore-coat composition, as well as the presence or absence of an additional outer layer known as the exosporium (Robinow, 1951). Exosporium ("EX", Fig. 46.1), most commonly studied in spores of *Bacillus* and *Clostridium* species, is a complex, paracrystalline shell which, under electron microscopic examination, appears separated from the exterior of the spore coat by a significant gap (Gerhardt and Ribi, 1964; Matz et al., 1970; Robinow, 1951). Referred to as a "balloon-like" structure (Robinow, 1951), the exosporium contains lipids, carbohydrates, proteins, and glycoproteins which impart structural, enzymatic, and physical characteristics to the spore surface (Boydston et al., 2006; Charlton et al., 1999; Liu et al., 2004; Matz et al., 1970; Redmond et al., 2004;

Steichen et al., 2003; Sylvestre et al., 2002; Todd et al., 2003). The role of the exosporium in overall spore fitness is not well understood (Basu et al., 2007; Kang et al., 2005), particularly given observations in certain animal models that spores stripped of their exosporium remain as infectious as those with intact exosporia (Bozue et al., 2007b; Giorno et al., 2007). A well-examined surface component of the exosporium is an immunodominant glycoprotein called *Bacillus* collagen-like antigen, or BclA (Steichen et al., 2003; Sylvestre et al., 2002). BclA is filamentous, forms the "hair-like" nap that is the dominant feature of the spore's outer surface ("HN", Fig. 46.1) (Hachisuka et al., 1966), and contains unique oligosaccharides in certain species of the *B. cereus* family (Daubenspeck et al., 2004; Leoff et al., 2009). Like the exosporium as a whole, the importance of this spore-surface appendage remains a subject of study and debate (Boydston et al., 2006; Bozue et al., 2007a, b; Brahmbhatt et al., 2007; Oliva et al., 2008; Sylvestre et al., 2002). General and analyte-specific methods have been used to detect carbohydrates and coat glycoproteins using MS (Callahan et al., 2009; Daubenspeck et al., 2004; Stump et al., 2005; Waller et al., 2005). In total, this complex, interwoven architecture enables bacterial spores to initiate rapid growth after resisting attack and prolonged stasis (Christopher et al., 1997; Vreeland et al., 2000). Improved methods of molecular extraction will aid the growing application of MS toward the analysis of spore fine structure.

46.1.3 Sporulation, Germination and Sample Heterogeneity

Aside from physico-chemical intractability as an experimental sample, a batch of spores is also heterogenous in compositional structure. Variability stems from association with components according to the developmental stage of spore genesis or spurious outgrowth, natural structural non-homogeneity of formed spores, and the degree of separation achieved during purification. Much of the understanding about sporulation and germination stems from pioneering genetic and microscopic research on *B. subtilis* about which concepts are compared to other bacteria where possible. The following description serves to illustrate some of the many sources of heterogeneity that can complicate spore sample preparation for molecular analysis.

Sporulation is controlled by a network of membrane-bound kinases and transcriptional regulators that respond to internal and external signaling in a highly regulated terminal stress response to starvation (Guerout-Fleury et al., 1996; Piggot, 2002; Stragier and Losick, 1996). In *B. subtilis*, germination requires nearly 8 h for completion (Piggot, 2002), necessitating careful attention to timing in collecting samples for analysis. Figure 46.2 summarizes the main steps of endosporulation in the *B. cereus* group of bacteria. Sporulation begins with asymmetrical division of the vegetative cell cytoplasm into large and small compartments known respectively as the mother cell and forespore. The division of cytoplasms coincides with a simultaneous division of gene regulation, as mother cell and forespore activate different transcriptional regulons using compartment-specific sigma factors (Driks, 1999; Losick et al., 1986; Piggot, 2002). A primary set of proteins expressed in

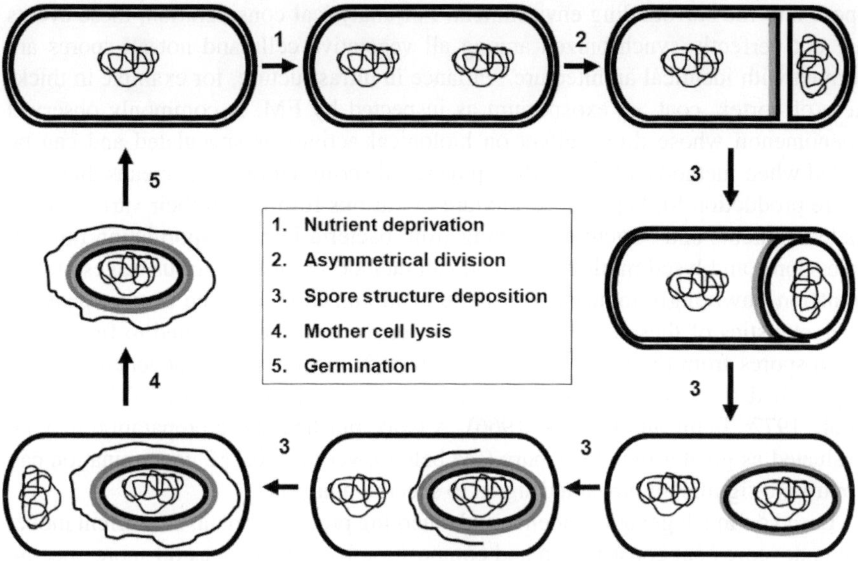

Fig. 46.2 Shown is the common five stage life-cycle for *Bacillus* class spore formers. In the absence of sufficient nutrients and the presence of oxygen (1), vegetative bacilli undergo sporulation. The process begins with an asymmetrical cell division (2), followed by a layering of proteins onto the developing endospore (3). Ultimately, the mother cell lyses and the spore is released (4). The metabolically inert yet highly resistant spore remains dormant until it becomes exposed to a nutrient-rich environment, such as a host. Specific nutrients trigger the process of germination (5), in which the spore rehydrates and the vegetative bacillus reemerges. Adapted from Driks (1999, 2004). Sample homogeneity is notably influenced by the timing of harvest, population heterogeneity, and degree of isolation achieved during purification

the forespore are small, acid soluble proteins (SASPs), which sequester DNA to protect the genome of the future spore from damage during dormancy. SASPs also serve as an amino acid depot for hydrolysis and use during germination (Paidhungat et al., 2000; Piggot, 2002). During sporulation, the forespore imports large deposits of dipicolinic acid (DPA) and divalent cations, principally calcium and magnesium, from the mother cell. These ions add protection by driving dehydration of the core (Camp and Losick, 2009). The dividing septum eventually engulfs the forespore, thereby creating the outer membrane of the future spore (Driks, 1999; Piggot, 2002). Peptidoglycan is deposited between these two membranes of the forespore and thickens into the cortex (Popham, 2002). Assembly of the cortex further promotes dehydration of the forespore, which presumably shuts down spore-specific protein synthesis (Meador-Parton and Popham, 2000; Popham et al., 1999). The proteinaceous spore coat, consisting of proteins synthesized in the mother cell cytoplasm, is assembled atop the outer membrane of the forespore (Driks, 1999). Construction of the exosporium occurs concomitantly with that of the coat (Beaman et al., 1971; Ohye and Murrell, 1973), both of which gradually envelop the forespore while still within the mother cell cytoplasm. Eventually, the mother cell lyses, releasing the spore into the surrounding environment. For analytical consideration, these events are not perfectly synchronized among all vegetative cells and not all spores are formed with identical architecture. Variance in ultrastructure, for example in thickness of cortex, coat, or exosporium as inspected by EM, is commonly observed phenomenon whose direct effect on biological activity is speculated and can be tested when methods allow for their physical discrimination. This means that each spore production lot begins as a mixture of various forms with their various associated contents and structures, ranging from bacteria in early sporulation through free spores and lysed mother cells, and can include spores in various stages of germination unwittingly induced by sample handling. In theory, the distinct physical characteristics of these different life stage forms could be exploited to fractionate clean spores from the heterogeneous mixture, but in practice homogeneity is only approached through consecutive purifications (Dean and Douthit, 1974; Prentice et al., 1972; Tamir and Gilvarg, 1966). A twice purified spore preparation may be evaluated as greater than 99% pure (see below), yet even 0.1 % contamination can contribute significant unwanted signal to spectral analysis.

Based on much genetic evidence and emerging proteomic data, the current model for endospore coat construction and composition lists as many as 60 spore-specific proteins of various function (Fenselau et al., 2007; Huang et al., 2004; Jagtap et al., 2006; Kuwana et al., 2002; Lai et al., 2003; Liu et al., 2004, 2008). These proteins serve roles in morphogenesis such as scaffolding, anchoring, nucleation and transporters for material deposition, as germinant receptors and lytic enzymes used in germination, for resistance to outside stresses, and for expression of virulence, among other spore functions. Spore coat assembly is believed to be assisted through a complex network of protein–protein interactions guided by key morphogenetic proteins (Henriques et al., 1995; Zilhao et al., 2004). Aside from these morphogenetic proteins, most protein components of the spore coat appear to display some degree of functional redundancy. Though not directly tested per se, some protein–protein interactions are proposed to be substantially nonspecific (Driks, 1999). The ability to better extract these proteins from their associated components for direct measurement will add to our understanding of spore biology.

Spore germination occurs when conditions favorable to vegetative growth are signaled by the presence of specific germinants. These are nutrients, such as sugars, single amino acids (particularly L-alanine) or purine nucleosides (particularly inosine), which serve as indicators of a nutrient-rich environment (Setlow, 2003). Germinants are bound by germinant receptors, a family of enzymes recognizing an overlapping array of molecules whose activation is the triggering event of germination (Cabrera-Martinez et al., 2003; Liu et al., 2004, 2008; Lybarger, 2008; Scott and Ellar, 1978). Receptor activation is essentially irreversible, with unsuccessful germination proceeding even if the germinants are subsequently removed (Cabrera-Martinez et al., 2003). Through a series of events and interactions still being deciphered, the spore releases a deposit of hydrogen ions stored in the core, which triggers the release of DPA and associated divalent cations, leading to rapid rehydration and activation of the spore core (Foster and Johnstone, 1990). These early events coincide with the loss of the heat-resistance and refractility, both of which are prominent and distinguishing spore features, and becomes evident in a matter of seconds (Popham et al., 1996). DPA release and core rehydration subsequently initiates hydrolysis of the spore cortex and spore coat, which allows further hydration of the spore and restoration of metabolic activity to the reawakening cell (Moir et al., 2002). Although conditions that foster germination inside a host organism are being characterized in detail, it has been found that germination can occur outside the host in the open environment, even for *B. anthracis* (Saile and Koehler, 2006; Setlow, 2003). Important for spore sample preparation are the facts that laboratory manipulations such as pressure or trace biochemicals (e.g., amino acids) can trigger germination that becomes abortive in the absence of sufficient nutrient to support complete vegetative outgrowth. Partially germinated spores, shed fragments, and ghosts of completed germination increase the micro-complexity of analytes within a spore lot. Many of spore-specific proteins have been directly observed by MS, but invariably with proteins ostensibly assigned to mother cell or germination processes (Giorno et al., 2007; Huang et al., 2004; Jagtap et al., 2006; Kuwana et al., 2002; Lai et al., 2003; Liu et al., 2004; Ryzhov et al., 2000). Differentiation among these proteomes will become clear with improved fractionation and further experimentation.

While current knowledge of spore structure and spore biology is particularly detailed for *B. subtilis*, greater understanding will come from physico-chemical studies that include MS, not only of true endospores but of resistant, spore-like forms from other organisms as well. This brief overview of the endospore lifecycle shows that content heterogeneity is a significant concern for experimental design and consideration. A phased culture becomes more heterogeneous as sporulation progresses, and once complete, not all spores of a given lot will initiate and proceed equally through germination, which occurs rapidly. Even twice-purified spore lots are observed to be heterogeneous in fine structure, stability profiles, and germination characteristics (Cronin and Wilkinson, 2008; Hornstra et al., 2009; Rose et al., 2007; Xu et al., 2009). Heterogeneity in spore structure and activity is a perplexing and poorly understood phenomenon. Fortunately, the new methods for sample preparation can help by enabling inactivation, differential lysis/extraction, and sample work-up as discussed below.

46.2 Conventional Methods of Spore Sample Treatment, Manipulation and Evaluation

To date, spore biology has largely been investigated through genetics, microscopy, and phenomenological methods that require little or no extraction of molecular components for manipulation or analysis. Although modern molecular and cellular approaches have notably advanced the understanding of spore formation, spore resistance, and spore outgrowth, such studies necessarily focus on the spore-bearing mother cell for genetic manipulation rather than on direct manipulation and component analysis of the spore itself. As protein structure and activity is clearly essential for spore biology, greater understanding of sub-spore localities and interactions would be gained from wider application of proteomic and systems biology approaches. As noted in the introduction, suitable methods to extract proteins for analysis are still under-developed. An aspect of processing that is critical to some samples is the inclusion of dependable decontamination or inactivation of infectivity. More recently published investigations have begun the process of defining spore proteomes and identifying ionizable biomarkers in *B. subtilis*, other strains used as models for virulent *B. anthracis* (Giorno et al., 2007; Huang et al., 2004; Jagtap et al., 2006; Kuwana et al., 2002; Lai et al., 2003; Liu et al., 2004, 2008; Ryzhov et al., 2000), and *Clostridium difficile* which presents unique difficulties of contamination by cellular debris (Thorne et al., 2010). It is important that methods for decontamination of live, infectious spores do not also destroy the molecules being extracted for study by MS. The remainder of this chapter focuses on treatment, handling and analysis of *B. anthracis* (aka anthrax) spores as a case examples of the methods discussed.

First, a brief review of methods used to certify a spore preparation will indicate the typical quality of spores as reagent material for use in biological studies. Anthrax spores, whether produced in broth by shaker flask, by batch fermentation, or on solid media, are naturally heterogeneous as described above. After water washes to remove media, and subsequent density gradient centrifugation purifications to remove debris, the quality of a preparation can be determined by several measurements of purity and activity. Colony inspections are also used to visually estimate percent encapsulation, another phenotypic measure of viability and genetic purity as well as suggested virulence. Microscopic inspections on Petroff-Hauser counting chambers are used to enumerate a spore lot for refractility, total spore count, total counts of vegetative cells, the presence of debris and extent of clumping. Endotoxin (obtained from solutions or vessels) can be assayed to qualify a lot for animal use. An acceptable spore lot may contain >95% refractile spores with no observed vegetative chains, <1% clumps of spores or vegetative cells, yield of 10^9–10^{11} viable cfu/ml depending on production method, >95% encapsulation, and <2,500 EU endotoxin/10^{10} spores. These macroscopic and microscopic observations and enumerations, averaged from a practical number of measurements ($n < 10$) are therefore influenced by subjective bias. In addition, due to losses by subsequent purification and a common end purpose for animal infection rather than molecular analysis, spore preparations have not typically been prepared to a level of purity best suitable for MS measurement of content. The need for standardized spore preparations and quality control assays arises from basic medical biological research and from the concurrent development of several medical treatments against infection by anthrax which requires sound and reliable demonstration of protective or therapeutic efficacy in animal models of the disease. Thus, safe, efficient and dependable methods of spore sample treatment and analysis are also needed as MS gains use in the FDA regulatory environment.

46.2.1 Spore Inactivation and Models

The methods currently employed to inactivate anthrax spores derive in large part from decontamination procedures that were developed for safety (Table 46.1). These techniques have the advantage of achieving spore decontamination, typically measured by plate count or liquid turbidity (e.g., Abs 650 nm) and reported in multiple orders of magnitude commonly referred to as log ($\log_{10}$) reductions in survivability or log killing. The nature of these destructive effects lies in their alteration of chemical integrity and biomolecular content and in doing so inactivate biological activity. However, these same disruptive effects negatively impact their suitability as preparative steps for subsequent analysis of the macromolecules comprising the structure and contents of the spore under investigation.

Table 46.1 Comparison spore decontamination methods prior to sample preparation for analysis

Technique	Advantages	Disadvantages
Formaldehyde	Formaldehyde gas up to 8-$\log_{10}$ reduction in *B. anthracis* spores (Loshon et al., 1999; Power and Russell, 1990; Rogers et al., 2007; Sinski and Pannier, 1963)	High safety hazard requires sealed room, specialized equipment, training, procedures and monitoring. Caustic to rubbers and plastics, caustic and mutagenic to biological tissue (Alderson, 1985; Cosma et al., 1988)
Peroxides	Peroxide gas yields up to 8-$\log_{10}$ reduction in *B. anthracis* spores (Rogers et al., 2005)	High safety hazard requires sealed room, specialized equipment, training, procedures and monitoring. Caustic to rubbers and plastics as well as biological tissue
Chlorites	Results in 5-$\log_{10}$ reduction of *B. subtilis* spore viability when combined with peroxide (DeQueiroz and Day, 2008)	Safety precautions required. Results in degradation/loss of protein from spore coat (Cho et al., 2006)
Chlorines	Achieves 6-$\log_{10}$ reduction of *B. anthracis* spores (Buttner et al., 2004) Results in 6.2-$\log_{10}$ drop in *B. subtilis* spores when combined with UV (Zhang et al., 2006). Deactivated spores analyzed directly by quantitative PCR (Zhang et al., 2006)	Safety precautions required. Results in degradation/loss of protein from spore coat (Cho et al., 2006)
Ozone	When combined with UV, results in only 2.5-$\log_{10}$ reduction in *B. anthracis* spores (Jung et al., 2008) Best-shown results are 3-$\log_{10}$ decrease in *B. anthracis* viability (Aydogan and Gurol, 2006)	Specialized equipment and safety precautions required. Results in degradation/loss of protein from spore coat (Cho et al., 2006)

Table 46.1 (continued)

Technique	Advantages	Disadvantages
UV	Results in 6.2-$\log_{10}$ reduction in *B. subtilis* spores when combined with chlorine (Zhang et al., 2006). Irradiation up to 24 h results in 6-$\log_{10}$ (*B. thuringiensis*) to 8-$\log_{10}$ (*B. cereus*) reduction alone, 13- to 16-$\log_{10}$ reduction in combination with titanium oxide (Luna et al., 2008)	Results in only 3.3-$\log_{10}$ reduction when used alone (Zhang et al., 2006) Results in degradation/loss of protein from spore coat (Cho et al., 2006)
Gamma rays	Achieved 8-log drop in *B. anthracis* spore viability after 60 min of irradiation (Dang et al., 2001). Direct addition of spores to PCR mix increased sensitivity (Dauphin et al., 2008)	Efficiency affected by sample arrangement in chamber. Results in loss of protein structure/ELISA- and FACS-based recognition (Dang et al., 2001; Dauphin et al., 2008) Reduction in sensitivity by PCR (Dang et al., 2001)
Wet heat	Simple and routine for food industry, requiring short time at >100°C (Belliveau et al., 1992; Coleman et al., 2007)	Not suitable alone to inactivate dangerous spores. Loss of protein structure/ELISA- and FACS-based recognition and reduced sensitivity by PCR (Dang et al., 2001)
Pressurized steam (autoclave)	Achieved 8-$\log_{10}$ drop after as little as 20 min of treatment (Dang et al., 2001). Heating/boiling/autoclaving effective for harvesting DNA (Drago et al., 2002; Luna et al., 2003; Ryu et al., 2003)	Efficiency affected by sample arrangement in chamber. Results in loss of protein structure/ELISA- and FACS-based recognition and reduced sensitivity by PCR (Dang et al., 2001). Not suitable for MS
Bead-beater	Demonstration of sporicidal effect on *B. cereus* (Windeler and Walter, 1975; Zadik and Peretz, 2008). Effective spore coat dissolution using UDS (Moody et al., 2010; Pandey and Aronson, 1979)	Bead-beating alone is a less effective sterilizer of bacillus spores than high salt treatments (Haddad et al., 1997). Requires safety environment and procedures to mitigate effect of aerosolized sample, particularly for dangerous spores. Many of the chemicals used to aid extraction must be removed before MS
Sonication	Strongest results (3- to 5.8-$\log_{10}$ drops with *Clostridium*) seen only in combination with high temperature (Raso et al., 1998)	Requires safety environment and procedures to mitigate effect of aerosolized sample. Not suitable alone to inactivate for dangerous spores
Microwave oven	12-min exposure leads to as much as 9-$\log_{10}$ reduction in *B. subtilis* spores (Celandroni et al., 2004; Goldblith and Wang, 1967; Latimer and Matsen, 1977; Wang et al., 2003)	Limited release of proteins. Requires pre-calibration of conventional microwave or user-crafted wave guide (Celandroni et al., 2004). Not suitable alone to inactivate for dangerous spores

Table 46.1 (continued)

Technique	Advantages	Disadvantages
Focused microwave technology (FMT)	1 min exposure leads to 4-$\log_{10}$ killing of *B. anthracis* Sterne strain spores (this chapter); can simultaneously treat 12 separate samples	Sample volumes limited to 0.1 ml within 0.5 ml microfuge tube. Larger volumes not yet tested
Ultra high pressure cycling (PCT)	Demonstration killing of 8-$\log_{10}$ for *B. anthracis* Sterne spores in 20 min (this chapter)	Sample volume limited to 0.5 ml per tube with 0.5 ml proprietary reagent per PULSE tube and up to three pulse tubes per run (this chapter)

Note: Final material derived from any of these treatments requires confirmatory testing for sterility when processing spores from pathogenic strains such as *B. anthracis*

A few bacterial strains or species have served as avirulent model organisms for use outside biological containment to measure the effectiveness of inactivation techniques. A survey over just the past 3 years of studies related to spore inactivation techniques reveals the use several *Bacillus* genera species as surrogate models for anthrax spores including *B. subtilis* (Hilgren et al., 2009), *B. cereus* (Luksiene et al., 2009), *B. licheniformus* (Kim et al., 2009), *B. atrophaeus* (previously and generally known as *B. subtilis globigii*) (Mohan et al., 2009), *B. amyloliquefaciens* (Ratphitagsanti et al., 2009), *Geobacillus stearothermophilus* (Rogers et al., 2007) and *B. anthracis* Sterne strain (Khan et al., 2009). These endospore-forming species arc attractive alternatives to the use of virulent *B. anthracis* for obvious safety reasons, and have therefore received greater use. In fact, many of the results reported in Tables 46.1 and 46.2 regarding the advantages and disadvantages of various inactivation and disruption techniques were obtained using avirulent, anthrax-surrogate species of *Bacillus*. Comparative studies examining differences in susceptibilities between *Bacillus* spores and those of the fellow spore-former *Clostridia* demonstrate that many findings are broadly transferable and can provide a general guideline regarding the effectiveness of inactivation strategies (Block, 2004; Votava and Slitrova, 2009). Nevertheless, numerous studies have also shown that a comparison of the structure among different *Bacillus* species reveals, not surprisingly, some significant differences in their protein content (Giorno et al., 2007, 2009; Lai et al., 2003; Manzano et al., 2009). Given that differences in spore resistance properties are observed among different spore preparations and procedures even within a single isolate of spore-former (Rose et al., 2007), one can speculate endlessly about the meaning and consequence of differing inactivation strategies among *B. anthracis* surrogates of differing compositions. Therefore, there exists a strong rationale to assess the effectiveness of inactivation, as well as the thoroughness of disruptive extraction for the purposes of proteomic analyses, using a strain as closely related to virulent *B. anthracis* as possible. For this reason, the attenuated *B. anthracis* Sterne strain currently stands as the closest surrogate for the testing and

Table 46.2 Comparison of spore disruption/extraction techniques

Technique	Advantages	Disadvantages
SDS/DTT	SDS/DTT combination used to inactivate sample and disrupt spore coat proteins (*B. anthracis* Sterne) (Huang et al., 2004; Lai et al., 2003; Liu et al., 2008; Stump et al., 2005)	Suitable only on small scale. Safety risk of tube breach during boiling if biological threat sample not previously inactivated. Necessary removal of SDS before MS
French press	Used for proteomic analysis of exosporium proteins from *B. anthracis* Sterne (Giorno et al., 2007; Steichen et al., 2003)	Not suitable for dangerous spores unless pre-inactivated. Proper biocontainment and PPE required
Bead-beater	Demonstration of utility in proteomic studies involving *B. anthracis* Sterne strain (Jagtap et al., 2006; Liu et al., 2004). Effective for harvesting DNA from spores (Kabir et al., 2003; Kuske et al., 1998; Luna et al., 2003)	Report that the amount of bead-beating necessary to reliably release DNA from spores led to DNA shearing (Levi et al., 2003)
Sonication	Used in combination with SDS and boiling to achieve spore disruption (Lai et al., 2003).· Used to achieve release of spore coat proteins for M/S (Ryzhov et al., 2000), ELISA (Borthwick et al., 2005), and DNA (Belgrader et al., 1999, 2000; Luna et al., 2003)	Use of sonication only to release spore coat proteins did not result in spore inactivation (Borthwick et al., 2005). Not suitable for dangerous spores unless pre-inactivated
Microwave oven	Demonstration of rapid extraction of DNA from spores using microwaves (Aslan et al., 2008)	Requires pre-calibration of conventional microwave or user-crafted wave guide (Celandroni et al., 2004). Not suitable for dangerous spores unless pre-inactivated
Focused microwave technology (FMT)	Combined inactivation and extraction simplifies procedures to single tube	Extended exposure resulting in temperatures >55°C degrade sample similarly to conventional microwave treatment (this chapter)
Ultra high pressure cycling (PCT)	Combined inactivation and extraction simplifies procedures to single tube. Demonstration of spore protein extraction and analysis by MS (this chapter)	High killing and extraction requires both 45,000 psi and proprietary chemical solvent (this chapter)

development of these techniques. An ideal method would extend directly to the safe, dependable, simple and portable inactivation and extraction of highly virulent and infectious *B. anthracis* sample material in a contained laboratory or field situation.

46.2.1.1 Conventional Chemical Inactivation

The most effective and common methods for inactivating *B. anthracis* spores derive from the knowledge and practice of using chemical disinfectants and sterilants in medical, dental, industrial, laboratory, and environmental applications. Chemical spore disinfectants are classified by killing efficiency into Group A agents which are bactericidal and sporistatic (e.g., alcohols, phenols, biguanides, organic acids and esters, organomercurials, quaternary ammonium compounds), and Group B agents which are both bactericidal and sporicidal (e.g., aldehydes, chlorine compounds, iodine compounds, peroxides, peroxy acids) (Russell, 1990). From a safety perspective, these agents are effective and desirable because they covalently modify or decompose critical chemical constituents of the spore coat and/or core. The variety, utility, and effects of chemical sporicidal agents are well reviewed elsewhere (Russell, 1990). However, this earlier publication's scope did not include examination or discussion of the ways in which the modes of action of these chemicals conflict with the desire to preserve the integrity of target compounds such as proteins while also limiting additives for detailed molecular analysis including MS.

The U.S. Army has tested various formulations of bleach against biological select agents and toxins since 1947 (Smart, 2009). Physical plant, industrial hygiene and safety guidelines and practices since then have coalesced on compounds containing derivatives of aldehyde, peroxide, or chlorine to decontaminate materials, equipment, and rooms. Federal decontamination response to the bioterrorism incidences of October 2001 included bleach (sodium hypochlorite), chlorine dioxide, ethylene oxide, hydrogen peroxide, and peroxyacetic acid, methyl bromide, paraformaldehyde, and vaporized hydrogen peroxide, with chlorine dioxide finally being recommended by the EPA for lower harmful side effects (EPA, 2007). These compounds reliably demonstrate 5- to 8-$\log_{10}$ reductions in viable cfu counts when used either alone (Buttner et al., 2004; Loshon et al., 1999; Power and Russell, 1990; Rogers et al., 2005, 2007; Sinski and Pannier, 1963) or in combination (DeQueiroz and Day, 2008; Zhang et al., 2006). As previously mentioned, many of these disinfectants have destructive effects on the biological materials that remain. Formaldehyde and peroxides yield the most effective killing when applied at elevated temperatures (Power and Russell, 1990). This results in proven loss of protein structure and consequent ELISA- and FACS-based recognition (Dang et al., 2001). A principal toxic effect of these treatments occurs by damage to DNA (Loshon et al., 1999), which also compromises intended nucleic acid analysis. However, lower concentrations that leave proteins unperturbed are not sufficiently damaging to DNA and are therefore compromised in sporicidal activity. Treatment with ozone has also been shown to have undesirable impacts on the protein components of the spore (Cho et al., 2006) while achieving only modest decontamination (Aydogan and Gurol, 2006; Jung et al., 2008). Chlorites and chlorine compounds have also been shown to cause the degradation and loss of proteins from the spore coat (Cho et al., 2006), though these chemical treatments seem to leave the deactivated spores suitable for at least some analysis by PCR (Zhang et al., 2006).

46.2.1.2 Conventional Physical Inactivation

Physical treatment techniques have similar limitations. Steam autoclave treatment by the standardized sterilization cycle of 121°C at 15 psi for at least 15 min is used routinely in medical/dental/veterinary settings and microbiology industries because of its availability and dependability in killing microbes. However, such harsh treatment destroys spore structure, producing spore coat deformities ranging from deep lacerations to small clefts as was observed by transmission electron microscopy while developing the standard indicator strips containing *B. subtilis* and *B. stearothermophilus* spores (Fonzi et al., 1999; Palenik et al., 1999). For 50 years, exposure to ionizing radiation of materials and biological samples contaminated with highly dangerous infectious microbes including *B. anthracis* spores has remained a steadfast standard method of physical decontamination in the research setting (Woese, 1958a, b, 1959a, b). The U.S. Postal service now irradiates unopened mail to mitigate the possible risk of contamination with anthrax spores (Smart, 2009). UV and gamma irradiation can reduce viable cfu counts by as much as 6-$\log_{10}$ (Dauphin et al., 2008; Luna et al., 2008; Zhang et al., 2006), but have destructive effects on structural proteins (Cho et al., 2006; Dang et al., 2001; Dauphin et al., 2008) and are known to reduce the sensitivity of subsequent DNA analyses (Dang et al., 2001). Heat, administered in combination with other physical (commonly pressure) or chemical treatments, can be an effective decontaminator but also degrades proteins, hampering molecular analysis (Dang et al., 2001). Heat exposure similarly has an established role as a preparative treatment of spores for DNA analysis (Drago et al., 2002; Luna et al., 2003; Ryu et al., 2003), though at least one report indicates that intense heat treatments such as boiling reduce the sensitivity of subsequent PCR analysis (Dang et al., 2001). On the other hand, heat alone at atmospheric pressure and <120°C can require several hours to fully inactivate spores.

Sterile filtration is an alternative and accessible physical method for inactivating biological threat agent samples, and has been approved under safety review for small volume and few number samples on a per-protocol basis. However, spore treatment by sterile filtration is constrained for two principal reasons. While spore samples are popularly inactivated on a small scale by boiling with detergent plus reducing agent and chaotrope (such as SDS-PAGE sample buffer) for analysis by PAGE, an intervening clean-up stage must be employed before analysis by MS. The SDS-PAGE approach is not amenable to high throughput analysis without specialized gel plug processing equipment. Also, residual spore debris will impair ultrafiltration function and safety. For this reason, ultrafiltration, whether by syringe filter or vacuum pump, is not a suitable method for sterile processing bulk or numerous samples of highly dangerous spore samples.

One technology that appears capable of decontaminating spores in a way that allows for subsequent direct analysis of spore components by MS is microwave irradiation. Microwaves, which were originally generated using household microwave ovens, are a powerful decontamination technique capable of as much as a 9-$\log_{10}$ reduction in spore counts when used on spores produced by *B. anthracis* relatives such as *B. subtilis* and *B. cereus* (Celandroni et al., 2004; Goldblith and Wang, 1967; Latimer and Matsen, 1977; Wang et al., 2003). Studies using spores from the species *B. lichenformis* indicate that the mechanism of decontamination stems from the ability of microwave energy to rupture the outer spore coat and inner spore membrane structures of the spore (Kim et al., 2009). Microwaves have been used for efficient extraction of spore DNA (Aslan et al., 2008). A rapid diagnostic method has been developed for microwave-assisted acid-hydrolysis and release of SASPs for detection by MALDI-TOF-MS (Swatkoski et al., 2006). Uncontrolled microwave decontamination, however, is associated with protein and nucleic acid denaturation from very high and rapid thermal conditions generated within the spore. Laboratory-crafted wave guides developed to generate more uniform and controllable distribution of the microwave electric-field amplitude showed less change in the structural and/or molecular components of spores than that associated with equivalent killing by simple heat, including 10 times less expansion of cortex and significant lowering in released DPA (Celandroni et al., 2004). These findings are concordant with a reduction of treatment-induced abortive germination by microwaves as compared to temperature-equivalent thermal killing. Fortunately, this technology has been improved and commercialized and recently became available on laboratory-scale instrumentation (see below). Abbreviated herein as focused microwave energy technology (FMT), this technology platform represents a promising avenue for the development of improved spore-inactivation methodology as has recently been demonstrated with *B. anthracis* Sterne spores (Powell B.S., unpublished data and below).

Finally, the application of high pressure with high heat, or ultra-high pressure with mild heat, can effectively inactivate spores depending on conditions applied (Mills et al., 1998; Paredes-Sabja et al., 2007; Raso et al., 1998). Static hydrostatic pressure is used in the food industry because of its demonstrated reliability to kill vegetative bacteria and inactivate infectious spores (4-$\log_{10}$ killing) where thermal processing is either unreliable or imparts undesirable changes to food flavor and quality (Ahn and Balasubramaniam, 2007; Evrendilek et al., 2008; Gill and Ramaswamy, 2008; Manas et al., 2000; Rastogi et al., 2007). Static application of ultra-high pressure at 483 MPa (70,053 psi, 4,830 bar, 4,770 atm) and 50°C for 5 min inactivated *Clostridium* spores by >2.5 log, but this also generated both germination and partial outgrowth (Kalchayanand et al., 2004). While a 4-log reduction of spore load is acceptable for the food industry (Shearer et al., 2000), this level of killing is not sufficient to inactivate dangerous spores for scientific research where the safety objective is 100% sporicidal efficiency. Because high pressure induces spore germination (Clouston and Wills, 1969; Kalchayanand et al., 2004), a single application of ultra-high hydrostatic pressure followed by return to atmospheric pressure represents an obstacle for use in pre-analytical processing for proteomics and MS. Furthermore, most commercial equipment developed for static pressure inactivation of microbes is large, process-scale equipment designed for industrial use and not bench-scale research. The recent availability of lab scale instruments to achieve rapid, repetitive cycling from ambient to ultra-high pressure and back again, introduces pressure cycling as another promising technology platform for anthrax sample preparation, as discussed below.

46.2.2 Spore Extraction

The extraction of molecules from noninfectious, purified spore samples typically employs methods described below with the noted caveats. Methods to disrupt and lyse dangerous, infectious spores emanate from or are heavily influenced by the disruptive inactivation approaches just described. As mentioned above, these treatments do not completely inactivate the spore and frequently damage the desired analyte. Many also derive from popular methods effective on less sturdy cells and have demonstrated ability to remove at least a subset of proteins from the exterior of the spore coat, such as filtration and boiling in gel sample buffer. While the extent of inactivation achieved by exclusive use of these methods (i.e., without explicit inactivation) may be sufficient for non-infectious samples, the yields with regard to extraction are unknown or not determined. Moreover, thoroughness with respect to retention of structural integrity for the protein population liberated by these techniques is similarly not well studied. Thus, the completeness of extraction which underlies the creation and analysis of whole proteomes will become evident with further development and application of improved methods such as PCT and FMT introduced below. A comparison of spore disruption and extraction techniques is provided in Table 46.2.

46.2.2.1 Conventional Chemical Extraction

In principal, the general chemical properties of the spore have guided the development of methods for their dissolution for structural analysis. Spore coat proteins are very resistant to alkali chemicals but are susceptible to dissolution by reducing agents because of high disulfide cross-linkage. The cortex contains acidic proteins which, together with the membranes, are alkali soluble and can be dissociated with detergents, while the nucleic acid protective proteins of the core are soluble in acid (Russell, 1990). Standard extraction treatments exploit these chemistries to disrupt the protein–protein interactions upon which spore structural integrity depends and to liberate proteins for proteomic analysis (Aronson and Fitz-James, 1976; Huang et al., 2004; Lai et al., 2003; Liu et al., 2004, 2008). A typical spore lysis buffer contains detergents (SDS, sodium deoxycholate, Triton X-100 and/or Nonidet P-40) combined with sulfhydryl reducing agent (DTT, BME, TCEP) and salts such as NaCl, buffered with 50 mM Tris (pH 8.0). Even with sulfhydryl reduction, the stripping of proteins from the spore coat can yield large protein aggregates that require further treatment using denaturants such as urea or guanidine to achieve separation of the constituent parts for further examination. Each of these extraction treatments damages spore chemistry to some degree, and such effects should be kept in mind when preparing and samples for MS. Interestingly, microwave-accelerated, hot acid site-specific cleavage has been developed as a proteomic tool for rapid analysis of microorganisms including *Bacillus* spores (Fenselau et al., 2007). To date, the lists of exosporium and spore coat proteins identified through these general methods vary from one study to the next (Aronson and Fitz-James, 1976; Huang et al., 2004; Lai et al., 2003; Liu et al., 2004, 2008), suggesting differences in efficiencies of extraction between the methods. An effort toward standardization of spore extraction methods, as proposed here, will aid comparisons of findings. Of critical note, we emphasize that methods developed with attenuated experimental model spores for transfer to samples from dangerous organisms must be tested empirically to demonstrate safety and utility on the target organism. For example, extraction methods developed for *B. anthracis* Sterne strain spores, employed as a veterinary vaccine and surrogate for fully virulent *B. anthracis* Ames strain spores, are not identical to the Ames strain and may not fully inactivate the spore or extract components similarly. Few data are reported for the subsequent direct transfer of surrogate methods and confirmation with highly virulent spore preparations.

46.2.2.2 Conventional Physical Extraction

Mechanical disruption techniques offer an alternative approach toward simultaneous decontamination and disassembly of spores. Sonic disruption has been used for harvesting spore DNA (Belgrader et al., 1999, 2000; Lai et al., 2003; Luna et al., 2003), but it achieves limited release of spore proteins (Borthwick et al., 2005; Ryzhov et al., 2000). Studies centered on the treatment of *B. anthracis*, *B. cereus*, *B. thuringiensis*, and *B. subtilis* have all demonstrated varying utility of this approach to unlock the components of the spore for further analysis. In addition to a comparatively modest sporicidal effect by sonication (Borthwick et al., 2005; Haddad et al., 1997; Windeler and Walter, 1975; Zadik and Peretz, 2008), the method presents increased safety risk of aerosolization in working with dangerous spore samples. Experiments using *Clostridium* spores indicate that reliable decontamination through sonication can only be achieved in combination with heating to elevated temperatures (Raso et al., 1998), which damages spore components of interest (Dang et al., 2001) and thereby undermines the utility of the technique. Another mechanical technique that has been employed in many studies involving *Bacillus* spores is pulverization. This is commonly achieved with glass or zirconium beads in a commercial or ad hoc bead-beating apparatus or dental amalgamator (Qbiogene Bio 101 Fast Prep; Mini Bead-Beater; Tissue Lyser; Bead Mill Homogenizer; Wig-L-Bug Grinding Mill). Bead-beating will inactivate spores (Windeler and Walter, 1975; Zadik and Peretz, 2008), though not as effectively as other methods (Haddad et al., 1997). This technique has been used as an effective means of spore disruption in several proteomic studies, typically in the presence of chemical extraction agents such as UDS (Jagtap et al., 2006; Liu et al., 2004; Moody et al., 2010; Pandey and Aronson, 1979). Bead-beating has successfully been applied for DNA-based detection assays (Kabir et al., 2003; Kuske et al., 1998; Luna et al., 2003; Ulrich et al., 2006), but there is indication that the extent of treatment necessary to ensure reliable release causes shearing of the spore DNA (Levi et al., 2003). More importantly, when working with pathogenic spore specimens, the bead beater method requires increased safety measures including additional physical containment and personal protective equipment. Anecdotal descriptions of tube failures from beaters employing brass bead's to process arthropod vectors of disease in remote location highlights the risk of pulverizing infectious sample whether in the field or the laboratory. An added scientific precaution about pulverization is that pathogenic spores require extended treatment to achieve full decontamination, which generates excessive heat and will degrade protein analytes.

Hydrostatic pressure has also been examined as a possible tool for sample preparation as the administration of intense pressure provides a means of disrupting and/or reordering biological structures at a molecular level. Hydrostatic pressure technologies stand as a potential improvement over other fluidic pressure-based technologies such as a French pressure cell or microfluidizer, which was previously used as a means of harnessing *B. anthracis* spore proteins for analysis (Giorno et al., 2007; Steichen et al., 2003). Unlike the mechanism of a French pressure cell or microfluidizer, hydrostatic pressure treatment does not expose a sample to shearing forces which have the potential to damage biomolecules such as proteins and DNA that are the object of study. Instead, the application of hydrostatic pressure changes thermodynamic conditions that otherwise maintain intermolecular structure and organization under normal conditions. Hydrophobic interactions are especially sensitive to pressure and are theorized as a key to effective cellular and subcellular disruption by ultra-high pressure (Gross et al., 2008a). Thus far, hydrostatic pressure has been examined primarily as a tool for decontaminating *Clostridium* strains (Gould and Sale, 1970; Mills et al., 1998; Paredes-Sabja et al., 2007; Raso et al., 1998; Sale et al., 1970). While the extent of killing achieved with hydrostatic pressure is impressive, successful efforts to this point have required higher than desirable temperatures (Mills et al., 1998; Paredes-Sabja et al., 2007; Raso et al., 1998). Beyond heat-associated damage, an unavoidable problem with extended hydrostatic pressure for proteomics sample extraction is its gratuitous induction of germination which rapidly and substantially alters the profile of biomolecules intended for analysis.

46.2.3 Sample Harvest and Work-Up for Mass Spectrometry

Spore sample purity is critical for generating interpretable spectra and to protect the instrument from contamination by sample, extraction or matrix residue. Spore matrix is especially sticky and can contaminate an instrument, causing increased service or repair. In general, partially formed or lysed spores and abortively germinated spores are largely removed using ultracentrifugation. Typical protocols for spore isolation include a harvest and bulk cleaning stage followed by one or two purification steps using density gradient centrifugation. Initial low speed centrifugation with water is used to remove much of the contaminants and chemicals remaining from the spore production process. Subsequent rounds of equilibrium density gradient centrifugation, using material such as Hypaque (Omnipaque) or Nycodenz, separates whole spores from vegetative cells, debris and other undesired products described above (Dean and Douthit, 1974; Prentice et al., 1972; Tamir and Gilvarg, 1966). However, very high purity can require several consecutive purifications. For extremely dirty source material, such as soil samples, spores can be extracted with Chelex 100 beads to bind abundant contaminants (Willard et al., 1998), followed by concentration and purification by density gradient centrifugation as described above.

Large-scale isolation by continuous flow centrifugation is useful for concentrating spores from milk fermentation lots (Agoston et al., 2009). Spore samples intended for detailed molecular analysis such MS, should receive at least two density gradient centrifugation purifications to exceed 99% purity. Two considerations belie this apparent achievement. First, the common optical evaluation of purity, as performed by inspection of morphology under the light microscope, is subjective and most likely overestimates homogeneity. The observation of mother cell-associated proteins in spore specific proteomic surveys (Huang et al., 2004; Jagtap et al., 2006; Kuwana et al., 2002; Lai et al., 2003; Liu et al., 2004) cannot yet be eliminated as supporting evidence of contamination. Second, small variation (e.g., 1%) in the macroscopic morphological characteristics of a spore population does not necessarily equate to an insignificant effect on mass spectral sample ionization or detection. Purified spores are typically then stored indefinitely in water or with a preservative, commonly 0.1% phenol.

Finally, gel electrophoresis is a useful and popular method for protein sample work-up preceding MS as has been reviewed elsewhere (Weiss and Gorg, 2009; Zimny-Arndt et al., 2009; Thorne et al., 2010). A principal advantage of one and two dimensional SDS-PAGE is that they provide analyte fractionation and supportive data concurrent with the removal of biological background and sample matrix. However this procedure in and of itself does not answer the need for inactivation as described above and pre-separation treatment remains a concern for handling infectious samples. Mass spectrometry has been employed with gel electrophoresis fractionation for analysis of whole spore proteomes (Giorno et al., 2007; Huang et al., 2004; Kuwana et al., 2002; Lai et al., 2003), of target specific fractions such as glycoproteins and small acid soluble proteins (Callahan et al., 2009; Daubenspeck et al., 2004; Stump et al., 2005; Waller et al., 2005), and for inspection of spore pathology proteomes (Mukhopadhyay et al., 2009; Seo et al., 2008).

46.2.4 Need for Standardized, Portable, and Robust Methods

The variety of spores as subjects of scientific inquiry is great, and the techniques used for inactivation and disruption is also varied, often being specific for species, sample quality or molecular target type. The use of chemical and/or physical methods to dismantle the layers of the spore, with no clear agreement as to which method or combinations of methods yields the most comprehensive collection of spore components, has yielded a wealth of differing results (Aronson and Fitz-James, 1976; Borthwick et al., 2005; Giorno et al., 2007, 2009; Huang et al., 2004; Jagtap et al., 2006; Kuwana et al., 2002; Lai et al., 2003; Liu et al., 2004, 2008; Ryzhov et al., 2000; Steichen et al., 2003). The use of these different approaches has had the advantage of extracting different subsets of spore components and thereby has contributed complementary information in assembling the current view of spore content and structure. This non-uniform approach also has some disadvantages. When a given technique achieves the isolation of only a subset of the larger collection of protein species known to be accessible on the exterior or near-exterior of the spore, uncertainty arises over the applicability or superiority of one approach compared to another regarding completeness and consistency of coverage. Naturally, target-focused techniques bias the creation and analysis of spore proteomes, both across species and among preparations. Nevertheless, such limitations have not precluded the use of traditional chemical and physical inactivation/disruption techniques for MS investigations of anthrax spores reported so far. Importantly, popular spore inactivation/ extraction methods such as bead-beating with UDS buffer and/or boiling in SDS-PAGE sample buffer embody a safety risk and impart contaminants to the sample that interfere with direct MS analysis. Although the common method of removing chaotrope and detergents from sample by SDS-PAGE has worked well for high abundance and soluble proteins, alternate approaches are needed to observe less tractable proteins. The desire to include low abundance markers for proteomic and metabolomic analyses of spores and spore-based phenomena will ultimately drive the creation and use of more comprehensive and standardized approaches. A technique that achieves the most complete representation of a spore population at a molecular level, and does so in a safe, simple and reproducible fashion across different preparations and species of spore-forming bacteria, will unlock the potential of MS for spore analysis.

The two techniques described next are new approaches for a simplified, safe and more universal inactivation and extraction of spores. Because of both activities are accomplished in a single tube by well controlled instrumentation, either method may be capable to answer the growing need for a common or reference approach to spore pre-treatment and handling of various spore sample preparations. Each technique is simple and brief, robust, safe, and can be standardized.

46.3 New Spore Inactivation and Extraction Methods Using PCT and FMT

Ultra-high pressure cycling technology (PCT) and focused microwave technology (FMT) are new technology platforms for biological and chemical sample treatment recently adapted for the preparation of spore samples for MS (Powell et al., unpublished data). Each method has a distinct advantage but both can provide inactivation and extraction in a single tube. These methods represent an alternative approach to prepare anthrax spore extracts for MS since the conventional methods of decontamination by gamma irradiation for transferring sample material out of the containment suite were found to be of tenuous utility and impractical application. In particular, the standard safety-approved methods of autoclave, gamma irradiation or formaldehyde/formalin treatment were found to be unsuitable as they reduced proteome output, presumably due to destroyed protein covalent structure. These new approaches are being developed in an effort to understand variation in virulence between different isolates or strains of *B. anthracis* and to control variability in biological activity among different spore batches prepared by standardized production procedures. A role for structural content heterogeneity is hypothesized and may be discovered using MS. These approaches are aiding the effort to detect and measure spore proteins and are outlined below. Since these technologies have so far shown themselves to be robust, safe, and simple, using instrumentation with relatively small bench foot prints and standard power requirement, they are each readily adaptable for sample decontamination and extraction in a field application. The methods are also transferable for the spectral exploration of many other forms of microbial spores. An advantage offered by pressure cycling is its capability to apply differential regimens of pressure treatment for differential cell lysis and extraction, as has been demonstrated for the isolation and study of mitochondria from whole cells (Gross et al., 2008a, b). The ability to differentially lyse and extract spores, vegetative bacteria, or infected host cells will improve pre-analytical sample preparation for better MS and analysis of spore structure and biology.

46.3.1 Equipment, Materials and Methods

46.3.1.1 PCT

Barocyclers, invented and manufactured by Pressure BioSciences Corporation (South Easton, MA, USA), permit programmed pressure cycling technology (PCT) through rapid, controlled cycles of ultra-high hydrostatic pressure onto a sample. PCT can alternate at intervals of nearly 5 s between ambient and target pressures of up to 45,000 psi (362 atm, 310 MPa) at defined temperature (8–55°C) to manipulate the thermodynamics of biomolecular interactions in a precise, reproducible, convenient and safe manner. Dry or dissolved sample added to extraction liquid is contained within a pressure-responsive, air tight and structurally stable tube that is submerged in a water-filled pressure chamber. The following method has served to completely inactivate and extract purified *B. anthracis* Sterne strain spores by PCT in a single treatment using automated instrumentation. Sample pressure chamber in two available instruments (NEP3229 and NEP2320) is controlled by an external water bath (Neslab RTE-140, or Thermocube), preheated and maintained at 55°C for spore sample treatment. Sample treatment occurs within a specialized PULSE tube (catalog no. FT-500), whose walls and caps are composed of high-density polymer, and which is engineered with a pressure responsive movable "ram" that compresses the liquid sample and contents while exerting no net pressure on the PULSE tube walls. Hydrophobic molecules and hydrophobic regions of the spore sample experience higher volume ratio compression than ionic components of the sample and water, which is generally estimated to collapse 7–8% at the operating pressure of 45,000 psi. An aqueous suspension of spores (500 μl), twice-purified by density gradient centrifugation and stored at 4°C, is combined 1:1 with ProteoSolve LRS Reagent A inside the PULSE tube. Using a PULSE tube tool, the movable ram is adjusted until sample volume touches the screw cap to remove excess air. The PULSE tube screw cap is closed to hand tightness using the same wrench tool and a wire Lawrence clip is fixed across both ends. Prior to processing, the instrument is programmed for 99 cycles at 55°C, each cycle comprising 5 s at 45,000 psi followed by 5 s at ambient pressure. An abbreviated run of 2–3 cycles is used to exchange heated fluid into the sample chamber and to insure the program operates as expected. Tubes are lowered into the pre-warmed and pre-flushed chamber individually, lid secured, then processing initiated by selecting "RUN." The sample is recovered immediately after the run, which lasts about 20 min as piston movement adds about 10 s per cycle. The PULSE tubes are opened inside a chemical fume hood using the PULSE tube tool. The sample is transferred to two microfuge tubes, a portion from which spore viability is tested and the remainder for evaporation with lid off to near dryness. Viability is tested by plating dilutions of the pressure-treated sample and untreated control on nutrient agar and grown at 37°C 5% CO_2 for up to 2 weeks with periodic inspection for cfu. Efficiency of killing is calculated from recorded values of colony counts using the following equation.

$$\text{Efficiency} = \frac{\text{Volume of treated sample} \times \text{cfu/ml observed}}{\text{Volume of untreated control} \times \text{cfu/ml observed}}$$

We have routinely achieved 8-$\log_{10}$ decrease in spore viability, which is equivalent to the sporicidal efficiency of autoclave treatment (Dang et al., 2001). In addition to spore inactivation, PCT with Reagent A simultaneously extracts spore sample for analysis. A distinct advantage of PCT is its utility for subsequent sample manipulation for protein analysis. Concentrated sample is transferred to a new PULSE tube with buffer and additives for a hold at medium high pressure to concurrently rehydrate the sample and accomplish rapid proteolysis. For rapid trypsin digestion, 0.05 M sodium bicarbonate buffer pH 8.5 and TPCK-treated trypsin (50:1) are added to concentrated sample (dry or wet) within the PULSE tube, and the ram is adjusted to the top of the tube as described above. Optimum conditions for trypsinolysis should be determined empirically, since several variables come into play, including temperature and duration, percent organic solvent (including residual Reagent A), the amount of sample, and degree of digestion desired. Importantly, the reaction pressure is kept below 20,000 psi to avoid trypsin inactivation (Lopez-Ferrer et al., 2008). A programmed run of 50 s at 19,000 psi, 10 s ambient, for 30 cycles at 50°C successfully yields sample ready for typical LC-MS/MS tandem MS. We found this procedure to provide the same coverage of spore proteins as observed for standard 18 h trypsinolysis. Other chapters in this volume contain additional information about using PCT to enhance enzymatic digestions.

We found that the combination of high pressure, 55°C temperature, and powerful solvents such as fluorinated alcohols contained in Reagent A result in partial stripping of the spore coat. Spore material is solubilized and rapid denaturation leads to complete inactivation of spores independent of the degree of their heterogeneity. The dried, PCT-extracted molecules are safe from degradation by endogenous enzymatic activity, thereby preserving the extracted spore material for safe transport and analysis at later time and another location without need for cold chain custodial care.

Beyond aiding the extraction and analysis proteins, PCT is useful for preparing multiple classes of analytes, including nucleic acids, carbohydrates and lipids, using non-aqueous solvent systems (Gross et al., 2008a). PCT has also been developed for differential treatment and isolation of cell types and organelles (Gross et al., 2008b). This aspect of PCT can be exploited for differential extraction and preparation of sub-spore components and particularly useful for recovery of spores from infected host tissue.

46.3.1.2 FMT

Instrumentation invented and manufactured by CEM Corporation (NC, USA) provides focused microwave technology (FMT) which bathes a reaction chamber with single mode electromagnetic (em) radiation of 2,450 MHz (12.2 cm wavelength). This causes sample molecules to rotate synchronously while aligning their dipole moments to the EM field of each passing microwave. Since microwave irradiation is much lower in frequency than x-ray, UV, or even infrared em, it causes molecules to rotate and quickly increase in energy without breaking covalent bonds. Polar attractions which normally help organize macromolecular structure and intermolecular associations reorganize in a convenient and safe manner. Dose-dependent extractions of spores have been generated by FMT using the Discover-S class microwave synthesizer with Synergy^{TM} software and onboard controls, 12 × 0.5 ml microfuge tube reactor cup and thermocouple probe. The use of a thermocouple temperature probe and a specialized cooling attachment increase controllability of sample treatment. Purified spore sample (500 μl) is dispensed into safe-lock cap microfuge tubes, cap closed, and arranged around the outside of the sample cup holder filled with water to the level of sample within the sample tubes. A reference tube of identical diluent is placed, cap open, in the center position and receives a thermocouple probe for automated temperature control. The instrument lid is shut and is programmed for 125 W, <95°C. Energy is automatically dosed according to the temperature setting, but the spore inactivation run typically lasts <5 min. The sample is recovered, plated for efficiency of killing, and is digested by trypsin for MS as described above. This method achieves 4-$\log_{10}$ decrease in spore viability, although fewer proteins were identified here than with a duplicate sample processed by PCT. Microwave-accelerated proteolysis can also be used for convenient upstream sample treatment and has gained recent popularity in preparing sample for MS (Reddy et al., 2009). Of relevance to spore analysis, this technology has proven useful for extracting protein from hyperthermophilic archaea which also resist conventional sample preparation methods (Lee et al., 2009) and has been adapted for automated sample processing in high throughput proteomics (Hahn et al., 2009).

Acknowledgements We thank Chunqin Li, Jeffery Enama, and Cale Whitworth and for laboratory assistance, and appreciate critical reading of the manuscript by Adam Driks, Stephen Little, Donald Chabot, Harry Hines and Alexander Lazarev. Opinions, interpretations, conclusions, and recommendations are those of the authors and are not necessarily endorsed by the U.S. Army. This work was performed under NIAID/MRMC interagency agreement Y1-AI-2663-01 A120 B.9.

References

Adye, J.C., and Powelson, D.M. (1961). Microcyst of Myxococcus xanthus. Chemical composition of the wall. J Bacteriol *81*, 780–785.

Agoston, R., Soni, K.A., McElhany, K., Cepeda, M.L., Zuckerman, U., Tzipori, S., Mohacsi-Farkas, C., and Pillai, S.D. (2009). Rapid concentration of Bacillus and Clostridium spores from large volumes of milk, using continuous flow centrifugation. J Food Prot *72*, 666–668.

Ahn, J., and Balasubramaniam, V.M. (2007). Physiological responses of Bacillus amyloliquefaciens spores to high pressure. J Microbiol Biotechnol *17*, 524–529.

Alderson, T. (1985). Formaldehyde-induced mutagenesis: A novel mechanism for its action. Mutat Res *154*, 101–110.

Aronson, A.I., and Fitz-James, P. (1976). Structure and morphogenesis of the bacterial spore coat. Bacteriol Rev *40*, 360–402.

Aslan, K., Previte, M.J., Zhang, Y., Gallagher, T., Baillie, L., and Geddes, C.D. (2008). Extraction and detection of DNA from Bacillus anthracis spores and the vegetative cells within 1 min. Anal Chem *80*, 4125–4132.

Auerbach, S., and Wright, G.G. (1955). Studies on immunity in anthrax. VI. Immunizing activity of protective antigen against various strains of Bacillus anthracis. J Immunol *75*, 129–133.

Aydogan, A., and Gurol, M.D. (2006). Application of gaseous ozone for inactivation of Bacillus subtilis spores. J Air Waste Manag Assoc *56*, 179–185.

Baker, D., and Torrey, J.G. (1980). Characterization of an effective actinorhizal microsymbiont, Frankia sp. AvcI1 (Actinomycetales). Can J Microbiol *26*, 1066–1071.

Basu, S., Kang, T.J., Chen, W.H., Fenton, M.J., Baillie, L., Hibbs, S., and Cross, A.S. (2007). Role of Bacillus anthracis spore structures in macrophage cytokine responses. Infect Immun *75*, 2351–2358.

Beaman, T.C., Pankratz, H.S., and Gerhardt, P. (1971). Paracrystalline sheets reaggregated from solubilized exosporium of Bacillus cereus. J Bacteriol *107*, 320–324.

Belgrader, P., Hansford, D., Kovacs, G.T., Venkateswaran, K., Mariella, R., Jr., Milanovich, F., Nasarabadi, S., Okuzumi, M., Pourahmadi, F., and Northrup, M.A. (1999). A minisonicator to rapidly disrupt bacterial spores for DNA analysis. Anal Chem *71*, 4232–4236.

Belgrader, P., Okuzumi, M., Pourahmadi, F., Borkholder, D.A., and Northrup, M.A. (2000). A microfluidic cartridge to prepare spores for PCR analysis. Biosens Bioelectron *14*, 849–852.

Belliveau, B.H., Beaman, T.C., Pankratz, H.S., and Gerhardt, P. (1992). Heat killing of bacterial spores analyzed by differential scanning calorimetry. J Bacteriol *174*, 4463–4474.

Biebl, H., Schwab-Hanisch, H., Sproer, C., and Lunsdorf, H. (2000). Propionispora vibrioides, nov. gen., nov. sp., a new gram-negative, spore-forming anaerobe that ferments sugar alcohols. Arch Microbiol *174*, 239–247.

Block, C. (2004). The effect of Perasafe and sodium dichloroisocyanurate (NaDCC) against spores of Clostridium difficile and Bacillus atrophaeus on stainless steel and polyvinyl chloride surfaces. J Hosp Infect *57*, 144–148.

Borthwick, K.A., Love, T.E., McDonnell, M.B., and Coakley, W.T. (2005). Improvement of immunodetection of bacterial spore antigen by ultrasonic cavitation. Anal Chem *77*, 7242–7245.

Boydston, J.A., Yue, L., Kearney, J.F., and Turnbough, C.L., Jr. (2006). The ExsY protein is required for complete formation of the exosporium of Bacillus anthracis. J Bacteriol *188*, 7440–7448.

Bozue, J., Cote, C.K., Moody, K.L., and Welkos, S.L. (2007a). Fully virulent Bacillus anthracis does not require the immunodominant protein BclA for pathogenesis. Infect Immun *75*, 508–511.

Bozue, J., Moody, K.L., Cote, C.K., Stiles, B.G., Friedlander, A.M., Welkos, S.L., and Hale, M.L. (2007b). Bacillus anthracis spores of the bclA mutant exhibit increased adherence to epithelial cells, fibroblasts, and endothelial cells but not to macrophages. Infect Immun *75*, 4498–4505.

Brahmbhatt, T.N., Janes, B.K., Stibitz, E.S., Darnell, S.C., Sanz, P., Rasmussen, S.B., and O'Brien, A.D. (2007). Bacillus anthracis exosporium protein BclA affects spore germination, interaction with extracellular matrix proteins, and hydrophobicity. Infect Immun *75*, 5233–5239.

Bryantseva, I.A., Gorlenko, V.M., Kompantseva, E.I., Tourova, B.B.K., and Osipov, G.A. (2000). Alkaliphilic heliobacterium Heliorestis baculata sp. nov. and emended description of the genus Heliorestis. Arch Microbiol *174*, 283–291.

Buttner, M.P., Cruz, P., Stetzenbach, L.D., Klima-Comba, A.K., Stevens, V.L., and Cronin, T.D. (2004). Determination of the efficacy of two building decontamination strategies by surface sampling with culture and quantitative PCR analysis. Appl Environ Microbiol *70*, 4740–4747.

Cabrera-Martinez, R.M., Tovar-Rojo, F., Vepachedu, V.R., and Setlow, P. (2003). Effects of overexpression of nutrient receptors on germination of spores of Bacillus subtilis. J Bacteriol *185*, 2457–2464.

Callahan, C., Fox, K., and Fox, A. (2009). The small acid soluble proteins (SASP alpha and SASP beta) of Bacillus weihenstephanensis and Bacillus mycoides group 2 are the most distinct among the Bacillus cereus group. Mol Cell Probe *23*, 291–297.

Camp, A.H., and Losick, R. (2009). A feeding tube model for activation of a cell-specific transcription factor during sporulation in Bacillus subtilis. Genes Dev *23*, 1014–1024.

Celandroni, F., Longo, I., Tosoratti, N., Giannessi, F., Ghelardi, E., Salvetti, S., Baggiani, A., and Senesi, S. (2004). Effect of microwave radiation on Bacillus subtilis spores. J Appl Microbiol *97*, 1220–1227.

Charlton, S., Moir, A.J., Baillie, L., and Moir, A. (1999). Characterization of the exosporium of Bacillus cereus. J Appl Microbiol *87*, 241–245.

Cheng, L.W., Onisko, B., Johnson, E.A., Reader, J.R., Griffey, S.M., Larson, A.E., Tepp, W.H., Stanker, L.H., Brandon, D.L., and Carter, J.M. (2008). Effects of purification on the bioavailability of botulinum neurotoxin type A. Toxicology *249*, 123–129.

Cho, M., Kim, J.H., and Yoon, J. (2006). Investigating synergism during sequential inactivation of Bacillus subtilis spores with several disinfectants. Water Res *40*, 2911–2920.

Christopher, G.W., Cieslak, T.J., Pavlin, J.A., and Eitzen, E.M., Jr. (1997). Biological warfare. A historical perspective. JAMA *278*, 412–417.

Clouston, J.G., and Wills, P.A. (1969). Initiation of germination and inactivation of Bacillus pumilus spores by hydrostatic pressure. J Bacteriol *97*, 684–690.

Coleman, W.H., Chen, D., Li, Y.-W., Cowan, A.E., and Setlow, P. (2007). How moist heat kills spores of Bacillus subtilis. J Bacteriol *189*, 8458–8466.

Collins, M.D., Lawson, P.A., Willems, A., Cordoba, J.J., Fernandez-Garayzabal, J., Garcia, P., Cai, J., Hippe, H., and Farrow, J.A. (1994). The phylogeny of the genus Clostridium: Proposal of five new genera and eleven new species combinations. Int J Syst Bacteriol *44*, 812–826.

Constable, F.L. (1959). Psittacosis elementary bodies. Nature *184*, 473–474.

Cosma, G.N., Jamasbi, R., and Marchok, A.C. (1988). Growth inhibition and DNA damage induced by benzo[a]pyrene and formaldehyde in primary cultures of rat tracheal epithelial cells. Mutat Res *201*, 161–168.

Cote, C.K., Bozue, J., Twenhafel, N., and Welkos, S.L. (2009). Effects of altering the germination potential of Bacillus anthracis spores by exogenous means in a mouse model. J Med Microbiol *58*, 816–825.

Cronin, U.P., and Wilkinson, M.G. (2008). Bacillus cereus endospores exhibit a heterogeneous response to heat treatment and low-temperature storage. Food Microbiol *25*, 235–243.

Dang, J.L., Heroux, K., Kearney, J., Arasteh, A., Gostomski, M., and Emanuel, P.A. (2001). Bacillus spore inactivation methods affect detection assays. Appl Environ Microbiol *67*, 3665–3670.

Daubenspeck, J.M., Zeng, H., Chen, P., Dong, S., Steichen, C.T., Krishna, N.R., Pritchard, D.G., and Turnbough, C.L., Jr. (2004). Novel oligosaccharide side chains of the collagen-like region of BclA, the major glycoprotein of the Bacillus anthracis exosporium. J Biol Chem *279*, 30945–30953.

Dauphin, L.A., Newton, B.R., Rasmussen, M.V., Meyer, R.F., and Bowen, M.D. (2008). Gamma irradiation can be used to inactivate Bacillus anthracis spores without compromising the sensitivity of diagnostic assays. Appl Environ Microbiol *74*, 4427–4433.

Dean, D.H., and Douthit, H.A. (1974). Buoyant density heterogeneity in spores of Bacillus subtilis: Biochemical and physiological basis. J Bacteriol *117*, 601–610.

DeQueiroz, G.A., and Day, D.F. (2008). Disinfection of Bacillus subtilis spore-contaminated surface materials with a sodium hypochlorite and a hydrogen peroxide-based sanitizer. Lett Appl Microbiol *46*, 176–180.

Dickinson, D.N., La Duc, M.T., Haskins, W.E., Gornushkin, I., Winefordner, J.D., Powell, D.H., and Venkateswaran, K. (2004). Species differentiation of a diverse suite of Bacillus spores by mass spectrometry-based protein profiling. Appl Environ Microbiol *70*, 475–482.

Dixon, T.C., Meselson, M., Guillemin, J., and Hanna, P.C. (1999). Anthrax. N Engl J Med *341*, 815–826.

Drago, L., Lombardi, A., Vecchi, E.D., and Gismondo, M.R. (2002). Real-time PCR assay for rapid detection of Bacillus anthracis spores in clinical samples. J Clin Microbiol *40*, 4399.

Driks, A. (1999). Bacillus subtilis spore coat. Microbiol Mol Biol Rev *63*, 1–20.

Driks, A. (2004). From rings to layers: Surprising patterns of protein deposition during bacterial spore assembly. J Bacteriol *186*, 4423–4426.

EPA, U.S. (2007). Anthrax Spore Decontamination Using Bleach (Sodium Hypochlorite) (United States Environmental Protection Agency).

Errington, J. (1993). Bacillus subtilis sporulation: Regulation of gene expression and control of morphogenesis. Microbiol Rev *57*, 1–33.

Errington, J. (2003). Regulation of endospore formation in Bacillus subtilis. Nat Rev Microbiol *1*, 117–126.

Errington, J. (2010). From spores to antibiotics via the cell cycle. Microbiology *156*, 1–13.

Evrendilek, G., Koca, N., Harper, J., and Balasubramaniam, V. (2008). High-pressure processing of Turkish white cheese for microbial inactivation. J Food Prot *71*, 102–108.

Fellows, P.F., Linscott, M.K., Ivins, B.E., Pitt, M.L., Rossi, C.A., Gibbs, P.H., and Friedlander, A.M. (2001). Efficacy of a human anthrax vaccine in guinea pigs, rabbits, and rhesus macaques against challenge by Bacillus anthracis isolates of diverse geographical origin. Vaccine *19*, 3241–3247.

Fenselau, C., Russell, S., Swatkoski, S., and Edwards, N. (2007). Proteomic strategies for rapid analysis of microorganisms. Eur J Mass Spectrom *13*, 35–39.

Fonzi, M., Montomoli, E., Gasparini, R., Devanna, D., and Fonzi, L. (1999). Morpho-structural variations of bacterial spores after treatment in steam vacuum assisted autoclave. Bull Group Int Rech Sci Stomatol Odontol *41*, 124–130.

Foster, S.J., and Johnstone, K. (1990). Pulling the trigger: The mechanism of bacterial spore germination. Mol Microbiol *4*, 137–141.

Friedlander, A.M., and Little, S.F. (2009). Advances in the development of next-generation anthrax vaccines. Vaccine *27*, 28–32.

GenomesOnline (2009). Genomes Online database v3.0.

Gerhardt, P., and Marquis, R.E. (1990). Spore Thermoresistance Mechanisms (Washington, DC, American Society for Microbiology).

Gerhardt, P., and Ribi, E. (1964). Ultrastructure of the exosporium enveloping spores of Bacillus cereus. J Bacteriol *88*, 1774–1789.

Gill, A.O., and Ramaswamy, H.S. (2008). Application of high pressure processing to kill Escherichia coli O157 in ready-to-eat meats. J Food Prot *71*, 2182–2189.

Giorno, R., Bozue, J., Cote, C., Wenzel, T., Moody, K.S., Mallozzi, M., Ryan, M., Wang, R., Zielke, R., Maddock, J.R., *et al.* (2007). Morphogenesis of the Bacillus anthracis spore. J Bacteriol *189*, 691–705.

Giorno, R., Mallozzi, M., Bozue, J., Moody, K.S., Slack, A., Qiu, D., Wang, R., Friedlander, A., Welkos, S., and Driks, A. (2009). Localization and assembly of proteins comprising the outer structures of the Bacillus anthracis spore. Microbiology *155*, 1133–1145.

Goldblith, S.A., and Wang, D.I. (1967). Effect of microwaves on Escherichia coli and Bacillus subtilis. Appl Microbiol *15*, 1371–1375.

Gould, G.W., and Sale, A.J. (1970). Initiation of germination of bacterial spores by hydrostatic pressure. J Gen Microbiol *60*, 335–346.

Gross, V., Carlson, G., Kwan, A.T., Smejkal, G., Freeman, E., Ivanov, A.R., and Lazarev, A. (2008a). Tissue fractionation by hydrostatic pressure cycling technology: The unified sample preparation technique for systems biology studies. J Biomol Tech *19*, 189–199.

Gross, V., Lazarev, A., Lawrence, N., and Schumacher, R. (2008b). Isolation of mitochondria from cell cultures by PCT for proteomic analysis. BioTechniques *45*, 99–100.

Guerout-Fleury, A.M., Frandsen, N., and Stragier, P. (1996). Plasmids for ectopic integration in Bacillus subtilis. Gene *180*, 57–61.

Hachisuka, Y., Kojima, K., and Sato, T. (1966). Fine filaments on the outside of the exosporium of Bacillus anthracis spores. J Bacteriol *91*, 2382–2384.

Haddad, A.J., Girard, B., Jr., Bouclin, R., Valois, M., and Landry, R.G. (1997). Effectiveness of salt versus glass bead sterilizers. J Can Dent Assoc *63*, 448–453.

Hahn, H.W., Rainer, M., Ringer, T., Huck, C.W., and Bonn, G.K. (2009). Ultrafast microwave-assisted in-tip digestion of proteins. J Proteome Res *8*, 4225–4230.

Henriques, A.O., Beall, B.W., Roland, K., and Moran, C.P., Jr. (1995). Characterization of cotJ, a sigma E-controlled operon affecting the polypeptide composition of the coat of Bacillus subtilis spores. J Bacteriol *177*, 3394–3406.

Henriques, A.O., and Moran, C.P., Jr. (2007). Structure, assembly, and function of the spore surface layers. Annu Rev Microbiol *61*, 555–588.

Hess, A., Hollander, R., and Mannheim, W. (1979). Lipoquinones of some spore-forming rods, lactic-acid bacteria and actinomycetes. J Gen Microbiol *115*, 247–252.

Hildebrandt, P.K., Conroy, J.D., McKee, A.E., Nyindo, M.B., and Huxsoll, D.L. (1973). Ultrastructure of Ehrlichia canis. Infect Immun *7*, 265–271.

Hilgren, J., Swanson, K.M., Diez-Gonzalez, F., and Cords, B. (2009). Susceptibilities of Bacillus subtilis, Bacillus cereus, and avirulent Bacillus anthracis spores to liquid biocides. J Food Prot *72*, 360–364.

Holt, S.C., and Leadbetter, E.R. (1969). Comparative ultrastructure of selected aerobic spore-forming bacteria: A freeze-etching study. Bacteriol Rev *33*, 346–378.

Hornstra, L.M., Ter Beek, A., Smelt, J.P., Kallemeijn, W.W., and Brul, S. (2009). On the origin of heterogeneity in (preservation) resistance of Bacillus spores: Input for a "systems" analysis approach of bacterial spore outgrowth. Int J Food Microbiol *134*, 9–15.

Huang, C.M., Foster, K.W., DeSilva, T.S., Van Kampen, K.R., Elmets, C.A., and Tang, D.C. (2004). Identification of Bacillus anthracis proteins associated with germination and early outgrowth by proteomic profiling of anthrax spores. Proteomics *4*, 2653–2661.

Jagtap, P., Michailidis, G., Zielke, R., Walker, A.K., Patel, N., Strahler, J.R., Driks, A., Andrews, P.C., and Maddock, J.R. (2006). Early events of Bacillus anthracis germination identified by time-course quantitative proteomics. Proteomics *6*, 5199–5211.

Jung, Y.J., Oh, B.S., and Kang, J.W. (2008). Synergistic effect of sequential or combined use of ozone and UV radiation for the disinfection of Bacillus subtilis spores. Water Res *42*, 1613–1621.

Kabir, S., Rajendran, N., Amemiya, T., and Itoh, K. (2003). Real-time quantitative PCR assay on bacterial DNA: In a model soil system and environmental samples. J Gen Appl Microbiol *49*, 101–109.

Kalb, S.R., Lou, J., Garcia-Rodriguez, C., Geren, I.N., Smith, T.J., Moura, H., Marks, J.D., Smith, L.A., Pirkle, J.L., and Barr, J.R. (2009). Extraction and inhibition of enzymatic activity of botulinum neurotoxins/A1, /A2, and /A3 by a panel of monoclonal anti-BoNT/A antibodies. PloS One *4*, e5355.

Kalchayanand, N., Dunne, C.P., Sikes, A., and Ray, B. (2004). Germination induction and inactivation of Clostridium spores at medium-range hydrostatic pressure treatment. Innovative Food Sci Emerg Technol *5*, 277–283.

Kane, M.D., and Breznak, J.A. (1991). Acetonema longum gen. nov. sp. nov., an H_2/CO_2 acetogenic bacterium from the termite, Pterotermes occidentis. Arch Microbiol *156*, 91–98.

Kang, T.J., Fenton, M.J., Weiner, M.A., Hibbs, S., Basu, S., Baillie, L., and Cross, A.S. (2005). Murine macrophages kill the vegetative form of Bacillus anthracis. Infect Immun *73*, 7495–7501.

Khan, S.A., Sung, K., Nawaz, M.S., Cerniglia, C.E., Tamplin, M.L., Phillips, R.W., and Kelley, L.C. (2009). The survivability of Bacillus anthracis (Sterne strain) in processed liquid eggs. Food Microbiol *26*, 123–127.

Kim, S.Y., Shin, S.J., Song, C.H., Jo, E.K., Kim, H.J., and Park, J.K. (2009). Destruction of Bacillus licheniformis spores by microwave irradiation. J Appl Microbiol *106*, 877–885.

Klobutcher, L.A., Ragkousi, K., and Setlow, P. (2006). The Bacillus subtilis spore coat provides "eat resistance" during phagocytic predation by the protozoan Tetrahymena thermophila. Proc Natl Acad Sci USA *103*, 165–170.

Kuhner, C.H., Frank, C., Griesshammer, A., Schmittroth, M., Acker, G., Gossner, A., and Drake, H.L. (1997). Sporomusa silvacetica sp, nov., an acetogenic bacterium isolated from aggregated forest soil. Int J Syst Bacteriol *47*, 352–358.

Kuske, C.R., Banton, K.L., Adorada, D.L., Stark, P.C., Hill, K.K., and Jackson, P.J. (1998). Small-Scale DNA sample preparation method for field PCR detection of microbial cells and spores in soil. Appl Environ Microbiol *64*, 2463–2472.

Kuwana, R., Kasahara, Y., Fujibayashi, M., Takamatsu, H., Ogasawara, N., and Watabe, K. (2002). Proteomics characterization of novel spore proteins of Bacillus subtilis. Microbiology *148*, 3971–3982.

Laaberki, M.H., and Dworkin, J. (2008). Role of spore coat proteins in the resistance of Bacillus subtilis spores to Caenorhabditis elegans predation. J Bacteriol *190*, 6197–6203.

Lacey, J. (1971). Thermoactinomyces sacchari sp. nov., a thermophilic actinomycete causing bagassosis. J Gen Microbiol *66*, 327–338.

Lai, E.M., Phadke, N.D., Kachman, M.T., Giorno, R., Vazquez, S., Vazquez, J.A., Maddock, J.R., and Driks, A. (2003). Proteomic analysis of the spore coats of Bacillus subtilis and Bacillus anthracis. J Bacteriol *185*, 1443–1454.

Latimer, J.M., and Matsen, J.M. (1977). Microwave oven irradiation as a method for bacterial decontamination in a clinical microbiology laboratory. J Clin Microbiol *6*, 340–342.

Lee, A.M., Sevinsky, J.R., Bundy, J.L., Grunden, A.M., and Stephenson J.L., Jr. (2009). Proteomics of Pyrococcus furiosus, a hyperthermophilic archaeon refractory to traditional methods. J Proteome Res *8*, 3844–3851.

Leoff, C., Saile, E., Rauvolfova, J., Quinn, C.P., Hoffmaster, A.R., Zhong, W., Mehta, A.S., Boons, G.J., Carlson, R.W., and Kannenberg, E.L. (2009). Secondary cell wall polysaccharides of Bacillus anthracis are antigens that contain specific epitopes which cross-react with three pathogenic Bacillus cereus strains that caused severe disease, and other epitopes common to all the Bacillus cereus strains tested. Glycobiology *19*, 665–673.

Levi, K., Higham, J.L., Coates, D., and Hamlyn, P.F. (2003). Molecular detection of anthrax spores on animal fibres. Lett Appl Microbiol *36*, 418–422.

Little, S.F., and Knudson, G.B. (1986). Comparative efficacy of Bacillus anthracis live spore vaccine and protective antigen vaccine against anthrax in the guinea pig. Infect Immun *52*, 509–512.

Liu, H., Bergman, N.H., Thomason, B., Shallom, S., Hazen, A., Crossno, J., Rasko, D.A., Ravel, J., Read, T.D., Peterson, S.N., *et al.* (2004). Formation and composition of the Bacillus anthracis endospore. J Bacteriol *186*, 164–178.

Liu, Y.T., Lin, S.B., Huang, C.P., and Huang, C.M. (2008). A novel immunogenic spore coat-associated protein in Bacillus anthracis: Characterization via proteomics approaches and a vector-based vaccine system. Protein Expr Purif *57*, 72–80.

Logan, N.A., Forsyth, G., Lebbe, L., Goris, J., Heyndrickx, M., Balcaen, A., Verhelst, A., Falsen, E., Ljungh, A., Hansson, H.B., *et al.* (2002). Polyphasic identification of Bacillus and Brevibacillus strains from clinical, dairy and industrial specimens and proposal of Brevibacillus invocatus sp. nov. Int J Syst Evol Microbiol *52*, 953–966.

Lopez-Ferrer, D., Petritis, K., Hixson, K.K., Heibeck, T.H., Moore, R.J., Belov, M.E., Camp, D.G., 2nd, and Smith, R.D. (2008). Application of pressurized solvents for ultrafast trypsin hydrolysis in proteomics: Proteomics on the fly. J Proteome Res *7*, 3276–3281.

Loshon, C.A., Genest, P.C., Setlow, B., and Setlow, P. (1999). Formaldehyde kills spores of Bacillus subtilis by DNA damage and small, acid-soluble spore proteins of the alpha/beta-type protect spores against this DNA damage. J Appl Microbiol *87*, 8–14.

Losick, R., Youngman, P., and Piggot, P.J. (1986). Genetics of endospore formation in Bacillus subtilis. Annu Rev Genet *20*, 625–669.

Luksiene, Z., Buchovec, I., and Paskeviciute, E. (2009). Inactivation of food pathogen Bacillus cereus by photosensitization in vitro and on the surface of packaging material. J Appl Microbiol *107*: 2037–2046.

Luna, V.A., Cannons, A.C., Amuso, P.T., and Cattani, J. (2008). The inactivation and removal of airborne Bacillus atrophaeus endospores from air circulation systems using UVC and HEPA filters. J Appl Microbiol *104*, 489–498.

Luna, V.A., King, D., Davis, C., Rycerz, T., Ewert, M., Cannons, A., Amuso, P., and Cattani, J. (2003). Novel sample preparation method for safe and rapid detection of Bacillus anthracis spores in environmental powders and nasal swabs. J Clin Microbiol *41*, 1252–1255.

Lybarger, S., Fisher, N., and Hanna, P. (2008). Spore Germination: Immediate Early Stage of Anthrax. Paper presented at: 5th Natl RCE Meeting (Chicago, IL).

Lyons, A.J., and Pridham, T.G. (1971). Streptomyces torulosus sp. n., an unusual knobby-spored taxon. Appl Microbiol *22*, 190–193.

Manas, P., Pagan, R., Raso, J., Sala, F.J., and Condon, S. (2000). Inactivation of Salmonella Enteritidis, Salmonella Typhimurium, and Salmonella Senftenberg by ultrasonic waves under pressure. J Food Prot *63*, 451–456.

Manzano, M., Giusto, C., Iacumin, L., Cantoni, C., and Comi, G. (2009). Molecular methods to evaluate biodiversity in Bacillus cereus and Bacillus thuringiensis strains from different origins. Food Microbiol *26*, 259–264.

Matz, L.L., Beaman, T.C., and Gerhardt, P. (1970). Chemical composition of exosporium from spores of Bacillus cereus. J Bacteriol *101*, 196–201.

Mazanec, K., Kocur, M., and Martinec, T. (1965). Electron microscopy of ultrathin sections of Sporosarcina ureae. J Bacteriol *90*, 808–816.

McCaul, T.F., and Williams, J.C. (1981). Developmental cycle of Coxiella burnetii: Structure and morphogenesis of vegetative and sporogenic differentiations. J Bacteriol *147*, 1063–1076.

Meador-Parton, J., and Popham, D.L. (2000). Structural analysis of Bacillus subtilis spore peptidoglycan during sporulation. J Bacteriol *182*, 4491–4499.

Mills, G., Earnshaw, R., and Patterson, M.F. (1998). Effects of high hydrostatic pressure on Clostridium sporogenes spores. Lett Appl Microbiol *26*, 227–230.

Miyashita, N., Kanamoto, Y., and Matsumoto, A. (1993). The morphology of Chlamydia pneumoniae. J Med Microbiol *38*, 418–425.

Mohan, A., Dunn, J., Hunt, M.C., and Sizer, C.E. (2009). Inactivation of Bacillus atrophaeus spores with surface-active peracids and characterization of formed free radicals using electron spin resonance spectroscopy. J Food Sci *74*, M411–M417.

Moir, A., Corfe, B.M., and Behravan, J. (2002). Spore germination. Cell Mol Life Sci *59*, 403–409.

Moody, K.L., Driks, A., Rother, G.L., Cote, C.K., Brueggemann, E.E., Hines, H.B., Friedlander, A.M., and Bozue, J. (2010). Processing, assembly and localization of a Bacillus anthracis spore protein. Microbiology *156*, 174–183.

Mukhopadhyay, S., Akmal, A., Stewart, A.C., Hsia, R.C., and Read, T.D. (2009). Identification of Bacillus anthracis spore component antigens conserved across diverse Bacillus cereus sensu lato strains. Mol Cell Proteomic *8*, 1174–1191.

Nicholson, W.L., Munakata, N., Horneck, G., Melosh, H.J., and Setlow, P. (2000). Resistance of Bacillus endospores to extreme terrestrial and extraterrestrial environments. Microbiol Mol Biol Rev *64*, 548–572.

Ohye, D.F., and Murrell, W.G. (1973). Exosporium and spore coat formation in Bacillus cereus T. J Bacteriol *115*, 1179–1190.

Oliva, C.R., Swiecki, M.K., Griguer, C.E., Lisanby, M.W., Bullard, D.C., Turnbough, C.L., Jr., and Kearney, J.F. (2008). The integrin Mac-1 (CR3) mediates internalization and directs Bacillus anthracis spores into professional phagocytes. Proc Natl Acad Sci USA *105*, 1261–1266.

Ormerod, J.G., Kimble, L.K., Nesbakken, T., Torgersen, Y.A., Woese, C.R., and Madigan, M.T. (1996). Heliophilum fasciatum gen. nov. sp. nov. and Heliobacterium gestii sp. nov.: Endospore-forming heliobacteria from rice field soils. Arch Microbiol *165*, 226–234.

Ovcinnkov, N.M., and Delektorskij, V.V. (1972). Current concepts of the morphology and biology of Treponema pallidum based on electron microscopy. Br J Vener Dis *47*, 315–328.

Paidhungat, M., Setlow, B., Driks, A., and Setlow, P. (2000). Characterization of spores of Bacillus subtilis which lack dipicolinic acid. J Bacteriol *182*, 5505–5512.

Palenik, C.J., Burke, F.J., Coulter, W.A., and Cheung, S.W. (1999). Improving and monitoring autoclave performance in dental practice. Br Dent J *187*, 581–584.

Pandey, N.K., and Aronson, A.I. (1979). Properties of the Bacillus subtilis spore coat. J Bacteriol *137*, 1208–1218.

Paredes-Sabja, D., Gonzalez, M., Sarker, M.R., and Torres, J.A. (2007). Combined effects of hydrostatic pressure, temperature, and pH on the inactivation of spores of Clostridium perfringens type A and Clostridium sporogenes in buffer solutions. J Food Sci *72*, M202–M206.

Piggot, P.J., and Losick, R. (2002). Sporulation Genes and Intercompartmental Regulation. In Bacillus subtilis and Its Closest Relatives: From Genes to Cells, J.A.H.A.L. Sonenshein, and R. Losick, eds. (Washington, DC, American Society for Microbiology), pp. 483–518.

Popham, D.L. (2002). Specialized peptidoglycan of the bacterial endospore: The inner wall of the lockbox. Cell Mol Life Sci *59*, 426–433.

Popham, D.L., Gilmore, M.E., and Setlow, P. (1999). Roles of low-molecular-weight penicillin-binding proteins in Bacillus subtilis spore peptidoglycan synthesis and spore properties. J Bacteriol *181*, 126–132.

Popham, D.L., Helin, J., Costello, C.E., and Setlow, P. (1996). Muramic lactam in peptidoglycan of Bacillus subtilis spores is required for spore outgrowth but not for spore dehydration or heat resistance. Proc Natl Acad Sci USA *93*, 15405–15410.

Popov, V.L., Shatkin, A.A., Pankratova, V.N., Smirnova, N.S., von Bonsdorff, C.H., Ekman, N.R., Morttinen, A., and Saikku, P. (1991). Ultrastructure of Chlamydia pneumoniae in cell culture. FEMS Microbiol Lett *68*, 129–134.

Power, E.G., and Russell, A.D. (1990). Sporicidal action of alkaline glutaraldehyde: Factors influencing activity and a comparison with other aldehydes. J Appl Bacteriol *69*, 261–268.

Prentice, G.A., Wolfe, F.H., and Clegg, L.F. (1972). The use of density gradient centrifugation for the separation of germinated from nongerminated spores. J Appl Bacteriol *35*, 345–349.

Raso, J., Palop, A., Pagan, R., and Condon, S. (1998). Inactivation of Bacillus subtilis spores by combining ultrasonic waves under pressure and mild heat treatment. J Appl Microbiol *85*, 849–854.

Rastogi, N.K., Raghavarao, K.S., Balasubramaniam, V.M., Niranjan, K., and Knorr, D. (2007). Opportunities and challenges in high pressure processing of foods. Crit Rev Food Sci Nutr *47*, 69–112.

Ratphitagsanti, W., Ahn, J., Balasubramaniam, V.M., and Yousef, A.E. (2009). Influence of pressurization rate and pressure pulsing on the inactivation of Bacillus amyloliquefaciens spores during pressure-assisted thermal processing. J Food Prot *72*, 775–782.

Reddy, P.M., Hsu, W.Y., Hu, J.F., and Ho, Y.P. (2010). Digestion completeness of microwave-assisted and conventional trypsin-catalyzed reactions. J Am Soc Mass Spectrom 21:421–424.

Redmond, C., Baillie, L.W., Hibbs, S., Moir, A.J., and Moir, A. (2004). Identification of proteins in the exosporium of Bacillus anthracis. Microbiology *150*, 355–363.

Rieu-Lesme, F., Fonty, G., and Dore, J. (1995). Isolation and characterization of a new hydrogen-utilizing bacterium from the rumen. FEMS Microbiol Lett *125*, 77–82.

Robinow, C.F. (1951). Observations on the structure of Bacillus spores. J Gen Microbiol *5*, 439–457.

Rode, P.M., and Smith, L.D. (1971). Taxonomic implications of spore fine structure in Clostridium bifermentans. J Bacteriol *105*, 349–354.

Rogers, J.V., Choi, Y.W., Richter, W.R., Rudnicki, D.C., Joseph, D.W., Sabourin, C.L., Taylor, M.L., and Chang, J.C. (2007). Formaldehyde gas inactivation of Bacillus anthracis, Bacillus subtilis, and Geobacillus stearothermophilus spores on indoor surface materials. J Appl Microbiol *103*, 1104–1112.

Rogers, J.V., Sabourin, C.L., Choi, Y.W., Richter, W.R., Rudnicki, D.C., Riggs, K.B., Taylor, M.L., and Chang, J. (2005). Decontamination assessment of Bacillus anthracis, Bacillus subtilis, and Geobacillus stearothermophilus spores on indoor surfaces using a hydrogen peroxide gas generator. J Appl Microbiol *99*, 739–748.

Rose, R., Setlow, B., Monroe, A., Mallozzi, M., Driks, A., and Setlow, P. (2007). Comparison of the properties of Bacillus subtilis spores made in liquid or on agar plates. J Appl Microbiol *103*, 691–699.

Russell, A.D. (1990). Bacterial spores and chemical sporicidal agents. Clin Microbiol Rev *3*, 99–119.

Ryu, C., Lee, K., Yoo, C., Seong, W.K., and Oh, H.B. (2003). Sensitive and rapid quantitative detection of anthrax spores isolated from soil samples by real-time PCR. Microbiol Immunol *47*, 693–699.

Ryzhov, V., Hathout, Y., and Fenselau, C. (2000). Rapid characterization of spores of Bacillus cereus group bacteria by matrix-assisted laser desorption-ionization time-of-flight mass spectrometry. Appl Environ Microbiol *66*, 3828–3834.

Sadoff, H.L. (1973). Comparative aspects of morphogenesis in three prokaryotic genera. Annu Rev Microbiol *27*, 133–153.

Saile, E., and Koehler, T.M. (2006). Bacillus anthracis multiplication, persistence, and genetic exchange in the rhizosphere of grass plants. Appl Environ Microbiol *72*, 3168–3174.

Sale, A.J., Gould, G.W., and Hamilton, W.A. (1970). Inactivation of bacterial spores by hydrostatic pressure. J Gen Microbiol *60*, 323–334.

Scott, I.R., and Ellar, D.J. (1978). Metabolism and the triggering of germination of Bacillus megaterium. Use of L-[3H]alanine and tritiated water to detect metabolism. Biochem J *174*, 635–640.

Seo, G.M., Jung, K.H., Kim, S.J., Kim, J.C., Yoon, J.W., Oh, K.K., Lee, J.H., and Chai, Y.G. (2008). Bacillus anthracis spores influence ATP synthase activity in murine macrophages. J Microbiol Biotechnol *18*, 778–783.

Setlow, P. (1995). Mechanisms for the prevention of damage to DNA in spores of Bacillus species. Annu Rev Microbiol *49*, 29–54.

Setlow, P. (2003). Spore germination. Curr Opin Microbiol *6*, 550–556.

Setlow, P. (2006). Spores of Bacillus subtilis: Their resistance to and killing by radiation, heat and chemicals. J Appl Microbiol *101*, 514–525.

Severson, K.M., Mallozzi, M., Bozue, J., Welkos, S.L., Cote, C.K., Knight, K.L., and Driks, A. (2009). Roles of the Bacillus anthracis spore protein ExsK in exosporium maturation and germination. J Bacteriol *191*, 7587–7596.

Shearer, A.E., Dunne, C.P., Sikes, A., and Hoover, D.G. (2000). Bacterial spore inhibition and inactivation in foods by pressure, chemical preservatives, and mild heat. J Food Prot *63*, 1503–1510.

Sinski, J.T., and Pannier, W.L. (1963). Decontamination of class III hood or animal holding system with formaldehyde. Tech Man US Army Biol Lab *86*, 1–11.

Sly, L.I., Taghavi, M., and Fegan, M. (1999). Phylogenetic position of Chitinophaga pinensis in the Flexibacter-Bacteroides-Cytophaga phylum. Int J Syst Bacteriol *49 Pt 2*, 479–481.

Smart, J.K. (2009). History of Decontamination (United States Army Solider and Biological Chemical Command).

Steichen, C., Chen, P., Kearney, J.F., and Turnbough, C.L., Jr. (2003). Identification of the immunodominant protein and other proteins of the Bacillus anthracis exosporium. J Bacteriol *185*, 1903–1910.

Sternbach, G. (2003). The history of anthrax. J Emerg Med *24*, 463–467.

Stragier, P., and Losick, R. (1996). Molecular genetics of sporulation in Bacillus subtilis. Annu Rev Genet *30*, 297–241.

Stump, M.J., Black, G., Fox, A., Fox, K.F., Turick, C.E., and Matthews, M. (2005). Identification of marker proteins for Bacillus anthracis using MALDI-TOF MS and ion trap MS/MS after direct extraction or electrophoretic separation. J Sep Sci *28*, 1642–1647.

Swatkoski, S., Russell, S.C., Edwards, N., and Fenselau, C. (2006). Rapid chemical digestion of small acid-soluble spore proteins for analysis of Bacillus spores. Anal Chem *78*, 181–188.

Sylvestre, P., Couture-Tosi, E., and Mock, M. (2002). A collagen-like surface glycoprotein is a structural component of the Bacillus anthracis exosporium. Mol Microbiol *45*, 169–178.

Tamir, H., and Gilvarg, C. (1966). Density gradient centrifugation for the separation of sporulating forms of bacteria. J Biol Chem *241*, 1085–1090.

Thorne, N.C., Shah, H.N., and Gharbia, S.E., eds. (2010). Isolation and Preparation of Spore Proteins and Subsequent Characterisation by Electrophoresis and Mass Spectrometry. In Mass Spectrometry for Microbial Proteomics, H.N. Shah and S.E. Gharbia, eds. (John Wiley & Sons Ltd. Chichester, UK).

Todd, S.J., Moir, A.J., Johnson, M.J., and Moir, A. (2003). Genes of Bacillus cereus and Bacillus anthracis encoding proteins of the exosporium. J Bacteriol *185*, 3373–3378.

Ulrich, M.P., Christensen, D.R., Coyne, S.R., Craw, P.D., Henchal, E.A., Sakai, S.H., Swenson, D., Tholath, J., Tsai, J., Weir, A.F., *et al.* (2006). Evaluation of the Cepheid GeneXpert system for detecting Bacillus anthracis. J Appl Microbiol *100*, 1011–1016.

Veening, J.W., Smits, W.K., and Kuipers, O.P. (2008). Bistability, epigenetics, and bet-hedging in bacteria. Annu Rev Microbiol *62*, 193–210.

Vesselinova, N.I., and Ensign, J.C. (1996). Flagellar proteins of motile spores of Actinomycetes. J Ind Microbiol Biotechnol *16*, 377–382.

Votava, M., and Slitrova, B. (2009). Comparison of susceptibility of spores of Bacillus subtilis and Czech strains of Clostridium difficile to disinfectants. Epidemiol Mikrobiol Imunol *58*, 36–42.

Vreeland, R.H., Rosenzweig, W.D., and Powers, D.W. (2000). Isolation of a 250 million-year-old halotolerant bacterium from a primary salt crystal. Nature *407*, 897–900.

Waller, L.N., Stump, M.J., Fox, K.F., Harley, W.M., Fox, A., Stewart, G.C., and Shahgholi, M. (2005). Identification of a second collagen-like glycoprotein produced by Bacillus anthracis and demonstration of associated spore-specific sugars. J Bacteriol *187*, 4592–4597.

Wang, J.C., Hu, S.H., and Lin, C.Y. (2003). Lethal effect of microwaves on spores of Bacillus spp. J Food Prot *66*, 604–609.

Weiss, W., and Gorg, A. (2009). High-Resolution Two-Dimensional Electrophoresis. In Proteomics, J. Reinders, and A. Sickmann, eds. (Totowa, NJ, Humana), pp. 13–32.

Willard, J.M., Lee, D.A., and Holland, M.M. (1998). Recovery of DNA for PCR amplification from blood and forensic samples using a chelating resin. Methods Mol Biol *98*, 9–18.

Williams, M.A., and Rittenberg, S.C. (1956). Microcyst formation and germination in Spirillum lunatum. J Gen Microbiol *15*, 205–209.

Windeler, A.S., Jr., and Walter, R.G. (1975). The sporicidal activity of glass bead sterilizers. J Endod *1*, 273–275.

Woese, C. (1959a). Further studies on the ionizing radiation inactivation of bacterial spores. J Bacteriol *77*, 38–42.

Woese, C. (1959b). Radiation destruction of the plaque-forming ability of spores of lysogenic Bacillus megaterium. Radiat Res *10*, 370–379.

Woese, C.R. (1958a). Comparison of the x-ray sensitivity of bacterial spores. J Bacteriol *75*, 5–8.

Woese, C.R. (1958b). Interpretation of inactivadon kinetics of spores of Bacillus megatherium. Arch Biochem Biophys *74*, 28–45.

Xu, H., He, X., Gou, J., Lee, H.Y., and Ahn, J. (2009). Kinetic evaluation of physiological heterogeneity in bacterial spores during thermal inactivation. J Gen Appl Microbiol *55*, 295–299.

Zadik, Y., and Peretz, A. (2008). The effectiveness of glass bead sterilizer in the dental practice. Refuat Hapeh Vehashinayim *25*, 36–39, 75.

Zhang, Y.J., Liu, W.J., and Zhang, L. (2006). Synergistic disinfection of Bacillus subtilis spores by UV irradiation and chlorine. Huan Jing Ke Xue *27*, 329–332.

Zilhao, R., Serrano, M., Isticato, R., Ricca, E., Moran, C.P., Jr., and Henriques, A.O. (2004). Interactions among CotB, CotG, and CotH during assembly of the Bacillus subtilis spore coat. J Bacteriol *186*, 1110–1119.

Zhilina, T.N., Garnova, E.S., Tourova, T.P., Kostrikina, N.A., and Zavarzin, G.A. (2001). Halonatronum saccharophilum gen. nov. sp. nov.: A new haloalkaliphilic bacterium of the order Haloanaerobiales from Lake Magadi. Microbiology *70*, 64–72.

Zimny-Arndt, U., Schmid, M., Ackermann, R., and Jungblut, P.R. (2009). Classical Proteomics: Two-Dimensional Electrophoresis/MALDI Mass Spectrometry. In Mass Spectrometry of Proteins and Peptides, M.S. Lipton, and L. Pasa-Tolic, eds. (Totowa, NJ, Humana Press), pp. 65–92.

Part XVI
Novel Approaches in Sample Preparation and LC-MS Analysis

Chapter 47
New Developments in LC-MS and Other Hyphenated Techniques

Mikhail E. Belov, Ruwan Kurulugama, Daniel Lopez-Ferrer, Yehia Ibrahim, and Erin Baker

Abstract Extensive challenges faced by analytical chemists in studying real world complex samples such as biological body fluids, tissue samples, environmental and geological samples have led to the exponential growth in analytical techniques in the late twentieth century. This vast array of different analytical techniques can be categorized into two major areas: sample separation and mass spectrometry analysis. Current state-of-the-art sample separation methods include gas chromatography, high performance liquid chromatography, ultra high pressure liquid chromatography, solid phase extraction, capillary electrophoresis, capillary zone electrophoresis and gas phase separation techniques such as ion mobility spectrometry. The current trend in sample separation is to combine multiple techniques that utilize different separation mechanisms to maximize the separation. The most widely used combinations include two-dimensional gas chromatography, strong cation exchange or weak cation exchange chromatography followed by reversed-phase liquid chromatography, two-dimensional reversed-phase liquid chromatography, liquid chromatography followed by ion mobility spectrometry and two-dimensional electrophoresis techniques. The introduction of atmospheric pressure ionization techniques such as electrospray ionization and matrix assisted laser desorption ionization and the variations of these two techniques have exponentially increased the utility of mass spectrometry in complex sample analysis. Mass spectrometry itself has tremendously improved over the years in terms of sensitivity, detection limits and dynamic range capabilities. Currently, mass spectrometers can attain zeptomolar detection limits with five orders of magnitude dynamic range.

Keywords Reversed-phase HPLC · Strong cation exchange · Capillary zone electrophoresis · Ion mobility spectrometry · High-field asymmetric waveform ion mobility spectrometry

---

M.E. Belov (✉)
Biological Science Division, Pacific Northwest National Laboratory, Richland, WA 99352, USA
e-mail: mikhail.belov@pnl.gov

A.R. Ivanov, A.V. Lazarev (eds.), *Sample Preparation in Biological Mass Spectrometry*, DOI 10.1007/978-94-007-0828-0_47,

Abbreviations

AGC	automated gain control
AMT	accurate mass and time
BSA	Bovine serum albumin
CE	capillary electrophoresis
CID	collision induced dissociation
cIEF	capillary isoelectric focusing
COFRADIC	combined fractional diagonal chromatography
CV	compensation voltage
CZE	capillary zone electrophoresis
DESI	desorption ESI
DMA	differential mobility analysis
DMS	differential mobility spectrometry
DTIMS	drift tube ion mobility spectrometry
EOF	electroosmotic flow
ESI	electrospray ionization
FAI	field amplified injection
FAIMS	high-field asymmetric waveform ion mobility spectrometry
FASS	field amplified sample stacking
FDR	false discovery rate
FTICR	Fourier Transform Ion Cyclotron Resonance
GC	gas chromatography
HPLC	high performance LC
HSA	human serum albumin
IGOT	isotope-coded glycosylation-site-specific tagging
IMAC	immobilized metal affinity chromatography
IMS	ion mobility spectrometry
ITP	isotachophoresis
LC	liquid chromatography
MALDI	matrix-assisted laser desorption ionization
MRM	multiple reaction monitoring
MS	mass spectrometry
OMS	overtone mobility spectrometry
PDADMAC	polydiallyl dimethylammonium chloride
PDMS	polydimethylsiloxane
PETG	polyethleneterephthalate
PMMA	polymethylmethacrylate
PNGase F	peptide-N-glycosidase F
PPV	positive predictive value
PSMA	polystyrene maleic anhydride
PTM	post-translational modification
qTOF	quadrupole time-of-flight
RP-HPLC	reversed-phase high performance liquid chromatography
RSD	relative standard deviation

SCX	strong cation exchange
SEM	scanning electron microscope
SIM	selected ion monitoring
SNR	Signal-to-noise ratio
SPE	solid phase extraction
SRM	selected reaction monitoring
tITP	transient isotachophoresis
TOFMS	time-of-flight MS
TWIG	travelling wave ion guide
TWIMS	traveling wave ion mobility spectrometry
WAX	weak anion exchange
μCZE	microfludic chip capillary zone electrophoresis

47.1 Introduction

Until the middle of the twentieth century, analytical separations were largely carried out by classical methods such as precipitation, distillation, and extraction. Due to the simplicity of these methods, only simple samples could be evaluated, ultimately limiting the information obtained from these studies. The development of a technique to quickly separate a complex mixture of analytes by transporting them in a mobile phase through a fixed immiscible stationary phase for qualitative and quantitative analysis was a welcome scientific finding. This technique, termed chromatography in 1903 by the Russian botanist Mikhail Tswett, established a direct approach to separate and analyze different components that makeup simple and complex mixtures. In the last 50 years, chromatography applications have grown explosively owing not only to the development of several new types of chromatographic techniques but also to the expanding need of scientists for better methods to characterize complex mixtures (Smith, 1988; Robards et al., 1994; Scott, 1995).

Since chromatography offers a quick way to separate, purify, and rapidly analyze mixtures, automation of certain sample preparation and analysis steps is possible. To further increase the throughput of analysis and enhance detection of analytes, coupling chromatography to other techniques such as mass spectrometry (MS) has attracted vast interest. During the twentieth century, gas chromatography (GC) combined with MS served as the preferred tool for separating mixtures (Greenway and Simpson, 1980; McLafferty, 1992; Kitson et al., 1996). However, the applicability of GC-MS was restricted because GC itself was only able to analyze volatile, low molecular weight species, which make up ~15% of all existing compounds. This critical limitation caused interest in liquid chromatography (LC) to grow. LC's great advantage over GC is its ability to sample aqueous mixtures and nonvolatile compounds including amino acids, proteins, nucleic acids, hydrocarbons, pesticides, carbohydrates, drugs, and many other aqueous mixtures that are of prime interest to industrial, research and public applications. Initial attempts to use LC separations were very slow due to long columns with large diameters, and packing material sizes exceeding 10 μm. Attempts were made to speed up the procedure; however, it wasn't until the late 1960s that the technology for producing smaller particle size packing was developed. After this development, high pressure systems were also implemented and this combination resulted in the technology being renamed to high performance LC (HPLC) to separate the new procedures from the initial basic methods (Snyder and Kirkland, 1997).

Currently, HPLC is the most widely used method of all analytical separation techniques due to its ability to separate molecules based on different properties such as hydrophobicity, size, and ionic characteristics just by changing the column packing and mobile phase solvent. The combination of HPLC with MS has matured into a vital analytical technique, but initially many challenges were faced in interfacing the two methods, such as dealing with both large liquid solvent volumes and vacuum requirements (Tomer and Parker, 1989; Garcia and Bacelo, 1993; Caprioli, 1994). The invention of electrospray ionization (ESI) in the early 1990s addressed the interface problem and allowed ions to be easily transferred from the HPLC column into the mass spectrometer for detection (Whitehouse et al., 1985). By performing HPLC-MS jointly, individual shortcomings of each method were addressed. Improvements included the ability to identify and quantify compounds that are not fully resolved with HPLC and drastic reduction in ionization suppression that ultimately limits detection of low abundance analytes in complex mixtures by MS. Today HPLC-MS continues to rapidly evolve towards higher speed, efficiency, and sensitivity driven by the emerging needs to prepare and evaluate numerous samples in areas of drug development (Lee, 2002), proteomics (Covey et al., 1991; Griffin et al., 1991), food safety (Andersen et al., 2008), environmental protection (Reiser and Fogiel, 1994; Molina et al., 1995), and pharmacokinetics (Hsieh and Korfmacher, 2006). Further areas are also of interest for HPLC-MS studies and, as the techniques continue to mature, additional applications become apparent. The other widely used liquid (or condensed) phase separation approach is capillary zone electrophoresis (CZE). This technology has been widely applied to the analysis of intact proteins (Haselberg et al., 2007), metabolites (Ptolemy and Britz-McKibbin, 2006), polysaccharides and carbohydrates (Mao et al., 2002). Readily amenable to miniaturization, CZE has been implemented in microfludic chips (μCZE) and efficiently coupled to MS (Lion et al., 2003). In this chapter, we will give a brief overview of different hyphenated condensed phase separation approaches, including multidimensional HPLC-MS and μCZE-MS, and will highlight the benefits of coupling HPLC-MS with gas phase separation techniques, such as ion mobility spectrometry (IMS) and field asymmetric waveform ion mobility spectrometry (FAIMS).

47.2 Multidimensional LC Separations in Proteomics

47.2.1 Bottom-Up Proteomics

In the field of proteomics, one of the most commonly used strategies is referred to as a "bottom-up" approach (Aebersold and Mann, 2003; Domon and Aebersold, 2006).

This approach aims at identification and characterization of the protein complement of a biological cell, tissue or biological fluid by an initial enzymatic digestion of proteins into peptides, which are then separated using various capillary liquid chromatographic separations and detected using mass spectrometry (LC-MS). This process is based on a "divide and conquer" strategy, where the goal is to separate complex peptide mixtures into distinct fractions, detect and identify peptides with high confidence and then map the identified peptides to protein classes, i.e. to obtain high coverage of the proteome. An enormous dynamic range of protein abundances in complex biological samples, such as blood plasma, requires high sensitivity of the analytical detection methods used in proteomics that is determined by the efficiencies of condensed phase separation stages and the MS interface (Qian et al., 2006, 2008). One of the key components in enhancing the sensitivity of condensed phase separations coupled to MS has been the development of new electrospray ionization (ESI) sources (Davis et al., 1995; Wilm et al., 1996). Initial sensitivity improvement of HPLC-MS due to the interfacing of the two techniques via a micro ESI source has been further augmented at reduced ESI flow rates (Kelly et al., 2008; Marginean et al., 2009). Decreasing flow rates from μL/min to nL/min range has been shown to enhance the ionization efficiency and ion transmission from an ion source to mass spectrometer (Marginean et al., 2007). Given a constant back pressure, lower LC flow rates are attained at smaller column diameters or longer column length. Shen et al. have shown a dramatic increase in proteome coverage when using 87-cm long columns packed with 3-μm C18-bonded porous (300 Å pores) silica particles. It was demonstrated that nano-LC-nano-ESI-MS sensitivity increased linearly with a decrease in the flow rates in the range of 20–400 nL/min, while separation peak capacity exceeded 1000 (Shen et al., 2001a, b, 2002, 2005). However, one of the limitations of reduced flow rates is an increased sample loading time. The latter can be reduced by the coupling of on-line solid phase extraction (SPE) to a nano-HPLC. This coupling enabled a 400-fold increase in the sample volume loading capacity and a 10-fold increase in the sample mass loading at a total injection and loading time of only a few minutes (Shen et al., 2004).

Broad characterization of a proteome poses formidable challenges even if highly efficient reversed-phase high performance liquid chromatography (RP-HPLC) separation is employed. Reignier and coworkers have estimated that a combination of strong cation exchange (SCX) and RP separations can potentially yield 2000–10,000 peptide peaks (Hsieh et al., 1996). This metric has placed multidimensional separations on the forefront of proteomic research. In 1999, Yates and coworkers published a seminal work, which took advantage of the orthogonality between SCX and RP separations. They applied this 2D separation approach in combination with tandem MS to *Saccharomyces cerevisiae* ribosome, leading to the identification of a novel protein component of the yeast and human 40S subunit (Link et al., 1999). The schematic diagram of the SCX-RP approach is shown in Fig. 47.1. In the subsequent publications, the method was further improved and evaluated with complex biological samples (Washburn et al., 2001; Wolters et al., 2001). To perform 2D separations, a biphasic column was constructed with 5 μm Partishpere (Whatman, Clifton, NJ) packing material and an RP column packed with 5 μm

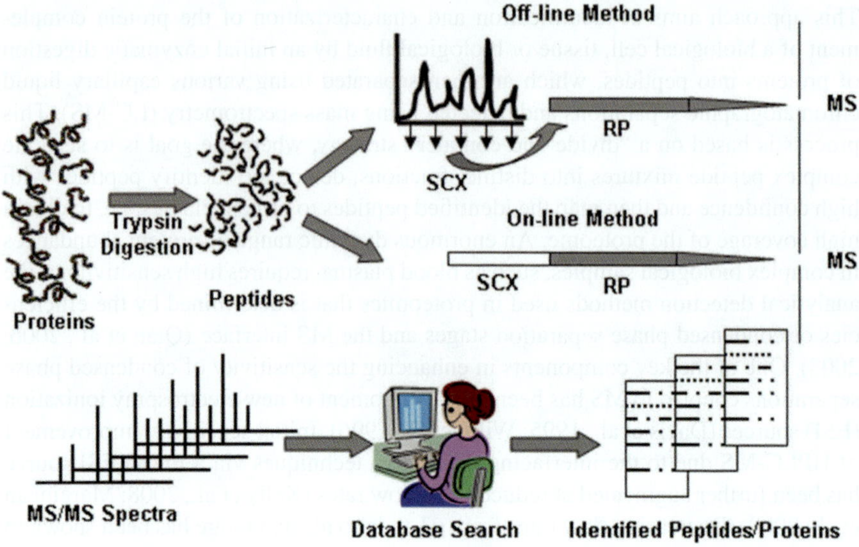

Fig. 47.1 Flow diagram of the two-dimensional separation of protein complexes. Typically, a denatured protein complex is digested with trypsin and the acidified peptide mixture is loaded onto the strong-cation exchange (SCX) column. A discrete fraction of peptides is displaced from the SCX column to the reversed-phase (RP) column and eluted into a mass spectrometer (MS) operating in tandem MS mode. This iterative process is repeated to obtain the fragmentation patterns of peptides in the original mixture. The program SEQUEST is used to correlate the tandem MS spectra of fragmented peptides to amino acid sequences using nucleotide databases

Zorbax Eclipse XDB (Hewlett Packard, Palo Alto, CA) C_{18} particles. The first column separates peptides according to their net charge, while the second separation is based on peptide hydrophobicities. The approach was applied for separation of peptides from the proteome of *S. cerevisiae* strain BJ5460 (Washburn et al., 2001). A total of 1484 proteins were detected and identified, including low abundance proteins such as transcription factors and protein kinases. Tri-phasic columns using RP separation followed by the SCX and then RP separation have been employed to further improve protein depth of coverage in multidimensional separations (Motoyama et al., 2006). These columns, denoted as RP-SCX-RP, are capable of simultaneous sample clean-up procedures which are performed on-line with analyte separations.

One of the main drawbacks of the on-line SCX-RP method is the lack of continuous salt gradient for efficient separation of peptides in the SCX dimension. The on-line 2D SCX-RP separation is typically performed using discrete iterative steps in the SCX salt gradient followed by the RP organic gradients. Alternatively, it is viable to perform the SCX offline by collecting distinct fractions followed by online RP-MS analysis of each SCX fraction (Peng et al., 2003; Lopez-Ferrer et al., 2004). One of the advantages of such an approach is that organic solvents such as acetonitrile can be used in the first dimension without any interference with the

RP-HPLC gradient. This provides better peptide recovery while the organic solvent can be easily removed by drying down the SCX fractions in a SpeedVac. The method can also be further improved if volatile salts such as ammonium formate or ammonium acetate are used. Thus, there would be practically no need for desalting even at high salt concentrations. A number of alternative 2D methodologies have been developed to further increase proteome coverage. To name only a few, these are the use of anion exchange combined with RP separations (Pieper et al., 2003), peptide isoelectric focusing (Cargile et al., 2004a, b), monolithic nano-capillary columns (Ivanov et al., 2003b), multimode pressure-assisted capillary electrochromatography/capillary electrophoresis (Ivanov et al., 2003a), and affinity chromatography (Qian et al., 2008) combined with RP separations. These systems have been successfully applied for the separation of complex mixtures. More recently, the orthogonal separation of peptides using both high and low pH RP separations has been shown to exhibit similar peptide coverage as that obtained using the traditional SCX-RP combination (Gilar et al., 2005a, b; Manadas et al., 2009).

For sensitive detection of peptides and confident protein identifications, it is highly desirable that multidimensional separations are interfaced to a high-resolution MS. One of the efficient proteomic pipelines, the accurate mass and time (AMT) tag approach, has been developed at Pacific Northwest National Laboratory (Lipton et al., 2002; Smith et al., 2002). Shown in Fig. 47.2, the AMT is a two step approach that employs retention time information from high peak capacity separations and accurate mass measurements of high resolution MS to uniquely identify peptide profiles in complex biological samples (Masselon et al., 2005). Using multidimensional LC separations and MS/MS measurements in the first step, an extensive database (i.e., peptide library) for a given proteome is generated, yielding a list of peptide sequences, normalized RP elution times and accurate peptide masses. The use of an Orbitrap MS (Makarov, 2000) and Fourier Transform Ion Cyclotron Resonance (FTICR) MS (Marshall, 2000) augments ion trap-generated tentative precursor ion sequences with accurate mass information for highly confident peptide identifications. Once the peptide library is compiled, there is no longer need to acquire tandem MS spectra or perform multidimensional separation steps, as highly accurate mass measurements and reproducible retention times are sufficient to correctly identify a peptide among all possible sequences in the database.

47.2.2 Post-translational Modifications (PTMs)

Other interesting hyphenated approaches are based on a combination of multiple chromatographic separations and specific enrichments of certain peptides from a complex pool of species generated after proteome digestion. Gevaert et al. developed COFRADIC (combined fractional diagonal chromatography) based on diagonal electrophoresis/chromatography (Gevaert et al., 2003, 2006; Vandekerckhove et al., 2004). This novel technology uses straightforward reactions to create shifts

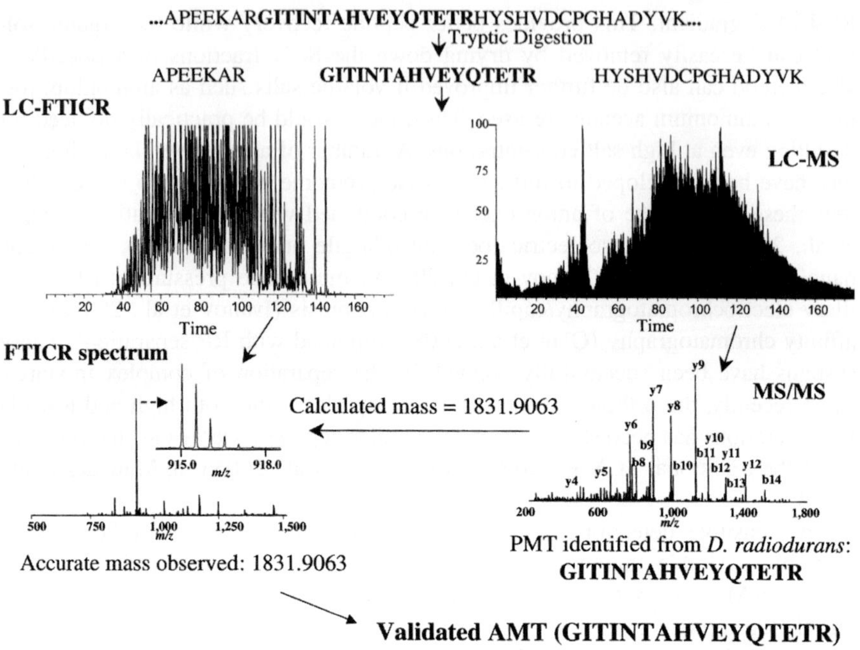

Fig. 47.2 Experimental strategy defining the accurate mass and time (AMT) tag approach as illustrated in identification of a peptide from elongation factor Tu (EF-Tu) within the *D. radiodurans* proteome. Sample digested with trypsin was analyzed sequentially with RPLC-ion trap tandem MS and RPLC-FTICR MS instruments. In this example, a tryptic peptide, or a putative mass and time (PMT) tag, was identified by MS/MS and the accurate mass was calculated based on its sequence (i.e. 1831.9063 Da) and its elution time recorded. The same proteome sample was then analyzed under the same conditions using a high-field FTICR MS. An accurate mass and time (AMT) tag was established when a peptide eluting at the same time and corresponding to the calculated mass (within 1 ppm) of the PMT identified in the first analysis. (Reproduced with permission from Smith et al. (2002). Copyright 2010 ©, with permission from the John Wiley and Sons, Inc.)

in the retention times of certain peptides by changing their overall hydrophobicity. RP runs are then performed to select specific peptide classes (e.g., N-terminal peptides, Met-containing peptides, etc.) (Neville et al., 1997; Gevaert et al., 2005; Villen et al., 2007; Gevaert and Vandekerckhove, 2009; Ghesquiere et al., 2009). In the case of methionine-containing peptides, the species of interest are first isolated and an RP-HPLC run is performed to collect a number of fractions. The fractions are then oxidized by the addition of hydrogen peroxide in such a way that methionine is transformed to methionine sulfoxide. When the modified and unmodified fractions are separated on the same column, the retention time of the oxidized peptides will be shifted, as oxidation produces a dipole moment which causes peptides to be more hydrophilic. A similar strategy has been developed for phosphopeptides and cysteine containing peptides. The success of the method for global proteomics depends on two key points: (1) the targeted peptides have to be an accurate representation of all the proteins in a proteome, and (2) reactions used to modify peptide hydrophobicities have to be complete and specific.

Other strategies for enrichment of specific peptides have also been developed. The isolation and enrichment of phosphopeptides is critically important because of the key role this post-translational modification (PTM) plays in the understanding of cell biology, including signal transduction and regulation of enzyme activities in many intracellular biological processes. The common strategy for phosphopeptides analysis is to use immobilized metal affinity chromatography (IMAC) (Villen and Gygi, 2008). This technique has been widely applied after peptide fractionation on a SCX column, which provides an initial enrichment of the phosphopeptides (Larsen et al., 2005; Thingholm et al., 2008a, b). An IMAC stationary phase typically consists of metal ions such as Fe^{3+}, Ga^{3+} or Al^{3+} chelated to a given substrate (Larsen et al., 2005; Thingholm et al., 2008a, b). The negative polarity of the phosphate group causes the phosphopeptides to be more neutral or negatively charged than the unmodified peptides, which are typically positively charged at a pH commonly used with SCX resins. Another phase that has proven to be efficient in the isolation of phosphopeptides is TiO_2 (Thingholm et al., 2008b), where peptides are loaded into an acid buffer (pH $\sim$2.9) and eluted using an alkaline buffer. The main difference between TiO_2 and IMAC is that the former has higher affinity for the mono-phosphorylated peptides, while IMAC is more specific for di- and multi-phosphorylated species. There is a variety of protocols for each resin to optimize the isolation of certain classes of phosphopeptides and also to attain sequential enrichment using both resins.

Another important PTM, which has drawn close attention due to its implication in various types of cancers, is glycosylation. Zhang et al. have reported on the technique of protein quantitation using stable isotope labeling and automated tandem MS (Zhang et al., 2003). The approach is based on conjugation of glycoproteins to a solid support using hydrazine chemistry, stable isotope labeling of glycopeptides and the specific release of formerly N-glycosylated peptides via peptide-N-glycosidase F (PNGase F). Figure 47.3 shows the specific capture of glycoproteins by hydrazine resin. Kaji et al. have developed a strategy for large-scale identification of N-glycosylated proteins from complex biological samples (Kaji et al., 2003). The approach, termed isotope-coded glycosylation-site-specific tagging (IGOT), is based on the lectin column-mediated affinity capture of a set of glycopeptides generated by tryptic digestion of protein mixtures, followed by peptide-N-glycosidase-mediated incorporation of a stable isotope tag, O-18, specifically into the N-glycosylation site. The O-18-tagged peptides are then identified by multi-dimensional LC-MS. The application of this method to the characterization of N-linked high-mannose and/or hybrid-type glycoproteins from an extract of *Caenorhabditis elegans* proteins allowed the identification of 250 glycoproteins, including 83 putative transmembrane proteins, with the simultaneous determination of 400 unique N-glycosylation sites. For detailed information on detection and identification of phosphorylated and glycosylated proteins, the reader is referred to excellent reviews (Blom et al., 2004; Wuhrer et al., 2007).

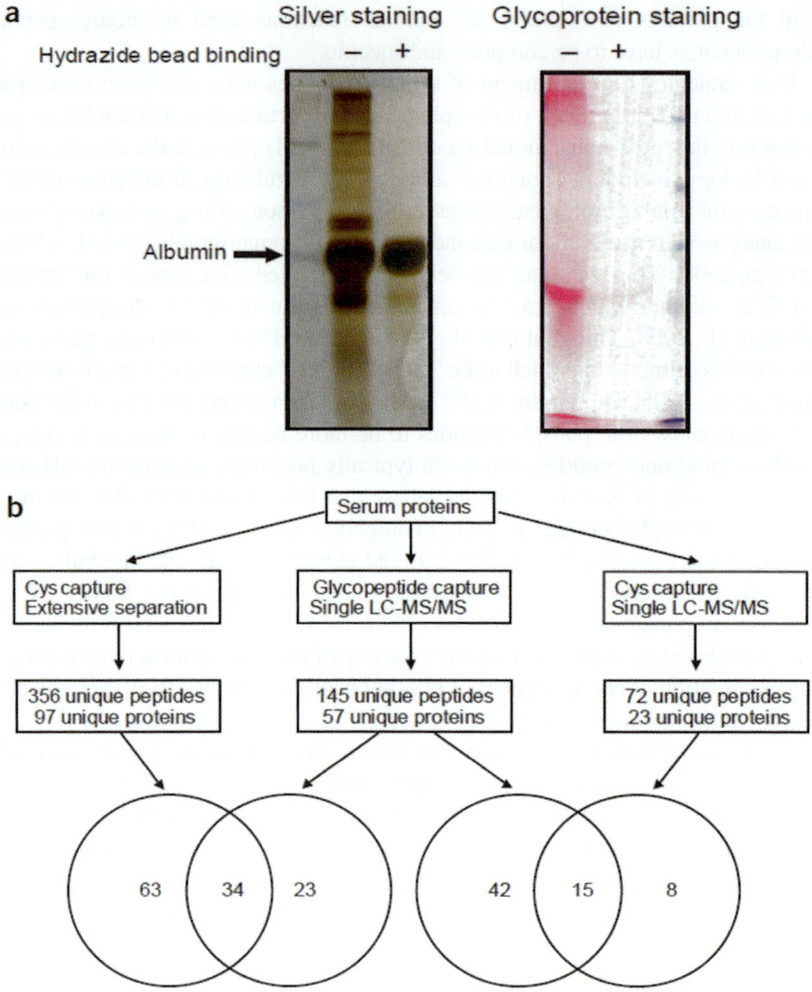

Fig. 47.3 Isolation of glycoproteins from serum. (**a**) Specific capture of serum glycoproteins by hydrazine resin. Total and glycoprotein protein staining results before (–) and after (+) the capture. (**b**) Comparison of the total number of proteins or peptides isolated and identified from serum samples by either cysteine-reactive tag or glycopeptides capture method. (Reproduced with permission from Zhang et al. (2003). Copyright 2010 ©, with permission from the Nature Publishing Group)

47.2.3 Top-Down Proteomics

A number of limitations inherent with the bottom-up strategy hamper its ability to provide a more comprehensive survey of biological systems. The most apparent obstacle in bottom-up proteomics analyses is that complete sequence coverage of proteins is rarely achieved, especially in the case of global and large-scale proteome analyses. As a result, important information with respect to the native proteins such as site-specific mutations and PTMs that are often critical for understanding protein function and regulation may be lost and cannot be examined in a full spectrum. Moreover, attributing certain peptide identifications to a specific protein is often challenging because of the presence of highly homologous proteins and protein isoforms in proteome samples, which also hinders accurate quantitation. In complex organisms, alternative splicing can lead to a significantly increased protein repertoire, e.g., up to three-fourth of human genes have at least one variant (Johnson et al., 2003; Resch et al., 2004; Stamm et al., 2005). Therefore, the use of short peptides as proxy markers for genes is inadequate and often misleading (Godovac-Zimmermann et al., 2005). Given these limitations of a bottom-up strategy, approaching analysis from the top-down, i.e., analyzing individual proteins directly by MS or MS/MS, has been increasingly pursued (Reid and McLuckey, 2002; Kelleher, 2004; Bogdanov and Smith, 2005). It is also becoming more and more popular to combine top-down and bottom-up strategies to obtain a more complete picture of the proteome (VerBerkmoes et al., 2002). In one application of the approach, Kelleher and coworkers have reported on detection and identification of 22 yeast proteins in a single RPLC-tandem MS run. Using anion exchange chromatography, proteins were fractionated prior to online RPLC runs, which resulted in detection of 231 metabolically (N-14/N-15) labeled protein pairs from *S. cerevisiae*. In another example, Sharma et al. have explored a combination of weak anion exchange (WAX) fractionation and on-line RPLC separation using a 12 T FTICR mass spectrometer for the detection of intact proteins from a *Shewanella oneidensis* MR-1 cell lysate (Sharma et al., 2007). A total of 715 intact proteins were detected, and the combined results from the WAX fractions and the unfractionated cell lysate were aligned using LC-MS features to facilitate protein abundance measurements. Protein identifications and post-translational modifications were assigned by comparing intact protein mass measurements to proteins identified in peptide MS/MS analysis of an aliquot of the same fraction.

Modern proteomics relies on implementation of a combination of efficient separation approaches and MS to fully characterize complex proteolytic digests. Selection of a particular strategy for multidimensional separation is mainly driven by the scope of biochemical research. Though multidimensional separations yield an improvement in proteome coverage, their practical use is hindered by the long analysis time, reduced reproducibility and the need for maintenance. Therefore, further developments aimed at increasing sample throughput, analysis reproducibility and the ability to efficiently isolate and enrich certain post-translationally modified proteins are essential for leveraging proteomics capabilities in high throughput screening of clinical samples.

47.3 Liquid Chromatography/Ion Mobility Spectrometry/Mass Spectrometry

The current state-of-the-art procedures for complex biological sample analysis involve multiple condensed phase separation and fractionation methods prior to mass spectrometry (MS) detection. These separation techniques are essential for detecting low abundance analytes from a complex biological sample. However, when thousands of samples are needed to be analyzed, e.g. in large scale clinical studies, the use of multiple slow condensed phase separation steps becomes impractical. In recent years, fast gas phase separation approaches, such as IMS, have emerged as an alternative separation step for complex sample analysis. The timescale of IMS measurements is such that it ideally fits between slower condensed phase separation and fast MS detection, representing a "nested", partially orthogonal to MS, separation approach. Herewith, we will primarily focus on hyphenated approaches where IMS is combined with LC and MS for biochemical applications.

The use of ion mobility as a chromatographic technique was described in early 1970s (Cohen and Karasek, 1970). Theory of IMS and historical perspectives have been previously discussed (Karasek, 1974; Revercomb and Mason, 1975). The basis of this technique is that different ions with varying collision cross-sections have different mobilities through an inert buffer gas under a constant weak electric field (Karasek, 1974; Revercomb and Mason, 1975). The conventional IMS experiments are performed in a drift tube, which is filled with a buffer gas (usually He or N_2) and consists of a series of ring electrodes providing a constant electric field. Ions traverse the drift tube at a constant velocity governed by the electric field strength and ion mobility. Mobility of an ion (K) is a physical characteristic of a charged molecule, which is governed by (Revercomb and Mason, 1975):

$$K = \frac{3}{16}\frac{Ze}{N}\frac{1}{\Omega_0}\sqrt{\frac{2\pi}{k_B T}\left(\frac{1}{m}+\frac{1}{M}\right)} \tag{47.1}$$

where Ze is the analyte charge, N is the number density of the buffer gas, m and M are the masses of the drift gas and analyte, respectively, k_B is Boltzmann's constant, T is the absolute temperature of the buffer gas, and Ω_0 is the diffusion collision integral. In practice, ion mobility is measured as a ratio of the ion velocity, v_D, to the electric field strength, E,

$$K = \frac{v_\mathrm{D}}{E} = \frac{L^2}{t_\mathrm{D} V} \tag{47.2}$$

where L is the drift tube length, V is the voltage across the drift tube, and t_D is the drift time of the analyte ion. Ion mobilities are typically reported as the reduced mobility values where temperature and pressure of the buffer gas are normalized to the standard values for comparing data sets across different instruments. The reduced mobility (K_0) is given by the following equation (Revercomb and Mason, 1975):

$$K_0 = K\frac{273.2}{T}\frac{P}{760} \tag{47.3}$$

where T and P correspond to the temperature (in K) and pressure (in Torr) in the drift tube, respectively. As evident from Eqs. (47.1) and (47.2), IMS is capable of characterizing molecular structures as well as separating components in a mixture.

Besides the conventional drift tube ion mobility technique (DTIMS), there are several other methods that employ the mobility of ions in gas phase as a separation method. These include: (1) differential mobility analysis (DMA) (de la Mora et al., 1998), in which ions migrate through a flowing gas under the influence of an electric field, (2) high-field asymmetric waveform ion mobility spectrometry (FAIMS) (Guevremont, 2004a; Kolakowski and Mester, 2007; Krylov et al., 2007) in which ions are separated based on differences between high- and low-field mobilities, (3) traveling wave ion mobility spectrometry (TWIMS) (Giles et al., 2004; Pringle et al., 2007) where a moving step wave applied to the drift tube facilitate the movement of ions, and (4) recently developed overtone mobility spectrometry (OMS) (Kurulugama et al., 2009; Valentine et al., 2009b), where a segmented drift tube with two or more distinct field settings is used to filter gas phase ions. These new approaches are becoming powerful tools for characterizing different classes of biomolecules and also for simplifying the analysis of complex mixtures through the removal of chemical noise.

47.3.1 Liquid Chromatography/Drift Tube Ion Mobility Spectrometery/Mass Spectrometry

DTIMS-MS has been primarily used for structural studies of gaseous molecules and clusters (Clemmer and Jarrold, 1997), gas phase reaction kinetic studies (Bowers et al., 1993; Wyttenbach et al., 2003; Moision and Armentrout, 2006), separation of closely related molecules such as isomers and conformers as well as to simplify analysis of complex mixtures.(Srebalus Barnes et al., 2002; Clowers et al., 2005; Koeniger et al., 2006a; Plasencia et al., 2008). IMS has been extensively used with various ion sources such as electrospray ionization (ESI), desorption ESI (DESI), matrix-assisted laser desorption ionization (MALDI), and radioactivity-based ionization, glow discharge and photoionization (Guharay et al., 2008). In recent years, the capability of the hyphenated high-performance liquid chromatography (HPLC)-IMS-TOFMS approach has been increasingly explored and a number of advances in instrument development and data processing have enabled the use of this technique in high-throughput proteomic applications (McLean and Russell, 2005; Taraszka et al., 2005; Valentine et al., 2006a; Baker et al., 2007a; Clowers et al., 2008). The utility of HPLC-IMS-TOFMS technology was initially described by Clemmer and coworkers in 2001 (Valentine et al., 2001). Since then several groups have implemented and further developed similar technologies using custom-built and commercial instruments (Kanu et al., 2008). Figure 47.4 shows an IMS-TOFMS instrument constructed by Clemmer's group (Isailovic et al., 2008). This instrument is capable of performing two-dimensional IMS separations in addition to the conventional HPLC-IMS-TOFMS experiments. The two-dimensional separation is

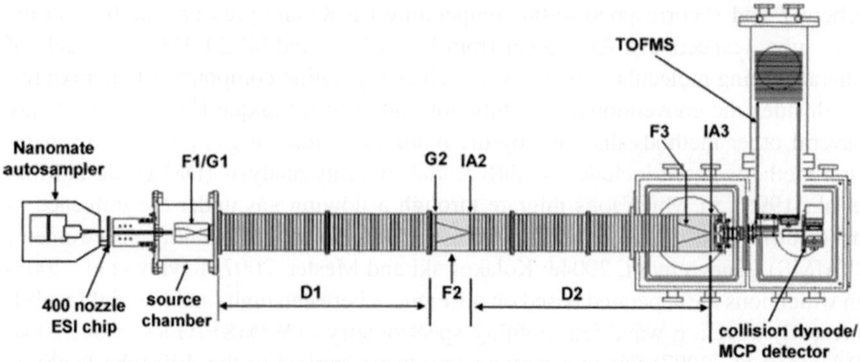

Fig. 47.4 Schematic diagram of the NanoMate-IMS-IMS-TOFMS instrument developed by Clemmer and coworkers. (Reproduced with permission from Isailovic et al. (2008). Copyright 2010 ©, with permission from the American Chemical Society)

obtained by collisionally activating parent ions at the end of the first drift tube so that the gas phase structures of these ions are altered while no dissociation is induced, and the activated intact ions are then separated in the second drift tube. This method was reported to improve the IMS peak capacity by at least a factor of 10 and in some favorable cases up to 60–80 (Koeniger et al., 2006b; Merenbloom et al., 2007; Kurulugama et al., 2008). The two-dimensional IMS technology has been used to perform high-throughput analysis of human plasma proteome from 70 individuals (Valentine et al., 2009a). Plasma samples were depleted of the most abundant proteins using affinity chromatography, digested with trypsin, purified with solid phase extraction (SPE) and then directly infused with a NanoMate autosampler (Advion Biosciences, Ithaca, NY) into the IMS-IMS-TOFMS analyzer. Although this technology is at a very early stage of development, the concept of ultra high-throughput proteome analysis has a multitude of applications (Valentine et al., 2009a). In another experiment, the IMS-IMS-TOFMS instrument was used to study N-linked serum glycans from cirrhosis and liver cancer patients. IMS profiles have been reported to identify the two study groups (Isailovic et al., 2008). Russell and coworkers have estimated the maximum deviation of the structure from the linear trend line between IMS and MS to be about 11% based on 14 individual data sets that included peptides from enzymatic digestions. This information provides an estimation of the orthogonality between IMS and MS measurements. The variation in peak capacity based on the drift gas used has also been studied (Ruotolo et al., 2002, 2004). Using a low resolution IMS-MS instrument, Valentine et al. have estimated the peak capacity for the two-dimensional LC-IMS and three-dimensional LC-IMS-MS separation to be ~900–1200 and $\sim 3.7\text{–}4.6\times 10^5$, respectively (Valentine et al., 2001). McLean et al. have compiled a table to represent the peak capacity for different multidimensional techniques (McLean et al., 2005). Based on that table, LC-FTICR-MS generates the highest peak capacity followed by the HPLC-IMS-MS and MALDI-IMS-MS techniques. However, MALDI-IMS-MS produced the

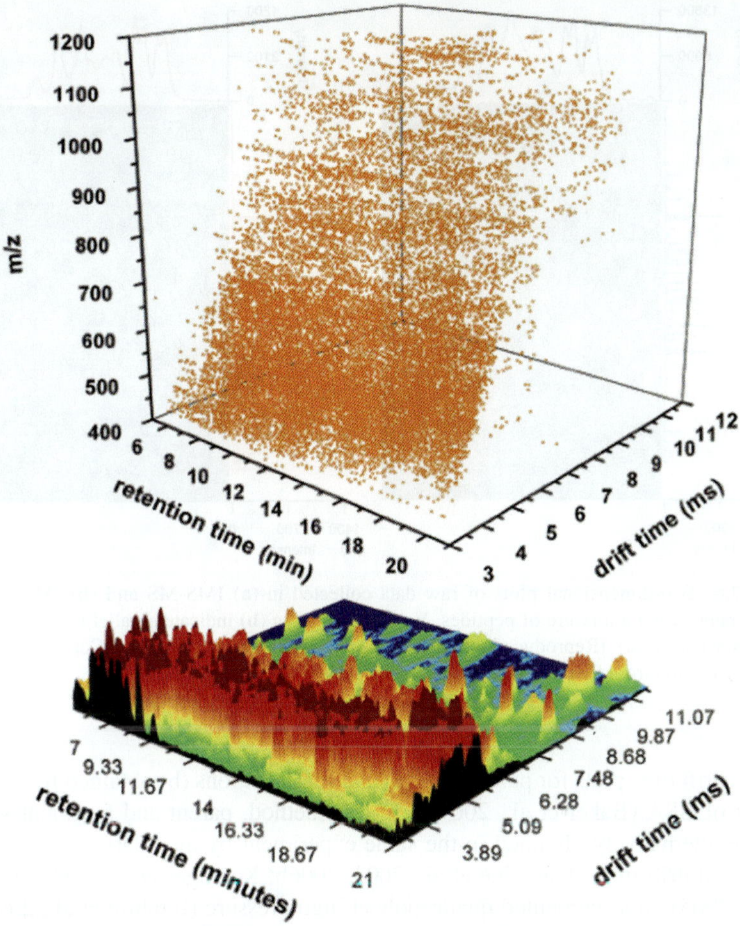

Fig. 47.5 Three-dimensional peak positions obtained from the 6000 most intense features observed from the LC-IMS-TOFMS analysis of human plasma proteome. (Reproduced with permission from Valentine et al. (2006a). Copyright 2010 ©, with permission from the American Chemical Society)

highest peak capacity production per second, which was calculated to be 2.8×10^6. Figure 47.5 shows a typical three-dimensional retention time, drift time, and *m/z* plot and a two-dimensional retention time versus drift time plot obtained for a human plasma sample (Valentine et al., 2006a). The three-dimensional plot was created using the 6000 most intense features observed in each of the 10 strong cation exchange (SCX) fractions. The two-dimensional plot was obtained for a single SCX fraction.

To perform tandem MS with IMS, collision induced dissociation (CID) can be carried out at the back of the drift tube using a parallel (also called time aligned parallel or multiplexed) fragmentation method. Figure 47.6 shows two-dimensional *m/z*

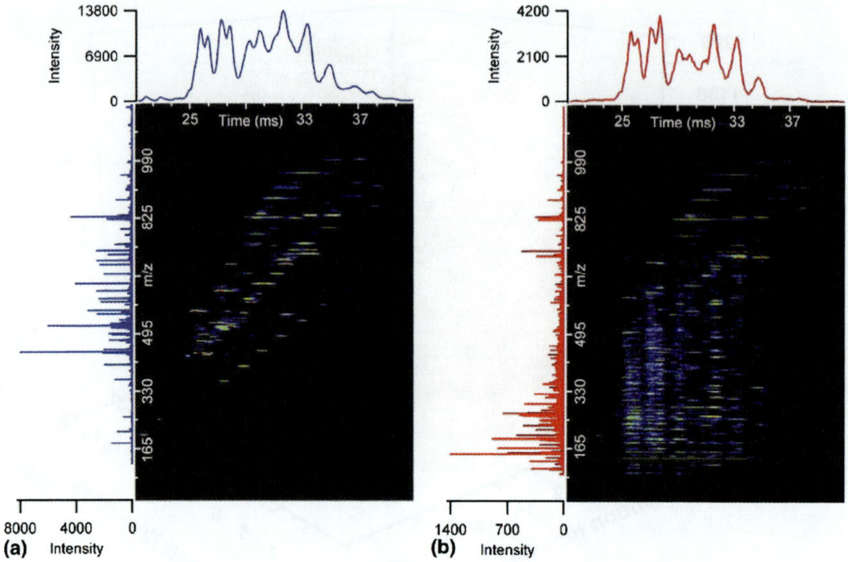

Fig. 47.6 Two-dimensional plots of raw data collected in (**a**) IMS-MS and (**b**) IMS-CID-MS experiments using a mixture of peptides. Vertical streaks in (**b**) indicate parallel fragmentation of precursor ions in (**a**). (Reproduced with permission from Baker et al. (2007b). Copyright 2010 ©, with permission from Elsevier B.V.)

versus drift time plots for parent ions (a) and fragment ions (b) obtained for a tryptic digest of BSA (Baker et al., 2007b). In this method, parent and fragment spectra can be alternatively obtained in the same experiment by modulating the field in a split-field drift tube (Valentine et al., 2003, 2006b; Koeniger et al., 2005; McLean et al., 2005), in a segmented quadrupole at high pressure (Ibrahim et al., 2009), or in a differentially pumped region as described below. Baker et al. have performed CID between the conductance limiting orifice behind the second short quadrupole and before the first octopole maintained at a pressure of $\sim 4\times10^{-2}$ Torr (Baker et al., 2007b). Here, a drift time dependent field gradient was used for efficient fragmentation of precursor ions with different charge states and different mobilities. Since the parent ions are fragmented at the back of the drift tube, the corresponding fragment ions have the same drift time signature as their parent ion. Therefore, fragmentation mass spectra obtained at high mass accuracy and resolution would yield high confidence identifications of the parent ions. However, the practical use of reversed-phase LC separation followed by IMS-CID-MS analysis for peptide identifications remains challenging due to inadequate data processing tools. When the parent ions are completely or partially overlapped in the drift domain, the resulting CID mass spectra include fragment ions from multiple precursors (as shown in Fig. 47.6), making it difficult to perform database searches for peptide identifications. Belov and coworkers have recently developed a novel algorithm for deconvolution of the overlapping (multiplexed) CID-MS spectra obtained from the multiple parent ions unresolved in the IMS domain. The algorithm is based on the efficient use of genomic information, IMS profiles and accurate mass measurements for both the precursors and fragment ions (Ibrahim et al., 2009).

For conventional DTIMS experiments, a packet of ions must be introduced into the drift tube to initiate the experiment. The early designs (and the contemporary IMS instruments operating at atmospheric pressures) made use of Bradbury-Nielsen gates to introduce ion packets into an IMS instrument. This limits the instrument duty cycle to <1%. The invention of ion funnel by Smith and Shaffer (2000) and its subsequent use in IMS-MS instrumentation dramatically improved the ion utilization in the ESI-IMS and IMS-TOF interfaces (Tang et al., 2005). Ion funnel was redesigned to be used for efficient trapping and releasing of ions at pressures of ~1–5 Torr (Ibrahim et al., 2007; Clowers et al., 2008). Ibrahim et al. have used an optimized configuration of an ion funnel trap (IFT) operated in automated gain control (AGC) mode in conjunction with an orthogonal TOFMS instrument to improve LC-MS sensitivity in analysis of complex bacterial proteome (Ibrahim et al., 2008). Despite high charge capacity of the IFT device ($\sim 10^7$ charges), large incoming ion currents observed in LC-ESI experiments with complex biological mixtures were found to fill the trap to its capacity in ~1–3 ms, limiting the duty cycle of IMS separation to 3–5%. To improve the IMS-TOFMS duty cycle (and the sensitivity), Belov et al. have developed a multiplexing technique, where multiplexing implies the capability of introducing multiple ion packets into the drift tube on a time scale of a single IMS separation (Belov et al., 2007). IMS-TOF signals were encoded with a pseudo-random binary sequence and the original signal vector was then successfully reconstructed using an inverse matrix transform. Since ions were accumulated in the IFT between adjacent IMS releases, the duty cycle of the multiplexing scheme was found to be >50%, which resulted in 7–10 fold improvements in signal-to-noise ratios (SNR). In the following development, Belov et al. have gone a step further and accounted for signal variability in LC-ESI experiments by developing a dynamic multiplexing scheme (Belov et al., 2008). The premise for the dynamic multiplexing is to track the dynamic source function (such as that of LC-ESI) with the IMS-TOFMS detection system by dynamically readjusting ion accumulation times and signal encoding sequences to ensure high dynamic range and high mass accuracy of IMS-TOFMS measurements throughout LC separation. This approach was initially demonstrated in analysis of reversed-phase LC fractions of human blood plasma samples using a NanoMate ESI coupled to an IMS-TOFMS system. The dynamic multiplexing IMS-TOF technology has been recently implemented with online LC separations of complex biological samples (Belov et al., 2009). Figure 47.7 shows the coupling of a custom built 88 cm long drift tube to a commercially available TOFMS (Belov et al., 2008). Here, the IFT is used to trap and release ion packets into the drift tube, and a rear ion funnel is used to recollimate the dispersed ion packets into a compact ion beam before entering to the MS. The exit grid of the IFT was used to modulate ion packet introduction into the IMS drift tube. A segmented quadrupole placed immediately after the rear funnel can be used to perform efficient CID. For this instrument the IMS resolving power was measured to be around 60–80, while the resolving power for the TOFMS is around 8000.

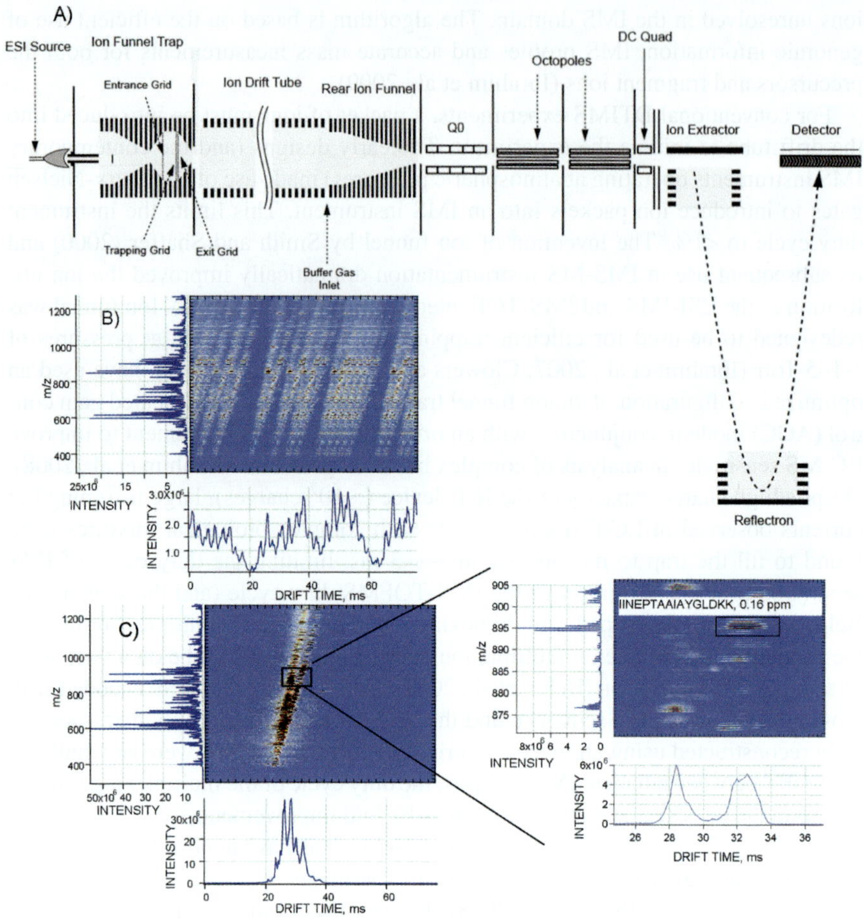

Fig. 47.7 (**a**) Schematic diagram of the IMS-TOFMS instrument constructed at Pacific Northwest National Laboratory (PNNL). Ion packet accumulation and multiplexing are accomplished by modulating the entrance, trapping and exit grids of the ion funnel trap. Multiplexed IMS-CID is performed inside the segmented quadrupole (Q0) at a pressure of ~200 mTorr. (**b**) and (**c**) IMS-TOFMS signals detected and reconstructed in experiments with a 0.5 mg/mL depleted human blood plasma sample. (**b**) 5-bit encoded IMS-TOFMS signal from a human blood plasma fraction, (**c**) reconstructed IMS-TOFMS signal from the 2D map in (**b**). The inset shows one of the identified peptides from the depleted human blood plasma (Reproduced with permission from Belov et al. (2008). Copyright 2010 ©, with permission from the American Chemical Society)

Synapt HDMS (Waters Corporation, Milford, USA) mass spectrometer is the first commercial mass spectrometer to incorporate pulsed IMS technology. Figure 47.8 shows the schematic diagram of a Synapt HDMS instrument (Pringle et al., 2007). Following ionization in a Z-spray ESI source, ions are collisionally focused by a travelling wave ion guide (TWIG) into a quadrupole analyzer. The mass selected ions are then trapped in the trap TWIG and separated according to their mobilities in the IMS TWIG and guided by transfer TWIG prior to introduction into the

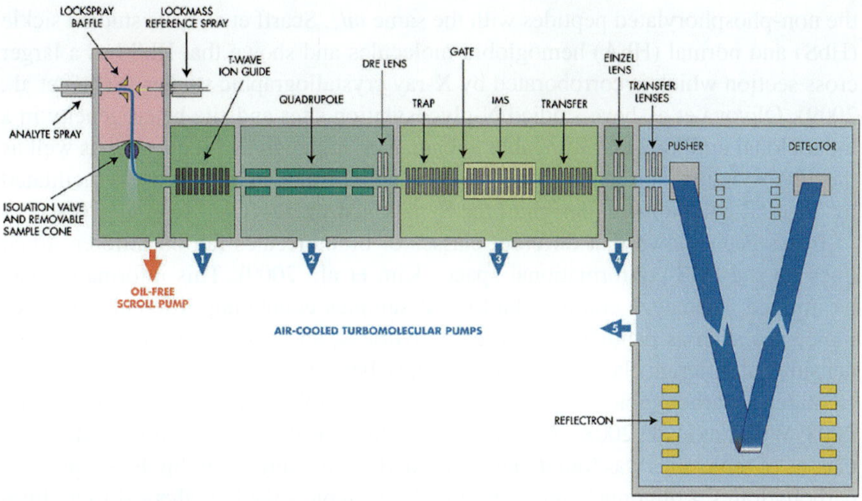

Fig. 47.8 Schematic diagram of Synapt HDMS (Waters Corporation, Milford, USA). This instrument incorporates a recently developed ion mobility separation method called TWIMS. (Reproduced with permission from Pringle et al. (2007). Copyright 2010 ©, with permission from Elsevier B.V.)

TOFMS. This instrument can be operated in a number of different configurations, so that both trap and transfer region can be used as collision cells for tandem MS analysis. The resolving power of a Synapt was shown to be ~10–15. The latest version of this instrument (Synapt G2) reportedly has an IMS resolving power of about 45 (Giles et al., 2009). An increase in the resolving power was achieved by increasing the buffer gas pressure in the IMS TWIG. However, to minimize the undesirable CID at the entrance to the drift cell a short region filled with Helium buffer gas has been added to the system. For the basic principles of TWIG theory and operation, the reader is referred to the comprehensive articles (Giles et al., 2004; Pringle et al., 2007; Shvartsburg and Smith, 2008). Synapt IMS-MS instrument has been used in various studies and applications such as detection of impurities in drugs, monoclonal antibody analysis, isomer separation, self assembly of macromolecules, post translational modifications (PTMs), peptide fragmentation, protein folding, structural studies of proteins, lipids and protein complexes, and identification of protein variants (Eckers et al., 2007; Garcia et al., 2008; Atmanene et al., 2009; Benassi et al., 2009; Chan et al., 2009; Thalassinos et al., 2009; Williams et al., 2009; Yamagaki and Sato, 2009). Garcia et al. have studied the structural variants of a- and b-type peptide fragment ions generated by CID using TWIMS/tandem MS techniques (Garcia et al., 2008). The use of IMS in conjunction with tandem MS has provided evidence for the formation of stable macrocyclic structures for several b-type ions. Additionally, different fragmentation patterns have been observed for isomeric a- or b-type ions with different mobilities. Thalassinos et al. have used Synapt instrument to separate phosphorylated peptides from other species (Thalassinos et al., 2009). Phosphorylated peptides were shown to have smaller cross sections compared to the non-phosphorylated peptides with the same *m/z*. Scarff et al. have studied sickle (HbS) and normal (HbA) hemoglobin molecules and shown that HbS has a larger cross section which is corroborated by X-ray crystallographic studies (Scarff et al., 2009). Olivova et al. have studied N-glycosylation sites and site heterogeneity in a monoclonal antibody and were able to characterize the glycan sequences as well as the glycosylation sites using a two-step sequential fragmentation process facilitated by the dual-collision-cell design of the Synapt instrument (Olivova et al., 2008).

It has been shown that different classes of biomolecules fall on different trend lines on the IMS conformational space (Kim et al., 2009). This information can be utilized to analyze complex biological samples containing different classes of molecules such as peptides, lipids, carbohydrates, and nucleic acids. The average densities of different classes of molecules have been determined as follows: oligonucleotides > carbohydrates > peptides > lipids (Ruotolo et al., 2007; Olivova et al., 2008; Williams et al., 2008). At a given *m/z* value, the average drift time for different classes of molecules was found to increase in the following order, lipids > peptides > carbohydrates > oligonucleotides. Figure 47.9a depicts the hypothetical trend lines on conformation space where different classes of biomolecules can be observed. The plot is presented for singly charged ions that are produced by matrix-assisted laser desorption ionization (MALDI) (Fenn et al., 2009). Figure 47.9b shows collision cross sections with respect to *m/z* values for different classes of molecules. The plot consists of data from 96 oligonucleotides, 192 carbohydrates, 610 peptides and 53 lipids. All species correspond to singly charged ions generated by MALDI. Figure 47.9c and d show the separation of lipids from peptides and peptides from carbohydrates, respectively in the same sample using MALDI-IMS-MS techniques. The advantage of using IMS prior to MS analysis for samples with multiple types of molecules is two-fold. First, IMS separation results in improved SNR due to dispersion of the chemical noise. Second, drift time information can be efficiently used as an additional criterion for reducing false discovery rate (FDR) in database searches. McLean et al. have shown that the use of drift time information in conjunction with mass measurements improves the protein confidence level assignments by a factor of 10–100 over that when using only MS data (McLean, 2009).

47.3.2 Liquid Chromatography/Field Asymmetric Ion Mobility Spectrometry/Mass Spectrometry

FAIMS, also referred to as differential mobility spectrometry (DMS), is widely used as an atmospheric pressure gas phase separation technique in conjunction with MS (Eiceman et al., 2002; Eiceman et al., 2004; Guevremont, 2004a, b; Shvartsburg et al., 2006; Kolakowski and Mester, 2007; Krylov et al., 2007). The device consists of two parallel electrodes that may have planar or cylindrical arrangements. A flow of gas between the two electrodes carries the ions through the device while they experience an asymmetric waveform. Typical gases used are air, nitrogen or mixture of nitrogen with other gases (such as Helium, SF_6 and CO_2) (Eiceman et al., 2004)

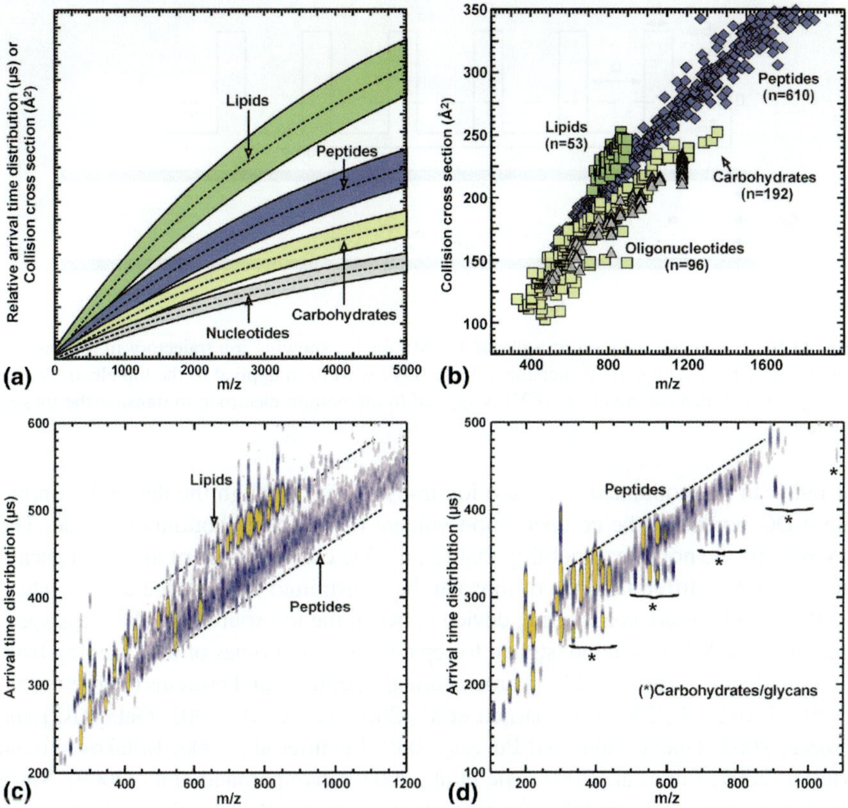

Fig. 47.9 A hypothetical representation of the conformation space for different classes of biomolecules (**a**), a plot of collision cross sections as a function of *m/z* for different classes of molecules (**b**), a 2D plot of drift time versus *m/z* obtained for the simultaneous analysis of lipids and peptides from a thin tissue section (**c**), and a 2D plot of drift time versus *m/z* obtained for the simultaneous analysis of peptides, glycans and in-source decay carbohydrate fragments from trypsin and PNGase F treated glycoprotein ribonuclease (**d**). (Reproduced with permission from McLean (2009). Copyright 2010 ©, with permission from Elsevier B.V.)

or dopants (Barnett et al., 1999a). Gas mixtures and dopants have been shown to improve the separation power of FAIMS.

As show in Fig. 47.10 the asymmetric waveform is applied to the top electrode such that (Purves and Guevremont, 1999; Guevremont, 2004a):

$$E_{\max} \times t_1 + E_{\min} \times t_2 = 0 \qquad (47.4)$$

where $E_{\max}$ represent the high field applied for duration t_1 and $E_{\min}$ is the low field applied for duration of t_2. If the mobility of a positive ion is higher in the high field than in the low field, the ion will drift toward the bottom electrode. After multiple cycles the ion will eventually be lost to the bottom electrode. The trajectory (path)

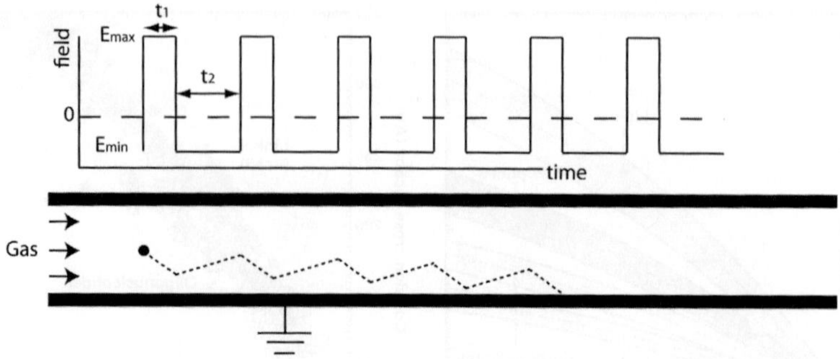

Fig. 47.10 Schematic diagram of a planar FAIMS device showing ion trajectory in response to variations of the asymmetric dispersion voltage (DV) waveform applied to the top electrode. An appropriate compensation voltage (CV) is applied to the bottom electrode to transmit the ions of interest

of ions can be modulated to ensure ion transmission through the device by applying a DC potential, the compensation voltage (CV), to the bottom electrode. The CV reflects the nonlinear mobility that is a unique characteristic of an ion. By scanning the CVs, different types of ions can be transmitted through the device. Thus, FAIMS can be used as a filtering device between the ion source and the mass spectrometer. FAIMS was demonstrated to separate different types of ions ranging from small molecules, amino acids, pharmaceuticals, peptides and proteins (Barnett et al., 1999b; Ells et al., 2000a, b; Barnett et al., 2002; Cui et al., 2003; Gabryelski and Froese, 2003; Guevremont and Purves, 2005; Levin et al., 2006; Kolakowski and Mester, 2007; Liu et al., 2007; Mie et al., 2007; Champarnaud et al., 2009). Since the demonstration of capability for separating isomers such as leucine and isoleucine (Barnett et al., 1999a) FAIMS use in LC-MS method developments where interferences are problematic has increased (Kapron and Thakur, 2006; Kapron et al., 2006; Hatsis et al., 2007; Liu et al., 2007; Wu et al., 2007; Xia et al., 2008; Klaassen et al., 2009). Kapron et al. have demonstrated the removal of metabolite interference from drug analysis using LC-MS/MS (Kapron et al., 2005). The N-oxide metabolite which co-eluted with the target drug underwent in-source conversion to the drug, resulting in incorrect quantitation by SRM (selected reaction monitoring) mode with a triple-quadrupole MS. By utilizing a CV difference of $\sim$11 V between the drug and the metabolite, FAIMS was shown to remove the N-oxide metabolite that co-eluted with the drug in LC-MS analysis. The removal of interference is most useful for targeted analysis, where FAIMS can be set up as a narrow band filter prior to MS detection. Figure 47.11a shows an example of interferences present in LC-MS

Fig. 47.11 (**a**) The representative LC/MS chromatograms of the pharmaceutical analyte (*top*) and internal standard (*bottom*) in rat serum. Data were acquired at the analyte LLOQ. (**b**) The representative LC/FAIMS-MS chromatogram of the analyte (*top*) and the internal standard (*bottom*) at the LLOQ in rat serum using FAIMS CV of –35 V (Reproduced with permission from Klaassen et al. (2009). Copyright 2010 ©, with permission from John Wiley and Sons, Inc.)

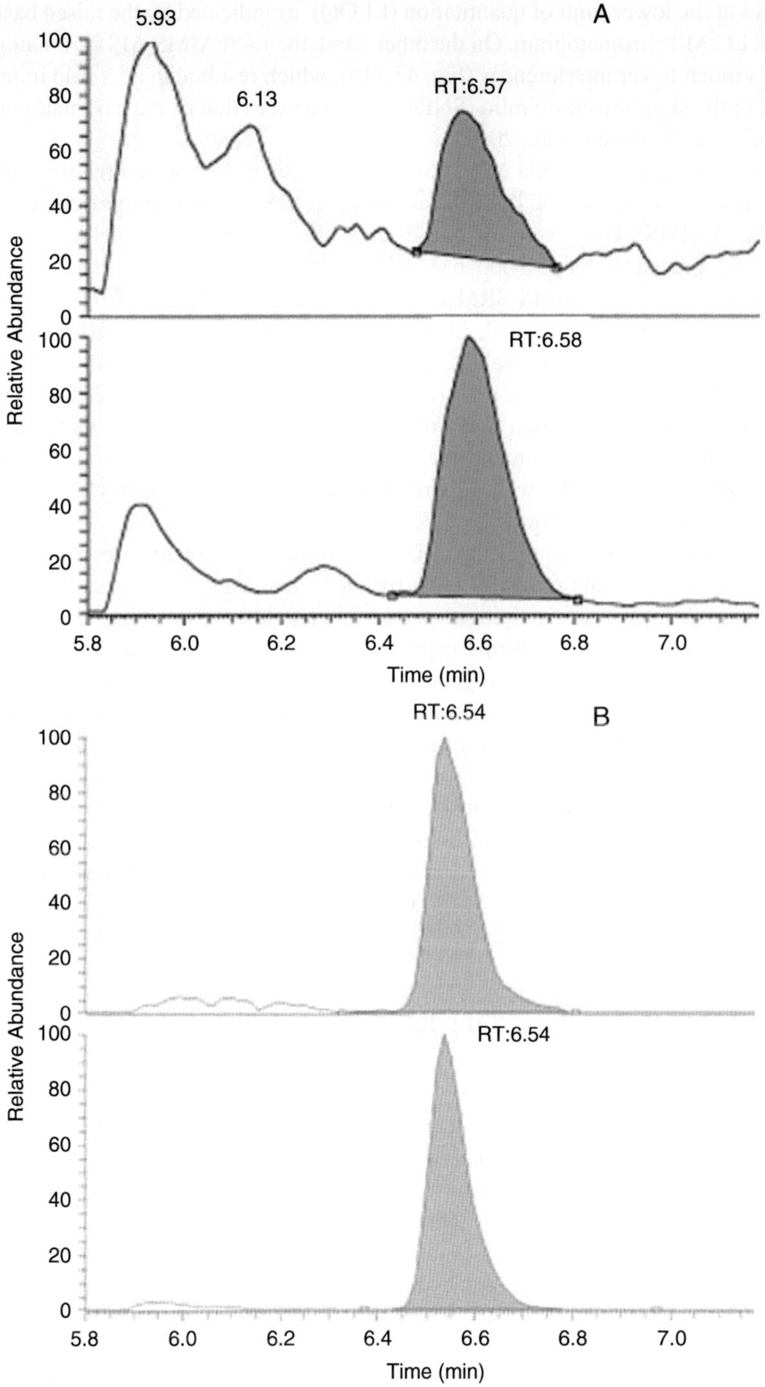

Fig. 47.11 (continued)

studies at the lower limit of quantitation (LLOQ), as indicated by the raised baseline in the LC/MS chromatogram. On the other hand, the LC/FAIMS-MS chromatogram shows much fewer interferences (Fig. 47.11b), which resulted in a 18-fold improvement in the signal-to-noise ratio (SNR) as compared to that of the chromatogram in Fig. 47.11a (Klaassen et al., 2009).

Wu et al. (2007) and Xia et al. (2008) systematically evaluated the ruggedness, accuracy and precision of LC-MS and LC-MS/MS methods employed with and without FAIMS. The results of Xia et al. (2008) showed a good level of accuracy and precision for 1–1000 nM calibration standards in rat plasma. At the 1 nM level, the LC-FAIMS-SRM exhibited accuracy and RSD of 97 and 18.6%, respectively, versus 143% and 33.6% observed with the LC-SRM method. The ruggedness of FAIMS was evaluated with more than 240 injections over 16 h with an RSD for the response area ratio (analyte/internal standard) of 3.72–5.78% using the three analytes. Although performed to simultaneously monitor three analytes in LC-FAIMS-SRM experiment, the use of FAIMS in multiple reaction monitoring (MRM) experiments, where more than one transition per analyte is typically monitored, is yet to be shown.

In conclusion, we would emphasize that hyphenated techniques such as LC-IMS-MS have become powerful tools for biological sample analysis. Proteomics, metabolomics, glycomics, lipidomics and other fields are evolving at a fast pace and new instrumental platforms capable of providing information on molecular sequences and structures are in great demand. IMS-TOF has a number of features that favorably position this technology at the forefront of biochemical research. These include improved sample coverage, extended dynamic range, low limits of detection and reduced chemical noise, the ability to obtain structural information and separate conformers and isobaric species. Although a number of technological challenges still remain to be addressed, LC-IMS-MS has gone through the infancy period and has been shown to have an increasing impact on biochemical and clinical research.

47.4 Microfluidic Capillary Electrophoresis/Mass Spectrometry

Despite being a powerful front-end approach for "bottom-up" proteomics, reversed-phase LC has limitations in analyses of larger peptides and intact proteins due to their adverse interaction with stationary phase and denaturation of the proteins in organic mobile phases. This hinders LC analyses of intact proteins and results in a decrease in the achievable resolution and peak capacity. On the other end of hydrophobicity scale are hydrophilic intracellular metabolites, which are poorly retained on the stationary phase and often remain undetected in LC analyses. Capillary electrophoresis (CE) represents a viable alternative to LC in analyses of both the intact proteins and hydrophilic metabolites. As to protein analysis, CE can be performed in aqueous buffers without organic modifiers and stationary phase to avoid protein conformational changes and eliminate protein degradation during analysis. In addition, high peak capacities typically attainable in CE enable efficient separation of large proteins based on their electrophoretic mobilities in background electrolytes (Tsukagoshi et al., 2005; Liu et al., 2006). Since first reports on coupling CE to MS (Olivares et al., 1987; Smith et al., 1988), this hyphenated approach has been increasingly used for characterization of intact proteins and their isoforms, as well as metabolites. The reader is referred to excellent reviews and references therein on capillary CE (Hernandez-Borges et al., 2004; Simpson and Smith, 2005; Ptolemy and Britz-McKibbin, 2006; Haselberg et al., 2007; Gaspar et al., 2008; Mischak et al., 2009; Ramautar et al., 2009) and microfludic CE systems (Lion et al., 2003; Peng et al., 2008). Herein, we will focus on miniaturization of CE, integration of different features, such as sample injection, separation, ionization, and interfacing microfluidic CE devices to MS. Most microfludic devices are coupled to MS through a nano-ESI interface, which requires stable reproducible flow rates. Therefore, the emphasis of this report will be on the capillary zone electrophoresis (CZE) capable of generating a solution bulk flow by electroosmosis.

47.4.1 Theoretical Background of Miniaturization

To better understand the effect of miniaturization on the efficiency of liquid phase separations, one needs to consider the behavior of physical parameters when a system is scaled down in size (Manz et al., 1990; Becker and Locascio, 2002). The underlying assumptions for microfludic systems are that their transport phenomena are controlled by diffusion and the flow regimes are strictly laminar. Diffusion is described by Fick's law as follows:

$$\frac{\partial n}{\partial t} = D\Delta n \tag{47.5}$$

where n is the density of concentration of particles, D is the diffusion coefficient, and Δ is the Laplace operator.

In the limit of low Reynold's numbers corresponding to the laminar flow, the diffusion coefficient of a spherical particle of radius r is given by:

$$D = \frac{k_B T}{6\pi \eta r} \tag{47.6}$$

where k_B is the Boltzmann's coefficient, T is the thermodynamic temperature, and η is the viscosity of the medium.

In a one-dimensional system, the mean squared displacement of a particle, $< x^2 >$, over its residence time in the capillary, t, is governed by:

$$< x^2 >= 2Dt \tag{47.7}$$

The parameter $< x^2 >$ also describes the variance of the migration zone width, i.e. the diffusion-limited broadening of the peak of a single component and is invoked to determine the capillary separation efficiency in terms of the number of theoretical plates, N, as follows:

$$N = \frac{L^2}{< x^2 >} = \frac{L^2}{2Dt} \tag{47.8}$$

where L is the length of the capillary.

The residence time of a charged particle in the capillary is determined by its electrophoretic mobility, μ_{CE}, and electric field strength, E, as follows:

$$t = \frac{L}{\mu_{\mathrm{CE}} E} \tag{47.9}$$

Substituting Eq. (47.9) into Eq. (47.8) results in

$$N = \frac{\mu_{\mathrm{CE}}}{2D} LE = \frac{\mu_{\mathrm{CE}}}{2D} U \tag{47.10}$$

Therefore, the number of theoretical plates (i.e., separation efficiency) is proportional to the applied voltage, U, across the capillary.

In practice, the number of theoretical plates, N, produced with capillary zone electrophoresis is often substantially below that expected under isothermal conditions dominated by the axial molecular diffusion. Even if symmetric peaks are generated, that would exclude migrational dispersion due to conductivity differences between the sample zone and bulk buffer, and analyte adsorption to the capillary walls, the parameter N has often been a fraction of that anticipated. Knox et al. have shown that peak broadening can be attributed to Joule heating (Knox and McCormack, 1994a, b). The optimum electric field in the capillary, E_{opt}, was found to be governed by (Knox and McCormack, 1994a):

$$E_{\mathrm{opt}} = \frac{5533.8 D^{1/3}}{d_{\mathrm{c}}} \tag{47.11}$$

where d_{c} is the capillary diameter.

Accounting for Eq. (47.10) results in:

$$N = \frac{2766.9}{D^{2/3}} \frac{L}{d_{\mathrm{c}}} E_{\mathrm{opt}} \propto \frac{L}{d_{\mathrm{c}}} \tag{47.12}$$

Therefore, the number of theoretical plates under the optimum conditions is inversely proportional to the capillary diameter.

Substituting Eq. (47.11) into Eq. (47.9) derives an expression of particle's residence time in the capillary through capillary length and diameter:

$$t \propto L d_{\mathrm{c}} \tag{47.13}$$

The combination of Eqs. (47.12) and (47.13) determines the underlying relationship which drives minituarization of capillary electrophoresis:

$$\frac{N}{t} \propto \frac{1}{d_c{}^2} \tag{47.14}$$

This means that the number of theoretical plates per unit time is proportional to the reciprocal of the capillary diameter squared, implying that the performance of capillary electrophoresis systems in chemical and biological separations drastically improves with a decrease in the capillary diameter.

47.4.2 Design of Microfluidic Chips for Electrophoretic Separations

A number of excellent reviews have described the properties of materials and technological processes used in fabrication of microfludic chips and the reader is referred to these publications for detailed information (Becker and Gartner, 2000; Becker and Locascio, 2002; Ng et al., 2002; Rossier et al., 2002). The most commonly used polymers for microfabrication are polymethylmethacrylate (PMMA), polydimethylsiloxane (PDMS) and polyethleneterephthalate (PETG). Several important features need to be considered in choosing the material for a microfluidic system.

Since electroosmotic pumping is the most common method used to generate the flow in a microfluidic device, the substrate material should have good electrical insulting properties to ensure the electric field gradient is established across the fluidic channel and not through the substrate.

The second consideration is a Joule heating that could be significant in systems employing electroosmotic flow (EOF). The number of theoretical plates, N, has been shown to strongly depend on the capillary temperature, T (Knox and McCormack, 1994b):

$$N = 1.851 \times 10^{18} r u_s \exp\frac{\left(\frac{1713}{T}\right)}{T} \tag{47.15}$$

where r is the radius of a diffusing molecule and u_s is the linear velocity of a sample. An increase in the capillary temperature from 20 to 55°C at a constant sample velocity results in more than a two-fold decrease in the expected plate number.

The third consideration is the microchannel charge, whose polarity and density varies drastically for different polymer materials. Surface charge establishes electroosmotic velocity, which determines the electromigration velocity of an ion, u_s:

$$u_{EO} = \frac{\varepsilon_0 \varepsilon \varsigma E}{\eta} \tag{47.16}$$

$$u_s = (\mu_{EO} + \mu_{EP})E \tag{47.17}$$

where u_{eo} is the electroosmotic velocity, ε_0 and ε are the dielectric permittivities of vacuum and medium, respectively, ζ is the zeta potential, μ_{EO} and μ_{EP} are the electroosmotic and electrophoretic mobilities. Separation of positively charged peptides and proteins using capillary electrophoresis in a microfludic format and interfacing the microfluidic chip to MS through an electrospray ionization (ESI) interface require suppression of analyte adsorption to the capillary walls and generation of a stable reproducible bulk flow in the range of 20–500 nL/min. Modifications of the capillary wall surface, including cationic polymer grafting process or dynamic coating (Dolnik, 2004; Kitagawa et al., 2006) are essential for controlling the microchannel charge.

As a detailed description of techniques for manufacturing microfludic devices is not the scope of this review, we would only briefly list several of the most popular approaches. In hot embossing (or imprinting), one approach employs a silicon stamp, which is made by transferring an image to a photomask or high resolution transparency (Martynova et al., 1997). A silicon wafer is coated with a masking material such as silicon dioxide, and then spin-coated with a layer of photoresist and subsequently exposed to UV light through the transparency image aligned to the wafer flat. The image is then transferred to the exposed masking layer by etching in a solution of e.g., hydrofluoric acid for silicon dioxide, and the result is a raised 3D image of the channels. In another approach (McCormick et al., 1997) a photomask is prepared to define the separation channel layout and a silicon master is wet-etched to produce a "negative" image of the desired channel layout on the silicon wafer. Subsequent to this step, a nickel electroform is produced from the etched silicon master; this electroform corresponds to a "positive" image of the microchip channels. Then a second electroform, which is a replica of the original silicon master, is prepared using the first one as a template. This electroform is further used to fabricate microchannels in plastic substrates by imprinting or injection molding.

A significant advancement in microfluidic system development was the introduction of elastomeric polymer molding techniques, known as soft photolithography (Duffy et al., 1998). Figure 47.12 shows the fabrication scheme of enclosed microchannels in oxidized PDMS. As with the methods described above, a positive relief of photoresist is fabricated on a silicon wafer. Following mounting glass posts on the wafer, which define reservoirs for analytes and buffers, a pre-polymer of PDMS is cast onto the silicon wafer and cured at 65°C for 1 h. The polymer replica of the master with negative relief of channels is then peeled away from the silicon wafer and the glass posts are removed. The PDMS replica and a flat slab are further oxidized in a plasma discharge (1) to obtain the conformal contact between PDMS surfaces and form an irreversible seal, and (2) to introduce silanol (SiOH) groups onto the surface of the polymer and support EOF in neutral and basic aqueous solutions.

Other approaches include laser photoablation (Roberts et al., 1997), X-ray lithography (Ford et al., 1998), plasma etching (Rossier et al., 2002; Wu et al., 2002), and UV pattering (Lee et al., 1995). Many of these fabrication approaches require the use of a master and are suitable for only a limited number of polymers and plastics.

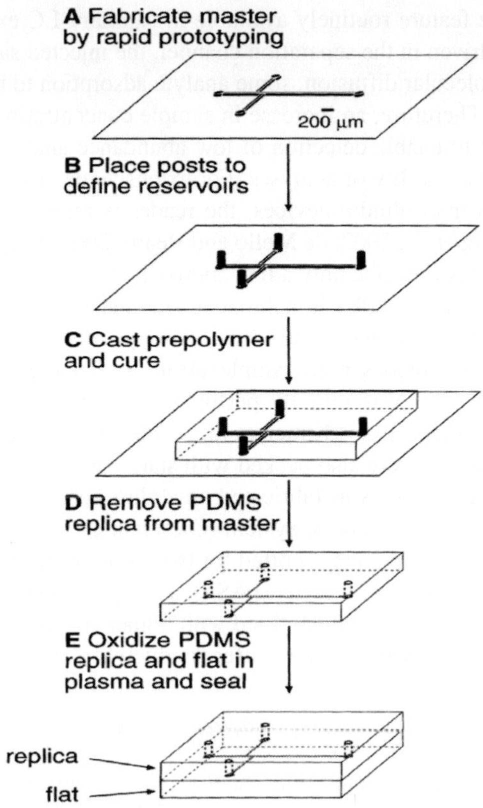

Fig. 47.12 Schematic diagram of fabrication of enclosed microchannels in oxidized PDMS. (**a**) A high-resolution transparency with CAD design of the channels was used as a mask in photolithography to produce a positive relief of photoresist on a slicon wafer. (**b**) Glass posts were placed on the wafer to define reservoirs for analytes and buffers. (**c**) A pre-polymer of PDMS was then cast onto the silicon wafer and cured at 65°C for 1 h. (**d**) The polymer replica of the master containing a negative relief of channels was peeled away from the silicon wafer, and the glass posts were removed. (**e**) The PDMS replica and a flat slab of PDMS were oxidized in a plasma discharge for 1 min. Plasma oxidation had two effects. First, when two oxidized PDMS surfaces were brought into conformal contact, an irreversible seal formed between them. This seal defined the channels as four walls of oxidized PDMS. Second, silanol (SiOH) groups introduced onto the surface of the polymer ionize in neutral or basic aqueous solutions and support EOF in the channels. (Reproduced with permission from Duffy et al. (1998). Copyright 2010 ©, with permission from the American Chemical Society)

47.4.3 Sample Pretreatment for Microfluidic Analyses

There are several reasons for pre-concentration of samples analyzed with capillary zone electrophoresis (CZE) in microfluidic format. First, low loading capacity of microfluidic devices limits the method's sensitivity in peptide and protein analyses. The total sample volume to be injected into a separation channel is practically constrained to 30–50 nL. Second, CZE doesn't provide temporal and spatial focusing of the analytes, the feature routinely available in capillary LC experiments. When electrokinetically driven in the separation channel, the injected sample plug further disperses due to molecular diffusion, some analyte adsorption to the capillary walls and Joule heating. Therefore, an increase in sample concentration on a chip would be highly desirable to enable detection of low abundance analytes and to improve reproducibility and reliability of analyses. For extensive details of sample pretreatment techniques in microfluidic devices, the reader is referred to comprehensive reviews (Lichtenberg et al., 2002; de Mello and Beard, 2003; Sueyoshi et al., 2008). Herein, we will briefly describe only a few approaches.

Solid-phase extraction (SPE) is a broad technique where analyte is retained by a chromatographic stationary phase and subsequently eluted in a more concentrated form. SPE provides both sample clean-up and pre-concentration. In a microfluidic format, SPE was initially performed by coating channel walls with a high affinity stationary phase (Kutter et al., 2000). To increase surface interaction area, microchannels were also packed with stationary phase material, so that a 330-pL chromatographic bed was fabricated on a glass substrate as part of an electroosmotically pumped microfluidic system (Oleschuk et al., 2000). Concentration enhancement up to 500-fold were reported for two fluorescent dyes.

In Field Amplified Sample Stacking (FASS), a sample prepared in a low conductivity buffer is injected into a channel filled with higher conductivity running buffer (Mikkers et al., 1979b). Assuming low buffer conductivity, the relative conductivity, γ, is defined as:

$$\gamma = c_\mathrm{r}/c_\mathrm{s} = \sigma_\mathrm{r}/\sigma_\mathrm{s} = E_\mathrm{s}/E_\mathrm{r} \tag{47.18}$$

where c_r and c_s are the concentrations of the running buffer and sample buffer, respectively, σ_r and σ_s are the conductivities of the two buffers, and E_r and E_s are the field strengths in the running buffer and sample, respectively. When voltage is applied, electric field strength is higher in the lower conductivity buffer. Since electrophoretic velocity is proportional to the electric field, analytes migrate rapidly through the sample zone, decelerate abruptly at the running buffer interface and "stack" into a narrow sample band. Generally, pre-concentration efficiency increases with an increase in γ and levels off at higher γ-values. Differences in conductivity and ionic strength between sample and running buffer result in differences in EOF, being much higher in the sample than in the running buffer. This causes additional hydrodynamic flow and broadening of the sample plug (Lichtenberg et al., 2002). Field Amplified Injection (FAI) was first demonstrated by Jacobson and Ramsey (Jacobson and Ramsey, 1995) and yielded a concentration enhancement for dansylated amino acids of up to 14-fold. The difference between FASS and FAI in the microfluidic format is illustrated in Fig. 47.13a (Lichtenberg et al., 2002). While FASS works for both negatively and positively charged ions, FAI has an electrophoretic bias depending on the EOF direction. FAI was combined with protein pre-concentration using a porous silica membrane between adjacent microchannels that allowed the passage of buffer ions and excluded larger migrating analytes (Foote et al., 2005). Figure 47.13b shows optimized arrangement of the pre-concentration device where a pre-concentrator and an injector were separated by a few millimeters

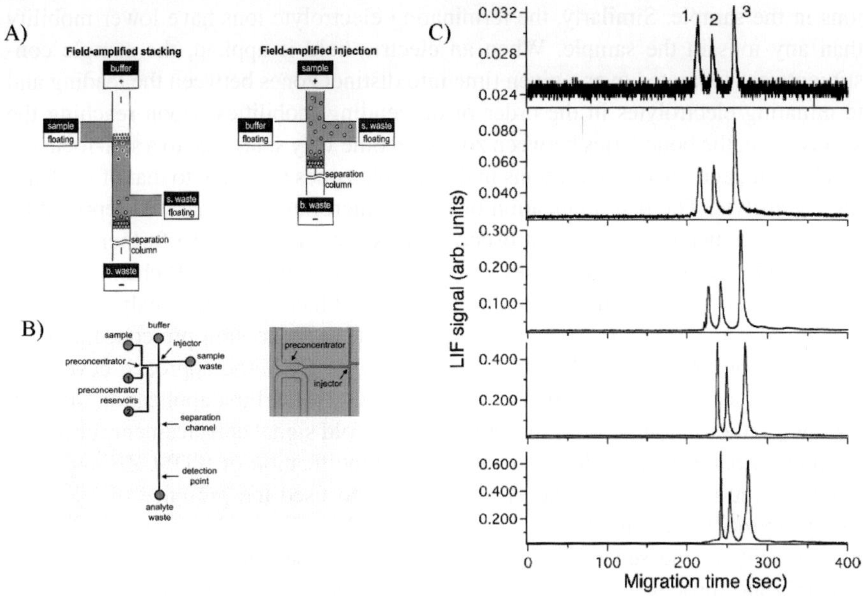

Fig. 47.13 (**a**) While Field Amplified Sample Stacking enables pre-concentration of both anions and cations in volumetrically defined plugs; the Field Amplified Injection enriches only the ionic species at the front boundary of the injection plug. (Reproduced with permission from Lichtenberg et al. (2002). Sample pretreatment on microfabricated devices. *Talanta*, 2002, 56, 233–266. Copyright 2002 ©, with permission from Elsevier). (**b**) Schematic of microchip layout for side-channel pre-concentration and the microscopic image of the pre-concentrator and injector. Channel width at half-depth, wide channel ~220 μm, narrow channels ~50 μm; injector length, 100 μm; pre-concentrator length, 0.5 mm; length of narrow channel between pre-concentrator and injector, 1.4 mm. (**c**) Capillary gel electrophoresis separations of α-lactalbumin (*1*), trypsin inhibitor (*2*), and carbonic anhydrase (*3*), following pre-concentration in the sample injector. Starting with the top electropherogram, pre-concentration times were 0, 1, 2, 3, and 4 min. Buffers, CE-SDS protein sample and run buffer (Bio-Rad), separation length, 7 mm; separation field strength, 300 V/cm. (Reproduced with permission from Foote et al. (2005). Copyright 2010 ©, with permission from the American Chemical Society)

to provide a relatively long sample plug and minimize the possibility of mistiming the loading and injection steps. Sample pre-concentration was achieved by applying a voltage gradient of ~1 kV/cm between the sample and pre-concentrator reservoirs. Leakage of analytes through the membrane was attributed to imperfect bonding or breakdown of the silicate structure at higher voltages (>2 kV). Loading without pre-concentration was performed by applying the voltage gradient between the sample reservoir and sample waste reservoir. Figure 47.13c illustrates capillary gel electrophoresis results with three proteins at different pre-concentration times. A signal enhancement of up 600-fold was obtained after 8 min of pre-concentration.

Isotachophoresis (ITP) is a technique where a sample is introduced between leading and terminating electrolytes (Mikkers et al., 1979a). The leading electrolyte is such that its ions have higher mobility than any of the same charge ions in the sample. Similarly, the terminating electrolyte ions have lower mobility than any ions in the sample. When an electric field is applied, the sample constituents get separated over a given time into distinct zones between the leading and terminating electrolytes in the order of descending mobilities. Upon reaching the steady state, the boundaries between zones become very sharp due to a self-focusing mechanism, and the concentrations in the sample zones are equal to that of the leading electrolyte. The first application of ITP in microfluidic format was reported by Walker et al. using Raman spectroscopy (Walker et al., 1998). To further improve on chip ITP resolution, Kaniansky et al. (2000) combined an ITP channel with a CZE channel. Under optimum conditions, 10^4-fold increase in the analyte ion concentration was observed at a baseline resolution. In on-column pre-concentration, the migration mode gradually changes from ITP to CZE, the approach developed and termed as transient ITP (tITP) by Foret et al. (1992). First application of tITP-CZE with microfludic devices reported up to 10-fold signal enhancement for model peptides (Zhang et al., 1999). An interesting modification of tITP-CZE approach has been reported by Vreeland et al. (2003) who used the presence of hydroxyl ions proximal to the sample zone for "destacking" a sample plug. Since the buffering capacity in the separation channel determines the number of hydroxyl ions that migrate through the separation channel to the cathode, the concentration of buffering cations determines how much of the separation channel (or separation time) is used for ITP versus CZE. A two-order of magnitude improvement in the limit of detection was observed with eTag reporter molecules under the tITP-CZE conditions.

A fundamentally novel approach which employs electrokinetic trapping enabled by a nanofluidic filter has been recently reported by Wang et al. (2005). Pre-concentration of proteins and peptides was achieved without any physical barriers or reagents because of (1) the energy barrier for charged biomolecules generated by the induced space charge in a separation microchannel, and (2) a faster nonlinear EOF for sample deliveries driven by the induced space charge in the microchannel. The mechanism of pre-concentration in the nano-filter device is shown in Fig. 47.14. Once a strong electric field is applied across the nano-filter, ion depletion region no longer maintains its electroneutrality. By applying an additional field along the microfluidic channel in the anodic side, a non-liner electroosmosis of the second kind is generated, which facilitates fast accumulation of the biomolecules in front of the induced space charge layer. Following the removal of the electric field across the nano-filter, analytes were separated in the microfluidic channel and then detected at a signal enhancement factor of 10^6.

Despite major advances in sample pre-concentration in microfludic devices, enormous challenges still remain to be addressed in analysis of "real world" biological samples, and particularly when high reproducibility and reliable quantitation are required. These difficulties are mostly related to the fact that conductivities of biological samples can be controlled only to an extent. This leads to limited reproducibility of pre-concentration efficiency from sample to sample and affects quantitation results.

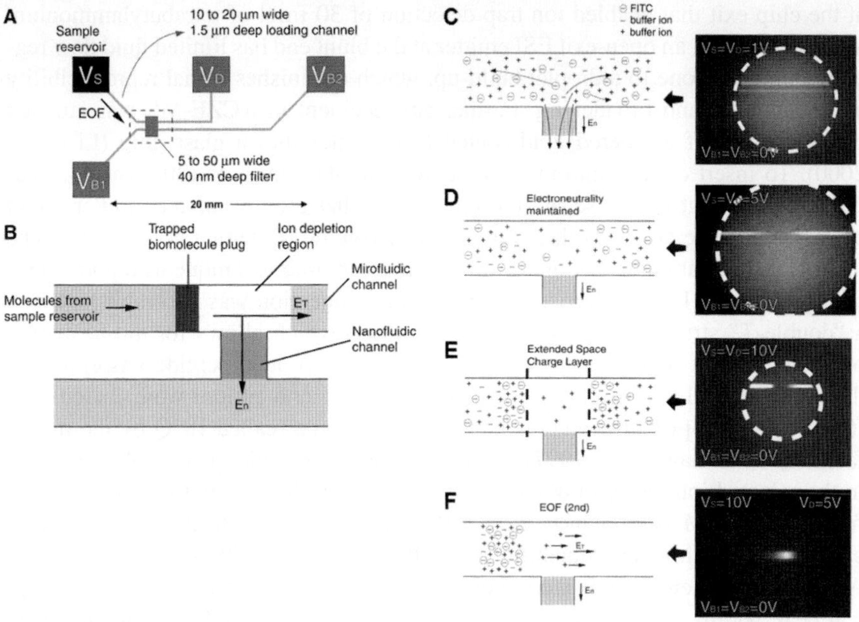

Fig. 47.14 A nanofluidic protein concentration device. (**a**) The layout of the device. (**b**) A schematic diagram that shows concentration mechanism. Once proper voltages are applied, the trapping region and the depletion region are formed as indicated in the drawing. The E_T specifies the electric field applied across the ion depletion region, while E_n specifies the cross nanofilter electric field. *Panels* (**c–f**) show the mechanism of pre-concentration in the nanofilter device. (**c**) No concentration polarization is observed when a small electrical field (E_n) is applied across the nanofilter. (**d**) With an increase in E_n, the transport of ions becomes diffusion-limited and generates the ion depletion zone. (**e**) Once a strong E_n field is applied, the nanochannel develops an induced space charge layer, where electroneutrality is no longer maintained. (**f**) By applying an additional E_T field along the microfluidic channel in the anodic side (from V_S to V_D in **a**), a nonlinear electrokinetic flow (called electroosmosis of the second kind) is induced, which results in fast accumulation of biomolecules in front of the induced space charge layer. (Reproduced with permission from Wang et al. (2005). Copyright 2010 ©, with permission from the American Chemical Society)

47.4.4 Microfluidic CZE-MS Interface

Complete characterization of complex biological samples with microfluidic (μ) CZE and MS detection remains a formidable challenge mainly due to the complexity of an ESI interface on a microchip. Comprehensive reviews of chip-based microfludic devices coupled with ESI-MS have been recently reported and the reader is referred to these publications for additional details (Issaq et al., 2004; Sung et al., 2005; Lazar et al., 2006). First demonstration of ESI feasibility with a microfludic chip was published by Ramsey and Ramsey (1997), who designed an ESI emitter on a blunt end of the microchip. ESI flow was supported by the electroosmotic pumping (see Eq. 12) and an estimated flow rate of ~100 nL/min was generated at the chip exit that enabled ion trap detection of 30 fmol of tetrabutylammonium iodide. However, an open-exit ESI emitter at the blunt end has limited fluid flow regulation and is prone to a droplet build-up, which diminishes signal reproducibility and results in band broadening. Further advancement in μCZE-MS was attained by integration of a tapered gold-coated ESI emitter into a glass chip (Li et al., 2000). To insert emitter into the microchip, a flat-bottom hole (200 μm i.d.) was drilled at the exit of the separation channel so that the end of the emitter could be coupled to the device with minimum dead volume (Bateman et al., 1997). The microchip was fabricated using a silica micromachining technique as reported earlier by Li et al. (Li and Harrison, 1997). Sample injection was accomplished with a "double-T" structure (Effenhauser et al., 1993) which allows for the formation of geometrically defined sample plugs. To prevent protein/peptide adsorption to the microchannel walls, both the separation channel and emitter were coated with [(acryloylamino)propyl] trymethylammonium chloride (called BCQ by the manufacturer) (Bateman et al., 1997). When coupled to a triple-quadrupole operating in the selected ion monitoring mode (SIM), concentration detection limits ranging from 3 to 45 nM were achieved for different peptides. Replicate injections indicated migration time reproducibility of 3% RSD and 6–13% RSD on the peak areas. Analyses of protein digests were typically achieved in less than 1.5 min with typical peak widths of 1.8–2.5 s. Though this approach and similar techniques (Gao et al., 2001) show high sensitivity and robustness, the drawback of these chips will always be variable dead volume with subsequent variability of band broadening in real applications.

In another implementation of a μCZE-ESI-MS interface, a miniature coupled microsprayer employing a micro liquid junction at the exit of the separation channel was developed (Deng et al., 2001; Wachs and Henion, 2001; Zhang et al., 2001). Detailed views of the chip and the microsprayer are shown in Fig. 47.15. The microsprayer is assisted with a make-up flow and a nebulizing gas to enhance droplet desolvation and ionization efficiency. In addition, the interface functions as a buffer zone to minimize nebulizer effect on separation quality. One of the objectives of this work was to develop a stand-alone ESI interface, which could be optimized outside of the microfluidic device and then be coupled to different chips. The glass microchip equipped with a microsprayer was interfaced to both a triple quadrupole MS and quadrupole time-of-flight (qTOF) MS. The measured CE separation efficiency for SIM CE/MS electropherograms was determined to be 2860 plates for small molecules, such as carnitines and acrylcarnitines extracted from human urine. In this design, however, the microsprayer sensitivity is affected by dilution due to the pneumatic flow. In addition, sample carry-over would be problematic for high throughput applications with biological samples.

An alternative approach would be to fabricate an ESI tip from the material at the outlet of the microchannel. Svedberg et al. reported on fabrications of ESI emitter tips directly from the end of microchips that were made of polycarbonate or poly(methyl methacrylate) with injection molding (Svedberg et al., 2003). ESI emitter tips were coated with either gold or graphite particles and ESI voltage was directly applied to the tips. No CZE experiments were performed with

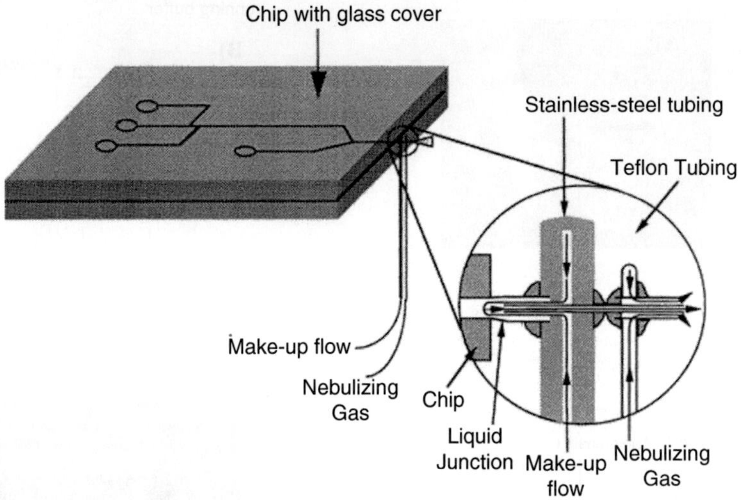

Fig. 47.15 Schematic drawing of the glass CE chip with associated sample, waste and separation buffer reservoirs. Also shown are the points of application of high voltage for sample loading in the double T (see inset). In this process, the sample reservoir was grounded and +6 and +4 kV were applied to the waste reservoir and the microsprayer, respectively. The *arrow* shows the direction of the sample flow. Electrophoresis operation mode. In this process, the separation reservoir was grounded and +4 kV was applied to the microsprayer. The sample and waste reservoirs were maintained at the same potential. The nanofabricated glass chip contained a capillary electrophoresis channel with dimensions of 20 μm (depth) × 60 μm (width) × 4.2 cm (length). (Reproduced with permission from Deng et al. (2001). Copyright 2010 ©, with permission from the American Chemical Society)

these chips and direct infusion results indicated a limit of detection of ~0.5 fmol for peptide mixtures. Thorlund et al. (2005) have reported on a PDMS microchip which integrated sample injection, CZE separation and ESI emitter in a single substrate. ESI emitter was coated with graphite powder and operated as a sheathless CE and ESI electrode. The microchip was interfaced to a TOFMS and evaluated with a mixture of peptides. When measured with CE-MS, electrophoretic mobilities of peptides were observed to be lower than those determined using UV detection. Possible explanation for that was a low grade of oxidation of the separation channel in the vicinity of the emitter. The authors also observed a spray signal decline during the injection step, which could lead to a distorted sample plug where the sample was concurrently directed toward the emitter and the waste reservoir. Our experience of applying ESI voltage directly to a microfabricated emitter indicated poor signal reproducibility from run to run mostly due to an electrolysis that could occur at the ESI voltage contact. An alternative design in Fig. 47.16a,b shows the scanning electron microscope (SEM) images of a nano-ESI tip incorporated into a PMMA microfludic chip, and the schematic diagram of the chip layout, respectively. Though the microchip was designed for a single-usage to eliminate sample carry-over for clinical analyses, the studies revealed that the microchip performance

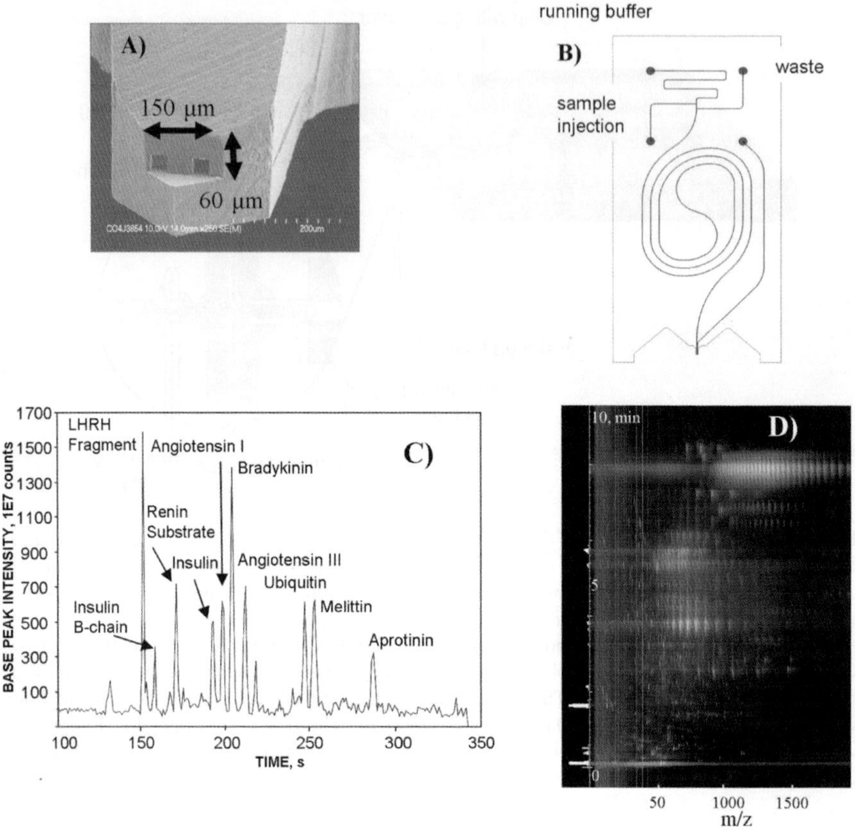

Fig. 47.16 The design and performance of a microfluidic CZE chip developed at Predicant Biosciences. (**a**) The SEM image of the electrospray emitter shows separation and side-arm ESI microchannels mating at the micromachined tip. (**b**) The PMMA microchip was fabricated by electroplating and hot embossing. The separation channel length was 16 cm, the channel width and the height were 30 μm each. In the CZE separation mode, –7 kV was applied to the running buffer well, while the ESI buffer voltage was +2 kV. The separation channel was positively coated with a dual polymer layer to prevent protein/peptide adsorption to the microchannel walls and maintain an EOF flow rate of ∼300 nL/min. (**c**) The signal obtained with a mixture of peptides and lower molecular weight proteins using μCE separation on the positively coated PMMA chip coupled to a high-resolution TOFMS. (**d**) The 2D map of μCZE-TOFMS separation of a lower molecular weight peptidome sample (MW <50 kDa) extracted from human blood serum

remained highly reproducible over tens of experiments. Two channels mating at the ESI tip represent the separation and ESI channels (Bousse and Stults, 2004). The latter was used to supply a spray voltage to the tip through a buffer solution. The separation channel was coated with a dual layer of polystyrene maleic anhydride (PSMA) and polydiallyl dimethylammonium chloride (PDADMAC) to prevent protein and peptide adsorption to the capillary walls and to generate EOF in the direction of the ESI emitter. Using "double-T" junction approach, a total sample amount of 30 nL was injected onto the separation channel. The EOF-generated flow rate was measured with a neutral marker to be ~300 nL/min. No coating was introduced into the ESI side channel to prevent sample flow to the ESI well. The microchip was coupled to a high-resolution high-mass-accuracy TOFMS instrument and rigorously characterized with a variety of biological samples. Figure 47.16c illustrates the typical performance of the microchip in analysis of a mixture of peptides and small proteins. The developed μCZE-MS exhibited a peak capacity of ~70 in 5 min long separations, and a concentration limit of detection of ~1 nM. In hundreds of experiments with mixtures of peptides and proteins using one microchip per experiment, variability in migration times was found to be ~3% RSD, while peak area reproducibility was better than ~20% RSD. Following optimization with model peptides and proteins, μCZE-TOFMS approach was then applied to analysis of serum samples collected from patients who were diagnosed with early stages of prostate cancer. Sample preparation protocol included removal of high-abundance high-molecular-weight proteins using ultra filtration with 30 and 50 kDa molecular weight cut-off membranes Minicon YM30 and YM50 (Millipore, Billerica, MA). Human blood serum samples were acidified with 1% formic acid and inspected with one-dimensional gel electrophoresis. The experiments revealed that acidification gave rise to disruption of non-covalent binding of lower molecular weight compounds to high-abundance proteins, such as human serum albumin (HSA) and immunoglobulin G (IgG), while ultra filtration accomplished 99.8% reduction in HSA and IgG concentrations. More than 70% of lower molecular weight components were recovered, desalted and concentrated using an SPE column, and then reconstituted in a CZE compatible buffer (isopropanol) for the following separation and MS detection. A protease inhibitor, such as pepstatin, was added to the serum samples to suppress endogenous protease activities. Figure 47.16d shows a typical 2D μCZE-TOFMS map of a blood serum peptidome sample. The results of hundreds of μCZE-TOFMS were processed using principal component analysis approaches and the developed pattern recognition algorithm, and then cross-validated with control serum samples. As a result, differentially expressed CZE-MS features were found to result in a two-fold increase in the positive predictive value (PPV) as compared to that available with the prostate-specific antigen (PSA) test.

In conclusion, we would like to emphasize that μCE-MS holds an enormous potential for proteomic applications. Several important features favorably position this approach for intact protein analysis. First, sample pre-concentration, separation, ionization and detection have been demonstrated in one device using model systems. Second, small microchannel dimensions and a relatively short length enable high-peak-capacity fast separations for high throughput studies. Third, a combination of orthogonal separation approaches, e.g. capillary isoelectric focusing (cIEF) and CZE (Herr et al., 2003) on a single chip drastically increases peak capacity of a microchip in protein separations. Fourth, low cost of plastic chip fabrication allows for mass production of single-use separation devices and complete elimination of sample carry-over, the feature of particular importance in clinical applications. Further developments are still needed to integrate sample preparation, pre-concentration, multi-dimensional CE separations, efficient ionization and sensitive MS detection with "real" biological samples. Reproducibility, robustness and high sensitivity will certainly be key metrics in commercialization of μCE-MS technology. The progress in these areas will undoubtedly define the success of microfludic chips in proteomics, glycomics and metabolomics applications.

47.5 Conclusions

Hyphenated condensed and gas phase separation approaches combined with mass spectrometry detection have been increasingly used in biochemical applications. Enormous complexity and variability of "real world" biological samples impose a strong need for highly efficient, high-throughput, robust, reproducible separations prior to signal detection. In the paradigm shifting to large-scale clinical studies, separation peak capacity per unit time, sensitivity, dynamic range, and accurate mass measurements are becoming key metrics for the successful approaches. Though the techniques reviewed in this chapter partially address this challenge, further enhancements are needed to reliably detect low abundance components, such as kinases or interleukins, in complex biological matrices. The future progress in biochemical analyses will strongly depend on reproducibility of sample preparation techniques and the efficiency of interfaces between the coupled separation/detection methods. It is becoming increasingly clear that only a combination of highly efficient separation/detection approaches in a single technology platform would be able to address the daunting challenges of life sciences, such as early diagnosis and prognosis of cancers and infectious diseases, and also help in understanding protein–protein interactions in biological systems and contribute to the development of pharmaceuticals and drugs for curing diseases.

Acknowledgements Portions of this research were supported by the U.S. Department of Energy (DOE), Office of Biological and Environmental Research, the Environmental Molecular Science Laboratory (EMSL) and the NIH National Center for Research Resources (Grant RR018522). This work was performed in the Environmental Molecular Science Laboratory (EMSL), a DOE national scientific user facility at the Pacific Northwest National Laboratory (PNNL). PNNL is operated by Battelle for the DOE under contract DE-AC05-76RL0 1830.

References

Aebersold, R., and Mann, M. (2003). Mass spectrometry-based proteomics. Nature *422*, 198–207.

Andersen, W.C., Turnipseed, S.B., Karbiwnyk, C.M., Clark, S.B., Madson, M.R., Gieseker, C.M., Miller, R.A., Rummel, N.G., and Reimschuessel, R. (2008). Determination and confirmation of melamine residues in catfish, trout, tilapia, salmon, and shrimp by liquid chromatography with tandem mass spectrometry. J Agric Food Chem *56*, 4340–4347.

Atmanene, C., Wagner-Rousset, E., Malissard, M., Chol, B., Robert, A., Corvaia, N., Dorsselaer, A.V., and Beck, A. (2009). Extending mass spectrometry contribution to therapeutic monoclonal antibody lead optimization: Characterization of immune complexes using noncovalent ESI-MS. Anal Chem *81*, 6364–6373.

Baker, E.S., Clowers, B.H., Li, F.M., Tang, K., Tolmachev, A.V., Prior, D.C., Belov, M.E., and Smith, R.D. (2007a). Ion mobility spectrometry–mass spectrometry performance using electrodynamic ion funnels and elevated drift gas pressures. J Am Soc Mass Spectrom *18*, 1176–1187.

Baker, E.S., Tang, K.Q., Danielson, W.F., Prior, D.C., and Smith, R.D. (2007b). Simultaneous fragmentation of multiple ions using IMS drift time dependent collision energies. J Am Soc Mass Spectrom *19*, 411–419.

Barnett, D.A., Ells, B., Guevremont, R., and Purves, R.W. (1999a). Separation of leucine and isoleucine by electrospray ionization–high field asymmetric waveform ion mobility spectrometry–mass spectrometry. J Am Soc Mass Spectrom *10*, 1279–1284.

Barnett, D.A., Guevremont, R., and Purves, R.W. (1999b). Determination of parts-per-trillion levels of chlorate, bromate, and iodate by electrospray ionization/high-field asymmetric waveform ion mobility spectrometry/mass spectrometry. Appl Spectrosc *53*, 1367–1374.

Barnett, D.A., Ells, B., Guevremont, R., and Purves, R.W. (2002). Application of ESI-FAIMS-MS to the analysis of tryptic peptides. J Am Soc Mass Spectrom *13*, 1282–1291.

Bateman, K.P., White, R.L., and Thibault, P. (1997). Disposable emitters for on-line capillary zone electrophoresis nanoelectrospray mass spectrometry. Rapid Commun Mass Spectrom *11*, 307–315.

Becker, H., and Gartner, C. (2000). Polymer microfabrication methods for microfluidic analytical applications. Electrophoresis *21*, 12–26.

Becker, H., and Locascio, L.E. (2002). Polymer microfluidic devices. Talanta *56*, 267–287.

Belov, M.E., Buschbach, M.A., Prior, D.C., Tang, K.Q., and Smith, R.D. (2007). Multiplexed ion mobility spectrometry-orthogonal time-of-flight mass spectrometry. Anal Chem *79*, 2451–2462.

Belov, M.E., Clowers, B.H., Prior, D.C., Danielson, W.F., Liyu, A.V., Petritis, B.O., and Smith, R.D. (2008). Dynamically multiplexed ion mobility time-of-flight mass spectrometry. Anal Chem *80*, 5873–5883.

Belov, M.E., Danielson, W.F., Ibrahim, Y., Prior, D.C., Baker, E.S., Zhao, R., Lopez-Ferrer, D., Petritis, B.O., and Smith, R.D. (2009). On-line Chromatography/Dynamically Multiplexed Ion Mobility Time-of-Flight Mass Spectrometry for High Throughput Proteomics. Paper Presented at Proceedings of the 57th ASMS Conference on Mass Spectrometry (Philadelphia, PA).

Benassi, M., Corilo, Y.E., Uria, D., Augusti, R., and Eberlin, M.N. (2009). Recognition and resolution of isomeric alkyl anilines by mass spectrometry. J Am Soc Mass Spectrom *20*, 269–277.

Blom, N., Sicheritz-Ponten, T., Gupta, R., Gammeltoft, S., and Brunak, S. (2004). Prediction of post-translational glycosylation and phosphorylation of proteins from the amino acid sequence. Proteomics *4*, 1633–1649.

Bogdanov, B., and Smith, R.D. (2005). Proteomics by FTICR mass spectrometry: Top down and bottom up. Mass Spectrom Rev *24*, 168–200.

Bousse, L., and Stults, J.T. (2004). Multi-channel microfluidic chip for electrospray ionization, United States Patent 6803568.

Bowers, M.T., Kemper, P.R., Vonhelden, G., and Vankoppen, P.A.M. (1993). Gas-phase ion chromatography – transition-metal state selection and carbon cluster formation. Science *260*, 1446–1451.

Caprioli, R.M. (1994). LC-MS and CE-MS (New York, NY, Wiley).

Cargile, B.J., Bundy, J.L., Freeman, T.W., and Stephenson, J.L., Jr. (2004a). Gel based isoelectric focusing of peptides and the utility of isoelectric point in protein identification. J Proteome Res *3*, 112–119.

Cargile, B.J., Talley, D.L., and Stephenson, J.L., Jr. (2004b). Immobilized pH gradients as a first dimension in shotgun proteomics and analysis of the accuracy of pI predictability of peptides. Electrophoresis *25*, 936–945.

Champarnaud, E., Laures, A.M.F., Borman, P.J., Chatfield, M.J., Kapron, J.T., Harrison, M., and Wolff, J.C. (2009). Trace level impurity method development with high-field asymmetric waveform ion mobility spectrometry: Systematic study of factors affecting the performance. Rapid Commun Mass Spectrom *23*, 181–193.

Chan, Y.T., Li, X., Soler, M., Wang, J.L., Wesdemiotis, C., and Newkome, G.R. (2009). Self-assembly and traveling wave ion mobility mass spectrometry analysis of hexacadmium macrocycles. J Am Chem Soc *131*, 16395–16397.

Clemmer, D.E., and Jarrold, M.F. (1997). Ion mobility measurements and their applications to clusters and biomolecules. J Mass Spectrom *32*, 577–592.

Clowers, B.H., Dwivedi, P., Steiner, W.E., Hill, H.H., and Bendiak, B. (2005). Separation of sodiated isobaric disaccharides and trisaccharides using electrospray ionization-atmospheric pressure ion mobility-time of flight mass spectrometry. J Am Soc Mass Spectrom *16*, 660–669.

Clowers, B.H., Ibrahim, Y.M., Prior, D.C., Danielson, W.F., Belov, M.E., and Smith, R.D. (2008). Enhanced ion utilization efficiency using an electrodynamic ion funnel trap as an injection mechanism for ion mobility spectrometry. Anal Chem *80*, 612–623.

Cohen, M.J., and Karasek, F.W. (1970). Plasma chromatography – a new dimension for gas chromatography and mass spectrometry. J Chromatogr Sci *8*, 330–337.

Covey, T., Huang, E., and Henion, J. (1991). Tryptic peptides via LC-MS and CID. Anal Chem *63*, 1193–1200.

Cui, M., Ding, L.Y., and Mester, Z. (2003). Separation of cisplatin and its hydrolysis products using electrospray ionization high-field asymmetric waveform ion mobility spectrometry coupled with ion trap mass spectrometry. Anal Chem *75*, 5847–5853.

Davis, M.T., Stahl, D.C., Hefta, S.A., and Lee, T.D. (1995). A microscale electrospray interface for on-line, capillary liquid chromatography/tandem mass spectrometry of complex peptide mixtures. Anal Chem *67*, 4549–4556.

de la Mora, J.F., de Juan, L., Eichler, T., and Rosell, J. (1998). Differential mobility analysis of molecular ions and nanometer particles. TRAC, Trends Anal Chem *17*, 328–339.

de Mello, A.J., and Beard, N. (2003). Dealing with 'real' samples: Sample pre-treatment in microfluidic systems. Lab Chip *3*, 11 N–19 N.

Deng, Y., Henion, J., Li, J.J., Thibault, P., Wang, C., and Harrison, D.J. (2001). Chip-based capillary electrophoresis/mass spectrometry determination of carnitines in human urine. Anal Chem *73*, 639–648.

Dolnik, V. (2004). Wall coating for capillary electrophoresis on microchips. Electrophoresis *25*, 3589–3601.

Domon, B., and Aebersold, R. (2006). Mass spectrometry and protein analysis. Science *312*, 212–217.

Duffy, D.C., McDonald, J.C., Schueller, O.J.A., and Whitesides, G.M. (1998). Rapid prototyping of microfluidic systems in poly(dimethylsiloxane). Anal Chem *70*, 4974–4984.

Eckers, C., Laures, A.M.F., Giles, K., Major, H., and Pringle, S. (2007). Evaluating the utility of ion mobility separation in combination with high-pressure liquid chromatography/mass spectrometry to facilitate detection of trace impurities in formulated drug products. Rapid Commun Mass Spectrom *21*, 1255–1263.

Effenhauser, C.S., Manz, A., and Widmer, H.M. (1993). Glass chips for high-speed capillary electrophoresis separations with submicrometer plate heights. Anal Chem *65*, 2637–2642.

Eiceman, G.A., Krylov, E.V., Nazarov, E.G., and Miller, R.A. (2004). Separation of ions from explosives in differential mobility spectrometry by vapor-modified drift gas. Anal Chem *76*, 4937–4944.

Eiceman, G.A., Nazarov, E.G., Miller, R.A., Krylov, E.V., and Zapata, A.M. (2002). Micro-machined planar field asymmetric ion mobility spectrometer as a gas chromatographic detector. Analyst *127*, 466–471.

Ells, B., Barnett, D.A., Purves, R.W., and Guevremont, R. (2000a). Detection of nine chlorinated and brominated haloacetic acids at part-per-trillion levels using ESI-FAIMS-MS. Anal Chem *72*, 4555–4559.

Ells, B., Froese, K., Hrudey, S.E., Purves, R.W., Guevremont, R., and Barnett, D.A. (2000b). Detection of microcystins using electrospray ionization high-field asymmetric waveform ion mobility mass spectrometry/mass spectrometry. Rapid Commun Mass Spectrom *14*, 1538–1542.

Fenn, L.S., Kliman, M., Mahsut, A., Zhao, S.R., and McLean, J.A. (2009). Characterizing ion mobility-mass spectrometry conformation space for the analysis of complex biological samples. Anal Bioanal Chem *394*, 235–244.

Foote, R.S., Khandurina, J., Jacobson, S.C., and Ramsey, J.M. (2005). Preconcentration of proteins on microfluidic devices using porous silica membranes. Anal Chem *77*, 57–63.

Ford, S.M., Kar, B., McWhorter, S., Davies, J., Soper, S.A., Klopf, M., Calderon, G., and Saile, V. (1998). Microcapillary electrophoresis devices fabricated using polymeric substrates and X-ray lithography. J Microcolumn Sep *10*, 413–422.

Foret, F., Szoko, E., and Karger, B.L. (1992). On-column transient and coupled column isotachophoretic preconcentration of protein samples in capillary zone electrophoresis. J Chromatogr *608*, 3–12.

Gabryelski, W., and Froese, K.L. (2003). Rapid and sensitive differentiation of anomers, linkage, and position isomers of disaccharides using high-field asymmetric waveform ion mobility spectrometry (FAIMS). J Am Soc Mass Spectrom *14*, 265–277.

Gao, J., Xu, J., Locascio, L.E., and Lee, C.S. (2001). Integrated microfluidic system enabling protein digestion, peptide separation, and protein identification. Anal Chem *73*, 2648–2655.

Garcia, J.F., and Bacelo, D. (1993). LC-MS interfacing systems. J High Resolut Chromatogr *16*, 633–641.

Garcia, I.R., Giles, K., Bateman, R.H., and Gaskell, S.J. (2008). Studies of peptide a- and b-type fragment ions using stable isotope labeling and integrated ion mobility/tandem mass spectrometry. J Am Soc Mass Spectrom *19*, 1781–1787.

Gaspar, A., Englmann, M., Fekete, A., Harir, M., and Scmitt-Kopplin, P. (2008). Trends in CE-MS 2005–2006. Electrophoresis *29*, 66–79.

Gevaert, K., Goethals, M., Martens, L., Van Damme, J., Staes, A., Thomas, G.R., and Vandekerckhove, J. (2003). Exploring proteomes and analyzing protein processing by mass spectrometric identification of sorted N-terminal peptides. Nat Biotechnol *21*, 566–569.

Gevaert, K., and Vandekerckhove, J. (2009). Reverse-phase diagonal chromatography for phosphoproteome research. Methods Mol Biol *527*, 219–227, ix.

Gevaert, K., Van Damme, P., Ghesquiere, B., and Vandekerckhove, J. (2006). Protein processing and other modifications analyzed by diagonal peptide chromatography. Biochim Biophys Acta *1764*, 1801–1810.

Gevaert, K., Van Damme, P., Martens, L., and Vandekerckhove, J. (2005). Diagonal reverse-phase chromatography applications in peptide-centric proteomics: Ahead of catalogue-omics? Anal Biochem *345*, 18–29.

Ghesquiere, B., Colaert, N., Helsens, K., Dejager, L., Vanhaute, C., Verleysen, K., Kas, K., Timmerman, E., Goethals, M., Libert, C., *et al.* (2009). In vitro and in vivo protein-bound tyrosine nitration characterized by diagonal chromatography. Mol Cell Proteomics *8*, 2642–2652.

Gilar, M., Olivova, P., Daly, A.E., and Gebler, J.C. (2005a). Orthogonality of separation in two-dimensional liquid chromatography. Anal Chem *77*, 6426–6434.

Gilar, M., Olivova, P., Daly, A.E., and Gebler, J.C. (2005b). Two-dimensional separation of peptides using RP-RP-HPLC system with different pH in first and second separation dimensions. J Sep Sci *28*, 1694–1703.

Giles, K., Pringle, S.D., Worthington, K.R., Little, D., Wildgoose, J.L., and Bateman, R.H. (2004). Applications of a travelling wave-based radio-frequency only stacked ring ion guide. Rapid Commun Mass Spectrom *18*, 2401–2414.

Giles, K., Gilbert, T., Green, M., and Scott, G. (2009). Enhancements to the Ion Mobility Performance of a Travelling Wave Separation Device. Paper presented at Proceedings of the 57th ASMS Conference on Mass Spectrometry (Philadelphia, PA).

Godovac-Zimmermann, J., Kleiner, O., Brown, L.R., and Drukier, A.K. (2005). Perspectives in spicing up proteomics with splicing. Proteomics *5*, 699–709.

Greenway, A.M., and Simpson, C.F. (1980). GC-MS combinations. Rev Sci Instrum *13*, 1131–1147.

Griffin, P.R., Coffman, J.A., Hood, L.E., and Yates, J.R.I. (1991). Proteins by LC-MS/MS using ESI. Int J Mass Spectrom Ion Processes *111*, 131–149.

Guevremont, R. (2004a). High-field asymmetric waveform ion mobility spectrometry: A new tool for mass spectrometry. J Chromatogr A *1058*, 3–19.

Guevremont, R. (2004b). High-field asymmetric waveform ion mobility spectrometry (FAIMS). Can J Anal Sci Spectrosc *49*, 105–113.

Guevremont, R., and Purves, R. (2005). FAIMS, a new technology for the study of protein structure. Faseb J *19*, A767–A767.

Guharay, S.K., Dwivedi, P., and Hill, H.H. (2008). Ion mobility spectrometry: Ion source development and applications in physical and biological sciences. IEEE Trans Plasma Sci *36*, 1458–1470.

Haselberg, R., de Jong, G.J., and Somsen, G.W. (2007). Capillary electrophoresis-mass spectrometry for the analysis of intact proteins. J Chromatogr A *1159*, 81–109.

Hatsis, P., Brockman, A.H., and Wu, J.T. (2007). Evaluation of high-field asymmetric waveform ion mobility spectrometry coupled to nanoelectrospray ionization for bioanalysis in drug discovery. Rapid Commun Mass Spectrom *21*, 2295–2300.

Hernandez-Borges, J., Neususs, C., Cifuentes, A., and Pelzing, M. (2004). On-line capillary electrophoresis-mass spectrometry for the analysis of biomolecules. Electrophoresis *25*, 2257–2281.

Herr, A.E., Molho, J.I., Drouvalakis, K.A., Mikkelsen, J.C., Utz, P.J., Santiago, J.G., and Kenny, T.W. (2003). On-chip coupling of isoelectric and free solution electrophoresis for multidimensional separations. Anal Chem *75*, 1180–1187.

Hsieh, Y., and Korfmacher, W.A. (2006). Increasing speed and throughput when using HPLC-MS/MS systems for drug metabolism and pharmacokinetic screening. Curr Drug Metab *7*, 479–489.

Hsieh, Y.L., Wang, H., Elicone, C., Mark, J., Martin, S.A., and Regnier, F. (1996). Automated analytical system for the examination of protein primary structure. Anal Chem *68*, 455–462.

Ibrahim, Y., Belov, M.E., Tolmachev, A.V., Prior, D.C., and Smith, R.D. (2007). Ion funnel trap interface for orthogonal time-of-flight mass spectrometry. Anal Chem *79*, 7845–7852.

Ibrahim, Y.M., Belov, M.E., Liyu, A.V., and Smith, R.D. (2008). Automated gain control ion funnel trap for orthogonal time-of-flight mass spectrometry. Anal Chem *80*, 5367–5376.

Ibrahim, Y., Prior, D.C., Baker, E.S., Smith, R.D., and Belov, M.E. (2010). Characterization of an ion mobility-multiplexed collision induced dissociation tandem time-of-flight mass spectrometry approach. Int J Mass Spectrom *293*, 34–44.

Isailovic, D., Kurulugama, R.T., Plasencia, M.D., Stokes, S.T., Kyselova, Z., Goldman, R., Mechref, Y., Novotny, M.V., and Clemmer, D.E. (2008). Profiling of human serum glycans associated with liver cancer and cirrhosis by IMS-MS. J Proteome Res *7*, 1109–1117.

Issaq, H.J., Janini, G.M., Chan, K.C., and Veenstra, T.D. (2004). Sheathless electrospray ionization interfaces for capillary electrophoresis-mass spectrometric detection – advantages and limitations. J Chromatogr A *1053*, 37–42.

Ivanov, A.R., Horvath, C., and Karger, B.L. (2003a). High-efficiency peptide analysis on monolithic multimode capillary columns: Pressure-assisted capillary electrochromatography/capillary electrophoresis coupled to UV and electrospray ionization-mass spectrometry. Electrophoresis *24*, 3663–3673.

Ivanov, A.R., Zang, L., and Karger, B.L. (2003b). Low-attomole electrospray ionization MS and MS/MS analysis of protein tryptic digests using 20-mu m-i.d. polystyrene–divinylbenzene monolithic capillary columns. Anal Chem *75*, 5306–5316.

Jacobson, S.C., and Ramsey, J.M. (1995). Microchip electrophoresis with sample stacking. Electrophoresis *16*, 481–486.

Johnson, J.M., Castle, J., Garrett-Engele, P., Kan, Z.Y., Loerch, P.M., Armour, C.D., Santos, R., Schadt, E.E., Stoughton, R., and Shoemaker, D.D. (2003). Genome-wide survey of human alternative pre-mRNA splicing with exon junction microarrays. Science *302*, 2141–2144.

Kaji, H., Saito, H., Yamauchi, Y., Shinkawa, T., Taoka, M., Hirabayashi, J., Kasai, K., Takahashi, N., and Isobe, T. (2003). Lectin affinity capture, isotope-coded tagging and mass spectrometry to identify N-linked glycoproteins. Nat Biotechnol *21*, 667–672.

Kaniansky, D., Masar, M., Bielcikova, J., Ivanyi, F., Eisenbeiss, F., Stanislawski, B., Grass, B., Neyer, A., and Johnck, M. (2000). Capillary electrophoresis separations on a planar chip with the column-coupling configuration of the separation channels. Anal Chem *72*, 3596–3604.

Kanu, A.B., Dwivedi, P., Tam, M., Matz, L., and Hill, H.H. (2008). Ion mobility-mass spectrometry. J Mass Spectrom *43*, 1–22.

Kapron, J., and Thakur, R.A. (2006). Developing robust LC-MS methods for drug urinalysis using FAIMS and H-SRM. LC GC Europe, 27–28.

Kapron, J., Wu, J., Mauriala, T., Clark, P., Purves, R.W., and Bateman, K.P. (2006). Simultaneous analysis of prostanoids using liquid chromatography/high-field asymmetric waveform ion mobility spectrometry/tandem mass spectrometry. Rapid Commun Mass Spectrom *20*, 1504–1510.

Kapron, J.T., Jemal, M., Duncan, G., Kolakowski, B., and Purves, R. (2005). Removal of metabolite interference during liquid chromatography/tandem mass spectrometry using high-field asymmetric waveform ion mobility spectrometry. Rapid Commun Mass Spectrom *19*, 1979–1983.

Karasek, F.W. (1974). Plasma chromatography. Anal Chem *46*, A710–A720.

Kelleher, N.L. (2004). Top-down proteomics. Anal Chem *76*, 196A–203A.

Kelly, R.T., Page, J.S., Zhao, R., Qian, W.J., Mottaz, H.M., Tang, K., and Smith, R.D. (2008). Capillary-based multi nanoelectrospray emitters: Improvements in ion transmission efficiency and implementation with capillary reversed-phase LC-ESI-MS. Anal Chem *80*, 143–149.

Kim, H.I., Kim, H., Pang, E.S., Ryu, E.K., Beegle, L.W., Loo, J.A., Goddard, W.A., and Kanik, I. (2009). Structural characterization of unsaturated phosphatidylcholines using traveling wave ion mobility spectrometry. Anal Chem *81*, 8289–8297.

Kitagawa, F., Kubota, K., Sueyoshi, K., and Otsuka, K. (2006). One-step immobilization of cationic polymer onto a poly(methyl methacrylate) microchip for high-performance electrophoretic analysis of proteins. Sci Technol Adv Mater *7*, 558–565.

Kitson, F.G., Larsen, B.S., and McEwen, C.N. (1996). A Practical Guide to Gas Chromatography and Mass Spectrometry (San Diego, CA, Academic).

Klaassen, T., Szwandt, S., Kapron, J.T., and Roemer, A. (2009). Validated quantitation method for a peptide in rat serum using liquid chromatography/high-field asymmetric waveform ion mobility spectrometry. Rapid Commun Mass Spectrom *23*, 2301–2306.

Knox, J.H., and McCormack, K.A. (1994a). Temperature effects in capillary electrophoresis. 1: Internal capillary temperature and effect upon performance. Chromatographia *38*, 207–214.

Knox, J.H., and McCormack, K.A. (1994b). Temperature effect in capillary electrophoresis. 2: Some theoretical calculations and predictions. Chromatographia *38*, 215–221.

Koeniger, S.L., Valentine, S.J., Myung, S., Plasencia, M., Lee, Y.J., and Clemmer, D.E. (2005). Development of field modulation in a split-field drift tube for high-throughput multidimensional separations. J Proteome Res *4*, 25–35.

Koeniger, S.L., Merenbloom, S.I., and Clemmer, D.E. (2006a). Evidence for many resolvable structures within conformation types of electrosprayed ubiquitin ions. J Phys Chem B *110*, 7017–7021.

Koeniger, S.L., Merenbloom, S.I., Valentine, S.J., Jarrold, M.F., Udseth, H.R., Smith, R.D., and Clemmer, D.E. (2006b). An IMS-IMS analogue of MS-MS. Anal Chem *78*, 4161–4174.

Kolakowski, B.M., and Mester, Z. (2007). Review of applications of high-field asymmetric waveform ion mobility spectrometry (FAIMS) and differential mobility spectrometry (DMS). Analyst *132*, 842–86.

Krylov, E.V., Nazarov, E.G., and Miller, R.A. (2007). Differential mobility spectrometer: Model of operation. Int J Mass Spectrom *266*, 76–85.

Kurulugama, R.T., Valentine, S.J., Sowell, R.A., and Clemmer, D.E. (2008). Development of a high-throughput IMS-IMS-MS approach for analyzing mixtures of biomolecules. J Proteomic *71*, 318–331.

Kurulugama, R.T., Nachtigall, F.M., Lee, S., Valentine, S.J., and Clemmer, D.E. (2009). Overtone mobility spectrometry: Part 1. Experimental observations. J Am Soc Mass Spectrom *20*, 729–737.

Kutter, J.P., Jacobson, S.C., and Ramsey, J.M. (2000). Solid phase extraction on microfluidic devices. J Microcolumn Sep *12*, 93–97.

Larsen, M.R., Thingholm, T.E., Jensen, O.N., Roepstorff, P., and Jorgensen, T.J. (2005). Highly selective enrichment of phosphorylated peptides from peptide mixtures using titanium dioxide microcolumns. Mol Cell Proteomic *4*, 873–886.

Lazar, I.M., Grym, J., and Foret, F. (2006). Microfabricated devices: A new sample introduction approach to mass spectrometry. Mass Spectrom Rev *25*, 573–594.

Lee, M.S. (2002). LC/MS Applications in Drug Development (New York, NY, Wiley).

Lee, K.Y., LaBianca, N., Rishton, S.A., Zolgharnain, S., Gelorme, J.D., Shaw, J., and Chang, T.H.P. (1995). Micromachining applications of a high resolution ultrathick photoresist. J Vac Sci Technol B *13*, 3012–3016.

Levin, D.S., Miller, R.A., Nazarov, E.G., and Vouros, P. (2006). Rapid separation and quantitative analysis of peptides using a new nanoelectrospray-differential mobility spectrometer-mass spectrometer system. Anal Chem *78*, 5443–5452.

Li, P.C.H., and Harrison, D.J. (1997). Transport, manipulation, and reaction of biological cells on-chip using electrokinetic effects. Anal Chem *69*, 1174–1178.

Li, J.J., Kelly, J.F., Chemushevich, I., Harrison, D.J., and Thibault, P. (2000). Separation and identification of peptides from gel-isolated membrane proteins using a microfabricated device for combined capillary electrophoresis/nanoelectrospray mass spectrometry. Anal Chem *72*, 599–609.

Lichtenberg, J., de Rooij, N.F., and Verpoorte, E. (2002). Sample pretreatment on microfabricated devices. Talanta *56*, 233–266.

Link, A.J., Eng, J., Schieltz, D.M., Carmack, E., Mize, G.J., Morris, D.R., Garvik, B.M., and Yates, J.R., 3rd (1999). Direct analysis of protein complexes using mass spectrometry. Nat Biotechnol *17*, 676–682.

Lion, N., Rohner, T.C., FDayon, L., Arnaud, I.L., Damoc, E., Youhnovski, N., Wu, Z.-Y., Roussel, C., Josserand, J., Jensen, H., *et al.* (2003). Microfluidic systems in proteomics. Electrophoresis *24*, 3533–3562.

Lipton, M.S., Pasa-Tolic, L., Anderson, G.A., Anderson, D.J., Auberry, D.L., Battista, K.R., Daly, M.J., Fredrickson, J., Hixson, K.K., Kostandarithes, H., *et al.* (2002). Global analysis of the *Deinococcus radiodurans* proteome by using accurate mass tags. Proc Natl Acad Sci USA *99*, 11049–11054.

Liu, S.R., Gao, L., Pu, Q.S., Lu, J.J., and Wang, X.J. (2006). Comprehensive protein profiling by multiplexed capillary zone electrophoresis using cross-linked polyacrylamide coated capillaries. J Proteome Res *5*, 323–329.

Liu, X., Zhao, Y.Y., Chan, K., Hrudey, S.E., Li, X.F., and Li, J.J. (2007). Analysis of nitrosamines by capillary electrospray-high-field asymmetric waveform ion mobility spectrometry-MS with programmed compensation voltage. Electrophoresis *28*, 1327–1334.

Lopez-Ferrer, D., Martinez-Bartolome, S., Villar, M., Campillos, M., Martin-Maroto, F., and Vazquez, J. (2004). Statistical model for large-scale peptide identification in databases from tandem mass spectra using SEQUEST. Anal Chem *76*, 6853–6860.

Makarov, A. (2000). Electrostatic axially harmonic orbital trapping: A high-performance technique of mass analysis. Anal Chem *72*, 1156–1162.

Manadas, B., English, J.A., Wynne, K.J., Cotter, D.R., and Dunn, M.J. (2009). Comparative analysis of OFFGel, strong cation exchange with pH gradient, and RP at high pH for first-dimensional separation of peptides from a membrane-enriched protein fraction. Proteomics *9*, 5194–5198.

Manz, A., Graber, N., and Widmer, H.M. (1990). Miniaturized total chemical analysis systems – a novel concept for chemical sensing. Sens Actuat B *1*, 244–248.

Mao, W.J., Thanawiroon, C., and Linhardt, R.J. (2002). Capillary electrophoresis for the analysis of glycosaminoglycans and glycosaminoglycan-derived oligosaccharides. Biomed Chromatogr *16*, 77–94.

Marginean, I., Kelly, R.T., Moore, R.J., Prior, D.C., LaMarche, B.L., Tang, K., and Smith, R.D. (2009). Selection of the optimum electrospray voltage for gradient elution LC-MS measurements. J Am Soc Mass Spectrom *20*, 682–688.

Marginean, I., Kelly, R.T., Page, J.S., Tang, K., and Smith, R.D. (2007). Electrospray characteristic curves: In pursuit of improved performance in the nanoflow regime. Anal Chem *79*, 8030–8036.

Marshall, A.G. (2000). Milestones in Fourier transform ion cyclotron resonance mass spectrometry technique development. Int J Mass Spectrom *200*, 331–356.

Martynova, L., Locascio, L., Gaitan, M., Kramer, G.W., Christensen, R.G., and McCrehan, W.A. (1997). Fabrication of plastic microfluid channels by imprinting methods. Anal Chem *69*, 4783–4789.

Masselon, C., Pasa-Tolic, L., Tolic, N., Anderson, G.A., Bogdanov, B., Vilkov, A.N., Shen, Y., Zhao, R., Qian, W.J., Lipton, M.S., *et al.* (2005). Targeted comparative proteomics by liquid chromatography-tandem Fourier ion cyclotron resonance mass spectrometry. Anal Chem *77*, 400–406.

McCormick, R.M., Nelson, R.J., AlonsoAmigo, J., Benvegnu, H.H., and Hooper, H.H. (1997). Microchannel electrophoretic separations of DNA in injection-molded plastic substrates. Anal Chem *69*, 2626–2630.

McLafferty, F.W. (1992). State-of-the-art GC/MS. Chem Tech *22*, 182–189.

McLean, J.A. (2009). The mass–mobility correlation redux: The conformational landscape of anhydrous biomolecules. J Am Soc Mass Spectrom *20*, 1775–1781.

McLean, J.A., and Russell, D.H. (2005). New vistas for mass spectrometry-based proteomics and biotechnology: Rapid two-dimensional separations using gas-phase electrophoresis/ion mobility-mass spectrometry. Am Biotechnol Lab *23*, 18, 20–21.

McLean, J.A., Ruotolo, B.T., Gillig, K.J., and Russell, D.H. (2005). Ion mobility-mass spectrometry: A new paradigm for proteomics. Int J Mass Spectrom *240*, 301–315.

Merenbloom, S.I., Bohrer, B.C., Koeniger, S.L., and Clemmer, D.E. (2007). Assessing the peak capacity of IMS-IMS separations of tryptic peptide ions in He at 300 K. Anal Chem *79*, 515–522.

Merenbloom, S.I., Koeniger, S.L., Valentine, S.J., Plasencia, M.D., and Clemmer, D.E. (2006). IMS-IMS and IMS-IMS-IMS/MS for separating peptide and protein fragment ions. Anal Chem *78*, 2802–2809.

Mie, A., Jornten-Karlsson, M., Axelsson, B.O., Ray, A., and Reimann, C.T. (2007). Enantiomer separation of amino acids by complexation with chiral reference compounds and high field asymmetric waveform ion mobility spectrometry: Preliminary results and possible limitations. Anal Chem *79*, 2850–2858.

Mikkers, F.E.P., Everaerts, F.M., and Peek, J.A.F. (1979a). Isotachophoresis – concepts of resolution, load capacity and separation efficiency. J Chromatogr *168*, 293–315.

Mikkers, F.E.P., Everaerts, F.M., and Verheggen, T.P.E.M. (1979b). High-performance zone electrophoresis. J Chromatogr *169*, 11–20.

Mischak, H., Coon, J.J., Noval, J., Weissinger, E.M., Schanstra, J.P., and Dominiczak, A.F. (2009). Capillary electrophoresis-mass spectrometry as a powerful tool in biomarker discovery and clinical diagnosis: An update of recent developments. Mass Spectrom Rev *28*, 703–724.

Moision, R.M., and Armentrout, P.B. (2006). The special five-membered ring of proline: An experimental and theoretical investigation of alkali metal cation interactions with proline and its four- and six-membered ring analogues. J Phys Chem A *110*, 3933–3946.

Molina, C., Honing, M., and Barcelo, D. (1995). Organophosphorus pesticides in water by SPE and ESI-LC-MS. Anal Chem *66*, 4444–4449.

Motoyama, A., Venable, J.D., Ruse, C.I., and Yates, J.R. (2006). Automated ultra-high-pressure multidimensional protein identification technology (UHP-MudPIT) for improved peptide identification of proteomic samples. Anal Chem *78*, 5109–5118.

Neville, D.C., Rozanas, C.R., Price, E.M., Gruis, D.B., Verkman, A.S., and Townsend, R.R. (1997). Evidence for phosphorylation of serine 753 in CFTR using a novel metal-ion affinity resin and matrix-assisted laser desorption mass spectrometry. Protein Sci *6*, 2436–2445.

Ng, J.M.K., Gitlin, I., Stroock, A.D., and Whitesides, G.M. (2002). Components for integrated poly(dimethylsiloxane) microfluidic systems. Electrophoresis *23*, 3461–3473.

Oleschuk, R.D., Shultz-Lockyear, L.L., Ning, Y.B., and Harrison, D.J. (2000). Trapping of bead-based reagents within microfluidic systems: On-chip solid-phase extraction and electrochromatography. Anal Chem *72*, 585–590.

Olivares, J.A., Nguyen, N.T., Yonker, C.R., and Smith, R.D. (1987). Online mass spectrometric detection for capillary zone electrophoresis. Anal Chem *59*, 1230–1232.

Olivova, P., Chen, W.B., Chakraborty, A.B., and Gebler, J.C. (2008). Determination of N-glycosylation sites and site heterogeneity in a monoclonal antibody by electrospray quadrupole ion-mobility time-of-flight mass spectrometry. Rapid Commun Mass Spectrom *22*, 29–40.

Peng, J.M., Elias, J.E., Thoreen, C.C., Licklider, L.J., and Gygi, S.P. (2003). Evaluation of multidimensional chromatography coupled with tandem mass spectrometry (LC/LC-MS/MS) for large-scale protein analysis: The yeast proteome. J Proteome Res *2*, 43–50.

Peng, Y.Y., Pallandre, A., Tran, N.T., and Taverna, M. (2008). Recent innovations in protein separation on microchips by electrophoretic methods. Electrophoresis *29*, 157–178.

Pieper, R., Gatlin, C.L., Makusky, A.J., Russo, P.S., Schatz, C.R., Miller, S.S., Su, Q., McGrath, A.M., Estock, M.A., Parmar, P.P., *et al.* (2003). The human serum proteome: Display of nearly 3700 chromatographically separated protein spots on two-dimensional electrophoresis gels and identification of 325 distinct proteins. Proteomics *3*, 1345–1363.

Plasencia, M.D., Isailovic, D., Merenbloom, S.I., Mechref, Y., and Clemmer, D.E. (2008). Resolving and assigning N-linked glycan structural isomers from ovalbumin by IMS-MS. J Am Soc Mass Spectrom *19*, 1706–1715.

Pringle, S.D., Giles, K., Wildgoose, J.L., Williams, J.P., Slade, S.E., Thalassinos, K., Bateman, R.H., Bowers, M.T., and Scrivens, J.H. (2007). An investigation of the mobility separation of some peptide and protein ions using a new hybrid quadrupole/travelling wave IMS/oa-ToF instrument. Int J Mass Spectrom *261*, 1–12.

Ptolemy, A.S., and Britz-McKibbin, P. (2006). Sample preconcentration with chemical derivatization in capillary electrophoresis – capillary as preconcentrator, microreactor and chiral selector for high-throughput metabolite screening. J Chromatogr A *1106*, 7–18.

Purves, R.W., and Guevremont, R. (1999). Electrospray ionization high-field asymmetric waveform ion mobility spectrometry-mass spectrometry. Anal Chem *71*, 2346–2357.

Qian, W.J., Jacobs, J.M., Liu, T., Camp, D.G., 2nd, and Smith, R.D. (2006). Advances and challenges in liquid chromatography-mass spectrometry based proteomic profiling for clinical applications. Mol Cell Proteomics *5*, 1727–1744.

Qian, W.J., Kaleta, D.T., Petritis, B.O., Jiang, H., Liu, T., Zhang, X., Mottaz, H.M., Varnum, S.M., Camp, D.G., 2nd, Huang, L., *et al.* (2008). Enhanced detection of low abundance human plasma proteins using a tandem IgY12-SuperMix immunoaffinity separation strategy. Mol Cell Proteomics *7*, 1963–1973.

Ramautar, R., Somsen, G.W., and de Jong, G.J. (2009). CE-MS in metabolomics. Electrophoresis *30*, 276–291.

Ramsey, R.S., and Ramsey, J.M. (1997). Generating electrospray from microchip devices using electroosmotic pumping. Anal Chem *69*, 1174–1178.

Reid, G.E., and McLuckey, S.A. (2002). 'Top down' protein characterization via tandem mass spectrometry. J Mass Spectrom *37*, 663–675.

Reiser, R.W., and Fogiel, A.J. (1994). LC-MS of herbicides using ESI. Rapid Commun Mass Spectrom *8*, 252–257.

Resch, A., Xing, Y., Modrek, B., Gorlick, M., Riley, R., and Lee, C. (2004). Assessing the impact of alternative splicing on domain interactions in the human proteome. J Proteome Res *3*, 76–83.

Revercomb, H.E., and Mason, E.A. (1975). Theory of plasma chromatography/gaseous electrophoresis – a review. Anal Chem *47*, 970–983.

Robards, K., Haddad, P.R., and Jackson, P.E. (1994). Principles and Practice of Modern Chromatographic Methods (New York, NY, Academic).

Roberts, M.A., Rossier, J.S., Bercier, P., and Girault, H. (1997). UV laser machined polymer substrates for the development of microdiagnostic systems. Anal Chem *69*, 2035–2042.

Rossier, J.S., Vollet, C., Carnal, A., Lagger, G., Gobry, V., Girault, H.H., Michel, P., and Reymond, F. (2002). Plasma etched polymer microelectrochemical systems. Lab Chip *2*, 145–150.

Ruotolo, B.T., Gillig, K.J., Stone, E.G., and Russell, D.H. (2002). Peak capacity of ion mobility mass spectrometry: Separation of peptides in helium buffer gas. J Chromatogr B *782*, 385–392.

Ruotolo, B.T., Hyung, S.J., Robinson, P.M., Giles, K., Bateman, R.H., and Robinson, C.V. (2007). Ion mobility-mass spectrometry reveals long-lived, unfolded intermediates in the dissociation of protein complexes. Angew Chem Int Edit *46*, 8001–8004.

Ruotolo, B.T., McLean, J.A., Gillig, K.J., and Russell, D.H. (2004). Peak capacity of ion mobility mass spectrometry: The utility of varying drift gas polarizability for the separation of tryptic peptides. J Mass Spectrom *39*, 361–367.

Scarff, C.A., Patel, V.J., Thalassinos, K., and Scrivens, J.H. (2009). Probing hemoglobin structure by means of traveling-wave ion mobility mass spectrometry. J Am Soc Mass Spectrom *20*, 625–631.

Scott, R.P.W. (1995). Techniques and Practice of Chromatography (New York, NY, Marcel Dekker).

Sharma, S., Simpson, D.C., Tolic, N., Jaitly, N., Mayampurath, A.M., Smith, R.D., and Pasa-Tolic, L. (2007). Proteomic profiling of intact proteins using WAX-RPLC 2-D separations and FTICR mass spectrometry. J Proteome Res *6*, 602–610.

Shen, Y., Tolic, N., Masselon, C., Pasa-Tolic, L., Camp, D.G., 2nd, Hixson, K.K., Zhao, R., Anderson, G.A., and Smith, R.D. (2004). Ultrasensitive proteomics using high-efficiency on-line micro-SPE-nanoLC-nanoESI MS and MS/MS. Anal Chem *76*, 144–154.

Shen, Y., Zhao, R., Belov, M.E., Conrads, T.P., Anderson, G.A., Tang, K., Pasa-Tolic, L., Veenstra, T.D., Lipton, M.S., Udseth, H.R., *et al.* (2001a). Packed capillary reversed-phase liquid chromatography with high-performance electrospray ionization Fourier transform ion cyclotron resonance mass spectrometry for proteomics. Anal Chem *73*, 1766–1775.

Shen, Y., Zhang, R., Moore, R.J., Kim, J., Metz, T.O., Hixson, K.K., Zhao, R., Livesay, E.A., Udseth, H.R., and Smith, R.D. (2005). Automated 20 kpsi RPLC-MS and MS/MS with chromatographic peak capacities of 1000–1500 and capabilities in proteomics and metabolomics. Anal Chem *77*, 3090–3100.

Shen, Y., Zhao, R., Berger, S.J., Anderson, G.A., Rodriguez, N., and Smith, R.D. (2002). High-efficiency nanoscale liquid chromatography coupled on-line with mass spectrometry using nanoelectrospray ionization for proteomics. Anal Chem *74*, 4235–4249.

Shen, Y.F., Tolic, N., Zhao, R., Pasa-Tolic, L., Li, L.J., Berger, S.J., Harkewicz, R., Anderson, G.A., Belov, M.E., and Smith, R.D. (2001b). High-throughput proteomics using high efficiency multiple-capillary liquid chromatography with on-line high-performance ESI FTICR mass spectrometry. Anal Chem *73*, 3011–3021.

Shvartsburg, A.A., and Smith, R.D. (2008). Fundamentals of traveling wave ion mobility spectrometry. Anal Chem *80*, 9689–9699.

Shvartsburg, A.A., Li, F.M., Tang, K.Q., and Smith, R.D. (2006). High-resolution field asymmetric waveform ion mobility spectrometry using new planar geometry analyzers. Anal Chem *78*, 3706–3714.

Simpson, D.C., and Smith, R.D. (2005). Combining capillary electrophoresis with mass spectrometry for applications in proteomics. Electrophoresis *26*, 1291–1305.

Smith, R.M. (1988). Gas and Liquid Chromatography in Analytical Chemistry (New York, NY, Wiley).

Smith, R.D., and Shaffer, S.A. (2000). Method and apparatus for directing ions and other charged particles generated at near atmospheric pressures into a region under vacuum, United States Patent 6107628.

Smith, R.D., Anderson, G.A., Lipton, M.S., Pasa-Tolic, L., Shen, Y.F., Conrads, T.P., Veenstra, T.D., and Udseth, H.R. (2002). An accurate mass tag strategy for quantitative and high-throughput proteome measurements. Proteomics *2*, 513–523.

Smith, R.D., Olivares, J.A., Nguyen, N.T., and Udseth, H.R. (1988). Capillary zone electrophoresis mass-spectrometry using electrospray ionization interface. Anal Chem *60*, 436–441.

Snyder, L.R., and Kirkland, J.J. (1997). Practical HPLC Method Development 2nd edn (New York, NY, Wiley).

Srebalus Barnes, C.A., Hilderbrand, A.E., Valentine, S.J., and Clemmer, D.E. (2002). Resolving isomeric peptide mixtures: A combined HPLC/ion mobility-TOFMS analysis of a 4000-component combinatorial library. Anal Chem *74*, 26–36.

Stamm, S., Ben-Ari, S., Rafalska, I., Tang, Y.S., Zhang, Z.Y., Toiber, D., Thanaraj, T.A., and Soreq, H. (2005). Function of alternative splicing. Gene *344*, 1–20.

Sueyoshi, K., Kitagawa, F., and Otsuka, K. (2008). Recent progress of online sample preconcentration techniques in microchip electrophoresis. J Sep Sci *31*, 2650–2666.

Sung, W.-C., Makamba, H., and Chen, S.-H. (2005). Electrokinetic-driven microfluidic system in poly(dimethylsiloxane) for mass spectrometry detection integrating sample injection, capillary electrophoresis, and electrospray emitter on-chip. Electrophoresis *26*, 1783–1791.

Svedberg, M., Pettersson, A., Nilsson, S., Bergquist, J., Nyholm, L., Nikolaeff, F., and Markides, K. (2003). Sheathless electrospray from polymer microchips. Anal Chem *75*, 3934–3940.

Tang, K., Shvartsburg, A.A., Lee, H.N., Prior, D.C., Buschbach, M.A., Li, F.M., Tolmachev, A.V., Anderson, G.A., and Smith, R.D. (2005). High-sensitivity ion mobility spectrometry/mass spectrometry using electrodynamic ion funnel interfaces. Anal Chem *77*, 3330–3339.

Taraszka, J.A., Kurulugama, R., Sowell, R.A., Valentine, S.J., Koeniger, S.L., Arnold, R.J., Miller, D.F., Kaufman, T.C., and Clemmer, D.E. (2005). Mapping the proteome of *Drosophila melanogaster*: Analysis of embryos and adult heads by LC-IMS-MS methods. J Proteome Res *4*, 1223–1237.

Thalassinos, K., Grabenauer, M., Slade, S.E., Hilton, G.R., Bowers, M.T., and Scrivens, J.H. (2009). Characterization of phosphorylated peptides using traveling wave-based and drift cell ion mobility mass spectrometry. Anal Chem *81*, 248–254.

Thingholm, T.E., Jensen, O.N., Robinson, P.J., and Larsen, M.R. (2008a). SIMAC (sequential elution from IMAC), a phosphoproteomics strategy for the rapid separation of monophosphorylated from multiply phosphorylated peptides. Mol Cell Proteomics *7*, 661–671.

Thingholm, T.E., Larsen, M.R., Ingrell, C.R., Kassem, M., and Jensen, O.N. (2008b). TiO(2)-based phosphoproteomic analysis of the plasma membrane and the effects of phosphatase inhibitor treatment. J Proteome Res *7*, 3304–3313.

Thorslund, S., lindberg, P., Andren, P.E., Nikolaeff, F., and Bergquist, J. (2005). Electrokinetic-driven microfluidic system in poly(dimethylsiloxane) for mass spectrometry detection integrating sample injection, capillary electrophoresis, and electrospray emitter on-chip. Electrophoresis *26*, 4674–4683.

Tomer, K., and Parker, C. (1989). Biochemical applications of liquid chromatography-mass spectrometry. J Chromatogr *492*, 189–221.

Tsukagoshi, K., Jinno, N., and Nakajima, R. (2005). Development of a micro total analysis system incorporating chemiluminescence detection and application to detection of cancer markers. Anal Chem *77*, 1684–1688.

Valentine, S.J., Koeniger, S.L., and Clemmer, D.E. (2003). A split-field drift tube for separation and efficient fragmentation of biomolecular ions. Anal Chem *75*, 6202–6208.

Valentine, S.J., Kulchania, M., Barnes, C.A.S., and Clemmer, D.E. (2001). Multidimensional separations of complex peptide mixtures: A combined high-performance liquid chromatography/ion mobility/time-of-flight mass spectrometry approach. Int J Mass Spectrom *212*, 97–109.

Valentine, S.J., Kurulugama, R.T., Bohrer, B.C., Merenbloom, S.I., Sowell, R.A., Mechref, Y., and Clemmer, D.E. (2009a). Developing IMS-IMS-MS for rapid characterization of abundant proteins in human plasma. Int J Mass Spectrom *283*, 149–160.

Valentine, S.J., Plasencia, M.D., Liu, X.Y., Krishnan, M., Naylor, S., Udseth, H.R., Smith, R.D., and Clemmer, D.E. (2006a). Toward plasma proteome profiling with ion mobility-mass spectrometry. J Proteome Res *5*, 2977–2984.

Valentine, S.J., Sevugarajan, S., Kurulugama, R.T., Koeniger, S.L., Merenbloom, S.I., Bohrer, B.C., and Clemmer, D.E. (2006b). Split-field drift tube/mass spectrometry and isotopic labeling techniques for determination of single amino acid polymorphisms. J Proteome Res *5*, 1879–1887.

Valentine, S.J., Stokes, S.T., Kurulugama, R.T., Nachtigall, F.M., and Clemmer, D.E. (2009b). Overtone mobility spectrometry: Part 2. Theoretical considerations of resolving power. J Am Soc Mass Spectrom *20*, 738–750.

Vandekerckhove, J., Van Damme, P., Staes, A., Martens, L., Van Damme, J., Demol, H., De Groot SGhesquiere, B., Timmerman, E., Thomas, G.R., Goethals, M., *et al.* (2004). COFRADIC (TM): A highly versatile and quantitative gel-free proteome technology, suited for global analysis of a variety of post-translational protein modifications. Mol Cell Proteomic *3*, S304–S304.

VerBerkmoes, N.C., Bundy, J.L., Hauser, L., Asano, K.G., Razumovskaya, J., Larimer, F., Hettich, R., and Stephenson, J.L. (2002). Integrating “top-down” and “bottom-up” mass spectrometric approaches for proteome analysis of *Shewanella oneidensis*. J Proteome Res *1*, 239–252.

Villen, J., Beausoleil, S.A., Gerber, S.A., and Gygi, S.P. (2007). Large-scale phosphorylation analysis of mouse liver. Proc Natl Acad Sci USA *104*, 1488–1493.

Villen, J., and Gygi, S.P. (2008). The SCX/IMAC enrichment approach for global phosphorylation analysis by mass spectrometry. Nat Protoc *3*, 1630–1638.

Vreeland, W.N., Williams, S.J., Barron, A.E., and Sassi, A.P. (2003). Tandem isotachophoresis-zone electrophoresis via base-mediated destacking for increased detection sensitivity in microfluidic systems. Anal Chem *75*, 3059–3065.

Wachs, T., and Henion, J. (2001). Electrospray device for coupling microscale separations and other miniaturized devices with electrospray mass spectrometry. Anal Chem *73*, 632–638.

Walker, P.A., Morris, M.D., Burns, M.A., and Johnson, B.N. (1998). Isotachophoretic separations on a microchip. Normal Raman spectroscopy detection. Anal Chem *70*, 3766–3769.

Wang, Y.C., Stevens, A.L., and Han, J.Y. (2005). Million-fold preconcentration of proteins and peptides by nanofluidic filter. Anal Chem *77*, 4293–4299.

Washburn, M.P., Wolters, D., and Yates, J.R., 3rd (2001). Large-scale analysis of the yeast proteome by multidimensional protein identification technology. Nat Biotechnol *19*, 242–247.

Whitehouse, C.M., Dreyer, R.N., Yamashita, M., and Fenn, J.B. (1985). Electrospray interface for LC-MS. Anal Chem *57*, 675–679.

Williams, J.P., Bugarcic, T., Habtemariam, A., Giles, K., Campuzano, I., Rodger, P.M., and Sadler, P.J. (2009). Isomer separation and gas-phase configurations of organoruthenium anticancer complexes: Ion mobility mass spectrometry and modeling. J Am Soc Mass Spectrom *20*, 1119–1122.

Williams, J.P., Giles, K., Green, B.N., Scrivens, J.H., and Bateman, R.H. (2008). Ion mobility augments the utility of mass spectrometry in the identification of human hemoglobin variants. Rapid Commun Mass Spectrom *22*, 3179–3186.

Wilm, M., Shevchenko, A., Houthaeve, T., Breit, S., Schweigerer, L., Fotsis, T., and Mann, M. (1996). Femtomole sequencing of proteins from polyacrylamide gels by nano-electrospray mass spectrometry. Nature *379*, 466–469.

Wolters, D.A., Washburn, M.P., and Yates, J.R., 3rd (2001). An automated multidimensional protein identification technology for shotgun proteomics. Anal Chem *73*, 5683–5690.

Wu, Z.Y., Xanthopoulos, N., Reymond, Rossier, J.S., and Girault, H.H. (2002). Polymer microchips bonded by O-2-plasma activation. Electrophoresis *23*, 782–790.

Wu, S.T., Xia, Y.Q., and Jemal, M. (2007). High-field asymmetric waveform ion mobility spectrometry coupled with liquid chromatography/electrospray ionization tandem mass spectrometry (LOESI-FAIMS-MSMS) multi-component bioanalytical method development, performance evaluation and demonstration of the constancy of the compensation voltage with change of mobile phase composition or flow rate. Rapid Commun Mass Spectrom *21*, 3667–3676.

Wuhrer, M., Catalina, M.I., Deelder, A.M., and Hokke, C.H. (2007). Glycoproteomics based on tandem mass spectrometry of glycopeptides. J Chromatogr B *849*, 115–128.

Wyttenbach, T., Paizs, B., Barran, P., Breci, L., Liu, D., Suhai, S., Wysocki Vicki, H., and Bowers Michael, T. (2003). The effect of the initial water of hydration on the energetics, structures, and H/D exchange mechanism of a family of pentapeptides: An experimental and theoretical study. J Am Chem Soc *125*, 13768–13775.

Xia, Y.Q., Wu, S.T., and Jemal, M. (2008). LC-FAIMS-MS/MS for quantification of a peptide in plasma and evaluation of FAIMS global selectivity from plasma components. Anal Chem *80*, 7137–7143.

Yamagaki, T., and Sato, A. (2009). Peak width-mass correlation in CID MS/MS of isomeric oligosaccharides using traveling-wave ion mobility mass spectrometry. J Mass Spectrom *44*, 1509–1517.

Zhang, B., Foret, F., and Karger, B.L. (2001). High-throughput microfabricated CE/ESI-MS: Automated sampling from a microwell Plate. Anal Chem *73*, 2675–2681.

Zhang, B., Liu, H., Karger, B.L., and Foret, F. (1999). Microfabricated devices for capillary electrophoresis-electrospray mass spectrometry. Anal Chem *71*, 3258–3264.

Zhang, H., Li, X.J., Martin, D.B., and Aebersold, R. (2003). Identification and quantification of N-linked glycoproteins using hydrazide chemistry, stable isotope labeling and mass spectrometry. Nat Biotechnol *21*, 660–666.

Chapter 48
Reversed-Phase HPLC of Peptides – A Valuable Sample Preparation Tool in Bottom-Up Proteomics: Separation Selectivity in Single and Multi-dimensional Separation Modes

Ravi C. Dwivedi, Vic Spicer, and Oleg V. Krokhin

Abstract Reversed-phase HPLC is a leading tool for peptide fractionation in proteomics. Various combinations of stationary and mobile phases are employed depending on the purpose of the separation and the detection procedure. This diversity results in significant changes in separation selectivity, which are currently not well understood. Our ongoing work, focused on collection of peptide retention data and the development of sequence specific algorithms for peptide retention prediction, reveals these differences and helps define sets of conditions where predictive models can be considered independent from chromatographic platform used. We describe our observations on peptide separation selectivity variations for a wide range of RP-HPLC conditions: C18 sorbents of different pore sizes, using trifuoroacetic/formic/heptafluorobutyric acids as ion-pairing modifiers, RP separation at pH 10, and RP separation with an alternative perfluorinated chemistry of the bonded phase. Both the charge distribution within the peptide chain and the ion-pair formation mechanisms were found to be major factors in determining the peptide separation selectivity in RP-HPLC.

Keywords Peptide fractionation · Peptide retention prediction · Reversed-phase HPLC · Separation selectivity

48.1 Introduction

The new field of proteomics was conceived following major developments in the mass spectrometry of biological compounds – matrix-assisted laser desorption ionization (MALDI) and electrospray ionization (ESI) (Karas and Hillenkamp, 1988; Fenn et al., 1989). The introduction of these techniques made the analysis of proteins/peptide mixtures a routine procedure for many laboratories worldwide (Cox

---

O.V. Krokhin (✉)
Department of Internal Medicine, Manitoba Centre for Proteomics and Systems Biology, University of Manitoba, Winnipeg, Canada R3E 3P4
e-mail: krokhino@cc.umanitoba.ca

A.R. Ivanov, A.V. Lazarev (eds.), *Sample Preparation in Biological Mass Spectrometry*, DOI 10.1007/978-94-007-0828-0_48,

and Mann, 2007; Cravatt et al., 2007). However, the growing complexity of proteomic samples revealed the limitations of mass spectrometry alone to resolve all components of analyzed mixtures. The traditional gel-based approach for reducing sample complexity and the visualization of protein composition appeared to be very labor intensive, and tended to be supplanted by high-throughput gel-free procedures (Lambert et al., 2005). Therefore, at this stage peptide HPLC (mostly reversed-phase (RP) and cation-exchange) became the cornerstone component of modern analytical chemistry of peptides and proteins. The development of new chromatographic approaches has proved to be very important for the deeper study of protein composition and interactions, both fundamental questions of proteomic research.

RP HPLC of peptides is a leading tool for peptide fractionation in proteomics. Typically, it employs water-acetonitrile gradient systems with the addition of volatile formic/acetic/trifluoroacetic (TFA) acids as mobile phase modifiers. These ingredients create minimal or no interference with the peptide ionization techniques (except for a limited detection sensitivity for TFA-based eluents in ESI-MS). Column effluent can be transferred directly to the mass spectrometer via an ESI source and analyzed on-line. Hydrophilic interaction liquid chromatography (HILIC) also employs volatile eluent components (Alpert, 1990; Yoshida, 2004), however this method has yet to find a wider application in proteomics. For the off-line LC-MALDI-MS approach, the influence of all listed ion-pairing modifiers on ionization efficiency is negligible. Moreover, when the alternative selectivity of RP high pH separation is required, volatile eluent ingredients like ammonium formate do not affect MALDI-MS detection sensitivity. In case of strong cation-exchange (SCX) chromatography, the high salt content in the mobile phase limits the application of the procedure to only the first dimension in a two-dimensional separation scheme.

Historically SCX separation mode has been the method of choice as a first dimension peptide separation in proteomics for many years. SCX separation selectivity is determined mainly by peptide charge, and was found to be significantly different compared to the RP-HPLC mode. This provides sufficient orthogonality with the RP-HPLC at acidic pH – the standard separation protocol used with ESI-MS analysis (Gilar et al., 2005). Another reason for its popularity is the compatibility of eluent systems, which allows combining both SCX and RP separation media into a compact separation device amenable for on-line 2D fractionation of complex peptide mixtures. The term MudPIT (multidimensional protein identification technology) was coined by Yates and co-workers (Link et al., 1999; Washburn et al., 2001; Wolters et al., 2001) to describe this separation procedure, which is now widely used by proteomic researchers. SCX columns have a high loading capacity, however the optimization of MudPIT is dependent on sample concentration and complexity. The use of SCX separations in off-line mode gives the advantage of sample manipulation and separation optimization between dimensions. Since the separation phases of SCX and RP column are different, it offers choices of buffers, gradient time, and column composition, thereby increasing the possibility for deeper analysis of proteome (peptide identification). However, presence of high salt in SCX eluates compels us to look for cleaner options for peptide elution.

Recent trends in high-throughput bottom-up proteomics also show the growing popularity of high pH RP-HPLC as the first dimension in a 2D (two-dimensional) separation scheme (Delmotte et al., 2007; Dowell et al., 2008; Dwivedi et al., 2008; Gilar et al., 2005, 2007; Schley et al., 2006; Toll et al., 2005). It provides higher separation efficiency compared to SCX and increases number of identified peptides/proteins, despite having lower separation orthogonality with low pH RP. Another advantage of the RP-LC mode compared to cation-exchange chromatography is the ability to accurately predict peptide retention. Peptide retention prediction can be used along with MS information as a constraint to harden protein identification (Palmblad et al., 2002; Strittmatter et al., 2004), and to reduce analysis time by constructing inclusion/exclusion lists for time-consuming MS/MS analysis (Krokhin et al., 2006). Historically, predictive models were first created for the RP mode (Browne et al., 1982; Guo et al., 1986; Meek, 1980; Sakamoto et al., 1988; Sasagawa et al., 1982; Su et al., 1981); the complexity of the ion-exchange separation mechanism has so far precluded the pursuit of a predictive model for SCX. It has also been shown that the fundamental approaches for predictive model development can be easily transplanted from acidic (trifluoroacetic acid) to basic conditions (Dwivedi et al., 2008; Spicer et al., 2007). Thus, retention prediction for both dimensions can significantly increase the effectiveness of all retention time prediction applications in proteomics. All of these aspects collectively make RP-HPLC the method of choice for peptide fractionation in proteomics for the foreseeable future.

The growing application of peptide retention prediction fuelled the development of new predictive models. While this problem was the subject of intensive development in 1980s and 1990s in the chromatographic community, many proteomics-derived models were recently developed (Baczek et al., 2005; Gorshkov et al., 2006; Klammer et al., 2007; Krokhin, 2004, 2006; Petritis et al., 2006; Pfeifer et al., 2007; Put et al., 2006; Shinoda et al., 2006). Mass spectrometry detection can provide massive peptide sequence-retention datasets; for example, Petritis et al. (2006) used a data set of ~345,000 peptides to optimize their models using an analytical neural network approach. We used data sets of at least ~5000 peptides for the development of our sequence specific retention calculator (SSRCalc) models (Krokhin, 2006). These were created for the sets of the most popular RP-HPLC conditions applied in proteomics: 100 and 300 Å pore size C18 sorbents with trifluoroacetic, 100 Å with formic acid as ion-pairing modifier, and C18 RP separation at pH 10. Unlike previous models, we recognized the differences in separation selectivity provided by switching between the different mobile phases and ion-pairing modifiers. This recognition is critical for proper application of the models, data sets collection and understanding the fundamental RP-HPLC peptide separation mechanisms.

Early developments in peptide retention prediction were hindered by the limited number peptides used to build the models: Meek (1980) reported a model optimization using 25 peptides, while Guo et al. (1986) used 58 peptides. Their reported correlation values of 0.98–0.99 R^2 gave the impression that the predictive problem could be solved using simple additive models and straightforward optimization procedures. But for the construction of the proper model, the number peptides in the training data set should be significantly larger than the number of parameters (i.e. >>20 for additive models, which is the number of retention coefficients for all naturally occurring amino acids). Recent application of these classical models to real proteomics data using least-squares regression optimization gave rather disappointing results (Palmblad et al., 2002). On the other hand, the collection of exhaustive data sets does not guarantee the quality of the model. It was established previously (Houghten and DeGraw, 1987), and confirmed in recent studies, that for an adequate description of peptide retention its sequence ordering should be taken into consideration. Thus our SSRCalc models have corrections related to the position of the residue (distance from N- and C-termini), nearest-neighbor effects, specific role of proline residues and propensity of the peptide to form helical structures (Krokhin, 2006). First two corrections are directly connected to the proper understanding of ion-pairing separation mechanism of peptide RP HPLC. The effect of various ion-pairing modifiers on the separation is very well known (Mahoney, 1982; Guo et al., 1987), however the real sequence-specific features were discovered and introduced into prediction models only recently. Thus, the quintessential view of additive models on ion-pairing mechanism can be found in early report by (Guo et al., 1987): "In general, each positive charge, whether originating from a Lys, Arg or His side chain, or from N-terminal amino group, exerts an equal effect on peptide retention". However, current sequence-specific models distinguish positively charged groups depending on the nature of neighboring amino acids (Dwivedi et al., 2008; Krokhin, 2006).

Another common problem in attempting to use retention prediction models in proteomics is the incorrect choice of chromatographic conditions and their applicable prediction algorithms. Partially, this is a consequence of insufficient emphasis of this issue in the literature – some papers give the impression that once the model has been developed it can cover all variety of RP-HPLC conditions. In-fact, even switching between sorbents of 100 and 300 Å pore size can provide differences in separation selectivity, and variations of eluent pH or the ion-pairing modifier may change selectivity drastically. There are two different approaches to provide optimal performance of the models. The first is to follow precisely the conditions that were used for collecting the optimization data set. This approach assumes existence of the models for variety of chromatographic conditions, as SSRCalc does (http://hs2.proteome.ca/SSRCalc/SSRCalc32.html) (Dwivedi et al., 2008; Krokhin, 2006). The second assumes that prediction model could be developed for each set of experimental conditions (Klammer et al., 2007); in this case part of the data set (or confident identifications) obtained from single LC-MS experiment is used for adjusting (re-optimizing) the model, which then can be applied for the rest of the data.

This chapter describes variations in separation selectivity for the most popular modes of peptide RP-HPLC applied in proteomics and defines the groups of conditions within which retention prediction models can be considered transferable. Optimization of retention prediction algorithms for different RP separation systems also allows selectivity comparison based on the values of the retention coefficient for different amino acid residues. Variations in these values demonstrate clearly which residues are responsible for selectivity variation; one can use this knowledge for the effective development of applied separation techniques. These observations were made based on the ongoing collection of extensive peptide sequence-retention data sets and the development of SSRCalc retention prediction model versions for these data sets.

48.2 Experimental

This chapter uses various peptide sequence-retention time data sets obtained over the years for optimization/testing of the SSRCalc models. They were collected in both 1D (one-dimensional) and 2D RP-LC modes with MALDI-MS or ESI-MS detection. A general summary of these conditions is shown in Table 48.1. It should be noted that the two-dimensional scheme allows us to collect retention data simultaneously for two sets of chromatographic conditions. Together with high-throughput MS/MS identification it can effectively generate massive data sets; we show a ~15,000 peptide identification in the whole cell lysate sample that was collected in just 80 h of mass spectrometer time (Dwivedi et al., 2008). However, dealing with 1D LC-MALDI-MS and known protein mixtures produces more confident identifications.

48.2.1 Tryptic Digestion: Sample Preparation for LC Fractionation

All chemicals used: dithiothreitol, iodoacetamide, trifluoroacetic acid (TFA), formic acid (FA), heptafluorobutyric acid (HFBA), ammonia, 2,5-dihydroxybenzoic acid (DHB) were obtained from Sigma Chemicals (St. Louis, MO, USA). The six peptides used for the standard mixture preparation: LGGGGGGDGSR, LGGGGGGDFR, LLGGGGDFR, LLLGGDFR, LLLLDFR, LLLLLDFR – were custom synthesized by BioSynthesis Inc. (Lewisville, TX). Sequencing-grade modified trypsin (Promega, Madison, WI, USA) was used for digestion. Mixtures of commercially available proteins and proteins purified at Manitoba Centre for Proteomics and Systems Biology were used to generate complex peptide mixtures for 1D fractionation in HPLC-MALDI-MS mode as described elsewhere (Krokhin, 2006). Sample preparation included reduction (10 mM dithiothreitol, 30 min, 57°C), alkylation (50 mM iodoacetamide, 30 min in the dark at room temperature), dialysis (100 mM NH_4HCO_3, 6 h, 7 kDa molecular weight cut-off, Pierce), and

Table 48.1 The list of RP HPLC conditions used to generate data for SSRCalc models

Model/reference	Mode 1D/2D	Column	Eluent conditions	# of peptides	Model accuracy
1. 300 Å-TFA (Krokhin, 2006)	1D LC-MALDI	150 μm×150 mm, Vydac TP218, 5 μm	4 μL/min, 0.1% TFA, 0.66% acetonitrile per min gradient	~5300	~0.979
2. 100 Å-TFA (Krokhin, 2006)	1D LC-MALDI	300 μm×150 mm, PepMap100, 3 μm	3 μL/min, 0.1% TFA, 0.75% acetonitrile per min gradient	~5200	~0.976
3. 100 Å-pH 10 (Dwivedi et al., 2008)	1D LC-MALDI	150 μm×150 mm, XTerra, 5 μm	3 μL/min, 20 mM ammonium formate pH 10, 0.75% acetonitrile per min gradient	~3600	~0.969
4. 100 Å-FA (Dwivedi et al., 2008)	1D LC-MALDI	300 μm×150 mm, PepMap100, 3 μm	3 μL/min, 0.1% FA, 0.75% acetonitrile per min gradient	~3550	~0.966
2D LC-ESI. 100 Å Fluophase-TFA – 100 Å FA/HFBA					
5. 100 Å Fluophase-TFA (Krokhin, unpublished)	First dimension, off-line	1 mm×100 mm, Fluophase RP, 5 μm	150 μL/min, 0.1% TFA, 0.75% acetonitrile per min gradient	~1900	~0.945
Using 100 Å-TFA model	Second dimension LC-ESI	100 μm×150 mm, Luna C18(2), 3 μm	400 nL/min, 0.1% FA/0.005 HFBA, 0.5% acetonitrile per min gradient	~1900	~0.976
2D LC-ESI. 100 Å-pH 10 – 100 Å FA					
(Dwivedi et al., 2008) using 100 Å-pH 10 model	First dimension, off-line	1 mm×100 mm, XTerra, 5 μm	150 μL/min, 20 mM ammonium formate pH 10, 1% acetonitrile per min gradient	~15,000	~0.948
(Dwivedi et al., 2008), using 100 Å-FA model	Second dimension LC-ESI	100 μm×150 mm, Luna C18(2), 5 μm	500 nL/min, 0.1% FA, 0.33% acetonitrile per min gradient	~15,000	~0.945

trypsin digestion (1/50 enzyme/substrate weight ratio, 12 h, 37°C). The peptide mixtures were diluted prior injection to provide ~2 pmol/injection sample load. Off-line HPLC-MALDI fractionations were performed using 3 μL/min flow rate with linear 0.75% acetonitrile per minute gradients. Column effluent was mixed on-line with MALDI matrix solution, deposited on MALDI target using robotic sample deposition device (30 s each fraction), air dried and submitted to MALDI-MS and MS/MS analyses. For 2D Fluophase-TFA–100 Å FA/HFBA the equimolar mixture of ~40 proteins was prepared (~50 μg of protein in total), digested and fractionated in first dimension on Fluophase RP column. Fractions collected in first dimension were dried, re-suspended in buffer A for the second dimension and subjected for injection (~0.5–1 pmol of each protein).

48.2.2 LC-Systems, Chromatographic Conditions

In the first dimension, separations were performed on an Agilent 1100 Series system (Agilent Technologies, Wilmington, DE, USA) with UV detection at 214 nm and linear water/acetonitrile gradients (Table 48.1). A micro-flow version (3–4 μL/min) of the same system was used for HPLC-MALDI-MS experiments in off-line mode. A splitless nano-flow Tempo LC system (Eksigent, Dublin, CA, USA) and a nano-flow LC-Packings/Dionex (Dionex, Sunnyvale, CA, USA) system were used for the second dimension separation for the 100 Å-pH 10–100 Å FA and 100 Å Fluophase-TFA–100 Å FA/HFBA experiments with electrospray ionization. The list of RP materials include: Vydac 218 TP C18 (5 μm; Grace Vydac, Hesperia, CA, USA), PepMap 100 (3 μm; Dionex, Sunnyvale, CA, USA), XTerra C18 (5 μm; Waters, Milford, MA, USA), Luna C18(2) (3 and 5 μm; Phenomenex, Torrance, CA, USA), Fluophase RP – perfluorinated C6 phase (5 μm; Thermo Fisher Scientific, Waltham, MA, USA). Details of the eluent compositions are listed in Table 48.1, with references to our previous work.

48.2.3 Mass Spectrometry and Peptide Identification

Three different mass spectrometers with orthogonal quadrupole/time-of-flight configuration were used. In LC-MALDI-MS experiments, dried chromatographic fractions (DHB matrix) were analyzed by single mass spectrometry (MS) with *m/z* range 560–5000 Da, and by tandem mass spectrometry (MS/MS) in the Manitoba/Sciex prototype MALDI quadrupole/TOF (time-of-flight, QqTOF) mass spectrometer (Loboda et al., 2000). QStar Pulsar (for the 100 Å Fluophase-TFA–100 Å FA/HFBA experiment) and QStar Elite (for the 100 Å-pH 10–100 Å FA experiment) mass spectrometers (both, SCIEX, Toronto, ON, Canada) were used in standard data-dependent acquisition mode for peptide identification in 2D-HPLC-ESI-MS experiments. The X!Tandem (GPM) search engine was used for MS/MS peptide identification (Craig and Beavis, 2004).

48.3 Results and Discussion

48.3.1 Comparison of Separation Selectivity: Retention Time–Retention Time (t_R–t_R) Plots and Separation Orthogonality

Answering the question of the applicability of peptide retention prediction models is equivalent to comparing separation selectivity of peptides under two different sets of chromatographic conditions. It is conceivable that if peptide retention under conditions A are completely identical to conditions B, the model developed for conditions A will be applicable to B and vice versa. The comparison of separation selectivity is typically achieved by plotting dependencies of retention times in one set of conditions versus another (t_R–t_R plot) for the same compounds. These dependencies often referred as orthogonality plots (Gilar et al., 2005). Figure 48.1 shows conceptual examples of such plots, which highlight definition of orthogonality of two chromatographic systems. On panel a, the t_R–t_R correlation with R^2-value 1 indicates an identical separation selectivity (or no separation orthogonality). Any peptide retention prediction model will produce identical results for both conditions A and B

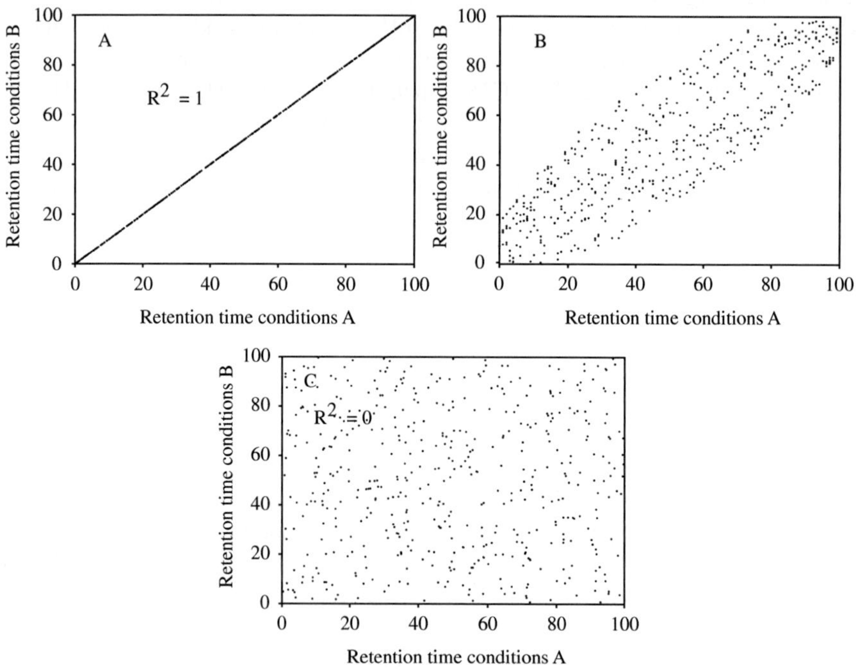

Fig. 48.1 t_R–t_R plots and separation orthogonality. (**a**) Identical selectivity (no orthogonality); (**b**) an intermediate case typical when two RP HPLC systems being compared; (**c**) completely orthogonal separations

independently of which data were used as a training set. Panel c shows an example of completely different separation selectivity: data points on t_R–t_R plot distributed randomly (perfect separation orthogonality). In this case, the cross-application of predictive models will provide no correlation. Panel b shows an intermediate case typical when two RP-HPLC conditions have been compared. Indeed, independent of buffers conditions (pH, ion-pairing modifiers) the separation in these systems is based on hydrophobic interactions. Peptide retention is mostly determined by the presence of the most hydrophobic residues: Trp, Phe, Leu, Ile, Tyr, Val. Thus t_R–t_R plots for two reversed-phase conditions will be scattered around a diagonal line, while the amplitude of deviations will be determined by specific factors: pH of the eluent, nature of ion-pairing modifiers, pore size of the sorbent, differences in chemistry of functional groups.

This chapter deals with the cases of substantial differences in separation selectivity, when development of specific prediction algorithms is required. The applicability of SSRCalc models within groups of stationary phases with nearly identical characteristics was recently studied. We showed that t_R–t_R plots for nine 100 Å pore size C18 sorbents with 0.1% TFA as the ion-pairing additive exhibit 0.9904–0.9995 correlations, indicating that the same prediction model can be used for all these phases (Spicer et al., 2007). These results also show that the chosen frequency of the fractionation (30 s at 0.75% acetonitrile per minute gradient) does not significantly affect the accuracy of retention time assignment, and, as a consequence, the resulting t_R–t_R correlations. Interestingly, the lowest correlations in the group of 100 Å pore size matrices were found for the C18 sorbent with embedded polar groups, highlighting the effect of sorbent chemistry on separation selectivity.

Figure 48.2 highlights the importance of having separate prediction models when the t_R–t_R plot shows a correlation significantly lower than 1. Panels a and b show the current accuracy of SSRCalc models for 100 Å-TFA and 300 Å-TFA conditions, respectively. Both plots were generated for the data sets of ~5000 peptides used as respective training sets for these models. Tryptic digests of the same set of known proteins were used to generate this data using off-line LC-MALDI-MS. Therefore, there is a significant overlap in the sequences detected in both systems. Figure 48.2c shows t_R–t_R plot for 100 Å-TFA and 300 Å-TFA peptides. The observed correlation of 0.978 indicates a significant difference in separation selectivity created by switching the pore size of the stationary phase. The steric effects of pore size mostly provide such difference: smaller peptides can easily penetrate inside the pores and realize their hydrophobic potential (Krokhin, 2006). The application of 100 Å-TFA SSRCalc model to 300 Å data indicate decrease in prediction accuracy from ~0.98 to ~0.96. A similar result was obtained when the 300 Å model was applied to 100 Å data (not shown here). This example clearly demonstrates that a change in chromatographic conditions required model adjustments.

However, it should be noted that the need to use an alternative algorithm might not be as obvious when the accuracy of the model itself is low. The traditional additive approach to retention prediction typically gives R^2-value correlations ~0.9–0.92, and provides very similar results when applied to both conditions discussed above. In other words, the peptide retention prediction algorithm should be

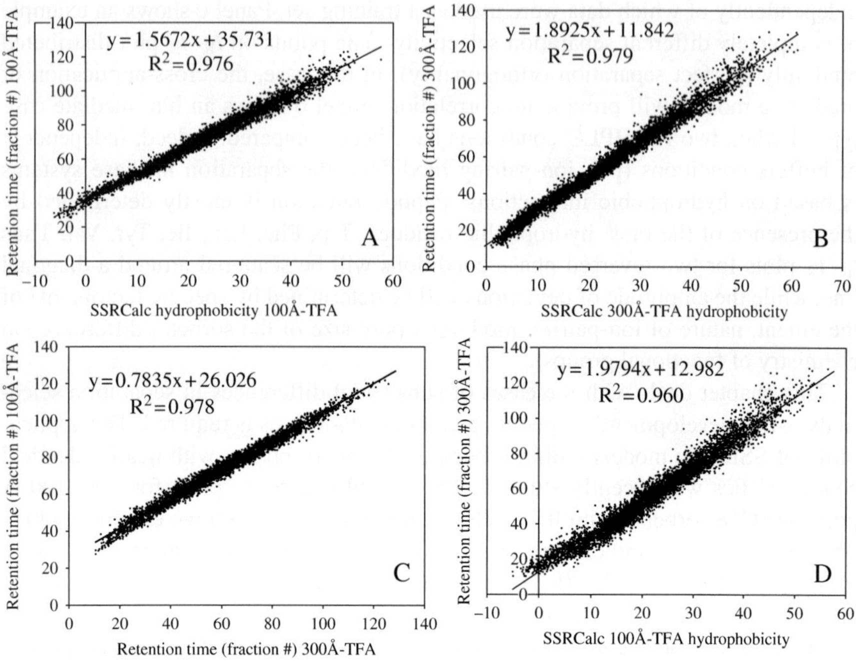

Fig. 48.2 Current accuracy of 100 Å-TFA and 300 Å-TFA models (for training sets) and their cross-applicability. (**a, b**) Retention time vs. hydrophobicity plots for respective 100 Å-TFA and 300 Å-TFA SSRCalc models; (**c**) t_R–t_R plot 100 Å-TFA vs. 300 Å-TFA conditions; (**d**) application of 100 Å-TFA SSRCalc to 300 Å data

accurate enough to see the difference in correlation on retention time–hydrophobicity plots provided by minor change in the separation conditions. Not surprisingly, these differences between 100 and 300 Å sorbents were first highlighted and studied following introduction of sequence-specific models like SSRCalc.

48.3.2 Comparison of Separation Selectivity for the Data Sets Generated Under Identical LC Conditions

It is conceivable that the use of identical mobile-stationary phase combinations will show no change in selectivity. However, there is another potential hurdle to overcome in this regard: differences in column size and flow rates. As we showed in the experimental section, modern proteomic techniques provide a variety of ways to separate and detect peptides. Thus, the majority of our data was collected using LC-MALDI-MS in micro-flow (3–4 μL/min) mode and with a column I.D. 150 and 300 μm. The most popular on-line LC-ESI-MS techniques typically apply sub-μL/min flow rates and 75–100 μm columns. In two-dimensional schemes peptides are first separated on larger columns (1 mm I.D. in our case) at higher flow rates and resulting fractions subsequently analyzed in micro- or nano-flow modes. The fraction number in the first dimension can be used as the equivalent of retention time. Will a significant change in column size and or flow rate affect peptide separation selectivity? Answering this question one needs to consider that peptides represent a group of "irregular" compounds from the point of view of linear-solvent-strength (LSS) theory (Snyder and Dolan, 2006). In other words, the slopes *S* in fundamental LSS equation ($\log k = \log k_w - S \times \phi$; where *k* is the retention factor at an organic solvent volume fraction ϕ (ϕ = ACN%/100) and k_w is the retention factor at $\phi = 0$) are different for different peptides. This may provide a change in separation selectivity or even retention order when different sizes of columns, flow rates and gradient slopes are employed (Glaich et al., 1986).

To explore this, two sets of peptide sequences–retention data collected using the same mobile/stationary phase combination, were compared. The XTerra pH 10 data set (~3620 peptides) in off-line LC-MALDI-MS mode (0.75% acetonitrile per minute gradient, 30 s fractions) was generated using purified proteins from different organisms including few human subjects (Table 48.1). The whole cell lysate of NK-type cells (~15,000 identifications) was fractionated in off-line 2D mode using identical pH 10 XTerra conditions in the first dimension, but different column size and flow rate (1% acetonitrile per minute gradient, 1 min fractions). Finding redundant (human) peptides in both sets should clarify if a significant change in column size and flow rate impacts separation selectivity. The t_R–t_R plot for 94 identical analytes found in both data sets shows an almost identical separation selectivity (R^2-value ~0.996, not shown here). This justifies the applicability of the SSRCalc algorithm that was developed using a micro-flow system to a system that employs a 1 mm I.D. column and 150 μL/min flow rate. Another important consequence of this finding is that the data sets can be aligned using the retention of identical peptides, allowing for concatenation and further development of more accurate prediction model.

It should be noted that the influence of different slopes *S* might not be visible when discrete fractions are used for MS acquisition. This might affect accuracy of the retention time assignment and, as a consequence also affects the monitoring of small variations in selectivity. Examination of peptide separation with continuous monitoring of elution profiles in ESI mode for different gradients and column sizes should provide a definitive answer to this question.

48.3.3 Separation Selectivity and Orthogonality for the Systems with Significant Variations in Mobile/Stationary Phase Chemistry

Figure 48.3a–e shows the t_R–t_R plots for systems with significant variations in selectivity. Four different sorbent-mobile phase combinations used for this comparison: 100 Å-TFA, 100 Å-FA, 100 Å Fluophase-TFA and 100 Å-pH 10. All sorbents share the same pore size and RP chemistry, except for perfluorinated C6 Fluophase RP.

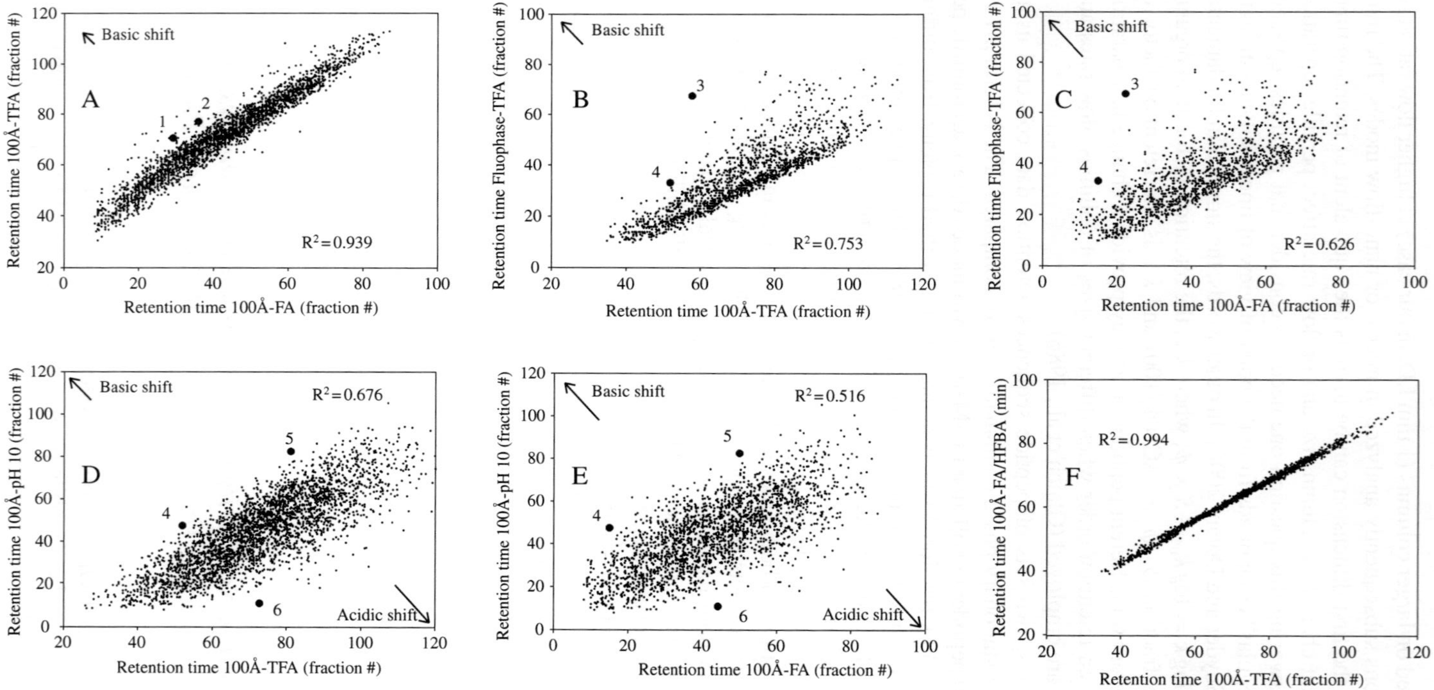

Fig. 48.3 t_R–t_R plots and separation orthogonality for various RP HPLC conditions. (**a**) 100 Å-TFA vs. 100 Å-FA; (**b**) Fluophase RP-TFA vs. 100 Å-TFA; (**c**) Fluophase RP-TFA vs. 100 Å-FA; (**d**) 100 Å-pH 10 vs. 100 Å-TFA; (**e**) 100 Å-pH 10 vs. 100 Å-FA; (**f**) 100 Å-FA/HFBA vs. 100 Å-TFA. Several examples of peptides exhibiting basic/acidic shifts are given: 1 – VHPFHVLR, 2 – HLHFTLVK, 3 – GHHEAELKPLAQSHATK, 4 – HKIPIK, 5 – LVPGWTKPITIGR, 6 – WEREEDEEQVDEEWR

We see t_R–t_R correlations ranging from ~0.94 (100 Å-TFA vs. 100 Å-FA) to ~0.52 (100 Å-FA vs. 100 Å-pH 10). As mentioned above, since peptide retention in all systems is determined mostly by hydrophobicity, most of the data points tend to concentrate close to a diagonal line. It appeared, however, that factors that determine the deviations are very similar for all of these examples. Identification and study of these factors helps to understand the underlying principles of RP-HPLC separation mechanisms.

(1) Panel e 100 Å-FA vs. 100 Å-pH 10 is the most extreme example of variation in separation selectivity. Correlation of just ~0.52 shows that these two systems are orthogonal enough to be used effectively in 2D RP-HPLC-ESI-MS for high-throughput proteomics applications as was demonstrated recently (Delmotte et al., 2007; Dwivedi et al., 2008; Gilar et al., 2005). Analysis of this plot shows that peptides with largest positive (above the diagonal line) and negative deviations are mostly basic and acidic, respectively. These are caused by dramatic changes in the charge distribution within peptide chains. Thus, residues most affected are those that are capable of carrying positive/negative charge. Aspartic and glutamic acids are neutral and moderately hydrophobic at pH 2 (FA conditions). But at pH 10, they are negatively charged and hydrophilic. Therefore peptide species with extremely low pI values (large number of acidic residues) retain stronger at acidic conditions and concentrate below the diagonal line. The opposite is true for the basic residues Arg, Lys and His – they are charged and hydrophilic at pH 2, and neutral/hydrophobic at pH 10. Similarly, basic peptides tend to retain stronger at pH 10 conditions (above the diagonal). We refer to these deviations shown in Fig. 48.3e as acidic and basic shifts.

(2) A similar dependence 100 Å-TFA vs. 100 Å-pH 10 (Fig. 48.3d) shows better correlations ~0.68 with the basic shift being not as profound as in the previous example. The reason for that becomes clear by analyzing the t_R–t_R plot in panel a. It shows ~0.94 correlation between FA and TFA conditions. There is a large number of basic peptides among peptides with largest positive deviation on this plot. This effect is not as profound compared to change of pH, but still visible. In other words, basic residues (Arg, Lys, His) contribute more to the peptide retention under TFA conditions compared to formic acid conditions. This is the consequence of the effect produced by the nature of the ion-pairing reagent. Trifluoroacetate anions are more hydrophobic compared to formate (Dai and Carr, 2005). This, in-turn, increases the apparent hydrophobicity of the basic peptides and provides the basic shift in Fig. 48.3a. Therefore TFA conditions can be placed between FA and pH 10 in terms of a stronger contribution of basic amino acids to the overall peptide hydrophobicity. This gradual change will become more evident when optimized retention coefficients in SSRCalc models will be compared in the next section.

It should be noted that comparison of separation selectivity between TFA and FA did not provide a clear picture as in the previous example of basic–acidic conditions. In the latter case, a dramatic swing in charge state dominates the factors affecting selectivity. But the situation observed in Fig. 48.3a is not as obvious. As we mentioned, one of the factors is related to hydrophobicity of an ion-pairing agent, which affects directly the hydrophobicity of basic residues. Some other factors are still not very well understood. A better understanding of the mechanisms responsible for the formation of peptide helical structures upon RP interaction (Krokhin, 2006) will improve this situation. Changes in the ion-pairing environment affect stabilization of helical structures upon interaction with C18 phase, therefore these processes as well cause additional alterations in selectivity.

(3) Figure 48.3b and c shows a comparison of separation selectivity on the perfluorinated sorbent (TFA as an ion-pairing modifier) with a traditional C18 phase. The first obvious difference is that there is no acidic shift on these plots, while basic peptides are retained much stronger when Fluophase-TFA conditions are applied (similar to basic shift in Fig. 48.3e). The charge state and hydrophobicity of Asp and Glu remain the same, since acidic eluent conditions were used. For basic residues, however, despite an unchanged charge, there is an apparent hydrophobicity increase causing a substantial basic shift. This is the consequence of much stronger interaction between TFA anions and perfluorinated functional groups of the stationary phase. Stronger interactions of the ion-pairing modifier cause stronger retention (larger hydrophobicity) of the residues involved in ion-pairing formation (Arg, Lys, His). Similar to the 3d–3e pair, the basic shift is more profound when formic acid conditions are used for comparison (Fig. 48.3c). It is interesting to note that a different underlying mechanism (stronger ion-pairing interactions with the stationary phase instead of a charge difference) provides a similar effect on the t_R–t_R plot. It shows once again the importance of ion-pairing interactions and respectively all ionogenic residues in determining the variation in separation selectivity between different RP-HPLC systems.

Earlier studies of the perfluorinated phase (Gilar et al., 2008) showed the influence of cation-exchange interactions on separation selectivity. Positively charged species may interact with residual silanol groups on the sorbent matrix providing a similar positive shift for basic peptides. Indeed, when a separation on Fluophase is performed in formic acid conditions, we found an increased retention for very basic peptides compared to TFA. These peptides, however, showed significantly deteriorated peak shapes under formic acid conditions: some of them were found in 10–15 consecutive fractions (5–7.5 min peak width). The use of TFA exhibited a much better peak shape, indicating that most likely the cation-exchange interactions with silica matrix are effectively suppressed. However, in discussing the separation selectivity and orthogonality for Fluophase RP stationary phase, the possibility of the minor influence of cation-exchange on silanol groups should be kept in mind.

Our initial goal in studying the separation selectivity of Fluophase RP was to evaluate its potential as first dimension separation in 2D-LC for high throughput analyses. The t_R–t_R plots (Fig. 48.3b and c) prove that it provides lower orthogonality compared to the pH 10 separation conditions. The absence of acidic shift will have negative impact on the number of identified proteins in bottom-up proteomics experiments. However, its application might be useful when such a specific change in separation selectivity is required (the increased retention of basic peptides).

48.3.4 *Separation Selectivity for the System with Mixed Ion-Pairing Modifiers: FA/HFBA*

As we mentioned above, trifluoroacetic acid reduces ESI-MS detection sensitivity. Formic and acetic acids are typically used as ion-pairing modifiers in ESI-MS detection mode, despite the lower separation efficiency they provide in RP-HPLC. The application of low concentration heptafluorobutyric acid together with FA is often recommended as an alternative solution; it maintains both a good shape for the chromatographic peaks and good detection sensitivity. So we ask, what separation selectivity is observed in this case, and which SSRCalc model will be most applicable? Figure 48.3f shows t_R–t_R plot for 100 Å-TFA–100 Å-FA/HFBA pair. A strong correlation of ~0.9942 suggests that separation selectivity for these two systems is very similar, and the SSRCalc model for 0.1% TFA can be applied for 0.1% FA/0.005% HFBA conditions. This finding is consistent with our previous results, when we reported a very similar prediction accuracy of SSRCalc 300 Å-TFA being applied to 300 Å-FA/HFBA conditions (Krokhin et al., 2004).

48.3.5 *The Effect of Separation Mode on Peptide Hydrophobicity, Expressed in Acetonitrile Percentage Units*

The discussion in previous sections concentrated on the orthogonality of separation modes or relative changes in separation selectivity. There are, however, overall changes in peptide hydrophobicity that can be expressed in the acetonitrile concentration required for elution of peptides from a reversed-phase column. To test these differences in detail, we designed a peptide retention standard (Krokhin and Spicer, 2009) consisting of six species covering a wide range of hydrophobicities: ~4.0–29.6% acetonitrile when applied to 100 Å-TFA system as shown in Fig. 48.4a. This mixture contains six designed peptide sequences: LGGGGGGDGSR, LGGGGGGDFR, LLGGGGDFR, LLLGGDFR, LLLLDFR, LLLLLDFR (P1–P6, listed in the order of increased hydrophobicity). Figure 48.4 shows a UV chromatograms of this mixture using all RP LC combinations described: 100 Å-TFA; 300 Å-TFA; 100 Å-FA; 100 Å-pH 10, perfluorinated RP-TFA and 100 Å-FA/HFBA. A 1%/min acetonitrile gradient starting from 0%, identical column size 1×100 mm and flow-rate of 150 μL/min were used for all separations. The retention time difference between the first (P1) and the last (P6) peaks on the chromatograms gives a direct indication of the ACN% scale covered.

Figure 48.4 illustrates the differences in apparent peptide hydrophobicity provided by switching ion-pairing modifiers, pore sizes and reversed-phase chemistry. Thus, changing from 100 to 300 Å (a and b) slightly decreases the retention times with no apparent change in relative selectivity. Conversely, the application of FA instead of TFA decreases the retention significantly (a and c) to the point where peptide P1 elutes under isocratic conditions of 0% acetonitrile and exhibits a poor peak shape. At pH 10 conditions (Fig. 48.4d), when peptides are affected by both acidic and basic shifts, the relative hydrophobicity remains unchanged. Peptide

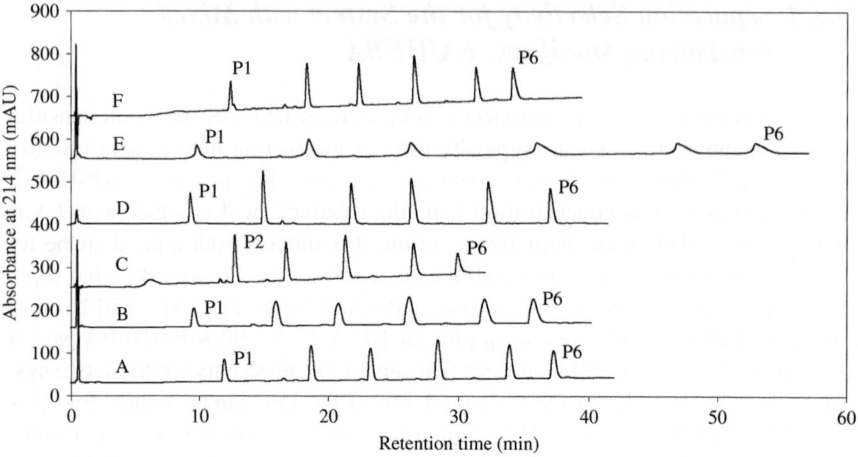

Fig. 48.4 Chromatographic separations of the 6-peptide standard mixture using 1 mm × 100 mm columns size, 150 μL/min flow-rate, and 1% acetonitrile per min gradient. (**a**) 100 Å (Luna C18(2)) 0.1% TFA, (**b**) 300 Å (Vydac TP218) 0.1% TFA, (**c**) 100 Å (Luna C18(2)) FA, (**d**) 100 Å (XTerra) pH 10, (**e**) Fluophase RP 0.1% TFA, (**f**) 100 Å (Luna C18(2)) 0.1% FA/0.005% HFBA

retention on perfluorinated RP phase with TFA as ion-pairing modifier shows only positive basic shift, resulting in a dramatic increase in hydrophobicity expressed in acetonitrile percentage scale (Fig. 48.4e: ~44% difference between the first and last peak instead of ~26%). Application of the FA/HFBA eluent system exhibits results very close to the TFA retention when applied with 100 Å C18 sorbent (Fig. 48.4a and f).

48.3.6 Differences in Apparent Hydrophobicity of the Residues Revealed During Optimization of SSRCalc Algorithms

Empirical observations of the selectivity change made for different t_R–t_R plots helps at initial stages of optimization of the prediction models. Adjusting SSRCalc to new conditions always starts with optimization of parameters for the most affected residues (Krokhin, 2006). Thus, for new pH 10 models, the retention coefficients for Asp, Glu, Arg, Lys, His were optimized first (Dwivedi et al., 2008). This allows for a directed, time efficient optimization process. The current accuracy of SSRCalc models for training sets and selected real samples is given in Table 48.1. While it reached ~0.98 correlation for the better-established TFA models, the latest Fluophase-TFA model is still at a 0.95 level. Nevertheless, the current optimized values of retention coefficients provide a clear picture of how amino acid hydrophobicity varies for different conditions. Figure 48.5 shows variation in R_c values for all four systems studied. As we mentioned above, the hydrophobic properties of the majority of residues remain basically unchanged (only Trp, Leu, Val and Ala are shown), while

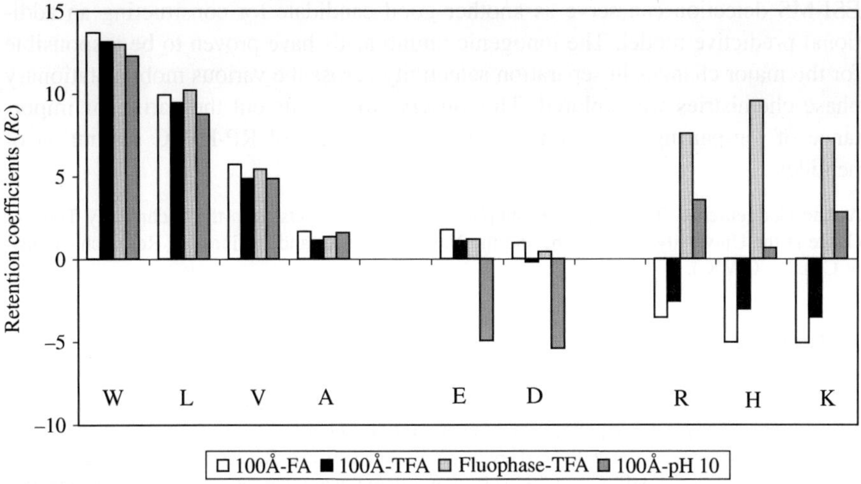

Fig. 48.5 Optimized retention coefficients (R_c) for respective SSRCalc models

the residues capable of carrying positive/negative charge account for most of the differences. Analysis of these changes illustrates why particular acidic or basic shifts appeared on t_R–t_R plots: slightly higher R_c values for basic residues in the TFA model compared to FA provide the small basic shift in Fig. 48.3a. This is also the reason why the basic shift is not so profound in Fig. 48.3b compared to Fig. 48.3c, and in Fig. 48.3d compared to Fig. 48.3e. Fluophase RP-TFA exhibits a dramatic positive shift for the basic residues, while at pH 10 both acidic and basic residues strongly alter the hydrophobicity. Overall, the four systems under consideration can be arranged in the order of increasing retention for basic residues and decreasing retention for acidic residues as follows: TFA – FA – Fluophase RP – pH 10.

48.4 Conclusions

The development of peptide retention prediction models for various RP-HPLC systems helps to determine the major factors affecting separation selectivity, and organize groups of chromatographic conditions where prediction models can be considered transferable. The application to three eluent systems (trifluoroacetic-based, formic acid-based and pH 10) using 100 Å C18 columns requires significant adjustment of the models. When these three eluent systems are used with 300 Å sorbents, the algorithms should again be re-optimized. This was demonstrated on TFA-based eluents, but there is ample evidence to suggest this will be necessary on the formic acid and pH 10 systems as well, giving a total of six predictive models. But since smaller pore-sized materials are the most popular for reversed-phase peptide fractionation, we expect the three 100 Å predictive models will find major application in proteomics. The acetic acid conditions commonly used with

ESI-MS detection can serve as another good candidate for constructing an additional predictive model. The ionogenic amino acids have proven to be responsible for the major changes in separation selectivity across the various mobile/stationary phase chemistries we explored. This observation points out the particular importance of ion-pairing formation on the overall picture of RP-HPLC separation of peptides.

Acknowledgements This work was supported in part by grants from the Technology Transfer Office at the University of Manitoba and the Natural Sciences and Engineering Research Council of Canada (O.V.K.).

References

Alpert, A.J. (1990). Hydrophilic-interaction chromatography for the separation of peptides, nucleic acids and other polar compounds. J Chromatogr *499*, 177–196.

Baczek, T., Wiczling, P., Marszall, M., Heyden, Y.V., and Kaliszan, R. (2005). Prediction of peptide retention at different HPLC conditions from multiple linear regression models. J Proteome Res *4*, 555–563.

Browne, C.A., Bennett, H.P., and Solomon, S. (1982). The isolation of peptides by high-performance liquid chromatography using predicted elution positions. Anal Biochem *124*, 201–208.

Cox, J., and Mann, M. (2007). Is proteomics the new genomics? Cell *130*, 395–398.

Craig, R., and Beavis, R.C. (2004). TANDEM: Matching proteins with tandem mass spectra. Bioinformatics *20*, 1466–1467.

Cravatt, B.F., Simon, G.M., and Yates, J.R., 3rd. (2007). The biological impact of mass-spectrometry-based proteomics. Nature *450*, 991–1000.

Dai, J., and Carr, P.W. (2005). Role of ion pairing in anionic additive effects on the separation of cationic drugs in reversed-phase liquid chromatography. J Chromatogr A *1072*, 169–184.

Delmotte, N., Lasaosa, M., Tholey, A., Heinzle, E., and Huber, C.G. (2007). Two-dimensional reversed-phase × ion-pair reversed-phase HPLC: An alternative approach to high-resolution peptide separation for shotgun proteome analysis. J Proteome Res *6*, 4363–4373.

Dowell, J.A., Frost, D.C., Zhang, J., and Li, L. (2008). Comparison of two-dimensional fractionation techniques for shotgun proteomics. Anal Chem *80*, 6715–6723.

Dwivedi, R.C., Spicer, V., Harder, M., Antonovici, M., Ens, W., Standing, K.G., Wilkins, J.A., and Krokhin, O.V. (2008). Practical implementation of 2D HPLC scheme with accurate peptide retention prediction in both dimensions for high-throughput bottom-up proteomics. Anal Chem *80*, 7036–7042.

Fenn, J.B., Mann, M., Meng, C.K., Wong, S.F., and Whitehouse, C.M. (1989). Electrospray ionization for mass spectrometry of large biomolecules. Science *246*, 64–71.

Gilar, M., Jaworski, A., Olivova, P., and Gebler, J.C. (2007). Peptide retention prediction applied to proteomic data analysis. Rapid Commun Mass Spectrom *21*, 2813–2821.

Gilar, M., Olivova, P., Daly, A.E., and Gebler, J.C. (2005). Orthogonality of separation in two-dimensional liquid chromatography. Anal Chem *77*, 6426–6434.

Gilar, M., Yu, Y.Q., Ahn, J., Fournier, J., and Gebler, J.C. (2008). Mixed-mode chromatography for fractionation of peptides, phosphopeptides, and sialylated glycopeptides. J Chromatogr A *1191*, 162–170.

Glaich, J.L., Quarry, M.A., Vasta, J.F., and Snyder, L.R. (1986). Separation of peptide mixtures by reversed-phase gradient elution. Use of flow rate changes for controlling band spacing and improving resolution. Anal Chem *58*, 280–285.

Gorshkov, A.V., Tarasova, I.A., Evreinov, V.V., Savitski, M.M., Nielsen, M.L., Zubarev, R.A., and Gorshkov, M.V. (2006). Liquid chromatography at critical conditions: Comprehensive approach to sequence-dependent retention time prediction. Anal Chem *78*, 7770–7777.

Guo, D., Mant, C.T., Taneja, A.K., Hodges, R.S. (1986). Prediction of peptide retention times in reversed-phase high-performance liquid chromatography II. Correlation of observed and predicted peptide retention times and factors influencing the retention times of peptides. J Chromatogr *359*, 519–532.

Guo, D.C., Mant, C.T., and Hodges, R.S. (1987). Effects of ion-pairing reagents on the prediction of peptide retention in reversed-phase high-performance liquid chromatography. J Chromatogr *386*, 205–222.

Houghten, R.A., and DeGraw, S.T. (1987). Effect of positional environmental domains on the variation of high-performance liquid chromatographic peptide retention coefficients. J Chromatogr *386*, 223–228.

Karas, M., and Hillenkamp, F. (1988). Laser desorption ionization of proteins with molecular masses exceeding 10,000 daltons. Anal Chem *60*, 2299–2301.

Klammer, A.A., Yi, X., MacCoss, M.J., and Noble, W.S. (2007). Improving tandem mass spectrum identification using peptide retention time prediction across diverse chromatography conditions. Anal Chem *79*, 6111–6118.

Krokhin, O.V. (2006). Sequence-specific retention calculator. Algorithm for peptide retention prediction in ion-pair RP-HPLC: Application to 300- and 100-A pore size C18 sorbents. Anal Chem *78*, 7785–7795.

Krokhin, O.V., Craig, R., Spicer, V., Ens, W., Standing, K.G., Beavis, R.C., and Wilkins, J.A. (2004). An improved model for prediction of retention times of tryptic peptides in ion pair reversed-phase HPLC: Its application to protein peptide mapping by off-line HPLC-MALDI MS. Mol Cell Proteomics *3*, 908–919.

Krokhin, O.V., and Spicer, V. (2009). Peptide retention standards and hydrophobicity indexes in reversed-phase high-performance liquid chromatography of peptides. Anal Chem *81*, 9522–9530.

Krokhin, O.V., Ying, S., Cortens, J.P., Ghosh, D., Spicer, V., Ens, W., Standing, K.G., Beavis, R.C., and Wilkins, J.A. (2006). Use of peptide retention time prediction for protein identification by off-line reversed-phase HPLC-MALDI MS/MS. Anal Chem *78*, 6265–6269.

Lambert, J.P., Ethier, M., Smith, J.C., and Figeys, D. (2005). Proteomics: From gel based to gel free. Anal Chem *77*, 3771–3787.

Link, A.J., Eng, J., Schieltz, D.M., Carmack, E., Mize, G.J., Morris, D.R., Garvik, B.M., and Yates, J.R., 3rd (1999). Direct analysis of protein complexes using mass spectrometry. Nat Biotechnol *17*, 676–682.

Loboda, A.V., Krutchinsky, A.N., Bromirski, M., Ens, W., and Standing, K.G. (2000). A tandem quadrupole/time-of-flight mass spectrometer with a matrix-assisted laser desorption/ionization source: Design and performance. Rapid Commun Mass Spectrom *14*, 1047–1057.

Mahoney, W.C. (1982). Isolation of denatured proteins and peptides by high-performance liquid chromatography. Effect of different perfluorinated acids, column length and large-pore supports. Biochim Biophys Acta *704*, 284–289.

Meek, J.L. (1980). Prediction of peptide retention times in high-pressure liquid chromatography on the basis of amino acid composition. Proc Natl Acad Sci USA *77*, 1632–1636.

Palmblad, M., Ramstrom, M., Markides, K.E., Hakansson, P., and Bergquist, J. (2002). Prediction of chromatographic retention and protein identification in liquid chromatography/mass spectrometry. Anal Chem *74*, 5826–5830.

Petritis, K., Kangas, L.J., Yan, B., Monroe, M.E., Strittmatter, E.F., Qian, W.J., Adkins, J.N., Moore, R.J., Xu, Y., Lipton, M.S., *et al.* (2006). Improved peptide elution time prediction for reversed-phase liquid chromatography-MS by incorporating peptide sequence information. Anal Chem *78*, 5026–5039.

Pfeifer, N., Leinenbach, A., Huber, C.G., and Kohlbacher, O. (2007). Statistical learning of peptide retention behavior in chromatographic separations: A new kernel-based approach for computational proteomics. BMC Bioinformatics *8*, 468.

Put, R., Daszykowski, M., Baczek, T., and Vander Heyden, Y. (2006). Retention prediction of peptides based on uninformative variable elimination by partial least squares. J Proteome Res *5*, 1618–1625.

Sakamoto, Y., Kawakami, N., and Sasagawa, T. (1988). Prediction of peptide retention times. J Chromatogr *442*, 69–79.

Sasagawa, T., Okuyama, T., Teller, D.C. (1982). Prediction of peptide retention times in reversed-phase high-performance liquid chromatography during linear gradient elution. J Chromatogr *240*, 329–340.

Schley, C., Altmeyer, M.O., Swart, R., Muller, R., and Huber, C.G. (2006). Proteome analysis of *Myxococcus xanthus* by off-line two-dimensional chromatographic separation using monolithic poly-(styrene-divinylbenzene) columns combined with ion-trap tandem mass spectrometry. J Proteome Res *5*, 2760–2768.

Shinoda, K., Sugimoto, M., Yachie, N., Sugiyama, N., Masuda, T., Robert, M., Soga, T., and Tomita, M. (2006). Prediction of liquid chromatographic retention times of peptides generated by protease digestion of the *Escherichia coli* proteome using artificial neural networks. J Proteome Res *5*, 3312–3317.

Snyder, L.R., and Dolan, J.W. (2006). High-Performance Gradient Elution: The Practical Application of the Linear-Solvent-Strength Model (New York, NY, Wiley).

Spicer, V., Yamchuk, A., Cortens, J., Sousa, S., Ens, W., Standing, K.G., Wilkins, J.A., and Krokhin, O.V. (2007). Sequence-specific retention calculator. A family of peptide retention time prediction algorithms in reversed-phase HPLC: Applicability to various chromatographic conditions and columns. Anal Chem *79*, 8762–8768.

Strittmatter, E.F., Kangas, L.J., Petritis, K., Mottaz, H.M., Anderson, G.A., Shen, Y., Jacobs, J.M., Camp, D.G., 2nd, and Smith, R.D. (2004). Application of peptide LC retention time information in a discriminant function for peptide identification by tandem mass spectrometry. J Proteome Res *3*, 760–769.

Su, S.J., Grego, B., Niven, B., and Hearn, M.T.W. (1981). High performance liquid chromatography of amino acids, peptides and proteins XXXVII. Analysis of group retention contributions for peptides separated by reversed phase high performance liquid chromatography. J Liq Chromatogr *4*, 1745–1764.

Toll, H., Oberacher, H., Swart, R., and Huber, C.G. (2005). Separation, detection, and identification of peptides by ion-pair reversed-phase high-performance liquid chromatography-electrospray ionization mass spectrometry at high and low pH. J Chromatogr A *1079*, 274–286.

Washburn, M.P., Wolters, D., and Yates, J.R., 3rd. (2001). Large-scale analysis of the yeast proteome by multidimensional protein identification technology. Nat Biotechnol *19*, 242–247.

Wolters, D.A., Washburn, M.P., and Yates, J.R., 3rd. (2001). An automated multidimensional protein identification technology for shotgun proteomics. Anal Chem *73*, 5683–5690.

Yoshida, T. (2004). Peptide separation by hydrophilic-interaction chromatography. J Biochem Biophys Methods *60*, 265–280.

Chapter 49
Applications of Microfluidic Devices with Mass Spectrometry Detection in Proteomics

Xiuli Mao and Iulia M. Lazar

Abstract Mass spectrometry (MS) was first used as a detection tool for microfluidic devices in the late nineties (Figeys et al., Anal Chem 69:3153–3160, 1997; Ramsey and Ramsey, Anal Chem 69:1174–1178, 1997; Xue et al., Anal Chem 69:426–430, 1997), and since then, significant efforts have been invested in the further development of efficient microfluidic-MS interfaces and the exploration of microfluidic-MS applicability in biomolecular analysis. Microfluidic devices can be coupled to MS through both electrospray ionization (ESI) and matrix-assisted laser desorption ionization (MALDI) interfaces. Both conventional, such as capillary electrophoresis (CE)/high performance liquid chromatography (HPLC) liquid sheath, liquid junction and micro/nano-ESI interfaces, as well as novel interfaces based on unique features enabled by microfabrication, have been developed. Current protein identification protocols proceed through a series of steps that are generally time-consuming and labor-intensive. The development of microfluidic devices with MS detection has facilitated protein analysis by minimizing sample consumption and enabling high-throughput, automatic sample manipulations. In this chapter, the applicability of microfluidic chips with MS detection in proteomics research is reviewed. Applications that focus on protein analysis, including sample pretreatment, proteolytic digestion and separation are described. Recent progress on cell culture and lysis, protein quantitation, and analysis of protein posttranslational modifications is discussed.

Keywords Microfluidic device · Mass spectrometry · Proteomics · Sample preparation

Abbreviations

2D	two dimensional
BSA	bovine serum albumin

I.M. Lazar (✉)
Department of Biological Sciences, Virginia Polytechnic Institute and State University, Blacksburg, VA 24061, USA
e-mail: malazar@vt.edu

A.R. Ivanov, A.V. Lazarev (eds.), *Sample Preparation in Biological Mass Spectrometry*, DOI 10.1007/978-94-007-0828-0_49,

CD	compact disc
CE	capillary electrophoresis
CEC	capillary electrochromatography
CGE	capillary gel electrophoresis
Cu(II)-IMAC	copper(II)-immobilized metal affinity chromatography
EOF	electroosmotic flow
ESI	electrospray ionization
EWOD	electro-wetting-on-dielectric
HPLC	high performance liquid chromatography
IEF	isoelectric focusing
IMAC	ion immobilized affinity chromatography
iTRAQ	isobaric tags for relative and absolute quantitation
MALDI	matrix-assisted laser desorption ionization
MEKC	micellar electrokinetic chromatography
MS	mass spectrometry
PMMA	poly(methyl methacrylate)
RP	reversed phase
SCX	strong cation exchange
SPE	solid phase extraction
TOFMS	time of flight mass spectrometry

49.1 Introduction

Microfluidic or 'lab on a chip' devices are versatile bioanalytical platforms with broad applicability in chemical and biological analysis. Microfluidic devices have been developed for applications that necessitate rapid and low-sample-consumption analysis, including DNA sequencing/sizing, PCR, SNPs genotyping, sample preparation and multidimensional separations (Armani et al., 2009; Inoue et al., 2009; Kim et al., 2009; Shi, 2006; Tia and Herr, 2009). More recently, the development of complex microfluidic bioassays for proteomics, drug screening and biodefense has been demonstrated, as well (West et al., 2008). The introduction of mass spectrometry as a detection tool for microchip devices has further expanded the functionality and applicability of microfluidic platforms, especially in proteomics. Microfluidic devices have been coupled with mass spectrometry detection via both ESI and MALDI, which are two of the most widely used soft ionization techniques in proteomics (Lazar 2006a; Lazar 2007). This interfacing was greatly facilitated by the compatibility between the typical microfluidic and optimal ESI/MALDI source flow rates (i.e., nL-μL/min). Microchip ESI can be generated directly from the chip edge, from chip-integrated ESI emitters, or from chip-inserted capillary ESI emitters via conventional liquid junction, liquid sheath or nano-ESI approaches. Strategies for improving the stability and efficiency of ESI have been recently reviewed (Sung et al., 2005). As sample ionization in MALDI sources usually occurs in the vacuum, interfacing of MALDI-MS with microchip devices is not always straightforward.

Microfluidic chips were initially connected with MALDI-MS off-line, separations being accomplished by depositing the analytes onto a separate MALDI target plate. Thereafter, a number of microfluidic systems that integrated MALDI targets into the microfluidic substrate have been developed, to enable tasks such as sample cleanup or proteolytic reactions to be seamlessly combined with target preparation (Devoe and Lee, 2006). Further development of the centrifugal and digital microfluidic devices facilitated direct and automatic sample preparation and delivery for MALDI-MS detection (Gustafsson et al., 2004; Wheeler et al., 2005). Digital or droplet-based microfluidics involves the generation and manipulation of discrete droplets at specific locations on an array of microelectrodes integrated on the chip (Belder, 2005; Jensen and Lee, 2004; Teh et al., 2008). Droplet-based microfluidic platforms have enabled controlled and rapid mixing of fluids in droplet microreactors ranging from nL to fL volumes, to greatly decrease the reaction time scales. These digital systems were rendered to be programmable and reconfigurable, to enable automatic sample preparation for MALDI-MS (Fair, 2007). The various interfaces and strategies for coupling microfluidic devices with MS detection have been thoroughly reviewed (Foret and Kusy, 2006; Koster and Verpoorte, 2007; Lazar et al., 2006a; Lazar 2007; Mogensen et al., 2004; Uchiyama et al., 2004; Zamfir, 2007), therefore, the authors will not discuss the topic in this chapter.

Ever since inception, proteomics has been referred to as the study of the full set of proteins encoded by the genome of an organism (Nedelkov and Nelson, 2006). Basic tasks in proteomics involve qualitative identifications of proteins and quantitative comparisons of proteomes to unravel the biological processes that control cellular life. The traditional technology for proteomic sample analysis involves the separation of proteins from cellular extracts by two dimensional (2D) gel electrophoresis, followed by proteolytic digestion, nano-LC separation and MS detection. The MS raw data are subjected to database searching for enabling the identification of the detected proteins. The complexity of proteomic samples (hundreds of thousands of peptide components), as well as the broad range of protein/peptide concentrations (spanning over 3–10 orders of magnitude), make protein analysis a challenging task. Proper sample preparation and high peak capacity separations are needed to enable sensitive MS detection. The advantages put forward by miniaturized devices such as low sample consumption, rapid separation capabilities, integration, automatic operation and high-throughput analysis, make microfluidic platforms ideal for proteomic explorations. For example, the tryptic digestion of proteins in gel or in solution at 37°C may take several hours. On-chip digestion in microreactors packed with immobilized trypsin beads, however, can be accomplished in a few seconds. The peptide components recovered from such enzyme microreactors can be further processed on the same microfluidic platform to enable efficient MS detection. Most conventional techniques used in protein analysis, such as solid phase extraction-SPE (Chen et al., 2007; Chung et al., 2005; Dahlin et al., 2005; Ekstrom et al., 2006; Kutter et al., 1999; Lee et al., 2008a; Liu et al., 2009a; Peterson et al., 2003; Ramsey and Collins, 2005; Verpoorte, 2003; Yang et al., 2005; Yu et al., 2001, 2002), two-dimensional gel electrophoresis (Das et al., 2007; Yang et al., 2009a), proteolytic digestion (Fan and Chen, 2007; Jankovicova et al., 2008; Jiang and Lee, 2001;

Krenkova et al., 2009; Lazar et al., 2001; Le Nel et al., 2008; Lee et al., 2008b; Liu et al., 2006; Qu et al., 2004; Slentz et al., 2003; Slovakova et al., 2005; Wu et al., 2004; Xu et al., 2008), isoelectric focusing-IEF (Sommer and Hatch, 2009), capillary electrophoresis (Fonslow and Yates, 2009), capillary electrochromatography-CEC (Huo and Kok, 2008), and liquid chromatography (Srbek et al., 2007; Lazar et al., 2003a; Lazar et al., 2006b; Xie et al., 2004a,b, 2005; Yin et al., 2005) have been integrated within a microfluidic chip format. The achievements enabled so far by miniaturized sample analysis have driven us to further explore and promote the critical role of microfluidic systems in proteomic studies.

49.2 Sample Delivery to MS

The analysis of relatively simple protein or protein digest mixtures was accomplished by direct infusion-ESI-MS or MALDI-MS. Initially, samples were infused into the ESI source directly from the chip edge (Figeys and Aebersold, 1998; Ramsey and Ramsey, 1997; Xue et al., 1997). Sample flows were generated via electroosmotic flow (EOF) or syringe pumping at 100–200 nL/min, and detection from low nM concentrations of myoglobin solutions was achieved. Thereafter, a multichannel chip was developed for high-throughput infusion analysis of standard protein digest or yeast proteomic cellular extracts separated on a 2D-gel (Figeys et al., 1998). The multiple sample delivery channels were fabricated in a glass substrate, and the sample flows were driven by EOF generated in a capillary inserted in the chip, that also facilitated the coupling of the chip to the ESI-MS source. Low fmol–pmol sample amounts/spot were detected. This work represented the first report of high-throughput protein analysis on a microfluidic chip. Other integrated microfluidic chips for sample infusion were demonstrated for protein glycosylation analysis, as well (Bindila et al., 2004; Zamfir et al., 2004; Zamfir et al., 2005; Zhang and Chelius, 2004; Zhang and Williamson Brian, 2005). While such simple sample delivery devices can be fabricated in a variety of materials (glass, polymeric or silicon), and can be designed to accommodate most conventional ESI sources (liquid sheet, liquid junction and nano-ESI) to enable rapid, reduced sample consumption analysis, they are useful mainly for the investigation of rather simple sample mixtures. Complex samples will be better processed by chips that enable high peak capacity separations prior to MS detection.

As an alternative to ESI-MS, digital microfluidic devices enable the manipulation of droplets in an automated and high-throughput manner to facilitate fast MALDI-MS analysis. Protein sample and matrix droplets can be loaded directly onto the MALDI target plate at specific locations by electro-wetting-on-dielectric (EWOD) operation, mixed, dried and analyzed by MS (Wheeler et al., 2004). The EWOD device allows the manipulation of droplets as small as 0.5 μL, of both sample and matrix solutions. However, only solutions with a large contact angle on the surface can be driven by such a mechanism, which limits the choice of materials for the fabrication of the EWOD device, and applicability to the analysis of only certain analytes.

49.3 Sample Preparation for MS

Sample preparation is one of the most critical steps in ensuring efficient MS detection. So far, microfluidic devices have been developed for almost all protein sample preparation steps including cell culture and lysis, protein extraction from cells, purification, desalting and enrichment.

49.3.1 Cell Culture and Lysis

Microfluidic devices provide new opportunities for advancing new cell culture strategies, to facilitate both basic cell biology research and tissue engineering efforts. Various microfluidic cell culture systems have been proposed for cell patterning, passive and active cell handling, and controlling the cell culture environment (El-Ali et al., 2005; Meyvantsson and Beebe, 2008). For example, an integrated microfluidic chip with pumps and valves was developed for performing long-term culture and monitoring of small populations of bacteria (Balagadde et al., 2005). The device allowed semi-continuous, planktonic growth in 6×16 nL microreactors. Each reactor consisted of a growth chamber, i.e., a fluidic loop – 11.5 mm in circumference, 10 μm high and 140 μm wide, an integrated peristaltic pump and micromechanical valves for supplying medium, remove waste and recover cells. Automated, real-time, noninvasive measurement of cell density and morphology was enabled by monitoring the cells by optical microscopy. Another example of a microfluidic chip designed for rapid stimulation and lysis of cells is shown in Fig. 49.1 (El-Ali et al., 2005). The chip comprises two cell treatment zones and three primary inlet ports. The inlet ports were used for loading the cell stream, the cell stimulation medium, and the gas necessary for the formation of a segmented gas-liquid flow that was used to enhance mixing. The two chip zones were used for cell stimulus and cell lysis and were controlled by integrated thermoelectric heaters and coolers at 37 and 4°C, respectively. Three types of cells, i.e., Jurkat E6-1, SKW6.4 and U937, were stimulated with various reagents and lysed with 1% triton X-100, 150 mM NaCl, 10 mM β-glycerophosphate, 10 mM $Na_2P_2O_7$, 10 mM NaF, 1 mM Na_3VO_4, 10 μg/mL leupeptin, 10 μg/mL pepstatin and 10 μg/mL chymostatin. Cell lysates were removed from the chip and analyzed by immunoblotting. For proteomic studies, the capability to perform cell culture and lysis within a microfluidic environment will facilitate not only the rapid manipulation of cells, but will also enable on-line sample processing by other integrated protein analysis units.

49.3.2 Protein Extraction from Cells

Protein extraction from cells is the first step in proteomic analysis. Standard extraction protocols involve multiple steps and the use of numerous buffer systems. Figeys *and coworkers* proposed a proteomic microreactor which can process minute

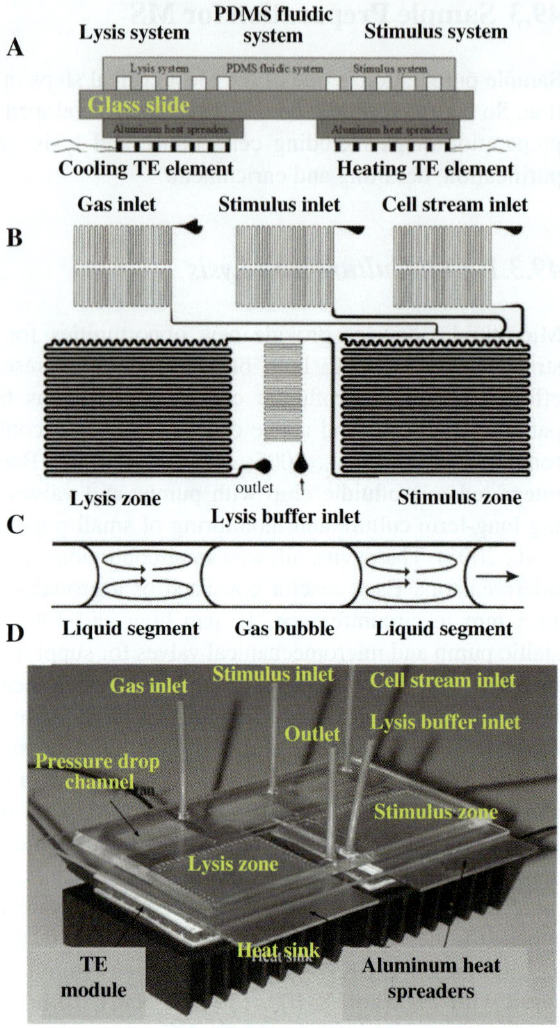

Fig. 49.1 Cell stimulus and lysis microfluidic device. (**a**) Device schematic with thermal control consisting of thermoelectric elements attached to the back of the chip through aluminum heat spreaders. (**b**) Schematic of the fluidic layout incorporating two main zones, for cell stimulus and cell lysis. (**c**) Recirculation in the segmented flow that creates enhanced convective mixing. (**d**) Photograph of the device with temperature control setup (reprinted with permission from El-Ali et al., 2005)

amounts of protein prior to MS analysis. The reactor contained a small bed of packed strong cation exchange (SCX) material, and was used to identify ubiquitinated proteins in human cells (Ethier et al., 2006; Vasilescu et al., 2007). Cells were loaded and lysed directly on the proteomic microreactor at low pH, such that positively charged proteins can be adsorbed on the SCX beads. Adsorbed proteins were then digested with trypsin and the released peptides were eluted and analyzed by MS. The chemical and enzymatic treatments were performed in only 50 nL effective volume, which resulted in an increased number of generated peptides per unit volume and improved detection limits. As little as 440 pg protein lysate was required for protein identifications, that is ∼10 times less than the amounts used by current

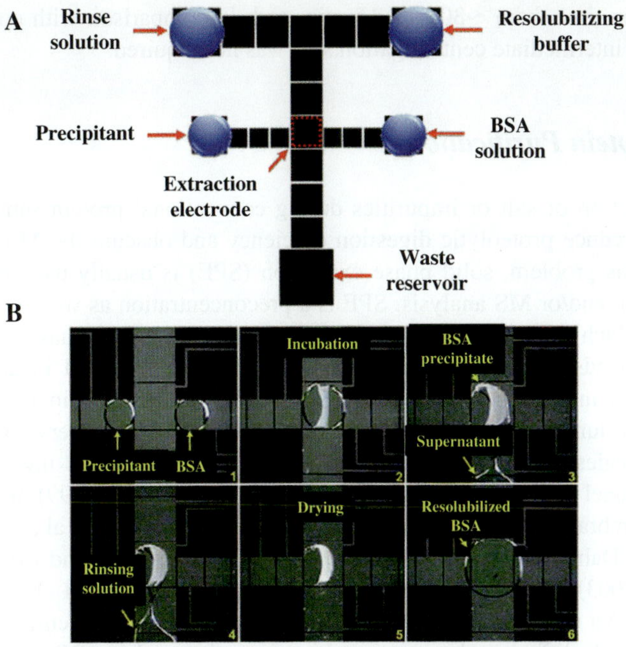

Fig. 49.2 Digital microfluidic device. (**a**) Device schematic showing the four reagent reservoirs, the waste reservoir, and the extraction electrode. (**b**) Movie frames depicting the extraction and purification of BSA (50 mg/mL) in 20% TCA (precipitant), as well as the washing with 70:30 v/v chloroform/acetonitrile rinse solution. The final frame shows the precipitated protein being redissolved in a droplet of 100 mM borate buffer with 1% SDS (reprinted with permission from Jebrail and Wheeler, 2009)

gel-free methodologies. Furthermore, only 300 cells were needed to perform the whole analysis.

Wheeler et al. presented a digital microfluidics-based automated protocol for extracting proteins from heterogeneous fluids by precipitation and re-solubilization (Jebrail and Wheeler, 2009). The chip shown in Fig. 49.2a has a double-cross geometry, with 1×1 mm and 1.5×1.5 mm actuation electrodes, 2.5×2.5 mm and 3×3 mm reservoir electrodes, and 40 μm inter-electrode gaps. The process of droplet operation for sample, precipitants, rinse and re-solubilization solutions is shown in Fig. 49.2b. The droplets containing sample and a precipitant are dispensed from their reservoirs and merged on the extraction electrode. The combined droplets are allowed to incubate until the protein precipitates from the solution, after which the supernatant is actuated away from the extraction electrode to the waste reservoir. The precipitate is then washed, dried and re-dissolved by dispensing and driving droplets of rinse and re-solubilization solution across the extraction electrode. This platform was used for protein extraction from complex mixtures including fetal bovine serum and cell lysates. The method was operated at droplet sizes of about ~100–300 nL and 80% protein recovery. Microfluidic operation reduced the extraction time from ∼30 to ∼15 min, and, in comparison with conventional methods, an intermediate centrifugation step was not required.

49.3.3 Protein Purification

The introduction of salt or impurities during conventional protein sample preparation may reduce proteolytic digestion efficiency and obscure the MS signal. To overcome this problem, solid phase extraction (SPE) is usually used before protein digestion and/or MS analysis. SPE is a preconcentration as well as a clean-up process in which analytes of interest are retained on a stationary phase, while undesired compounds are washed out. The analytes are later eluted in an adequate buffer system in a concentrated form. SPE is well suited to miniaturization, as very small volumes are necessary for sample elution and recovery (Yang et al., 2005). Strategies for SPE implementation on microfluidic chips involve the use of microchannel surface coated microstructures (Kutter et al., 1999), hydrophobic polymer membranes, reversed-phase (RP) microparticles (Chen et al., 2007; Chung et al., 2005; Dahlin et al., 2005; Ekstrom et al., 2006; Ramsey and Collins, 2005; Verpoorte, 2003), and in situ polymerized monoliths (Peterson et al., 2003; Yang et al., 2005; Yu et al., 2001, 2002). For protein purification and enrichment, SPE techniques with different chromatographic properties such as RP, ion exchange, or affinity binding to isolate specific subsets of molecules, have been used, followed by subsequent proteolytic digestion or MS analysis. For example, a nL-scale (3.5 nL) SPE device with C18 particles packed in a weir structured channel was developed for the detection of low-abundance gel-separated proteins by MALDI-MS (Chen et al., 2007). Fifty nanoliters of matrix solution were needed to elute the proteins from the SPE column, and the sequence coverage for bovine serum albumin (BSA), 5 fmol tryptic digest loaded on the SPE column, was 51%, which was better than that obtained from 25 fmol of the same sample processed by conventional methods.

Another widely used method for the fabrication of SPE elements involves the use of microchannels packed with porous polymeric monoliths. An enrichment factor of $\sim 10^3$ for green fluorescent protein was demonstrated for an SPE device prepared by in situ copolymerization of butyl methacrylate with ethylene dimethacrylate, or 2-hydroxyethyl methacrylate and [2-(methacryloyloxy) ethyl]trimethylammonium chloride with ethylene dimethacrylate, by photoinitiation (Yu et al., 2001).

An integrated high-throughput sample preparation compact disc (CD) with MALDI-MS detection was developed by Gyros Corp, as shown in Fig. 49.3a (http://www.gyros.com). There are 96 parallel microstructures on the CD, each microstructure containing a 10 nL RP chromatography column. The inset in Fig. 49.3a shows the individual microstructure elements. Sample and washing eluent solutions are driven by centrifugal force. The protein digest samples are loaded into the sample reservoir (see inset) by capillary action. Upon rotation, the sample solution passes through the RP column, hydrophobic peptides are retained on the column, and the liquid solution is removed from the system to waste. A wash buffer is loaded into a common distribution channel and a well-defined volume chamber. The disc is

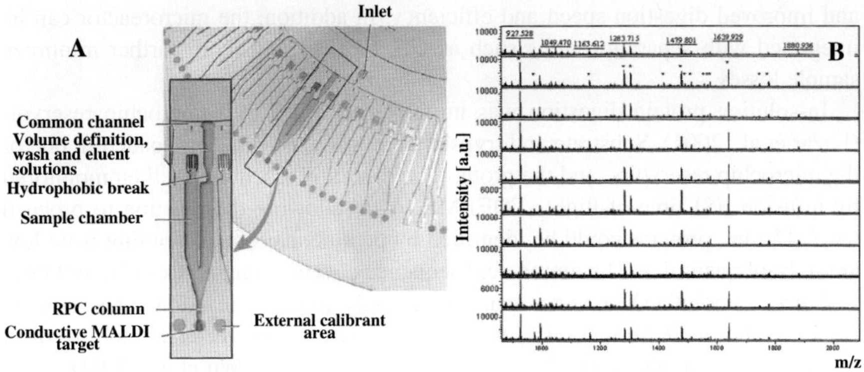

Fig. 49.3 MALDI-MS centrifugal device. (**a**) Image of the Gyrolab MALDI sample preparation CD. (**b**) MALDI mass spectra of a tryptic digest of BSA from 10 different CD microstructures. The *asterisks* indicate BSA fragment positions (reprinted with permission from Gustafsson et al., 2004)

rotated at a speed that will empty the distribution channel but will not allow the wash solution to pass through a hydrophobic break zone. A further increase in rotation speed will allow the well-defined volume of wash solution to pass through the hydrophobic break and wash the RP column, and be then discarded as waste. Next, a well-defined volume of elution/MALDI matrix solution is loaded and passed through the RP column, to elute the peptides and carry them to the MALDI target zone. The reproducibility of the CD device was demonstrated with 1 fmol tryptic digests of BSA-see ten MALDI spectra in Fig. 49.3b generated from 10 different CD processing lines (Gustafsson et al., 2004). The limit of detection for proteolytic digests was 50 amol, and the success rate of the CD technology in protein identification was estimated to be twice as high as that obtained with C18 ZipTips and standard MALDI steel targets.

Digital microfluidics, involving the manipulation of droplets by an electrically controlled driving mechanism such as EWOD, also provides an excellent platform for parallel sample preparation for MALDI-MS. An integrated digital microfluidic chip for multiplexed proteomic sample preparation and MALDI-MS detection has been reported (Moon et al., 2006; Wheeler et al., 2005). The sample droplets were transported to target areas by EWOD operation, and dried. After impurities were removed by rinsing droplets, matrix-containing droplets were delivered to the sample, dried, and the sample was further analyzed in situ by MALDI-MS. This design allowed multiple sample preparation steps on the MALDI target plate, and automatic manipulation of sample solution droplets as small as 70 nL.

49.4 Proteolytic Digestion

Proteolytic digestion is a critical step in shotgun and bottom-up proteomic studies, and is one of the most successful miniaturized protein analysis processes. The miniaturized size of the reactor results in significantly reduced sample consumption and improved digestion speed and efficiency. In addition, the microreactor can be integrated with separation units such as CE, CEC or HPLC, to further minimize sample losses.

In-solution protein digestion was initially performed in a microchip reservoir (Lazar et al., 2001). Substrate and trypsin were mixed off-line and loaded in one of the microchip reservoirs, and the proteolytic digestion reaction was then monitored by infusion ESI time-of-flight (TOF) MS. After allowing the reaction to proceed for 7–15 min, proteins could be identified by peptide mass fingerprinting from low amol–fmol quantities. The amino acid sequence coverage ranged from 70 to 95%.

Since immobilized enzymes can significantly improve reaction rates, storage stability, and decrease autolysis, enzymes (proteases) have been immobilized on microchannel surfaces (Lee et al., 2008b; Liu et al., 2006; Wu et al., 2004), polymeric membrane microstructures (Jiang and Lee, 2001), stable silica gel-derived microchannels (Qu et al., 2004), microparticles in a microchannel (Jankovicova et al., 2008; Le Nel et al., 2008; Slentz et al., 2003; Slovakova et al., 2005; Wang et al., 2000), and monolithic materials (Fan and Chen, 2007; Krenkova et al., 2009; Xu et al., 2008). The samples were driven through these microreactors by a syringe pump or by gas pressure.

For example, a tryptic digestion microreactor was prepared by immobilizing trypsin on a 100 μm wide, 100 μm deep and 4 cm long (400 nL volume) poly(methyl methacrylate) (PMMA) channel surface. The proteolytic digestion of cytochrome C was accomplished in 24 s by driving the sample solution through the channel at 1 μL/min (Lee et al., 2008b). The shortcoming of such a bioreactor is that the sample must be driven at low flow rates to ensure sufficient time for diffusion, for the sample to reach the channel surface with immobilized enzymes. To overcome this issue, a multilayer assembled microchip was proposed (Liu et al., 2006). Positively charged chitosan and negatively charged hyaluronic acid were multilayer-assembled on the surface of a poly(ethylene terephthalate) microchip to form a microstructured, biocompatible network for trypsin immobilization. Low fmol amounts of proteins at a concentration of 0.5 ng/μL could be identified after 5 s digestion time.

The protease can be also immobilized on particles that are packed in microchannel bioreactors. For example, trypsin-loaded beads were packed in a large channel (~800 μm wide, 150 μm deep and 15 mm long), which was connected with a double-T microchannel structure to further enable CE separation and detection with ESI-MS, as shown in Fig. 49.4 (Wang et al., 2000). Protein samples were driven through the digestion microreactor with a syringe pump at flow rates of 0.5–60 μL/min. BSA could be completely digested at room temperature in ~3–6 min, while small proteins such as melittin were consumed within 5 s. High amino acid sequence coverage was obtained for standard proteins.

Other refined devices with even higher digestion rate and low sample consumption were developed. Immobilized trypsin on a silica-coated fiberglass core inserted in a microchip channel reduced the digestion time to <10 s for even larger proteins such as bovine serum albumin and cytochrome C (Liu et al., 2009b). The achievable sequence coverage for proteins digested in the "in-channel fiber bioreactor" ranged

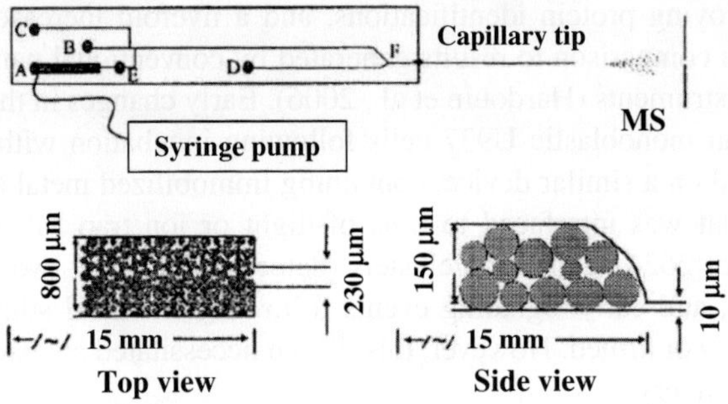

Fig. 49.4 Microchip integrated enzyme microreactor and capillary electrophoresis separation system. The microreactor is 800 µm wide, 150 µm deep and 15 mm long. The channels on the chip are 230 µm wide and 10 µm deep near each fluid reservoir, then become 30 µm wide and 10 µm deep for the rest of the length (reprinted with permission from Wang et al., 2000)

from 45% (BSA) to 77% (cytochrome C). Recently, magnetic beads with immobilized trypsin that could be easily handled in magnetic fields have attracted attention for rapid proteolytic digestion (Jankovicova et al., 2008; Le Nel et al., 2008; Slovakova et al., 2005). When compared to a batch wise system, kinetic studies of the proteolysis of a model peptide in a PDMS microchannel loaded with grafted trypsin magnetic beads, showed a 100-fold increase in digestion speed (Slovakova et al., 2005). Alternatively, multiple proteases were immobilized on porous polymer monoliths to enable high efficiency digestion of high-molecular weight proteins (Krenkova et al., 2009). Even though the monolith was fabricated in a capillary, which was coupled with MS off-line through MALDI, or on-line through ESI, the process could be easily implemented on a microfluidic chip.

49.5 Separation

49.5.1 One-Dimensional Separations

Various types of electrically driven separation methods have been implemented on the chip for protein analysis, including CE, CEC and IEF (Dolnik and Liu, 2005; Lazar et al., 2003b; Oleschuk et al., 2000; Sommer and Hatch, 2009). For all these separations, miniaturization reduced analysis times to a few seconds or minutes, and sample consumption to pL volumes. Most importantly, however, for proteomic applications, nano-LC systems with ESI-MS detection have been integrated on the chip (Brambilla et al., 2009; Lazar et al., 2002, 2006b; Levkin et al., 2007; Yin et al., 2005). A PMMA microfluidic chip with a sample enrichment column, HPLC column, and ESI spray tip was developed and made commercially available in 2006 (www.agilent.com). The chip was demonstrated for the analysis of standard samples and proteins isolated from plasma with detection limits of ~1–5 fmol (Fortier et al., 2005; Yin et al., 2005). The reproducibility and sensitivity of this chip were tested for improving protein identifications, and a fivefold increase in sensitivity was obtained in comparison to results generated by conventional nano-LC-MS/MS with ion trap instruments (Hardouin et al., 2006). Early changes in the phosphoproteome of human monoblastic U937 cells following incubation with phorbol ester was investigated on a similar device, containing immobilized metal affinity enrichment media, that was interfaced to time-of-flight or ion trap MS (Ghitun et al., 2006). A total of 2624 peptide ion clusters (detected in at least two replicate runs) were identified, and early signaling events following chemical stimulation of the U937 cells were confirmed. However, this device necessitated an external pump for mobile phase delivery.

A stand-alone HPLC chip that integrated enrichment and HPLC columns with particle packing or monolithic material, has been developed (Srbek et al., 2007; Xie et al., 2004a,b, 2005). The chip was fabricated by photolithography from silicon wafers and parylene, and comprised three electrolysis-based electrochemical pumps for loading the sample and delivering the solvent gradient, platinum electrodes for providing current to the pumps and establishing the electrospray potential, a low-volume static mixer, a separation column packed with silica-based RP material, integrated frits for bead capture, and an ESI nozzle (Fig. 49.5a). Pressures in excess of 250 psi were generated. The chip was used to perform the analysis of a BSA digest by LC-MS/MS (1.2 cm separation column) with gradient elution at a flow rate of 80 nL/min. The base peak chromatogram of the BSA tryptic digest (~600 fmol), and the pump current setting for the entire cycle of analysis, are provided in Fig. 49.5b. The sequence coverage for BSA was 53%. The chromatographic resolution was close to that provided by a commercial nano-LC system, but the total cycle time was significantly reduced as a result of the minimal volume between the pumps and the separation column. Alternatively, a fully integrated LC device, driven by a multichannel electroosmotic pumping and valving system was described by Lazar

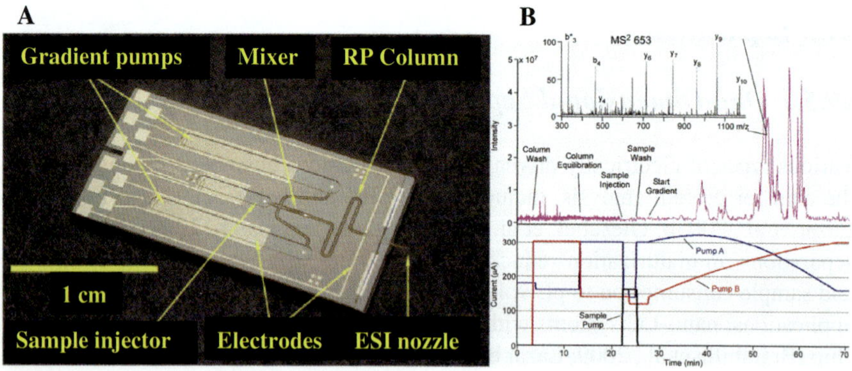

Fig. 49.5 Integrated LC chip driven by electrolysis based electrochemical pumps. (**a**) Photograph of the LC chip. (**b**) Base peak chromatogram (*upper panel*) illustrating the separation of a 600 fmol BSA digest, and pump current settings (*lower panel*) for the entire cycle of analysis. The *inset* represents the tandem mass spectrum of the peptide HLVDEPQNLIK with $m/z = 653.3$ (reprinted with permission from Xie et al., 2005)

(Lazar et al., 2002, 2003a, 2006b). Pressures in excess of 80 psi and flow rates in the ~20 nL/min to 1 μL/min range were generated with microchips containing four hundred pumping channels (~1.5 μm deep). The applicability of this device to complex sample processing is described later in the chapter (Dawoud et al., 2007).

49.5.2 Multi-dimensional Separations

The performance of the multi-dimensional separation systems is heavily affected by the sample separation rate and the effectiveness of the sample transfer between dimensions. The separation efficiency is affected by sample leakage and the method used for sample injection into the second dimension. The microfluidic format ensures zero dead volume, automated/programmable sample transfer between dimensions, and significantly improves reproducibility and separation efficiency. Two dimensional separations using IEF and CE (Herr et al., 2003; Wang et al., 2004), IEF followed by capillary gel electrophoresis-CGE (Das et al., 2007; Griebel et al., 2004; Li et al., 2004; Liu et al., 2008; Xu et al., 2004; Yang et al., 2009b), or CGE followed by micellar electrokinetic chromatography-MEKC (Osiri et al., 2008) have been used for protein separations. The peak capacity of 2D MEKC-CE separations was estimated to be 4500 (Ramsey et al., 2003). A multidimensional microfluidic system combining IEF and parallel SDS gel electrophoresis was described for protein separations (Liu et al., 2008). The design of the microfluidic chip is shown in Fig. 49.6. The sample was injected under negative pressure, separated first by IEF in the first dimension and CGE in the second dimension. The

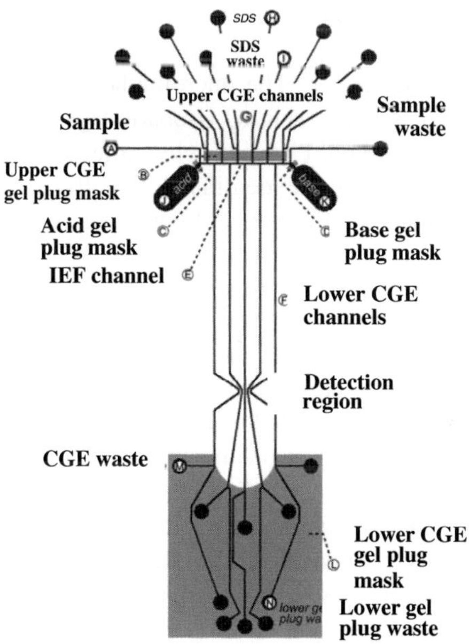

Fig. 49.6 Schematic diagram of an on-line IEF-CGE microfluidic 2D separation device (reprinted with permission from Liu et al., 2008)

separation of an *E. coli* cell lysate was accomplished on this microfluidic chip within 10 min with good reproducibility. Even though the microfluidic device was not connected to MS detection, it provides a very effective way for conducting 2D protein separations in a microfluidic format.

A proof-of-concept three-dimensional electrochromatography system including tryptic digestion, copper(II)-immobilized metal affinity chromatography (Cu(II)-IMAC), and RP capillary electrochromatography was integrated in a microchip format (Slentz et al., 2003). The tryptic and Cu(II)-IMAC columns were packed with particles, while the RP separation column was prepared from collocated monolithic support structures. The Cu(II)-IMAC column was used to select histidine-containing peptides, and peptide identification occurred via fluorescence detection. Multidimensional HPLC-chip/MS was also demonstrated for the analysis of the human nucleolar proteome (Vollmer et al., 2006). The analysis was conducted with off-line SCX chromatography followed by microfraction collection and HPLC chip-MS. Two hundred and six proteins, matched by 2,024 unique tryptic peptides, were identified in the analysis.

49.6 Protein Quantitation

The development of large-scale proteomic quantitation protocols has become a major driving force in proteomics. Isotopic labeling coupled to mass spectrometry is the dominant approach in quantitative proteomics. Stable, nearly chemically equivalent isotopes with a mass difference that can be resolved by MS or MS/MS, are used to label protein or peptide samples for comparison. The mass difference can be used as a marker to identify related peptides, while the intensity/area of the peptide peaks can be used for relative quantitation (Abu-Farha et al., 2009). A microfluidic LC/MS system that performs quantitative analysis of complex proteomic cellular extracts was described. The microfluidic platform was used to explore protein differential expression in iTRAQ (isobaric tags for relative and absolute quantitation)-labeled MCF-7 breast cancer cells cultured in the presence of estradiol and tamoxifen, and enabled the identification of 40–50 proteins and of several previously reported human putative cancer biomarkers (cathepsin D precursor, cytoskeletal keratins 8, 18, and 19, heat shock cognate 71 kDa protein, calreticulin precursor, heterogeneous nuclear ribonucleoprotein A/B, UDP-glucose dehydrogenase, and alpha-enolase) (Armenta et al., 2009).

The main drawbacks of stable isotope labeling quantitation methods are high cost, time-consuming sample preparation, and capability to perform comparisons only between labeled samples that are processed simultaneously. Incomplete labeling, or protein precipitation due to experimental conditions that are ideal for promoting the labeling reaction but not for maintaining the proteins in solution, or changes in chromatographic affinity and retention properties, are additional problems that may affect the quantitation process. On the contrary, label-free and signal intensity-based protein quantitation approaches benefit from not having to label the sample, and enable comparisons between multiple samples that are processed independently. Quantitation accuracy, however, is inferior to the one obtained with label-based approaches. The two major label-free protein quantitation strategies are based on: (a) spectral counting (i.e., the protein abundance is inferred from the total number of tandem mass spectra that match the corresponding peptides), and (b) peptide intensity or area measurements (i.e., the protein abundance is inferred from the intensity or the area of the corresponding peptides in a mass spectrum or a chromatogram). Recently, peak-intensity based, label-free absolute quantitation of carbonic anhydrase II in human serum was performed by an HPLC-chip coupled to a triple quadrupole mass spectrometer (Callipo et al., 2009). An internal standard peptide with sequence GGPLEGTYR was selected for the carbonic anhydrase II specific peptide GGPLDGTYR to perform quantitation via multiple reaction monitoring (MRM)-MS. Calibration was performed based on peak area ratios for GGPLDGTYR and GGPLEGTYR. The quantitation limit for carbonic anhydrase II was estimated to be ~2 fmol/mL. A similar method was used for the simultaneous qualitative characterization and relative quantitation of bioactive lupin proteins (extracts of lupin cultivars) based on the intensities of peptide precursor ions from data-dependent MS analysis with an ion trap instrument (Brambilla et al., 2009).

49.7 Microfluidic Posttranslational Modifications Analysis

49.7.1 Phosphorylation Analysis

Protein phosphorylation is involved in a number of biological regulatory mechanisms that control cell division, growth, differentiation, apoptosis and metabolism. Protein phosphorylation occurs mainly on serine, threonine, and tyrosine residues in eukaryotic cells, and/or on histidine, glutamic acid and aspartic acid residues in prokaryotic cells. Phosphorylated proteins are present in very low copy numbers, and only a small proportion of any given protein is phosphorylated at any given time (Lazar, 2009). In addition, multiple sites on a protein can be phosphorylated, and the MS detection of phosphorylated proteins/peptides is made difficult by the phospho group itself that imparts a negative charge to the peptide, reducing detection efficiency in commonly used (+) mode ESI-MS. As a result, enrichment steps are essential for the analysis of phosphorylated proteins by MS. Enrichment by ion immobilized affinity chromatography (IMAC) is one of the most widely used methods in phosphorylation analysis. An integrated microfluidic system comprising a large channel, a CE separation channel, and a low dead volume ESI-MS interface, was described for the rapid analysis of trace-level target proteins/peptides (2 pmol). Immobilized metal affinity chromatography beads were packed in the large channel to enable the affinity selection of phosphopeptides prior to electrophoretic separation and MS detection on a quadrupole/time-of-flight instrument. Samples were loaded on the chip by an autosampler (Li et al., 2002). In another example, a polyimide device integrating a precolumn loaded with Ga(III) IMAC beads, a separation channel packed with C18 particles, sample ports, frits, and a triangular shape ESI tip, was used for the trace level analysis of phosphoproteins from cell extracts. The combination of IMAC selection prior to microfluidic separation facilitated the identification of phosphopeptides present at only 1.5 fmol in complex protein digests, with 25-fold enrichment compared to non-IMAC enriched samples. This chip was demonstrated for differential phosphoproteome analysis for identifying early signaling events in human monoblast lymphoma U937 cells after chemical stimulation (Ghitun et al., 2006). A similar application was reported with a TiO_2 or TiO_2/C18 instead of a Ga(III) precolumn for phosphoprotein analysis from whole cell lysates (Mohammed et al., 2008). IMAC enrichment has been also integrated on a MALDI target plate for on-line phosphorylation analysis (Ibanez et al., 2007; Qiao et al., 2007). TiO_2 particles were sintered to the MALDI target plate, and the phosphorylated peptides were captured and enriched on the plate surface (Ibanez et al., 2007; Qiao et al., 2007).

An interesting phosphoprotein analysis approach involved an in situ tagging reaction in the ESI Taylor cone of a microfluidic device. Phosphopeptide tagging reactions by dinuclear zinc(II) complexes were performed on a microfluidic device with a dual-channel microsprayer. The device consisted of a polyimide microchip with two microchannels that connected at the tip of the microchip. In situ reactions were achieved inside the Taylor cone with the dual-channel microsprayer, both with the tag synthesized chemically before the experiments, and with the tag electrogenerated by in situ oxidation of a zinc electrode. Two solutions of different physicochemical properties can be mixed inside the Taylor cone and used to selectively tag target molecules (Prudent et al., 2008).

49.7.2 Glycosylation Analysis

Glycosylation plays an important role in many biological processes including immune defense, protein stability, cell growth, and cell-cell adhesion. Analysis of glycosylation is more complicated than that of phosphorylation due to the complex glycan structures and the presence of multiple glycosylation sites. So far, direct sample infusion (Zamfir et al., 2004, 2005; Zhang and Chelius, 2004; Zhang and Williamson Brian, 2005), capillary electrophoresis (Mao et al., 2006), lectin affinity column for sample preparation (Bedair and Oleschuk, 2006), and chip HPLC (Chu et al., 2009) have been used for miniaturized analysis of glycosylation followed by MS detection. A sheath-flow nanoelectrospray interface that coupled a microchip with ESI/MS was used to analyze the glycopeptides of RNase B. Tandem mass spectra were generated to further elucidate the glycan structure of five glycopeptides (Mao et al., 2006). Detection limits were in the low fmol range. In another example, N-linked glycans were released from human serum without derivatization, and analyzed on a microfluidic chip with graphitized carbon packing for enrichment and separation (Chu et al., 2009). Two hundred glycans including high mannose, fucosylated complex, and fucosylated-sialylated complex glycans were identified by this chip HPLC-MS system.

49.8 Integrated Microfluidics/MS for Protein Analysis

The ability to integrate biological and/or chemical lab operations into a small chip, which ultimately enables automatic, rapid, high-throughput, and low sample consumption analysis, is the most important attribute of microfluidic systems. For proteomics studies, an ideal microfluidic device should be able to perform sample extraction, clean-up (desalting, purification and enrichment), proteolytic digestion, separation and detection. Up to date, partial integration of some of these steps has been accomplished, e.g., sample enrichment, separation and detection (Dahlin et al., 2005), or proteolytic digestion, HPLC separation and detection. Most designs, however, are still at the proof-of-concept stage. For real world applications, the most widely used device is the integrated chip-HPLC, which commonly integrates an enrichment column, an RP-HPLC column, and an ESI tip (www.agilent.com). Such a chip was demonstrated for the analysis of numerous proteomic samples involving ion trap and time-of-flight MS detection (see previous examples). For example, a protein extract from rat plasma digested with trypsin and analyzed by the chip-HPLC enabled the reproducible identification of more than 600 peptide ions in ten replicate runs (Fortier et al., 2005). Another type of integrated HPLC chip involves the use of electrolysis-based electrochemical pumps for driving the fluids on the chip (Xie et al., 2005). In addition to the pumps, the system integrates an enrichment column, a separation column and an ESI sprayer (see discussion in the previous sections of the chapter).

A microfluidic device (Fig. 49.7a) integrating several EOF pumps (Lazar et al., 2002), an EOF valve, an LC separation channel and an ESI emitter was used for neutral loss ion mapping, data-dependent triple-play and neutral loss analysis, and in situ dephosphorylation followed by LC separation and MS detection of α-casein

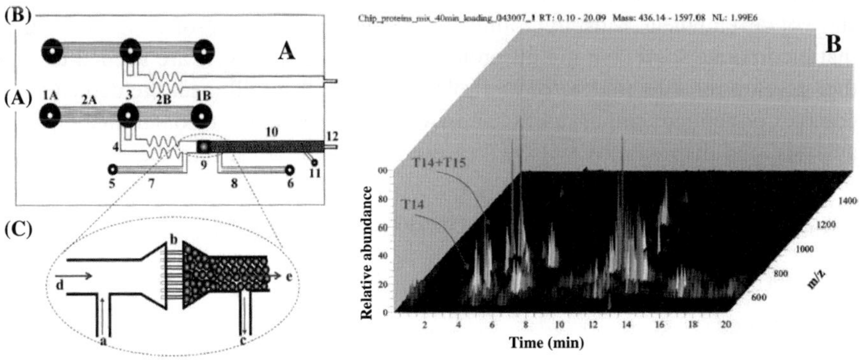

Fig. 49.7 Microfluidic platform for the analysis of phosphoproteins. (**a**) Schematic representation of the microfluidic chip including an EOF-driven LC system (A/A) and infusion pump (A/B). The inset (A/C) illustrates an expanded view of the separation channel head/preconcentrator. (**b**) 2D-view chromatogram of a microfluidic LC data dependent triple-play MS analysis of a protein mixture digest solution (0.5 μM). Phosphorylated peptides were detected from low intensity signals in the presence of multiple, intense, nonphosphorylated peptides (reprinted with permission from Dawoud et al., 2007)

tryptic digest solutions (Dawoud et al., 2007). Two sample processing units, one for performing EOF-driven sample infusion, and one for performing EOF-driven microfluidic LC separations, were fabricated on a 3"×1.5" glass chip. The chip was interfaced to MS by incorporating in the chip fused-silica ESI emitters. The sample infusion unit was used to explore protein phosphorylation in relatively simple mixtures, while the LC separation unit was used for further analysis of more complex samples. The LC unit also enabled the analysis of peptides prior and after in situ dephosphorylation. Figure 49.7b shows a 3D view chromatogram of the data-dependent triple-play MS analysis of a tryptic digest of ten bovine proteins. Twenty three peptides were detected for all proteins, and phosphorylated peptides were detected from very low intensity signals (T14 and T15 as shown in Fig. 49.7b). Generally, the microfluidic device enabled the completion of an analysis from only a few microliters of sample, in ∼10–15 min, and the detection of low fmol amounts of sample.

49.9 Conclusions

Numerous strategies for coupling miniaturized, integrated analysis platforms with ESI or MALDI mass spectrometry detection have been proposed and demonstrated for a variety of proteomic sample preparation needs. The advantage of integrating on a microfabricated platform the protein sample preparation process revolves around the benefits enabled by miniaturization, i.e., speed of analysis, low sample consumption, automatic/high-throughput manipulations and sensitive detection. Some of these benefits were captured by commercially available microfluidic devices that have been used as versatile sample inlet systems for various types of mass spectrometers. Despite these accomplishments, microfluidics MS is still at an early stage of development. Most microfluidics-MS systems are proof-of-concept designs that incorporate some, but not all protein analysis steps. Efforts are needed to further develop reliable, integrated and robust systems that will meet the demands of complex proteomic sample analysis. Low-cost, high-throughput, contamination-free analysis, and disposability will be key ingredients for advancing these devices in biomedical and clinical research.

Acknowledgment This work was supported by NCI/NIH grant 1R21CA126669-01A1.

References

Abu-Farha, M., Elisma, F., Zhou, H.J., Tian, R.J., Zhou, H., Asmer, M.S., and Figeys, D. (2009). Proteomics: From technology developments to biological applications. Anal Chem *81*, 4585–4599.

Armani, M., Rodriguez-Canales, J., Gillespie, J., Tangrea, M., Erickson, H., Emmert-Buck, M.R., Shapiro, B., and Smela, E. (2009). 2D-PCR: A method of mapping DNA in tissue sections. LOC *9*, 3526–3534.

Armenta, J.M., Dawoud, A.A., and Lazar, I.M. (2009). Microfluidic chips for protein differential expression profiling. Electrophoresis *30*, 1145–1156.

Balagadde, F.K., You, L.C., Hansen, C.L., Arnold, F.H., and Quake, S.R. (2005). Long-term monitoring of bacteria undergoing programmed population control in a microchemostat. Science *309*, 137–140.

Bedair, M., and Oleschuk, R.D. (2006). Lectin affinity chromatography using porous polymer monolith assisted nanoelectrospray MS/MS. Analyst *131*, 1316–1321.

Belder, D. (2005). Microfluidics with droplets. Ang Chem Int Edn *44*, 3521–3522.

Bindila, L., Froesch, M., Lion, N., Vukeli, Z., Rossier, J.S., Girault, H.H., Peter-Katalini, J., and Zamfir, A.D. (2004). A thin chip microsprayer system coupled to Fourier transform ion cyclotron resonance mass spectrometry for glycopeptide screening. Rapid Commun Mass Spectrom *18*, 2913–2920.

Brambilla, F., Resta, D., Isak, I., Zanotti, M., and Arnoldi, A. (2009). A label-free internal standard method for the differential analysis of bioactive lupin proteins using nano HPLC-Chip coupled with ion trap mass spectrometry. Proteomics *9*, 272–286.

Callipo, L., Foglia, P., Gubbiotti, R., Samperi, R., and Lagana, A. (2009). HPLC-CHIP coupled to a triple quadrupole mass spectrometer for carbonic anhydrase II quantification in human serum. Anal Bioanal Chem *394*, 811–820.

Chen, W.Z., Shen, J., Yin, X.F., and Yu, Y.N. (2007). Optimization of microfabricated nanoliter-scale solid-phase extraction device for detection of gel-separated proteins in low abundance by matrix-assisted laser desorption/ionization mass spectrometry. Rapid Commun Mass Spectrom *21*, 35–43.

Chu, C.S., Ninonuevo, M.R., Clowers, B.H., Perkins, P.D., An, H.J., Yin, H.F., Killeen, K., Miyamoto, S., Grimm, R., and Lebrilla, C.B. (2009). Profile of native N-linked glycan structures from human serum using high performance liquid chromatography on a microfluidic chip and time-of-flight mass spectrometry. Proteomics *9*, 1939–1951.

Chung, W.J., Kim, M.S., Cho, S., Park, S.S., Kim, J.H., Kim, Y.K., Kim, B.G., and Lee, Y.S. (2005). Microaffinity purification of proteins based on photolytic elution: Toward an efficient microbead affinity chromatography on a chip. Electrophoresis *26*, 694–702.

Dahlin, A.P., Bergstrom, S.K., Andren, P.E., Markides, K.E., and Bergquist, J. (2005). Poly(dimethylsiloxane)-based microchip for two-dimensional solid-phase extraction-capillary electrophoresis with an integrated electrospray emitter tip. Anal Chem *77*, 5356–5363.

Das, C., Zhang, J., Denslow, N.D., and Fan, Z.H. (2007). Integration of isoelectric focusing with multi-channel gel electrophoresis by using microfluidic pseudo-valves. LOC *7*, 1806–1812.

Dawoud, A.A., Sarvalyal, H.A., and Lazar, I.M. (2007). Microfluidic platform with mass spectrometry detection for the analysis of phosphoproteins. Electrophoresis *28*, 4645–4660.

Devoe, D.L., and Lee, C.S. (2006). Microfluidic technologies for MALDI-MS in proteomics. Electrophoresis *27*, 3559–3568.

Dolnik, V., and Liu, S. (2005). Applications of capillary electrophoresis on microchip. J Sep Sci *28*, 1994–2009.

Ekstrom, S., Wallman, L., Hok, D., Marko-Varga, G., and Laurell, T. (2006). Miniaturized solid-phase extraction and sample preparation for MALDI MS using a microfabricated integrated selective enrichment target. J Proteome Res *5*, 1071–1081.

El-Ali, J., Gaudet, S., Günther, A., Sorger, P.K., and Jensen, K.F. (2005). Cell stimulus and lysis in a microfluidic device with segmented gas–liquid flow. Anal Chem *77*, 3629–3636.

Ethier, M., Hou, W.M., Duewel, H.S., and Figeys, D. (2006). The proteomic reactor: A microfluidic device for processing minute amounts of protein prior to mass spectrometry analysis. J Proteome Res *5*, 2754–2759.

Fair, R.B. (2007). Digital microfluidics: Is a true lab-on-a-chip possible? Microfluidics Nanofluidics *3*, 245–281.

Fan, H.Z., and Chen, G. (2007). Fiber-packed channel bioreactor for microfluidic protein digestion. Proteomics *7*, 3445–3449.

Figeys, D., and Aebersold, R. (1998). Nanoflow solvent gradient delivery from a microfabricated device for protein identifications by electrospray ionization mass spectrometry. Anal Chem *70*, 3721–3727.

Figeys, D., Gygi, S.P., McKinnon, G., and Aebersold, R. (1998). An integrated microfluidics tandem mass spectrometry system for automated protein analysis. Anal Chem *70*, 3728–3734.

Figeys, D., Ning, Y.B., and Aebersold, R. (1997). A microfabricated device for rapid protein identification by microelectrospray ion trap mass spectrometry. Anal Chem *69*, 3153–3160.

Fonslow, B.R., and Yates, J.R., III (2009). Capillary electrophoresis applied to proteomic analysis. J Sep Sci *32*, 1175–1188.

Foret, F., and Kusy, P. (2006). Microdevices in Mass Spectrometry. Paper presented at: 17th International Mass Spectrometry Conference (Prague, Czech Republic).

Fortier, M.H., Bonneil, E., Goodley, P., and Thibault, P. (2005). Integrated microfluidic device for mass spectrometry-based proteomics and its application to biomarker discovery programs. Anal Chem *77*, 1631–1640.

Ghitun, M., Bonneil, E., Fortier, M.H., Yin, H.F., Killeen, K., and Thibault, P. (2006). Integrated microfluidic devices with enhanced separation performance: Application to phosphoproteome analyses of differentiated cell model systems. J Sep Sci *29*, 1539–1549.

Griebel, A., Rund, S., Schonfeld, F., Dorner, W., Konrad, R., and Hardt, S. (2004). Integrated polymer chip for two-dimensional capillary gel electrophoresis. LOC *4*, 18–23.

Gustafsson, M., Hirschberg, D., Palmberg, C., Jornvall, H., and Bergman, T. (2004). Integrated sample preparation and MALDI mass spectrometry on a microfluidic compact disk. Anal Chem *76*, 345–350.

Hardouin, J., Duchateau, M., Joubert-Caron, R., and Caron, M. (2006). Usefulness of an integrated microfluidic device (HPLC-Chip-MS) to enhance confidence in protein identification by proteomics. Rapid Commun Mass Spectrom *20*, 3236–3244.

Herr, A.E., Molho, J.I., Drouvalakis, K.A., Mikkelsen, J.C., Utz, P.J., Santiago, J.G., and Kenny, T.W. (2003). On-chip coupling of isoelectric focusing and free solution electrophoresis for multidimensional separations. Anal Chem *75*, 1180–1187.

Huo, Y., and Kok, W.T. (2008). Recent applications in CEC. Electrophoresis *29*, 80–93.

Ibanez, A.J., Muck, A., and Svatos, A. (2007). Metal-chelating plastic MALDI (pMALDI) chips for the enhancement of phosphorylated-peptide/protein signals. J Proteome Res *6*, 3842–3848.

Inoue, A., Han, A., Makino, K., Hosokawa, K., and Maeda, M. (2009). SNP genotyping of unpurified PCR products by sandwich-type affinity electrophoresis on a microchip with programmed autonomous solution filling. LOC *9*, 3297–3302.

Jankovicova, B., Rosnerova, S., Slovakova, M., Zverinova, Z., Hubalek, M., Hernychova, L., Rehulka, P., Viovy, J.L., and Bilkova, Z. (2008). Epitope Mapping of Allergen Ovalbumin Using Biofunctionalized Magnetic Beads Packed in Microfluidic Channels: The First Step Towards Epitope-Based Vaccines. Paper presented at: 22nd International Symposium on Microscale Bioseparations and Methods for Systems Biology (Berlin, Germany).

Jebrail, M.J., and Wheeler, A.R. (2009). Digital microfluidic method for protein extraction by precipitation. Anal Chem *81*, 330–335.

Jensen, K., and Lee, A. (2004). The science & applications of droplets in microfluidic devices. LOC *4*, 31 N–32 N.

Jiang, Y., and Lee, C.S. (2001). On-Line Coupling of Micro-enzyme Reactor with Micro-membrane Chromatography for Protein Digestion, Peptide Separation, and Protein Identification Using Electrospray Ionization Mass Spectrometry. Paper presented at: 14th International Symposium on Microscale Separations and Analysis (Boston, MA).

Kim, J., Johnson, M., Hill, P., and Gale, B.K. (2009). Microfluidic sample preparation: Cell lysis and nucleic acid purification. Integr Biol *1*, 574–586.

Koster, S., and Verpoorte, E. (2007). A decade of microfluidic analysis coupled with electrospray mass spectrometry: An overview. LOC *7*, 1394–1412.

Krenkova, J., Lacher, N.A., and Svec, F. (2009). Highly efficient enzyme reactors containing trypsin and endoproteinase LysC immobilized on porous polymer monolith coupled to MS suitable for analysis of antibodies. Anal Chem *81*, 2004–2012.

Kutter, J.P., Jacobson, S.C., and Ramsey, J.M. (1999). Solid Phase Extraction on Microfluidic Devices. Paper presented at: 21st International Symposium on Capillary Chromatography and Electrophoresis (Park City, UT).

Lazar, I.M. (2007). Microfluidic Devices with Mass Spectrometry Detection. In Handbook of Capillary and Microchip Electrophoresis and Associated Microtechniques, 3rd Edition, J.P. Landers, ed. (CRC Press, New York, NY), pp 1459–1506 13th Dec 2007.

Lazar, I.M. (2009). Recent advances in capillary and microfluidic platforms with MS detection for the analysis of phosphoproteins. Electrophoresis *30*, 262–275.

Lazar, I.M., Grym, J., and Foret, F. (2006a). Microfabricated devices: A new sample introduction approach to mass spectrometry. Mass Spectrom Rev *25*, 573–594.

Lazar, I.M., and Karger, B.L. (2003a). Microchip Integrated Separation Systems for Proteomic Applications. 51th Conference on Mass Spectrometry and Allied Topics (Montreal, QC).

Lazar, I.M., and Karger, B.L. (2002). Multiple open-channel electroosmotic pumping system for microfluidic sample handling. Anal Chem 74, 6259–6268.

Lazar, I.M., Li, L.J., Yang, Y., and Karger, B.L. (2003b). Microfluidic device for capillary electrochromatography-mass spectrometry. Electrophoresis *24*, 3655–3662.

Lazar, I.M., Ramsey, R.S., and Ramsey, J.M. (2001). On-chip proteolytic digestion and analysis using "wrong-way-round" electrospray time-of-flight mass spectrometry. Anal Chem *73*, 1733–1739.

Lazar, I.M., Trisiripisal, P., and Sarvaiya, H.A. (2006b). Microfluidic liquid chromatography system for proteomic applications and biomarker screening. Anal Chem *78*, 5513–5524.

Le Nel, A., Krenkova, J., Kleparnik, K., Smadja, C., Taverna, M., Viovy, J.L., and Foret, F. (2008). On-chip tryptic digest with direct coupling to ESI-MS using magnetic particles. Electrophoresis *29*, 4944–4947.

Lee, E.Z., Huh, Y.S., Jun, Y.-S., Won, H.J., Hong, Y.K., Park, T.J., Lee, S.Y., and Hong, W.H. (2008a). Removal of bovine serum albumin using solid-phase extraction with in-situ polymerized stationary phase in a microfluidic device. J Chromatogr, A *1187*, 11–17.

Lee, J., Musyimi, H.K., Soper, S.A., and Murray, K.K. (2008b). Development of an automated digestion and droplet deposition microfluidic chip for MALDI-TOF MS. J Am Soc Mass Spectrom *19*, 964–972.

Levkin, P.A., Eeltink, S., Stratton, T.R., Brennen, R., Robotti, K., Yin, H., Killeen, K., Svec, F., and Frechet, J.M.J. (2007). Monolithic Porous Polymer Stationary Phases in Polyimide Chips for the Fast High-Performance Liquid Chromatography Separation of Proteins and Peptides. Paper presented at: 31st International Symposium on Capillary Chromatography (Albuquerque, NM).

Li, J.J., LeRiche, T., Tremblay, T.L., Wang, C., Bonneil, E., Harrison, D.J., and Thibault, P. (2002). Application of microfluidic devices to proteomics research – identification of trace-level protein digests and affinity capture of target peptides. Mol Cell Proteomic *1*, 157–168.

Li, Y., Buch, J.S., Rosenberger, F., DeVoe, D.L., and Lee, C.S. (2004). Integration of isoelectric focusing with parallel sodium dodecyl sulfate gel electrophoresis for multidimensional protein separations in a plastic microfluidic network. Anal Chem *76*, 742–748.

Liu, J., Chen, C.-F., Tsao, C.-W., Chang, C.-C., Chu, C.-C., and De Voe, D.L. (2009a). Polymer microchips integrating solid-phase extraction and high-performance liquid chromatography using reversed-phase polymethacrylate monoliths. Anal Chem *81*, 2545–2554.

Liu, J.K., Yang, S., Lee, C.S., and Devoe, D.L. (2008). Polyacrylamide gel plugs enabling 2-D microfluidic protein separations via isoelectric focusing and multiplexed sodium dodecyl sulfate gel electrophoresis. Electrophoresis *29*, 2241–2250.

Liu, T., Wang, S., and Chen, G. (2009b). Immobilization of trypsin on silica-coated fiberglass core in microchip for highly efficient proteolysis. Talanta *77*, 1767–1773.

Liu, Y., Lu, H.J., Zhong, W., Song, P.Y., Kong, J.L., Yang, P.Y., Girault, H.H., and Liu, B.H. (2006). Multi layer-assembled microchip for enzyme immobilization as reactor toward low-level protein identification. Anal Chem *78*, 801–808.

Mao, X.L., Chu, I.K., and Lin, B.C. (2006). A sheath-flow nanoelectrospray interface of microchip electrophoresis MS for glycoprotein and glycopeptide analysis. Electrophoresis *27*, 5059–5067.

Meyvantsson, I., and Beebe, D.J. (2008). Cell culture models in microfluidic systems. Annu Rev Anal Chem *1*, 423–449.

Mogensen, K.B., Klank, H., and Kutter, J.P. (2004). Recent developments in detection for microfluidic systems. Electrophoresis *25*, 3498–3512.

Mohammed, S., Kraiczek, K., Pinkse, M.W. H., Lemeer, S., Benschop, J.J., and Heck, A.J. R. (2008). Chip-based enrichment and NanoLC-MS/MS analysis of phosphopeptides from whole lysates. J Proteome Res *7*, 1565–1571.

Moon, H., Wheeler, A.R., Garrell, R.L., Loo, J.A., and Kim, C.J. (2006). An integrated digital microfluidic chip for multiplexed proteomic sample preparation and analysis by MALDI-MS. LOC *6*, 1213–1219.

Nedelkov, D., and Nelson, R.W. (2006). New and Emerging Proteomic Techniques (Totowa, NJ, Humana Press).

Oleschuk, R.D., Shultz-Lockyear, L.L., Ning, Y.B., and Harrison, D.J. (2000). Trapping of bead-based reagents within microfluidic systems: On-chip solid-phase extraction and electrochromatography. Anal Chem *72*, 585–590.

Osiri, J.K., Shadpour, H., Park, S., Snowden, B.C., Chen, Z.Y., and Soper, S.A. (2008). Generating high peak capacity 2-D maps of complex proteomes using PMMA microchip electrophoresis. Electrophoresis *29*, 4984–4992.

Peterson, D.S., Rohr, T., Svec, F., and Frechet, J.M.J. (2003). Dual-function microanalytical device by in situ photolithographic grafting of porous polymer monolith: Integrating solid-phase extraction and enzymatic digestion for peptide mass mapping. Anal Chem *75*, 5328–5335.

Prudent, M., Rossier, J.S., Lion, N., and Girault, H.H. (2008). Microfabricated dual sprayer for on-line mass tagging of phosphopeptides. Anal Chem *80*, 2531–2538.

Qiao, L., Roussel, C., Wan, J.J., Yang, P.Y., Girault, H.H., and Liu, B.H. (2007). Specific on-plate enrichment of phosphorylated peptides for direct MALDI-TOF MS analysis. J Proteome Res *6*, 4763–4769.

Qu, H.Y., Wang, H.T., Huang, Y., Zhong, W., Lu, H.J., Kong, J.L., Yang, P.Y., and Liu, B.H. (2004). Stable microstructured network for protein patterning on a plastic microfluidic channel: Strategy and characterization of on-chip enzyme microreactors. Anal Chem *76*, 6426–6433.

Ramsey, J.D., and Collins, G.E. (2005). Integrated microfluidic device for solid-phase extraction coupled to micellar electrokinetic chromatography separation. Anal Chem *77*, 6664–6670.

Ramsey, J.D., Jacobson, S.C., Culbertson, C.T., and Ramsey, J.M. (2003). High-efficiency, two-dimensional separations of protein digests on microfluidic devices. Anal Chem *75*, 3758–3764.

Ramsey, R.S., and Ramsey, J.M. (1997). Generating electrospray from microchip devices using electroosmotic pumping. Anal Chem *69*, 1174–1178.

Shi, Y. (2006). DNA sequencing and multiplex STR analysis on plastic microfluidic devices. Electrophoresis *27*, 3703–3711.

Slentz, B.E., Penner, N.A., and Regnier, F.E. (2003). Protein proteolysis and the multi-dimensional electrochromatographic separation of histidine-containing peptide fragments on a chip. J Chromatogr A *984*, 97–107.

Slovakova, M., Minc, N., Bilkova, Z., Smadja, C., Faigle, W., Futterer, C., Taverna, M., and Viovy, J.L. (2005). Use of self assembled magnetic beads for on-chip protein digestion. LOC *5*, 935–942.

Sommer, G.J., and Hatch, A.V. (2009). IEF in microfluidic devices. Electrophoresis *30*, 742–757.

Srbek, J., Eickhoff, J., Effelsberg, U., Kraiczek, K., van de Goor, T., and Coufal, P. (2007). Chip-based nano-LC-MS/MS identification of proteins in complex biological samples using a novel polymer microfluidic device. J Sep Sci *30*, 2046–2052.

Sung, W.C., Makamba, H., and Chen, S.H. (2005). Chip-based microfluidic devices coupled with electrospray ionization-mass spectrometry. Electrophoresis *26*, 1783–1791.

Teh, S.Y., Lin, R., Hung, L.H., and Lee, A.P. (2008). Droplet microfluidics. LOC *8*, 198–220.

Tia, S., and Herr, A.E. (2009). On-chip technologies for multidimensional separations. LOC *9*, 2524–2536.

Uchiyama, K., Nakajima, H., and Hobo, T. (2004). Detection method for microchip separations. Anal Bioanal Chem *379*, 375–382.

Vasilescu, J., Zweitzig, D.R., Denis, N.J., Smith, J.C., Ethier, M., Haines, D.S., and Figeys, D. (2007). The proteomic reactor facilitates the analysis of affinity-purified proteins by mass spectrometry: Application for identifying ubiquitinated proteins in human cells. J Proteome Res *6*, 298–305.

Verpoorte, E. (2003). Beads and chips: New recipes for analysis. LOC *3*, 60 N–68 N.

Vollmer, M., Horth, P., Rozing, G., Coute, Y., Grimm, R., Hochstrasser, D., and Sanchez, J.C. (2006). Multi-dimensional HPLUMS of the nucleolar proteome using HPLC-chip/MS. J Sep Sci *29*, 499–509.

Wang, C., Oleschuk, R., Ouchen, F., Li, J.J., Thibault, P., and Harrison, D.J. (2000). Integration of immobilized trypsin bead beds for protein digestion within a microfluidic chip incorporating capillary electrophoresis separations and an electrospray mass spectrometry interface. Rapid Commun Mass Spectrom *14*, 1377–1383.

Wang, Y.C., Choi, M.N., and Han, J.Y. (2004). Two-dimensional protein separation with advanced sample and buffer isolation using microfluidic valves. Anal Chem *76*, 4426–4431.

West, J., Becker, M., Tombrink, S., and Manz, A. (2008). Micro total analysis systems: Latest achievements. Anal Chem *80*, 4403–4419.

Wheeler, A.R., Moon, H., Bird, C.A., Loo, R.R. O., Kim, C.J., Loo, J.A., and Garrell, R.L. (2005). Digital microfluidics with in-line sample purification for proteomics analyses with MALDI-MS. Anal Chem *77*, 534–540.

Wheeler, A.R., Moon, H., Kim, C.J., Loo, J.A., and Garrell, R.L. (2004). Electrowetting-based microfluidics for analysis of peptides and proteins by matrix-assisted laser desorption/ionization mass spectrometry. Anal Chem *76*, 4833–4838.

Wu, H.L., Zhai, J.J., Tian, Y.P., Lu, H.J., Wang, X.Y., Jia, W.T., Liu, B.H., Yang, P.Y., Xu, Y.M., and Wang, H.H. (2004). Microfluidic enzymatic-reactors for peptide mapping: Strategy, characterization, and performance. LOC *4*, 588–597.

Xie, J., Miao, Y.N., Shih, J., He, Q., Liu, J., Tai, Y.C., and Lee, T.D. (2004a). An electrochemical pumping system for on-chip gradient generation. Anal Chem *76*, 3756–3763.

Xie, J., Miao, Y.N., Shih, J., Tai, Y.C., and Lee, T.D. (2005). Microfluidic platform for liquid chromatography-tandem mass spectrometry analyses of complex peptide mixtures. Anal Chem *77*, 6947–6953.

Xie, J., Shih, J., He, Q., Pang, C.L., Tai, Y.C., Miao, Y., Lee, T.D., and Ieee (2004b). An Integrated LC-ESI Chip with Electrochemical-Based Gradient Generation. Paper presented at: 17th IEEE International Conference on Micro Electro Mechanical Systems (Maastricht, The Netherlands).

Xu, A.H., Sluszny, C., and Yeung, E.S. (2004). Prototype for Integrated Two-Dimensional Gel Electrophoresis for Protein Separation. Paper presented at: 25th International Symposium on Chromatography (Paris, France).

Xu, X.J., Wang, X.Y., Liu, Y., Liu, B.H., Wu, H.L., and Yang, P.Y. (2008). Trypsin entrapped in poly(diallyldimethylammonium chloride) silica sol-gel microreactor coupled to matrix-assisted laser desorption/ionization time-of-flight mass spectrometry. Rapid Commun Mass Spectrom *22*, 1257–1264.

Xue, Q.F., Foret, F., Dunayevskiy, Y.M., Zavracky, P.M., McGruer, N.E., and Karger, B.L. (1997). Multichannel microchip electrospray mass spectrometry. Anal Chem *69*, 426–430.

Yang, S., Liu, J., Lee, C.S., and DeVoe, D.L. (2009a). Microfluidic 2-D PAGE using multifunctional in situ polyacrylamide gels and discontinuous buffers. LOC *9*, 592–599.

Yang, S., Liu, J.K., Lee, C.S., and Devoe, D.L. (2009b). Microfluidic 2-D PAGE using multifunctional in situ polyacrylamide gels and discontinuous buffers. LOC *9*, 592–599.

Yang, Y.N., Li, C., Lee, K.H., and Craighead, H.G. (2005). Coupling on-chip solid-phase extraction to electrospray mass spectrometry through an integrated electrospray tip. Electrophoresis *26*, 3622–3630.

Yin, N.F., Killeen, K., Brennen, R., Sobek, D., Werlich, M., and van de Goor, T.V. (2005). Microfluidic chip for peptide analysis with an integrated HPLC column, sample enrichment column, and nanoelectrospray tip. Anal Chem *77*, 527–533.

Yu, C., Davey, M.H., Svec, F., and Frechet, J.M.J. (2001). Monolithic porous polymer for on-chip solid-phase extraction and preconcentration prepared by photoinitiated in situ polymerization within a microfluidic device. Anal Chem *73*, 5088–5096.

Yu, C., Xu, M.C., Svec, F., and Frechet, J.M. J. (2002). Preparation of monolithic polymers with controlled porous properties for microfluidic chip applications using photoinitiated free-radical polymerization. J Polymer Sci Polymer Chem *40*, 755–769.

Zamfir, A., Vakhrushev, S., Sterling, A., Niebel, H.J., Allen, M., and Peter-Katalinic, J. (2004). Fully automated chip-based mass spectrometry for complex carbohydrate system analysis. Anal Chem *76*, 2046–2054.

Zamfir, A.D. (2007). Recent advances in sheathless interfacing of capillary electrophoresis and electrospray ionization mass spectrometry. J Chromatogr A *1159*, 2–13.

Zamfir, A.D., Lion, N., Vukelic, Z., Bindila, L., Rossier, J., Girault, H.H., and Peter-Katalinic, J. (2005). Thin chip microsprayer system coupled to quadrupole time-of-flight mass spectrometer for glycoconjugate analysis. LOC *5*, 298–307.

Zhang, S., and Chelius, D. (2004). Characterization of protein glycosylation using chip-based infusion nanoelectrospray linear ion trap tandem mass spectrometry. J Biomol Tech *15*, 120–133.

Zhang, S., and Williamson Brian, L. (2005). Characterization of protein glycosylation using chip-based nanoelectrospray with precursor ion scanning quadrupole linear ion trap mass spectrometry. J Biomol Tech *16*, 209–219.

Index

A.R. Ivanov, A.V. Lazarev (eds.), *Sample Preparation in Biological Mass Spectrometry*, DOI 10.1007/978-94-007-0828-0,

F

Q

T